About the Cover

Wildlife Can Thrive in Urban Habitats

The coyote (*Canis latrans*) is an extremely adaptable species that can thrive in locations as diverse as forests, deserts, grasslands, and wetlands. It also does well in rural, suburban, and even urban settings. Coyotes are omnivores and their diet is as diverse as the habitats in which they live. They eat small mammals, rabbits, insects, rodents, flowers, and fruits as well as young, immature, and adult white-tailed deer, and sometimes livestock.

There is a long history of coyotes being hated and feared by people because they are predators on wildlife, livestock, and pets. Cash bounties have been paid for the pelts of coyotes for decades. While the hunting season for most animals is typically a few weeks each year, more than 30 states in the United States have a 12-month hunting season for coyotes. Despite this, coyote populations have thrived. While habitat alteration by people has resulted in a decrease in many species, this is not the case for coyotes. They exist both far from humans and close by. Many major cities in the United States have had news stories about coyotes living within their boundaries. There is reportedly one coyote living in New York City's Central Park. San Francisco's Candlestick Park and Chicago's Lincoln Park have also had coyote sightings. Just type "coyote" and the name of a nearby urban park into any search engine and you will see what we mean.

The cover of the fourth edition of our book features a female coyote with car headlights and pavement in the background, indicating that this is an urban coyote. The photograph was taken in the San Francisco Bay area as part of the Urban Coyote Initiative (www.urbancoyoteproject.com) to inform the public about the fascinating lives of coyotes, their role in the environment, and the need to appreciate and embrace wild animals in urban settings. Thus, the coyote serves as an excellent example of the complexity of environmental science, which focuses on understanding the interactions of people and the natural environment.

Jaymi Heimbuch/Minden Pictures

Environmental Science
for the AP® Course

Fourth Edition

Andrew Friedland
Dartmouth College

Rick Relyea
Rensselaer Polytechnic Institute

Executive Vice President and General Manager: Charles Linsmeier
Vice President, Social Sciences and High School: Shani Fisher
Executive Program Director: Ann Heath
Senior Executive Program Manager: Yolanda Cossio
Program Manager: Heidi Bamatter
Editorial Assistant: Calyn Clare Liss
Senior Marketing Manager: Thomas Menna
Marketing Assistant: Brianna DiGeronimo
Senior Media Editor: Justin Perry
Senior Director, Content Management Enhancement: Tracey Kuehn
Senior Managing Editor: Lisa Kinne
Senior Content Project Manager: Edward Dionne
Senior Workflow Project Manager: Paul Rohloff
Production Supervisor: Robert Cherry
Director of Design, Content Management: Diana Blume
Senior Design Services Manager: Natasha Wolfe
Interior Designer: Kevin Kall
Senior Cover Design Manager: John Callahan
Art Manager: Matthew McAdams
Illustration Coordinator: Janice Donnola
Illustrations: Joseph BelBruno
Text Permissions Editor: Michael McCarty
Senior Media Permissions Manager: Christine Buese
Photo Research: Lisa Passmore, Lumina Datamatics, Inc.
Senior Director, Digital Production: Keri deManigold
Senior Media Project Manager: Elton Carter
Composition: Lumina Datamatics, Inc.
Printing and Binding: Transcontinental Printing
Banners (lights and the coyote from the cover image): Jaymi Heimbuch/Minden Pictures

Library of Congress Control Number: 2022946782
ISBN-13: 978-1-319-40928-9
ISBN-10: 1-319-40928-8

3 4 5 6 7 8 29 28 27 26 25 24

Printed in Canada

Bedford, Freeman & Worth Publishers
120 Broadway
New York, NY 10271
highschool.bfwpub.com/catalog

Macmillan is committed to lessening our company's impact on the environment. The Macmillan family of publishing houses in the USA achieved a 50% reduction in our 2020 CO_2 emissions as compared to our 2010 baseline. Based on this progress, and further mitigation via the purchase of carbon offsets, Macmillan USA has been carbon neutral for the last 8 years commencing in 2013.
https://sustainability.macmillan.com/

To Katie, Jared, and Ethan

—AJF

To Christine, Isabelle, and Wyatt

—RAR

Brief Contents

Contents

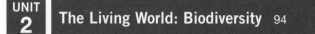

About the Authors

Andrew Friedland is the Richard and Jane Pearl Professor in Environmental Studies at Dartmouth College. He was the founding chair of the AP® Test Development Committee (College Board®) for Environmental Science. He has a strong interest in high school science education, and in the early years of AP® environmental science he participated in many trainer and teacher workshops. For more than 15 years, Andy has been a guest lecturer at various Advanced Placement® Institutes for Secondary Teachers and in high school APES classrooms. He also served on the College Board AP® Environmental Science Curriculum Development and Assessment Committee.

Andy has taught introductory environmental science and energy courses at Dartmouth as well as courses in forest ecosystems and global change, and soil science. He created an online introductory environmental science course that is accessible through edX.org and YouTube.

Andy received a B.A. degree in both biology and environmental studies, and a Ph.D. in earth and environmental science, from the University of Pennsylvania. For more than four decades, he has investigated the effects of air pollution on the cycling of carbon, nitrogen, and lead in high-elevation forests of New England and the Northeast. During the last decade, he has examined the impact of increased demand for wood as a fuel, and the subsequent effect on carbon stored deep in forest soils.

Andy has served on panels for the National Science Foundation, USDA Forest Service, and Science Advisory Board of the Environmental Protection Agency. He has authored or coauthored 80 peer-reviewed publications and one other book, *Writing Successful Science Proposals*, Third Edition (Yale University Press). In 2015, he was named a Fellow of the American Association for the Advancement of Science.

Andy is passionate about saving energy and at his home he has installed a 4 kW photovoltaic tracker that follows the sun during the day.

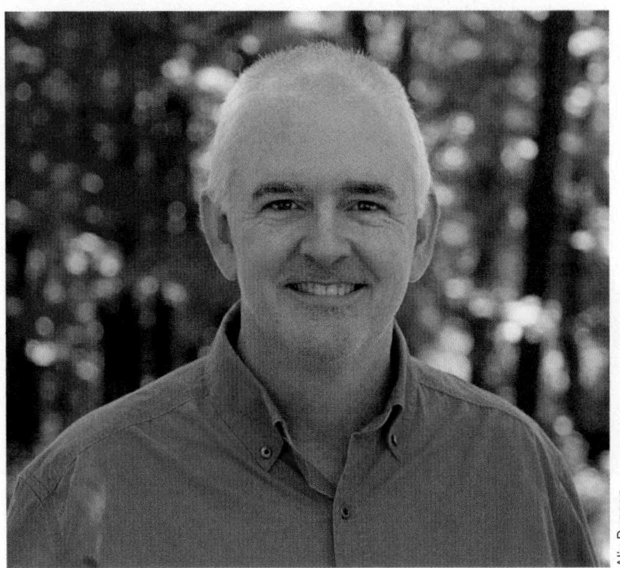

Rick Relyea is the David Darrin Senior '40 Endowed Chair in Biological Sciences at Rensselaer Polytechnic Institute. Rick teaches courses in ecology, evolution, and animal behavior at the undergraduate and graduate levels. He received a B.S. in environmental forest biology from the State University of New York College of Environmental Science and Forestry, an M.S. in wildlife management from Texas Tech University, and a Ph.D. in ecology and evolution from the University of Michigan.

Rick is recognized throughout the world for his work in the fields of ecology, evolution, animal behavior, and ecotoxicology. He has served on multiple scientific panels for the National Science Foundation and has been an associate editor for the journals of the Ecological Society of America. For three decades, he has conducted research on a wide range of topics, including predator–prey interactions, phenotypic plasticity, eutrophication of aquatic habitats, sexual selection, disease ecology, long-term dynamics of populations and communities, and pesticide impacts on aquatic ecosystems. He has authored more than 200 scientific articles, and has presented research seminars throughout the world. Rick was a professor at the University of Pittsburgh for 15 years, where he was named the Chancellor's Distinguished Researcher in 2005 and received the Tina and David Bellet Teaching Excellence Award in 2014. In 2014, he moved to Rensselaer Polytechnic Institute.

Rick has a strong interest in high school education. High school science teachers conduct research in his laboratory and he offers summer workshops for high school teachers in the fields of ecology, evolution, and ecotoxicology. Rick also works to bring cutting-edge research experiments into high school classrooms.

Rick's commitment to the environment extends to his personal life. He lives in a home constructed with a passive solar building design and equipped with active solar panels on the roof.

Content Advisory Board

This book is the product of collective intelligence. For this edition, we were fortunate to collaborate closely with an expert content advisory board throughout the development process. The content advisory board understands the needs of the environmental science teacher and student. They provided helpful direction on how to make the content relevant, engaging, and appropriate for a high school classroom. We extend gratitude and admiration to each of these talented educators for their enduring contributions to the teaching of environmental science.

© Jennifer Campbell Photography

Hemalatha Bhaskaran

James M. Bennett High School

Hema has taught at James M. Bennett High School since 2004. She is passionate about providing meaningful educational experiences for all her students. Hema was recognized by CBT as the State's Environmental Teacher of the Year, she received the Presidential Award for Excellence in Science Teaching (2019), was named Wicomico County Teacher of the Year (2020-2021) and was one of seven finalists for Maryland Teacher of the Year (2020-2021). She promotes diversity beyond the classroom through Solutions for Multicultural Achievement and Recognition Team. She engages in collaborative leadership by serving on JMB's ILT, Youth Environmental Action Summit's steering committee and CBF's Teacher Environmental Literacy Leadership.

Cambria Carmody

Suzanne Carmody

Widefield High School

Suzanne has been teaching secondary science for 29 years, 20 of those years at the high school level. She has taught geology and AP® Environmental Science for the last 10 years. Suzanne has been an AP® reader since 2015 and has taught Saturday study sessions. Suzanne enjoys cheering her daughters on at sporting events as well as gardening, camping, and kayaking.

Amy Fassler

Amy Fassler

Marshfield High School

Amy teaches AP® Environmental Science at Marshfield High School in Marshfield, Wisconsin. For the past 19 years her passion for environmental science has inspired many students to become more environmentally aware and pursue additional studies in the field. She coaches the Science Olympiad Team at MHS and serves as a mentor for a science research team. Amy is active in professional development at the state and national level and serves as a College Board consultant. Amy has been involved in development of the AP® Environmental Science Course and Exam Description, and has served on the test development committee and leadership team for scoring the AP® Environmental Science Exam.

Highpoint Pictures, Inc.

Jim Lehner

Taft School

Jim has served as a reader, table leader, and question leader for the AP® Environmental Science exam over the past 18 years and will serve as assistant chief reader starting in 2023. Jim also served on the APES Test Development Committee from 2016 to 2022 and has led many APES summer institutes over the last decade. In his spare time, Jim loves learning new things about the Universe and playing genuine country music on his Telecaster.

Courtney Mayer

Courtney Mayer

Courtney Mayer taught science courses for 23 years, including AP® Environmental Science in San Antonio, Texas. She is a College Board consultant for AP® Environmental Science in the Southwestern Region and has presented all over the United States and in China. Courtney has been a reader and table leader at the AP® Environmental Science reading where she scored free-response questions for thousands of students. She was honored in 2011 with the College Board's AP® Award for helping to promote and continue the mission set by the College Board. Courtney has published 10 books, all around science topics, particularly AP® Environmental Science.

Kristi Schertz

Kristi Schertz

Saugus High School

Kristi Schertz is an experienced professional development leader, curriculum developer, and coach for science, AP® courses, literacy, and technology. She is an AP® consultant and travels the country presenting at summer institutes and workshops for teachers. She is currently in her 23rd year teaching and 16th year teaching AP® Environmental Science. She has been an AP® reader since 2011 as well as a mentor to new science and AP® teachers. In addition, she is a published author of AP® practice tests and environmental science test-prep guides. She also writes a popular blog site for AP® science teachers.

Joni Villarreal

Antonio (Tony) Villarreal

Sharyland Advanced Academic Academy

Tony has served as a Reader, college board consultant, and question writer, and has led AP® Summer Institutes each summer since 2016. He has also recorded AP® Daily and AP® Daily Live videos for AP® Classroom. In addition to teaching AP® Environmental Science for 15 years, Tony also teaches Dual Biology and Honors Chemistry. He served as a lecturer at University of Texas Rio Grande Valley and as an adjunct faculty member at South Texas College. Tony was a Consultant/Coach for National Math & Science Initiative, where he conducted teacher training and mentoring and student study sessions all over Texas and the United States.

Erin Foley

Alyson Wasko

Montclair High School

Alyson has served as a reader and table leader for the AP® Environmental Science exam since 2011, and has been teaching the course for 13 years. She completed her undergraduate work in biology at Syracuse University, and masters work in both education (MA) and geoscience (MS) from Montclair State University. In addition to teaching, Alyson serves as the assistant marching band director and student coalition advisor at Montclair High School.

Acknowledgments

We would like to thank the many people at Bedford, Freeman, and Worth who helped guide us through the publication process of this book. They have taught us a great deal and have been crucial to our book becoming greatly appreciated by so many people. We especially want to acknowledge: Ann Heath, Yolanda Cossio, Heidi Bamatter, Rebecca Kohn, Joseph BelBruno, Edward Dionne, Diana Blume, Natasha Wolfe, Fred Burns, Christine Buese, Lisa Passmore, Calyn Clare Liss, Matt McAdams, Justin Perry, Thomas Menna, and Brianna DiGeronimo. Thanks also to all the people who helped with aspects of the program, including Jabin Burnworth, Andrew Milbauer, Aaron Stoler, and Leah Kohn. We thank David Courard-Hauri, Ross Jones, and Susan Weisberg for contributions to the first edition of this book.

We also wish to convey our appreciation to the dozens of colleagues, teachers, students, and reviewers who constantly challenged us to write a clear, correct, and scientifically and philosophically balanced textbook.

High School Reviewers and Focus Group Participants

Cynthia Ahmed, *Signature School, IN*
Timothy Allen, *Thomas A. Edison Preparatory High School, OK*
Julie Back, *Kecoughtan High School, VA*
Maureen Bagwell, *Collierville High School, TN*
Fredrick Baldwin, *Kendall High School, NY*
Lisa Balazs, *Indian Springs School, AL*
Debra Bell, *Montgomery High School, TX*
Melinda Bell, *Flagstaff Arts and Leadership Academy, AZ*
Karen Benton, *South Brunswick High School, NJ*
Richard Benz, *Wickliffe High School, OH*
Hemalatha Bhaskaran, *Wicomico County Public Schools, MD*
Cindy Birkner, *Webber Township High School, IL*
Christine Bouchard, *Milford Public Schools, CT*
Gail Boyarsky, *East Chapel Hill High School, NC*
Rebecca Bricen, *Johnsonburg High School, PA*
Deanna Brunlinger, *Elkhorn Area High School, WI*
Kevin Bryan, *Woodrow Wilson Senior High School, CA*
Jabin Burnworth, *Manchester Junior Senior High School, IN*
Tanya Bunch, *Carter High School, TN*
Diane Burrell, *Starr's Mill High School, GA*
Teri Butler, *New Hanover High School, NC*
Charles Campbell, *Russellville High School, AR*
Suzanne Carmody, *Widefield High School, CO*
Sande Caton, *Concord High School, DE*
Andrea Charles, *West Side Leadership Academy, IN*
Linda Charpentier, *Xavier High School, CT*
Blanca Ching, *Fort Hamilton High School, NY*
Ashleigh Coe, *Bethesda–Chevy Chase High School, MD*
Bethany Colburn, *Randolph High School, MA*
Jonathan D. Cole, *Holmdel High School, NJ*
Robert Compton, *Walled Lake Northern High School, MI*
Ann Cooper, *Osceola High School, AR*
Thomas Cooper, *The Walker School, GA*
Joyce Corriere, *Hampton High School, VA*
Stephanie Crow, *Milford High School, MI*
Stephen Crowley, *Winooski High School, VT*
Linda D'Apolito, *Trinity School, NY*
Brygida DeRiemaker, *Eisenhower High School, MI*
Chand Desai, *Martin Luther King Magnet High School, TN*
Michael Douglas, *Bronx Prep Charter School, NY*
Nancy Dow, *A. Crawford Mosley High School, FL*
Nat Draper, *Deep Run High School, VA*
Denis DuBay, *Leesville High School, NC*
John Dutton, *Shaw High School, OH*
Heather Earp, *West Johnston High School, NC*
Kim Eife, *Academy of Notre Dame, PA*
Brian Elliot, *San Dimas High School, CA*
Christina Engen, *Crescenta Valley High School, CA*
Mary Anne Evans, *Allendale Columbia School, NY*
Kay Farkas, *Rush-Henrietta High School, NJ*
Tim Fennell, *LASA at LBJ High School, TX*
Michael Finch, *Greene County Tech High School, AR*
Robert Ford, *Fairfield College Preparatory School, CT*
Paul Frisch, *Fox Lane High School, NY*
Bob Furhman, *The Covenant School, VA*
Nivedita (Nita) Ganguly, *Oak Ridge High School, TN*
Mike Gaule, *Ladywood High School, MI*
Billy Goodman, *Passaic Valley High School, NJ*
Amanda Graves, *Mt. Tahoma High School, WA*
Barbara Gray, *Richmond Community High School, VA*
Jack Greene, *Logan High School, UT*
Julie Quinn Kiernan, *Cretin-Durham Hall, NC*
Jeannie Kornfeld, *Hanover High School, NH*
Jen Kotkin, *St. Philip's Academy, NJ*
Pat Kretzer, *Timber Creek High School, FL*
Michelle Krug, *Coral Springs High School, FL*
Jim Kuipers, *Chicago Christian High School, IL*
Claire Kull, *Career Center, NC*
Jay Kurima, *O. D. Wyatt High School, TX*
Tom LaHue, *Aptos High School, CA*
Cathy Larson, *Patuxent High School, MD*
Michael Lauer, *Danville High School, KY*
Sonia Laureni, *West Orange High School, NJ*
Amy Lawson, *Naples High School, FL*

Jim Lehner, *The Taft School, CT*
Dr. Avon Lewis, *Lexington High School, MA*
Marie Lieberman, *Ravenscroft School, NC*
John Ligget, *Conestoga High School, PA*
Ann Linsley, *Bellaire High School, TX*
Mark Little, *Broomfield High School, CO*
Leyana Lloyd, *Washington Senior Academy, GA*
Larry Lollar, *Alice High School, TX*
Stephanie Longfellow, *Deltona High School, FL*
Leslie Lopez, *Round Rock High School, TX*
Sue Ellen Lyons, *Holy Cross School, LA*
Theresa Lyster, *Camden County High School, GA*
John F. Madden, *Ashley Hall School, SC*
Jeremy Magee, *Sandy High School, OR*
Mike Mallon, *James I. O'Neill High School, NY*
Scott Martin, *Deer Creek High School, OK*
Kristi Martinez, *Eastlake High School, WA*
Christeena Mathews, *The Philadelphia High School for Girls, PA*
Courtney Mayer, *Winston Churchill High School, TX*
Monica Maynard, *Schurr High School, CA*
James McAdams, *Center Grove High School, IN*
Kristen McClellan, *Grand Junction High School, CO*
Sandy McDonough, *North Salem Middle/High School, NY*
Diane Medford, *Los Alamos High School, NM*
Andrew Milbauer, *Poudre High School, CO*
Leslie Miller, *Flintridge Sacred Heart Academy, CA*
Lonnie Miller, *El Diamante High School, CA*
Melody Mingus, *Breckinridge County High School, KY*
Myra Morgan, *National Math & Science Initiative, AP® Environmental Consultant*
Tammy Morgan, *Lake Placid High School, NY*
David Moscarelli, *Ponaganset High School, RI*
Terri Mountjoy, *Greene County Career Center, OH*
Bill Mulhearn, *Archmere Academy, DE*
Sharna Murphy, *Millikan High School, CA*
Jeanine Musgrove, *Oakton High School, VA*
Anna Navarro, *Veterans Memorial High School, TX*
Barbara Nealon, *Southern York County School District, PA*
Dara Nix-Stevenson, *American Hebrew Academy, NC*
Bennett O'Connor, *Dallas ISD, TX*
Robert Oddo, *Horace Greeley High School, NY*
Kate Oitzinger, *El Molino High School, CA*
Paul Olson, *Redwood High School, CA*
Janet Ort, *Hoover High School, AL*
Roger Palmer, *Bishop Dunne High School, TX*
Annetta Pasquarello, *Triton Regional High School, NJ*
Lynn Paulsen, *Mayde Creek High School, TX*
Nicole Peffley, *Cinco Ranch High School, TX*
Judy Perrella, *Academy of the Holy Names, FL*
Carolyn Phillips, *Southeastern High School, IL*
Pam Phillips, *Hayden High School, AL*
Alanna Piccillo, *Palisade High School, CO*
Jenny Ramsey, *Charlotte Christian School, NC*
Susan Ramsey, *VASS, VA*
Cristen Rasmussen, *Costa Mesa High School, CA*
Alesa Rehmann, *Coral Shores High School, FL*
Mark Reilly, *Jeffersonville High School, IN*
Kimbell Reitz, *Penn High School, IN*
Cheryl Rice, *Howard High School, MD*
Sharon Riley, *Springfield High School, OH*
Chris Robson, *Ironwood Ridge High School, AZ*
James Rodewald, *Shaker High School, NY*
Kurt Rogers, *Northern Highlands Regional High School, NJ*
Kris Rohrbeck, *Almont High School, MI*
David Rouby, *Hall High School, AR*
Rebecca Rouch, *East Bay High School, FL*
Jennifer Roy, *TrekNorth Junior & Senior High School, MN*
Reva Beth Russell, *Lehi High School, UT*
Sheila Scanlan, *Highland High School, AZ*
Kristi Schertz, *Saugus High School, CA*
Greg Schiller, *James Monroe High School, CA*
Amy Schwartz, *Aragon High School, CA*
Kristin Shapiro, *Plano East Senior High School, TX*
Shashi Sharma, *Henry Snyder High School, NJ*
Tonya Shires, *Edgewood High School, MD*
Pamela Shlachtman, *South Dade Senior High School, FL*
Julie Smiley, *Winchester Community High School, IN*
Amy Snodgrass, *Central High School, AR*
Bill Somerlot, *New Albany High School, OH*
Anne Soos, *Stuart Country Day School of the Sacred Heart, NJ*
Joan Stevens, *Arcadia High School, CA*
Marianne Strickhart, *Henry Snyder High School, NJ*
Timothy Strout, *Jericho High School, NY*
Robert Summers, *A+ College Ready, AL*
Jeff Sutton, *The Harker School, CA*
Dave Szaroleta, *Salesianum School, DE*
Kristen Thomson, *Saratoga High School, CA*
James Timmons, *Carrboro High School, NC*
Thomas Tokarski, *Woodlands High School, NY*
Susan Tully, *Salem Academy Charter School, MA*
Debra Tyson, *Brooke Pointe High School, VA*
Melissa Valentine, *Elizabeth Seton High School, MD*
Dirk Valk, *McKeel Academy, FL*
Gene Vann, *Head-Royce School, CA*
Rebecca Van Tassell, *Herron High School, IN*
Ashley Veenema, *Lebanon High School, NH*
Marc Vermeire, *Friday Harbor High School, WA*
Antonio Villarreal, *Sharyland Independent School District, TX*

Naomi Volain, *Springfield Central High School, MA*
Betty Walden, *Merritt Island High School, FL*
Craig Wallace, *North Oldham High School, KY*
Abbie Walston, *North Haven High School, CT*
Alyson Wasko, *Montclair High School, NJ*
Annette Weeks, *Battle Ground High School, WA*
Pamela Weghorst, *Ardrey Kell High School, NC*
Matthew Wells, *Cypress Lakes High School, TX*
Michelle Whitehurst, *Powhatan High School, VA*
Jane Whitelock, *Easton High School, MD*
Laurie Whitesell, *Eli Whitney Middle School, OK*
Robert Whitney, *Westview High School, CA*
Carol Widegren, *Lincoln Park High School, IL*
Sarrah Williams, *Hamden Hall Country Day School, CT*
Robert Willis, *Lakeside High School, GA*

College Reviewers

M. Stephen Ailstock, PhD, *Anne Arundel Community College*
Deniz Z. Altin-Ballero, *Georgia Perimeter College*
Daphne Babcock, *Collin County Community College District*
Jay L. Banner, *University of Texas at San Antonio*
James W. Bartolome, *University of California, Berkeley*
Ray Beiersdorfer, *Youngstown State University*
Grady Price Blount, *Texas A&M University, Corpus Christi*
Dr. Edward M. Brecker, *Palm Beach Community College, Boca Raton*
Anne E. Bunnell, *East Carolina University*
Ingrid C. Burke, *Colorado State University*
Anya Butt, *Central Alabama Community College*
John Callewaert, *University of Michigan*
Kelly Cartwright, *College of Lake County*
Mary Kay Cassani, *Florida Gulf Coast University*
Young D. Choi, *Purdue University Calumet*
John C. Clausen, *University of Connecticut*
Richard K. Clements, *Chattanooga State Technical Community College*
Thomas Cobb, *Bowling Green State University, OH*
Stephen D. Conrad, *Indiana Wesleyan University*
Terence H. Cooper, *University of Minnesota, Saint Mary's Winona Campus*
Douglas Crawford-Brown, *University of North Carolina at Chapel Hill*
Wynn W. Cudmore, *Chemeketa Community College*
Katherine Kao Cushing, *San Jose State University*
Maxine Dakins, *University of Idaho*
Robert Dennison, *Heartland Community College*
Michael Denniston, *Georgia Perimeter College*
Roman Dial, *Alaska Pacific University*
Robert Dill, *Bergen Community College*
Michael L. Draney, *University of Wisconsin, Green Bay*
Anita I. Drever, *University of Wyoming*
James Eames, *Loyola University. New Orleans*
Kathy Evans, *Reading Area Community College*
Mark Finley, *Heartland Community College*
Dr. Eric J. Fitch, *Marietta College*
Karen F. Gaines, *Northeastern Illinois University*
James E. Gawel, *University of Washington, Tacoma*
Carri Gerber, *Ohio State University Agricultural Technical Institute*
Julie Grossman, *Saint Mary's University of Minnesota, Saint Mary's Winona Campus*
Lonnie J. Guralnick, *Roger Williams University*
Sue Habeck, *Tacoma Community College*
Hilary Hamann, *Colorado College*
Dr. Sally R. Harms, *Wayne State College*
Floyd Hayes, *Pacific Union College*
Keith R. Hench, *Kirkwood Community College*
William Hopkins, *Virginia Tech*
Richard Jensen, *Hofstra University*
Sheryll Jerez, *Stephen F. Austin State University*
Shane Jones, *College of Lake County*
Caroline A. Karp, *Brown University*
Erica Kipp, *Pace University, Pleasantville/Briarcliff*
Christopher McGrory Klyza, *Middlebury College*
Frank T. Kuserk, *Moravian College*
Matthew Landis, *Middlebury College*
Kimberly Largen, *George Mason University*
Larry L. Lehr, PhD, *Baylor University*
Zhaohui Li, *University of Wisconsin, Parkside*
Thomas R. MacDonald, *University of San Francisco*
Robert Stephen Mahoney, *Johnson & Wales University*
Bryan Mark, *Ohio State University, Columbus Campus*
Paula J.S. Martin, *Juniata College*
Robert J. Mason, *Tennessee Temple University*
Michael R. Mayfield, *Ball State University*
Alan W. McIntosh, *University of Vermont*
Dr. Kendra K. McLauchlan, *Kansas State University*
Patricia R. Menchaca, *Mount San Jacinto Community College*
Dr. Dorothy Merritts, *Franklin and Marshall College*
Bram Middeldorp, *Minneapolis Community and Technical College*
Tamera Minnick, *Mesa State College*
Mark Mitch, *New England College*
Ronald Mossman, *Miami Dade College, North*
William Nieter, *St. John's University*
Mark Oemke, *Alma College*
Victor Okereke, PhD, PE, *Morrisville State College*
Duke U. Ophori, *Montclair State University*
Chris Paradise, *Davidson College*
Dr. Clayton A. Penniman, *Central Connecticut State University*
Christopher G. Peterson, *Loyola University Chicago*
Craig D. Phelps, *Rutgers, The State University of New Jersey, New Brunswick*
F. X. Phillips, PhD, *McNeese State University*
Rich Poirot, *Vermont Department of Environmental Conservation*
Bradley R. Reynolds, *University of Tennessee, Chattanooga*
Amy Rhodes, *Smith College*
Marsha Richmond, *Wayne State University*
Sam Riffell, *Mississippi State University*
Jennifer S. Rivers, *Northeastern Illinois University*
Ellison Robinson, *Midlands Technical College*
Bill D. Roebuck, *Dartmouth Medical School*
William J. Rogers, *West Texas A&M University*

Thomas Rohrer, *Central Michigan University*
Aldemaro Romero, *Arkansas State University*
William R. Roy, *University of Illinois at Urbana-Champaign*
Steven Rudnick, *University of Massachusetts, Boston*
Heather Rueth, *Grand Valley State University*
Eleanor M. Saboski, *University of New England*
Seema Sah, *Florida International University*
Shamili Ajgaonkar Sandiford, *College of DuPage*
Robert M. Sanford, *University of Southern Maine*
Nan Schmidt, *Pima Community College*
Jeffery A. Schneider, *State University of New York at Oswego*
Bruce A. Schulte, *Georgia Southern University*
Eric Shulenberger, *University of Washington*
Michael Simpson, *Antioch University New England*
Annelle Soponis, *Reading Area Community College*
Douglas J. Spieles, *Denison University*
David Steffy, *Jacksonville State University*
Christiane Stidham, *State University of New York at Stony Brook*
Peter F. Strom, *Rutgers, The State University of New Jersey, New Brunswick*
Kathryn P. Sutherland, *University of Georgia*
Christopher M. Swan, *University of Maryland, Baltimore County*
Melanie Szulczewski, *University of Mary Washington*
Jamey Thompson, *Hudson Valley Community College*
John A. Tiedemann, *Monmouth University*
Conrad Toepfer, *Brescia University*
Todd Tracy, *Northwestern College*
Steve Trombulak, *Middlebury College*
Zhi Wang, *California State University, Fresno*
Jim White, *University of Colorado, Boulder*
Rich Wolfson, *Middlebury College*
C. Wesley Wood, *Auburn University*
David T. Wyatt, *Sacramento City College*

Getting the Most from This Book

Congratulations!

We are delighted that you have made the decision to take AP® Environmental Science. You are about to embark on a journey in which you will gain an appreciation for and understanding of the many ways in which humans depend on and impact the natural world. We will introduce you to the environmental science concepts and systems underlying policy issues that exist in the real world and help you think critically about our environment. Our mission is to present information about the world's leading environmental science issues, so that you can (1) understand them and (2) objectively evaluate the best strategies for sustaining both people and the natural world.

As with any AP® course, your path requires curiosity and commitment, and we are confident that you will find it to be highly interesting and personally rewarding. We wrote this fourth edition with the dual goals of helping you understand environmental science and to perform your very best on the exam at the end of the school year. This edition has been organized to precisely align with the AP® Environmental Science curriculum. Every item on the College Board's curriculum is covered thoroughly in the text, in an order and depth that will give you the confidence you need on exam day.

We break down the topics into nine major units. Each unit begins with an interesting opening case study that foreshadows the topics covered in the unit and ends with a closing case study that teaches you how the ideas discussed can be used to pursue real-world environmental solutions. Within each unit, the topics are further broken down into short, easy-to-read modules to keep you on pace to complete the material by the end of the school year.

As you read through the units and modules throughout the year, you will encounter hundreds of AP® Practice Questions. We also provide a full-length Cumulative AP® Practice Exam at the end of the book and additional questions online. In addition, you will have lots of opportunities to learn new math, data analysis, and graphing skills that will be helpful in preparing for the exam. We have also included AP® Exam Tips in every module, which are written by AP® Environmental Science teachers who have years of experience in grading the exam. These tips give you a critical inside look at what is expected on the exam and the best strategies for answering exam questions and avoiding common misconceptions.

Throughout the book, you will find photographs and figures that clearly communicate the information in the text. All of the information for a given unit is then brought together into a two-page visual representation to help you understand the "big picture" of how all the concepts fit together.

Together, we are confident that the fourth edition of our book is going to provide you with a highly valuable understanding of environmental science issues that will serve you well throughout your life. It will also provide you with outstanding preparation for the AP® exam.

We wish you all the best in the coming year!

Andrew Friedland, Rick Relyea

Get the most from your book's organization and pacing

Your AP® Environmental Science adventure begins here! This book has been created with YOU in mind. It is packed with features to help you learn effectively and do well on the AP® Environmental Science Exam.

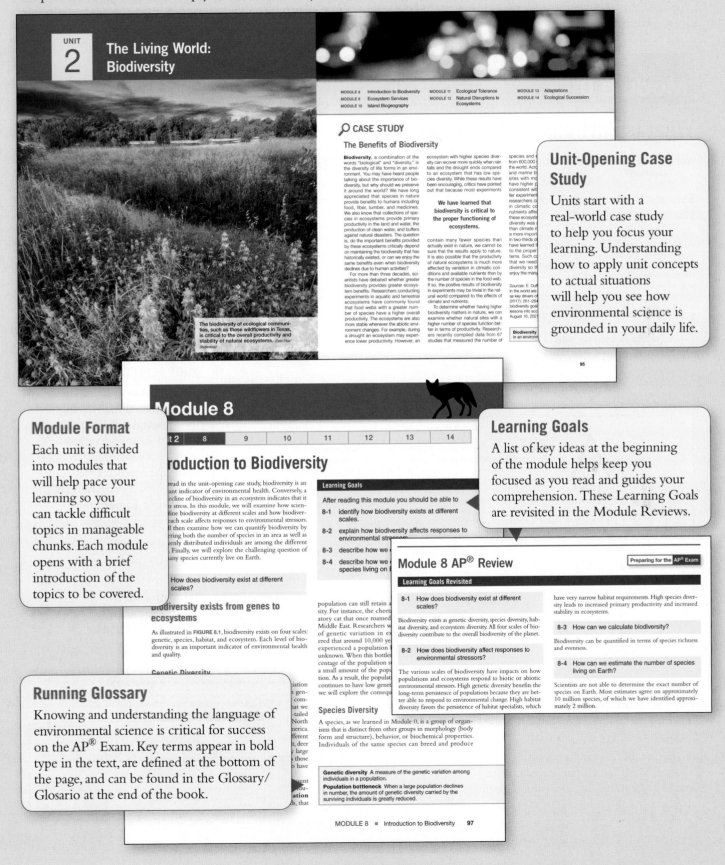

UNIT 2 The Living World: Biodiversity

MODULE 8 Introduction to Biodiversity
MODULE 9 Ecosystem Services
MODULE 10 Island Biogeography
MODULE 11 Ecological Tolerance
MODULE 12 Natural Disruptions to Ecosystems
MODULE 13 Adaptations
MODULE 14 Ecological Succession

CASE STUDY

The Benefits of Biodiversity

Biodiversity, a combination of the words "biological" and "diversity," is the diversity of life forms in an environment. You may have heard people talking about the importance of biodiversity, but why should we preserve it around the world? We have long appreciated that species in nature provide benefits to humans including food, fiber, lumber, and medicines. We also know that collections of species in ecosystems provide primary productivity in the land and water, the production of clean water, and buffers against natural disasters. The question is, do the important benefits provided by these ecosystems critically depend on maintaining the biodiversity that has historically existed, or can we enjoy the same benefits even when biodiversity declines due to human activities?

For more than three decades, scientists have debated whether greater biodiversity provides greater ecosystem benefits. Researchers conducting experiments in aquatic and terrestrial ecosystems have commonly found that food webs with a greater number of species have a higher overall productivity. The ecosystems are also more stable whenever the abiotic environment changes. For example, during a drought an ecosystem may experience lower productivity. However, an ecosystem with higher species diversity can recover more quickly when rain falls and the drought ends compared to an ecosystem that has low species diversity. While these results have been encouraging, critics have pointed out that because most experiments

We have learned that biodiversity is critical to the proper functioning of ecosystems.

contain many fewer species than actually exist in nature, we cannot be sure that the results apply to nature. It is also possible that the productivity of natural ecosystems is much more affected by variation in climatic conditions and available nutrients than by the number of species in the food web. If so, the positive results of biodiversity in experiments may be trivial in the natural world compared to the effects of climate and nutrients.

To determine whether having higher biodiversity matters in nature, we can examine whether natural sites with a higher number of species function better in terms of productivity. Researchers recently compiled data from 67 studies that measured the number of

The biodiversity of ecological communities, such as these wildflowers in Texas, is critical to the overall productivity and stability of natural ecosystems. (Dean Fikar/Shutterstock)

95

Unit-Opening Case Study

Units start with a real-world case study to help you focus your learning. Understanding how to apply unit concepts to actual situations will help you see how environmental science is grounded in your daily life.

Module 8

Unit 2 | 8 | 9 | 10 | 11 | 12 | 13 | 14

Introduction to Biodiversity

Module Format

Each unit is divided into modules that will help pace your learning so you can tackle difficult topics in manageable chunks. Each module opens with a brief introduction of the topics to be covered.

Learning Goals

A list of key ideas at the beginning of the module helps keep you focused as you read and guides your comprehension. These Learning Goals are revisited in the Module Reviews.

Learning Goals

After reading this module you should be able to

8-1 identify how biodiversity exists at different scales.

8-2 explain how biodiversity affects responses to environmental stressors.

8-3 describe how we

8-4 describe how we species living on

How does biodiversity exist at different scales?

Biodiversity exists from genes to ecosystems

As illustrated in FIGURE 8.1, biodiversity exists on four scales: genetic, species, habitat, and ecosystem. Each level of biodiversity is an important indicator of environmental health and quality.

Species Diversity

A species, as we learned in Module 0, is a group of organisms that is distinct from other groups in morphology (body form and structure), behavior, or biochemical properties. Individuals of the same species can breed and produce

Genetic diversity A measure of the genetic variation among individuals in a population.

Population bottleneck When a large population declines in number, the amount of genetic diversity carried by the surviving individuals is greatly reduced.

MODULE 8 ■ Introduction to Biodiversity 97

Module 8 AP® Review

Preparing for the AP® Exam

Learning Goals Revisited

8-1 How does biodiversity exist at different scales?

Biodiversity exists as genetic diversity, species diversity, habitat diversity, and ecosystem diversity. All four scales of biodiversity contribute to the overall biodiversity of the planet.

8-2 How does biodiversity affect responses to environmental stressors?

The various scales of biodiversity have impacts on how populations and ecosystems respond to biotic or abiotic environmental stressors. High genetic diversity benefits the long-term persistence of populations because they are better able to respond to environmental change. High habitat diversity favors the persistence of habitat specialists, which have very narrow habitat requirements. High species diversity leads to increased primary productivity and increased stability in ecosystems.

8-3 How can we calculate biodiversity?

Biodiversity can be quantified in terms of species richness and evenness.

8-4 How can we estimate the number of species living on Earth?

Scientists are not able to determine the exact number of species on Earth. Most estimates agree on approximately 10 million species, of which we have identified approximately 2 million.

Running Glossary

Knowing and understanding the language of environmental science is critical for success on the AP® Exam. Key terms appear in bold type in the text, are defined at the bottom of the page, and can be found in the Glossary/Glosario at the end of the book.

Emphasize the Big Ideas and build essential skills with all the Science Practices

Use these features spiraled throughout the text to help you hone the skills that are central to the study of environmental science.

Module 0

What Is Environmental Science?

0-1 What is environmental science and why is it important?

Environmental science is important if you live on this planet

Stop reading for a moment and look up to observe your surroundings. Consider the air you breathe, the heating or cooling system that keeps you at a comfortable temperature, and the natural or artificial light that helps you see. Our **environment** is the sum of all the conditions surrounding us that influence life. These conditions include living organisms as well as nonliving components such as soil, temperature, and water. The influence of humans is an important

Learning Goals

After reading this module you should be able to

- **0-1** describe the field of environmental science and discuss its importance.
- **0-2** identify ways in which humans have altered and continue to alter our environment.
- **0-3** explain the four "Big Ideas" in environmental science.
- **0-4** describe the scientific method and justify how it is used to design and evaluate information in environmental science.

Module 0

What is Environmental Science all about? Module 0 offers context for what to expect in this fascinating and interconnected course, while introducing the Big Ideas that will come up time and time again throughout your learning.

Practice your science skills

1. **Text Analysis:** Identify one ecological claim about the benefit of biodiversity made by the author.
2. **Concept Explanation:** Explain why an increase in ecosystem biodiversity will also increase ecosystem resilience.
3. **Scientific Experiments:** Identify a research method used to determine the importance of biodiversity in nature.

As we have just learned, biodiversity is an important part of any ecosystem because of the role it plays in making ecosystems productive and stable over time. In this unit, we will examine biodiversity in depth, including biodiversity from the level of genes to ecosystems. We will also see how this diversity allows ecosystems to provide us with a number of important services. Then we explore how the number of species in an ecosystem is the net result of some species colonizing places they can tolerate and other species going extinct due to unsuitable biotic or abiotic conditions. Given that many ecosystems experience various disruptions over time, we examine how short- and long-term disruptions can vary in magnitude and alter the composition of species in different ecosystems. Whether an ecosystem is frequently disrupted or not, species can evolve traits to persist in a given ecosystem as a result of natural selection. Over long periods of time, the initial species that colonize an area are eventually replaced by other species that possess traits that are better suited to the changing environmental conditions. Collectively, the modules in this unit will provide an excellent discussion of how we obtain biodiversity in different ecosystems and the consequences of doing so.

Unit-Opening "Practice your science skills"

After reading the unit-opening case studies, try answering these thoughtful questions that test your understanding and assess your ability to draw upon the Science Practices; particularly **Text Analysis** and **Scientific Experiments**.

Mathematical Routines

It is a challenge for students to develop mastery of the **Mathematical Routines** science practice. Spend time with the "Do the Math" and "Practice Math and Graphing" features to gain confidence in answering math problems on the exam.

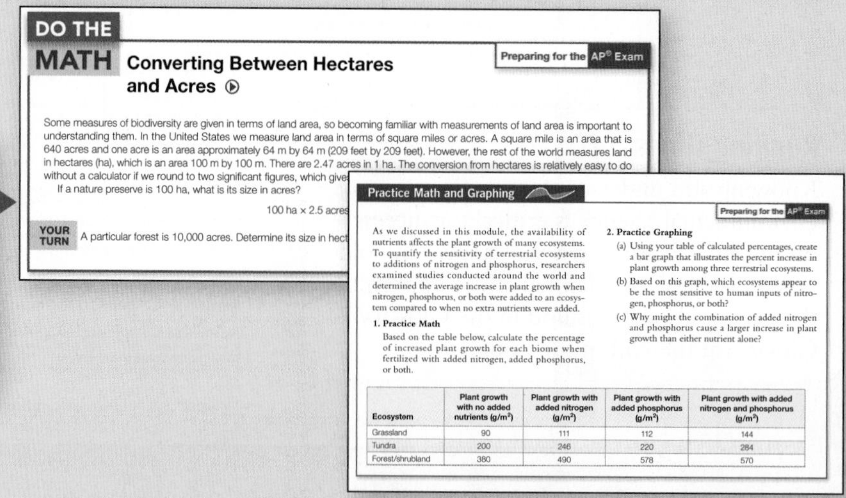

DO THE MATH **Converting Between Hectares and Acres**

Preparing for the AP® Exam

Some measures of biodiversity are given in terms of land area, so becoming familiar with measurements of land area is important to understanding them. In the United States we measure land area in terms of square miles or acres. A square mile is an area that is 640 acres and one acre is an area approximately 64 m by 64 m (209 feet by 209 feet). However, the rest of the world measures land in hectares (ha), which is an area 100 m by 100 m. There are 2.47 acres in 1 ha. The conversion from hectares is relatively easy to do without a calculator if we round to two significant figures, which give

If a nature preserve is 100 ha, what is its size in acres?

100 ha × 2.5 acres

YOUR TURN A particular forest is 10,000 acres. Determine its size in hect

Practice Math and Graphing

Preparing for the AP® Exam

As we discussed in this module, the availability of nutrients affects the plant growth of many ecosystems. To quantify the sensitivity of terrestrial ecosystems to additions of nitrogen and phosphorus, researchers examined studies conducted around the world and determined the average increase in plant growth when nitrogen, phosphorus, or both were added to an ecosystem compared to when no extra nutrients were added.

1. **Practice Math**

   Based on the table below, calculate the percentage of increased plant growth for each biome when fertilized with added nitrogen, added phosphorus, or both.

2. **Practice Graphing**
   - (a) Using your table of calculated percentages, create a bar graph that illustrates the percent increase in plant growth among three terrestrial ecosystems.
   - (b) Based on this graph, which ecosystems appear to be the most sensitive to human inputs of nitrogen, phosphorus, or both?
   - (c) Why might the combination of added nitrogen and phosphorus cause a larger increase in plant growth than either nutrient alone?

Ecosystem	Plant growth with no added nutrients (g/m^2)	Plant growth with added nitrogen (g/m^2)	Plant growth with added phosphorus (g/m^2)	Plant growth with added nitrogen and phosphorus (g/m^2)
Grassland	90	111	112	144
Tundra	200	246	220	284
Forest/shrubland	380	490	578	570

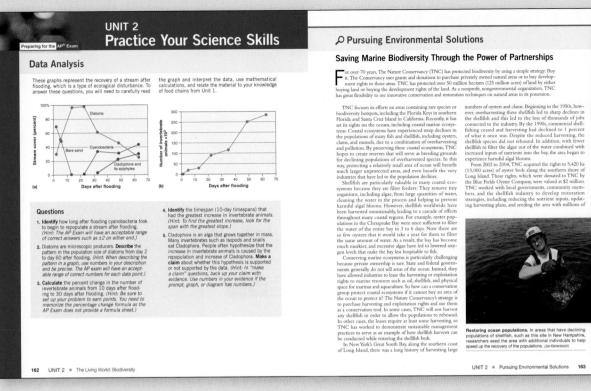

Preparing for the AP® Exam

UNIT 2
Practice Your Science Skills

Data Analysis

These graphs represent the recovery of a stream after flooding, which is a type of ecological disturbance. To answer these questions, you will need to carefully read the graph and interpret the data, use mathematical calculations, and relate the material to your knowledge of food chains from Unit 1.

Stream cover (percent)
100%
80
60
40
20
0
Diatoms
Bare sand
Cyanobacteria
Cladophora and its epiphytes
0 10 20 30 40 50 60 70
Days after flooding
(a)

Number of invertebrate animals ×10^3
300
250
200
150
100
50
0
0 10 20 30 40 50 60 70
Days after flooding
(b)

Questions

1. **Identify** how long after flooding cyanobacteria took to begin to repopulate a stream after flooding. *(Hint: The AP Exam will have an acceptable range of correct answers such as ±2 on either end.)*
2. Diatoms are microscopic producers. **Describe** the pattern in the population size of diatoms from day 2 to day 60 after flooding. *(Hint: When describing the pattern in a graph, use numbers in your description and be precise. The AP exam will have an acceptable range of correct numbers for each data point.)*
3. **Calculate** the percent change in the number of invertebrate animals from 10 days after flooding to 30 days after flooding. *(Hint: Be sure to set up your problem to earn points. You need to memorize the percentage change formula as the AP Exam does not provide a formula sheet.)*
4. **Identify** the timespan (10-day timespans) that had the greatest increase in invertebrate animals. *(Hint: To find the greatest increase, look for the span with the greatest slope.)*
5. Cladophora is an alga that grows together in mass. Many invertebrates such as isopods and snails eat Cladophora. People often hypothesize that the increase in invertebrate animals is caused by the repopulation and increase of Cladophora. **Make a claim** about whether this hypothesis is supported or not supported by the data. *(Hint: In "make a claim" questions, back up your claim with evidence. Use numbers in your evidence if the prompt, graph, or diagram has numbers.)*

162 UNIT 2 ■ The Living World: Biodiversity

Pursuing Environmental Solutions

Saving Marine Biodiversity Through the Power of Partnerships

For over 70 years, The Nature Conservancy (TNC) has protected biodiversity by using a simple strategy: Buy it. The Conservancy uses grants and donations to purchase privately owned natural areas or to buy development rights to those areas. TNC has protected over 50 million hectares (125 million acres) of land by either buying land or buying the development rights of the land. As a nonprofit, nongovernmental organization, TNC has great flexibility to use innovative conservation and restoration techniques on natural areas in its possession.

TNC focuses its efforts on areas containing rare species or biodiversity hotspots, including the Florida Keys in southern Florida and Santa Cruz Island in California. Recently, it has set its sights on the oceans, including coastal marine ecosystems. Coastal ecosystems have experienced steep declines in the populations of many fish and shellfish, including oysters, clams, and mussels, due to a combination of overharvesting and pollution. By preserving these coastal ecosystems, TNC hopes to create reserves that will serve as breeding grounds for declining populations of overharvested species. In this way, protecting a relatively small area of ocean will benefit much larger unprotected areas, and even benefit the very industries that have led to the population declines.

Shellfish are particularly valuable in many coastal ecosystems because they are filter feeders: They remove tiny organisms, including algae, from large quantities of water, cleaning the water in the process and helping to prevent harmful algal blooms. However, shellfish worldwide have been harvested unsustainably, leading to a cascade of effects throughout many coastal regions. For example, oyster populations in the Chesapeake Bay were once sufficient to filter the water of the entire bay in 3 to 6 days. Now there are so few oysters that it would take a year for them to filter the same amount of water. As a result, the bay has become much murkier, and excessive algae have led to lowered oxygen levels that make the bay less hospitable to fish.

Conserving marine ecosystems is particularly challenging because private ownership is rare. State and federal governments generally do not sell areas of the ocean. Instead, they have allowed industries to lease the harvesting or exploitation rights to marine resources such as oil, shellfish, and physical space for marinas and aquaculture. So how can a conservation group protect coastal ecosystems if it cannot buy an area of the ocean to protect it? The Nature Conservancy's strategy is to purchase harvesting and exploitation rights and use them as a conservation tool. In some cases, TNC will not harvest any shellfish in order to allow the populations to rebound. In other cases, the leases require at least some harvesting, so TNC has worked to demonstrate sustainable management practices to serve as an example of how shellfish harvests can be conducted while restoring the shellfish beds.

In New York's Great South Bay, along the southern coast of Long Island, there was a long history of harvesting large numbers of oysters and clams. Beginning in the 1950s, however, overharvesting these shellfish led to sharp declines in the shellfish and this led to the loss of thousands of jobs connected to the industry. By the 1990s, commercial shellfishing ceased and harvesting had declined to 1 percent of what it once was. Despite the reduced harvesting, the shellfish species did not rebound. In addition, with fewer shellfish to filter the algae out of the water combined with increased inputs of nutrients into the bay, the area began to experience harmful algal blooms.

From 2002 to 2004, TNC acquired the rights to 5,420 ha (13,000 acres) of oyster beds along the southern shore of Long Island. These rights, which were donated to TNC by the Blue Fields Oyster Company, were valued at $2 million. TNC worked with local governments, community members, and the shellfish industry to develop restoration strategies, including reducing the nutrient inputs, updating harvesting plans, and seeding the area with millions of

Restoring ocean populations. In areas that have declining populations of shellfish, such as this site in New Hampshire, researchers seed the area with additional individuals to help speed up the recovery of the populations. *(Joe Klementovich)*

UNIT 2 ■ Pursuing Environmental Solutions 163

Unit-Closing "Practice Your Science Skills"

At the end of each Unit, we close with another opportunity to review and test your mastery of the science practices. Here, you'll find:

- **Data Analysis**, where you describe the patterns and trends found in data sets;
- **Environmental Solutions**, where you read about real-world solutions to environmental problems, answer critical thinking questions and use evidence to describe unintended consequences of those solutions, and more;
- and **Concept Explanation**. These Science Applied features explain the concepts and processes for current issues in environmental science related to the unit's topics. You'll have an opportunity to complete a practice FRQ related to the feature's coverage.

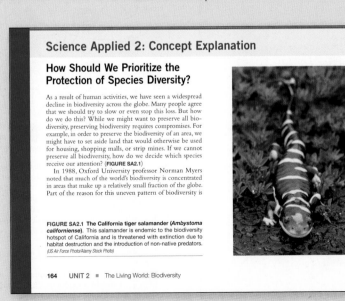

Science Applied 2: Concept Explanation

How Should We Prioritize the Protection of Species Diversity?

As a result of human activities, we have seen a widespread decline in biodiversity across the globe. Many people agree that we should try to slow or even stop this loss. But how do we do this? While we might want to preserve all biodiversity, preserving biodiversity requires compromises. For example, in order to preserve the biodiversity of an area, we might have to set aside land that would otherwise be used for housing, shopping malls, or strip mines. If we cannot preserve all biodiversity, how do we decide which species receive our attention? (**FIGURE SA2.1**)

In 1988, Oxford University professor Norman Myers noted that much of the world's biodiversity is concentrated in areas that make up a relatively small fraction of the globe. Part of the reason for this uneven pattern of biodiversity is

FIGURE SA2.1 The California tiger salamander (*Ambystoma californiense*). This salamander is endemic to the biodiversity hotspot of California and is threatened with extinction due to habitat destruction and the introduction of non-native predators. *(US Air Force Photo/Alamy Stock Photo)*

164 UNIT 2 ■ The Living World: Biodiversity

Practice for the exam from day one

This book is your ultimate study tool. Use all the AP® practice opportunities at the module, unit, and book level to give you the confidence you will need on exam day.

AP® Exam Tip

You may be asked to design an experiment on the AP® Environmental Science exam. To do this you must start with a question, and then determine the variables you would test and measure to answer the question.

- The *independent* variable is the one you manipulate or change in your experimental design.
- The *dependent* variable is the one that you would measure at the end of your experiment.

AP® Exam Tips

Use these tips, found in each module, to help you focus on key content that you should know for the exam and avoid common misconceptions.

AP® Practice Questions

Preparing for the AP® Exam

Multiple-Choice Questions

1. Which of the following is a major factor in defining a biome?
   (a) proximity to the ocean
   (b) abundance of prey
   (c) longitude position
   (d) latitude position
2. In which biome does permafrost play a factor in growth of vegetation?
   (a) tundra
   (b) taiga
   (c) temperate grassland
   (d) deciduous forest
3. In which biome is plant growth primarily constrained by precipitation?
   (a) taiga
   (b) temperate seasonal forest
   (c) temperate grassland
   (d) tropical rainforest
4. Which biome has the highest soil nutrient levels?
   (a) tropical rainforest
   (b) taiga
   (c) shrubland
   (d) temperate seasonal forest
5. Scientists predict that the planet will continue to experience warmer temperatures over the next century. Given this prediction, which shift in biome boundary seems likely?
   (a) Alpine tundra biomes will expand to include a larger area of mountain tops.
   (b) Taiga biomes will shift toward the poles.
   (c) Tropical rainforest biomes will shift toward the equator.
   (d) Temperate rainforest biomes will shift toward the equator.

Use the data below to answer questions 6 & 7:

Rainfall (mm)	Biome A	Biome B
May	50	171
June	53	98
July	50	77

Temperature (°C)	Biome A	Biome B
May	12	28
June	14	28
July	16	

6. Given the data above, it is most l
   (a) tropical rainforest.
   (b) tundra.
   (c) hot desert.
   (d) temperate seasonal forest.

Free-Response Question

Barn owls are predators of small prey, as illustrated in the food web below.

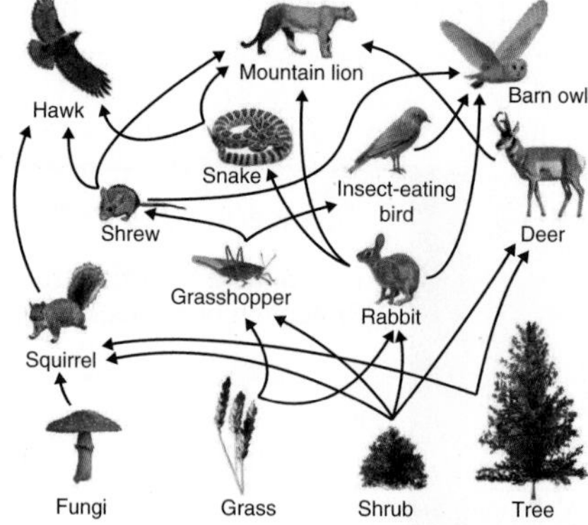

(a) Use the food web to answer the following:
   (i) **Identify** a primary producer. (1 pt.)
   (ii) **Identify** a primary consumer. (1 pt.)
   (iii) **Identify** a detritivore. (1 pt.)
   (iv) **Identify** a secondary consumer. (1 pt.)
(b) **Identify** an organism that is a decomposer. (1 pt.)
(c) **Identify** what the arrows in a food web/chain represent. (1 pt.)
(d) **Explain** how much energy moves from one trophic level to the next. (1 pt.)
(e) **Explain** how humans consuming food at a lower trophic level would impact ecosystems and human land use. (2 pts.)
(f) **Explain** how human removal of a species could impact a food web. (1 pt.)

Module AP® Practice Questions

At the end of each module, AP® Practice Questions, which are structured like the questions you'll see on the exam, help you build confidence as you sharpen your test-taking skills.

UNIT 2 Review

Key Terms to Remember

Biodiversity (p. 95)
Genetic diversity (p. 97)
Population bottleneck (p. 97)
Species diversity (p. 98)
Habitat diversity (p. 98)
Specialists (p. 98)
Generalists (p. 98)
Ecosystem diversity (p. 99)
Species richness (p. 102)
Species evenness (p. 102)
Ecosystem services (p. 105)
Provision (p. 106)
Aquaculture (p. 109)
Island biogeography (p. 113)
Species-area curve (p. 113)
Ecological tolerance (Fundamental niche) (p. 122)
Realized niche (p. 123)
Geographic range (p. 123)
Mass extinction (p. 126)
Periodic disruption (p. 129)
Episodic disruption (p. 129
Random disruption (p. 129)
Resistance (p. 130)
Resilience (p. 130)
Intermediate disturbance hypothesis (p. 134)
Evolution (p. 138)
Microevolution (p. 138)
Macroevolution (p. 138)
Evolution by artificial selection (p. 139)
Evolution by natural selection (p. 140)
Fitness (p. 141)
Adaptation (p. 141)
Evolution by random processes (p. 141)
Allopatric speciation (p. 145)
Sympatric speciation (p. 146)
Genetically modified organism (GMO) (p. 148)
Ecological succession (p. 150)
Primary succession (p. 150)
Pioneer species (p. 150)
Secondary succession (p. 151)
Climax community (p. 152)
Keystone species (p. 156)
Indicator species (p. 157)
Endemic species (p. 165)
Biodiversity hotspots (p. 165)

Unit 2 AP® Environmental Science Practice Exam Preparing for the AP® Exam

Section 1: Multiple-Choice Questions

1. After a severe drought, the productivity in an ecosystem took many years to return to pre-drought conditions. This observation indicates that the ecosystem has
(a) high resilience.
(b) low resilience.
(c) high resistance.
(d) low resistance.

2. Intertidal communities are frequently disturbed by storms that generate large waves. Following a period of intermediate wave disturbance, a group of researchers examined the species richness of north and south Pacific intertidal communities. Several weeks after the disturbance, they found that northern communities had returned to their original state whereas southern communities were still recovering. This result suggests that northern intertidal communities
(a) have greater resilience.
(b) have greater resistance.
(c) follow predictions of the intermediate disturbance hypothesis.
(d) experience more frequent disturbance.

Questions 3–5 refer to the following table:

Researchers in North America collected data on tree abundance for a group of forest communities in the three states shown in the table below.

Total percent state-wide cover

Species	Connecticut	Pennsylvania	Georgia
Red maple	25	40	20
Black oak	20	20	40
White pine	15	10	10
Eastern hemlock	20	10	0
Black cherry	20	20	30

3. According to the data, in which state(s) would the lowest percentage of white pine trees be found?
(a) Connecticut and Georgia
(b) Pennsylvania and Georgia
(c) Pennsylvania
(d) Georgia

168 UNIT 2 ▪ The Living World: Biodiversity

4. Which statement best describes the forest communities within each state represented in the table?
(a) Pennsylvania has the highest species richness; Connecticut has the highest species evenness.
(b) Connecticut and Pennsylvania have equal species richness; Georgia has the highest species evenness.
(c) Georgia and Pennsylvania have equal species richness; Georgia has the highest species evenness.
(d) Connecticut and Pennsylvania have equal species richness; Connecticut has the highest species evenness.

5. According to the data in the table, in which state would the forest communities be least vulnerable to environmental disturbances?
(a) Georgia, because it has the highest species richness, which increases the resistance of an ecosystem
(b) Connecticut, because it has the highest species evenness, which increases the resistance of an ecosystem
(c) Pennsylvania, because it has the highest species evenness, which increases the resistance of an ecosystem
(d) Connecticut, because it has the highest species richness, which increases the resistance of an ecosystem

6. A dramatic decline in genetic diversity could best be explained by which of the following?
(a) genetic drift, because genotypes are lost quickly within large populations
(b) the founder effect, because species need to migrate to other areas and are always able to do so quickly
(c) mutations, because random changes in genotypes reduces genetic diversity
(d) the bottleneck effect, because the number of individuals in the gene pool are rapidly reduced

For questions 7–9, use the following answer choices. Choices may be used once, more than once, or not at all.

(a) allopatric speciation
(b) sympatric speciation
(c) fundamental niche
(d) realized niche

7. Two fish species in a single lake evolve from a single ancestor that once lived in the lake. This is a description of a(n) ________.

8. The range of abiotic and biotic conditions under which a species actually lives is the definition of a(n) ________.

9. A stream divides a single turtle lated populations that evolv describes a(n)________.

Use the photo below to answer questions 10 & 11:

(Cavan Images/Alamy Stock Photo)

10. Which sequence of secondary succession would be likely to occur after the disturbance shown in the photo above?
(a) bare soil, lichens, mosses, grasses, deciduous trees
(b) bare rock, lichens, mosses, grasses, shrubs, mixed shade- and sunlight-tolerant trees
(c) bare soil, grasses and wildflowers, shrubs, sunlight-tolerant trees, shade-tolerant trees
(d) bare rock, grasses and wildflowers, lichens, mosses, shrubs, coniferous trees

11. Which statement about ecological succession is correct?
(a) Secondary succession begins in a community lacking soil.
(b) Terrestrial succession is not influenced by competition for limiting resources such as available soil moisture, sunlight, and nutrients.
(c) In forest succession, more shade-tolerant trees replace less shade-tolerant trees.
(d) Forest fires and hurricanes lead to primary succession because a soil base still exists.

12. The theory of island biogeography suggests that species richness is increased by which of the following factors?
(a) larger islands, because there are more available niches
(b) smaller islands, because there are more available resources
(c) larger islands, because there is increased competition for space
(d) smaller islands, because there are more niche specialists

Unit AP® Practice Exams

At the end of each unit, a comprehensive set of AP® Practice Exam questions address content across the unit. These multiple-choice and free-response questions match the style and scope you can expect to see on the exam in May.

Cumulative AP® Practice Exam

A complete practice exam at the end of the book matches the format of the actual AP® Environmental Science Exam.

Cumulative AP® Environmental Science Practice Exam

Section 1: Multiple-Choice Questions

1. Which biome has the greatest amount of nutrients in the soil?
(a) tropical rainforest
(b) temperate seasonal forest
(c) taiga
(d) tundra

2. Suppose there is an ecosystem with several plant species and several generalist and specialist consumers that eat those plants. Which of the following best describes what will most likely happen to the ecosystem following a severe drought that kills most of the plant species?
(a) The specialist species will adapt to consume other plants that they did not originally consume.
(b) The generalist species in the area will be advantaged over the specialist species and outcompete the specialists.
(c) The specialist species will be advantaged over the generalist species and outcompete the generalists.
(d) The generalist species will begin to prey on the specialist species as a way to adapt and survive.

4. The graphic below depicts the ways in which phosphorus cycles through the global ecosystem. Which arrow is not part of the phosphorus cycle?

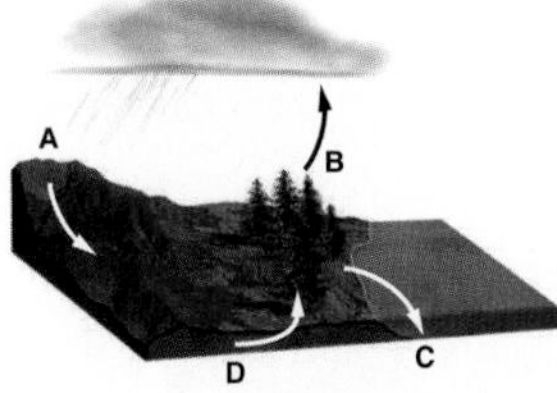

(a) A
(b) B
(c) C
(d) D

3. The four letters on the map below represent four distinct biomes. Which biome is most likely to develop the thickest soil O-horizon?

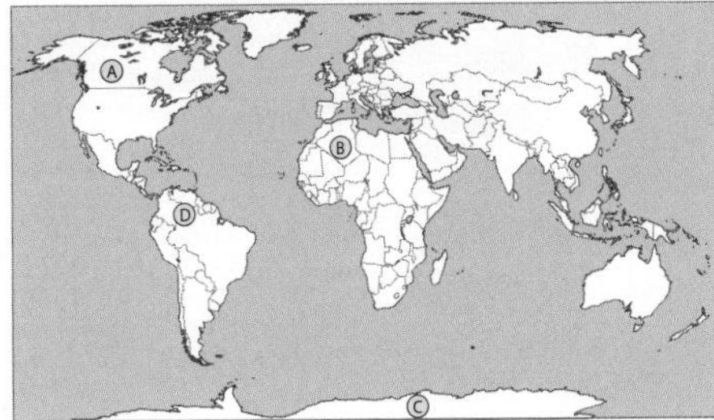

(a) biome A
(b) biome B
(c) biome C
(d) biome D

Cumulative AP® Environmental Science Practice Exam Exam-1

Use the stunning visuals to bring Environmental Science to life

The figures, photographs, graphs, and brand new two-page visual representation features will help you understand and remember the Big Ideas and important concepts that will be on the exam.

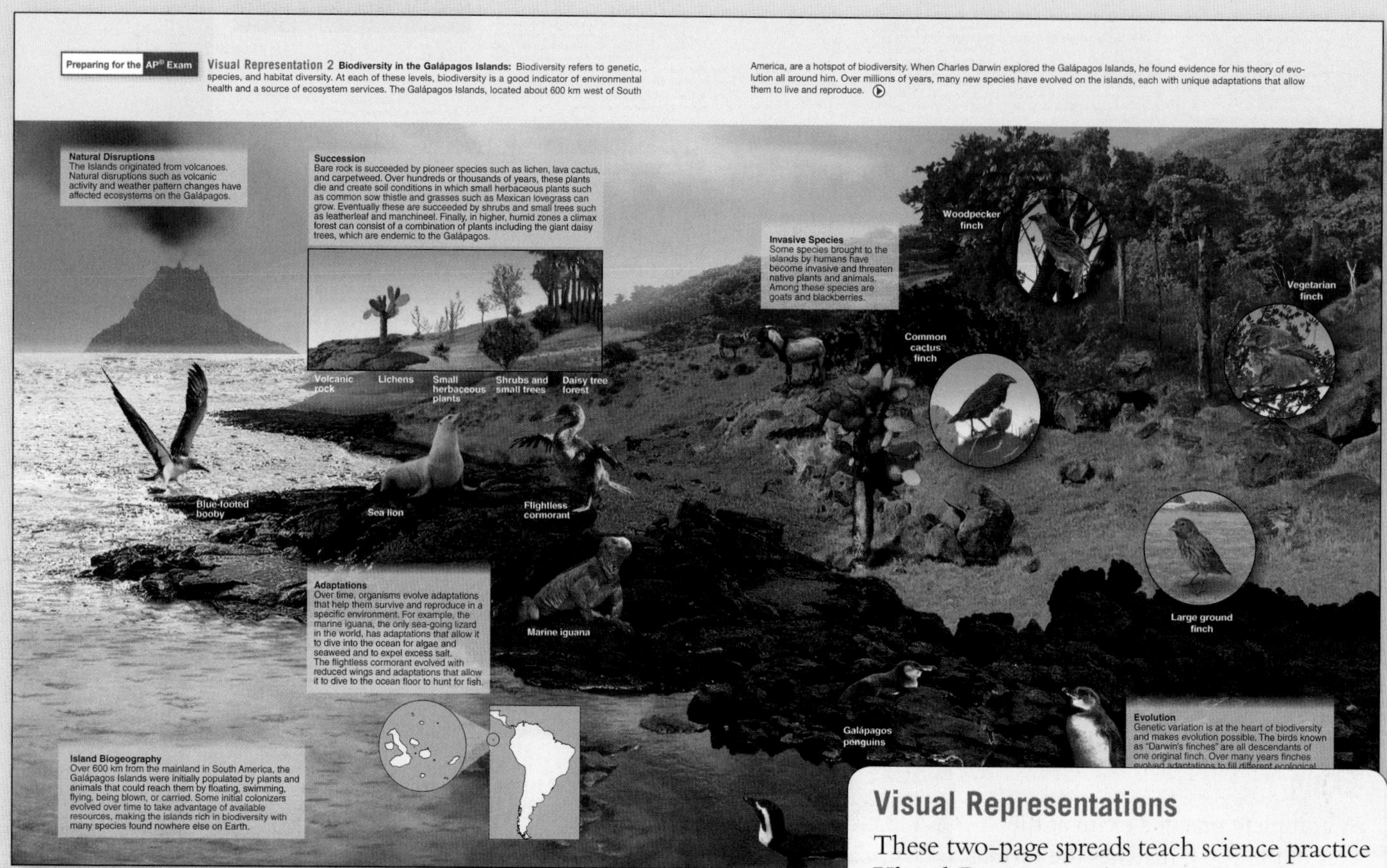

Preparing for the AP® Exam
Visual Representation 2 Biodiversity in the Galápagos Islands: Biodiversity refers to genetic, species, and habitat diversity. At each of these levels, biodiversity is a good indicator of environmental health and a source of ecosystem services. The Galápagos Islands, located about 600 km west of South America, are a hotspot of biodiversity. When Charles Darwin explored the Galápagos Islands, he found evidence for his theory of evolution all around him. Over millions of years, many new species have evolved on the islands, each with unique adaptations that allow them to live and reproduce.
Natural Disruptions
The Islands originated from volcanoes. Natural disruptions such as volcanic activity and weather pattern changes have affected ecosystems on the Galápagos.
Succession
Bare rock is succeeded by pioneer species such as lichen, lava cactus, and carpetweed. Over hundreds or thousands of years, these plants die and create soil conditions in which small herbaceous plants such as common sow thistle and grasses such as Mexican lovegrass can grow. Eventually these are succeeded by shrubs and small trees such as leatherleaf and manchineel. Finally, in higher, humid zones a climax forest can consist of a combination of plants including the giant daisy trees, which are endemic to the Galápagos.
Volcanic rock
Lichens
Small herbaceous plants
Shrubs and small trees
Daisy tree forest
Invasive Species
Some species brought to the islands by humans have become invasive and threaten native plants and animals. Among these species are goats and blackberries.
Woodpecker finch
Vegetarian finch
Common cactus finch
Blue-footed booby
Sea lion
Flightless cormorant
Adaptations
Over time, organisms evolve adaptations that help them survive and reproduce in a specific environment. For example, the marine iguana, the only sea-going lizard in the world, has adaptations that allow it to dive into the ocean for algae and seaweed and to expel excess salt. The flightless cormorant evolved with reduced wings and adaptations that allow it to dive to the ocean floor to hunt for fish.
Marine iguana
Large ground finch
Galápagos penguins
Evolution
Genetic variation is at the heart of biodiversity and makes evolution possible. The birds known as "Darwin's finches" are all descendants of one original finch. Over many years finches
Island Biogeography
Over 600 km from the mainland in South America, the Galápagos Islands were initially populated by plants and animals that could reach them by floating, swimming, flying, being blown, or carried. Some initial colonizers evolved over time to take advantage of available resources, making the islands rich in biodiversity with many species found nowhere else on Earth.

Visual Representations

These two-page spreads teach science practice **Visual Representations** and show the interconnectedness of the topics within each unit.

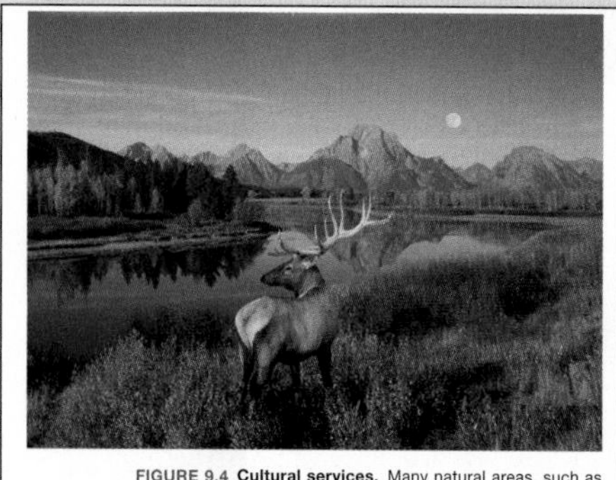

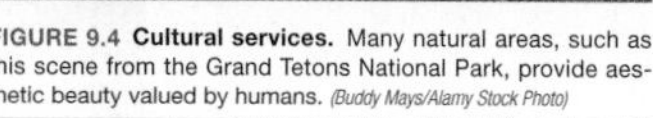

FIGURE 9.4 **Cultural services.** Many natural areas, such as this scene from the Grand Tetons National Park, provide aesthetic beauty valued by humans. *(Buddy Mays/Alamy Stock Photo)*

Exposed rocks
Lichens and mosses
Annual wild-flowers
Perennial wildflowers and grasses
Shrubs
Aspen, cherry, and young pine forest
Beech and maple broadleaf forest
Time

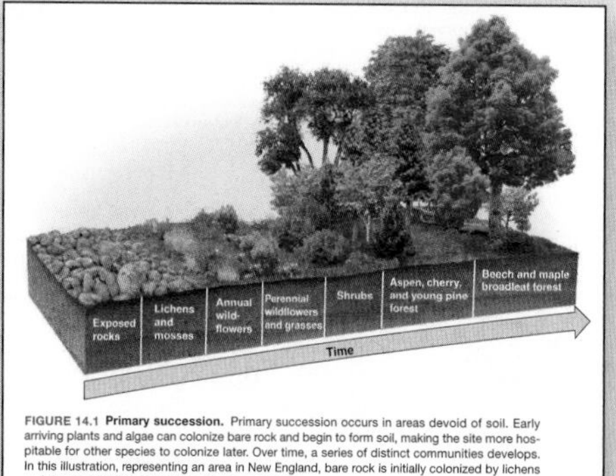

FIGURE 14.1 **Primary succession.** Primary succession occurs in areas devoid of soil. Early arriving plants and algae can colonize bare rock and begin to form soil, making the site more hospitable for other species to colonize later. Over time, a series of distinct communities develops. In this illustration, representing an area in New England, bare rock is initially colonized by lichens and mosses and later by grasses, shrubs, and trees.

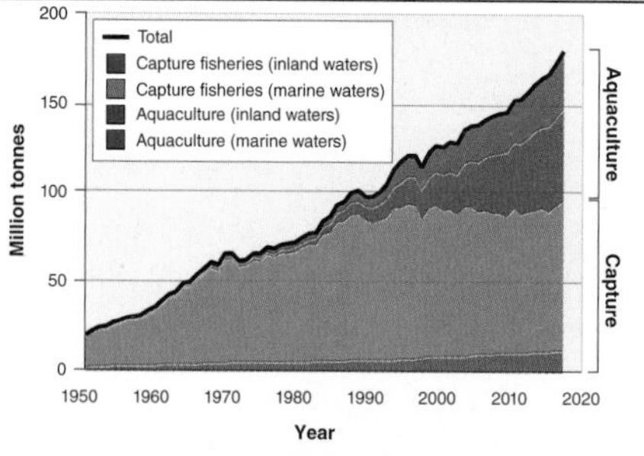

Total
Capture fisheries (inland waters)
Capture fisheries (marine waters)
Aquaculture (inland waters)
Aquaculture (marine waters)
Million tonnes
200
150
100
50
0
1950
1960
1970
1980
1990
2000
2010
2020
Year
Aquaculture
Capture

FIGURE 9.6 **Fish and shellfish production over six decades.** The amount of fish and shellfish captured from inland and marine waters has slowly increased since 1990 while the amount grown using aquaculture has grown rapidly. *(Data from http://www.fao.org/state-of-fisheries-aquaculture.)*

Photos, Illustrations, and Graphs

The visuals in this text have been carefully chosen and drawn to help you comprehend key concepts and practice analyzing data that are presented in a variety of formats.

Access everything you need for this course online

Our digital platform includes all of the resources you need in one convenient place.

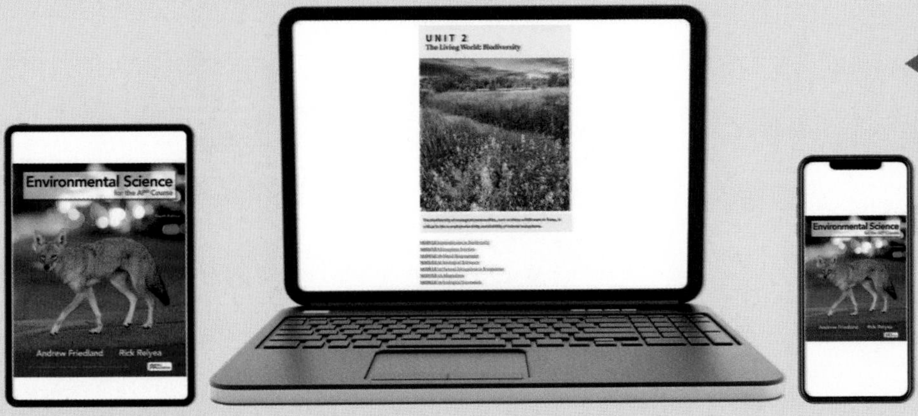

E-Book

The interactive, mobile-ready e-book allows you to read and reference the text when you are online and offline. All offline highlights and notes sync when you connect to the Internet.

Cover image: Jaymi Heimbuch/Minden Pictures; credits for the devices (l-r): Lemberg Vector studio/Shutterstock, Cigdem Simsek/Alamy Stock Photo, Alem Omerovic/EyeEm/Getty Images

Online Homework System

Created and supported by educators, every homework problem contains hints, targeted feedback, and solutions to ensure you get the help you need with course content.

Environmental Science

for the AP® Course

MODULE 0 What Is Environmental Science?

Decreased air pollution during the pandemic. This view of the India Gate monument in New Delhi during a typical day in 2019 (left) and in 2020 during the Covid pandemic (right) illustrates one positive effect of less travel and fuel use during the pandemic. *(Ruhani Kaur/Bloomberg via Getty Images; Yawar Nazir/Getty Images)*

CASE STUDY

The Global Pandemic Was a Global Environmental Science Event Too

The Covid-19 pandemic was the most important and devastating global health crisis in 100 years. It was also a profoundly important environmental science event. The pandemic was caused by SARS coronavirus disease 2019 (thus the name Covid-19), which is a zoonotic disease, meaning that it spreads from animals to humans and from humans to animals. We will examine animals and their interactions within biological systems further in Units 1 and 2. Rabies, malaria, Lyme disease, and salmonella are other zoonotic diseases, which you will encounter in Unit 8. In general, cases of zoonotic diseases have increased in recent years because the human population has grown exponentially, and greater numbers of humans have had more contact with animals that transmit zoonotic diseases. This greater contact has occurred because the growing human population has encroached upon animal habitats and has had more engagement in hunting, fishing, trapping, and consuming those animals.

Covid-19 was also an important environmental science event because it greatly affected birth rates, death rates, and life expectancy numbers for the human population. These are all important topics in environmental science that we will cover in Unit 3. In addition to the tragedy of premature deaths and illnesses affecting people of all ages from Covid-19, this pandemic has resulted in a decreased population growth rate, which is the difference between the birth rate and the death rate. Because of an increase in the death rate among individuals in many age groups, the growth rate of the human population has gone down. It also appears that as of 2022 the birth rate — the rate of births per thousand people in the population — has slowed following the pandemic. No one knows for how long this trend will continue. But for now, changes in both of these parameters have led to a slower growth rate of the human population.

During the pandemic, people consumed fewer materials and less fuel, and traveled less, all of which ended up leading to decreased emissions of carbon dioxide and other air pollutants.

Finally, although the Covid-19 virus introduced pain, suffering and death to millions of people around the world, disproportionately so to lower income populations and people of color, there was a great deal of knowledge gained from it. We learned about the role of humans on the natural environment by watching what happened when our impact was reduced. During the pandemic, people consumed fewer materials and less fuel, and traveled less, all of which ended up leading to decreased emissions of carbon dioxide and other air pollutants. Carbon dioxide is the most important greenhouse gas, and its increase in the atmosphere contributes to global climate change. So, reducing carbon dioxide emissions contributes to mitigating global climate change, which we will examine further throughout the text. Changes from the pandemic even contributed to the return of wildlife to locations that they had been driven from decades earlier. For example, many national parks and other tourist destinations received fewer visitors in 2020, resulting in a return to more natural conditions in those locations, allowing a return of wildlife and restoration of the biological systems. At Yosemite National Park, deer, bobcat, and bear were observed in areas normally crowded with human visitors and those areas have returned to more natural conditions. So, a pandemic resulting from the spread of a zoonotic disease resulted in changes in the growth rate of the human population, and it also led to reduced consumption and travel by humans, which led to changes in air pollution and wildlife patterns. These are just a few of many reasons why the Covid-19 pandemic was an important environmental science event, as well as an important global health event.

Sources: T. Rumea and S.M. Didar-Ul Islamb, Environmental effects of COVID-19 pandemic and potential strategies of sustainability, *Heliyon* 6:9 (2020): e04965; T. Mohn, The traffic tradeoff, *New York Times*, June 4, 2020.

Practice your science skills

1. **Scientific Experiments:** The Covid-19 pandemic was an important environmental science event in 2019. Identify a testable hypothesis for a scientific investigation that includes an environmental science aspect of the Covid-19 pandemic post 2020.
2. **Scientific Experiments:** Identify and describe a research method used to test the hypothesis proposed in the response to question 1.
3. **Scientific Experiments:** Explain one modification to the research method described in the answer to question 2 that might alter the results of the scientific experiment.

Humans are dependent on Earth resources such as air, water, and soil for our existence. However, we have altered the planet in many ways, both large and small, that has potentially put these resources at risk. The study of environmental science can help us understand how we have changed the planet and identify ways of responding to those changes. Four "Big Ideas" in environmental science help us put the vast amount of information in context. They contain underlying assumptions and overarching principles of this course and they relate to energy transfer, interactions between Earth systems, interactions between different species and the environment, and sustainability. Because there is so much information available to us and much of it is new, we need to use scientific inquiry and investigation and the scientific method to evaluate this information.

Module 0

What Is Environmental Science?

0-1 What is environmental science and why is it important?

Learning Goals

After reading this module you should be able to

0-1 describe the field of environmental science and discuss its importance.

0-2 identify ways in which humans have altered and continue to alter our environment.

0-3 explain the four "Big Ideas" in environmental science.

0-4 describe the scientific method and justify how it is used to design and evaluate information in environmental science.

Environmental science is important if you live on this planet

Stop reading for a moment and look up to observe your surroundings. Consider the air you breathe, the heating or cooling system that keeps you at a comfortable temperature, and the natural or artificial light that helps you see. Our **environment** is the sum of all the conditions surrounding us that influence life. These conditions include living organisms as well as nonliving components such as soil, temperature, and water. The influence of humans is an important part of the environment as well. The environment we live in determines how healthy we are, how fast we grow, how easy it is to move around, and even how much food we can obtain. One environment may be strikingly different from another—a hot, dry desert versus a cool, humid tropical rainforest, or a coral reef teeming with marine life versus a crowded city street.

We are about to begin the study of **environmental science**, the field of study that looks at interactions among human systems and those found in nature. By "system" we mean any set of interacting components that influence one another by exchanging energy, materials, or information. We have already seen that a change in one part of a system—for example, a pandemic disease in the human population such as Covid-19—can cause changes elsewhere in the system, such as in the populations of wildlife returning to locations in Yosemite National Park.

An environmental system may be completely human-made, like a subway system, or it may be natural, like weather. The scope of an environmental scientist's work can vary from looking at a small population of individuals, to multiple populations that make up a species, to a community of interacting species, or to even larger systems, such as ecosystems or the global climate system. Some environmental scientists are interested in local or regional problems, such as the impact of fewer cars on the road in a particular city during the pandemic. Other environmental scientists work on global issues, such as species extinction and climate change.

Many environmental scientists study a specific type of natural system known as an ecosystem. An **ecosystem** is a particular location on Earth with interacting **biotic**, or living, and **abiotic**, or nonliving, components.

As you embark on this new field of study, you should recognize that environmental science is different from **environmentalism**, which is a social movement that seeks to protect the environment through lobbying, activism, and education. An environmentalist is a person who participates in environmentalism. In contrast, an environmental scientist, like any scientist, follows the process of observation, hypothesis testing, and field and laboratory research. We'll learn more about the process of science later in this module and throughout this book.

So, what then does the study of environmental science actually include? As **FIGURE 0.1** shows, environmental science encompasses topics from many scientific disciplines, such as chemistry, biology, and Earth science. Environmental science is itself a subset of the broader field known as

Environment The sum of all the conditions surrounding us that influence life.

Environmental science The field of study that looks at interactions among human systems and those found in nature.

Ecosystem A particular location on Earth with interacting biotic and abiotic components.

Biotic Living.

Abiotic Nonliving.

Environmentalism A social movement that seeks to protect the environment through lobbying, activism, and education.

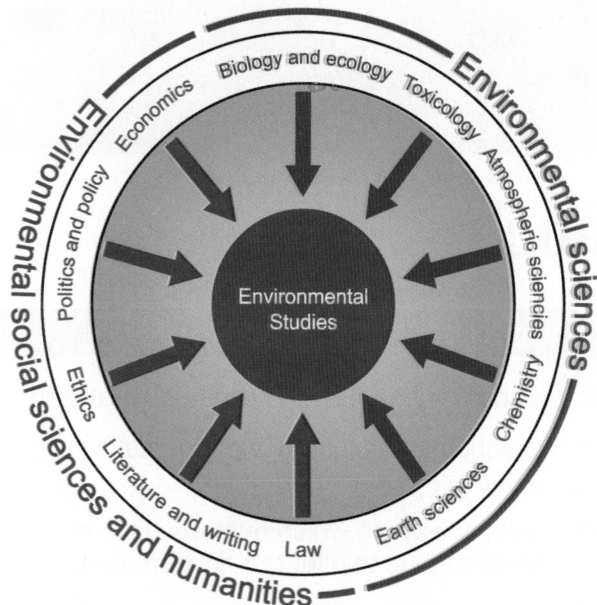

FIGURE 0.1 Environmental studies. The study of environmental science uses knowledge from many disciplines.

environmental studies, which is the field of study that includes environmental science and additional subjects such as environmental policy, economics, literature, and ethics.

We have seen that environmental science is a profoundly interdisciplinary field. It is also a rapidly growing area of study. As human activities continue to affect the environment, environmental science can help us understand the consequences of our interactions with our planet and help us to make better decisions for ourselves and our surroundings.

0-2 What are the ways in which humans have altered and continue to alter our environment?

Humans impact natural systems in many ways

Think of the last time you walked in a wooded area. Did you notice any dead or fallen trees? Chances are that even if you did, you were not aware that living and nonliving components were interacting all around you. Perhaps an insect pest killed the tree you saw and many others of the same species. Over time, dead trees in a forest lose moisture. The increase in dry wood makes the forest more vulnerable to intense wildfires. But the process doesn't stop there. Wildfires trigger the germination of certain tree seeds, some of which lie dormant until after a fire. And so, what began with the activity of insects leads to a transformation of the forest. In this way, living factors interact with nonliving factors to influence the future of the forest. All of these factors are part of a system.

Environmental studies The field of study that includes environmental science and additional subjects such as environmental policy, economics, literature, and ethics.

Systems vary in size and a large system may contain many smaller systems within it. **FIGURE 0.2** shows an example of complex, interconnecting systems that operate at multiple space and time scales: the fisheries of the North Atlantic.

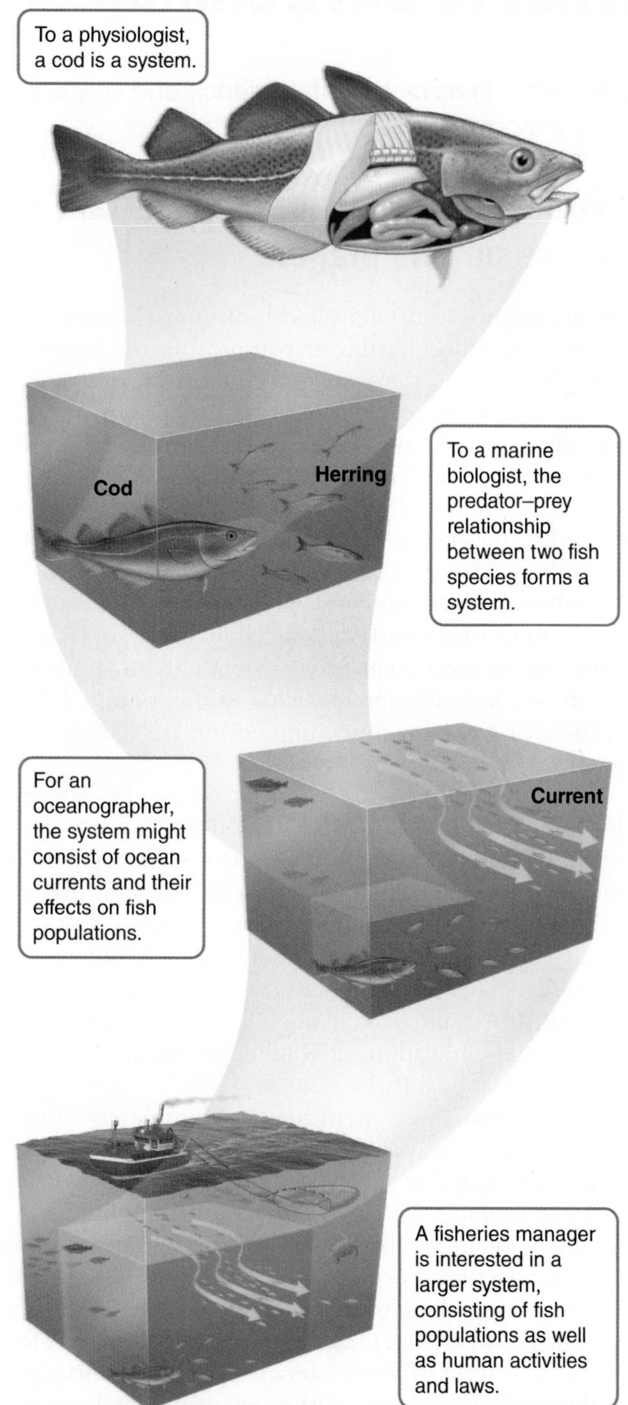

To a physiologist, a cod is a system.
Cod
Herring
To a marine biologist, the predator–prey relationship between two fish species forms a system.
For an oceanographer, the system might consist of ocean currents and their effects on fish populations.
Current
A fisheries manager is interested in a larger system, consisting of fish populations as well as human activities and laws.

FIGURE 0.2 Systems within systems. The boundaries of an environmental system may be defined by the researcher's point of view. Physiologists, marine biologists, oceanographers, and fisheries managers would all describe the North Atlantic Ocean fisheries system differently.

A physiologist who wants to study how codfish survive in the cold waters of the North Atlantic must consider all the biological adaptations of the cod that enable it to be part of one system. In this case, the fish and its internal organs are the system being studied. In the same environment, a marine biologist might study the predator–prey relationship between cod and herring. That relationship constitutes another system, which includes two fish species and the environment they live in. At an even larger scale, an environmental scientist might examine a system that includes all of these systems as well as people, fishing technology, policy, and law. The global environment is composed of both small-scale and large-scale systems.

Humans manipulate the systems in their environment more profoundly than any other species. We convert land from its natural state into urban, suburban, and agricultural areas. We change the chemistry of our air, water, and soil, both intentionally—for example, by adding fertilizers—and unintentionally—for example, by our activities that generate pollution. Even where we don't manipulate the environment directly, the simple fact that there are so many of us affects our surroundings.

Humans and our direct ancestors (other members of the genus *Homo*) have lived on Earth for about 2.5 million years. During this time, and especially during the last 10,000 to 20,000 years, we have shaped and influenced our environment. As tool-using, social animals, we have continued to develop a capacity to directly alter our environment in substantial ways. *Homo sapiens*—genetically modern humans, or you and me—evolved to be successful hunters; historically, when we entered a new environment we often hunted large animal species to extinction. In fact, early humans are thought to be responsible for the extinction of mammoths, mastodons, giant ground sloths, and many types of birds. Roughly a century ago, hunting in North America led to the extinction of the passenger pigeon (*Ectopistes migratorius*) and nearly caused the loss of the American bison (*Bison bison*) **FIGURE 0.3**.

But the picture isn't all bleak. Human activities and manipulation of the natural environment have also created opportunities for certain species to thrive. For example, for thousands of years Native Americans on the Great Plains used fire to capture animals for food. The fires they set kept trees from encroaching on the plains, which in turn created an opportunity for an entire ecosystem to develop. Because of human activity, this ecosystem—the tallgrass prairie—is now home to numerous unique species. And in China, the Giant Panda (*Ailuropoda melanoleuca*) is no longer endangered. While it is now considered vulnerable, this is an improvement due to active human efforts to expand habitat critical to the panda.

During the last two centuries, the rapid and widespread development of technology, coupled with dramatic human population growth, has substantially increased both the rate and the scale of our global environmental impact. Modern cities with electricity, running water, sewer systems, Internet connections, and public transportation systems have improved human well-being, but they have come at a cost. Because cities cover land that was once undisturbed by people, species that relied on the specific conditions of that undisturbed land had to adapt or relocate, or they went extinct. Coyotes are a species that have adapted well and can be found in a variety of habitats that have been modified by humans, including cities, as illustrated by the cover of this textbook! Human-induced changes in climate—for example, in patterns of temperature and precipitation—also affect the health of natural systems on a global scale. Current changes in land use and climate are rapidly outpacing the rate at which natural systems can evolve. Some species have not "kept up" and can no longer compete in the human-modified environment.

As the global population continues to grow, so does our effect on the environment. Four thousand people can live in

(a)

(b)

FIGURE 0.3 Extinct species and those on the brink. (a) The passenger pigeon was overhunted and went extinct. (b) The American bison was overhunted and on the brink of extinction but recovered. *(a: Florilegius/Alamy Stock Photo; b: Eastcott Momatiuk/Getty Images)*

a relatively small area with only minimal effects on the environment. But when roughly 4 million people live in a modern city like Los Angeles, their combined activity will cause environmental damage that will inevitably pollute the water, air, and soil as well as introduce other adverse consequences.

0-3 What are the four "Big Ideas" in Environmental Science?

The AP® environmental science course contains four "Big Ideas"

This book contains many illustrations of the big ideas in environmental science and you will encounter them many times in each unit (see the Module 0 Visual Representation feature on pages 10–11). Here we briefly describe each one.

Big Idea 1: Energy Transfer

Energy is contained within all objects and it is neither created nor destroyed, but it can change from one form to another. Understanding the conversion and transfer of energy among different components of systems (whether natural or created by humans) is integral to the study of environmental science. Whenever energy flows through systems, it becomes more and more unusable—it degrades. The consequences of this, including waste and pollution, are important in environmental science.

Big Idea 2: Interactions Between Earth Systems

Earth is one large system with many connections among its individual components. Natural systems both change over time as well as when you move from one location on Earth to another. As humans and natural factors influence these systems, they inevitably change. The change can occur in the numbers of organisms as well as in the pools of energy or nutrients. For example, carbon is an element that cycles between the atmosphere, land, and water. As part of a biogeochemical cycle, carbon exists in many different forms, including as carbon dioxide (CO_2) and organic molecules that make up the bodies of every organism. With increased human activities, there is at present substantially more carbon dioxide in the atmosphere than in the past, which contributes to global climate change. The ability to withstand change—and survive or thrive—depends on many biotic and abiotic aspects of systems that vary over space and time.

Sustainability Using Earth's resources in a way that does not jeopardize future generations from engaging in similar activities.

Big Idea 3: Interactions Between Different Species and the Environment

Species have existed on Earth for 4.5 billion years and have interacted with one another for most of that time. Humans and early human-like species have existed for at least 2.5 million years as outlined above. However, it is only in the last 10,000 years or so that human population growth and technology have begun to have an impact on Earth. Both the sheer number of people and the technology being utilized influence the type and extent of the environmental impact. Through manipulation of land and water, primarily through agriculture, and through the combustion of fossil fuels, humans have had a substantial impact on resources and terrestrial and aquatic pollution.

Big Idea 4: Sustainability

Sustainability means being able to do something now without jeopardizing the ability of future generations to engage in similar activities. As evidence of human impact on virtually all natural systems on Earth becomes more apparent, the need for sustainability becomes more and more important, thus its unifying theme throughout this course. Both efficient and wise use of resources is needed in conjunction with the expenditure of those resources to provide shelter, food, health care and other needs to humans worldwide. In fact, our survival depends on it. We must consider the social, cultural, and economic factors when attempting to achieve an equitable and sustainable allocation of resources.

Aspects of these foundational Big Ideas are present in every module and probably every page of this book. By understanding these ideas and how they influence or inform every topic in this book, you will gain a fuller appreciation of environmental science. Because there is so much information available in the field of environmental science, and because a good fraction of this information is relatively new, it is essential to assess it for its quality and integrity. We need to use scientific inquiry and investigation and the scientific method to evaluate this information. Therefore, an understanding of the scientific method is also essential for your study of environmental science.

0-4 What is the scientific method and how is it used to design and evaluate information in environmental science?

The scientific method is an important process in environmental science

During the past century, humans have learned a lot about the impact of their activities on the natural world. Scientific inquiry has provided great insights into the challenges we are facing and has suggested ways to address those challenges. For example, a hundred years ago, we did not appreciate how significantly or rapidly humans could alter the chemistry of the atmosphere by burning fossil fuels. Nor did we understand the effects of many common materials, such as lead and mercury, on human health. Much of our knowledge comes from the work of researchers who study a particular problem or situation in order to understand why it occurs and to determine how we can fix or prevent it from occurring. In this section we will look at the process scientists use to ask and answer questions about the environment.

The Scientific Method

To investigate the natural world, scientists have to be as objective and methodical as possible. They must conduct their research in such a way that other researchers can understand how their data were collected and agree on the validity of their findings. To do this, scientists follow a process known as the **scientific method**, which is an objective way to explore the natural world, draw inferences from it, and predict the outcome of certain events, processes, or changes. The scientific method is used in some form by scientists in all parts of the world and is a generally accepted way to conduct science, often called "Western" or "peer-reviewed science."

As we can see in **FIGURE 0.4**, the scientific method has a number of steps, including observing and questioning, forming hypotheses, collecting data, interpreting results, and disseminating findings.

Observing and Questioning

Scientists are constantly observing both the natural world and the world engineered by humans, and are asking questions about what they see, hear, or smell. Often their observations relate to other things that they encounter. For example, we know that air pollution causes respiratory problems and can contribute to increased mortality in humans. So let's say that scientists studying air pollution observed that there were fewer cars on the road during the pandemic. This observation may have led them to question if fewer cars driving on roads correlated with lower concentrations of certain air pollutants. Such observing and questioning is where the process of scientific research begins.

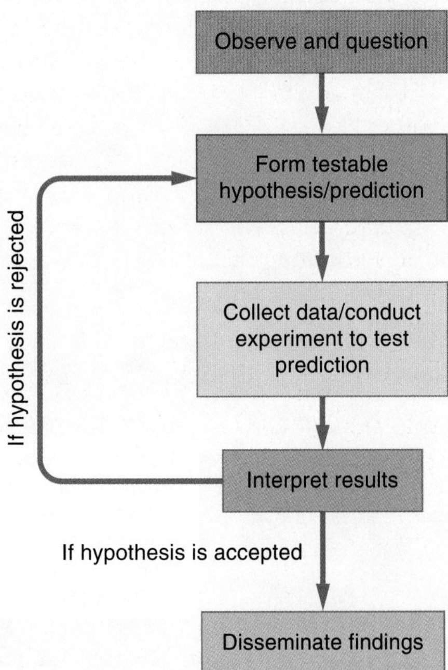

FIGURE 0.4 The scientific method. In an actual investigation, a researcher might reject a hypothesis and investigate further with a new hypothesis, several times if necessary, depending on the results of the experiment.

Forming Hypotheses and Designing the Experiment

Observing and generating questions leads a scientist to formulate a hypothesis. A **hypothesis** is a testable conjecture about how something works. It may be an idea, a proposition, a possible mechanism of interaction, or a statement about an effect. So an air pollution scientist might hypothesize that when there are fewer cars traveling the roads each day, the concentration of an air pollutant emitted by cars, say particulate matter (also known as soot), might be lower than when there are more cars on the roads each day.

What makes a hypothesis testable? We can test our idea about the relationship between the number of cars traveling and the concentration of particulate matter in the air by conducting an experiment. In this case, we would need to find two parallel highways, separated by a good distance so air pollution from one highway cannot travel to the other highway. Then we can vary the number of cars that travel on each highway and measure particulate matter adjacent to each highway. Consider this statement: "Highways that have fewer cars traveling on them will have less particulate matter in the air adjacent to them." The statement is a testable hypothesis because we can test it: It speculates that there is an interaction between something that cars release when they operate and measured concentrations of particulate matter. We can collect samples of air from along the two highways and determine the concentrations of particulate matter in the air. In a laboratory, we can directly measure air emitted from the exhaust pipe of cars, and make direct determinations that particulate matter is released from the exhaust system of cars. It also must be something that can be proven to be false. These are examples of testing hypotheses.

AP® Exam Tip

Make sure you understand the definition of a testable hypothesis *and* can also provide examples of non-testable hypotheses and explain why they aren't testable.

Scientific method An objective method to explore the natural world, draw inferences from it, and predict the outcome of certain events, processes, or changes.

Hypothesis A testable conjecture about how something works.

Preparing for the AP® Exam

Visual Representation 0 **The Four Big Ideas of Environmental Science:** Four Big Ideas shape the study of environmental science: Energy Transfer, Interactions Between Earth Systems, Interactions Between Different Species and the Environment, and Sustainability. You will return to these ideas throughout your

Sustainable solutions to energy needs include capturing geothermal energy and solar energy. Rooftop gardens help manage stormwater runoff, reduce pollution, and provide green space in urban settings. ENG STB EIN
Burning fossil fuels for transportation, heating and cooling, electricity, and manufacturing releases carbon and other pollutants into the atmosphere. ENG ERT
Urban waterways become polluted through industrial discharge, wastewater, runoff, and trash. ERT EIN
Some species such as pigeons thrive in urban areas where food is plentiful and there are few or no natural predators. EIN
Sustainable methods of transportation such as biking and walking are more possible in urban areas where people are able to live near work and school. STB

Energy Transfer (ENG): All ecological processes involve energy conversions and transfers. Energy cannot be created; as energy flows through systems it becomes less useful.

Interactions Between Earth Systems (ERT): Earth is one large, interconnected system which changes over time and space. Both biotic and abiotic components interact to influence the different systems within the biosphere.

exploration of environmental science. As your knowledge grows, you will continue to appreciate how the interplay of these Big Ideas helps us understand environmental problems and formulate solutions.

Human activity threatens many species. The California condor was nearly extinct in the 1980s, but captive breeding programs enabled it to make a comeback. STB EIN
Fires are important for nutrient cycling in some areas. Fire management policy must consider ways to protect people while allowing natural processes to occur. ERT EIN
Preserving green space in both urban and rural areas provides many benefits to wildlife, humans, and the overall health of the environment. STB EIN
Nitrogen and phosphorus runoff from fertilizer pollutes rural waterways, with adverse consequences for aquatic organisms. ERT EIN
More sustainable farming methods such as drip irrigation cause less environmental harm and improve crop productivity. ENG STB EIN
Interactions Between Different Species and the Environment (EIN): Humans have had a substantial impact on natural systems and the environment. Agriculture, technology, use of fossil fuels, and population growth have all contributed to human manipulation of natural systems.
Sustainability (STB): Solutions to problems created by human manipulation of natural systems will enable future generations to engage in similar activities. Sustainable methods involve resource management, a combination of conservation and development, and an understanding of cultural, social, and economic factors.

In each testable hypothesis, we are measuring **variables**, which are any categories, conditions, factors, or traits that differ in the natural world or in experimental situations. An **independent variable** is a variable that is not dependent on other factors that you are trying to study. In our example of cars and particulate matter, the number of cars travelling the highway each day is the independent variable. A **dependent variable** is a variable that is dependent on other factors that you are measuring. In our example, the concentration of particulate matter in the air is the dependent variable.

It is almost always easier to prove something wrong than to prove it is true beyond doubt. In this case, scientists use a null hypothesis. A **null hypothesis** is a prediction that there is no difference between the groups or conditions that are being compared. The statement "particulate matter concentrations in the air have no relationship with the number of cars that travel along the highway" is an example of a null hypothesis. So by generating a null hypothesis, scientists examine how easy or difficult it is to accept or reject the null hypothesis. If the null hypothesis is accepted, then they conclude that the pollutant, chemical, or process in question did not affect whatever factor they were measuring. They then move on to a different question or concern.

When designing experiments, the investigator must always include a **control group**, a group that experiences exactly the same conditions as the experimental group, except for the single variable under study. Sometimes in environmental science, a natural event occurs that acts as an experimental treatment in an ecosystem and is called a **natural experiment**.

Variable Any categories, conditions, factors, or traits that differ in the natural world or in experimental situations.

Independent variable A variable that is not dependent on other factors.

Dependent variable A variable that is dependent on other factors.

Null hypothesis A prediction that there is no difference between the groups or conditions that are being compared.

Control group In a scientific investigation, a group that experiences exactly the same conditions as the experimental group, except for the single variable under study.

Natural experiment A natural event that acts as an experimental treatment in an ecosystem.

Replication The data collection procedure of taking repeated measurements.

Sample size (*n*) The number of times a measurement is replicated in data collection.

Accuracy How close a measured value is to the actual or true value.

Precision How close the repeated measurements of a sample are to one another.

Uncertainty An estimate of how much a measured or calculated value differs from a true value.

Collecting Data

Scientists typically take several, repeated sets of measurements of what they believe to be the same sample—a procedure called **replication**. The number of times a measurement is replicated in data collection is the **sample size** (sometimes referred to as *n*). A sample size that is too small can cause misleading results. For example, to replicate our study on the relationship between number of cars traveling and particulate matter concentrations in the air, a scientist might look for 10 highways with lots of cars traveling on them and 10 highways with very few cars travelling on them. They would need to collect data to obtain this information. Once data are collected, they can be analyzed, which may involve summing or averaging the data, or conducting other types of numerical analyses. "Do the Math: Range of Electric Vehicles" provides a data averaging exercise.

Proper procedures—testable hypotheses, appropriate sample size, careful measurements—yield results that are accurate and precise. They also help us determine the possible relationship between our measurements or calculations and the true value. **Accuracy** refers to how close a measured value is to the actual or true value. For example, an environmental scientist might put a known concentration of particulate matter in an environmental chamber and repeatedly make measurements to see how close or accurate the measured values are to the "known" concentration. If the measured value is close to the true value, it is an accurate measurement. **Precision** is how close the repeated measurements of a sample are to one another. In the same example, if the scientist measured particulate matter in the environmental chamber five times and obtained five results that were very similar to one another, the estimates would be precise. **Uncertainty** is an estimate of how much a measured or calculated value differs from a true value. In some cases, it represents the likelihood that additional repeated measurements will fall within a certain range. Looking at **FIGURE 0.5**, we see that high accuracy and high precision is the most desirable result.

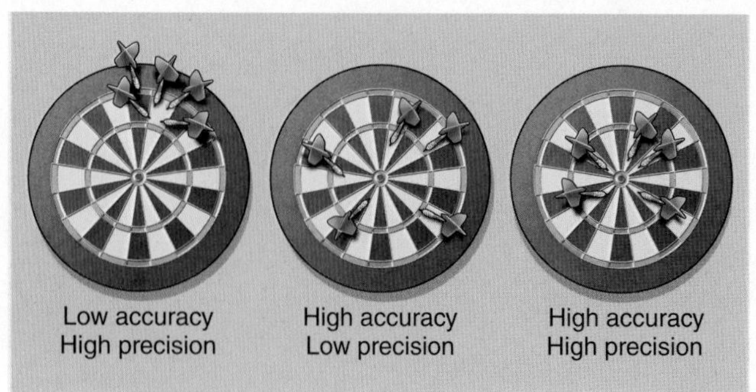

FIGURE 0.5 Accuracy and precision. Accuracy refers to how close a measured value is to the actual or true value. Precision is how close repeated measurements of the same sample are to one another.

DO THE MATH Range of Electric Vehicles

Preparing for the AP® Exam

All-electric vehicles, which run on batteries only, are becoming more common. The range that a given vehicle can travel without being charged is a function of the efficiency of the vehicle and the storage capacity of the vehicle. The mileage ranges of five common electric vehicles are: 226 miles, 249 miles, 305 miles, 238 miles, and 313 miles. What is the average mileage range of these five electric vehicles? Round to the nearest whole number.

Add the mileage ranges together and divide by the number of electric vehicles:

$$226 \text{ miles} + 249 \text{ miles} + 305 \text{ miles} + 238 \text{ miles} + 313 \text{ miles} = 1{,}331 \text{ miles}$$

$$1{,}331 \text{ miles} \div 5 \text{ vehicles} = 266 \text{ miles average range per vehicle}$$

YOUR TURN Five common hybrid-electric vehicles (which are powered by both an internal combustion engine and batteries) have mileage ranges (distance they can travel on a full tank of gasoline + a fully charged battery) as follows: 409 miles, 702 miles, 513 miles, 583 miles, and 492 miles.

What is the average mileage range of these five hybrid-electric vehicles? Round to the nearest whole number.

Interpreting Results

We have followed the steps in the scientific method from making observations and asking questions, to forming a hypothesis, to collecting data. What happens next? Once results have been obtained, analysis of data begins. A scientist may use a variety of techniques to assist with data analysis, including summaries, graphs, charts, and diagrams.

As data analysis proceeds, scientists begin to interpret their results. This process normally involves two types of reasoning: inductive and deductive. **Inductive reasoning** is the process of making general statements from specific facts or examples. If the scientist who measured particulate matter concentrations along highways made a statement about all highways, they would be using inductive reasoning. It might be reasonable to make such a statement if the highways that were sampled were representative of all highways. **Deductive reasoning** is the process of applying a general statement to specific facts or situations. For example, if we know that, in general, greater numbers of cars traveling on highways leads to higher concentrations of particulate matter in the air along highways, we might attribute high particulate matter concentrations on, say, a particular highway in Texas, to the number of cars. However, that particular attribution might be incorrect, if for example there is a factory near that highway in Texas that emitted particulate matter. Without additional observations and measurements (how close is the factory to the highway? does it emit particulate matter? does the wind blow from the factory to the highway?), and possibly experimentation, the observer would have no way of knowing the source of the particulate matter with any degree of certainty.

The most careful scientists always maintain multiple working hypotheses—that is, they entertain many possible explanations for their results. They accept or reject certain hypotheses based on what the data show or do not show. Eventually, they determine that certain explanations are the most likely, and they begin to generate conclusions based on their results.

Disseminating Findings

A hypothesis is never confirmed by a single experiment. That is why scientists not only repeat their experiments themselves, but also present papers at conferences and publish the results of their investigations. This dissemination of scientific findings allows other scientists to repeat the original experiment and verify or challenge the results. The process of science involves ongoing discussion among scientists, who frequently disagree about hypotheses, experimental conditions, results, and the interpretation of results. Two investigators may even obtain different results from similar measurements and experiments, as might happen if there was a highway near a factory and each investigator interpreted the source of particulate matter differently. Only when the same results are obtained over and over by different investigators can we begin to trust that those results are valid. In the meantime, the disagreements and discussion about contradictory findings are a valuable part of the scientific process. They help scientists refine their research to arrive at more consistent, reliable conclusions.

AP® Exam Tip

The free-response section of the AP® Environmental Science Exam regularly asks students to design an experiment that covers each step of the scientific method. Be sure you know how to do a lab write-up using the scientific method.

Inductive reasoning The process of making general statements from specific facts or examples.

Deductive reasoning The process of applying a general statement to specific facts or situations.

Like any scientist, you should always read reports of "exciting new findings" with a critical eye. Question the source of the information, consider the methods or processes that were used to obtain the information, evaluate the journal or location where the information was published, and draw your own conclusions. This process, essential to all scientific endeavors, is known as critical thinking.

A hypothesis that has been repeatedly tested and confirmed by multiple groups of researchers and has reached wide acceptance becomes a **theory**. Current theories about the role of cars along highways and particulate matter concentrations adjacent to those highways are derived from decades of research and studies. Notice that this sense of theory is different from the way we might use the term in everyday conversation (such as, "But that's just a theory!"). To be considered a theory, a hypothesis must be consistent with a large body of experimental results. A theory cannot be contradicted by any replicable tests.

Scientists work under the assumption that the world operates according to fixed, knowable laws. We accept this assumption because it has been successful in explaining a vast array of natural phenomena and continues to lead to new discoveries. When the scientific process has generated a theory that has been tested multiple times, we can call that theory a natural law. A natural law is a theory to which there are no known exceptions, and which has withstood rigorous testing. Familiar examples include the law of gravity and the laws of thermodynamics. These theories are accepted as facts by the scientific community, but they remain subject to revision if contradictory data are found.

Theory A hypothesis that has been repeatedly tested and confirmed by multiple groups of researchers and has reached wide acceptance.

First law of thermodynamics A theory with no known exception that states that energy is neither created nor destroyed but it can change from one form to another.

Second law of thermodynamics A theory with no known exception that states that when energy is transformed, the quantity of energy remains the same, but its ability to do work diminishes.

Theories Without Exceptions

Some theories have no known exceptions. Two theories concerning energy fall in this category. The laws of thermodynamics are among the most significant principles in all of science. In environmental science, they describe energy transformation from fuels to useful work, and they dictate why energy transformations result in wasted heat and pollution.

Impounded water behind a dam is quiet, still, and unmoving. It doesn't seem like it contains a lot of energy. But in fact, that water contains a great deal of potential energy. If you release that water by opening a gate in the dam, the water rushes out. The potential energy of the impounded water becomes the kinetic energy of the water rushing through the gates of the dam. This is an illustration of the **first law of thermodynamics** a theory that states that energy is neither created nor destroyed but it can change from one form to another. The **second law of thermodynamics** is a theory that states that when energy is transformed, the quantity of energy remains the same, but its ability to do work diminishes. Examples of the second law are abundant in environmental science. A full tank of gasoline in an automobile contains a certain quantity of energy. But not all of that energy in gasoline is transformed into motion of the car. Some energy is converted to waste heat and noise, because any time there is a conversion of energy from one form to another, some of that energy will be lost. We will explore these very important theories with no known exceptions many times throughout this book.

Module 0 AP® Review

Preparing for the AP® Exam

Learning Goals Revisited

0-1 What is environmental science and why is it important?

Environmental science is the study of the connections between the natural world and humans. It is important because it allows us to understand the way that humans impact the natural world.

0-2 What are the ways in which humans have altered and continue to alter our environment?

Some of the major ways humans impact the natural world include contributing to the clearing of land, extinction of animals and contributing to greenhouse gases that accelerate climate change. Studying environmental science also allows us to identify how changes in our activities might decrease our impact on Earth and increase the likelihood of achieving sustainability.

0-3 What are the four "Big Ideas" in Environmental Science?

The four Big Ideas in environmental science, which will appear throughout this book, are energy transfer, interactions between Earth systems, interactions between different species and the environment, and sustainability.

0-4 What is the scientific method and how is it used to design and evaluate information in environmental science?

The scientific method is an objective process used to explore the natural world, draw inferences from it, and predict the outcome of events, processes, or changes. The scientific method is a primary tool used in environmental science. The first and second laws of thermodynamics are important theories that have no known exceptions and relate to energy and energy transfer.

MODULE 0 Review

Key Terms to Remember

Environment (p. 5)
Environmental science (p. 5)
Ecosystem (p. 5)
Biotic (p. 5)
Abiotic (p. 5)
Environmentalism (p. 5)
Environmental studies (p. 6)
Sustainability (p. 8)
Scientific method (p. 9)
Hypothesis (p. 9)
Variable (p. 12)
Independent variable (p. 12)
Dependent variable (p. 12)
Null hypothesis (p. 12)
Control group (p. 12)
Natural experiment (p. 12)
Replication (p. 12)
Sample size (*n*) (p. 12)
Accuracy (p. 12)
Precision (p. 12)
Uncertainty (p. 12)
Inductive reasoning (p. 13)
Deductive reasoning (p. 13)
Theory (p. 14)
First law of thermodynamics (p. 14)
Second law of thermodynamics (p. 14)

UNIT 1 The Living World: Ecosystems

Grapes used for fine wines grow best in only a few regions of the world that have a similar climate. This vineyard is in Pinhão, Portugal. *(Colors Hunter - Chasseur de Couleurs/Getty Images)*

CASE STUDY

Growing Grapes to Make Fine Wines

Wine making has its origin in the Mediterranean region, in places such as Egypt, Greece, Italy, France, and Spain. As the European nations began to colonize other parts of the world, they brought grape vines with them. Today wine is made throughout the world, but the regions known for the finest wines are the Mediterranean, California, Chile, South Africa, and southwestern Australia. What is it about these regions that favors the production of great wines?

Wine making critically depends on having proper growing conditions for grapes. The best conditions are mild, moist winters and hot, dry summers. Mild winters are important because temperatures that fall below freezing can damage the grape vines. Hot, dry summers are important because they are moderately stressful for the plants, which causes them to create the perfect balance of sugars and acids in the grapes. The dry climate also reduces outbreaks of many grape vine diseases that are more prevalent in humid environments.

The five regions with the best growing conditions are all situated at northern and southern latitudes between 30° and 50° next to the ocean, and typically on the western side of continents. Their similarity in geographic position causes them to have air and water currents that produce comparable climates that provide similar growing conditions. Because of this, these regions also contain plants that look quite similar. Although the plant species in these five locations are not closely related, they consist of drought-tolerant grasses, wildflowers, and shrubs. In short, the plants that grow naturally in these areas, much like grapes, are well adapted to the local climate.

Because wine grapes grow best under a narrow range of climatic conditions, global climate change is causing great concern among winegrowers.

Because wine grapes grow best under a narrow range of climatic conditions, global climate change is causing great concern among winegrowers. Scientists recently evaluated the current and future climates in the wine region of California, where 90 percent of U.S. wine is produced. They predicted that many vineyards would have to shift northward to Oregon and Washington by 2040 because the ideal climate for wine growing will shift northward. Similarly, winemakers in France have experienced a decade of exceptionally hot and dry summers, which has made it difficult to grow the unique varieties of French wine grapes. In 2016, scientists reported that warming temperatures in France caused the date of grape harvesting to be 2 weeks earlier than the average harvest date for the past 400 years. In contrast, English wineries, which are located farther north and have not previously had ideal conditions for growing wine grapes, are now producing some of their best wines because the current English climate is starting to resemble the historic French climate. Moreover, new research predicts that temperatures in England will continue to increase by another 2°C by the year 2100. As winemakers face the reality of global warming and changing climates, they will have to decide whether to move their vineyards to more hospitable climates, modify the growing conditions by adding irrigation, or plant varieties of grapes that are more tolerant to the changing climate, and therefore produce different types of wines.

The story of wine grapes illustrates the point that different regions of the world contain distinct climates and that these climates affect the species that can live in each region. It further demonstrates that when these climates change, we can expect changes in the species that live in the regions and changes in the way humans use these ecosystems.

Sources: E. Asimov, How climate change impacts wine, *The New York Times*, October 14, 2019, at https://www.nytimes.com/interactive/2019/10/14/dining/drinks/climate-change-wine.html?searchResultPosition=1; S. Chaudhuri, Climate change uncorks British wine production, *Wall Street Journal*, December 1, 2016, at https://www.wsj.com/articles/britains-wine-production-could-be-boosted-by-climate-change-1480619748; B. Cook, et al., Climate change decouples drought from early wine grape harvests in France, *Nature Climate Change* 6 (2016):715–719; L. Hannah et al., Climate change, wine and conservation, *Proceedings of the National Academy of Sciences* 110 (2013): 6907–6912.

Practice your science skills

1. **Concept Application:** Explain how anthropogenic greenhouse gases are contributing to global climate change in the Mediterranean region.
2. **Environmental Solutions:** Make a claim by citing an environmental problem associated with adding irrigation to grow grapes in a warmer climate.
3. **Environmental Solutions:** Justify why shifting vineyards northward to Oregon and Washington by 2040 would be economically advantageous to those states.

As we have seen with wine grapes, annual patterns of temperature and precipitation help to determine the types of plants and animals that can live in aquatic and terrestrial ecosystems around the world. These annual patterns represent a region's **climate**, which is the average weather that occurs in a given region over a long period of time—typically over several decades. We can contrast climate with the concept of **weather**, which is the short-term conditions of the atmosphere in a local area that include temperature, humidity, clouds, precipitation, and wind speed. In this unit, we will examine how differences in climate around the world favor different suites of plants and animals on land and then consider the different suites of organisms that live in different types of water bodies. With this foundation, we will then explore how water, nutrients, and energy move through ecosystems.

Climate The average weather that occurs in a given region over a long period of time.

Weather The short-term conditions of the atmosphere in a local area that include temperature, humidity, clouds, precipitation, and wind speed.

Module 1

Unit 1	1	2	3	4	5	6	7

Introduction to Ecosystems

As we begin our exploration of environmental science, we need to appreciate that ecosystems are comprised of nonliving and living elements that interact, as first discussed in Module 0. The nonliving elements include abiotic conditions, such as the range of temperatures, humidity, rainfall, and nutrients that are present in different regions of the world. The living, or biotic, elements of an ecosystem include the many species of plants, animals, fungi, and bacteria that interact in different ways, which can affect the species' likelihood of obtaining resources and persisting in an ecosystem. In this module, we examine the characteristics of ecosystems, including how we determine their boundaries. We then examine the science of **community ecology**, which is the study of interactions among species. Species interact in many ways, including the struggle that species experience when there are limited resources, the consumption of one species by another, and the neutral or positive impacts that some species experience when interacting. As we will see, many species live in a **symbiosis**, which means that the two species are living in a close and long-term association with one another in an ecosystem.

Learning Goals

After reading this module you should be able to

1-1 explain how we define ecosystem boundaries.

1-2 describe how competing species respond to limited resources.

1-3 identify which species interactions involve one species consuming another species.

1-4 describe which species interactions cause neutral or positive effects on both species.

1-5 explain how invasive species represent novel species interactions.

1-1 How do we define ecosystem boundaries?

Ecosystem boundaries are defined by biotic and abiotic components or defined by humans

The characteristics of any given ecosystem are highly dependent on the climate that exists in that location on Earth. For example, ecosystems in the dry desert of Death Valley, California, where temperatures may reach 50°C (120°F), are very different from those on the continent of Antarctica, where temperatures may drop as low as −85°C (−120°F). Similarly, water can range from being immeasurable in deserts to being the defining feature of the ecosystem in lakes and oceans. On less extreme scales, small differences in precipitation and the ability of the soil to retain water can favor different terrestrial ecosystem types. Regions with greater quantities of water in the soil can support trees, whereas regions with less water in the soil can support only grasses.

The biotic and abiotic components of an ecosystem provide the boundaries that distinguish one ecosystem from another. Some ecosystems have well-defined boundaries, whereas others do not. A cave, for example, is a well-defined ecosystem (**FIGURE 1.1**). It contains identifiable biotic components, such as animals and microorganisms that are specifically adapted to live in a cave environment, as well as distinctive abiotic components, including temperature, salinity, and water that flows through the cave as an underground stream. Roosting bats fly out of the cave each night and consume insects. When the bats return to the cave and defecate, their feces provide energy that passes through the relatively few animal species that live in the cave. In many caves, for example, small invertebrate animals consume bat feces and are in turn consumed by cave salamanders.

The cave ecosystem is relatively easy to study because its boundaries are clear. With the exception of bats feeding

Community ecology The study of interactions among species.

Symbiosis Two species living in a close and long-term association with one another in an ecosystem.

FIGURE 1.1 A cave ecosystem. Cave ecosystems, such as this one in Tanzania with emerging bats, typically have distinct boundaries and are home to highly adapted species. *(Michele Menegon/ardea.com)*

outside the cave, the cave ecosystem is easily defined as everything from the point where the stream enters the cave to the point where it exits. Likewise, many aquatic ecosystems, such as lakes, ponds, and streams, are relatively easy to define because the ecosystem's boundaries correspond to the boundaries between land and water. Knowing the boundaries of an ecosystem makes it easier to identify the system's biotic and abiotic components and to trace the cycling of energy and matter through the system.

In most cases, however, determining where one ecosystem ends and another begins is difficult. For this reason, ecosystem boundaries are often subjective. Environmental scientists might define a terrestrial ecosystem as the range of a particular species of interest, such as the area where wolves roam, or they might define it by using topographic features, such as two mountain ranges enclosing a valley. The boundaries of some managed ecosystems, such as national parks, are set according to administrative rather than scientific criteria. Yellowstone National Park, for example, was once managed as its own ecosystem until scientists began to realize that many species of conservation interest, such as grizzly bears (*Ursus arctos horribilis*), spent time both inside and outside the park, despite the park's massive area of 898,000 ha (2.2 million acres). To manage these species effectively, scientists had to think much more broadly; they had to include nearly 20 million ha (50 million acres) of public and private land outside the park. This larger region, shown in **FIGURE 1.2a**, was named the Greater Yellowstone Ecosystem. As the name suggests,

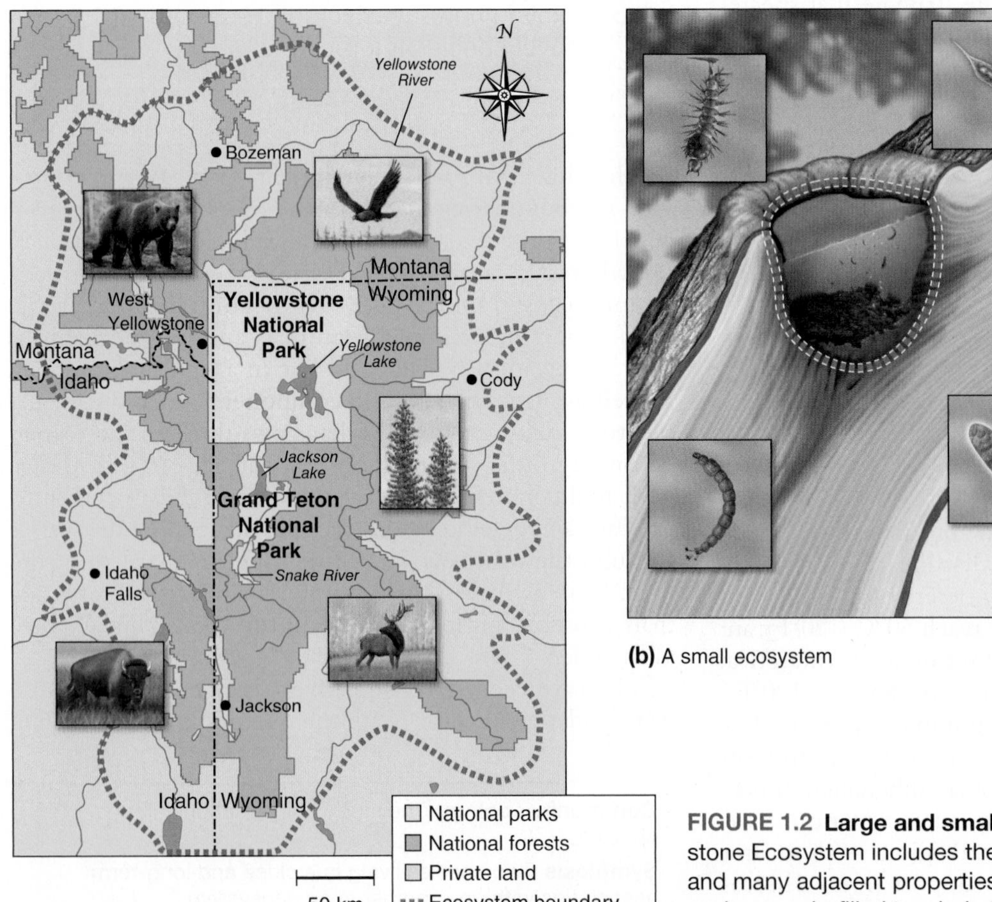

FIGURE 1.2 Large and small ecosystems. (a) The Greater Yellowstone Ecosystem includes the land within Yellowstone National Park and many adjacent properties. (b) Some ecosystems are very small, such as a rain-filled tree hole that houses a diversity of microbes and aquatic insects.

the actual ecosystem extends well beyond the administrative boundaries of the park.

Not all ecosystems are as vast as the Greater Yellowstone Ecosystem. Some can be quite small, such as a water-filled hole in a fallen tree trunk, illustrated in Figure 1.2b. Such tiny ecosystems include all the physical and chemical components necessary to support a diverse set of species, although the species are typically very small as well, including microbes, mosquito larvae, and other insects.

Although it is helpful to divide locations on Earth into distinct ecosystems, it is important to remember that each ecosystem interacts with surrounding ecosystems through the exchange of energy and matter. Organisms such as bats—which fly in and out of caves—move across ecosystem boundaries, as do chemical elements, such as carbon or nitrogen dissolved in water. As a result, changes in any one ecosystem can ultimately have far-reaching effects on the global environment.

The combination of all ecosystems on Earth forms the **biosphere**, which is the region of our planet where life resides. The biosphere is a 20-km (12-mile) thick layer around Earth between the deepest ocean bottom and the highest mountain peak.

1-2 How do competing species respond to limited resources?

Competition for limited resources between species can lead to resource partitioning

Within ecosystems, species interact in a number of different ways. One of the best-known type of species interaction is that caused by individuals competing for limited resources.

Competition

Competition is the struggle of individuals, either within or between species, to obtain a shared limited resource. In a classic experiment demonstrating competition, the Russian biologist Georgy Gause examined how different species of single-cell eukaryotes, in the genus *Paramecium* affected each other's population growth. **FIGURE 1.3** shows the results of his experiments. When the two species—*P. caudatum* and *P. aurelia*—were grown in separate laboratory cultures, each species thrived and reached a relatively high population size within 10 days. However, when the two species were grown together, *P. aurelia* continued to thrive but *P. caudatum* declined to extinction. Gause's observations, combined with additional experiments with other organisms, led researchers to formulate the **competitive exclusion principle**, which states that two species competing for the same limiting resource cannot coexist. Under a given set of

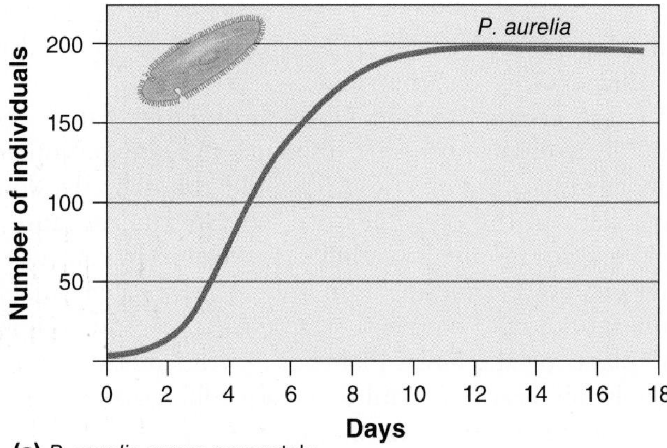

(a) *P. aurelia* grown separately

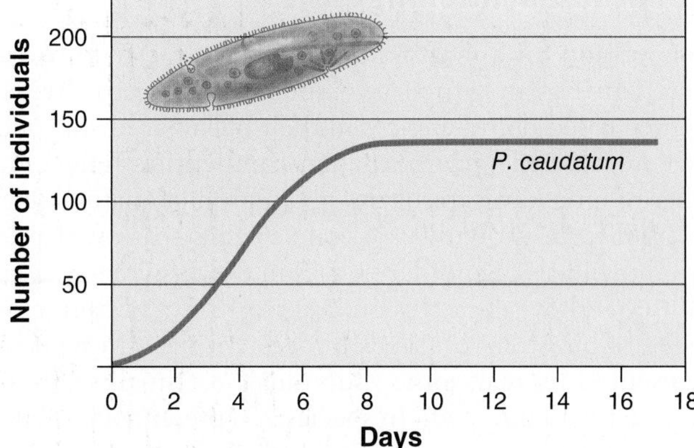

(b) *P. caudatum* grown separately

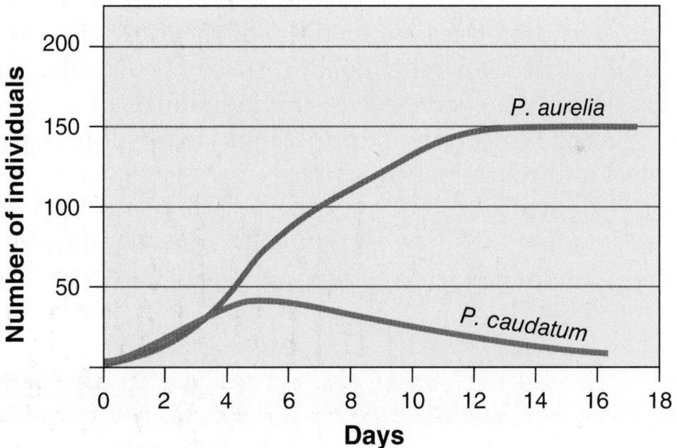

(c) *P. aurelia* and *P. caudatum* grown together

FIGURE 1.3 Competition for a limiting resource. When Gause grew two species of Paramecium separately, both achieved large population sizes. However, when the two species were grown together, *P. aurelia* continued to grow well, while *P. caudatum* declined to extinction. These experiments demonstrated that two species competing for the same limiting resource cannot coexist. *(Data from Gause, 1934.)*

Biosphere The region of our planet where life resides.

Competition The struggle of individuals, either within or between species, to obtain a shared limiting resource.

Competitive exclusion principle The principle stating that two species competing for the same limiting resource cannot coexist.

environmental conditions, when two species have the same realized niche, one species will perform better and will drive the other species to extinction.

We see competition at work throughout nature. For example, many plant species influence the distribution and abundance of other plant species. Goldenrods are the dominant plant in the old fields of New England because of their superior competitive ability; they can grow taller than other wildflowers and obtain more of the available sunlight. Similarly, the wild oat plant (*Avena fatua*) can outcompete crop plants on the Great Plains of North America because its seeds ripen earlier, permitting oat seedlings to start growing before the other species.

Resource Partitioning

Competition for a limiting resource can lead to **resource partitioning**, in which two species evolve to divide a resource based on differences in their behavior or morphology. In evolutionary terms, when competition reduces the ability of individuals to survive and reproduce, natural selection will favor individuals of one species that overlap less with individuals of another species in the resources they use. FIGURE 1.4 shows how this process works. Let's imagine two species of birds that eat seeds of different sizes. In species 1, represented by teal, some individuals eat small seeds and others eat medium seeds. In species 2, represented by yellow, some individuals eat medium seeds and others eat large seeds. As a result, some individuals of both species compete for medium seeds. This overlap is represented by green. If species 1 is the better competitor for medium seeds, then evolution will favor individuals of species 2 that specialize in eating only the large seeds; any individuals of species 2 that continue to compete for medium seeds will have poor survival and reproduction. After several generations, species 2 will evolve to feed mostly on large seeds. This process of resource partitioning results in reduced competition between the two species.

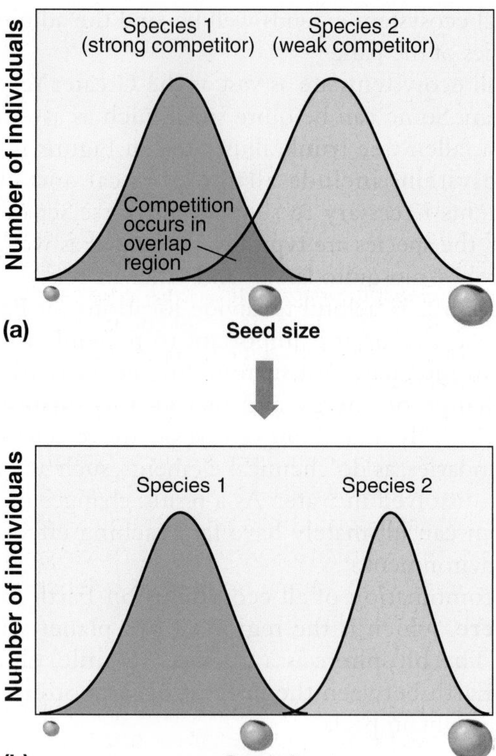

FIGURE 1.4 The evolution of resource partitioning. (a) When two species overlap in their use of a limiting resource, selection favors those individuals of each species whose use of the resource overlaps the least with that of the other species. (b) Over many generations, the two species can evolve to reduce their overlap and thereby partition their use of the limiting resource.

AP® Exam Tip

You may be asked to design an experiment on the AP® Environmental Science exam. To do this you must start with a question, and then determine the variables you would test and measure to answer the question.

- The *independent* variable is the one you manipulate or change in your experimental design.
- The *dependent* variable is the one that you would measure at the end of your experiment.

Resource partitioning When two species evolve to divide a resource based on differences in their behavior or morphology.

Species in nature reduce resource overlap in several ways. One strategy is temporal resource partitioning, a process in which two species utilize the same resource but at different times. For example, wolves and coyotes that live in the same territory are active at different times of day to reduce the overlap in the times that they are hunting for similar prey. Similarly, some plants reduce competition for pollinators by flowering at different times of the year.

If two species reduce competition by using different habitats, they are exhibiting spatial resource partitioning. For example, desert plant species have evolved a variety of different root systems that reduce competition for water and soil nutrients; some have very deep roots while others have shallow roots. For example, black grama grass (*Bouteloua eriopoda*) has shallow roots that extend over a large area to capture rainwater, whereas tarbush (*Flourensia cernua*) sends roots deep into the ground to tap deep sources of water.

A third method of reducing resource overlap is by morphological resource partitioning, which is the evolution of differences in body size or shape. For example, Charles Darwin observed morphological resource partitioning among the finches of the Galápagos Islands. While the original finch species that came to the islands from South America was a seed eater, over time this one species evolved into 14 unique species on the islands, with some species evolving beaks that were well

FIGURE 1.5 Morphological resource partitioning. On the Galápagos islands, a single species of seed-eating finch has evolved into multiple species of different finches that have different beak shapes that are well suited to particular diets. The large ground finch (*Geospiza magnirostris*) eats large seeds, the small ground finch (*G. fuliginosa*) eats small seeds, the woodpecker finch (*G. pallida*) eats insects, and the vegetarian finch (*Platyspiza crassirostris*) eats plant buds.

suited to eating large or small seeds, others evolving beaks that were well suited to eating insects, and still others that were well suited to eating plant buds, as shown in **FIGURE 1.5**.

This morphological resource partitioning allows for reduced competition among the species of finches that live on the Galápagos Islands.

1-3 Which species interactions involve one species consuming another species?

Interactions involving one species consuming another include predation, parasitism, and herbivory

Species interactions that involve one species consuming another includes predation, parasitism, and herbivory.

Predation

One of the most dramatic species interactions occurs between predators and their prey. **Predation** is an interaction in which one animal typically kills and consumes another animal. Predators include African lions that eat gazelles and great horned owls (*Bubo virginianus*) that eat small rodents. Organisms of all sizes may be predators, and their effects on their prey vary widely.

One specialized type of predator is the group known as **parasitoids**, which is a specialized type of predator that lays eggs inside other organisms—referred to as its host. When the eggs hatch, the parasitoid larvae slowly consume the host from the inside out, eventually leading to the host's death. Parasitoids include certain species of wasps and flies.

Predators can play an important role in controlling the abundance of prey. In Sweden, red foxes prey on several species of hares and grouse. In the 1970s and 1980s, the fox population was reduced by a disease called mange. With fewer foxes around, the foxes' prey increased in abundance; grouse populations doubled in size, and hare populations increased sixfold. Similarly, researchers have monitored the population of moose (*Alces alces*) and wolves (*Canis lupus*) on Isle Royale in Lake Superior since the 1950s. In the late 1970s, the wolf population experienced an outbreak of a disease caused by canine virus and the wolf population plummeted by approximately 70 percent. This reduced population resulted in much less wolf predation on the moose, which allowed the moose population to more than triple in size over the subsequent 15 years. With this rapid growth, the moose experienced intense competition among members of its own species for the plants that they eat and, as a result, the moose began to starve and their population plummeted in the mid-1990s. As the wolves continued to decline during the past decade, the growth of the plants rebounded, and the moose population has once again increased to high numbers. As you can see from these examples, predators can play a critical role in their ecosystems by regulating prey populations.

To avoid being eaten or harmed by predators, many prey species have evolved defenses. These defenses may be behavioral, morphological, or chemical, or may simply mimic another species' defense. Animal prey commonly use behavioral defenses, such as hiding and reduced movement, to attract less attention from predators. Other prey species have evolved impressive morphological defenses, as shown in **FIGURE 1.6**, including camouflage to help them hide from

Predation An interaction in which one animal typically kills and consumes another animal.

Parasitoid A specialized type of predator that lays eggs inside other organisms – referred to as its host.

(a)

(b)

FIGURE 1.6 Prey morphological defenses. Predation and herbivory has favored the evolution of fascinating defenses. (a) The camouflage of the satanic leaf-tailed geckos makes it difficult for predators to see it. (b) Sharp spines protect this porcupine from predators. *(a: Nature Picture Library/Alamy Stock Photo; b: bucky_za/Getty Images)*

(a)

(b)

FIGURE 1.7 Prey chemical defenses. (a) The poison dart frog (*Epipedobates bilinguis*) has a toxic skin. (b) This nontoxic frog (*Allobates zaparo*) mimics the appearance of the poison dart frog. *(a: Morley Read/Alamy; b: Luis Louro/Alamy Stock Photo)*

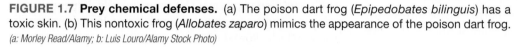

predators and spines to help deter predators. For example, the satanic leaf-tailed gecko (*Uroplatus phantasticus*) lives on the island of Madagascar and has evolved superb camouflage to match the twig leaves in its environment. Many animals, such as porcupines, stingrays, and puffer fish, have spines that deter predator attack. Similarly, many plants have evolved spines that deter herbivores from grazing on their leaves and fruits.

Chemical defenses are another common mechanism of protection from predators. Several species of insects, frogs, and plants emit chemicals that are toxic or distasteful to their predators. For example, the poison dart frog shown in **FIGURE 1.7a** produces a toxin on its skin that is lethal to many species of predators. In fact, many toxic prey are brightly colored, and predators learn to recognize and avoid consuming them. In some cases, prey have not evolved toxic defenses of their own, but instead have evolved to mimic the physical characteristics of other prey species that do possess chemical defenses (Figure 1.7b). By mimicking species that possess toxic defenses, these nontoxic species can fool predators into not attacking them.

Parasitism

Parasitism is an interaction in which one organism lives on or in another organism—referred to as the host. Because parasites typically consume only a small fraction of their host, a single parasite rarely causes the death of its host. Parasites include tapeworms that live in the intestines of animals as well as the protists that live in the bloodstream of animals and cause malaria.

Parasites that cause disease in their host are called **pathogens**. Pathogens include viruses, bacteria, fungi, protists, and wormlike organisms called helminths. These organisms cause many of the most well-known diseases in the world, ranging from the common cold to some forms of cancer.

Parasitism An interaction in which one organism lives on or in another organism, referred to as the host.

Pathogen A parasite that causes disease in its host.

Herbivory An interaction in which an animal consumes plants or algae.

Herbivory

Herbivory is an interaction in which an animal consumes plants or algae. The gazelles of the African plains are well-known herbivores, as are the various species of deer, rabbits, goats, and sheep. In aquatic systems, herbivores include sea urchins in the ocean and tadpoles in ponds that consume various species of algae. When herbivore populations increase, they can have dramatic effects on the abundance of plants and algae as well as the species composition that remains. One way to assess the effect of herbivores is by excluding herbivores from an area and seeing how the ecosystem responds. For example, when deer are excluded from an area by putting up a high fence, scientists commonly observe an extraordinary increase in plants, especially in those species of plants that deer prefer to consume (**FIGURE 1.8**). As in the case of predators and prey, many species of plants and algae have evolved defenses against herbivores that include sharp spines and distasteful chemicals.

FIGURE 1.8 Herbivory. When fences are erected to exclude herbivory by deer, plant growth typically increases dramatically, especially for species that deer prefer to eat. *(Jean-Louis Martin RGIS/CEFE-CNRS)*

FIGURE 1.9 Mutualism. Acacia trees and *Pseudomyrmex* ants have each evolved adaptations that enhance their mutualistic interaction. The trees' thorns serve as nest sites for the ants (inset) and provide them with food in the form of nectar produced in specialized nectaries (arrow). In return, the ants protect the trees from herbivores and competing plants. *(a: Oxford Scientific/Getty Images; b: Alex Wild Photography)*

1-4 Which species interactions cause neutral or positive effects on both species?

Species interactions that cause neutral or positive effects include mutualisms and commensalisms

Some species interactions can be quite beneficial to the participants. In this section, we will example two such positive interactions: mutualisms and commensalisms.

Mutualisms

One of the most ecologically important interactions is the relationship between plants and their pollinators, which include birds, bats, and insects. The plants depend on the pollinators for their reproduction, and the pollinators depend on the plants for food. In some cases, one pollinator species might visit many species of plants, while in others, a range of pollinator species may visit many plant species. In still other cases, one animal species pollinates only one plant species, and that plant is pollinated only by that animal species. For example, there are about 900 species of fig trees, and almost every species of fig is pollinated by a unique species of fig wasp.

An interaction between two species that increases the chances of survival or reproduction for both species is called a **mutualism**. Each species in a mutualistic interaction is ultimately assisting the other species in order to benefit itself. If the benefit is too small, the interaction will no longer be worth the cost of helping the other species. Under such conditions, natural selection will favor individuals that no longer engage in the mutualistic interaction.

One well-studied example of plant-animal mutualism is the interaction between acacia trees and several species of *Pseudomyrmex* ants in Central America. The acacia tree supplies the ants with food and shelter, and the ants protect the tree from herbivores and competitors. The ants eat nectar produced by the tree in special structures called nectaries, and they live in the tree's large thorns, which have soft cores that the ants can easily hollow out (**FIGURE 1.9**). The ants protect the tree by attacking anything that falls on it—for example, by stinging intruding insects that might eat the acacia's leaves and destroying vines that might shade the tree's branches. Through natural selection over generations of close association, the ants and the trees have both evolved traits that make this mutualism work. The ants have evolved several behavioral adaptations: living in the hollowed-out thorns of the tree, extensive patrolling of branches and leaves, and attacking all foreign animals and plants regardless of whether they are useful to the ants as food. For its part, the acacia tree has evolved specialized thorns, year-round leaf production, and modified leaves and nectaries that provide food for the ants.

Two other well-known mutualistic relationships are those found in coral reefs and in lichens. Coral reefs consist of tiny coral animals that build the reefs, and they provide a home to particular species of algae while the algae provide the coral with sugars produced via **photosynthesis**, which is the process of using solar energy to convert water (H_2O) and carbon dioxide (CO_2) into glucose ($C_6H_{12}O_6$) and oxygen (O_2). Their relationship is critical; if the algae die, the coral will die as well. Similarly, many of the lichens that grow on the surfaces of rocks and trees are made up of an alga, one or two

Mutualism An interaction between two species that increases the chances of survival or reproduction for both species.

Photosynthesis The process by which plants and algae use solar energy to convert carbon dioxide (CO_2) and water (H_2O) into glucose ($C_6H_{12}O_6$) and oxygen (O_2).

FIGURE 1.10 Lichens. Lichens, such as this *Hypogymnia physodes*, are composed of one or two fungi and an alga that are tightly linked together in a mutualistic interaction. The fungi provides nutrients to the alga, and the alga provides carbohydrates for the fungus via photosynthesis. *(Duncan Shaw/Science Source)*

species of fungus, and often bacteria living in close association. The fungus provides many of the nutrients the alga needs, and the alga provides carbohydrates for the fungus via photosynthesis (**FIGURE 1.10**).

Commensalisms

Sometimes species can interact in a way that benefits one species but has no effect on the other species. For example, tree branches can serve as perch locations and nest sites for birds. The birds benefit from the presence of the tree because it provides them perches to search for food and a place to raise their offspring. However, the survival and reproduction of the tree is not affected by the presence of the bird. An interaction between two species in which one species benefits and the other species is neither harmed nor helped is called **commensalism**. Commensalisms are common in nature. For example, many species of fish use coral reefs as hiding places to avoid predators. The fish receive a major benefit from the presence of the coral, but the species of coral that construct the reefs are typically neither harmed nor benefited by the presence of the fish.

Interactions among species are important in determining which species can live in an ecosystem. **TABLE 1.1** summarizes these interactions and the effects they have on each of the interacting species, whether positive (+), negative (–), or neutral (0). Competition for a limited resource has a negative effect on both competing species. In contrast, predation, parasitism, and herbivory each has a positive effect on the consumer, but a negative effect on the organism being consumed. Mutualism has positive effects on both interacting species. Commensalism has a positive effect on one species and no effect on the other species.

Commensalism An interaction between two species in which one species benefits and the other species is neither harmed nor helped.

Native species A species that lives in its historical range, typically where it has lived for thousands or millions of years.

Exotic species A species living outside its historical range. *Also known as* **alien species.**

TABLE 1.1 Interactions between species and their effects

Type of interaction	Species 1	Species 2
Competition	–	–
Predation	+	–
Parasitism	+	–
Herbivory	+	–
Mutualism	+	+
Commensalism	+	0

1-5 How do invasive species represent novel species interactions?

Invasive species represent novel species interactions due to a lack of evolutionary history

As we have discussed, it is common to observe prey species that have evolved defenses against their predators and plants that have evolved defenses against herbivores. Such evolution takes place over a very long time. As a result, species are often not defended when a new species is introduced to a region where it can act as a predator, parasitoid, parasite, or competitor. Thus, interactions among species can have very different impacts on an ecosystem when the species is native versus exotic.

Native species are species that live in their historical range, typically where they have lived for thousands or millions of years. In contrast, **exotic species**, also known as **alien species**, are species that live outside their historical range. For example, honeybees (*Apis mellifera*) were introduced to North America in the 1600s to provide a source of honey for European colonists. Red foxes, now abundant in Australia, were introduced there in the 1800s for the purpose of fox hunts, which were popular in Europe at the time. Because neither the honeybee nor the red fox was native to their new land, they are living outside their historical range as exotic species.

During the past several centuries, humans have frequently moved animals, plants, and pathogens around the world. Some species are also moved accidentally. For example, rats that stowed away in shipping containers ended up on distant oceanic islands. Because these islands did not previously have rats or other ground predators, there had never been any natural selection against nesting on the ground. As a result, numerous island bird species evolved to nest on the ground. When rats arrived in places such as Hawaii, they found the eggs and hatchlings from ground nests as an easy source of food, resulting in a high rate of extinction in ground-nesting birds. Similar accidental movements have occurred for many pathogens, including exotic fungi that were introduced to North America nearly a century ago and have since killed nearly all American elm (*Ulmus americana*) and American chestnut (*Castanea dentata*) trees in eastern North America. Similarly, an exotic protist that causes avian malaria has driven many species of Hawaiian birds to extinction.

In most cases, exotic species fail to establish successful populations when they are introduced to a new region because they are not well suited to the biotic or abiotic conditions of the new region. For the small percentage of introductions that are successful, exotic species can live in their new surroundings and have no negative effects on the native species. In other cases, however, the exotic species rapidly increase in population size and cause harmful effects on native species. An exotic species that spreads rapidly across large areas and causes harm is called an **invasive species**. The rapid spread of invasive species is possible because invasive species, which have natural enemies in their native regions that act to control their population, often have no natural enemies in the regions where they are introduced. Some of the best-known examples of invasive exotic species in North America are rats, zebra mussels (*Dreissena polymorpha*) and kudzu vines (*Pueraria lobata*), which you can see in **FIGURE 1.11**.

In this module, we have examined the boundaries of ecosystems and how species in these ecosystems interact in several different ways, such as competition, predation, herbivory, parasitism, and mutualisms. In the next module, we will examine how terrestrial species are distributed around the planet in response to different environmental conditions, including temperature and precipitation.

FIGURE 1.11 Invasive species. The kudzu vine was introduced to the United States where it was planted along roadsides and on farms to reduce erosion. Where there is abundant sunlight and no large mammals to eat it, the vine can grow rapidly and climb up trees, completely shading the trees and eventually killing them. *(morgan hill/Alamy Stock Photo)*

Invasive species A species that spreads rapidly across large areas and causes harm.

Module 1 AP® Review

Preparing for the AP® Exam

Learning Goals Revisited

1-1 How do we define ecosystem boundaries?

Ecosystem boundaries distinguish one ecosystem from another. Although boundaries can be well-defined, often they are not. Boundaries are commonly defined either by topographic features, such as mountain ranges, or are subjectively set by administrative criteria rather than biological criteria.

1-2 How do competing species respond to limited resources?

Competition is an interaction between two species that share a limiting resource. Over time, competition for a resource can cause natural selection to favor those individuals that have reduced overlap in resource use and this can lead to spatial, temporal, or morphological resource partitioning.

1-3 Which species interactions involve one species consuming another species?

Predation is an interaction in which animals partially or entirely consume another animal. Predators can affect the abundance of prey populations and cause the evolution of antipredator defenses in prey populations. Parasitism is an interaction in which one organism lives on or in another organism. Those parasites that can cause disease in their hosts are known as pathogens. Herbivory is an interaction in which animals consume plants or algae. In some cases, herbivores can have dramatic effects on plant and algal communities by removing the most palatable species.

1-4 Which species interactions cause neutral or positive effects on both species?

Mutualisms are interactions that benefit two interacting species by increasing the chances of survival or reproduction for both. One of the most common mutualisms is the interaction between flowering plants and their pollinators. A second well-known mutualism is between acacia trees and the ants that defend the trees in exchange for food and a place to live. Commensalisms are interactions in which one species benefits but the other is neither harmed nor helped. Examples include birds perching on trees and marine fish using coral reefs for protection from predators.

1-5 How do invasive species represent novel species interactions?

Given that the evolution of strategies against competitors, predators, parasites, and herbivores requires long periods of time, the introduction of exotic species can have substantial impacts on the native species in an ecosystem. While some species have been intentionally introduced, others have been accidentally introduced.

AP® Practice Questions

Preparing for the AP® Exam

Multiple-Choice Questions

1. Resource partitioning
   (a) can occur through morphological differences between competing species.
   (b) can cause the extinction of a competing species.
   (c) is not the result of behavioral changes.
   (d) does not occur among competing predators.

2. The interaction between bees and sunflowers is an example of
   (a) predation. (c) mutualism.
   (b) parasitism. (d) commensalism.

3. Pathogens are a type of
   (a) mutualist. (c) predator.
   (b) parasite. (d) herbivore.

4. Which interaction can harm both species involved?
   (a) competition (c) parasitism
   (b) predation (d) commensalism

5. Species that spread rapidly across large areas, causing harm to local species are called
   (a) exotic species. (c) alien species.
   (b) invasive species. (d) native species.

Free-Response Question

Ecosystem boundaries are defined by many factors.

(a) **Identify** one abiotic factor defining ecosystems. (1 pt.)
(b) **Identify** one biotic factor defining ecosystems (1 pt.)
(c) **Explain** why the boundary of an aquatic ecosystem is easier to define than a terrestrial ecosystem. (1 pt.)
(d) **Explain** two different ways scientists and administrators define the boundaries of an ecosystem. (2 pts.)
(e) **Describe** two different ways ecosystems interact with each other. (2 pts.)
(f) **Identify** the following about global ecosystems:
   (i) The region of the planet where life resides. (1 pt.)
   (ii) The range where life on the planet resides. (1 pt.)
(g) **Explain** why the ecosystem boundary in the image to the right extends beyond the Yellowstone national park in the center. (1 pt.)

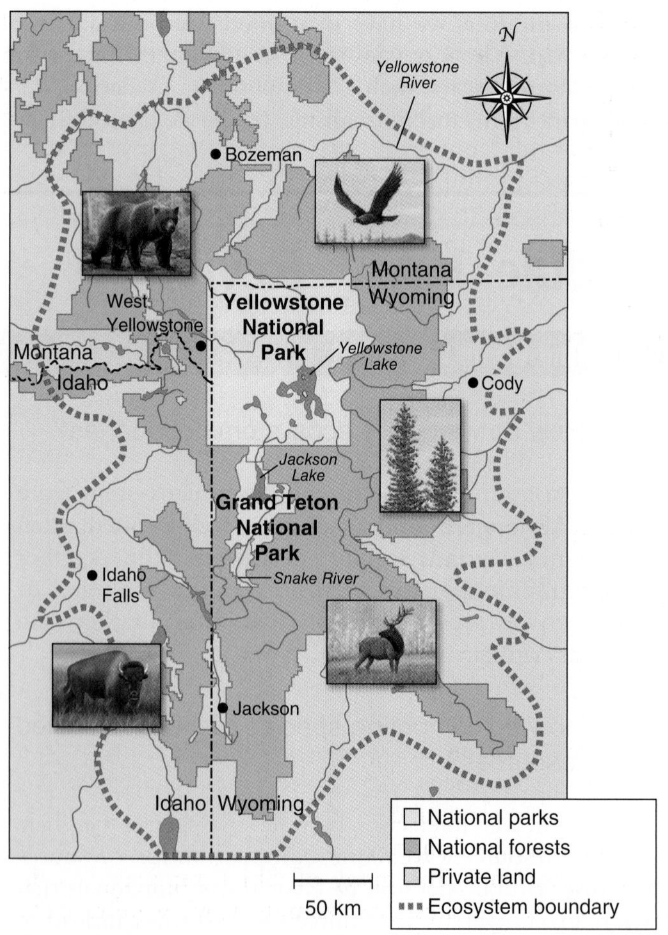

The Greater Yellowstone Ecosystem

Module 2

Unit 1 | 1 | 2 | 3 | 4 | 5 | 6 | 7

Terrestrial Biomes

In the previous module, we examined the many ways in which species interact, including as predators, competitors, and parasites. Such interactions occur in all regions of the world, yet the actual species that we find in different regions can vary a great deal. The plants and animals that are found in a particular region of the world is called a **biome**. A major factor that determines whether a species can survive and reproduce in a particular terrestrial region depends on the annual patterns of temperature and precipitation. For example, only species that are well adapted to hot and dry conditions can survive in deserts. A very different set of species survives in cold, snowy places. In this module, we will examine how differences in the annual patterns of temperature and precipitation help determine the distribution of terrestrial plants and animals that live in each region and how scientists categorize these regions. These differences in patterns of precipitation and temperature have major impacts on the distribution of natural resources, including the use of the biomes for grazing animals, growing crops, and harvesting lumber. We will also examine how the boundaries of terrestrial biomes have changed in the past and will continue to change in the future.

Learning Goals

After reading this module you should be able to

2-1 explain how we define terrestrial biomes.

2-2 describe the information contained in climate diagrams.

2-3 identify the nine terrestrial biomes.

2-4 describe the causes of changing boundaries of terrestrial biomes.

2-1 How do we define terrestrial biomes?

Terrestrial biomes are defined by the dominant plant growth forms

The annual patterns of temperature and precipitation represent a region's climate, which as we learned in the unit-opening case study, is the average weather that occurs in a given region over a long period—typically over several decades. We can contrast climate with the concept of weather, which is the short-term conditions of the atmosphere in a local area that include temperature, humidity, clouds, precipitation, and wind speed. In categorizing terrestrial biomes, we focus on how different climates affect the distribution of plants and animals.

Scientists have long recognized that organisms possess distinct growth forms, many of which represent adaptations to local climates. For example, if we were to examine the plants from all the deserts of the world, we would find many cactus-like species. Plants that look like cacti in North American deserts are indeed members of the cactus family, but those in the Kalahari Desert are members of the euphorbia family. These two distantly related families of plants look similar because they have evolved similar adaptations to hot, dry environments—including the ability to store large amounts of water in their tissues and a waxy coating to reduce water loss (**FIGURE 2.1**).

Despite these common adaptations to desert conditions, these two plant families have distinctive flowers and spines, and only the euphorbia family produces a milky sap. These differences help to confirm that while the two groups may look similar, genetically they are not closely related. Instead, exposure to similar selective pressures over long periods of time has favored the evolution of many similar adaptations. We will discuss this in more detail in Unit 2. As a result, scientists can categorize terrestrial regions of the world into **terrestrial biomes**, which are geographical regions of land categorized by a particular combination of average annual temperature, annual precipitation, and distinctive plant growth forms that are adapted to that climate. In Module 3 we will examine **aquatic biomes**, which are

Biome The plants and animals that are found in a particular region of the world.

Terrestrial biome A geographic region of land categorized by a particular combination of average annual temperature, annual precipitation, and distinctive plant growth forms.

Aquatic biome An aquatic region characterized by a particular combination of salinity, depth, and water flow.

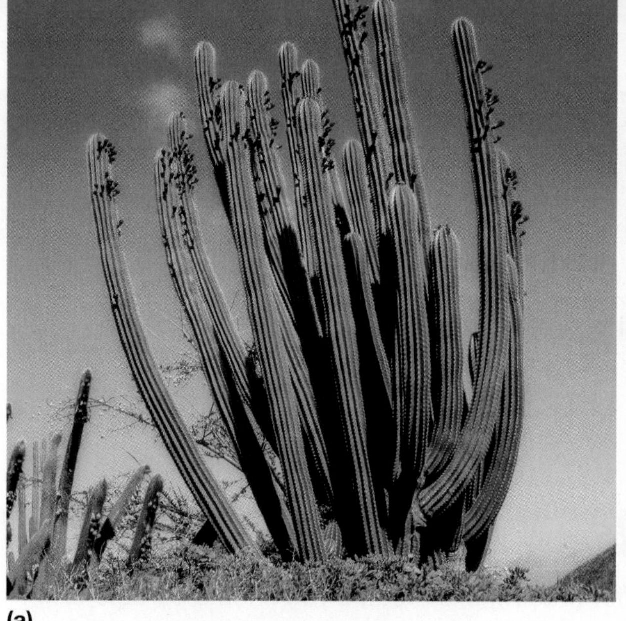

(a)

(b)

FIGURE 2.1 Similar growth forms. Cacti and euphorbia plants are not closely related, but they have evolved many similar adaptations that allow them to live in hot, dry environments. (a) Organ pipe cactus (*Stenocereus thurberi*) in Arizona. (b) Euphorbia (*E. damarana*) in Namibia. *(a: Dhoxax/Shutterstock; b: Patti Murray/Earth Scenes/Animals Animals)*

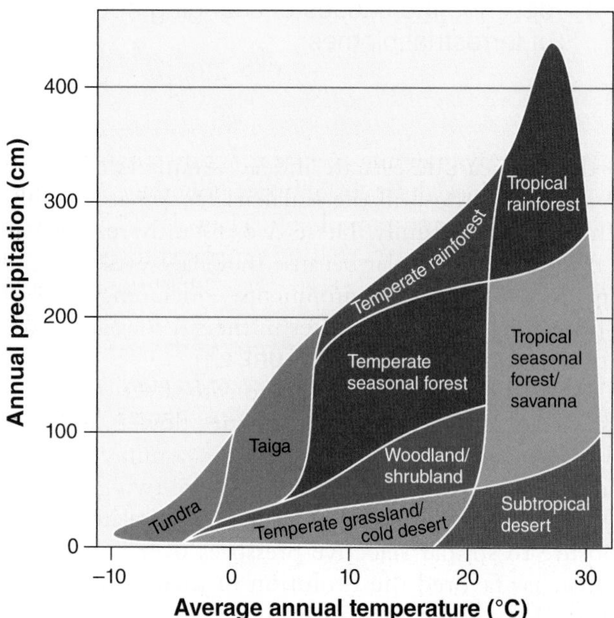

FIGURE 2.2 Locations of the world's biomes. Tundra and taiga biomes generally exist when the average annual temperature is less than 5°C (41°F). In contrast, temperate biomes exist in regions that experience average annual temperatures between 5°C and 20°C (41°F and 68°F). Tropical biomes exist in regions that experience average annual temperatures above 20°C.

aquatic regions categorized by a particular combinations of salinity, depth, and water flow.

FIGURE 2.2 shows the range of terrestrial biomes on Earth in the context of precipitation and temperature.

Habitat An area where a particular species lives in nature.

AP® Exam Tip

Make sure you know the difference between a habitat and a biome. A habitat is more specific than a biome. A biome is a large area of similar ecosystems, and can contain countless habitats. A habitat is a specific area that an organism requires within a biome, but it isn't the entire biome.

For example, tundra and taiga biomes have average annual temperatures below 5°C (41°F), whereas temperate biomes have average annual temperatures between 5°C and 20°C (68°F), and tropical biomes have average annual temperatures above 20°C. Within each of these temperature ranges, we can observe a wide range of precipitation. **FIGURE 2.3** shows the distribution of biomes around the world.

Note that although terrestrial biomes are categorized by plant growth forms, the animal species living in different biomes are often quite distinctive as well. For example, rodents inhabiting deserts around the world have a number of adaptations for hot, dry climates, including highly efficient kidneys that allow very little water loss through urination. For both plants and animals, biomes refer to large geographical areas that are home to large numbers of species. In contrast, a **habitat** is an area where a particular species lives in nature. A habitat is a subset of a biome. For example, the forests of the northeastern United States represent a biome, but this biome contains multiple terrestrial habitats including agricultural fields, abandoned fields, young forests with sapling trees, and mature forests.

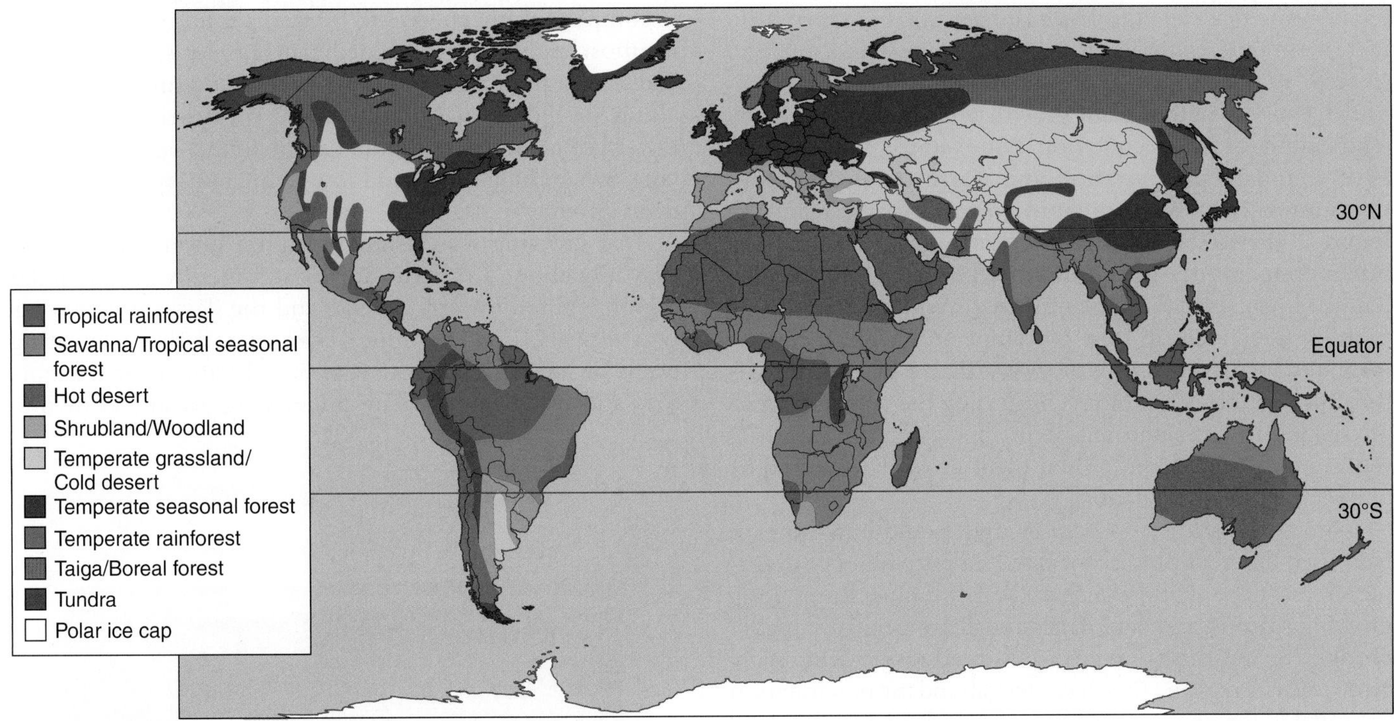

FIGURE 2.3 Biomes. Biomes are categorized by particular combinations of average annual temperature and annual precipitation.

2-2 How can we describe the information contained in climate diagrams?

Climate diagrams illustrate patterns of annual temperature and precipitation

Climate diagrams, such as those shown in **FIGURE 2.4**, are a helpful way to visualize regional patterns of temperature and precipitation. By graphing average monthly temperature and precipitation, these diagrams illustrate how the conditions in a biome vary during a typical year. They also indicate when the temperature is warm enough for plants to grow—that is, the months when it is above 0°C (32°F), known as the growing season. In Figure 2.4a, we can see that the growing season is mid-March through mid-October.

In addition to the growing season, climate diagrams can illustrate the seasonal combinations of precipitation and

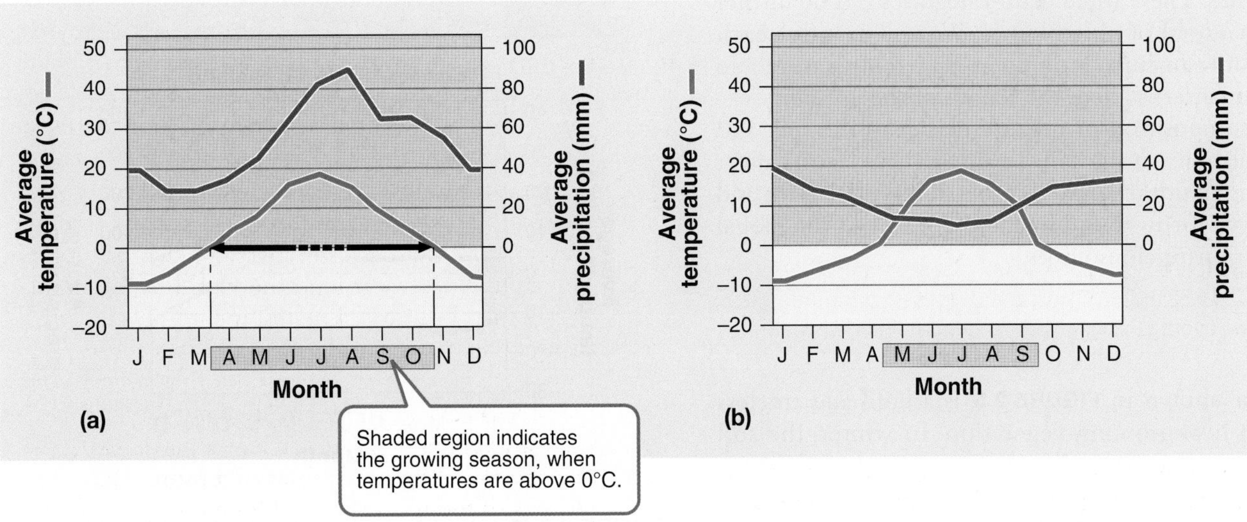

FIGURE 2.4 Climate diagrams. Climate diagrams display monthly temperature and precipitation values, which help determine the plants and animals that can live in the biome.

temperature that can constrain plant growth. For every 10°C increase in temperature, plants need 20 mm of precipitation to grow, so the two *y* axes are scaled in this way to allow us to determine whether plant growth is more constrained by temperature or by precipitation. For example, some biomes have abundant precipitation and warm temperatures in the summer, which favors rapid growth of large plants. In Figure 2.4b, we see a different scenario in which there are warm temperatures in the summer but very little precipitation. In this situation, plant growth will be constrained more by precipitation than by temperature. Thus, climate diagrams are an efficient way to visualize a lot of information about a terrestrial biome, including the annual pattern of changes in temperature and precipitation, the length of the growing season, and when plant growth is limited by temperature or precipitation.

Climate diagrams also help us understand how humans have used different biomes for different purposes. For example, areas of the world that have warm temperatures, long growing seasons, and abundant rainfall are generally highly productive and therefore are well suited to growing many crops. Warm regions that have less abundant precipitation are suitable for growing grains such as wheat and for grazing domesticated animals, including cattle and sheep. Colder regions are often best suited to grow forests that will be harvested for lumber.

2-3 What are the nine terrestrial biomes?

Terrestrial biomes range from tundra to tropical rainforests to deserts

We can divide the major terrestrial biomes into three main categories that are related to latitudes and altitudes on Earth: tundra and taiga biomes, temperate biomes, and tropical biomes. These three main categories can be further divided into a total of nine biomes. We will examine each of these biomes in turn, looking at how temperature and precipitation patterns affect the length of the growing season, the decomposition of organic matter in the soil, and nutrient availability in the soil. Together, these factors affect the global distributions of the biomes, the typical plant and animal growth forms found in each biome, and the global distribution of natural resources.

Tundra

The **tundra**, shown in FIGURE 2.5, is a cold and treeless biome, with low-growing vegetation. In winter, the soil is completely frozen. Arctic tundra is found in the northernmost regions of the Northern Hemisphere in Russia, Canada, Scandinavia, and Alaska. Antarctic tundra is found along the edges of Antarctica and on nearby islands. At lower latitudes, alpine tundra can be found on high mountains, where high winds and low temperatures prevent trees from growing.

The growing season in the tundra is very short, usually only about 4 months during summer, when the polar region is tilted toward the Sun and the days are very long. During this time the upper layer of soil thaws, creating pools of standing water that are ideal habitat for mosquitoes and other insects. The underlying subsoil, known as

Tundra A cold and treeless biome with low-growing vegetation.

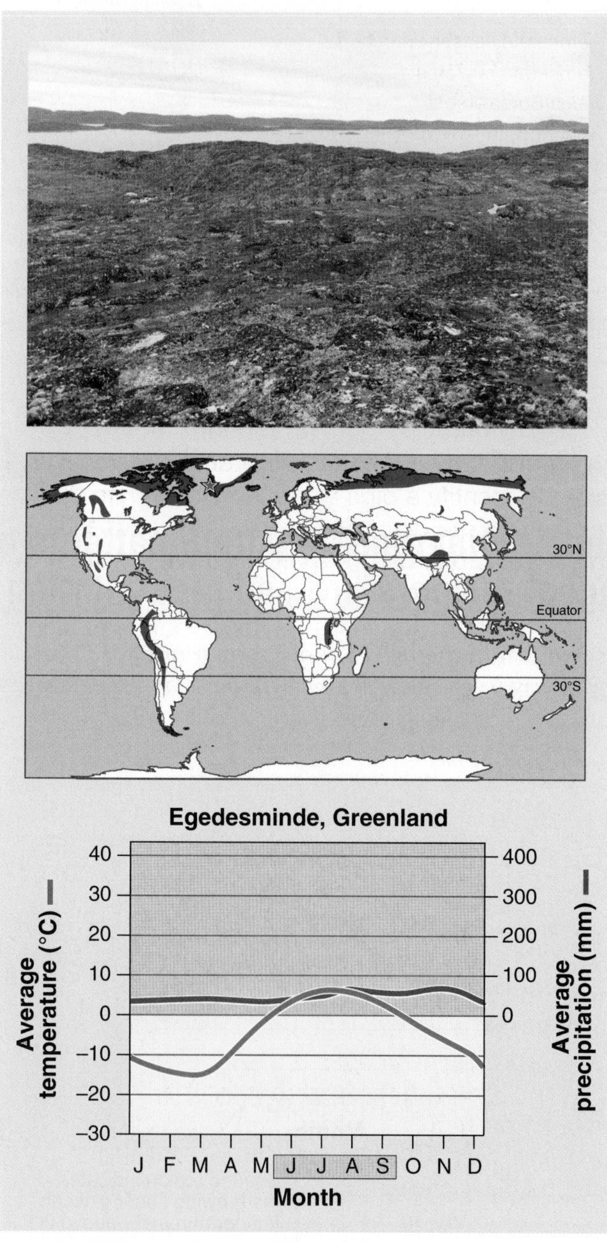

FIGURE 2.5 **Tundra biome.** The tundra is cold and treeless, with low-growing vegetation. *(Panther Media GmbH/Alamy)*

permafrost, is an impermeable, permanently frozen layer of soil that prevents water from draining and roots from penetrating. Permafrost, combined with the cold temperatures and a short growing season, prevents deep-rooted plants such as trees from living in the tundra.

While the tundra receives little precipitation, there is enough to support some plant growth. The characteristic plants of this biome, such as small woody shrubs, mosses, heaths, and lichens, can grow in shallow, waterlogged soil and can survive short growing seasons and bitterly cold winters. At these cold temperatures, chemical reactions occur slowly, and as a result, dead plants and animals decompose slowly. This slow rate of decomposition results in the accumulation of organic matter in the soil over time with relatively low levels of soil nutrients. Common animals in the Arctic tundra include muskoxen (*Ovibos moschatus*), Arctic foxes (*Vulpes lagopus*), and polar bears (*Ursus maritimus*).

The major human impact on the tundra biome is the threat of warming global temperatures melting the permafrost, which would then decompose dead plants and animals at a faster rate than usual. Such decomposition would release additional greenhouse gases. We will discuss this in more detail in Unit 9.

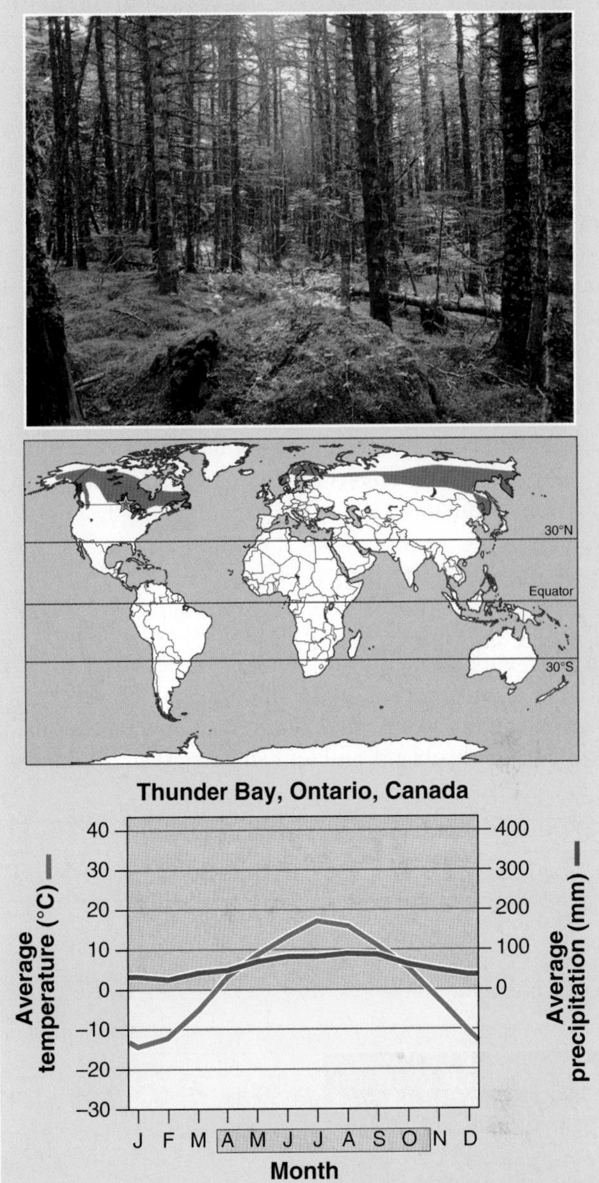

FIGURE 2.6 Taiga biome. Also known as the boreal forest biome, it is made up primarily of coniferous evergreen trees that can tolerate cold winters and short growing seasons. *(Bill Brooks/Alamy)*

AP® Exam Tip

On the AP® Environmental Science exam you may be asked to identify a biome based on a climate diagram. Realize that the diagram has two *y* axes, one for temperature and one for precipitation. Look at temperature first. Identify if there is a winter period where it goes below freezing. This will narrow down your options for terrestrial biomes. Next, look at precipitation. Does this place look wet, average, or dry? This will make it easier to determine each terrestrial biome.

Taiga

The **taiga** biome, also referred to as a **boreal forest** biome, shown in **FIGURE 2.6**, consists primarily of coniferous (cone-bearing) evergreen trees that can tolerate cold winters and short, cool growing seasons in the summer. Evergreen trees appear green year-round because they drop only a fraction of their needles each year. Taigas are found between about 50° N and 60° N in Europe, Russia, and North America. This subarctic biome has a very cold climate, and plant growth is more constrained by temperature than by precipitation.

As in the tundra, cold temperatures and relatively low precipitation make decomposition in taigas a slow process. In addition, the waxy needles of evergreen trees contain compounds that are resistant to decomposition. As a result of the slow rate of decomposition and the low nutrient content of the needles, taiga soils are covered in a thick layer of organic material, but the soils are poor in nutrients.

While the taiga biome generally receives a bit more precipitation than the tundra biome, the climate characteristics

Permafrost An impermeable, permanently frozen layer of soil.

Taiga A forest biome made up primarily of coniferous evergreen trees that can tolerate cold winters and short growing seasons. *Also known as* **boreal forest.**

of taigas—cold temperatures, low precipitation, and nutrient-poor soil—determine the species of plants that can survive in them. In addition to coniferous trees such as pine, spruce, and fir, some deciduous trees, such as birch, maple, and aspen, can also be found in this biome. The needles of coniferous trees can tolerate below-freezing conditions, but the deciduous trees drop all their leaves in autumn before the subfreezing temperatures of winter can damage them. When the weather warms, the deciduous trees produce new leaves and grow rapidly. Common animal species include beavers (*Castor Canadensis*), brown bears (*Ursus arctos*), and wolverines (*Gulo gulo*).

Because taigas have poor soils and short growing seasons, they are poorly suited for agriculture. However, they serve as an important source of trees for pulp, paper, and building materials. As a result, many taigas have been extensively logged. Other threats to taigas include mining and the extraction of oil and gas.

Temperate Rainforest

Moving to the mid-latitudes, we find that the climate is more temperate, with average annual temperatures between 5°C and 20°C (41°F and 68°F). A range of temperate biomes exists in this area including temperate rainforest, temperate seasonal forest, woodland/shrubland, and temperate grassland/cold desert.

Temperate rainforest A coastal biome typified by moderate temperatures and high precipitation.

The **temperate rainforest** is a coastal biome typified by moderate temperatures and high precipitation, and is shown in **FIGURE 2.7**. The temperate rainforest can be found in relatively narrow areas around the world. Temperate rainforests exist along the west coast of North America from northern California to Alaska, in southern Chile, on the east coast of Australia and in neighboring Tasmania, and on the west coast of New Zealand. Ocean currents along these coasts help to moderate temperature fluctuations, and ocean water provides a source of water vapor. The result is relatively mild summers and winters, compared with other biomes at similar latitudes, and a nearly 12-month growing season. In the temperate rainforest, winters are rainy and summers are foggy.

The combination of mild temperatures and high precipitation supports the growth of very large trees. In North America, the most common temperate rainforest trees are coniferous species, including fir, spruce, cedar, and hemlock as well as some of the world's tallest trees: the coastal redwoods (*Sequoia sempervirens*). These immense trees can live hundreds to thousands of years and achieve heights of 90 m (295 feet) and diameters of 8 m (26 feet). Because many of these large tree species are attractive sources of lumber, much of this biome has been logged and subsequently converted into single-species tree plantations.

As we have already seen, coniferous trees produce needles that are slow to decompose. The relatively cool temperatures in the temperate rainforest also favor slow decomposition, although it is not nearly as slow as in taiga and tundra. The nutrients released are rapidly taken up by the trees or carried

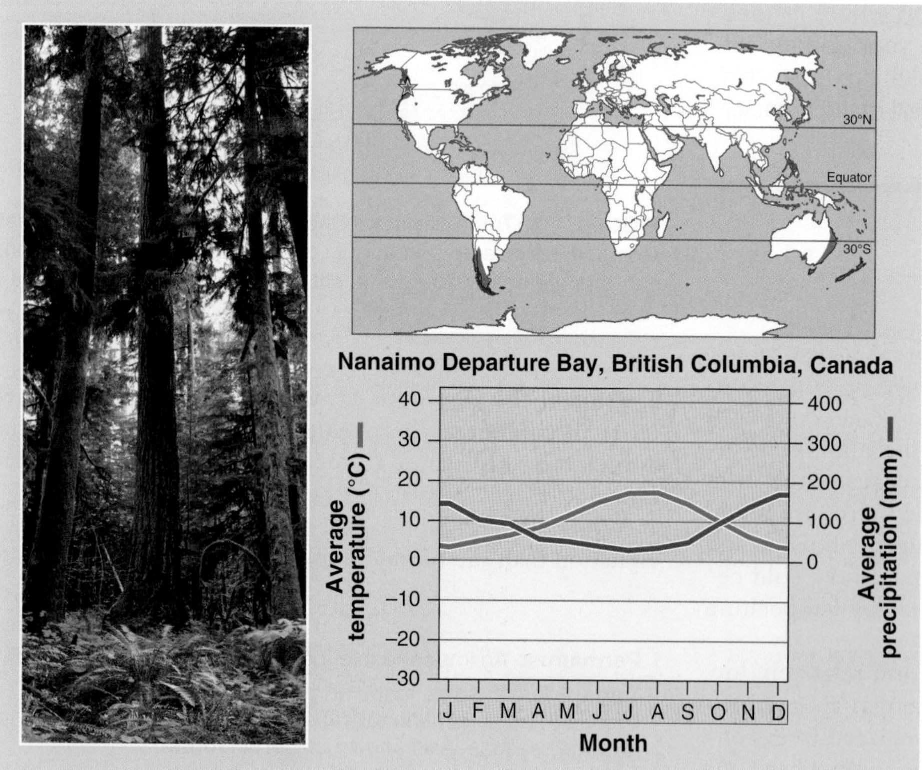

FIGURE 2.7 Temperate rainforest biome. Temperate rainforests have moderate mean annual temperatures and high precipitation that supports the growth of very large trees. *(BGSmith/Shutterstock)*

down through the soil by the abundant rainfall through the process of leaching, which leaves the soil low in nutrients. Ferns and mosses, which can survive in nutrient-poor soil, are commonly found living under the enormous trees. Animals that are characteristic of the temperate rainforest in North America include the black-tailed deer (*Odocoileus hemionus columbianus*), the Pacific giant salamander (*Dicamptodon ensatus*), and the Pacific treefrog (*Pseudacris regilla*).

Temperate Seasonal Forest

The **temperate seasonal forest**, shown in **FIGURE 2.8**, is a biome with warm summers and cold winters with over 1 m (39 inches) of annual precipitation. It is found in the eastern United States, Japan, China, Europe, Chile, and eastern Australia. Away from the moderating influence of the ocean, these forests experience much warmer summers and colder winters than temperate rainforests. They are dominated by broadleaf deciduous trees such as beech, maple, oak, and hickory, although some coniferous tree species may also be present. Because of the predominance of deciduous trees, these forests are also called temperate deciduous forests.

The warm summer temperatures in temperate seasonal forests favor rapid decomposition. Because the leaves shed by broadleaf trees are more readily decomposed than the needles of coniferous trees, the soils of temperate seasonal forests generally contain more nutrients than those of taigas. Their higher soil fertility, combined with their longer growing season, means that temperate seasonal forests have greater plant productivity than taigas. Common animal species include white-tailed deer (*Odocoileus virginianus*), red foxes (*Vulpes vulpes*), and gray squirrels (*Sciurus carolinensis*).

Because temperate seasonal forests are so productive, they have historically been one of the first biomes to be converted to agriculture on a large scale. When European settlers arrived in North America, they cleared large areas of the eastern forests for agriculture, although much of this has since grown back.

Shrubland

Shrubland, also referred to as a **woodland** and illustrated in **FIGURE 2.9**, is a biome characterized by hot, dry summers and mild, rainy winters. This biome is found on the coast of southern California (where it is called chaparral), in southern South America (matorral), in southwestern Australia (mallee), in southern Africa (fynbos), and in a large region surrounding the Mediterranean Sea (maquis). There is a 12-month growing season, but plant growth is constrained by high temperatures and low precipitation in summer and by cool temperatures and higher precipitation in winter.

Temperate seasonal forest A biome with warm summers and cold winters with over 1 m (39 inches) of annual precipitation.

Shrubland A biome characterized by hot, dry summers and mild, rainy winters. *Also known as* **woodland**.

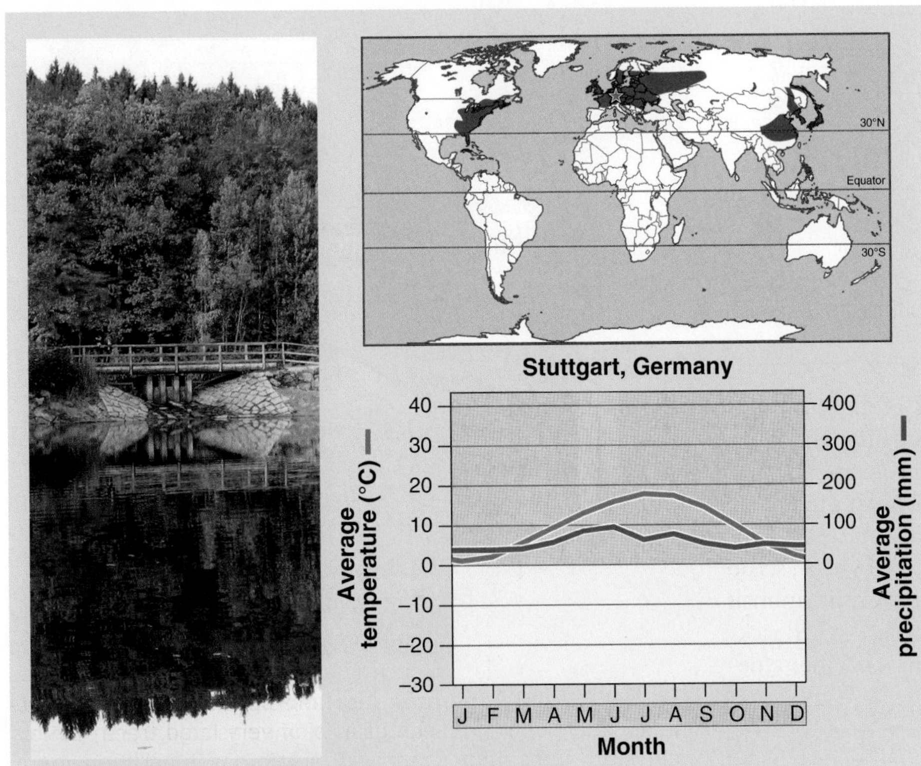

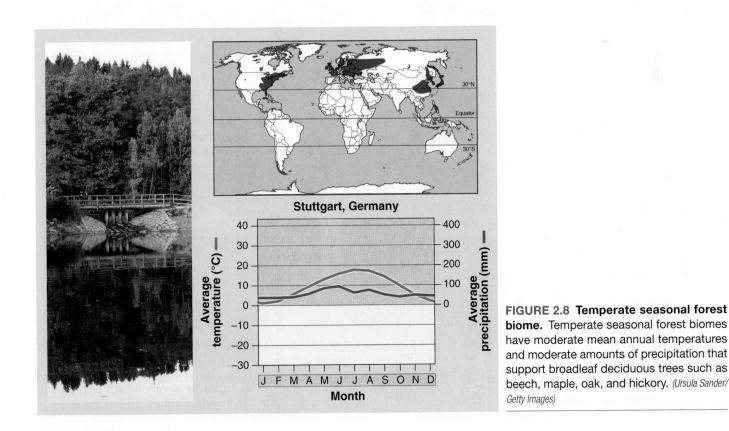

FIGURE 2.8 Temperate seasonal forest biome. Temperate seasonal forest biomes have moderate mean annual temperatures and moderate amounts of precipitation that support broadleaf deciduous trees such as beech, maple, oak, and hickory. *(Ursula Sander/Getty Images)*

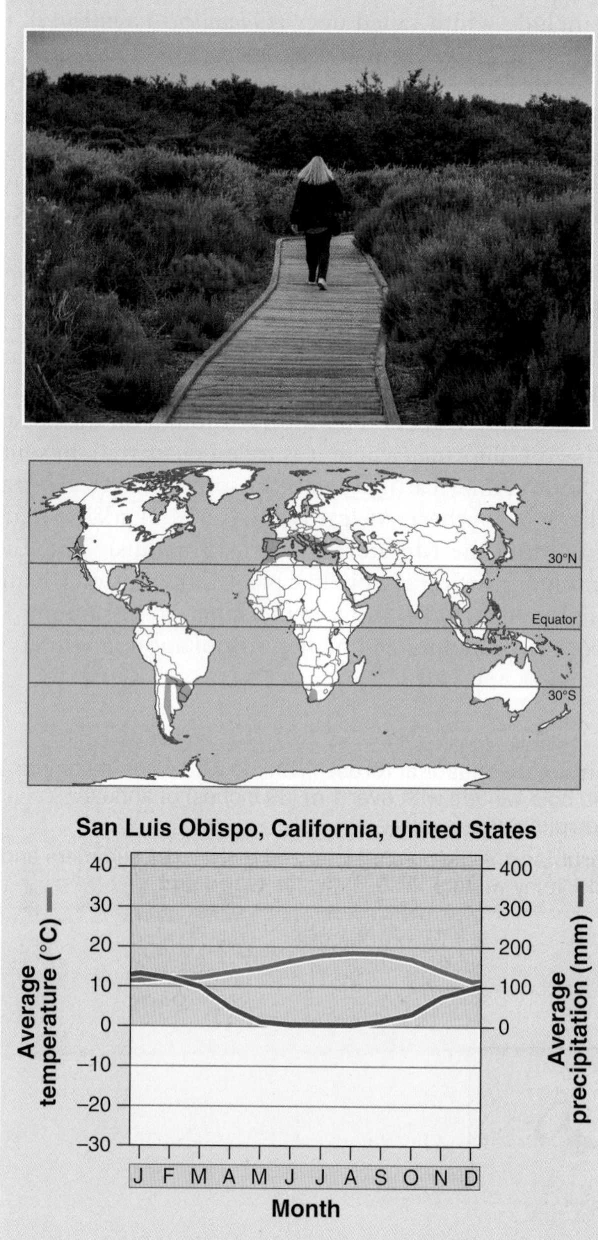

FIGURE 2.9 Woodland/shrubland biome. The woodland/shrubland biome is characterized by hot, dry summers and mild, rainy winters. *(George Rose/Getty Images)*

The hot, dry summers of the shrubland biome favor the natural occurrence of wildfires. Plants of this biome are well adapted to both fire and drought. Many plants quickly resprout after a fire, while others produce seeds that open only upon exposure to the intense heat of a fire. Typical plants of this biome include drought-resistant shrubs such as yucca, scrub oak, and sagebrush. In North America, characteristic animals include California quail (*Callipepla californica*), black-tailed jackrabbits (*Lepus californicus*), coyotes, and southern pacific rattlesnakes (*Crotalus oreganus helleri*).

Temperate grassland A biome characterized by cold, harsh winters, and hot, dry summers. *Also known as* **cold desert**.

Soils in this biome are low in nutrients because of leaching by the winter rains. As a result, the major agricultural uses of this biome are for grazing by animals and growing drought-tolerant, deep-rooted crops, such as grapes. In fact, this biome has the ideal conditions for growing grapes for making wine. The major threats from humans are increased human development and an increased frequency of fire that is making it difficult for many native species to survive in this biome.

Temperate Grassland

The **temperate grassland** biome, also referred to as a **cold desert**, is shown in **FIGURE 2.10**. It is characterized

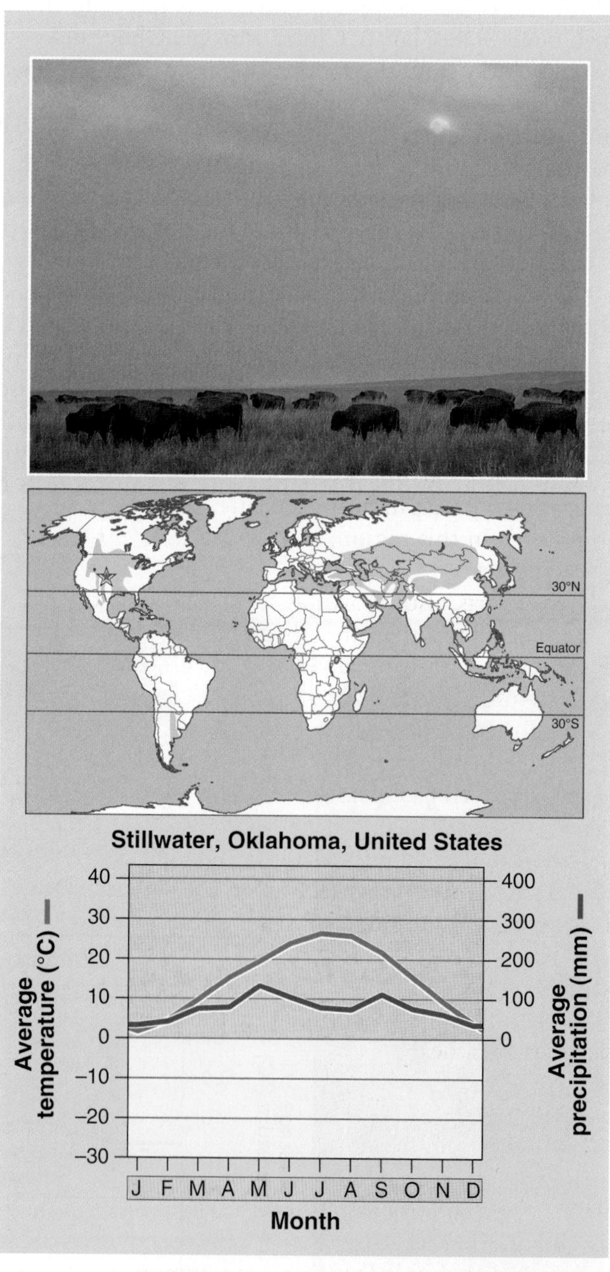

FIGURE 2.10 Temperate grassland/cold desert biome. The temperate grassland/cold desert biome has cold, harsh winters and hot, dry, summers that support grasses and nonwoody flowering plants. *(AP Photo/Brandi Simons)*

by cold, harsh winters and hot, dry summers. Temperate grasslands are found in the Great Plains of North America (where they are called prairies), in South America (pampas), and in central Asia and eastern Europe (steppes). As in the shrubland biome, plant growth is constrained by insufficient precipitation in summer and cold temperatures in winter. Fires are common, as the dry and frequently windy conditions fan flames ignited by lightning. Although estimates vary, it is thought that, historically, large wildfires occurred in this biome every few years, sometimes burning as much as 10,000 ha (nearly 25,000 acres) in a single fire.

Typical plants of temperate grasslands include grasses and nonwoody flowering plants. These plants are generally well adapted to wildfires and frequent grazing by animals. Their deep roots store energy to enable quick regrowth. Within this biome, the amount of rainfall determines which plants can survive in a region. In the North American prairies, for example, nearly 1 m (39 inches) of rain falls per year on the eastern edge of the biome, supporting grasses that can grow up to 2.5 m (8 feet) high. Although these tallgrass prairies receive sufficient rainfall for trees to grow, frequent wildfires keep trees from encroaching. In fact, the American Indians are thought to have intentionally kept the eastern prairies free of trees by using controlled burning. To the west, annual precipitation drops to 0.5 m (20 inches), favoring the growth of grasses less than 0.5 m (20 inches) tall. These shortgrass prairies are simply too dry to support trees or tall grasses. Farther west, closer to the Rocky Mountains, annual precipitation continues to decline to 0.25 m (10 inches). In this region, the short grass prairie gives way to cold desert.

Cold deserts, also known as temperate deserts, have even sparser vegetation than shortgrass prairies. Cold deserts are distinct from subtropical deserts in that they have much colder winters and do not support the characteristic plant growth forms of hot deserts, such as cacti. Typical animals of the North American prairie include bison, greater prairie chickens (*Tympanuchus cupido*), and the prairie kingsnake (*Lampropeltis calligaster*).

The combination of a relatively long growing season and rapid decomposition that adds large amounts of nutrients to the soil makes temperate grasslands very productive. More than 98 percent of the tallgrass prairie in the United States has been converted to agriculture, while the less productive shortgrass prairie is predominantly used for growing wheat and grazing cattle.

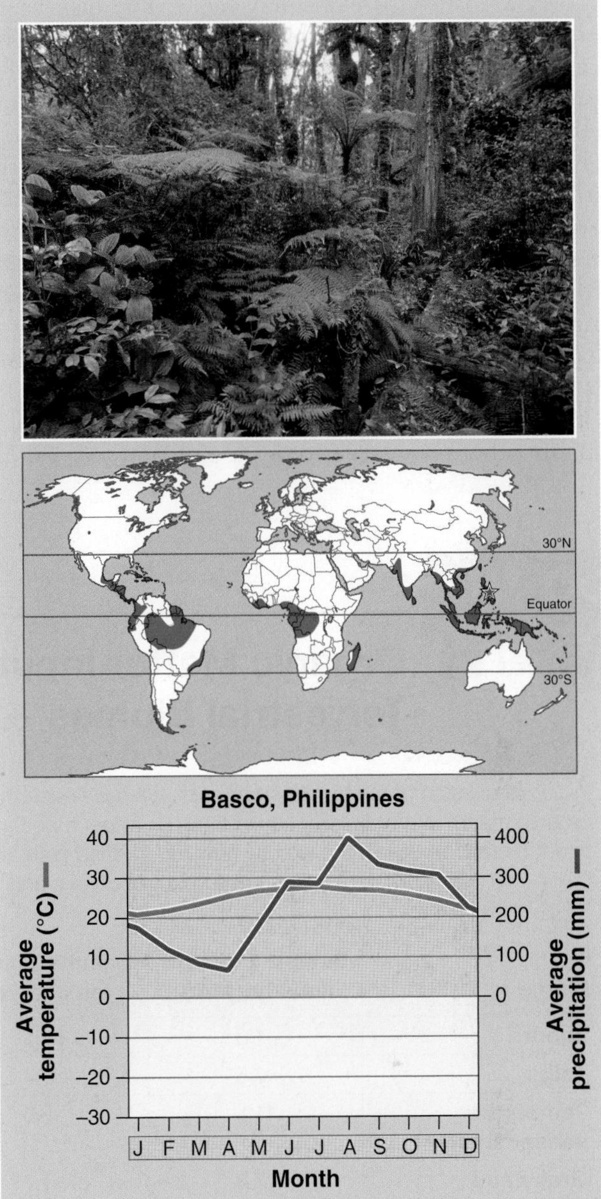

FIGURE 2.11 Tropical rainforest biome. Tropical rainforests are warm and wet, with little seasonal temperature variation. These forests are highly productive with several distinctive layers of vegetation. *(Doug Weschler/Earth Scenes/Animals Animals)*

Tropical Rainforest

In the tropics, average annual temperatures exceed 20°C. Here we find the tropical biomes: tropical rainforests, savannas, and hot deserts.

The **tropical rainforest** biome, shown in **FIGURE 2.11**, is a warm and wet biome found between 20° N and 20° S of the equator. This biome is found in Central and South America, Africa, Southeast Asia, and northeastern Australia. It is also found on large tropical islands, where the oceans provide a constant source of atmospheric water vapor.

Precipitation occurs frequently, although there are seasonal patterns in precipitation. Because of the warm temperatures and abundant rainfall, productivity is high,

Tropical rainforest A warm and wet biome found between 20° N and 20° S of the equator, with little seasonal temperature variation and high precipitation.

and decomposition is extremely rapid. The lush vegetation takes up nutrients quickly, leaving few nutrients to accumulate in the soil. Because of its high productivity, approximately 24,000 ha (59,500 acres) of tropical rainforest are cleared each year for agriculture. However, the high rate of decomposition causes the soils to lose their fertility quickly, so they have relatively little undecomposed organic matter (humus). As a result, farmers growing crops on tropical soils often have to keep moving to newly deforested areas. See "Do the Math: Organic Matter Inputs Among Terrestrial Biomes" to apply your knowledge and calculate the average amount of dead plant material that falls in each biome.

Tropical rainforests contain more biodiversity per hectare than any other terrestrial biome. As much as two-thirds of Earth's terrestrial species are found in this biome. These forests have several distinctive layers of vegetation. Large trees form a forest canopy that shades the underlying vegetation. Several layers of successively shorter trees make up the subcanopy, also known as the understory. Attached to the trunks and branches of the trees are epiphytes, plants that hold small pools of water that support small aquatic ecosystems far above the forest floor. Numerous species of woody vines (also called lianas) are rooted in the soil, but climb up the trunks of trees and often into the canopy. Common animals in tropical rainforests around the world include jaguars (*Panthera onca*), orangutans, and red-eyed treefrogs (*Agalychnis callidryas*). The dominant human threat to tropical rainforests is deforestation due to expanding forestry and agriculture.

DO THE MATH Organic Matter Inputs Among Terrestrial Biomes

Preparing for the AP® Exam

In our discussion of the terrestrial biomes, we have seen that each biome differs in the productivity of plants and in the amount of organic matter left to decompose in the soil. To get a better understanding of these differences, scientists set out containers in different biomes to determine exactly how much dead plant material actually falls to the ground. Using their data, shown in the table below, we can calculate the average amount of dead plant material falling in each biome in units of metric tons per hectare per year.

Biome	Sample 1 (metric tons/ha/year)	Sample 2 (metric tons/ha/year)	Sample 3 (metric tons/ha/year)	Sample 4 (metric tons/ha/year)	Sample 5 (metric tons/ha/year)
Tundra	1.3	1.1	1.7	1.5	1.9
Taiga	7.5	7.4	7.6	7.5	7.5
Temperate seasonal forest	11.1	12.5	10.4	9.9	13.2
Grassland	7.3	7.7	7.4	7.6	7.5
Tropical rainforest	29.0	30.0	31.0	35.0	25.0

To calculate the average amount of dead plant litter falling in the tundra biomes, for example, we sum the five values and divide by the number of samples

(1.3 metric tons/ha/year + 1.1 metric tons/ha/year + 1.7 metric tons/ha/year + 1.5 metric tons/ha/year
+ 1.9 metric tons/ha/year) ÷ 5
= 1.5 metric tons/ha/year

We can also express this number in terms of kilograms per square meter per year:

$$1.5 \text{ metric tons/ha/year} \times 1{,}000 \text{ kg/metric tons} = 1{,}500 \text{ kg/ha/year}$$

$$1{,}500 \text{ kg/ha/year} \times 1 \text{ ha}/10{,}000 \text{ m}^2 = 0.15 \text{ kg/m}^2\text{/year}$$

YOUR TURN

1. Calculate the average amount of dead plant material that fall in the other biomes and convert each into kilograms per square meter per year.
2. Explain the reasons for the differences among biomes. How do these relative differences in the amount of dead plant material in each biome compare to the amount of organic matter in the soil of each biome?

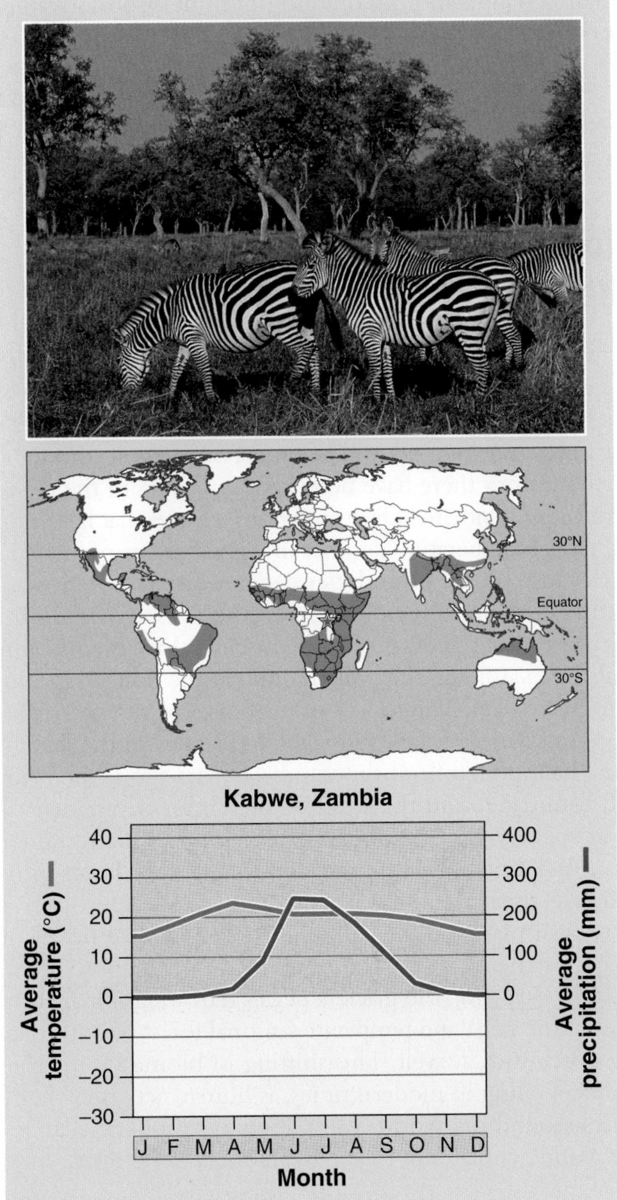

FIGURE 2.12 Tropical seasonal forest/savanna biome. Tropical seasonal forest and savannas have warm temperatures and distinct wet and dry seasons. Vegetation ranges from dense stands of shrubs and trees to relatively open landscapes dominated by grasses and scattered deciduous trees. *(John Warburton-Lee/DanitaDelimont.com)*

Savanna

The **savanna** biome (also known as a **tropical seasonal forest**), shown in **FIGURE 2.12**, is marked by warm temperatures and distinct wet and dry seasons. Most precipitation only occurs during summer. The trees drop their leaves during the dry season as an adaptation to survive the drought conditions and produce new leaves during the wet season. Thus these forests are also called tropical deciduous forests.

Savannas are common in much of Central America, on the Atlantic coast of South America, in southern Asia, in northwestern Australia, and in sub-Saharan Africa. Areas with moderately long dry seasons support dense stands of shrubs and trees. In areas with the longest dry seasons, the tropical seasonal climate leads to the formation of savannas, relatively open landscapes dominated by grasses and scattered deciduous trees. Common plants in this biome include acacia and baobab trees. Grazing and fire discourage the growth of many smaller woody plants and keep the savanna landscape open. The presence of trees and a warmer average annual temperature distinguish savannas from grasslands.

The warm temperatures of the savanna biome promote decomposition, but the low amounts of precipitation constrain plants from using the soil nutrients that are released. As a result, the soils of this biome are fairly fertile and can be farmed. Their fertility has resulted in the conversion of large areas of tropical seasonal forest and savanna into agricultural fields and grazing lands. For example, over 99 percent of the savanna of Pacific Central America and the Atlantic coast of South America has been converted to human uses, including agriculture and grazing. Common animals in this biome from around the world include large grazing animals such as several species of gazelles and zebras (*Equus* spp.), and large predators such as African lions and cheetahs.

Hot Desert

The **hot desert** biome, also known as a subtropical desert, shown in **FIGURE 2.13**, is located at roughly 30° N and 30° S, and is characterized by hot temperatures, extremely dry conditions, and sparse vegetation. Also known as subtropical deserts, this biome includes the Mojave Desert in the southwestern United States, the Sahara in Africa, the Arabian Desert of the Middle East, and the Great Victoria Desert of Australia. Cacti, euphorbia plants, and succulent plants are well adapted to this biome. To prevent water loss, the leaves of desert plants may be small, nonexistent, or modified into spines, and the outer layer of the plant is thick, with few pores for water and air exchange. Most photosynthesis occurs along the plant stem, which stores water so that photosynthesis can continue even during very dry periods. To protect themselves from herbivores, desert plants have developed defense mechanisms such as spines to discourage grazing, a concept first addressed in Module 1. Desert-adapted animals around the world include multiple species of tortoises, camels, and roadrunners.

Savanna A biome marked by warm temperatures and distinct wet and dry seasons. *Also known as* **tropical seasonal forest**.

Hot desert A biome located at roughly 30° N and 30° S, and characterized by hot temperatures, extremely dry conditions, and sparse vegetation.

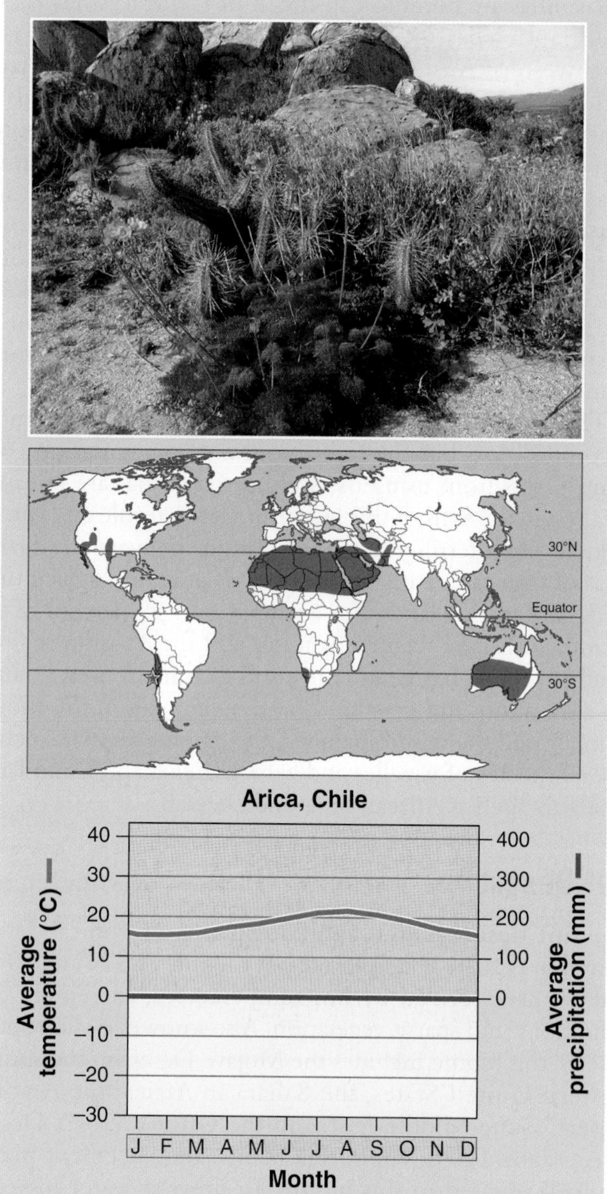

FIGURE 2.13 Hot desert biome. Subtropical deserts have hot temperatures, extremely dry conditions, and sparse vegetation. *(Rose Deakin/TopFoto)*

When rain does fall, the desert landscape is transformed. Annual plants — those that live for only a few months, reproduce, and die — grow rapidly during periods of rain. In contrast, perennial plants — those that live for many years — experience spurts of growth when it rains, but then exhibit little growth during the rest of the year. The slow overall growth of perennial plants in deserts makes them particularly vulnerable to disturbance, and they have long recovery times. The major threats to this biome include climate change and the draining of underground water for human use.

2-4 What are the causes of changing boundaries of terrestrial biomes?

Biome boundaries shift as climates change

As we have seen, terrestrial biomes have boundaries that are determined by modern climates that vary around the world. An important point to realize is that these boundaries are not fixed over long periods of time. For example, throughout Earth's history there have been major changes in the climates of the continents, including numerous ice ages that have caused substantial temperature declines in many regions of the world.

Because biome boundaries depend on plant responses to temperature and precipitation, the boundaries have shifted during the ice ages. Researchers have documented these shifts by examining the pollen found at the bottom of North American lakes. The pollen of terrestrial plants is carried by the wind, where it lands in lakes and settles to the lake bottom. That pollen from thousands of years ago is then stored and found deep in the mud of the lakes. Given that each plant species makes a distinctively shaped pollen grain and scientists can determine the age of the various layers of mud in a lake, they can then determine when different species of terrestrial plants were present. In doing so, we now know that after the last ice age in North America (approximately 10,000 years ago), the glaciers began retreating north and the trees of the taiga and temperate seasonal forest began expanding northward as well. This shifting of biome boundaries is also happening in modern times, as human activities alter climates around the world. We will discuss modern changes in the world's climates in much greater detail in Unit 9.

In this module, we have examined the diversity of terrestrial biomes in relation to global patterns of temperature and precipitation. We have also seen that the boundaries of these biomes can change over time, particularly due to human impacts of global climates. In the next module, we will explore the diversity of aquatic biomes on Earth.

AP® Exam Tip

A hypothesis should indicate both variables and directionality. Directionality means that you are describing an increase or decrease. On the exam, don't merely say that changing one factor will impact another. That is too vague. Give variables direction whenever possible.

Module 2 AP® Review

Preparing for the AP® Exam

Learning Goals Revisited

2-1 How do we define terrestrial biomes?

Terrestrial biomes are categorized by the dominant plant forms that exist in a region.

2-2 How can we describe the information contained in climate diagrams?

Climate diagrams illustrate monthly patterns of temperature and precipitation during the year. They also illustrate the growing season of a biome and the months during which plants are more constrained by temperature or precipitation.

2-3 What are the nine terrestrial biomes?

The nine terrestrial biomes are tundra, taiga, temperate rainforests, temperate seasonal forests, shrublands, temperate grasslands/cold deserts, tropical rainforests, savannas, and hot deserts.

2-4 What are the causes of changing boundaries of terrestrial biomes?

Given that biome boundaries are determined by climate, the changing climates in Earth's history have caused biome boundaries to shift over time. Current and future climate changes will continue to cause biome boundaries to shift.

Practice Math and Graphing

Preparing for the AP® Exam

Answer the following questions. Be sure to show all your work.

1. Practice Math

To create climate graphs, we need to know the typical precipitation for a biome. Scientists do this by selecting a location somewhere within a biome and then using weather data from that location to determine the average precipitation over multiple years. The two tables below (Table 2 on page 42) show data collected for Nashville, Tennessee, for 3 years. Use the data in **TABLE 1** to calculate the average precipitation for each month.

TABLE 1 Precipitation

Month	Year 1 (mm)	Year 2 (mm)	Year 3 (mm)
January	91	90	95
February	89	98	101
March	95	99	103
April	91	93	104
May	142	135	134
June	99	93	105
July	87	89	97
August	81	84	87
September	72	101	94
October	82	83	78
November	94	105	105
December	83	125	95

(continues)

2. Practice Graphing

(a) Using the average precipitation data from your answers to the "Practice Math" problem above and **TABLE 2** on average monthly temperatures, create a climate diagram by arranging the months on the *x* axis, plotting the average temperature on the left-side *y* axis, and plotting the average precipitation on the right-side *y* axis. To ensure that the two *y* axes are properly aligned with each other—to show whether plant growth is more constrained by temperature or precipitation—the range of the *y* axis for temperature should be from 0 to 70°C and the range of the *y* axis for precipitation should be from 0 to 140 mm.

(b) Based on your climate diagram, what is the growing season for this location?

(c) Based on your climate diagram, is the biome more limited by temperature or precipitation?

TABLE 2 Temperature

Month	Average temperature (°C)
January	8.3
February	11.1
March	16.1
April	21.7
May	25.6
June	30.0
July	31.7
August	31.7
September	27.8
October	22.2
November	15.6
December	9.4

AP® Practice Questions

Preparing for the AP® Exam

Multiple-Choice Questions

1. Which of the following is a major factor in defining a biome?
   (a) proximity to the ocean
   (b) abundance of prey
   (c) longitude position
   (d) latitude position

2. In which biome does permafrost play a factor in growth of vegetation?
   (a) tundra
   (b) taiga
   (c) temperate grassland
   (d) deciduous forest

3. In which biome is plant growth primarily constrained by precipitation?
   (a) taiga
   (b) temperate seasonal forest
   (c) temperate grassland
   (d) tropical rainforest

4. Which biome has the highest soil nutrient levels?
   (a) tropical rainforest
   (b) taiga
   (c) shrubland
   (d) temperate seasonal forest

5. Scientists predict that the planet will continue to experience warmer temperatures over the next century. Given this prediction, which shift in biome boundary seems likely?
   (a) Alpine tundra biomes will expand to include a larger area of mountain tops.
   (b) Taiga biomes will shift toward the poles.
   (c) Tropical rainforest biomes will shift toward the equator.
   (d) Temperate rainforest biomes will shift toward the equator.

Use the data below to answer questions 6 & 7:

Rainfall (mm)	Biome A	Biome B
May	50	171
June	53	98
July	50	77

Temperature (°C)	Biome A	Biome B
May	12	28
June	14	28
July	16	29

6. Given the data above, it is most likely that Biome A is
   (a) tropical rainforest.
   (b) tundra.
   (c) hot desert.
   (d) temperate seasonal forest.

7. Given the data above, Biome B is most likely a
   (a) temperate seasonal forest.
   (b) shrubland.
   (c) tropical rainforest.
   (d) hot desert.

Free-Response Question

Biomes are defined by many factors.

(a) (i) **Identify** which temperate biome listed is dominated by broadleaf deciduous trees. (1 pt.)
  (ii) **Describe** the climate conditions in a temperate deciduous forest. (1 pt.)

(b) (i) **Identify** which biome contains succulent plants with spines for protection. (1 pt.)
  (ii) **Describe** a threat currently faced by the biome in part b (i). (1 pt.)

(c) (i) **Propose a solution** to the threat identified in part b (ii). (1 pt.)
  (ii) **Identify** a drawback to the solution proposed in part c (i). (1 pt.)

(d) **Explain** why boundaries of biomes have shifted during the past century. (1 pt.)

Scientists are measuring the productivity of biomes over time, looking at the amount of plant material produced by each biome. Use the table below to answer the following questions:

Biome	Sample 1 (metric tons/ha/year)	Sample 2 (metric tons/ha/year)	Sample 3 (metric tons/ha/year)	Sample 4 (metric tons/ha/year)	Sample 5 (metric tons/ha/year)
Tundra	1.3	1.1	1.7	1.5	1.9
Taiga	7.5	7.4	7.6	7.5	7.5
Tropical rainforest	29.0	30.0	31.0	35.0	25.0

(e) **Identify** the independent variable in the data collected here. (1 pt.)

(f) **Identify** the dependent variable in the data collected here. (1 pt.)

(g) **Identify** a possible conclusion to be made from the data collected from these biomes. (1 pt.)

Module 3

Unit 1	1	2	3	4	5	6	7

Aquatic Biomes

Learning Goals

After reading this module you should be able to

3-1 identify the major freshwater biomes.

3-2 identify the major marine biomes.

In the previous module, we took a tour of the nine terrestrial biomes, which are categorized by climate and dominant plant growth forms. In this module, we examine the aquatic biomes of the world, which are categorized by physical characteristics such as salinity, depth, and water flow. Although temperature is an important factor in determining which species can survive in a particular aquatic habitat, it is not a factor used to categorize aquatic biomes. These biomes provide several important ecological functions, including serving as important drinking water sources for humans. Aquatic biomes fall into two broad categories: freshwater and marine. Freshwater biomes include streams, rivers, lakes, and wetlands. Marine biomes, also known as saltwater biomes, include shallow marine areas such as estuaries and coral reefs as well as the open ocean. As was the case of terrestrial biomes, scientists do not categorize aquatic biomes based on the presence of different animal species,

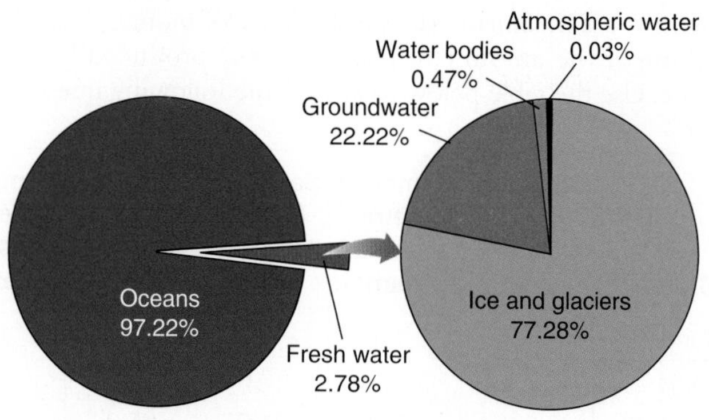

FIGURE 3.1 Distribution of water on Earth. Fresh water represents less than 3 percent of all water on Earth, and only about three-fourths of that fresh water is surface water. Most of that surface water is frozen as ice and in glaciers. Therefore, less than 1 percent of all water on the planet is accessible for use by humans. *(Data from R.W. Christopherson,* Geosystems, *7th ed. Pearson/Prentice Hall, 2009.)*

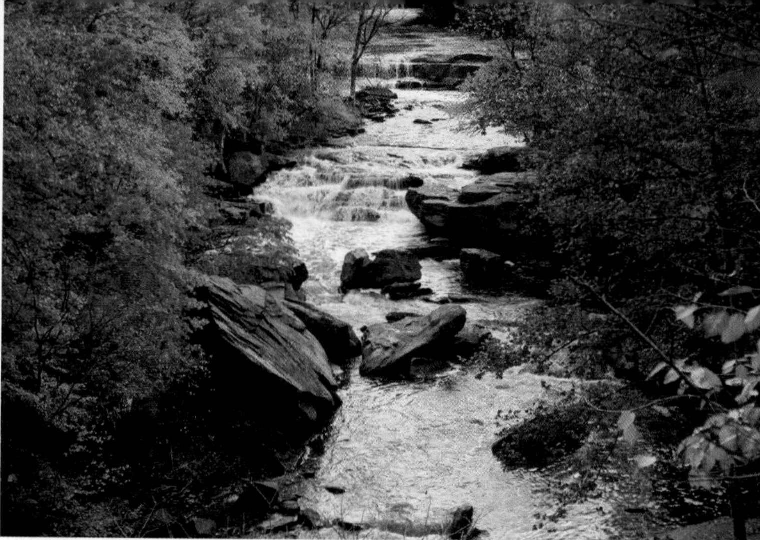

FIGURE 3.2 Streams and rivers. Streams and rivers are freshwater aquatic biomes that are characterized by flowing water. This photo shows Berea Falls on the Rocky River near Cleveland, Ohio. *(Jim West/Alamy Stock Photo)*

but the environmental conditions present in each aquatic biome—such as temperature, salinity, nutrient availability or turbidity (i.e., water cloudiness)—are major determinants of which species of plants and animals can live in the biome. For example, most species of fish and shellfish are specially adapted to live in either fresh water or salt water, but few species can live under both conditions.

Nearly 70 percent of Earth's surface is covered by water. As **FIGURE 3.1** shows, this water is found in five main repositories, not including the water vapor and precipitation contained in the atmosphere. The vast majority of Earth's water—more than 97 percent—is found in the oceans as salt water. The remainder—less than 3 percent—is fresh water, the type of water that can be consumed by humans. Of this small percentage of Earth's water that is fresh water, nearly three-fourths is aboveground, though mostly in the form of ice and glaciers, and therefore generally not available for human consumption. Only one-fourth resides underground in the form of groundwater. This means that available fresh water is less than 1 percent of all water on Earth.

3-1 What are the major freshwater biomes?

Freshwater biomes have low salinity

Freshwater biomes are categorized as streams and rivers, lakes and ponds, or freshwater wetlands. Freshwater biomes are widely distributed throughout the world, from the Arctic to the tropics.

Freshwater biomes Categorized as streams and rivers, lakes and ponds, or freshwater wetlands.

Streams and Rivers

Streams and rivers are characterized by flowing fresh water that may originate from underground springs or as runoff from rain or melting snow (**FIGURE 3.2**). Streams (also called creeks) are typically narrow and carry relatively small amounts of water. Rivers are typically wider and carry larger amounts of water. It is not always clear, however, at what point a particular stream, as it combines with other streams, becomes large enough to be categorized as a river.

As water flow changes, biological communities also change. Most streams and many rapidly flowing rivers have few plants or algae. Instead, inputs of organic matter from terrestrial biomes, such as fallen leaves, provide the energy that is required by organisms. This organic matter is consumed by insect larvae and crustaceans such as crayfish, which then become prey for predators such as fish. As fast-moving streams combine to form rivers, the water flow typically slows, sediments and organic material settle to the bottom, and rooted plants and algae are better able to grow and provide an alternative energy source for the herbivores.

Fast-moving streams and rivers typically have stretches of turbulent water called rapids, where water and air are mixed together. This mixing allows large amounts of atmospheric oxygen to dissolve into the water. Such high-oxygen environments support fish species such as trout and salmon that need large amounts of oxygen to support their high-energy swimming. Slower-moving rivers experience less mixing of air and water. These lower-oxygen environments favor species such as catfish that can better tolerate low-oxygen conditions. Some common fish living in streams and rivers include several species of catfish, and sturgeon. Today, the major threats to streams and rivers are excess nutrients and pollutants, which we will discuss in detail in Unit 8.

FIGURE 3.3 **Lakes and ponds.** Lakes, such as Lake George in New York State, are characterized by standing water and a central zone of water that is too deep for emergent vegetation. *(Frank Paul/Alamy)*

Lakes and Ponds

Lakes and ponds contain standing water, at least some of which is too deep to support emergent vegetation (plants that are rooted to the bottom and emerge above the water's surface). Lakes are larger than ponds, but as with streams and rivers, there is no clear point at which a pond is considered large enough to be categorized as a lake (FIGURE 3.3). Common fish in ponds and lakes include several species of sunfish, bass, perch, and pike.

As FIGURE 3.4 shows, lakes and ponds can be divided into several distinct zones. The **littoral zone** is the shallow area of soil and water in lakes and ponds near the shore where most algae and emergent plants such as cattails grow. Most photosynthesis occurs in this zone. In the zone of open water in lakes and ponds as deep as the sunlight can penetrate, we can find the **limnetic zone**, where rooted plants can no longer survive. As a result, floating algae called **phytoplankton** are the only photosynthetic organisms in this zone. The plants and algae together make a substantial contribution to the global production of oxygen and the removal of carbon dioxide. The **profundal zone** is a region of water where sunlight does not reach, below the limnetic zone in very deep lakes. As a result, plants and algae cannot survive in the profundal zone, so nutrients are not easily taken up. Bacteria decompose the detritus that reaches the profundal zone, but they consume oxygen in the process. As a result, dissolved oxygen concentrations can become so low that there is an insufficient amount of oxygen to support many large organisms. The muddy bottom of a lake, pond, or ocean beneath the limnetic and profundal zones is called the **benthic zone**.

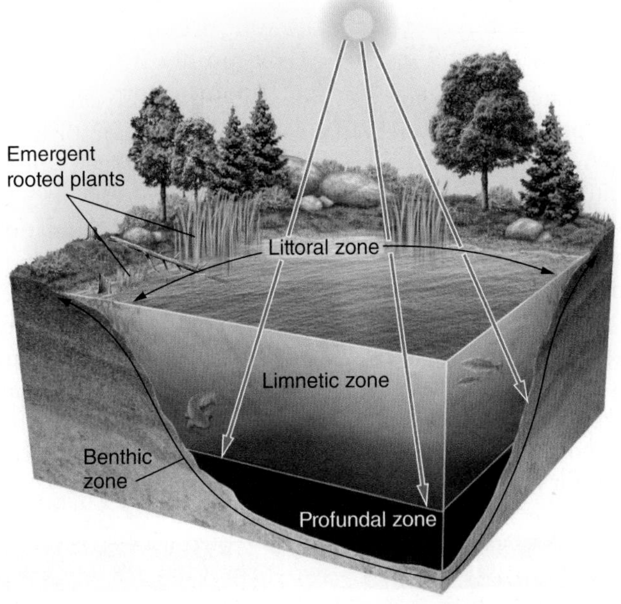

FIGURE 3.4 **Lake zones.** The littoral zone consists of shallow water with emerging, rooted plants whereas the limnetic zone is the deeper water where plants do not emerge. The deepest water, where oxygen can be limiting because little sunlight penetrates to allow photosynthesis by producers, is the profundal zone. The sediments that lie beneath the littoral, limnetic, and profundal zones constitute the benthic zone.

Lakes are further classified by their level of fertility. Lakes that have low amounts of phytoplankton due to low amounts of nutrients in the water—such as phosphorus and nitrogen—are called **oligotrophic** lakes. Given the low concentrations of algae, oligotrophic lakes can look very clear for tens of meters down from the surface. In contrast, lakes with a moderate level of fertility are called **mesotrophic** lakes, and lakes with a high level of fertility are called **eutrophic** lakes. Eutrophic lakes can be so fertile that their very high concentrations of algae make the water quite turbid, resembling pea soup, such that you may only be able to see a half meter down from the water's surface. A current concern for many oligotrophic lakes is that human activities are causing increased nutrient inputs, which is causing them to become less clear due to the increased growth of the algae.

Littoral zone The shallow zone of soil and water in lakes and ponds near the shore where most algae and emergent plants such as cattails grow.

Limnetic zone A zone of open water in lakes and ponds as deep as the sunlight can penetrate.

Phytoplankton Floating algae.

Profundal zone A region of water where sunlight does not reach, below the limnetic zone in very deep lakes.

Benthic zone The muddy bottom of a lake, pond, or ocean beneath the limnetic and profundal zones.

Oligotrophic Describes a lake with a low level of phytoplankton due to low amounts of nutrients in the water.

Mesotrophic Describes a lake with a moderate level of fertility.

Eutrophic Describes a lake with a high level of fertility.

FIGURE 3.5 Freshwater wetlands. Freshwater wetlands have soil that is saturated or covered by fresh water for at least part of the year and are characterized by particular plant communities. (a) In this swamp in southern Illinois, bald cypress trees emerge from the water. (b) This marsh in south central Wisconsin is characterized by cattails, sedges, and grasses growing in water that is not acidic. (c) This bog in northern Wisconsin is dominated by sphagnum moss as well as shrubs and trees that are adapted to acidic conditions. *(a–c: Lee Wilcox)*

Freshwater Wetlands

Freshwater wetlands are aquatic biomes that are submerged or saturated by water for at least part of each year, but shallow enough to support emergent vegetation. They support species of plants that are specialized to live in submerged or saturated soils.

Freshwater wetlands include swamps, marshes, and bogs. Swamps are wetlands that contain emergent trees, such as the Great Dismal Swamp in Virginia and North Carolina and the Okefenokee Swamp in Georgia and Florida (**FIGURE 3.5a**). Marshes are wetlands that contain primarily nonwoody vegetation, including cattails and sedges (Figure 3.5b). Bogs, in contrast, are very acidic wetlands that typically contain sphagnum moss and spruce trees (Figure 3.5c).

Freshwater wetlands are among the most productive biomes on the planet, and they provide several critical services that benefit humans and biodiversity. For example, wetlands can take in large amounts of rainwater and release it slowly into the groundwater or into nearby streams, thus reducing the severity of floods and droughts. Wetlands also filter pollutants from water, recharging the ground with clean water. Many bird species depend on wetlands during migration or breeding. As many as one-third of all endangered bird species in the United States spend some part of their lives in wetlands, even though this biome makes up only 5 percent of the nation's land area. More than half of the freshwater wetland area in the United States has been drained for agriculture or development or to eliminate breeding grounds for mosquitoes and various disease-causing organisms. As a result, we now have much less habitat for those species that rely on freshwater wetlands and we have a reduced ability to filter contaminants, which is important since freshwater biomes provide a vital source of water for drinking, agriculture, and industries.

Freshwater wetland An aquatic biome that is submerged or saturated by water for at least part of each year, but shallow enough to support emergent vegetation.

AP® Exam Tip

FRQs often ask for solutions to water issues. Restoring wetlands is a good solution as they reduce flooding by providing a place for water infiltration. They can filter out certain pollutants to protect water quality. When using this as a solution, clearly explain *how* wetlands are a solution. If mentioning that they can filter out pollutants, name the pollutant.

3-2 What are the major marine biomes?

Marine biomes have high salinity

Marine biomes contain salt water and can be categorized as estuaries/salt marshes, mangrove swamps, intertidal zones, coral reefs, and the open ocean.

Estuaries and Salt Marshes

Estuaries are areas along the coast where the fresh water of rivers mixes with salt water from the ocean. Because rivers carry large amounts of nutrient-rich organic material, estuaries are extremely productive places for plants and algae, and the abundant plant life helps filter contaminants out of the water. Found along the coasts in temperate climates, we can find salt marshes within the estuaries. Like freshwater marshes, **salt marshes** contain nonwoody emergent vegetation (**FIGURE 3.6**). Estuaries and salt marshes provide important habitat for spawning fish and shellfish; two-thirds of marine fish and shellfish species spend their larval stages in this biome. Like many freshwater wetlands, salt marshes have suffered from being filled in, being developed to support growing human populations, and being contaminated by pollutants that arrive from the rivers that supply water. This is having negative impacts on many commercially important species of fish and shellfish.

Mangrove Swamps

Mangrove swamps occur along tropical and subtropical coasts and, like freshwater swamps, contain trees whose roots are submerged in water (**FIGURE 3.7**). Unlike most trees, however, mangrove trees are salt tolerant. They often grow in estuaries, but they can also be found along shallow coastlines that lack inputs of fresh water. The trees help to protect those coastlines from erosion and storm damage. Falling leaves and trapped organic material produce a nutrient-rich environment. Like salt marshes, mangrove swamps provide sheltered habitat for commercially important fish and shellfish. Nearly one-third of the world's mangrove swamps have been destroyed, primarily for human habitation or to make space to grow crops such as rice and rubber trees.

FIGURE 3.6 Estuary and salt marsh. Estuaries experience a mixture of fresh water and salt water where rivers empty into oceans. Salt marshes are commonly found in the estuaries of temperate regions. This salt marsh is in Plum Island Sound in Massachusetts. *(© 2006, 2007 Jerry and Marcy Monkman/www.ecophotography.com)*

FIGURE 3.7 Mangrove swamp. Salt-tolerant mangrove trees, such as these in Everglades National Park, are important in stabilizing tropical and subtropical coastlines and in providing habitat for marine organisms. *(© T. & S. Allofs/Biosphoto)*

Intertidal Zones

The **intertidal zone** is the narrow band of coastline that exists between the levels of high tide and low tide (**FIGURE 3.8**). Intertidal zones range from steep, rocky areas to broad, sloping mudflats. Environmental conditions in this biome are relatively stable when submerged during high tide. However, conditions can become quite harsh during low tide when organisms are exposed to direct sunlight, high temperatures, and desiccation (i.e., drying out). Moreover, waves crashing onto shore can make it a challenge for organisms to hold on to rocks and not get washed away. Intertidal zones are home to a wide variety of organisms that have adapted to these conditions, including barnacles, sponges, algae, mussels, crabs, and sea stars. The main threats of human impacts are pollution, including trash, chemical pollution, and oil spills.

Estuary An area along the coast where the fresh water of rivers mixes with salt water from the ocean.

Salt marsh Found along the coast in temperate climates, a marsh containing nonwoody emergent vegetation.

Mangrove swamp A swamp that occurs along tropical and subtropical coasts, and contains salt-tolerant trees with roots submerged in water.

Intertidal zone The narrow band of coastline that exists between the levels of high tide and low tide.

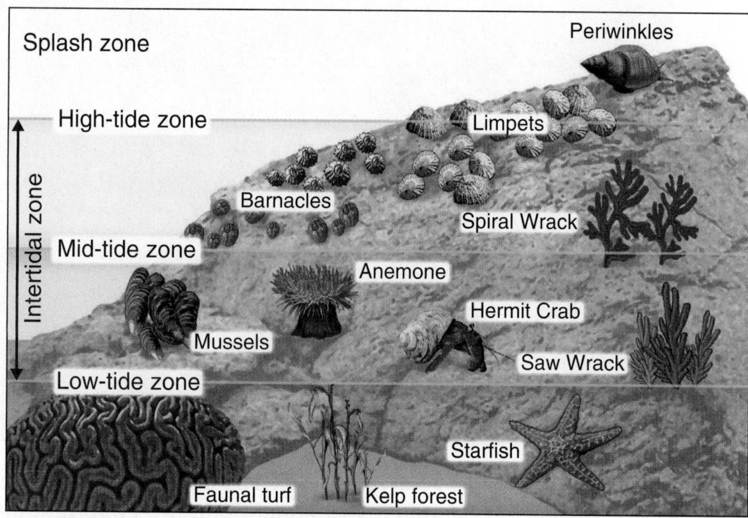

FIGURE 3.8 Intertidal zone. Organisms that live in the area between high and low tide, such as sea anemones, barnacles, and sea stars, must be highly tolerant of the harsh, desiccating conditions that occur during low tide.

Coral Reefs

Coral reefs, which are found in warm, shallow waters beyond the shoreline in tropical regions, represent Earth's most diverse marine biome. Corals are tiny animals that secrete a layer of limestone (calcium carbonate) to form an external skeleton. The animal living inside this tiny skeleton is essentially a hollow tube with tentacles that draw in plankton and detritus for food. Corals live in water that is relatively poor in nutrients and food, which is possible because of their relationship with single-celled algae that live within the tissues of the corals. When a coral digests the food it captures, it releases CO_2 and nutrients. The algae use the CO_2 during photosynthesis to produce sugars and the nutrients stimulate the algae to release their sugars to the coral. You might recall our treatment of corals in our discussion of mutualisms in Module 1. The coral gains energy in the form of sugars, and the algae obtain CO_2, nutrients, and a safe place to live within the coral's tiny limestone skeleton. But this association with photosynthetic algae means that corals can live only in shallow waters where sunlight can penetrate.

Although each individual coral is tiny, most corals live in vast colonies. As individual corals die and decompose, their limestone skeletons remain. Over time, these skeletons accumulate and develop into coral reefs, which can become quite massive. The Great Barrier Reef of Australia, for example, covers an area of 2,600 km^2 (1,600 $miles^2$). A tremendous diversity of other organisms, including fish and invertebrates, use the structure of the reef as both a refuge in which to live and a place to find food. At the Great Barrier Reef there are more than 400 species of coral, 1,500 species of tropical fish, and 200 species of birds.

Coral reefs are currently facing a wide range of challenges, including pollutants and sediments that make it difficult for the corals to survive. Coral reefs also face the growing problem of **coral bleaching**, a phenomenon in which the algae inside the corals die, causing the corals to turn white (**FIGURE 3.9**). Without the algae, the corals soon

FIGURE 3.9 Coral reef. This coral reef is in American Samoa in the South Pacific. The image on the left was taken prior to coral bleaching in December 2014. The image on the right was taken 2 months later, after coral bleaching occurred. *(© The Ocean Agency/XL Catlin Seaview Survey/Richard Vevers)*

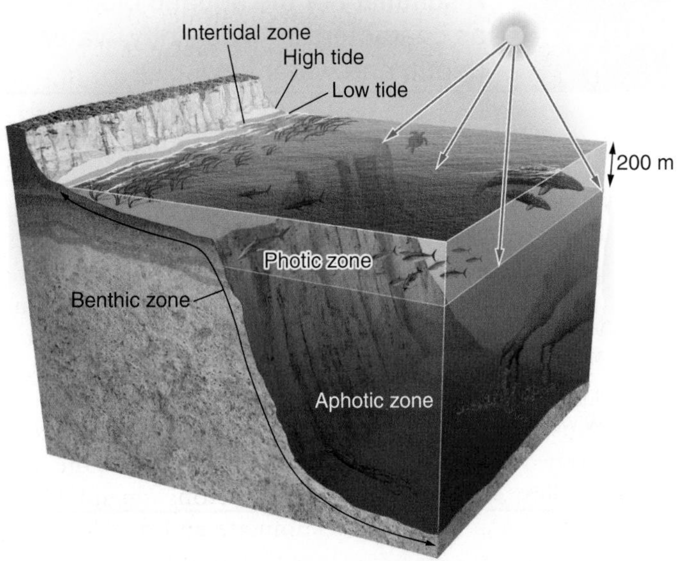

FIGURE 3.10 The open ocean. The open ocean can be separated into several distinct zones.

FIGURE 3.11 Making light in the aphotic zone. Several species that live in the deep ocean have evolved the ability to generate light to help them see and locate food. This bobtail squid (*Rossia macrosoma*) lives in the deep waters of the Mediterranean and off the coast of western Europe. *(Andrey Nekrasov/Getty Images)*

die as well, losing their vibrant colors and resulting in the reef turning white. Scientists believe that the algae are dying from a combination of disease and environmental changes, including lower ocean pH and abnormally high water temperatures. Coral bleaching is a serious problem; without the corals, the entire coral reef biome is endangered.

The Open Ocean

The **open ocean** contains deep ocean water that is located away from the shoreline where sunlight can no longer reach the ocean bottom. The exact depth of penetration by sunlight depends on a number of factors, including the amounts of sediment and algae suspended in the water, but it generally does not exceed 200 m (approximately 650 feet).

Like a pond or lake, the ocean can be divided into zones. These zones are shown in **FIGURE 3.10**. The upper layer of ocean water that receives enough sunlight to allow photosynthesis is the **photic zone**, where phytoplankton produce a large amount of the planet's oxygen and take up a large amount of carbon dioxide. The deeper layer of water that lacks sufficient sunlight for photosynthesis is the **aphotic zone** and the ocean floor is called the benthic zone, which is similar to the benthic zone of lakes and ponds.

In the photic zone, algae are the major photosynthetic organisms. They form the base of a food web that includes tiny zooplankton, fish, and whales. In the aphotic zone, because of the lack of light, there are no plants or algae and therefore no photosynthesis. However, there are some species of bacteria in the deep ocean that can use the energy contained in the bonds of methane and hydrogen sulfide. These bacteria use **chemosynthesis**, which is a process used by some bacteria generate energy using methane and hydrogen sulfide. These bacteria form the base of a deep-ocean food web that includes animals such as tube worms. The aphotic zone also contains a variety of organisms that can generate their own light to help them feed in the dark waters. These organisms include several species of crustaceans, jellyfish, squid, and fish (**FIGURE 3.11**).

Coral reef Represents Earth's most diverse marine biome, and are found in warm, shallow waters beyond the shoreline in tropical regions.

Coral bleaching A phenomenon in which algae inside corals die, causing the corals to turn white.

Open ocean Deep-ocean water, located away from the shoreline where sunlight can no longer reach the ocean bottom.

Photic zone The upper layer of ocean water in the ocean that receives enough sunlight for photosynthesis.

Aphotic zone The deeper layer of ocean water that lacks sufficient sunlight for photosynthesis.

Chemosynthesis A process used by some bacteria to generate energy with methane and hydrogen sulfide.

As you can appreciate, there are many different types of aquatic biomes and we can categorize them in terms of salinity, water depth, and water flow. Now that we have examined all of the aquatic biomes, we can see a summary of the biomes and their characteristics in **TABLE 3.1**.

In this module, we have toured the diversity of saltwater and freshwater aquatic biomes and seen that they are primarily categorized by salinity, flow, and depth. In the next module, we will consider how aquatic and terrestrial biomes play important roles in the cycling of carbon and nitrogen, two elements that are key to the organisms that live in every biome.

TABLE 3.1 A summary of the aquatic biomes as characterized by salinity, water depth, and water flow

	Freshwater (low salinity)	Marine (high salinity)
Flowing water	Streams and rivers	Estuaries/salt marshes (temperate regions) Mangrove swamps (tropical regions)
Standing water, deep water	Ponds and lakes	Open ocean
Standing water, shallow water	Freshwater wetlands	Coral reefs (tropical regions)
Fluctuating water depths		Intertidal zones

Module 3 AP® Review

Preparing for the AP® Exam

Learning Goals Revisited

3-1 What are the major freshwater biomes?

The major freshwater biomes are streams and rivers, ponds and lakes, and freshwater wetlands.

3-2 What are the major marine biomes?

The major marine biomes are estuaries/salt marshes, mangrove swamps, intertidal zones, coral reefs, and the open ocean.

AP® Practice Questions

Preparing for the AP® Exam

Multiple-Choice Questions

1. Which ecosystem experiences harsh conditions due to fluctuations from tides?
   (a) coral reef
   (b) open ocean
   (c) intertidal zone
   (d) ponds and lakes
2. Most of the photosynthesis in lakes and ponds occurs in the
   (a) benthic zone.
   (b) littoral zone.
   (c) limnetic zone.
   (d) profundal zone.
3. Which is an important ecosystem service provided by wetlands?
   (a) flood control
   (b) food production though agriculture
   (c) habitat for pollutants
   (d) releasing of carbon into the atmosphere
4. Using the image below, identify which zone in this freshwater system contains the photic and aphotic zones.

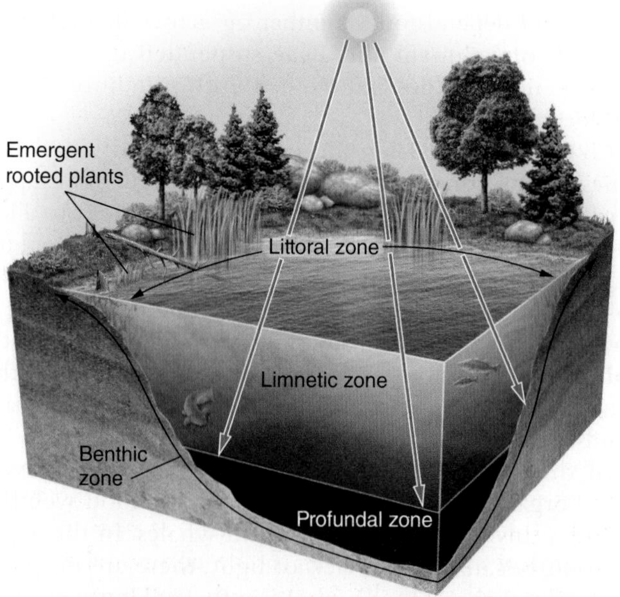

   (a) littoral zone
   (b) limnetic zone
   (c) benthic zone
   (d) profundal zone

5. Which of the following aquatic ecosystems contains the aphotic zone?
(a) coral reefs
(b) streams and rivers
(c) freshwater wetlands
(d) open ocean

Free-Response Question

The table below shows the amounts of nitrogen and phosphorus found in three different lakes.

Lake	Nitrogen (ppm)	Phosphorus (ppm)
A	330	45
B	50	9
C	116	18

(a) **Identify** the condition of nutrient levels found in lake A. (1 pt.)
(b) **Describe** one consequence of nutrient levels remaining as they are in Lake A. (1 pt.)
(c) **Describe** human activities that could cause the high nutrient levels of Lake A. (1 pt.)
(d) **Propose a solution** to the human activities identified in part (c). (1 pt.)
(e) **Justify** the solution proposed in part (d) by identifying an additional benefit to the method besides water nutrient reduction. (1 pt.)
(f) **Describe** a nutrient category that would be typical in Lake B. (1 pt.)
(g) The water cycle moves nutrients through the biosphere.
  (i) **Identify** the process in which plants release water back into the atmosphere. (1 pt.)
  (ii) **Describe** a way that humans affect the water cycle. (1 pt.)
  (iii) **Describe** an ecological disadvantage to the problem from part g (i). (1 pt.)
(h) **Explain** which nutrient is a limiting factor in water ecosystems. (1 pt.)

Module 4

Unit 1	1	2	3	4	5	6	7

The Carbon and Nitrogen Cycles

Learning Goals

After reading this module you should be able to

4-1 explain how carbon cycles within ecosystems.

4-2 describe how nitrogen cycles within ecosystems.

Thus far in Unit 1, we have examined how the biotic and abiotic conditions help to determine the distribution of plant and animal species in terrestrial and aquatic biomes. Given that all species are comprised of matter—including carbon, nitrogen, phosphorus, and water—these atoms and molecules are critical for the functioning of species in ecosystems. Other elements are also required, although in much smaller amounts, such as calcium, magnesium, potassium, and sulfur. An interesting insight is that when it comes to matter, all ecosystems combined represent a closed system, meaning these atoms and molecules are not lost or gained but instead they cycle within and among ecosystems in different forms. The specific chemical forms that elements take determine how they cycle within the biosphere.

Because the movement of matter within and between ecosystems involves cycles of biological, geological, and chemical processes, these cycles are known as **biogeochemical cycles**. To keep track of the movement of matter in biogeochemical cycles, we refer to the components of the biogeochemical cycle that contain the matter—including air, water, and organisms—as **reservoirs**. Each reservoir can serve as a "source" of the element when atoms and molecules leave a reservoir and as a "sink" when atoms

Biogeochemical cycle The movements of matter within and between ecosystems involving cycles of biological, geological, and chemical processes.

Reservoirs The components of the biogeochemical cycle that contain the matter, including air, water, and organisms.

and molecules get stored in the reservoir. The movement of the atoms and molecules between different reservoirs occurs through biotic and abiotic processes. In this module, we examine the cycles of carbon and nitrogen in several different forms with a focus on the major reservoirs where carbon and nitrogen are stored on Earth as well as the important biotic and abiotic processes that cause these elements to cycle among the reservoirs. In Module 5, we will examine the cycling of phosphorus and water.

4-1 How does carbon cycle within ecosystems?

The carbon cycle moves carbon between air, water, and land

Carbon is the most important element in living organisms; it makes up about 20 percent of their total body weight. Carbon is the basis of the long chains of organic molecules that form the membranes and walls of cells, constitute the backbones of proteins, and store energy for later use. Other than water, there are few molecules in the bodies of organisms that lack carbon. To understand the **carbon cycle**, which is the movement of carbon around the biosphere among reservoir sources and sinks, we need to examine the flows that move carbon and the major reservoirs that contain carbon.

The Carbon Cycle

FIGURE 4.1 illustrates the seven processes that drive the carbon cycle: photosynthesis, respiration, exchange, sedimentation, burial, extraction, and combustion. These processes can be categorized as either fast or slow. The fast part of the cycle involves processes that are associated with living organisms that hold carbon for a relatively short period of time. It also includes the exchange of CO_2 between the air and water and the combustion of organic carbon, which releases CO_2 into the atmosphere. The slow part of the cycle involves carbon that is held in rocks, in soils, or as petroleum hydrocarbons (the materials we use as fossil fuels). Carbon may be stored in these forms for millions of years.

Carbon cycle The movement of carbon around the biosphere among reservoir sources and sinks.

Aerobic respiration The process by which cells convert glucose and oxygen into energy, carbon dioxide, and water.

Steady state When a system's inputs equal outputs, so that the system is not changing over time.

Photosynthesis and Respiration

Let's take a closer look at Figure 4.1, which depicts the processes of the carbon cycle. When plants and algae use photosynthesis, whether on land or in the water, they use solar energy to convert carbon dioxide (CO_2) and water (H_2O) into glucose ($C_6H_{12}O_6$) and oxygen (O_2). A portion of this energy is consumed by herbivores and subsequently by predators of those herbivores.

These organisms return a portion of their carbon as CO_2 when they use **aerobic respiration**, whereby cells convert glucose and oxygen into energy, carbon dioxide, and water. Additional carbon is returned after organisms die. When organisms die, carbon that was part of the live biomass reservoir becomes part of the dead biomass reservoir. Decomposers break down the dead material, which returns CO_2 to the water or air via respiration and thus continues the cycle.

Exchange, Sedimentation, and Burial

As you can see on the left side of Figure 4.1, carbon is exchanged between the atmosphere and the ocean. The amount of carbon released from the ocean into the atmosphere roughly equals the amount of atmospheric CO_2 that diffuses into ocean water. Some of the CO_2 dissolved in the ocean enters the food web via photosynthesis by algae.

Another portion of the CO_2 dissolved in the ocean combines with calcium ions in the water to form calcium carbonate ($CaCO_3$), a compound that can precipitate out of the water and form limestone and dolomite rock via sedimentation and burial. Although sedimentation is a very slow process, the small amounts of calcium carbonate sediment formed each year have accumulated over millions of years to produce the largest carbon reservoir in the slow part of the carbon cycle.

A small fraction of the organic carbon in the dead biomass reservoir is buried and incorporated into ocean sediments before it can decompose into its constituent elements. This organic matter becomes fossilized and, over millions of years, some of it may be transformed into fossil fuels. The amount of carbon removed from the food web by this slow process is roughly equivalent to the amount of carbon returned to the atmosphere by weathering of rocks containing carbon (such as limestone) and by volcanic eruptions. Thus, the slow part of the carbon cycle is in **steady state**, which means that inputs equal outputs, so that the system is not changing over time.

Extraction and Combustion

The final processes in the carbon cycle are extraction and combustion, as are shown in the middle of Figure 4.1. The extraction of fossil fuels by humans is a relatively recent phenomenon that began when human society started to rely on coal, oil, and natural gas as energy sources. Extraction by itself does not alter the carbon cycle; it is the subsequent step of combustion that makes the difference. Combustion of fossil fuels by humans and the natural combustion of

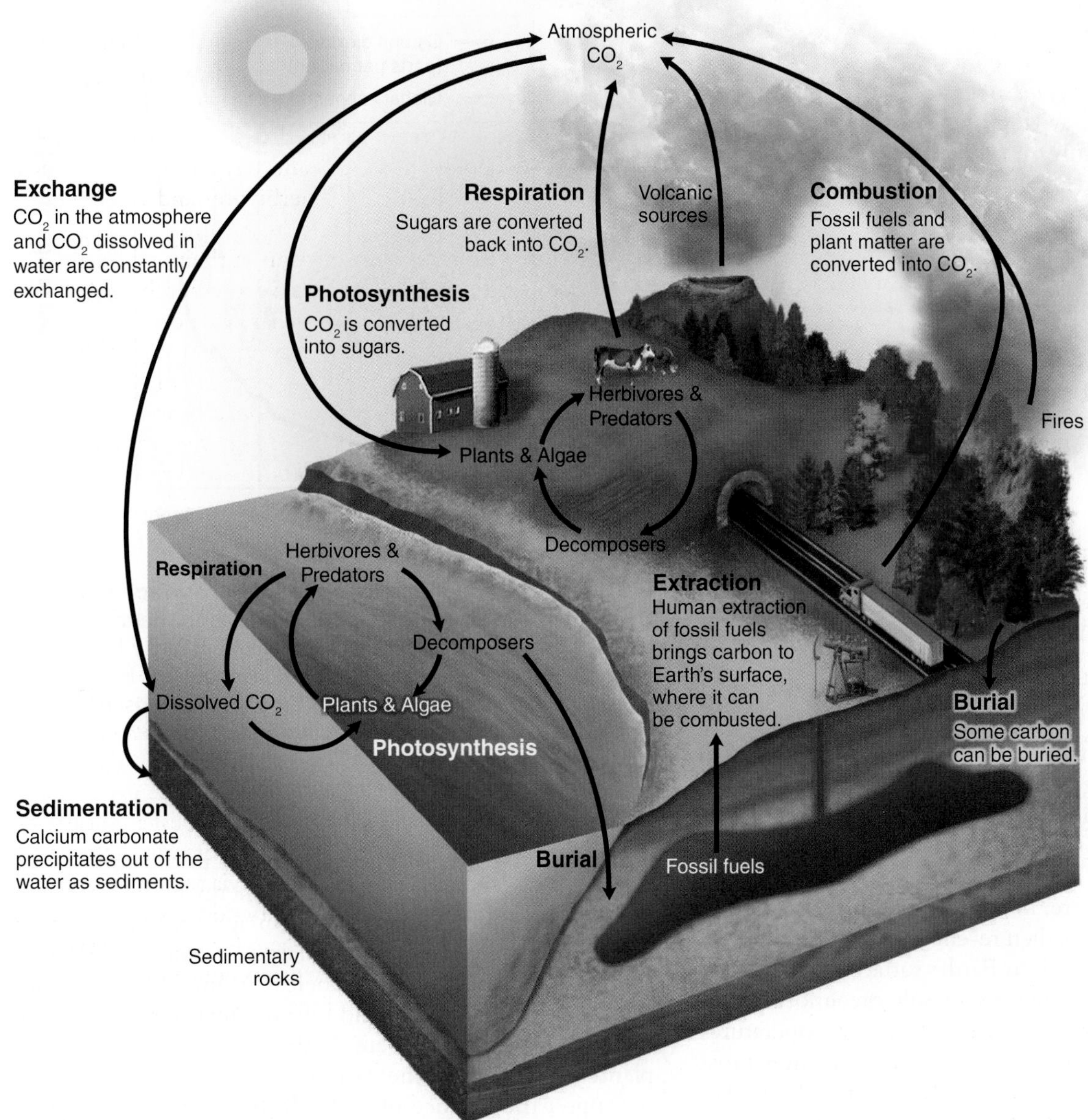
Atmospheric CO_2
Exchange
CO_2 in the atmosphere and CO_2 dissolved in water are constantly exchanged.
Respiration
Sugars are converted back into CO_2.
Volcanic sources
Combustion
Fossil fuels and plant matter are converted into CO_2.
Photosynthesis
CO_2 is converted into sugars.
Herbivores & Predators
Fires
Plants & Algae
Decomposers
Respiration
Herbivores & Predators
Extraction
Human extraction of fossil fuels brings carbon to Earth's surface, where it can be combusted.
Decomposers
Dissolved CO_2
Plants & Algae
Burial
Some carbon can be buried.
Photosynthesis
Sedimentation
Calcium carbonate precipitates out of the water as sediments.
Burial
Fossil fuels
Sedimentary rocks

FIGURE 4.1 The carbon cycle. Producers take up carbon from the atmosphere via photosynthesis and pass it on to consumers and decomposers. Some inorganic carbon sediments precipitate out of the water to form sedimentary rock while some organic carbon may be buried and become fossil fuels. Respiration by organisms returns carbon to the atmosphere and water. Combustion of fossil fuels and other organic matter returns carbon to the atmosphere.

carbon by fires or volcanoes release carbon into the atmosphere as CO_2 or into the soil as ash.

As you can see, combustion, respiration, and decomposition operate in very similar ways: All three processes cause organic molecules to be broken down to produce CO_2, water, and energy. However, respiration and decomposition are biotic processes, whereas combustion is an abiotic process.

Human Impacts on the Carbon Cycle

In the absence of human disturbance, the exchange of carbon between Earth's surface and atmosphere is in a steady state. Carbon taken up by photosynthesis eventually ends up in the soil. Decomposers in the soil gradually release that carbon at roughly the same rate it is added. Similarly, the gradual movement of carbon into the buried or fossil fuel reservoirs is offset by the slow processes that release it.

AP® Exam Tip

A very important part of learning about cycles is to know how humans have impacted each cycle.

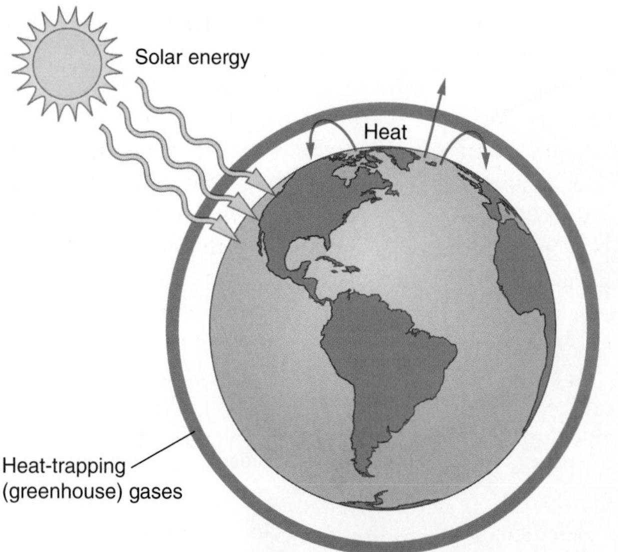

FIGURE 4.2 The Earth-surface energy balance. As Earth's surface is warmed by the Sun, it radiates heat outward. Heat-trapping gases absorb the outgoing heat and reradiate some of it back to Earth. Without these greenhouse gases, Earth would be much cooler.

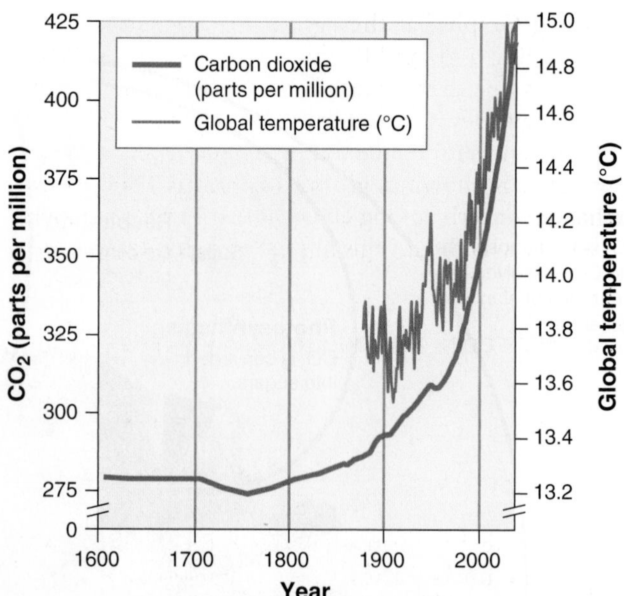

FIGURE 4.3 Changes in average global surface temperature and in atmospheric CO_2 concentrations. Earth's average global surface temperature has increased steadily for at least the past 100 years. Carbon dioxide concentrations in the atmosphere have varied over geologic time, but have risen steadily since 1960. *(Data from http://data.giss.nasa.gov/gistemp/graphs_v3/ and http://www.esrl.noaa.gov/gmd/ccgg/trends/#mlo_full)*

A major current concern is that the combustion of CO_2 is leading to the warming of our planet. As **FIGURE 4.2** shows, our thick planetary atmosphere contains many gases. When light energy from the Sun hits Earth's surface, the energy is absorbed and a portion is emitted as heat. Some of the gases in the Earth's atmosphere, known as **greenhouse gases**, absorb this heat. The gases then re-emit this absorbed heat in all directions, including back to Earth's surface, which causes the planet to become warmer. As a result, greenhouse gases help keep Earth's surface within the range of temperatures at which life can flourish. The gas that contributes most to warming of the atmosphere is carbon dioxide (CO_2). During most of the history of life on Earth, greenhouse gases have been present in the atmosphere at fairly constant concentrations for long periods of time. For example, before the Industrial Revolution, the atmospheric concentration of carbon dioxide had changed very little for hundreds of years, as you can see in **FIGURE 4.3**. As we have discussed, during this time carbon entering any of the various reservoirs was balanced by carbon leaving these reservoirs.

Since the Industrial Revolution, human activities have had a major influence on carbon cycling. Largely due to the combustion, or burning, of fossil fuels, humans have been moving fossilized carbon into the atmosphere at a much faster rate than carbon leaves the atmosphere through the processes of sedimentation and burial. As a result, atmospheric carbon concentrations have increased and this has upset the balance between Earth's carbon reservoirs and the atmospheric reservoir. The excess CO_2 in the atmosphere acts to increase the retention of heat energy in the biosphere.

Tree harvesting is another human activity that can affect the carbon cycle. Trees store a large amount of carbon in their wood, both above and below ground. The destruction of forests by cutting and burning increases the amount of CO_2 in the atmosphere. Unless enough new trees are planted to recapture the carbon, the destruction of forests will upset the balance of CO_2. To date, large areas of forest, including tropical forests as well as North American and European temperate forests, have been converted into pastures, grasslands, and croplands. In addition to destroying a great deal of biodiversity, this destruction of forests has added large amounts of carbon to the atmosphere. The increases in atmospheric carbon due to human activities have been partly offset by an increase in carbon absorption by the ocean. While some regions of the world have experienced increased reforestation, such as the northeastern United States, the loss of forest remains a global concern.

As a result of combusting fossil fuels and tree harvesting, atmospheric CO_2 concentrations have dramatically increased during the past 200 years. From 1600 to 1800, CO_2 concentrations were steady at approximately 280 parts per million (ppm). Since 1800, CO_2 concentrations have rapidly increased to approximately 420 ppm today, as you can see in Figure 4.3. In this figure, you can also see that during roughly the same period, global temperatures have displayed an overall increase, as we would expect given that CO_2 is a greenhouse gas. (Note that this graph has two *y* axes. See the appendix

Greenhouse gases Gases in Earth's atmosphere that trap heat near the surface.

"Reading Graphs" at the end of the book to learn more about reading a graph like this one.) The current scientific evidence is that the increase in atmospheric CO_2 during the last two centuries is caused by human activities. The increase in global temperatures due to humans producing more greenhouse gases is known as **global warming**. Global warming is a major concern among environmental scientists and policy makers. We will discuss this in much greater detail in Unit 9.

4-2 How does nitrogen cycle within ecosystems?

The nitrogen cycle includes many chemical transformations

Nitrogen is used to form amino acids, the building blocks of proteins, and nucleic acids, the building blocks of DNA and RNA. Because so much of it is required, nitrogen is often a limiting nutrient for plants and algae. A **limiting nutrient** is a nutrient required for the growth of an organism but available in a lower quantity than other nutrients. The presence of a limiting nutrient, such as nitrogen, constrains the growth of plants and algae. Adding other nutrients, such as water or phosphorus, will not improve their growth.

The Nitrogen Cycle

The **nitrogen cycle** is the movement of nitrogen around the biosphere among reservoir sources and sinks. As nitrogen moves through an ecosystem, it experiences many chemical transformations. **FIGURE 4.4** shows the five major transformations in the nitrogen cycle: nitrogen fixation, nitrification, assimilation, mineralization, and denitrification. Unlike the carbon cycle, most reservoirs hold nitrogen for relatively short periods of time.

Nitrogen Fixation

While Earth's atmosphere is 78 percent nitrogen by volume, making it the largest nitrogen reservoir, the vast majority of that nitrogen is in the form of nitrogen gas, which most plants and algae cannot use. However, **nitrogen fixation** is a process that converts nitrogen gas (N_2) in the atmosphere into forms of nitrogen that plants and algae can use. As you can see in Figure 4.4, nitrogen fixation can occur through biotic or abiotic processes.

In the biotic process, a few species of bacteria can convert N_2 gas directly into ammonia (NH_3), which is rapidly converted to ammonium (NH_4^+), a form that is readily used by plants and algae. Nitrogen-fixing organisms include cyanobacteria (also known as blue-green algae) and certain bacteria that live within the roots of legumes, which include plants such as peas, beans, and a few species of trees. Nitrogen-fixing organisms use the fixed nitrogen to synthesize their own tissues, then excrete any excess. Cyanobacteria, which are primarily aquatic organisms, excrete excess ammonium ions into the water, where they can be taken up by aquatic plants and algae. Nitrogen-fixing bacteria that live within plant roots excrete excess ammonium ions into the plant's root system; the plant, in turn, supplies the bacteria with sugars it produces via photosynthesis.

Nitrogen fixation can also occur through two abiotic pathways. N_2 can be fixed in the atmosphere by lightning or during combustion processes such as fires and the burning of fossil fuels. These processes convert N_2 into nitrate (NO_3^-), which is carried to Earth's surface in precipitation, where it is then usable by plants. Humans have developed techniques for nitrogen fixation into ammonia or nitrate to be used in plant fertilizers. Although these processes require a great deal of energy, humans now fix more nitrogen than is fixed in nature. The development of synthetic nitrogen fertilizers has led to increases in crop yields, particularly for crops such as corn that require large amounts of nitrogen.

Nitrification

Another step in the nitrogen cycle is **nitrification**, which is the conversion of ammonium (NH_4^+) into nitrite (NO_2^-) and then into nitrate (NO_3^-). These conversions are conducted by specialized species of bacteria. Although nitrite is not used by most plants and algae, nitrate is readily used.

Assimilation

Once plants or algae take up nitrogen in the form of ammonia, ammonium, nitrite, or nitrate, they incorporate the element into their tissues in a process called **assimilation**. When herbivores feed on the plants and algae, some of the nitrogen is assimilated into the tissues of the herbivores while the rest is eliminated as waste products.

Mineralization

Eventually, organisms die and their tissues decompose. In a process called **mineralization**, fungal and bacterial decomposers break down the organic matter found in dead bodies and waste products and convert these organic compounds back into inorganic compounds such as ammonium (NH_4^+). Because this process produces ammonium, the process of mineralization is sometimes called **ammonification**. The ammonium

Global warming The increase in global temperatures due to humans producing more greenhouse gases.

Limiting nutrient A nutrient required for the growth of an organism but available in a lower quantity than other nutrients.

Nitrogen cycle The movement of nitrogen around the biosphere among reservoir sources and sinks.

Nitrogen fixation The process that converts nitrogen gas in the atmosphere (N_2) into forms of nitrogen that plants and algae can use.

Nitrification The conversion of ammonia (NH_4^+) into nitrite (NO_2^-) and then into nitrate (NO_3^-).

Assimilation A process by which plants and algae incorporate nitrogen into their tissues.

Mineralization The process by which fungal and bacterial decomposers break down the organic matter found in dead bodies and waste products and convert these organic compounds back into inorganic compounds, such as inorganic ammonium (NH_4^+). *Also known as* **ammonification**.

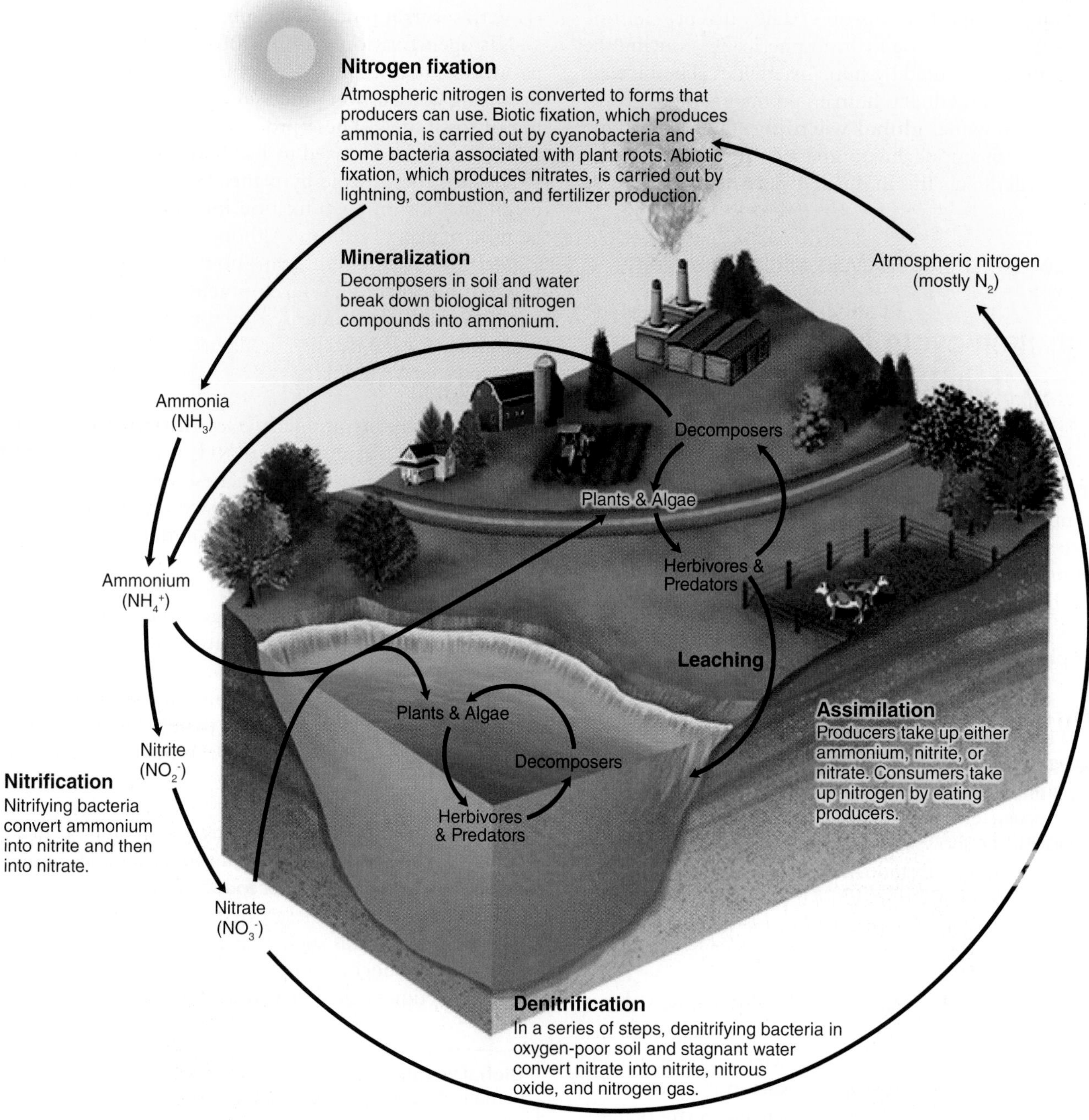

Nitrogen fixation
Atmospheric nitrogen is converted to forms that producers can use. Biotic fixation, which produces ammonia, is carried out by cyanobacteria and some bacteria associated with plant roots. Abiotic fixation, which produces nitrates, is carried out by lightning, combustion, and fertilizer production.
Mineralization
Decomposers in soil and water break down biological nitrogen compounds into ammonium.
Atmospheric nitrogen (mostly N_2)
Ammonia (NH_3)
Decomposers
Plants & Algae
Herbivores & Predators
Ammonium (NH_4^+)
Leaching
Plants & Algae
Decomposers
Herbivores & Predators
Assimilation
Producers take up either ammonium, nitrite, or nitrate. Consumers take up nitrogen by eating producers.
Nitrite (NO_2^-)
Nitrification
Nitrifying bacteria convert ammonium into nitrite and then into nitrate.
Nitrate (NO_3^-)
Denitrification
In a series of steps, denitrifying bacteria in oxygen-poor soil and stagnant water convert nitrate into nitrite, nitrous oxide, and nitrogen gas.

FIGURE 4.4 The nitrogen cycle. The nitrogen cycle moves nitrogen from the atmosphere and into soils through several fixation pathways, including the production of fertilizers by humans. In the soil, nitrogen can exist in several forms. Denitrifying bacteria release nitrogen gas back into the atmosphere.

produced by this process can either be taken up by plants and algae in the ecosystem or be converted into nitrite (NO_2^-) and nitrate (NO_3^-) through the process of nitrification.

Denitrification The conversion of nitrate (NO_3^-) in a series of steps into the gases nitrous oxide (N_2O) and, eventually, nitrogen gas (N_2), which is emitted into the atmosphere.

Anaerobic An environment that lacks oxygen.

Denitrification

The final step that completes the nitrogen cycle is **denitrification**, which is the conversion of nitrate (NO_3^-) in a series of steps into the gases nitrous oxide (N_2O) and, eventually, nitrogen gas (N_2), which is emitted into the atmosphere. Denitrification is conducted by specialized bacteria that live under **anaerobic** conditions, which means the environment lacks oxygen. Such environments include waterlogged soils or the bottom sediments of

TABLE 4.1	A summary of the five transformations of nitrogen that occur in the nitrogen cycle
Nitrogen fixation	Nitrogen fixation converts N_2 from the atmosphere. Biotic processes convert N_2 to ammonia (NH_3), whereas abiotic processes convert N_2 to nitrate (NO_3^-).
Nitrification	Nitrifying bacteria convert ammonium (NH_4^+) into nitrite (NO_2^-) and then into nitrate (NO_3^-).
Assimilation	Plants and algae take up either ammonium (NH_4^+) or nitrate (NO_3^-). Herbivores assimilate nitrogen by eating plants and algae.
Mineralization/ Ammonification	Decomposers in soil and water break down biological nitrogen compounds into ammonium (NH_4^+).
Denitrification	In a series of steps, denitrifying bacteria in oxygen-poor soil and stagnant water convert nitrate (NO_3^-) into nitrous oxide (N_2O) and eventually nitrogen gas (N_2).

oceans, lakes, and swamps. These bacteria do not live under **aerobic** conditions, which are environments with abundant oxygen.

As you can see, the nitrogen cycle is a fairly complicated cycle because of the many transformations of nitrogen that take place. A summary of these processes can be found in TABLE 4.1.

Human Impacts on the Nitrogen Cycle

Nitrogen is a limiting nutrient in most terrestrial ecosystems, so excess inputs of various forms of nitrogen can have consequences in these ecosystems. For example, nitrate is readily transported through the soil with water through **leaching**, a process in which dissolved molecules are transported through the soil via groundwater. In addition, adding nitrogen to soils in fertilizers ultimately increases atmospheric concentrations of nitrogen in regions where the fertilizer is applied. This nitrogen can be transported through the atmosphere and deposited by rainfall in natural ecosystems that have adapted over time to a particular level of nitrogen availability. The added nitrogen can alter the distribution or abundance of species in those ecosystems.

In one study of nine different terrestrial ecosystems across the United States, scientists added nitrogen fertilizer to some plots and left other plots unfertilized as controls. They found that adding nitrogen reduced the number of species by up to 48 percent because some species that could survive under low-nitrogen conditions could no longer compete against larger plants that thrived under high-nitrogen conditions. Other studies have documented cases in which plant communities that have grown on low-nitrogen soils for millennia are now experiencing changes in their species composition as a result of nitrogen being deposited from the atmosphere. An influx of fixed nitrogen due to human activities has favored colonization by new species that are better adapted to soils with higher fertility.

In this module, we have examined the reservoirs and processes that cause carbon and nitrogen to cycle around the planet. We have also considered the many ways in which human activities can impact these reservoirs and processes. In the next module, we build on this knowledge by examining the cycling of phosphorus and water, which also play important roles in aquatic and terrestrial biomes.

Aerobic An environment with abundant oxygen.

Leaching A process in which dissolved molecules are transported through the soil via groundwater.

Module 4 AP® Review

Preparing for the AP® Exam

Learning Goals Revisited

4-1 How does carbon cycle within ecosystems?

Carbon cycles through ecosystems as a result of seven processes: photosynthesis, respiration, exchange, sedimentation, burial, extraction, and combustion.

4-2 How does nitrogen cycle within ecosystems?

Nitrogen cycles through ecosystems as a result of five processes: nitrogen fixation, nitrification, assimilation, mineralization, and denitrification.

AP® Practice Questions

Preparing for the AP® Exam

Multiple-Choice Questions

1. Which process in the carbon cycle is considered a slow part of the cycle?
   (a) sedimentation
   (b) respiration
   (c) photosynthesis
   (d) combustion

2. The largest carbon reservoir is found in
   (a) oceans.
   (b) the atmosphere.
   (c) living organisms.
   (d) fossil fuels.

3. Which process is causing a major human impact on the carbon cycle?
   (a) sedimentation
   (b) combustion
   (c) photosynthesis
   (d) respiration

4. Which process in the nitrogen cycle involves converting nitrogen gas (N_2) into ammonia (NH_3)?
   (a) fixation
   (b) nitrification
   (c) assimilation
   (d) mineralization

5. Which of the following processes is also known as ammonification?
   (a) nitrogen fixation
   (b) nitrification
   (c) mineralization
   (d) denitrification

6. Which organism and process shows the correct pairing?
   (a) photosynthesis: animals
   (b) nitrification: viruses
   (c) denitrification: bacteria
   (d) photosynthesis: bacteria

Free-Response Question

Nitrogen is crucial for sustaining life in both terrestrial and aquatic ecosystems. See the steps of the nitrogen cycle on the diagram below:

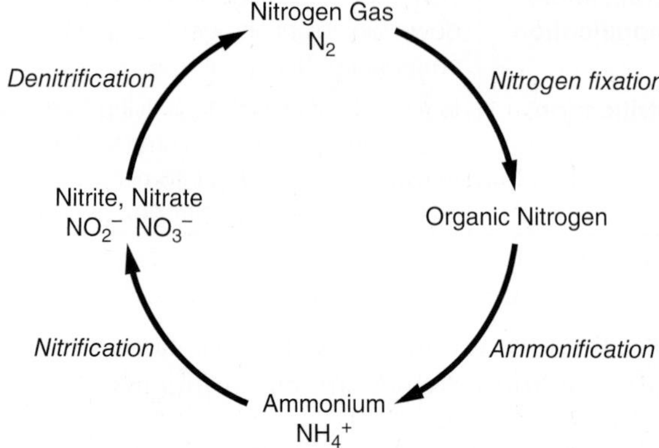

(a) **Identify** which organisms fix nitrogen. (2 pts.)
(b) **Identify** which organisms do nitrification. (2 pts.)
(c) **Describe** the following steps in the nitrogen cycle:
  (i) nitrogen fixation (1 pt.)
  (ii) ammonification (1 pt.)
(d) **Identify** an anthropogenic effect on the nitrogen cycle. (2 pts.)
(e) Human sources of nitrogen can lead to acid precipitation. **Propose a solution** to reduce sources of acid precipitation. (2 pts.)

Module 5

Unit 1	1	2	3	4	5	6	7

The Phosphorus and Hydrologic (Water) Cycles

In the previous module, we examined how carbon and nitrogen cycles through ecosystems in different forms as it moves between reservoirs via biotic and abiotic processes. We examined how these reservoirs can serve as both sinks and sources for these two elements and how human activities can alter the transformations of the elements, thereby upsetting the amount of carbon and nitrogen in each reservoir. In this module, we will expand on these biogeochemical cycles as we examine the cycling of phosphorus and water within ecosystems. We will see that these two important chemical components have their own distinct reservoirs and their own unique transformations. We will also examine the ways in which human activities are altering the distribution of phosphorus and water among the different reservoirs and the consequences of these activities.

Learning Goals

After reading this module you should be able to

5-1 explain how phosphorus cycles within ecosystems.

5-2 describe how water cycles within ecosystems.

5-1 How does phosphorus cycle within ecosystems?

The phosphorus cycle moves between land and water

Plants and animals need phosphorus for many biological processes. Phosphorus is a major component of DNA and RNA as well as ATP (adenosine triphosphate), the molecule cells use for energy transfer. In plants, phosphorus is a limiting nutrient second only to nitrogen in its importance for successful agricultural yields. Thus phosphorus, like nitrogen, is commonly added to soils in the form of fertilizer. As we will see, phosphorus is even more limiting in aquatic ecosystems.

The Phosphorus Cycle

The **phosphorus cycle** is the movement of phosphorus around the biosphere among sources and sinks. As shown in **FIGURE 5.1**, the phosphorus cycle primarily operates between land and water. The major reservoir for phosphorus is the collection of rocks and sediments comprised of phosphorus-containing minerals. In contrast to the carbon and nitrogen cycles, the phosphorus cycle has no gas phase, although phosphorus does enter the atmosphere in very small amounts when dust is dissolved in rainwater or sea spray. An important consequence of the lack of a gas phase is that it limits the movement of phosphorus from the ocean back to terrestrial and freshwater environments, which makes it a limiting nutrient in these ecosystems. Unlike nitrogen, which can be transformed into many different compounds, phosphorus rarely changes form; it is typically found in the form of phosphate (PO_4^{3-}). There are five processes that drive the phosphorus cycle: assimilation, mineralization, sedimentation, geologic uplift, and weathering.

Assimilation and Mineralization

The biotic processes that affect the phosphorus cycle are not complex. Plants and animals on land and in the water take up inorganic phosphate and assimilate the phosphorus into their tissues as organic phosphorus. The waste products and eventual dead bodies of these organisms are decomposed by fungi and bacteria, which causes the mineralization of organic phosphorus back to inorganic phosphate.

Sedimentation, Geologic Uplift, and Weathering

The abiotic processes of the phosphorus cycle involve movements between the water and the land. In water, phosphorus is not very soluble; so much of it precipitates out of solution in the form of phosphate-laden sediments in the ocean. You can see this process in the left corner of Figure 5.1. Over time, geologic forces can lift these ocean layers up and they become mountains. The phosphate rocks in the mountains

Phosphorus cycle The movement of phosphorus around the biosphere among reservoir sources and sinks.

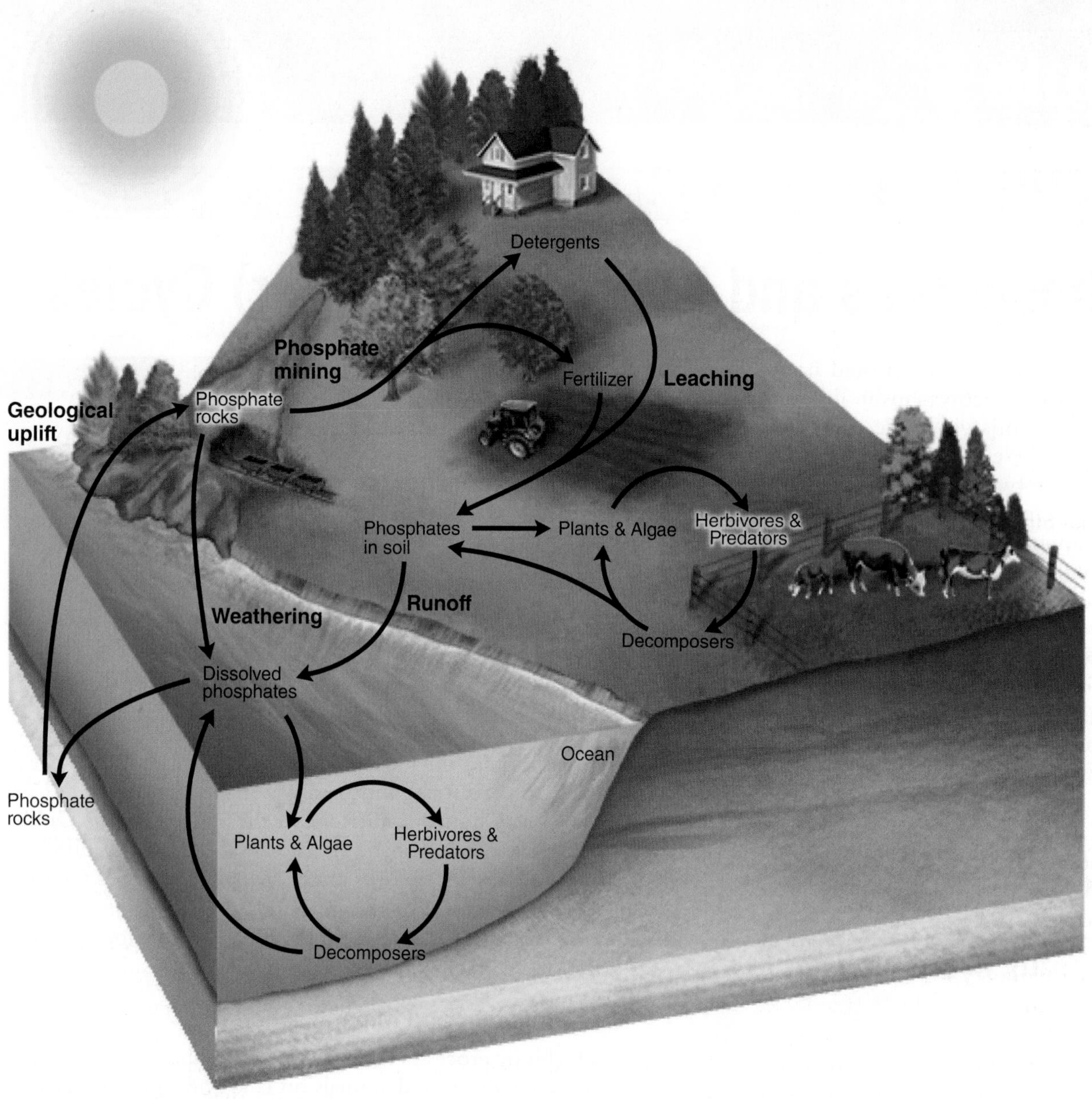

FIGURE 5.1 The phosphorus cycle. The phosphorus cycle begins with the weathering or mining of phosphate rocks and use of phosphate fertilizer, which releases phosphorus into the soil and water. This phosphorus can be used by producers and subsequently moves through the food web. In water, phosphorus can precipitate out of solution and form sediments, which over time are transformed into new phosphate rocks.

are slowly weathered by natural forces including rainfall and this weathering brings phosphorus to terrestrial and aquatic habitats. Phosphorus is tightly held by soils, so it is not easily leached from soils into water bodies. Because so little phosphorus leaches into water bodies and because much of what enters water precipitates out of solution, very little dissolved phosphorus is naturally available in streams, rivers, and lakes. As a result, phosphorus is a limiting nutrient in many aquatic systems.

Human Impacts on the Phosphorus Cycle

Humans have had a dramatic effect on the phosphorus cycle. For example, humans mine the phosphate sediments that are found in the geologically uplifted mountains to produce fertilizer. When these fertilizers are applied to lawns, gardens, and agricultural fields, excess phosphorus can leach into water bodies. Because aquatic systems are commonly limited by a low availability of phosphorus, even

FIGURE 5.2 Algal bloom. When excess phosphorus enters waterways, it can stimulate a sudden and rapid growth of algae that turns the water bright green, like this area along the Susquehanna River in Pennsylvania. Some species of blooming algae can release toxins that are harmful to people, pets, and wildlife. The algae eventually die, and the resulting increase in decomposition can reduce dissolved oxygen to levels that are lethal to fish and shellfish. *(Michael P. Gadomski/Science Source)*

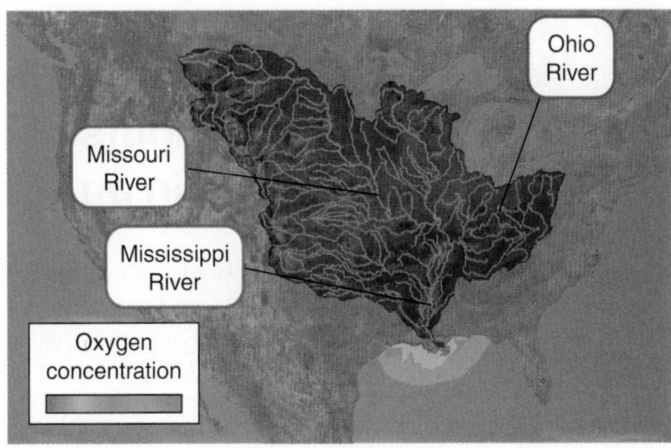

FIGURE 5.3 Dead zone. The Mississippi River drains water from more than one-third of the United States and, in doing so, carries phosphorus to the Gulf of Mexico. In the Gulf, this phosphorus fertilizes the algae, causing it to rapidly grow and subsequently die. The dead algae then decompose, which consumes most of the oxygen in the water. This results in a dead zone that cause some species to leave the area and other less mobile species to die. *(Source: https://coastalscience.noaa.gov/news/noaa-forecasts-very-large-dead-zone-for-gulf-of-mexico/)*

small inputs of leached phosphorus into these systems can greatly increase the growth of algae. Phosphorus inputs can cause a rapid increase in the algal population of a waterway, known as an **algal bloom**, that can quickly increase the biomass of algae in the ecosystem (**FIGURE 5.2**). Some types of algae, such as the cyanobacteria, can produce dangerous toxins when they bloom. These toxins can harm people, pets, domesticated animals, and wildlife species that drink the water. These toxic algal blooms commonly result in the closing of beaches and the shutdown of intake pipes for public water.

After the algae bloom, there is a second major impact on water quality. As the algae die, their decomposition consumes large amounts of oxygen. As a result, the water becomes low in oxygen, or **hypoxic**. When oxygen concentrations become so low that it kills fish and other aquatic animals, we refer to it as a **dead zone**. Dead zones occur around the world and one of the largest occurs where the Mississippi River empties into the Gulf of Mexico. As you can see in **FIGURE 5.3**, the Mississippi River drains about one-third of the continental United States, carrying a large amount of nutrients to the Gulf of Mexico from farms, suburbs, and urban areas. In the Gulf, the algae grow and then die, causing a massive dead zone that negatively impacts the fish and shellfish in the region.

A second major source of phosphorus in waterways is from the use of household detergents. From the 1940s through the 1990s, laundry detergents in the United States contained phosphates to make clothes cleaner. As a result, the water discharged from washing machines contained phosphorus, which inadvertently fertilized streams, rivers, and lakes. Because ecological dead zones triggered by excess phosphorus cause substantial environmental and economic damage, manufacturers stopped adding phosphates to laundry detergents in 1994 and 17 states have imposed partial or total bans on phosphates in dishwashing detergents since 2010. In Europe, the European Union passed a ban on phosphates in laundry detergents starting in 2013 and a ban on phosphates in dishwashing detergents starting in 2017.

In addition to causing algal blooms, increases in phosphorus concentrations can alter plant communities. For example, southern Florida has experienced agricultural expansion to such an extent that the water that flows through the Everglades has elevated phosphorus concentrations. This has altered the ecosystem of the Everglades. For instance, over time, the cattail plant has become more common and sawgrass has declined. As a result, animals that depend on sawgrass are now experiencing reduced food and habitat.

See "Do the Math: Measuring Inputs of Nitrogen and Phosphorus in an Urban Environment" on page 62 to apply your knowledge of the nitrogen and phosphorus cycles and calculate the percentage of various phosphorus sources that enters the Mississippi River from the city of St. Paul.

Algal bloom A rapid increase in the algal population of a waterway.

Hypoxic Low in oxygen.

Dead zone When oxygen concentrations become so low that it kills fish and other aquatic animals.

DO THE MATH Measuring Inputs of Nitrogen and Phosphorus in an Urban Environment

Preparing for the AP® Exam

As we have seen in our discussions of nutrients throughout this unit, nutrients can come from multiple sources. Scientists recently quantified the various sources of nitrogen and phosphorus that enter the Mississippi River from St. Paul, Minnesota. Using the data below listing the amount of nitrogen coming into the river, we can calculate the percentages of each source.

Source	Nitrogen (kg/km^2/year)	Phosphorus (kg/km^2/year)
Atmospheric deposition	1,300	40
Household pet waste	2,100	280
Residential fertilizer	3,800	0
Weathering	0	10
County compost	200	30

To calculate the percentage of each nitrogen source, we can begin by calculating the sum of all sources:

$$\text{Total nitrogen} = 1{,}300\ kg/km^2/year + 2{,}100\ kg/km^2/year + 3{,}800\ kg/km^2/year + 0\ kg/km^2/year + 200\ kg/km^2/year$$
$$= 7{,}400\ kg/km^2/year$$

We can then calculate each percentage by dividing each source by the total of all nitrogen sources (rounded to two significant digits):

Percent nitrogen from atmospheric deposition = $1{,}300\ kg/km^2/year \div 7{,}400\ kg/km^2/year \times 100\% = 18\%$
Percent nitrogen from household pet waste = $2{,}100\ kg/km^2/year \div 7{,}400\ kg/km^2/year \times 100\% = 28\%$
Percent nitrogen from residential fertilizer = $3{,}800\ kg/km^2/year \div 7{,}400\ kg/km^2/year \times 100\% = 51\%$
Percent nitrogen from weathering = $0\ kg/km^2/year \div 7{,}400\ kg/km^2/year \times 100\% = 0\%$
Percent nitrogen from county compost = $200\ kg/km^2/year \div 7{,}400\ kg/km^2/year \times 100\% = 3\%$

YOUR TURN Using the phosphorus data from the table, calculate the percentage of each phosphorus source that enters the Mississippi River from the city of St. Paul.

5-2 How does water cycle within ecosystems?

The hydrologic cycle moves water through the biosphere

Water is essential to life. It makes up over one-half of a typical mammal's body weight, and no organism can survive without it. Water allows essential molecules to move within and between cells, draws nutrients into the leaves of trees, dissolves and removes toxic materials, and performs many other critical biological functions. On a larger scale, water is the primary agent responsible for dissolving and transporting the chemical elements necessary for living organisms. The movement of water around the biosphere among reservoir sources and sinks, is known as the **hydrologic cycle**. As it cycles, water moves through source and sink reservoirs in its liquid, gas, and solid phases. The largest reservoir of water is in the world's oceans while the ice found in the polar ice caps and glaciers and the water present in soils serve as much smaller reservoirs.

Hydrologic cycle The movement of water around the biosphere among reservoir sources and sinks.

Transpiration The release of water from leaves into the atmosphere during photosynthesis.

The Hydrologic Cycle

FIGURE 5.4 shows how the hydrologic cycle works. Heat from the Sun causes water to evaporate from oceans, lakes, and soils. Solar energy also provides the energy for photosynthesis, during which plants release water from their leaves into the atmosphere—a process known as **transpiration**. The water vapor that enters the atmosphere eventually cools and forms clouds, which, in turn, produce precipitation in the form of rain, snow, and hail. Some precipitation falls back into the ocean and some falls on land.

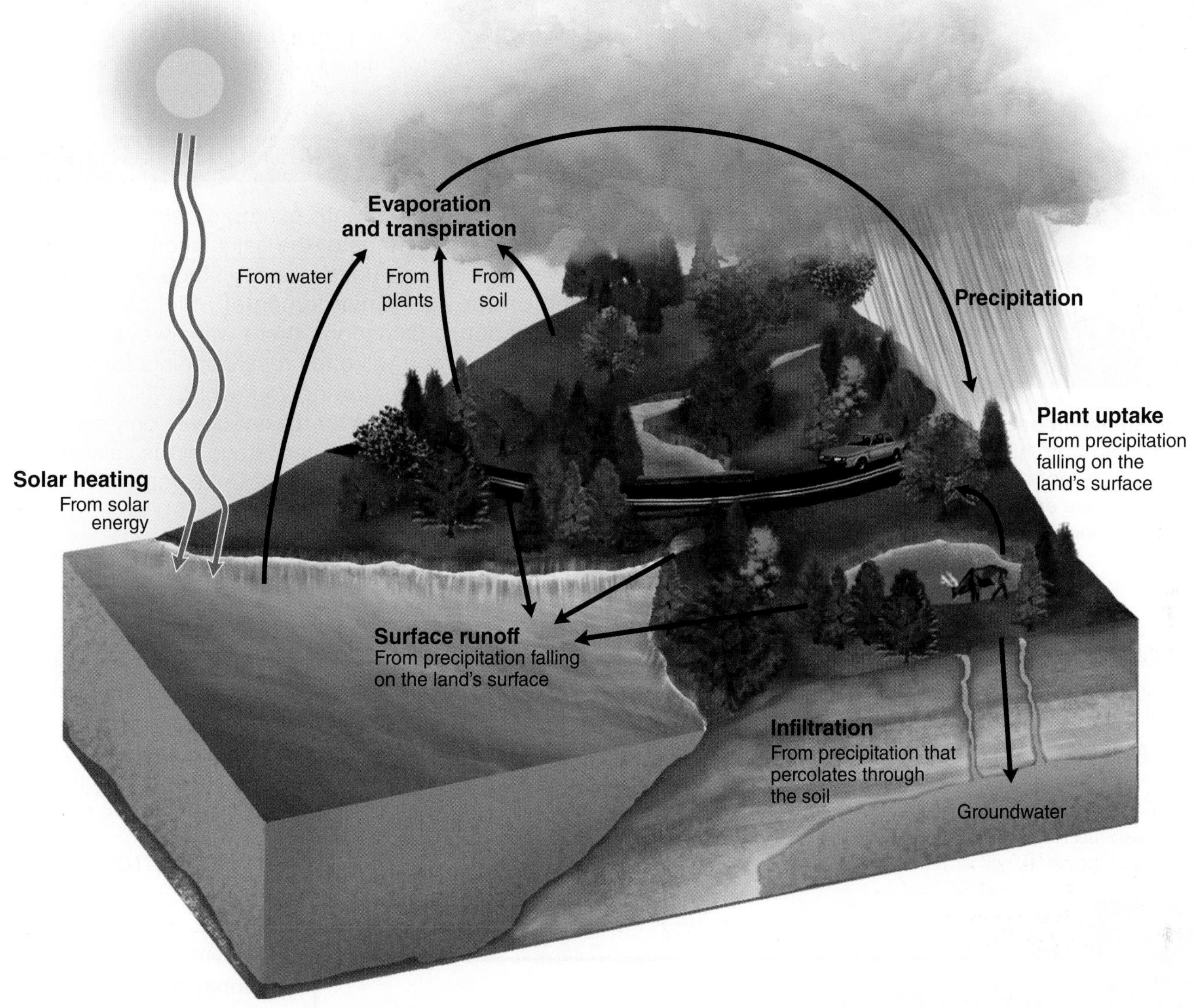

FIGURE 5.4 The hydrologic cycle. Water moves from the atmosphere to Earth's surface and back to the atmosphere.

When water falls on land, it may take one of three distinct routes. First, it may return to the atmosphere by evaporation or, after being taken up by plant roots, by transpiration. The combined amount of evaporation and transpiration is called **evapotranspiration**. Alternatively, water can be absorbed by the soil and percolate down into the groundwater. Finally, water can move as **runoff** across the land surface and into streams and rivers, eventually reaching the ocean, which is the ultimate reservoir of water on Earth's surface. As water in the ocean evaporates, the cycle begins again. While the hydrologic cycle describes the movement of water between Earth and the atmosphere, we have seen that it also plays a key role in moving many other elements that are dissolved in the water, including carbon, nitrogen, and phosphorus.

Human Activities and the Hydrologic Cycle

Because Earth is a closed system with respect to matter, water never leaves our planet. Nevertheless, human activities can alter the hydrologic cycle in several ways. For example, harvesting trees from a forest can reduce evapotranspiration by reducing plant biomass. If evapotranspiration decreases, then runoff or percolation will increase. On a moderate or steep slope, most water will leave the land surface as runoff.

Evapotranspiration The combined amount of evaporation and transpiration.

Runoff Water that moves across the land surface and into streams and rivers.

FIGURE 5.5 Human impacts on the hydrologic cycle. When forests are cleared from hillsides, more water enters the soils and there are fewer roots to hold the soil in place. As a result, heavy rains can cause erosion and flooding, such as this village in South Sulawesi, Indonesia in 2021. *(Moh Niaz Sharief/ZUMAPRESS.com)*

At the same time, harvesting the trees results in fewer roots that can hold the soil on the mountainside. Thus, clear-cutting a mountain slope can lead to massive erosion and flooding, as you can see in **FIGURE 5.5**.

Similarly, paving over land surfaces to build roads, businesses, and homes reduces the amount of percolation that can take place in a given area, increasing runoff and evaporation. Humans can also alter the hydrologic cycle by diverting water from one area to another to provide water for drinking, irrigation, and industrial uses. In "Do the Math: Raising Mangoes" you can apply some basic math skills to get a sense of the decisions involved in planting trees to reforest Haiti and reduce the impacts of humans on the hydrologic cycle.

AP® Exam Tip

Notice in the Do the Math example below, that units are used in every stage of the calculation, not just the answer. Always use units in every stage of the calculation on the AP® Environmental Science exam. In Free-Response Questions there will always be math problems, and you will often receive points for the correct answer *and* for showing your work, which is called the "set up." The points for the set up are only earned if there are units written for every stage of the calculation. If one number lacks the correct units you will not get these points.

In this module, we have learned about the major reservoirs and processes that are responsible for the cycling of phosphorus and water around the planet. We also examined how human activities are affecting these processes and the consequences of doing so in terrestrial and aquatic ecosystems. In the next module, we will consider the productivity of ecosystems and evaluate which ecosystems are the most and least productive.

DO THE MATH Raising Mangoes

Preparing for the AP® Exam

In Haiti, farmers are being encouraged to plant mango trees to help reduce runoff and increase the uptake of water by the soil and the trees. Consider a group of Haitian farmers that decides to plant mango trees. Mango saplings cost $10 each. Once the trees become mature, each tree will produce $75 worth of fruit per year. A village of 225 people decides to pool its resources and set up a community mango farm. Their goal is to generate a per capita income of $300 per year for everyone in the village.

1. How many mature trees will the village need to meet the goal?

Total annual income desired:

$$\$300/\text{person} \times 225 \text{ persons} = \$67{,}500$$

Number of trees needed to produce $67,500 in annual income:

$$\$67{,}500 \div \$75/\text{tree} = 900 \text{ trees}$$

2. Each tree requires 25 m^2 of space. How many hectares must the village set aside for the mango farm?

$$900 \text{ trees} \times 25 \text{ m}^2/\text{tree} = 22{,}500 \text{ m}^2 = 2.25 \text{ ha}$$

YOUR TURN Each tree requires 20 L of water per day during the 6 hot months of the year (180 days). The water must be pumped to the farm from a nearby stream. How many liters of water are needed each year to water the farm of 900 trees?

Module 5 AP® Review

Preparing for the AP® Exam

Learning Goals Revisited

5-1 How does phosphorus cycle within ecosystems?

The phosphorus cycle involves a large pool of phosphorus in rock that is formed by the precipitation of phosphate onto the ocean floor. When geologic forces cause these sediments to uplift and become mountains, the phosphorus-containing rock can weather and provide phosphorus to algae, plants, and animals. When these organisms die and decompose, some of the phosphorus is recycled in the food web and the rest is returned to the ocean.

5-2 How does water cycle within ecosystems?

The major reservoir for water is the ocean and the cycling of water is driven by the energy of the Sun. Water evaporates from water bodies and the land and this evaporated water eventually drops back to Earth as precipitation, where it can percolate into the soil, be taken up by plants, or run off the surface of the soil and head back to the ocean.

Practice Math and Graphing

Preparing for the AP® Exam

As we discussed in this module, the availability of nutrients affects the plant growth of many ecosystems. To quantify the sensitivity of terrestrial ecosystems to additions of nitrogen and phosphorus, researchers examined studies conducted around the world and determined the average increase in plant growth when nitrogen, phosphorus, or both were added to an ecosystem compared to when no extra nutrients were added.

1. Practice Math

Based on the table below, calculate the percentage of increased plant growth for each biome when fertilized with added nitrogen, added phosphorus, or both.

2. Practice Graphing

(a) Using your table of calculated percentages, create a bar graph that illustrates the percent increase in plant growth among three terrestrial ecosystems.

(b) Based on this graph, which ecosystems appear to be the most sensitive to human inputs of nitrogen, phosphorus, or both?

(c) Why might the combination of added nitrogen and phosphorus cause a larger increase in plant growth than either nutrient alone?

Ecosystem	Plant growth with no added nutrients (g/m^2)	Plant growth with added nitrogen (g/m^2)	Plant growth with added phosphorus (g/m^2)	Plant growth with added nitrogen and phosphorus (g/m^2)
Grassland	90	111	112	144
Tundra	200	246	220	284
Forest/shrubland	380	490	578	570

AP® Practice Questions

Preparing for the AP® Exam

Multiple-Choice Questions

1. Phosphorus
   (a) is a limiting nutrient in many aquatic systems.
   (b) has an important gaseous phase.
   (c) is easily lost from soils due to leaching.
   (d) is often produced by volcanic eruptions.

2. Hypoxia is a direct result of adding too much phosphorus to the natural cycle. Hypoxia is
   (a) too high oxygen concentration in water.
   (b) too low oxygen concentration in water.
   (c) too high algae blooming in water.
   (d) too low algae blooming in water.

3. What is the largest reservoir in the water cycle?
   (a) lakes
   (b) oceans
   (c) soils
   (d) polar ice caps and glaciers

4. Human construction of buildings and pavement affect which of the following cycles the most?
   (a) phosphorus cycle
   (b) nitrogen cycle
   (c) carbon cycle
   (d) hydrologic cycle

5. Haitian farmers are buying mango tree saplings for $10 each, and the full-grown trees produce $75 of fruit each year. If a farmer wishes to earn $1,500 per year when the trees are grown, how much will the farmer have to spend on saplings?
   (a) $100
   (b) $150
   (c) $200
   (d) $750

Free-Response Question

Biogeochemical cycles are systems composed of abiotic and biotic processes.

(a) **Describe** a biological process by which phosphorus is taken into tissues. (1 pt.)
(b) **Describe** a biological process by which phosphorus is removed from tissues. (1 pt.)
(c) **Describe** one abiotic process by which phosphorus moves through water and land. (1 pt.)
(d) **Explain** why phosphorus tends to be the limiting factor in aquatic ecosystems. (1 pt.)
(e) **Identify** two sources of excess phosphorus introduced by humans. (2 pts.)
(f) **Explain** how excess phosphorus can alter plant communities. (1 pt.)
(g) The water cycle allows nutrients to pass through abiotic and biotic phases. **Identify** two processes through which the water cycle moves nutrients through abiotic and biotic systems. (2 pts.)
(h) **Propose a solution** for mitigating the amount of phosphorus ending up in aquatic systems. (1 pt.)

Module 6

Unit 1	1	2	3	4	5	6	7

Primary Productivity

In the previous two modules, we examined how matter cycles through ecosystems in the forms of carbon, nitrogen, phosphorus, and water. An important message in this discussion was the interconnectedness of the various reservoirs of matter through biotic and abiotic processes. In this module, we are going to transition from the movement of matter to the movement of energy in ecosystems. We will see many similarities, such as the interconnectedness between the biotic and abiotic components. However, we will also see that as energy moves through an ecosystem, much of the available energy is not efficiently moved from the Sun to plants and algae, herbivores, and carnivores. Much of the energy is lost as heat. The movement of energy helps determine which species can grow and reproduce in the ecosystem. By understanding the

Learning Goals

After reading this module you should be able to

6-1 describe how photosynthesis and respiration affect energy flow.

6-2 calculate gross primary productivity and net primary productivity.

6-3 explain why primary productivity has a low efficiency.

6-4 explain why some ecosystems are much more productive than others.

processes that determine energy movement, environmental scientists can better understand how human activities alter these processes and understand how to reduce such impacts. We will then consider how organisms use photosynthesis and respiration to capture and release energy. In the subsequent module, we will examine how this energy is converted into the biomass of herbivores and carnivores.

6-1 How do photosynthesis and respiration affect energy flow?

Photosynthesis captures energy while respiration releases energy

To understand how ecosystems function and how best to protect and manage them, ecosystem ecologists study the processes by which energy flows across an ecosystem. We can begin to trace this energy flow by looking at the processes of photosynthesis and cellular respiration.

Photosynthesis

Nearly all energy that powers ecosystems comes from the Sun as solar energy, which is a form of kinetic energy. Plants, algae, and some bacteria that use the Sun's energy to produce usable forms of energy, such as sugars, are called **producers**, or **autotrophs**. As you can see in the top half of **FIGURE 6.1**, these producers use photosynthesis, which means they use solar energy to convert carbon dioxide (CO_2) and water (H_2O) into glucose ($C_6H_{12}O_6$) and oxygen (O_2). Glucose is a form of potential energy that can be used by a wide range of organisms. The photosynthesis process also produces oxygen (O_2) as a waste product. That is why plants and other producers are beneficial to our atmosphere; they produce the oxygen we need to breathe.

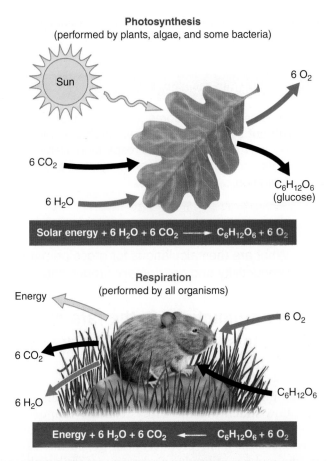

Photosynthesis
(performed by plants, algae, and some bacteria)

Sun

$6\ O_2$

$6\ CO_2$

$6\ H_2O$

$C_6H_{12}O_6$ (glucose)

Solar energy + $6\ H_2O + 6\ CO_2 \longrightarrow C_6H_{12}O_6 + 6\ O_2$

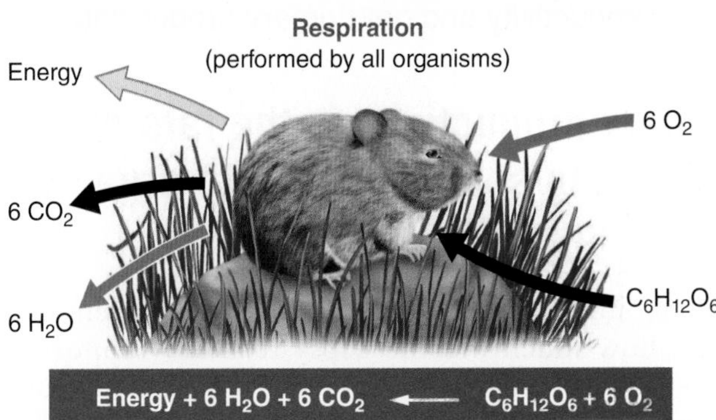

FIGURE 6.1 Photosynthesis and respiration. Photosynthesis is the process by which producers use solar energy to convert carbon dioxide and water into glucose and oxygen. Respiration is the process by which organisms convert glucose and oxygen into water and carbon dioxide, releasing the energy needed to live, grow, and reproduce. All organisms, including producers, perform respiration.

Cellular Respiration

Producers use the glucose they produce by photosynthesis to store energy and to build structures such as leaves, stems, and roots. Herbivores can eat the tissues of producers and gain energy from the chemical energy contained in those tissues. They do this through **cellular respiration**, a process by which cells unlock the energy of chemical compounds. We first learned about aerobic respiration in Module 4. This process, which is shown in the bottom half of Figure 6.1, is the opposite of photosynthesis; cells convert glucose and oxygen into energy, carbon dioxide, and water. In essence, organisms conducting aerobic respiration run photosynthesis backward to recover the solar energy stored in glucose. Some organisms, such as bacteria that live in the mud underlying a swamp where oxygen is not available, conduct **anaerobic respiration**, a process by which cells convert glucose into energy in the absence of oxygen. Anaerobic respiration does not provide as much energy as aerobic respiration.

Many organisms—including producers—carry out aerobic respiration to fuel their own metabolism and growth. Thus, producers both produce and consume oxygen. When the Sun is shining and photosynthesis occurs, producers generate more oxygen through photosynthesis than they consume through respiration. At night, when producers only respire, they consume oxygen without generating it. Overall, producers photosynthesize more than they respire. The net effect is that producers release oxygen into the air and store carbon in their tissues. This is one of the reasons that scientists worry about the loss of forests around the world because it results in less carbon being stored in the trees and more carbon dioxide accumulating in the atmosphere. Since

Producers Plants, algae, and some bacteria that use the Sun's energy to produce usable forms of energy, such as sugars. *Also known as* **autotrophs**.

Cellular respiration The process by which cells unlock the energy of chemical compounds.

Anaerobic respiration The process by which cells convert glucose into energy in the absence of oxygen.

carbon dioxide is a greenhouse gas, this can contribute to global warming, as discussed in Module 4.

AP® Exam Tip

All organisms complete some form of cellular respiration. Without oxygen it is called anaerobic respiration. With oxygen it is called aerobic respiration. Remember that plants and algae do both photosynthesis and cellular respiration.

6-2 What are the calculations for gross primary productivity and net primary productivity?

Primary productivity is the rate of converting solar energy into organic compounds over time

The amount of energy available in an ecosystem determines how much life the ecosystem can support. For example, the amount of sunlight that reaches a lake surface determines how many producers can live in the lake. In turn, the number of producers determines the number of herbivores the lake can support, and the size of the herbivore population determines the number of carnivores the lake can support.

Gross versus Net Primary Productivity

To quantify the rate of capturing energy in an ecosystem, environmental scientists measure the ecosystem's productivity. We define **primary productivity** as the rate of converting solar energy into organic compounds over a period of time. We can further refine this definition by clarifying the impacts of photosynthesis and respiration. For example, the **gross primary productivity (GPP)** of the ecosystem is a measure of the total amount of solar energy that producers in tan ecosystem capture via photosynthesis over a given amount of time. Note that the term *gross*, as used here, indicates the total amount of energy captured by producers. GPP does not subtract the energy that is lost when the producers respire. In contrast, we can quantify the energy captured by producers in an ecosystem—minus the energy producers use for respiration—as the ecosystem's **net primary productivity (NPP)**:

$$\text{net primary productivity (NPP)} = \text{gross primary productivity (GPP)} - \text{respiration by producers (R)}$$

You can think of GPP and NPP in terms of a paycheck from your summer job: GPP is the total amount your employer pays you whereas NPP is the actual amount you take home after taxes are deducted. In this analogy, the taxes removed from your paycheck represent the energy removed from GPP due to respiration.

GPP is essentially a measure of how much photosynthesis is occurring over some amount of time. Determining GPP is a challenge for scientists because a plant rarely photosynthesizes without simultaneously respiring. However, if we can determine the rate of photosynthesis and the rate of respiration, we can use this information to calculate GPP. Knowing the value of GPP allows us to compare the GPP of different ecosystems and to examine how different human activities—such as fertilizer inputs into aquatic biomes—cause changes in GPP.

Measuring Gross versus Net Primary Productivity

We can determine the rate of photosynthesis by measuring the compounds that participate in the reaction. So, for example, we can measure the rate at which CO_2 is taken up during photosynthesis and the rate at which CO_2 is produced during respiration. A common approach to measuring GPP is to first measure the production of CO_2 in the dark. Because no photosynthesis occurs in the dark, this measure eliminates CO_2 uptake by photosynthesis. Next, we measure the uptake of CO_2 in sunlight. This measure gives us the net movement of CO_2 when respiration and photosynthesis are both occurring. By adding the amount of CO_2 produced in the dark to the amount of CO_2 taken up in the sunlight, we can determine the gross amount of CO_2 that is taken up during photosynthesis:

$$CO_2 \text{ taken up during photosynthesis} = CO_2 \text{ taken up in sunlight} + CO_2 \text{ produced in the dark}$$

We can see an example of this type of experiment in **FIGURE 6.2**, which illustrates the CO_2 concentrations in a sealed chamber containing a leaf that is either placed in sunlight or in the dark. In sunlight, the leaf takes in CO_2 for photosynthesis and emits a smaller amount of CO_2 for respiration. The net increase in CO_2 is a measure of NPP. In the dark, the leaf takes in no CO_2, since it cannot photosynthesize, but it continues to emit a small amount of CO_2 for respiration (R). In this way, we can derive the GPP of an ecosystem by summing NPP and R. We quantify GPP and NPP in units of kilocalories of energy per square meter per year (kcal/m^2/year). We can also quantify them in units

Primary productivity The rate of converting solar energy into organic compounds over a period of time.

Gross primary productivity (GPP) The total amount of solar energy that producers in an ecosystem capture via photosynthesis over a given amount of time.

Net primary productivity (NPP) The energy captured by producers in an ecosystem minus the energy producers respire.

Biomass The total mass of all living matter in a specific area.

Standing crop The amount of biomass present in an ecosystem at a particular time.

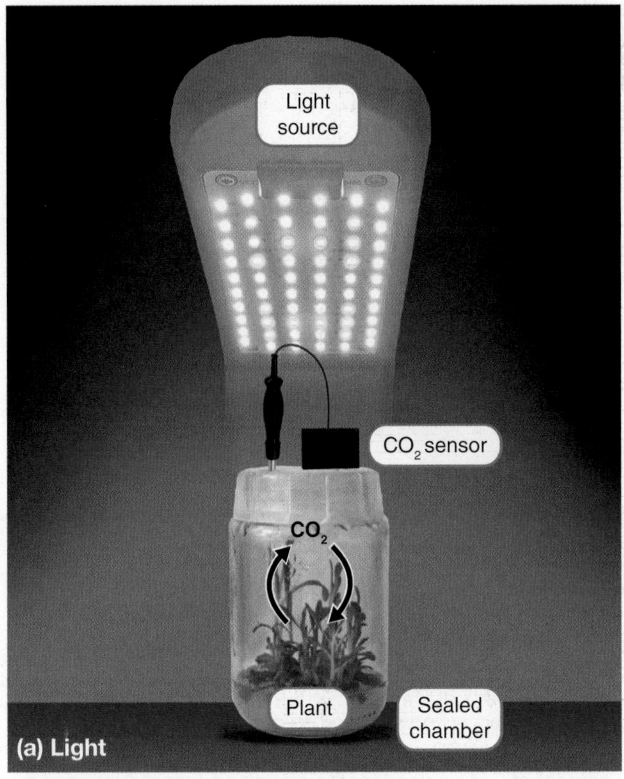

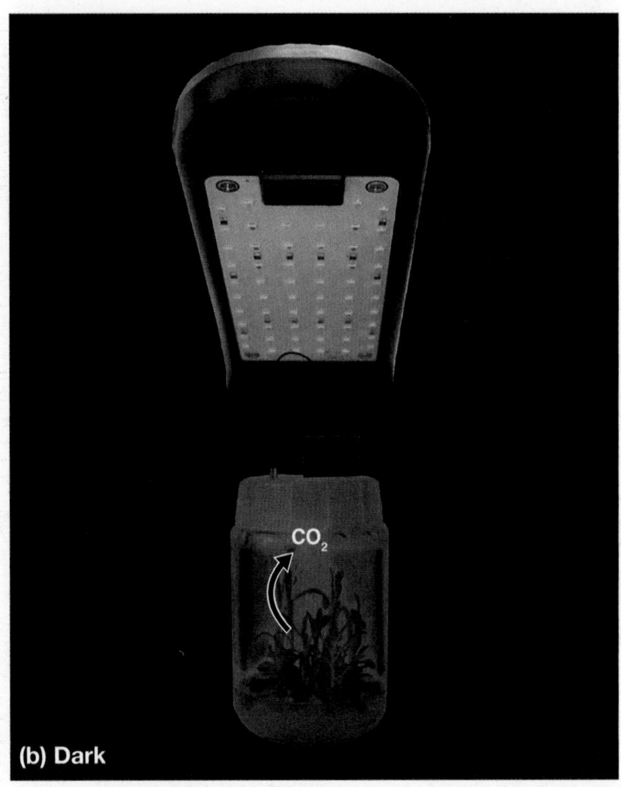

FIGURE 6.2 Quantifying gross and net primary productivity. Researchers place a leaf into a sealed chamber in which they can carefully measure the concentration of CO_2. (a) When the leaf is exposed to light, it takes in a large amount of CO_2 as it photosynthesizes while also emitting a small amount of CO_2 as it respires. The difference in CO_2 concentration represents the net primary productivity (NPP) of the leaf. (b) When the leaf is in the dark, it takes in no CO_2 for photosynthesis, but it continues to emit CO_2 as it respires. Thus, the increase in CO_2 concentration is a measure of respiration (R). By quantifying NPP and R in the experiment, we can calculate GPP as the sum of NPP and R.

of kilograms of carbon taken up per square meter per year (kg C/m^2/year). You can practice these calculations in "Do the Math: Calculating NPP, GPP, and R."

The net primary productivity of an ecosystem establishes the rate at which **biomass**—the total mass of all living matter in a specific area—is produced over a given amount of time. The amount of biomass present in an ecosystem at a particular time is its **standing crop**. It is important to differentiate standing crop, which measures the amount of energy in a system at a given time, from productivity, which measures

DO THE MATH Calculating NPP, GPP, and R

Preparing for the AP® Exam

As we have seen, we can calculate either GPP, NPP, and R using the following equation:

$$NPP = GPP - R$$

Using this equation, we can determine the value of any single variable if we know the values of the other two variables. The two most commonly measured variables in natural ecosystems are GPP and R. In a temperate seasonal forest, GPP can be approximately 13,000 kcal/m^2/year and R can be 8,000 kcal/m^2/year. Based on this information, what is the NPP for the forest?

$$NPP = GPP - R$$

$$NPP = 13{,}000 \text{ kcal/m}^2\text{/year} - 8{,}000 \text{ kcal/m}^2\text{/year}$$

$$NPP = 5{,}000 \text{ kcal/m}^2\text{/year}$$

YOUR TURN If a tropical rainforest has an NPP of 8,500 kcal/m^2/year and a GPP of 21,000 kcal/m^2/year, how much energy is used for the rainforest for respiration (R)?

the rate of energy production over a span of time. For example, slow-growing forests have low productivity; the trees add only a small amount of biomass through growth and reproduction each year. However, the standing crop of long-lived trees—the biomass of trees that has accumulated over hundreds of years—is quite high. In contrast, the high growth rates of algae living in the ocean make them extremely productive. But because herbivores eat these algae so rapidly, the standing crop of algae at any time is relatively low.

6-3 Why does primary productivity have low efficiency?

Primary productivity is not an efficient process

Converting sunlight into chemical energy is not an efficient process. As **FIGURE 6.3** shows, only about 1 percent of the total amount of solar energy that reaches the producers in an ecosystem—the sunlight on a pond surface, for example—is converted into chemical energy via photosynthesis. Most of that solar energy is lost from the ecosystem as heat that returns to the atmosphere. Some of the lost energy consists of wavelengths of light that producers cannot absorb. Those wavelengths are either reflected from the surfaces of producers or pass through their tissues.

The NPP of ecosystems ranges from 25 to 50 percent of GPP. Given that only 1 percent of the available solar energy is converted into chemical energy, which is measured as GPP, this means that as little as 0.25 percent of the solar energy striking the planet is represented by NPP. Clearly, it takes a lot of energy to conduct photosynthesis!

Let's look at the math. Recall from Figure 6.3 that on average, of the 1 percent of the Sun's energy that is captured by a producer, about 60 percent is used to fuel the producer's respiration. The remaining 40 percent can be used to support the producer's growth and reproduction. A grassland in North America, for example, might have a GPP of 6,000 kcal/m^2/year and lose 3,600 kcal/m^2/year to respiration by plants. Because NPP = GPP − R, the NPP of the forest is 2,400 kcal/m^2/year. In this example, NPP is 40 percent of GPP.

In aquatic ecosystems, the low efficiency of converting solar energy to plants and algal tissues is further impeded by the water absorbing the various wavelengths of sunlight. As you may know, sunlight is comprised of many different wavelengths of light, including red, green, and blue light, as illustrated in **FIGURE 6.4a**. A fundamental property of water is that it absorbs long wavelengths of light more than it absorbs short wavelengths of light. As a result, in the upper 1 m of water, much of the red light is absorbed. Therefore, algae and plants living deeper

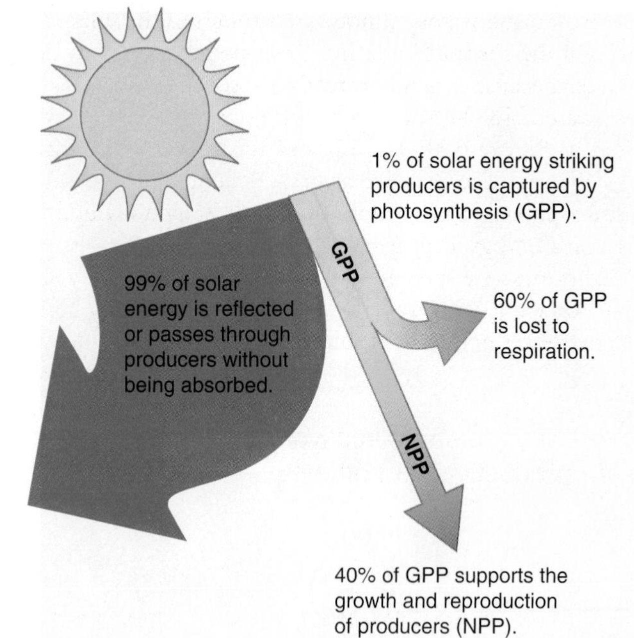
1% of solar energy striking producers is captured by photosynthesis (GPP).
GPP
99% of solar energy is reflected or passes through producers without being absorbed.
60% of GPP is lost to respiration.
NPP
40% of GPP supports the growth and reproduction of producers (NPP).

FIGURE 6.3 Gross and net primary productivity. Producers typically capture only about 1 percent of available solar energy via photosynthesis. This is known as gross primary productivity, or GPP. About 60 percent of GPP is typically used for respiration. The remaining 40 percent of GPP is used for the growth and reproduction of the producers. This is known as net primary productivity, or NPP.

(a)

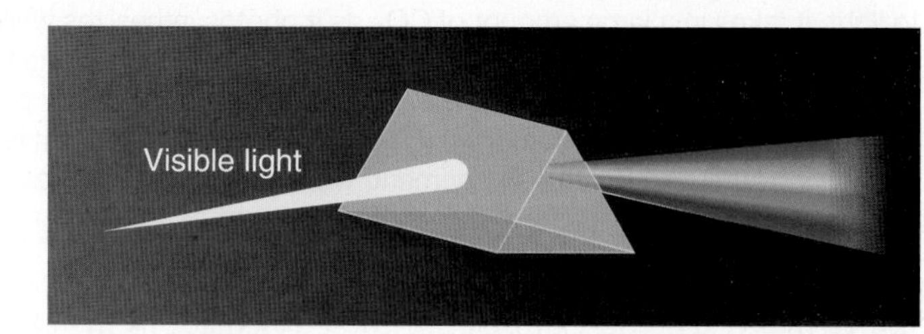
Visible light

(b)

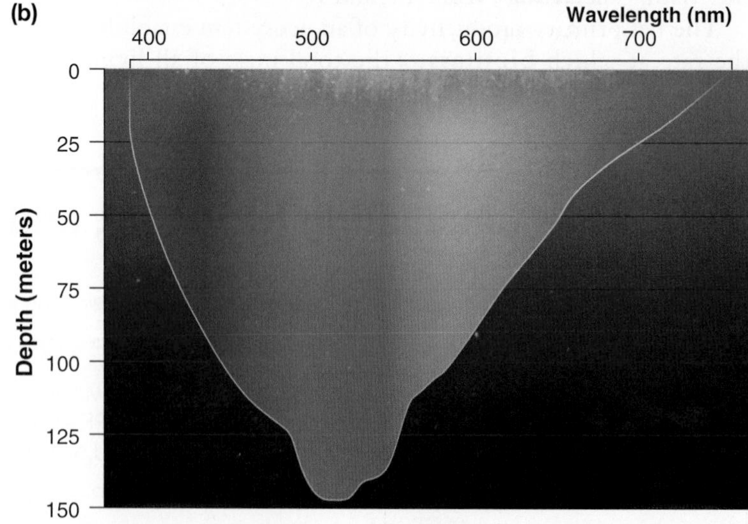
Wavelength (nm)
400
500
600
700
Depth (meters)
0
25
50
75
100
125
150

FIGURE 6.4 The wavelengths of visible light. The white light that we can see is comprised of several different wavelengths of light. (a) When we pass white light through a glass prism, the different wavelengths get separated based on whether the color is composed of short or long wavelengths. (b) Similar to the effects of the prism, the different wavelengths of light also penetrate water to different degrees, with the blue wavelengths penetrating deeper into the water than the red wavelengths.

than 1 m in the water cannot access much of the energy from red light. In contrast, blue light can penetrate much deeper in water because it is not as readily absorbed by the water. In clear water, blue light can penetrate down to 100 m, as illustrated in Figure 6.4b. In response to different intensities of light energy, algae that live at greater depths have evolved to conduct photosynthesis more efficiently, which includes using additional energy-capturing pigments other than chlorophyll. In addition, some species of algae can migrate up toward the water surface to capture light energy during the day and then sink to greater depths to obtain nutrients at night.

6-4 Why are some ecosystems much more productive than others?

Some ecosystems are much more productive than others

Measurement of NPP allows us to compare the productivity of different ecosystems, as shown in **FIGURE 6.5**. As you can see, producers grow best in ecosystems where they have plenty of sunlight, lots of available water and nutrients, and warm temperatures, such as tropical rainforests and salt marshes, which are the most productive ecosystems on Earth. Conversely, producers grow poorly in the cold regions of the Arctic, dry deserts, and the dark regions of the deep sea. In general, the greater the productivity of an ecosystem, the more herbivores can be supported.

Measuring NPP is also a useful way to measure change in an ecosystem. For example, after a drastic change alters an ecosystem, such as a hurricane devastating a forest or a drought occurring in the grape-growing regions of the world, the amount of stored energy (NPP) tells us whether the new system is more or less productive than the previous system.

In this module, we have considered the importance of photosynthesis and respiration in helping us to determine the productivity of aquatic and terrestrial ecosystems. We have also seen that primary producers have a very low efficiency in converting available sunlight into usable energy. Finally, we have seen that some ecosystems are much more productive than others for a variety of reasons, including differences in temperature and precipitation. In the next module, we will examine how the energy captured by producers moves up through the series of species that consume the producers and are, in turn, consumed by predators.

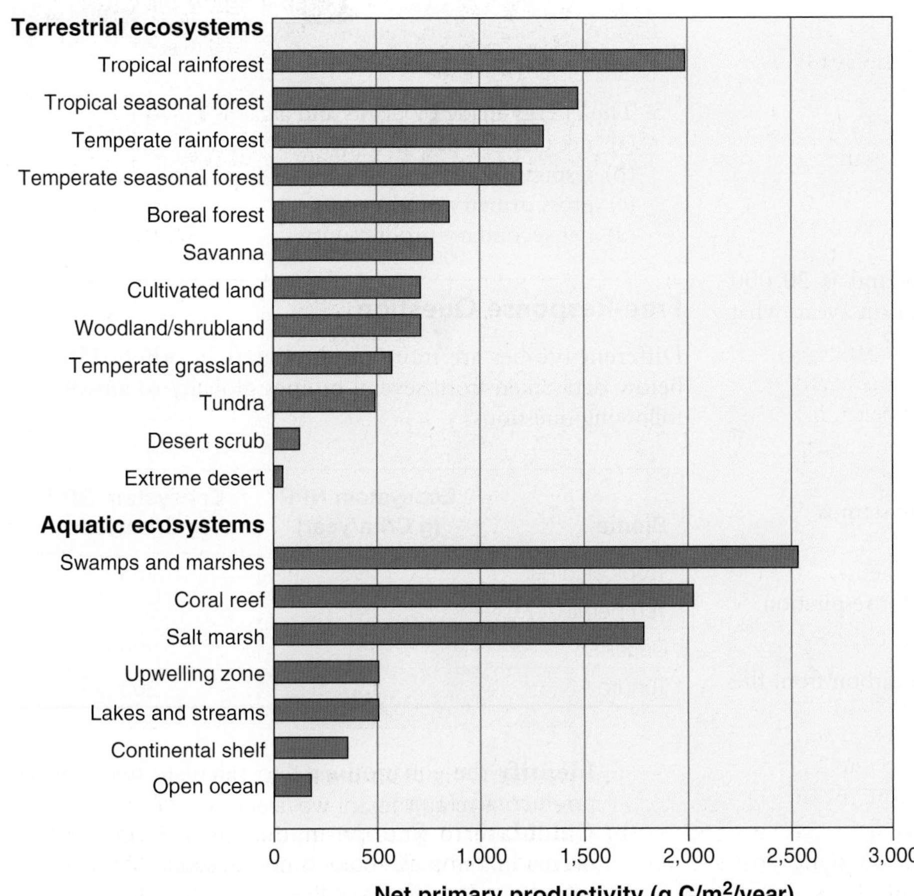

FIGURE 6.5 Net primary productivity. Net primary productivity varies among ecosystems. Productivity is highest where temperatures are warm and water and solar energy are abundant. As a result, NPP varies tremendously among different areas of the world. *(Data from R. H. Whittaker and G. E. Likens, Primary production: The biosphere and man,* Human Ecology *1 (1973): 357–369.)*

Module 6 AP® Review

Preparing for the AP® Exam

Learning Goals Revisited

6-1 How do photosynthesis and respiration affect energy flow?

Photosynthesis captures the energy of the Sun to convert CO_2 and water into carbohydrates. Respiration, whether aerobic or anaerobic, unlocks the chemical energy stored in the cells of organisms.

6-2 What are the calculations for gross primary productivity and net primary productivity?

Ecosystem productivity can be quantified by measuring the total amount of solar energy that producers capture, which is gross primary productivity, or by measuring the total amount of solar energy captured minus the amount of energy used for respiration, which is net primary productivity.

6-3 Why does primary productivity have low efficiency?

The low efficiency occurs because producers can only capture about 1 percent of all available solar energy. The remainder is lost as heat to the atmosphere, including those wavelengths of light that are not usable by the producers.

6-4 Why are some ecosystems much more productive than others?

Producers have the highest rates of primary productivity in ecosystems with warm temperatures, plenty of available water, and high concentrations of nutrients.

AP® Practice Questions

Preparing for the AP® Exam

Multiple-Choice Questions

1. Autotrophs
   (a) use photosynthesis.
   (b) can survive without oxygen.
   (c) are herbivores.
   (d) do not respire.

2. If gross primary productivity in a wetland is 20,000 kcal/m^2/year and respiration is 12,000 kcal/m^2/year, what is the net primary productivity of the wetland?
   (a) 8,000 kcal/m^2/year
   (b) 32,000 kcal/m^2/year
   (c) 12,000 kcal/m^2/year
   (d) 24,000 kcal/m^2/year

3. The gross primary productivity of an ecosystem is
   (a) the total amount of biomass.
   (b) the total energy captured by photosynthesis.
   (c) the energy captured after accounting for respiration.
   (d) the energy available to herbivores.

4. Which is the biotic process that removes carbon from the atmosphere?
   (a) respiration
   (b) photosynthesis
   (c) nitrification
   (d) greenhouse effect

5. The energy made by plants and algae is called
   (a) net primary productivity.
   (b) respiration.
   (c) gross primary productivity.
   (d) net secondary productivity.

Free-Response Question

Different biomes are more productive than others. Use the below data taken from several biomes globally to answer the following questions.

Biome	Ecosystem NPP (g C/ha/year)	Ecosystem GPP (g C/ha/year)
Tropical forest	1,098	1,355
Temperate forest	741	1,087
Deserts	151	229
Tundra	130	208

(a) **Identify** the equation used to calculate net primary productivity. (1 pt.)
(b) **Calculate** to find the amount of respiration that occurs in a tropical forest. Show all work. (2 pts.)

(c) **Calculate** to find the amount of respiration that occurs in the tundra. Show all work. (2 pts.)
(d) **Explain** why there is a difference in the net primary productivity of tropical forests and tundras. (1 pt.)
(e) **Calculate** the net primary productivity of deserts and tundra and compare the difference. Show all work. (2 pts.)
(f) **Explain** why deserts and tundra both have net primary productivity. (1 pt.)
(g) **Explain** how humans affect tropical forests and how human involvement impacts net primary productivity. (1 pt.)

Module 7

Unit 1	1	2	3	4	5	6	7

Trophic Levels, Energy Flow and the 10% Rule, Food Chains, and Food Webs

As we discussed earlier in this unit, ecosystems are comprised of a diversity of organisms including animals, plants, algae, fungi, and bacteria. We have also seen how these organisms interact as predators, prey, parasites, competitors, and mutualists. Thus, we have considered how pairs of species interact, but given the large number of species that live in most terrestrial and aquatic biomes, the interactions among species can be vast and complex. To help make sense of this complexity, we can think about species functioning at different levels in an ecosystem. In the previous module, we discussed how producers capture energy from the Sun and use it for respiration, growth, and reproduction. In this module we will examine how energy and matter are moved from the producers to the herbivores and predators. We will also focus on the low efficiency of energy movement from one organism to another. Finally, we will consider how a large number of interacting species can cause interesting impacts when a species is added or removed from the ecosystem.

Learning Goals

After reading this module you should be able to

7-1 describe how energy and matter move through trophic levels in an ecosystem.

7-2 explain how low ecological efficiency causes energy to decrease at higher trophic levels.

7-3 explain why food webs experience feedback loops.

7-1 How do energy and matter move through trophic levels in an ecosystem?

Ecosystems depend on energy flowing through and matter cycling around trophic levels

To understand how ecosystems function and how best to protect and manage them, ecologists study the processes that move energy and matter within an ecosystem. To understand energy relationships, we need to look at the way energy flows through an ecosystem and matter cycles around an ecosystem.

To appreciate the movement of energy and matter, we can think about how producers are eaten by herbivores, how herbivores are eaten by predators, and how some predators are eaten by other predators. In the previous module, we saw that producers in terrestrial, freshwater, and shallow marine biomes make their own food by capturing the energy of the Sun by photosynthesis. We also learned from Module 3, that in the deep ocean producers make their food through the process of chemosynthesis. Regardless of which process the producers use to make their food, they serve as the foundational group of organisms that are able to transfer energy and matter to consumers.

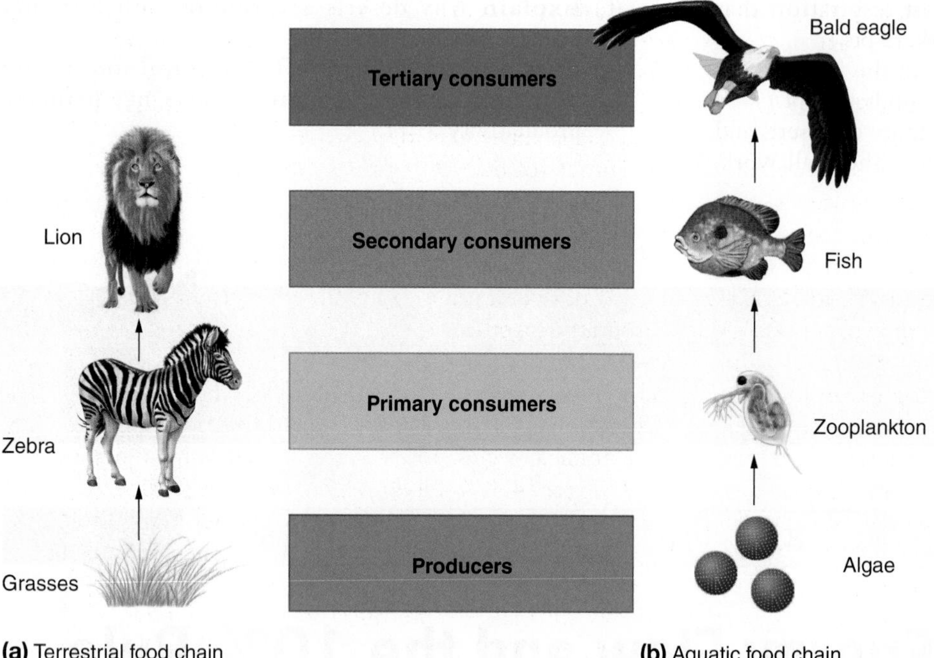

FIGURE 7.1 Simple food chains. A simple food chain that links producers and consumers in a linear fashion illustrates how energy and matter move through the trophic levels of an ecosystem. (a) An example of a terrestrial food chain. (b) An example of an aquatic food chain.

In contrast to the producers, **consumers**, also known as **heterotrophs**, are incapable of photosynthesis and therefore must obtain their energy by consuming other organisms. In **FIGURE 7.1**, we can see that consumers in terrestrial and aquatic ecosystems fall into different categories. Consumers that eat producers are called **herbivores** or **primary consumers**. Primary consumers include a variety of familiar plant- and alga-eating animals, such as zebras, grasshoppers, tadpoles, and zooplankton. Consumers that eat other consumers are called **carnivores**. Carnivores that eat primary consumers are called **secondary consumers**. Secondary consumers include creatures such as lions, hawks, and rattlesnakes. Carnivores that eat secondary consumers are called **tertiary consumers**. As you can see on the right side of Figure 7.1, animals such as bald eagles can be tertiary consumers. The algae (producers) living in lakes convert sunlight into glucose, zooplankton (primary consumers) eat the algae, fish (secondary consumers) eat the zooplankton, and eagles (tertiary consumers) eat the fish.

Consumer An organism that is incapable of photosynthesis and must therefore obtain its energy by consuming other organisms. *Also known as* **heterotroph**.

Herbivore A consumer that eats producers. *Also known as* **primary consumer**.

Carnivore A consumer that eats other consumers.

Secondary consumer A carnivore that eats primary consumers.

Tertiary consumer A carnivore that eats secondary consumers.

Trophic levels The successive levels of organisms consuming one another.

Food chain The sequence of consumption from producers through tertiary consumers.

AP® Exam Tip

Trophic levels are a good way to understand the distribution of energy in an ecosystem. The AP® Environmental Science exam will only ask you to describe the trophic level of an organism after showing you a food chain, so you won't be expected to memorize any food chains for the exam. The most important thing to understand is the loss of energy as you move up the trophic pyramid.

The successive levels of organisms consuming one another are known as **trophic levels** (from the Greek word *trophe*, which means "nourishment"). The sequence of consumption from producers through tertiary consumers is known as a **food chain**, where energy moves from producers, which are the lowest trophic level, up through primary, secondary, and tertiary consumers, which is typically the highest trophic level. A food chain helps us visualize how energy and matter move between trophic levels.

Not all organisms fit neatly into a single trophic level. Some organisms, called omnivores, operate at several trophic levels. Omnivores include grizzly bears, which eat berries and fish, and the Venus flytrap (*Dionaea muscipula*), which can photosynthesize as well as digest insects that become trapped in its leaves.

Each trophic level eventually produces dead individuals and waste products. Three groups of organisms feed on this dead organic matter: scavengers, detritivores, and decomposers. **Scavengers** are organisms, such as vultures, that consume dead animals. **Detritivores** are organisms, such as dung beetles, that specialize in breaking down dead tissues and waste products (referred to as detritus) into smaller particles. These particles can then be further processed by **decomposers**, which are the fungi and bacteria that complete the breakdown process by converting organic matter into small elements and molecules that can be recycled back into the ecosystem. Without scavengers, detritivores, and decomposers, there would be no way of recycling organic matter, and the world would rapidly fill up with dead plants and animals.

To visualize how energy and matter are moving among the trophic levels, we can combine the food chain consisting of producers and their consumers with a food chain consisting of scavengers, detritivores, and decomposers that break down dead organic matter. You can see an example of this in **FIGURE 7.2**. On the left side of the figure, we see that producers are eaten by primary consumers, which are then eaten by secondary consumers. However, we also see that all of these trophic groups produce waste products and dead individuals, which collectively create a source of dead organic matter. As the dead organic matter is broken down, shown on the right side of the figure, nutrients such as nitrogen and phosphorus are released and subsequently taken up by the producers. A portion of the energy and matter is also used for the respiration, growth, and reproduction of the scavengers, detritivores, and decomposers. These consumers of dead organic matter can then be consumed by secondary consumers, thereby completing the cycling of matter in the ecosystem.

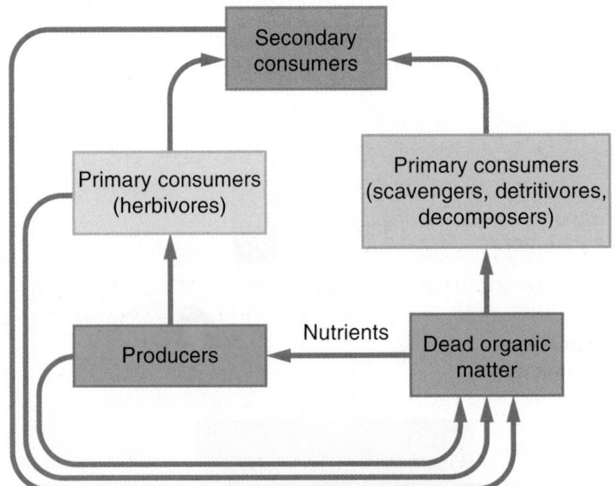

FIGURE 7.2 The cycling and conservation of matter. On the left side of the figure, some of the matter flows from producers to primary consumers and then to secondary consumers. The remaining matter, consisting of waste and dead individuals, collectively becomes the dead organic matter in the food chain. On the right side of the figure, the dead organic matter is broken down, releasing nutrients that can be taken up by the producers and providing nutrients that allows the growth and reproduction of scavengers, detritivores, and decomposers. These organisms can then be eaten by the secondary consumers.

As we can see in these examples, the matter that we discussed in our modules on biogeochemical cycles—such as carbon, nitrogen, and phosphorus—is cycling through the food chain from the producers to the consumers and then through the scavengers, detritivores, and decomposers. Thus, the species in a food chain serve important roles in biogeochemical cycles, which we discussed in Modules 4 and 5 (see Figures 4.1, 4.4, and 5.1). At the same time, the biogeochemical cycles are essential for species living in ecosystems because they help cycle these elements into and out of the food chains. Note that there is a conservation of matter because the matter is not lost from the ecosystem. For example, carbon enters the food chain when plants take up CO_2 to make glucose via photosynthesis. A portion of the glucose is used for the respiration of the plants, which releases CO_2 to the atmosphere, and the remainder is allocated to growth and reproduction. Plant reproduction is either eaten by a consumer, or the plant eventually dies and decomposes, which ultimately releases the remaining carbon back into the abiotic portion of the carbon cycle. Through all of these flows, the carbon has not been lost from the system; it has simply changed forms. Thus, there is a conservation of matter. In contrast, much of the energy available in each trophic level is lost as respiration and heat. As a result, much of the energy that flows through an ecosystem is ultimately lost from the ecosystem as heat that moves into the atmosphere. This is why we say that matter cycles through an ecosystem whereas energy flows through an ecosystem.

7-2 How does low ecological efficiency cause energy to decrease at higher trophic levels?

The flow of energy from one trophic level to the next has a low efficiency

An interesting pattern that we can observe in ecosystems around the world is that there is commonly much more energy in producers than in primary consumers, and much

Scavenger An organism that consumes dead animals.

Detritivore An organism that specializes in breaking down dead tissues and waste products into smaller particles.

Decomposers Fungi and bacteria that complete the breakdown process by converting organic matter into small elements and molecules that can be recycled back into the ecosystem.

FIGURE 7.3 The flow of energy in the Serengeti ecosystem of Africa. The Serengeti ecosystem has more plants than herbivores, and more herbivores than carnivores. *(Stu Porter/Alamy Stock Photo)*

more energy in primary consumer than in secondary consumers. To illustrate this point, consider the Serengeti Plain in East Africa (**FIGURE 7.3**). Producers, such as grasses and acacia trees, absorb energy directly from the Sun. That energy spreads throughout an ecosystem as plants are eaten by primary consumers, such as gazelles, and those animals are subsequently eaten by secondary consumers, such as cheetahs. There are hundreds of millions of plants, and millions of primary consumers, such as zebras and wildebeests, but there are far fewer secondary consumers, such as lions (*Panthera leo*) and cheetahs (*Acinonyx jubatus*). In accordance with the second law of thermodynamics, when one organism consumes another, not all the energy in the consumed organism is transferred to the consumer.

Ecological Efficiency and Trophic Pyramids

One reason that not all the energy gets transferred to a higher trophic level is that some of the energy of the lower trophic level is not in a usable form. For example, some parts of plants are not digestible and are excreted by primary consumers. Similarly, secondary consumers such as owls consume the muscles and organs of rodents, but they cannot digest the bones and hair, so the owls regurgitate these undigestible rodent parts.

A second reason that not all the energy gets transferred to a higher trophic level is how the digestible energy gets used by organisms. A substantial portion of this energy is used to power the consumer's day-to-day activities, including moving, eating, and (for birds and mammals) maintaining a constant body temperature. That energy is ultimately lost as heat. Any remaining energy may be converted into consumer biomass for growth and reproduction. Given that a portion of energy from a lower trophic level is not digestible and a portion of the digestible energy is used for the consumer's day-to-day activities, there is only a small percentage of the energy remaining that can be devoted to growth and reproduction of the consumers.

The proportion of consumed energy that can be passed from one trophic level to another is referred to as **ecological efficiency**. Ecological efficiencies are low; they range from 5 to 20 percent and average about 10 percent across all ecosystems. In other words, of the total biomass available at a given trophic level, only about 10 percent can be converted into energy at the next higher trophic level, a process known as **the 10% rule**. We can represent the distribution of biomass, numbers, or energy among trophic levels using a **trophic pyramid**, like the one of the Serengeti ecosystem shown in **FIGURE 7.4**. Trophic pyramids tend to have similar proportions of energy or biomass across ecosystems. Most of the energy and biomass are found at the producer level, and they commonly decrease as we move up the pyramid. The 10 percent efficiency of energy flow between adjacent trophic levels helps to determine the population sizes of the various species within each trophic level.

Ecological efficiency The proportion of consumed energy that can be passed from one trophic level to another.

The 10% rule Of the total biomass available at a given trophic level, only about 10 percent can be converted into energy at the next higher trophic level.

Trophic pyramid A representation of the distribution of biomass, numbers, or energy among trophic levels.

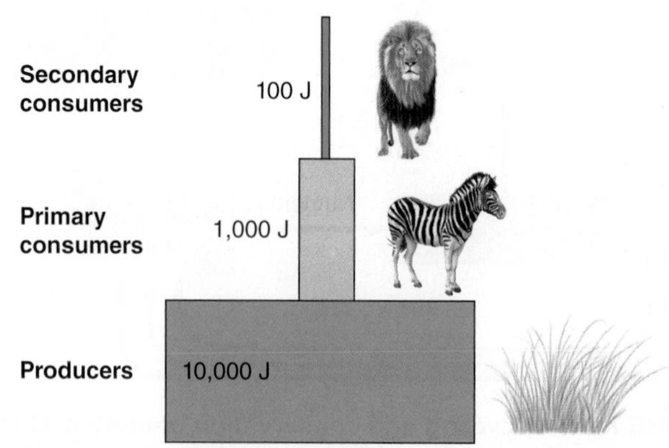

FIGURE 7.4 Trophic pyramid for the Serengeti ecosystem. This trophic pyramid represents the amount of energy that is present at each trophic level, measured in joules (J). While this pyramid assumes 10 percent ecological efficiency, actual ecological efficiencies range from 5 to 20 percent across different ecosystems. For most ecosystems, graphing the numbers of individuals or biomass within each trophic level would produce a similar pyramid.

Ecological Efficiency Applied to Humans

The principle of ecological efficiency also has implications for the human diet. For example, if all humans were to act only as primary consumers—that is, become vegetarians—we would harvest much more energy from any given area of land or water. How would this work?

Suppose a hectare of cropland could produce 1,000 kg of soybeans. This food could feed humans directly. Or, if we assume 10 percent ecological efficiency, it could be fed to cattle to produce approximately 100 kg of meat. In terms of biomass, there would be 10 times more food available for humans acting as primary consumers by eating soybeans than for humans acting as secondary consumers by eating beef. However, 1 kg of soybeans contains about 2.5 times as many calories as 1 kg of beef. Therefore, 1 ha of land would produce 25 times more calories when used for soybeans than when used for beef. When we act as secondary consumers, we obtain far less energy than when we act as primary consumers.

7-3 Why do food webs experience feedback loops?

Food webs contain multiple, interconnected food chains with feedback loops

Species in natural ecosystems are rarely connected in such a simple, linear fashion as suggested by a food chain diagram. A more realistic type of model is known as a **food web**, which is a model of how energy and matter,—such as carbon, nitrogen, and phosphorus—move through two or more interconnected food chains, such as the Serengeti food web shown in **FIGURE 7.5**. Food webs nicely illustrate one

Food web A model of how energy and matter move through two or more interconnected food chains.

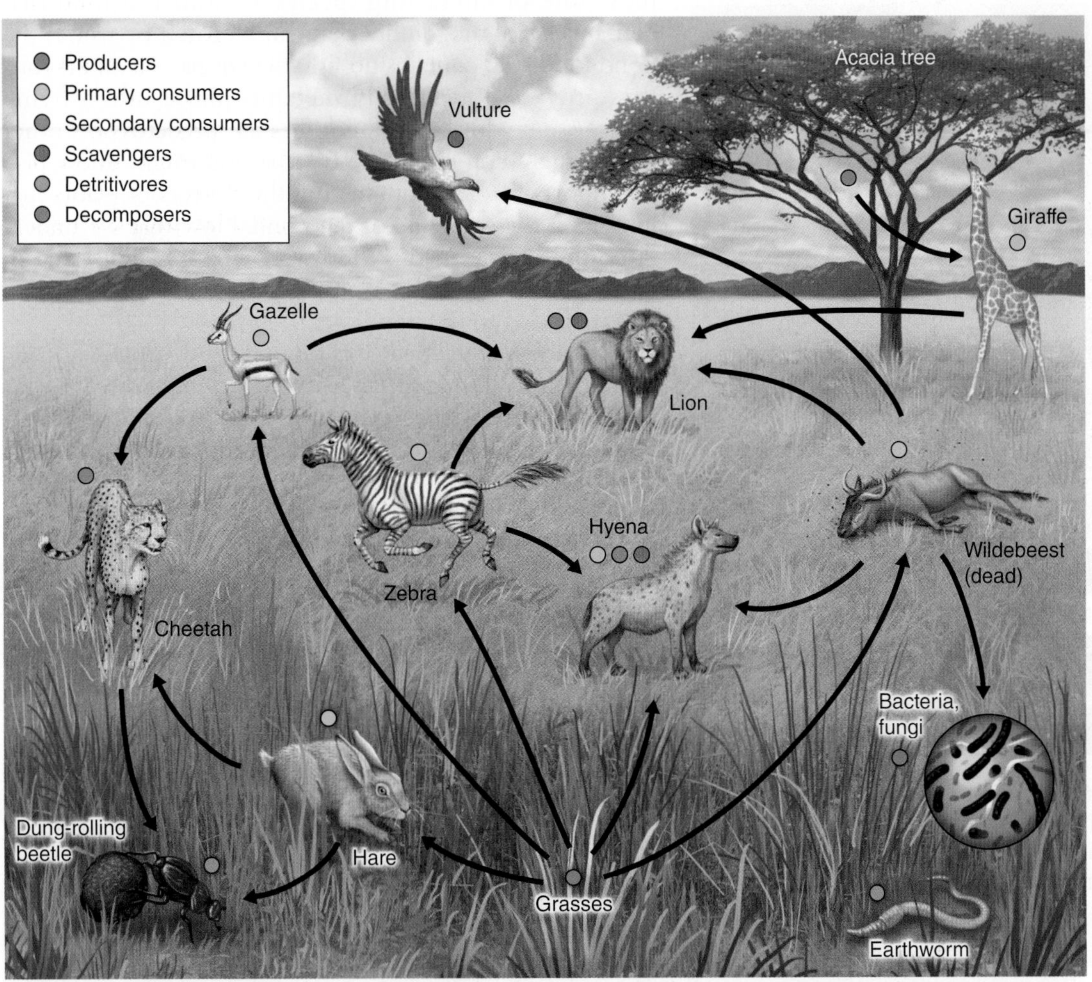

FIGURE 7.5 A terrestrial food web. Food webs are more realistic representations of trophic relationships than simple food chains. They include scavengers, detritivores, and decomposers, and they recognize that some species feed at multiple trophic levels. Arrows indicate the direction of energy movement. This is a real but somewhat simplified food web; in an actual ecosystem, many more organisms are present and many more pathways of energy flow among them.

of the most important concepts of ecology: All species in an ecosystem are connected to one another.

Food webs also occur in aquatic biomes. In contrast to the simplified food chain in Figure 7.1, we can see the much more complex and realistic food web illustrated in **FIGURE 7.6**. In this aquatic food web, we can see a large number of species in a lake that interact by consuming lower trophic levels. The producers are the plants that live in the shallow water, the attached algae that live on the lake bottom, and the phytoplankton that are present throughout the water. These producers are eaten by several different primary consumers, including crickets that eat the plants and snails that eat attached algae and bacteria. Secondary consumers include the redwing blackbirds, which eat the crickets, and the bluegill sunfish, which eat the snails. Finally, largemouth bass eat the sunfish, which makes them a tertiary consumer. You might also notice that some species in a food web do not fit neatly into a single trophic level. For example, great blue herons feed on snails, which makes the herons a secondary consumer, but they also feed on bluegill sunfish, which makes the herons a tertiary consumer. Species feeding at multiple trophic levels is a common observation in food webs.

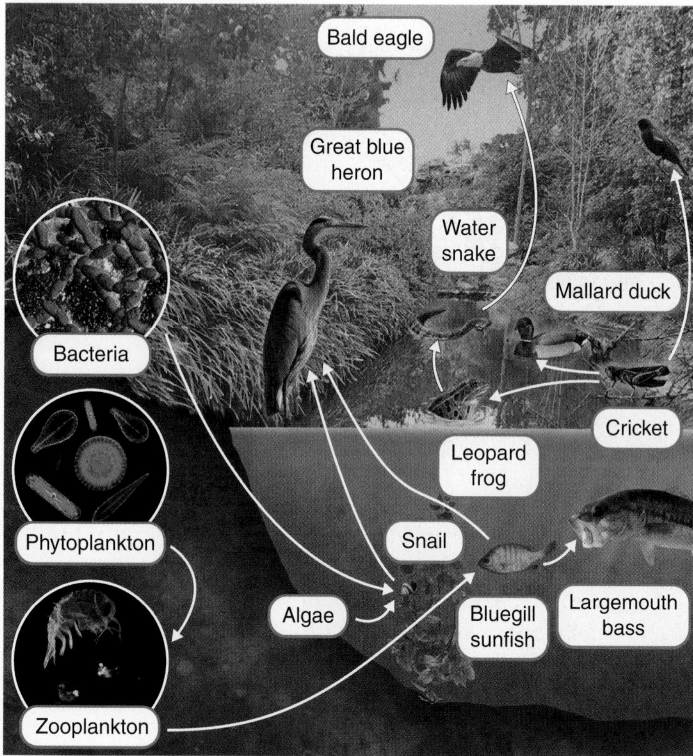

FIGURE 7.6 An aquatic food web. The food web in a lake includes producers such as the phytoplankton, attached algae, and plants living along the shoreline. It also includes primary consumers such as snails and zooplankton, secondary consumers such as the bluegill sunfish, and tertiary consumers such as largemouth bass and eagles.

Feedback Loops

Given that species are interconnected in a food web, any changes to the abundance of a species has the potential to impact the other species in the web. For example, if lions on the Serengeti become unusually abundant, they can dramatically reduce the abundance of gazelles. With fewer gazelles, there will be more plants left uneaten. In this simple scenario, the lions have a negative direct effect on the gazelles, but they have a positive indirect effect on the plants. As a result, fluctuations in the abundance of one species can cause positive feedback loops on some species and negative feedback loops on other species.

With this perspective, we can consider a bit more complex suite of interacting species. Consider a pond that contains larval dragonflies that can be eaten by fish. If the pond food web contains no fish, the aquatic, larval dragonflies grow and eventually metamorphose into a flying adult. As an adult, the dragonflies are predators on other insects including bees. The bees, as you might guess, play an important role in pollinating the flowers of plants that live along the shoreline of the pond. Now that we understand the species in this food web, which is illustrated in **FIGURE 7.7a**, how might the dragonflies be impacted when fish are added to the pond?

When fish are added to the pond food web, they consume the larval dragonflies, thereby having a direct negative effect on the species as shown by the solid arrow. In doing so, the fish also cause several indirect positive and negative effects on the other species, as shown by the dashed arrows. For example, the consumption of larval dragonflies indirectly causes a reduction in the number of adult dragonflies that metamorphose from the water. With fewer adult dragonflies, there is reduced consumption of the bees, and a resulting increase in the number of bees visiting the flowers along the shoreline. You can see the resulting impact of adding fish to the pond on the number of pollinator visits in Figure 7.7b. Thus, the presence or absence of just one species, in this case a predatory fish, has widespread indirect impacts on all of the species in the food web. This occurs because all of these species are interconnected. Indeed, it is this interconnectedness that allows the flow of energy and the cycling of matter in aquatic and terrestrial biomes that we have discussed throughout this entire unit.

(a)

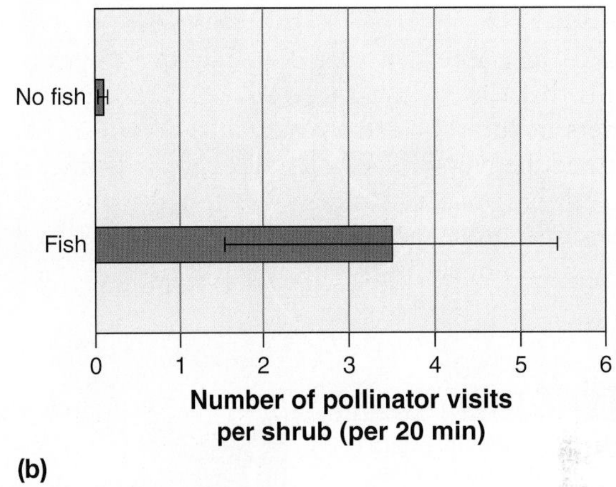

(b)

FIGURE 7.7 **Feedback loops in a food web.** Given that species in a food web are interconnected, changes in the abundance of one species can have large impacts on many other species. (a) When a predatory fish is added to a pond, it has a direct negative impact on aquatic, larval dragonflies. It also has indirect negative effects on adult dragonflies, positive indirect effects on pollinators such as bees, and a positive indirect effect on the pollination of flowers. (b) When we examine the number of pollination visits on the flowers along the pond's shoreline, we see that the presence of fish in the water has a large positive effect on the number of pollinators visiting the plants. Error bars are standard errors. *(Data from Knight, T. M., et al. 2005. Trophic cascades across ecosystems.* Nature *237: 880–883.)*

AP® Exam Tip

Remember data never proves a hypothesis. When designing an experiment on the exam, make sure to mention the value of repeated trials in your conclusion statement.

In this final module of Unit 1, we examined how the energy and matter captured by producers moves up through several trophic levels and the fact that the energy conversion is relatively low at 10 percent. We also saw that species are arranged in food webs with multiple connections, which causes changes in the abundance of a given species to have direct and indirect feedback loops on other species in the food web. For a summarizing visual of how many concepts from this unit are interconnected, turn to the Unit 1 Visual Representation feature on pages 80–81.

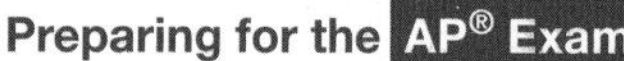

Visual Representation 1 The Yellowstone Ecosystem: The Yellowstone Ecosystem is a system of interdependent components, processes, and relationships. The intersection of a coyote's life with each of these components in the Yellowstone ecosystem illustrates the interconnectedness of the living and non living world.

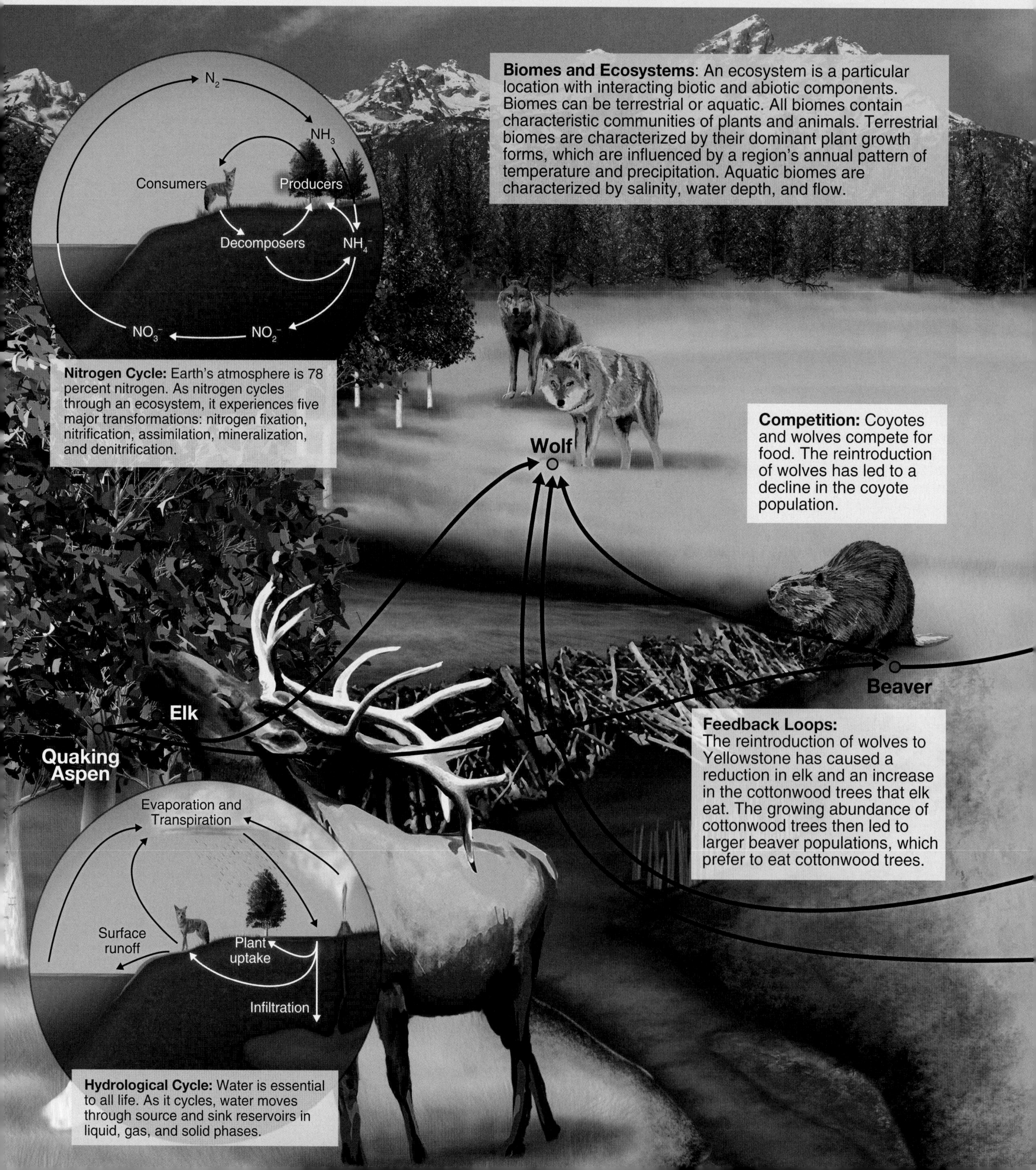

Biomes and Ecosystems: An ecosystem is a particular location with interacting biotic and abiotic components. Biomes can be terrestrial or aquatic. All biomes contain characteristic communities of plants and animals. Terrestrial biomes are characterized by their dominant plant growth forms, which are influenced by a region's annual pattern of temperature and precipitation. Aquatic biomes are characterized by salinity, water depth, and flow.

N_2

NH_3

Consumers

Producers

Decomposers

NH_4^-

NO_3^-

NO_2^-

Nitrogen Cycle: Earth's atmosphere is 78 percent nitrogen. As nitrogen cycles through an ecosystem, it experiences five major transformations: nitrogen fixation, nitrification, assimilation, mineralization, and denitrification.

Wolf

Competition: Coyotes and wolves compete for food. The reintroduction of wolves has led to a decline in the coyote population.

Beaver

Elk

Quaking Aspen

Feedback Loops: The reintroduction of wolves to Yellowstone has caused a reduction in elk and an increase in the cottonwood trees that elk eat. The growing abundance of cottonwood trees then led to larger beaver populations, which prefer to eat cottonwood trees.

Evaporation and Transpiration

Surface runoff

Plant uptake

Infiltration

Hydrological Cycle: Water is essential to all life. As it cycles, water moves through source and sink reservoirs in liquid, gas, and solid phases.

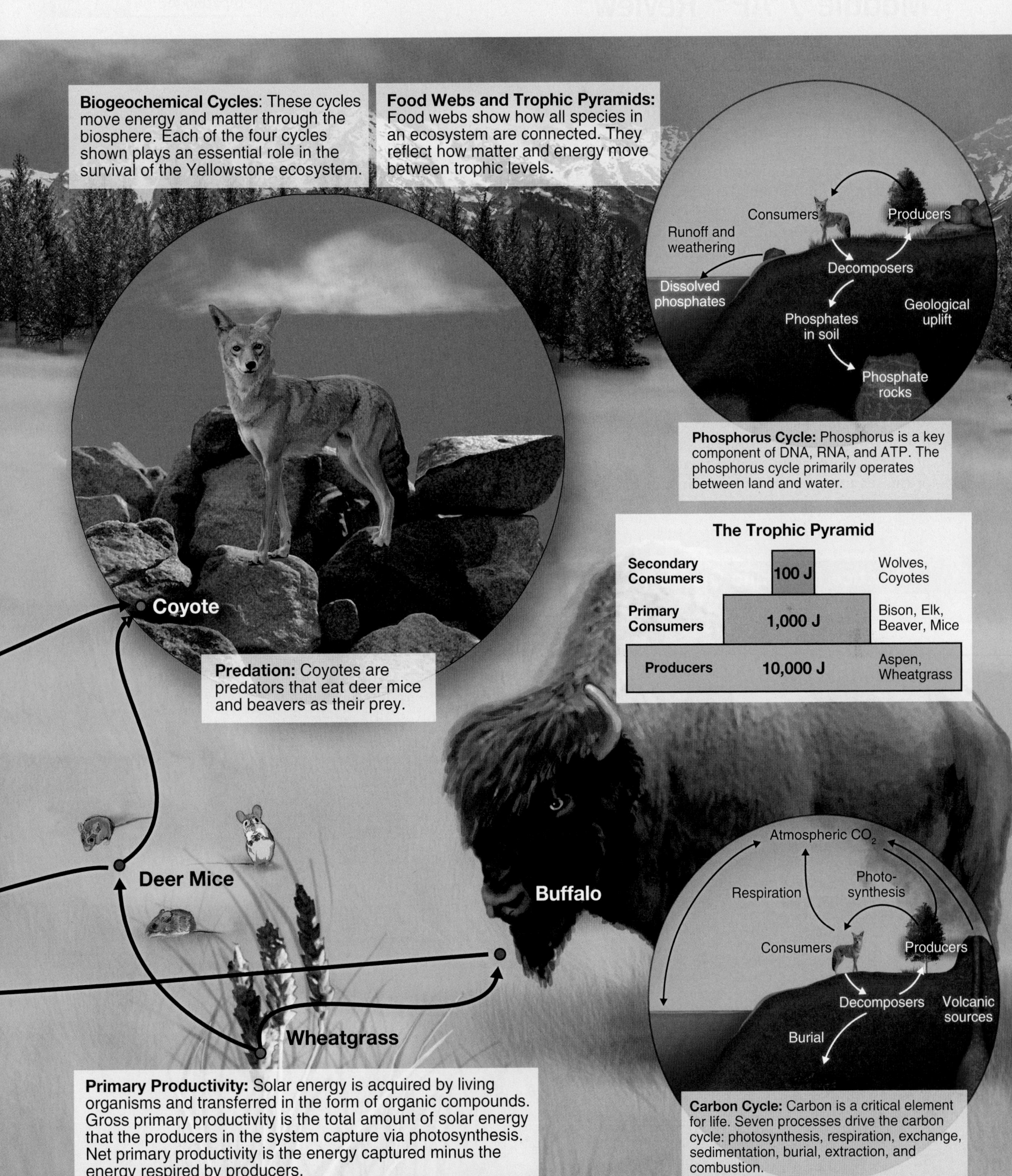
Biogeochemical Cycles: These cycles move energy and matter through the biosphere. Each of the four cycles shown plays an essential role in the survival of the Yellowstone ecosystem.
Food Webs and Trophic Pyramids: Food webs show how all species in an ecosystem are connected. They reflect how matter and energy move between trophic levels.
Consumers
Producers
Runoff and weathering
Decomposers
Dissolved phosphates
Geological uplift
Phosphates in soil
Phosphate rocks
Phosphorus Cycle: Phosphorus is a key component of DNA, RNA, and ATP. The phosphorus cycle primarily operates between land and water.
The Trophic Pyramid
Secondary Consumers
100 J
Wolves, Coyotes
Primary Consumers
1,000 J
Bison, Elk, Beaver, Mice
Producers
10,000 J
Aspen, Wheatgrass
Coyote
Predation: Coyotes are predators that eat deer mice and beavers as their prey.
Deer Mice
Buffalo
Wheatgrass
Atmospheric CO_2
Respiration
Photo-synthesis
Consumers
Producers
Decomposers
Volcanic sources
Burial
Primary Productivity: Solar energy is acquired by living organisms and transferred in the form of organic compounds. Gross primary productivity is the total amount of solar energy that the producers in the system capture via photosynthesis. Net primary productivity is the energy captured minus the energy respired by producers.
Carbon Cycle: Carbon is a critical element for life. Seven processes drive the carbon cycle: photosynthesis, respiration, exchange, sedimentation, burial, extraction, and combustion.

Module 7 AP® Review

Preparing for the AP® Exam

Learning Goals Revisited

7-1 How do energy and matter move through trophic levels in an ecosystem?

Energy and matter move from producers to primary, secondary, and tertiary consumers. While energy is ultimately lost as heat to the atmosphere, matter such as carbon, phosphorus, and nitrogen cycle through the ecosystem as the producers and consumers are broken down by scavengers, detritivores, and decomposers. This cycling of matter among trophic groups plays a key role in the biogeochemical cycles of ecosystems.

7-2 How does low ecological efficiency cause energy to decrease at higher trophic levels?

On average, only about 10 percent of the energy available in one trophic level is transferred to the next highest trophic level, which is referred to as ecological efficiency. Part of the reason for such low efficiency is that a substantial portion of the biomass in each trophic level is not digestible. Of the energy that can be digested, a portion is used for the day-to-day activities of the consumer, with a small portion remaining for the growth and reproduction of the consumer. This low efficiency also has implications for humans acting as primary versus secondary consumers.

7-3 Why do food webs experience feedback loops?

Food webs are two or more interacting food chains that represent the flows of energy matter in an ecosystem. When food webs have a new species added or an existing species' numbers reduced, the positive and negative feedback loops can have widespread effects on the other species in the food web because they are interconnected.

Practice Math and Graphing

Preparing for the AP® Exam

Answer the following questions. Be sure to show all your work.

1. Practice Math

In Cedar Bog Lake in Minnesota, a scientist measured the productivity of producers, primary consumers, and secondary consumers. Using the values in the table below, calculate the percent energy transferred from producers to primary consumers and the percent energy transferred from primary consumers to secondary consumers. How do these values compare to the 10% rule?

Trophic level	Productivity (cal/cm^2/year)
Producers	111.3
Primary consumers	14.8
Secondary consumers	3.1

2. Practice Graphing

Based on the food web shown on the next page, researchers collected data on the abundance of wolves, elk, woody plants, and beavers in two different locations.

Using a bar graph, plot the abundance of wolves, elk, and beavers in each of the three locations.

Location	Species	Number present
A	Wolf	2
B	Wolf	6
C	Wolf	10
A	Elk	30
B	Elk	20
C	Elk	10
A	Beaver	2
B	Beaver	10
C	Beaver	20

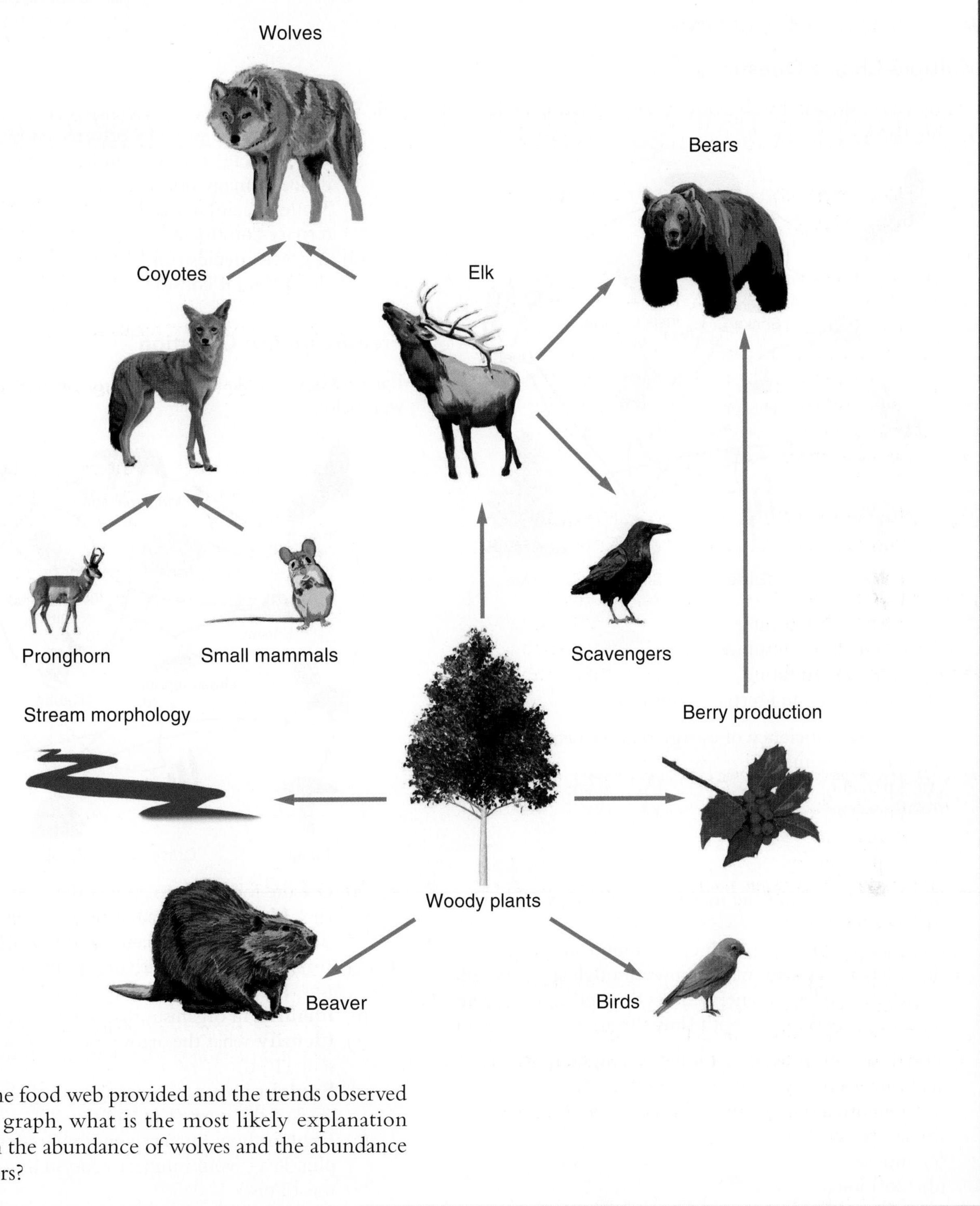

Given the food web provided and the trends observed in your graph, what is the most likely explanation between the abundance of wolves and the abundance of beavers?

AP® Practice Questions

Multiple-Choice Questions

1. Based on the food web below, what is the trophic level of the hawk?

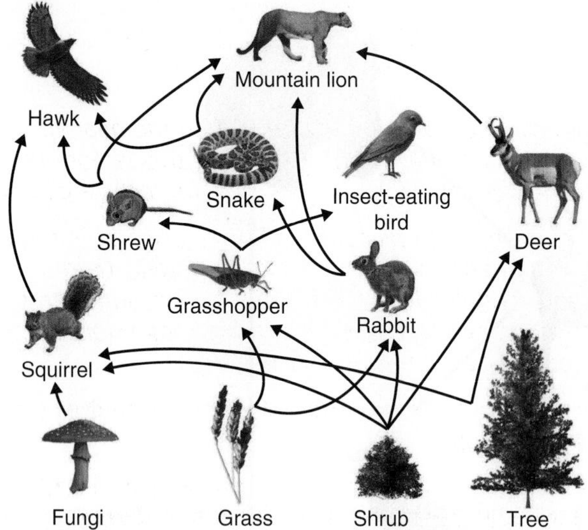

   (a) a primary consumer
   (b) a secondary consumer
   (c) a tertiary consumer
   (d) a secondary and tertiary consumer

2. The average efficiency of energy transfer between trophic levels is approximately
   (a) 1 percent.
   (b) 4 percent.
   (c) 10 percent.
   (d) 40 percent.

3. Which is true about the movement of energy and matter in ecosystems?
   (a) Energy and matter both cycle around an ecosystem.
   (b) Energy cycles but matter flows through an ecosystem.
   (c) Matter cycles but energy flows through an ecosystem.
   (d) Energy and matter both flow through an ecosystem.

4. Which organism would typically be considered a secondary consumer?
   (a) cyanobacteria
   (b) rabbits
   (c) moose
   (d) scorpions

5. Why do food webs experience feedback loops?
   (a) Producers are eaten by primary consumers.
   (b) Each species can have indirect effects on the abundance of many other species.
   (c) Some species act as both secondary consumers and tertiary consumers.
   (d) A given species can have a direct negative effect on the species it consumes.

Free-Response Question

Barn owls are predators of small prey, as illustrated in the food web below.

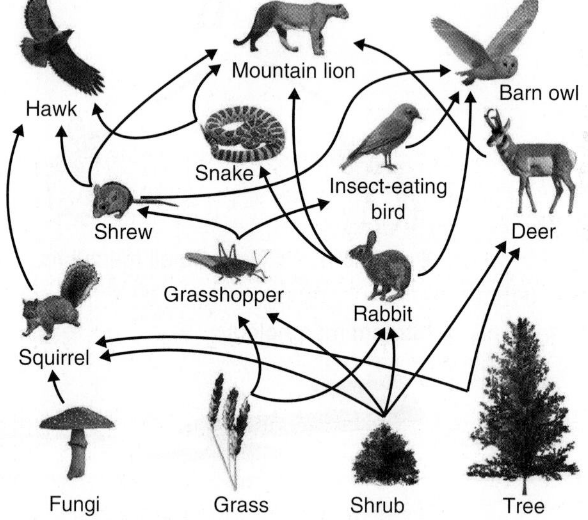

(a) Use the food web to answer the following:
   (i) **Identify** a primary producer. (1 pt.)
   (ii) **Identify** a primary consumer. (1 pt.)
   (iii) **Identify** a detritivore. (1 pt.)
   (iv) **Identify** a secondary consumer. (1 pt.)

(b) **Identify** an organism that is a decomposer. (1 pt.)

(c) **Identify** what the arrows in a food web/chain represent. (1 pt.)

(d) **Explain** how much energy moves from one trophic level to the next. (1 pt.)

(e) **Explain** how humans consuming food at a lower trophic level would impact ecosystems and human land use. (2 pts.)

(f) **Explain** how human removal of a species could impact a food web. (1 pt.)

UNIT 1
Practice Your Science Skills

Preparing for the AP® Exam

Data Analysis

This graph represents the average annual precipitation and temperature for terrestrial biomes. To answer the following questions, carefully read the graph and use your knowledge of terrestrial biomes and biogeochemical cycles.

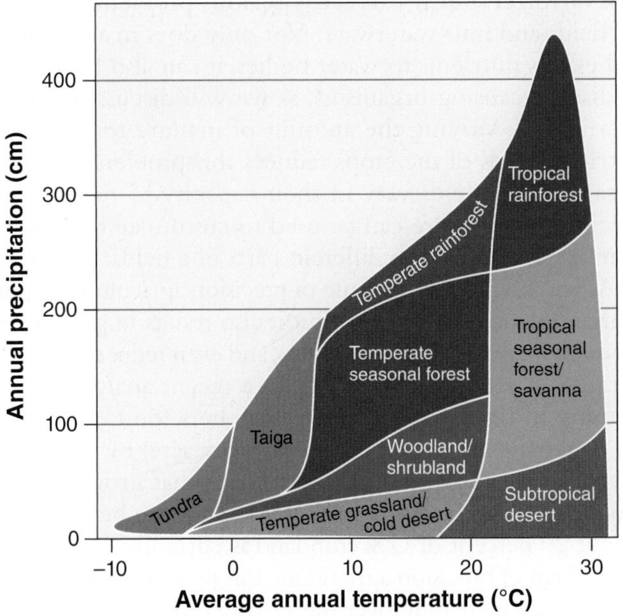

Questions

1. Based on the graph, **identify** the range of the air temperature in a woodland/shrubland biome. *(Hint: On the AP® exam, you need to be very precise in your measurements. Graders will have an acceptable range such as ±2 on either end of the answer. Many students will mistakenly find the nearest labeled number on the x axis, but this is not precise. You must also provide units to receive points.)*
2. Based on the graph, **identify** the range of annual precipitation in a temperate rainforest. *(Hint: You must be very precise in your measurements. Graders will have an acceptable range, for example around ±10 cm on either end.)*
3. Based on the graph, **identify** a biome that exists in polar regions.
4. The rate of nitrogen cycling is slower in the boreal forest than the temperate rainforest. **Explain** why. *(Hint: This question requires you to use your knowledge of biogeochemical cycles from Unit 1 along with this graph to answer the question. The task verb "explain" requires more information than "describe." You must include steps in a process, cause and effect, or more detail. One way to remember to do this is to add "because" or "leads to" in your answer to make sure it is complete.)*
5. Based on the graph, **make a claim** about which biome would contain a cactus. *(Hint: You must use your knowledge about dominant plants in a biome from Unit 1 along with data from the graph to answer this question. The task verb "make a claim" requires evidence to back up a statement. If your prompt, diagram, graph, or chart has quantitative data (numbers), you must use numbers in your evidence. If the prompt or diagram has qualitative data or prose, you can use that instead of numbers.)*

Pursuing Environmental Solutions

The Practice of Precision Agriculture

As we have discussed in this unit, nutrients such as phosphorus and nitrogen make their way into water bodies and subsequently cause harmful algal blooms and dead zones. Much of this nutrient input comes from agricultural fertilizer that is not taken up by crops and washes into local streams, rivers, and lakes when it rains. Not only is this movement of fertilizer harmful to the environment, but it also represents a loss to farmers who have purchased the fertilizer and used their time and equipment to distribute it over the fields. As a result, reducing the runoff of excess nutrients from agricultural fields represents a major global challenge.

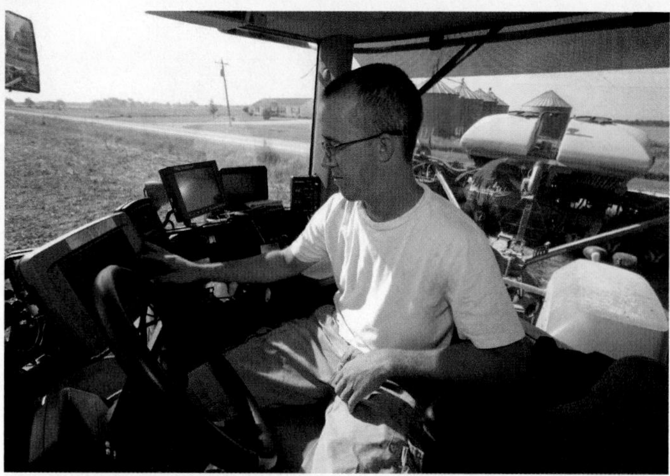

Precision agriculture. A farmer in Indiana plants soybeans using a GPS-equipped tractor that automatically steers itself. Such tractors can be used to plant crops and tailor the amount of fertilizer or pesticides applied to different areas of a field. *(United Soybean Board)*

One reason nutrients wash off fields is that the soil fertility can vary across a field so that crops in some portions of a field need more fertilizer than crops in other portions of the same field. However, farmers traditionally spread fertilizer evenly across a field. In a perfect world, a farmer could customize fertilizer applications across the field, adding high amounts only where high amounts are needed. Such an approach should use less fertilizer, improve the growth of the crops, and result in less unused fertilizer washing off the fields and into water bodies.

A customized approach to fertilizing crops, known as precision agriculture, is becoming a solution to excess nutrient runoff thanks to two technological advances. The first advance has been the ability to assess variation in crop growth across large fields. This is commonly done by using satellite photos that quantify areas of a field where crops are growing well, which is a sign of high fertility, and areas where crops are growing poorly, which is a sign of low fertility. Using this information, combined with data on how the type of soil varies across the field, a digital map can be created to tell the farmer which areas of the field need more or less fertilizer.

The second advance is the use of Global Positioning System (GPS) technology, which is the same technology used on modern cell phones to tell a person where they are located on Earth. By combining a digital map of where fertilizer needs to be applied with GPS coordinates of where the tractor is located in the field, farmers can customize their fertilizer application rates as they move across the field. For very large fields that are fertilized by small planes, the same principles can be used. Such applications can increase the efficiency of fertilizer use by the crops. For example, using precision agriculture allows a 300 percent increase in the efficiency of nitrogen uptake by crops. As a result, there is much less fertilizer wasted and running off the fields into streams, rivers, and lakes.

In addition to these two advances, other technological advances have been essential to the development of precision farming. These advances include dramatic improvements in software, huge increases in the ability of computers to analyze large amounts of data from satellite images across thousands of square kilometers, and advancements in soil sensors that are placed directly in fields to detect variation in nutrients and water, which complements the crop growth data obtained from the satellite images. One of the most fascinating advances is the ability to have GPS-equipped tractors automatically steer themselves as a farmer drives across large fields. Such automated steering not only helps control where different amounts of fertilizer are applied, but also helps the tractor minimize overlapping fertilizer applications as it moves back and forth across the field.

Precision agriculture is not only valuable for applying commercial fertilizers, but also for applying manure and irrigating the fields. As animal waste, manure is a natural fertilizer, but it can cause serious problems if it runs off fields and into waterways. Not only does manure runoff add excess nutrients to water bodies, it can also be a source of disease-causing organisms, as we will discuss further in Module 52. Varying the amount of manure to match the nutrient needs of the crops reduces this problem. Similarly, because large fields vary in their capacity to retain water, precision agriculture can be used to customize the amount of irrigation applied to different parts of a field.

As you might guess, the use of precision agriculture requires a larger financial investment, but it also results in greater crop production, reduced fertilizer costs, and even reduced fuel costs because of reduced tractor use. In a recent analysis of corn farming in the United States, researchers found that farmers could save up to \$62 per ha (\$25 per acre) by using these advanced technologies. In larger farms that grow crops on thousands of hectares, the total cost savings can be substantial. About 20 percent of U.S. crop land is currently farmed with some form of precision agriculture; this percentage is expected to continue increasing over the next decade. The investment in precision agriculture has increased crop production, saved farmers money, and helped the environment by reducing the human impacts on biogeochemical cycles.

Critical Thinking Questions

1. Given that crop-eating pests can also vary in number across a farm, how might farmers use precision technologies for applying pesticides?
2. With the environmental benefits provided by precision agriculture, how might governments motivate farmers to adopt these technologies?

References

Cost Savings from Precision Agriculture Technologies on U.S. Corn Farms. May 2, 2016. https://www.ers.usda.gov/amber-waves/2016/may/cost-savings-from-precision-agriculture-technologies-on-us-corn-farms/.

Hedley, C. 2014. The role of precision agriculture for improved nutrient on farms. *Journal of the Science of Food and Agriculture* 95:12–19.

Itzhaki, R. Artificial intelligence and precision farming: The dawn of the next agricultural revolution. *Forbes*, January 7, 2021.

Science Applied 1: Concept Explanation

Reversing Human Impacts on a Salty Lake

Located between the deserts of the Great Basin and the mountains of the Sierra Nevada, California's Mono Lake is an unusual site (**FIGURE SA1.1**). It is characterized by eerie towers of limestone rock known as tufa, glassy waters, unique animal species, and frequent dust storms. Mono Lake is a terminal lake, which means that water flows into it but does not flow out. As water moves through the mountains and desert soil, it picks up salt and other minerals, which it deposits in the lake. As the water evaporates, these minerals are left behind. Over time, evaporation has caused a buildup of salt concentrations so high that the lake became as salty as the ocean. **FIGURE SA1.2** summarizes these inputs to and outputs from the Mono Lake system.

While the lake is too salty for fish and many other lake species to survive, there are a few species that are specially adapted to live there, including the Mono brine shrimp (*Artemia monica*) and larvae of the Mono Lake alkali fly (*Ephydra hians*). The shrimp and the fly larvae consume microscopic algae, millions of tons of which grow in the lake each year. In turn, large flocks of migrating birds, such as sandpipers, gulls, and flycatchers, use the lake as a stop-over, feeding on the brine shrimp and fly larvae. The lake is an oasis on the migration route for these birds and they have come to depend on its food and water resources. The health of Mono Lake is therefore critical for many species.

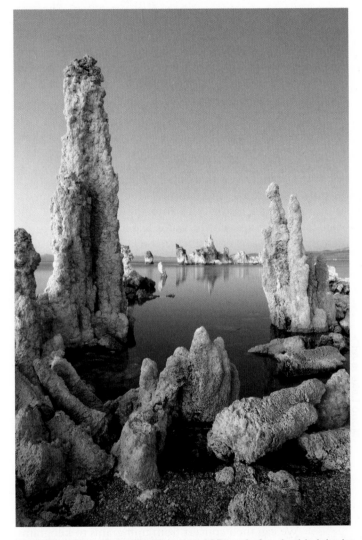

FIGURE SA1.1 The tufta towers of Mono Lake. In this lake in California, water enters the lake carrying minerals. As the water evaporates, the minerals are left behind to accumulate, which makes the lake very salty. *(Rachid Dahnoun/Getty Images)*

What happens when humans altered the natural water cycle?

In 1941, the City of Los Angeles built an aqueduct to divert water from the streams feeding Mono Lake so it could be used by residents of the city. With less stream water feeding the lake, the surface of the lake dropped 14 m (45 feet) by 1982. The lake now had half as much water and the salinity of the water doubled to more than twice that of the ocean. The salt killed algae in the lake; without the algae to eat, the brine shrimp also died. Most migrating birds stayed away, and newly exposed land bridges allowed coyotes from the desert to prey on the colonies of nesting birds that remained. These changes underscored the interconnectedness of the Mono Lake food web.

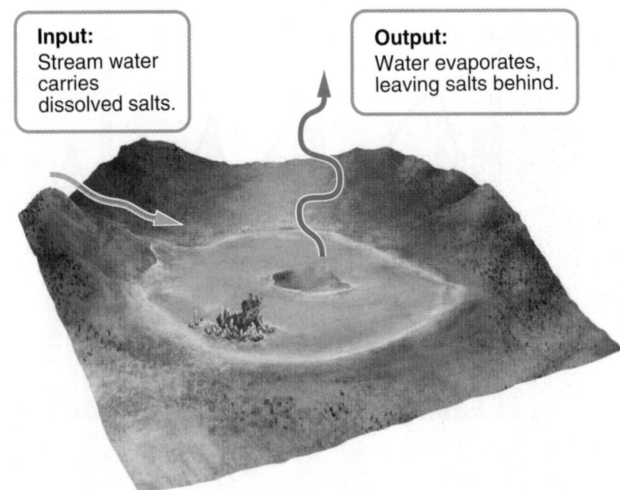

FIGURE SA1.2 The Mono Lake System. In this terminal lake system, inputs are from stream water while outputs are evaporated water only. In contrast, all salts that enter the lake from the surrounding streams remain in the lake.

How can we reduce human impacts on the natural water cycle?

Just when it appeared that Mono Lake would never recover, circumstances changed. In 1994, after years of litigation led by the National Audubon Society and tireless work by environmentalists and environmental scientists, the Los Angeles Department of Water and Power agreed to reduce the amount of water it diverted and to allow the lake to refill to about two-thirds of its historical depth. To decrease the amount of water diverted from Mono Lake, the Los Angeles residents had to reduce their water consumption. The city converted grass lawns requiring a great deal of water to native shrubs that are drought-tolerant, and it imposed new rules requiring low-flow showerheads and water-saving toilets. With this increase in lake water, the brine shrimp were thriving, and many birds returned to Mono Lake.

Over the past two decades, the region has experienced drought years and wet years as part of the natural water cycle. In years with more abundant water, Los Angeles is allowed to divert more stream water away from the lake. As a result, Mono Lake now contains more water, but has remained at about 4 m (12 feet) below the targeted goal for the past 20 years. In **FIGURE SA1.3**, you can see how annual changes in snow and rain caused above-average and below-average amounts of runoff into the lake and how the changes in runoff caused changes in the lake depth.

Where is the missing salt of Mono Lake?

To understand how water diversion alters water depth and salt accumulation, ecosystem scientists had to examine the flows of water and salts into the lake from streams and the loss of water due to evaporation out of the lake

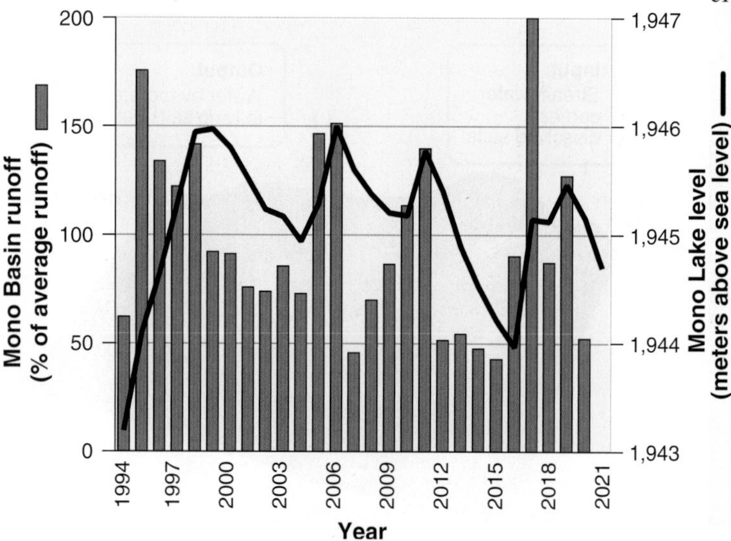

FIGURE SA1.3 Mono lake runoff and lake levels. Over the past three decades, the runoff of water from the streams into the lake has been highly variable as the region experienced wet and dry years. Years of high runoff into the lake result in higher lake levels.

FIGURE SA1.4 Research at Mono Lake. Scientists collect water sample at Mono Lake to monitor the concentrations of salt and nutrients in the water. *(Henry Bortman/NASA)*

(**FIGURE SA1.4**). The scientists made a series of observations and drew conclusions that helped them account for the movement of water, but they could not account for the movement of salt. In short, they discovered that more salt came into the lake than existed in the water of the lake. The question then arose, where is the missing salt?

Given that Mono Lake is a terminal lake, water enters the lake from streams, underground springs, and from precipitation, but water does not flow out. In a typical year before Los Angeles began diverting water, the lake level did not rise or fall. Given that this lake has no outlets, water outputs must equal water inputs. This suggests that the amount of water historically entering the lake was being balanced by the amount of water leaving the lake by evaporation. If this was not the case, the lake would eventually either dry up or overflow its banks.

While the balance of water in the lake is straightforward, the balance of salt in the lake is much more interesting. Streams and underground springs contain some amount of salt that varies seasonally. The salt concentration of the water flowing into Mono Lake averages about 50 mg per liter, which is equivalent to 50 parts per million.

To calculate the total amount of salt that historically entered Mono Lake each year, we can multiply the concentration of salt in the inflowing water (50 mg per liter) by the number of liters of water flowing into the lake, prior to being diverted by the City of Los Angeles (120 billion liters per year):

$$50 \text{ mg/L salt} \times 120 \text{ billion L/year} = 6 \text{ trillion mg salt/year}$$

We can then convert from milligrams to kilograms:

$$6 \text{ trillion mg salt/year} \times \frac{1 \text{ million kg}}{1 \text{ trillion mg}} = 6 \text{ million kg salt/year}$$

This is the annual input of salt to Mono Lake. Scientists estimate that no water has flowed out of the Mono Lake basin since it was formed about 120,000 years ago. If we assume that Earth's climate hasn't changed significantly over that time and that water inputs to Mono Lake have not changed drastically over that period, how much salt should be in the water of Mono Lake?

At the historic input rate of 6 million kg salt/year, the total amount of salt in the water of Mono Lake today should be:

$$6 \text{ million kg/year} \times 120{,}000 \text{ years}$$
$$= 720 \text{ billion kg of dissolved salt}$$

However, the lake today only contains about 285 billion kg of dissolved salt, which is much less salt than predicted. This made scientists wonder about the fate of the huge amount of unaccounted salt that entered the lake over the past 120,000 years.

The lake's tall tufa towers, prominently featured in the photograph, hold the answer. There is a limit to the amount of salt that can remain dissolved in water. As the concentrations of salt (primarily calcium) increase over time, the water reaches this limit. When the water can hold no more dissolved salt, the calcium precipitates out of solution by combining with carbonate ions to form calcium carbonate, which is a form of limestone. Over thousands of years, the precipitating calcium carbonate created the tufa towers. In this way, the excess salt has been removed from the lake water, but not from the Mono Lake system as a whole. If we account for the amount of salt in the water and in the tufa towers, we can balance the amount of salt coming into the lake and leaving the lake water.

The research on Mono Lake is an excellent example of how environmental scientists make observations and develop hypothesis for how the natural world works. By examining the inputs and outputs of both salt and water, they were able to determine how human diversion of essential water has a cascading effect on water levels, salt concentrations, and the consequences to the algae and animals that depend on the lake. The story also demonstrates a key principle of environmental science: A change in any one factor often has both expected and unexpected effects.

Questions

1. How did Los Angeles inadvertently conduct an experiment at Mono Lake?
2. What can be done to further increase the depth to the management goal of 6,391 feet above sea level, given that the City of Los Angeles diverts more stream water in years of higher runoff?
3. What is the reason for the discrepancy between the two calculations of salt content in Mono Lake?

Practice AP® Free-Response Question

Water that flows into Mono Lake contains a much smaller concentration of salt than the water already in the lake. This inflowing water tends to stratify, or float on top of existing water, because fresh water is less dense that salt water. As salt from the lower layer dissolves into the upper layer, nutrients from the bottom of the lake also rise to the surface. This exchange of nutrients is critical for the growth of algae in the surface waters. Recent research suggests that the reduction of water diversion from Mono Lake had unexpected results.

In 1995, the reduction of stream diversions from Mono Lake, combined with greater than average quantities of fresh water from snowmelt runoff, led to a rapid rise in lake water level. The large volume of fresh water from streams led to a long-term stratification of the lake, with fresh water on the surface and salt water on the bottom. Relative to data taken before the initial stream diversions in 1941, the current stratification has reduced the rate at which nutrients rise from the bottom of the lake. Long-term projections, based on mathematical models, suggest that the current degree of stratification will persist for decades.

(a) **Identify** three potential consequences of reduced lake mixing. (3 pts.)

(b) **Describe** two adaptive management strategies that could reduce lake stratification in Mono Lake. (3 pts.)

(c) **Identify** the chemical property of water that allows salt to dissolve. (2 pts.)

(d) **Explain** why the mixing of salt water with fresh water would be considered an example of increased entropy. (2 pts.)

References

California water wars: Mono Lake's unlikely victory. *ABC News*, July 1, 2019. https://www.abc10.com/article/news/mono-lake-tufa/103-cc9531b5-2a0e-41d0-b4ed-4e8824bc180f.

Mono Lake: State of the Lake (2021), https://www.monolake.org/learn/stateofthelake/.

L. Sahagun, Once teetering, Mono Lake is revived by heavy rains, snow, *Los Angeles Times*, January 10, 2017. http://www.latimes.com/local/california/la-me-mono-lake-storm-20170109-story.html.

UNIT 1

Review

Key Terms to Remember

Climate (p. 18)
Weather (p. 18)
Community ecology (p. 19)
Symbiosis (p. 19)
Biosphere (p. 21)
Competition (p. 21)
Competitive exclusion principle (p. 21)
Resource partitioning (p. 22)
Predation (p. 23)
Parasitoid (p. 23)
Parasitism (p. 24)
Pathogen (p. 24)
Herbivory (p. 24)
Mutualism (p. 25)
Photosynthesis (p. 25)
Commensalism (p. 26)
Native species (p. 26)
Exotic species (Alien species) (p. 26)
Invasive species (p. 27)
Biome (p. 29)
Terrestrial biome (p. 29)
Aquatic biome (p. 29)
Habitat (p. 30)
Tundra (p. 32)
Permafrost (p. 33)
Taiga (Boreal forest) (p. 33)
Temperate rainforest (p. 34)
Temperate seasonal forest (p. 35)
Shrubland (Woodland) (p. 35)
Temperate grassland (Cold desert) (p. 36)
Tropical rainforest (p. 37)
Savanna (Tropical seasonal forest) (p. 39)
Hot desert (p. 39)
Freshwater biomes (p. 44)
Littoral zone (p. 45)
Limnetic zone (p. 45)
Phytoplankton (p. 45)
Profundal zone (p. 45)
Benthic zone (p. 45)
Oligotrophic (p. 45)
Mesotrophic (p. 45)
Eutrophic (p. 45)
Freshwater wetland (p. 46)
Estuary (p. 47)
Salt marsh (p. 47)
Mangrove swamp (p. 47)
Intertidal zone (p. 47)
Coral reef (p. 48)
Coral bleaching (p. 48)
Open ocean (p. 49)
Photic zone (p. 49)
Aphotic zone (p. 49)
Chemosynthesis (p. 49)
Biogeochemical cycle (p. 51)
Reservoirs (p. 51)
Carbon cycle (p. 52)
Aerobic respiration (p. 52)
Steady state (p. 52)
Greenhouse gases (p. 54)
Global warming (p. 55)
Limiting nutrient (p. 55)
Nitrogen cycle (p. 55)
Nitrogen fixation (p. 55)
Nitrification (p. 55)
Assimilation (p. 55)
Mineralization (Ammonification) (p. 55)
Denitrification (p. 56)
Anaerobic (p. 56)
Aerobic (p. 57)
Leaching (p. 57)
Phosphorus cycle (p. 59)
Algal bloom (p. 61)
Hypoxic (p. 61)
Dead zone (p. 61)
Hydrologic cycle (p. 62)
Transpiration (p. 62)
Evapotranspiration (p. 63)
Runoff (p. 63)
Producers (Autotrophs) (p. 67)
Cellular respiration (p. 67)
Anaerobic respiration (p. 67)
Primary productivity (p. 68)
Gross primary productivity (GPP) (p. 68)
Net primary productivity (NPP) (p. 68)
Biomass (p. 69)
Standing crop (p. 69)
Consumer (Heterotroph) (p. 74)
Herbivore (Primary consumer) (p. 74)
Carnivore (p. 74)
Secondary consumer (p. 74)
Tertiary consumer (p. 74)
Trophic levels (p. 74)
Food chain (p. 74)
Scavenger (p. 75)
Detritivore (p. 75)
Decomposers (p. 75)
Ecological efficiency (p. 76)
The 10% rule (p. 76)
Trophic pyramid (p. 76)
Food web (p. 77)

Unit 1 AP® Environmental Science Practice Exam **Preparing for the AP® Exam**

Section 1: Multiple-Choice Questions

1. Which is an example of a biotic component of an ecosystem?
   (a) water
   (b) sunlight
   (c) fungi
   (d) air

2. Which of the following is a product of photosynthesis?
   (a) glucose
   (b) sulfur
   (c) carbon dioxide
   (d) nitrates

3. Which could be a cause of decreased evapotranspiration?
   (a) increased precipitation
   (b) decreased runoff
   (c) increased percolation
   (d) increased vegetation

For questions 4 & 5, select from the following choices:

(a) producers (c) secondary consumers
(b) primary consumers (d) tertiary consumers

4. At which trophic level are eagles that consume fish that consume zooplankton that eat algae?

5. At which trophic level do organisms use a process that produces oxygen as a waste product?

6. Which statement about precision agriculture is false?
   (a) It focuses on areas where crops are growing poorly.
   (b) It uses satellite imagery and GPS technology.
   (c) It prevents all fertilizer runoff to local streams and rivers.
   (d) It increases the efficiency of fertilizer application.

7. Which food chain from the Serengeti Plain is in the correct sequence from lowest to highest trophic level?
   (a) shrubs–gazelles–cheetahs–decomposers
   (b) shrubs–decomposers–cheetahs–gazelles
   (c) gazelles–decomposers–cheetahs–shrubs
   (d) decomposers–cheetahs–shrubs–gazelles

8. Roughly what percentage of incoming solar energy is converted into chemical energy by producers?
   (a) 99 (c) 50
   (b) 80 (d) 1

9. The net primary productivity of an ecosystem is 1 kg C/m^2/year, and the energy needed by the producers for their own respiration is 1.5 kg C/m^2/year. The gross primary productivity of such an ecosystem would be
   (a) 0.5 kg C/m^2/year
   (b) 1.5 kg C/m^2/year
   (c) 2.0 kg C/m^2/year
   (d) 2.5 kg C/m^2/year

10. A particular plot of land can produce 700 kg of beef per hectare. Beef sells for \$4/kg. If that land is converted to producing corn, which sells for \$0.15/kg, approximately how much will the farmer make selling corn? Assume 10 percent ecological efficiency.
    (a) \$100 (c) \$1,000
    (b) \$500 (d) \$3,000

11. Which statement about the carbon cycle is TRUE?
    (a) Carbon transfer from photosynthesis is in steady state with respiration and death.
    (b) The majority of dead biomass is accumulated in sedimentation.
    (c) Combustion of carbon is equivalent in mass to sedimentation.
    (d) Most of the carbon entering the oceans is from terrestrial ecosystems.

12. Human interaction with the nitrogen cycle is primarily due to
    (a) the leaching of nitrates into terrestrial ecosystems.
    (b) the fixation of nitrogen for fertilizers.
    (c) the acceleration of the nitrification process in aquatic ecosystems.
    (d) the decreased assimilation of ammonium and nitrates.

13. Small inputs of this substance, commonly a limiting factor in aquatic ecosystems, can result in algal blooms and dead zones.
    (a) dissolved carbon dioxide
    (b) sulfur
    (c) dissolved oxygen
    (d) phosphorus

14. Which describes a key difference between photosynthesis and aerobic respiration?
    (a) Photosynthesis generates organic compounds whereas aerobic respiration generates inorganic compounds.
    (b) Photosynthesis releases energy whereas aerobic respiration consumes energy.
    (c) Aerobic respiration has a higher energy efficiency than photosynthesis.
    (d) Photosynthesis consumes chemical energy whereas aerobic respiration generates chemical energy.

15. Why do scientists use dominant plant growth forms to categorize terrestrial biomes?
    (a) Plants with similar growth forms are always closely related genetically.
    (b) Different plant growth forms indicate climate differences, whereas different animal forms do not.
    (c) Plants from similar climates evolve different adaptations.
    (d) Similar plant growth forms are found in climates with similar temperatures and amounts of precipitation.

16. Which statement about tundras and boreal forests is correct?
    (a) Both are characterized by slow plant growth, so there is little accumulation of organic matter.
    (b) Tundras are warmer than boreal forests.
    (c) Boreal forests have shorter growing seasons than tundras.
    (d) Boreal forests have larger dominant plant growth forms than tundras.

17. Which statement about temperate biomes is correct?
    (a) Similar biomes can be found along lines of longitude.
    (b) Temperate rainforests receive the most precipitation, whereas cold deserts receive the least precipitation.
    (c) Temperate rainforests can be found in eastern South America.
    (d) Taiga biomes are adapted to frequent fires.

18. Which statement about tropical biomes is correct?
    (a) Tropical rainforests have the highest precipitation.
    (b) Savannas are characterized by the densest forests.
    (c) Tropical rainforests have the slowest rates of decomposition due to high rainfall.
    (d) Subtropical deserts have the highest species diversity.

19. Which statement about aquatic biomes is correct?
    (a) They are characterized by dominant plant growth forms.
    (b) They can be categorized by temperature and precipitation.
    (c) Lakes contain littoral zones and intertidal zones.
    (d) Freshwater wetlands have emergent plants in their deepest areas, whereas ponds and lakes do not.

20. The limnetic zone is defined as the area of a lake where
    (a) algae and emergent plants grow.
    (b) sunlight does not reach.
    (c) the thermocline exists.
    (d) phytoplankton are the only photosynthetic organisms.

21. Which biome provides ecosystem services that include reducing the severity of floods and filtering pollutants from the water?
    (a) freshwater wetlands
    (b) temperate grassland
    (c) coral reef
    (d) open ocean

22. In the coniferous forests of Oregon, eight species of woodpeckers coexist. Four species select their nesting sites based on tree diameter, while the fifth species nests only in fir trees that have been dead for at least 10 years. The sixth species also nests in fir trees, but only in live or recently dead trees. The two remaining species nest in pine trees, but each selects trees of different sizes. This pattern is an example of
    (a) resource partitioning.
    (b) commensalism.
    (c) predator-mediated competition.
    (d) a mutualism.

23. Which is an example of commensalism?
    (a) oxpeckers (a type of bird) that live on a rhino and eat parasites off the rhino's back
    (b) cuckoos, which lay their eggs in the nest of another bird species that then raises the cuckoo's young
    (c) cattle egrets eat insects that are disturbed when the cattle forage
    (d) protozoa living inside termites, helping termites to digest wood

For question 24, use the diagram below that shows the distributions of two barnacle species.

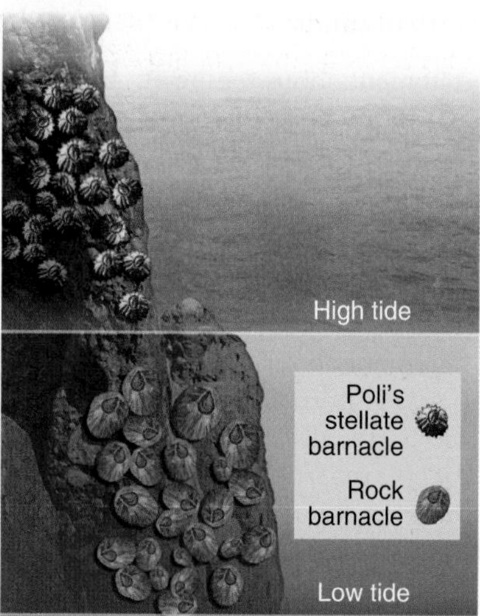

24. When researchers remove rock barnacles, the Poli's stellate barnacles expand their distribution downward into the intertidal zone. However, when researchers remove Poli's stellate barnacles, the rock barnacles do not expand their distribution upward. Which is the best interpretation of this result?
    (a) Rock barnacles reside in deeper water due to competition from Poli's stellate barnacles.
    (b) Poli's stellate barnacles reside in shallower water due to competition from rock barnacles.
    (c) Poli's stellate barnacles reside in shallower water to avoid desiccation.
    (d) Rock barnacles reside in deeper water to avoid deep-water predators.

For questions 25 & 26, use the graph below:

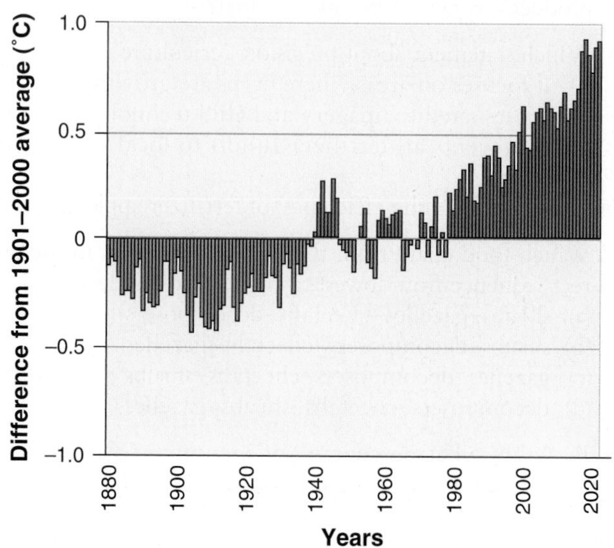

25. In the graph above, approximately which year did temperatures begin to be above the average?
    (a) 1880
    (b) 1920
    (c) 2000
    (d) 1940

26. How much above the average were temperatures in 2010?
    (a) −0.2
    (b) 0.2
    (c) 0.7
    (d) 1.0

For questions 27 & 28, use the text below:

European starlings are birds that are native to England. In the continental United States, they displace native birds such as woodpeckers and chickadees from using nesting sites. They are much more aggressive in seeking and finding nesting sites than native species. As a result, their numbers have grown exponentially.

27. Which of the following identifies the author's main claim?
    (a) European starlings are native species to the United States.
    (b) European starlings are competitors of native species in the United States.
    (c) Woodpeckers are an invasive species in the United States.
    (d) Native species outcompete the invasive starlings in the United States.

28. Which of the following is evidence of the author's main claim?
    (a) Woodpeckers and chickadees compete for nesting sites with each other.
    (b) European starlings are native to England.
    (c) European starling populations have grown exponentially in the United States.
    (d) Starlings are not as aggressive in competition for nests.

Section 2: Free-Response Questions

1. Use the diagram below to answer the following questions:

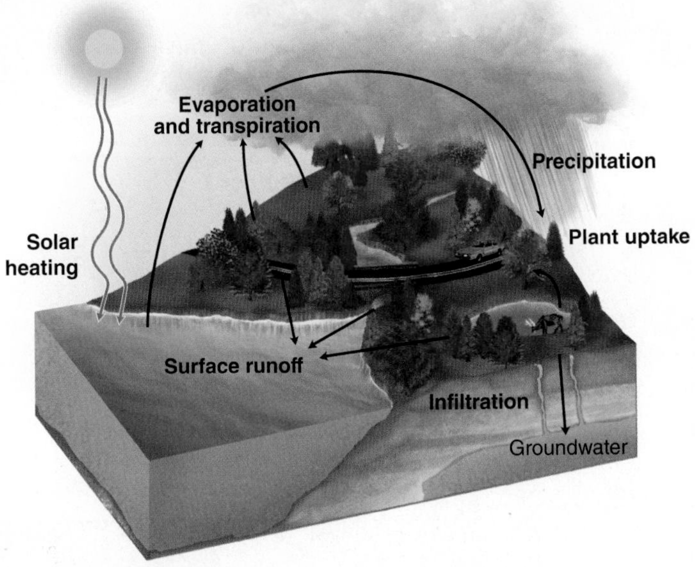

(a) **Identify** the process in the diagram that allows water to pass through the ground surface into bodies of water. (1 pt.)

(b) **Identify** the process in the diagram that filters water that can be used for drinking purposes. (1 pt.)

(c) **Explain** one way humans interfere with the hydrologic cycle. (1 pt.)

(d) **Explain** what role water plays in the other biogeochemical cycles. (2 pts.)

(e) Scientists have been measuring the effect of deforestation on water availability in ecosystems. They set up a lab experiment to see how water moves through land that has abundant tree roots compared to land without tree roots.

(i) **Identify** the independent variable in this experiment. (1 pt.)

(ii) **Identify** the dependent variable in this experiment. (1 pt.)

(iii) **Explain** a possible hypothesis tested through this experiment. (1 pt.)

(f) **Explain** the effect of deforestation on the hydrologic cycle in regards to nitrogen and phosphorus. (2 pts.)

2. (a) Marine biomes are different from freshwater biomes in many different ways.

(i) **Describe** one key difference between marine and freshwater biomes. (1 pt.)

(ii) **Describe** how the key difference described in part (i) affects living organisms. (1 pt.)

(b) **Identify** how an estuary biome differs from an open ocean biome or a river biome. (1 pt.)

(c) Mangrove swamps are a type of saltwater biome.

(i) **Describe** how a mangrove swamp is impacted by humans. (1 pt.)

(ii) **Make a claim** for the importance of mangrove biomes. (2 pts.)

(iii) **Justify** the claim made in question c part (i). (1 pt.)

Use the following image to answers parts d-f:

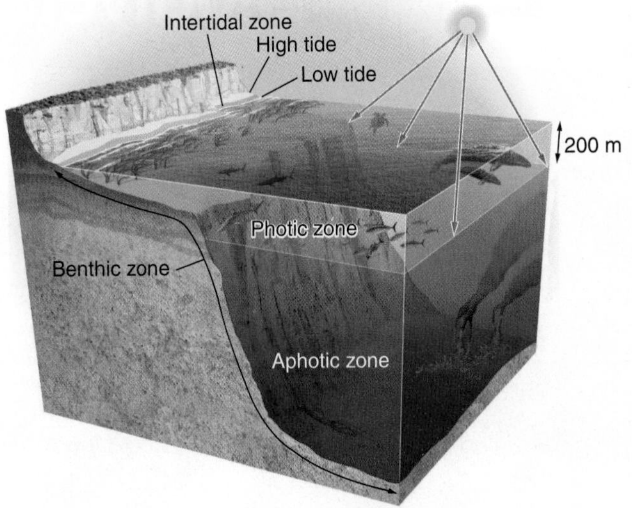

(d) **Identify** where biodiversity would typically be highest. (1 pt.)

(e) **Explain** which zone would be impacted the most by an increase in runoff of fertilizers. (1 pt.)

(f) **Identify** one way that humans could decrease the impact to the zone in question c, part (i). (1 pt.)

3. (a) Energy in ecosystems passes through living organisms.

(i) **Identify** the main source of energy for most ecosystems. (1 pt.)

(ii) **Explain** how much efficiency exists as energy moves from one trophic level to the next. (1 pt.)

(b) Besides Joules or kcal, **identify** another method in which energy can be measured in an ecosystem. (2 pts.)

(c) Use the diagram below for the following questions.

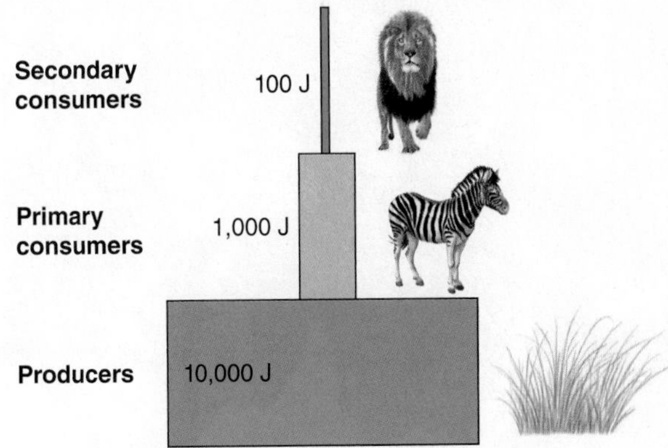

(i) **Calculate and show work** for energy lost from producers to consumers. (2 pts.)

(ii) **Calculate and show work** for the percent of energy remaining from producers to secondary consumers. (2 pts.)

(iii) **Calculate and show work** to determine how much energy would be available if there were another trophic level above the lion. (2 pts.)

UNIT 2

The Living World: Biodiversity

The biodiversity of ecological communities, such as these wildflowers in Texas, is critical to the overall productivity and stability of natural ecosystems. *(Dean Fikar/Shutterstock)*

MODULE 8	Introduction to Biodiversity	**MODULE 11**	Ecological Tolerance	**MODULE 13**	Adaptations
MODULE 9	Ecosystem Services	**MODULE 12**	Natural Disruptions to Ecosystems	**MODULE 14**	Ecological Succession
MODULE 10	Island Biogeography				

CASE STUDY

The Benefits of Biodiversity

Biodiversity, a combination of the words "biological" and "diversity," is the diversity of life forms in an environment. You may have heard people talking about the importance of biodiversity, but why should we preserve it around the world? We have long appreciated that species in nature provide benefits to humans including food, fiber, lumber, and medicines. We also know that collections of species in ecosystems provide primary productivity in the land and water, the production of clean water, and buffers against natural disasters. The question is, do the important benefits provided by these ecosystems critically depend on maintaining the biodiversity that has historically existed, or can we enjoy the same benefits even when biodiversity declines due to human activities?

For more than three decades, scientists have debated whether greater biodiversity provides greater ecosystem benefits. Researchers conducting experiments in aquatic and terrestrial ecosystems have commonly found that food webs with a greater number of species have a higher overall productivity. The ecosystems are also more stable whenever the abiotic environment changes. For example, during a drought an ecosystem may experience lower productivity. However, an ecosystem with higher species diversity can recover more quickly when rain falls and the drought ends compared to an ecosystem that has low species diversity. While these results have been encouraging, critics have pointed out that because most experiments contain many fewer species than actually exist in nature, we cannot be sure that the results apply to nature. It is also possible that the productivity of natural ecosystems is much more affected by variation in climatic conditions and available nutrients than by the number of species in the food web. If so, the positive results of biodiversity in experiments may be trivial in the natural world compared to the effects of climate and nutrients.

We have learned that biodiversity is critical to the proper functioning of ecosystems.

To determine whether having higher biodiversity matters in nature, we can examine whether natural sites with a higher number of species function better in terms of productivity. Researchers recently compiled data from 67 studies that measured the number of species and ecosystem productivity from 600,000 sample locations around the world. Across terrestrial, freshwater, and marine biomes, they found that sites with more species did indeed have higher productivity, which was consistent with the results of the earlier experiments. In addition, when the researchers considered how variation in climatic conditions and available nutrients affected the productivity of these ecosystems, they found that biodiversity was a more important factor than climate in half of the studies and a more important factor than nutrients in two-thirds of the studies. In short, we have learned that biodiversity is critical to the proper functioning of ecosystems. Such conclusions make it clear that we need to protect Earth's biodiversity so that we can continue to enjoy the many benefits that it provides.

Sources: E. Duffy et al., Biodiversity effects in the world are common and as strong as key drivers of productivity, *Nature* 549 (2017): 261–264; D. Obura, New global biodiversity goals must take these key lessons into account, *The Conversation*, August 16, 2021.

Biodiversity The diversity of life forms in an environment.

Practice your science skills

1. **Text Analysis:** Identify one ecological claim about the benefit of biodiversity made by the author.
2. **Concept Explanation:** Explain why an increase in ecosystem biodiversity will also increase ecosystem resilience.
3. **Scientific Experiments:** Identify a research method used to determine the importance of biodiversity in nature.

As we have just learned, biodiversity is an important part of any ecosystem because of the role it plays in making ecosystems productive and stable over time. In this unit, we will examine biodiversity in depth, including biodiversity from the level of genes to ecosystems. We will also see how this diversity allows ecosystems to provide us with a number of important services. Then we explore how the number of species in an ecosystem is the net result of some species colonizing places they can tolerate and other species going extinct due to unsuitable biotic or abiotic conditions. Given that many ecosystems experience various disruptions over time, we examine how short- and long-term disruptions can vary in magnitude and alter the composition of species in different ecosystems. Whether an ecosystem is frequently disrupted or not, species can evolve traits to persist in a given ecosystem as a result of natural selection. Over long periods of time, the initial species that colonize an area are eventually replaced by other species that possess traits that are better suited to the changing environmental conditions. Collectively, the modules in this unit will provide an excellent discussion of how we obtain biodiversity in different ecosystems and the consequences of doing so.

Module 8

Unit 2 | 8 | 9 | 10 | 11 | 12 | 13 | 14

Introduction to Biodiversity

As we read in the unit-opening case study, biodiversity is an important indicator of environmental health. Conversely, a rapid decline of biodiversity in an ecosystem indicates that it is under stress. In this module, we will examine how scientists define biodiversity at different scales and how biodiversity at each scale affects responses to environmental stressors. We will then examine how we can quantify biodiversity by considering both the number of species in an area as well as how evenly distributed individuals are among the different species. Finally, we will explore the challenging question of how many species currently live on Earth.

Learning Goals

After reading this module you should be able to

8-1 identify how biodiversity exists at different scales.

8-2 explain how biodiversity affects responses to environmental stressors.

8-3 describe how we can calculate biodiversity.

8-4 describe how we estimate the number of species living on Earth.

8-1 How does biodiversity exist at different scales?

Biodiversity exists from genes to ecosystems

As illustrated in **FIGURE 8.1**, biodiversity exists on four scales: genetic, species, habitat, and ecosystem. Each level of biodiversity is an important indicator of environmental health and quality.

Genetic Diversity

Genetic diversity is a measure of the genetic variation among individuals in a population. Large populations generally have a high amount of variation in their genetic composition and this can occur for a variety of reasons that we discuss in detail in Module 13. For example, the white-tailed deer (*Odocoileus virginianus*) lives throughout much of North America, Central America, and into northern South America. Across this wide geographic range, the deer live in different habitats, including fields, forests, and swamps. As a result, deer herds living in colder climates have evolved relatively large bodies to help conserve heat, while other herds, such as those living in the warm climate of Florida, have evolved to have much smaller bodies that more readily lose heat.

If a large population declines in number, the amount of genetic diversity carried by the surviving individuals is greatly reduced, which is known as a **population bottleneck**. If the population subsequently rebounds, that population can still retain a low amount of genetic diversity. For instance, the cheetah (*Acinonyx jubatus*) is a predatory cat that once roamed across much of Africa and the Middle East. Researchers who have examined the amount of genetic variation in existing cheetahs have discovered that around 10,000 years ago the cheetah population experienced a population bottleneck for reasons that are unknown. When this bottleneck occurred, only a small percentage of the population survived and it could only carry a small amount of the population's original genetic variation. As a result, the population of cheetahs that exists today continues to have low genetic variation. In the next section, we will explore the consequences.

Species Diversity

A species, as we learned in Module 0, is a group of organisms that is distinct from other groups in morphology (body form and structure), behavior, or biochemical properties. Individuals of the same species can breed and produce

Genetic diversity A measure of the genetic variation among individuals in a population.

Population bottleneck When a large population declines in number, the amount of genetic diversity carried by the surviving individuals is greatly reduced.

(d) Ecosystem diversity

(c) Habitat diversity

(b) Species diversity

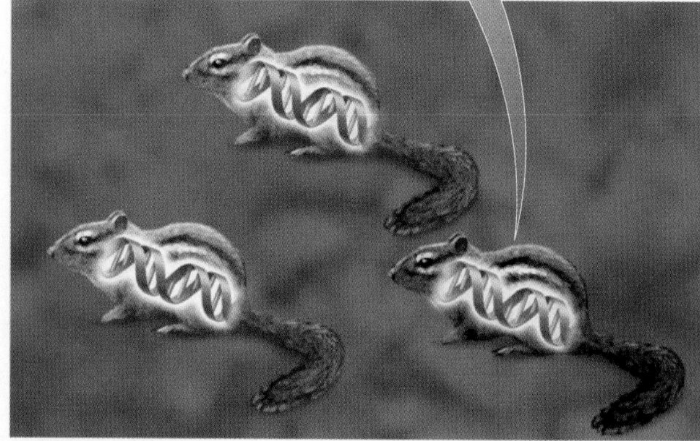

(a) Genetic diversity

FIGURE 8.1 Levels of biodiversity. Biodiversity exists at four scales. (a) Genetic diversity is the variety of genes among individuals of a species. (b) Species diversity is the variety of species within an ecosystem. (c) Habitat diversity is the variety of habitats present in an ecosystem. (d) Ecosystem diversity is the variety of ecosystems within a region.

viable and fertile offspring. This last requirement is important because sometimes individuals from different species can mate, but they typically produce offspring that do not survive or cannot breed. Based on this definition, **species diversity** indicates the number of species in a region or in a particular ecosystem. For example, in our discussion of terrestrial biomes in Module 2, we learned that the tundra and taiga biomes have a low species diversity of plants and animals while tropical rainforests have a very high species diversity. Similarly, a temperate deciduous forest typically has dozens of tree and shrub species, but a commercial timber plantation typically has just one or a few species of trees that are grown for their high-value lumber (**FIGURE 8.2**). In "Do the Math: Converting hectares and acres" you can see how scientists examine species diversity in terms of land area.

Habitat Diversity

Habitat diversity is the variety of habitats that exist in a given ecosystem. Some species are **specialists**, which means that they only live under a narrow range of biotic or abiotic conditions. Examples include herbivores that only consume a single plant species, such as the koala (*Phascolarctos cinereus*) that live among eucalyptus trees and feed only on eucalyptus leaves. In contrast, other species are **generalists**, which means they can live under a wide range of biotic or abiotic conditions. White-tailed deer are an example of a generalist because they can live in a wide range of climates throughout North and South America and feed on a large number of different plants. For example, consider a region that has a mixture of farms growing crops, farms with pastures for cattle, abandoned farms with small trees and shrubs, and forests with large trees. Each of these habitats support different suites of specialist and generalist plant and animal species. The combination of the four habitats supports many more species. As a result, a greater diversity of habitats supports a greater diversity of species.

AP® Exam Tip

When mentioning a "habitat" on the AP® Environmental Science Exam, you should always specify which type of habitat and name a particular organism or group of organisms that live there. Forests provide habitat (this won't earn credit). Forests provide nesting habitats for some species of birds (this will earn credit).

(a)

(b)

FIGURE 8.2 Species diversity and ecosystems. (a) Natural forests contain a high diversity of tree species. (b) In forest plantations, in which a single tree species has been planted for lumber and paper products, species diversity is low. *(a: Paul Biggins/Alamy Stock Photo; b: Brent Waltermire/Alamy)*

We can see this pattern in the number of bird species living along streams that differ in depth and the types of plants that are present along the shoreline. When researchers examine the number of bird species living along 27 streams in Vermont that supply water to Lake Champlain, they found that the average stream contained only 17 species of birds, but the entire collection of streams contained 101 species of birds. The reason is that different types of birds preferred streams with different habitats. For example, fish-eating birds congregated at deeper streams that have very little agriculture along the shoreline, wading birds congregated in shallow streams, and insect-eating birds congregated in streams surrounded by meadows and forests. Having a diverse suite of habitats is therefore important to maintaining a high diversity of species.

Ecosystem Diversity

At the largest scale, we can think of **ecosystem diversity**, which is the variety of ecosystems that exist in a given region. For example, along the coasts of China, we can find a mixture of temperate forest ecosystems, river ecosystems, and the ocean ecosystem. As we saw in the case of habitat diversity, a larger diversity of different ecosystems allows for more species to live in a given area. Similarly, Los Angeles County in California has the highest diversity of birds in the United States as a result of the many different habitats in the area including oak woodlands, chaparral, deserts, urban environments, and the ocean.

AP® Exam Tip

You are required to show all work for math problems on the AP® Environmental Science Exam. You will not get full credit for a correct answer unless you show each step, no matter how simple. Remember to include units. Develop the habit of showing your work from the beginning of the year, starting with the math practice in this module.

Species diversity The number of species in a region or in a particular ecosystem.

Habitat diversity The variety of habitats that exist in a given ecosystem.

Specialists Species that only live under a narrow range of biotic or abiotic conditions.

Generalists Species that can live under a wide range of biotic or abiotic conditions.

Ecosystem diversity The variety of ecosystems that exist in a given region.

DO THE MATH

Converting Between Hectares and Acres

Preparing for the AP® Exam

Some measures of biodiversity are given in terms of land area, so becoming familiar with measurements of land area is important to understanding them. In the United States we measure land area in terms of square miles or acres. A square mile is an area that is 640 acres and one acre is an area approximately 64 m by 64 m (209 feet by 209 feet). However, the rest of the world measures land in hectares (ha), which is an area 100 m by 100 m. There are 2.47 acres in 1 ha. The conversion from hectares is relatively easy to do without a calculator if we round to two significant figures, which gives us 2.5 acres in 1 ha.

If a nature preserve is 100 ha, what is its size in acres?

$$100 \text{ ha} \times 2.5 \text{ acres} = 250 \text{ acres}$$

YOUR TURN A particular forest is 10,000 acres. Determine its size in hectares.

8-2 How does biodiversity affect responses to environmental stressors?

Increased biodiversity improves how populations and ecosystems respond to environmental stressors

The various scales of biodiversity impact how populations and ecosystems respond to biotic or abiotic environmental stressors. In this section we will explore how genetic diversity, species diversity, and habitat diversity provide benefits to populations and ecosystems.

Consequences of Genetic Diversity

High genetic diversity benefits the long-term persistence of populations because they are better able to respond to environmental change compared to populations with lower genetic diversity. For example, if a population of fish possesses high genetic diversity for disease resistance, at least some individuals are likely to carry genes that allow them to survive whatever diseases move through the population. When there is low genetic diversity, there is an increased likelihood that the population will decline more when exposed to a disease. In the case of the cheetahs, which experienced a population bottleneck 10,000 years ago, their low genetic diversity has made them particularly susceptible to a pathogen that kills up to 70 percent of cheetahs that are held in captivity.

Genetic diversity can also be very important in domesticated animals and plants. In the case of agricultural crops, centuries of plant breeding have produced hundreds of different varieties of crops such as corn, rice, and wheat. When a new disease appears that can kill one crop variety, researchers can look to other varieties that contain unique genetics that are resistant to the new disease. As we will discuss later (Module 59), environmental scientists are very concerned that over the past century the number of genetically distinct varieties of important domesticated plant and animal species have been rapidly declining as farmers focus on a small number of highly productive varieties. For example, apples originated as a wild plant in Kazakhstan and the original varieties have been bred into thousands of different varieties. In 1900, U.S. farmers grew 8,000 varieties of apples, but today more than 95% of these varieties are extinct (**FIGURE 8.3**).

Consequences of Habitat Diversity for Specialist and Generalist Species

When a region contains a high diversity of habitats, there is an increase in the total number of species. This is because

FIGURE 8.3 Genetic diversity in domesticated crops. Centuries of plant breeding has produced a tremendous diversity of genetic varieties of domesticated animals and plants, including more than 8,000 genetic varieties of apples. Today, most of these varieties no longer exist as farmers have focused on growing just the most productive varieties. *(simo bogdanovic/Alamy Stock Photo)*

each habitat contains some unique specialist species while multiple habitats often contain the same generalist species. So, what happens when we start to lose habitats due to human activities? Given that specialist species are only found in a single habitat type while generalists are found across multiple habitat types, the loss of a single habitat type generally leads to the loss of the specialist species, but not the loss of the generalists since they can live in other habitat types. However, as habitat loss continues and multiple habitats are destroyed, this too eventually leads to the loss of the generalist species. Habitat loss can also lead to reduced numbers of any species that require large, continuous habitats, such as large migrating herds of buffalo or gazelles, or large predators such as wolves and mountain lions that roam over several kilometers of land in search of prey.

Consequences of Species Diversity

As we discussed in the unit-opening case study, scientists have discovered that having a higher species diversity results in ecosystems that are more productive and resilient. That is, they have greater primary productivity (as discussed in Module 6) and they are better able to either resist the impact of disturbances—such as drought, hurricanes, or fires—or they are better able to rapidly recover from disturbances. To see the impacts of species diversity on productivity, researchers have examined soil fungi that live in close proximity to plant roots and act as mutualists with the plants. The fungi search for phosphorus in the soil and provide some of this important nutrient to the plants. In exchange, the plant roots provide the fungi with sugars produced through photosynthesis. When researchers conducted an experiment in which they manipulated the number of species of fungi in the soil, they found that soil with a higher diversity of fungi caused the plants to increase the biomass of their roots and their shoots (i.e., stems and leaves), as shown in **FIGURE 8.4**.

Scientists have also examined how species diversity affects the stability of ecosystems. An excellent example of this comes from an experiment in which scientists manipulated the number of grassland plants in Minnesota and then monitored how much the abundance of herbivores, predators, and parasites varied over 11 years as the ecosystem experienced environmental change including years of droughts and years of abundant rainfall. As you can see in **FIGURE 8.5**, the researchers found that increasing the number of plant species resulted in greater stability in the number of herbivore species and the number of predator and parasite species.

Environmental scientists often focus on species diversity as a critical environmental indicator. The number of frog species, for example, is used as an indicator of regional environmental health because frogs are exposed to both the water and the air in their ecosystems. A decrease in the number of frog species in a particular ecosystem may be an indicator of environmental problems in that location. Species losses in several ecosystems can indicate environmental problems on a larger scale.

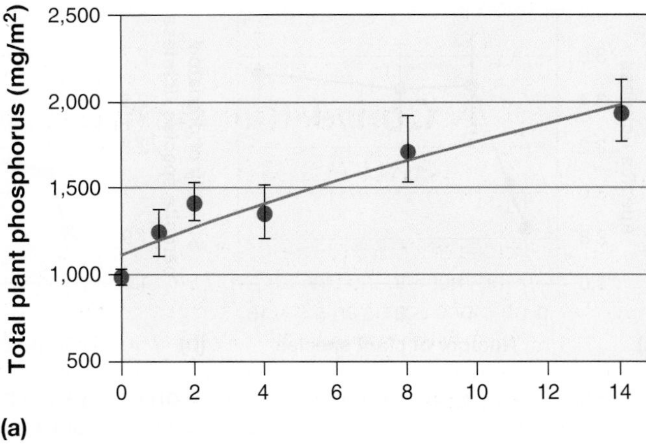

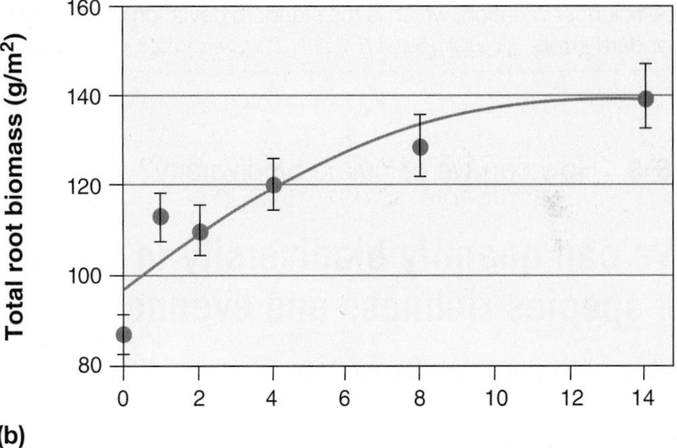

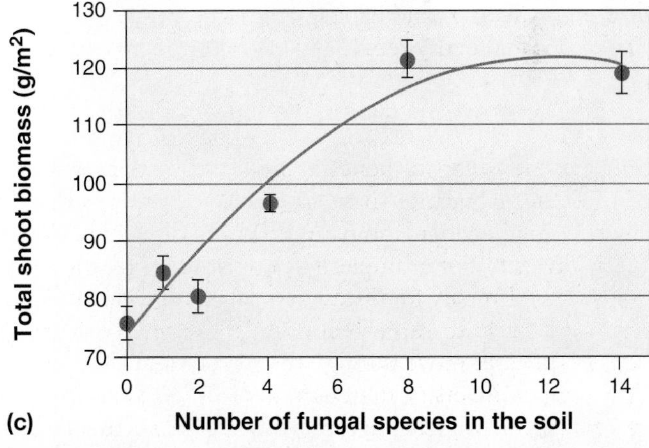

FIGURE 8.4 The effects of fungal species diversity on plant growth. When researchers manipulated the number of fungal species in the soil, which function as mutualists with plants, they found that increased fungal diversity leads to increased plant growth. The fungi take up phosphorus from the soil and pass it to the plant roots. As the number of soil fungi increases, it causes (a) an increase in the amount of phosphorus found in the plants, (b) an increase in the biomass of the plant roots, and (c) an increase in the biomass of the plant shoots, which includes the stems and leaves. *(Data from van der Heijden, A. G. A., et al. 1998. Mycorrhizal fungal diversity determines plant biodiversity, ecosystem variability, and productivity.* Nature *396: 69–72.)*

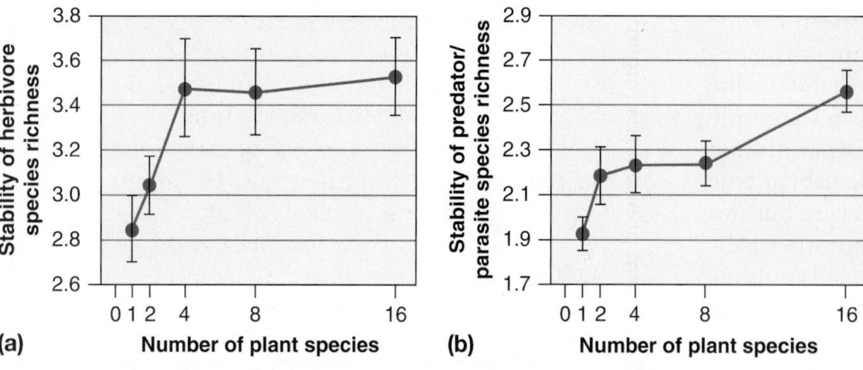

FIGURE 8.5 The effect of species richness on ecosystem stability. Researchers manipulated the number of plant species in a Minnesota grassland and then monitored the herbivores, predators, and parasites over 11 years of environmental variation. Increasing the number of plant species resulted in greater stability of the number of (a) herbivore species and (b) predator and parasite species. Stability is defined as the inverse of the coefficient of variation, which is the standard deviation divided by the mean. Error bars are standard errors. *(Data from Haddad, N. M. 2011. Plant diversity and the stability of foodwebs.* Ecology Letters *14: 42–46.)*

8-3 How can we calculate biodiversity?

We can quantify biodiversity in terms of species richness and evenness

When we want to understand the diversity of species in an area and how this diversity differs among locations due to natural causes or human activities, scientists have developed two measures: species richness and species evenness. The number of different species in a given area of an ecosystem, such as a pond, the canopy of a tree, or a plot of grassland, is known as **species richness**. Species richness is used to give an approximate sense of the biodiversity in a given area of an ecosystem. However, we may also want to know the **species evenness**, which is the relative proportion of individuals within the different species in a given area. Species evenness tells us whether a particular ecosystem is numerically dominated by one species or whether all of its species have similar abundances. An ecosystem has high species evenness if its species are all represented by similar numbers of individuals. An ecosystem has low species evenness if one species is represented by many individuals whereas other species are represented by only a few individuals. A location with low species evenness is considered to be less diverse than a location with high species evenness.

Given these two components of species diversity, scientists evaluating the biodiversity of an area must consider both species richness and species evenness. Consider the two forest communities, Community 1 and Community 2, shown in **FIGURE 8.6**. Both forests contain 20 trees that are

FIGURE 8.6 Measures of species diversity. Species richness and species evenness are two different measures of species diversity. Although both communities contain the same number of species, Community 1 has a more even distribution of species and is therefore more diverse than Community 2. Percentages of the four tree species are provided below each community.

distributed among four species. In Community 1, each species is represented by 5 individuals. In Community 2, one species is represented by 14 individuals and each of the other three species is represented by 2 individuals. Although the species richness of the two forests is identical, the four species are more evenly represented in Community 1. That forest therefore has greater species evenness and is considered to be more diverse. Because species richness or evenness often declines after a human disturbance, knowing the species richness and species evenness of an ecosystem gives environmental scientists a baseline they can use to determine how much that ecosystem has changed.

AP® Exam Tip

When taking the AP® Environmental Science Exam make certain you describe vocabulary words. Sometimes using the word isn't enough. For example, you could write "Species richness, or the number of different species in an area, is one way to measure biodiversity." Note: If you can't remember the vocabulary term, you can receive credit by describing the word instead.

8-4 How can we estimate the number of species living on Earth?

It is difficult to estimate the number of species on Earth

A short walk through the woods, a corner lot, or a city park makes one thing clear: Life comes in many forms. A small plot of untended land or a tiny pond contains dozens, perhaps hundreds, of different kinds of plants and animals visible to the naked eye as well as thousands of different kinds of microscopic organisms. In contrast, a carefully tended lawn or a commercial timber plantation usually supports only a few types of grasses or trees. The total number of organisms in the plantation or lawn may be the same as the number in the pond or in the untended plot, but the number of species will be far smaller.

As we have noted, estimating the number of species in any given place is the most common measure of biodiversity, but estimating the total number of species on Earth is a challenge. Many species are easy to find, such as the birds or small mammals you might see in your neighborhood. Others are not so easy to find. Some species are active only at night, live in inaccessible locations such as the deep ocean, or cannot be seen without a microscope. To date, scientists have named approximately 2 million species, which means the total must be larger than that. While 2 million described species is an impressive number, it is equally impressive that the rate of discovering new species remains very high. Currently, a staggering 15,000 to 18,000 new species are discovered each year from around the world. This highlights the fact that we still have a long way to go in knowing the biodiversity of our planet.

Because insects contain more species than most other groups, scientists reason that if we could get a good estimate for the number of insect species in the world, we would have a much better sense of the total number of species. In one study, researchers fumigated the canopies of a single tree species in the tropical rainforest and then collected all the dead insects that fell from the trees onto a tarp on the ground. From this collection, they counted the number of beetle species that fed on only the one tree species they fumigated. By multiplying this number of beetle species by the total number of tropical tree species, they estimated that in the tropics there were perhaps 8 million species of beetles that feed on a single species of tree. Because beetles make up about 40 percent of all insect species, and because insect species in the forest canopy tend to be about twice as numerous as insect species on the forest floor, the researchers suggested that a reasonable estimate for the total number of tropical insect species might be 30 million. More recent work has indicated that this number is probably too high. Current estimates for the total number of species on Earth range between 5 million and 100 million, but most scientists estimate that there are about 10 million total species on Earth.

In this module, we have explored the important role that biodiversity plays in providing resilience from the levels of genes to ecosystems. In the next module, we will examine how all of this biodiversity works together to provide life-supporting resources such as clean water, timber, fisheries, and agricultural crops that are critical to humans.

Species richness The number of different species in a given area.

Species evenness The relative proportion of individuals within the different species in a given area.

Module 8 AP® Review

Preparing for the AP® Exam

Learning Goals Revisited

8-1 How does biodiversity exist at different scales?

Biodiversity exists as genetic diversity, species diversity, habitat diversity, and ecosystem diversity. All four scales of biodiversity contribute to the overall biodiversity of the planet.

8-2 How does biodiversity affect responses to environmental stressors?

The various scales of biodiversity have impacts on how populations and ecosystems respond to biotic or abiotic environmental stressors. High genetic diversity benefits the long-term persistence of populations because they are better able to respond to environmental change. High habitat diversity favors the persistence of habitat specialists, which have very narrow habitat requirements. High species diversity leads to increased primary productivity and increased stability in ecosystems.

8-3 How can we calculate biodiversity?

Biodiversity can be quantified in terms of species richness and evenness.

8-4 How can we estimate the number of species living on Earth?

Scientists are not able to determine the exact number of species on Earth. Most estimates agree on approximately 10 million species, of which we have identified approximately 2 million.

AP® Practice Questions

Preparing for the AP® Exam

Multiple-Choice Questions

1. How many species are estimated to exist on Earth?
   (a) 2 million
   (b) 8 million
   (c) 10 million
   (d) 30 million

Use the table below to answer questions 2–4:

Two savanna communities both contain 5 plant species. Researchers collected data for each community, which are shown in the table below.

Species	Community 1	Community 2
A	5 individuals	10 individuals
B	5 individuals	12 individuals
C	5 individuals	1 individual
D	5 individuals	1 individual
E	5 individuals	1 individual

2. What is the species richness of these communities?
   (a) 5
   (b) 10
   (c) 12
   (d) 25

3. Which statement best describes the two communities?
   (a) Community 1 has the same species evenness as Community 2.
   (b) Community 2 has a higher species richness.
   (c) Community 1 has a higher species evenness.
   (d) Community 2 has a lower species richness.

4. Which statement best explains the biodiversity of these communities?
   (a) Community 1 has the highest biodiversity because it has the highest species richness.
   (b) Community 1 has the same biodiversity as Community 2.
   (c) Community 2 has the highest biodiversity because it has the highest number of individuals.
   (d) Community 1 has the highest biodiversity because it has the highest species evenness.

5. Which statement best describes an environmental problem that can result from the loss of habitat diversity?
   (a) Generalist species will be lost more quickly than specialist species.
   (b) Specialist and generalist species are likely to be similarly impacted.
   (c) Specialist species will be lost more quickly than generalist species.
   (d) Neither specialist nor generalist species are likely to be impacted.

6. Which of the following best explains how ecosystems would be affected by an increase in species richness?
   (a) An increase in species richness would decrease the stability in an ecosystem because the chance of a species being lost during a disruption would increase.
   (b) An ecosystem with increased species richness would be less likely to recover from disruptions because with more species, there is a greater chance that many of them will be lost.

(c) An increase in species richness would increase primary productivity in an ecosystem, because fewer species will be present to compete with one another.

(d) An increase in species richness would be more likely to recover from disruptions, because more individuals would be present to reproduce.

Free-Response Question

Researchers set out to determine if there is a relationship between climate warming in the tundra and biodiversity. They collected species richness data for groups of vegetation at three different elevations in Denali National Park in Alaska. Their findings are listed in the table below.

Species richness of vegetation in Denali National Park study area

	Total richness at elevation		
Species type	**300 meters**	**700 meters**	**1200 meters**
Vascular Plants	153	217	276
Mosses	133	170	196
Lichens	82	126	191

(Data from https://esajournals.onlinelibrary.wiley.com/doi/10.1002/ecs2.1848)

(a) Using the data in the table, **identify** the species richness of mosses at an elevation of 1200 meters. (1 pt.)

(b) **Describe** the general relationship between species richness and elevation for the data in Table 1. (1 pt.)

(c) **Describe** how climate warming would affect the species richness in the tundra ecosystem. (1 pt.)

(d) **Explain** why ecosystems with higher species richness are better able to recover from disturbances. (2 pts.)

(e) For the researchers' experiment described above,

(i) **identify** the independent variable in the investigation. (1 pt.)

(ii) **describe** a research question for the investigation. (1 pt.)

(iii) **identify** one variable that would have been held constant in the investigation. (1 pt.)

(f) Another research team looks at the study data and predicts that an elevation of zero meters would be most affected by climate warming, and species richness would be lowest. **Make a claim** to support or refute the researcher's prediction, and justify the claim using evidence. (2 pts.)

Module 9

Unit 2	8	9	10	11	12	13	14

Ecosystem Services

In the previous module, we examined how biodiversity exists at different scales and how maintaining high biodiversity helps populations, species, and ecosystems resist the harmful impacts of environmental stressors. A critical question that environmental scientists investigate is whether the planet's natural life-support systems are being degraded by stressors caused by human activities. Ecosystems provide **ecosystem services**, which are the processes by which life-supporting resources such as clean water, timber, fisheries, and agricultural crops are produced. Ecosystem services may benefit people directly, for example by providing the food that we eat or the water that we drink. They may also benefit us indirectly, for example by providing a diversity of conditions, nutrients, and species, all of which make the ecosystem healthier for all organisms, including humans. Although we often take a healthy ecosystem for granted, we tend to notice when an ecosystem is degraded or stressed because it is unable to provide the same services such as water purification or produce the same goods such as fruit or firewood.

Learning Goals

After reading this module you should be able to

9-1 explain the four categories of ecosystem services.

9-2 describe how human activities can disrupt ecosystem services.

Ecosystem services The processes by which life-supporting resources such as clean water, timber, fisheries, and agricultural crops are produced.

9-1 What are the four categories of ecosystem services?

Ecosystem services include provisions, regulating services, support systems, and cultural services

Given that we rely on a relatively small number of species for our essential needs, why should we care about the millions of other species that live in various ecosystems? To understand the value of entire ecosystems, we need to consider both intrinsic values and instrumental values.

Many people believe that ecosystems have intrinsic value—that is, that ecosystems are valuable independent of any benefit to humans. These beliefs may grow out of religious or philosophical convictions. People who believe that ecosystems are inherently valuable may argue that we have a moral obligation to preserve them. They may equate the obligation of protecting ecosystems with our responsibility toward people or animals that might need our help to survive. People who argue that ecosystems are valuable independent of any benefit to humans generally believe that environmental policy and the protection of ecosystems should be driven by this intrinsic value.

As mentioned in the module introduction, ecosystems also have ecosystem services, of which we will consider four categories: provisions, regulating services, support systems, and cultural services.

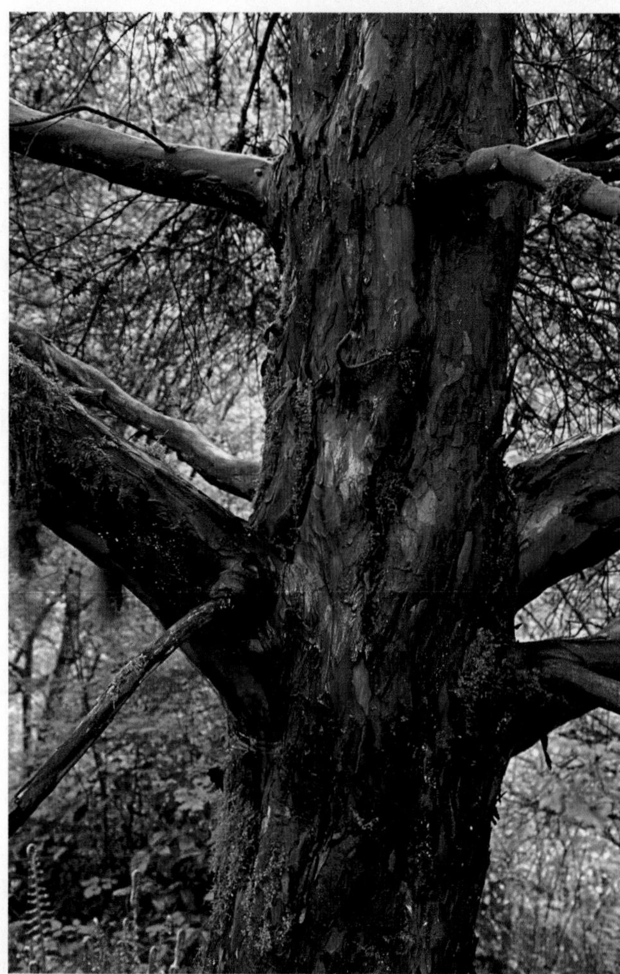

FIGURE 9.1 Provisions. Scientists discovered that the bark of the Pacific yew contains a chemical that has anticancer properties. *(Ray Pfortner)*

Provisions

Goods produced by ecosystems that humans can use directly are called **provisions**. Examples include lumber, food crops, medicinal plants, natural rubber, and furs. Of the top 150 prescription drugs sold in the United States, about 70 percent come from natural sources. For example, Taxol, a potent anticancer drug, was originally discovered by a botanist in the bark of the Pacific yew (*Taxus brevifolia*), a rare tree that grows in forests of the Pacific Northwest (**FIGURE 9.1**). Once approved by the FDA, the synthetic version of this single drug has had annual sales of over $1.5 billion and has helped more than a million cancer patients. There is no way to estimate the potential value of natural pharmaceuticals that have yet to be discovered, but currently more than 800 natural chemicals have been identified as having potential uses to improve human health. Therefore, our best strategy may be to preserve as much biodiversity as we can to improve our chances of finding the next critical drug.

Regulating Services

Natural ecosystems help to regulate environmental conditions. For example, humans currently add about 8 gigatons of carbon to the atmosphere annually (1 gigaton = 1 trillion kilograms), but only about 4 gigatons of carbon remain in the atmosphere. The rest is removed by natural ecosystems, such as tropical rainforests and oceans, which provide us with more time to address climate change than we would otherwise have. As we have already learned in Unit 1, ecosystems also are important in regulating nutrient and hydrologic cycles.

An important aspect of regulating services is the ability of an ecosystem to be resilient, ensuring that it will continue to exist in its current state in the face of environmental change. A resilient ecosystem also ensures it can continue to provide benefits to humans. You may recall that in the previous module we discussed the importance of maintaining high species diversity for maintaining ecosystem resilience. For example, several different species may perform similar functions in an ecosystem but differ in their susceptibility to disturbance. If a pollutant kills one plant species that contains nitrogen-fixing bacteria but does not kill other plant species that contain nitrogen-fixing bacteria, the ecosystem can still continue to fix nitrogen, so there is built-in resilience when we maintain the natural level of biodiversity (**FIGURE 9.2**).

Provision A good produced by an ecosystem that humans can use directly.

FIGURE 9.2 Species diversity as a component of resilience. The Zumwalt Prairie ecosystem in Oregon contains a high diversity of grasses and wildflowers, including many species of nitrogen-fixing wildflowers. If one nitrogen-fixing species is eliminated, the lost function can be compensated for by another nitrogen-fixing species. *(Dennis Frates/Alamy)*

Support Systems

Natural ecosystems provide numerous support services that would be extremely costly for humans to generate. One example is pollination of food crops (**FIGURE 9.3**). The American Institute of Biological Sciences estimates that crop pollination in the United States by native species of bees and other insects, hummingbirds, and bats is worth roughly $3 billion in added food production. In addition to providing habitat for animals that pollinate crops, ecosystems offer natural pest control services because they provide habitat for predators that prey on agricultural pests. Although organic farmers, who rarely use synthetic pesticides, gain the most from these pest controls, conventional agriculture benefits as well.

Healthy ecosystems also filter harmful pathogens and chemicals from water, leaving humans with water that requires relatively little treatment prior to drinking. Without these water-filtering services, humans would have to

FIGURE 9.3 Support systems. Pollinators such as this honeybee on a cherry tree play an essential role in ensuring the pollination of food crops. *(Steffan & Alexandria Sailer/Ardea/Earth Scenes/Animals Animals)*

build many new water treatment facilities that use expensive filtration technologies. New York City, for example, draws its water from naturally clean reservoirs in the Catskill Mountains. But residential development and tourism in the area have threatened to increase contamination of the reservoirs with silt and chemicals. Building a filtration plant adequate to address these problems would cost $6 billion to $8 billion. For this reason, New York City and the U.S. Environmental Protection Agency have been working to protect sensitive regions of the Catskills so that the reservoirs continue to serve as an important support system.

AP® Exam Tip

Organisms can provide multiple different ecosystem services. On the AP® Environmental Science Exam your description of the service is necessary to defend your statement.

You should write, *"Bees provide support services by pollinating our crops."*

Merely writing, *"Bees provide provisioning services"* is too vague, and won't receive credit.

Cultural Services

Ecosystems provide cultural or aesthetic benefits to many people. The awe-inspiring beauty of nature has instrumental value because it provides an aesthetic benefit for which people are willing to pay (**FIGURE 9.4**). Similarly, scientific funding agencies may award grants to scientists for research that explores biodiversity with no promise of any economic gain. Nevertheless, the research itself has instrumental value because the scientists and others benefit from these activities by gaining knowledge. While intellectual gain and aesthetic satisfaction may be difficult to quantify, they can be considered cultural services that have instrumental value.

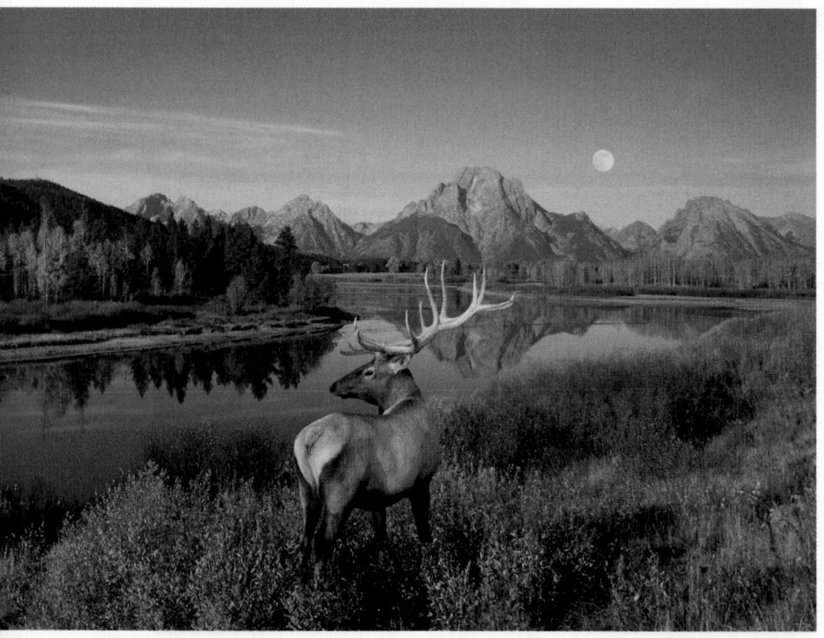

FIGURE 9.4 Cultural services. Many natural areas, such as this scene from the Grand Tetons National Park, provide aesthetic beauty valued by humans. *(Buddy Mays/Alamy Stock Photo)*

The Monetary Value of Ecosystem Services

Most economists believe that the instrumental values of an ecosystem can be assigned monetary values, and they are beginning to incorporate these values into their calculations of the economic costs and benefits of various human activities. However, assigning a dollar value is easier for some categories of ecosystem services than for others. Recently, a team of scientists and economists attempted to estimate the total value of ecosystem services to the human economy. They considered replacement value—the cost to replace the services provided by natural ecosystems. They also looked at other factors, such as how property values were affected by location relative to these services—for example, oceanfront housing. Finally, they considered how much time or money people were willing to spend to use these services—for example, whether they were willing to pay a fee to visit a national park. Using this method, researchers estimated that ecosystem services were worth over $125 trillion per year, which is about twice the entire global economy. This underscores the importance of preserving ecosystem services to maintain a sustainable planet.

AP® Exam Tip

When answering an economic question on the exam, make sure to use language that indicates economic costs such as "money," "jobs," or "tourism," to name a few. Implying economic impact without using the correct terminology will not score you points.

9-2 How can human activities disrupt ecosystem services?

Human activities can disrupt ecosystem services with economic and ecological consequences

Because biodiversity helps determine the services that ecosystems can provide, we would expect declines in genetic diversity, species diversity, and habitat diversity to be associated with declines in ecosystem services. While we will discuss declines in biodiversity in much more detail in Module 59, here we will focus on how human activities can impact ecosystem services and have economic and ecological consequences.

To assess the state of ecosystem services, teams of scientists from around the world have worked together to conduct the United Nations' Millennium Ecosystem Assessment. As part of this assessment, scientists examine trends in 24 ecosystem functions that represent the four categories of ecosystem services, including food production, fish and shellfish production, water purification, and pollination. The trends are examined in light of the past 50 years, as the human population has more than doubled and caused a dramatic increase in the consumption of ecosystem services. Of these 24 different ecosystem functions, 15 were found to be declining or used at a rate that cannot be sustained. In some cases, an ecosystem function has improved, but this improvement has come at the cost of a decline in other functions. If we want to improve ecosystem functions, we need to improve the fate of the species and ecosystems that provide these services.

FIGURE 9.5 A cost of increased food production. Increased food production has led to an increase in the clearing of land. On this farm in Brazil, the diverse rainforest was cleared for crop production. *(Frontpage/Shutterstock)*

Food Production

Around the world, the production of food has increased faster than the growth of the human population and this activity has disrupted natural ecosystem services. However, there are substantial regional differences in food production with large increases occurring in East Asia and much smaller increases occurring in sub-Saharan Africa. As we will discuss in upcoming modules, the reasons for the rapid growth in food production are linked to new varieties of crops, increased irrigation, and increased production of synthetic fertilizers to provide an additional source of fixed nitrogen for plant growth. As a result, the increase in food production has caused impacts on other ecosystem services. For example, as we mentioned in Module 4, the rapid growth in the use of synthetic nitrogen has dramatically increased the amount of nitrogen moving through terrestrial and aquatic ecosystems and has altered the composition of species that are favored in these biomes. The increase in food production has also led to increased clearing of land for crops in some regions of the world, which results in less land containing the natural biodiversity of the region (**FIGURE 9.5**). Land clearing also results in less carbon being locked up in the crops compared to the natural plant biomass that was cleared, which results in more carbon dioxide going into the atmosphere as a greenhouse gas.

Fish and Shellfish Production

We can also examine the growth of fish and shellfish production and how it has altered ecosystem services. In doing so, we need to think about two sources of fish and shellfish: the capture of wild animals from the ocean and the farming of fish, shellfish, and seaweed, which is known as **aquaculture**. We will elaborate more on aquaculture in a later module. In the case of fishing for wild animals, there has been a slow, continuous growth in marine fish captures, with a 14 percent increase from 1990 to 2018 according to a 2020 report by the Food and Agriculture Organization of the United Nations. During this same time period, there has been a massive 527 percent increase in the farming of marine fish and shellfish. You can see the growth of freshwater and marine fish production in **FIGURE 9.6**.

The increase in wild-caught fish and shellfish is the result of increased fishing effort and technological advances. Commercial fishing has driven down the abundances of many fish species to the point where the populations have collapsed and are now too scarce to pursue. These population declines have caused commercial fishing operations to shift to other species

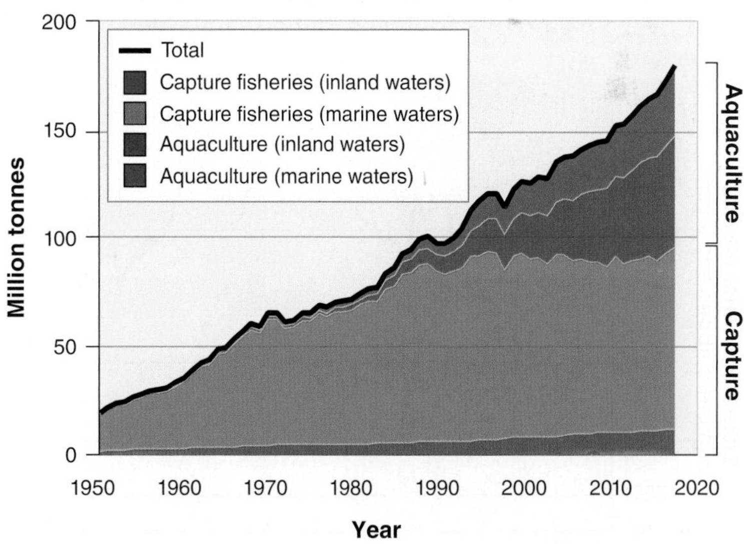

FIGURE 9.6 Fish and shellfish production over six decades. The amount of fish and shellfish captured from inland and marine waters has slowly increased since 1990 while the amount grown using aquaculture has grown rapidly. *(Data from http://www.fao.org/state-of-fisheries-aquaculture.)*

Aquaculture The farming of fish, shellfish, and seaweed.

of fish, many of which have populations that are now also declining. The encouraging news is that some nations have regulated the amount of commercial fishing for a given species of fish or shellfish, such as multiple species of wild salmon in Alaska, and this has allowed the long-term sustainable harvest of these species. Unfortunately, commercial fishing for many other species is not effectively regulated around the world. In addition, the growth of aquaculture operations around the world has led to the loss of natural coastal habitats in tropical countries. We will have much more to say about sustainable fishing and aquaculture in Module 34.

Water Availability

The supply of fresh water for humans as an ecosystem service has increased globally over the past 50 years, but this varies a great deal by region. In many parts of the world, including North Africa, the Middle East, Mexico, and the American West, water is being used faster than it can be naturally replenished as a result of the increasing demands of a growing human population. In 2021, for example, several reservoirs in the American West experienced their lowest water levels ever recorded as a result of increased drought and record high temperatures. The largest reservoir, Lake Mead in Nevada and Arizona, supplies water to 40 million people across four states and serves to generate electricity for the region (**FIGURE 9.7a**). In 2021, a water emergency was declared as Lake Mead dropped to a record low, falling 43 m (140 feet) and containing only 35 percent of the water that it held when initially filled in the 1930s (Figure 9.7b). Such declines are causing the states to reduce their water consumption to allow the water supply to rebound.

In addition to having available water, it is also important that the available water is clean water. The greatest human need for water is for agriculture, and this has led to declines in water quality in many parts of the world due to various pollutants related to agriculture, including excess nutrients, pesticides, and animal waste. We will discuss these water pollution issues, and their solutions, in Unit 8.

Pollination Services

As we noted earlier, pollination of flowers by insects, hummingbirds, and bats provides a critical ecosystem support service for humans. These pollinators also are responsible for the sexual reproduction of plants in nature. One large concern is the global decline in bees, which is happening due to a combination of habitat loss, parasites, and pesticide use. Habitat loss is important because natural habitats provide nesting areas for bees and provide natural flowers that can sustain bee populations with pollen and nectar when a nearby agricultural crop is not flowering. Parasites of bees appear to be increasingly common, and this may be due to a decline in bee health due to pesticides in the environment. Because pesticides are designed to kill insects that eat crops, pesticides can have the unintended consequence of weakening or killing beneficial insects such as bees. Moreover, bees are exposed to multiple pesticides in the environment, with the average beehive in the United States containing six different agricultural chemicals. The most complete data on bee populations is from the United States and Europe and both regions are observing declines in both wild species of

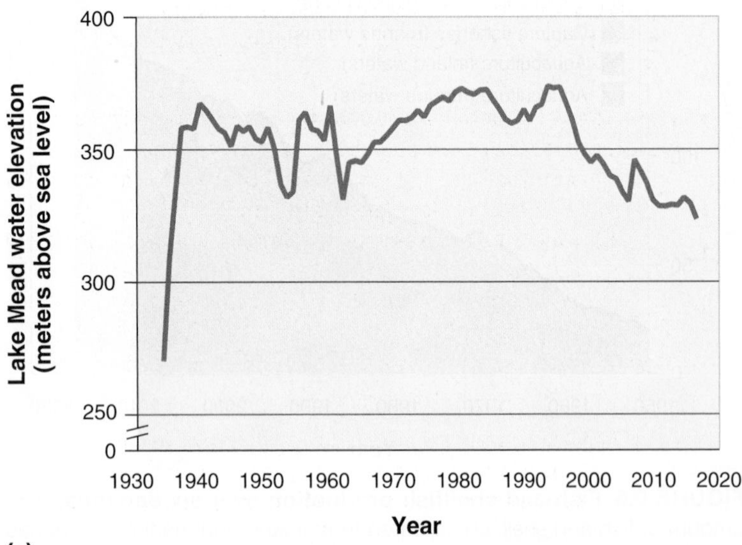

FIGURE 9.7 Declining water availability. Lake Mead is a human-made reservoir created by damming the Colorado River with the Hoover Dam. (a) Since 2000, the lake has steadily declined in depth due to water use and climate change. (b) In 2021, the lake had dropped 43 m (140 feet) to only 35 percent of its volume. The white rocks along the sides of the lake show the high-water mark of past decades. *(b: BRIDGET BENNET/REUTERS/Newscom)*

bees and in the domesticated honeybees that are raised by people. These declines may be happening in other regions of the world as well, and this is causing a great deal of concern for the conservation of biodiversity and future crop production.

As we have seen in this module, ecosystems provide a number of important services to humans, but several of them have been declining over recent decades. In subsequent modules, we will explore a variety of ways in which humans can continue to rely on these ecosystem services in a more sustainable manner.

Module 9 AP® Review

Preparing for the AP® Exam

Learning Goals Revisited

9-1 What are the four categories of ecosystem services?

Ecosystems can have intrinsic values, which are independent of any benefit to humans, or they can have instrumental values, which provide a benefit to humans and can be assigned a monetary value. Instrumental values include provisions, regulating services, support systems, and cultural services.

9-2 How can human activities disrupt ecosystem services?

Recent assessments of ecosystem functions have found that more than half of those assessed are either declining or used at a rate that cannot be sustained. In some cases, such as food production or fish and shellfish production, there has been an increase over the past several decades but this has come at the cost of other functions, including the availability of clean, fresh water and the support system provided by pollinators.

AP® Practice Questions

Preparing for the AP® Exam

Multiple-Choice Questions

1. The intrinsic value of an ecosystem
   (a) is the total monetary worth of its features.
   (b) is based on the goods the ecosystem produces.
   (c) considers the potential benefits that could be discovered in the ecosystem.
   (d) is the value it has independent of any benefit to humans.

2. A cultural ecosystem service of forests could best be described by which statement?
   (a) harvesting timber to produce homes for humans
   (b) scenic overlooks for hikers to enjoy
   (c) water and air purification by trees and soil
   (d) habitats for predators of agricultural pests

3. Which is an example of a provision service?
   (a) Many important pharmaceuticals have their origin in nature.
   (b) Ecosystems provide habitats for native pollinator species.
   (c) A rainforest with high diversity can store more carbon than most crops.
   (d) High diversity can increase the aesthetic value of nature.

Use the passage below to answer questions 4 & 5:

In the United States, California grows more produce for food than any other individual state. According to the U.S. Department of Agriculture (USDA), in 2020 California produced $49 billion in revenue, which accounted for 13.5% of all of the food production in the U.S. In contrast, the United Nations reports that in June 2021 there were 47 countries listed as "low-income food deficit countries" in the world, including Nicaragua, which imported $275 million in agricultural products from the United States.

4. According to the passage, which statement is correct about food production as an ecosystem service?
   (a) The increase in food production is not keeping up with human population growth.
   (b) Increases in food production vary among different regions of the world.
   (c) Increased food production has allowed the use of less synthetic fertilizer.
   (d) Increased food production has been done without increased clearing of land.

5. According to the passage, which location would be at the highest risk if it experienced a long drought?
   (a) California, because its supporting services would be severely affected by drought.
   (b) The United States, because its regulating services would be severely affected by drought.
   (c) Central America, because its cultural services would be severely affected by drought.
   (d) Nicaragua, because its provisions would be severely affected by drought.

6. Which of the following provides the best explanation regarding the importance of pollination as an ecosystem service?
   (a) Increased crop production has led to increased pollinator populations, which has decreased food production everywhere in the world.
   (b) Increased land clearing has led to less pollinator habitat, which has increased food production globally.
   (c) In some regions of the world, bee populations have been declining, which will decrease global food production.
   (d) Pesticides remove harmful insects and help improve pollination by bees, which will increase global food production.

Free-Response Question

In California's Sacramento Valley, farmers have been planting wildflowers as cover crops to increase the habitat for native pollinator species such as monarch butterflies, as well as ladybugs, which are natural predators of crop-eating aphids.

(a) **Identify** a supporting ecosystem service provided by the wildflowers. (1 pt.)

(b) **Describe** how the ecosystem service identified in part (a) benefits humans. (1 pt.)

(c) **Describe** one way in which human activities can impact ecosystem services. (1 pt.)

(d) **Explain** one way that the increase in wildflowers in the pecan orchards could have unintended consequences on ecosystem services. (1 pt.)

(e) In 2019, United States pecan farms made up 396,000 acres of land, and produced 265 million pounds of pecans.

   (i) **Calculate** how many hectares of land were used in the United States to produce pecans in 2019, if there are 2.47 acres in a hectare. Show all work. (2 pts.)

   (ii) Aphids have the potential to cause crop loss of up to 75% in pecan orchards. **Calculate** how many hectares of pecan orchard would remain unaffected if this were to take place. Show all work. (2 pts.)

   (iii) In 2019, the U.S. pecan industry produced $465 million worth of pecans. **Calculate** the value per hectare of the pecan crops that could be saved by using wildflowers to encourage natural pest control of the aphids. Show all work. (2 pts.)

Module 10

Unit 2	8	9	10	11	12	13	14

Island Biogeography

In the prior modules, we examined the different types of biodiversity and how biodiversity plays a key role in providing essential ecosystem functions. We also know from previous modules that species are not evenly distributed; some locations have high diversity while others have low diversity. There are a number of factors that affect species diversity, including the environmental conditions found in terrestrial biomes at different latitudes, as discussed in Module 2. In this module, we examine how species diversity differs among islands at similar latitudes, as a result of two other important factors: island area and distance to the mainland. As we will see, the insights gained from studying species diversity on islands has important implications to conserving biodiversity around the world.

Learning Goals

After reading this module you should be able to

10-1 describe how island biogeography affects which species live on islands.

10-2 identify what determines the number of species on islands.

10-3 describe why some species on islands are particularly vulnerable to invasive species.

10-1 How does island biogeography affect which species live on islands?

Island biogeography affects the number of species living on islands and their ecological relationships

The study of how species are distributed and interacting on islands is known as **island biogeography**. Island biogeography began with the recognition that larger islands contained more species than smaller islands.

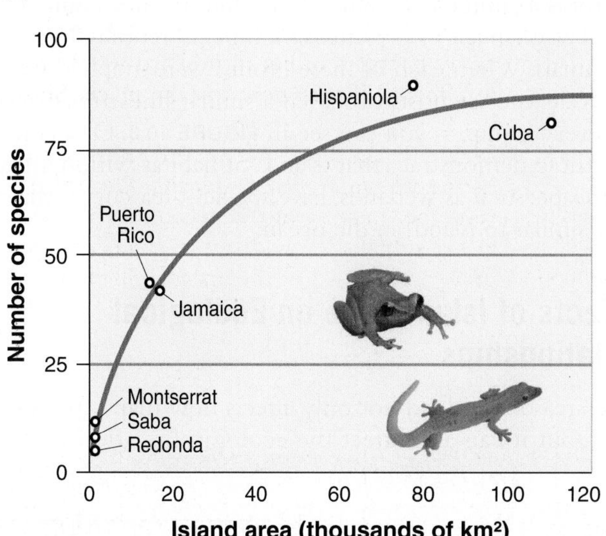

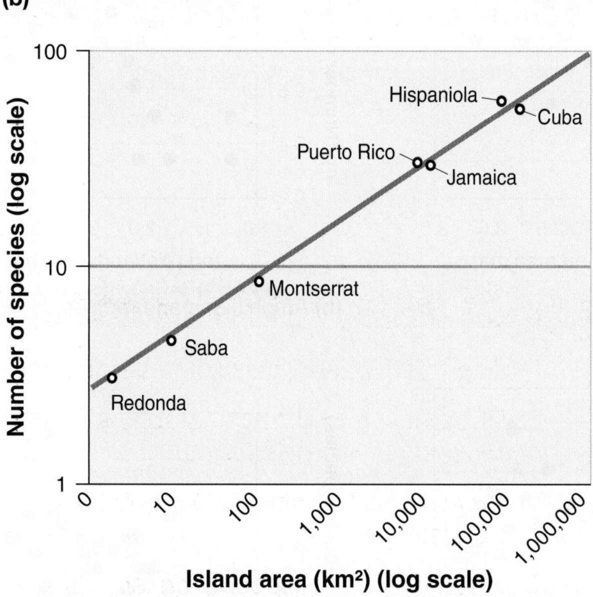

FIGURE 10.1 A species-area curve for amphibians and reptiles. When researchers plot the number of species on different islands against the areas of the islands, they find a positive relationship. In this example of amphibian and reptile species living on islands in the West Indies, (a) there is a positive plateauing relationship. (b) If the same data are plotted on a log scale, the values fall along a straight line. *(Data from MacArthur, R. H., and E. O. Wilson. 1967.* The Theory of Island Biogeography. *Princeton University Press.)*

The Species-area Curve for Islands

One of the most repeatable patterns in studies of biodiversity is that larger islands contain more species. For example, when researchers examined the number of amphibians and reptiles living on islands in the West Indies, they discovered a plateauing relationship, as shown in **FIGURE 10.1a**. This relationship is called the **species-area curve** because it is a description of how the number of species on an island increases with the area of the island. Even more striking is that when we graph the same data using log scales of both variables, the relationship becomes a straight line, as shown in Figure 10.1b.

It turns out that this relationship is quite common among different types of species living on islands. For example, when researchers examined the number of bird species living on the islands around the Philippines, Malaysia, and New Guinea, they once again observed a linear relationship between island area and species richness (**FIGURE 10.2**).

Given that the relationship is a straight line on a log scale, we can describe the relationship using the general equation for any line, which we typically write as:

$$y = mx + b$$

FIGURE 10.2 A species-area curve for birds. When researchers plotted the number of birds living on islands around Malaysia, the Philippines, and New Guinea, they discovered a linear relationship between island area and the number of species. *(Data from MacArthur, R. H., and E. O. Wilson. 1963. An equilibrium theory of insular zoogeography.* Evolution *17: 373–387.)*

Island biogeography The study of how species are distributed and interacting on islands.

Species-area curve A description of how the number of species on an island increases with the area of the island.

In this case, y is the dependent variable, x is the independent variable, m is the slope, and b is the y intercept. Based on this equation for any line, we can write a similar equation for the logged relationship between island area (A) and species richness (S):

$$\log S = z \log A + \log c$$

In this equation, z is the slope of the line and $\log c$ is a constant that represents the y intercept. Using this equation, researchers have examined the species-area curves for a very diverse suite of species and they made a striking discovery. The slopes of the lines all generally fall between 0.20 and 0.35, even when the areas of islands span a very large range (from 1 m^2 to thousands of km^2). This suggests that there are consistent processes in nature that determine how many species can live on islands of different sizes.

Given this consistency of species-area curves among different groups of organisms, it raises an important question: Why do larger islands contain more species than smaller islands? There are three reasons for this pattern. First, dispersing species that come from other islands or from a nearby mainland are more likely to find larger islands than smaller islands. Second, at any given latitude, larger islands can support more individuals of a given species than smaller islands. This matters because larger populations are less prone to extinction. Third, larger islands typically contain a wider variety of biotic and abiotic conditions that support a larger number of species. A wider range of environmental conditions also provides greater opportunities for new species to evolve on the island over time.

Species-area Curves for Isolated Habitats in a Terrestrial Landscape

While scientists initially focused on oceanic islands when looking for relationships between island area and species richness, we now know that the relationship also holds true for other types of isolated habitats, such as wetlands of different sizes scattered across a terrestrial landscape. For example, when scientists examined 30 wetlands in Ontario, they counted the number of species for plants, amphibians, reptiles, birds, and mammals. When each of these groups were graphed using a log scale, the scientists observed a similar linear relationship for every group, as you can see in **FIGURE 10.3**. Observations like these demonstrate that islands of habitat within a larger landscape, such as wetlands, have species-area curves that are very similar to islands in the ocean.

Effects of Island Size on Ecological Relationships

The area of an island not only affects how many species can exist, but it can also affect the ecological relationships that

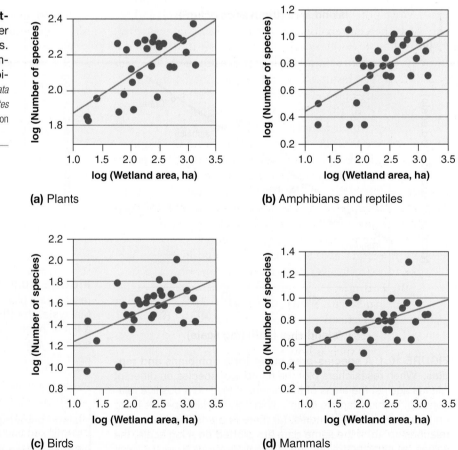

FIGURE 10.3 Species-area curves for wetlands. Scientists in Ontario counted the number of species living in wetlands of different sizes. When plotted on a log scale, they observed similar linear relationships for (a) plants, (b) amphibians and reptiles, (c) birds, and (d) mammals. *(Data from Findlay, C. S., and J. Houlahan. 1997. Anthropogenic correlates of species richness in southeastern Ontario wetlands.* Conservation Biology *11: 1000–1009.)*

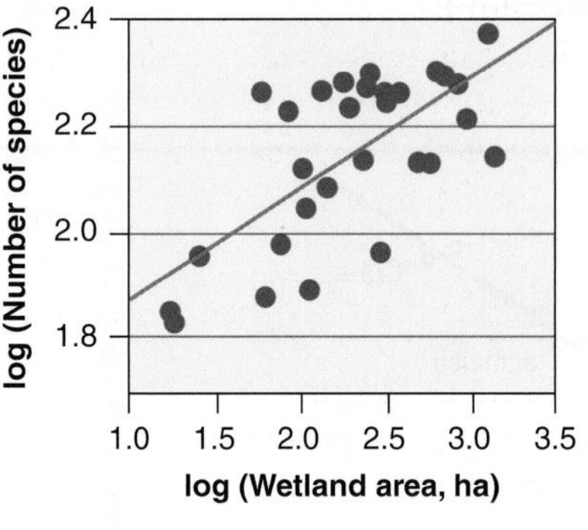

(a) Plants

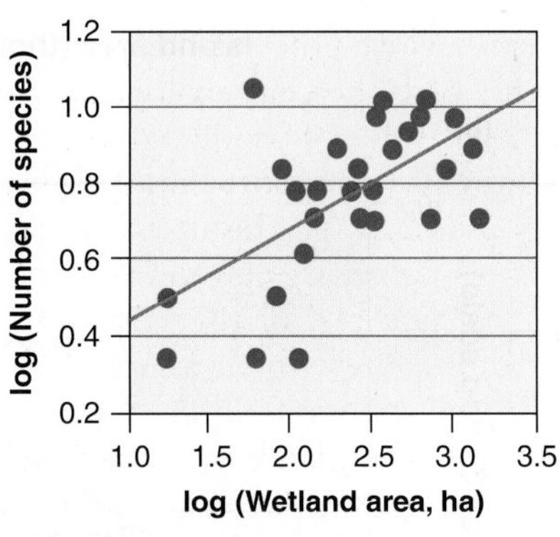

(b) Amphibians and reptiles

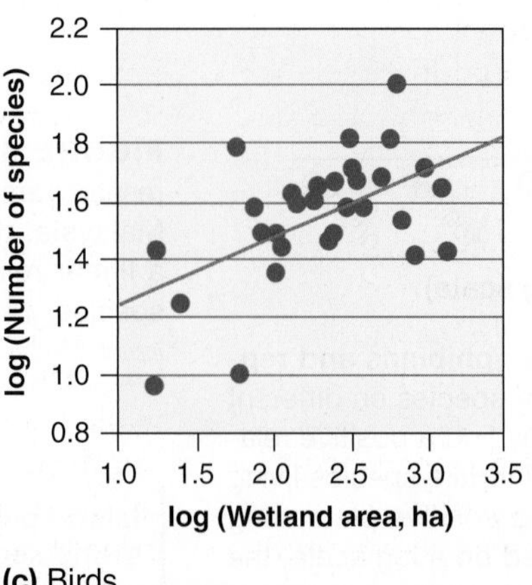

(c) Birds

log (Number of species)
1.4
1.2
1.0
0.8
0.6
0.4
0.2
1.0 1.5 2.0 2.5 3.0 3.5
log (Wetland area, ha)

(d) Mammals

(a)

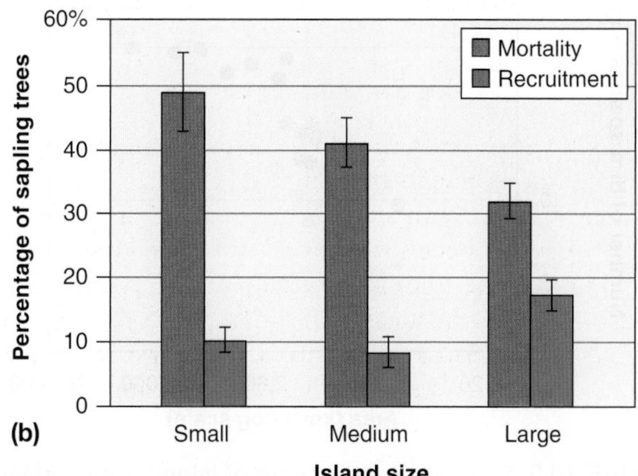

(b)

FIGURE 10.4 Effects of islands on ecological relationships. (a) When the Caroni River was dammed to form Lago Guri in Venezuela, a set of islands was created. On the smallest islands, the populations of primary consumers were too small to support predators, so the predators went extinct. (b) As a result of not having predators, the primary consumers on the smallest islands increased in number and this caused increased herbivory on the seedling and sapling trees, resulting in high sapling mortality and a lower percentage of seedlings surviving into the sapling stage (i.e., recruitment). Error bars are standard errors. *(Data from Terborgh, J., et al. 2006. Vegetation dynamics of predator-free land-bridge islands.* Journal of Ecology *94: 253–263.) (a: Peter Langer/DanitaDelimont.com)*

exist. As we noted in our discussion of trophic levels and the rule of 10 percent ecological efficiency in Module 7, small populations of producers and primary consumers make it difficult to support populations of secondary and tertiary consumers. As a result, we would expect smaller islands to not support large predators.

A nice test of this hypothesis was possible when the Caroni River in Venezuela was dammed to create a lake known as Lago Guri. As the water rose behind the dam, the surrounding landscape was flooded and the highest points in the forested landscape became islands in the lake, as you can see in **FIGURE 10.4a**. While the landscape originally contained large predators, the smallest islands did not contain enough primary consumers to support predator populations, so the predators on these islands went extinct. In contrast, the larger islands continued to support larger populations of primary consumers and this allowed predators to persist. This mixture of small islands without predators and large islands with predators meant that the two types of islands experienced very different community structures. The lack of predators on the small islands allowed the primary consumers such as howler monkeys (*Alouatta seniculus*), iguana lizards, and leaf-cutter ants to become much more abundant and consume more of the plants on the island. In Figure 10.4b, you can see the impact of this increased herbivory in terms of the higher mortality of sapling trees and the lower recruitment of sapling trees, which was defined by the researchers as the survival of seedling trees into the sapling stage. As you can see from this example, smaller islands not only have fewer species, but the species can also experience quite different ecological relationships, including in some cases a complete lack of large predators.

10-2 What determines the number of species on islands?

The number of species on islands depends on rates of colonization and extinction

We can now understand that the size of an island is one factor that greatly impacts the number of species that it can support. As scientists investigated the number of species living on islands, they noticed that distance between a habitat and a source of colonizing species is a second factor that affects species richness. For example, oceanic islands that are more distant from continents generally have fewer species than islands that are closer to continents. Distance matters because, while many species can disperse short distances, only a few can disperse long distances. In other words, if two islands are the same size and contain the same resources, the nearer island should accumulate more species than the farther island because it has a higher rate of immigration by new species.

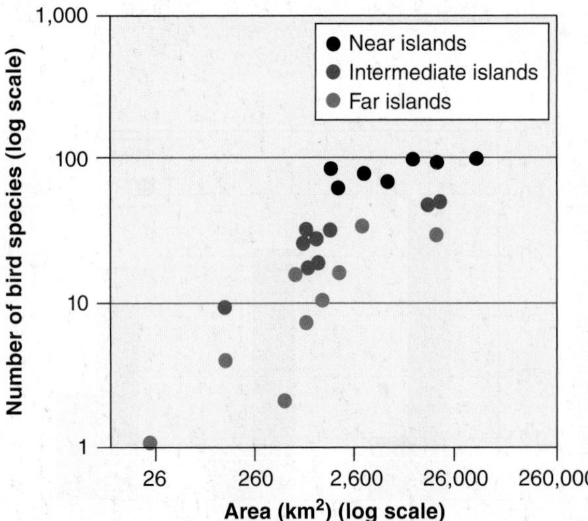

FIGURE 10.5 The combined effects of island area and island distance from a mainland. Researchers examined the number of bird species living on 25 islands in the South Pacific. Larger islands contained more species, but island distance from a mainland source of birds (New Guinea) also mattered. For any given island area, near islands contained more bird species than far islands. *(Data from MacArthur, R. H., and E. O. Wilson. 1963. An equilibrium theory of insular zoogeography.* Evolution *17: 373–387.)*

Observing the Natural Effect of Island Distance

We can see an example of how island area and distance to a mainland affect species richness by examining the number of bird species living on 25 islands in the South Pacific. In this region, island areas range from small to large; island distances range from close to far from New Guinea, which is a very large island and serves as a large potential source of colonizing bird species. You can see how island size and island distance affect the number of species in **FIGURE 10.5**. Just as we have seen in earlier studies, islands with larger areas contain more species of birds. However, for islands of a particular area, such as 2,600 km^2, those near to New Guinea had more bird species than those at intermediate distances or far from New Guinea. In short, the number of species on an island is determined by both island size and island distance.

The combined effects of island area and island distance can also be observed in the mammal species living on mountain tops in the American Southwest, which represent another type of habitat "island." These mountaintops contain alpine tundra and coniferous forests, which are surrounded by very different biomes including woodlands, grasslands, and deserts. As a result, the mammals living on the mountaintops are isolated on habitat islands, similar to birds living on oceanic islands. The nearest large source of these mountaintop mammals is the southern Rocky Mountains and the Mogollon Rim, which is a mountain range in northern Arizona. When researchers counted the mammal species on each mountain top, they found that larger mountain habitats contained more mammals (**FIGURE 10.6a**). However, they also found that mountaintops that were more distant from the Southern Rocky Mountains and Mogollon Rim contained fewer species of mammals (Figure 10.6b). Results such as these make a compelling case that island area and island distance combine to determine the number of species present.

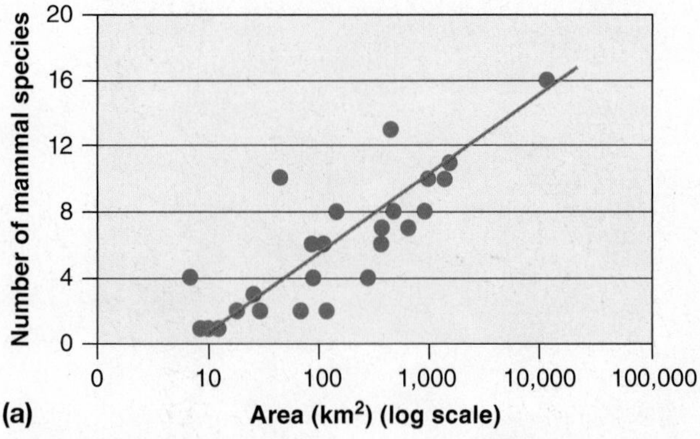

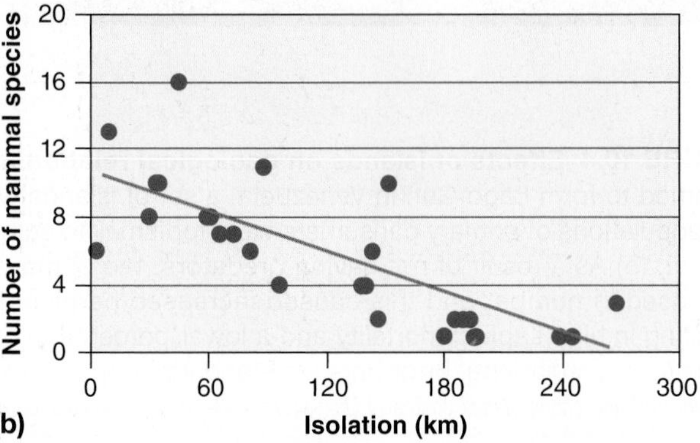

FIGURE 10.6 Effects of mountaintop area and distance from a mainland on mammal species richness. Mountaintops in the American Southwest are tundra and conifer habitats, which are isolated from each other by the presence of other habitats between the mountaintops. (a) Mountaintops with larger areas contain more mammal species. (b) Mountaintops that are closer to large mountain ranges also contain more mammal species. *(Data from Lomolino, M. V., et al. 1989. Island biogeography of montane forest mammals in the American Southwest.* Ecology *70: 180–194.)*

Experimentally Demonstrating the Effect of Island Distance

Thus far we have described patterns in nature and hypothesized that islands farther from a mainland have fewer species because they are less likely to be colonized from the mainland. It would be even more convincing if we could conduct a controlled experiment to show this mechanism actually operating. Several decades ago, researchers E. O. Wilson and Daniel Simberloff conducted an experiment that tested this hypothesis using small islands in Florida that typically contained a single mangrove tree. They began by visiting three islands that were near, intermediate, or far from mainland

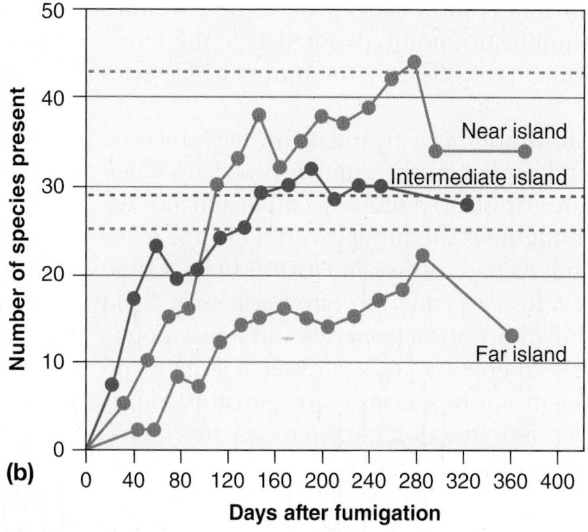

(a) (b)

FIGURE 10.7 Experimentally testing the effect of island distance. Researchers selected three islands off the coast of Florida and counted the number of invertebrates present. (a) They then constructed tents around each island and fumigated with a pesticide to remove all of the animals. For the following year, they revisited the islands to determine how many species had recolonized each island. (b) The dashed lines represent the original number of species on each island before the fumigation. *(Data from Simberloff, D. S., and E. O. Wilson. 1969. Experimental zoogeography of islands: The colonization of empty islands.* Ecology *50: 278–296.) (a: Daniel Simberloff)*

Florida. During these visits, they counted how many species of insects, spiders, and other arthropods were present on each island. Then they erected a tent around each island, as shown in **FIGURE 10.7a**, and fumigated the island with an insecticide to kill nearly every invertebrate animal on the island (e.g., insects and spiders). Their goal was to then revisit the islands every few weeks to determine how many species had arrived to colonize each island, which you can see in Figure 10.7b. In this figure, the horizontal dashed lines represent the number of invertebrates originally found on near, intermediate, and far islands. As shown in the solid lines, each of the three islands were rapidly recolonized by invertebrates, with the near island having the most species and the far island having the fewest. This classic experiment provided clear evidence that islands closer to a mainland have more species because they are more likely to be colonized.

A Model of Island Biogeography

Based on these observations and experiments, scientists have developed a model for predicting how many species an island will have based on island size and distance to the mainland. The model considers the rate of colonization by new species to the island and the rate of extinction by species already living on the island, as illustrated in **FIGURE 10.8**. Thus, the model begins with a colonization curve, shown in orange, and an extinction curve, shown in blue. When an island contains very few species, the rate of colonization by new species is expected to be very high. As the number of species on the island increases, the rate of colonization declines because there are few new species that can colonize. In addition, when an island contains very few species, the rate of extinction is expected to be low because there should be relatively little competition, predation, or parasitism occurring that could drive a species extinct. As the number of species on the island increase, it becomes much more likely that some species will be driven extinct, so the blue line curves upward. The key idea from this model is that the number of species that

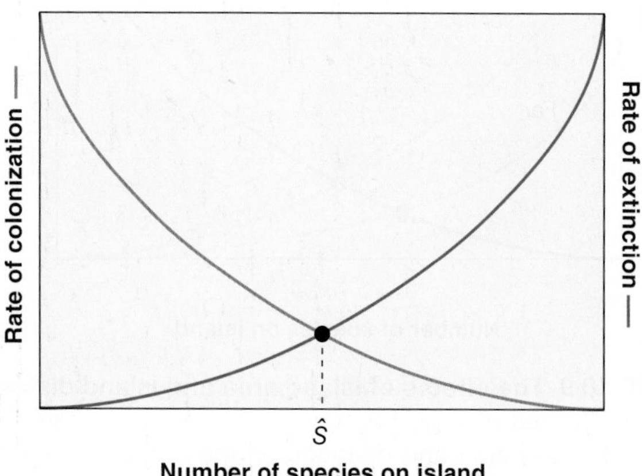

FIGURE 10.8 A model of island biogeography. The model includes the rate of colonization by new species to an island and the rate of extinction of species already living on the island. An equilibrium occurs where the two curves cross, denoted as $\hat{S}$, which is the number of species on the island when the rate of colonization is offset by the rate of extinction.

one should find on the island occurs where the two lines intersect; at this equilibrium point, denoted as $\hat{S}$, the number of species colonizing is offset by the number of species going extinct.

Using this basic model as a foundation, we can now imagine what would happen to the number of species when we change island area or island distance to the mainland. We can start by imagining how the model would change if we added a larger island. As you can see in **FIGURE 10.9a**, a large island would have a lower extinction rate because it could support larger populations of each species and larger populations are less likely to go extinct. As a result, a large island is predicted to contain more species than a small island, as we can see by where its extinction curve crosses the colonization curve, denoted as $\hat{S}_{Large}$.

We can also imagine how the model would change if we added a far island. As illustrated in Figure 10.9b, a far island would experience a lower rate of colonization because fewer species could disperse to it. As a result, a far island is predicted to contain fewer species than a near island, as we can see by where its colonization curve crosses the extinction curve, denoted as $\hat{S}_{Far}$.

We can now combine the four options of large versus small islands and near versus far island to have a complete overview of the model. As shown in **FIGURE 10.10**, the model predicts that large islands near the mainland should contain the highest number of species (denoted as $\hat{S}_{LN}$). In contrast, small islands far from the mainland should contain the lowest number of species (denoted as $\hat{S}_{SF}$). As you can see, the model predictions are quite consistent with researchers' observations of how many species live on islands in nature and consistent with the experimental test of island biogeography conducted on the small islands near Florida.

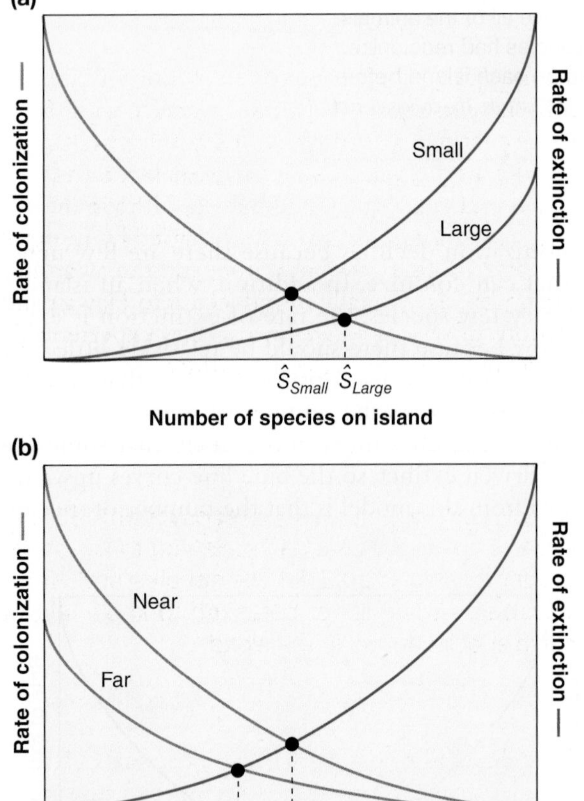

FIGURE 10.9 The effects of island area and island distance. Using the model of island biogeography, we can predict the effects of island area and distance on the number of species present when colonizations are offset by extinctions. (a) If we change island area, the model predicts that large islands will contain more species than small islands. (b) If we change island distance, the model predicts that far islands will contain fewer species than near islands.

Implications for Species Conservation

The research into island biogeography helps us understand the biodiversity found on oceanic islands and habitat islands within a terrestrial landscape. The insights also provide guidance for how to preserve biodiversity in setting aside habitats as national parks, national forests, and wildlife refuges. For example, based on the observations from the model, we can infer that setting aside larger tracts of land will allow us to protect more species than a smaller tract of land. Moreover,

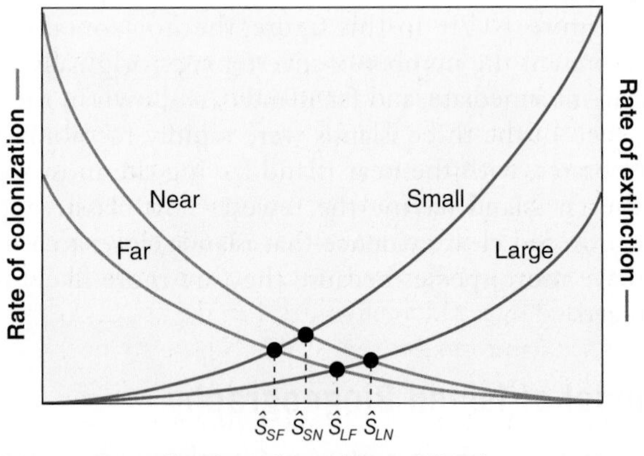

FIGURE 10.10 Modeling the combined effects of island area and island distance. The four combinations of large and small islands crossed with near and far islands produced four different equilibria for how many species should exist on an island. The highest number of species should live on islands that are large and near ($\hat{S}_{LN}$) while the lowest number of species should live on islands that are small and far ($\hat{S}_{SF}$).

a larger tract of land will contain more species than multiple smaller tracts that together comprise a similar total area. The research also informs us that protecting tracts of habitat islands closer to a large, protected habitat will protect more species than protecting tracts of habitat islands that are farther away. Thus, understanding island biogeography has real-world applications to conserving biodiversity around the world.

FIGURE 10.11 **Brown tree snake.** The brown tree snake was introduced to the island of Guam where it decimated the populations of several island species, including birds, bats, and lizards. *(John Mitchell/Science Source)*

10-3 Why are some species on islands particularly vulnerable to invasive species?

Some species on islands have evolved to be specialists and are vulnerable to invasive species

When it comes to the conservation of biodiversity, species living on islands are of high concern because they have often evolved to be specialists. Why is this the case? Species on islands are commonly much more isolated than species on the mainland and they interact with a smaller number of other species. For example, many species living on islands lack predators and do not come into contact with the many pathogens that exist on the mainland. As a result, these species have not evolved predator defenses. Many island species of birds, for instance, have evolved to build their nests on the ground since there are no nest predators on the island to avoid. Some island birds, such as the species that live on the islands of New Zealand, have historically lacked predators and have also lost the ability to fly, which is an energetically costly activity. Finally, the environments of islands may contain a relatively small diversity of food items, so island species can evolve to specialize on the types of food that are present. Under the pristine conditions of an isolated island, these specialized evolved responses can be beneficial to the survival and reproduction of the island species.

While being specialized can be beneficial under isolated island conditions, it can make these species highly vulnerable to invasive species. For example, numerous islands have been invaded by accidentally introduced mice and rats when humans visit and inhabit the islands. These rodents are generalists, allowing them to live under a wide range of conditions and eat a wide range of food items. Moreover, they can either consume the food required by the native specialists or even serve as predators on the island fauna. One of the best examples of an invasive predator wreaking havoc on native island species happened when the brown tree snake (*Boiga irregularis*) was accidentally introduced to the island of Guam as ships were using the island as a port for transporting goods after World War II (**FIGURE 10.11**). The snakes began to consume several species of native island animals. The population quickly grew to more than 2 million snakes, eventually causing the decline or extinction of nine species of birds, three species of bats, and several species of lizards. The native fauna did not evolve with tree snakes, so they had no defenses against this new predator. Similarly, the birds of the Hawaiian Islands evolved without the pathogen that causes avian malaria and is carried by mosquitoes. By the early 1900s, both mosquitoes and the avian malaria pathogen were accidentally introduced into Hawaii and the populations of several native bird species experienced dramatic declines, with several going extinct.

AP® Exam Tip

All species are native to some ecosystem. If the AP® Environmental Science Exam asks you to list an invasive species you should list the species and where it is a problem. You will not be asked to know all of the invasive species found in the world.

The vulnerability of island species becomes clear when we consider the percentage of extinct species that are island species. According to the Convention on Biological Diversity, there have been 724 animal extinctions in the past 400 years and nearly half of those have been island species. Moreover, it is estimated that at least 90 percent of extinct birds were birds living on islands. As you can see, the unique conditions that favor the evolution of specialists on islands makes them particularly vulnerable to the typically generalist invasive species. This knowledge, combined with our understanding of island biogeography, provides the insights we need to help conserve biodiversity on islands and continents throughout the world.

In this module, we learned how island size and distance from a mainland affect the number of species present and their ecological roles. We also learned that some species living on islands are particularly vulnerable to invasive species because they have not evolved any defenses against the invasive species. In the next module, we will explore a related topic by focusing on how species have distinctive biotic and abiotic conditions under which they can live.

Module 10 AP® Review

Preparing for the AP® Exam

Learning Goals Revisited

10-1 How does island biogeography affect which species live on islands?

For both oceanic islands and islands of habitat within a terrestrial landscape, islands with greater area contain more species. When plotted on a log scale, the relationship of land area to species is linear with a similar slope among many groups of plants and animals. Given the difference in number of species, small islands can contain fewer predators than large islands and therefore experience different ecological relationships, including predator–prey interactions.

10-2 What determines the number of species on islands?

The number of species on an island depends on both the area of the island and the distance from the island to a mainland source of species. Using a model of island biogeography, we can understand the combined effects of island area and distance on the colonization rates and extinctions rates of different islands and predict the number of species that will be present at equilibrium.

10-3 Why are some species on islands particularly vulnerable to invasive species?

Many species that are native to islands have evolved into specialists that can have a narrow diet and have not developed defenses against predators or pathogens. When invasive species are introduced, many of them are generalists that can consume the limited food of the native species or act as predators or pathogens that can decimate the island species.

AP® Practice Questions

Preparing for the AP® Exam

Multiple-Choice Questions

Use the graph below to answer questions 1–3:

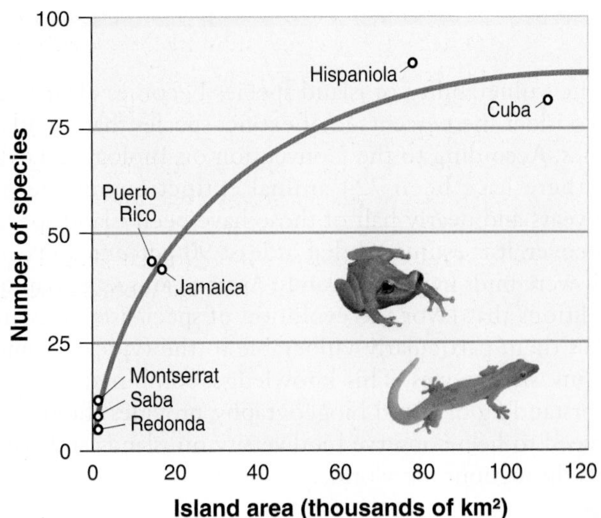

1. Which statement is the best description of the species-area curve?
   (a) It shows that species increase with island area.
   (b) It shows species richness peaks on islands of intermediate area.
   (c) The relationship is exponential when plotted on a log scale.
   (d) The slope of the logged relationship is commonly between 0.50 and 0.65.

2. According to the graph, which island is the largest in area?
   (a) Cuba
   (b) Hispaniola
   (c) Jamaica
   (d) Montserrat

3. The best explanation of why large islands contain more species is that
   (a) larger islands are easier to migrate to because they are always closer to the mainland.
   (b) larger islands can only support specialist species, but they have high species richness.
   (c) larger islands have more available niches in a larger habitat, which can support more species.
   (d) larger islands are farther away from mainlands, which means only the fittest species can migrate there.

4. Why do more distant islands have fewer species?
   (a) Distant islands are generally smaller.
   (b) Distant islands have less habitat diversity.
   (c) Distant islands can only be colonized by species with the best dispersal abilities.
   (d) Distant islands have greater habitat diversity.

5. Which statement is the best explanation of the model of island biogeography?
   (a) As an island increases in species richness, the extinction rate decreases.
   (b) As an island increases in species richness, the colonization rate increases.
   (c) As an island increases in species richness, the extinction rate increases.
   (d) As an island increases in species richness, the extinction rate and colonization rate both increase.

6. Why are many island species particularly vulnerable to invasive species?
   (a) Island species are often generalists with a varied diet of available food.
   (b) Island species are often specialists with highly evolved defenses against predators.
   (c) Island species are often generalists, but have no evolved defenses against predators.
   (d) Island species are often specialists with no evolved defenses against pathogens.

Free-Response Question

The below graph shows the number of bird species on islands of varying size and distance from a mainland.

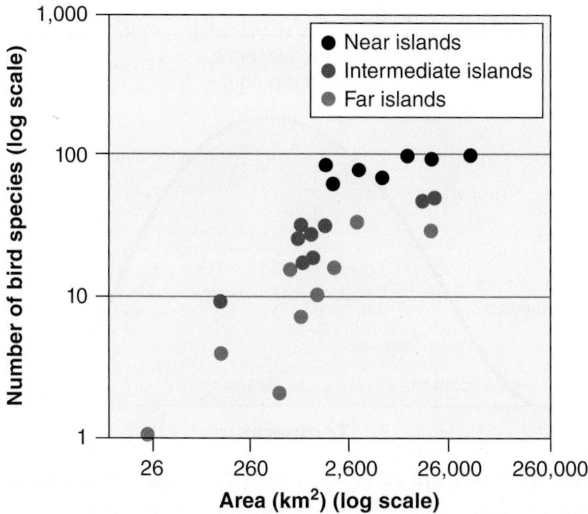

(a) **Identify** the maximum species richness of birds for the islands in the graph. (1 pt.)

(b) **Describe** the relationship between species richness and the "far islands" in the graph. (1 pt.)

(c) **Explain** why the smallest of the "far islands" has the lowest species richness. (2 pts.)

(d) **Describe** a potential ecological problem of having low species richness. (1 pt.)

(e) Students in an AP® Environmental Science class are designing an investigation to simulate the effects of the Theory of Island Biogeography. They draw a mainland and four islands on a large piece of paper, and simulate the migration of species from the mainland by dropping beans onto their paper. Their results are presented in the table below:

	Species that migrated	
Island	Trial 1	Trial 2
Large and close	10	11
Large and far	8	8
Small and close	5	4
Small and far	2	1

(i) **Identify** the students' most likely research question for this investigation. (1 pt.)

(ii) **Identify** the dependent variable for the students' simulation. (1 pt.)

(iii) **Describe** one aspect of the students' simulation that would have been held constant in the investigation. (1 pt.)

(iv) **Explain** a modification the students could make to their simulation that would increase the reliability of their study. (2 pts.)

Module 11

Unit 2 | 8 | 9 | 10 | 11 | 12 | 13 | 14

Ecological Tolerance

As you may recall from our discussion of terrestrial and aquatic biomes, different biomes contain species that are well adapted to the environmental conditions present in each biome. In this module, we explore how individuals and species vary in the ecological conditions that they can tolerate, including both biotic and abiotic conditions. We then examine how changes in the environment have caused species to shift their distributions and how this is likely to continue with future environmental changes. Species that cannot evolve in their tolerance or move to more hospitable environmental conditions will likely go extinct.

Learning Goals

After reading this module you should be able to

11-1 explain what determines where individuals and species can live.

11-2 describe how environmental change affects species.

11-3 explain how environmental change can cause species extinctions.

11-1 What determines where individuals and species can live?

Ecological tolerance determines where individuals and species can live

Every individual has an optimal environment in which it performs particularly well. Thus, all individuals exhibit a range of **ecological tolerance**, which is the suite of abiotic conditions under which a species can survive, grow, and reproduce (also known as the **fundamental niche**). Given that individuals of a particular species have performed well under a similar set of conditions, with some variation due to genetic diversity, we can also think about each species having a particular range of ecological tolerance. Thus, both individuals and species have limits to the abiotic conditions they can tolerate, such as extremes of temperature, humidity, salinity, sunlight, and water flow. **FIGURE 11.1** illustrates this concept using one environmental factor—temperature. As conditions move further away from the ideal, individuals may be able to survive, and perhaps even to grow, but not be able to reproduce. As conditions continue to move away from the ideal, individuals can only survive. If conditions move beyond the range of tolerance, individuals will die. We

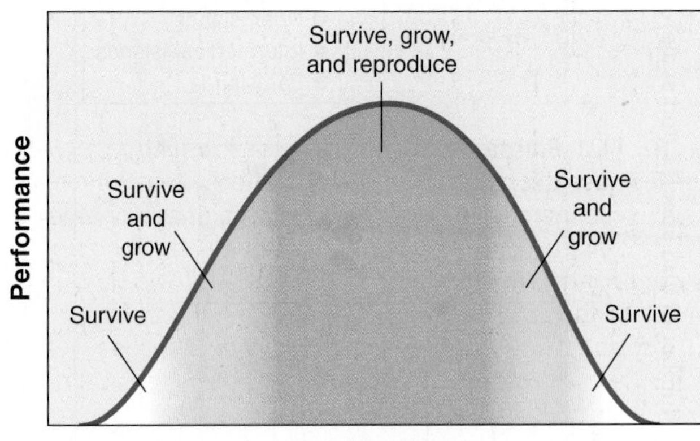

FIGURE 11.1 Range of ecological tolerance. All species have an ideal range of abiotic conditions, such as temperature, under which their members can survive, grow, and reproduce. Under more extreme conditions, their ability to perform these essential functions declines.

can see a similar scenario for the response to sunlight. Many plant species that live in open fields grow very well in full sunlight, but they have trouble performing photosynthesis in the deep shade of a forest. In contrast, small plants living along the forest floor are highly efficient at conducting photosynthesis in low light, but they perform poorly in the intense sunlight of an open field. Because the combination of abiotic conditions in a particular environment fundamentally determines whether a species can

Ecological tolerance The suite of abiotic conditions under which a species can survive, grow, and reproduce. *Also known as* **fundamental niche.**

(a) Red-winged blackbirds only

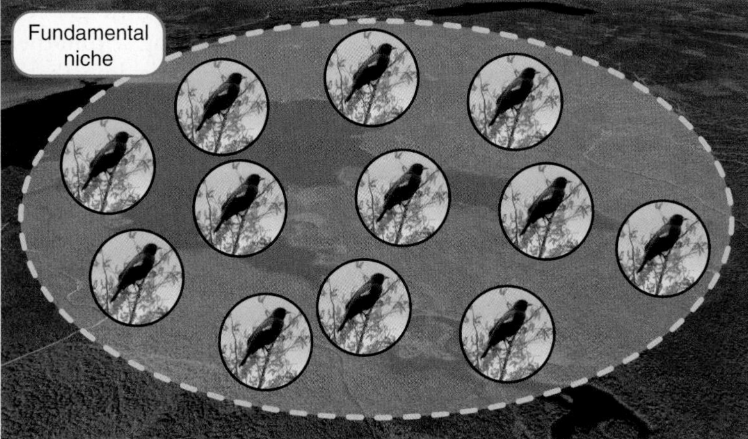

(b) Red-winged blackbirds and yellow-headed blackbirds

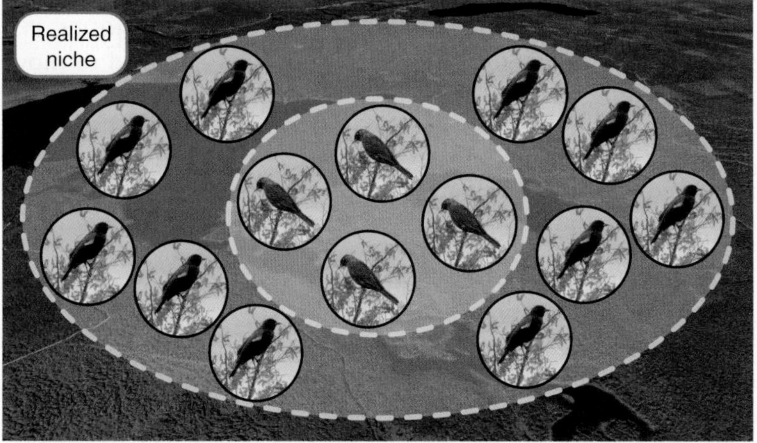

FIGURE 11.2 Fundamental versus realized niches. The fundamental niche includes the range of abiotic conditions under which a species can persist while the realized niche is the combination of abiotic conditions and biotic interactions. (a) The red-winged blackbird is able to build nests in emerging marsh vegetation over both shallow and deep water. This represents its fundamental niche. (b) If yellow-headed blackbirds arrive, they displace the red-winged blackbirds from the deep-water sites of the marsh. Thus, the realized niche of the red-winged blackbird becomes a subset of its fundamental niche.

persist there, ecological tolerance is also known as the fundamental niche of a species.

The fundamental niche establishes the abiotic limits for the persistence of a species. However, biotic factors can further limit the locations where a species can live. Common biotic limitations include the presence of competitors, predators, and diseases. For example, even if abiotic conditions are favorable for a plant species in a particular location, other plant species may be better competitors for water and soil nutrients. Those competitors might prevent the species from growing and thriving in that environment. Similarly, even if a small rodent can tolerate the temperature and humidity of a tropical forest, a deadly rodent disease might prevent the species from persisting in the forest. Therefore, biotic factors further narrow the fundamental niche to the conditions under which a species actually lives. The range of abiotic and biotic conditions under which a species actually lives is called its **realized niche**. Once we determine what contributes to the realized niche of a species, we have a better understanding of the **geographic range** of the species, or the areas of the world in which the species lives.

We can explore an example of the difference between fundamental and realized niches by considering the nesting sites that are chosen by the red-winged blackbird (*Agelaius phoeniceus*), which is a common bird throughout North America. The red-winged blackbird builds its nests in the vegetation that emerges from a marsh and it will build the nests over both shallow and deep marsh water, as illustrated in **FIGURE 11.2a**. However, the distribution of nests changes if yellow-headed blackbirds (*Xanthocephalus xanthocephalus*) arrive at the marsh to build their nests as well. The yellow-headed blackbirds prefer to build their nests over the deep water of the marsh and they aggressively chase away the red-winged blackbirds from the deep-water areas, as shown in Figure 11.2b. As a result, while the red-winged blackbird's fundamental niche includes the entire marsh, its realized niche is smaller and more restricted when it interacts with yellow-headed blackbirds.

AP® Exam Tip

When answering free-response questions, avoid statements such as, "it will kill all the animals and plants." Become familiar with the phrase "range of ecological tolerance" to answer how organisms will respond when conditions change. For example, if ocean temperatures increase, some organisms will not survive because the temperature exceeds their range of ecological tolerance.

11-2 How does environmental change affect species?

Environmental change can alter the distribution of species

Because species are adapted to particular environmental environments, we would expect that changes in environmental conditions would alter the distribution of species on Earth. For example, scientists have found evidence for the

Realized niche The range of abiotic and biotic conditions under which a species actually lives.

Geographic range Areas of the world in which a species lives.

relationship between environmental change and species distribution in layers of sediments that have accumulated over time at the bottom of modern lakes. Each sediment layer contains pollen from plants that lived in the region when the sediments were deposited. In some cases, this pollen record goes back thousands of years.

In much of northern North America, lakes formed 12,000 years ago at the end of the last Ice Age when temperatures warmed and the glaciers slowly retreated to the north. The retreating glaciers left behind a great deal of barren land, which was quickly colonized by plants, including trees. Some of the pollen produced by these trees fell into lakes and was buried in the lake sediments. Scientists can measure the ages of these sediment layers with carbon dating, which is a process that examines how different forms of carbon, known as isotopes, change over thousands of years. Carbon atoms in the atmosphere are 99 percent carbon-12 and 1 percent carbon-13. In addition, there is a rare form of carbon known as carbon-14, which is radioactive with a half-life of 5,730 years. When plants fix atmospheric carbon by photosynthesis, they take in these ratios of the three isotopes. When this fixed carbon ends up in lake sediments, the radioactive carbon-14 breaks down over time and this allows us to know how long ago the carbon was deposited in the lake sediments. Each tree species produces uniquely shaped pollen, so it is possible to determine when a particular tree species lived near a particular lake and how the entire community of plant species has changed over time. **FIGURE 11.3** shows pollen records for three tree species in North America. These pollen records suggest that changes in climatic conditions after the Ice Age produced substantial changes in the distributions of plants over time. Pine trees, spruce trees, and birch trees all moved north with the retreat of the glaciers. This allowed each tree species to continue living within their range of ecological tolerance. As the plants moved north, the animals that depend on these plants followed.

Because past changes in environmental conditions have led to changes in species distributions, it is reasonable to ask whether current and future changes in environmental conditions might also cause changes in species distributions. For example, as the global climate warms, some areas of the world are expected to receive less precipitation, while other areas are predicted to receive more. If our predictions of future environmental changes are correct, and if we have a good understanding of the ecological tolerance of many species, we should be able to predict how the distribution of species will change in the future. North American trees, for example, are expected to have more northerly distributions with future increases in global temperatures. As **FIGURE 11.4** shows, the loblolly pine (*Pinus taeda*) is expected to move from its far southern distribution and become common throughout the eastern half of the United States.

When we consider the impacts of current and future environmental changes, it is important to remember that species vary in their ability to move across the landscape as the environment changes. Some species are highly mobile at particular life stages: Adult birds and wind-dispersed seeds, for example, move across the landscape easily. Other organisms, such as the desert tortoise (*Gopherus agassizii*), are slow movers. Furthermore, the movements of many species may be impeded by obstacles built by humans, including roads and dams. It remains unclear how species that face these

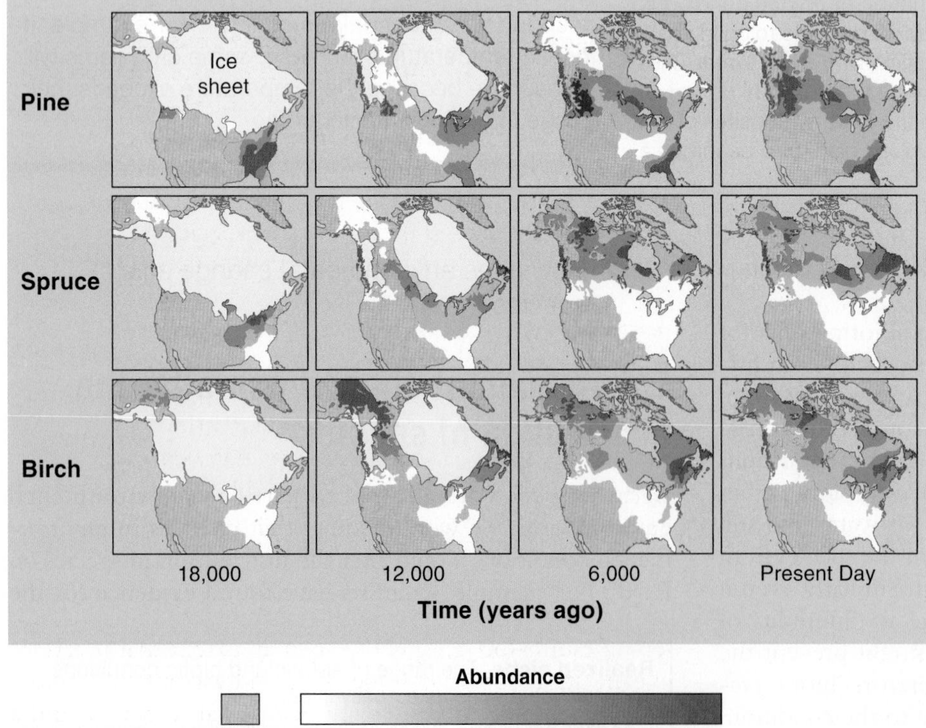

FIGURE 11.3 Changes in tree species distributions over time. Pollen recovered from lake sediments indicates that plant species moved north as temperatures warmed following the retreat of the glaciers, beginning about 12,000 years ago. Areas shown in color or white were sampled for pollen, whereas areas shown in gray were not sampled. *(Data from http://vemages.gsfc.nasa.gov//3453/boreal_model.gif.)*

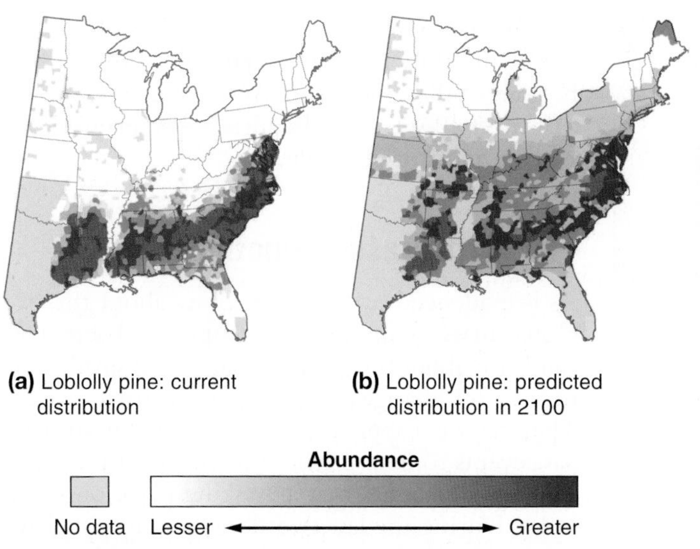

FIGURE 11.4 Predicting future species distributions. Based on our knowledge of the niche requirements of the loblolly pine tree, we can predict how the distributions might change as a result of future changes in environmental conditions. *(Data from https://www.fs.fed.us/nrs/atlas/tree/v3/131)*

FIGURE 11.5 Shifting biome boundaries with global climate change. On the southwest tip of Australia there is a small area of woodland/shrubland biome which is home to plants in the genus *Banksia*. Due to global climate change, the shifting environmental conditions in this region over the next 60 years are going to make it harder for the plants to survive, resulting in two-thirds of the species declining in abundance and one-fourth of the species going extinct. *(Phil Morley/AGE Fotostock)*

substantial challenges will shift their distributions as global climate change occurs or, if they cannot shift, how their populations may be impacted.

AP® Exam Tip

The AP® Environmental Science Exam includes various models such as data tables, graphs, and other infographics. Practice analyzing various graphs and data tables so that you are comfortable gathering data, recognizing trends, and making predictions.

11-3 How can environmental change cause species extinctions?

Environmental change beyond a species' ecological tolerance can cause species extinctions

Understanding the concept of ecological tolerance helps us understand how changes in the environment can cause species to start performing poorly and eventually going extinct. If environmental conditions change, any species that cannot adapt to the change or move to more favorable environments will eventually go extinct. Throughout Earth's history, there have been major environmental changes which have caused large number of extinctions. In fact, 99 percent of the species that have ever lived on Earth are now extinct.

There are several reasons why species might go extinct when there are substantial changes in the environmental conditions. First, there may be no favorable environment close enough to which they can move to match their ecological tolerance. For example, a small area of land on the southwest coast of Australia experiences a current climate that supports a woodland/shrubland biome (see Figure 2.9 on Page 36). This biome is surrounded by a large desert biome to the north and east and an ocean to the south and west. If scientists are right in predicting that the coming century will bring a hotter and drier climate to southwestern Australia, then the unique species of plants that live in this small biome will have nowhere to go where they can survive within their range of ecological tolerance. Scientists have closely investigated one plant genus in this biome (*Banksia*) that contains more than 100 species (**FIGURE 11.5**). Over the next 60 years, they predict that two-thirds of the species will decline in abundance and one-fourth of the species will go extinct.

In situations where there is an alternative favorable environment to which a species can move, it may already be occupied by other species against which the moving populations cannot successfully compete. For example, the predicted northern movement of the loblolly pine, shown in Figure 11.4b, might not happen if another pine tree species in the northern United States is a better competitor and prevents the loblolly pine from surviving in that area. There may also be herbivores that live in the north that will consume loblolly pine to such an extent that it cannot expand its geographic range. This underscores the importance of not only considering ecological tolerance and a species' fundamental niche, but also the importance of biotic

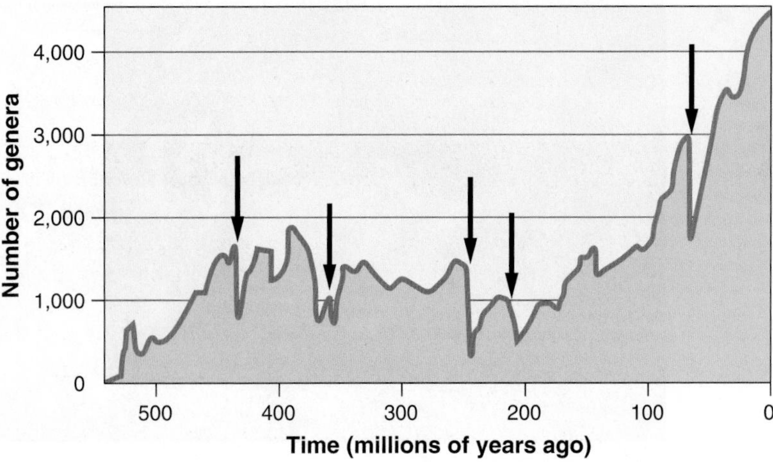

FIGURE 11.6 Mass extinctions. Five global mass extinction events have occurred since the evolution of complex life roughly 500 million years ago. *(Data from GreenSpirit, https://www.nationalgeographic.org/media/mass-extinctions)*

interactions further restricting where a species can live as its realized niche.

The Five Global Mass Extinctions

Throughout Earth's history, individual species have evolved and gone extinct. The fossil record has revealed five periods of global **mass extinction**, in which large numbers of species went extinct over relatively short periods of time. The times of these mass extinctions are shown in **FIGURE 11.6**. Note that because species are not always easy to discriminate in the fossil record, scientists count the number of genera, rather than species, that once roamed Earth but are now extinct. As we will see, several of these mass extinction events are associated with major changes in the environment around the planet that made it difficult for species to survive outside of their ecological tolerances.

The greatest mass extinction on record took place 251 million years ago when roughly 90 percent of marine species and 70 percent of land vertebrates went extinct. The cause of this mass extinction is not known. A better-known mass extinction occurred at the end of the Cretaceous period (65 million years ago), when roughly one-half of Earth's species, including the dinosaurs, went extinct. The cause of this mass extinction has been the subject of great debate, but there is now a general agreement that a large meteorite struck Earth and produced a dust cloud that circled the planet and blocked incoming solar radiation. This resulted in an almost complete halt to photosynthesis, and thus an almost total lack of food at the bottom of the food chain. Among the few species that survived was a small squirrel-sized primate that was the ancestor of humans.

Mass extinction A large number of species that went extinct over a relatively short period of time.

The Sixth Mass Extinction

As is evidenced by what we know about the five global mass extinctions, environmental scientists can learn about the effects of large and small environmental changes by studying historic events. They can then apply the lessons learned from these past events to help predict the effects of the environmental changes that are taking place on Earth today. During the last two decades, scientists have reached a consensus that we are currently experiencing a sixth global mass extinction of a magnitude within the range of the previous five mass extinctions. Given that we have poor data on the abundance of most species, especially the very small invertebrate animals, estimates of extinction rates vary widely. These estimates are extrapolated from extinction rates observed in a few well-described groups of organisms. For example, the Millennium Ecosystem Assessment, which is conducted by the United Nations, has estimated the extinction rate at 24 species per day. In contrast, the United Nations Convention on Biological Diversity has estimated the extinction rate at 150 species per day. What we know for sure is that there have been about 800 confirmed extinctions during the past 400 years, which represents a rate of about 2 species per year. These confirmed extinctions do not include any possible extinctions of species not yet discovered, making it is difficult to know the true extinction rate.

There is widespread agreement among scientists that the current mass extinction has human causes. These causes include a wide variety of environmental changes that are pushing species outside of their ecological tolerances, such as habitat destruction, overharvesting, introductions of invasive species, climate change, and emerging diseases. We'll use our new understanding of the importance of ecological tolerances when we examine all of these factors in detail in Unit 9. Because much of the current environmental change caused by human activities is both dramatic and sudden, many species may not be able to move or adapt in time to avoid extinction.

In this module, we have explored the idea that individuals and species have specific environmental conditions under which they perform well. Thus, the concept of ecological tolerance is not only useful for understanding why different species are found living in different biomes around the world, but also how human-caused environmental changes have the potential to reduce the abundance of many species, with some being driven to extinction.

Module 11 AP® Review

Preparing for the AP® Exam

Learning Goals Revisited

11-1 What determines where individuals and species can live?

Individuals and species have a range of abiotic conditions under which they can survive, grow, and reproduce. The suite of all suitable abiotic conditions, known as the fundamental niche, determines where individuals and species can potentially live. Biotic interactions including competition, predation, and pathogens can further restrict this range to a realized niche.

11-2 How does environmental change affect species?

Given that species are adapted to particular environmental conditions that determine their distributions on Earth, changes in the environment can alter the geographic range of species. We see examples of this in the shifting ranges of trees in North America since the last Ice Age. Scientists also predict future changes in the geographic ranges of species as they experience future environmental changes.

11-3 How can environmental change cause species extinctions?

When the environmental conditions experience more extreme changes, species can go extinct. This can happen because the changes are too rapid for species to respond, the species are unable to shift their distributions, or the attempt to shift the distribution is met with resistance by other species of predators, pathogens, or competitors that restrict the fundamental niche of the species that is trying to shift.

AP® Practice Questions

Preparing for the AP® Exam

Multiple-Choice Questions

1. The abiotic conditions under which a species can survive and reproduce is called its
   (a) species diversity.
   (b) realized niche.
   (c) geographic range.
   (d) fundamental niche.

2. Which best describes a factor that might prevent a species from shifting its geographic range in response to a changing climate?
   (a) a slow rate of environmental change
   (b) a highway built through a native species' habitat
   (c) a new tract of forest planted to combat the effects of deforestation
   (d) a new law preserving thousands of acres of forest from logging

3. How would the introduction of a predator species affect the distribution of a species?
   (a) The species will immediately go extinct.
   (b) The species will change its location, as its fundamental niche has changed.
   (c) The species will change its location, as its realized niche has changed.
   (d) The species will remain in its current location and outcompete the predator species.

4. How many mass extinction events have there been in Earth's history?
   (a) two
   (b) four
   (c) five
   (d) seven

5. Which is a significant cause of the current mass extinction event?
   (a) captive breeding programs
   (b) recreational fishing
   (c) laws protecting species' habitats
   (d) extreme weather events

Use the diagram below to answer questions 6 & 7:

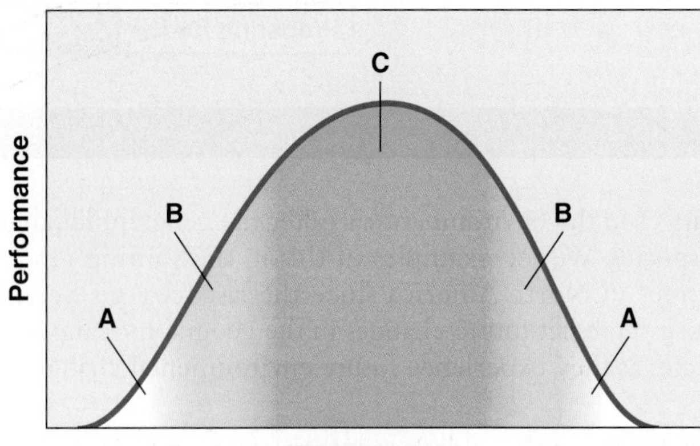

6. The areas on the graph marked "A" represent where a species is able to
   (a) only survive.
   (b) survive and grow.
   (c) survive and reproduce.
   (d) survive, grow, and reproduce.

7. The area on the graph marked "C" represents an organism's
   (a) range of tolerance.
   (b) fundamental niche.
   (c) realized niche.
   (d) geographic range.

Use the passage below to answer questions 8 & 9:

Researchers recently compiled data from 67 research studies that measured the number of species and ecosystem productivity from 600,000 sample locations around the world. Across terrestrial, freshwater, and marine biomes, they found that sites with more species had higher productivity, which was consistent with the results of experiments conducted earlier.

In addition, when the researchers considered how variation in climatic conditions and available nutrients affected the productivity of these ecosystems, they found that biodiversity was a more important factor than climate in half of the studies and a more important factor than nutrients in two-thirds of the studies.

In short, we have learned that biodiversity is critical to the proper functioning of ecosystems. Such conclusions make it clear that we need to protect Earth's biodiversity so that we can continue to enjoy the many benefits that it provides.

8. Which of the following best describes the author's conclusion in the passage?
   (a) Climate is the most important factor affecting the productivity in an ecosystem.
   (b) More species leads to a decrease in productivity in ecosystems.
   (c) Species richness has a larger effect on productivity in ecosystems than nutrients or climate.
   (d) Ecosystem productivity directly increases biodiversity of ecosystems.

9. Which of the following is evidence that supports the author's conclusion in the passage?
   (a) 67 research studies were conducted.
   (b) Terrestrial, freshwater, and marine biomes were investigated.
   (c) Two-thirds of the studies found that biodiversity affected productivity more than nutrients.
   (d) Sampling was conducted in 600,000 locations around the world.

Free-Response Question

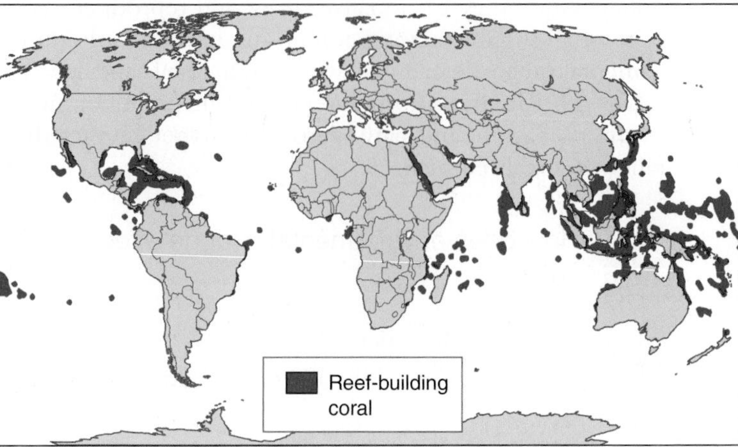

(a) Using the map above:
   (i) **Identify** the latitudinal range of the highest concentrations of reef-building coral on Earth. (1 pt.)
   (ii) **Describe** the relationship between the latitudinal range identified in part (i) and the range of tolerance for reef-building coral. (1 pt.)
   (iii) **Explain** how the global distribution of coral might change as a result of rising ocean temperatures. (1 pt.)

(b) **Identify** TWO factors that could determine the fundamental niche of reef-building coral. (2 pts.)

(c) **Describe** how the realized niche of reef-building coral might change with the introduction of a new predator species. (1 pt.)

(d) **Explain** why the geographical range of a species may change over time. (1 pt.)

(e) Many scientists agree that the current mass extinction event has human causes.
   (i) **Describe** TWO ways human activities are playing a role in the current mass extinction. (2 pts.)
   (ii) For one example provided in part (i), **propose a solution** that may slow the extinction rate of species. (1 pt.)

Module 12

Natural Disruptions to Ecosystems

In the previous module, we discussed how each species has particular ranges of ecological tolerance to different abiotic conditions and how biotic interactions can further restrict where species can live as part of their realized niche. We also saw that historic slow changes in environmental conditions of thousands of years can cause species to shift their geographic ranges as they strive to find suitable environmental conditions. At the ecosystem level, we have seen that flows of energy and matter in ecosystems are essential to the species that live in them. However, sometimes ecosystems experience major disruptions that alter how they operate. Disruptions can occur over time and space, and environmental scientists are often interested in how disruptions affect the flow of energy and matter through an ecosystem. More specifically, they are interested in whether an ecosystem can resist the impact of disruptions and whether a disrupted ecosystem can recover its original condition. In this module, we consider how environments can be disrupted and how this can affect species and ecosystems. We will also investigate how these ecosystems can bounce back to their pre-disturbance condition.

Learning Goals

After reading this module you should be able to

12-1 describe how natural disruptions vary over time and space.

12-2 identify how Earth's climate and sea level have changed.

12-3 explain how natural disruptions can affect habitats and animal migrations.

12-1 How do natural disruptions vary over time and space?

Natural disruptions operate on a range of scales

Environments can change a great deal and the rate of these changes can vary widely. For example, air temperature can plummet several degrees in a few hours when a cold front moves into an area, but the water temperature of a lake requires weeks or months to change by the same amount. In this section, we explore how environments naturally change over space and time. While our focus is on natural disruptions, disruptions also can be due to human activities, such as housing, agriculture, air pollution, forest clear-cutting, and the removal of entire mountaintops for coal mining. We will discuss human-caused disruptions in future modules.

Environmental Variation Over Time

Environments around the world can change substantially over days, weeks, years, and millennia. Some changes can be **periodic disruptions**, which means they occur regularly, such as the cycles of day and night or the daily and monthly cycles of the moon's effect on ocean tides. Other changes can be **episodic disruptions**, which means they occur somewhat regularly, such as cycles of high rain and low rain that occur every 5 to 10 years. Still other disruptions are **random disruptions**, which have no regular pattern, such as volcanic eruptions or hurricanes. When Hurricane Ida hit the Caribbean in 2021, for example, it caused widespread damage to homes, businesses, and ecosystems in Cuba, Louisiana, and surrounding areas (**FIGURE 12.1**). Such natural events can disrupt ecosystems even more than many human-caused disruptions.

Some environmental disruptions can cause large impacts on species and ecosystems, but the probability of it occurring twice at a particular location is low. For example, fires, hurricanes, tsunamis, and droughts happen every year somewhere in the world and they can completely change the ecosystem for many years, but they rarely hit the same location for multiple years in a row. Other large disruptions can occur at the same locations with some regularity. For

Periodic disruption Occurring regularly, such as the cycles of day and night or the daily and monthly cycles of the moon's effect on ocean tides.

Episodic disruption Occurring somewhat regularly, such as cycles of high rain and low rain that occur every 5 to 10 years.

Random disruption Occurring with no regular pattern, such as volcanic eruptions or hurricanes.

FIGURE 12.1 Ecosystem disruption. Large disturbances can have major effects on ecosystems, such as this area in Louisiana that was devastated by Hurricane Ida in 2021. *(Win McNamee/Getty Images)*

FIGURE 12.2 Fire as an ecosystem disruption. After many years of growth, forests accumulate dead leaves and branches on the forest floor. In many parts of the world, this accumulation of dry, dead organic matter makes the forest susceptible to forest fires every few decades. *(Cavan Images/Alamy Stock Photo)*

example, after a forest fire, coniferous forests have very little burnable material left on the ground. Each year after the fire, however, the forest floor continues to accumulate more conifer needles and dead branches until there is enough to feed another fire (**FIGURE 12.2**). Similarly, disease outbreaks can occur with regularity in nature as a host population increases in abundance and the hosts are then attacked by a disease-causing pathogen that can easily spread among densely populated hosts. As the pathogen kills a large number of hosts, the host population declines sharply and the survivors may now be immune, so the pathogen population declines sharply. The subsequent birth of new hosts that have no acquired immunity begins to build the host population back up until there are enough to allow the pathogen to spread rapidly once again. In this way, the cycling of host and pathogen populations occurs every few years.

Duration and Spatial Extent of Disruptions

When we consider the timing of disruptions and their spatial extent, an interesting pattern arises that is illustrated in **FIGURE 12.3**. As you can see in the figure, disruptions that have a relatively small spatial extent are also disruptions that occur over a short duration. For example, thunderstorms typically occur locally and may only last a few hours or less. In contrast, there are disruptions such as monsoons that can last for several weeks and impact large areas. Geologic changes in terrestrial and aquatic environments can take much longer. This includes processes such as the rising of mountains, volcanoes repeatedly erupting to form new islands, erosion cutting rivers into bedrock, and moving continents. In the next section, we will focus on other slowly changing environmental conditions, such as altered climate and sea level, which have changed over hundreds of thousands of years.

Ecosystem Resistance and Resilience

Not every ecosystem disruption is a disaster. For example, a low-intensity fire might kill some plant species, but at the same time it might benefit fire-adapted species that can use the additional nutrients released from the dead plants. So, although the population of a particular producer species might be diminished or even eliminated, the net primary productivity of all the producers in the ecosystem might remain the same. When this is the case, we say that the productivity of the system is resistant to disruptions. The **resistance** of an ecosystem is a measure of how much a disruption can affect the flows of energy and matter. When a disruption influences populations and communities but has no effect on the overall flows of energy and matter, we say that the ecosystem has high resistance.

When the flows of energy and matter of an ecosystem are affected by a disruption, environmental scientists often ask how quickly and how completely the ecosystem can recover its original condition. The rate at which an ecosystem returns to its original state after a disruption is termed **resilience**. You might recall that we first mentioned ecosystem resilience in our discussion of ecosystem services in Module 9. A highly resilient ecosystem returns to its original state relatively rapidly; a less resilient ecosystem does so more slowly. For example, imagine that a severe drought has eliminated half the species in an area. In a highly resilient ecosystem, the flows of energy and matter might return to normal in the following year. In a less resilient ecosystem, the flows of energy and matter might not return to their pre-drought conditions for many years.

An ecosystem's resilience often depends on specific interactions of the biogeochemical and hydrologic cycles. For example, as human activity has led to an increase in global atmospheric

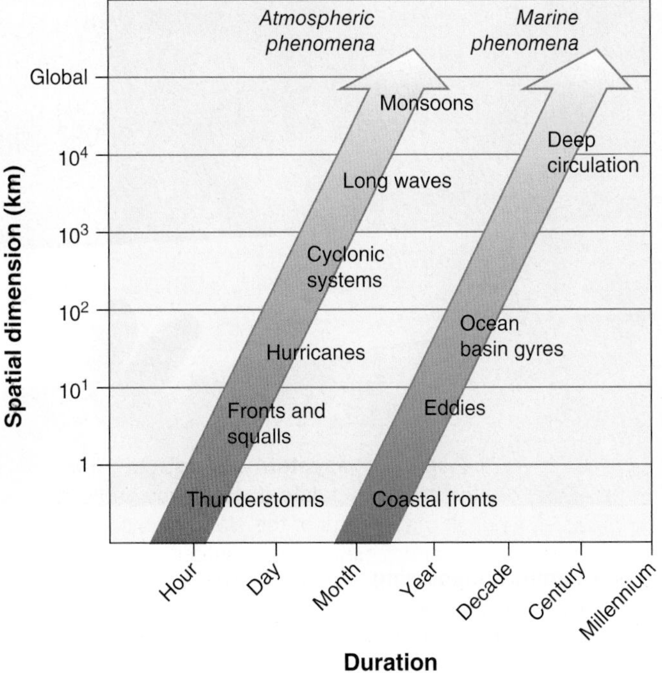

FIGURE 12.3 Ecosystem disruptions over space and time. Disruptions in marine environments and the atmosphere that have a short duration also generally impact a relatively small area. In contrast, disruptions that occur over a long duration generally impact a much larger area.

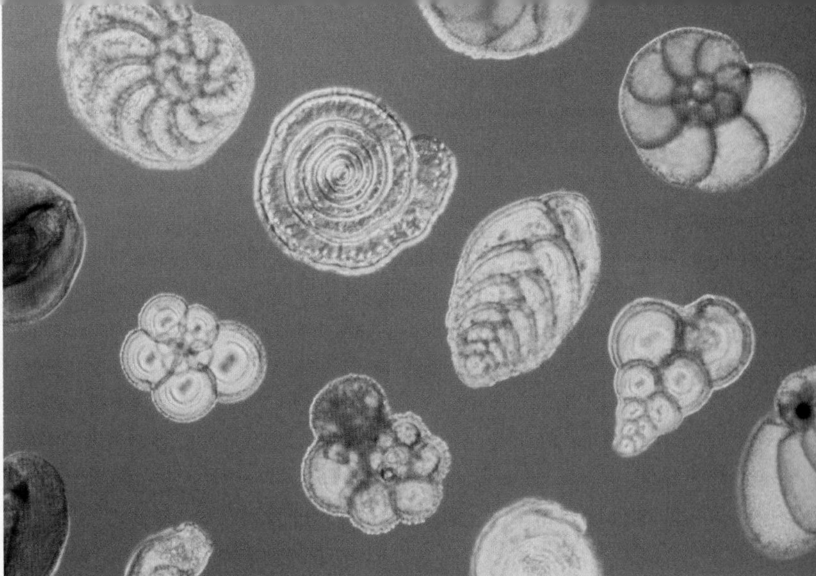

FIGURE 12.4 Estimating past temperatures using the ancient shells of foraminifera. The tiny shells of the protists become buried in layers of ocean sediments. By knowing the age of different ocean sediments and the preferred temperature of different species of foraminifera, scientists can indirectly estimate ocean temperature changes over time based on which species are found in each sediment layer. *(Astrid & Hanns-Frieder Michler/Science Source)*

CO_2 concentrations, terrestrial and aquatic ecosystems have increased the amount of carbon they absorb. In this way the carbon cycle as a whole has offset some of the changes that we might expect from increases in atmospheric CO_2 concentrations, including global climate change. Conversely, when a drought occurs, the soil may dry out and harden so much that when it eventually does rain, the soil cannot absorb as much water as it did before the drought. The soil changes in response to the drought, which leads to further drying and intensifies the drought damage. In this case, the hydrologic cycle does not relieve the effects of the drought; instead, a positive feedback loop in the system makes the situation worse.

12-2 How have Earth's climate and sea level changed?

Earth's climate and sea level have changed over geologic time

As we have noted, some environmental changes can only be observed when we can look back over long periods of time. Two excellent examples can be found in Earth's changes in climate and sea levels over hundreds of thousands of years. In this section, we examine what we know and, just as importantly, how we know it.

Earth's Changing Climates

Since nobody was measuring temperatures on Earth thousands of years ago, we must use indirect measurements to determine historic climates. Common indirect measurements include changes in the species composition of organisms that have been preserved over millions of years and chemical analyses of air bubbles formed in ice long ago that capture changes in greenhouse gases, which drive changes in temperatures.

Changing Species Compositions

One commonly used biological measurement is the change in species composition of a group of small protists, called foraminifera. Foraminifera are tiny, marine organisms with hard shells that resist decay after death (**FIGURE 12.4**). In

Resistance In an ecosystem, a measure of how much a disruption can affect the flows of energy and matter.

Resilience The rate at which an ecosystem returns to its original state after a disruption.

(a)

(b)

FIGURE 12.5 Estimating past greenhouse gas concentrations and past temperatures using ice cores. (a) Ice cores are extracted from very cold regions of the world such as this glacier on Mount Sajama in Bolivia. (b) Ice cores have tiny, trapped air bubbles of ancient air that can provide indirect estimates of greenhouse gas concentrations and global temperatures. *(a: ARCTIC IMAGES/ Alamy Stock Photo; b: Marc Steinmetz/VISUM/Redux)*

some regions of the ocean floor, the tiny shells have been building up in sediments for millions of years. The youngest sediment layers are near the top of the ocean floor whereas the oldest sediment layers are much deeper. Fortunately, different species of foraminifera prefer different water temperatures. As a result, when scientists identify the predominant species of foraminifera in a layer of sediment, they can infer the likely temperature of the ocean at the time the layer of sediment was deposited. By examining thousands of sediment samples, we can gain insights into temperature changes over millions of years.

Air Bubbles in Ancient Ice

Scientists can determine changes in greenhouse gas concentrations and temperatures over long periods of time by examining ancient ice. In cold areas such as Antarctica and at the top of the Himalayas, the snowfall each year eventually compresses to become ice. Similar to marine sediments, the youngest ice is near the surface and the oldest ice is much deeper. During the process of compression, the ice captures small air bubbles. These bubbles contain tiny samples of the atmosphere that existed at the time the ice was formed. Scientists have traveled to these frozen regions of the world to drill deep into the ice and extract long tubes of ice, which we call ice cores (**FIGURE 12.5**). Samples of ice cores can span up to 500,000 years of ice formation. Scientists determine the age of different layers in the ice core and then melt the ice from a piece associated with a particular time period. When the piece of ice melts, air bubbles are released and scientists measure the concentration of greenhouse gases in the air when the bubbles were trapped in the ancient ice.

AP® Exam Tip

Make certain you know that the long-term source for carbon dioxide levels in the atmosphere is ice cores. You should know that the long-term data for temperature comes for foraminifera exoskeletons found in ocean sediments.

Oxygen atoms in the air bubbles of ancient ice can also be used to determine temperatures from the distant past. Oxygen atoms occur in two forms, or isotopes: light oxygen, also known as oxygen-16 (^{16}O), contains 8 protons plus 8 neutrons. In contrast, heavy oxygen, also known as oxygen-18 (^{18}O), contains 8 protons plus 10 neutrons. Ice formed during a period of warmer temperatures contains a higher percentage of heavy oxygen than ice formed during colder temperatures. By examining changes in the percentage of heavy oxygen atoms from different layers of the ice core, we can indirectly estimate temperatures from hundreds of thousands of years ago.

FIGURE 12.6 charts historic temperatures and CO_2 concentrations. Looking at the blue line, we see that temperatures have changed dramatically over the past 400,000 years. Compared to modern-day temperatures, historic temperatures have seen 10°C decreases and 4°C increases. Most of these rapid shifts occurred during the onset of an ice age or during the transition from an ice age to a period of warm temperatures after an ice age. Because these changes occurred before humans could have had an appreciable

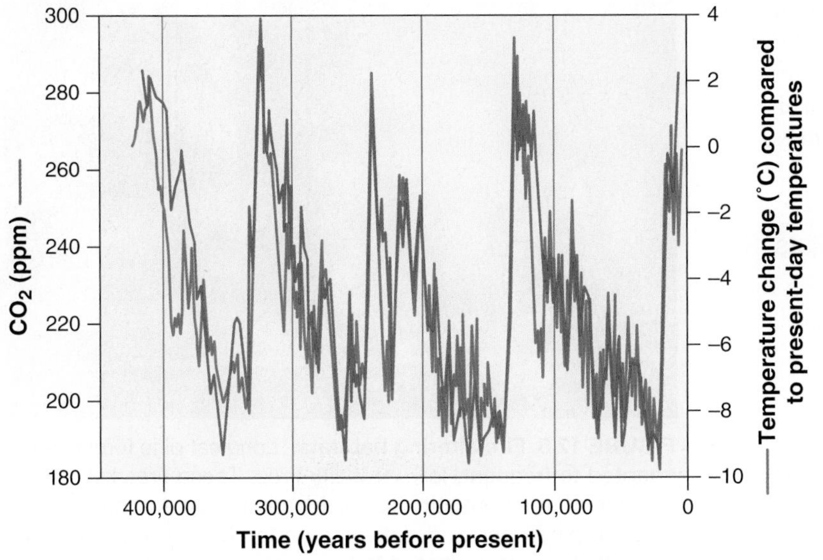

FIGURE 12.6 Historic temperature and CO_2 concentrations. Ice cores used to estimate historic temperatures and CO_2 concentrations indicate that the two factors vary together. *(Data from https://www.ncei.noaa.gov/products/paleoclimatology)*

effect on global systems, scientists suspect the changes were caused by small, regular shifts in the orbit of Earth. The path of the orbit, the amount of tilt on Earth's axis, and the position relative to the Sun all change regularly over hundreds of thousands of years. These changes alter the amount of sunlight that hits high northern latitudes in the winter, the amount of snow that can accumulate, and the way that snow and ice can reflect solar energy rather than absorbing it. These changes could give rise to fairly regular shifts in temperature over a long period of time.

An important insight from Figure 12.6 is the close correspondence between historic temperatures and CO_2 concentrations. But the graph does not tell us the nature of this relationship. Did periods of increased CO_2 cause increased temperature (as discussed in Module 4); did periods of increased temperature cause increased production of CO_2; or is another factor at work? Scientists have determined that the relationship between fluctuating levels of CO_2 and the temperature is complex and that both factors play a role. As we know, the increase of CO_2 in the atmosphere causes a greater capacity for warming through the greenhouse effect. However, when Earth experiences higher temperatures, the oceans warm and cannot hold as much CO_2 gas and, as a result, they release CO_2 into the atmosphere. What ultimately matters are the net movement of CO_2 between the atmosphere and the oceans and how these different feedback loops have worked together to affect global temperatures.

Earth's Changing Sea Levels

Much like temperature, scientists only have direct measurement of sea levels going back to the mid-1800s. However, they can indirectly determine changes in sea levels by examining fossils on land that contain aquatic species. The fossils tell us that water was present in a particular location and, by using carbon dating, we can determine the age of the fossils to tell us when the sea level had risen.

When scientists examine the data, they find that sea levels have varied a great deal during Earth's history, as shown in **FIGURE 12.7**. For example, during the peak of the last Ice Age, which was about 26,000 years ago, a large portion of Earth's water was in the form of massive ice sheets that expanded down from Canada to cover the Midwest, parts of the Northeast and Rocky Mountains, and Northern Europe. In the southern hemisphere, ice sheets expanded from Antarctica to cover parts of Chile and Argentina. With so much water being converted into ice, sea levels dropped by more than 120 m (400 feet). Following this Ice Age (21,000 to 11,700 years ago), the Earth warmed by 4°C; this caused the major ice sheets to melt and sea levels to rise an astounding 85 m (280 feet). Even more surprising is that after the temperature leveled off, for

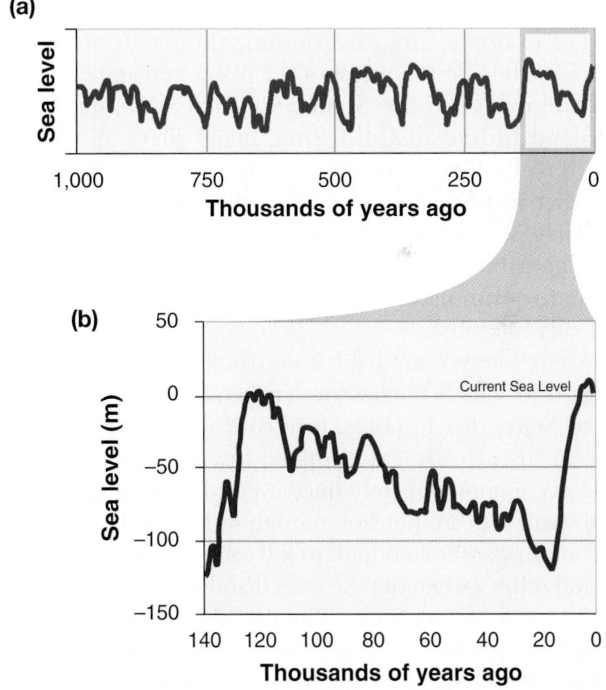

FIGURE 12.7 Fluctuating sea levels. Scientists can indirectly measure changing sea levels using dated fossils of aquatic organisms to determine changes over hundreds of thousands of years. (a) Over the past million years, sea levels have fluctuated by more than 130 m (420 feet) as the planet experienced periods of warming and cooling. (b) Inset showing the past 140,000 years shows how the peak of the most recent ice age (21,000 years ago) caused a decline in sea levels. The subsequent warming that occurred resulted in substantial melting of the glaciers and a sharp increase in sea levels of approximately 130 m (680 feet). *(Data from NOAA: PaleoClimatology.)*

the next 8,000 years the ice sheets continued to melt and sea levels continued to rise another 45 m (150 feet). Such variation in sea levels represent major ecosystem disruptions, particularly along coastal regions around the world that fluctuated between being terrestrial and marine biomes.

12-3 How can natural disruptions affect habitats and animal migrations?

Disruptions can cause large habitat changes and animal migrations

Environmental scientists are interested in natural environmental disruptions because they can produce important changes in habitats. Given that disruptions can occur over different time scales and over small or large areas, the type of disruption determines the scale of habitat change and whether it can cause animal migrations.

Large, Rare Disruptions

As we have seen, some disruptions such as tornadoes, hurricanes, and wildfires can create large areas of habitat destruction. For example, fires can consume thousands of hectares and completely destroy most of the plants and animals in the region. However, in fire-prone areas such as the shrubland/woodland biome in California, many plant species are adapted to frequent fires. As a result, the plants either have seeds that readily germinate after the fire or they readily sprout new shoots from their protected underground tissues. Thus, the habitat is dramatically altered because of the fire, but the fire stimulates plants to germinate and the habitat eventually recovers.

In some biomes, the habitat destruction is more selective in regard to which species are harmed. In the southeastern United States, for instance, forests of longleaf pine (*Pinus palustris*) historically experienced low-intensity fires that would occur approximately once every decade (**FIGURE 12.8**). These small fires are not hot enough to kill the longleaf pine trees, but they are hot enough to kill other species, such as oak trees and other species of pine trees that slowly try to encroach into the longleaf pine forest. The fires also allow the longleaf pine seed to germinate, as they come into contact with soil that has had all organic matter burned away. However, human campaigns to prevent fires since the 1920s have resulted in these longleaf pine forests being increasingly invaded by other species of trees that are not fire adapted, causing the savanna-like longleaf pine forests to be outcompeted by oaks and other pines. Thus, the disruption caused by natural fires was responsible for controlling the tree species in this habitat. Today, forest managers use controlled burns in longleaf pine forests to remove competing trees and maintain the savanna-like habitat that was historically present due to frequent disruptions by low-intensity fires.

FIGURE 12.8 **Fire altering habitats.** Longleaf pine forests are adapted to frequent, low-intensity fires. These fires kill other species of trees but stimulate the germination and growth of longleaf pine trees and a diverse understory of grasses and herbs that are also adapted to fire. *(William D. Boyer, UDSA Forest Service, Bugwood.org)*

Intermediate disturbance hypothesis The hypothesis that ecosystems experiencing intermediate levels of disturbance will favor a higher level of diversity of species than those with high or low disturbance levels.

Intermediate Levels of Disruption

Intermediate levels of ecosystem disruption can also play an important role in changing habitats by altering species diversity. The **intermediate disturbance hypothesis** states that ecosystems experiencing intermediate levels of disturbance will favor a higher diversity of species than those with high or low disturbance levels. The graph in **FIGURE 12.9a** illustrates this relationship between ecosystem disturbance and species diversity. Ecosystems in which disturbances are rare experience intense competition among species. Because of this, populations of only a few highly competitive species eventually dominate the ecosystem. In places where disturbances are frequent, only a few species with high population growth rates are able to counter the effects of frequent disturbance and prevent species extinction. At an intermediate level of disturbance, many more species are capable of persisting.

A classic example of the intermediate disturbance hypothesis can be found in marine algae that spend their lives attached to rocks along the rocky coast of New England. In these areas, common periwinkle snails (*Littorina littorea*) are present and they eat multiple species of marine algae. When few snails are present, they only eat a small amount of the algae, which represents a small disturbance. Under these conditions, only a few algal species—the most competitive species—dominate the rocks, as shown in Figure 12.9b. In areas containing high densities of snails, however, a lot more algae are consumed, which represents a high disturbance.

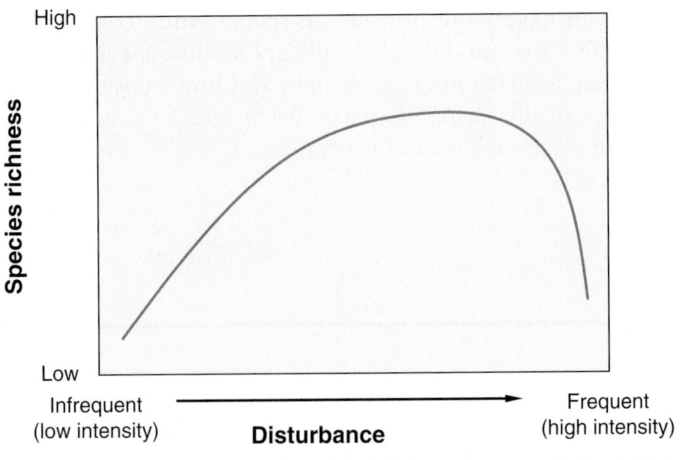

(a)

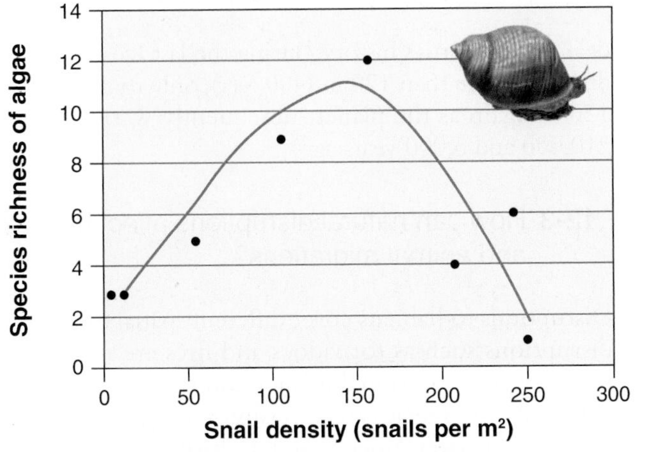

(b)

FIGURE 12.9 Intermediate disturbance hypothesis. (a) In general, we expect to see the highest species diversity at intermediate levels of disturbance. Rare disturbances favor the best competitors, which outcompete other species. Frequent disturbances eliminate most species except those that have evolved to live under such conditions. At intermediate levels of disturbance, species from both extremes can persist. (b) This shows an example of the intermediate disturbance in the number of algal species observed in response to different amounts of herbivory by marine snails. When few or many snails are present, there is a low diversity of algal species, but when an intermediate density of snails are consuming algae, the snails cause an intermediate amount of disturbance and a higher diversity of algal species can persist in the ecosystem. *(Data from Lubchenco, J. 1978. Plant species diversity in a marine intertidal community: Importance of herbivore food preference and algal competitive abilities.* American Naturalist *112: 23–39.)*

Under these conditions, only the most unpalatable algal species can persist. However, when snails were present at an intermediate density, this represents an intermediate disturbance to the algae; there is too much consumption by snails for the best competitors to dominate and the other species are never driven extinct. As a result, the intermediate density of snails favors 12 species of algae being able to persist on the rocks. As you can see in this example, the highest diversity of species can occur when ecosystems experience an intermediate frequency of disturbance.

Seasonal Disruptions Favoring Migrations

As we have mentioned, some disruptions are cyclic, such as seasonal changes in temperature that are associated with seasonal changes in food abundance. These seasonal disruptions are common causes of animal migrations in search of more abundant food. Short-term migrations include the movement of grazing animals up and down a mountainside during alternating periods of snowfall and snowmelt as they strive to find grasses to graze on. Well known long-term migrations include many North American bird species that fly south in the fall in search of more abundant food and then fly back north in the spring when the northern weather begins to warm and generate new food sources in the form of plants and insects.

One of the most famous annual migrations occurs on the Serengeti National Park and surrounding game reserves of Tanzania and Kenya. Here, the natural disruptions of seasonal changes in rain and drought cause millions of wildebeests (*Connochaetes taurinus*), zebras, and gazelles to migrate in a very large circular path to seek new habitats with food and water (**FIGURE 12.10**). As the seasons pass, the giant herds of animals eventually

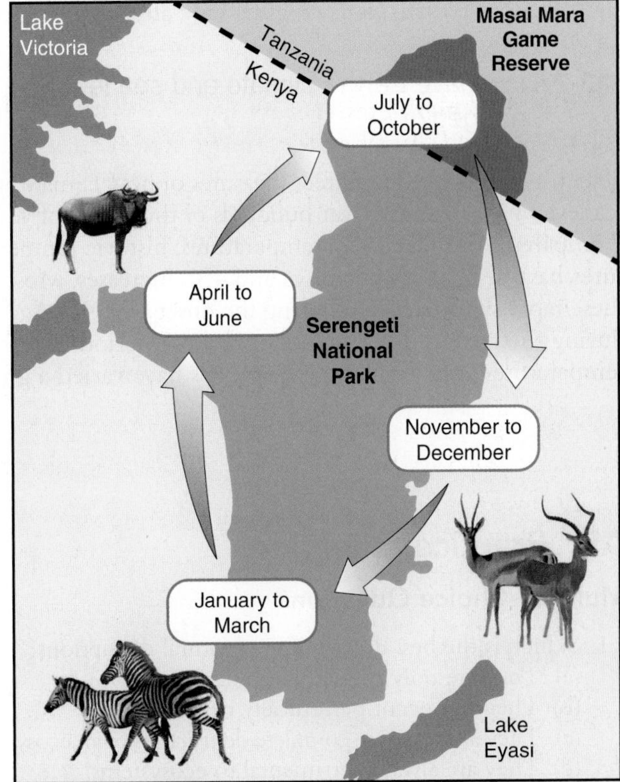

FIGURE 12.10 Environmental disruptions and animal migrations. Seasonal changes in precipitation cause herds of large grazing animals to migrate in a circular pattern among different regions around the Serengeti National Park. The herds of wildebeest, zebras, and gazelles conduct this migration as each habitat experiences seasonal reductions of plants and water.

return to their original habitat when the seasonal rains return, allowing the grasses and wildflowers to grow, which the migrating animals consume.

Throughout this module, we have seen that natural environmental disruptions occur over time and space and they can have large impacts on species and ecosystems. In the next two modules, we will explore how species can adapt to such environmental changes and how ecosystems can dramatically change as particular species of plants and animals replace each other over time.

Module 12 AP® Review

Preparing for the AP® Exam

Learning Goals Revisited

12-1 How do natural disruptions vary over time and space?

Natural disruptions can occur on the scale of days to years to millennia. Large disruptions last longer and commonly affect large areas, but they are relatively rare in any particular location. The resistance of an ecosystem is a measure of how much a disturbance can affect the flows of energy and matter. The resilience of an ecosystem is the rate at which an ecosystem returns to its original state after a disruption.

12-2 How have Earth's climate and sea level changed?

Scientists must rely on indirect measures of how climate and sea levels have changed over hundreds of thousands of years. Compared to modern-day temperatures, historic temperatures have seen 10°C decreases and 4°C increases. Most of these rapid shifts occurred during the onset of an Ice Age or during the transition from an Ice Age to a period of warm temperatures after an Ice Age. Sea levels have varied a great deal during Earth's history; during the last Ice Age, sea levels dropped more than 120 m (400 feet), only to rise more than 120 m again as the planet subsequently warmed between 210,000 and 3,000 years ago.

12-3 How can natural disruptions affect habitats and animal migrations?

Disruptions to habitats cover different spatial extents. Large disruptions such as tornadoes and fires are relatively rare for any given location, but they can dramatically alter the habitat. Some species are specially adapted to such disruptions and recover quickly. When disruptions are intermediate in frequency, we can see ecosystems that experience increased species diversity as dominant competitors are reduced, making way for other species in the ecosystem. Seasonal disruptions due to regular changes in the climate can also alter habitats. One common response to seasonal disruptions is animal migrations in search of more abundant food supplies.

AP® Practice Questions

Preparing for the AP® Exam

Multiple-Choice Questions

1. Which is the best description of natural disruptions?
   (a) They are only caused by humans.
   (b) They can occur periodically or randomly.
   (c) They always cause complete destruction of an ecosystem.
   (d) They are always detrimental to ecosystems.

2. An ecosystem that rapidly returns to its original state after a disturbance is
   (a) resistant.
   (b) resilient.
   (c) stable.
   (d) adaptable.

Use the graph below to answer questions 3-5:

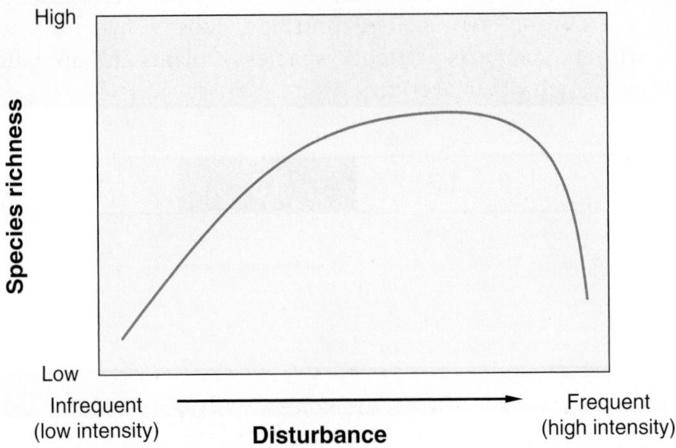

3. According to the graph, what effect will an intermediate level of disturbance have on an ecosystem?
   (a) An intermediate level of disturbance will increase runoff.
   (b) An intermediate level of disturbance will decrease primary productivity.
   (c) An intermediate level of disturbance will increase species diversity.
   (d) An intermediate level of disturbance will decrease biomass.

4. Which of the following best describes the relationship between species richness and disturbance?
   (a) As species richness increases, the level of disturbance increases to a maximum, and then decreases with more species richness.
   (b) As species richness increases, the level of disturbance decreases.
   (c) As the level of disturbance increases, species richness decreases.
   (d) As the level of disturbance increases, species richness increases to a maximum, and then decreases with further disturbance.

5. Which of the following best explains why an extreme level of disturbance will greatly affect species richness?
   (a) Ecosystems where disturbances are rare will have increased species richness, as they favor generalist species with a good ability to compete.
   (b) Ecosystems where disturbances are rare will have decreased species richness, as they favor specialist species with a good ability to compete.
   (c) Ecosystems where disturbances are frequent will have increased species richness, as they favor generalist species with a good ability to reproduce.
   (d) Ecosystems where disturbances are frequent will have decreased species richness, as they favor specialist species with a good ability to reproduce.

6. What is the relationship between a natural disruption's duration and its spatial coverage?
   (a) Disruptions with a large spatial coverage generally last longer.
   (b) Disruptions with a large spatial coverage generally last for shorter periods.
   (c) Disruptions that last longer generally have a small spatial coverage.
   (d) Disruptions that last longer generally have an intermediate spatial coverage.

7. Which is a measure of how much a disturbance can affect the flows of energy and matter in an ecosystem?
   (a) diversity
   (b) intensity
   (c) resistance
   (d) resilience

Free-Response Question

The East Asian-Australasian Flyway is an annual migration route for 61 species of shorebirds between Australia and 21 other countries. These shorebird species can travel seasonally to Australia from as far as Alaska and Siberia.

(a) **Identify** in which season shorebirds are most likely to travel to Australia. (1 pt.)

(b) **Describe** why shorebirds make this annual migration. (1 pt.)

(c) **Explain** how the migration of shorebirds may change over time if the average temperature in Arctic areas of Alaska and Siberia continue to increase. (1 pt.)

(d) **Describe** a potential ecological problem that could be caused by the change in shorebird migration explained in part (c). (1 pt.)

(e) The global average sea level has increased from approximately 2 inches in 1920 to 10 inches in 2020. **Calculate** the percent change in global average sea level from 1920 to 2020. Show all work. (2 pts.)

(f) A component of sea level rise is from the melting of ice sheets such as those in Greenland. It is estimated that 308 billion tons of ice have been lost from Greenland each year from 2002 to 2020. **Calculate** the rate of ice mass loss in tons per second. Show all work. (2 pts.)

(g) There are approximately 245 gallons of water contained in one ton of ice. **Calculate** the total amount of water added to the ocean per year in gallons as Greenland melts. (2 pts.)

Module 13

Unit 2	8	9	10	11	12	13	14

Adaptations

Throughout this unit, we have discussed the importance of biodiversity for the functioning and persistence of ecosystems. In this module, we will look at the processes of evolution, which is the source of biodiversity at the genetic, population, and ecosystem levels. Because the evolution of biodiversity depends on genetic diversity, we will begin with a reminder of how genetic diversity is created and then consider how humans and the natural world select from this variation to favor particular individuals that go on to reproduce in the next generation. We will also examine how several random processes can also cause evolution. With this foundation, we will then consider how natural and human processes have caused new species to arise over short and long timescales, including very rapid evolution as humans strive to dramatically modify species for a variety of reasons.

Learning Goals

After reading this module you should be able to

13-1 describe how populations can evolve adaptations in response to environmental changes.

13-2 explain how evolution can occur over short and long timescales.

13-1 How can populations evolve adaptations in response to environmental changes?

Populations can evolve by artificial selection, natural selection, or random processes

Earth's biodiversity is the product of **evolution**, which can be defined as a change in the genetic composition of a population over time. Evolution can occur at multiple levels. Evolution at the population level, such as the evolution of different varieties of apples or potatoes, is called **microevolution**. In contrast, when genetic changes give rise to new species, or to new genera, families, classes, or phyla—the larger categories of organisms into which species are organized—we call the process **macroevolution**. Among these many levels of macroevolution, the term speciation is restricted to the evolution of new species.

Evolution A change in the genetic composition of a population over time.

Microevolution Evolution at the population level.

Macroevolution Evolution that gives rise to new species, genera, families, classes, or phyla.

Genes and Genetic Variation

As you likely recall from an introductory biology class, to understand how genetic diversity is created we first need to understand genes. Genes are physical locations in chromosomes within each cell of an organism. A given gene has DNA that codes for particular traits, such as body size, but the DNA can take different forms known as alleles. The complete set of genes in an individual is called its genotype, which serves as the blueprint for the complete set of traits that organism may potentially possess. The traits that an individual actually exhibits is called its phenotype, including an individual's anatomy, physiology, and behavior. Phenotypes are the product of both the genotype of an individual and the environment of an individual. For example, your genotype may code for growing into a tall adult, but if your environment has limited food, you may end up as a shorter adult. Other phenotypes are less influenced by the environment. For instance, eye color is largely determined by the genes you carry rather than the environment in which you live. In either case, changes in genotypes can produce important changes in phenotypes.

You likely also recall that populations contain individuals that vary in their genotypes. This variation is caused by mutations, in which there are occasional mistakes in the copying process, which produces a random change in the genetic code. Environmental factors, such as ultraviolet radiation from the Sun, can also cause mutations. When mutations occur in cells responsible for reproduction, such as the eggs and sperm of animals, those mutations can be passed on to the next generation. Most mutations are detrimental to the performance of an organism, but sometimes a mutation improves an organism's survival or reproduction. If such a mutation is passed along to the next generation, it adds new genetic diversity to the population. Some mosquitoes, for example, possess a mutation that makes them less vulnerable to insecticides. In areas that are sprayed with insecticides, this mutation improves an individual mosquito's chance of surviving and reproducing.

Genetic diversity can also be created through recombination. In plants and animals, genetic recombination occurs as chromosomes are duplicated during reproductive cell division and a piece of one chromosome breaks off and attaches to another chromosome. This process does not create new genes, but it does bring together new combinations of alleles on a chromosome and can therefore produce novel traits. For example, the human immune system must battle a large variety of viruses and bacteria that regularly attempt to invade the body. Recombination in human cells allows new allele combinations to come together and provide new immune defenses that may prove to be effective against invading organisms.

AP® Exam Tip

Remember organisms don't choose to adapt. Adaptations only happen in populations over many generations. On the exam you should write, "Higher genetic diversity allows a population a greater likelihood to adapt to disturbance." Your answer shouldn't suggest an organism adapts within its lifetime.

Evolution by Artificial Selection

Evolution can occur in three ways: artificial selection, natural selection, and random processes. In this section we will look at artificial selection.

For centuries, humans have caused microevolution by breeding plants and animals for desirable traits. For example, all breeds of domesticated dogs belong to the same species as the gray wolf (*Canis lupus*), yet dogs exist in an amazing variety of sizes and shapes, ranging from toy poodles to Siberian huskies. **FIGURE 13.1** shows the relatedness among the wolf and different breeds of domestic dogs that were bred from the wolf by humans. Beginning with the domestication of wolves, dog breeders have selectively bred individuals that had particular qualities they desired, including body size, body shape, and coat color. After many generations of breeding, the selected traits became more and more exaggerated until breeders felt satisfied that the desired characteristics of a new dog breed had been achieved. As a result of this carefully controlled breeding, we have a tremendous variety of dog sizes, shapes, and coat colors today. In fact, for many breeds we even have a paper trail that documents how dog breeders have created new breeds. Despite this diversity of genotypes and phenotypes, dogs remain a single species: All dog breeds can still mate with one another and produce viable offspring.

When humans determine which individuals to breed, typically with a preconceived set of traits in mind, we call the process **evolution by artificial selection**. Artificial selection has produced numerous breeds of horses, cattle, sheep, pigs, and chickens with traits that humans find useful or aesthetically pleasing. Most of our modern agricultural crops are also the result of many years of artificial selection. For example, starting with a single species of wild mustard (*Brassica oleracea*), plant breeders have produced a variety of

Evolution by artificial selection The process in which humans determine which individuals to breed, typically with a preconceived set of traits in mind.

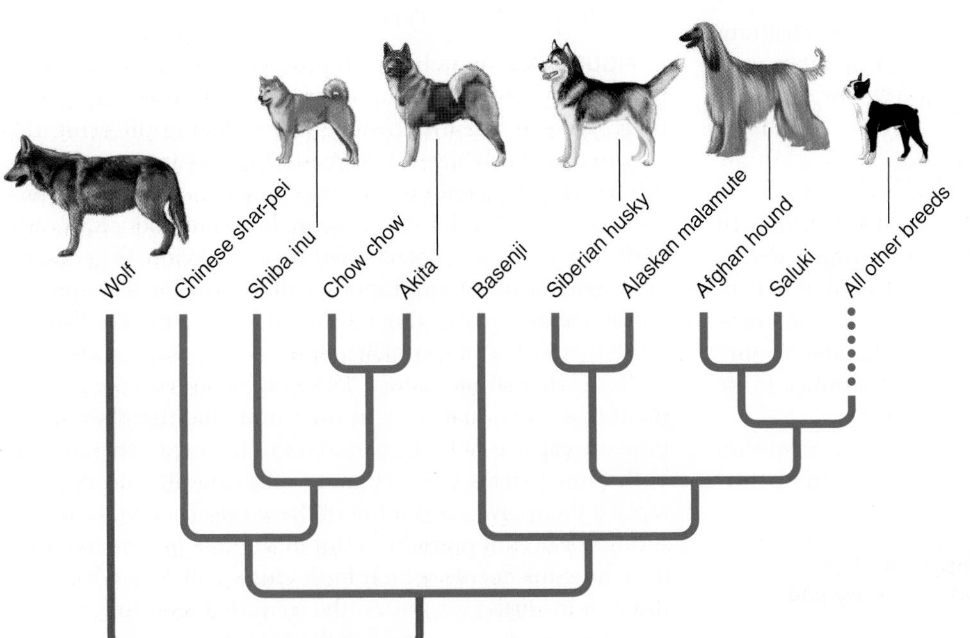

FIGURE 13.1 Artificial selection on animals. The diversity of domesticated dog breeds is the result of artificial selection on wolves. The wolf is the ancestor of the various breeds of dogs. It is shown at the same level as the dogs in this family tree because it is a species that is still alive today. *(Data from H.G. Parker et al. 2004. Science 304: 1160–1164.)*

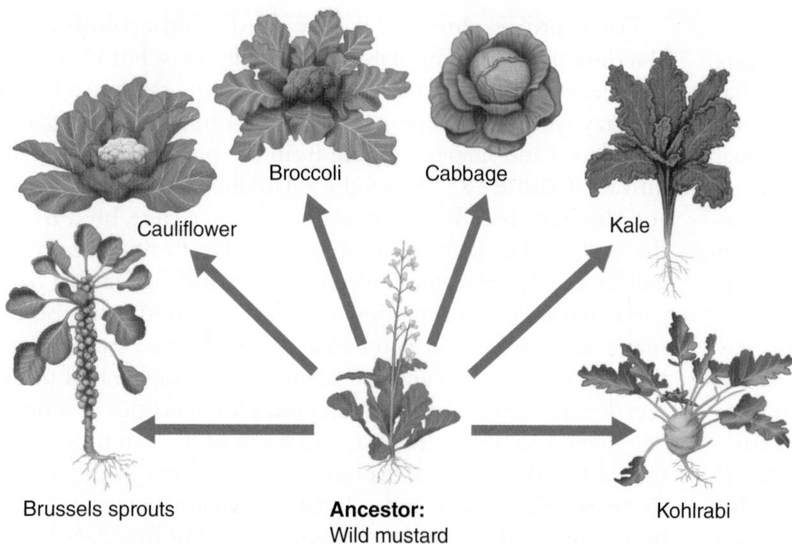

FIGURE 13.2 Artificial selection on plants. Plant breeders have produced a wide range of edible plants from a single species of wild mustard.

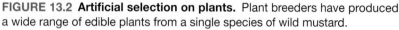

food crops, including cabbage, cauliflower, broccoli, Brussels sprouts, kale, and kohlrabi, shown in **FIGURE 13.2**.

As useful as artificial selection has been to humans, it can also produce a number of unintended results. For example, farmers often use herbicides to kill weeds. However, as we cover larger and larger areas with herbicides, there is an increasing chance that at least one weed will possess a mutation that allows it to survive the herbicide application. If that one mutant plant passes on its herbicide resistance to its offspring, we will have artificially selected for herbicide resistance in that weed. This process is occurring in many parts of the world where increased use of popular herbicides has led to the evolution of several species of herbicide-resistant weeds. A similar process has occurred in hospitals, where the use of antibiotics and antibacterial cleaners has caused artificial selection of harmful drug-resistant bacteria. These examples underscore the importance of understanding the mechanisms of evolution and the ways in which humans can either purposefully or inadvertently direct the evolution of organisms.

Evolution by Natural Selection

Evolution also takes place through natural mechanisms. In **evolution by natural selection**, the environment determines which individuals survive and reproduce. Members of a population naturally vary in their traits, and certain combinations of those traits make individuals better able to survive and reproduce. As a result, the genes that produce those traits are more common in the next generation.

Prior to the mid-nineteenth century, the idea that species could evolve over time had been suggested by a number of scientists and philosophers. However, the concept of evolution by natural selection did not become synthesized into a unifying theory until two scientists, Alfred Wallace (1823–1913) and Charles Darwin (1809–1882), independently put the various pieces together.

Evolution by natural selection The process in which the environment determines which individuals survive and reproduce.

Of the two scientists, Charles Darwin is perhaps the better known. At age 22, he became the naturalist on board HMS *Beagle*, a British survey ship that sailed around the world from 1831 to 1836. During his journey, Darwin made many observations of trait variation across a wide diversity of species, both wild species and the many varieties of domesticated plants and animals that humans had bred. In addition to observing living organisms, he found fossil evidence of a large number of extinct species. He recognized that organisms produce many more offspring than are needed to replace the parents, and that most of these offspring do not survive. Darwin questioned why, out of all the species that had once existed on Earth, only a small fraction had survived. Similarly, he wondered why, among all the offspring produced in a population in a given year, only a small fraction survived to the next year. During the decades following his voyage, he developed his ideas into a robust theory. His *On the Origin of Species by Means of Natural Selection*, published in 1859, changed the way people thought about the natural world. The key ideas of Darwin's theory of evolution by natural selection are the following:

- Individuals produce an excess of offspring.
- Not all offspring can survive.
- Individuals differ in their traits.
- Differences in traits can be passed on from parents to offspring.
- Differences in traits are associated with differences in the ability to survive and reproduce.

FIGURE 13.3 shows how this process works using the example of body size in a group of crustaceans known as amphipods. We can begin with parents producing offspring that vary in their body size. The largest offspring are consumed by fish because fish prefer to eat large prey rather than small prey. As a result, the smaller offspring are left to reproduce. Because body size is, in part, determined by an individual's genes, the next generation of amphipods will be smaller. This process continues over many generations and over time the fish will cause the evolution of smaller body sizes in amphipods.

Both artificial and natural selection begin with the requirement that individuals vary in their traits and that these variations are capable of being passed on to the next generation. In both cases, parents produce more offspring than necessary to replace themselves, and some of these offspring either do not survive or do not reproduce. But in the case of artificial selection, humans decide which individuals will breed, based on those individuals that possess the traits that tend toward some predetermined human goal, such as a curly coat or large body

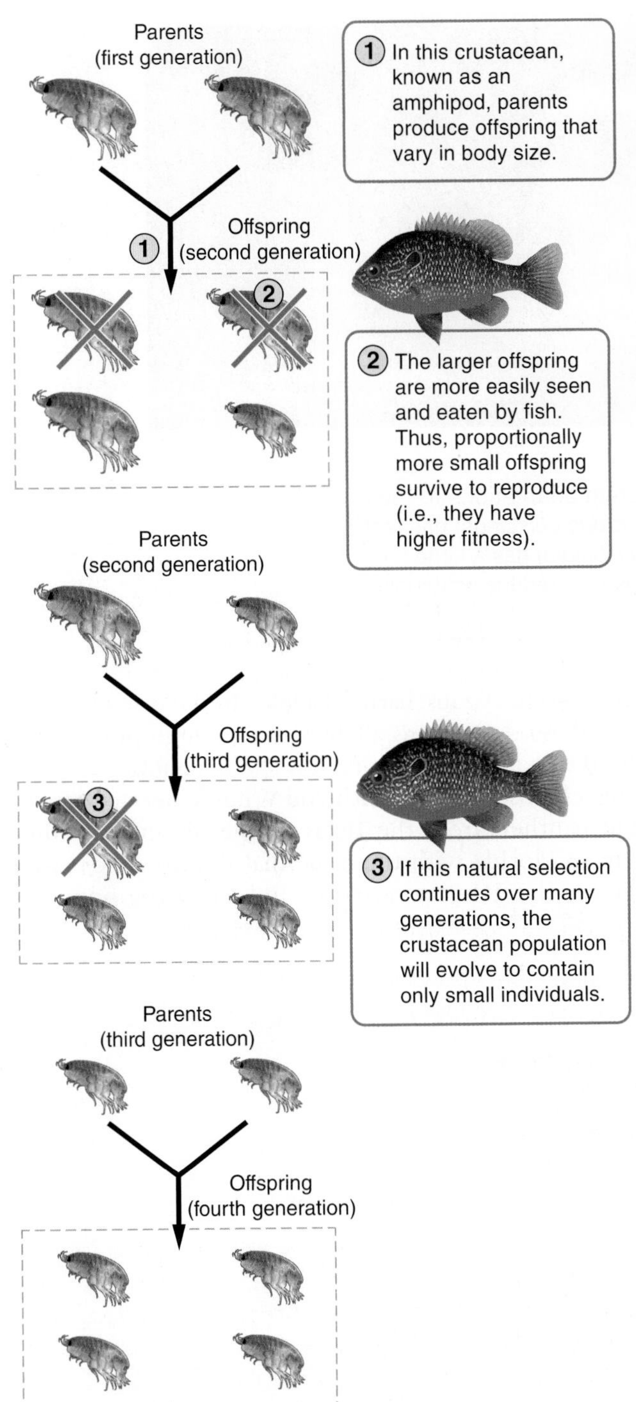
Parents
(first generation)
1 In this crustacean, known as an amphipod, parents produce offspring that vary in body size.
1 Offspring
(second generation)
2
2 The larger offspring are more easily seen and eaten by fish. Thus, proportionally more small offspring survive to reproduce (i.e., they have higher fitness).
Parents
(second generation)
Offspring
(third generation)
3
3 If this natural selection continues over many generations, the crustacean population will evolve to contain only small individuals.
Parents
(third generation)
Offspring
(fourth generation)

FIGURE 13.3 Evolution by natural selection. All species produce an excess number of offspring. Only those offspring with the fittest genotypes will pass on their genes to the next generation.

size. Natural selection does not select for specific traits that tend toward some predetermined goal. Rather, natural selection favors any combination of traits that improves an individual's **fitness**—its ability to survive and reproduce—as we saw in the case of the smallest amphipods surviving predation by fish. Traits that improve an individual's fitness are called **adaptations**. As we will see in the next section, adaptations can occur over both short and long timescales.

Natural selection can favor multiple solutions to a particular environmental challenge, as long as each solution improves an individual's ability to survive and reproduce. For example, while all plants living in the desert face the challenge of low water availability in the soil, different species have evolved different solutions to this common challenge. Some species have evolved the ability to store excess water during infrequent rains (**FIGURE 13.4**). Other species have evolved large taproots to draw water from deep in the soil. Still other species have evolved waxy or hairy leaf surfaces that reduce water loss. Each of these very different adaptations allows the plants to survive and reproduce in a dry, desert environment.

Evolution by Random Processes

Artificial and natural selection are important mechanisms of evolution, but we can also have **evolution by random processes**. As you may recall from an introductory biology class, these random processes alter the genetic composition of a population over time, but the changes are not related to differences in fitness among individuals. Thus, they are not adaptations. There are five random processes: mutation, gene flow, genetic drift, bottleneck effects, and founder effects. As we will see, these random processes can have important impacts on the genetic diversity of species and our understanding of how to keep species from going extinct.

Mutation

If a random mutation is not lethal, it can add to the genetic variation of a population. The larger the population, the more opportunities there will be for mutations to appear within it. As the number of mutations accumulates in the population over time, evolution occurs. If a mutation improves an individuals' fitness, then it can be favored by natural selection, which then becomes an adaptive process rather than a random process.

Gene Flow

Gene flow is the process by which individuals move from one population to another and thereby alter the genetic composition of both populations. Populations can experience an influx of migrating individuals with different alleles. The arrival of these individuals from adjacent populations alters the frequency of alleles in the population.

Gene flow can be helpful in bringing genetic variation to a population that lacks it. For example, the Florida panther is a subspecies of panther that once roamed throughout much of the southeastern United States and likely experienced gene flow with other subspecies of panthers. By 1995, the

Fitness An individual's ability to survive and reproduce.

Adaptation A trait that improves an individual's fitness.

Evolution by random processes The processes that alter the genetic composition of a population over time, but the changes are not related to differences in fitness among individuals.

(a)

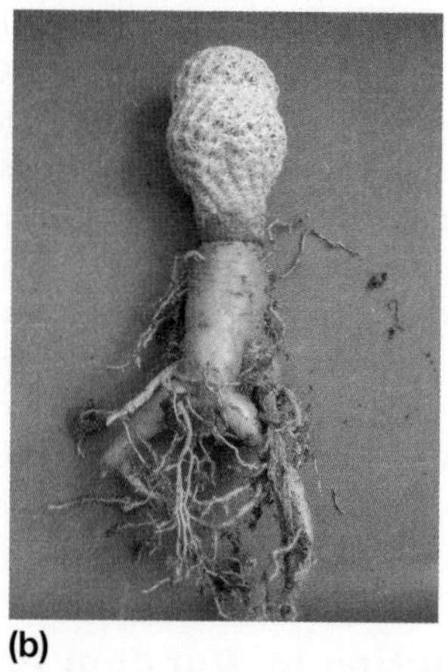
(b)

(c)

FIGURE 13.4 Adaptations. Desert plants have evolved several different adaptations to their desert environment. (a) The "Old Man of the Mountain" cactus (*Oreocereus celsianus*) has leaf hairs that reduce water loss. (b) Another species of cactus (*Mammillaria albiflora*) has a large taproot to draw water from deep in the soil. (c) The waxy outer layers of Aloe vera reduce water loss. *(a: FPI/Alamy; b: Ian Nartowicz; c: Scott Green/EyeEm/Getty Images)*

Florida panther only lived in southern Florida, occupying just 5 percent of its original habitat and the number of panthers declined to only about 30 individuals. Given that the Florida subspecies was isolated from other subspecies, it did not experience gene flow. As a result, shown in **FIGURE 13.5**, the small population had low genetic variation and the remaining individuals became very inbred, which causes individuals to express homozygous, harmful alleles. In the case of the panthers, these deleterious alleles caused a high prevalence of kinked tails, heart defects, and low sperm counts.

In response, the U.S. Fish and Wildlife Service captured eight panthers from the Texas subspecies and introduced them to Florida with the hope that this gene flow would increase the genetic variation and allow the population

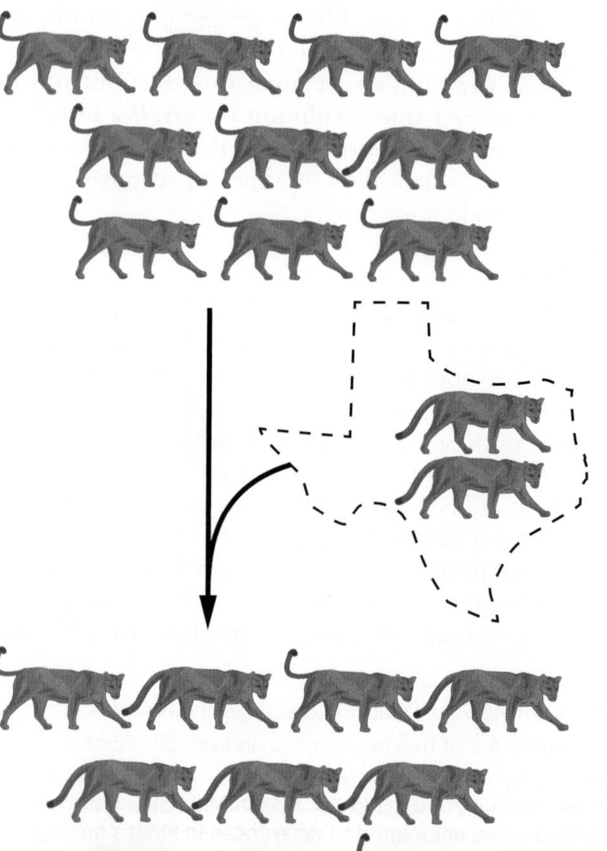

Florida
As the population of Florida panthers declined to very low numbers, the percentage of kinked tails increased to approximately 90 percent.
Texas
Several Texas panthers were brought to Florida.
Florida
After breeding with the Texas panthers, the Florida panther population was less inbred and the percentage of kinked tails declined.

FIGURE 13.5 Microevolution by gene flow. As the Florida panther declined in population size, the animals experienced low genetic variation and showed signs of inbreeding, which lead to kinky tails, heart defects, and low sperm counts. With the introduction of eight panthers from Texas, the Florida population experienced a decline in the prevalence of defects and a growth in population from 30 to 160 individuals.

to grow. By 2021, the Florida panther population had grown to more than 230 individuals and the prevalence of defects previously seen from inbreeding had declined. These improvements in the panther population underscore the importance of gene flow to providing beneficial genetic diversity to save declining populations.

Genetic Drift

Genetic drift is a change in the genetic composition of a population over time as a result of random mating. Like mutation and gene flow, genetic drift is a nonadaptive, random process. It can have a particularly important role in altering the genetic composition of small populations. In small populations, shown in **FIGURE 13.6a**, random mating among individuals can eliminate some of the individuals that carry unique traits simply because they did not find a mate in a given year. For example, imagine a small population of five mice, in which two individuals carry genes that produce black hair and three individuals carry genes that produce white hair. If, by chance, the individuals that carry the genes for black hair fail to find a mate, those genes will not be passed on. The next generation will be entirely white-haired, and the black-haired phenotype will be lost. In this case, the genetic composition of the population has changed, so the population has therefore evolved. The cause underlying this evolution is random; the failure to find a mate has nothing to do with hair color. In contrast, a large population that has the same proportion of black-haired mice, shown in Figure 13.6b, has a greater absolute number of mice. As a result, it is less likely that random mating events will cause all of the black-haired mice to not find a mate, so their genes are passed on to the next generation. In short, genetic drift is more likely to occur in a small population than in a large population.

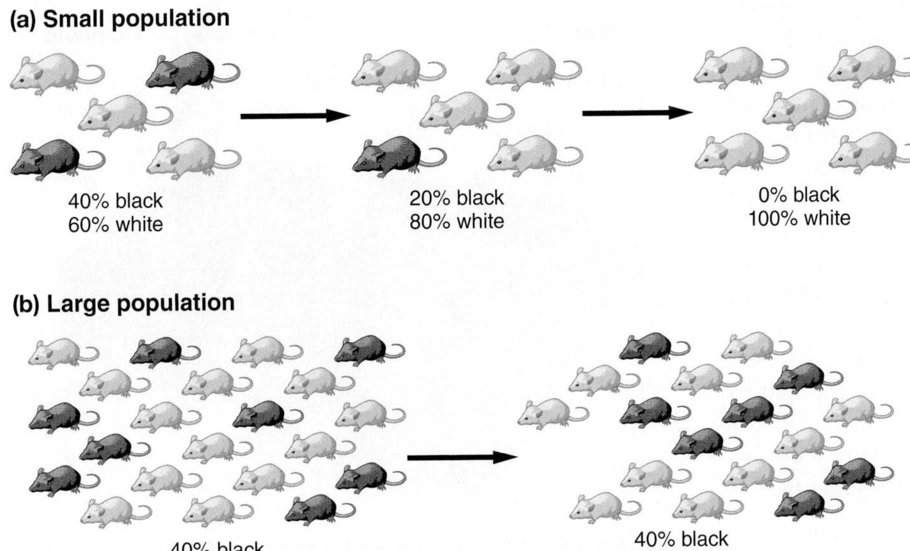

FIGURE 13.6 Microevolution by genetic drift. (a) In a small population, some less-common genotypes can be lost by chance as random mating among a small number of individuals can result in the less-common genotype not mating. As a result, the genetic composition can change over time. (b) In a large population, it is more difficult for the less-common genotypes to be lost by chance because the absolute number of these individuals is large. As a result, the genetic composition tends to remain the same over time in larger populations.

Bottleneck Effect

A drastic reduction in the size of a population that reduces genetic variation—known as a bottleneck effect—is another random process that can change a population's genetic composition. As discussed in Module 8, a population might experience a drastic reduction, or bottleneck, in its numbers for many reasons, including habitat loss, a natural disaster, harvesting by humans, or changes in the environment. **FIGURE 13.7** illustrates the

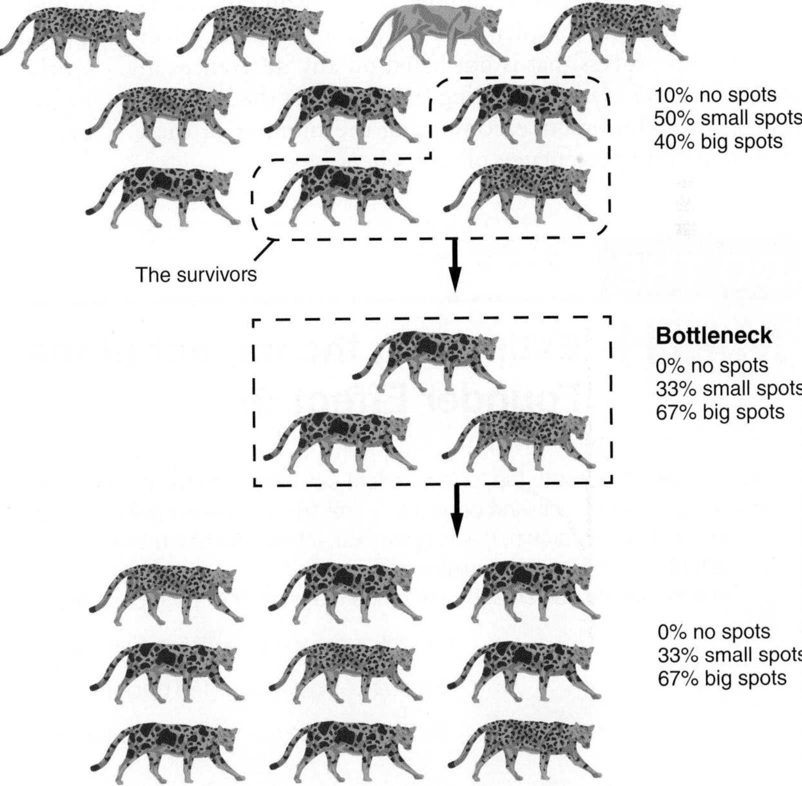

FIGURE 13.7 Microevolution by the bottleneck effect. If a population experiences a drastic decrease in size (goes through a "bottleneck"), some genotypes will be lost, and the genetic composition of the survivors will differ from the composition of the original group.

bottleneck effect using the example of cheetahs and the size of their spots. When the size of a population is reduced, the amount of genetic variation that can be present in the population is also reduced. With fewer individuals there are fewer unique genotypes remaining in the population, even if it subsequently rebounds to have more individuals.

Low genetic variation in a population can cause several problems, including increased risk of disease and low fertility. In addition, species that have been through a population bottleneck are often less able to adapt to future changes in their environment. In some cases, once a species has been forced through a bottleneck, the resulting low genetic diversity causes the population to decline. Such declines are thought to be occurring in a number of species today. The cheetah, for example, has relatively little genetic variation due to a bottleneck that appears to have occurred about 10,000 years ago.

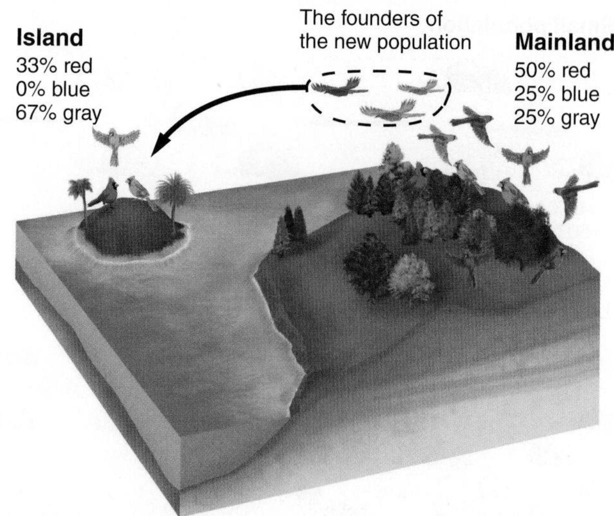

FIGURE 13.8 Microevolution by the founder effect. If a few individuals from a mainland population colonize an island, the genotypes on the island will represent only a subset of the genotypes present in the mainland population. As with the bottleneck effect, some genotypes will not be present in the new population.

Founder Effect

Imagine that a few individuals of a particular bird species happen to be blown off their usual migration route and land on a hospitable oceanic island, as illustrated in **FIGURE 13.8**. These few individuals will have been drawn at random from the mainland population, and the genotypes they possess are only a subset of those in the original mainland population. These colonizing individuals, or founders, will give rise to an island population that has a genetic composition very different from that of the original mainland population. A change in the genetic composition of a population as a result of descending from a small number of colonizing individuals is known as the founder effect. Thus, the founder effect is a random process that is not based on any differences in fitness. "Do the Math: Estimating the Impact of the Founder Effect" provides practice converting the number of each phenotype into percentages of each phenotype.

We can see another example of the founder effect in the Amish communities of Pennsylvania. The Amish population was founded by a relatively small number of individuals—about 200 people from Germany. This group happened to carry a mutation for Ellis-van Creveld syndrome, a condition that causes a variety of effects including extra fingers. The mutation is rare in humans around the world, but by chance the frequency of the mutation was higher in the early Amish colonists and has remained higher because the population is an isolated group with little gene flow from outside its community.

DO THE MATH Estimating the Impact of the Founder Effect

Preparing for the AP® Exam

As we have discussed, the founder effect can alter the phenotypes of a population in a newly colonized island compared to the phenotypes in the mainland population. Imagine a mainland population of rats that was comprised of 1,000 individuals with three unique coat colors: brown, black, and white. If there are 550 brown rats, 150 black rats, and 300 white rats, what is the percentage of each phenotype in the population?

To estimate the percentages, we divide the number of each phenotype by the total number of rats:

$$\text{Percent of brown rats} = (550 \text{ brown rats} \div 1{,}000 \text{ rats}) \times 100\% = 55\%$$

$$\text{Percent of black rats} = (150 \text{ black rats} \div 1{,}000 \text{ rats}) \times 100\% = 15\%$$

$$\text{Percent of white rats} = (300 \text{ white rats} \div 1{,}000 \text{ rats}) \times 100\% = 30\%$$

YOUR TURN If the founding population is comprised of 40 percent brown rats, 50 percent black rats, and 10 percent white rats, how many rats of each coat color would you expect in an island population of 200 rats?

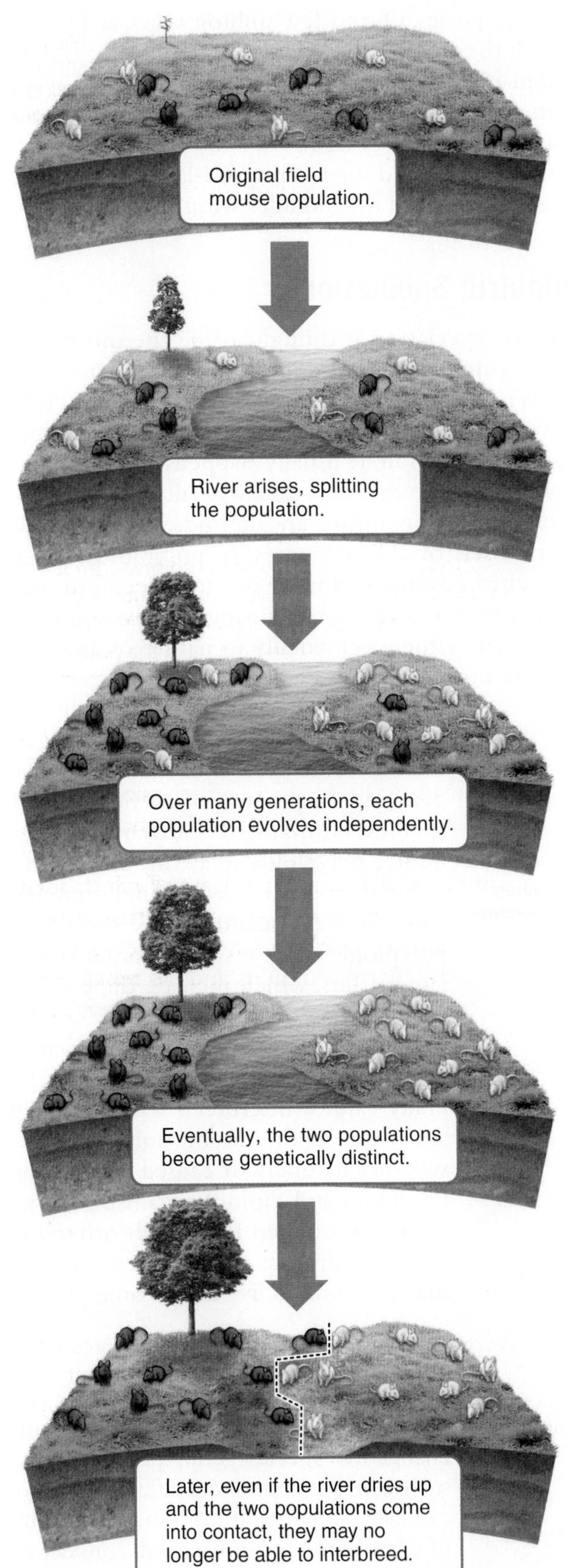

Original field mouse population.
River arises, splitting the population.
Over many generations, each population evolves independently.
Eventually, the two populations become genetically distinct.
Later, even if the river dries up and the two populations come into contact, they may no longer be able to interbreed.

FIGURE 13.9 **Allopatric speciation.** Geographic barriers can split populations. Natural selection may favor different traits in the environment of each isolated population, resulting in different adaptations. Over time, the two populations may become so genetically distinct that they are no longer capable of interbreeding.

AP® Exam Tip

When you have to calculate percentages on the exam make sure you always include units and "× 100." 0.55 is a ratio. 55% is a percentage. To convert them you need to show this in your set up.

13-2 How can evolution occur over short and long timescales?

Evolution can occur over short and long time scales by natural and artificial selection

As we have seen, microevolution is happening all around us, from the breeding of agricultural crops, to the unintentional evolution of drug-resistant bacteria in hospitals, to the bottleneck that reduced genetic variation in the cheetah. But how do we move from the evolution of genetically distinct populations of a species to the evolution of genetically distinct species? That is, how do we move from microevolution to macroevolution? In this section, we examine two common speciation processes and then examine how new species can evolve over short and long timescales.

Allopatric Speciation

One common way in which evolution creates new species is through geographic isolation, which means the physical separation of a group of individuals from others of the same species. The process of speciation that occurs with geographic isolation is known as **allopatric speciation** (from the Greek *allos*, meaning "other," and *patris*, meaning "fatherland"). As shown in FIGURE 13.9, geographic isolation can occur when a subset of individuals from a larger population colonizes a new area of habitat that is physically separated from that larger population. For example, a single large population of field mice might be split into two smaller populations as geographic barriers change over time. A river might change course and divide a large prairie into two halves, a large lake might split into two smaller lakes, or a new mountain range could rise. In such cases, the genetic composition of the isolated populations might diverge over time, either because of random processes—such as the founder effect—or because natural selection favors different adaptations on each side of the barrier.

If the two separated habitats differ in environmental conditions, such as temperature, precipitation, or the occurrence of predators, natural selection will favor different phenotypes

Allopatric speciation The process of speciation that occurs with geographic isolation.

in each of the habitats. If individuals cannot move between the populations, then over time the two geographically isolated populations will continue to become more and more genetically distinct. Eventually, the two populations will be separated not only by geographic isolation but also by reproductive isolation, which means the two populations of a species will evolve separately to the point that they can no longer interbreed and produce viable offspring. At this point, the two populations will become distinct species.

Allopatric speciation is thought to be responsible for the diversity of the group of birds known as Darwin's finches. When Charles Darwin visited the Galápagos Islands, located west of Ecuador, he noted a large variety of finch species, each of which seemed to live in different habitats or to eat different foods. Research on these birds has demonstrated that they all share a common ancestor that colonized the islands from the mainland long ago. **FIGURE 13.10** is a family tree for these finches that shows the relatedness patterns of the 13 species. Over a few million years, as Darwin discovered, the finches that were geographically isolated on different islands became genetically distinct and eventually became reproductively isolated. See the Visual Representation feature on pages 158–159 for a closer look at how Darwin's finches and the Galápagos Islands reveal the interconnectedness of all the modules within this unit.

Sympatric speciation The evolution of one species into two species, without any geographic isolation.

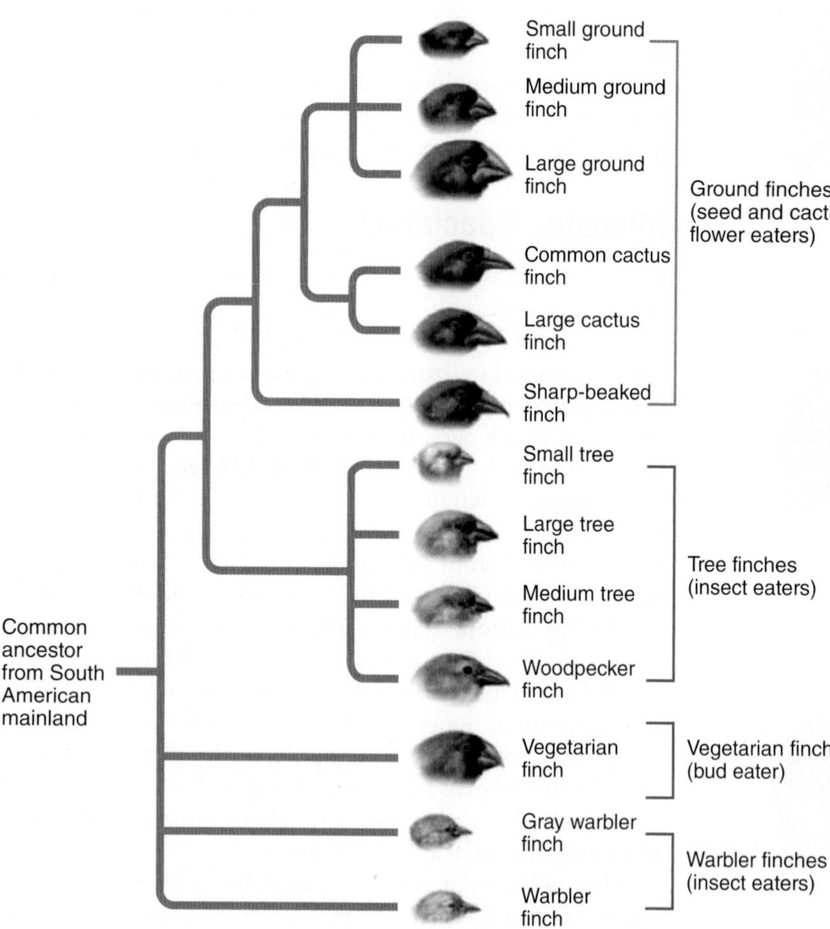

FIGURE 13.10 Allopatric speciation of Darwin's finches. In the Galápagos Islands, allopatric speciation has led to a large variety of finch species, all descended from a single species that colonized the islands from the South American mainland.

Sympatric Speciation

Allopatric speciation is thought to be the most common way in which evolution generates new species diversity. However, it is not the only way. **Sympatric speciation** is the evolution of one species into two species without any geographic isolation. It usually happens through a process known as polyploidy. As you may recall from a prior biology class, most organisms are diploid, which means they have two sets of chromosomes. In polyploidy, the number of chromosome sets increases to three, four, or even six. Such increases can occur during the division of reproductive cells, either accidentally in nature or as a result of deliberate human actions. Plant breeders, for example, have found several ways to interrupt the normal cell division process. Polyploid organisms include some species of snails and salamanders and 15 percent of all flowering plant species. Humans have discovered ways to induce polyploidy in a wide variety of agricultural crops such as bananas, strawberries, and wheat. As **FIGURE 13.11** shows for wheat, polyploidy often results in larger plants and larger fruits, which increase the commercial value of the domesticated crops.

The key feature of polyploid organisms is that once they become polyploid, they generally cannot interbreed with their diploid ancestors. At the instant polyploidy occurs, whether naturally or caused by humans, the polyploid and diploid organisms are reproductively isolated from each other and are therefore distinct species, even though they may continue to live in the same place.

The Pace of Evolution

How long does evolution take? A significant change in a species' genotype and phenotype, such as an adaptation to a different food source, can take anywhere from hundreds to millions of years. In this section, we will consider examples of rapid evolution by natural selection and even faster evolution by artificial selection.

Slow Evolution by Natural Selection

Sometimes evolution can occur relatively slowly, as in the case of the cichlid fish of

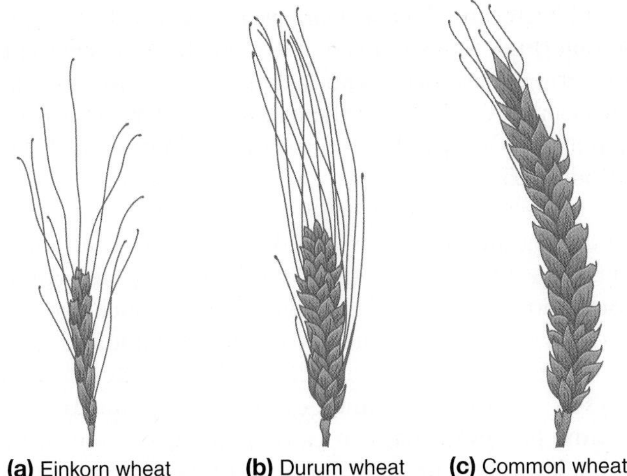

FIGURE 13.11 Sympatric speciation. Flowering plants such as wheat commonly form new species through the process of polyploidy, an increase in the number of sets of chromosomes beyond the normal two sets. (a) The ancestral einkorn wheat (*Triticum boeoticum*) has two sets of chromosomes and produces small seeds. (b) Durum wheat (*Triticum durum*), which is used to make pasta, was bred to have four sets of chromosomes and produces medium-sized seeds. (c) Common wheat (*Triticum aestivum*), which is used mostly for bread, was bred to have six sets of chromosomes and produces the largest seeds.

Lake Tanganyika, one of the African Great Lakes. You can see a variety of these species in **FIGURE 13.12**. Evidence indicates that the roughly 200 different species of cichlids in the lake evolved from a single ancestral species over a period of several million years. During this period, some cichlid species specialized to become insect eaters and others to become fish eaters, while still others evolved to eat invertebrates such as snails and clams.

Although the cichlids of Lake Tanganyika evolved relatively slowly in evolutionary terms, the pupfishes of the Death Valley region of California and Nevada evolved much more rapidly. In the 20,000 to 30,000 years since the large lakes of the region were reduced to isolated springs, several species of pupfish have evolved.

The ability of a species to survive an environmental change depends greatly on how quickly it evolves the adaptations needed to thrive and reproduce under the new conditions. If a species cannot adapt quickly enough, it will go extinct. This can happen when the rate of environmental change is faster than the rate at which evolution can respond. Slow rates of evolution can occur when a population has long generation times or when a population contains low genetic variation on which natural selection can act. For example, when pollutants are illegally dumped into streams, the environmental change occurs very quickly compared to the generation times of the animals living in the streams, so the animals in the stream die because they do not have sufficient time to evolve tolerance to the pollutant.

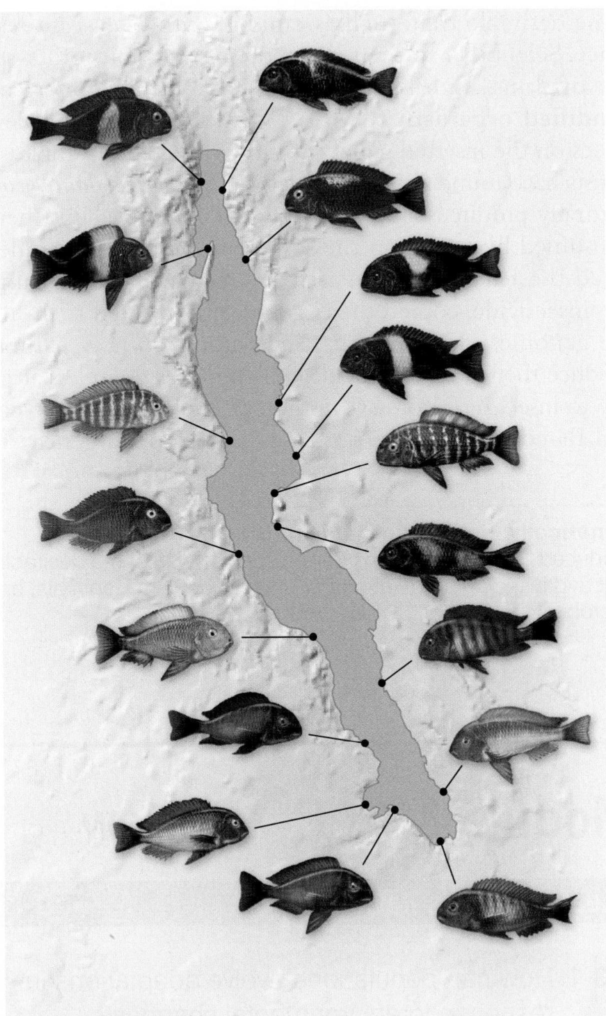

FIGURE 13.12 Rapid evolution. The cichlid fishes of Lake Tanganyika have evolved approximately 200 distinct and colorful species in the relatively short period since the lake formed in eastern Africa. The location where each species can be found in the lake is indicated by a corresponding black dot on the map.

Very Rapid Evolution by Artificial Selection

The pace of evolution can be incredibly fast when artificial selection is applied. Such rapid evolution is occurring in many species of commercially harvested fish, including the Atlantic cod (*Gadus morhua*). Intensive fishing over several decades has continually targeted the largest adults, selectively removing most of those individuals from the population and, therefore, also removing the genes that produce large adults. Because larger fish tend to reach sexual maturity later, the genes that code for a later onset of sexual maturity have also been removed. As a result, after just a few decades of intensive fishing, the Atlantic cod population has evolved to reach reproductive maturity at a smaller size and a younger age.

Evolution occurs even more rapidly in populations of genetically modified organisms. Using genetic engineering techniques, scientists can now copy genes from a species with some desirable trait, such as rapid growth or disease resistance. Scientists can then insert these genes into other species of plants, animals, or microbes to produce a **genetically modified organism (GMO)**. When a GMO reproduces, it passes on the inserted genes to its offspring. For example, scientists have found that a soil bacterium (*Bacillus thuringiensis*) naturally produces an insecticide as a defense against being consumed by insects in the soil. Plant breeders have identified the bacterial genes that are responsible for making the insecticide, copied those genes, and inserted them into the genomes of crop plants. Such crops can now naturally produce their own insecticide, which makes them less attractive to insect herbivores. Common examples include Bt-corn and Bt-cotton, so named because they contain genes from the soil bacterium *Bacillus thuringiensis*. As you might guess, inserting genes into an organism is a much faster way to produce desired traits than traditional plant and animal breeding, which can only select from the naturally available variation in a population. We will have much more to say about GMOs in Module 25.

Genetically modified organism (GMO) An organism produced by copying genes from a species with some desirable trait and inserting them into other species of plants, animals, or microbes.

The take-home message in our discussion of evolution and adaptations is that microevolution is happening all around us through both natural and artificial selection. Such evolution can occur slowly over time or occur very rapidly, depending on the processes causing the evolution. Moreover, the processes that can cause microevolution of populations are the same processes that can cause macroevolution of new species. However, macroevolution is often more difficult to observe because of the longer timescale required, unless it occurs through the rapid process of artificial selection. In the next module, we will explore how the species living in an area changes over time, with those species that arrive to an area first being replaced by those that arrive later.

Module 13 AP® Review

Preparing for the AP® Exam

Learning Goals Revisited

13-1 How can populations evolve adaptations in response to environmental changes?

Genetic variation helps to determine the traits that individuals express. Artificial and natural selection can act on genetic variation to produce changes in a population's phenotype over time in a process we call microevolution. When phenotypic changes improve the fitness of individuals, they are considered to be adaptations. Some genetic changes in a population are not due to selection, but rather due to random processes.

13-2 How can evolution occur over short and long timescales?

Through the process of macroevolution, new species can be created. In nature, new species can arise through either allopatric or sympatric speciation. There are a number of factors that can affect the pace of evolution including how quickly the environment changes, how much genetic variation exists in the population, the size of the population, and the generation time of the species. Evolution can happen even faster as a result of human breeding for polyploids or through the use of genetically modified organisms.

AP® Practice Questions

Preparing for the AP® Exam

Multiple-Choice Questions

1. A phenotype is best described by which of the following?
   (a) It is an adaptation that creates a new species.
   (b) It is the genes of a particular individual.
   (c) It is a result of genetic recombination.
   (d) It is the set of traits expressed in an individual.

2. Which is the best definition of an adaptation?
   (a) a mutation that creates a new species
   (b) a trait that improves an individual's fitness
   (c) a trait that is passed on to the next generation
   (d) a trait created by natural selection

3. The change in the genetic composition of a population over time due to random mating is called
   (a) the bottleneck effect.
   (b) gene flow.
   (c) genetic drift.
   (d) mutation.

Use the passage below to answer questions 4 & 5:

Spider monkeys are omnivorous animals that are found in tropical rain forests in many parts of the world. They commonly feed on flowers, seeds, and small insects. Spider monkeys don't have many predators, but jaguars and pumas are known to prey on them, as well as eagles and large snakes. They have fur ranging in colors from black, brown, golden, red, or tan. In a particular zoo, the small population of spider monkeys has a higher proportion of individuals with light, golden-brown fur than spider monkeys found in the wild.

4. If the monkeys were recently captured from the wild and if fur color is largely determined by genetics, what evolutionary process is at work?
   (a) the founder effect
   (b) the bottleneck effect
   (c) artificial selection
   (d) genetic drift

5. If the small zoo population of spider monkeys is later released onto an island where it can grow into a large population, which outcome is most likely?
   (a) The genetic diversity will be lower than the zoo population.
   (b) The genetic diversity will be lower than the original wild population.
   (c) The genetic diversity will be similar to the original wild population.
   (d) Genetic diversity will be higher than the original wild population.

6. Which would cause the most-rapid evolution?
   (a) artificial selection
   (b) geographic isolation
   (c) recombination
   (d) natural selection

7. The Bt-cotton is an example of
   (a) evolution through geographic isolation.
   (b) evolution through genetic modification.
   (c) evolution through natural selection.
   (d) macroevolution.

Free-Response Question

A team of researchers studying Atlantic salmon in a commonly fished area publishes their findings of the average size of five groups of fish marked and recaptured in the wild during two separate samples 10 years apart. Their results are in the table below:

	Average length (cm)	
Group	**Sample 1**	**Sample 2**
A	28.0	23.2
B	29.1	23.7
C	28.9	24.3
D	29.8	25.0
E	28.5	23.2

(Data from https://www.fws.gov)

(a) **Identify** the average length of Group C in the first sample of Atlantic salmon. (1 pt.)
(b) **Describe** the overall trend in the average length of salmon that changed over the 10 years of the study. (1 pt.)
(c) **Explain** how the body size of a fish could affect its fitness. (1 pt.)
(d) **Explain** whether this change in average length of salmon is due to artificial or natural selection. (2 pts.)

Students studying a local pond are interested to see if a change in pH affects algae growth. They collect three samples of pond water containing algae to bring back to their classroom, where they change the pH of the samples for the experiment.

(e) **Identify** the dependent variable in the students' experiment. (1 pt.)
(f) **Identify** a hypothesis for the students' experiment. (1 pt.)
(g) **Describe** a possible control for the students' experiment. (1 pt.)
(h) **Identify** an aspect of the students' experiment that would be held constant. (1 pt.)
(i) **Describe** a modification that could be made to the students' experiment that could increase the reliability of the results. (1 pt.)

Module 14

Ecological Succession

Throughout this unit, we have examined the biodiversity of ecosystems, including how species evolve and the ecosystem services they provide. However, the biodiversity of natural communities does not stay the same forever. Change in the species composition of communities over time is a perpetual natural process. In this module, we will look at how both terrestrial and aquatic communities change in their composition of species over time. We will also examine how these community changes affect species richness, biomass, and productivity of the ecosystems. Finally, we examine the special roles of some species that have disproportionally large effects on communities and other species that are indicators of human impacts on ecosystems.

Learning Goals

After reading this module you should be able to

14-1 describe how terrestrial ecosystems experience succession.

14-2 describe whether succession occurs in aquatic ecosystems.

14-3 identify how succession impacts species richness, biomass, and productivity.

14-4 explain the importance of keystone species and indicator species.

14-1 How do terrestrial ecosystems experience succession?

Terrestrial ecosystems can experience primary or secondary succession

Virtually every community experiences **ecological succession**, which is the predictable replacement of one group of species by another group of species over time. Depending on the community type, ecological succession can occur over time spans varying from weeks to centuries. In this section, we will discuss primary succession and secondary succession.

Primary Succession

Some terrestrial communities begin with bare rock and no soil. For example, a community may begin forming on newly exposed rock left behind after a glacial retreat, newly cooled lava after a volcanic eruption, or an abandoned parking lot. When succession begins with bare rock and no soil, we call it **primary succession**.

Ecological succession The predictable replacement of one group of species by another group of species over time.

Primary succession Ecological succession occurring on surfaces with bare rock and no soil.

Pioneer species In primary succession, species that can survive with little or no soil.

FIGURE 14.1 shows the process of primary succession in a temperate forest biome in New England. The bare rock can be colonized by **pioneer species** such as algae, lichens, and mosses — organisms that can survive with little or no soil. As these early-successional species grow, they excrete acids that allow them to take up nutrients directly from the rock. The resulting chemical alteration of the rock also makes it more susceptible to erosion. When the algae, lichens, and mosses die, they become the organic matter that mixes with minerals eroded from the rock to create new soil.

Over time, soil develops on the bare surface and it becomes a hospitable environment for plants with deep root systems. Mid-successional plants such as grasses and wildflowers are easily dispersed to such areas. These species are typically well adapted to exploiting open, sunny areas and are able to survive in the young, nutrient-poor soil. The lives and deaths of these mid-successional species gradually improve the quality of the soil by increasing its ability to retain nutrients and water. As a result, new species colonize the area and outcompete the mid-successional species. The type of community that

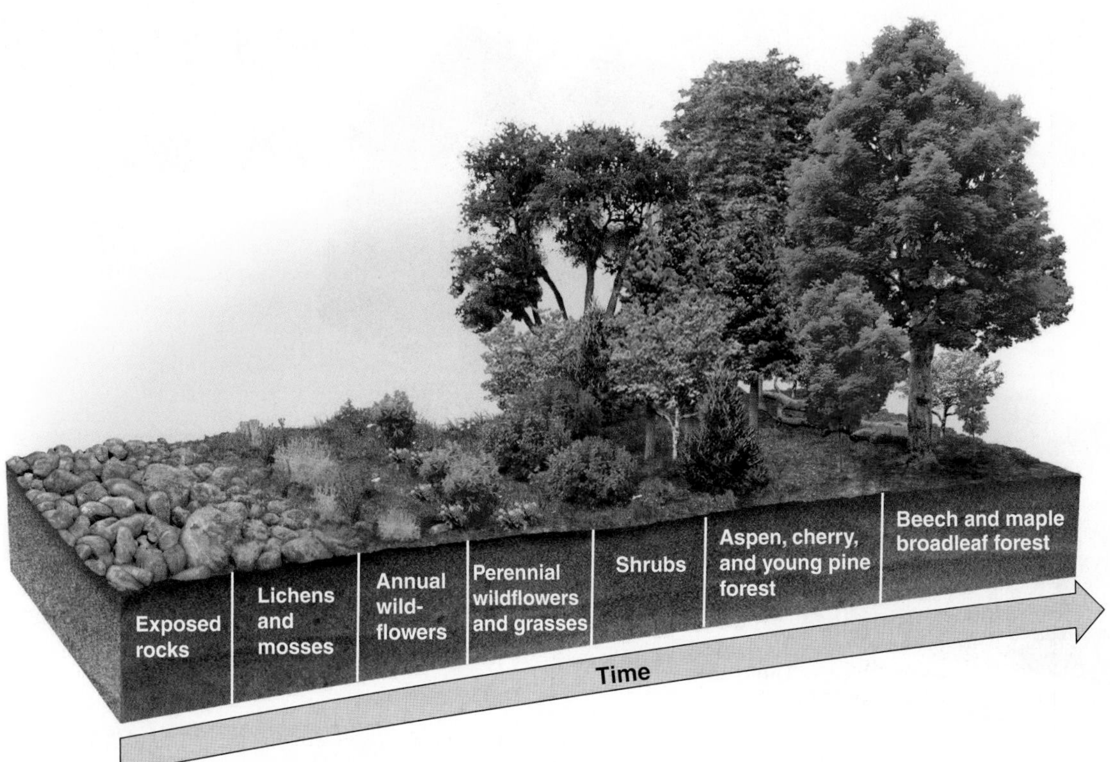

FIGURE 14.1 Primary succession. Primary succession occurs in areas devoid of soil. Early arriving plants and algae can colonize bare rock and begin to form soil, making the site more hospitable for other species to colonize later. Over time, a series of distinct communities develops. In this illustration, representing an area in New England, bare rock is initially colonized by lichens and mosses and later by grasses, shrubs, and trees.

eventually develops is determined by the temperature and rainfall of the region. In the United States, succession produces forest communities in the East, grassland communities in the Midwest, and shrubland communities or deserts in the Southwest.

Secondary Succession

Secondary succession occurs in areas that have been disturbed but have not lost their soil. Secondary succession follows an event, such as a forest fire or hurricane, that removes vegetation but leaves the soil mostly intact. Secondary succession also occurs on abandoned farm fields that have been plowed, causing all of the plants to be killed.

FIGURE 14.2 shows the process of secondary succession for a typical forest in New England. It usually begins with rapid colonization by plants that can easily disperse to the disturbed area. The first plants to arrive are typically grasses and wildflowers that have wind-borne seeds. As in primary succession, these species are eventually replaced by species that are better competitors for sunlight, water, and soil nutrients. In regions that receive sufficient rainfall, trees replace grasses and flowers. The first tree species to colonize the area are those that can disperse easily and grow rapidly. For example, the seeds of aspen trees are carried on the wind, and cherry tree seeds are carried by birds that consume the fruit and excrete the seeds on the ground. These early-succession trees can colonize new areas rapidly and grow well in full sunshine. As these trees increase in number and grow larger, however, they cause an increased amount of shade on the ground. Because they need full sunshine to grow, new seedlings of these tree species cannot persist. In contrast, species that are more shade tolerant, including many species of beech and maple, survive and grow well in the shade of the early-succession tree species. These shade-tolerant species grow up through the canopy and eventually outcompete the early-succession trees and dominate the forest community.

The process of secondary succession happens on many spatial scales, from small-scale disturbances such as a single treefall to huge areas cleared by natural disasters such as hurricanes or wildfires. The process is similar in both cases. For example, if a large tree falls and creates an opening for light to reach the forest floor, the seedlings of early-succession trees that grow rapidly in full sunlight outcompete other seedlings, and secondary succession begins again.

Secondary succession The succession of plant life that occurs in areas that have been disturbed but have not lost their soil.

FIGURE 14.2 Secondary succession. Secondary succession occurs where soil is present, but all plants have been removed. Early arriving plants set these areas on a path of secondary succession. Secondary succession in a New England forest begins with grasses and wildflowers, which are later replaced by trees.

Historically, ecologists described succession as having a final stage known as a **climax community**. In forests, for example, ecologists considered the oldest forests to be climax forests. However, it is now recognized that natural disturbances—such as fire, wind, and outbreaks of insect herbivores—and human-caused disturbances, such as logging—are a regular part of most communities. Thus, a late-successional stage is typically not final because at any moment it can be reset to an earlier stage by natural or human-caused disturbances.

14-2 Does succession occur in aquatic ecosystems?

Succession also occurs in aquatic ecosystems

Disturbances also create opportunities for succession in aquatic environments. For example, the rocky intertidal zone along the Pacific coast of North America is exposed to the air during low tide and lies under water during high tide. From time to time in this biome, major storms turn over rocks or clear their surfaces of living things. These bare rocks can then be colonized through the process of primary succession. Diatoms, which are also known as brown algae, and short-lived species of red and green algae are usually the first to arrive. Not long afterward, the rocks are colonized by barnacles and several species of long-lived red algae. This mid-successional stage develops rather quickly—in about 15 months. Finally, if the rock is not disturbed again for 3 years or more, the area may become dominated by several species of attached barnacles and mussels.

We also see succession in streams that experience major floods. When a stream floods, the fast-moving water can displace rocks and soil in ways that eliminate virtually all of the algae and invertebrate animals. However, the stream can be rapidly recolonized from unaffected upstream areas, with algae and animals traveling downstream via stream currents. In addition, many stream invertebrates are larval-stage insects that metamorphose into adult flying insects. As a result, disturbed areas of a stream can also be rapidly recolonized by adult flying insects that lay their eggs

Climax community Historically described as the final stage of succession.

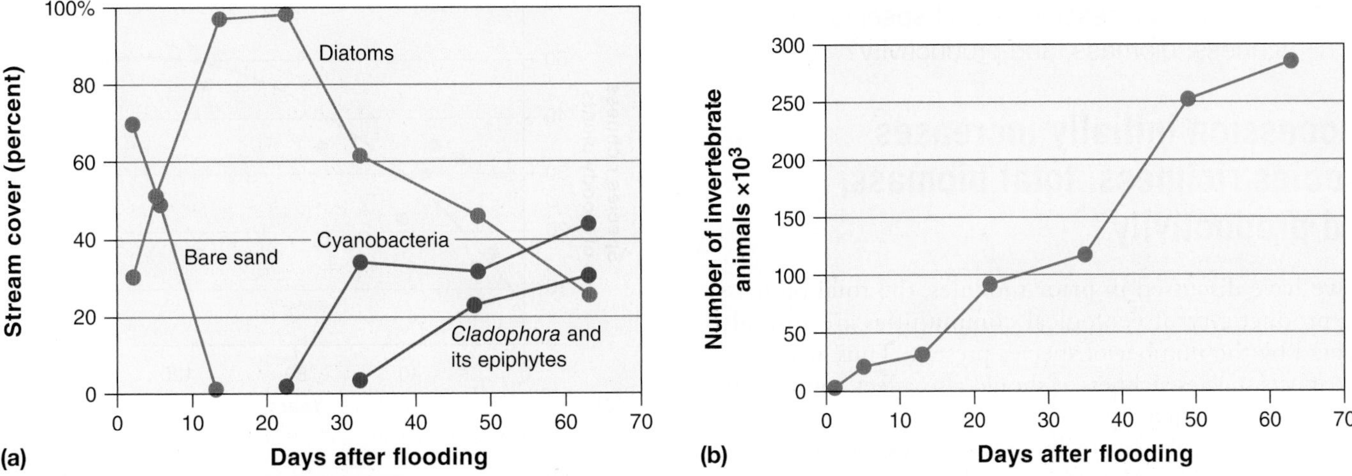

FIGURE 14.3 Succession in streams. When the Sycamore Creek in Arizona experienced a major flood, most of the algae and invertebrate animals were destroyed. (a) Monitoring of the river over the next 2 months found colonization of algae, with an initial influx of diatoms followed by cyanobacteria (i.e., blue-green algae) and green algae (i.e., Cladophora). As the stream bottom became covered in algae, the percentage of bare sand rapidly declined. (b) As the algae provided a source of food for invertebrate animals, the number of invertebrate animals quickly increased. *(Data from Fisher, S. G., et al. 1982. Temporal succession in a desert stream ecosystem following flash flooding.* Ecological Monographs *52: 93–110.)*

in the stream. We can see an example of this in Sycamore Creek in Arizona, which experienced a major flooding event that destroyed most of the organisms in the stream. When researchers monitored the creek over the subsequent 2 months, they found that the creek underwent a rapid succession of species, as you can see in **FIGURE 14.3**. Among the stream algae, diatoms arrived first, covering nearly 50 percent of the stream bottom after 5 days and more than 90 percent after 13 days. The diatoms began to then decline with the arrival of other types of algae, including cyanobacteria (i.e., blue-green algae) and green algae (i.e., Cladophora). Over the same time period, the number of invertebrate animals dramatically increased.

Shallow freshwater lakes and ponds experience a very different pattern of succession, as illustrated in **FIGURE 14.4**. Tens of thousands of years ago, a glacier may have carved out a lake or pond basin, scouring it of sediments and vegetation. Over time, algae and aquatic plants colonized the water body. The growth of plants and algae and the erosion of surrounding rock and soil slowly filled the basin with sediment and organic matter, making it increasingly shallow. This process, which may have taken hundreds or even thousands of years, eventually filled in the basin completely and made it a terrestrial habitat.

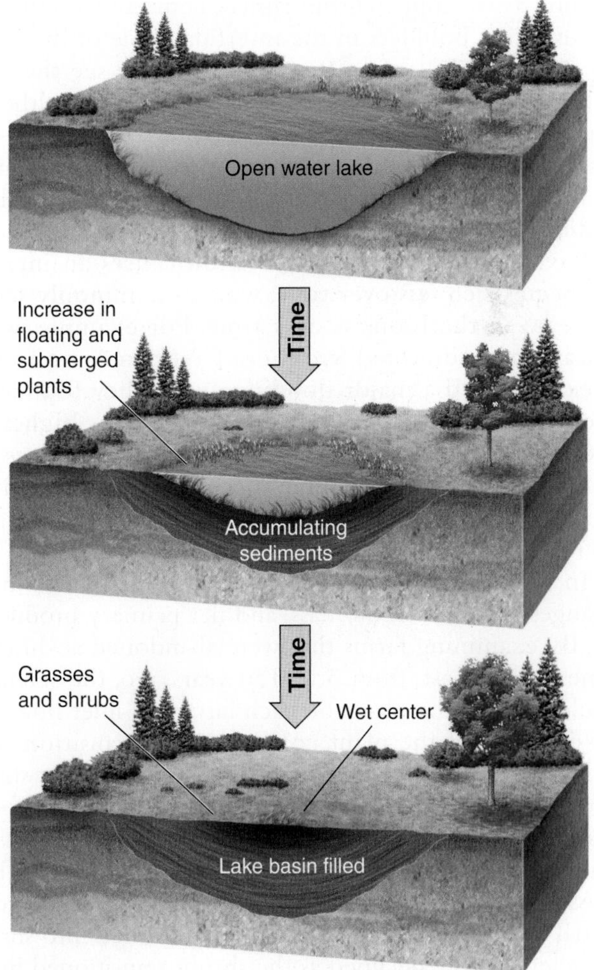

FIGURE 14.4 Succession in shallow lakes and ponds. Over a time span of hundreds to thousands of years, lakes and ponds are filled with sediments and slowly become terrestrial habitats.

14-3 How does succession impact species richness, biomass, and productivity?

Succession initially increases species richness, total biomass, and productivity

As we have discussed in prior modules, the total biomass and productivity of ecological communities are strongly affected by the number of species present. Thus, if succession affects species richness, it should also affect community biomass and productivity.

To examine patterns in species richness, we can look into the research results from a wide range of organisms and biomes, as shown in **FIGURE 14.5**. In all cases, we start with very few species. For example, if we examine the number of woody plant species that are present in abandoned farm fields in North Carolina and experiencing succession (Figure 14.5a), we see a continual increase over 100 years that begins to plateau. If we observe the number of bird species present in abandoned farm fields in Georgia (Figure 14.5b), we once again see an increase over 100 years, which plateaus and then shows a small decline as we approach the climax community. Finally, if we examine boulders in the intertidal zone of the ocean shoreline in California (Figure 14.5c), we see the same pattern of increasing species richness over time, although the process occurs much faster for these organisms—in less than 2 years—due to the much faster generation times of algae and invertebrates compared to woody plants and birds.

Given that ecological succession causes an increase in species richness over time, we also commonly see an increase in the biomass over time. For example, when researchers monitored Sycamore Creek in Arizona after it experienced a major flooding event, they found that the biomass increase over time was 3 times higher for the algae and nearly 10 times higher for the invertebrate animals. Thus, the large increase in species richness as the stream underwent succession also led to a large increase in biomass.

In the forests of China, researchers have examined changes in richness, biomass, and net primary productivity. By examining farms that were abandoned at different times in the past, from 3 to 150 years ago, the scientists could observe the plants on each farm and infer how succession caused the plant community to transition from fields dominated by grasses to fields dominated by shrubs and then trees. As you can see in **FIGURE 14.6**, they found that as the fields transitioned from grasses to forests, there was an increase in species richness and an increase in biomass. However, the productivity of the areas exhibited an initial large increase as the grasses transitioned into shrubs, but then sharply declined as the shrubs transitioned into a

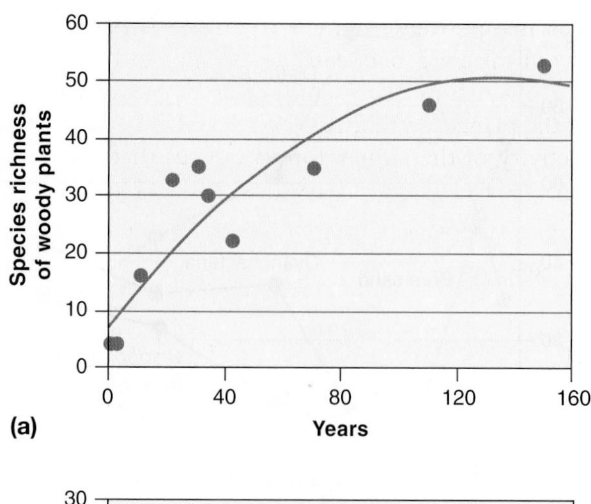

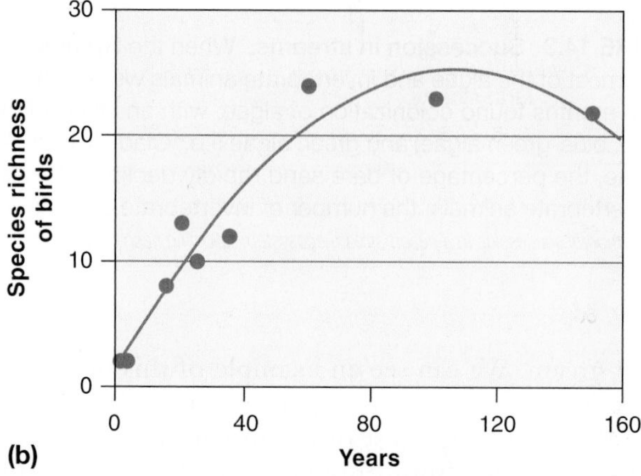

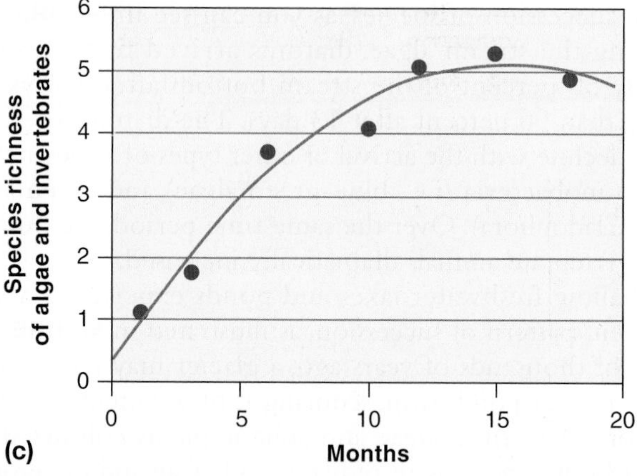

FIGURE 14.5 Effects of succession on species richness. Across many different communities, we commonly see a pattern of ecological succession causing an increase in species richness that can plateau and even have a small decline. The examples include (a) woody plant species growing in abandoned farm fields in North Carolina, (b) bird species in abandoned farm fields in Georgia, and (c) algae and invertebrates from the intertidal boulders of southern California. *(Data from (a) Oosting, H. J. 1942. An ecological analysis of the plant communities of Piedmont, North Carolina.* American Midland Naturalist *28: 1–126; (b) Johnston, D. W., and E. P. Odum. 1956. Breeding bird populations in relation to plant succession on the Piedmont of Georgia.* Ecology *37: 50–62; (c) Sousa, W. P. 1979. Disturbance in marine intertidal boulder fields: The nonequilibrium maintenance of species diversity.* Ecology *60: 1225–1239.)*

forest of mature trees. As we discussed in Module 6, this is why we distinguish between the amount of biomass present at a given time (i.e., standing crop) versus the rate at which the plants add biomass each year. In this case, the low productivity of the climax forests tells us that these forests add a small amount of new biomass each year, but the total biomass that they accumulate as large trees over a century (i.e., their standing crop) is quite large.

Species richness
80
60
40
20
0
Grasslands Shrubs Forest
3–5 15–25 30–40 50–60 80–100 80–100 150+
Forest age (Years)

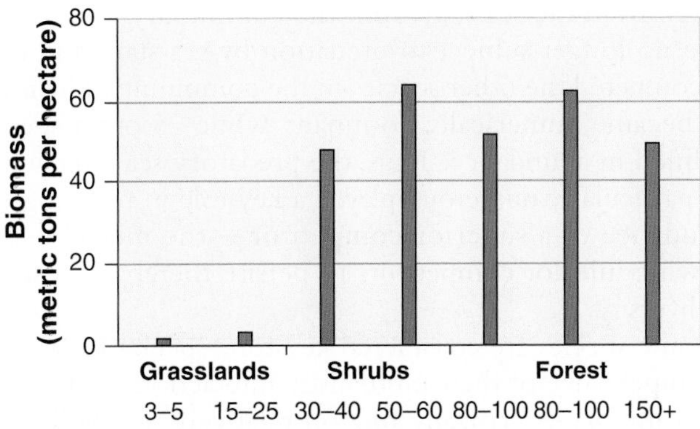

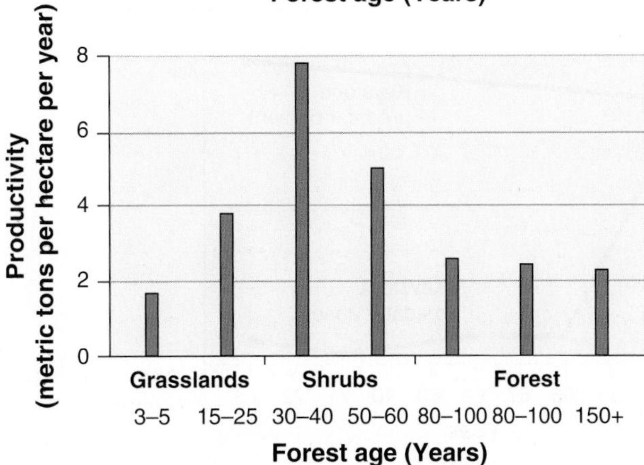

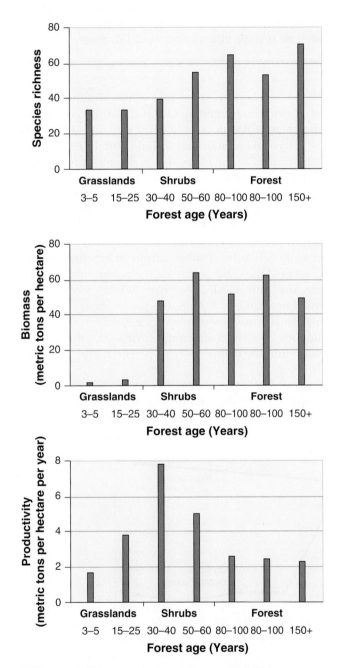

FIGURE 14.6 **Effects of succession on species richness, biomass, and net primary productivity.** By examining farm fields that had been abandoned at different times in the past, research could examine how the community changed. They observed (a) an increase in species richness, (b) an increase in biomass, and (c) an initial increase in productivity in the shrub stages, followed by a decline in the climax forest. *(Data from Wang, K., R. Shao, and Z. Shangguan. 2020. Changes in species richness and community productivity during succession on the Loess Plateau (China).* Polish Journal of Ecology *58: 549–558.)*

14-4 What is the importance of keystone species and indicator species?

Keystone species alter species composition while indicator species indicate ecosystem characteristics

As ecological communities undergo succession, they can contain certain species that play important roles. In this section, we will examine the roles of keystone species and indicator species.

Keystone Species

Some species that are not abundant can have very large effects on a community. The beaver is a prime example. Although they make up only a small percentage of the total biomass of the North American forest, beavers play a critical role in the forest community. They build dams that convert narrow streams into large ponds, thereby creating new habitat for pond-adapted plants and animals (**FIGURE 14.7**). These ponds also flood many hectares of forest, causing the trees to die and creating habitat for animals that rely on dead trees. Several species of woodpeckers and some species

FIGURE 14.7 **Beavers.** Beavers are an important species to the community because of the role they play in creating new pond and wetland habitat. For a species that is not particularly abundant, the beaver has a strong influence on the presence of other organisms in the community. *(a: Thomas & Pat Leeson/Science Source; b: Troy Harrison/Getty Images)*

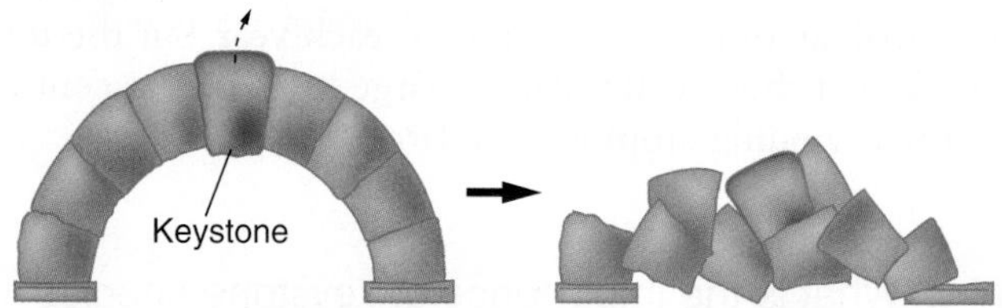

FIGURE 14.8 Keystone. Keystone species get their name from the keystone of an arch. Without the keystone in place, the arch would fall apart.

AP® Exam Tip

You may be asked to describe how a keystone species affects an ecosystem. Most students will answer that a keystone species has a significant role within the ecosystem. While this is correct, the response in incomplete. A complete description should note that keystone species play a more significant role in their ecosystems than their relative abundance would suggest.

of ducks make their nests in cavities that are carved into the dead trees. In short, one beaver can have a major impact on the entire community of plants and animals.

The beaver is an example of a **keystone species**, which is a species that is not very abundant but has large effects on an ecological community. The name keystone species is a metaphor that comes from architecture. As **FIGURE 14.8** shows, in a stone arch, the keystone is the single center stone that supports all the other stones. Without the keystone, the arch would collapse. Typically, the most abundant species and the major energy producers, while vital to the health of a community, are not keystone species. Keystone species typically exist in low numbers. They may be predators, prey, mutualistic species, or providers of some other essential service such as an important habitat. For example, alligators play an important role in their communities by digging deep "gator holes" in summer. These gator holes serve as critical sources of water for many other animals during dry months.

Keystone species A species that is not very abundant but has large effects on an ecological community.

The role of keystone predators was well demonstrated in a classic experiment conducted in intertidal communities off the coast of Washington State, as shown in **FIGURE 14.9**. These intertidal communities include mobile animal species such as sea stars and snails as well as dozens of other species that attach themselves to rocks, including mussels, barnacles, and algae. When sea stars are present, they prey on mussels (*Mytilus californianus*). This predation continually clears space where other species can attach to the rocks. When researchers removed the sea stars from the community, the mussels were no longer subject to predation by sea stars, and they outcompeted the other species in the community. The mussels became numerically dominant, while 25 other species declined in abundance. Thus, the predatory sea stars, while not particularly numerous, played a key role in reducing the abundance of a superior competitor—the mussel—and allowing inferior competitors to persist, thereby increasing biodiversity.

Some species are considered keystone species because of the importance of their mutualistic interactions. For example, some plants critically rely on relatively rare pollinator

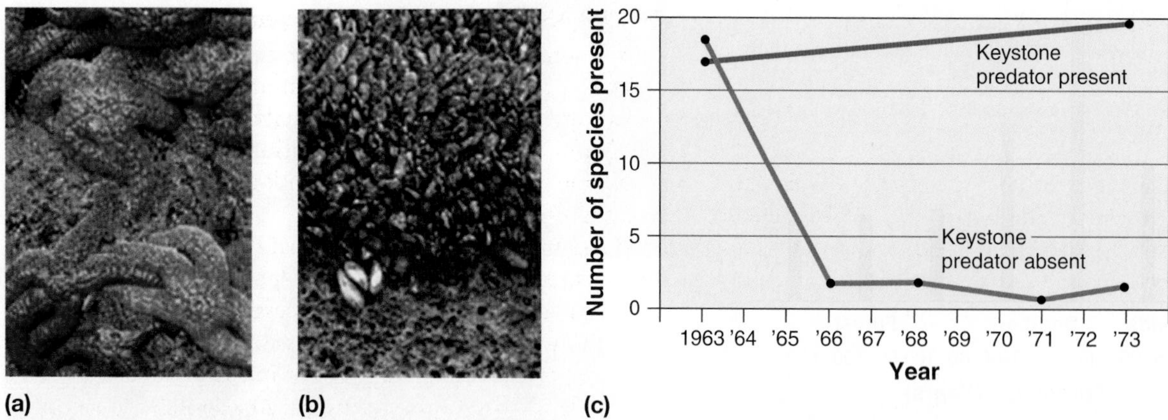

FIGURE 14.9 Keystone predators. (a) Sea stars are keystone predators in their rocky intertidal communities in Washington State. When sea stars are present, they consume mussels, which are strong competitors for space. This predation creates open spaces that inferior competitors can colonize. As a result, the diversity of species is high. (b) In the absence of sea stars, the mussels dominate the surfaces of the intertidal rocks, and the diversity of species declines dramatically. (c) Evidence of the role sea stars play in intertidal communities was discovered in an experiment in which the sea stars were allowed to be present in intertidal communities or were removed. *(a: blueeyes/Shutterstock; b: Alex L. Fradkin/Getty Images; c: Data from R. T. Paine, Intertidal community structure: Experimental studies on the relationship between a dominant competitor and its principal predator,* Oecologia *15 (1974): 93–120.)*

FIGURE 14.10 Indicator species. Indicator species indicate particular characteristics of an ecosystem. (a) Larval mayflies living in streams can indicate water that is uncontaminated by pollutants. (b) Red-cockaded woodpeckers are indicators of healthy old-growth forests of longleaf pine in the southeastern United States. *(a: Nature Picture Library/Alamy Stock Photo; b: Brady Beck Photography)*

species, which makes these pollinators a keystone species. On many South Pacific islands, a species of bat known as the flying fox (*Pteropus vampyrus*) is the only pollinator and seed disperser for hundreds of tropical plant species. Unfortunately, flying foxes have been hunted for food to near-extinction. Without the pollinating and seed-dispersing functions of the flying foxes, many plant species may become extinct, dramatically changing these island communities.

Certain soil fungi are another group of mutualists that serve as keystone species. These fungi are found on and in the roots of many plant species, where they increase the plants' ability to extract nutrients from the soil. As a result, they play a critical role in the growth of plant species, which in turn provide habitat and resources for other members of a forest or field community.

Indicator species

Sometimes we have a species that demonstrates a particular characteristic of an ecosystem, which we call **indicator species**. Indicator species are often used to quickly characterize when ecosystems have been negatively impacted by humans or when ecosystems are rebounding from some harmful impact (**FIGURE 14.10**). For instance, scientists monitor streams to determine how many species of mayflies are present. Certain species of mayflies are particularly sensitive to pollutants in streams, so finding a low number of mayfly species is an indicator of stream pollution. Similarly, the presence of red-cockaded woodpeckers (*Dryobates borealis*) in the southeastern United States is an indicator of a healthy old-growth forest of longleaf pine trees, since they prefer such trees for nesting. Even lichens growing attached to trees can be indicator species; they are particularly sensitive to air pollution, so researchers can indirectly assess air pollution by examining where lichens have declined in abundance.

Indicator species can also be important for suggesting the presence of harmful pathogens. For example, a number of pathogens can be present in drinking water that is contaminated by improperly treated sewage. Thus, we need to test for the presence of pathogens in water bodies or well water for drinking. However, it is not feasible to test for all of the many different pathogens that can exist in contaminated water. Instead, scientists have settled on using a species that indicates whether or not disease-causing pathogens are likely to be present. A good indicator for water that potentially contains pathogens from sewage is *Escherichia coli* (known more commonly as *E. coli*). Most strains of *E. coli* live naturally in humans and are not harmful, although certain strains can be deadly to people who are very young, very old, or possess weak immune systems. Given that *E. coli* is commonly found in human intestines, detecting *E. coli* in water is a reliable indicator that human waste has entered the water. This does not necessarily mean that the water is harmful for drinking or recreation, but the presence of *E. coli* does indicate an increased risk that wastewater pathogens are also in the water.

In this module, we have seen that ecosystems can experience dramatic transitions in their species composition over time as a result of ecological succession in both terrestrial and aquatic environments. Succession not only alters species composition and richness, but it also alters the biomass and productivity of ecosystems. It can also include keystone species and indicator species that play important roles in ecosystems. In the closing story for this unit, we will examine how conservation groups are developing partnerships to protect biodiversity while making use of ecosystem services in a sustainable way.

Indicator species A species that demonstrates a particular characteristic of an ecosystem.

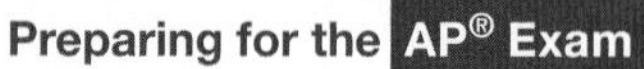

Visual Representation 2 **Biodiversity in the Galápagos Islands:** Biodiversity refers to genetic, species, and habitat diversity. At each of these levels, biodiversity is a good indicator of environmental health and a source of ecosystem services. The Galápagos Islands, located about 600 km west of South

Natural Disruptions
The Islands originated from volcanoes. Natural disruptions such as volcanic activity and weather pattern changes have affected ecosystems on the Galápagos.

Succession
Bare rock is succeeded by pioneer species such as lichen, lava cactus, and carpetweed. Over hundreds or thousands of years, these plants die and create soil conditions in which small herbaceous plants such as common sow thistle and grasses such as Mexican lovegrass can grow. Eventually these are succeeded by shrubs and small trees such as leatherleaf and manchineel. Finally, in higher, humid zones a climax forest can consist of a combination of plants including the giant daisy trees, which are endemic to the Galápagos.

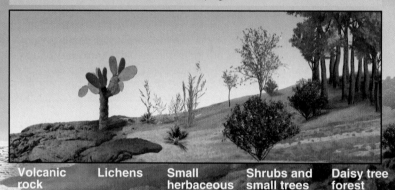

Volcanic rock
Lichens
Small herbaceous plants
Shrubs and small trees
Daisy tree forest

Blue-footed booby
Sea lion
Flightless cormorant

Adaptations
Over time, organisms evolve adaptations that help them survive and reproduce in a specific environment. For example, the marine iguana, the only sea-going lizard in the world, has adaptations that allow it to dive into the ocean for algae and seaweed and to expel excess salt. The flightless cormorant evolved with reduced wings and adaptations that allow it to dive to the ocean floor to hunt for fish.

Marine iguana

Island Biogeography
Over 600 km from the mainland in South America, the Galápagos Islands were initially populated by plants and animals that could reach them by floating, swimming, flying, being blown, or carried. Some initial colonizers evolved over time to take advantage of available resources, making the islands rich in biodiversity with many species found nowhere else on Earth.

America, are a hotspot of biodiversity. When Charles Darwin explored the Galápagos Islands, he found evidence for his theory of evolution all around him. Over millions of years, many new species have evolved on the islands, each with unique adaptations that allow them to live and reproduce.

Invasive Species
Some species brought to the islands by humans have become invasive and threaten native plants and animals. Among these species are goats and blackberries.

Evolution
Genetic variation is at the heart of biodiversity and makes evolution possible. The birds known as "Darwin's finches" are all descendants of one original finch. Over many years finches evolved adaptations to fill different ecological niches on the islands. For example, the large ground finch has a bill adapted for picking and crushing seeds. In contrast, the woodpecker finch has a sharp bill adapted for extracting and eating insects. The vegetarian finch has evolved adaptations to pull buds off trees. Darwin's finches continue to evolve today.

Module 14 AP® Review

Preparing for the AP® Exam

Learning Goals Revisited

14-1 How do terrestrial ecosystems experience succession?

In terrestrial communities, an area that begins with no soil undergoes primary succession whereas an area that begins with soil undergoes secondary succession.

14-2 Does succession occur in aquatic ecosystems?

In aquatic ecosystems, streams and intertidal habitats that are disturbed can become quickly recolonized by algae and invertebrate animals. Shallow lakes and ponds can slowly fill in with sediments over thousands of years to eventually become terrestrial habitats.

14-3 How does succession impact species richness, biomass, and productivity?

As ecological succession proceeds, we generally see an increase in species richness that plateaus or slightly declines, an increase in biomass that plateaus, and an initial increase in productivity that subsequently declines.

14-4 What is the importance of keystone species and indicator species?

Keystone species are not very abundant, but they have large effects on an ecological community as predators, sources of food, mutualistic species, or providers of other essential services such as an important habitat. Indicator species indicate characteristics of the ecosystem; they are often used to quickly characterize when ecosystems have been impacted by humans.

AP® Practice Questions

Preparing for the AP® Exam

Multiple-Choice Questions

Use the passage below to answer questions 1–3:

A newly-formed volcanic island in the Pacific Ocean is covered with freshly cooled lava. After several months, researchers observe that algae and lichens are growing on the lava rocks. After several more years, the researchers find that there are now larger populations of lichens as well as mosses growing on the rocks, and the rocks have begun to break apart. The researchers concluded that primary succession is occurring on the island.

1. According to the passage, which species are the first to colonize the freshly cooled lava?
   (a) mosses & algae
   (b) trees & shrubs
   (c) lichens & algae
   (d) grasses & lichens

2. According to the passage, what can be concluded about the species richness of the volcanic island over time?
   (a) After several years, the species richness of the new island is the same as it was when it was formed.
   (b) After several months, the species richness of the new island is decreasing.
   (c) After several months, the species richness of the new island is at its maximum.
   (d) After several years, the species richness of the new island has increased to include species of lichens, mosses, and algae.

3. According to the passage, what evidence do the researchers have to reach their conclusion?
   (a) Primary succession starts with bare soil.
   (b) Primary succession causes the number of species to initially decrease.
   (c) Primary succession occurs after forest fires.
   (d) Primary succession begins with colonization by algae, lichens, and mosses.

4. The process of succession in a shallow lake
   (a) results in a terrestrial ecosystem.
   (b) occurs rarely because disturbances are rare.
   (c) progresses fastest in very deep lakes.
   (d) has no climax species.

5. Which tree species is an early-succession species in North American forests?
   (a) beech
   (b) fir
   (c) aspen
   (d) maple

6. Which of the following best describes an environmental problem associated with a sudden decrease in the population of a keystone species in an ecosystem?
   (a) A sudden decrease in population of a keystone species is always caused by human activities.
   (b) The loss of a keystone species will cause a sharp increase in the primary production in an ecosystem.
   (c) A sudden decrease in the population of a keystone species will result in increased gene flow in a population.
   (d) The loss of a keystone species can have a cascading effect on the populations of other species in its food web.

7. Which describes an indicator species?
   (a) a species that has a large effect on other species in the community
   (b) a species that suggests the presence of human sewage
   (c) a species that serves as a pioneer species in succession
   (d) a species that serves as a climax species in succession

Use the diagram below to answer questions 8 & 9:

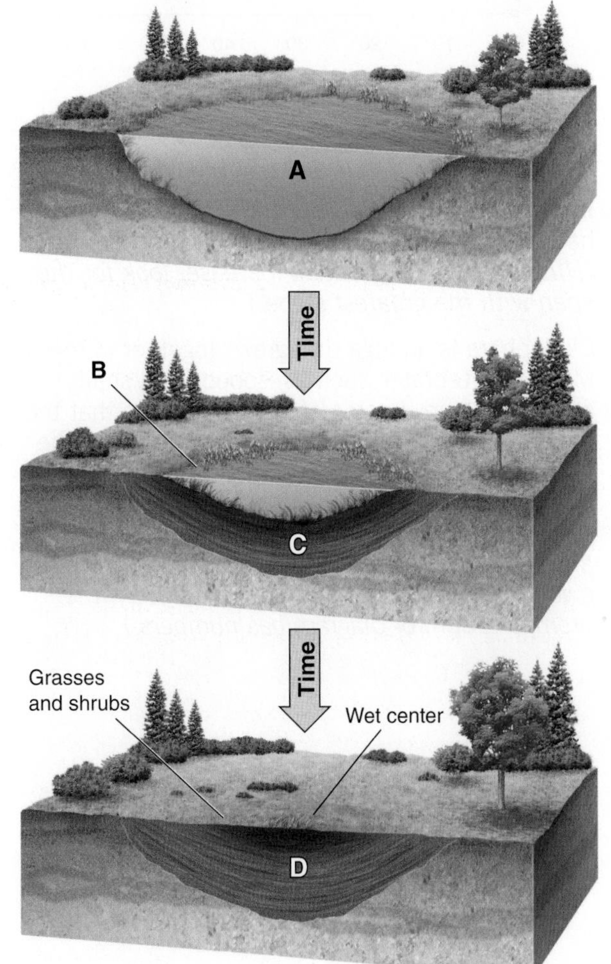

8. Which letter in the diagram represents the latest stage of aquatic succession?
   (a) A
   (b) B
   (c) C
   (d) D

9. Which of the following best explains the progression of events depicted in the diagram?
   (a) The process of succession in an aquatic ecosystem can occur over thousands of years as sediment slowly fills in a body of freshwater.
   (b) The process of primary succession occurs in a freshwater ecosystem after sediment has filled in a body of water.
   (c) Aquatic succession takes place over millions of years and gives rise to new aquatic species.
   (d) Aquatic succession takes place more quickly than terrestrial succession.

Free-Response Question

Many bats in the state of Pennsylvania play a critical part in maintaining the state's ecosystems. As the primary predators of night-flying insects, bats play a significant role in controlling insect populations. As such, they can be considered keystone species for their enormous economic value as biological control agents of insects, valued in the multi-millions of dollars each year.

(a) **Identify** the trophic level of these bats. (1 pt.)
(b) **Describe** how these bats can be considered a keystone species. (1 pt.)
(c) **Explain** how this ecosystem in Pennsylvania might be affected if a disease suddenly decreased the bat population. (2 pts.)

Researchers estimate that the loss of bats would raise agriculture costs by about \$75 per acre, in terms of the pesticides needed to replace them. In 2014, this was estimated to cost the state of Pennsylvania \$300 million.

(d) **Calculate** the number of acres of farmland affected by the loss of bats. Show all work. (2 pts.)
(e) **Calculate** the cost per hectare of farmland, if there are approximately 0.4 acres in one hectare. Show all work. (2 pts.)
(f) In 2014, the population of Pennsylvania was 12.79 million people. **Calculate** the cost per capita in Pennsylvania of the pesticides needed to replace the pest control service of the bats. Show all work. (2 pts.)

Preparing for the AP® Exam

UNIT 2
Practice Your Science Skills

Data Analysis

These graphs represent the recovery of a stream after flooding, which is a type of ecological disturbance. To answer these questions, you will need to carefully read the graph and interpret the data, use mathematical calculations, and relate the material to your knowledge of food chains from Unit 1.

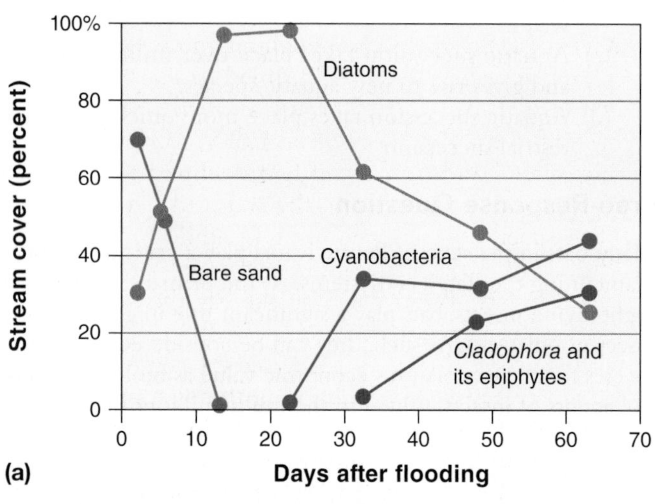

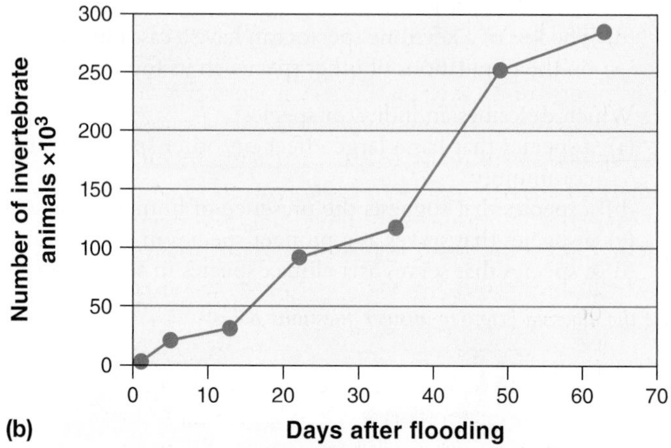

Questions

1. **Identify** how long after flooding cyanobacteria took to begin to repopulate a stream after flooding. *(Hint: The AP Exam will have an acceptable range of correct answers such as ±2 on either end.)*
2. Diatoms are microscopic producers. **Describe** the pattern in the population size of diatoms from day 2 to day 60 after flooding. *(Hint: When describing the pattern in a graph, use numbers in your description and be precise. The AP exam will have an acceptable range of correct numbers for each data point.)*
3. **Calculate** the percent change in the number of invertebrate animals from 10 days after flooding to 30 days after flooding. *(Hint: Be sure to set up your problem to earn points. You need to memorize the percentage change formula as the AP Exam does not provide a formula sheet.)*
4. **Identify** the timespan (10-day timespans) that had the greatest increase in invertebrate animals. *(Hint: To find the greatest increase, look for the span with the greatest slope.)*
5. Cladophora is an alga that grows together in mass. Many invertebrates such as isopods and snails eat Cladophora. People often hypothesize that the increase in invertebrate animals is caused by the repopulation and increase of Cladophora. **Make a claim** about whether this hypothesis is supported or not supported by the data. *(Hint: In "make a claim" questions, back up your claim with evidence. Use numbers in your evidence if the prompt, graph, or diagram has numbers.)*

Pursuing Environmental Solutions

Saving Marine Biodiversity Through the Power of Partnerships

For over 70 years, The Nature Conservancy (TNC) has protected biodiversity by using a simple strategy: Buy it. The Conservancy uses grants and donations to purchase privately owned natural areas or to buy development rights to those areas. TNC has protected over 50 million hectares (125 million acres) of land by either buying land or buying the development rights of the land. As a nonprofit, nongovernmental organization, TNC has great flexibility to use innovative conservation and restoration techniques on natural areas in its possession.

TNC focuses its efforts on areas containing rare species or biodiversity hotspots, including the Florida Keys in southern Florida and Santa Cruz Island in California. Recently, it has set its sights on the oceans, including coastal marine ecosystems. Coastal ecosystems have experienced steep declines in the populations of many fish and shellfish, including oysters, clams, and mussels, due to a combination of overharvesting and pollution. By preserving these coastal ecosystems, TNC hopes to create reserves that will serve as breeding grounds for declining populations of overharvested species. In this way, protecting a relatively small area of ocean will benefit much larger unprotected areas, and even benefit the very industries that have led to the population declines.

Shellfish are particularly valuable in many coastal ecosystems because they are filter feeders: They remove tiny organisms, including algae, from large quantities of water, cleaning the water in the process and helping to prevent harmful algal blooms. However, shellfish worldwide have been harvested unsustainably, leading to a cascade of effects throughout many coastal regions. For example, oyster populations in the Chesapeake Bay were once sufficient to filter the water of the entire bay in 3 to 6 days. Now there are so few oysters that it would take a year for them to filter the same amount of water. As a result, the bay has become much murkier, and excessive algae have led to lowered oxygen levels that make the bay less hospitable to fish.

Conserving marine ecosystems is particularly challenging because private ownership is rare. State and federal governments generally do not sell areas of the ocean. Instead, they have allowed industries to lease the harvesting or exploitation rights to marine resources such as oil, shellfish, and physical space for marinas and aquaculture. So how can a conservation group protect coastal ecosystems if it cannot buy an area of the ocean to protect it? The Nature Conservancy's strategy is to purchase harvesting and exploitation rights and use them as a conservation tool. In some cases, TNC will not harvest any shellfish in order to allow the populations to rebound. In other cases, the leases require at least some harvesting, so TNC has worked to demonstrate sustainable management practices to serve as an example of how shellfish harvests can be conducted while restoring the shellfish beds.

In New York's Great South Bay, along the southern coast of Long Island, there was a long history of harvesting large numbers of oysters and clams. Beginning in the 1950s, however, overharvesting these shellfish led to sharp declines in the shellfish and this led to the loss of thousands of jobs connected to the industry. By the 1990s, commercial shellfishing ceased and harvesting had declined to 1 percent of what it once was. Despite the reduced harvesting, the shellfish species did not rebound. In addition, with fewer shellfish to filter the algae out of the water combined with increased inputs of nutrients into the bay, the area began to experience harmful algal blooms.

From 2002 to 2004, TNC acquired the rights to 5,420 ha (13,000 acres) of oyster beds along the southern shore of Long Island. These rights, which were donated to TNC by the Blue Fields Oyster Company, were valued at $2 million. TNC worked with local governments, community members, and the shellfish industry to develop restoration strategies, including reducing the nutrient inputs, updating harvesting plans, and seeding the area with millions of

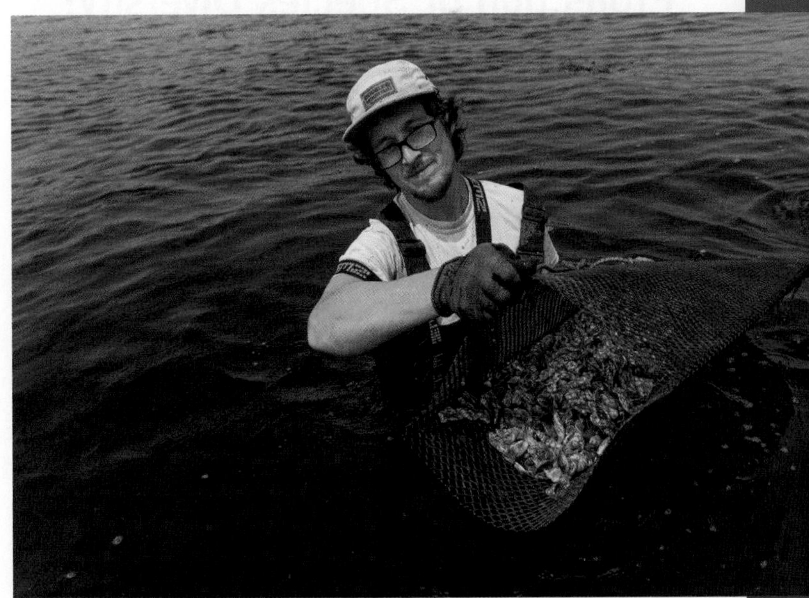

Restoring ocean populations. In areas that have declining populations of shellfish, such as this site in New Hampshire, researchers seed the area with additional individuals to help speed up the recovery of the populations. *(Joe Klementovich)*

small clams to give the population a jump start. Within a few years, the populations of shellfish began to show signs of recovery, yet the bay remains challenged by the excess inputs of nutrients. If the nutrient problem can be remedied, TNC hopes to engage in sustainable harvesting over part of this area and conduct research in the rest of it.

TNC has similar projects under way off the coasts of Virginia, North Carolina, and Washington State. In California, TNC has accumulated the rights to 10,000 ha (25,000 acres) of marine fisheries along the coast. Today, it has developed a collaborative program with local commercial fishermen to collectively harvest fish and shellfish using sustainable practices, which includes reporting their catches on an iPad app to help the TNC and its partners better assess fish abundance and fishing pressure. These fishermen can also post any information about areas of the ocean where they accidentally catch protected species, which helps other fishing boats avoid these areas. One result is that the amount of unintended species caught, known as "bycatch," has declined from between 15 and 20 percent to only 1 percent. In making these changes, TNC has found effective ways to continue the tradition of fishing that provides jobs to the local communities while working to ensure that the fish populations persist long into the future.

Critical Thinking Questions

1. What are the advantages of working to increase fish populations by leasing areas of the ocean and banning fishing in those areas versus leasing areas and working with local commercial fishing boats to change fishing practices?
2. What are some of the challenges in getting commercial fishing operations to change their practices?

References

California Fishermen, Once Blocked by Conservationists, Now Work with Them. *Market Place,* June 19, 2017. https://www.marketplace.org/2017/06/19/california-fishermen-once-blocked-conservationists-now-work-with-them/; Groundfish comeback brings to a close unprecedented collaboration between The Nature Conservancy, fishermen, and California coastal communities. The Nature Conservancy. September 19, 2019. https://www.nature.org/en-us/newsroom/groundfish-comeback-closes-unprecedented-collaboration/; The Nature Conservancy. 2002. *Leasing and Restoration of Submerged Lands: Strategies for Community-Based, Watershed-Scale Conservation*; Partnership Preserves Livelihoods and Fish Stocks. *New York Times,* November 27, 2011. https://www.nytimes.com/2011/11/28/science/earth/nature-conservancy-partners-with-california-fishermen.html?_r%EF%80%BD0.

Science Applied 2: Concept Explanation

How Should We Prioritize the Protection of Species Diversity?

As a result of human activities, we have seen a widespread decline in biodiversity across the globe. Many people agree that we should try to slow or even stop this loss. But how do we do this? While we might want to preserve all biodiversity, preserving biodiversity requires compromises. For example, in order to preserve the biodiversity of an area, we might have to set aside land that would otherwise be used for housing, shopping malls, or strip mines. If we cannot preserve all biodiversity, how do we decide which species receive our attention? (**FIGURE SA2.1**)

In 1988, Oxford University professor Norman Myers noted that much of the world's biodiversity is concentrated in areas that make up a relatively small fraction of the globe. Part of the reason for this uneven pattern of biodiversity is

FIGURE SA2.1 The California tiger salamander (*Ambystoma californiense*). This salamander is endemic to the biodiversity hotspot of California and is threatened with extinction due to habitat destruction and the introduction of non-native predators. *(US Air Force Photo/Alamy Stock Photo)*

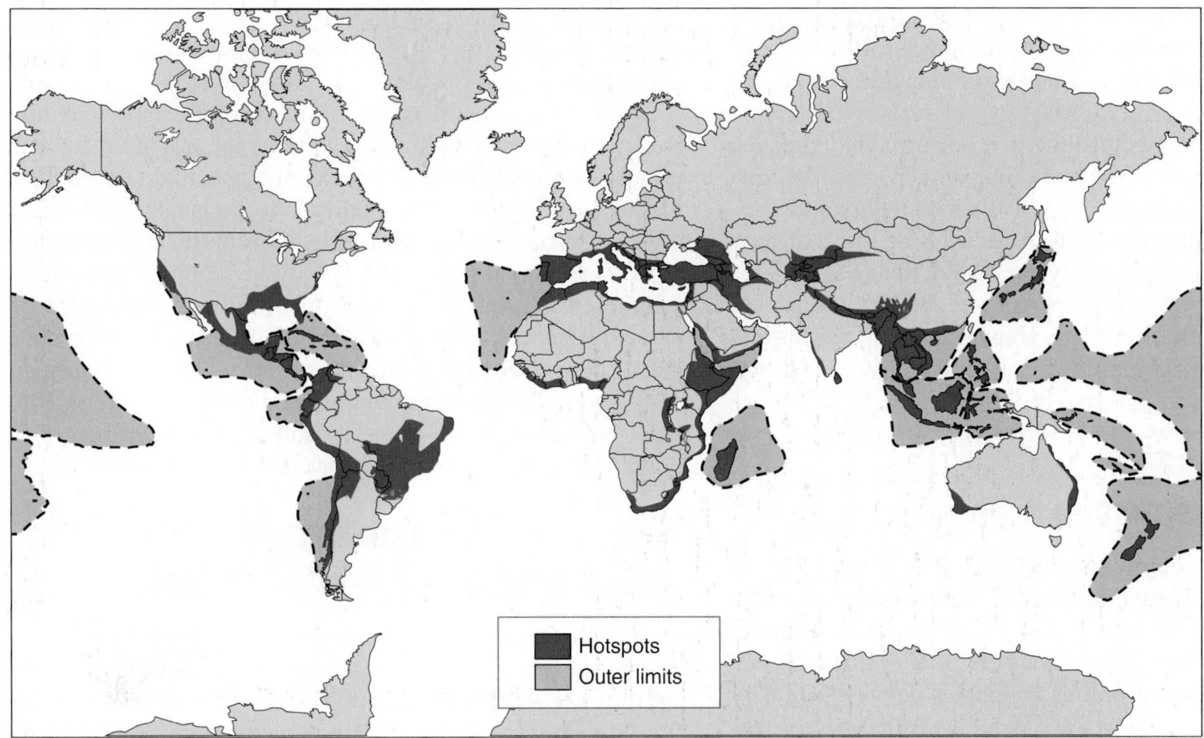

FIGURE SA2.2 Biodiversity hotspots. Conservation International has identified 34 biodiversity hotspots (shown in red) that have at least 1,500 endemic plant species and a loss of at least 70 percent of all vegetation. *(Data from Conservation International at https://www.conservation.org/priorities/biodiversity-hotspots.)*

that so many species are *endemic species*. **Endemic species** are those that live in a very small area of the world and nowhere else, often in isolated locations such as the Hawaiian Islands. Because these isolated areas are home to so many endemic species, they end up containing a high proportion of all the species found on Earth. Myers called these areas **biodiversity hotspots**.

Scientists originally identified 10 biodiversity hotspots, including Madagascar, western Ecuador, and the Philippines. Myers argued that these 10 areas needed immediate conservation attention because human activities there could have disproportionately large negative effects on the world's biodiversity. A year later, the group Conservation International adopted Myers's concept of biodiversity hotspots to guide its conservation priorities. As of 2022, Conservation International had identified the 36 biodiversity hotspots shown in **FIGURE SA2.2**. Although these hotspots collectively represent only 2.3 percent of the world's land area, more than 50 percent of all plant species and 42 percent of all vertebrate species are confined to these areas. As a result of this categorization, major conservation organizations have adjusted their funding priorities and are spending hundreds of millions of dollars to conserve these areas. What does environmental science tell us about the hotspot approach to conserving biodiversity?

What makes a hotspot hot?

Since Norman Myers initiated the idea of biodiversity hotspots, scientists have debated which factors should be considered most important when deciding where to focus conservation efforts. For example, most scientists agree that species richness is an important factor. For example, there are more than 1,300 bird species in the small nation of Ecuador—more than twice the number of bird species living in the United States and Canada. For this reason, protecting a habitat in Ecuador has the potential to save many more bird species than protecting the same amount of habitat in the United States and Canada. From this point of view, the choice to protect areas with a lot of species makes sense.

Identifying biodiversity hotspots is challenging, however, because scientists have not yet discovered and identified all the species on Earth. Because the distribution of plants is

Endemic species Species that live in a very small area of the world and nowhere else, often in isolated locations such as the Hawaiian Islands.

Biodiversity hotspots Isolated areas that are home to so many endemic species, that they contain a high proportion of all the species found on Earth.

typically much better known than that of animals, the most practical way to identify hotspots has been to locate areas containing high numbers of endemic plant species. Conservation International considers two criteria when determining whether an area qualifies as a hotspot. First, the area must contain at least 1,500 endemic plant species. By conserving plant diversity, the hope is that we will simultaneously conserve animal diversity, especially for those groups, such as insects, that are poorly cataloged. Second, the area must have lost more than 70 percent of the vegetation that contains those endemic plant species. In this way, high-diversity areas with a high level of habitat loss receive the highest conservation priority. High-diversity areas that are not being degraded receive lower conservation priority.

What else can make a hotspot hot?

The number of endemic species in an area is undoubtedly important in identifying biodiversity hotspots, but other scientists have argued that this criterion alone is not enough. They suggest that we also consider the total number of species in an area or the number of species currently threatened with extinction in an area. Would all three approaches identify similar regions of conservation priority? A recent analysis of birds suggests they would not. When scientists identified bird diversity hotspots using each of the three criteria—endemic species, total species richness, and threatened species—their results, shown in **TABLE SA2.1**, identified very different hotspot areas. Scientists using the three criteria to identify hotspots of mammal diversity reached the same conclusion.

As we can see, some areas of the world that have high species richness do not contain high numbers of endemic species. The Amazon, for example, has a high number of bird species, but not a particularly high number of endemic bird species compared with other more-isolated regions of the world, such as the islands of the Caribbean. Similarly, areas with high numbers of threatened or endangered species do not always have high numbers of endemic species. These findings highlight the critical problem of deciding whether conservation efforts should be focused on areas containing the greatest number of species, areas containing the greatest number of threatened species, or areas containing the greatest number of endemic species. All three approaches are reasonable.

In addition to considering species diversity, some scientists have argued that we should also consider the size of the human population in diverse areas. For example, we might expect that natural areas containing more people face a greater probability of being affected by human activities. Whereas the world has an average human population density of 42 people per square kilometer, the average hotspot has a human population of 73 people per square kilometer. Such places may be at a higher risk of degradation from human activities. This risk should be considered when determining priority areas for conservation, and it should motivate us to promote development that does not come at the cost of species diversity.

What are the costs and benefits of conserving biodiversity hotspots?

Conservation efforts that focus on regions with large numbers of species place a clear priority on preserving the largest number of species possible. However, people making these efforts do not explicitly consider the likelihood of succeeding in this goal, nor do they necessarily consider the costs associated with the effort. For example, there may be many ways of helping a species persist in an area, including buying habitat, entering into agreements with landowners not to develop their land, or removing threats such as invasive species. In the case of the California tiger salamander, for example, while eliminating invasive predators may not be feasible, it is possible to protect salamander habitat. In 2011, the U.S. Fish and Wildlife Service agreed to designate more than 20,000 ha (50,000 acres) of habitat as being critical for the salamander's persistence and in 2016 the agency announced their plan to help populations of the salamander to recover by working with public and private landowners to protect critical salamander habitat.

What about biodiversity coldspots?

The concept of biodiversity hotspots assumes that our primary goal is to protect the maximum number of species.

TABLE SA2.1 Biodiversity hotspots for birds, identified by three criteria

Rank	Total number of species	Number of endemic species	Number of threatened species
1	Andes	Andes	Andes
2	Amazon Basin	New Guinea and Bismarck Archipelago	Amazon Basin
3	Western Great Rift Valley	Panama and Costa Rica Highlands	Guyana Highlands
4	Eastern Great Rift Valley	Caribbean	Himalayas
5	Himalayas	Lesser Sunda Islands	Atlantic Coastal Forest, Brazil

That goal is admirable, but it could come at the cost of many important ecosystems that do not fall within hotspots. Yellowstone National Park, for example, has a relatively low diversity of species, yet it is one of the few places in the United States that contains remnant populations of large mammals, including wolves, grizzly bears, and bison. Does this mean that places such as Yellowstone National Park should receive decreased conservation attention?

Biodiversity coldspots also provide ecosystem services that humans value at least as much as species diversity. For example, wetlands in the United States are incredibly important for flood control, water purification, wildlife habitat, and recreation. Many wetlands, however, have relatively low plant diversity and, as a result, would not be identified as biodiversity hotspots. It is true that increased species richness leads to improved ecosystem services, but only as we move from very low species richness to moderate species richness. Moving from moderate species richness to high species richness generally does not further improve the functioning of an ecosystem. Since very high species diversity is not expected to provide any substantial improvement in ecosystem function, protecting more and more species produces diminishing returns in terms of protecting ecosystem services. Hence, if our primary goal is to preserve the functioning of the ecosystems that improve our lives, we do not necessarily need to preserve every species in those ecosystems.

How can we reach a resolution?

During the past two decades, it has become clear that scientists and policy makers need to set priorities for the conservation of biodiversity. No single criterion may be agreed upon by everyone. However, it is important to appreciate the bias of each approach and to consider the possible unintended consequences of favoring some geographic regions over others. Our investment in conservation cannot be viewed as an all-or-nothing choice of some areas over others. Instead, our decisions must consider the costs and benefits of alternative conservation strategies and incorporate current and future threats to both species diversity and ecosystem function. In this way, we can strike a balance between our desire to preserve Earth's species diversity and our desire to protect the functioning of Earth's ecosystems.

Questions

1. If you were in the role of a decision maker, how might you balance your desire to protect biodiversity around the world with the high cost of preserving biodiversity in any given location?
2. When identifying hotspots around the world, why is it important to simultaneously consider the biodiversity of different areas and the future human population size in each of these locations?
3. If you had to make a choice between protecting biodiversity hotspots and biodiversity coldspots, how would you make this decision?

Practice AP® Free-Response Question

Write your answer to each part clearly. Support your answers with relevant information and examples. Where calculations are required, show your work.

The Florida Reef is the only living coral barrier reef in the continental United States. Like other coral reefs on the planet, Florida Reef is home to many species of coral, fish, and other aquatic life forms. Also, like other coral reefs, its health is at risk due to a number of human impacts including those associated with global warming. However, Conservation International does not list this as a biodiversity hotspot.

(a) **Identify** two reasons why Florida Reef might not be internationally recognized as a hotspot. (3 pts.)

(b) **Describe** two ecosystem services provided by Florida Reef. (3 pts.)

(c) Researchers recently found that the disturbance generated by hurricanes, which cause high winds and damaging waves, is important for maintaining the diversity of coral reef ecosystems. While a long period of no hurricanes leads to declining coral reef diversity, massive hurricanes often put many coral reef species at risk of local extinction. **Make a claim** to explain this pattern of coral reefs with occasional modest hurricanes having the greatest diversity. (4 pts.)

References

Habel, J. C., et al. 2019. Final countdown for biodiversity hotspots. *Conservation Letters* 12:e12668.

Joppa, L. N., et al. 2011. Biodiversity hotspots house most undiscovered plant species. *PNAS* 108:13171–13176.

Kareiva, P., and M. Marvier. 2003. Conserving biodiversity coldspots. *American Scientist* 91:344–351.

Myers, N., et al. 2000. Biodiversity hotspots for conservation priorities. *Nature* 403:853–858.

Orme, C. D., et al. 2005. Global hotspots of species richness are not congruent with endemism or threat. *Nature* 436:1016–1019.

UNIT 2 Review

Key Terms to Remember

Biodiversity (p. 95)
Genetic diversity (p. 97)
Population bottleneck (p. 97)
Species diversity (p. 98)
Habitat diversity (p. 98)
Specialists (p. 98)
Generalists (p. 98)
Ecosystem diversity (p. 99)
Species richness (p. 102)
Species evenness (p. 102)
Ecosystem services (p. 105)
Provision (p. 106)
Aquaculture (p. 109)
Island biogeography (p. 113)
Species-area curve (p. 113)
Ecological tolerance (Fundamental niche) (p. 122)
Realized niche (p. 123)
Geographic range (p. 123)
Mass extinction (p. 126)
Periodic disruption (p. 129)
Episodic disruption (p. 129
Random disruption (p. 129)
Resistance (p. 130)
Resilience (p. 130)
Intermediate disturbance hypothesis (p. 134)
Evolution (p. 138)
Microevolution (p. 138)
Macroevolution (p. 138)
Evolution by artificial selection (p. 139)
Evolution by natural selection (p. 140)
Fitness (p. 141)
Adaptation (p. 141)
Evolution by random processes (p. 141)
Allopatric speciation (p. 145)
Sympatric speciation (p. 146)
Genetically modified organism (GMO) (p. 148)
Ecological succession (p. 150)
Primary succession (p. 150)
Pioneer species (p. 150)
Secondary succession (p. 151)
Climax community (p. 152)
Keystone species (p. 156)
Indicator species (p. 157)
Endemic species (p. 165)
Biodiversity hotspots (p. 165)

Unit 2 AP® Environmental Science Practice Exam

Preparing for the AP® Exam

Section 1: Multiple-Choice Questions

1. After a severe drought, the productivity in an ecosystem took many years to return to pre-drought conditions. This observation indicates that the ecosystem has
   (a) high resilience.
   (b) low resilience.
   (c) high resistance.
   (d) low resistance.

2. Intertidal communities are frequently disturbed by storms that generate large waves. Following a period of intermediate wave disturbance, a group of researchers examined the species richness of north and south Pacific intertidal communities. Several weeks after the disturbance, they found that northern communities had returned to their original state whereas southern communities were still recovering. This result suggests that northern intertidal communities
   (a) have greater resilience.
   (b) have greater resistance.
   (c) follow predictions of the intermediate disturbance hypothesis.
   (d) experience more frequent disturbance.

Questions 3–5 refer to the following table:

Researchers in North America collected data on tree abundance for a group of forest communities in the three states shown in the table below.

Total percent state-wide cover

Species	Connecticut	Pennsylvania	Georgia
Red maple	25	40	20
Black oak	20	20	40
White pine	15	10	10
Eastern hemlock	20	10	0
Black cherry	20	20	30

3. According to the data, in which state(s) would the lowest percentage of white pine trees be found?
   (a) Connecticut and Georgia
   (b) Pennsylvania and Georgia
   (c) Pennsylvania
   (d) Georgia

4. Which statement best describes the forest communities within each state represented in the table?
   (a) Pennsylvania has the highest species richness; Connecticut has the highest species evenness.
   (b) Connecticut and Pennsylvania have equal species richness; Georgia has the highest species evenness.
   (c) Georgia and Pennsylvania have equal species richness; Georgia has the highest species evenness.
   (d) Connecticut and Pennsylvania have equal species richness; Connecticut has the highest species evenness.

5. According to the data in the table, in which state would the forest communities be least vulnerable to environmental disturbances?
   (a) Georgia, because it has the highest species richness, which increases the resistance of an ecosystem
   (b) Connecticut, because it has the highest species evenness, which increases the resistance of an ecosystem
   (c) Pennsylvania, because it has the highest species evenness, which increases the resistance of an ecosystem
   (d) Connecticut, because it has the highest species richness, which increases the resistance of an ecosystem

6. A dramatic decline in genetic diversity could best be explained by which of the following?
   (a) genetic drift, because genotypes are lost quickly within large populations
   (b) the founder effect, because species need to migrate to other areas and are always able to do so quickly
   (c) mutations, because random changes in genotypes reduces genetic diversity
   (d) the bottleneck effect, because the number of individuals in the gene pool are rapidly reduced

For questions 7–9, use the following answer choices. Choices may be used once, more than once, or not at all.

(a) allopatric speciation
(b) sympatric speciation
(c) fundamental niche
(d) realized niche

7. Two fish species in a single lake evolve from a single ancestor that once lived in the lake. This is a description of a(n) ____________.

8. The range of abiotic and biotic conditions under which a species actually lives is the definition of a(n) ____________.

9. A stream divides a single turtle population into two isolated populations that evolve into two species. This describes a(n)____________.

Use the photo below to answer questions 10 & 11:

(Cavan Images/Alamy Stock Photo)

10. Which sequence of secondary succession would be likely to occur after the disturbance shown in the photo above?
    (a) bare soil, lichens, mosses, grasses, deciduous trees
    (b) bare rock, lichens, mosses, grasses, shrubs, mixed shade- and sunlight-tolerant trees
    (c) bare soil, grasses and wildflowers, shrubs, sunlight-tolerant trees, shade-tolerant trees
    (d) bare rock, grasses and wildflowers, lichens, mosses, shrubs, coniferous trees

11. Which statement about ecological succession is correct?
    (a) Secondary succession begins in a community lacking soil.
    (b) Terrestrial succession is not influenced by competition for limiting resources such as available soil moisture, sunlight, and nutrients.
    (c) In forest succession, more shade-tolerant trees replace less shade-tolerant trees.
    (d) Forest fires and hurricanes lead to primary succession because a soil base still exists.

12. The theory of island biogeography suggests that species richness is increased by which of the following factors?
    (a) larger islands, because there are more available niches
    (b) smaller islands, because there are more available resources
    (c) larger islands, because there is increased competition for space
    (d) smaller islands, because there are more niche specialists

13. Which combination of factors would result in the highest biodiversity?
    (a) a small island that is also close to a continent
    (b) a large island that is also close to a continent
    (c) a small island that is also distant from a continent
    (d) a large island that is also distant from a continent

14. Which is correct regarding keystone species?
    (a) Keystone species cannot be mutualists with other species.
    (b) Keystone species are often abundant in a community.
    (c) Keystone species are most likely to be commensalists.
    (d) Some keystone species are habitat engineers.

15. Which statement best describes the principles of island biogeography?
    (a) A larger protected area should contain fewer available niches.
    (b) Protected areas that are closer together should contain fewer species.
    (c) National parks can be thought of as islands of biodiversity.
    (d) A larger protected area will have fewer habitats.

Use the following passage to answer questions 16–18:

The giant kangaroo rat lives in semiarid climates such as the Carrizo Plain in the Central Valley region of California, where they prefer to forage during nighttime hours in the summer months when the temperature is more favorable. The giant kangaroo rats' burrows provide shelter for squirrels and lizards, and they forage on seeds and grasses in a distinctive circular pattern. Their predators include barn owls, kit foxes, and badgers.

16. The characteristics above best describe the giant kangaroo rats'
    (a) habitat.
    (b) fundamental niche.
    (c) range of tolerance.
    (d) realized niche.

17. The giant kangaroo rat is an endangered species that is essential to its ecosystem, providing shelter for animals and serving as prey for others, all while living in less than 2 percent of its original range. This important role in their community, despite their small population, makes the giant kangaroo rat an example of a(n)
    (a) commensalist.
    (b) keystone species.
    (c) invasive species.
    (d) niche generalist.

18. The giant kangaroo rat currently lives in small, isolated populations in the Central Valley of California. Which of the following best explains why the giant kangaroo rat is likely to have low genetic variation?
    (a) The populations experience sympatric speciation, because they have been geographically isolated from one another.
    (b) The populations experience genetic drift, because they have reduced gene flow between populations.
    (c) The populations experience artificial selection, because humans continue to fragment their habitats through agriculture.
    (d) The populations experience the founder effect, because they are separating themselves into new, smaller populations.

Read the following article written for a local newspaper to answer questions 19–21:

Neighbors Voice Opposition to Proposed Clear-Cutting

A heated discussion took place last night at the monthly meeting of the Fremont Zoning Board. Local landowner Julia Taylor has filed a request that her 150-acre woodland area be rezoned from residential to multiuse in order to allow her to remove all of the timber from the site.

"This is my land, and I should be able to use it as I see fit," explained Ms. Taylor. "In due course, all of the trees will return and everything will go back to the same as it is now. The birds and the squirrels will still be there in the future. I have to sell the timber because I need the extra revenue to supplement my retirement as I am on a fixed income. I don't see what all the fuss is about," she commented.

A group of owners of adjacent properties see things very differently. Their spokesperson, Ethan Jared, argued against granting a change in the current zoning. "Ms. Taylor has allowed the community to use these woods for many years, and we thank her for that. But I hope that the local children will be able to hike and explore the woods with their children as I have done with mine. Removing the trees in a clear-cut will damage our community in many ways, and it could lead to contamination of the groundwater and streams and affect many animal and plant species. Like the rest of us property owners, Ms. Taylor gets her drinking water from a well, and I do not think she has really looked at all the ramifications should her plan go through. We strongly oppose the rezoning of this land—it has a right to be left untouched."

After more than 2 hours of debate between Ms. Taylor and many of the local residents, the chair of the Zoning Board decided to research the points raised by the neighbors and report on his findings at next month's meeting.

19. According to the author, Ms. Taylor is concerned that which of the following ecosystem services provided by her wooded area will be eliminated by the proposed clear-cutting?
    (a) a cultural service
    (b) a provision
    (c) a regulating service
    (d) a supporting service

20. According to the author, Mr. Jared is interested in preventing the clear-cutting to preserve which ecosystem service the forest provides?
    (a) a cultural service
    (b) a provision
    (c) a regulating service
    (d) a supporting service

21. According to the author, in addition to issues surrounding ecosystem services, what additional problem may be caused by the clear-cutting project?
    (a) loss of space for housing projects
    (b) an increase in the population of Fremont
    (c) the loss of habitat for local species
    (d) the contamination of well water

Section 2: Free-Response Questions

1. The diagram below shows ecological succession in a forest ecosystem.

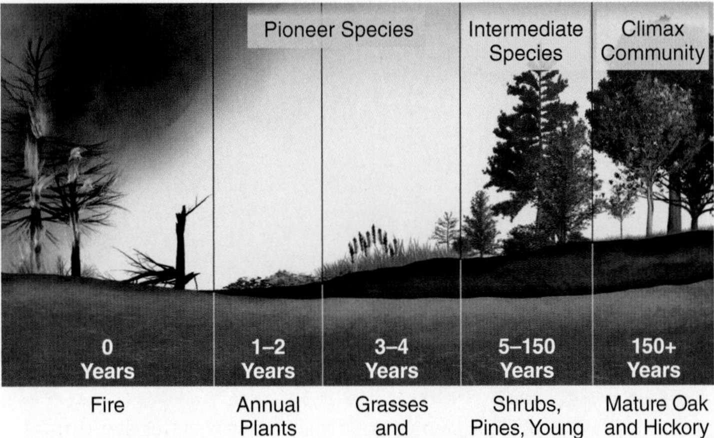

(a) **Identify** one pioneer species depicted in the diagram. (1 pt.)

(b) **Describe** how the type of succession depicted in the diagram is able to occur. (1 pt.)

(c) **Explain** how a supporting service of forests would be affected by the disturbance depicted in the diagram. (1 pt.)

(d) **Explain** one characteristic of a plant species that would increase its ability to recover from a drastic decline in their population size. (2 pts.)

Researchers studying the effects of the intensity of fires on the rate of succession isolated two plots of land in a forest ecosystem. One experiences a low-intensity fire and the other experiences a high-intensity fire.

(e) **Identify** a likely hypothesis for this investigation. (1 pt.)

(f) **Identify** the dependent variable for this investigation. (1 pt.)

(g) **Describe** a possible control for this investigation. (1 pt.)

(h) **Describe** how a modification could be made to improve the reliability of the results of this investigation. (1 pt.)

(i) A student analyzing the results of this study states that the shade-tolerant tree species are the first to repopulate after a fire. **Make a claim** using evidence that supports or refutes the student's statement. (1 pt.)

2. A team of scientists wants to observe the effects of disturbance on a temperate forest ecosystem. They create three forest plots, and after 6 months, they measure the number of plant species in each plot. Their data is listed in the table below.

Species diversity in three forest plots		
Plot	**Treatment**	**Number of species**
A	Undisturbed	17
B	100% clear-cut	3
C	50% clear-cut	25

(a) **Identify** the species richness of the plot where all of the trees were cut down. (1 pt.)

(b) **Describe** which plot is likely to have the highest resistance and resilience. (1 pt.)

(c) **Explain** why Plot C has a greater species diversity than Plot A. (2 pts.)

(d) **Explain** one ecological benefit of high genetic diversity. (2 pts.)

(e) **Identify** a regulating service of a forest. (1 pt.)

(f) Other than the loss of the regulating service you identified in part (e), **identify** an environmental problem that can be caused by clear-cutting a forest. (1 pt.)

(g) **Propose a solution** to mitigate the effects of clear-cutting. (1 pt.)

(h) **Describe** a potential disadvantage to the solution you proposed in part (g). (1 pt.)

3. Genetic variation within a population is the variation in the genotypes of the individuals that make up that population. Differences in genetic variation can influence an organism's ability to thrive in an ecosystem.

(a) **Identify** one process that increases genetic variation in a population. (1 pt.)

(b) **Describe** how the process you identified in part (a) increases genetic variation. (1 pt.)

(c) **Explain** how natural selection can affect genetic variation. (1 pt.)

The brown marmorated stink bug is a native species in Japan and China that was first found in Pennsylvania in 1998. In the years since, the stink bug has caused major economic damage to crops in the Middle Atlantic states. In 2010, the stink bug was responsible for $37 million in damage to the apple crop in the Mid-Atlantic states.

(d) **Describe** an environmental problem that could be caused by the overabundance of stink bugs in the Middle Atlantic states. (1 pt.)

(e) The apple crop of the Middle Atlantic states is grown on approximately 85,000 acres of land, which comprises 26 percent of the apple production in the United States.

(i) **Calculate** the damage the stink bugs caused to apple production per acre in 2010 in the Middle Atlantic states. (2 pts.)

(ii) **Calculate** how many acres of land are used in the United States to produce apples. Show all work. (2 pts.)

(iii) **Calculate** how many hectares of land are used to produce apples in the United States outside of the Middle Atlantic states, if there are 2.47 acres in one hectare. Show all work. (2 pts.)

UNIT 3

Populations

A former New England field is now a forest. *(Jerry and Marcy Monkman/EcoPhotography.com/Alamy)*

MODULE 15	Generalist and Specialist Species, *K*- and *r*-selected Species, and Survivorship Curves	**MODULE 16**	Carrying Capacity, Population Growth, and Resource Availability	**MODULE 18**	Human Population Dynamics and the Demographic Transition
		MODULE 17	Age Structure Diagrams and Total Fertility Rates		

CASE STUDY

New England Forests Contain a Variety of Species

In the early 1600s, European visitors to a region that is now called New England found large areas of what appeared to be undisturbed temperate seasonal forest. These forests contained a variety of tree species including maple, beech, pine, and hemlock. Over the next 200 years, Europeans cut down most of these trees to clear the land for farming and housing. They also built stone walls to demarcate their land, keep livestock contained, or as a means of disposal of the vast quantity of rocks they found in the soil when plowing their fields. This deforestation peaked in the 1800s, at which point up to 80 percent of all New England forests had been cleared. Between 1850 and 1950, however, many people abandoned their New England farms to take jobs in the growing textile industry in nearby cities. Others moved to the Midwest, where farmland was more fertile and considerably less expensive. The old stone walls that are visible in the middle of forested land in the New England countryside today are the only evidence that this forest was once farmland. They also tell a story about human migration from one continent to another and within North America.

What happened to the former farmland is a testament to the resilience of the forest ecosystem and the generalist and specialist species that exist within. The ecosystem transformations began shortly after the farmers left. Seeds of grasses and wildflowers that specialize in colonizing open, sunny, bare soils were carried to the abandoned fields by birds or blown there by the wind. Within a year, the fields were carpeted with a large variety of plant species. Eventually, a single group of plants — the goldenrods, which are specialists in colonizing bare, open soil — came to dominate the fields by growing taller and outcompeting other species of plants for sunlight. The other plant species remained in the fields, but they were not very abundant. Nevertheless, the dominance of the goldenrods was short-lived.

The complex interactions among populations of goldenrods, insects, and other species created an ever-changing ecosystem.

Goldenrods and other wildflowers play an important part in areas that were once agricultural fields, commonly called "old-fields," because they support a diverse group of plant-eating insects. Some of these herbivorous insects are generalists while others are specialists. The number of individuals of each insect species varies from year to year, and occasionally some species experience very large population increases, or outbreaks. One such species is a leaf beetle (*Microrhopala vittata*) that specializes in eating goldenrods. Periodic outbreaks of this species in the abandoned fields of New England dramatically reduced goldenrod populations. With fewer goldenrods, other plant species could compete and prosper.

The complex interactions among populations of goldenrods, insects, and other species created an ever-changing ecosystem. For example, as the leaf beetle population increased in the community, so did the populations of predators and parasites that fed on them. As these predators and parasites reduced the population of leaf beetles, the goldenrod population began to rebound. As the goldenrods surged, they once again caused other plant species to decline in numbers.

Over time, tree seeds arrived and tree seedlings began to grow, which changed the species composition of the old fields once again. One species in particular, the fast-growing white pine (*Pinus strobus*), eventually came to dominate. White pine is known for its ability to grow and thrive in many different environmental conditions including shady forest understory and open sunny old fields. This means that white pine can be present across many forest successional stages, from early stages as a colonist of recently disturbed old fields, to later stages as a shade-tolerant tree in the forest understory. In this sense the white pine is a forest succession generalist. And in the present example, the pine trees grew tall and cast so much shade that the goldenrods and other sunlight-loving understory plant species could not survive.

White pines dominated the old-field communities until humans began another harvest in the early 1900s, following another increase in industrialization and economic activity. Just as the reduction of goldenrod populations made room for other plant species, logging of the white pines made room for broadleaf tree species. Two of these broadleaf species, American beech (*Fagus grandifolia*) and sugar maple (*Acer saccharum*) are dominant in New England forests today. The New England fields that had been abandoned earlier were slowly transformed into communities that resembled the original forests of centuries ago, with a mix of pines, hemlocks, and broadleaf trees.

The story of the New England forests shows us that populations of specialists and generalists can increase or decrease dramatically over time. It also illustrates how species interactions within a community can alter their abundance and that both specialists and generalist species play a role in the communities that occur. Finally, it demonstrates a common theme in environmental science: that humans — especially when the human population grows, resource consumption increases, and people migrate from one location to another — can alter ecosystems repeatedly over time.

Sources: M. J. Duveneck, et al., Recovery dynamics and climate change effects to future New England forests, *Landscape Ecology* 32 (2017): 1385–1397; M. D. Abrams, Eastern white pine versatility in the presettlement forest, *Bioscience* 51 (2001): 967–979; W. P. Carson and R. B. Root, Herbivory and plant species coexistence: Community regulation by an outbreaking phytophagous insect, *Ecological Monographs* 70 (2000): 73–99; T. Wessels, *Reading the Forested Landscape* (Countryman Press, 2005).

Practice your science skills

1. **Concept Explanation:** Describe how the dominant population of goldenrods is an example of one stage of primary succession.
2. **Text Analysis:** Based on the keystone species material presented in Module 14, describe what evidence and data the author might use to make the claim that the leaf beetle (*Microrhopala vittata*) is a keystone species.
3. **Text Analysis:** The author describes the presence of stone walls in the middle of New England forests. Describe one bias the author might have about why these stone walls appear in the middle of a forest and suggest an experiment to assess that bias (Hint: It would be a very long experiment).

Changes that occur in a New England forest are a great illustration of some of the different populations that exist in the natural world as a result of plant, animal, and human population influences on ecosystems. This unit will start by examining many different plant and animal species and end by focusing specifically on one species, human beings. As we saw in the unit-opening account, there are distinct patterns of species occurrence over space and time, such as when a specialist dominates an ecosystem after disturbance by humans. Understanding the factors that generate these patterns can help us understand populations and how individual species perform certain functions and roles in ecosystems. These factors include the ways in which populations increase and decrease in size and how species interact in ecological communities. In this unit we will explore generalist and specialist species, the reproductive strategies each possesses, the way individuals in different species survive, and ultimately the carrying capacity of an ecosystem for a given species. As we explore animals and plants and then the human population, we will also examine population growth and resource availability, age structure diagrams, and other topics related specifically to human population dynamics. These topics help us understand why populations change over time, why some grow larger than others, and how the human population has behaved in the past and may change in the future. Because we are discussing human lives and human births and deaths, we must remember to be sensitive to the fact that the numbers we are discussing represent someone's parent or sibling or child. We are writing about human demographics in academic terms but these are very personal issues.

Module 15

Unit 3 | 15 | 16 | 17 | 18

Generalist and Specialist Species, *K*- and *r*-selected Species, and Survivorship Curves

In this module we will describe how populations change over time in reaction to a variety of ecosystem factors, including the impacts of and differences between generalist and specialist species and how each excels in specific types of habitats. We will also identify differences between *K*-selected species, which tend to be longer lived and provide intensive parental care, and *r*-selected species, which tend to have shorter lifespans and provide little or no parental care. Finally, we will explain survivorship curves, which show the likelihood of survival at any given time within an organism's lifetime for different species. These topics allow us to gain an appreciation for why some species thrive and others don't in certain ecosystems, and why the same species may exhibit different behaviors in other ecosystems.

Learning Goals

After reading this module you should be able to

15-1 identify the differences between generalist and specialist species.

15-2 describe the differences between *K*- and *r*-selected species.

15-3 explain survivorship curves.

15-1 What are the differences between generalist and specialist species?

Generalists exist under a broad range of conditions while specialists exist under a narrow range

As we saw in Unit 2, the evolution and adaptation of species over time combined with their current interactions affect the abundance of species in a particular location or habitat. The concept of ecological tolerance was also presented in Unit 2 and it states that there is a range of conditions under which organisms can exist. However, ecological tolerance can exist for both individual organisms and for a species. Because every species has an optimal environment in which it performs particularly well, there are both biotic and abiotic factors that contribute to species abundance and specific locations where a particular species exists. When we examine the realized niches of species in nature—the actual locations where they exist and thrive—we see that some species, known as niche generalists, can live under a very wide range of abiotic or biotic conditions. As we saw in Figure 11.1, this means they can survive under a variety of conditions, as exhibited by a broad distribution along the *x* axis, temperature. Other species are niche specialists. For these species, the distribution occurs across a narrow temperature range and displays a very narrow curve.

With niche generalists, there is roughly an equal distribution of individuals at a variety of values. For example, there are species of stickleback fish that can survive and reproduce under a wide range of temperatures, an abiotic factor. They are temperature generalists. There are other generalists that do well along a broad gradient of biotic and abiotic factors. Such gradients exist in environments like a successional forest community. When you go from an old field to an open forest to a closed canopy forest, there are changes in both biotic factors such as competition, and abiotic factors such as soil temperature. As you can recall from the Unit 3–opening case study, the white pine tree species was described as being very successful at colonizing recently cleared areas. But white pine can also do well in forests that are intermediate between young and mature, and even in some mature forests. So, we might then infer that white pine should be regarded as a forest succession generalist. You also read about the leaf beetle in the Unit 3 case study, and how they specialized in eating goldenrod plants almost exclusively. This beetle would then be categorized as a niche specialist with respect to food sources, a biotic factor. These differences between generalist and specialist species occur in larger

(a)

(b)

FIGURE 15.1 Generalists and specialists. (a) Some organisms, such as gray kangaroos, are niche generalists with broad diets and wide habitat preferences. (b) Other organisms, such as koalas, are niche specialists with narrow diets and highly specific habitat preferences. *(a: Sohns, Juergen & Christine/Animals Animals; b: Lourens Smak/Alamy)*

animals as well. Some mammals, such as gray kangaroos (*Macropus giganteus*) (**FIGURE 15.1a**), feed on numerous plant species and are generalists. Other niche specialists live under a very narrow range of conditions or feed on a small group of species. A good example of a mammal niche specialist is the koala (*Phascolarctos cinereus*) (Figure 15.1b), which feeds only on the leaves of eucalyptus trees.

In general, niche specialists persist quite well when environmental conditions remain relatively constant or if the food source they specialize on is abundant. But specialists are more vulnerable to reductions in numbers and possible extinction if conditions change because the loss of a favored habitat or food source leaves them with few alternatives for survival. In contrast, niche generalists should fare better under changing food conditions because they have a number of alternative food sources available to them. So, if one species of food source decreases in number, the niche generalist should have other food options to survive on. This is highly relevant today as global climate change is expected to cause shifts in temperature and precipitation, which might lead habitats to change and food abundance to increase or decrease. This has led many environmental scientists to suggest that in most cases, niche generalists should be more adaptable to and more resilient when encountering global change.

15-2 What are the differences between *K*- and *r*-selected species?

Species have different reproductive strategies

Populations change over the course of time in response to a variety of abiotic and biotic factors. We've seen what happens when generalist and specialist species are confronted with both stable and changing environments, and how those environments impact their success. Now, let's investigate this phenomenon a little further by exploring what we mean when we refer to the success of a given species. Generally, when an environmental scientist refers to the success of a species, they are speaking about the population size of that species. And population size most commonly increases through reproduction. We can define **population growth rate** (also known as **intrinsic growth rate**) as the number of offspring an individual can produce in a given time period, minus the deaths of the individual or its offspring during that same period. Under ideal conditions, with unlimited resources available, every population has a particular maximum potential for growth, which is called the

biotic potential. When organisms reproduce, they utilize a range of reproductive strategies in nature. Although not every species will exhibit all of these strategies, most will demonstrate a number of traits within one or both of these two groupings: *K*-selected or *r*-selected species.

K-selected Species

***K*-selected species** are species with a low intrinsic growth rate that causes the population to increase slowly until it reaches the carrying capacity of the environment. The **carrying capacity** is the limit to the number of individuals that can be supported by an existing habitat or ecosystem and is denoted as *K*. The population size of a *K*-selected species is largely determined by the carrying capacity (*K*), and their population fluctuations are usually small (**FIGURE 15.2**). Carrying capacity will be more fully described in Module 16.

K-selected species have certain traits in common. *K*-selected animals are typically large organisms that reach reproductive maturity relatively late, produce few, large offspring per reproductive event, and expend significant energy providing substantial parental care. Elephants, for example, are *K*-selected species that have populations that grow slowly; they do not become reproductively mature until they are 13 years old, breed once every 2 to 4 years, and produce only one calf at a time. However, once an elephant population approaches its carrying capacity, the population remains near the carrying capacity and does not widely fluctuate in size (except in cases of interference by human beings, such as poachers). Large mammals and most birds are *K*-selected species. Perhaps in reading this description you've also realized that you and we (the authors) are members of a *K*-selected species as well. For environmental scientists interested in biodiversity management, or protecting and restoring endangered species, the slow growth of *K*-selected species poses a challenge; in practical terms, it means that an endangered *K*-selected species cannot respond quickly to efforts to save it from extinction.

K-selected species usually have a low population growth rate and often have long life spans. They also live in relatively stable environments. As a result, their populations are relatively stable and remain close to the carrying capacity of the ecosystem. Elephants, in the absence of poachers, fall into this category. They live a long time and can heavily browse and graze an area, removing a vast amount of vegetation, before they move on to another area. Therefore, competition for resources is usually relatively high in the habitats of *K*-selected species, such as with elephants. High competition for resources also contributes to why *K*-selected species are often vulnerable to adverse impacts from invasive species. Very often the invasive species can compete for the same resources as the slow growing, long-lived *K*-selected species and reproduce more quickly, thereby exploiting more of the limited resource. For example, kudzu is an invasive vine native to China that has overgrown many habitats in the southeastern United States. It covers slow-growing *K*-selected tree species and shades them, blocking out sunlight and eventually killing the trees.

r-selected Species

At the opposite end of the spectrum from *K*-selected species, ***r*-selected species** have a high intrinsic growth rate and their populations typically increase rapidly. Such species typically reproduce often and produce large numbers of offspring. However, there are some *r*-selected species such as certain salmon species that reproduce only once and then die. It is this ability to have a high growth rate that allows *r*-selected species to rapidly surpass their carrying capacity and sometimes reach population larger than the environment's carrying capacity. This is a process known as **overshoot** and **dieback**, an increase and then a rapid decline in a population due to death (also known as **die-off**). Once overshoot happens, there are more individuals than the system can support and so a major dieback is unavoidable, as

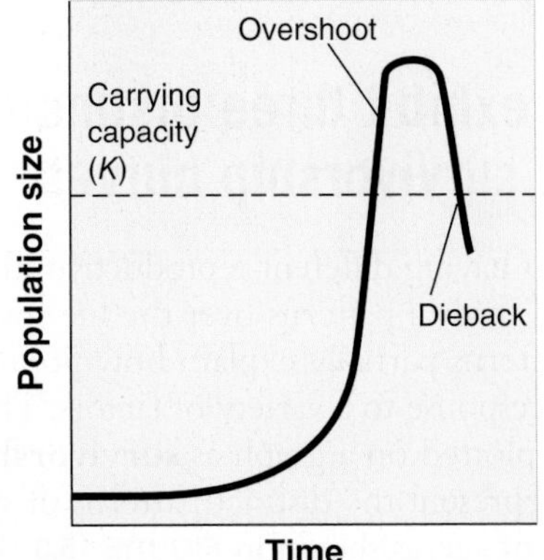

FIGURE 15.2 Population overshoot and dieback. Some populations experience an overshoot of the carrying capacity and a subsequent dieback.

Population growth rate The number of offspring an individual can produce in a given time period, minus the deaths of the individual or its offspring during the same period. *Also known as* **intrinsic growth rate.**

Biotic potential Under ideal conditions with unlimited resources available, every population has a maximum potential for growth.

***K*-selected species** A species with a low intrinsic growth rate that causes the population to increase slowly until it reaches the carrying capacity of the environment.

Carrying capacity The limit to the number of individuals that can be supported by an existing habitat or ecosystem, and is denoted as *K*.

***r*-selected species** A species that has a high intrinsic growth rate, and their population typically increases rapidly.

Overshoot When a population becomes larger than the environment's carrying capacity.

Dieback A rapid decline in a population due to death. *Also known as* **die-off.**

TABLE 15.1 Traits of *K*-selected and *r*-selected species

Trait	*K*-selected species	*r*-selected species
Life span	Long	Short
Time to reproductive maturity	Long	Short
Number of reproductive events	Few	Many (although in some cases 1)
Number of offspring	Few	Many
Size of offspring	Tend to be larger	Small
Parental care	Present	Absent
Population growth rate	Slow	Fast
Population regulation	Density dependent	Density independent
Population dynamics	Stable, near carrying capacity	Highly variable
Competition for resources	Relatively high	Low
Impact of invasive species	Generally high	Low

you can see in Figure 15.2. The name *r*-selected species refers to the fact that the growth rate is designated as *r* in population models.

Among animals, *r*-selected species tend to be small organisms with the potential for high population growth rates. They reach reproductive maturity relatively early, tend to reproduce frequently, produce many small offspring, and provide little or no parental care. House mice (*Mus musculus*), for example, become reproductively mature at 6 weeks of age, can breed every 5 weeks, and produce up to a dozen offspring at a time. Other *r*-selected organisms include small fish, many insect species, and certain plant species. Many organisms that humans consider to be pests, such as cockroaches, dandelions, and rats, are *r*-selected species.

Because *r*-selected species have a high growth rate and often have relatively short lifespans, their populations fluctuate widely above and below the carrying capacity of the ecosystem. This usually results in a wide range of resources being available to *r*-selected species and competition for those resources is usually low—until the population exceeds the carrying capacity. Relatively low competition for a wide variety of resources also contributes to why *r*-selected, often generalist, species are not usually vulnerable to adverse impacts from invasive species. Very often there is an ample supply of resources and therefore the invasive species, which is frequently also a *r*-selected species, is not competing for the same resources as the fast growing, short-lived *r*-selected species.

TABLE 15.1 summarizes the traits of *K*-selected and *r*-selected species. These two categories represent opposite ends of a wide spectrum of reproductive strategies. Most species fall somewhere in between these two extremes, and many exhibit combinations of traits from the two strategies. For example, tuna and redwood trees are both long-lived species that take a long time to reach reproductive maturity, which is a trait of a *K*-selected species. Once they do reach reproductive maturity, however, they produce millions of small offspring that receive no parental care, which is a trait of an *r*-selected species.

AP® Exam Tip

You should understand the different ways ecologists describe populations and be able to use these descriptors appropriately when writing for the AP® Environmental Science Exam. For example, be prepared to describe the differences between *K*-selected and *r*-selected reproductive strategies. Review Table 15.1 to compare and contrast the traits and vulnerabilities of these two categories of reproductive strategies.

15-3 What are survivorship curves?

Species exhibit three distinctly different survivorship curves

In addition to having different reproductive strategies, species have distinct survival patterns over the life span of individuals. These patterns partially explain how populations change over time in response to a variety of factors. The survival patterns can be plotted on a graph as **survivorship curves**, or graphs that represent the distinct patterns of species survival as a function of age, as shown in **FIGURE 15.3**. There are three basic types of survivorship curves. A **type I survivorship curve** has a pattern of survival over time in which there is high survival throughout most of the life span, but then individuals start to die in large numbers as they approach old age.

Survivorship curve A graph that represents the distinct patterns of species survival as a function of age.

Type I survivorship curve A pattern of survival over time in which there is high survival throughout most of the life span, but then individuals start to die in large numbers as they approach old age.

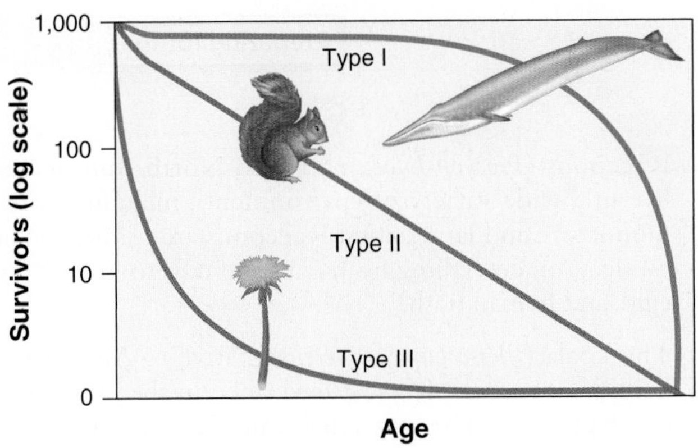

FIGURE 15.3 Survivorship curves. Different species have distinct patterns of survivorship over the life span. Species range from exhibiting excellent survivorship until old age (type I curve) to exhibiting a relatively constant decline in survivorship over time (type II curve) to having very low rates of survivorship early in life (type III curve). *K*-selected species tend to exhibit type I curves, whereas *r*-selected species tend to exhibit type III curves.

Species with type I curves include *K*-selected species such as elephants, whales, and humans, especially those in countries that are highly developed economically (often called "developed countries" such as the United States—we will define this term more fully in Module 17). In contrast, a **type II survivorship curve** is a pattern of survival over time in which there is a relatively constant decline in survivorship throughout most of the life span. Species with a type II curve include chipmunks and squirrels. Species with type II curves can also include *K*-selected species such as certain raptors and birds of prey that have equal chance of survivorship throughout their life span but provide a great deal of parental support for their young. A **type III survivorship curve** has low survivorship (a high death rate) early in life with few individuals reaching adulthood. However, those individuals that reach adulthood tend to live for a relatively long time. Species with type III curves include *r*-selected species such as mosquitoes, many fish, dandelions, and many frog species. For example, most frogs experience a very high death rate early in life (loss of many eggs and young, including in the tadpole stage), but once they reach the adult stage as a frog, they tend to live a relatively long life, given their abilities to blend in with their surroundings and escape capture through rapidly responding to threats by jumping.

The combination of the type of species, generalist or specialist, and reproductive strategy, *r*- or *K*-, and survivorship curves all contribute to the survival and success of a given species. When there is pressure on a population due to human activity, each of these factors will contribute to the ability of the species to respond and may be part of influencing whether or not the species survives.

In this module we described some of the differences that we see in species and how those differences lead to different reproductive strategies and how they influence survivorship and longevity. In the next module we will examine how these and other traits contribute to population size and growth and the carrying capacity of ecosystems.

Type II survivorship curve A pattern of survival over time in which there is a relatively constant decline in survivorship throughout most of the life span.

Type III survivorship curve A pattern of survival over time in which there is low survivorship (a high death rate) early in life with few individuals reaching adulthood.

Module 15 AP® Review

Preparing for the AP® Exam

Learning Goals Revisited

15-1 What are the differences between generalist and specialist species?

Species can be characterized as generalists and specialists. Generalist species have an advantage in habitats that are changing while specialist species have an advantage in habitats that remain relatively constant.

15-2 What are the differences between *K*- and *r*-selected species?

K-selected species are typically larger, have fewer offspring, and provide parental support. *r*-selected species are usually smaller, have many offspring, and provide no parental support. In actuality, many species have some properties of both *r*- and *K*-selection.

15-3 What are survivorship curves?

Species survivorship curves represent the probability of survival at different stages of life. Some species have a high likelihood of survival until they reach an older age (type I); some have an equal likelihood of survival throughout their lifetime (type II); and some have a low likelihood of survival at a young age (type III).

AP® Practice Questions

Multiple-Choice Questions

1. The biotic potential of a population
   (a) occurs at the population's carrying capacity.
   (b) depends on the limiting resources of the population.
   (c) increases as the population size increases.
   (d) occurs only under ideal conditions.

2. Which would be expected for a niche generalist?
   (a) many food sources
   (b) a narrow temperature range
   (c) limited adaptability to changing conditions
   (d) few suitable habitats

3. An *r*-selected species characteristically has
   (a) a type I survivorship curve.
   (b) few offspring.
   (c) a population near carrying capacity.
   (d) a fast population growth rate.

4. Which is true of a population overshoot?
   (a) It occurs when reproduction quickly responds to changes in food supply.
   (b) It is typically followed by a dieback.
   (c) It is common in *K*-selected species.
   (d) It occurs when a species stops growing after reaching the carrying capacity.

5. Which of the following is not common in *K*-selected species?
   (a) long time to reproductive maturity
   (b) little parental care
   (c) long lifespan
   (d) small number of offspring

6. Black bears (*Ursus americanus*) are a species of small bear living in North America. They have a life span of around 18 years, and have two to three cubs at a time. Gestation can last around 235 days (about 8 months) and the cubs remain with their mother until about 2 years. Given this description, and the figure shown below, black bears would likely fall under the

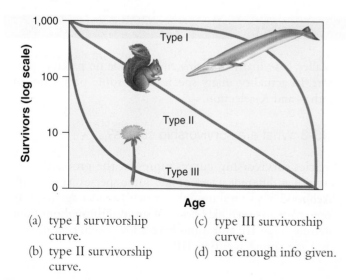

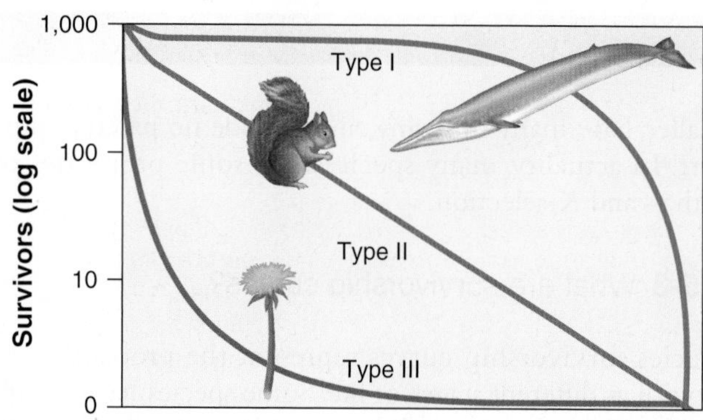

   (a) type I survivorship curve.
   (b) type II survivorship curve.
   (c) type III survivorship curve.
   (d) not enough info given.

7. Raccoons (*Procyon lotor*), native to North America, can live in a wide variety of environments, including forests, mountains, and large cities. Raccoons are omnivores and can feast on everything from fruit and nuts to insects, frogs, eggs, and human trash.

   The koala (*Phascolarctos cinereus*), native to Australia, are herbivorous marsupials that feed only on the leaves of the eucalyptus tree. Therefore, their range is restricted to habitats that support eucalyptus trees. Within this diet some koalas specialize even further and eat leaves from only one or two specific trees.

   Identify the claim of the author given the information above:
   (a) Raccoons are specialists only.
   (b) Koalas are generalists only.
   (c) Raccoons are generalists and koalas are specialists.
   (d) Raccoons are specialists and koalas are generalists.

Free-Response Question

The below table lists two species of lizard showing three aspects of their population and reproductive practices:

	Species A	Species B
Clutch size	Large	Small
Life expectancy	Short life	Long life
Survivorship type	Type II	Type I

(a) **Identify** the type of selection in species A. (1 pt.)
(b) **Identify** the type of selection in species B. (1 pt.)
(c) **Explain** which type of species might become an invasive species. (2 pts.)
(d) **Explain** which type of species might be a specialist species. (2 pts.)
(e) **Describe** which species would be at higher risk from habitat loss. (2 pts.)
(f) **Propose** a solution to reducing invasive species. (1 pt.)
(g) **Justify** the solution proposed in part (f). (1 pt.)

Module 16

Unit 3 | 15 | 16 | 17 | 18

Carrying Capacity, Population Growth, and Resource Availability

In this module we will examine carrying capacity, population growth, and resource availability. In order to explore those topics, we will first discuss how populations change and identify the different models or simplifications of the real world that describe population growth and decline. For short periods of time, populations in nature may grow rapidly without limits—that is until they become limited by a resource and the growth rate slows. The carrying capacity, or the maximum number of individuals that can be supported by the environment—a key topic in this module—is usually what limits population growth. If a population grows beyond its carrying capacity, we often see a dieback of the population, followed by an oscillation above and below the carrying capacity until a stable equilibrium is reached. Ultimately, when a population's resource base (its food supply, for example) shrinks, the carrying capacity will decrease and the population will decline well below the initial carrying capacity.

Scientists often use models to help them explain how systems behave and to predict how systems like populations might change in the future. To do this, they often use growth models that incorporate **density-dependent factors**—factors that influence an individual's probability of survival and reproduction in a manner that depends on the size of the population—and **density-independent factors**—those that have the same effect on an individual's probability of survival and reproduction at any population size—to explain and predict changes in population size. These models are important tools for environmental scientists, whether they are protecting an endangered condor population, managing a commercially harvested fish species, or controlling an insect pest. They are essential to understanding how many individuals in a population can exist and be supported by the surrounding environment. In this module we will examine two growth models and additional tools for understanding changes in population size, then we will use these models and tools to explore the relationships between carrying capacity, population growth, and resource availability.

Learning Goals

After reading this module you should be able to

16-1 explain the exponential growth model, and why it produces a J-shaped curve.

16-2 describe how the logistic growth model explains carrying capacity and produces an S-shaped curve.

16-3 justify how carrying capacity can have an impact on ecosystems.

16-4 identify how population can decline below carrying capacity.

16-1 What is the exponential growth model, and why does it produce a J-shaped curve?

The exponential growth model describes populations that continuously increase, and is the first step in explaining carrying capacity

Population growth models are mathematical equations that can be used to predict population size at any moment in time.

Density-dependent factor A factor that influences an individual's probability of survival and reproduction in a manner that depends on the size of the population.

Density-independent factor A factor that has the same effect on an individual's probability of survival and reproduction at any population size.

Population growth models Mathematical equations that can be used to predict population size at any moment in time.

Understanding the exponential and logistic growth models are essential to understanding and describing carrying capacity.

A population can initially grow very rapidly when its growth is not limited by scarce resources. We defined this population growth rate in Module 15 when we described *r*-selected species and how their populations grew rapidly and exceeded the carrying capacity of the system. Recall that under ideal conditions, with unlimited resources available, every population has a particular maximum potential for growth, called the biotic potential. We can also refer to it as the growth rate *r*. When food is abundant, individuals have a tremendous ability to reproduce. This is also called high **fecundity**, or ability to produce an abundance of offspring. For example, domesticated hogs (*Sus domestica*) can have litters of 10 piglets, and American bullfrogs (*Lithobates catesbeianus*) can lay up to 20,000 eggs. Under ideal conditions, the probability of an individual surviving also increases. Together, a high number of births and a low number of deaths produce a high population growth rate. When conditions are less than ideal due to limited resources, a population's growth rate will be lower than its biotic potential because individuals will produce fewer offspring (or forgo reproduction entirely) and the number of deaths will increase. The **exponential growth model** is a model that estimates a population's future size after a given period of time based on the biotic potential and the number of reproducing individuals currently in the population. The equation used in this model tells us the expected population size at a set time in the future under ideal conditions, when resources are not limiting and the population increases at its biotic potential.

When populations are not limited by resources, growth can be very rapid because more births occur with each unit of time. When graphed, the exponential growth model produces a **J-shaped curve**, as shown in **FIGURE 16.1**. The J-shape of the curve represents the change in a growing population over time. At first, the population is so small that it cannot increase rapidly because there are few individuals present to reproduce. As the population increases however, there are more individuals of reproductive age, and therefore the same per capita growth rate applies to a much larger number of individuals. This means that the number of new births during each unit of time increases from one time period to the next. The rate of change is always a fixed proportion, but the magnitude of change increases rapidly, leading to the J-shaped curve.

Fecundity The ability to produce an abundance of offspring.

Exponential growth model A growth model that estimates a population's future size after a period of time based on the biotic potential and the number of reproducing individuals currently in the population.

J-shaped curve The curve of the exponential growth model when graphed.

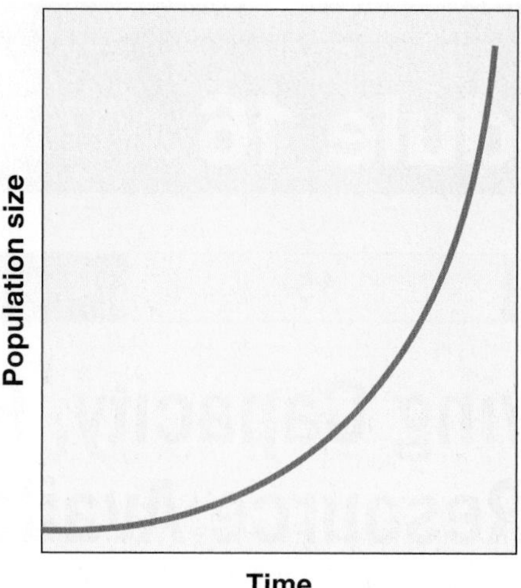

FIGURE 16.1 The exponential growth model. When populations are not limited by resources, their growth can be very rapid. More births occur over time, creating a J-shaped growth curve.

One way to think about exponential growth in a population is to compare it to the growth of a money market account that pays compound interest, where there is an account balance at some starting time and there is a known interest rate. Let's say you put \$100 in the account at an annual interest rate of 5 percent (not so common in recent years but nice for our example). After 1 year the new balance in the account would be \$105 (\$100 + 5% of 100, or \$5, = \$105). In the second year, the balance in the account would be \$105 plus 5 percent of \$105, so the amount added would be a little bit more than \$5. In fact, 25 cents more than if there had been a simple addition of \$5 to the account in the second year. So the balance on the account after year 2 would be \$110.25. That seems like a very small difference, but if you continue to compound the interest each year then after 10 years, the account would have not have (\$100 + \$5 + \$5 + \$5 + \$5 + \$5 + \$5 + \$5 + \$5 + \$5 + \$5 =) \$150 but rather \$162.89. After the twentieth year, the same 5 percent interest rate would produce a balance of \$265.33. If you can follow this example of compounding interest growth for a bank account, you will understand the power of exponential growth and how it can lead to relatively rapid increases in population.

Applying a constant annual rate of growth to an increasing amount, whether money in an account or a population of organisms, produces rapid growth over time. Exponential growth is density independent because the value will grow by the same percentage every year and the result is dependent on population size. The exponential growth model is an excellent starting point for understanding population growth. Indeed, there is solid evidence that real populations—even small ones—can grow exponentially, at least initially. However, no population can experience exponential growth indefinitely. Eventually, it will approach the carrying capacity of the environment, at which point its growth will slow and level off.

> **16-2** How does the logistic growth model explain carrying capacity and produce an S-shaped curve?

The logistic growth model describes populations that are limited by the carrying capacity

While the exponential growth model describes a continuously increasing population that grows at a fixed rate, as we just explained, populations do not experience exponential growth indefinitely. For this reason, ecologists have modified the exponential growth model to incorporate environmental limits on population growth, including limiting resources. The **logistic growth model** describes a population whose growth is initially exponential but slows as the population approaches the carrying capacity of the environment. When using the logistic growth model, environmental scientists define the carrying capacity with the letter *K*. As we can see in **FIGURE 16.2**, if a population starts out small, its growth can be very rapid. As the population size nears about one-half of the carrying capacity, however, population growth reaches an inflection point and begins to slow. As the population size approaches the carrying capacity, the population stops growing. This slowing and then stopping of population growth occurs because resources begin to limit population growth—the population becomes constrained by resource availability in the ecosystem. When graphed, the logistic growth model produces an **S-shaped curve**.

The logistic growth model is used to predict the growth of populations that are subject to density-dependent factors as the population grows, such as increased competition for food, water, or nest sites. Because density-independent factors such as hurricanes and floods are inherently unpredictable, the logistic growth model does not account for them. One of the assumptions of the logistic growth model is that the number of offspring that survive depends on the current population size and the carrying capacity of the environment. There are situations when the food supply or other density-dependent factors are plentiful at the time of mating and gestation but become limited at the time of birth of the offspring. For example, many species of mammals mate during the fall or winter, and the number of pregnant females in the population depends partially or fully on the food supply at the time of mating. In the following months, when offspring are born, there is a chance that food availability will not match the new population size. If there is less food available than needed to feed the offspring, the population will be larger than the environment's carrying capacity and as a result, there will be overshoot. A major environmental consequence of this overshoot is resource depletion because there will not be enough food for all the individuals in the population, and the population will experience a dieback.

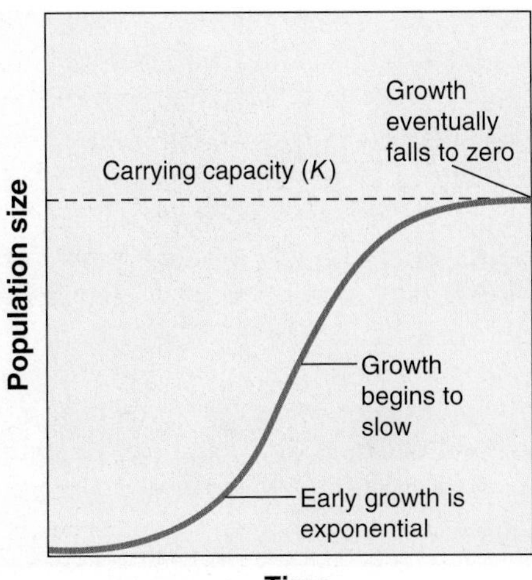

FIGURE 16.2 The logistic growth model. A small population initially experiences exponential growth. As the population becomes larger, however, resources become more scarce, and the growth rate slows. When the population size reaches the carrying capacity of the environment, growth stops. As a result, the pattern of population growth follows an S-shaped curve.

> **AP® Exam Tip**
>
> Remember any population, regardless of reproductive strategy, has a carrying capacity indicated by the letter *K*. Both *K*-strategists and *r*-strategists have a carrying capacity.

Density-Dependent Factors

Density-dependent factors were first illustrated in a now-classic experiment conducted by Russian biologist Georgy Gause in 1932. Gause monitored population growth in two species of *Paramecium* (a type of single-celled aquatic organism) living under ideal conditions in test tubes with a constant amount of food added each day. As the graph in **FIGURE 16.3a** shows, both species of *Paramecium* initially experienced rapid population growth, but over time the rate of growth began to slow. Eventually, the two population sizes reached a plateau and remained there for the rest of the experiment.

Gause suspected that *Paramecium* population growth was limited by food supply. To test this hypothesis, he conducted a second experiment in which he doubled the amount

> **Logistic growth model** A growth model that describes a population whose growth is initially exponential, but slows as the population approaches the carrying capacity of the environment.
>
> **S-shaped curve** The shape of the logistic growth model when graphed.

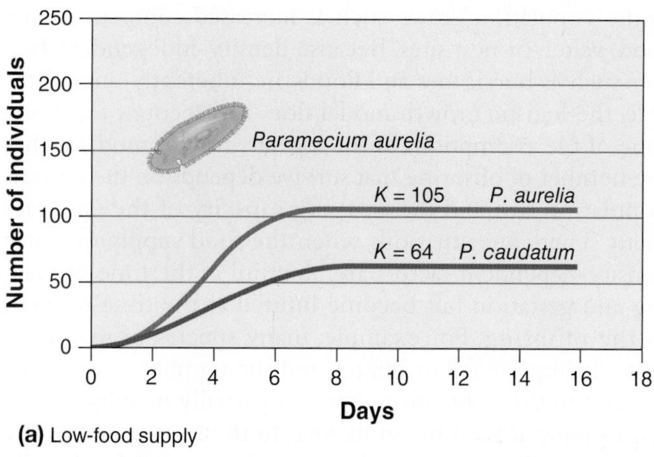

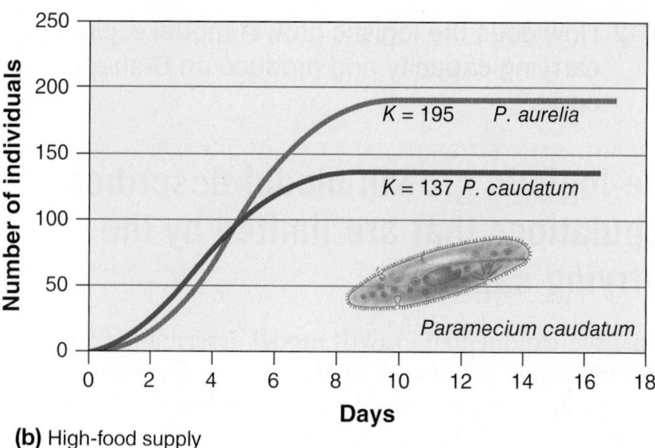

FIGURE 16.3 Gause's experiments. (a) Under low-food conditions, the population sizes of two species of *Paramecium* initially increased rapidly, but then leveled off as their food supply became limiting. (b) When twice as much food was provided, both species attained population sizes that were nearly twice as large, but they again leveled off. *(Data from Gause, 1932.)*

of food he added to the test tubes. Again, both species of *Paramecium* experienced rapid population growth early in the experiment, and the rate of growth, again, slowed over time. However, as Figure 16.3b shows, their maximum population sizes were approximately double those observed in the first experiment.

Gause's results confirmed that food is a limiting resource for *Paramecium*, and the implications of this study extend to many other organisms such as large mammals. A **limiting resource** is a resource that a population cannot live without and that occurs in quantities lower than the population would require to increase in size. If a limiting resource decreases, so does the size of a population that depends on it. For terrestrial plant populations, water and nutrients such as nitrogen and phosphorus are common limiting resources. For animal populations, food, water, and nest sites are common limiting resources.

Density-Independent Factors

We have seen that density-dependent factors, like food, affect an individual's probability of survival and expected number of offspring differently depending on whether the population is large or small. In contrast, density-independent factors have the same effect on an individual's probability of survival and reproduction rate regardless of the population size. A tornado, for example, can uproot and kill many trees in an area (**FIGURE 16.4**). However, a given tree's probability of being killed by the tornado does not depend on whether it resides in a forest with a high or low density of other trees. Other density-independent factors include hurricanes, floods, fires, volcanic eruptions, and other climatic events. An individual's likelihood of mortality increases during such an event regardless of whether the population happens to be at a low or high density.

Bird populations are often regulated by density-independent factors. For example, in the United Kingdom, a particularly cold winter can freeze the surfaces of ponds, making amphibians and fish inaccessible to wading birds such as herons. With their food supply no longer available, herons have an increased risk of starving to death, regardless of whether the heron population is at a low or a high density.

FIGURE 16.4 Path of destruction. This forest in Pennsylvania was damaged by a tornado. Natural disasters such as tornadoes can cause large declines in population sizes and therefore regulate plant and animal populations. Because the magnitude of the decline is not related to the density of the population prior to the disaster, these natural disasters represent density-independent factors in population regulation. *(Michael P. Gadomski/Science Source)*

Limiting resource A resource that a population cannot live without and that occurs in quantities lower than the population would require to increase in size.

16-3 How can carrying capacity have an impact on ecosystems?

Carrying capacity can impact ecosystem diebacks

Animal populations, and even the human population, can so severely transform an ecosystem that a resource becomes depleted and the carrying capacity of the ecosystem declines. Common examples typically involve populations that become so large that they completely exhaust the food supply or some other constraining factor. The reindeer (*Rangifer tarandus*) population on St. Paul Island in Alaska is a good example of this. As you can see in **FIGURE 16.5**, a small population of 25 reindeer was introduced to the island in 1910 to serve as a food source for the people living on the island. The island contained abundant vegetation for the reindeer and there were no large predators, so the population grew exponentially until it reached more than 2,000 individuals in 1938. However, the amount of vegetation on the island was not sufficient for such a large population of reindeer. The population subsequently crashed to only 8 animals by 1950, most likely because the reindeer ran out of food. After this dieback and a rebounding of the vegetation, 31 new reindeer were introduced to the island. Today, the St. Paul Tribal Government manages the reindeer population to avoid exponential increases followed by dieback so that the population provides a stable source of food for the people of the island. The most recent data from a few years ago indicated that the reindeer population was approximately 400 animals.

In situations where the population is not managed, or not managed properly, overpopulation can result in a severe depletion of a resource. In some cases, the loss is easily reversible, such as vegetation that regrows if it has been browsed too heavily like on St. Paul Island, but in other cases, such as loss of water from an aquifer or soil from an area of land, it may take many decades, or in some cases centuries, to recover. Thus, a reduction in carrying capacity can be temporary or in some cases permanent, depending on what change in resource availability caused the reduction.

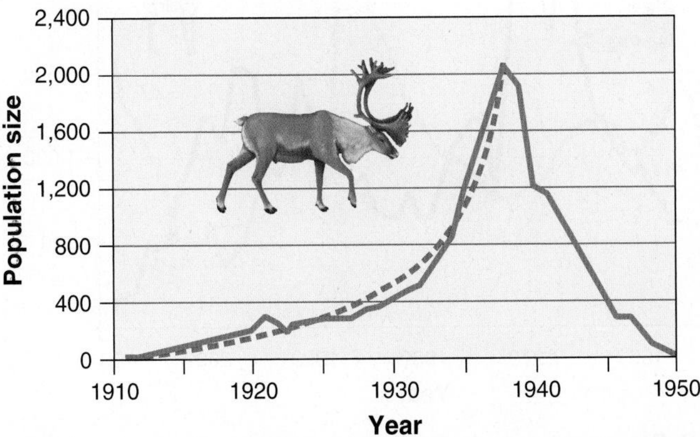

FIGURE 16.5 Growth and decline of a reindeer population. Humans introduced 25 reindeer to St. Paul Island, Alaska, in 1910. The population initially experienced rapid growth (blue line) that approximated a J-shaped exponential growth curve (orange line). In 1938, the population began to dieback, probably because the animals exhausted the food supply. *(Data from Scheffer, V. B. 1951. The rise and fall of a reindeer herd,* Scientific Monthly *73: 356–362.)*

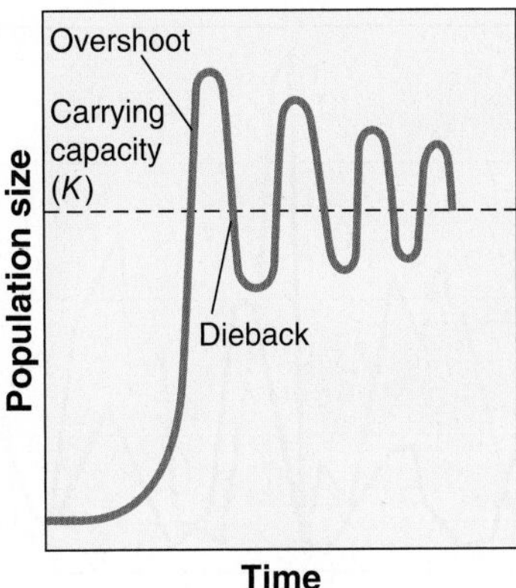

FIGURE 16.6 Population oscillations. Some populations experience recurring cycles of overshoots and diebacks that lead to a pattern of oscillations around the carrying capacity of their environment.

As we have just seen, diebacks can take a population well below the carrying capacity of the environment. In subsequent cycles of reproduction, however, the population may grow large again. **FIGURE 16.6** illustrates the dynamics of a population experiencing a recurring cycle of overshoot and dieback that causes it to oscillate around the carrying capacity. In many cases, these oscillations decline over time and approach the carrying capacity.

So far in this module we have considered what happens when populations are limited by resources such as food, water, and top soil. However, there are also situations where the carrying capacity is defined by the existence of a predator. A classic example is the relationship between snowshoe hares (*Lepus americanus*) and lynx (*Lynx canadensis*) that prey on them in North America. Trapping records from the Hudson's Bay Company, which purchased hare and lynx pelts for nearly 90 years in Canada, indicate that the populations of both species cycle over time. **FIGURE 16.7** shows how this interaction works. The lynx population peaks 1 or 2 years after the hare population peaks. As the hare population increases, it provides more prey for the lynx, and thus the lynx population begins to grow. As the hare population reaches a peak, food for hares becomes scarce, and the hare population dies back. The decline in hares leads to a subsequent decline in the lynx population. Because the low lynx numbers reduce predation, the hare population increases again.

There are other examples of population fluctuations connected to carrying capacity and predator-prey relationships such as when overpopulation leads to overgrazing and food shortages, increased competition for food and other

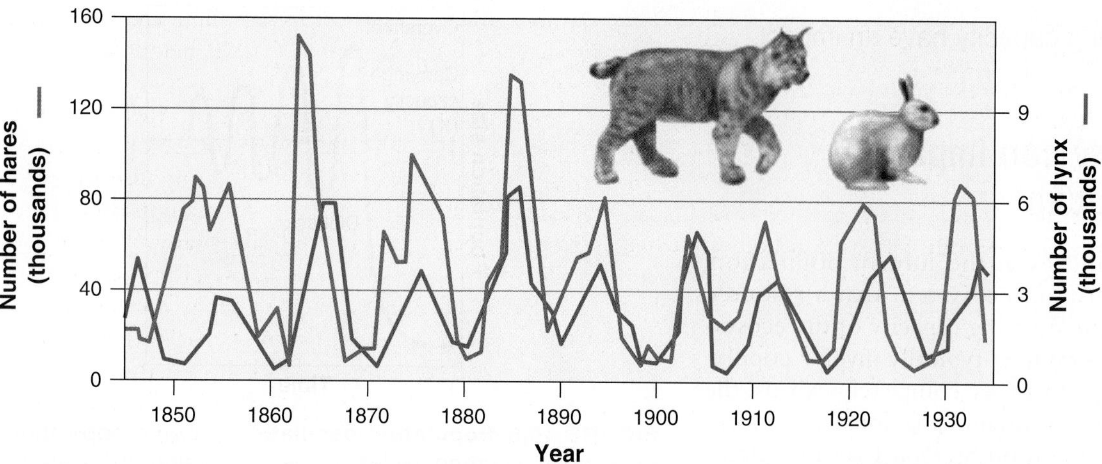

FIGURE 16.7 Population oscillations in lynx and hares. Both lynx and hares exhibit repeated oscillations of abundance, with the lynx population peaking 1 to 2 years after the hare population. When hares are not abundant, there is plenty of food, which allows the hare population to increase. As the hare population increases, there are more hares for lynx to eat, so then the lynx population increases. As the hare population becomes very abundant, they start to run out of food and the hare population dies back. As hares become less abundant, the lynx population subsequently dies back. With less predation and more food once again available, the hare population increases again, and the cycle repeats. *(Data from Hudson's Bay Company.)*

resources, or disease. In human populations, which we will consider in the next two modules, overpopulation can contribute to famine, conflict (war), and the spread of infectious diseases. There are frequent examples in recent decades of famine, conflict, and the spread of disease. For example, a drought in East Africa in 2011 and 2012 led to food shortages, famine, conflict, and the congregating of people in refugee camps. Because of crowded conditions, infectious diseases and malnutrition flourished, resulting in an estimated 260,000 deaths.

We can find one more example in the natural world as a result of a natural experiment in Michigan during the last century that was partially described in Unit 1. As you saw with the moose-wolf Isle Royale, Michigan, example from Module 1, wolves and moose have coexisted on Isle Royale for several decades and recorded estimates of their population sizes go back to 1959. However, prior to their coexistence, moose lived on the island without wolves and would go through population fluctuations as they over-browsed the vegetation and experienced diebacks due to food shortages. In the very cold winter of 1948, ice connected the mainland with Isle Royale and wolves crossed the ice bridge that formed after weeks of cold temperatures and populated the island. This started a predator control on moose, and ultimately led to a reduction in the moose population. Then, as we learned in Module 1, starting in the late 1970s and early 1980s, the wolf population declined sharply, probably as a result of a deadly canine virus. Its transmission was probably facilitated by the relatively large numbers of wolves at the time that allowed the spreading of the virus. As wolf predation on moose declined, the moose population grew rapidly until it ran out of food and experienced another large dieback that began in 1995. **FIGURE 16.8** shows the changes in both populations over a period of more than 50 years. By 2018, only two wolves

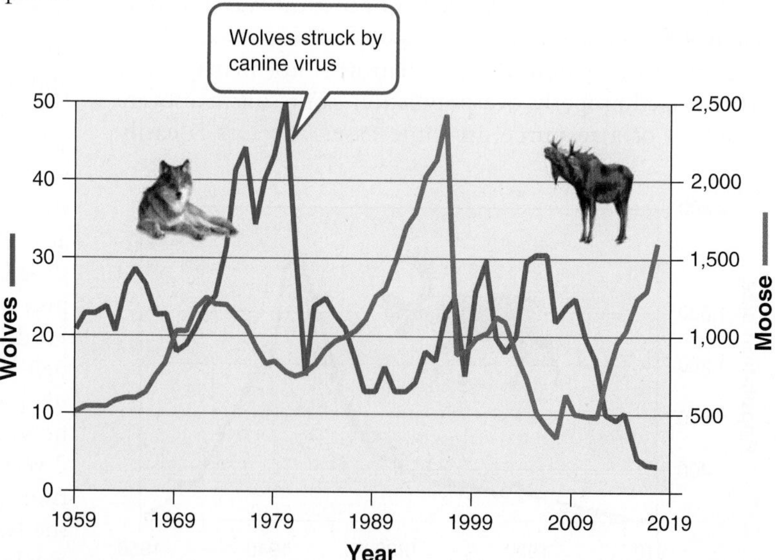

FIGURE 16.8 Predator control of prey populations. As the population of wolves on Isle Royale succumbed to a canine virus, their moose prey experienced a dramatic population increase. The moose population ultimately grew so large that they exceeded their carrying capacity and then experienced a large dieback. In 2018, only two wolves remained and the moose population began to once again increase. *(Data from R. O. Peterson and J. A. Vucetich, Ecological Studies of Wolves on Isle Royale: Annual Report 2016–2017, School of Forest Resources and Environmental Science, Michigan Technological University.)*

remained on Isle Royale. As a result, and after a great deal of debate among scientists and the public, the National Park Service reintroduced a dozen wolves to the island. Based on the most recent survey conducted in 2020, there were at least 14 wolves on Isle Royale. As a result, the moose population and many vegetation species browsed by moose will be relatively stable, for at least a period of time. The Isle Royale example shows how wolves, a predator, not only keep a limit on the moose population but they also prevent the moose from being limited by a resource, vegetation.

16-4 How can population decline below carrying capacity?

Populations change when a resource base shrinks

Organisms can be limited by food but they can also be limited by habitat. For example, a species of diving ducks called the common goldeneye nests in dead or dying trees and logs that are hollowed out by woodpeckers and other species. However, if something reduces the number of dead and dying trees, such as logging and forest removal, or if something causes the population of woodpeckers to decline so fewer hollowed out areas exist, then the number of suitable nesting sites for ducks declines as well. So there are situations where there is an abundant food supply to support large numbers of goldeneye ducks, but there are not enough nesting sites, and the population declines below what could be supported by the food supply.

As you can imagine, the change in nest sites could happen quickly if a forest removal operation took place during the course of a summer. And if the logging continued for many years, the decrease in nesting sites could continue. The return of nesting sites could take many years because most hollowed-out areas in trees come about when a tree is diseased or dying and its rotted areas contain insects that attract woodpeckers. You can see that in this example, the resource base—nesting sites—is heavily dependent on a number of other factors, and the time frame in which the resource base decreases or increases could vary widely.

Reductions in the resource base can adversely affect human populations as well as plants and animals. There are numerous relatively recent examples of situations involving humans where too many people crowding into a finite area led to a resource reduction. For example, the population of sub-Saharan Africa grew from 177 million people in 1950 to 657 million people 50 years later. This drastically reduced the amount of land and fresh water available to each person. For another example, we can look to western China, where overgrazing of the land by livestock in the previous decades led to soil degradation by the early 2000s (**FIGURE 16.9**). As a result, shepherds and farmers found it much more challenging to raise livestock and grow food. Neither of these examples led to outright famine and deaths of large numbers of humans, but they did lead to limitations and the unequal distribution of many resources.

FIGURE 16.9 Overgrazing by sheep. The adverse effects of too many people and too much livestock are apparent in this photograph of soil compaction and erosion from overgrazing of sheep. *(FLPA/Alamy Stock Photo)*

Module 16 AP® Review

Preparing for the AP® Exam

Learning Goals Revisited

16-1 What is the exponential growth model, and why does it produce a J-shaped curve?

Population growth models help us understand and describe population increases and decreases and why they might be fluctuating over time. Exponential growth models are the simplest because they assume unlimited resources, and therefore a constant growth rate can be maintained as the population size increases. This means that more individuals are added to the population each year than were added in the previous year.

16-2 How does the logistic growth model explain carrying capacity and produce an S-shaped curve?

Logistic growth models are more realistic because they incorporate a carrying capacity and the associated density-dependent factors that occur and eventually limit natural populations. Density dependent factors can regulate populations more strongly as populations grow, whereas density-independent factors can regulate populations at any population size.

16-3 How can carrying capacity have an impact on ecosystems?

By varying the assumptions of growth models, we see that a population can overshoot its carrying capacity and experience a dieback. This cycle of overshoot and dieback can happen multiple times, for a variety of reasons, including as a result of predator-prey relationships. The amount of resource per organism can be reduced by increases in the number of organisms or by reductions in the overall quantity of the actual resource.

16-4 How can population decline below carrying capacity?

The carrying capacity is the maximum population size of a particular species that can be sustained by a specific environment, but this does not mean that the population will always be at the carrying capacity. If the population overshoots—increases beyond what the environment can sustain—we often see a sharp dieback as competition for resources makes the population susceptible to food shortages, competition and famine, conflict, and disease. Changes in the availability of resources at a given time can also increase mortality and slow reproduction, leading to a negative growth rate and a population below carrying capacity.

AP® Practice Questions

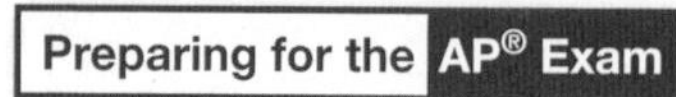

Multiple-Choice Questions

1. The J-shaped population curve is a classic representation of
   (a) a population at the carrying capacity.
   (b) declining populations.
   (c) exponential growth.
   (d) growth under limiting conditions.

2. Which is true about a population's carrying capacity?
   (a) It is denoted as C.
   (b) It is determined by a limiting resource.
   (c) It is controlled by density-independent factors.
   (d) It is always known.

3. Logistic growth models incorporate a time with almost unlimited growth and
   (a) a time with growth that approaches the carrying capacity.
   (b) a time with no growth.
   (c) nothing else.
   (d) a time with a rapid decrease in growth.

4. Which is true of a population overshoot?
   (a) It occurs when reproduction quickly responds to changes in food supply.
   (b) It is followed by a dieback.
   (c) It is most likely to be experienced by *K*-selected species.
   (d) It occurs when a species stops growing after reaching the carrying capacity.

5. What usually happens when a resource base decreases?
   (a) The population grows.
   (b) The carrying capacity increases.
   (c) Population growth decreases.
   (d) The predator population normally increases.

6. Which is an example of a density-independent factor?
   (a) limited water for plants
   (b) limited prey for predators
   (c) a fixed amount of grass for dairy cows
   (d) a hurricane

7. In the table shown below, the population goes up and down seasonally; this is known as:

Month	Population Size
January	55 individuals
February	58 individuals
March	63 individuals
April	63 individuals
May	65 individuals
June	65 individuals
July	58 individuals
August	58 individuals
September	58 individuals
October	58 individuals

(a) overshoot
(b) dieback
(c) oscillation
(d) carrying capacity

Free-Response Question

The graphs below show two different types of growth curves for a population.

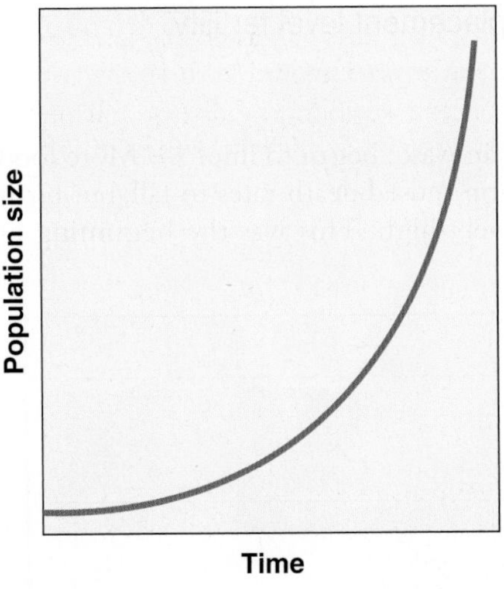

Graph A

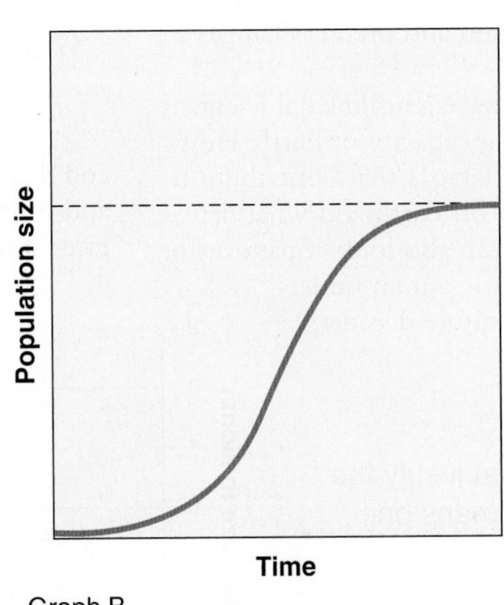

Graph B

(a) **Identify** which graph is logistic growth. (1 pt.)
(b) **Identify** which graph is exponential growth. (1 pt.)
(c) **Identify** the dashed line at the top of graph B. (2 pts.)
(d) **Describe** which population would be most affected by a seasonal drought. (2 pts.)
(e) **Explain** how a species from your answer to part (d) would respond to a seasonal drought. (2 pts.)
(f) **Make a claim** for which type of species is most affected by urbanization. (1 pt.)
(g) **Propose a solution** to the problem identified in part (f). (1 pt.)

Module 17

Unit 3 | 15 | 16 | 17 | 18

Age Structure Diagrams and Total Fertility Rates

The fundamental rules governing populations are essentially the same for plants, animals, and humans. However, humans have the ability to make choices and alter their environment to a much greater extent than plants and animals. In this module we will focus on population principles and applications as they pertain to humans. After explaining the factors that may limit carrying capacity, we will describe the specific social and cultural drivers of human population change. Then we will learn how to read and interpret age structure diagrams for the human population and discuss what they tell us about population change in the past and future. Finally, we will introduce the concepts of the total fertility rate and replacement level fertility, explore why fertility rates differ from one country to the next, and explain how these differences help us to explain and predict changes in their populations over time.

Although it may seem surprising, environmental scientists do not know the human carrying capacity of Earth. However, we are able to discuss the factors that contribute to the carrying capacity of humans on Earth and what drives human population changes. We can also look at patterns of past human population growth to gain an understanding of what might occur in future decades.

Learning Goals

After reading this module you should be able to

17-1 explain the factors that justify the carrying capacity of humans on Earth.

17-2 describe the social and cultural drivers of human population change.

17-3 describe and interpret age structure diagrams.

17-4 explain the factors that influence the total fertility rate.

17-5 describe replacement level fertility.

17-1 What are the factors that justify the carrying capacity of humans on Earth?

The total number of people the Earth can support is not known

Every 5 days, the global human population increases by roughly 1.3 million lives: 2.0 million infants are born and 750,000 people die. The human population has not always grown at this rate, however. As **FIGURE 17.1** shows, until a few hundred years ago the human population was relatively stable: Deaths and births occurred in roughly equal numbers. This situation changed about 400 years ago, when agricultural output increased and the separation and treatment of human waste began to improve. More food and improved sanitation caused death rates to fall, but birth rates remained relatively high. This was the beginning of

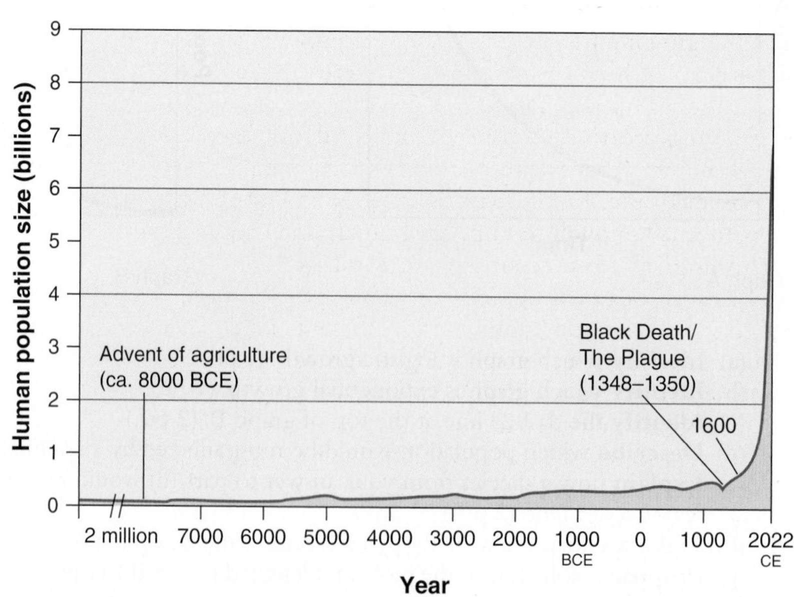

FIGURE 17.1 Human population growth. The global human population has grown more rapidly in the last 400 years than at any other time in history.

a period of rapid population growth—almost exponential growth—that has brought us to the current human population of 7.8 billion people.

As we saw earlier in this unit, under ideal conditions, all populations grow exponentially. In most cases, exponential growth slows or stops when an environmental limit is reached. The limiting factor can be a scarcity of resources such as food or water or an increase in predators, parasites, or diseases. Limiting resources determine the carrying capacity of a habitat. All animal and plant populations are ultimately constrained by the carrying capacity. Are human populations constrained by carrying capacity? You might think that the answer is an easy "yes" but in fact there has been considerable discussion about this topic within the scientific community for hundreds of years.

Environmental scientists have differing opinions on Earth's carrying capacity for humans. Some scientists believe we have already outgrown, or soon will outgrow, the available supply of food, water, timber, fuel, and other resources on which humans rely. One of the first proponents of the notion that the human population could exceed Earth's carrying capacity was English clergyman and professor, Thomas Malthus. In 1798, Malthus observed that the human population was growing exponentially while the food supply we rely on was growing linearly. In other words, the food supply increases by a fixed amount each year, while the human population increases in proportion to its own increasing size. Malthus hypothesized that the human population size would eventually exceed the food supply. **FIGURE 17.2a** shows this projection graphically.

A number of environmental scientists today subscribe to Malthus's view that humans will eventually reach the carrying capacity of Earth—or already have—after which the population will decline. Other scientists do not believe that Earth has a fixed carrying capacity for humans. They argue that the growing population of humans provides an increasing supply of intellect that leads to increasing amounts of innovation. Humans can alter Earth's carrying capacity by employing creativity—one of the fundamental ways in which humans differ from most other species on Earth.

For example, in the past whenever the food supply seemed small enough to limit the human population, major technological advances increased food production. This progression began thousands of years ago. The development of arrows made hunting more efficient, which allowed hunters to feed a larger number of people. Early farmers increased crop yields with hand plows and later with oxen- or horse-drawn plows. More recently, mechanical harvesters made farming even more efficient. Each of these inventions increased the planet's carrying capacity for humans, as Figure 17.2b shows.

The ability of humans to innovate in the face of challenges has led some scientists to expect that we will continue to make technological advances indefinitely. This expectation is reasonable, but it is not certain. Based on our history, should we assume that humans will continue to find ways to feed a growing population? Are there other limits to human population growth? And how do we know if we have exceeded Earth's carrying capacity? Answers to these questions are difficult to obtain but understanding drivers of human population change will help us begin to address them.

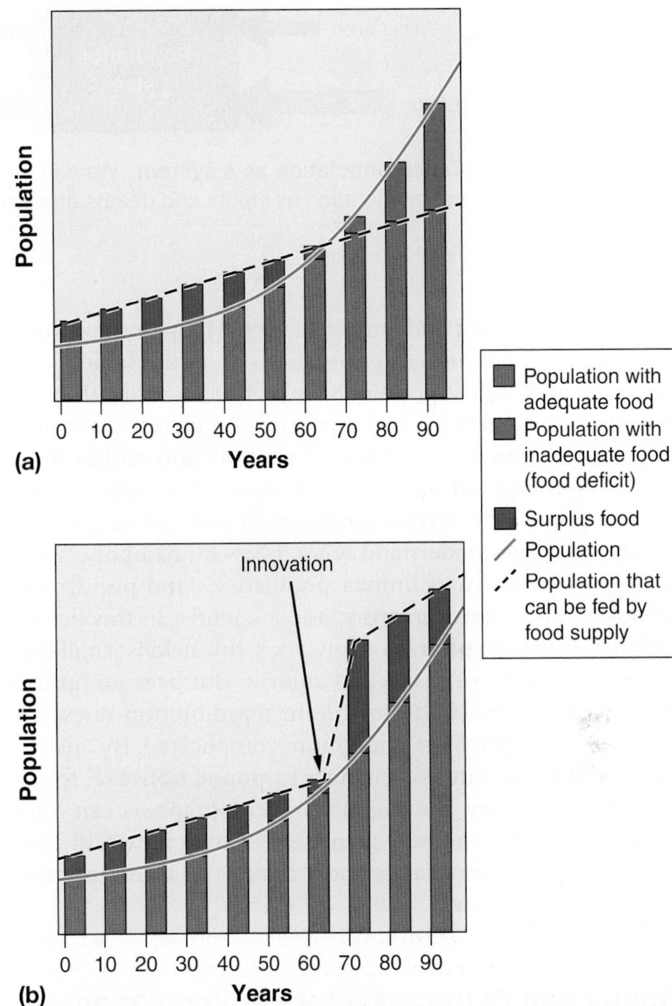

FIGURE 17.2 A theoretical model of food supply and population size. (a) In a theoretical 100-year period *without* significant improvements in agricultural technology, the human population grows exponentially, while the food supply grows linearly. Consequently, a food surplus is followed by a food deficit. (b) In a theoretical 100-year period *with* significant improvements in agricultural technology, the food supply increases suddenly. Consequently, there is a continuing food surplus.

17-2 What are the social and cultural drivers of human population change?

Many factors drive human population change

As we know from our study of plant and animal populations, a variety of factors influence the growth, reproduction,

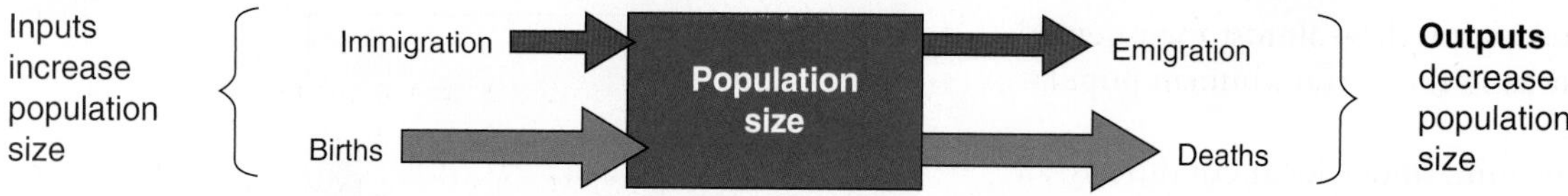

FIGURE 17.3 **The human population as a system.** We can think of the human population as a system, with births and immigration as inputs and deaths and emigration as outputs.

and success of plant and animal species. Many of these same factors influence human populations as well. Habitat size, resource availability, population size, birth and death rates, fertility, life expectancy, and migration are factors that influence population size on Earth as a whole and within individual countries. In order to understand the impact of the human population on the environment and carrying capacity, we must first understand what drives human population change. The study of human populations and population trends is called **demography**, and a scientist in this field is called a **demographer**. In many ways, this field is parallel to the studies done for plants and animals. But because human choice plays a much larger role in reproduction rates, the study of demography is often more complicated. By analyzing specific data such as changes in population size, fertility, life expectancy, and migration, demographers can offer insights—some of them surprising—into how and why human populations change and what can be done to influence rates of change.

Social and Cultural Factors in Population Change

We can view the human population as a system with inputs and outputs, like all biological systems. However, human population trends are influenced by a wide range of social and cultural factors that are not relevant to most other biological systems, such as access to health care, birth control and education, economic prosperity, and geopolitical conflicts.

Demography The study of human populations and population trends.

Demographer A scientist in the field of demography.

Immigration The movement of people into a country or region, from another country or region.

Emigration The movement of people out of a country or region.

Crude birth rate (CBR) The number of births per 1,000 individuals per year.

Crude death rate (CDR) The number of deaths per 1,000 individuals per year.

Net migration rate The difference between immigration and emigration in a given year per 1,000 people in a country.

Births, Deaths, and Migration

When demographers look at population trends in individual countries, they consider the inputs and outputs. As FIGURE 17.3 shows, inputs include both births and **immigration**, which is the movement of people into a country or region from another country or region. Outputs, or decreases, include deaths and **emigration**, which is the movement of people out of a country or region. When inputs to the population are greater than outputs, the growth rate is positive. Conversely, if outputs are greater than inputs, the growth rate is negative.

Demographers use specific measurements to determine yearly birth and death rates. The **crude birth rate (CBR)** is the number of births per 1,000 individuals per year. The **crude death rate (CDR)** is the number of deaths per 1,000 individuals per year. Worldwide, there were 19 births and 7 deaths per 1,000 people in 2020. Migration is not factored in for the global population because, even though people move from place to place, they do not leave Earth. Thus, in 2020, the global population increased by 12 people per 1,000 people. This rate can be expressed mathematically as a percentage (CBR and CDR are expressed per 1,000 so we divide by 10 to express as per 100):

$$\text{Global population growth rate in percent} = \frac{[\text{CBR} - \text{CDR}]}{10} = \frac{[19-7]}{10} = \frac{[12]}{10} \times 100 = 1.2\%$$

When we consider demographic trends for a specific country, we must look beyond its birth and death rates because a country may experience population growth, stability, or decline as a result of migration. **Net migration rate** is the difference between immigration and emigration in a given year per 1,000 people in a country. A positive net migration rate means there is more immigration than emigration, and a negative net migration rate means the opposite. For example, approximately 1 million people immigrate to the United States each year, and a much smaller number emigrate. With a U.S. population of 330 million, these rates are equal to 3 immigrants per 1,000 people.

To calculate the population growth rate for a single nation, immigration and emigration are taken into account:

$$\text{National population \% growth rate} = \frac{[(\text{CBR} + \text{immigration}) - (\text{CDR} + \text{emigration})]}{10}$$

A country with a relatively low CBR but a high immigration rate may still experience population growth. For example, currently, the United States is not replacing its population with births from the existing population, but it has a high net migration. As a result, the U.S. population will probably increase by 20 percent by 2050. Canada has a net migration rate of 9 per 1,000 and is also not replacing its population with births from the existing population. The current population in Canada of 36.7 million will probably increase by 28 percent by 2050. Therefore, both the United States and Canada will experience net population growth over the next few decades, but the growth will come from immigration rather than from births originating within the existing population. "Do the Math: Calculating Population Growth" on page 195 shows how this can happen.

In countries with a negative net migration rate and fewer births than deaths, the population actually decreases over time. Very few countries fit this model. One is the country of Georgia, in western Asia. In 2015, it had a growth rate of 0.2 percent, births were fewer than deaths, and a net migration rate of −5 per 1,000. Georgia is projected to have a 20 percent population decrease by 2050.

Life Expectancy

To understand more about the number of deaths in a human population system, demographers study the human life span. **Life expectancy** is the average number of years that an infant born in a particular year in a particular country can be expected to live, given the current average life span and death rate in that country. Life expectancy is generally higher in countries with better health care. A high life expectancy also tends to be a good predictor of high resource consumption rates and environmental impacts. **FIGURE 17.4** shows life expectancies around the world.

Life expectancy is reported in at least three different ways: for the overall population of a country, for males only, and for females only. For example, in 2020, global life expectancy was 73 years overall, 70 years for men, and 75 years for women. It is not known yet how the Covid-19 pandemic has affected global life expectancy, but it appears that life expectancy will decrease. In the United States, life expectancy was 79 years overall, 76 years for men, and 81 years for women. Already, population scientists in the United States have reported that the pandemic has resulted in a decrease in U.S. life expectancy by 1 to 2 years.

You may be wondering why global and U.S. life expectancies are shorter for men than women. In general, human males have higher death rates than human females, leading

> **Life expectancy** The average number of years that an infant born in a particular year in a particular country can be expected to live, given the current average life span and death rate in that country.

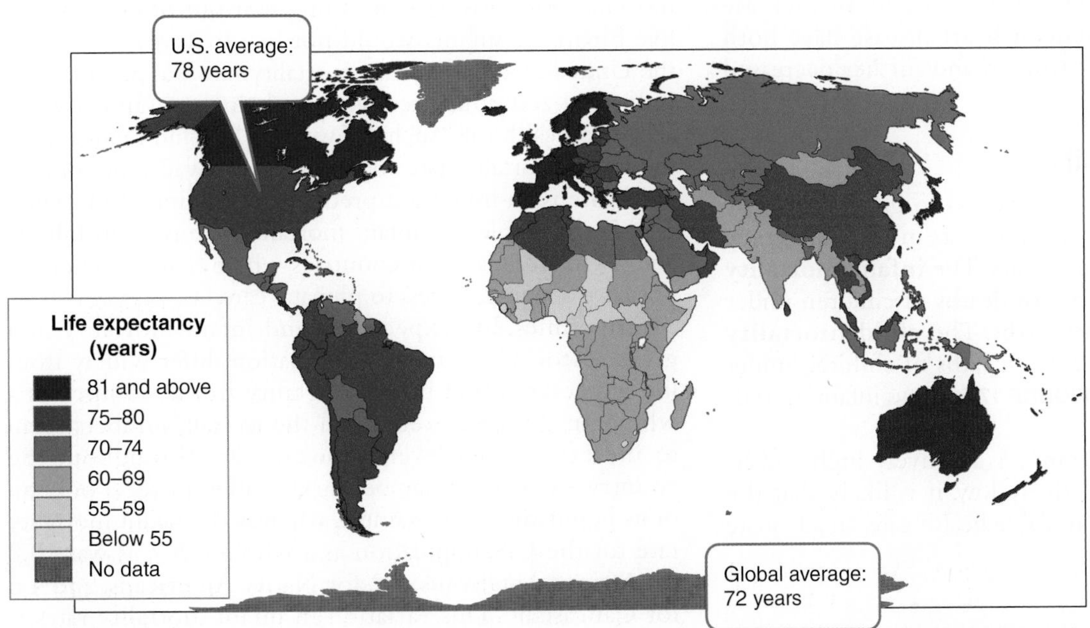

FIGURE 17.4 Average life expectancies around the world. Life expectancy varies significantly by continent and in some cases by country. *(Data from https://www.americashealthrankings.org; https://www.americashealthrankings.org/learn/reports/2015-annual-report/comparison-nations-2.)*

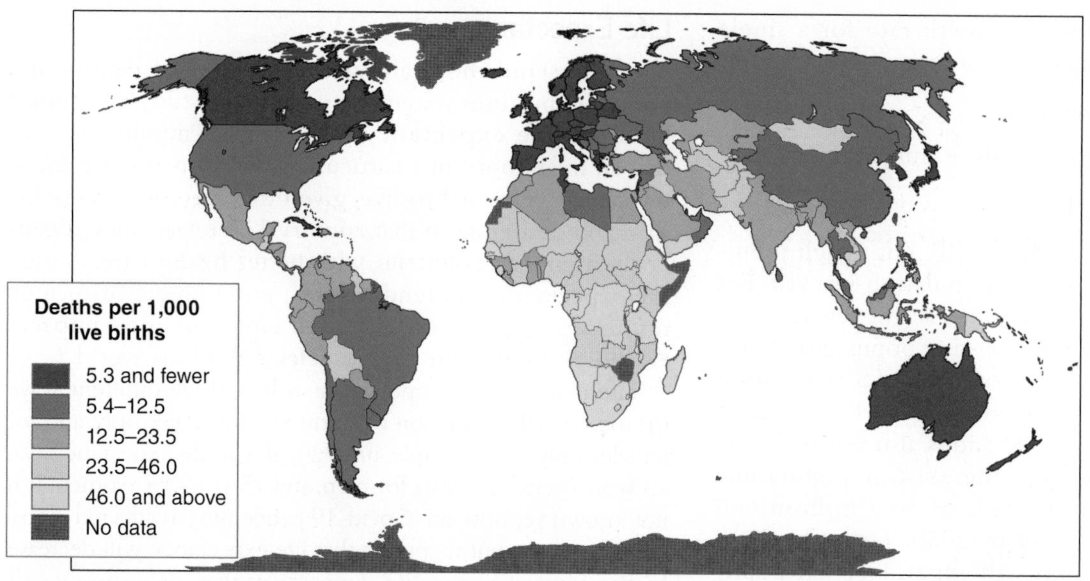

FIGURE 17.5 Infant mortality around the world. Infant mortality rates are lower in developed countries and higher in developing countries. *(Data from https://www.americashealthrankings.org/learn/reports/2015-annual-report/comparison-nations-2.)*

to a shorter life expectancy for men. In addition to biological factors, men have historically tended to face greater dangers in the workplace, made more hazardous lifestyle choices, and been more likely to die in wars. Cultural and employment norms have changed over time in some countries, however, and as more and more women enter the workforce and the armed forces, the life expectancy gap between men and women has decreased. A decline in alcohol and tobacco consumption and an increase in prevention and treatment of heart disease have both benefitted men more than women and further decreased the gap.

Infant and Child Mortality

The availability of health care, access to good nutrition, and exposure to pollutants are all factors in life expectancy, infant mortality, and child mortality. The **infant mortality** rate is defined as the number of deaths of children under 1 year of age per 1,000 live births. The **child mortality** rate is defined as the number of deaths of children under age 5 per 1,000 live births. **FIGURE 17.5** shows infant mortality rates around the world.

If a country's life expectancy is relatively high and its infant mortality rate is relatively low, it is likely that the country has a high level of available health care, an adequate food supply, potable drinking water—water that is safe to drink—good sanitation, and a moderate level of pollution. Conversely, if its life expectancy is relatively low and its infant mortality rate is relatively high, it is likely that the country's population does not have sufficient health care or sanitation and that potable drinking water and food are in limited supply. Pollution and exposure to other environmental hazards may also be high. In 2020, the global infant mortality rate was 31—meaning that out of every 1,000 live births, 31 infants would not live to 1 year of age. In the United States, the infant mortality rate was 5.7. In other industrialized countries with high income and reliable access to health care, such as Sweden (1.8) and France (3.5), the infant mortality rate was even lower. Availability of prenatal care is an important predictor of the infant mortality rate. For example, the infant mortality rate is 63 in Liberia and 29 in Bolivia, both countries where many women do not have adequate access to prenatal care.

Sometimes, life expectancy and infant mortality in a given sector of a country's population differ widely from life expectancy and infant mortality in the country as a whole. In this case even when the overall numbers seem to indicate a high level of health care throughout the country, the reality may be starkly different for a portion of its population. For example, whereas the infant mortality rate for the U.S. population as a whole is 5.7, it was 10.8 for African Americans, 8.2 for Native Americans, and 4.6 for Caucasians. This variation in infant mortality rates is related to socioeconomic status and varying degrees of access to adequate nutrition, prenatal care, and overall health care, as well as many other factors including exposure to pollution. Over the past few decades, the study of this topic

Infant mortality The number of deaths of children under 1 year of age per 1,000 live births.

Child mortality The number of deaths of children under age 5 per 1,000 live births.

DO THE MATH Calculating Population Growth

Preparing for the AP® Exam

Although today TFR is much lower, a decade ago, New Zealand had a population of 4.3 million people, a total fertility rate (TFR) of 2.1, and a net migration rate of 2 per 1,000. How many people would you expect New Zealand to gain in the following year as a result of migration? If the TFR stays the same for the next century, and the net migration rate stays the same as well, how will the population of New Zealand change?

$$\text{Net migration rate} = \frac{\text{number of immigrants/year}}{\text{number of people in the population}}$$

A TFR of 2.1 for a developed country suggests that the CBR is approximately equal to the CDR, the country is at replacement-level fertility and, therefore, the population is stable. The migration rate suggests that

$$\frac{2}{1{,}000} = \frac{x}{4{,}300{,}000}$$

and therefore that

$$x = 8{,}600 \text{ people/year}$$

So, 8,600 people are added to New Zealand each year by migration. If there is no growth due to biological replacement, then the rate of increase is

$$\frac{8{,}600 \text{ people/year}}{4{,}300{,}000 \text{ people}} = 0.002/\text{year} \times 100 = 0.2\%/\text{year}$$

YOUR TURN What will happen to the population in New Zealand if TFR drops below 2.1 and migration decreases to 0 per 1,000 per year?

has been given a name: **environmental justice**, which is the disproportionate exposure to environmental hazards experienced by people of color, recent immigrants, and people of lower socio-economic backgrounds. Environmental justice is both an academic field and a social movement. A number of studies have documented that there are greater numbers of landfills and hazardous waste sites, solid-waste incinerators, and other industries that generate pollutants in closer proximity to people of color than to Caucasians.

Aging and Disease

Even with a high life expectancy and a low infant mortality rate, a country may have a high crude death rate, in part because it has a large number of older individuals. The United States, for example, has a higher per capita income than Mexico, which is consistent with the higher life expectancy and lower infant mortality rate in the United States. At the same time, the United States has a much higher CDR, at 9 deaths per 1,000 people on average, than Mexico, which has 6 deaths per 1,000 people on average. This higher CDR results from the much larger elderly population in the United States, with 16 percent of its population aged 65 years or older, compared with the 7 percent of the population aged 65 or older in Mexico.

Disease is also a substantial regulator of human populations. According to the World Health Organization, infectious diseases — those caused by microbes that are transmissible from one person to another — are the second biggest killer worldwide after heart disease. In the past, tuberculosis and malaria were two of the infectious diseases responsible for the greatest number of human deaths. Until the last few years, the human immunodeficiency virus (HIV), which causes acquired immune deficiency syndrome (AIDS), was responsible for more deaths annually than either tuberculosis or malaria. In 2016, there were 1 million AIDS-related deaths. Since then, due to health care improvements, annual deaths have decreased. Between 1990 and 2015,

Environmental justice The study of the disproportionate exposure to environmental hazards experienced by people of color, recent immigrants and people of lower socio-economic backgrounds; and is both an academic field and a social movement.

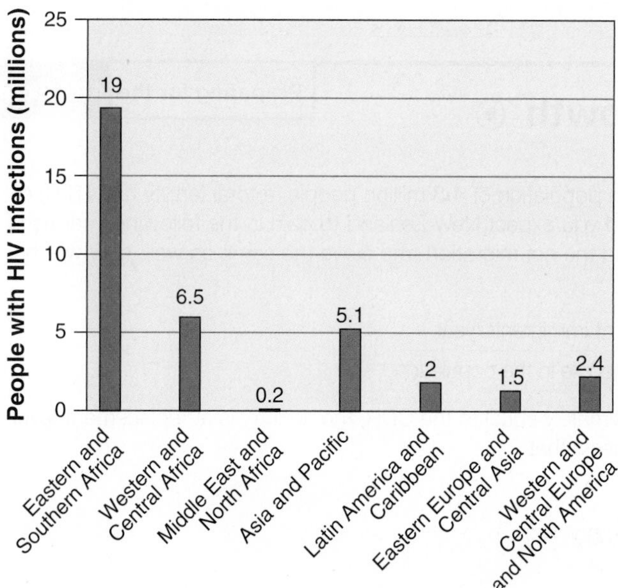

FIGURE 17.6 HIV infection worldwide. Worldwide, about 37 million people are living with HIV, two-thirds of them in sub-Saharan Africa, which includes all of Africa south of the Sahara (essentially all of Africa except North Africa). *(Data from http://www.unaids.org/sites/default/files/media_asset/2016-AIDS-data_en.pdf.)*

AIDS-related illnesses killed more than 30 million adults and children. Because HIV disproportionately infects people aged 15 to 49—the most productive years in a person's life span—HIV has had a more disruptive effect on society and life expectancy than other illnesses that affect the very young and the very old. In 2020, AIDS-related deaths decreased to fewer than 700,000, which is lower than deaths from tuberculosis (1.2 million) but higher than deaths from malaria (410,000). Covid-19 deaths worldwide have also been highly disruptive. In 2020, 1.9 million deaths occurred worldwide from Covid-19 and a total of more than 6 million deaths have occurred since the beginning of the pandemic through mid 2022.

As **FIGURE 17.6** shows, more than 37 million people were living with HIV in 2020, with approximately 26 million of them in Africa south of the Sahara. The annual number of deaths due to AIDS reached a peak of 2.1 million in 2005 but has decreased since then. We will talk more about AIDS and other infectious diseases in Unit 8.

Age structure diagram A visual representation of the number of individuals within specific age groups for a country, typically expressed for males and females.

Population pyramid An age structure diagram that is widest at the bottom and smallest at the top, typical of developing countries.

Developing countries Countries with relatively low levels of industrialization and income.

Developed countries Countries that have relatively high levels of industrialization and income.

17-3 How can we describe and interpret age structure diagrams?

Age structure diagrams describe how populations are distributed across age ranges and provide insights into future population changes

The crude birth rate and death rate, the life expectancy, and other factors determine how many people are alive on Earth or in a particular country. We can be even more specific by illustrating graphically how many people are currently alive within specific age ranges. The age structure of a population describes how its members are distributed across age ranges, usually in 5-year increments. An **age structure diagram**, examples of which are shown in **FIGURE 17.7**, is a visual representation of the number of individuals within specific age groups for a country, typically expressed separately for males and females. Each horizontal bar of the diagram represents a 5-year age group. The total area of all the bars in the diagram represents the size of the whole population. Most age-structure diagrams delineate three age ranges: pre-reproductive years (roughly ages 0–14), reproductive years (ages 15–44) and post-reproductive years (ages 45 and older). Later in this section you will see how understanding the fraction of the population that is below age 15 and greater than 65 is a useful metric to know. We can also make projections about what a population will look like in the future. Demographers use data on the current age structure of a population and this population's life expectancy to predict how rapidly or slowly a population will increase or decrease and what its size will be at a future point in time.

While every nation has a unique age structure, we can group countries very broadly into three categories based on how the population appears when graphed. A country with many more younger people than older people has an age structure diagram that is widest at the bottom and narrowest at the top, as shown in Figure 17.7a. This type of age structure diagram, called a **population pyramid**, is typical of developing countries—such as Venezuela, South Sudan, and Nigeria. **Developing countries** are those with relatively low levels of industrialization and income (sometimes defined as less than $3/day per capita). **Developed countries** have relatively high levels of industrialization and income. Developing countries also tend to have more children in the work force than developed countries. The wide base of the graph compared with the levels above it indicates that the population will grow because a large number of females aged 0 to 15 have yet to bear children. Even if each one of these future potential mothers has two children and no more, the total population size will grow simply because there are increasing numbers of females able to give birth in the younger age cohorts.

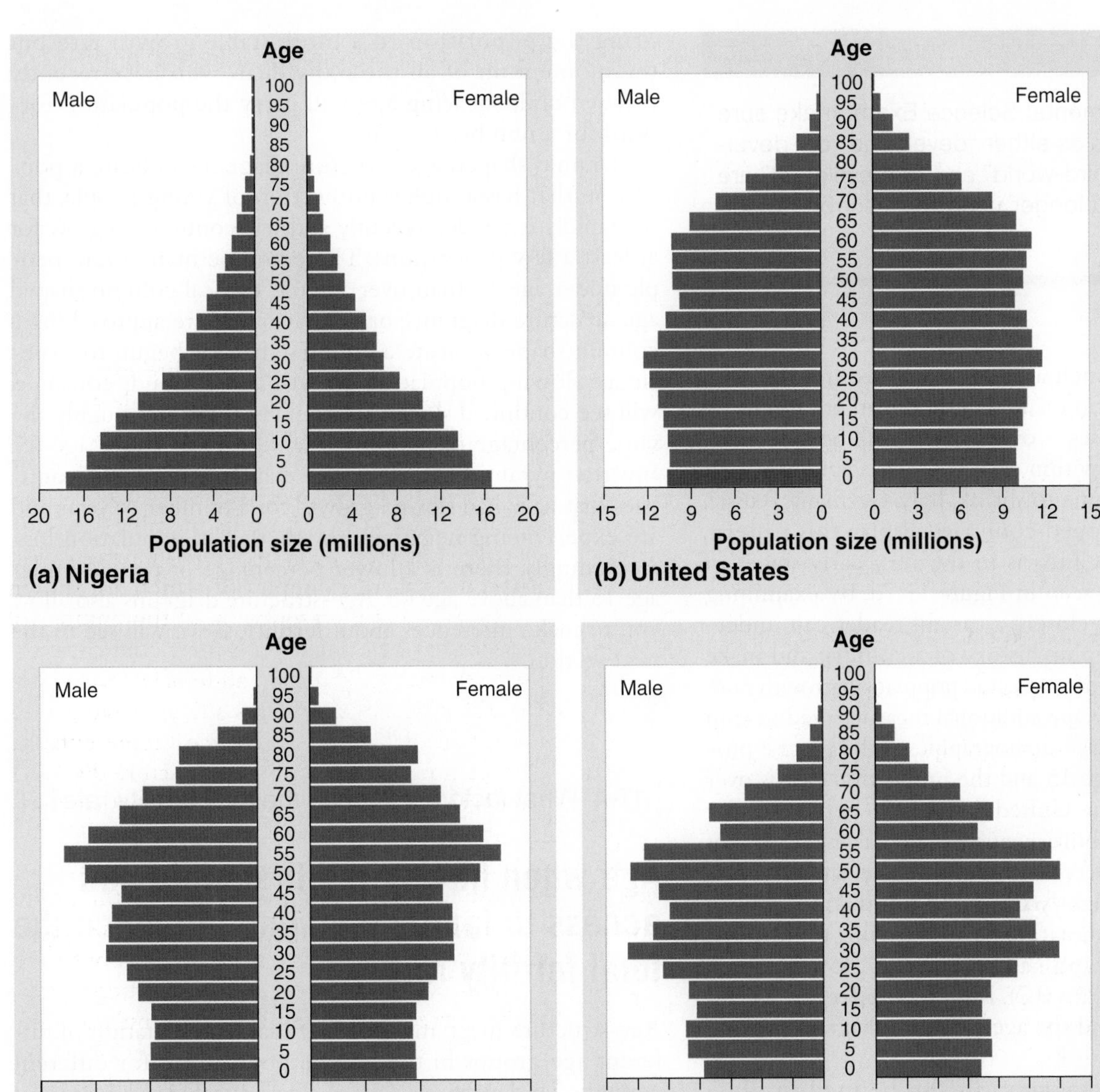

FIGURE 17.7 Age structure diagrams for four countries in 2020. The horizontal axis of the age structure diagram shows the population size in millions for males and females in each 5-year age group shown on the vertical axis. (a) A population pyramid illustrates a rapidly growing population (Nigeria). (b) A column-shaped age structure diagram indicates population stability (United States). (c) In some developed countries, the population is declining (Germany). (d) China's population control measures will eventually lead to a population decline. *(2020 data from https://www.census.gov/population/international/data/idb.)*

A country with little difference between the number of individuals in younger age groups and in older age groups has an age structure diagram that looks more like a vertical column, from age 0 through age 50, as shown in Figure 17.7b. Virtually every age-structure diagram begins to show tapering at the oldest age groups because of natural senescence (death) of human beings. If a country has few individuals in the younger age classes, we can deduce that it has slow population growth or is approaching no growth at all. The United States, Canada, Australia, Sweden, and many other developed countries have this type of age structure diagram. A number of developing countries that have recently lowered their growth rates should begin to show this pattern within the next 10 to 15 years.

A country with a greater number of older people than younger people has an age structure diagram that resembles

AP® Exam Tip

On the AP® Environmental Science Exam, make sure to describe countries as either "developed" or "developing." Terms like "third-world" and "undeveloped" are outdated and are no longer used to describe human populations.

an inverted pyramid. Such a country will exhibit decreasing growth rates over time and—although there might be some occasional increases—overall there will be a decreasing number of females within each younger age range. Such a population will continue to shrink. Italy, Germany, Russia, and a few other developed countries display this pattern, seen in Figure 17.7c. China is in the very early stages of showing this pattern as seen in Figure 17.7d. By examining age-structure diagrams closely, a careful reader can understand the past demographic history of a country and make educated guesses about the expected population growth rates in the coming decades. One additional measure used to gain a "snapshot" of a country's demographics is the precise proportion that is under age 15 and the proportion that is over age 65. For example, the United States is 18 percent under age 15 and 16 percent above age 65. Guatemala is 37 percent under age 15 and 5 percent over age 65. The United States is showing stable or slightly declining population growth (Figure 17.7b) while Guatemala is showing rapid population growth (no graph shown but the growth rate is 1.7 percent and the TFR is 2.7). Both the under age 15 and over age 65 numbers and the age structure diagram provide this information.

The age-structure diagram can also be used to illustrate how long a time it takes for social changes to affect a growing population. **Population momentum** is continued population growth after growth reduction measures have been implemented. It occurs because there are relatively large numbers of individuals at reproductive maturity in the population. Population momentum has been compared with the momentum of a long, heavy freight train, which takes longer to stop than a shorter, lighter freight train. It is the reason why a population keeps on growing after birth control policies or voluntary birth reductions have begun to lower the CBR of a country. Eventually, over several generations, those actions will bring the population to a more stable growth rate but the momentum of all the individuals who have recently reached child-bearing age will carry the population forward for a number of years.

Pyramid shaped age-structure diagrams indicate a population that has a higher proportion of young people, that has rapidly expanded recently and will continue to grow for at least a few generations. There will be many more people under age 15 than over age 65. Vertical column shaped age-structure diagrams, or diagrams that are approaching a column shape, indicate a country that has begun to experience slowing population growth and, if trends continue, will see continued slowing. Generally, there are roughly the same percentage of people under age 15 as above age 65. Inverted pyramid age-structure diagrams represent countries that have had slowed growth for a number of years and are experiencing negative growth, that is, population loss. Accordingly, there is a lower percentage of people under age 15 than above age 65. Age-structure diagrams also allow you to make inferences about fertility, as we will see in the next section.

Population momentum Continued population growth after growth reduction measures have been implemented.

Total fertility rate (TFR) An estimate of the average number of children that each woman in a population will bear throughout her childbearing years.

17-4 What factors influence the total fertility rate?

Age when having the first child and access to family planning influence the total fertility rate

Age-structure diagrams allow a visual representation of different age groups in the human populations for different countries and allow us to make an educated prediction of whether the population as a whole is increasing or decreasing. Now we will examine one additional human population parameter, total fertility rate, which provides another indication of whether or not a population is growing, influences age-structure diagrams, and plays an important role in overall population dynamics. We also explain how total fertility rate is influenced by access to family planning services, education, the age a woman has her first child, and other social and economic factors.

To understand more about the role that births play in population growth, demographers look at the **total fertility rate (TFR)**, an estimate of the average number of children that each woman in a population will bear throughout her childbearing years (between the onset of puberty and menopause). Typically, childbearing years are considered as from ages 15 to 49. In the United States in 2020, the TFR was 1.7, meaning that, on average, each woman who was alive in 2020 was expected during ages 15 and 49 to give birth to an average of 1.7 children. Note that, unlike crude birth rate and crude death rate, TFR is not calculated per

1,000 people. Instead, it is a measure of births per individual woman over her lifetime, for an average woman in the population or country.

Many factors influence TFR. The age at which a woman has her first child is strongly correlated with TFR: If a woman delays having her first child, there will be fewer years during which she can have children. This single act of delaying when the first child is born often leads to a decrease in the total number of children a woman has over her lifetime. Pursuing education and entering the workplace both take women out of the home, particularly in developing countries. And both of these activities directly correlate with delayed child rearing. These activities also lead to higher household income. As **FIGURE 17.8** illustrates for a variety of countries, total fertility rate correlates inversely with income—that is, countries that have higher TFR tend to have lower income per capita. Women with higher incomes and more education also tend to have more access to information about methods of birth control; they are more likely to interact with their partners as equals; and they may choose to practice family planning. **Family planning** is regulation of the number or spacing of offspring through the use of birth control. When women have the option to use family planning, crude birth rates and total fertility rates decrease. **FIGURE 17.9** shows how female education levels correlate with total fertility rates. In Ethiopia, for instance, women with a secondary school education or higher have a TFR of 1.5, whereas the TFR among women with no formal education is 5.8.

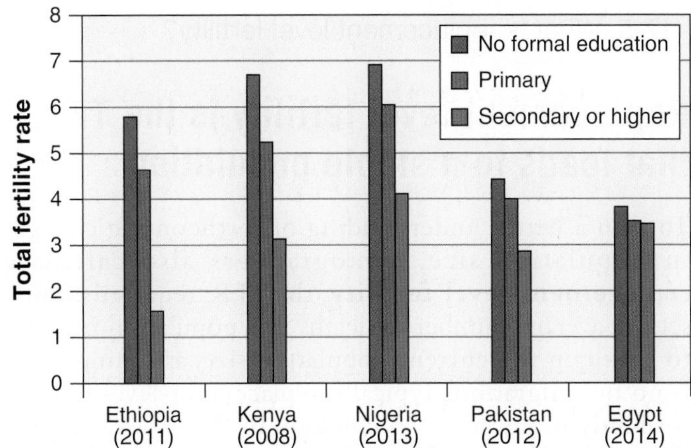

FIGURE 17.9 Total fertility rates for educated women and those lacking formal education. Fertility is strongly related to female education in many developing countries.

AP® Exam Tip

One of the more successful ways to reduce population growth without violating human rights is by providing women with access to education.

There have been many examples of effective family planning campaigns. In the 1980s, Kenya had one of the highest population growth rates in the world, and its TFR was almost 8. By 1990, its TFR was about 4—one-half of the previous rate. Kenya's government achieved these dramatic results by implementing an active family planning campaign. The campaign, which began in the 1970s, encouraged smaller families. Advertising directed toward both men and women emphasized that overpopulation led to both unemployment and harm to the natural environment. The ads also promoted the use of contraceptives. Today the TFR in Kenya is 3.5. There are also countries such as Japan where TFR is too low (1.3 in 2020) and advertising campaigns attempt to increase TFR!

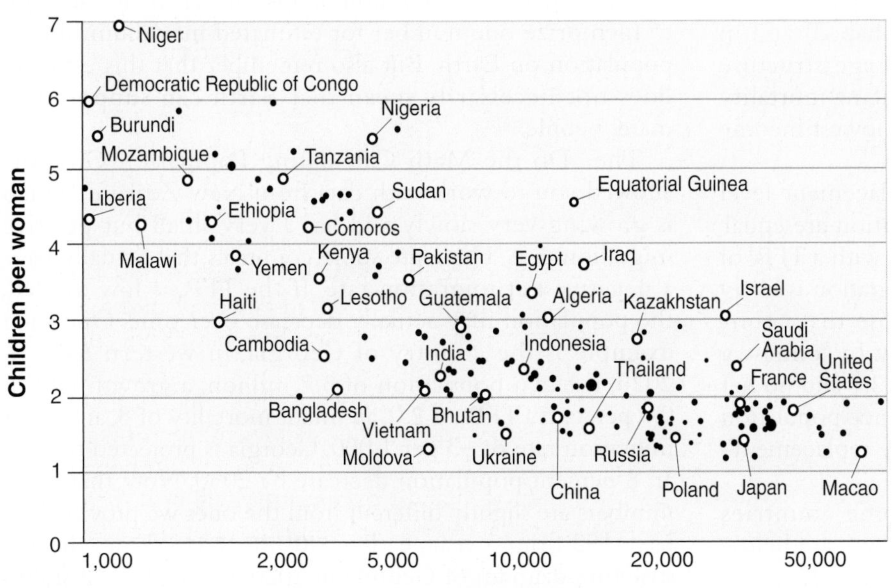

FIGURE 17.8 Total fertility rate and per capita income. Wealthier nations tend to have lower total fertility rates.

Family planning The regulation of the number or spacing of offspring through the use of birth control.

17-5 What is replacement level fertility?

Replacement level fertility is the TFR that leads to a stable population

To gain a better understanding of forthcoming changes in population size, demographers also calculate **replacement-level fertility**, the TFR required to offset the average number of deaths in a population in order to maintain the current population size, assuming there is no net migration. Typically, replacement-level fertility is slightly more than two children. In theory, exactly two children will replace the two parents who conceived them. In practice, the replacement level fertility—the number of children needed to replace two parents—is higher because a small percentage of children die before they are able to have children of their own, and some people cannot or choose not to have children. As we have seen and will explore further, the rate of death among infants and children depends on a number of factors including the level of health care, the availability of health care, and the economic status of each country.

In developed countries, we typically see a replacement-level fertility of about 2 or 2.1 (to allow for the occasional death before a woman has children). In developing countries—those with a slightly higher infant mortality and death rate of children and young adults—a TFR of greater than 2.1 is needed to achieve replacement-level fertility because a higher portion of women will die before they have children. The TFR for the world is 2.3. The TFR in the highest income countries is 1.6 (and age-structure diagrams are or are approaching column-shaped) and in the lowest income countries it is 4.6 (and age-structure diagrams are typically pyramid-shaped). Infant mortality in the highest income countries is 5 and in lowest income countries it is 50.

In a country where TFR is equal to replacement-level fertility, and where immigration and emigration are equal, the country's population is stable. A country with a TFR of less than 2.1 and no net increase from immigration is likely to experience a population decrease because that country's TFR is below replacement-level fertility. In contrast, a developed country with a TFR greater than 2.1 and no net decrease from emigration is likely to experience population growth because that country's TFR is above replacement-level fertility.

Around the world, TFR varies and some countries are above replacement level fertility and others are below it. Overall world population changes are the net result of all the individual country TFRs. **FIGURE 17.10** shows the current projections through the year 2100. Most demographers believe that the human population will be somewhere between 8 billion and 10 billion in 2050 and will stabilize between 7 billion and 11 billion by roughly 2100. Remember "10 billion people on Earth" if you want to memorize one number for estimated maximum human population on Earth. But also remember that this estimate does not necessarily mean that Earth can support this many people.

The "Do the Math Calculating Population Growth" allowed you to work with data from New Zealand, which is growing very slowly and has a very small but positive migration rate. There are some countries that actually have a negative net migration rate. If the TFR is low as well, the population may actually decrease over time. One such example is the country of Georgia, in western Asia. In 2020, it had a population of 3.7 million, a growth rate of 0.0 percent, a TFR of 2.0, an infant mortality of 8, and a net migration rate of −3 per 1,000. Georgia is projected to have an 8 percent population decrease by 2050. Note that these numbers are slightly different from the ones we provided on page 193 from 5 years earlier. **FIGURE 17.11a** shows the age structure diagram of Georgia in 2020. Figure 17.11b shows the age-structure diagram in 2050. As you might expect, by 2050, the country age structure diagram is approaching the look of a column, especially in the younger age groups. And the TFR for Georgia in 2050 will be lower than 2.0.

Replacement-level fertility The total fertility rate required to offset the average number of deaths in a population in order to maintain the current population size.

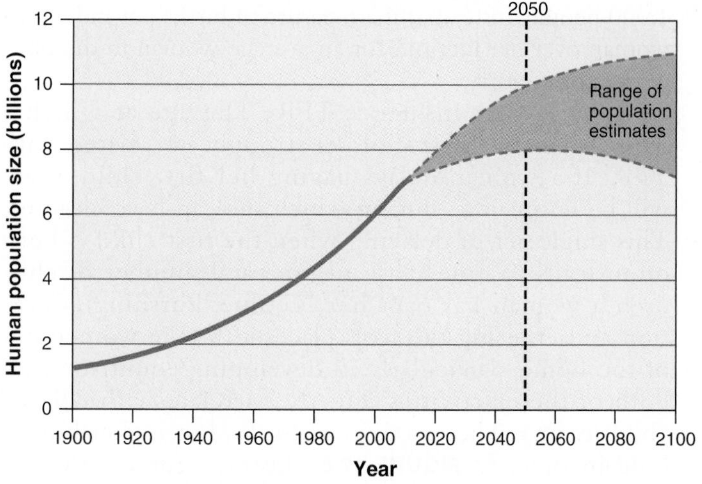

FIGURE 17.10 Projected world population growth. Demographers project that the global human population will be between 8 billion and 10 billion by 2050. By 2100, it is projected to be between 7 billion and 11 billion. The dashed lines represent estimated values. Ten billion people is a common mid-range estimate for total number of people on Earth. *(Data from Millennium Ecosystem Assessment, 2005; Population Reference Bureau [www.prb.org], 2020.)*

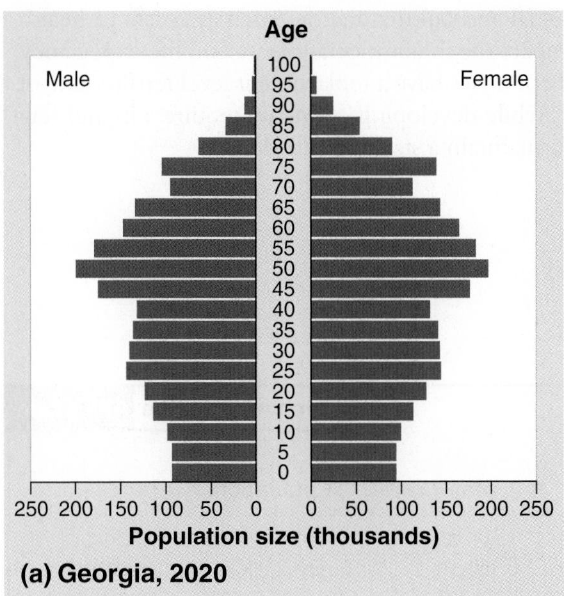

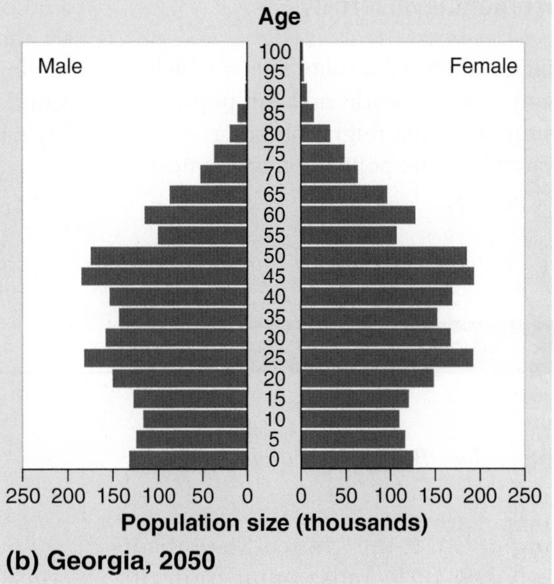

FIGURE 17.11 **Age-structure diagram for the country of Georgia.** (a) In 2020; (b) in 2050. Georgia is projected to have a column-shape by 2050.

Module 17 AP® Review

Preparing for the AP® Exam

Learning Goals Revisited

17-1 What are the factors that justify the carrying capacity of humans on Earth?

Scientists disagree about the human carrying capacity of Earth. Some believe that we have exceeded human carrying capacity on Earth. An early presenter of this view was Thomas Malthus. Others suggest that human ingenuity and innovation will allow an almost infinite increase in the carrying capacity for human existence.

17-2 What are the social and cultural drivers of human population growth?

Many factors drive human population growth, including crude birth rate, crude death rate, infant mortality, life expectancy and migration. These factors are influenced by both social and cultural drivers that vary by country, level of income and education, and regions of the world.

17-3 How can we describe and interpret age structure diagrams?

Age structure diagrams are a common way to represent the current age composition of a population and allow demographers to predict future changes in population age distributions. For example, age structure diagrams for countries with more young people look like a pyramid, and indicate these countries will have rapidly growing populations. Countries with age structure diagrams that look like vertical columns will experience slow growth or a stable population, while an age structure diagram that looks like an inverted pyramid is indicative of a shrinking population. Over time, as birth and death rates of a country change, the age structure diagram of that country will change as well.

17-4 What factors influence the total fertility rate?

The total fertility rate is the average number of children that a woman in a given population will have during her childbearing years. The TFR is influenced by many social and cultural factors, such as norms around sex and marriage, the opportunities for education and work that are available to women, and access to family planning services.

17-5 What is replacement level fertility?

Replacement level fertility is the total fertility rate at which births happen at the same rate as deaths and the population remains stable, assuming the net migration rate is zero. This fertility rate will vary from one population to the next based on social and cultural factors, such as access to health care, that impact the infant mortality rate and life expectancy. Developed countries have a replacement level fertility rate of around 2.1, while developing countries require a higher fertility rate to maintain a stable population.

Practice Math and Graphing

Preparing for the AP® Exam

Answer the following questions. Be sure to show all your work.

1. **Practice Math**

   At the beginning of 2020, the United States had a CBR of 12, a CDR of 9, and a population of 330 million. Based on births and deaths, and ignoring migration, how many people were expected to be in the United States at the end of 2020?

2. **Practice Graphing**

   The U.S. Census Bureau reports the following total population values for the United States over the past 6 decades (with some rounding).

Year	Population
1960	180 million
1970	200 million
1980	225 million
1990	250 million
2000	280 million
2010	310 million
2020	330 million

   (a) Graph the U.S. population over time with millions of people on the y axis and year on the x axis.
   (b) Calculate the percentage increase (relative to the 1960 population) in the U.S. population between 1960 and 2020.

AP® Practice Questions

Preparing for the AP® Exam

Multiple-Choice Questions

1. Thomas Malthus introduced
   (a) the study of human demographics.
   (b) the hypothesis that humans could exceed Earth's carrying capacity.
   (c) the concept of variation in replacement-level fertility due to differences in industrialization.
   (d) the age structure diagram.

2. A country has a net migration rate of 3 per 1,000, a CBR of 9 per 1,000, and a CDR of 11 per 1,000. What is the growth rate of this country?
   (a) 0.1 percent
   (b) 0.2 percent
   (c) 0.5 percent
   (d) 1 percent

3. Replacement level fertility in a developed country is roughly equal to
   (a) 1.5.
   (b) 1.8.
   (c) 2.1.
   (d) 2.4.

4. An age-structure diagram that is a pyramid shape indicates a country that is
   (a) experiencing more deaths than births.
   (b) growing rapidly.
   (c) not growing.
   (d) gradually decreasing in population size.

5. The total fertility rate of the world population over time is shown in the graph below. Explain the best cause of the pattern seen in total fertility in this time period.

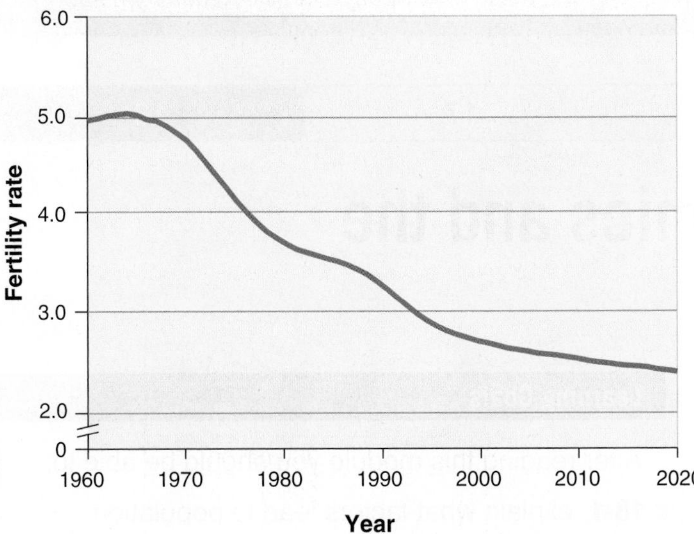

(a) TFR has decreased due to the decrease in available birth control.
(b) TFR has increased due to the education of women.
(c) TFR has decreased because of employment and education conditions have improved for women.
(d) TFR has decreased because of employment and education conditions for men.

Free-Response Question

Look at the age structure diagrams for country A and country B below and answer the questions that follow.

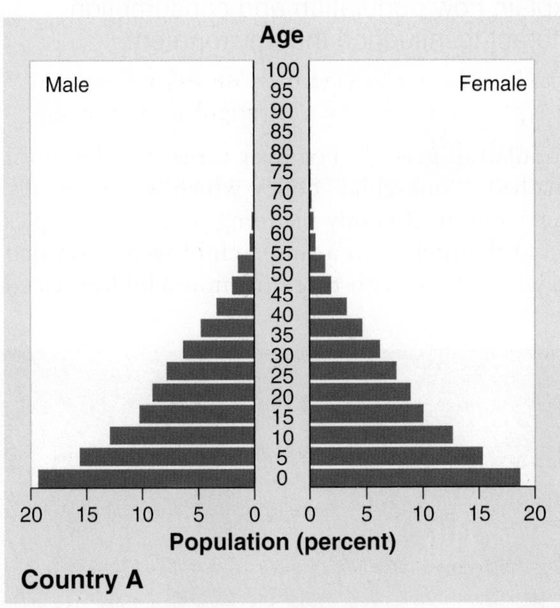

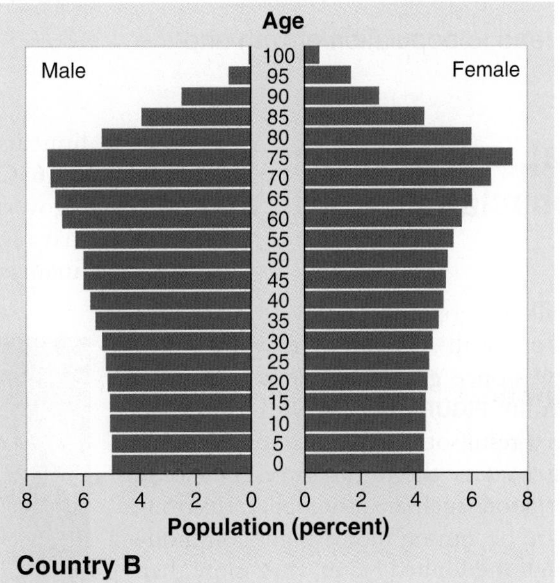

(a) **Explain** how country A differs from country B in terms of
  (i) the age structure of its population. (2 pts.)
  (ii) the total fertility rate of the country. (2 pts.)
  (iii) the life expectancy of the population. (2 pts.)
  (iv) the growth rate of the population. (2 pts.)
(b) **Describe** one socioeconomic feature of country A. (2 pts.)

Human Population Dynamics and the Demographic Transition

Demographic parameters discussed earlier in this unit such as crude birth rates and death rates, infant mortality, and total fertility rates all have interrelated impacts on human population dynamics. These population changes do have certain recognizable patterns and, in some cases, can be modeled and described with the demographic transition model. This model illustrates the differences between countries as they transition from one stage of economic development to another.

Learning Goals

After reading this module you should be able to:

18-1 explain what factors lead to population growth and decline.

18-2 calculate doubling times using the rule of 70.

18-3 describe the demographic transition.

18-4 explain how population and consumption interact to influence the environment.

18-1 What factors lead to population growth and decline?

Social, political, and economic factors contribute to population growth and decline

With more than 1.4 billion people—roughly 18 percent of the human population on Earth—China is the world's most populous nation and was once growing rapidly. Today it is below replacement fertility. **FIGURE 18.1** shows the crowded capital city, Beijing. As a result of its rapid economic development, it may one day become the world's largest economy. Once-scarce consumer goods such as automobiles, electronics, and refrigerators are becoming increasingly commonplace in China. Although the United States, with more than 330 million people, has historically been the world's largest consumer of resources and the greatest producer of many pollutants, China is rapidly surpassing the United States in both consumption and pollution. It is already the largest emitter of carbon dioxide and sulfur dioxide, and it consumes one-third of commercial fish and seafood. China is making considerable environmental gains and is also facing considerable environmental challenges as its affluence increases.

Limiting Population Growth in China

To manage the presence of humans on Earth in a sustainable way, we must address both population growth and resource consumption. China has already taken dramatic steps to limit its population growth. For over three decades, until 2016, China had a "one-child" policy which was essentially a government-enforced family planning program. Couples that restricted themselves to a single child were rewarded financially, while those with three or more children faced

FIGURE 18.1 There are 20 million people in the greater Beijing, China, area. China has a population of 1.4 billion. *(Peter Adams/Getty Images)*

sanctions, such as a 10 percent salary reduction. Chinese officials used numerous tools—many controversial—to meet population targets, including abortions, forced sterilizations, and the designation of certain pregnancies as "illegal."

China is one of only a few countries where government-mandated population control measures contributed to reducing population growth. After decades of having one of the world's highest fertility rates—for example, China's TFR in 1970 was 6.3—China now has a TFR of 1.5. In addition to the one-child policy, economic prosperity has also contributed to slowing population growth. While China currently has the largest population in the world, in a matter of years, India is projected to surpass it in population size.

Population decline is only one part of the picture, however. Even if China's population were to stop growing today, the country's resource consumption would continue to increase as standards of living improve. The middle class in China is over 500 million people, which is larger than the total population of every other country in the world other than China and India. Greater numbers of Chinese people are purchasing cars, home appliances, and other material goods that people in Western nations commonly own. These products require resources to produce and use. Manufacturing a refrigerator requires mining and processing raw materials such as steel and copper, producing plastic from oil, and using large quantities of electricity. Having a refrigerator in the home increases daily electricity demand. All of these processes generate carbon dioxide, air and water pollution, and other waste products.

A look at a typical Chinese city street is evidence of the country's growing affluence. Between 1985 and 2002, China's population increased 30 percent, but the number of motor vehicles used in China grew by over 500 percent, from 3 million to 20 million. According to current Chinese government estimates, China now has over 270 million vehicles. China is the second largest consumer of petroleum (after the United States), and concentrations of urban air pollutants, such as carbon monoxide and photochemical smog, are high relative to many other countries. A study in the *Proceedings of the National Academy of Sciences of the United States of America* documented that air pollution from China travels across the Pacific Ocean and has resulted in declining air quality in the western United States.

There is some good news, however. China already has higher fuel efficiency standards for cars than does the United States, and it is a leader in the manufacturing of renewable energy technologies. In 2017, China announced its intention to ban gasoline- and diesel-powered automobiles and recently it released a target date of 2035.

China's influence on the environment is dramatic because of its size and its increasing industrial activity. However, increasing industrial activity is occurring in many other parts of the world as human populations and consumption increase. The middle class in India may be larger than that in China.

Factors of Global Population Change

Human population growth and associated resource consumption have large impacts on the environment. In early 2022, Earth's human population was 7.8 billion and increasing by 225,000 people per day. At the same time, people in many parts of the world are using more resources than ever before. Almost all aspects of environmental science are affected by the numbers of people on Earth and their activities. This final module in Unit 3 describes how human populations grow and decline and explains transitions that have taken place in various countries.

TABLE 18.1 shows the general factors that contribute to population growth and decline. Many factors have been discussed earlier in this unit and the results appear predictable. For example, high crude birth rates lead to population growth and low crude birth rates lead to population decline. But some factors are less obvious. For example, access to adequate nutrition in developing countries might allow women to have a higher TFR, especially if they suffered from malnutrition or did not receive enough calories per day. However, in a developed country, access to adequate nutrition might correlate with access to education or better employment, which might lead to lower TFR. Infant mortality is a metric where the ultimate impact on population is sometimes less obvious. High infant mortality can directly lead to a population decline. But, when lower infant mortality is experienced in a country or region for a number of years, some families will deliberately choose to have fewer children. This might be because now they have the expectation that more of their children will live. They once had larger families because they thought that some of their children might die; now they no longer need to follow that practice.

There are a number of factors that limit human population on Earth. These include the carrying capacity of Earth, which we learned earlier might be fixed or expandable, depending on human ingenuity and inventiveness. As Malthus stated, in theory, eventually the human population growth should be limited by one factor. Some of these factors are density independent, as we saw in Module 16 when discussing biological populations. These include factors such as storms, fires, heat waves, and droughts, all of which are projected to increase with increasing global climate change. There is evidence that many of these are increasing already.

TABLE 18.1 Factors that contribute to population growth and decline

Population growth	Population decline
High crude birth rate	Low crude birth rate
High total fertility rate	Low total fertility rate
Low infant mortality	High infant mortality
Early marriage	Postponement of marriage
Lack of access to family planning	Access to family planning
Access to adequate nutrition (developing countries)	Access to adequate nutrition (developed countries)
Low overall death rate	High overall death rate

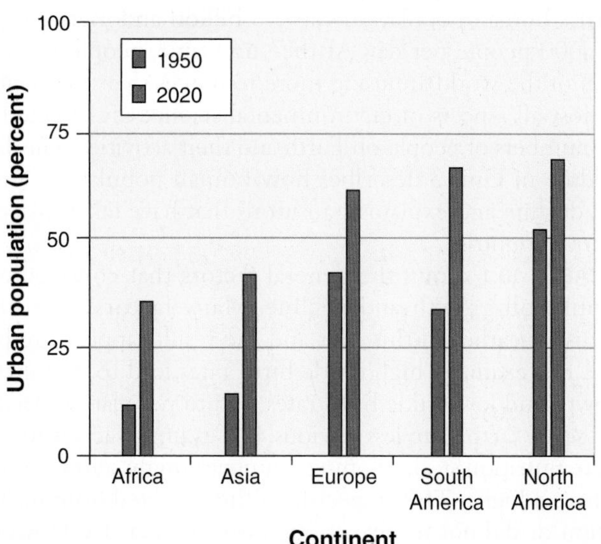

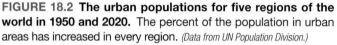

FIGURE 18.2 The urban populations for five regions of the world in 1950 and 2020. The percent of the population in urban areas has increased in every region. *(Data from UN Population Division.)*

TABLE 18.2 The largest 10 urban areas in the world in 2021

Rank	City, Country	Population in 2021 (millions)
1.	Tokyo, Japan	37.4
2.	Delhi, India	29.4
3.	Shanghai, China	26.3
4.	São Paulo, Brazil	21.9
5.	Mexico City, Mexico	21.7
6.	Al-Qahirah (Cairo), Egypt	20.5
7.	Dhaka, Bangladesh	20.3
8.	Mumbai, India	20.2
9.	Beijing, China	20.0
10.	Osaka, Japan	19.2

Note: Data are from 2021 and contain areas in and around the cities defined by the United Nations as "urban agglomerations." *(Data from World Population Review, 2021, https://worldpopulationreview.com/)*

There are also density-dependent factors that are similar to those described by Malthus. Examples of these are access to clean water, air, and food and protection from transmission of infectious diseases. Also, the size of the territory or ecosystem in which a population exists can limit a population if it is relatively small and may facilitate a larger population size if it is relatively large.

Largest Cities in the World

Over the past 70 years, more and more people are living in urban areas. **FIGURE 18.2** shows the urban population in five regions of the world in 1950 and 2020. The urban population in some regions has increased by a factor of 3. Today, more than 56 percent of the world lives in urban areas. In North America, almost 70 percent lives in urban areas and that percentage is expected to increase in the next 15 years. More than 75 percent of people in developed countries live in urban areas, and that number is expected to increase slightly over the next 15 years. In developing countries, 48 percent of people live in urban areas. This number is estimated to increase to 56 percent by 2030. **TABLE 18.2** shows that, of the 10 largest cities in the world, eight are in developing countries. Worldwide, almost 5 billion people are expected to live in urban areas by 2030.

Doubling time The number of years it takes a population to double.

Although no urban populations are expected to double anytime soon, calculating the rate of change for various populations is very useful. One standard calculation used by environmental scientists is the time it takes for a population to double, called the doubling time.

18-2 How can you calculate doubling time using the rule of 70?

The rule of 70 allows you to calculate the doubling time

When describing a country, sometimes we wish to characterize the rate at which its population is going to double in size. In Module 16, we described how the exponential growth model is used to calculate the rate of population change based on the growth rate, the original population size, and the number of years into the future we wanted to project. There is a quick calculation that can be made that is less precise, but it will give a reasonable approximation of the time it takes a population to double in size.

If we know the growth rate of a population and assume that growth rate is constant, we can calculate the number of years it takes for a population to double, which is known as its **doubling time**. Because growth rates may change in future years, we can never determine a country's doubling time with certainty. Therefore, we say that a population will double in a certain number of years if the growth rate remains constant during that time period.

The doubling time can be approximated mathematically using a formula called the rule of 70. The **rule of 70** dictates that by dividing the number 70 by the percentage population growth rate we can determine a population's doubling time:

$$\text{Doubling time (years)} = \frac{70}{\text{growth rate (expressed in \%)}}$$

Therefore, a population growing at 2 percent per year will double every 35 years:

$$\frac{70}{2} = 35 \text{ years}$$

Note that this is true of any population growing at 2 percent per year, regardless of the size of that population: at a 2 percent growth rate, population size will double—a population of 50,000 people will increase by 50,000 in 35 years, and a population of 50 million people will increase by 50 million in 35 years. See the "Do The Math: Comparing Population Growth in Two Countries" for practice calculations for growth rate and doubling time. The human population on Earth has doubled several times since 1600 but it is not expected to double again. As we illustrated in Figure 17.10, the human population is expected to stabilize at somewhere between 8 billion and 10 billion people by 2050.

AP® Exam Tip

You can use the rule of 70 to determine the percent growth of a population if you know doubling time. On the exam, you may be given the initial and final population to determine how many times it has doubled. Look at the years and you can figure out doubling time. If a population doubled twice in the span of 14 years you can establish the doubling time is 7 years. 70/doubling time = percent growth. 70/7years = 10% growth/year.

Rule of 70 A method which dictates that by dividing the number 70 by the percentage population growth rate we can determine a population's doubling time.

DO THE MATH Comparing Population Growth in Two Countries

Preparing for the AP® Exam

Hungary and Sweden are two European countries each with approximately 10 million people. When will the population of each country double? Hungary has a CBR of 9 and a CDR of 13. Sweden has a CBR of 11 and a CDR of 9. Ignore migration in and out of each country and calculate the national population growth rate of each country.

1. Hungary

$$(\text{CBR} - \text{CDR}) \div 10 = \frac{9-13}{10} = -0.4$$

So the growth rate is –0.4%.

The doubling time is $\frac{70}{-0.4} = -175$. Because this is a negative number, it means the time for the population to double will never occur.

At current birth and death rates, Hungary will never double in population.

2. Sweden

$$(\text{CBR} - \text{CDR}) \div 10 = \frac{11-9}{10} = 0.2$$

So the growth rate is 0.2%.

The doubling time is $\frac{70}{0.2} = 350$ years.

At current birth and death rates, Sweden will double in size in 350 years.

YOUR TURN The Dominican Republic has 10.7 million people and a national population growth rate of 1.5 percent. Paraguay has 6.8 million people and a national population growth rate of 1.5 percent. In how many years will each country double in size?

18-3 What is the demographic transition?

The demographic transition is a model that predicts population changes as countries develop

Human populations undergo change as a variety of natural and societal conditions vary over time. Some of these changes occur with specific patterns that can be described and explained. As we have seen, populations in specific countries and in rural and urban areas change over time and the population of Earth has also fluctuated over time. Demographers are interested in understanding the reasons behind fluctuations in population growth in the past and whether they apply to contemporary or future demographic issues. We will begin by looking at demographic transitions, and then move on to investigate the effect of these transitions on the environment, and the relationship between economic development and sustainability.

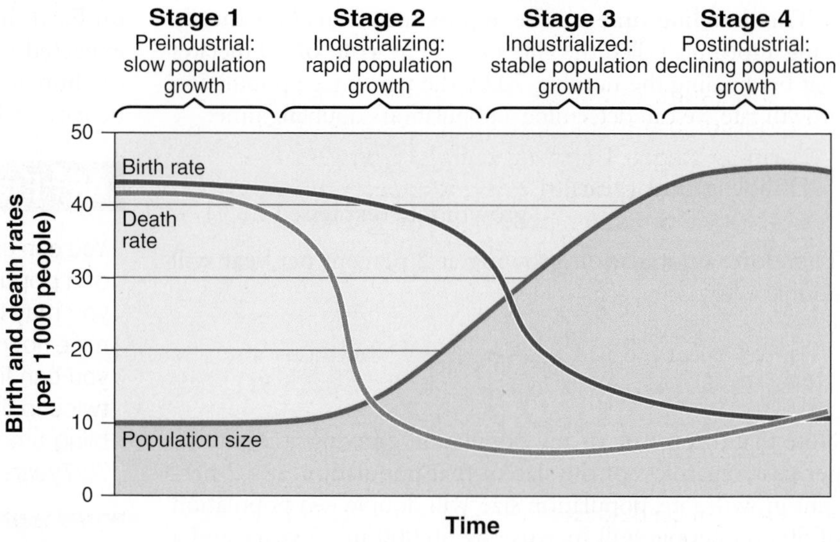

FIGURE 18.3 Demographic transition. The theory of demographic transition models the way that birth, death, and growth rates for a nation change with economic development. Stage 1 is a preindustrial period characterized by high birth rates and high death rates. In stage 2, as the society begins to industrialize, death rates drop rapidly, but birth rates do not change. Population growth is greatest at this point. In stage 3, birth rates decline for a variety of reasons. In stage 4, the population stops growing and sometimes begins to decline as birth rates drop below death rates.

The Theory of Demographic Transition

Historically, nations that have gone through similar processes of economic development have experienced similar patterns of population growth. Scientists who studied the population growth patterns of European countries in the early 1900s described a four-stage process they referred to as a demographic transition. The **theory of demographic transition** states that a country moves from high to lower birth and death rates as development occurs and that country moves from a preindustrial to an industrialized economic system. During this time, it undergoes a predictable shift in population growth. The four stages of the demographic transition model are shown in **FIGURE 18.3**.

The theory of demographic transition, while helpful as a learning tool, does not adequately describe the population growth patterns of many developing countries either today or during the last half-century. Both birth and death rates have declined rapidly in a number of developing countries for a variety of reasons that are not yet entirely understood. In some developing countries governments have taken measures to improve health care and sanitation and promote birth control, in spite of the country's poverty.

Despite the limitations of the demographic transition model, it is worth examining in more detail because it allows us to visualize a representation of the way some countries influence the environment as they undergo growth and development.

Stage 1: Slow Population Growth

Stage 1 represents a population that is nearly at steady state. The size of the population will not change very quickly because high birth rates and high death rates offset one another. In other words, CBR equals CDR. This pattern is typical of countries before they begin to modernize and was typical of the entire world until a few centuries ago. In stage 1 countries, life expectancy for adults is relatively short due to difficult and often dangerous living or working conditions. The infant mortality rate, which is the number of deaths among children under age 1 per 1,000 live births, is also high because of disease, lack of health care, and poor sanitation, all of which contribute disproportionately to deaths among the very young and very old. In a stage 1 developing economy, where most people are farmers, having numerous children is an asset. Children can do jobs such as collecting firewood, tending crops, watching livestock, and caring for younger siblings. With no social security system, parents also count on having many children so that some of them will be available to care for them when they are older.

Theory of demographic transition A theory that states that a country moves from high to lower birth and death rates as development occurs and that country moves from a preindustrial to an industrialized economic system.

> **AP® Exam Tip**
>
> For the exam, remember that in the four stages of demographic transition, only two stages experience growth. In stages 1 and 4 the population doesn't grow as birth rate and death rate are similar. It is only in stages 2 and 3 that the population will grow.

Western Europe and the United States were in stage 1 before the Industrial Revolution, which began in the late eighteenth century. Today, crude birth rates exceed crude death rates in almost every country in the world, so even the poorest nations have moved beyond stage 1. However, an increase in crude death rates due to war, famine, and diseases such as AIDS has pushed some countries back in the direction of stage 1. For example, more than a decade ago, Lesotho (CBR = 26, CDR = 28 in 2005) moved back into stage 1 as a result of its high crude death rate.

Stage 2: Rapid Population Growth

In stage 2, death rates decline while birth rates remain high and, as a result, the population grows rapidly. Oftentimes, this scenario arises as follows: As a country modernizes, better sanitation, clean drinking water, increased access to food and goods, and access to health care, including childhood vaccinations, all reduce the infant mortality rate and CDR. However, the CBR does not markedly decline. Couples continue to have large families because it takes at least one generation, if not more, for people to become fully aware of the decline in infant mortality and adjust to it. This is another example of population momentum that we first discussed in Module 17. In addition, there may be a desire from government officials to reduce the birth rate due to lower infant mortality, but it also takes time to implement educational systems and birth control measures to empower families to have fewer children.

A stage 2 country is in a state of imbalance: Births outnumber deaths. India is in stage 2 today (CBR = 20; CDR = 16). The U.S. population exhibited a stage 2 population structure in the early twentieth century when there were high birth rates and low death rates (CBR = 30; CDR = 17).

Stage 3: Stable Population Growth

A country enters stage 3 as its economy and educational system improve. As we saw earlier, when family income increases, people have fewer children. As a result, the CBR begins to fall. Stage 3 is typical of many developed countries.

As societies transition from developing to developed nations, having large numbers of children may become a financial burden rather than an economic benefit. When people are relatively affluent, they tend to spend more time pursuing education and they are more likely to have access to birth control and to choose to have smaller families. However, cultural, societal, and religious norms may also play a role in birth rates.

As birth rates and death rates decrease in stage 3, the system returns to a state where growth rates slow. Population size is relatively stable during this stage because birth rates and death rates are approaching equality. Uruguay and Albania are two countries that are in stage 3 right now.

FIGURE 18.4 Elderly populations. Some countries have very large elderly populations. Here, men and women gather at a senior residence in Matsuyama, Japan. *(Bloomberg/Getty Images)*

Stage 4: Declining Population Growth

Stage 4 is characterized by declining population size and often by a relatively high level of affluence and economic development. Japan, Germany, Portugal, and Italy are stage 4 countries, with the CBR below the CDR.

The declining population in stage 4 means a country will have fewer young people and a higher proportion of elderly people (**FIGURE 18.4**). This demographic shift can have important social and economic effects. With fewer people in the labor force and more people retired or working part-time, the ratio of dependent elderly to wage earners increases, and the costs of pension programs and social security services will increase the tax burden on each wage earner. There may be a shortage of health care workers to care for an aging population. Governments may encourage immigration as a source of additional workers. In some countries, such as Japan, the government provides economic incentives to encourage families to have more children in order to offset the demographic shift.

Recent studies on demographic shifts in highly developed countries suggest that the TFR actually increases after reaching a low point between 1.2 and 1.5. The reasons for this increase are unclear, but it appears that when a population becomes affluent and well educated, it becomes somewhat easier for families to raise children, and they choose to do so in slightly greater numbers. Such a pattern is occurring in Norway, Italy, the United States, and other developed nations.

18-4 How do population and consumption interact to influence the environment?

Population and consumption influence the environment differently, depending on the country

Both population size and the amount of resources each person uses are critical factors that determine the impact of humans on Earth. Every human exacts a toll on the environment by eating, drinking, generating waste, and consuming products. Even relatively simple foods such as beans and rice require energy, water, and mineral resources to produce and prepare. Raising and preparing meat requires even more resources: Animals require crops to feed them as well as water and energy resources. Building homes, manufacturing cars, and making clothing and consumer products all require energy, water, wood, steel, and other resources. These and many other human activities contribute to environmental degradation.

Both population and economic development contribute to the consumption of resources and to human impact on the environment. In this section we will examine how relationships among population size, economic development, and resource consumption influence the environment.

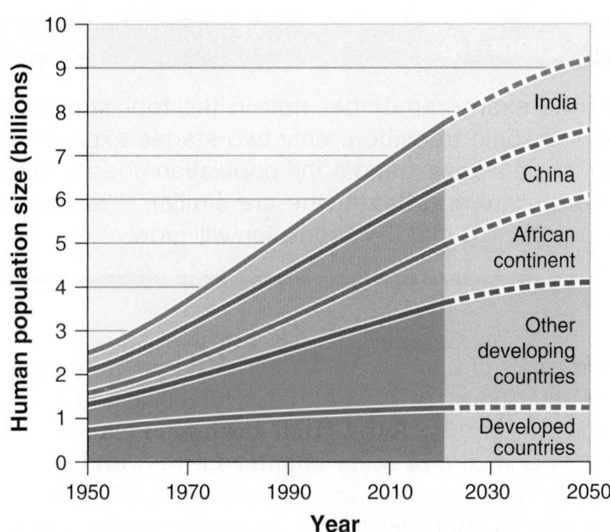

FIGURE 18.6 Population growth in specific developing nations and all developed nations. Population growth in the developed nations has mostly ceased, while that in developing nations is slowing, but is expected to continue until 2050 or beyond.

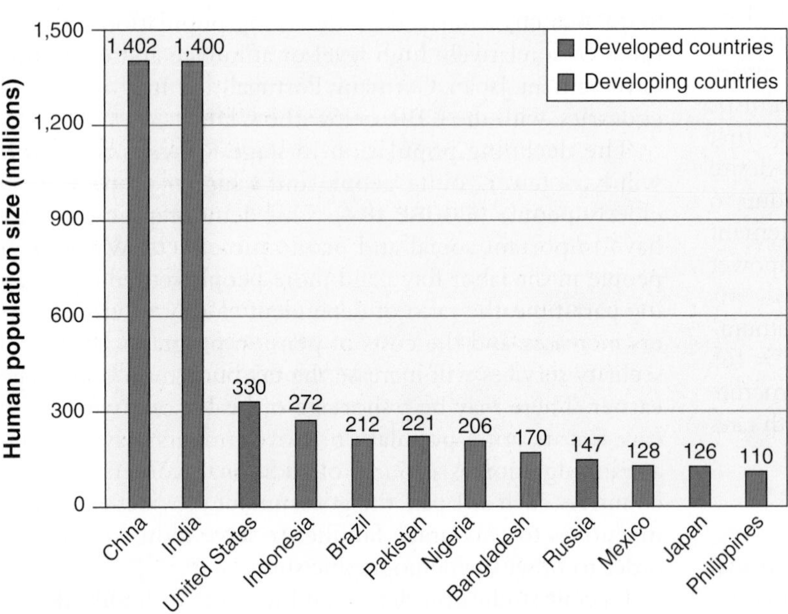

FIGURE 18.5 The 12 most populous countries in the world. China and India are by far the largest nations in the world. Only 3 of the 12 most populous countries are developed nations. *(Data from Population Reference Bureau, www.prb.org.)*

Resource Use

Of Earth's 7.8 billion human inhabitants, roughly 6.5 billion live in developing countries and 1.3 billion live in developed countries. As **FIGURE 18.5** shows, 9 of the 12 most populous nations on Earth are developing countries. **FIGURE 18.6** charts the relationship between population growth rate over time for certain developing nations and all of the developed nations grouped together. Populations in developing parts of the world have continued to grow relatively rapidly, at an average rate of 1.5 percent per year. At the same time, populations in the developed world have almost leveled off, with an average growth rate of 0.1 percent per year. Impoverished countries are increasing their populations more rapidly than are affluent countries.

Differences in resource use are striking in terms of how population and wealth affect the environment. Calculating the per capita ecological footprint for a country provides a way to measure the effect of affluence and consumption on the planet. This will be covered in Unit 5.

The IPAT Equation

The total environmental impact of 7.8 billion people is hard to appreciate and even more difficult to quantify. Some people consume large amounts of resources and have a negative impact

(a)

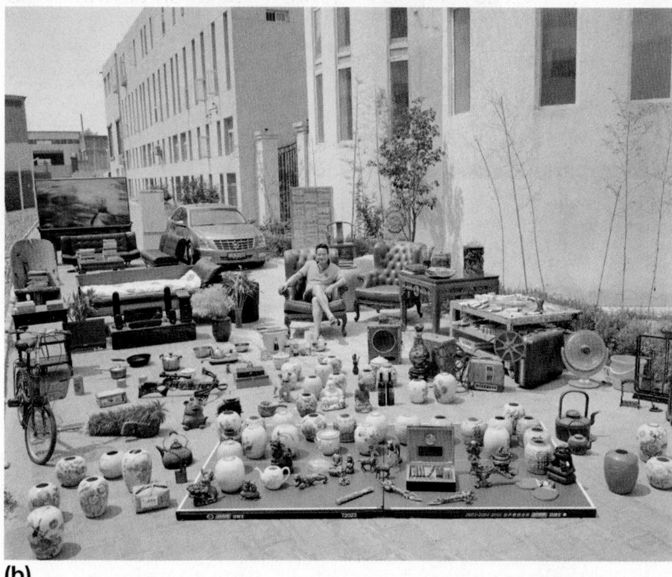

(b)

FIGURE 18.7 Material possessions. Most families in developing countries have few possessions compared with their counterparts in developed countries. In each of these photos, members of a family are shown outside their homes with all their possessions. (a) This family lives in a rural region of Tibet. (b) This individual lives in an urban part of China, where more development has occurred. *(a & b: Huang Quingjun/http://www.huangqinglun.com/)*

on environmental systems, while others live much more lightly on the land (**FIGURE 18.7**). Living lightly can be intentional, as when people in the developed world make an effort to live "green," or sustainably. But it can also be unintentional, as when poverty prevents people from acquiring material possessions or building homes.

To estimate the impact of human lifestyles on Earth, environmental scientists Barry Commoner, Paul Ehrlich, and John Holdren developed the **IPAT equation**

$$\text{impact} = \text{population} \times \text{affluence} \times \text{technology}$$

Although it is written mathematically, the IPAT equation is a conceptual representation of the three major factors that influence environmental impact. It cannot be solved mathematically, but rather it is written out in a math-like formula to emphasize the three factors that contribute to environmental impact. In this context impact is the overall environmental effect of a human population multiplied by affluence, multiplied by technology. It is useful to look at each of these factors individually.

Population has a straightforward effect on impact. All else being equal, two people consume twice as much as one. Therefore, when we compare two countries with similar economic circumstances, the one with more people is likely to have a larger impact on the environment.

Affluence is created by economic opportunity and does not have as simple a relationship to impact as population does. One person in a developed country can have a greater impact than two or more people in a developing country. A family of four in the United States that owns two large sport utility vehicles and lives in a spacious home with a lawn and swimming pool uses a much larger share of Earth's resources than a Bangladeshi family of four living in a two-room apartment and traveling on bicycles and buses. The more affluent a society or individual is, the higher the environmental impact.

The effect of technology is even more complicated. Technology can both degrade the environment and create solutions to minimize our impact on the environment. For example, the manufacturing of chlorofluorocarbons (CFCs) resulted in safe and effective refrigeration and air conditioning that was beneficial to human health, yet CFCs led to ozone destruction in the stratosphere. In contrast, the hybrid electric car helps to reduce the impact of the automobile on the environment because it has greater fuel efficiency than a conventional internal combustion vehicle of the same size and weight. The IPAT equation originally used the term *technology*, but some scientists now use the term *destructive technology* to differentiate it from beneficial technologies such as the hybrid electric car. For a summarizing visual of how many concepts from this unit are interconnected, turn to the Unit 3 Visual Representation feature on pages 212-213. In the next unit, we will begin to investigate Earth systems and resources.

IPAT equation A conceptual representation of the three major factors that influence environmental **I**mpact: **P**opulation of humans, **A**ffluence, **T**echnology.

Preparing for the AP® Exam

Visual Representation 3 **Population dynamics in Japan and Tanzania:** Japan and Tanzania present a significant contrast in human population dynamics. Much of Japan's population is older and the population growth rate is negative. Women, on average, give birth to slightly more than one child. Japan is an affluent country with an advanced economy and a large urban population. Tanzania's population is younger, and

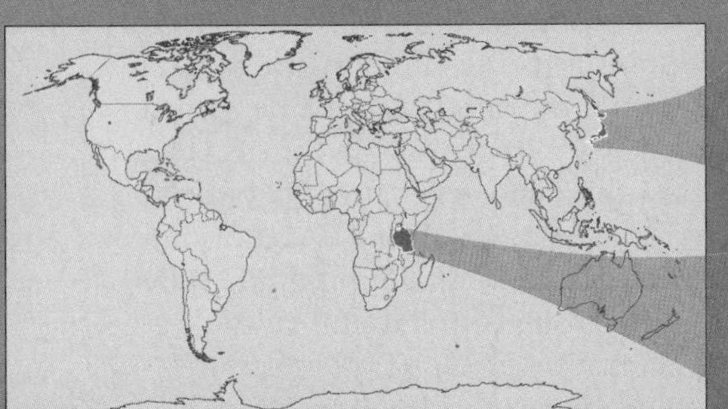

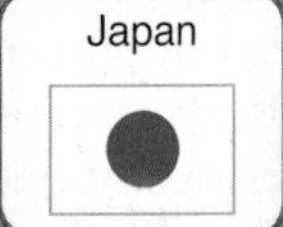

Japan and Tanzania:
Japan is an East Asian island nation in the northwest Pacific Ocean. The United Republic of Tanzania is in East Africa with the Indian Ocean to the east and borders with eight other African nations. Japan and Tanzania are separated by more than 7,000 miles and present very different demographics.

Crude birth rate (CBR) is the number of births per 1,000 individuals per year.

Crude death rate (CDR) is the number of deaths per 1,000 individuals per year.

National population % growth rate:

$$\frac{[(\text{CBR} + \text{immigration}) - (\text{CDR} + \text{emigration})]}{10}$$

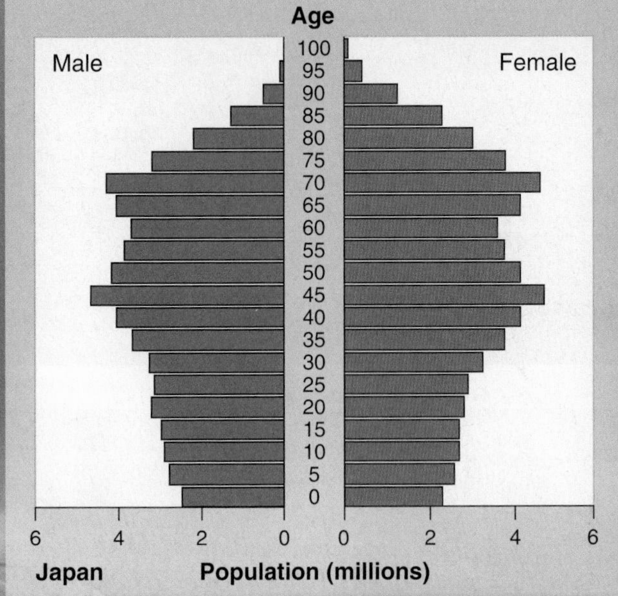

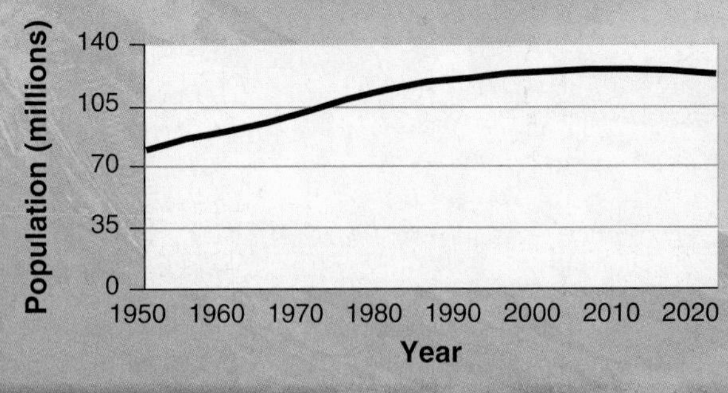

Japan's population is aging rapidly. The crude birth rate (CBR) is lower than the crude death rate (CDR). Total fertility is low and life expectancy is high.

Population growth rate = **−0.4%**
TFR = **1.3**
CBR = **7 / 1,000**
CDR = **11 / 1,000**
Infant mortality = **1.8 / 1,000**
Life expectancy = **84**
Urban population = **92%**
Migration = **0 migrants / 1,000**

https://www.livepopulation.com/country/japan.html

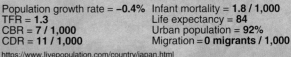

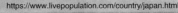

Carrying capacity (*K*):
The growth of the Sika deer population in the Nara area shows a pattern similar to the pattern Gause found in his limiting resource experiments. Even though these deer are fed extensively by tourists, the population growth has slowed over time and settled at carrying capacity (*K*).

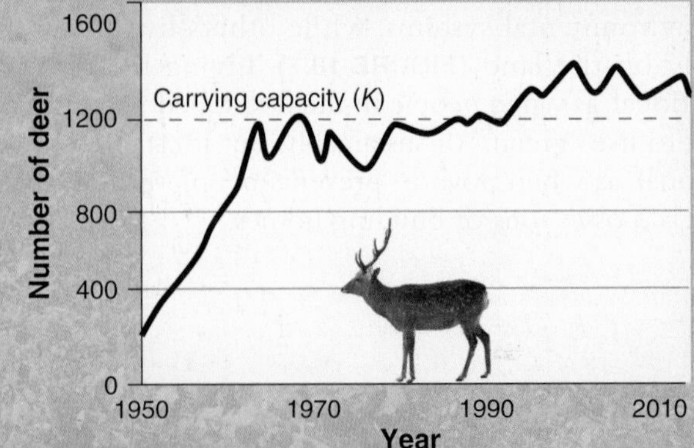

Data from Harumi Torii and Shirow Tatsuzawa, *Sika Deer in Nara Park: Unique Human-Wildlife Relations*; Hokaido University, 2009.

the population growth rate is positive. Women on average give birth to slightly more than four children. Tanzania is a developing country and most of the population lives in rural areas. Case studies in wildlife populations from both countries provide examples of exponential growth, logistic growth, and the limits of carrying capacity which apply to non-human populations.

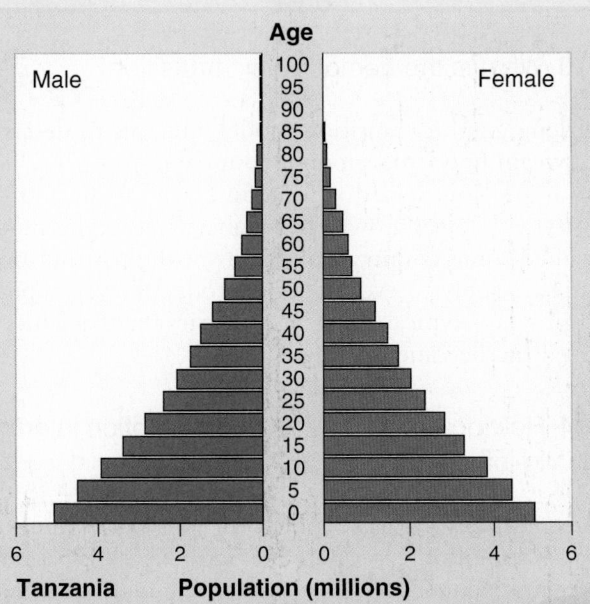

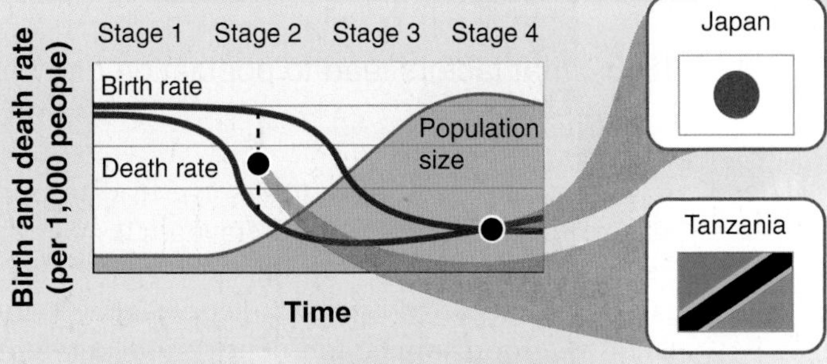

Demographic transition:
The theory of demographic transition says that a country moves from high to lower birth and death rates as development occurs and that country moves from a pre-industrial to an industrialized economic system. Tanzania is in stage 2, where death rates decline while birth rates remain high. As a result, population is growing rapidly. Japan is in stage 4 with declining population size and a relatively high level of economic development and affluence.

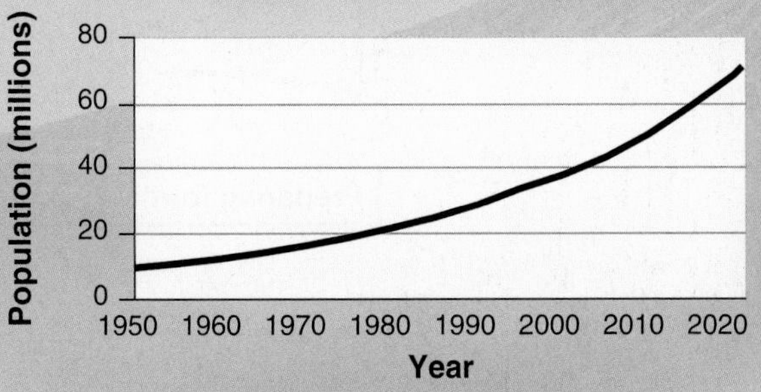

Doubling time is the number of years it takes a population to double.

The rule of 70 is a mathematical approximation of doubling time.

$$\text{Doubling time (years)} = \frac{70}{\text{\% Population growth rate}}$$

Tanzania's population is young and growing. The crude birth rate (CBR) is higher than the crude death rate (CDR). Total fertility rate (TFR) is relatively high and life expectancy is moderate.

Population growth rate = **2.9%**
TFR = **4.5**
CBR = **34 / 1,000**
CDR = **5 / 1,000**
Infant mortality = **32 / 1,000**
Life expectancy = **70**
Urban population = **35%**
Migration = **0 migrants / 1,000**

https://www.livepopulation.com/country/tanzania.html

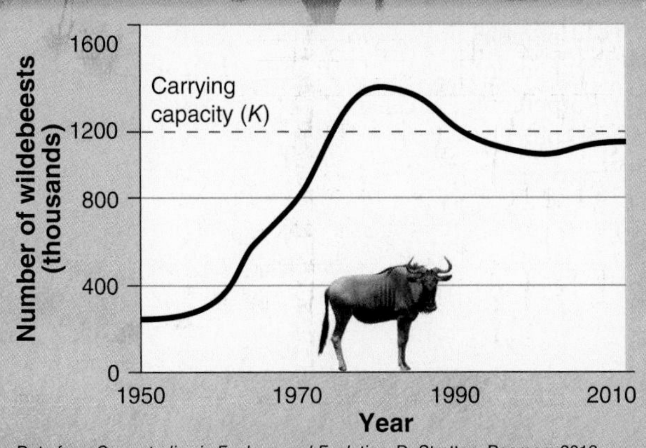

Data from *Case studies in Ecology and Evolution*, D. Stratton, Pearson, 2012.

Logistic growth (S-shaped curve):
The recovery of the wildebeest population in East Africa after 90 percent were killed by a virus shows logistic growth. Initially the population grew exponentially. Eventually growth slowed and then, with some yearly deviations, populations settled at a stable number, the carrying capacity (*K*).

Module 18 AP® Review

Preparing for the AP® Exam

Learning Goals Revisited

18-1 What factors lead to population growth and decline?

Many social and cultural factors can lead to population growth and decline. Government policies can influence fertility rates that affect population change. Population change can be affected by both density-dependent factors such as the food supply and density-independent factors such as extreme weather events and fires.

18-2 How can you calculate doubling time using the rule of 70?

The rule of 70 allows a quick calculation based on the growth rate of a population to determine when a population will double in size.

18-3 What is the demographic transition?

The demographic transition model attempts to describe and explain how a developing country moves from high growth and low wealth to become a more-developed country with lower growth and higher wealth. This model described some country transitions in the past but does not always apply to every country today. Slower population growth is typical of many countries that have industrialized and become wealthier.

18-4 How do population and consumption interact to influence the environment?

Those countries that have industrialized typically increase their consumption of resources, with adverse impacts on the environment. The IPAT equation is a conceptual representation of the three major factors that influence environmental impact: population, affluence, and technology.

AP® Practice Questions

Preparing for the AP® Exam

Multiple-Choice Questions

1. Doubling time can be approximated mathematically by
   (a) $\frac{70}{\%\ \text{growth rate}}$.
   (b) $\frac{10\%}{\text{growth rate}}$.
   (c) 2% × growth rate.
   (d) $\frac{\text{CBR} - \text{CDR}}{10}$.
2. According to the theory of demographic transition, countries move through growth stages in which order?
   (a) stable growth, rapid growth, slow growth, declining growth
   (b) rapid growth, slow growth, stable growth, declining growth
   (c) slow growth, declining growth, rapid growth, stable growth
   (d) slow growth, rapid growth, stable growth, declining growth
3. Which of the following likely reduces fertility rate the most?
   (a) birth control for men
   (b) government ad campaigns for birth control
   (c) decreasing income of populations
   (d) increased education of women
4. The IPAT equation was developed to estimate
   (a) the affluence of a country's population.
   (b) the rate of demographic transition.
   (c) the impact of human lifestyles on Earth.
   (d) the gross domestic product of a country.
5. Approximately what percentage of the population of developed countries lives in urban areas?
   (a) 44 percent
   (b) 50 percent
   (c) 75 percent
   (d) 86 percent

Free-Response Question

The graph below shows how a country progresses through demographic transition.

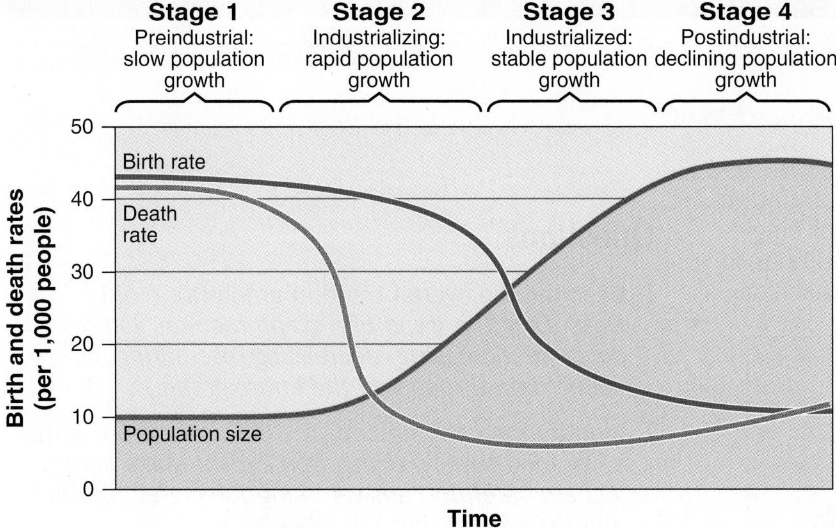

(a) **Identify** the stage of demographic transition with high levels of elderly people. (2 pts.)
(b) **Identify** an economic effect of having more elderly people in the population. (2 pts.)
(c) **Describe** the changes occurring in fertility rate from stage 2 to stage 3. (2 pts.)
(d) **Explain** the type of age-structure diagram seen in stage 1. (1 pt.)
(e) **Propose a solution** that a government might implement to slow its population growth that could be utilized by a country undergoing a demographic transition. (1 pt.)
(f) **Justify** the previous solution in part (e) by describing two other advantages. (2 pts.)

Preparing for the AP® Exam

UNIT 3 Practice Your Science Skills

Data Analysis

Russian biologist Georgy Gause monitored population growth in two species of Paramecium (a type of single-celled aquatic organism) living under ideal conditions in test tubes with a constant amount of food added each day.

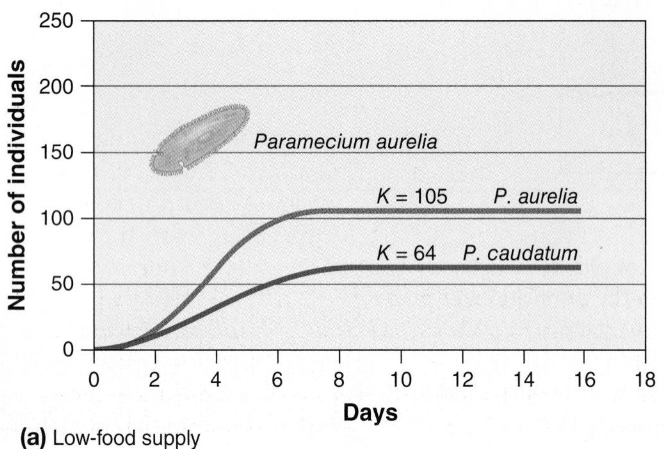

(a) Low-food supply

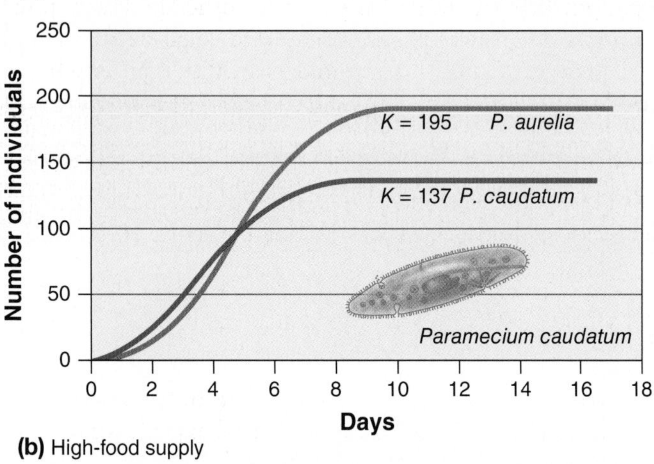

(b) High-food supply

Questions

1. **Describe** the overall trend in graph (a). *(Hint: Describing the trend of a graph requires you to describe increasing, decreasing, oscillating, or stable data throughout the entire graph.)*
2. **Identify** the carrying capacity of *P. caudatum* with a low food supply. *(Hint: Don't rush! Make sure you are careful to answer using the correct graph and the correct line.)*
3. **Identify** the independent variable in this experiment. *(Hint: The independent variable is the variable that is manipulated by the scientist.)*
4. **Describe** the range of days in which *P.aurelia* grew exponentially in graph (a). *(Hint: If this type of question is on an FRQ, the AP® readers will have an acceptable range that would earn points.)*
5. A student hypothesizes that food supply is a limiting factor and that Paramecium populations would increase with increased food. **Make a claim** defending or refuting this hypothesis based on the information in these graphs. *(Hint: A "claim" question requires you to back up your claim with evidence. If you have numbers, use them as evidence.)*

Pursuing Environmental Solutions

Gender Equity and Population Control in Kerala

India is the world's second most populous country and will probably overtake China as the most populous by 2025 or 2026. Since the middle of last century, India has tried various methods of population control, steadily reducing its growth rate to the current 1.4 percent. Although this growth rate would suggest below-replacement-level fertility, India's population will continue to increase for some time because of population momentum; it has a large number of young people. The Population Reference Bureau estimates that India's population will not stabilize until 2050 or later.

Students in Kerala, India. Literacy rates similar to those in developed countries may be a factor in the population stabilization that has occurred in the state of Kerala, India. *(Sreekesh Raveendran Nair)*

Decades ago, India attempted to enforce nationwide population control through sterilization, but massive protests against this coercive approach led to changes in policy. Since the 1970s, India has emphasized family planning and reproductive health, although each Indian state chooses its own approach to population stabilization. Some states have had tremendous success in lowering growth rates. Kerala, in southwestern India, is one of those states.

The state of Kerala is about twice the size of Connecticut. But Connecticut has a population of roughly 3.5 million people while Kerala has a population of 35 million. Kerala's population density is about 860 people per square kilometer (2,200 people per square mile), roughly 2 times that of India as a whole, and 23 times that of the United States. At one point, Kerala's population was growing even faster than the population in the rest of India, in large part because the state government had implemented an effective health care system that decreased infant and adult mortality rates.

However, for more than 40 years, Kerala's birth rate, like its mortality rate, has fallen to levels similar to those in North America and other industrialized countries. The current total fertility rate in Kerala is about 1.8—roughly the same as the United States and lower than the TFR of 2.2 for India as a whole. As a result, Kerala's population has stabilized. How did the state's population stabilize despite its poverty?

A special combination of social and cultural factors seems to be responsible for Kerala's sustainable population growth. Kerala has emphasized "the three Es" in its approach: education, employment, and equality. In addition to its accessible health care system, Kerala has good schools that support a literacy rate of over 95 percent, the highest in India. Furthermore, unlike rates in the rest of India, male and female literacy rates in Kerala are almost equal. In other parts of India and in most other developing countries, women attend school, on average, only half as long as men do. As we have learned, higher education levels for women lead to greater female empowerment and increased use of family planning. In Kerala, 58 percent of women use contraceptives, compared with 48 percent in the rest of India, and many women delay childbirth and join the workforce before deciding to start families. Kerala has a strong matriarchal tradition in which women are highly valued and their education is encouraged. Education and empowerment allow these women to be in a better position to make decisions regarding family size.

The World Bank estimates that if the rest of the developing world had followed Kerala's lead 30 years ago and equalized education for men and women, current TFR throughout the developing world would be close to replacement-level fertility. Evidence for this view can be found in the reduced fertility rates of other relatively poor countries where women and men have equal status, such as Sri Lanka and Cuba. In fact, gender equality is an important component of a number of United Nations programs today.

Critical Thinking Questions

1. Describe some typical ways countries have tried to lower birth rates.
2. Identify some other innovative ways to lower birth rates.
3. Can achieving replacement-level fertility sometimes cause difficulties? If so, identify and explain some problems that might occur.
4. Describe what is preventing the model used to promote slower population growth in Kerala from being used elsewhere in the world.

References

Franke, R. W., and B. H. Chasin. 2005. Kerala: Radical reform as development in an Indian state. In *The Anthropology of Development and Globalization: From Classical Political Economy to Contemporary Neoliberalism*, ed. M. Edelman and A. Haugerud. Blackwell.

Soni, P. Highest literacy (most educated) rate in India state wise. *Business Insider India*. Dec. 18, 2020.

Thulaseedharan, J. V. 2018. Contraceptive use and preferences of young married women in Kerala, India. *Journal of Contraception*. 9:1–10. doi: 10.2147/OAJC.S152178

Science Applied 3: Concept Explanation

How Can We Control Overabundant Animal Populations?

The growing human population has caused population declines in many animal species. At the same time, human alteration of the landscape has caused other species to flourish. Increases in the population of these species have had many consequences. How have these increases happened, and what, if anything, should we do about them?

What causes animal population explosions?

The white-tailed deer is a high-profile example of dramatic animal population growth in the United States. Before European colonists arrived, its densities were approximately 4 to 6 deer per square kilometer (10 to 15 deer per square mile). In many areas today, however, deer densities are up to 20 times higher. In New York State, for example, unregulated hunting and the clearing of preferred deer habitats for farming had nearly eliminated the deer population by 1900. The New York State population has since grown exponentially to roughly 1 million deer.

This population growth can be explained by processes we have discussed throughout Unit 3. For example, white-tailed deer prefer to live in early-successional fields containing wildflowers and shrubs. Abandoned farms that have undergone succession during the twentieth century, such as the old fields of New England that we described in the unit-opening case study, are prime deer habitat. Such increases in habitat and food supply have increased the carrying capacity of the environment for deer. With a higher carrying capacity, density-dependent controls on the population are smaller, and reproduction increases.

In addition to providing deer with habitat and food, humans have eliminated many of the deer's major predators, including wolves and cougars, as we discussed in Module 16. During much of the twentieth century, regulated hunting held deer populations in check, taking the place of natural predators that had largely been eliminated. Over the last few decades, however, the number of deer hunters has steadily declined, allowing the deer population to increase. In short, a higher carrying capacity and a lower mortality by both predators and hunters have allowed deer populations to grow exponentially and exceed their historic carrying capacity (**FIGURE SA3.1**).

FIGURE SA3.1 Deer overpopulation. As humans have provided increased food and habitat for white-tailed deer and simultaneously removed their major predators, deer populations have increased to the point that they exceed the historic carrying capacity. This photo shows deer in a yard of a home in Pennsylvania. *(Philippe Gerber/Getty Images)*

The white-tailed deer is just one of many species whose populations are exploding throughout North America and around the world. As in the case of deer, these population increases are due to predator extinctions, reduced hunting, increased habitat, and increased food supplies provided by agriculture and by people who feed them. In the southeastern United States, some varieties of domesticated European and Russian hogs (*Sus scrufa*) have been allowed to be free ranging and this has allowed feral populations to evolve and spread. Over the past three decades, feral hogs have spread to at least 35 states. Wild horses (*Equus ferus*) are another species that is not native to the United States but has a growing population that is a perpetual challenge to control in the American West (**FIGURE SA3.2**).

Large herbivorous mammals are not the only animals that experience overpopulation as the result of human activity. The tens of thousands of monk parakeets (*Myiopsitta monachus*) in Florida are the offspring of individuals that have escaped from the pet trade since the 1960s. Populations of Canada geese (*Branta canadensis*) have grown along with the number of agricultural fields and golf courses on which they can feed. Other former pets, including dogs and cats, quickly establish feral populations when released into the wild.

What effects do overabundant species have on communities?

Populations that exceed their historic carrying capacity can cause widespread problems. Large populations of herbivore species such as deer, hogs, and horses can overgraze the

FIGURE SA3.2 Overabundant horses. Wild horses in the American West were introduced more than a century ago. Without major predators, their population has increased to a point where they are impacting the habitat needed by native species. How to best control horse populations is a controversial issue with many different ethical and scientific perspectives. *(Cliff Owen/AP Images)*

plants in their community, leaving little food for other herbivorous species. Hogs are also causing major problems with rooting up the soil. In Australia, grazing by the overabundant kangaroos removes many of the plants needed by other herbivores, including several species of endangered animals that depend on these plants for food and habitat.

These large animal populations also have direct effects on humans. Herbivorous mammals consume agricultural crops, landscape plants, and valuable wild plant species. In the United States, for example, the deer population so greatly exceeds the natural carrying capacity of the land and eats so many tree seedlings and saplings that it is now difficult to regenerate several valuable forest tree species, including northern red oak (*Quercus rubra*) and sugar maple. Deer also consume suburban landscape plants leading to many sad and angry homeowners. In New York State it has been estimated that deer damage $60 million worth of agricultural crops each year. Wild hogs across the United States cause $1.5 billion in crop damage each year.

Wildlife overabundance increases wildlife-human encounters, such as deadly collisions of vehicles with deer or kangaroos. For example, white-tailed deer in the United States cause roughly 1.3 million vehicle collisions per year, with an average cost of about $4,000 per accident. Approximately 200 people die each year from these collisions.

Over-abundant animal populations can also have less appreciated impacts. For example, most people are surprised to know that even house cats can have a major impact on native animal populations. Researchers in 2014 estimated that domestic cats annually kill approximately 1 billion to 4 billion birds and 6 billion to 20 billion mammals.

How can we control overabundant species?

Controlling the population of any species involves issues that are ecological, political, moral, and ethical. From an ecological perspective, it is clear how to control a wildlife population: reduce the available food and habitat to lower the carrying capacity, compensate for the missing predators by killing individuals in the population, or slow the population's ability to reproduce. On a local scale, it can be helpful to discourage people from feeding wildlife. On a larger scale, however, reducing food and habitat is often the least viable option because it would typically require restricting the access of large herbivores to millions of hectares of land.

Compensating for absent predators by culling animal herds through hunting can be an effective approach, but it is often met with resistance from members of the public. For example, in 2009 the Australian government ordered the army to shoot 6,000 kangaroos to reduce an overabundant population living on a large military compound. The objective was to reduce the amount of grazing done by the kangaroo herd and therefore protect rare plants and the insects that live on those plants. This decision was opposed by animal rights activists and by many Australians who consider the kangaroo a national symbol that should be protected. In other cases, such as overabundant deer in suburban areas of the United States, shooting the animals is not a viable approach for reasons of safety. In short, culling a herd through shooting is an increasingly unpopular strategy. Moreover, most populations that are reduced in number rapidly bounce back because of a sudden increase in the abundance of food for the remaining individuals.

A strategy that has more public support is the use of birth control. This approach is common in cities, where many feral cats are trapped, neutered, and returned to the wild. Researchers are also investigating ways to give animals contraceptive drugs to reduce their populations over time. The drugs being developed differ in their modes of action, ranging from chemicals that prevent animals from making critical reproductive hormones to chemicals that prevent sperm from fertilizing an egg. Contraceptive strategies are

FIGURE SA3.3 Administering contraceptives to wild animals. Researchers deliver the PZP contraceptive drug to large herbivores, such as wild horses, using a dart gun. *(Anne Sherwood/The New York Times/Redux)*

currently being tested on a variety of species, including deer, kangaroos, elephants, and pigeons (**FIGURE SA3.3**).

Reducing animal overpopulation with contraceptives has widespread public appeal because it eliminates the need to kill animals. Getting the contraceptive drugs into the animals, however, presents substantial challenges. Drugs can be administered to some animals via dart guns. For others, such as monk parakeets, researchers are working on ways to administer medications in food. Some contraceptive drugs last for a single breeding season and require a new dose each season, which can be difficult and expensive. The cost of giving an animal a single dose of a contraceptive drug ranges from \$100 to \$500 per animal. However, recent tests with a new contraceptive drug, known as PZP (porcine zona pellucida), suggest that the birth control effects on deer may last for up to 5 years, which may make the approach more economically viable. At the same time, researchers are developing less labor-intensive methods of delivering the drugs to animals, including setting up feeders with facial recognition software to ensure that the drug is given to the targeted species and not to other species.

The issue of overabundant wildlife highlights a number of other environmental issues. The underlying causes of animal overpopulation are typically associated with human population increases. These human-caused wildlife population explosions can, in turn, have substantial negative effects on humans and their environment. Although there are several potential solutions to the problem of animal overpopulation, successful solutions must take into account the public's affection for wildlife, its opposition to killing animals, and its frequent lack of knowledge about the negative effects of overabundant species.

Questions

1. If humans are responsible for environmental changes that cause increases in the populations of native species, under what conditions should we work to reverse these population increases?
2. How might the options to control overabundant populations differ in rural versus urban areas?

Practice AP® Free-Response Question

(a) **Identify** two density-dependent factors that can influence the population size of white-tailed deer. (2 pts.)

(b) **Identify** two density-independent factors that can influence the population size of white-tailed deer. (2 pts.)

(c) **Describe** two ways that a large number of deer in a relatively small area of land could slow succession or maybe even reverse it. (2 pts.)

(d) Consider a population of 100,000 deer with a crude birth rate of 40 and a crude death rate of 20. **Calculate** the population growth rate in percent and identify the number of years it will take for this population to double. (2 pts.)

(e) **Identify** two different strategies for reducing deer populations and describe one benefit and one consequence of each. (2 pts.)

References

Foderaro, L. 2013. A kinder, gentler way to thin the deer herd. *New York Times*, July 5. (http://www.nytimes.com/2013/07/06/nyregion/providing-birth-control-to-deer-in-an-overrun-village.html)

Klein, A. 2016. Cont-roo-ception: Hormone implants bring kangaroos under control. *New Scientist*, June 20. https://www.newscientist.com/article/2094401-cont-roo-ceptionhormone-implants-bring-kangaroosunder-control/

Masters, B. 2017. Can fertility control keep wild horse herds in check? *National Geographic*, February 8. https://www.nationalgeographic.com/adventure/features/environment/wild-horses-part-three/

Nordstrum, A. 2014. Can wild pigs ravaging the U.S. be stopped? *Scientific American*, October 21. https://www.scientificamerican.com/article/can-wild-pigs-ravagingthe-u-s-be-stopped/

UNIT 3

Review

Key Terms to Remember

Population growth rate (Intrinsic growth rate) (p. 176)
Biotic potential (p. 177)
K-selected species (p. 177)
Carrying capacity (p. 177)
r-selected species (p. 177)
Overshoot (p. 177)
Dieback (Die-off) (p. 177)
Survivorship curve (p. 178)
Type I survivorship curve (p. 178)
Type II survivorship curve (p. 179)
Type III survivorship curve (p. 179)
Density-dependent factor (p. 181)
Density-independent factor (p. 181)
Population growth models (p. 181)
Fecundity (p. 182)
Exponential growth model (p. 182)
J-shaped curve (p. 182)
Logistic growth model (p. 183)
S-shaped curve (p. 183)
Limiting resource (p. 184)
Demography (p. 192)
Demographer (p. 192)
Immigration (p. 192)
Emigration (p. 192)
Crude birth rate (CBR) (p. 192)
Crude death rate (CDR) (p. 192)
Net migration rate (p. 192)
Life expectancy (p. 193)
Infant mortality (p. 194)
Child mortality (p. 194)
Environmental justice (p. 195)
Age structure diagram (p. 196)
Population pyramid (p. 196)
Developing countries (p. 196)
Developed countries (p. 196)
Population momentum (p. 198)
Total fertility rate (TFR) (p. 198)
Family planning (p. 199)
Replacement-level fertility (p. 200)
Doubling time (p. 206)
Rule of 70 (p. 207)
Theory of demographic transition (p. 208)
IPAT equation (p. 211)

Unit 3 | AP® Environmental Science Practice Exam | Preparing for the AP® Exam

Section 1: Multiple-Choice Questions

1. Which is true regarding limiting resources?
   (a) A limiting resource is an important density-independent factor.
   (b) A limiting resource is common to all members of a population.
   (c) Abundance of predators is an example of a limiting resource.
   (d) The number of reproducing individuals in a population is a limiting resource.

2. Which characteristic will species with a type III survivorship curve most likely exhibit?
   (a) slow population growth rate
   (b) density-dependent population regulation
   (c) little to no parental care
   (d) large body size

3. Which of the following will lead to an increase in the carrying capacity of the global human population?
   (a) consuming food at a higher trophic level
   (b) decrease in technological advancements
   (c) increase in per capita consumption rate
   (d) lower consumption of meat products

4. In Vietnam, there were 16 births and 6 deaths out of every 100 people in the year 2011. Using the rule of 70, estimate the doubling time for Vietnam's population.
   (a) 0.7 years
   (b) 7 years
   (c) 0.14 years
   (d) 1.4 years

5. Which of the following is most likely to describe a population in a developing country?
   (a) Replacement level fertility is less than 2.1.
   (b) Infant mortality is at 10 per 1,000 live births or more.
   (c) Immigration is very high.
   (d) The age structure will be shaped like pyramid.

Use the diagrams below to answer questions 6–8:

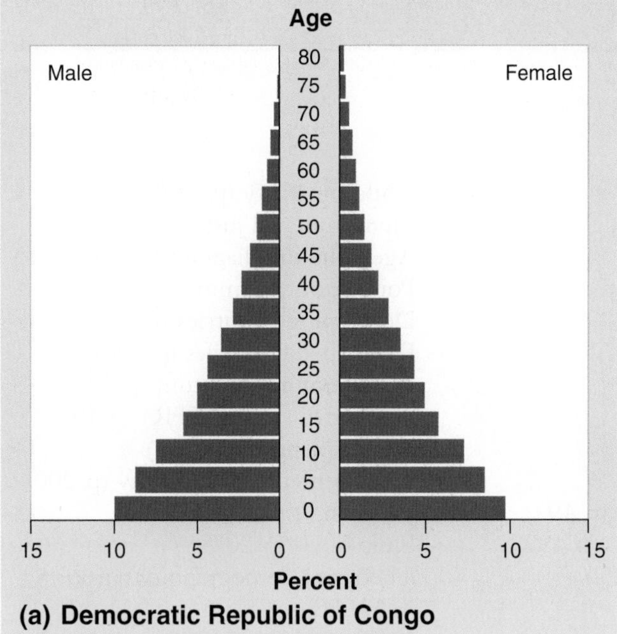

(a) Democratic Republic of Congo

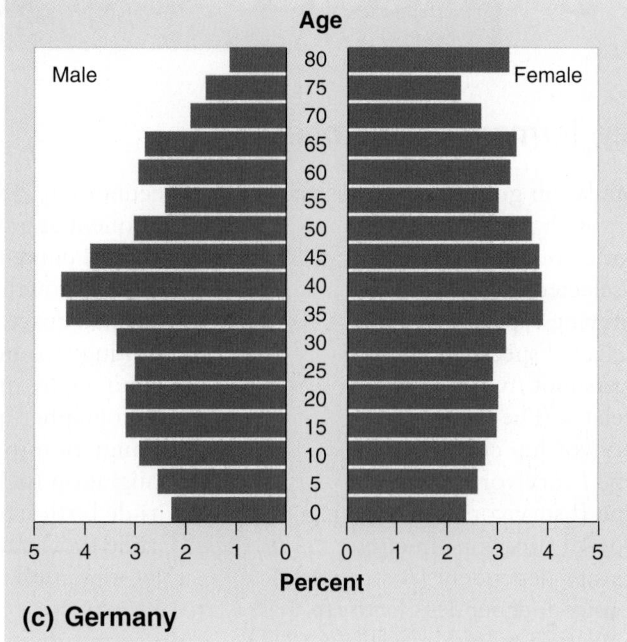

(c) Germany

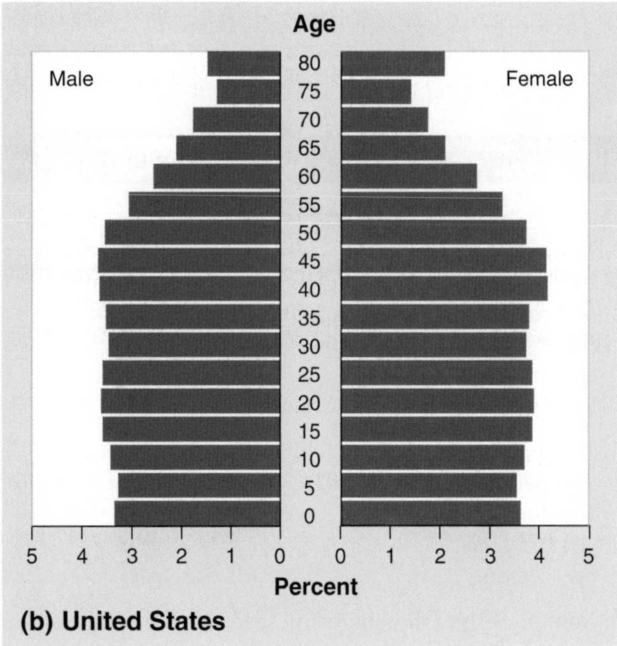

(b) United States

6. Which of the countries above is likely to have the highest growth rate?
   (a) Democratic Republic of Congo
   (b) United States
   (c) Germany
   (d) unable to determine

7. Which stage of demographic transition is Germany likely in?
   (a) stage 1
   (b) stage 2
   (c) stage 3
   (d) stage 4

8. Which country in the diagram above would have a crude birth rate below the crude death rate?
   (a) Democratic Republic of Congo
   (b) United States
   (c) Germany
   (d) all would have a CBR below CDR

9. World Bank records indicate that in 2010, the total fertility rate in Puerto Rico was 1.6 and population size was 4,000,000. If net migration rate was −1.0, then the best estimate of the population in 2012 is
   (a) 2,500,000.
   (b) 3,500,000.
   (c) 4,000,000.
   (d) 4,500,000.

10. A developing country with a total fertility rate of 4.0 and a net migration rate of −3.5 is likely to have an age structure
    (a) shaped like a pyramid.
    (b) shaped like an inverted pyramid.
    (c) shaped like a column.
    (d) that is variable depending on the demographic.

11. Which statement is true under the theory of demographic transition?
    (a) Improved sanitation and health care lead to reduced population growth.
    (b) Population momentum holds birth rates stable while death rates increase.
    (c) Density independent factors slow economic growth.
    (d) Education reduces the growth rate of a population.

12. Increasing the gross domestic product of a nation ultimately leads to which of the following?
    (a) a decrease in pollution
    (b) an increase in birth and death rates
    (c) an increase in population size
    (d) a decrease in educational opportunities

13. Which country is most likely to be in the second stage of demographic transition?
    (a) Argentina
    (b) Colombia
    (c) Philippines
    (d) Brazil

14. A city has a CBR of 17 and a CDR of 15. Ignoring immigration and emigration, when will the population of this city double?
    (a) 20 years
    (b) 35 years
    (c) 200 years
    (d) 350 years

Use the text below to answer questions 15 & 16:

The 2020 U.S. growth rate of 0.7 percent per year during the preceding decade was the slowest decade of growth since the 1930s.

In the absence of sufficient immigration, the total U.S. population is expected to decline. On the surface, this could seem to have some positive impacts such as reducing environmental stresses and other unwanted consequences of overcrowding.

15. Which of the following is one of the author's claims?
    (a) Immigration in the United States is keeping the growth rate high.
    (b) There may be positive environmental impacts to a declining population.
    (c) U.S. population is currently on the rise.
    (d) The U.S. population is declining but is expected to rise.

16. Which of the following would be an environmental impact of a declining population?
    (a) less production of CO_2 from vehicles
    (b) less resources available for fuel and production
    (c) more habitat loss for human living spaces
    (d) more water pollution due to more intensive farming practices

Use the graph below to answer questions 17–19:

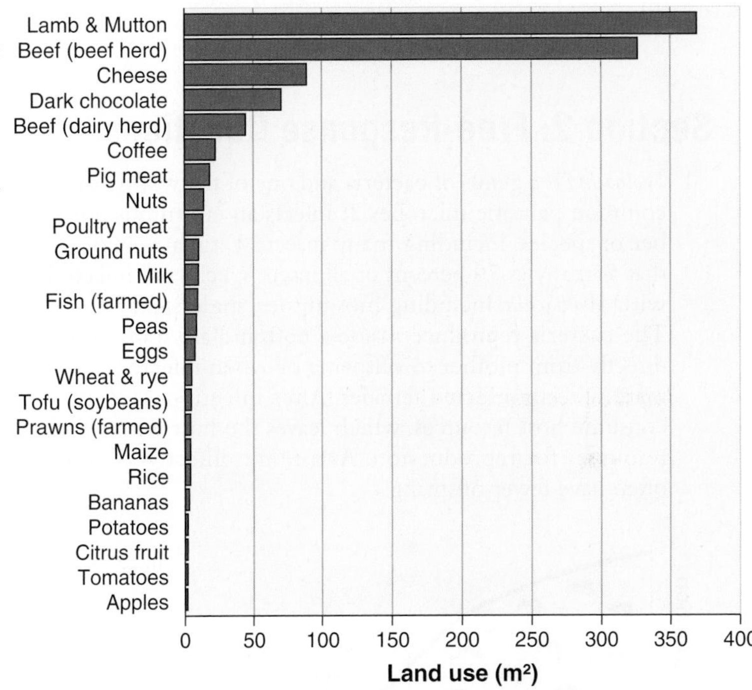

17. A more developed country, like the United States, consumes more beef than a less developed country even though a more developed country has a lower population. How much more land is required for beef than poultry meat?
    (a) 320 m^2
    (b) 550 m^2
    (c) 12 m^2
    (d) 120 m^2

18. Which of the following foods would a country in stage 2 of demographic transition most likely consume more of?
    (a) Lamb and mutton
    (b) Rice
    (c) Beef
    (d) Cheese

19. What conclusion could be made from the data in this graph?
    (a) Populations consuming meat require much more land than those consuming plants and vegetables.
    (b) Populations consuming plants require much more land than those consuming meat.
    (c) A population's food consumption plays no impact on land use.
    (d) If a population is using land to grow more meat, it is likely in stage 1 of demographic transition.

20. As a country moves into stage 2 of the demographic transition, you would expect replacement level fertility to ____________ and crude death rate to ____________.
(a) increase; decrease
(b) decrease; increase
(c) increase; increase
(d) decrease; decrease

21. If a population of 3.5 million people has a growth rate of 2.7 percent per year in 2010, what year would the population reach 14 million people?
(a) 2062
(b) 2036
(c) 2088
(d) 2050

Section 2: Free-Response Questions

1. *Wolbachia* is a genus of bacteria and one of the world's most common parasitic microbes. It infects an enormous number of species, including many insects. Estimates indicate that as many as 70 percent of all insect species are infected with *Wolbachia*, including mosquitoes, spiders, and mites. The bacteria reproduce inside a host and are transmitted directly from mother to offspring or when infected males mate with uninfected females. After infection, *Wolbachia* consume host resources, which leaves the host with fewer resources for reproduction. As a result, infected females often have fewer offspring.

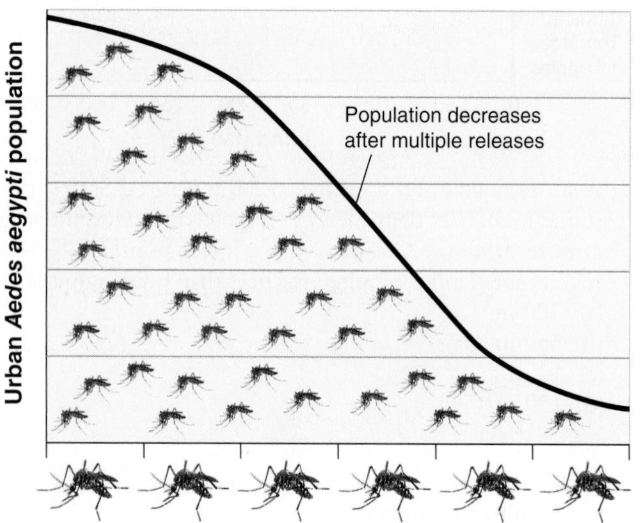

A group of scientists are studying the *Wolbachia* for population studies by releasing *Wolbachia* infected male mosquitoes into the populations.The data in the graph to the left show the results of the release.

(a) **Identify** the possible hypothesis being tested by this experiment? (1 pt.)
(b) **Identify** the independent variable in the experiment conducted. (1 pt.)
(c) **Describe** the population trajectory of *Wolbachia* in a population of hosts. (2 pts.)
(d) Based on your answer to part (c), **explain** one possible way that natural selection is altering the interaction between *Wolbachia* and its hosts? (2 pts.)
(e) **Describe** how the presence of nearby host populations might alter the spread of *Wolbachia*. (2 pts.)
(f) **Describe** the expected result of the experiment releasing infected males into the population. (2 pts.)

2. Country Y and country Z are two nations in different stages of demographic transition. Country Y is in stage two, while Country Z is in stage four.

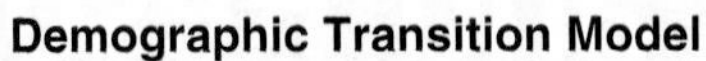

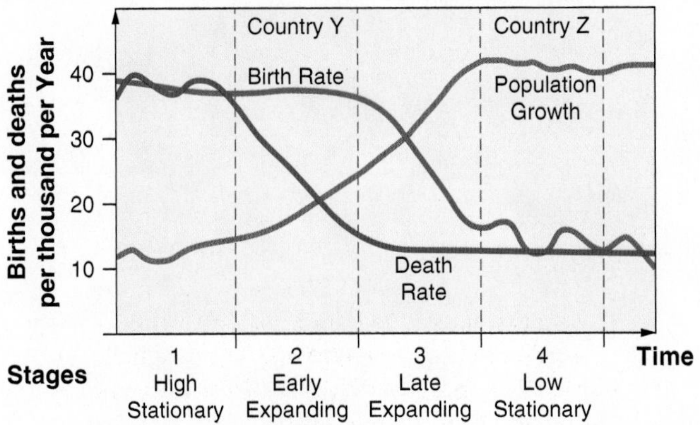

(a) **Describe** what an age structure diagram for country Y and country Z will look like. (2 pts.)

(b) **Explain** the differences between country Y and Z in terms of the following:

(i) CBR, CDR, and the rate of population growth (2 pts.)

(ii) the age structure of the population (1 pt.)

(c) **Explain** the social and economic implications of a high immigration rate into country Z. (2 pts.)

(d) **Explain** one example of an actual country that could be represented by country Y and one example that could be represented by country Z. (2 pts.)

(e) **Describe** the impact country Y and country Z have on the environment and how the environmental impacts will differ. (1 pt.)

3. By 2012, China's one-child policy had drastically reduced total fertility rate to 1.4 and led to a reduction in population growth rate to 0.50 percent. One problem associated with this is the development of an inverted population pyramid. Care for the elderly requires a substantial amount of energy and resources. China's elderly primarily rely on their families for health care.

(a) Given China's one-child policy, **describe** two examples of how China's growing elderly population will affect future population growth. (2 pts.)

(b) **Describe** one potential benefit and one potential problem associated with the emigration of elderly individuals from rural to urban areas. (2 pts.)

(c) **Calculate** the doubling time of the population above. (2 pts.)

(d) The total population size of China was 1,400,000,000 in 2012, with a crude birth rate of 12 births per 1,000 individuals. The number of individuals that emigrated from China was 700,000 and the number of individuals that immigrated to China was 200,000. **Calculate** the crude death rate. (2 pts.)

(e) **Describe** what the total fertility rate would be if all families in China obeyed the one-child policy rule. (1 pt.)

(f) **Explain** an alternative method to slow the fertility rate of a country. (1 pt.)

UNIT

4

Earth Systems and Resources

The eruption of Mount Saint Helens in 1980. After weeks of warning tremors, the mountain experienced an earthquake, which caused a landslide followed by the eruption of ash and gases. *(Jim Sugar/Getty Images)*

CASE STUDY

The Eruption of Mount Saint Helens: 40 Years Later

In the state of Washington stands Mount St. Helens. Mount St. Helens is a large mountain that was formed by volcanoes, along with the other mountains in the Cascade Range, which had remained quiet for more than a hundred years. All of this changed on May 18, 1980, when the mountain experienced an explosion of rocks and soil moving at 482 km/hour (300 mph), destroying 388 km^2 (150 mi^2) of forest, and killing 57 people. The emerging ash and gases lasted for 9 hours and rose 24 km (15 miles) into the air. The winds carried the ash across eleven states to the east and southeast, including as far east as Nebraska. This eruption stimulated a major increase in research on the processes that cause eruptions. Looking back 40 years, we now have a much better understanding of these processes and the long-term effects of the eruption on the ecological communities.

The first sign that the mountain was becoming active again occurred in March of 1980, when geologists detected a number of small earthquakes and gas emerging from the mountain. They also noticed that the mountain was starting to develop a bulge on its surface. We now know that this bulge was the pressure of the underlying volcano that was being held back by a thick layer of ice, rocks, and soil. On the morning of the eruption, there was a much larger earthquake and the shaking of the land caused a massive landslide down the north side of the mountain. In fact, it was the largest landslide ever observed. This landslide removed the thick layer that was holding back the volcano, so at 8:32 AM Mount St. Helens erupted, blowing away the north face of the mountain. It is estimated that the mountain released 540 million tons of ash, which would fill a football field with ash stacked 150 miles high. Although many volcanoes also release lava, this eruption did not have flows of lava coming down the mountain.

The emerging ash and gases lasted for 9 hours and rose 24 km (15 miles) into the air.

The eruption also impacted the water bodies in the area. The ash that dropped into the nearby Columbia River caused the river's depth to go from 12 m (39 feet) to only 4 m (13 feet). Nearby Spirit Lake received large deposits of ash and fir trees — 200 years old and 55 m (180 feet) tall — that were blasted into the lake by the explosion and pushed all of the water out of the lake. The amount of debris was so large that it dammed the outlet of the lake. As the lake refilled with water over time, this dam posed a risk of failing and flooding towns that are located downstream. As a result, engineers had to build a tunnel to help water drain out of the lake. Today, the trees are still in the lake, but the original tunnel is failing so plans are being made to build a new and improved tunnel.

The landscape around the north side of the mountain was described as looking like Mars. The explosion of heat, rocks, and ash scoured and scorched a massive area around the mountain. It was expected that it would take decades to centuries for the area to undergo primary succession, a process we discussed in Module 14. Interestingly, the large field of ash did not have to wait for plant seeds to arrive to initiate ecological succession. Researchers discovered that a few plants were still alive under the snow and ash layer and these plants were able to poke up through the layers and jump-start the process of succession. As a result, some climax species were

Mount St. Helens 40 years after the eruption. In this view of the north face of the mountain, we can see the remnants of the massive landslide and the crater caused by combined effects of the landslide and eruption. In the foreground, we can see the diversity of plants that are now growing where there was once a forest of trees. *(Roman Khomlyak/Shutterstock)*

present early in the succession process while some early-succession species did not show up until much later. In addition, the scarce surviving plants attracted small mammals and birds, which likely carried additional seeds into the area in their digestive systems. Researchers also discovered that pocket gophers, which were in their underground burrows at the time of the eruption, below a layer of snow and soil, had survived and were able to dig out of the volcanic materials. In doing so, they mixed the volcanic ash, which contained no organic matter, with the underlying soil; this made the ground more hospitable for plants to grow. They also learned that it was better to leave the dead trees lying on the ground rather than remove them, because the trees helped prevent erosion of the soil. Today, the once destroyed forests are a thicket of willows, alders, huckleberries, and young fir trees.

In 1980, volcano research was still in its infancy. Over the past 40 years, however, Mount St. Helens has become a living laboratory for hundreds of environmental scientists who are working to understand the process that cause volcanoes around the world, to predict when they will erupt, and to forecast how the processes of ecological succession will subsequently unfold. The mountain continues to have small releases of ash, gas, and lava, and nobody can be sure when a major eruption may happen again. When it does, the researchers will be ready.

Sources: C. Apple, Eruption! *The Spokesman Review*, May 18, 2020, https://www.spokesman.com/stories/2020/may/17/eruption-mt-st-helens/; S. Nash, Making sense of Mount St. Helens, *BioScience* 60 (2010): 571–575; A. Orlando, 40 years ago: Lessons from the eruption of Mount St. Helens, *Discover Magazine*, August 13, 2020, https://www.discovermagazine.com/planet-earth/40-years-ago-lessons-from-the-eruption-of-mount-st-helens.

Practice your science skills

1. **Text Analysis:** Describe the author's perspective and assumptions about the future of volcanic research.
2. **Environmental Solutions:** Describe one potential environmental solution (other than building a new and improved tunnel) to control the volcanic ash sediment and plant debris flowing downstream.
3. **Environmental Solutions:** Describe a consequence for the potential environmental solution proposed in question 2 to control the sediment flowing downstream.

The story of Mount Saint Helens is an excellent example of how Earth systems collectively impact the biotic and abiotic conditions in a region through a combination of physical, chemical, and biological processes. In this unit, we will explore these processes, including how the movement of the Earth's surface causes earthquakes and volcanoes, how soils are formed and eroded, and the movement of water over the land. We will also explore how the Earth's orientation and the Sun's differential warming of latitudes drives air currents in the atmosphere and water currents in the oceans, which combine to determine the planet's various climates and biomes. Given that these processes help determine the distribution of species diversity, the many human uses of nature, and the many harmful impacts on nature, the topics covered in this unit are essential to your understanding of environmental science.

Module 19

Unit 4	19	20	21	22	23

Plate Tectonics

Thus far we have discussed species, communities, and ecosystems and how they are interconnected in ways that sustain nature and the ecosystem services upon which humans rely. In this unit, we place our focus at the scale of the biosphere to understand the various systems of Earth, including the different layers inside our planet, the formation of soils, the composition of our atmosphere, and global patterns of air circulation that drives global weather. In this first module we focus on how the surface of Earth is comprised of moving plates that, over geologic time, have dramatically altered the landscape of oceans and continents, substantially changed the biodiversity, and even today produce geologic events that impact our lives.

Learning Goals

After reading this module you should be able to

19-1 describe the layers of Earth.

19-2 explain how tectonic plate movement produces divergent, convergent, and transform boundaries.

19-3 identify how the global distribution of plate boundaries predicts the locations of geologic events.

19-1 What are the layers of Earth?

Earth is comprised of layers from the core to the tectonic plates on the surface

Approximately 4.6 billion years ago, Earth was formed from cosmic dust in the solar system that consolidated into a planet. Early Earth was a hot, molten sphere and, for a period of time, additional debris from the formation of the solar system bombarded Earth. As this molten material slowly cooled, the elements within it separated into layers according to their mass. Heavier elements such as iron sank toward Earth's center, and lighter elements such as silica floated toward its surface. Some gaseous elements left the solid planet and became part of Earth's atmosphere. Although asteroids occasionally strike Earth today, the bombardment phase of planet formation has largely ceased and the elemental composition of Earth has stabilized. In other words, the elements and minerals that were present when the planet formed—and which are distributed unevenly around the globe—are all that we have today.

Because Earth's elements settled into place according to their mass, the inside of the planet is characterized by distinct layers. If we could slice into Earth, as shown in **FIGURE 19.1**, we would see concentric layers composed of various materials. The innermost zone of Earth's interior is the planet's

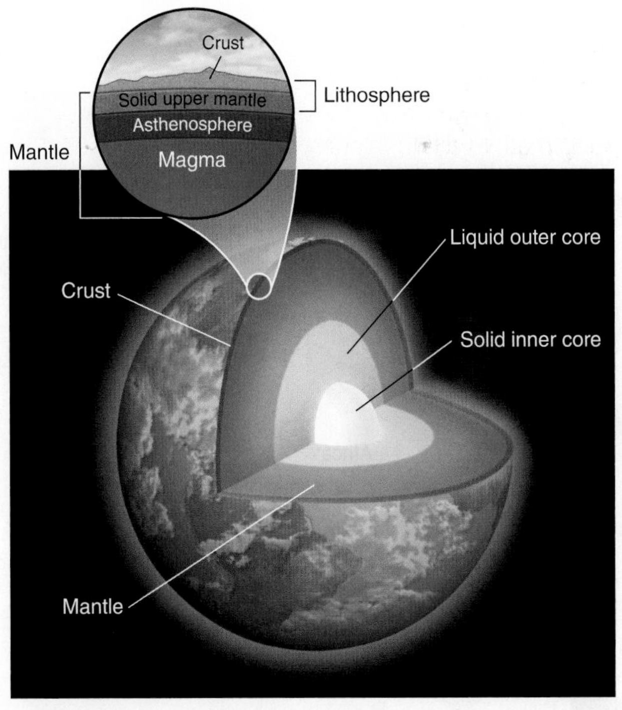

FIGURE 19.1 Earth's layers. Earth is composed of concentric layers, with a solid inner core that is surrounded by a liquid outer core and then the mantle. As shown in the inset, the mantle is comprised of three layers: magma, the asthenosphere, and the solid upper mantle.

core, over 3,000 km (1,860 miles) below Earth's surface. The core is a dense mass largely made of nickel and some iron. The inner core is solid, and the outer core is liquid. Above the core is the **mantle**, which itself is comprised of three layers. The innermost portion of the mantle contains molten rock, or **magma**, that slowly circulates. In the upper mantle is the **asthenosphere**, which is composed of semi-molten, flexible rock, and then the solid upper mantle. Above the solid upper mantle is the crust, which is a chemically distinct layer. Together, the solid upper mantle and the crust form the **lithosphere** (from the Greek word *lithos*, which means "rock"). The lithosphere is approximately 100 km (60 miles) thick. Over the crust lies the thin layer of soil that allows life to exist on the planet. The crust and overlying soil provide most of the chemical elements that make up life, along with elements from the atmosphere. In this module our primary focus is on the lithosphere, which is made up of numerous plates that are slowly moved by the flowing magma below it.

Core The innermost zone of Earth's interior, composed mostly of iron and nickel. It includes a liquid outer layer and a solid inner layer.

Mantle The layer of Earth above the core, containing magma, the asthenosphere, and the solid upper mantle.

Magma Molten rock.

Asthenosphere The layer of Earth located in the outer part of the mantle, composed of semi-molten rock.

Lithosphere The outermost layer of Earth, including the solid upper mantle and crust.

19-2 How does tectonic plate movement produce divergent, convergent, and transform boundaries?

Tectonic plate movement creates major geographic features on Earth

Prior to the 1900s, scientists believed that the major features of Earth—such as continents and oceans—were fixed in place. In this section we will see that the continents and ocean floors are slowly moving.

The Theory of Plate Tectonics

In 1912, a German meteorologist named Alfred Wegener published a revolutionary hypothesis proposing that the world's continents had once been joined in a single landmass, which he called "Pangaea." His evidence included observations of identical rock formations on both sides of the Atlantic Ocean, as shown in **FIGURE 19.2a**. The positions of these formations suggested that a single supercontinent broke up into separate landmasses. Fossil evidence also suggested that a single large continent existed in the past. For example, Figure 19.2b shows fossils of a single species that have been found on different continents that are currently separated by oceans. The evidence from rocks and fossils collectively suggest that the continents have indeed moved over long periods of time.

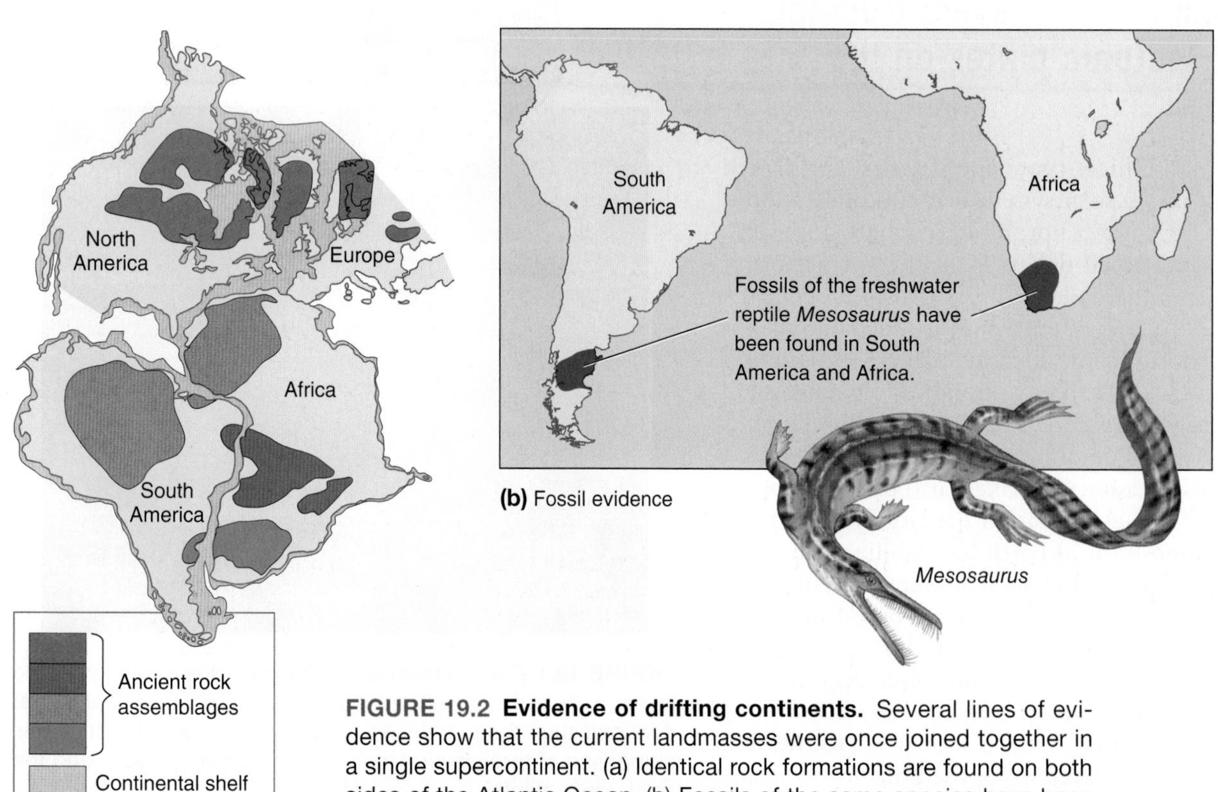

FIGURE 19.2 Evidence of drifting continents. Several lines of evidence show that the current landmasses were once joined together in a single supercontinent. (a) Identical rock formations are found on both sides of the Atlantic Ocean. (b) Fossils of the same species have been collected from different continents.

But how is this possible? Many scientists resisted the idea that Earth's lithosphere could move laterally. However, following the publication of Wegener's hypothesis, scientists found additional evidence that Earth's landmasses had existed in several different configurations over time. Pivotal to this discovery was the work of geologist Marie Tharp. In the 1950s, many scientists believed the oceans were flat on the bottom, but as Tharpe mapped the ocean bottom, she called attention to the fact that the ocean was filled with underwater mountain ranges and massive canyons. This led to the theory of **plate tectonics**, which states that Earth's lithosphere is divided into plates, most of which are in constant motion and create mountain ranges and canyons, both on land and in the oceans. Today, scientists fully appreciate that the lithosphere consists of a number of tectonic plates.

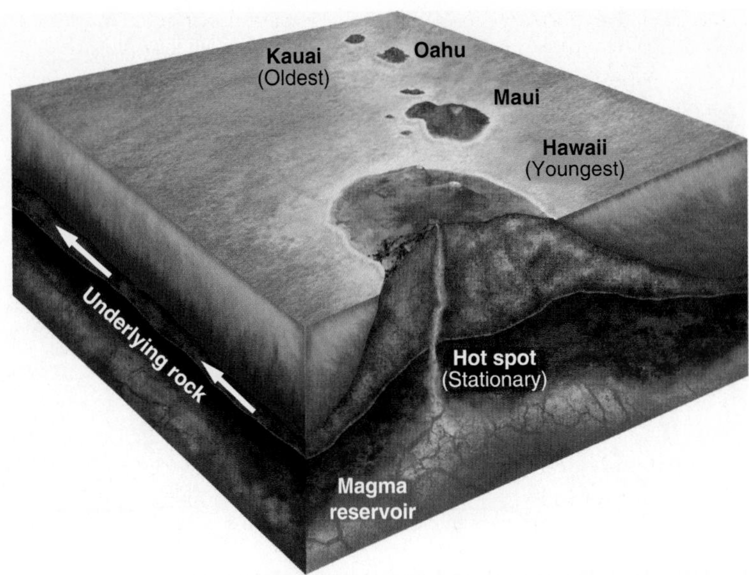

FIGURE 19.3 Plate movement over a hot spot. As the Pacific Plate moves over a hot spot, a series of volcanic eruptions that occurred over several million years led to formation of the Hawaiian Islands.

Earthquakes and Volcanoes

Although the rate of plate movement is too slow for us to notice, geologic activity provides vivid evidence that the plates are in motion. This includes the forces that produce earthquakes and volcanoes.

Earthquakes

Although the plates are always in motion, their slow movement is not always smooth. Imagine rubbing two rough and jagged rocks past each other. The rocks would resist that movement and get stuck together. The rock along a fault is also jagged and thus resists movement, but the mounting pressure eventually overcomes the resistance and the plates give way, slipping quickly. This is an **earthquake**, which is a sudden movement of Earth's crust caused by a release of potential energy from the movement of tectonic plates. The plates can move up to several meters in just a few seconds. The epicenter of an earthquake is the exact point on the surface of Earth directly above the location where the rock ruptures.

Earthquakes occur many times a day throughout the world, but most are so small that humans do not feel them. The magnitude of an earthquake is reported on the Richter scale, a measure of the largest ground movement that occurs during an earthquake. The Richter scale measures the intensity of earthquake force on a logarithmic scale. Thus, a value increases by a factor of 10 for each unit increase. For example, a magnitude 7.0 earthquake, which causes serious damage, is 10 times greater than a magnitude 6.0 earthquake and 1,000 times greater than a magnitude 4.0 earthquake, which only some people can feel or notice. Worldwide, there may be as many as 800,000 small earthquakes of magnitude 2.0 or less per year, but an earthquake of magnitude 8.0 or more occurs approximately once every year.

Volcanoes

One of the critical consequences of Earth's formation and elemental composition is that the planet remains very hot at its center. The high temperature of Earth's outer core and mantle is thought to be the result of the radioactive decay of various isotopes of elements such as potassium, uranium, and thorium, which release heat. The heat causes plumes of hot magma to well upward from the mantle. These plumes produce **hot spots**, which are places where molten material from the mantle reaches the lithosphere.

As a plate moves over a geologic hot spot, heat from the rising mantle plume melts the crust, forming a **volcano**, which is a vent in Earth's surface that emits ash, gases, and molten lava. Over time, as the plate moves past the hot spot, it can leave behind a trail of volcanic islands. As shown in **FIGURE 19.3**, the Hawaiian Islands were formed by a series of volcanic eruptions over several million years as the Pacific Plate traveled northwest over a geologic hot spot.

When earthquakes or volcanoes occur in the ocean, they can cause a **tsunami**, which is a series of waves in the ocean caused by seismic activity in the ocean or an undersea volcano that causes a massive displacement of water. As those

Plate tectonics The theory that the lithosphere of Earth is divided into plates, most of which are in constant motion.

Earthquake A sudden movement of Earth's crust caused by a release of potential energy from the movement of tectonic plates.

Hot spot In geology, a place where molten material from Earth's mantle reaches the lithosphere.

Volcano A vent in the surface of Earth that emits ash, gases, or molten lava.

Tsunami A series of waves in the ocean caused by seismic activity or an undersea volcano that causes a massive displacement of water.

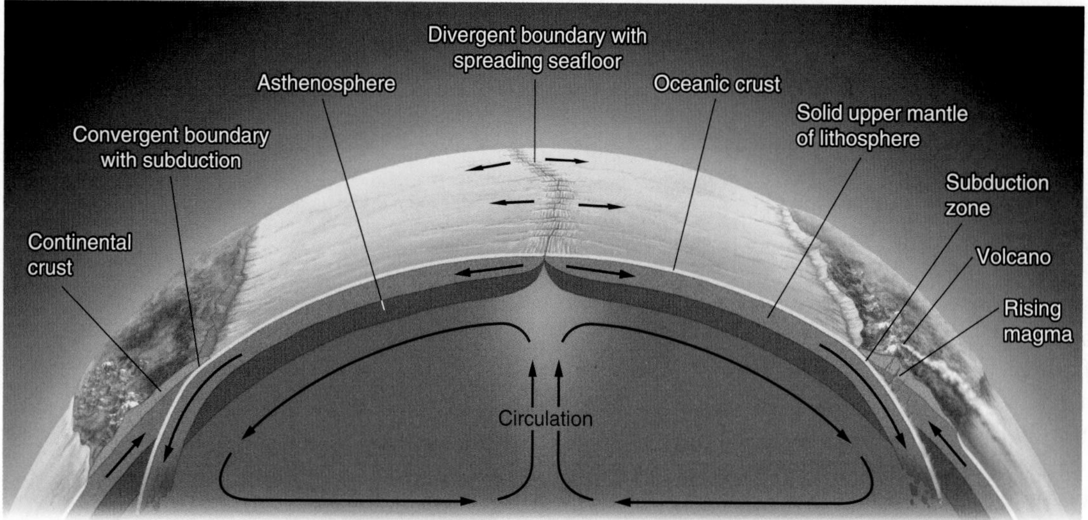

FIGURE 19.4 Divergent and convergent plate movement. Circulation in the mantle causes oceanic plates to spread apart as new rock rises to the surface at spreading zones. Where oceanic and continental plate margins come together, older oceanic crust is subducted.

waves approach land, their height becomes greater and they can cause substantial—sometimes catastrophic—damage when they reach land.

Divergent, Convergent, and Transform Boundaries

It is helpful to think of the plates in two categories: oceanic and continental. Oceanic plates lie primarily beneath the oceans, whereas continental plates lie beneath landmasses. The crust of oceanic plates is dense and rich in iron, while the crust of continental plates generally contains more silicon dioxide, which is much less dense than iron. The continental plates are therefore lighter and typically rise above the oceanic plates. Oceanic and continental plates "float" on top of the denser material beneath them and their slow movements are driven by the circulation of magma that occurs in Earth's mantle.

Divergent boundary An area below the ocean where tectonic plates move away from each other.

Seafloor spreading Caused by a divergent boundary, in which rising magma forms new oceanic crust on the seafloor at the boundaries between those plates.

Convergent boundary An area where one plate moves toward another plate and collides.

Subduction The process in which the edge of an oceanic plate moves downward beneath the continental plate and is pushed toward the center of Earth.

Island arc A chain of islands formed by volcanoes as a result of two tectonic plates coming together and experiencing subduction.

Collision zone An area where two continental plates are pushed together and the colliding forces push up the crust to form a mountain range.

Adjacent tectonic plates interact to form three different types of boundaries: divergent boundaries, convergent boundaries, and transform boundaries. You can see divergent and convergent boundaries in **FIGURE 19.4**. As the tectonic plates below the ocean move away from each other, we observed a **divergent boundary** in which rising magma forms new oceanic crust on the seafloor at the boundaries between those plates. Divergent boundaries in the ocean floor cause **seafloor spreading**, where rising magma forms new oceanic crust on the seafloor at the boundaries between those plates, as shown in the center of the figure. These locations of divergent boundaries can also produce volcanoes, earthquakes, and rift valleys. For example, the Great Rift Valley in eastern Africa was caused by a divergence of plates that subsequently filled in with water to become the Rift Valley lakes that have very high biodiversity, such as Lake Tanganyika that we discussed in Module 13.

There also are **convergent boundaries**, which occur as one plate moves toward another plate and collides. There are two outcomes when plates collide. In the case where an oceanic plate meets a continental plate, the edge of the oceanic plate moves downward beneath the continental plate and is pushed toward the center of Earth in a process known as **subduction**. Subduction happens because the heavier oceanic plate slides underneath the lighter continental plate. This is best seen on the right side of Figure 19.4. Subduction can also occur when two oceanic plates collide. In areas where subduction occurs, we can see **island arcs**, which are chains of islands formed by volcanoes as a result of two tectonic plates coming together and experiencing subduction.

A different outcome is observed when two continental plates are pushed together, which we call a **collision zone**. In this case, the collision of the two continental plates of similar densities does not result in subduction. Instead, the colliding forces push up the crust to form a mountain range.

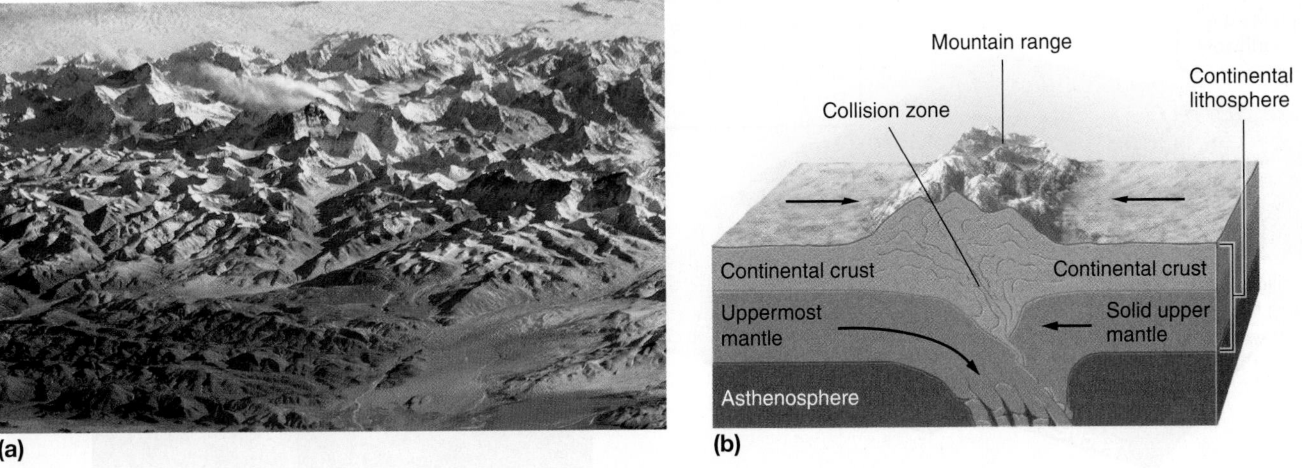

FIGURE 19.5 Collision of two continental plates. (a) This photo shows Mount Everest and Makalu in the Himalayas from space. (b) The Himalayas, which include the highest mountains on Earth, was formed when the collision of two continental plates forced the margins of both plates upward. *(a: NASA)*

A nice example of this is when the Indo-Australian plate collided with the Eurasian plate, beginning about 40 million years ago, and gave rise to the Himalayan mountain range (**FIGURE 19.5**). Convergent boundaries also give rise to volcanoes, earthquakes, and island arcs.

We can see the locations of the plates and the many divergent and convergent boundaries in **FIGURE 19.6**. In this figure, the arrows indicate the direction that each plate is slowly moving, the red lines indicate divergent boundaries, and the blue lines indicate convergent boundaries.

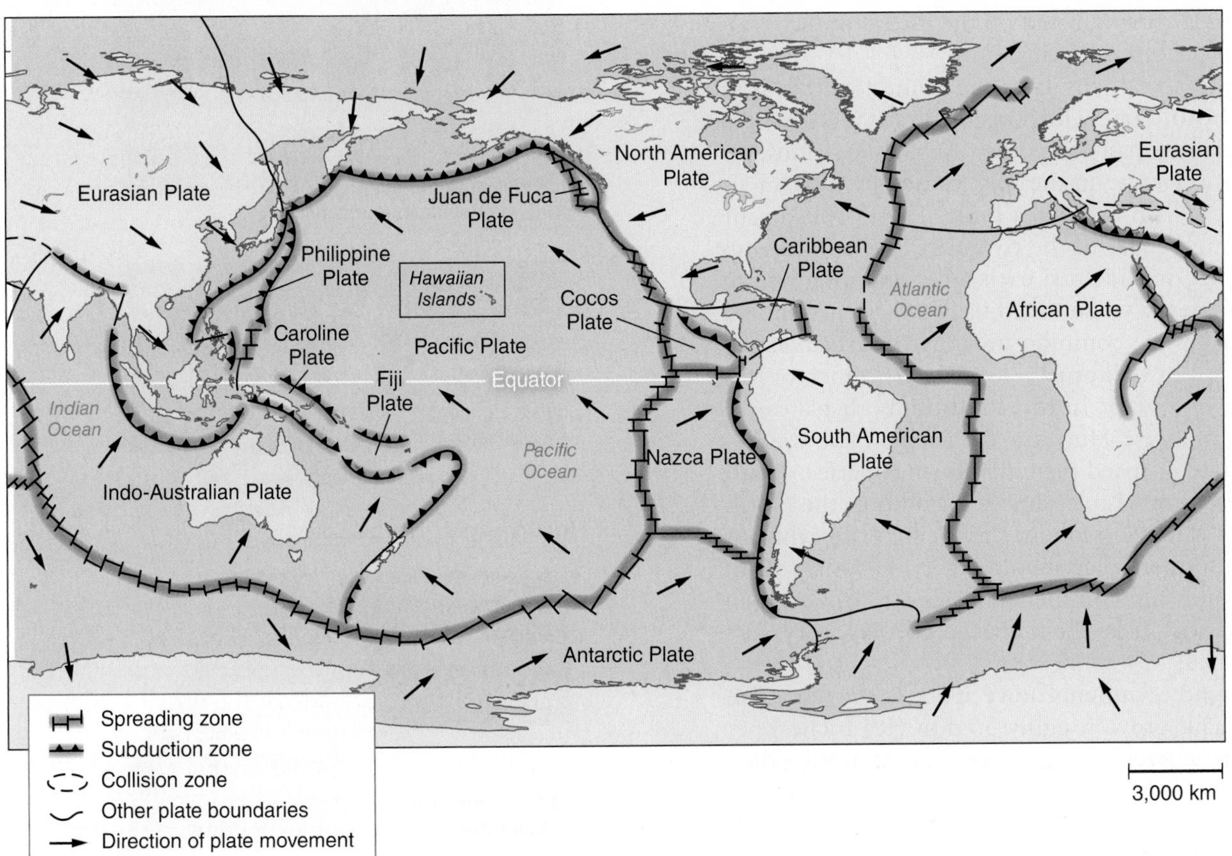

FIGURE 19.6 Tectonic plates of the world. Earth is covered with tectonic plates, most of which are in constant motion. The arrows indicate the direction of plate movement. New lithosphere is added at spreading zones and older lithosphere is recycled into the mantle at subduction zones.

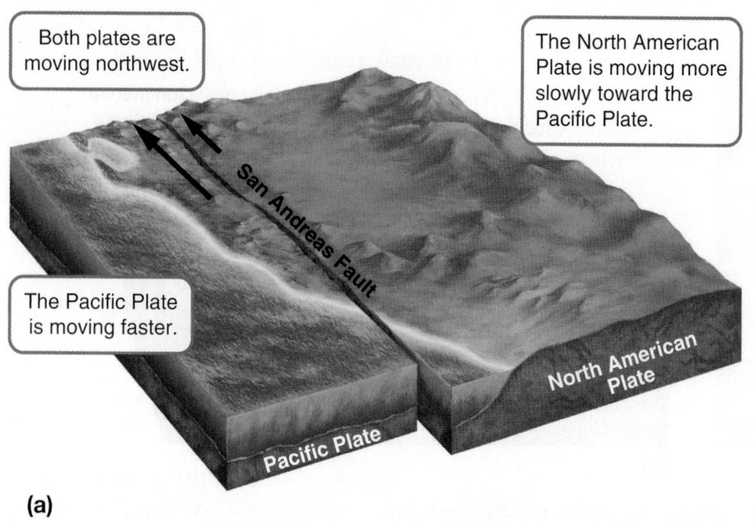
Both plates are moving northwest.
The North American Plate is moving more slowly toward the Pacific Plate.
San Andreas Fault
The Pacific Plate is moving faster.
North American Plate
Pacific Plate
(a)

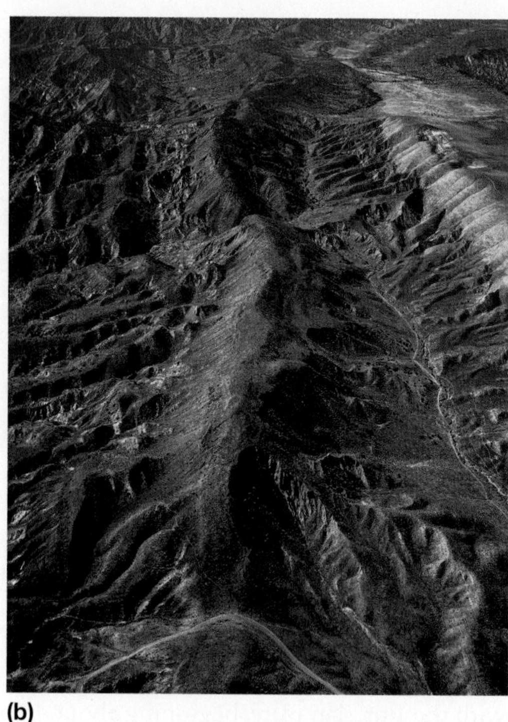
(b)

FIGURE 19.7 A fault in California. (a) The San Andreas Fault in California is an example of a transform boundary. Although plates are moving north and west, the Pacific Plate is moving at a greater speed than the North America Plate. The North America Plate is also moving toward the Pacific Plate. (b) In this photo of the San Andreas Fault, you can see the results of seismic activity. *(b: DEA/PUBBLI AER FOTO/Getty Images)*

You should be able to predict whether a boundary is divergent or convergent simply based on the direction that each of two adjacent plates are moving.

The third way in which plates can come together is at a **transform boundary**, which occurs in locations where plates move sideways past each other. At such locations, we commonly see a fracture in the rock caused by movement in Earth's crust, which is called a **fault**. These faults occur because the rough edges of the two plates are building up energy as they try to slide past each other and then release a great deal of energy when they finally experience movement. Earthquakes are common in fault zones such as the San Andreas Fault in California (**FIGURE 19.7**), which occurs at a transform boundary. In this location, both plates are moving north and west. However, the Pacific Plate is moving at a faster rate of speed than the North American Plate, which is one cause of seismic activity. In addition, the North American Plate is moving toward the Pacific Plate, which is another reason for periodic seismic activity. The main point is that the oceanic and continental plates are constantly in motion in ways that make them diverge, converge, or transform (**FIGURE 19.8**).

Most plates and continents move at about the same rate as your fingernails grow: roughly 36 mm (1.4 inches) per year. While this movement is far too slow to notice on a

Transform boundary An area where tectonic plates move sideways past each other.

Fault A fracture in rock caused by a movement of Earth's crust.

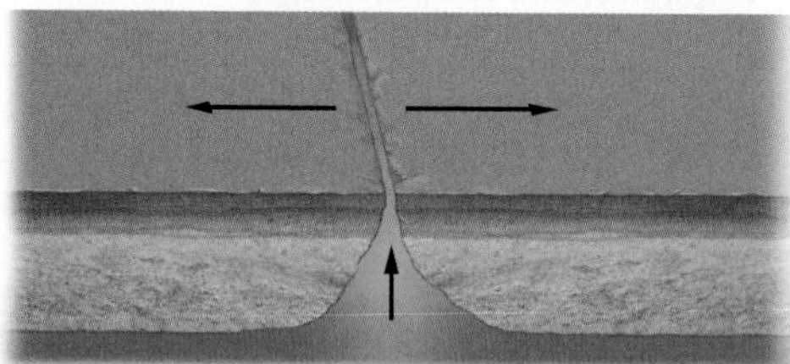
(a) Divergent plate boundary

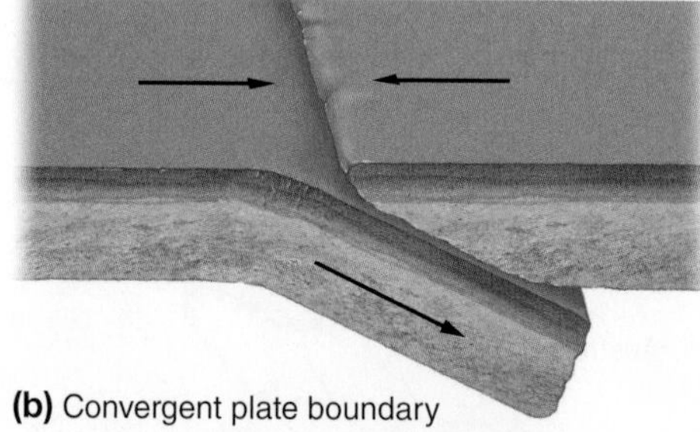
(b) Convergent plate boundary

(c) Transform fault boundary

FIGURE 19.8 Types of plate boundaries. (a) At divergent plate boundaries, plates move apart. (b) At convergent plate boundaries, plates collide. (c) At transform fault boundaries, plates slide past each other.

daily basis, distance moved can be large over long periods of time. Think how often you've needed to clip your nails over the course of your life! For instance, the two plates underlying the Atlantic Ocean have spread apart and come together twice over the past 500 million years, causing Europe and Africa to collide with North America and South America and separate from them again. The "Do the Math: Plate Movement" below shows how we can calculate the time it takes for plates to move a given distance.

Consequences of Plate Movement on Biodiversity

Because the plates move, continents on those plates slowly move over the surface of Earth. As the continents have moved, their climates have changed and geographic barriers were formed or removed. As a result, species evolved and adapted, or slowly or rapidly went extinct. In some places, as the plates moved, a continent that straddled two plates broke apart, creating two separate smaller continents or islands in different climatic regions. We can see an example of this in the case of Australia and Antarctica, which split into separate continents and moved to areas of the world that have very different climates. As you may recall from our discussion of allopatric speciation in Module 13, species that become separated can take different evolutionary paths and over time evolve into two or more separate species. The fossil record tells us how species adapted to the changes that took place over geologic time. Climate scientists and ecologists can use this information to anticipate how species will adapt to the relatively rapid climate changes happening on Earth today.

AP® Exam Tip

Be specific in your language when you answer free-response questions on the AP® Environmental Science Exam. For example, plate tectonics is a dynamic process. When describing plate movement, indicate that plates are currently in motion. Qualifying terms such as time, position, or direction are often necessary to earn points. For example, earthquakes occur when plates move past each other quickly or suddenly. For full points, the qualifying term "quickly" or "suddenly" must be included in the answer.

DO THE MATH Plate Movement

Preparing for the AP® Exam

If two cities lie on different tectonic plates, and those plates are moving so that the cities are approaching each other, how many years will it take for the two cities to be situated adjacent to each other?

Los Angeles is 630 km (380 miles) southeast of San Francisco. The plate under Los Angeles is moving northward at about 36 mm per year relative to the position of the plate under San Francisco. Given this average rate of plate movement, how long will it take for Los Angeles to be located next to San Francisco?

The distance traveled is 630 km, and the net distance moved is 36 mm per year. We can use this formula to determine the answer

$$\text{time} = \text{distance} \div \text{rate}$$

$$630 \text{ km} = 630{,}000 \text{ m} = 630{,}000{,}000 \text{ mm}$$

$$\frac{630{,}000{,}000 \text{ mm}}{36 \text{ mm/year}} = 17{,}500{,}000 \text{ years}$$

We could also put these dimensional relationships together as follows and then cancel units that occur in both the numerator and denominator. We are left with an answer in numbers of years

$$630 \text{ km}\left(\frac{1{,}000 \text{ m}}{1 \text{ km}}\right)\left(\frac{1{,}000 \text{ mm}}{1 \text{ m}}\right)\left(\frac{1 \text{ year}}{36 \text{ mm}}\right) = 17{,}500{,}000 \text{ years}$$

Thus, it will take about 18 million years for Los Angeles to be located alongside San Francisco.

YOUR TURN How long will it take for a plate that moves at 20 mm per year to travel the distance of one football field? Note that a football field is 91.44 m (100 yards) long.

19-3 How does the global distribution of plate boundaries predict the locations of geologic events?

Plate boundaries determine the location of many volcanoes, island arcs, earthquakes, hot spots, and faults

Given that volcanoes and earthquakes can be produced at plate boundaries, it is perhaps not surprising that many of their locations map closely to plate boundaries. Sometimes volcanoes and earthquakes are observed in close proximity to where tectonic activity is high. **FIGURE 19.9** shows one example, in which earthquake locations and volcanoes form a circle of tectonic activity, called the "Ring of Fire," around the Pacific Ocean. While many hot spots and island arcs also occur along plate boundaries, we can see on the map that some hot spots and island arcs, such as the Hawaiian Islands, are not located on plate boundaries.

Environmental and Human Impacts of Earthquakes and Volcanoes

The surfacing and sinking of the lithosphere are ongoing processes. When humans live in close proximity to areas of seismic or volcanic activity, however, the results can be dramatic and devastating. For earthquakes, even a moderate amount of Earth movement can be disastrous. Moderate earthquakes (defined as magnitudes 5.0 to 5.9) can cause collapsed structures and buildings, fires, contaminated water supplies, ruptured dams, and deaths. Loss of life is more often a result of the proximity of large population centers to the epicenter than of the magnitude of the earthquake itself. The quality of building construction in the affected area is also an important factor in the amount of damage that occurs. For example, a magnitude 7.2 earthquake in Haiti killed more than 2,000 people and left many more homeless in 2021. This followed an earlier earthquake in Haiti in 2010 that killed more than 100,000 people. In that case, the epicenter was near the populated capital of Port-au-Prince where many buildings were probably not built to withstand a large earthquake. In 2017, two earthquakes of magnitudes 8.1 and 7.1 occurred within two weeks in Mexico. A series of smaller earthquakes continued for days afterwards and over 300 people were killed. Many of the victims were trapped under collapsed buildings (**FIGURE 19.10**). In 2021, the most powerful earthquake in the United States in the past 5 decades hit Alaska's Aleutian Islands with a magnitude of 8.7. This earthquake occurred at a convergent boundary, where the Pacific Plate experiences subduction under the North American plate. Fortunately, the epicenter was relatively deep underground and the area affected was not heavily populated, resulting in only one death.

Extra safety precautions are needed when dangerous materials are used in areas of earthquake activity. For example, nuclear power plants are designed to withstand

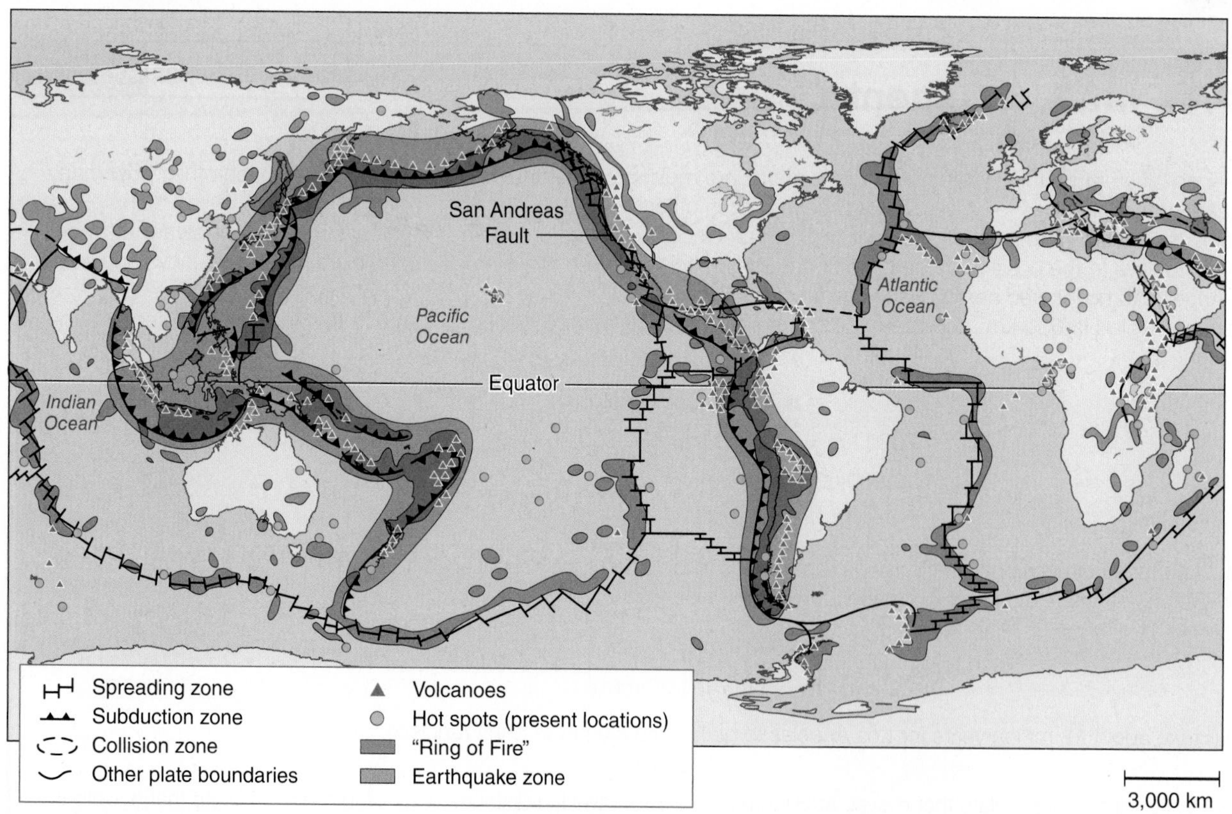

FIGURE 19.9 Locations of earthquakes and volcanoes. A "Ring of Fire" circles the Pacific Ocean along plate boundaries. Other zones of seismic and volcanic activity, including hot spots, are also shown on this map.

FIGURE 19.10 Earthquake damage in Mexico. This photo shows earthquake damage in September 2017 in Mexico City, Mexico, after an earthquake of magnitude 7.1 on the Richter scale. *(MARIO VAZQUEZ/Getty Images)*

FIGURE 19.11 A volcanic eruption. This eruption of the Mexico's Popocatépetl volcano occurred in 2020. *(www.webcamsdemexico.com)*

significant ground movement and are programmed to shut down if movement occurs above a certain threshold. The World Nuclear Association estimates that 20 percent of nuclear power plants operate in areas of significant earthquake activity. Between 2004 and 2009, nuclear power plants in Japan shut down their operation four times because of ground movement that exceeded the threshold. In 2011, seismic activity in Japan led to a devastating major earthquake and tsunami that caused the second-worst nuclear power plant accident ever to occur, as we will see in Module 38.

AP® Exam Tip

Not all earthquakes produce tsunamis. For an earthquake to create a tsunami there must be a shift of the plates that displaces a large amount of water.

Volcanoes, when active, can be equally disruptive and harmful to human life. Active volcanoes are not distributed randomly over Earth's surface; 85 percent of them occur along plate boundaries. As we have seen, volcanoes can also occur over hot spots. Depending on the type of volcano, an eruption may eject cinders, ash, dust, rock, or lava into the air (**FIGURE 19.11**). Volcano eruptions can result in loss of life, habitat destruction and alteration, reduction in air quality, and many other environmental consequences.

The world gained a new awareness of the impact of volcanoes when eruptions from a volcano in Iceland disrupted air travel to and from Europe in April 2010. Ash from the eruption entered the atmosphere in a large cloud and prevailing winds spread it over a wide area. The ash contained small particles of silicon dioxide, which have the potential to damage airplane engines. Air travel was suspended in many parts of Europe, and millions of travelers were stranded in what was the greatest travel disruption ever to have been caused by a volcano in modern times.

In this module we have seen that the surface of our planet is comprised of plates that are continuously moving and interacting with each other as the magma circulates below. This plate movement has resulted in continent movement over long periods of time as well as areas of concentrated volcanic and earthquake activity, which can produce mountain ranges, island arcs, and rifts that shape the geography of Earth. Volcanoes and earthquakes also have impacts on biodiversity as well as having major impacts on humans that live in close proximity to where these events happen.

Module 19 AP® Review

Preparing for the AP® Exam

Learning Goals Revisited

19-1 What are the layers of Earth?

The center of Earth has a solid core that is surrounded by a liquid core. Outside of these layers is the magma, the asthenosphere, and the lithosphere.

19-2 How does tectonic plate movement produce divergent, convergent, and transform boundaries?

Tectonic plates are continuously moving as the underlying magma circulates and they interact at convergent, divergent, and transform boundaries. The movement of these plates over millions of years has caused dramatic changes in biodiversity around the world. The planet remains very hot at its center and this heat causes plumes of hot magma to well upward from the mantle to produce volcanoes and hot spots. Earthquakes occur where jagged edges of plates rub together and the mounting pressure causes the plates to ultimately slip. When this happens, the plates can move up to several meters in just a few seconds and cause tremendous destruction to human property and lives.

19-3 How does the global distribution of plate boundaries predict the locations of certain geologic events?

Earthquakes and volcanoes are concentrated in the areas around plate boundaries. Volcanoes can also be found away from these boundaries where there are hot spots of magma flowing up to the surface of Earth, such as the hot spot that formed the Hawaiian Islands.

AP® Practice Questions

Preparing for the AP® Exam

Multiple-Choice Questions

1. Which layer of Earth is composed primarily of liquid nickel and iron?
   (a) the outer core
   (b) the inner core
   (c) the asthenosphere
   (d) the mantle

2. Which of the following statements best describes the process of subduction?
   (a) Subduction is the result of a hot spot moving near a plate boundary.
   (b) Subduction occurs when one plate passes under another.
   (c) Subduction occurs when oceanic plates diverge and form volcanoes.
   (d) Subduction is the process that occurs in transform boundaries that results in earthquakes.

3. The Hawaiian Islands were formed
   (a) at a divergent plate boundary.
   (b) at a hot spot.
   (c) at a convergent plate boundary.
   (d) at a transform fault.

4. How far will a plate travel in 60,000 years if it moves at a net rate of 25 mm/year?
   (a) 24 m
   (b) 1,500 m
   (c) 3,000 m
   (d) 4,800 m

5. An earthquake of magnitude 7.0 on the Richter scale is how much greater than an earthquake of magnitude 5.0?
   (a) 2
   (b) 5
   (c) 10
   (d) 100

Free-Response Question

The figure below shows a convergent tectonic boundary in the Pacific Northwest region of the United States.

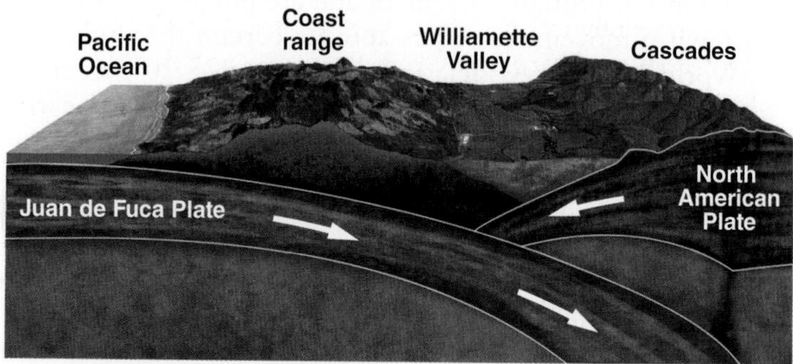

(a) **Identify** the mountain range produced by this tectonic plate boundary. (1 pt.)
(b) **Describe** the tectonic process occurring to the Juan de Fuca Plate at this convergent boundary. (1 pt.)
(c) **Explain** why volcanic mountains are formed at this type of plate boundary. (1 pt.)
(d) **Describe** an environmental problem that can occur as a result of a volcanic eruption at this plate boundary. (1 pt.)
(e) The Juan de Fuca Plate is moving toward the North American Plane at an average of 11.8 feet per 100 years.
   (i) **Calculate** the average rate of movement of the Juan de Fuca Plate in millimeters per year (mm/yr) if there are 30.5 centimeters in a foot. Show all work. (2 pts.)
   (ii) At its fastest, the Juan de Fuca Plate moves 13.8 feet per 100 years, or 42 mm/yr. **Calculate** the percent difference between the plate's fastest rate and its average rate from (i). (2 pts.)
   (iii) **Calculate** how many years it would take for the Juan de Fuca Plate to travel 100 meters. Show all work. (2 pts.)

Module 20

Soil Formation, Erosion, Composition, and Its Properties

In the previous module we explored the layers of Earth and how the rocky lithosphere moves as tectonic plates float above liquid magma. In this module we examine how different types of rocks are formed in lithosphere and how they are broken down by physical and chemical processes. These broken-down rocks, in turn, form the foundation of soils, which are formed as a mixture of minerals from rocks and organic matter from decomposing organisms. Since the organic matter lies above the rocks, soils develop distinctive layers that differ in their composition. Soils are a resource that can take tens of thousands of years to fully develop, yet human activities can lead to the rapid erosion of soils, which degrades the ability to grow plants and retain the essential ecosystem functions that soils provide.

Learning Goals

After reading this module you should be able to

20-1 describe the formation of igneous, sedimentary, and metamorphic rocks.

20-2 identify the processes that break down rocks.

20-3 explain how soils are formed.

20-4 describe the causes of soil erosion.

20-5 identify the properties that affect soil productivity.

20-1 How are igneous, sedimentary, and metamorphic rocks formed?

The three major types of rock are formed by different combinations of heat and pressure

While magma is the original source of all rock, there are three major ways in which the rocks we see at Earth's surface can form: directly from molten magma; by compression of sediments; and by exposure of rocks and other Earth materials to high temperatures and pressures. In this section we will examine the processes that produce three distinct rock types: igneous, sedimentary, and metamorphic, as shown in **FIGURE 20.1**.

Igneous Rocks

Igneous rocks form directly from magma, which we discussed in Module 19. Igneous rocks are classified by their chemical composition as basaltic or granitic. Basaltic rock is a dark rock that contains minerals with high concentrations of iron, magnesium, and calcium. It is the dominant rock type in the crust of oceanic plates. In contrast, granitic rock is a lighter-colored rock made up of the minerals feldspar, mica, and quartz, and contains elements such as silicon, aluminum, potassium, and calcium. It is the dominant rock type in the crust of continental plates. When granitic rock breaks down due to weathering, it forms

Igneous rock Rock formed directly from magma.

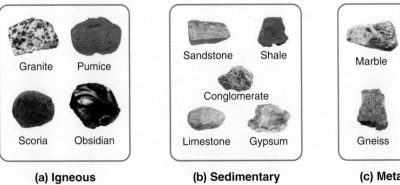

Granite Pumice
Scoria Obsidian

(a) Igneous

(b) Sedimentary

Marble Slate
Gneiss Quartzite

(c) Metamorphic

FIGURE 20.1 Three major types of rocks. Rocks fall into three categories depending on how they are formed. (a) Igneous rocks are formed by heat. (b) Sedimentary rocks are formed by pressure. (c) Metamorphic rocks are formed by a combination of heat and pressure.

sand. Soils that develop from granitic rock tend to be more permeable to water than those that develop from basaltic rock, but both types of rock can form fertile soil.

The formation of igneous rock often brings to the surface rare elements and metals that humans find economically valuable, such as lanthanum, which is a scarce metal used in the batteries of electric cars. When rock cools, it is subject to stresses that cause it to break. Cracks that occur when rocks cool, known as fractures, can occur in any kind of rock. Water from the surface of Earth running through fractures may dissolve valuable metals, which may precipitate out in the fractures to form concentrated deposits called veins. These deposits are important sources of gold- and silver-bearing ores as well as rare metals such as tantalum, which is used to manufacture electronic components of cell phones.

Sedimentary Rocks

Sedimentary rocks form when sediments such as muds, sands, or gravels are compressed by overlying sediments. Sedimentary rock formation occurs over long periods when environments such as sand dunes, mudflats, lake beds, or areas prone to landslides are buried and the overlying materials create pressure on the materials below. The resulting rocks can be uniform in composition, such as sandstones and mudstones that formed from ancient oceanic or lake environments. Alternatively, they may be highly heterogeneous, such as conglomerate rocks formed from mixed cobbles, gravels, and sands.

Sedimentary rocks hold the fossil record that provides a window into our past. When layers of sediment containing plant or animal remains are compressed over eons, those organic materials may be preserved as fossils.

Sedimentary rock Rock that forms when sediments such as muds, sands, or gravels are compressed by overlying sediments.

Metamorphic rock Rock that forms when sedimentary rock, igneous rock, or other metamorphic rock is subjected to high temperature and pressure.

Rock cycle The geologic cycle governing the constant formation, alteration, and destruction of rock material that results from tectonics, weathering, and erosion, among other processes.

Metamorphic Rocks

Metamorphic rocks form when sedimentary rocks, igneous rocks, or other metamorphic rocks are subjected to high temperatures and pressures. The pressures that form metamorphic rock cause profound physical and chemical changes in the rock. These pressures can be exerted by overlying rock layers or by tectonic processes such as continental collisions, which cause extreme horizontal pressure and distortion. Metamorphic rocks include stones such as slate and marble as well as anthracite, which is a type of coal. Metamorphic rocks have long been important as building materials in human civilizations because they are structurally strong and visually attractive.

The Rock Cycle

The three different modes of rock formation are interlinked through the **rock cycle**, which governs the constant formation, alteration, and destruction of rock material that results from tectonics, weathering, and erosion, among other processes. The rock cycle is the slowest of all of Earth's cycles. Environmental scientists are most often interested in the part of the rock cycle that occurs at or near Earth's surface.

FIGURE 20.2 shows the processes of the rock cycle. Rock forms when magma from Earth's interior reaches the surface,

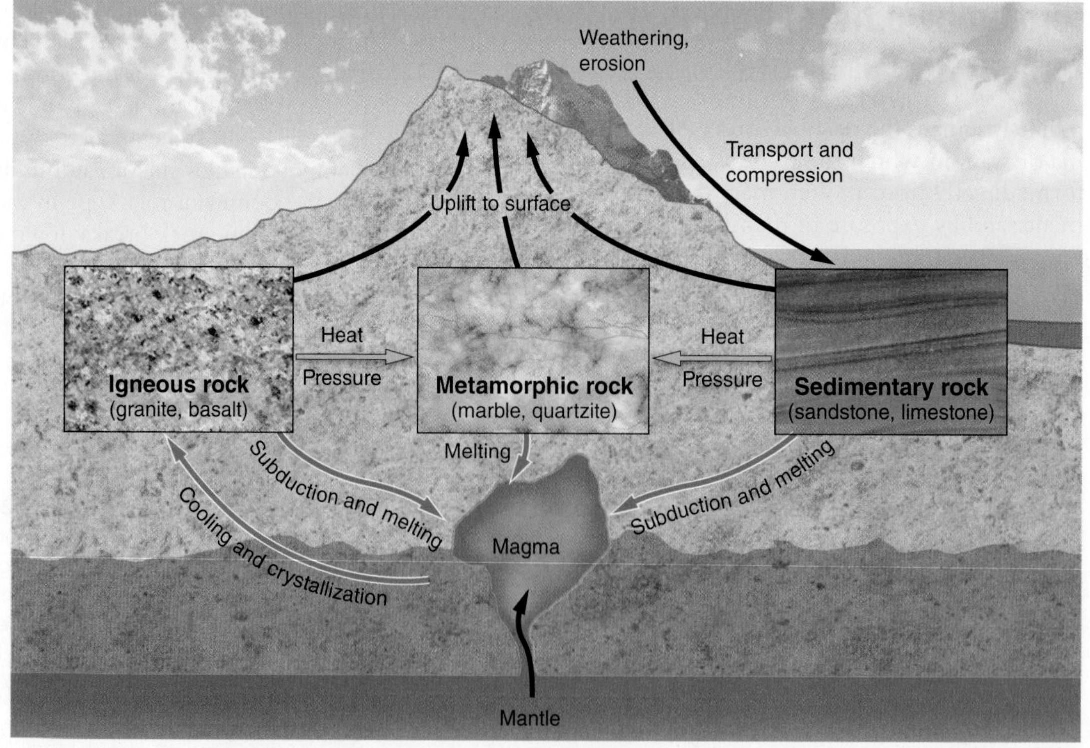

FIGURE 20.2 The rock cycle. The rock cycle slowly but continuously forms new rock and breaks down old rock over millions of years. Three types of rock are created in the rock cycle: Igneous rock is formed from magma; sedimentary rock is formed by the compression of sedimentary materials; and metamorphic rock is created when rocks are subjected to high temperatures and pressures.

cools, and hardens. Once at Earth's surface, rock masses are broken up, moved, and deposited in new locations by processes such as weathering and erosion, which is caused by water, wind, and changing temperatures. We will discuss weathering and erosion in much more detail later in the module. New sedimentary rock may be formed from the deposited material; if there is heat and pressure, new metamorphic rock may be formed from the deposited material. Eventually, the rock is subducted into the mantle, where it melts and becomes magma again. The rock cycle slowly but continuously breaks down rock and forms new rock over millions of years.

20-2 What are the processes that break down rocks?

Rock is broken down by weathering and erosion

We have seen that rock forms beneath Earth's surface under intense heat, pressure, or both. When rock is exposed at Earth's surface, however, it begins to break down through the processes of weathering and erosion. These processes are components of the rock cycle, returning chemical elements and rock fragments to the crust by depositing them as sediments through the hydrologic cycle. This physical breakdown and chemical alteration of rock begins the cycle all over again, as shown in Figure 20.2. Without this part of the rock cycle, elements would never be recycled and the precursors of soils would not be present. In this section we examine the processes of weathering and erosion.

Weathering

Weathering occurs when rock is exposed to air, water, certain chemical compounds, or biological agents such as plant roots, lichens, and burrowing animals. There are two major categories of weathering—physical and chemical—that work in combination to degrade rocks.

Physical weathering is the mechanical breakdown of rocks and minerals. Physical weathering can be caused by water, wind, or variations in temperature such as seasonal freeze-thaw cycles. When water works its way into cracks or fissures in rock, it can remove loose material and widen the cracks, as illustrated in **FIGURE 20.3a**. When water freezes in the cracks, the water expands, and the pressure of its expansion can force rock to break. Different responses to temperature can cause two minerals within a rock to expand and contract differently, which also results in splitting or cracking.

Biological agents can also cause physical weathering. Plant roots can work their way into small cracks in rocks and pry them apart, as illustrated in Figure 20.3b. Burrowing animals may also contribute to the breakdown of rock material, although their contributions are usually minor. However it occurs, physical weathering exposes more

Physical weathering The mechanical breakdown of rocks and minerals.

(a)

(b)

FIGURE 20.3 Physical weathering. (a) Water can work its way into cracks in rock, where it can wash away loose material. When the water freezes and expands, it can widen the cracks. (b) Growing plant roots can force rock sections apart. *(a: Walt Anderson; b: Bruce Hamms/Alamy)*

(a)

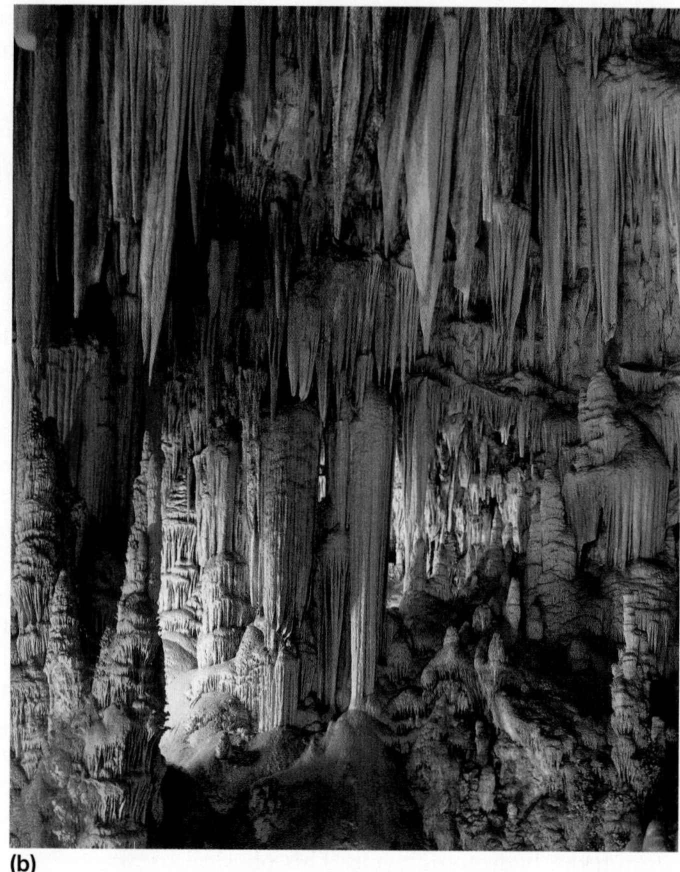
(b)

FIGURE 20.4 Chemical weathering. Chemical weathering helps to break down rocks and minerals by chemical reactions. (a) Lichens living on a gravestone produce weak acids that cause chemical weathering of the rock. (b) Water that contains carbonic acid wears away limestone, sometimes forming spectacular caves. *(a: Catherine Paffey/Alamy Stock Photo; b: mauritius images/Thonig)*

surface area and makes rock more vulnerable to further degradation. By producing more surface area for weathering processes to act on, physical weathering increases the rate of chemical weathering. **Chemical weathering** is the breakdown of rocks and minerals by chemical reactions, the dissolving of chemical elements from rocks, or both these processes. It releases essential nutrients from rocks, making them available for use by plants and other organisms.

Chemical weathering occurs most rapidly on newly exposed minerals, known as primary minerals. It alters primary minerals to form secondary minerals and the ionic forms of their constituent chemical elements. For example, when feldspar—a primary mineral found in granitic rock—is exposed to natural acids in rain, it forms clay particles as a secondary mineral and releases ions such as potassium, an essential nutrient for plants. Lichens can break down rock in a similar way by producing weak acids.

Chemical weathering The breakdown of rocks and minerals by chemical reactions, the dissolving of chemical elements from rocks, or both these processes.

Their effects can commonly be seen on soft gravestones and masonry, as shown in **FIGURE 20.4a**. You might recall that we mentioned the effect of lichens on rocks in our discussion of primary succession in Module 14.

Rocks that contain compounds that dissolve easily, such as calcium carbonate, tend to weather quickly. Rocks that contain compounds that do not dissolve readily are often the most resistant to chemical weathering. In examining element cycles in Module 5, we noted that weathering of rocks is an important part of the phosphorus cycle.

Depending on the starting chemical composition of rock and the pH of the water that comes in contact with it, hundreds of different chemical reactions can take place. For example, when carbon dioxide in the atmosphere dissolves in water vapor it creates carbonic acid. When waters containing carbonic acid flow into rocks that are rich in limestone, they dissolve the limestone (which is composed of calcium carbonate) and leave behind spectacular cave systems (Figure 20.4b).

Some chemical weathering is the result of human activities. For example, sulfur emitted into the atmosphere from fossil fuel combustion combines with oxygen to form sulfur dioxide. That sulfur dioxide reacts with water vapor in

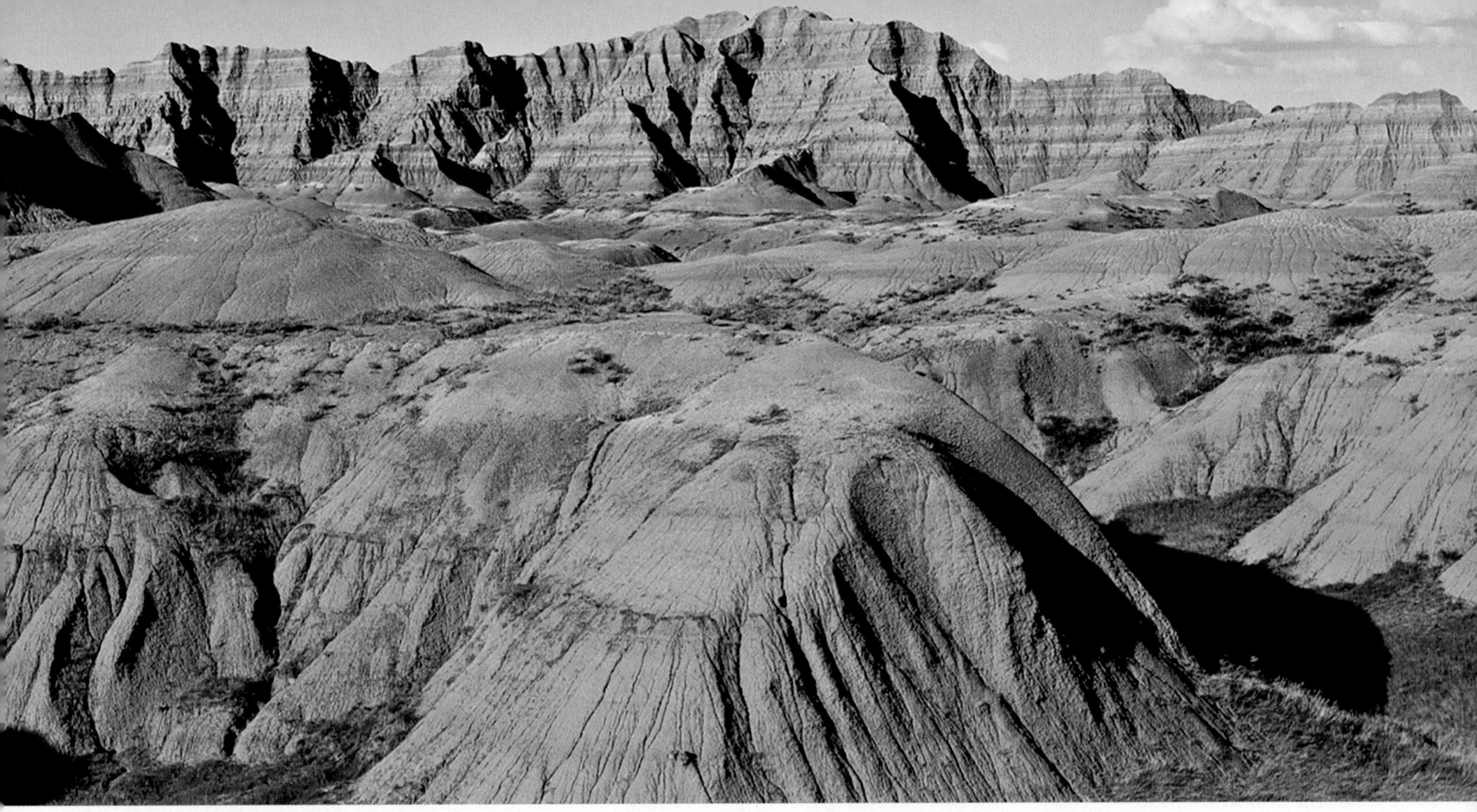

FIGURE 20.5 Erosion. Some erosion, such as the erosion that created these formations in the Badlands of South Dakota, occurs naturally as a result of water, glaciers, or wind. The Badlands are the result of the erosion of softer sedimentary rock types, such as shales and clays. Harder rocks, including many types of metamorphic and igneous rocks, are more resistant to erosion. *(welcomia/Shutterstock)*

the atmosphere to form sulfuric acid. Similarly, emissions of nitrous oxides and water vapor in the atmosphere form nitric acid. As a result, rain and snow can contain high amounts of sulfuric acid and nitric acid, which we call **acid precipitation** or **acid rain**. Similar to carbonic acids dissolving limestone and creating caves, acid precipitation is responsible for the rapid degradation of many old statues, gravestones, and other limestone and marble structures. We will discuss acid precipitation in much more detail in Module 46.

Chemical weathering, due to natural processes or human-caused acid precipitation, can contribute additional elements to an ecosystem. Knowing the rate of weathering helps researchers assess how rapidly soil fertility can be renewed in an ecosystem. In addition, because the chemical reactions involved in the weathering of certain granitic rocks consume carbon dioxide from the atmosphere, weathering can actually reduce atmospheric carbon dioxide concentrations.

Erosion

We have seen that physical and chemical weathering results in the breakdown and chemical alteration of rock. **Erosion** is the physical removal of rock fragments (sediment, soil, rock, and other particles) from a landscape or ecosystem. Erosion is usually the result of two mechanisms. In the first mechanism, wind, water, and ice move soil and other materials down a slope under the force of gravity. In the second mechanism, living organisms, such as animals that burrow under the soil, cause erosion. After eroded material has traveled a certain distance from its source, it accumulates in an area of deposition.

Erosion is a natural process: Streams, glaciers, and wind-borne sediments continually carve, grind, and scour rock surfaces (**FIGURE 20.5**). In many places, however, human land use contributes substantially to the rate of erosion. Poor land use practices such as deforestation, overgrazing, unmanaged construction activity, and road building can create and accelerate rock erosion. Furthermore, erosion usually leads to deposition of the eroded material somewhere else, which may cause additional environmental problems. As we will see later in this module, erosion is also an important process that impacts soils.

Acid precipitation Precipitation high in sulfuric acid and nitric acid. *Also known as* **Acid rain.**

Erosion The physical removal of rock fragments from a landscape or ecosystem.

AP® Exam Tip

Do not use the terms "weathering" and "erosion" interchangeably. Weathering is the breakdown of big rocks into smaller rocks. Erosion is the movement of soil. On the AP® Environmental Science Exam the correct answer may only be one of these terms, not both of them.

20-3 How are soils formed?

Soils form as a result of parent material, climate, topography, organisms, and time

Soil has a number of functions that benefit organisms and ecosystems. As you can see in **FIGURE 20.6**, soil is a medium for plant growth. It also serves as the primary filter of water, as water moves from the atmosphere into rivers, streams, and groundwater. Thus, protecting soil helps protect water quality to ensure clean water for humans and other organisms. Soil also contributes greatly to biodiversity by providing habitat for a wide variety of living organisms—from bacteria, algae, and fungi to insects and other animals. Soil and the organisms within it filter chemical compounds deposited by air pollution and by household sewage systems; some of these materials remain in the soil and some are released to the atmosphere or into groundwater. In this section we investigate the formation and properties of soil.

To appreciate the role of soil in ecosystems, we need to understand how and why soil forms and what happens to soil when humans alter it. It takes hundreds to thousands of years for soil to form. Soil is the result of physical and chemical weathering of rocks, as previously discussed, and the gradual accumulation of detritus from the biosphere. We can determine the specific properties of a soil if we know its parent rock type, the amount of time it has been forming, and its associated biotic and abiotic components.

FIGURE 20.7 shows the stages of soil development from rock to mature soil. The processes that form soil work in two directions simultaneously. The breakdown of rocks and primary minerals by weathering provides the raw material for soil from below. The deposition of organic matter from organisms and their wastes contributes to soil formation from above. What we normally think of as "soil" is a mix of these mineral and organic components. A poorly developed (young) soil has substantially less organic matter and fewer

FIGURE 20.6 Ecosystem services provided by soil. Soil serves as a medium for plant growth, as a habitat for other organisms, and as a recycling system for organic wastes. Soil also helps to filter and purify water.

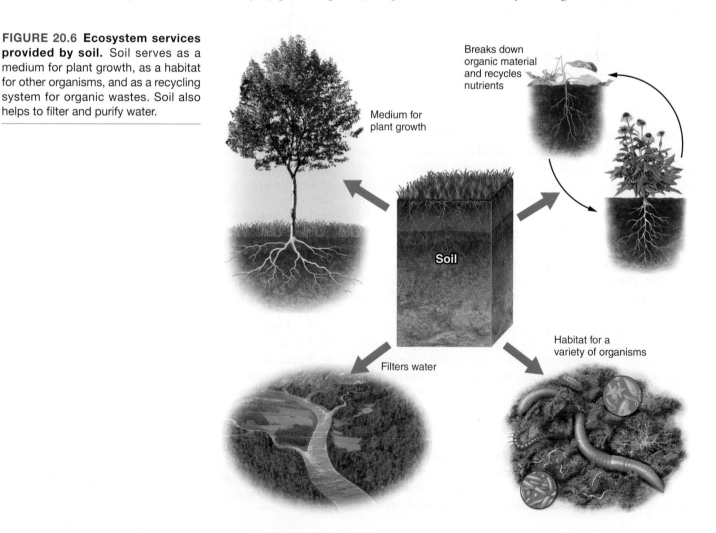

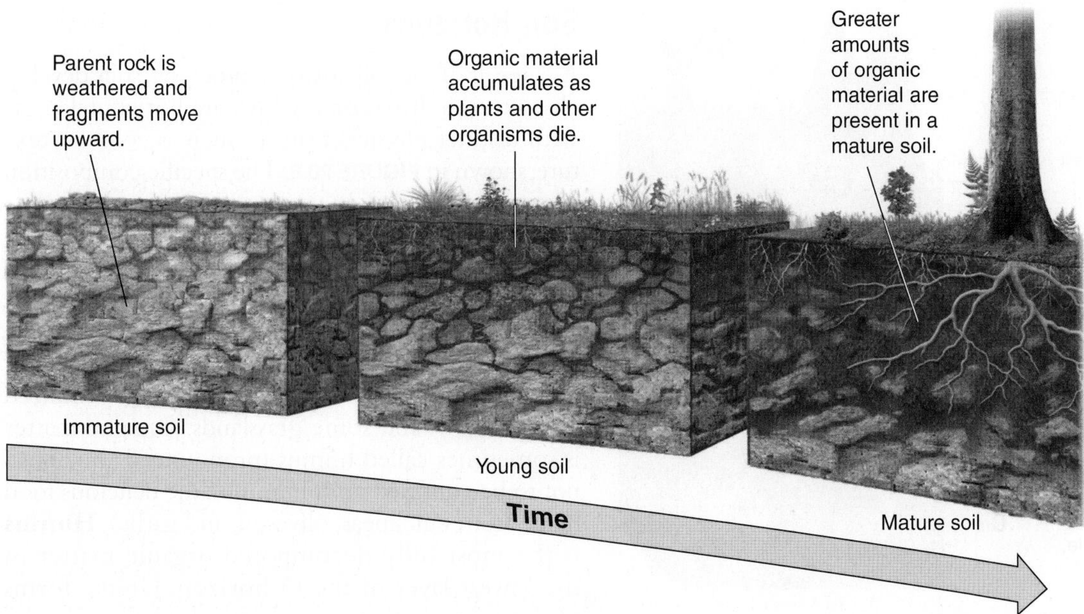
Parent rock is weathered and fragments move upward.
Organic material accumulates as plants and other organisms die.
Greater amounts of organic material are present in a mature soil.
Immature soil
Young soil
Time
Mature soil

FIGURE 20.7 Soil formation. Soil is a mixture of organic and inorganic matter. The breakdown of rock and primary minerals from the parent material provides the inorganic matter. The organic matter comes from organisms and their wastes.

nutrients than a more developed (mature) soil. Very old soils may also be nutrient poor because over time plants remove many essential nutrients and water leaches away others. Five factors combine to determine the properties of soils: parent material, climate, topography, organisms, and time.

Parent Material

A soil's **parent material** is the underlying rock material from which the inorganic components of a soil are derived. Different soil types arise from different parent materials. For example, a parent material composed of quartz sand will give rise to a soil that is nutrient poor, such as those along the Atlantic coast of the United States. By contrast, a soil that has calcium carbonate as its parent material will contain an abundant supply of calcium, have a high pH, and may also support high agricultural productivity. Such soils are found in the area surrounding Lake Champlain in Vermont and northern New York, as well as many other locations.

Climate

Climate influences soil formation in a number of ways. The long-term effects of temperature, humidity, and water affect soil development. Soils do not develop when temperatures are below freezing because decomposition of organic matter and water movement are both extremely slow in frozen and nearly frozen soils. Therefore, soils at high latitudes of the Northern Hemisphere are composed largely of organic material in an undecomposed state, as we saw in Module 2 in our discussion of taiga and tundra biomes. In contrast, soil development in the humid tropics is accelerated by rapid weathering of rock and soil minerals, leaching of nutrients, and decomposition of organic detritus. Climate also has an indirect effect on soil formation because it affects the type of vegetation that develops, and therefore the type of detritus left after the vegetation dies.

Topography

Topography—the surface slope and arrangement of a landscape—is another factor in soil formation. Soils that form on steep slopes are constantly subjected to erosion and, on occasion, to more drastic mass movements of material as happens in landslides. In contrast, soils that form at the bottoms of steep slopes may continually accumulate material from higher elevations and become quite deep.

Organisms

Many organisms influence soil formation. Plants remove nutrients from soil and excrete organic acids that speed chemical weathering. Animals that tunnel or burrow—for example, earthworms, gophers, voles, and field mice—mix the soil, uniformly distributing organic and mineral matter. Collectively, soil organisms act as recyclers of organic matter. In the process of using dead organisms and wastes as an energy source, soil organisms break down organic detritus and release mineral nutrients and other materials that benefit plants.

Parent material The underlying rock material from which the inorganic components of a soil are derived.

Time

The final factor that determines the properties of a soil is the amount of time during which the soil has developed. As soils age, they develop a variety of characteristics. The grassland soils that support much of the food crop and livestock feed production in the United States are relatively old soils. Because they have had continual inputs of organic matter for hundreds of thousands of years from the grassland and prairie vegetation growing above them, they have become deep and fertile. Other soils that are equally old, but with less productive communities above them and perhaps greater quantities of water moving through them, can become relatively infertile.

O horizon: Organic matter in various stages of decomposition
A horizon (topsoil): Zone of overlying organic material mixed with underlying mineral material
E horizon: Zone of leaching of metals and nutrients; occurs in some soils beneath either the O horizon or the A horizon
B horizon (subsoil): Zone of accumulation of metals and nutrients
C horizon (subsoil): Least-weathered portion of the soil profile, similar to the parent material
Horizons
O
A
E
B
C
0
<.1 m
.3 m
1 m
1.3 m

FIGURE 20.8 Soil horizons. All soils have horizons, or layers, which vary depending on soil-forming factors such as climate, organisms, and parent material. Most soils have either an O or A horizon and usually not both. Some soils that have an O horizon also have an E horizon.

Soil Horizons

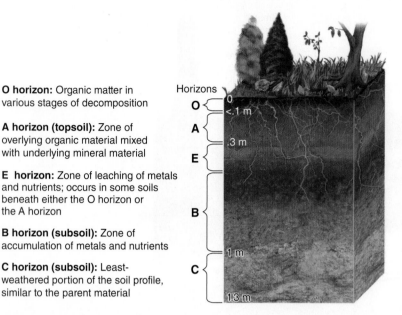

As a result of the soil-forming processes, soils develop characteristic **horizons**, which are horizontal layers with distinct physical features such as color or texture, shown in FIGURE 20.8. The specific composition of those horizons—in terms of the minerals, organic matter, and overall fertility—depends largely on climate, vegetation, and parent material. At the surface of many soils is a layer known as the **O horizon**, composed of organic detritus such as leaves, needles, twigs, and even animal bodies, all in various stages of decomposition. The O horizon is most pronounced in forest soils and some grasslands. Organic matter is sometimes called humus (pronounced "hu-mus," not to be confused with hummus, the delicious food made from chickpeas, olive oil, and garlic). **Humus** is the most fully decomposed organic matter in the lowest layer of the O horizon. Unlike forms of decomposing organic matter higher up in the O horizon, humus does not contain recognizable plant or animal components. Humus and "humic material" may also occur in smaller concentrations in the mineral soil horizons below the O horizon.

In a soil that is mixed, either naturally or by human agricultural practices, the top layer is the **A horizon**, also known as **topsoil**, a zone of organic material (including humus) and minerals that have been mixed together. In some acidic soils, an **E horizon**—a zone of leaching—forms under the O horizon or, less often, the A horizon. When an E horizon is present, iron, aluminum, and dissolved organic acids from the overlying horizons are transported through and removed from the E horizon and then deposited in the B horizon, where they accumulate. When an E horizon is present, it always occurs above the B horizon. The **B horizon**, commonly known as subsoil, is composed primarily of mineral material with very small amounts of organic matter, including humus. If nutrients are present in the subsoil, they will be in the B horizon. The **C horizon**—the least-weathered soil horizon—occurs beneath the B horizon and is similar to the parent material.

Horizon A horizontal layer in a soil defined by distinctive physical features such as color and texture.

O horizon The organic horizon at the surface of many soils, composed of organic detritus in various stages of decomposition.

Humus The most fully decomposed organic matter in the lowest section of the O horizon.

A horizon Frequently the top layer of soil, a zone of organic material and minerals that have been mixed together. *Also known as* **Topsoil**.

E horizon A zone of leaching, or eluviation, found in some acidic soils under the O horizon or, less often, the A horizon.

B horizon Commonly known as subsoil, a soil horizon is composed primarily of mineral material with very little organic matter.

C horizon The least-weathered soil horizon, which always occurs beneath the B horizon and is similar to the parent material.

AP® Exam Tip

Be cautious not to refer to soil as "dirt" on the exam. Dirt lacks living organisms, and it isn't alive. Soil is what is required for ecosystems to develop and is what humans need to grow their own food.

20-4 What causes soil erosion?

Soil erodes by wind and water, as well as increases in human activities

Human activity has dramatic effects on soils. For centuries, land use for agriculture, forestry, and other human activities

(a)

(b)

FIGURE 20.9 Soil erosion by water. (a) Small-scale erosion in a cornfield after a rainstorm. (b) Large-scale erosion as a mudslide engulfs the village of Conchita, CA. *(a: Tim McCabe/USDA Natural Resources Conservation Service; b: Mark Reid, USGS. Public domain)*

has led to significant soil degradation, resulting in a reduced ability for soils to support plant growth. One of the major causes of soil degradation is soil erosion, which occurs when topsoil is disturbed, as occurs in a plowed field, or when vegetation is removed or a forest is cleared. These activities lead to erosion by water or wind.

Erosion by Water

When the topsoil is disturbed or the vegetation is removed, the topsoil lacks a complex system of plant roots that can hold onto the soil particles. As a result, large rainstorms can cause the rapid removal of large amounts of soil. You can see examples of this on building construction sites, where barriers to erosion are put in place, and in many farm fields after a flood (**FIGURE 20.9a**). Even more devastating is erosion that occurs after there has been logging on steep slopes. Once again, the logging causes the network of tree roots to die and no longer hold onto the soil particles. During large rainstorms, the erosion process is assisted by the force of gravity, leading to massive landslides of soil sliding down mountainsides and being deposited into the valleys below (Figure 20.9b). Such landslides can bury homes and cause considerable economic damage. Moreover, while topsoil loss can happen rapidly—in as little as a single growing season—it can take centuries for the lost topsoil to be replaced.

Erosion by Wind

The erosion of soils can also occur by wind. Similar to water, wind can have a large impact on soils that have had their natural vegetation removed. In the midwestern United States in the 1920s and 1930s, for example, the amount of conversion from native grasslands to wheat fields increased dramatically. Wheat fields are more susceptible to soil erosion by wind than are native grasslands because the wheat stems are removed from the land during harvesting, so the bare soil is exposed to more of the wind. As a result, this conversion of grasslands to wheat fields produced many more hectares of land susceptible to erosion. In the early 1930s, a severe and prolonged drought caused a decade of crop failures. With few crop plants remaining alive to hold the soil in place, the drought led to massive dust storms as winds picked up the parched topsoil and blew it away (**FIGURE 20.10**). One dust storm, on April 14, 1935, was so severe that the Sun was entirely blocked out in the southern Great Plains, leading to the name "Black Sunday." Topsoil

FIGURE 20.10 Soil erosion by wind. This photo shows a farmhouse in Stratford, Texas, just prior to being hit by a massive dust storm in 1935. Poor agricultural practices combined with prolonged droughts can produce severe dust storms that carry away topsoil and rob the land of its fertility. *(National Oceanic and Atmospheric Administration/Science Source)*

from these dust storms traveled as far as Washington, D.C., and earned the southern Great Plains the nickname of "the Dust Bowl." The decade of dust storms and the resulting losses of topsoil led to large-scale human migration out of the region and impacted the economy of the entire nation. Today, improved farming practices have reduced the susceptibility of soil to erosion by wind, making major dust storms in the Great Plains and elsewhere much less likely.

20-5 What properties affect soil productivity?

Soils have different physical, chemical, and biological properties that affect productivity

Soils with different properties serve different functions for humans. For example, some soil types are good for growing crops and others are more suited for building a housing development. In addition, different soils around the world differ in their productivity as a result of differences in water-holding capacity and nutrient availability. To understand and classify soil types, we need to understand the physical, chemical, and biological properties of soils. In this section we examine how the role of different sizes of soil particles determines these soil properties.

Physical Properties of Soil

The physical properties of soil refer to physical characteristics such as size and weight. Sand, silt, and clay are mineral particles of different sizes. The smallest particles are clay, intermediate size particles are silt, and the largest particles are sand (**FIGURE 20.11a**). These differences in particle size affects the size of the air spaces between particles, which we refer to as **porosity**. Large sand particles have large air spaces between particles than the much smaller clay particles, so sandy soil has a higher porosity than clay soils.

Soil scientists categorize different kinds of soil by the percentages of sand, silt, and clay they contain. Figure 20.11b plots those percentages on a soil texture diagram that allows us to identify and compare soil types. Each side of the triangle lists the percentages of sand, silt, and clay. A point in the middle of the "loam" category (the red dot in the figure) represents a soil that contains 40 percent sand, 40 percent silt, and 20 percent clay. We can determine this by following the lines from that point to the scales on each of the three sides of the triangle. The sum of sand, silt, and clay will always be 100 percent. Conversely, in the laboratory, a soil scientist can determine the percentage of each component in a soil sample and then plot the results. The name for each soil type, for example "loam," follows from the percentage of the three components in the soil.

We can determine the percent of clay, silt, and sand in a sample of soil by filling a clear jar half full of soil and then fill the rest of the jar with water and a small amount of dishwashing detergent. We then put a lid on the jar, shake it vigorously, and then let the soil settle. The sand particles are the heaviest, so they settle to the bottom, followed by a layer of silt particles and then an upper layer of clay particles

Porosity The size of the air spaces between particles.

(a)

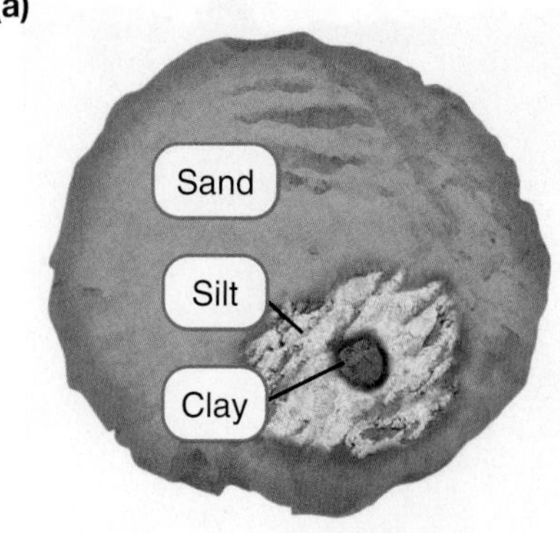

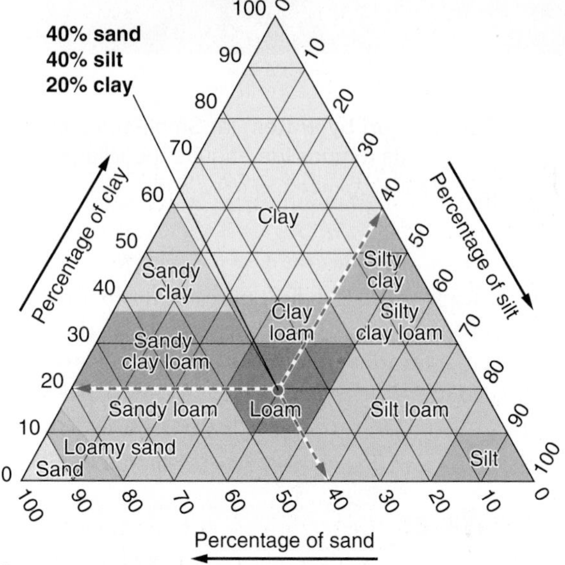

(b) Soil texture chart

FIGURE 20.11 Soil properties. Soils consist of a mixture of clay, silt, and sand. (a) The relative sizes of sand, silt, and clay. (b) The relative proportions of these particles determine the texture of the soil.

AP® Exam Tip

Practice using the soil properties chart, like the one in Figure 20.11. Remember that the long sides of the triangle each represent 100%. If you are doing it correctly, all three lines will intersect. Practice using sides of a piece of paper, index cards, or rulers on the diagram.

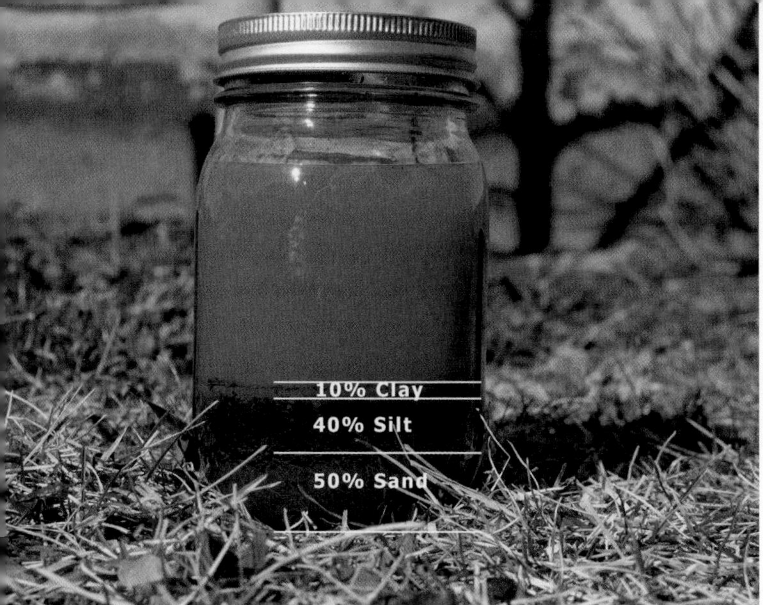

FIGURE 20.12 Testing for soil texture. If we fill a jar half full of soil, and then add water and a small amount of dishwashing detergent and then mix the contents, the heavy sand particles will settle first, followed by the silt particles and then the clay particles. By measuring the height of each layer, we can calculate the percent of sand, silt, and clay. *(Science Source/Science Source)*

(**FIGURE 20.12**). If we measure the thickness of each layer we can estimate the percent of sand, silt, and clay in the sample and then use the soil texture triangle to determine the name of the soil.

The **water holding capacity** of soil—the amount of water a soil can hold against the draining force of gravity—depends on its texture, shown in **FIGURE 20.13**. Soils with a high water holding capacity have low water **permeability**, which is the ability of water to move through the soil. Sand particles—the largest of the three components—pack together loosely. Water can move easily between the particles, making sand quick to drain and quick to dry out. Soils with a high proportion of sand are also easy for roots to penetrate, making sandy soil somewhat advantageous for growing plants such as carrots and potatoes. Clay particles—the smallest of the three components—pack together much more tightly than sand particles. As a result, there is less pore space in a soil dominated by clay, and water and roots cannot easily move through it. Silt particles are intermediate both in size and in their ability to drain or retain water.

Knowing the soil texture helps farmers in a particular climate determine how much to fertilize and irrigate their field. For example, the most fertile and productive agricultural soil is a mixture of sand, silt, and clay, which we know from the soil texture triangle in Figure 20.11 is characterized as a loam. This mixture promotes balanced water drainage and retention. In natural ecosystems, various plants have adapted to growing in wet, intermediate, and dry environments, so there are plants that thrive in soils of all textures.

Given that soil texture determines the permeability of water, it can have a strong influence on how the physical environment responds to environmental pollution. For example, the ground water of western Long Island in New York State has been contaminated over the years by toxic chemicals discharged from local industries. One major reason for the contamination is that Long Island is dominated by sandy soils that readily allow surface water to drain into the groundwater. While soil usually serves as a filter that removes pollutants from the water moving through it, sandy soils are so permeable that pollutants move through them quickly and therefore are not filtered effectively.

In contrast, clay is particularly useful where a potential contaminant needs to be contained. For example, many modern landfills are lined with clay, which helps keep the contaminants from leaching into the soil and groundwater beneath the landfill. We will discuss this in much more detail in Module 50.

Chemical Properties of Soil

Chemical properties are also important in determining how a soil functions. Clay particles contribute the most to

Water holding capacity The amount of water a soil can hold against the draining force of gravity.

Permeability The ability of water to move through the soil.

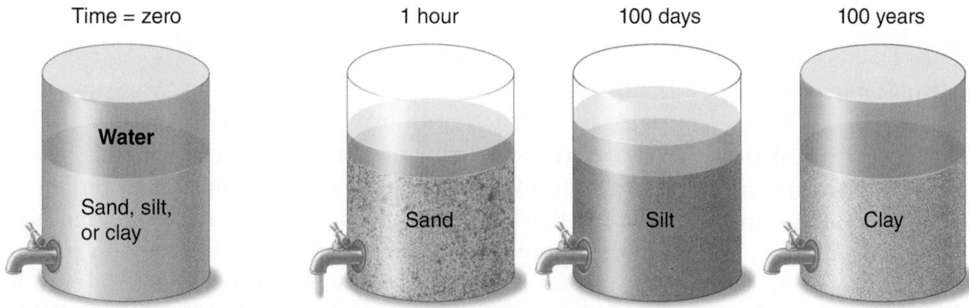

FIGURE 20.13 Soil water holding capacity and permeability. The water holding capacity and permeability of soil depends on its composition of soil particles. Sand has large, loosely packed particles, so it is very permeable to water and holds very little. Clay drains much more slowly and holds much more water.

the chemical properties of a soil because of their ability to attract positively charged mineral ions, referred to as cations. Because clay particles have a negative electrical charge, cations are adsorbed—held on the surface—by the particles. The cations can be subsequently released from the particles and used as nutrients by plants.

The ability of a particular soil to adsorb and release cations is called its **cation exchange capacity (CEC)**, sometimes referred to as the nutrient holding capacity. The overall CEC of a soil is a function of the amount and types of clay particles present. Soils with high CECs have the potential to provide essential cations to plants and therefore are desirable for agriculture. If a soil is more than 20 percent clay, however, its water retention becomes too great for most crops as well as many other types of plants. In such waterlogged soils, plant roots are deprived of oxygen. Thus, there is a trade-off between CEC and water permeability.

The relationship between soil bases and soil acids is another important soil chemical property. Calcium, magnesium, potassium, and sodium are collectively called soil bases because they can neutralize or counteract soil acids such as aluminum and hydrogen. Soil acids are generally detrimental to plant nutrition, while soil bases tend to promote plant growth. With the exception of sodium, all the soil bases are essential for plant nutrition. **Base saturation** is the proportion of soil bases to soil acids, expressed as a percentage.

Because of the way they affect nutrient availability to plants, CEC and base saturation are important determinants of overall ecosystem productivity. If a soil has a high CEC, it can retain and release plant nutrients. If it has a relatively high base saturation, its clay particles will hold important plant nutrients such as calcium, magnesium, and potassium. A soil with both high CEC and high base saturation is likely to support high productivity. Understanding how different soils affect the availability of nutrients is key to identifying what types of fertilizers to use, whether for growing your lawn, growing vegetables in your garden, or growing crops in a large field. Testing for nutrients can be done with home-testing kits or by bringing a soil sample to a local Agricultural Extension Office, which are available in many counties.

Biological Properties of Soil

As we have seen, a diverse group of organisms lives in the soil. **FIGURE 20.14** shows a representative sample. Three groups of organisms account for 80 to 90 percent of the biological activity in soils: fungi, bacteria, and protozoans (a diverse group of single-celled organisms). Rodents and

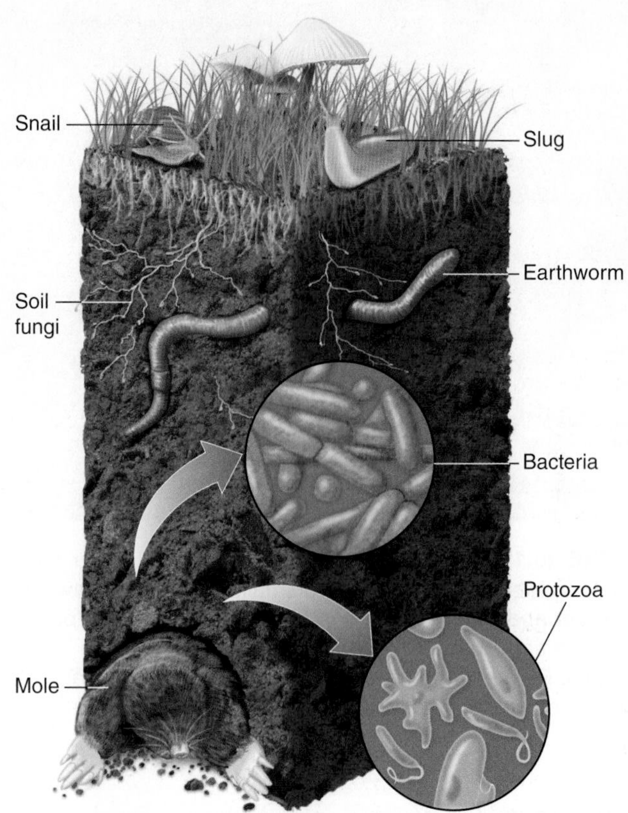

FIGURE 20.14 Soil organisms. Bacteria, fungi, and protozoans account for 80 to 90 percent of soil organisms. Also present are snails, slugs, insects, earthworms, and rodents.

earthworms contribute to soil mixing and the breakdown of large organic materials into smaller pieces. Earthworms are one of many organisms that contribute to humus formation in soils. Some soil organisms, such as snails and slugs, are herbivores that eat plant roots as well as the aboveground parts of plants. However, the majority of soil organisms are detritivores, which consume dead plant and animal tissues and recycle the nutrients they contain. Some soil bacteria also fix nitrogen, which, as we saw in Module 4, is essential for plant growth. The presence or absence of these important organisms can affect both water permeability and nutrient availability of the soils, which in turn affects decisions about fertilizing and irrigating the soil.

In this module we examined the various ways in which rocks are formed and are subsequently weathered. The weathered rocks form the parent material for soils, which are formed by a suite of physical, chemical, and biological processes that produce soil horizons and make different soils possess different properties, including differences in particle size, porosity, and permeability. We also saw that soils can be eroded by wind and water, particularly when impacted by human activities. In subsequent modules we will see how the atmosphere and wind patterns brings different climates to these soils to determine the plants and animals that can exist in different biomes.

Cation exchange capacity (CEC) The ability of a particular soil to adsorb and release cations.

Base saturation The proportion of soil bases to soil acids, expressed as a percentage.

Module 20 AP® Review

Preparing for the AP® Exam

Learning Goals Revisited

20-1 How are igneous, sedimentary, and metamorphic rocks formed?

Igneous rocks form directly from magma. Sedimentary rocks form when sediments such as muds, sands, or gravels are compressed by overlying sediments. Metamorphic rocks form when sedimentary rocks, igneous rocks, or other metamorphic rocks are subjected to high temperatures and pressures.

20-2 What are the processes that break down rocks?

Rocks are broken down by physical weathering, which can be caused by water, wind, or variations in temperature. It can also be caused by the activity of organisms. Rocks are also broken down by chemical weathering, such as the reactions caused by weak acids and the dissolving of minerals.

20-3 How are soils formed?

Soil formation begins with the breakdown of rocks that serve as the parent material. It is also influenced by the climate of an area because climate determines the range of temperatures and precipitation that can affect weathering and leaching of nutrients. Additional factors include topography, organisms, and the amount of time that has passed for these processes to occur. Soil horizons are horizontal layers with distinct physical features such as color or texture.

20-4 What causes soil erosion?

While erosion occurs naturally, human activities can dramatically increase the amount of erosion. Water can cause erosion when plants have been removed from an area, leaving no roots to hold onto the soil particles. Wind can also cause erosion, particularly when land has been plowed and droughts prevent many plants from growing and holding the soil particles.

20-5 What properties affect soil productivity?

Physical, chemical, and biological properties affect the productivity of soils. Soils with more clay particles have a greater water-holding capacity and a higher nutrient capacity. However, having too much clay can make it hard for plants to extract the water and nutrients from the soil.

AP® Practice Questions

Preparing for the AP® Exam

Multiple-Choice Questions

1. Which rock is formed at high temperatures and pressures?
   (a) igneous
   (b) sandstone
   (c) sedimentary
   (d) metamorphic

2. Which statement is the best explanation of acid precipitation?
   (a) Acid precipitation causes physical weathering, which can release key soil nutrients from breaking up rock.
   (b) Acid precipitation causes chemical weathering, which can release key soil nutrients from dissolving minerals in rock.
   (c) Acid precipitation causes chemical weathering, which can consume key soil nutrients from breaking up rock.
   (d) Acid precipitation causes physical weathering, which can consume key soil nutrients from dissolving minerals in rock.

3. What are the five primary soil formation factors?
   (a) climate, parent material, pH, organisms, topography
   (b) parent material, topography, organisms, time, latitude
   (c) parent material, climate, pH, latitude, altitude
   (d) climate, parent material, topography, organisms, time

4. Which is the correct order of soil horizons starting at the surface?
   (a) A, B, C, E, O
   (b) O, A, B, C, E
   (c) O, E, A, B, C
   (d) O, A, E, B, C

5. What type of soil would be best to add to the bottom of a constructed pond, where the goal is to have as little leakage of water as possible?
   (a) mostly clay
   (b) mostly silt
   (c) mostly sand
   (d) equal amounts of sand, silt, and clay

6. Which of the following, if added to soil, would lower the base saturation?
   (a) sodium
   (b) potassium
   (c) magnesium
   (d) aluminum

7. Using the soil texture triangle, what is the soil type that contains 30 percent clay, 30 percent silt, and 40 percent sand?
   (a) silt loam
   (b) clay loam
   (c) sandy loam
   (d) aluminum

Free-Response Question

The soil triangle below shows a soil pyramid with the texture of two soil samples marked at points "A" and "B." Both samples were collected from a depth of 0.25 meters (0.82 feet) in two different locations.

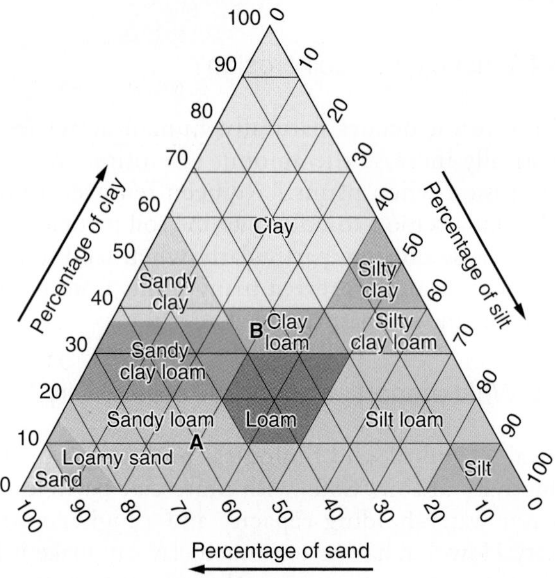

(a) **Identify** the percentage of silt in Sample A. (1 pt.)
(b) **Describe** which soil sample would be best for plants that prefer quick-draining soils. (1 pt.)
(c) **Describe** from which soil horizon the samples were most likely collected. (1 pt.)
(d) **Explain** how the water-holding capacity of soil can inform decisions about watering plants. (1 pt.)
(e) A farmer decides that soil Sample B would be the better choice of soil for planting crops. Using the data in the soil triangle, **justify** whether you support or refute this farmer's claim. (2 pts.)
(f) A group of AP® Environmental Science students design an experiment to test the cation exchange capacity (CEC) of different soil particles. Their results are shown in the table below:

Soil sample	Clay content	Cation exchange capacity (cmolc/kg)
A	0%	12
B	8%	16
C	16%	20
D	24%	25

(Data from http://nmsp.cals.cornell.edu/publications/factsheets/factsheet22.pdf)

(i) **Identify** the independent variable for this experiment. (1 pt.)
(ii) **Identify** a likely hypothesis for this experiment. (1 pt.)
(iii) **Identify** the control for this experiment. (1 pt.)
(iv) **Describe** a modification that could be made to the students' experiment that could improve their results. (1 pt.)

Module 21

Unit 4	19	20	21	22	23

Watersheds

In our prior modules, we explored the ecosystem services provided by nature and how disturbances can alter these services. In Module 20 we discussed the importance of soils in affecting the permeability of water and the availability of nutrients. Nowhere are these topics more important than in the consideration of **watersheds**, which is all the land in an area that drains into a particular stream, river, lake, or wetland. You can see an example illustrated in **FIGURE 21.1**, where all of the water between two mountain ranges flows into a river. As a result what happens on the land has tremendous importance to what happens to the water body that receives the draining water.

Learning Goals

After reading this module you should be able to

21-1 describe the characteristics of a watershed.

21-2 identify the impacts humans have on watersheds.

River

FIGURE 21.1 Watershed. A watershed is the area of land that drains into a particular body of water.

21-1 What are the characteristics of a watershed?

Watersheds are characterized by their area, length, slope, soils, and vegetation

Watersheds have a number of characteristics that affect the movement of water and nutrients. In this section we will explore how area, length, slope, soil type, and vegetation type all characterize a watershed.

Area and Length

The area of the watershed can vary a great deal, from a few hectares to thousands of hectares that drain water from multiple states in the United States. For example, the Mississippi River drains water from nearly one-third of the United States and carries all of this water to the Gulf of Mexico.

The length of a watershed is measured along the main flow of the water in a stream or river that drains the watershed from the beginning to the outlet of the watershed. For a watershed of a given size, a greater watershed length means that there will be a longer travel time

Watershed All the land in an area that drains into a particular stream, river, lake, or wetland.

FIGURE 21.2 The effect of slope on water flow and erosion. On steep slopes, such as this railway embankment, steep slopes cause water to move downhill much faster, which causes substantial erosion. *(American Society of Civil Engineers)*

to move water to the outlet of the watershed. As you might imagine, the area and length of a watershed combined play a large role in the amount of water, the materials carried by the water, and the time required to move the water from the land area to the water body at the outlet of the watershed.

Slope

The slope of a watershed is the steepness of the land. When the land has a very gentle slope, such as an open field that gradually slants down to a stream, the movement of water on the surface of the soil or under the soil moves at a slow speed. An important result of this slow flow is that the water causes very little erosion, so the stream receives very little sediment. In contrast, when the land has a steep slope, such as a tall mountain, the water moves at a much faster speed (**FIGURE 21.2**). The higher speed of the water produces a much larger force, allowing the water to displace soil and even large rocks down the steep slope and depositing them at the bottom. We saw an example of such massive erosion in Module 20 when we discussed mudslides that occur on steep mountains and how the mud can rapidly add sediments to streams, turning the water from clear to brown.

Soil Type

While the slope of the watershed affects water velocity, the soil type in the watershed also affects the land and water. As we noted in our discussion of soils, sandy soils allow water to rapidly permeate, which means less water remains to run along the surface of the soil. In contrast, soils containing high amounts of clay are much less permeable, so the water from a rainstorm is not able to quickly soak into the ground. Instead, more of the water runs along the soil's surface. In addition, because silt and clay particles are much smaller and lighter in mass than sand particles, the water running along the soil surface can carry these soil particles into water bodies at the end of the watershed. The result is that the water body turns brown with sediments, which can prevent photosynthesis by producers since light cannot penetrate deeply into the water. The sediments can also smother the eggs of fish and bury the rocky habitats that are used by many animals.

Vegetation Type

The final characteristic of a watershed is the vegetation. As we mentioned in our discussion of soils, plants play a key role in holding onto soils with their roots. When plants are removed, the watershed can experience much more erosion, especially on steep slopes. We have seen examples of this in plowed fields that experience moderate erosion and logged mountains that can experience massive landslides. The presence of plants also facilitates water percolating into the soil through the many pathways made by roots. Plants also take up large amounts of nutrients from the soils and assimilate the nutrients in their tissues. Thus, the presence of vegetation—combined with the effects of watershed area, length, slope, and soil type—plays a key role in the amount of water that leaves a watershed and the amount of soil sediments that are carried away.

AP® Exam Tip

The AP® Environmental Science Exam often asks students to propose ways to improve water quality in a watershed. Restoring vegetation or forest cover to the watershed is an easy solution because it decreases erosion of soil into surface waters, decreasing that impact on water quality. Make certain to give directionality to your answer. Restoring vegetation and forest cover would be *increase* forest cover, which would *decrease* erosion of topsoil. Ultimately, this would *improve* water quality.

21-2 What impacts do humans have on watersheds?

Humans impact watersheds by altering water flow and inputting excess nutrients and soil

Watersheds are found all over the world, ranging from pristine regions almost untouched by humankind to areas that are heavily populated by people. The impacts of human activities that affect watersheds are widespread, including the building of dams, mining for minerals, and an increase in the impermeable surfaces that don't let water percolate into the soil—as discussed in Module 5—and cause the water to rapidly run off. In this section we consider two well-studied watersheds with a focus on the impacts that humans have on altering water flow and inputting excess nutrients and soils.

The Hubbard Brook Watersheds

One of the most thorough studies of how disturbance affects the movement of water and nutrients has been ongoing for nearly 60 years in the Hubbard Brook ecosystem of New Hampshire. Since 1962, investigators have monitored the hydrological and biogeochemical cycles of six watersheds at Hubbard Brook, ranging in area from 12 to 43 ha (30 to 106 acres). The soil in each watershed is underlain by impenetrable bedrock, so there is no deep percolation of water; all precipitation that falls on the watershed leaves either by evapotranspiration or by runoff. Scientists measure precipitation throughout each watershed, and track the amount of water and nutrients leaving the system through a stream gauge at the bottom of the stream that drains a given watershed (**FIGURE 21.3**).

Researchers at Hubbard Brook investigated the effects of logging an entire watershed and the subsequent suppression of plant regrowth. The researchers cut down the forest in one watershed and used herbicides to suppress the regrowth of vegetation for several years. A nearby watershed that was not logged served as a control (**FIGURE 21.4**). The concentrations of nitrate in stream water were similar in the two watersheds before the logging occurred. Within 6 months after the cutting, the logged watershed showed significant increases in stream nitrate concentrations. With this information, the researchers were able to determine that when trees are no longer present to take up nitrate from the soil, nitrate leaches out of the soil and ends up in the stream that drains the watershed. These nitrates fertilize the streams and can result in unusually high growth of plants and algae that alters the normal functioning of stream ecosystems.

Continued monitoring of Hubbard Brook over five decades after the experiment began discovered that when ecological succession was allowed to happen and the trees grew back, the growth of the trees and other plants caused a 90 percent decline in the nitrates running off the mountains and into the stream. This study demonstrated the consequences of not allowing new vegetation to grow when a forest is cut. It also highlighted the importance of plant succession in regulating the cycling of nutrients and keeping excess nutrients out of the watershed's water bodies.

The Chesapeake Bay Watershed

Thus far we have considered individual watersheds in mountainous areas that drain into a nearby water body. However, watersheds can be much larger in area. For example, the Chesapeake Bay receives water from a massive area of land from New York to Virginia, covering 17 million ha

FIGURE 21.3 Measuring water and nutrient flow from a watershed. At the Hubbard Brook ecosystem in New Hampshire, the forest grows on top of impermeable rock, so any nutrients leaving the watershed are carried by the stream water. To measure the amount of water and nutrients leaving individual watersheds, researchers built small dams at the end of the watersheds. *(USDA Forest Service, Northern Research Station)*

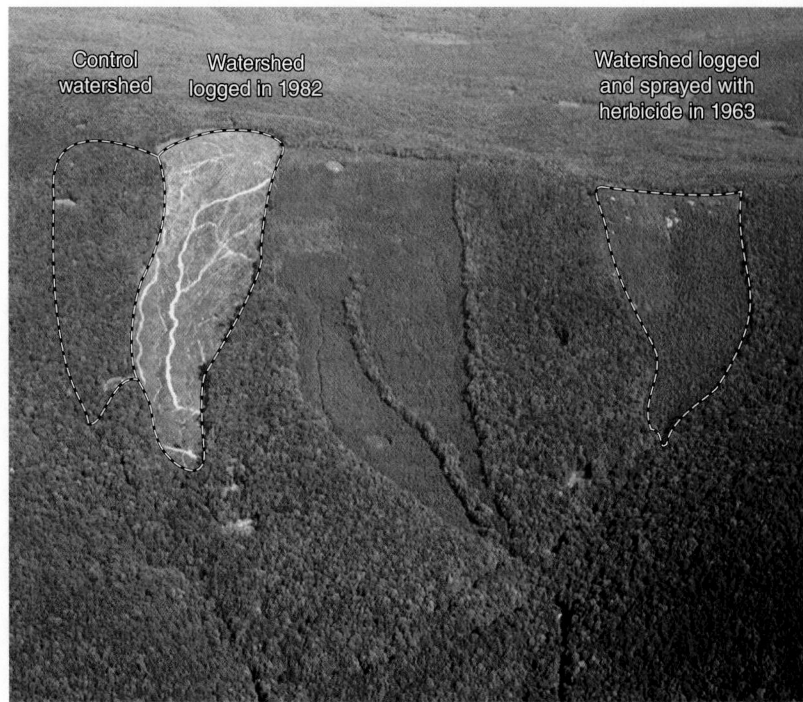

FIGURE 21.4 The Hubbard Brook watershed in New Hampshire. In the Hubbard Brook ecosystem, researchers clear-cut one watershed to determine the importance of trees in retaining soil nutrients. They compared nutrient runoff in the clear-cut watershed with that in a control watershed that was not clear-cut. (The two other watersheds shown in the photo received other experimental treatments.) *(US Forest Service, Northern Research Station)*

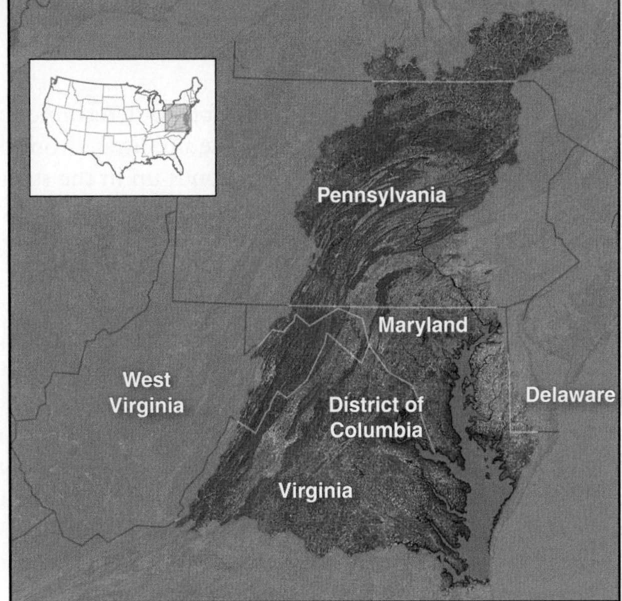

FIGURE 21.5 The Chesapeake Bay watershed. This watershed spans parts of six states and the District of Columbia. As a result, the bay receives water from a very large area that contains substantial human activities that can affect water quality.

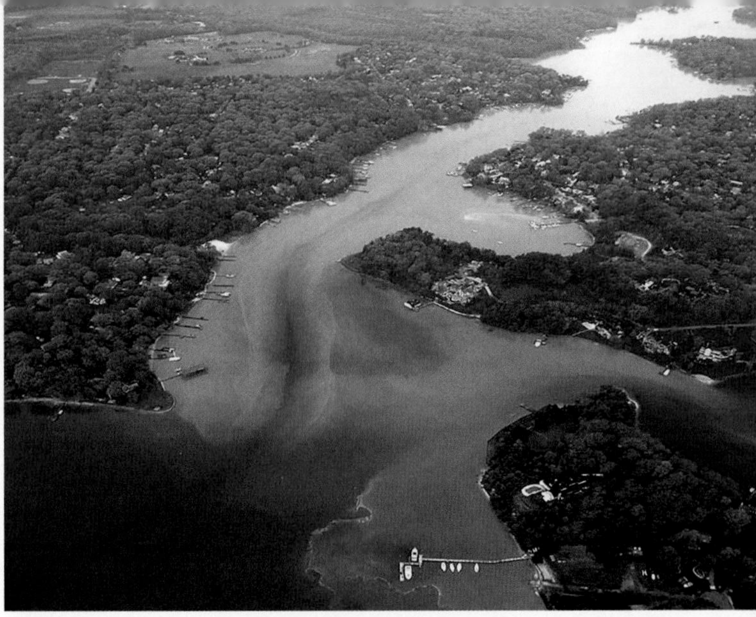

FIGURE 21.6 Inputs of sediments from a watershed. After heavy rains, the Chesapeake Bay receives a large amount of sediment from the surrounding watershed. This sediment has harmful effects on the plants and animals living in the bay. *(David Wallace P.E./Underwood & Associates)*

(41 million acres), as you can see in **FIGURE 21.5**. Numerous streams exist throughout the watershed and they feed into major rivers that release nearly 193 billion liters (51 billion gallons) of water into the bay each day. This includes the Susquehanna, Potomac, Patuxent, Rappahannock, York, and James rivers. The watershed contains forests, farms, suburbs, and major cities such as Baltimore, Richmond, and Washington, D.C.

The bay is an estuary that is a mixture of fresh water from the rivers and salt water from the ocean. As we discussed in Module 3, estuaries provide a number of important ecosystem services. For example, the bay and the organisms it contains filter a large amount of water, which removes many contaminants. It also serves as a buffer against large ocean storms moving onshore and as a buffer against floods on land since it can hold a great volume of water. The bay is also famous for its seafood, including several species of oysters and crabs. Finally, it is also a major tourist attraction. In short, the Chesapeake Bay provides substantial ecosystem services to the region, with an annual value of $33 billion.

Given that this large area of land drains billions of liters of water from millions of hectares of land, the water also delivers contaminants, nutrients, and sediments to the Chesapeake Bay. The contaminants include pesticides that are applied to lawns and farm fields, as well as pharmaceuticals that people take, which pass through their bodies and through sewage-treatment plants that ultimately deliver treated water to the rivers in the watershed.

There is also an enormous amount of nutrients that come into the bay; scientists estimate that the bay receives 272 million kg (600 million pounds) of nitrogen and 14 million kg (30 million pounds) of phosphorus each year. These excess nutrients come from agricultural fields, fertilized lawns, improperly treated sewage, and the waste of domesticated animals such as cattle and hogs. As we mentioned in our discussion of biogeochemical cycles (Modules 4 and 5), such large inputs of nutrients lead to algal blooms in the bay. Moreover, as the blooming populations of algae begin to die, their decomposition consumes much of the oxygen out of the water, leading to hypoxia and dead zones where many species of fish and shellfish cannot survive.

Increased sediments are also an issue in the Chesapeake Bay. Sediments are soils washed away from fields and forests as well as soils washed away from the banks of streams and the ocean shoreline. According to current estimates, 8 billion kg (18 billion pounds) of sediments come into the bay each year. The tiniest soil particles stay suspended in the water, making the water cloudy, and preventing sunlight from reaching the grasses that historically have been abundant in the bay (**FIGURE 21.6**). These grasses are important because they serve as a habitat for fish and blue crabs (*Callinectes sapidus*). As a result, the blue crab population has experienced substantial declines over several decades.

Because the Chesapeake Bay watershed is so large, cleaning it up has required a sustained and monumental effort. In 2000, the states surrounding the Chesapeake Bay formed a partnership along with multiple federal departments to develop the Chesapeake Bay Action Plan. This plan outlines a series of goals to reduce the impacts of nutrients, sediments, and chemicals coming into the bay. Many of the Action Plan's goals are being met, including a reduction in nitrogen, an increase in water clarity, and an increase in the crab population. However, the Chesapeake Bay serves as an excellent reminder that to understand human impacts on water, we need to consider impacts on the entire watershed.

In this module we examined the importance of understanding watersheds as regions of land that drain water into nearby water bodies, thereby altering water quality. We saw that the movement of water, soil, and nutrients depends on the characteristics of the watershed. These inputs are also substantially impacted by human activities, underscoring our need to think about human impacts as not simply local, but affecting areas far from the sources of the human activities. With this focus on water movement, in the next module we will explore patterns of precipitation and temperature around the world and the processes that determine these patterns.

Module 21 AP® Review

Preparing for the AP® Exam

Learning Goals Revisited

21-1 What are the characteristics of a watershed?

Watersheds are characterized by their area, length, slope, soils, and vegetation. Steeper slopes cause water to move at higher speeds, which causes greater erosion. The type of soil affects whether water can permeate or must run over the soil's surface. Vegetation is important for holding onto soil particles to reduce erosion, increasing the permeability of water, and taking up nutrients from the soil.

21-2 What impacts do humans have on watersheds?

Human impacts have been demonstrated in several well-studied watersheds. Removing all of the trees causes increased water flow and increased nutrient loss, although ecological succession can eventually reverse these impacts. Some watersheds are very large, such as the Chesapeake Bay watershed, which experiences large inputs of contaminants and sediments that harm the ecological community in the bay, and excess nutrients that contribute to algal blooms, hypoxia, and dead zones.

AP® Practice Questions

Preparing for the AP® Exam

Multiple-Choice Questions

1. What role does vegetation play in a watershed?
   (a) increases erosion
   (b) decreases water permeability
   (c) increases landslides
   (d) decreases erosion

2. What is the effect of a steeper slope in a watershed?
   (a) increased water speed, decreased erosion
   (b) decreased water speed, decreased erosion
   (c) decreased water speed, increased erosion
   (d) increased water speed, increased erosion

3. When the Hubbard Brook watershed was logged, nutrients in the stream initially ________ and then ________ as ecological succession occurred over five decades.
   (a) increased; increased
   (b) increased; decreased
   (c) decreased; increased
   (d) decreased; decreased

4. Which is the best description of the human impacts on the Chesapeake Bay watershed?
   (a) Human activities increase the rate of ecological succession in the watershed.
   (b) Nutrients released from human activities improve water quality in the bay.
   (c) Eroded soil sediments decrease primary productivity in the bay.
   (d) Human use of pesticides has not had an impact on the watershed

5. Which of the following is a regulating ecosystem service provided by the Chesapeake Bay?
   (a) flood control
   (b) tourism
   (c) pharmaceuticals
   (d) food

Free-Response Question

The figure below shows a watershed that drains into a river.

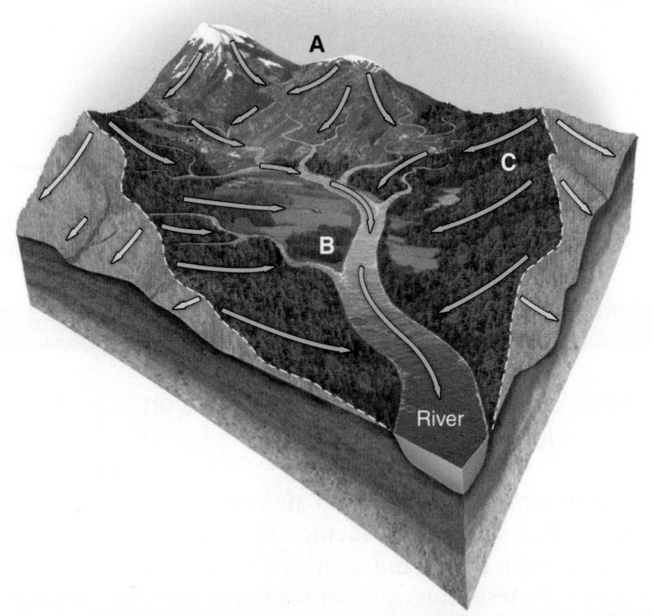

(a) **Identify** the position in the diagram where water would flow into the watershed from melting snow. (1 pt.)
(b) **Describe** one position in the diagram where water would flow at a fast rate toward the river. (1 pt.)
(c) Besides the speed of water flow, **explain** one factor that may influence how water in the diagram drains into the river. (2 pts.)
(d) **Identify** one way that humans can negatively impact watersheds. (1 pt.)
(e) **Explain** how the human impact identified in (d) can affect ecosystems. (2 pts.)
(f) At position B, an area of forest was clear-cut for the construction of riverfront homes:
   (i) **Describe** a potential environmental problem that could be caused by this project. (1 pt.)
   (ii) **Propose a solution** to the problem described in part (i). (1 pt.)
   (iii) **Describe** an additional benefit of the solution proposed in part (ii). (1 pt.)

Module 22

Earth's Atmosphere, Global Wind Patterns, Solar Radiation, and Earth's Seasons

In this unit so far, we have examined Earth's systems, including the formation of rocks and soils and the role of watersheds, and now we move on to learn about the creation of climates by the movement of air and precipitation. In this module we explore why incoming solar radiation (also known as **insolation**), which is the main source of energy on Earth, differs with latitude and how it changes with the seasons. With this knowledge, we then focus on the planet's atmosphere, starting with an exploration of atmospheric gases and the different atmospheric layers above Earth's surface. We then consider how solar radiation and the properties of air combine to cause air to circulate in the atmosphere. These air currents bring warm and cold air to different regions. They also move water vapor around the planet, which causes some regions to receive more precipitation than others. In doing so, the air currents determine the locations of different climates, the location of terrestrial biomes, and the prevailing directions of wind across the globe.

Learning Goals

After reading this module you should be able to

22-1 explain why the amount of solar radiation varies with latitude and the seasons.

22-2 identify the major gases and layers in Earth's atmosphere.

22-3 describe how the properties of air determine patterns of air circulation.

22-4 explain what drives atmospheric convection currents.

22-5 explain how the Coriolis effect alters global wind directions.

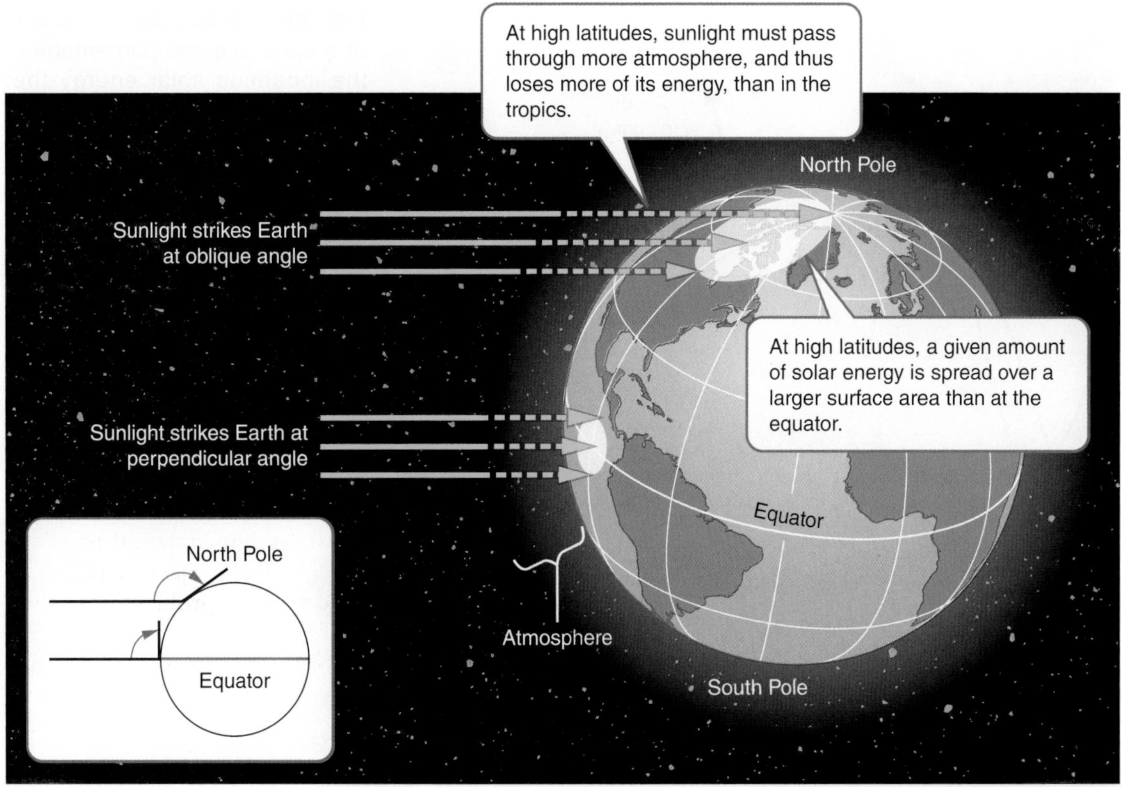
At high latitudes, sunlight must pass through more atmosphere, and thus loses more of its energy, than in the tropics.
North Pole
Sunlight strikes Earth at oblique angle
At high latitudes, a given amount of solar energy is spread over a larger surface area than at the equator.
Sunlight strikes Earth at perpendicular angle
Equator
North Pole
Equator
Atmosphere
South Pole

FIGURE 22.1 Differential heating of Earth. Tropical regions near the equator receive more solar energy than mid-latitude and polar regions, where the Sun's rays strike Earth's surface at an oblique angle.

22-1 Why does the amount of solar radiation vary with latitude and the seasons?

Solar radiation varies due to Earth's curvature and tilted axis

To understand why different biomes and climates are located in different locations on Earth, we need to first understand how the amount of solar radiation is distributed across latitudes and across the seasons.

Effects of Latitude

As the Sun's energy passes through the atmosphere and strikes land and water, it warms the planet's surface. But this warming does not occur evenly across the planet. This uneven warming pattern has three primary causes.

The first cause of unequal warming is variation in the angle at which the Sun's rays strike Earth. As we can see in **FIGURE 22.1**, in the region nearest to the equator—the tropics—the Sun strikes at a perpendicular, or right, angle. In the mid-latitude and polar regions, the Sun's rays strike at a more oblique angle. As a result, the Sun's rays travel a relatively short distance through the atmosphere to reach Earth's surface in the tropics but they must travel a longer distance through the atmosphere to reach Earth's surface near the poles. Because solar energy is lost as it passes through the atmosphere, more solar energy reaches the equator than mid-latitude and polar regions.

The second cause of the uneven warming of Earth is variation in the amount of surface area over which the Sun's rays are distributed. As you now know, the Sun's rays strike Earth at different angles in different places on the globe, and the angle can change with the time of year. When the Sun's rays strike near the equator, the solar energy is distributed over a smaller surface area than near the poles. Thus, regions near the equator receive more solar energy per square meter than mid-latitude and polar regions. You can replicate this phenomenon by shining a flashlight onto a round object, such as a basketball, in a dark room. If you shine the light perpendicular to the surface of the ball, you will create a small circle of bright light. If you shine the flashlight at an oblique angle, you will create a large oval pool of dim light because the light is distributed over a larger area.

Finally, the third cause of the uneven warming of Earth is that some areas of Earth reflect more solar energy than others. The percentage of incoming sunlight that is reflected from a

Insolation Incoming solar radiation, which is the main source of energy on Earth.

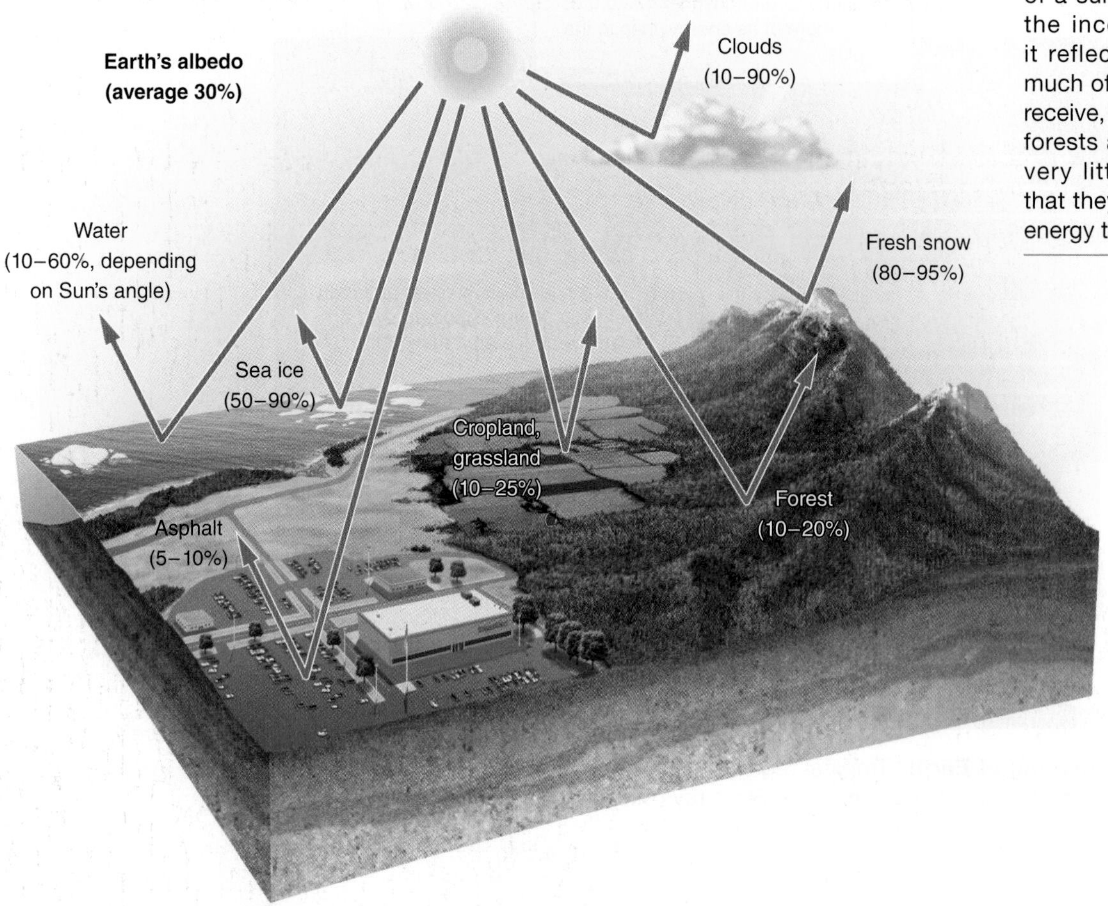

FIGURE 22.2 Albedo. The albedo of a surface is the percentage of the incoming solar energy that it reflects. Snow and ice reflect much of the solar energy that they receive, but darker objects such as forests and asphalt paving reflect very little energy, which means that they absorb most of the solar energy that strikes them.

surface is called its **albedo**. The higher the albedo of a surface, the more solar energy it reflects and the less it absorbs. A white surface has a higher albedo than a black surface, so a white surface tends to stay cooler. **FIGURE 22.2** shows albedo values for various surfaces on Earth. Although Earth has an average albedo of 30 percent, tropical regions with dense green foliage have albedo values of 10 to 20 percent, whereas the snow-covered polar regions have values of 80 to 95 percent.

Effects of Seasons

As we have just discussed, differences in the amount of solar energy striking various latitudes on Earth depend on the angle of the Sun's rays. For the same reason, the amount of solar energy reaching various latitudes shifts over the course of the year. Because Earth's axis of rotation is tilted 23.5°, Earth's orbit around the Sun causes most regions of the world to experience seasonal changes in temperature and precipitation. Specifically, when the Northern Hemisphere is tilted toward the Sun, the Southern Hemisphere is tilted away from the Sun, and vice versa.

Albedo The percentage of incoming sunlight reflected from a surface.

FIGURE 22.3 helps us visualize how this works. The Sun's rays strike the equator directly twice a year: during the March equinox, on March 20 or 21, and again during the September equinox, on September 22 or 23. On those days, all regions of Earth except those nearest the poles receive 12 hours of daylight and 12 hours of darkness. For the 6 months between the March and September equinoxes, the Northern Hemisphere tilts toward the Sun, experiencing more hours of daylight than darkness. The opposite is true in the Southern Hemisphere. On June 20 or 21, the Sun is directly above the Tropic of Cancer at 23.5° N latitude. On this day — the June solstice — the Northern Hemisphere experiences more daylight hours than on any other day of the year. For the 6 months between the September and March equinoxes, the Northern Hemisphere tilts away from the Sun, experiencing fewer hours of daylight than darkness. On December 21 or 22 — the December solstice — the Sun is directly over the Tropic of Capricorn at 23.5° S latitude. On this day, the Northern Hemisphere experiences its shortest day of the year, and the Southern Hemisphere experiences its longest day of the year.

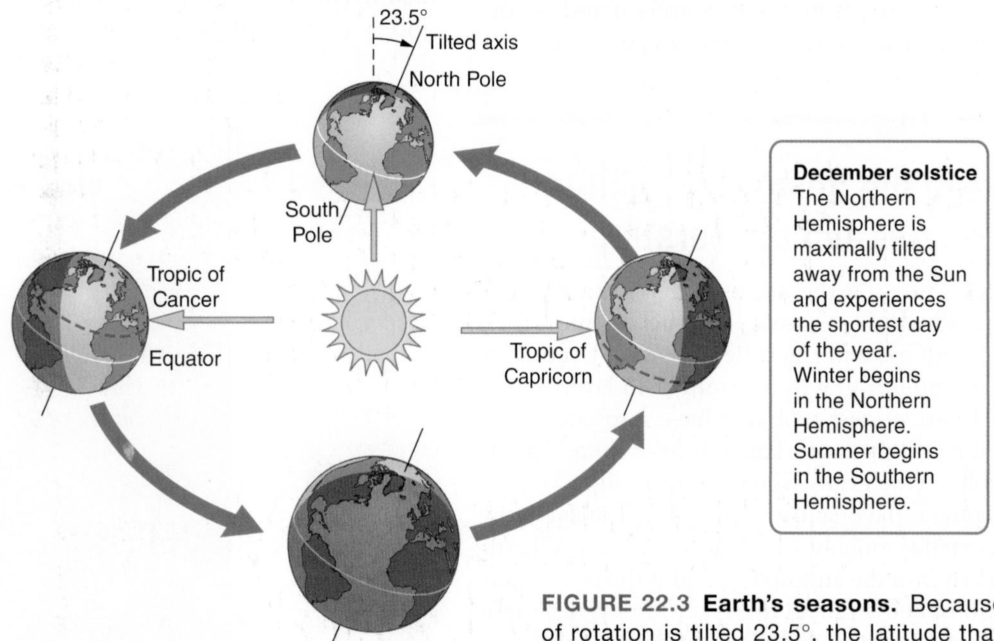

March equinox
The Sun is directly overhead at the equator and all regions of Earth receive 12 hours of daylight and 12 hours of darkness. Spring begins in the Northern Hemisphere. Fall begins in the Southern Hemisphere.
23.5°
Tilted axis
North Pole
South Pole
June solstice
The Northern Hemisphere is maximally tilted toward the Sun and experiences the longest day of the year. Summer begins in the Northern Hemisphere. Winter begins in the Southern Hemisphere.
Tropic of Cancer
Equator
Tropic of Capricorn
December solstice
The Northern Hemisphere is maximally tilted away from the Sun and experiences the shortest day of the year. Winter begins in the Northern Hemisphere. Summer begins in the Southern Hemisphere.
September equinox
The Sun is directly overhead at the equator and all regions of Earth receive 12 hours of daylight and 12 hours of darkness. Fall begins in the Northern Hemisphere. Spring begins in the Southern Hemisphere.

FIGURE 22.3 **Earth's seasons.** Because Earth's axis of rotation is tilted 23.5°, the latitude that receives the most direct rays of the Sun and the most hours of daylight changes throughout the year as Earth orbits the Sun. Thus, Earth's tilt produces predictable seasons. This diagram illustrates the pattern of seasons in the Northern Hemisphere.

22-2 What are the major gases and layers in Earth's atmosphere?

To understand how the atmosphere contributes to different climates around Earth, we also need to understand the layers of gas in the atmosphere.

The atmosphere is dominated by nitrogen and oxygen gases

The atmosphere of our planet extends to approximately 10,000 km (6,200 miles) above the surface of Earth. You may be surprised to know that the atmosphere is comprised primarily of nitrogen and oxygen, as illustrated in **FIGURE 22.4**. In fact, these two gases make up 99 percent of all gases in the atmosphere. All other gases are present in very small amounts, but the role they play is anything but small. For example, greenhouse gases such as carbon dioxide, methane, and nitrous oxide all are present at low concentrations, yet we know that these scarce gases help make the planet much warmer. These relatively rare gases naturally make our planet 33 degrees C (59 degrees F) warmer than it would be without natural levels of greenhouse gases in the atmosphere. Thus, you can also understand the concerns of scientists when they assert that even small increases in greenhouse gases, due to human activities, can have a large effect on global temperature.

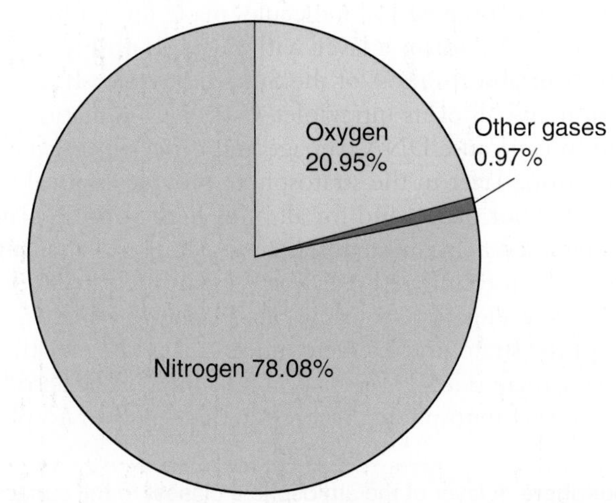

FIGURE 22.4 **The composition of gases in the atmosphere.** Nitrogen and oxygen comprise approximately 99 percent of the atmospheric gases. The remaining 1 percent may be small, but it includes greenhouse gases that play a large role in warming our planet.

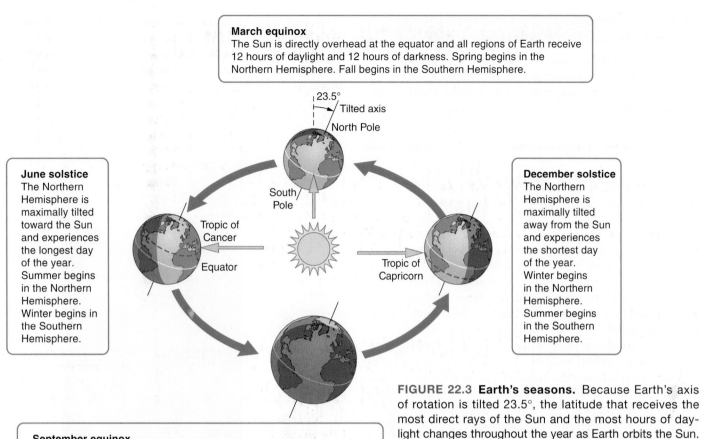

AP® Exam Tip

Make sure to distinguish between nitrogen oxides and nitrogen. Nitrogen oxides are an air pollutant. Nitrogen is the most abundant gas in our atmosphere and is not an air pollutant. Always write nitrogen oxides as NOx, or the specific nitrogen compound.

The layers of the atmosphere differ in mass, pressure, and temperature

As **FIGURE 22.5** shows, Earth's atmosphere consists of five layers of gases. The pull of gravity on the gas molecules keeps these layers of gases in place. Because each layer of gas has mass and the pull of gravity is stronger in layers closer to Earth, the layers closest to Earth have a greater mass. This causes the layers closest to Earth to have more densely packed molecules, which causes higher air pressure.

The atmospheric layer closest to Earth's surface, the **troposphere**, extends roughly 16 km (10 miles) above Earth. It is the densest layer of the atmosphere and is the layer where most of the atmosphere's nitrogen, oxygen, and water vapor occur. The troposphere experiences a great deal of circulation of liquids and gases, and is the layer where Earth's weather occurs. Air temperature in the troposphere decreases with distance from Earth's surface and varies with latitude. Temperatures can fall as low as −52°C (−62°F) near the top of the troposphere. In 2021, scientists made a fascinating new discovery about the troposphere: during the past 40 years, the continued impacts of global warming have been expanding the height of the troposphere by 50 to 60 m per decade.

Above the troposphere is the **stratosphere**, which extends roughly 16 to 50 km (10–31 miles) above Earth's surface. The stratosphere is less dense than the troposphere. **Ozone**, a pale blue gas composed of molecules made up of three oxygen atoms (O_3), forms a layer within the stratosphere. This ozone layer absorbs most of the Sun's ultraviolet-B (UV-B) radiation and all of its ultraviolet-C (UV-C) radiation. UV radiation can cause DNA damage and cancer in organisms, so the ozone layer in the stratosphere provides critical protection for our planet and for all of us living on the planet. The upper layers in the stratosphere absorb the UV radiation and convert it to infrared radiation, which is released as heat. Because UV radiation from the Sun reaches the higher altitudes of the stratosphere, where much of the UV radiation is absorbed, there is less UV radiation remaining to be absorbed in the lower stratosphere. As a result, the upper stratosphere is warmer than the lower stratosphere, spanning the range of approximately 0°C to −60°C.

Beyond the stratosphere are the mesosphere, thermosphere, and exosphere, respectively. The atmospheric pressure and density in each of these layers continues to decrease as we move toward the outer layers of the atmosphere. The

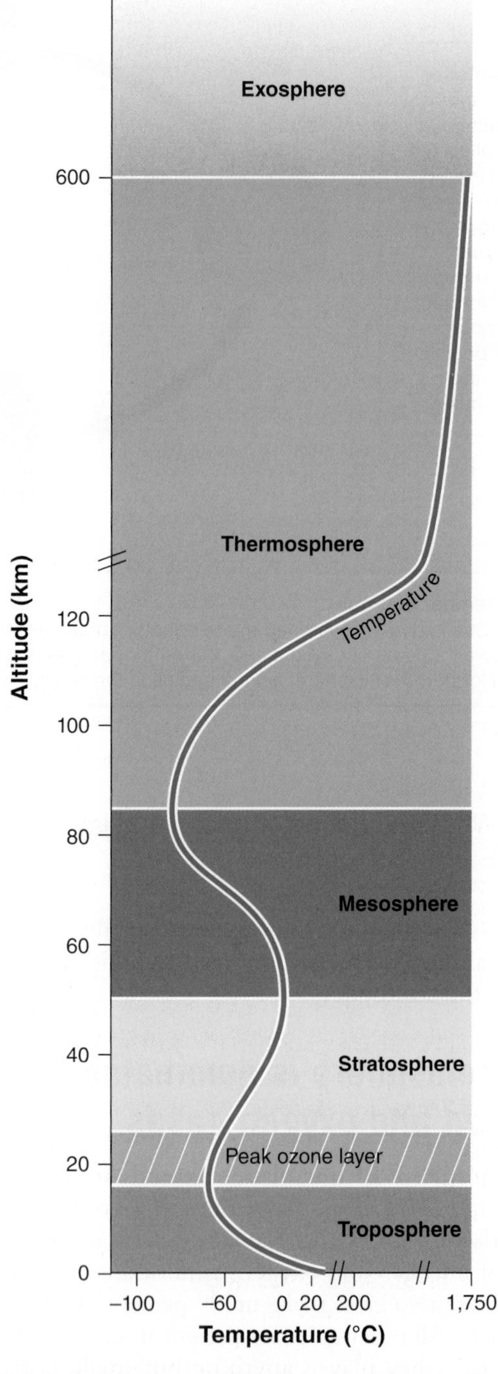

FIGURE 22.5 The layers of Earth's atmosphere. The troposphere is the atmospheric layer closest to Earth. Because the density of air decreases with altitude, the troposphere's temperature also decreases with altitude. Temperature increases with altitude in the stratosphere because the Sun's UV-B and UV-C rays warm the upper part of this layer. Temperatures in the thermosphere can reach 1,750°C (3,182°F).

Troposphere A layer of the atmosphere closest to the surface of Earth, extending up to approximately 16 km (10 miles).

Stratosphere The layer of the atmosphere above the troposphere, extending roughly 16 to 50 km (10–31 miles) above the surface of Earth.

Ozone A pale blue gas composed of molecules made up of three oxygen atoms (O_3).

FIGURE 22.6 Northern lights. The glowing, moving lights that are visible at high latitudes in both hemispheres are the product of solar radiation energizing the gases of the thermosphere. *(Stephen Mcsweeny/Shutterstock)*

mesosphere extends roughly 50 to 85 km (31–53 miles) above the surface of Earth. It has a temperature range of 0 to −90°C and is the layer where most meteors burn up in the atmosphere.

The **thermosphere** extends 85 to 600 km (53–375 miles) above the surface of Earth and is particularly important to organisms on Earth's surface because of its ability to block harmful X-ray radiation and some UV radiation from reaching our planet, with the closer ozone layer absorbing much of the remaining UV radiation. This absorption of radiation is what causes the temperature in the upper thermosphere to reach nearly 2,000°C. The thermosphere is also interesting because it contains charged gas molecules that, when hit by solar energy, begin to glow and produce light, in the same way that a light bulb glows when electricity is applied. Because this interaction between solar energy and gas molecules is driven most intensely by magnetic forces at the North Pole and South Pole, the best places to view the phenomenon are at high latitudes. In the northern United States, Canada, and northern Europe, these glowing gases are known as the northern lights, or aurora borealis. In Australia and southern South America, they are called the southern lights, or aurora australis (**FIGURE 22.6**). The outermost layer of the atmosphere is the **exosphere**, which extends from 600 to 10,000 km (375–6,200 miles) above the surface of Earth. This is the layer in which satellites orbit the planet. The temperature range varies with altitude, from 0 to 1,700°C.

22-3 How do the properties of air determine patterns of air circulation?

Air circulates in the atmosphere as a result of changing density, water vapor capacity, and temperature

Air has four properties that determine how it circulates in the atmosphere: density, water vapor capacity, adiabatic heating or cooling, and latent heat release.

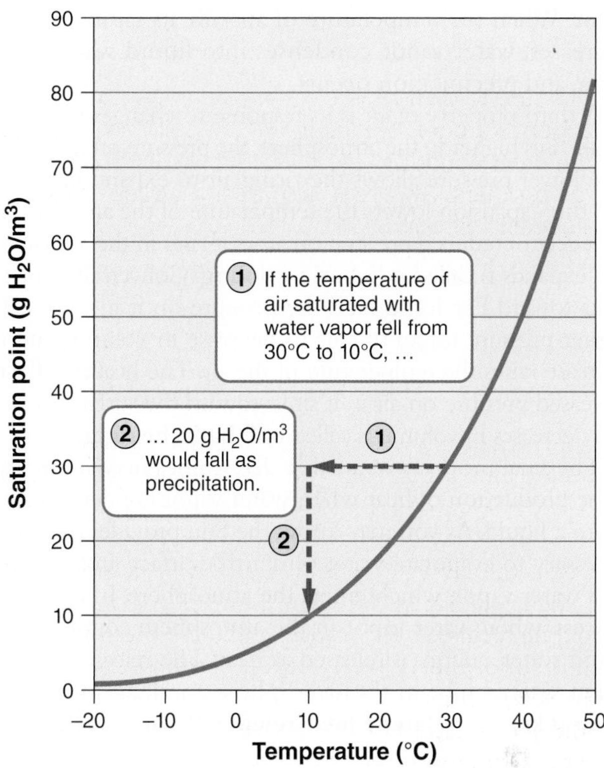

FIGURE 22.7 The saturation point of air. When air cools and its saturation point drops, water vapor condenses into liquid water that forms clouds. These clouds are ultimately the source of precipitation.

The first property is air density, which is the mass of all molecules in the air in a given volume. The density of air determines its movement: Less dense air rises, whereas denser air sinks. At a constant atmospheric pressure, warm air has a lower density than cold air. Because of this density difference, warm air rises, whether in a room in your house or in the atmosphere.

The second property that determines how air circulates is its capacity to contain water vapor. For example, warm air is not only less dense than cold air, but it also has a higher capacity for water vapor. In regions of the world that receive substantial amounts of precipitation, hot summer days are associated with high humidity because the warm air contains a lot of water vapor. The maximum amount of water vapor that can be in the air at a given temperature is called its **saturation point**. **FIGURE 22.7** shows the relationship between the temperature of air and its saturation

Mesosphere The layer of the atmosphere above the stratosphere, extending roughly 50 to 85 km (31–53 miles) above the surface of Earth.

Thermosphere The layer of the atmosphere above the mesosphere, extending 85 to 600 km (53–375 miles) above the surface of Earth.

Exosphere The outermost layer of the atmosphere, which extends from 600 to 10,000 km (375– 6,200 miles) above the surface of Earth.

Saturation point The maximum amount of water vapor in the air at a given temperature.

point. When the temperature of air falls, its saturation point decreases, water vapor condenses into liquid water, clouds form, and precipitation occurs.

A third property of air is its response to changes in pressure. As air rises higher in the atmosphere, the pressure on it decreases. The lower pressure allows the rising air to expand in volume, and this expansion lowers the temperature of the air. The cooling effect of reduced pressure on air as it rises in the atmosphere and expands is called **adiabatic cooling**. Conversely, when air sinks toward Earth's surface, the pressure on it increases. The higher pressure forces the air to decrease in volume, and this decrease raises the temperature of the air. The heating effect of increased pressure on air as it sinks toward the surface of Earth and decreases in volume is called **adiabatic heating**.

The final property of air that determines how it circulates is the production of heat when water vapor condenses from a gas to a liquid. As you may know, the Sun provides the energy necessary to evaporate water on Earth's surface and convert it into water vapor, which enters the atmosphere. In the reverse process, when water vapor in the atmosphere condenses into liquid water, energy is released as heat. The release of energy when water vapor in the atmosphere condenses into liquid water is known as **latent heat release**. Because of latent heat release, whenever water vapor condenses in the atmosphere, the air will become warmer and rise.

22-4 What drives atmospheric convection currents?

Atmospheric convection currents are driven by solar radiation at the equator

A major contributor to the different climates of the world is atmospheric convection currents. **Atmospheric convection currents** are global patterns of air movement that are initiated by the unequal heating of Earth, with regions near the equator receiving more solar radiation than regions near the poles. These air currents are called convection currents because they involve the movement of air that absorbs and releases heat as the air moves up into the atmosphere and then back to Earth. Now that we understand the four properties of air, we can look at how these properties contribute to atmospheric convection currents. In the next section we will consider how these atmospheric currents are deflected by Earth's rotation.

FIGURE 22.8 shows how atmospheric currents develop. The warming of humid air at Earth's surface decreases the

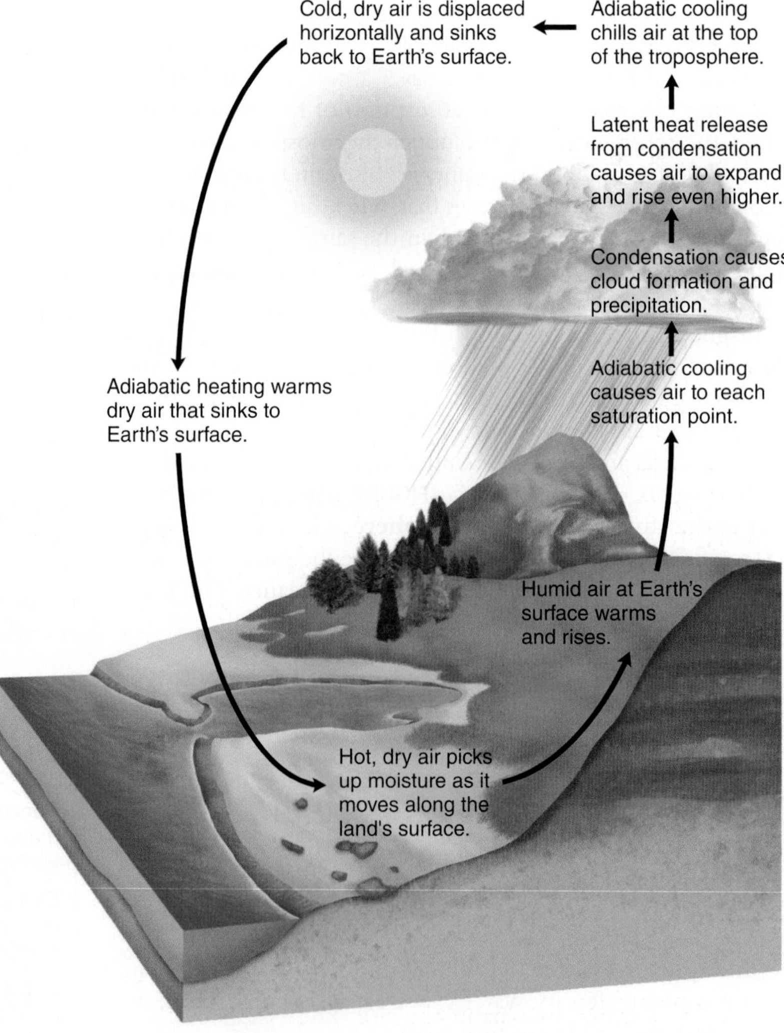

Cold, dry air is displaced horizontally and sinks back to Earth's surface.
Adiabatic cooling chills air at the top of the troposphere.
Latent heat release from condensation causes air to expand and rise even higher.
Condensation causes cloud formation and precipitation.
Adiabatic cooling causes air to reach saturation point.
Adiabatic heating warms dry air that sinks to Earth's surface.
Humid air at Earth's surface warms and rises.
Hot, dry air picks up moisture as it moves along the land's surface.

FIGURE 22.8 Atmospheric currents. Warming at Earth's surface causes air to rise up into the atmosphere where it experiences lower pressures, adiabatic cooling, and latent heat release. The cool air near the top of the atmosphere is then displaced horizontally before it sinks back to Earth. As it sinks, the air experiences adiabatic heating and then moves horizontally along the surface of Earth to complete the cycle.

Adiabatic cooling The cooling effect of reduced pressure on air as it rises higher in the atmosphere and expands.

Adiabatic heating The heating effect of increased pressure on air as it sinks toward the surface of Earth and decreases in volume.

Latent heat release The release of energy when water vapor in the atmosphere condenses into liquid water.

Atmospheric convection current Global patterns of air movement that are initiated by the unequal heating of Earth.

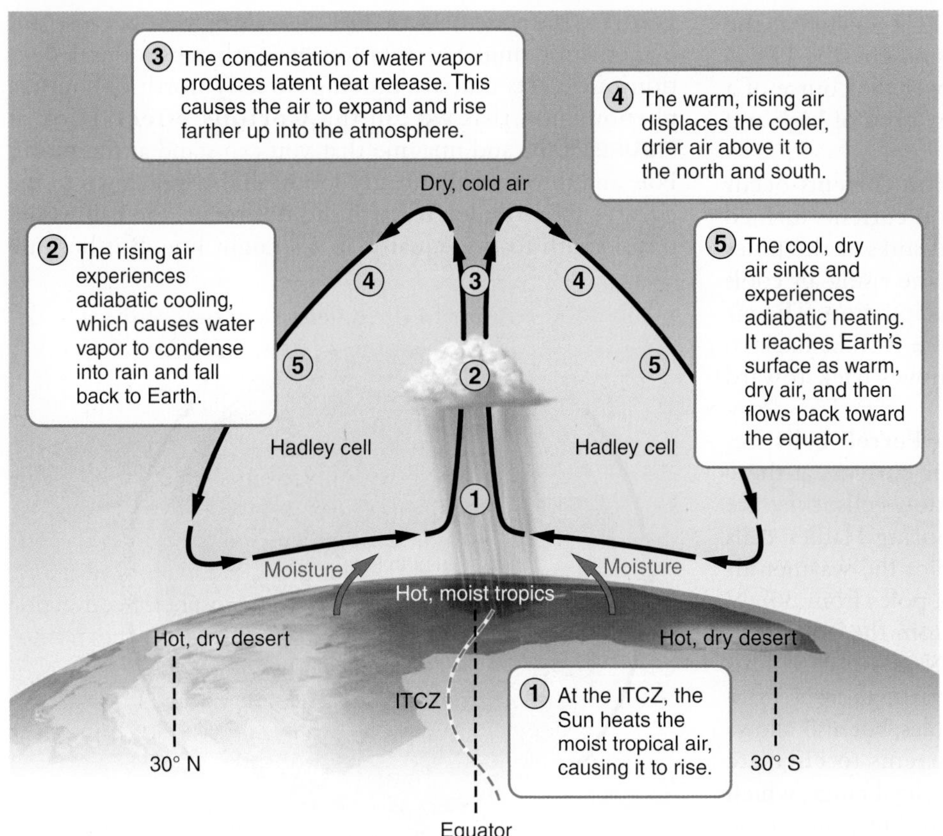

3 The condensation of water vapor produces latent heat release. This causes the air to expand and rise farther up into the atmosphere.
4 The warm, rising air displaces the cooler, drier air above it to the north and south.
Dry, cold air
2 The rising air experiences adiabatic cooling, which causes water vapor to condense into rain and fall back to Earth.
5 The cool, dry air sinks and experiences adiabatic heating. It reaches Earth's surface as warm, dry air, and then flows back toward the equator.
Hadley cell
Hadley cell
Moisture
Moisture
Hot, moist tropics
Hot, dry desert
Hot, dry desert
ITCZ
1 At the ITCZ, the Sun heats the moist tropical air, causing it to rise.
30° N
30° S
Equator

FIGURE 22.9 Hadley cells. Hadley cells are atmospheric convection currents that operate between the equator and 30° N and 30° S. Solar energy warms humid air in the tropics. The warm air rises and eventually cools below its saturation point. The water vapor it contains condenses into clouds and precipitation. The air, which now contains little moisture, sinks to Earth's surface at approximately 30° N and 30° S. As the air descends, it is warmed by adiabatic heating. This descent of hot, dry air causes desert environments to develop at those latitudes.

density of the air. As a result, the air begins to rise and, as it does, it begins to experience lower atmospheric pressures and adiabatic cooling. The cooling causes the air to reach its saturation point, which leads to condensation that causes cloud formation and precipitation. Condensation also causes latent heat release, which offsets some of the adiabatic cooling and becomes a strong driving force to make the air expand further and rise higher. Collectively, these processes cause air to rise continuously from Earth's surface near the equator, forming a river of air flowing upward into the troposphere.

As the air reaches the top of the troposphere, it is chilled by adiabatic cooling. This air now contains relatively little water vapor. As warmer air continually rises from below, the cold, dry air at the top of the troposphere is displaced horizontally and eventually begins to sink back to Earth's surface. As the air sinks, it experiences higher atmospheric pressures, and its reduction in volume causes adiabatic heating. By the time the air reaches Earth's surface, it is hot and dry. This hot and dry air circulates back to the starting point near the equator and picks up moisture along the way.

Atmospheric convection currents on Earth are found at specific locations. For example, **Hadley cells** are convection currents that cycle between the equator and approximately 30° N and 30° S. In a Hadley cell, illustrated in **FIGURE 22.9**, the greater warming of the tropics causes the air near the equator to expand and rise. This rising air cools and the condensed water falls as precipitation over the tropics, which is why tropical rainforests are located near the equator. After being chilled by adiabatic cooling near the top of the troposphere, the cold dry air is displaced horizontally north and south of the equator. This air descends back to Earth at approximately 30° N and 30° S. As it sinks, adiabatic heating causes the air to warm. As a result, regions at 30° N and 30° S are typically hot, dry deserts. Once the warm, dry air reaches Earth, it flows back toward the equator to complete the cycle.

The circulation of the Hadley cells is driven by the intense solar energy that strikes Earth near the equator. The latitude that receives the most intense sunlight, which causes the ascending branches of the two Hadley cells to converge, is called the **intertropical convergence zone (ITCZ)**. Because humid air rises at this latitude and precipitation falls, the ITCZ is typified by dense clouds and intense thunderstorm activity.

The latitude of the ITCZ is not fixed. Because Earth's axis of rotation is tilted, the area receiving the most intense sunlight shifts between approximately 23.5° N and 23.5° S

Hadley cell A convection current in the atmosphere that cycles between the equator and 30° N and 30° S.

Intertropical convergence zone (ITCZ) The latitude that receives the most intense sunlight, which causes the ascending branches of the two Hadley cells to converge.

as Earth orbits the Sun. Because the ITCZ occurs at the latitudes that receive the most intense sunlight, the ITCZ moves north and south of the equator over the course of a year. As a result, the tropics experience seasons of high and low precipitation.

Another set of atmospheric convection currents occur near the poles. **Polar cells** are convection currents formed by air that rises at 60° N and 60° S and sinks at the poles (90° N and 90° S). At 60° N and 60° S, the rising air cools and the water vapor condenses into precipitation. The air dries as it moves toward the poles, where it sinks back to Earth's surface. At the poles, the air moves back toward 60° N and 60° S, completing the cycle.

A third convection current, known as **Ferrell cells**, lies between Hadley cells and polar cells. Air currents at these latitudes do not form distinct convection cells; they are driven by the circulation of the neighboring Hadley cells and polar cells. At Earth's surface, some of the warmer air from the Hadley cells moves toward the poles from 30° N and 30° S, and some of the cooler air from the polar cells moves toward the equator from 60° N and 60° S. This movement not only helps to distribute warm air away from the tropics and cold air away from the poles, but also allows a wide range of warm and cold air currents to circulate between 30° and 60° N or S. In this latitudinal range, which includes most of the United States, wind direction can be quite variable at Earth's surface.

Collectively, these convection currents slowly move the warm air of the tropics toward the mid-latitude and polar regions. Because these currents determine the patterns of temperature and precipitation around the world, they are largely responsible for the locations of terrestrial biomes, including rainforests, deserts, and grasslands on Earth.

22-5 How does the Coriolis effect alter global wind directions?

Earth's rotation causes the Coriolis effect, which deflects global wind patterns

The atmospheric convection currents cause air to move directly north or south. However, the actual air currents at Earth's surface are deflected to the east or west because our planet is rotating. The rotation of Earth causes the deflection of objects that are moving directly north or south, a phenomenon that we call the **Coriolis effect**. Look at **FIGURE 22.10a** and imagine that you can stand at the North Pole and throw a ball directly south, all the way down to the equator (0° latitude). If Earth did not rotate, the ball would travel south to the equator in a straight line. But because

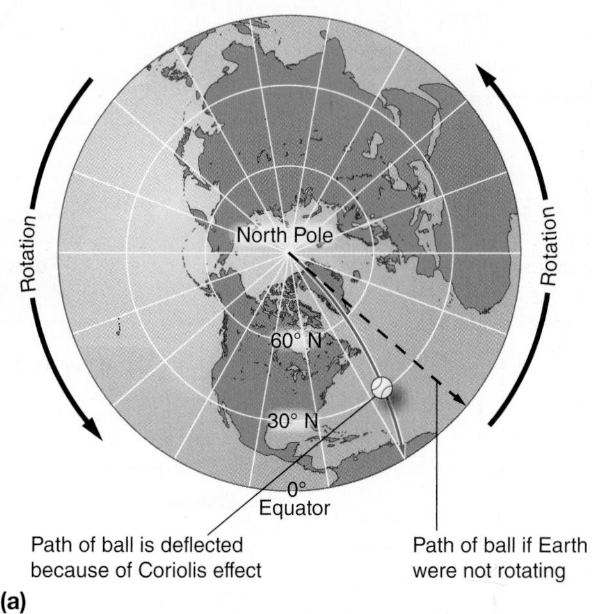

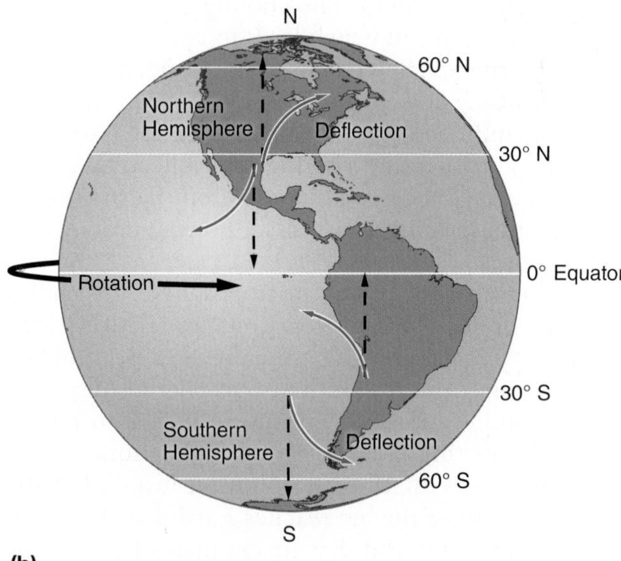

FIGURE 22.10 The Coriolis effect. (a) A ball thrown from the North Pole toward the equator would be deflected to the west by the Coriolis effect. (b) The different rotation speeds of Earth at different latitudes cause a deflection in the paths of traveling objects.

Polar cell A convection current in the atmosphere, formed by air that rises at 60° N and 60° S and sinks at the poles, 90° N and 90° S.

Ferrell cell A convection current in the atmosphere that lies between Hadley cells and polar cells.

Coriolis effect The deflection of an object's path due to the rotation of Earth.

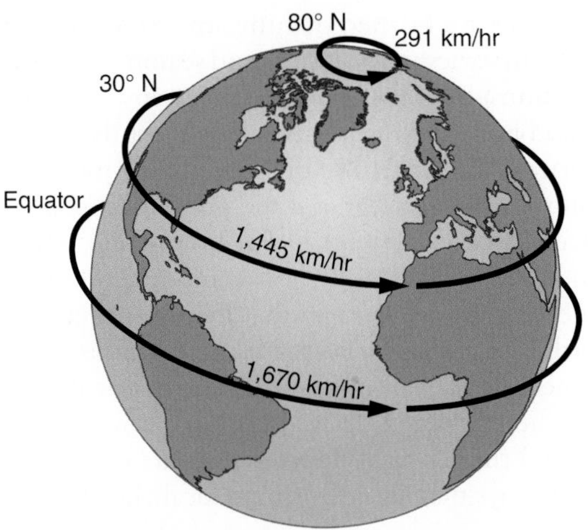

FIGURE 22.11 The speed of Earth's rotation. Because all locations on Earth complete one revolution every 24 hours, and because Earth has a greater circumference near the equator than near the poles, its speed of rotation is much faster at the equator than near the poles.

Earth rotates to the east while the ball is traveling through the air, the ball will land in a location that is west of its intended target. This happens because Earth's surface beneath the ball is moving faster and faster as the ball moves toward the equator. In other words, the path of the ball is deflected with respect to a given location on the globe because Earth is rotating. Now imagine you are standing at 30° N or 30° S latitude. As shown in Figure 22.10b, throwing a ball toward the equator results in a deflection of the ball's path to the west. However, throwing a ball toward the nearest pole results in a deflection of the ball's path to the east.

The deflection of objects moving north or south occurs because the planet's surface moves much faster at the equator than at higher latitudes. The planet's circumference is 40,000 km (25,000 miles) at the equator, but only 7,000 km (4,350 miles) at 80° N or 80° S latitude. Now imagine traveling all the way around Earth in 24 hours at the equator versus at 80° N latitude. **FIGURE 22.11** will help you visualize this journey. At the equator, you must travel 40,000 km in 24 hours; at 80° N latitude you must travel only 7,000 km. To make the trips in 24 hours, you must travel much faster at the equator. Indeed, Earth's surface moves at 1,670 km per hour (1,038 miles per hour) at the equator but only 291 km per hour (181 miles per hour) at 80° N or 80° S.

If Earth did not rotate, the air in the three types of convection cells that we discussed earlier would simply move directly north or south and cycle back again. For example, the air in a Hadley cell sinks to Earth's surface at approximately 30° latitude and travels along Earth's surface toward the equator. However, the Coriolis effect produced by Earth's rotation deflects the path of atmospheric convection currents that move toward or away from the equator, as we can see in **FIGURE 22.12**. Because Earth's speed of rotation is faster near the equator, the path of the air current moving toward the equator is deflected to the west, just as we saw in our example of throwing a ball toward the equator. This deflection causes the Hadley cell north of the equator to produce prevailing winds along Earth's surface that come from the northeast (called northeast trade winds), whereas the cell south of the equator produces prevailing winds that come from the southeast (called southeast trade winds).

The Coriolis effect also explains the prevailing wind directions in the mid-latitudes (between 30° and 60°) where we find the Ferrell cells. These winds can be quite variable as a result of the mixing of air currents from the adjacent Hadley cells and polar cells. Closer to 30°, air tends to move along Earth's surface toward the poles. If Earth were not rotating, this air would move straight north in the Northern Hemisphere and straight south in the Southern Hemisphere. Because Earth is rotating faster at 30° than

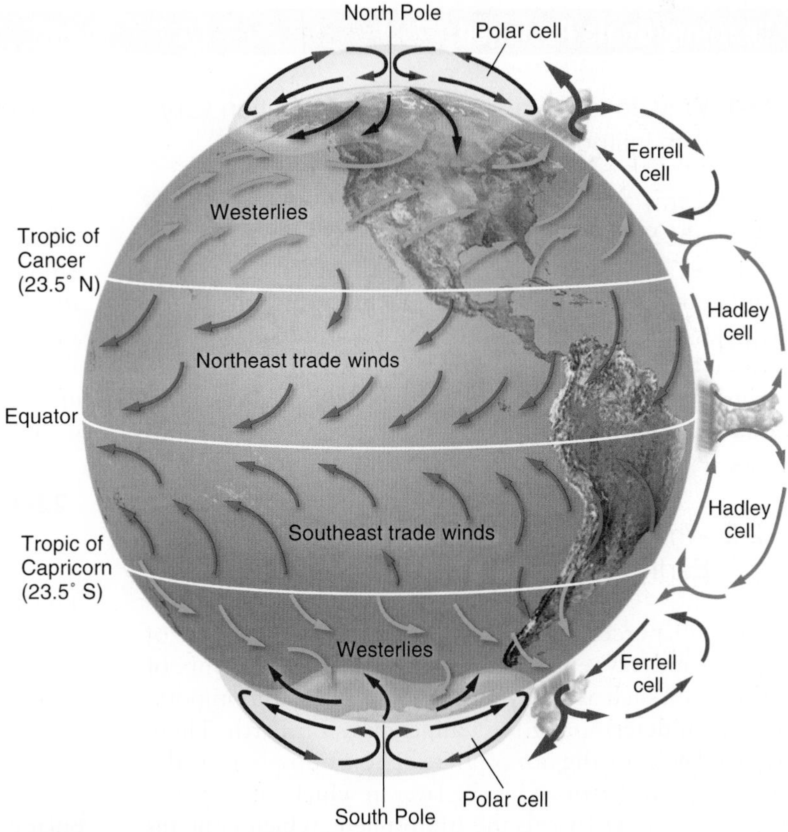

FIGURE 22.12 Prevailing wind patterns. Prevailing wind patterns around the world are produced by a combination of atmospheric convection currents and the Coriolis effect. This global map shows air circulation when the Sun is directly over the equator (e.g., spring and fall equinox).

at 60°, this air movement toward the poles is deflected to the east. As a result, the prevailing winds above 30° in the Northern Hemisphere come from the southwest, while in the Southern Hemisphere the prevailing winds above 30° are from the northwest. In both cases, these winds are called westerlies.

Finally, the Coriolis effect helps us understand the prevailing wind directions in the polar regions. Along Earth's surface, the polar cells move air away from the poles and toward 60° latitude. Since Earth is rotating faster at 60° than it is at 90°, the air movement is deflected to the west. As a result, polar winds come out of the northeast in the Northern Hemisphere and out of the southeast in the Southern Hemisphere. These winds are called easterlies.

In this module we learned that the amount of solar radiation striking Earth varies with latitude and seasons. We also learned that the atmosphere is comprised primarily of nitrogen and oxygen and how it exists in layers that affect the environment we experience on Earth. We discovered how the intense solar radiation near the equator and the properties of air combine to drive convection currents in the atmosphere, which determine the patterns of temperature and precipitation around the planet. Finally, we learned that these convection currents near Earth's surface are deflected by the Coriolis effect, which determines the prevailing wind directions at different latitudes. In the next module we will learn how air circulation patterns and the Coriolis effect determine ocean circulation patterns in ways that also determine local climates.

Module 22 AP® Review

Preparing for the AP® Exam

Learning Goals Revisited

22-1 Why does the amount of solar radiation vary with latitude and the seasons?

The equator receives more solar radiation because it receives the Sun's ray at a perpendicular angle, so the Sun's rays travel a shorter distance through the atmosphere and is distributed over a smaller area at the tropics than near the poles. The tropics also absorb more of the Sun's energy because they have lower albedo values than polar regions. Because Earth's axis of rotation is tilted 23.5°, Earth's orbit around the Sun causes most regions of the world to experience seasonal changes in temperature and precipitation.

22-2 What are the major gases and layers in Earth's atmosphere?

Nearly 99 percent of the atmosphere is comprised of nitrogen and oxygen. The remaining 1 percent is a mix of scarce gases, such as greenhouse gases, that play an important role in determining the temperatures on Earth. There are five layers of the atmosphere. The troposphere is the layer closest to Earth and is the layer in which our weather occurs. The next layer is the stratosphere, which contains ozone that absorbs most of the Sun's ultraviolet-B (UV-B) radiation and all of its ultraviolet-C (UV-C) radiation. The other layers are the mesosphere, thermosphere, and exosphere.

22-3 How do the properties of air determine patterns of air circulation?

Air density affects whether it rises or sinks in the atmosphere; less dense air rises and denser air sinks. Warm air can contain more water vapor than cold air. Air pressure affects how air warms and cools; reduced pressure leads to adiabatic cooling whereas increased pressure leads to adiabatic heating. When water vapor condenses into a liquid, there is latent heat release, causing the air to become warmer.

22-4 What drives atmospheric convection currents?

The four properties of air, combined with the most intense solar radiation near the equator, drives Hadley cells. Hadley cells rise near the equator and as the air cools, the air releases water as precipitation. The cool dry air sinks back to Earth near 30° N and 30° S latitudes and as it sinks it becomes warmer. The movement of the Hadley cells helps drive the circulation of air in Ferrell cells and polar cells.

22-5 How does the Coriolis effect alter global wind directions?

The rotation of Earth on its axis causes points closer to the equator to spin much faster than points near the poles. As a result, air movement from Hadley, Ferrell, and polar cells near the planet's surface are deflected east or west. The combination of the convection currents and the Coriolis effect produce winds that differ in direction at different latitudes.

AP® Practice Questions

Preparing for the AP® Exam

Multiple-Choice Questions

Use the diagram below to answer questions 1 & 2:

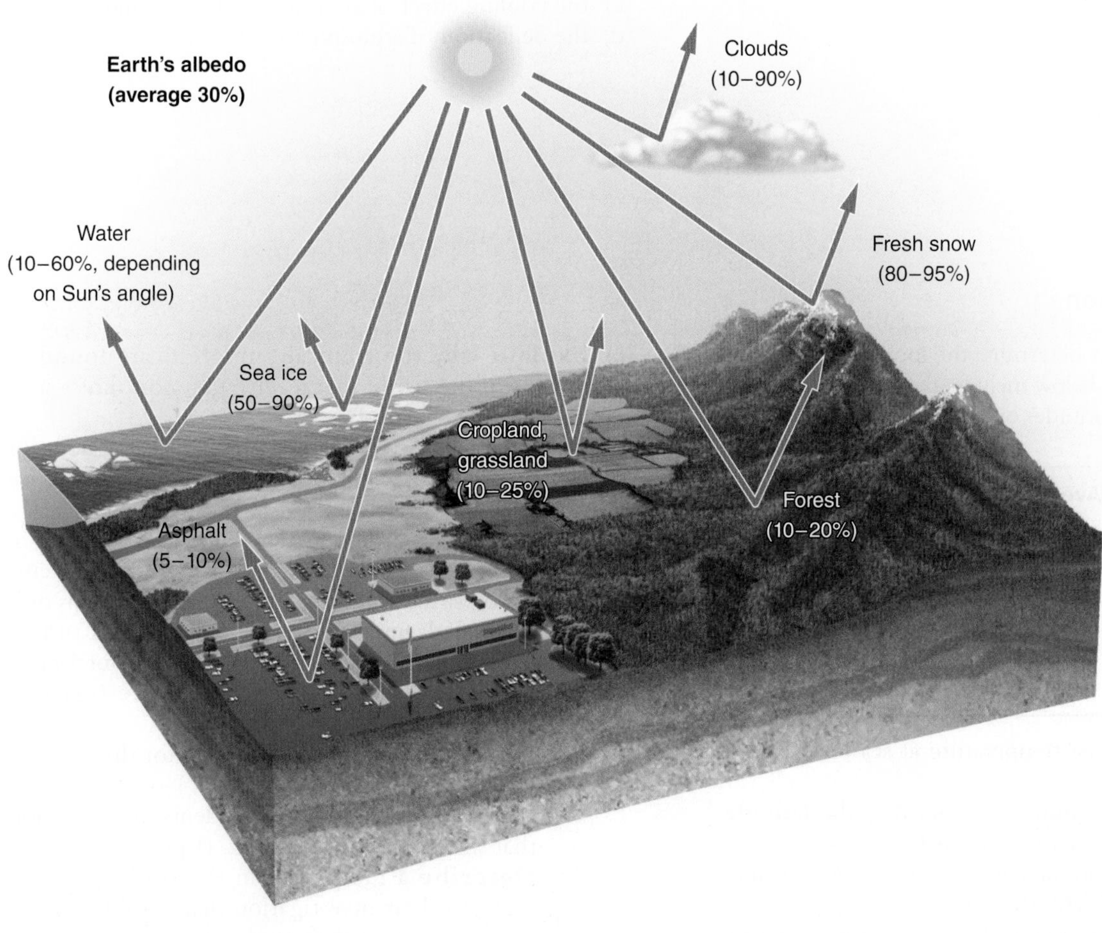

1. Which has the highest albedo?
   (a) soil
   (b) snow
   (c) water
   (d) pavement

2. What is the average albedo of water on Earth?
   (a) 25 percent
   (b) 30 percent
   (c) 40 percent
   (d) 90 percent

3. Summer in the Northern Hemisphere is warmer primarily because of
   (a) increased atmospheric absorption of solar radiation.
   (b) reduced albedo.
   (c) the tilt of Earth's axis.
   (d) increased cloud cover.

4. Which atmospheric layer contains the protective ozone layer?
   (a) exosphere
   (b) stratosphere
   (c) thermosphere
   (d) troposphere

5. Temperature change in adiabatic heating occurs due to
   (a) water vapor condensing.
   (b) a pressure increase.
   (c) absorption of sunlight.
   (d) a pressure decrease.

6. Latent heat release
   (a) increases air temperature.
   (b) decreases air temperature.
   (c) increases air pressure.
   (d) decreases air pressure.

7. Which of the following is a direct cause of vertical atmospheric convection currents?
   (a) Hadley cells
   (b) northern lights
   (c) polar cells
   (d) sun angle

8. The maximum amount of water vapor air can hold at a given temperature is called the
   (a) saturation point.
   (b) absolute humidity.
   (c) dew point.
   (d) vapor pressure.

9. The Coriolis effect is responsible for
   (a) the convergence of two ascending Hadley cells.
   (b) the location of atmospheric currents.
   (c) the cooling effect of air as it rises and expands.
   (d) the deflection of atmospheric currents.

Free-Response Question

A number of Earth's features determine the locations of biomes around the world. The table below includes the average temperature at sea level for various latitudes on earth.

Latitude	Average temperature at sea level (°C)
90°	–30
60°	–12
30°	18
0°	27

(a) **Identify** the average temperature at sea level of the equator. (1 pt.)

(b) **Identify** a biome commonly found at the latitude where average temperatures are 18°C. (1 pt.)

(c) **Describe** why the average temperature at 90° latitude is lower than at 60° latitude. (1 pt.)

(d) **Describe** how movement of the Intertropical Convergence Zone (ITCZ) during a year can influence seasonal forests in tropical regions. (1 pt.)

(e) **Explain** why the tropical rainforests are found in regions of the world that receive the most direct sunlight. (1 pt.)

(f) **Describe** one environmental problem that may occur as a result of increased melting of snow and ice on earth. (1 pt.)

(g) In their AP® Environmental Science class, a group of students is investigating the effects of color on the temperature of various surfaces. They set up heat lamps over several materials (ice, water, sand, soil), and took surface temperature measurements every 10 minutes for 1 hour.
   (i) **Identify** a hypothesis for this investigation. (1 pt.)
   (ii) **Identify** the dependent variable for this investigation. (1 pt.)
   (iii) **Identify** an aspect of the students' investigation that should be kept constant. (1 pt.)
   (iv) **Describe** a modification the students could make to their investigation that would increase the reliability of their study. (1 pt.)

Module 23

Unit 4 | 19 | 20 | 21 | 22 | 23

Earth's Geography and Climate, El Niño, and La Niña

In the previous modules, we saw how the unequal heating of Earth together with the Coriolis effect drives air currents around the world in ways that determine global climate. However, this is only part of the story of how climates are determined. The oceans can carry massive amounts of heat, so in this module we will learn about the factors that drive oceans to circulate in particular ways and how cool or warm ocean water affect the climates of the continents. We will also discover how geographic factors play a role in affecting global climates.

Learning Goals

After reading this module you should be able to

23-1 describe how ocean circulation affects weather and global climates.

23-2 explain how mountains interrupt ocean airflow to cause rain shadows.

23-3 identify the causes and consequence of El Niño and La Niña events.

23-1 How does ocean circulation affect weather and global climates?

Ocean currents are driven by unequal heating, gravity, wind, salinity, and the location of continents

The flow of ocean water is an important factor for global climates because it moves warm and cold waters to different parts of the world. For example, the Gulf Stream is a current of warm water that emerges from the Gulf of Mexico. The warm waters emit heat into the air above it and this warmed air helps to warm much of the eastern coast of the United States. In doing so, ocean currents affect the primary productivity found in different ocean regions as well as the climate of the adjacent continents. When warm ocean currents flow from the tropics to higher latitudes, the neighboring continents receive warm air from the ocean. In this section we will examine the major ocean currents that occur near the oceans and then consider currents that mix the surface water with much deeper water.

Unequal Heating and Gravity

Ocean currents are driven by a combination of temperature, gravity, prevailing winds, the Coriolis effect, salinity, and the locations of continents. As we have already observed, the tropics receive the most direct sunlight throughout the year, which makes tropical surface waters warm. Warm water, like warm air, expands and rises. This process raises the tropical water surface about 8 cm (3 inches) higher than mid-latitude waters. While this increase in water height might seem small, it is sufficient for the force of gravity to make ocean water flow away from the equator. As we will see, this movement away from the equator due to gravity combines with other forces to make the surface waters circulate.

Wind and the Coriolis Effect

As the warm tropical waters are pushed away from the equator by gravity, prevailing winds around the globe also play a role in determining the direction in which ocean surface water moves. In the Northern Hemisphere, for example, the trade winds near the equator push water from the northeast to the southwest. However, the Coriolis effect, which we learned about in Module 22, deflects this wind-driven current so that water actually moves from east to west at the equator. Similarly, when winds in northern mid-latitude regions push water from the southwest to the northeast, the Coriolis effect deflects this current so that water actually

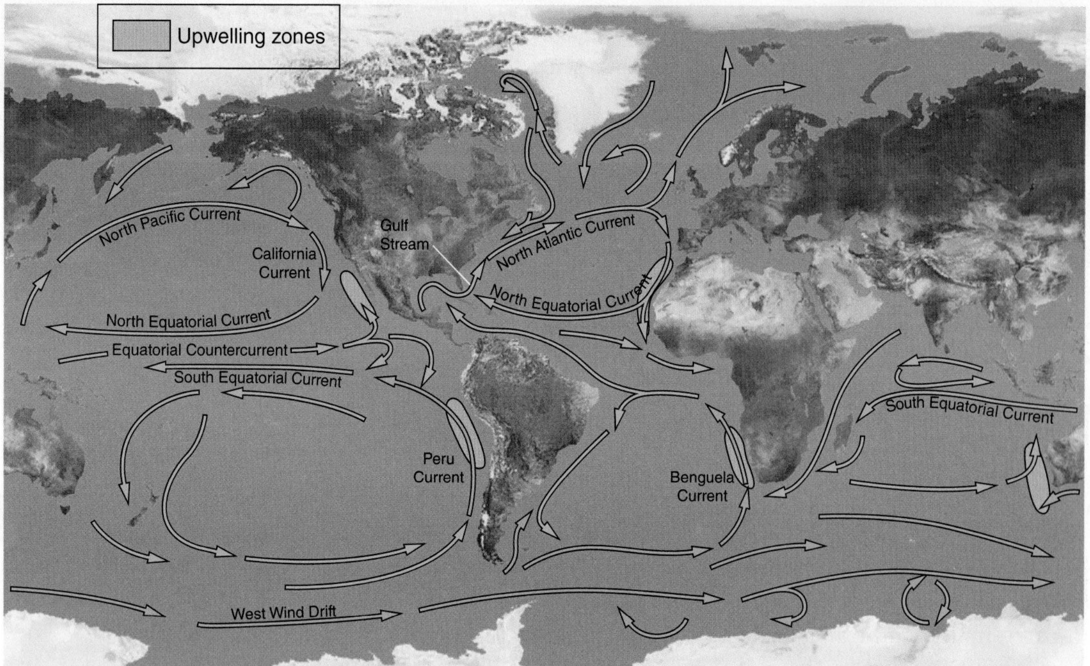

FIGURE 23.1 Oceanic circulation patterns. Oceanic circulation patterns are the result of differential heating, gravity, prevailing winds, the Coriolis effect, and the locations of continents. Each of the five major ocean basins contains a gyre driven by the trade winds in the tropics and the westerlies at mid-latitudes. The result is a clockwise circulation pattern in the Northern Hemisphere and a counterclockwise circulation pattern in the Southern Hemisphere. Along the west coasts of many continents, currents diverge and cause the upwelling of deeper and more fertile water.

moves from west to east. **FIGURE 23.1** shows the overall effect: Ocean surface currents between the equator and the mid-latitudes rotate in a clockwise direction in the Northern Hemisphere and in a counterclockwise direction in the Southern Hemisphere. These large-scale patterns of water circulation that move clockwise in the Northern Hemisphere and counterclockwise in the Southern Hemisphere are called **gyres**.

Gyres redistribute heat in the ocean, just as atmospheric convection currents redistribute heat in the atmosphere. Cold water from the polar regions moves along the west coasts of continents, and the cool air immediately above these waters brings cooler temperatures to the adjacent continents. For example, the California Current, which flows south from the North Pacific along the coast of California, causes coastal areas of California to have cooler temperatures than areas at similar latitudes on the east coast of the United States. Similarly, warm water from the tropics moves along the east coasts of continents, and the warm air immediately above these waters causes warmer temperatures on land.

Gyre A large-scale pattern of water circulation that moves clockwise in the Northern Hemisphere and counterclockwise in the Southern Hemisphere.

Upwelling The upward movement of ocean water toward the surface as a result of diverging currents.

Thermohaline circulation An oceanic circulation pattern that drives the mixing of surface water and deep water.

Upwelling

Ocean currents also help explain why some regions of the ocean support highly productive ecosystems. Along the west coasts of most continents, for example, the surface currents diverge, or separate from one another, causing deeper waters to rise and replace the water that has moved away. This upward movement of water toward the surface, shown as dark blue regions in Figure 23.1, is called **upwelling**. The deep waters bring nutrients from the ocean bottom that support large populations of producers. The producers, in turn, support large populations of fish that have long been important to commercial fisheries.

Deep Ocean Currents

Another oceanic circulation pattern, **thermohaline circulation**, drives the mixing of surface water and deep water. This process is crucial for moving heat and nutrients around the globe. Thermohaline circulation is driven by

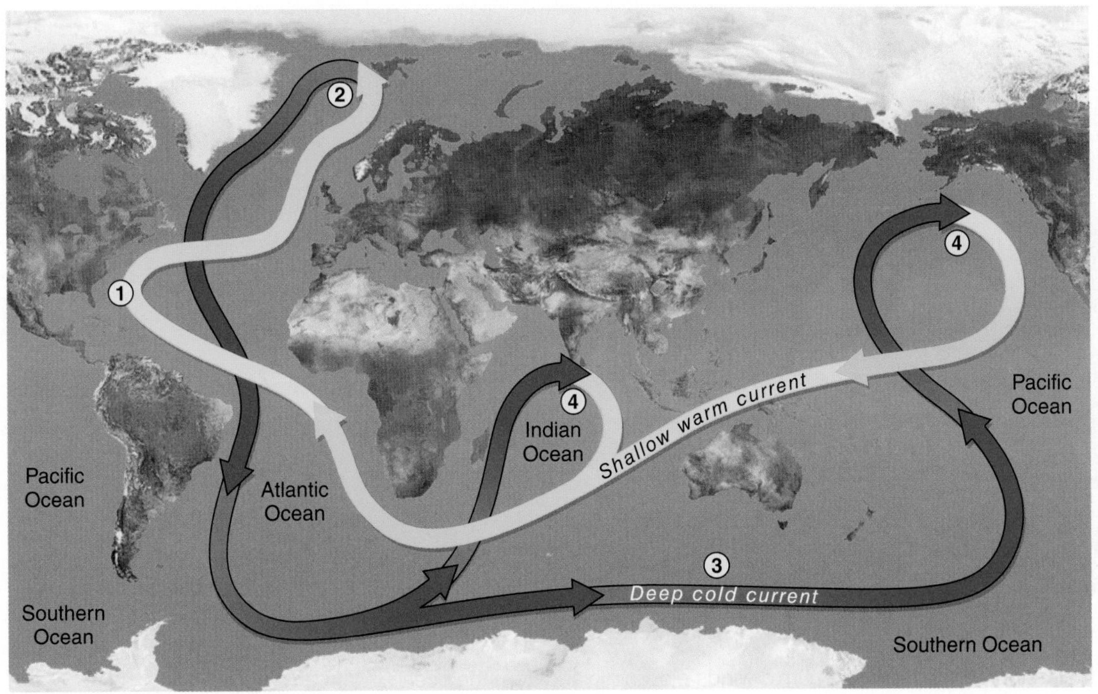

Pacific Ocean
Atlantic Ocean
Indian Ocean
Pacific Ocean
Southern Ocean
Southern Ocean
Shallow warm current
Deep cold current
1
2
3
4
4
1 Warm water flows from the Gulf of Mexico to the North Atlantic, where some of it freezes and evaporates.
2 The remaining water, now saltier and denser, sinks to the ocean bottom.
3 The cold water travels along the ocean floor, connecting the world's oceans.
4 The cold, deep water eventually rises to the surface and circulates back to the North Atlantic.

FIGURE 23.2 Thermohaline circulation. The sinking of dense, salty water in the North Atlantic drives a deep, cold current that moves slowly around the world.

surface waters that contain unusually large amounts of salt. As Figure 23.1 shows, warm currents flow from the Gulf of Mexico to the very cold North Atlantic. As the ocean water arrives in the North Atlantic, some of the water has evaporated, but the salt remains in the ocean. In addition, some of the ocean water freezes; given that ice is fresh water with no salt, the salt once again remains in the ocean water. Thus, evaporation and freezing both cause the remaining ocean water to have an increased salt concentration. This cold, salty water is relatively dense, so it sinks to the bottom of the ocean, mixing with deeper ocean waters. Two processes—the sinking of cold, salty water at high latitudes and the rising of warm water near the equator—create the movement necessary to drive a deep, cold current that slowly moves past Antarctica and northward to the northern Pacific Ocean, where it returns to the surface and then makes its way back to the Gulf of Mexico. This global round trip, shown in **FIGURE 23.2**, can take hundreds of years to complete. As you can see in the figure, thermohaline circulation helps to mix the water of all the oceans.

Ocean currents associated with thermohaline circulation affect the temperature of nearby landmasses, similar to how the major gyres affect surface waters. For example, the ocean current known as the Gulf Stream originates in the tropics near the Gulf of Mexico and flows toward western Europe. This movement of warm tropical waters brings warmer temperatures to latitudes that would otherwise be much colder. For instance, England's average winter temperature is 20°C (36°F) warmer than that of Newfoundland, Canada, which is located at a similar latitude. England receives warm currents from the Gulf Stream water while Newfoundland receives cold ocean currents from the North Atlantic.

Scientists are concerned that global warming could affect thermohaline circulation. If increased air temperatures accelerate the melting of glaciers in the Northern Hemisphere, the waters of the North Atlantic could become less salty and thus less likely to sink. Such a change could potentially shut down thermohaline circulation and stop the transport of warm water to western Europe, making it a much colder place.

23-2 How do mountains interrupt ocean airflow to cause rain shadows?

Rain shadows cause mountains to be dry on one side

Although many processes that affect weather and climate operate on a global scale, local features, such as mountain ranges, can also play a role. Air moving inland from the ocean often contains a large amount of water vapor. As shown in **FIGURE 23.3**, when this air meets the windward side of a mountain range—the side facing the wind—it rises and begins to experience adiabatic cooling, which we first learned about in Module 22. Because water vapor condenses as air cools, clouds form and precipitation falls. As is the case in Hadley cells that we discussed in the previous module, this condensation causes latent heat release, which helps to

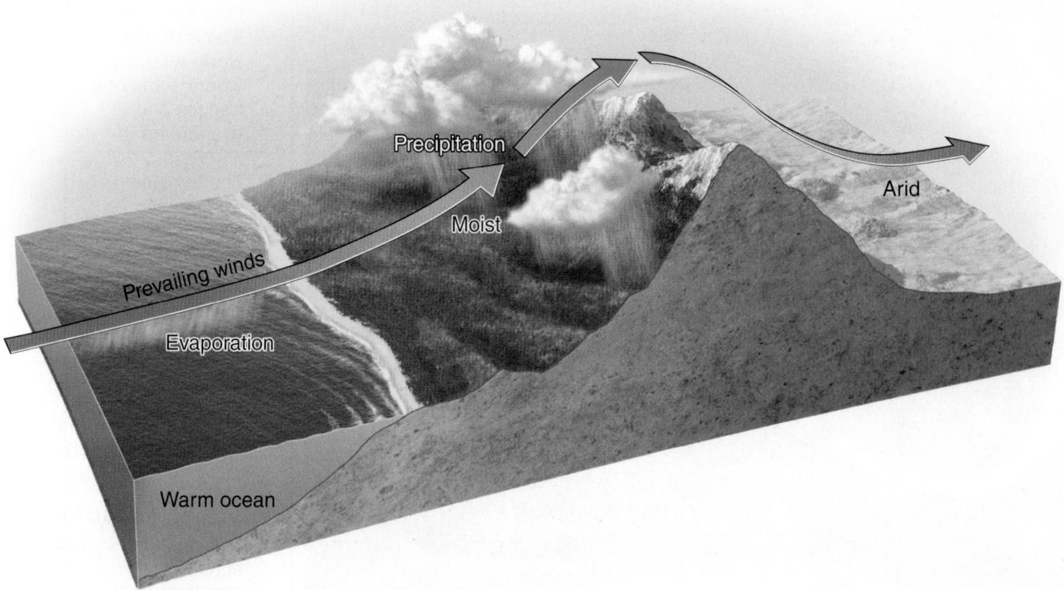

FIGURE 23.3 **Rain shadow.** Rain shadows occur where humid winds blowing inland from the ocean meet a mountain range. On the windward (wind-facing) side of the mountains, air rises and cools, and large amounts of water vapor condense to form clouds and precipitation. On the leeward side of the mountains, cold, dry air descends, warms via adiabatic heating, and causes much drier conditions.

accelerate the upward movement of the air. Thus, the presence of the mountain range causes large amounts of precipitation to fall on its windward side. The cold, dry air then travels over to the leeward side of the mountain range—the side not facing the wind—where it descends and experiences higher pressures, which cause adiabatic heating. This now warm, dry air produces arid conditions on the leeward side of the range. It is common to see lush vegetation growing on the windward side of a mountain range while very dry conditions prevail on the leeward side. The dry region formed on the leeward side of the mountain as a result of the humid winds from the ocean that cause precipitation on the windward side is known as a **rain shadow**. We can see an example of this along the western side of the United States, where the Sierra Nevada and Cascade mountain ranges impede the movement of warm ocean air, resulting in a rain shadow on the east side of the mountains where we find the Great Basin Desert and the Mojave Desert (FIGURE 23.4). Indeed, we can observe the rain shadow phenomenon all the way down the West Coast to southern California.

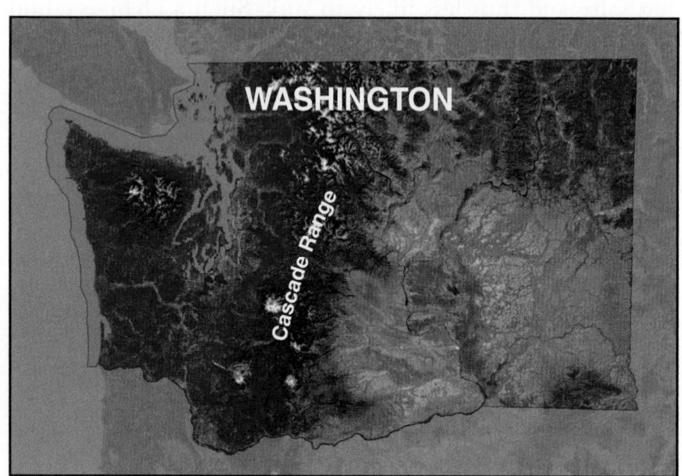

FIGURE 23.4 **Rain shadow in Washington State.** The Cascade Mountain range causes a rain shadow, with abundant precipitation on the western side of the mountains and much drier conditions on the eastern side of the mountains. This difference in precipitation results in temperate rainforests and boreal forests on the western side of the mountains, but temperate grasslands/cold deserts on the eastern side.

23-3 What are the causes and consequence of El Niño and La Niña events?

The El Niño–Southern Oscillation is caused by a shift in prevailing winds and ocean currents that alters global weather

Our discussion of ocean currents in the previous section focused on the ocean currents that exist in most years. In the southern Pacific Ocean, for example, trade winds from South America and the Coriolis effect cause the equatorial

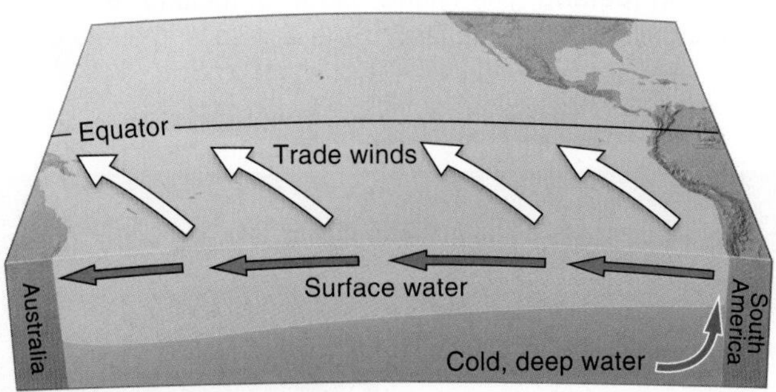

(a) Normal year

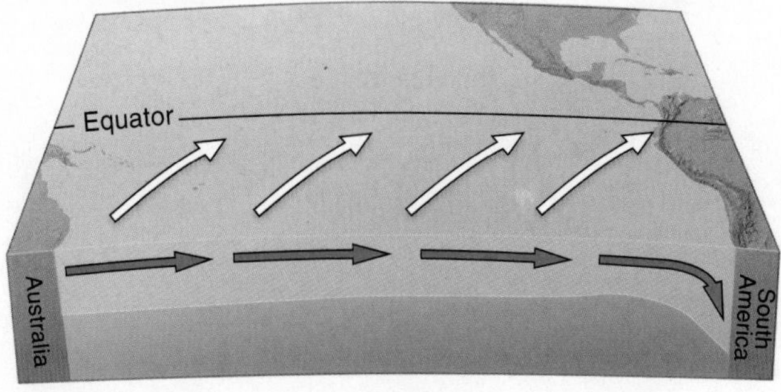

(b) El Niño year

FIGURE 23.5 The El Niño–Southern Oscillation. (a) In a normal year, trade winds push warm surface waters away from the coast of South America and promote the upwelling of water from the ocean bottom. (b) In an El Niño year, trade winds weaken or reverse direction, so warm waters build up along the west coast of Peru.

water to flow from east to west. As we have discussed, this current produces an upwelling of cold, nutrient-rich water along the west coast of South America, which produces large populations of fish.

Every 3 to 7 years, however, the tropical current moves in the opposite direction. Because this phenomenon often begins around the December 25 Christmas holiday, it has been named El Niño ("the baby boy"). However, since the cause of the phenomenon is due to a reversal of wind and water currents in the South Pacific, the preferred name for the phenomenon is the **El Niño–Southern Oscillation (ENSO)**.

FIGURE 23.5 shows this process in action. First, the trade winds that normally blow from South America and push the surface waters to the west weaken or reverse direction. This change in the wind allows warm equatorial water from the western Pacific to move eastward toward the west coast of South America. The movement of warm water and air toward South America suppresses upwelling off the coast of Peru. The effect can last from a few weeks to a few years.

Because ocean currents are so important to global climates, the ENSO has several consequences on global climates. As we have noted, it reduces the upwelling off the South American coast, which decreases productivity and dramatically reduces fish populations. It also has widespread effects around the world, including cooler and wetter conditions in the southeastern United States and unusually dry weather in the northern United States, Canada, southern Africa, and Southeast Asia. Such changes can have major impacts on crop production, with some ENSO events leading to poor crop yields and tens of thousands of people dying in developing countries. The ENSO event that occurred during 2015–2016 was the most severe in 30 years. Scientists estimate that nearly 100 million people experienced food shortages as a result.

After an El Niño event occurs, the trade winds that are caused will reverse back to their more typical direction, moving to the northwest from South America. However, the reversal in trade winds is stronger than usual, which is a phenomenon known as **La Niña**. As a result, regions that became hotter and dryer during an El Niño year are cooler and wetter during a La Niña year. Once we experience the oscillation of an El Niño year followed by a La Niña year, global climates become more typical for several years.

AP® Exam Tip

It will be helpful to learn how El Niño impacts your home location, but you won't be expected to know how El Niño impacts every part of the world. You will need to know that El Niño is the result of warm surface waters piling up on the Pacific Coast of Central and South America as a prevailing wind change. You will not need to explain why the wind patterns change in this area.

In this module we learned that global climates are not only determined by the Sun's energy or by air circulation but also by ocean circulation because the oceans are carrying vast amounts of heat. Ocean circulation is determined by several factors, including the locations of continents, and the impacts on continental climates are also affected by the rain shadows of mountain ranges. We have now reached the end of our unit on earth systems, and we bring many of the concepts together in the Unit 4 Visual Representation, which features a close look at the island nation of Iceland (pages 276–277).

Rain shadow A region with dry conditions found on the leeward side of a mountain range as a result of humid winds from the ocean causing precipitation on the windward side.

El Niño–Southern Oscillation (ENSO) A reversal of wind and water currents in the South Pacific.

La Niña Following an El Niño event, trade winds in the South Pacific reverse strongly, causing regions that were hot and dry to become cooler and wetter.

Preparing for the AP® Exam

Visual Representation 4 **Earth Systems in Iceland** The Island nation of Iceland reflects many features of Earth systems such as how plate tectonics shapes the landscape and soils, the role of ocean

Watersheds:
Iceland is a land of glaciers, rivers, and lakes, which means it is a land of watersheds. Lake Thingvallavatn in the southwest is Iceland's largest natural lake. It is located in a rift valley where the movement of tectonic plates has created a deep crack in the land that filled with water to create the lake. The lake is fed by a glacier 40 miles away through numerous underground channels. It can take 20–30 years for water from the glacier to reach the lake. Other sources reach the lake in a few months. Water flowing over igneous rocks and through Earth's mantle picks up nutrients which enrich the area surrounding the lake.

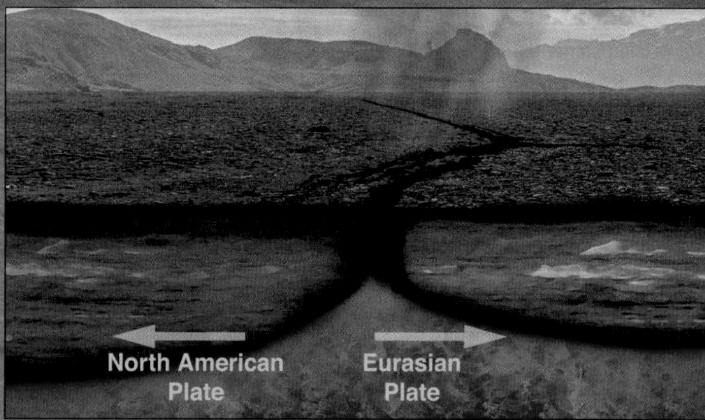

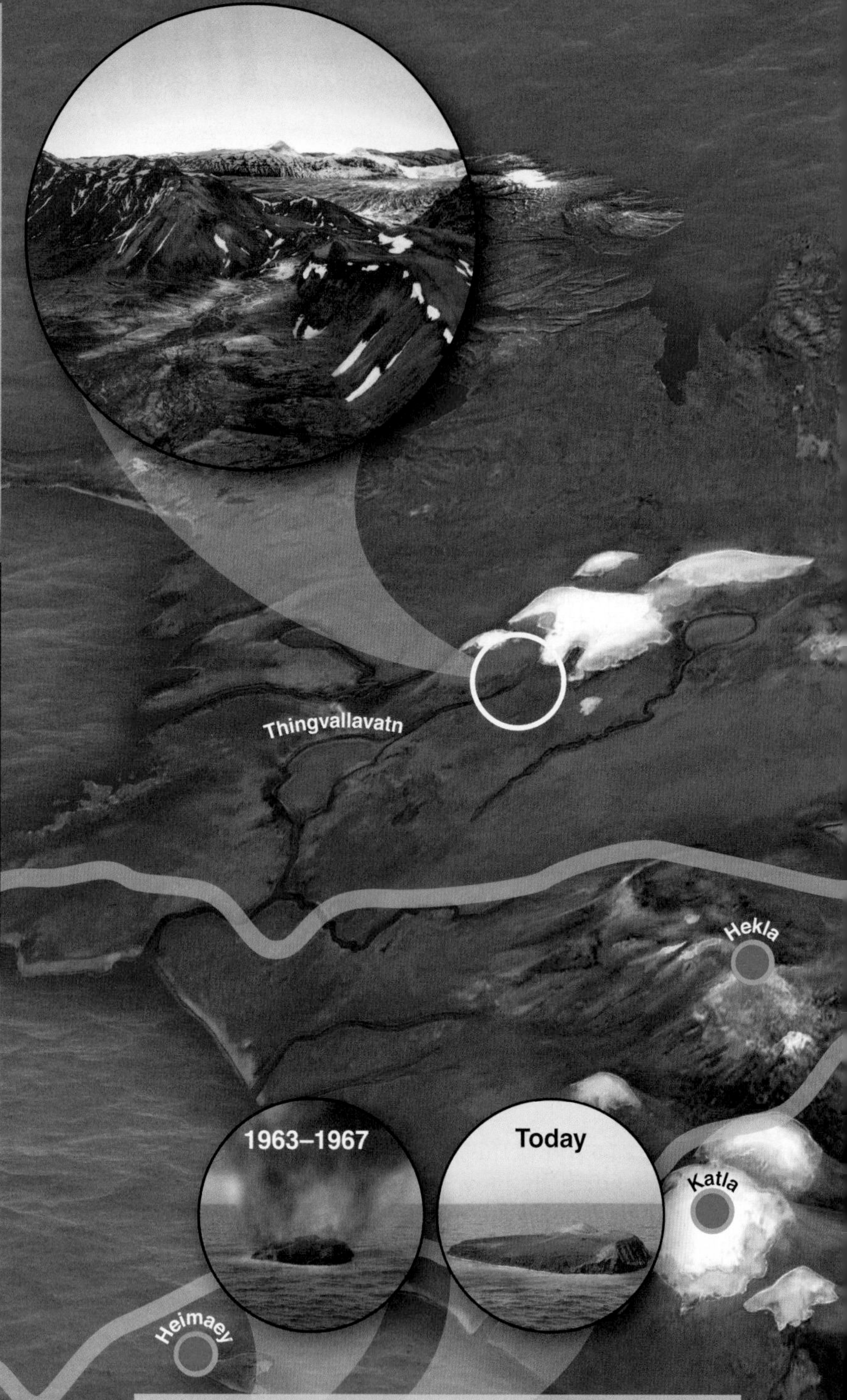

Plate tectonics:
This geographically young island sits on top of the mid-Atlantic ridge, which divides the North American and Eurasian tectonic plates. The two divergent tectonic plates are moving apart at about 2 cm per year, creating fissures. In some locations you can walk between the two plates.

Large active volcanoes

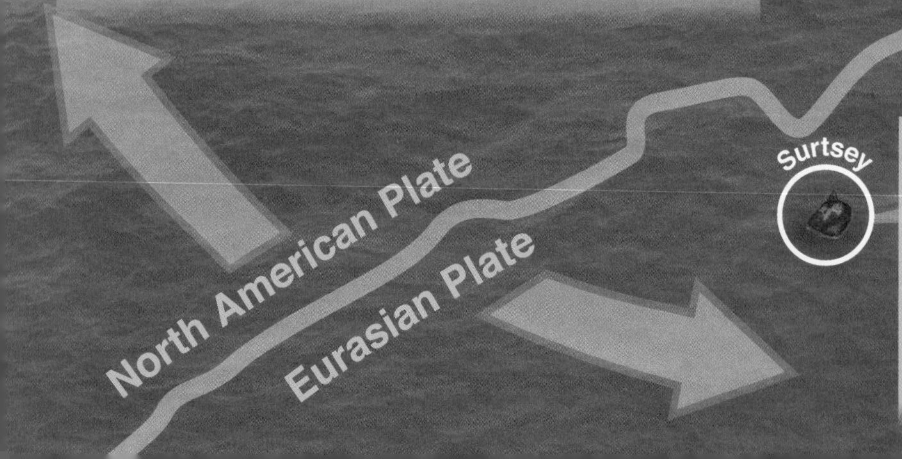

Thingvallavatn

Hekla

1963–1967

Today

Katla

Heimaey

Surtsey

North American Plate

Eurasian Plate

Divergent tectonic plates: Given that Iceland is located on top of two diverging tectonic plates, it is a geologically active island. The deep rift valleys have magma rising up, causing earthquakes and creating nearly 200 volcanoes on the island. From 1963 to 1967 volcanoes off the southern coast created the new island of Surtsey.

currents in climate, and the interconnectedness of water resources. Its northern location demonstrates the role of solar radiation on the seasons and climate. All these factors affect the ways in which Earth's resources support life.

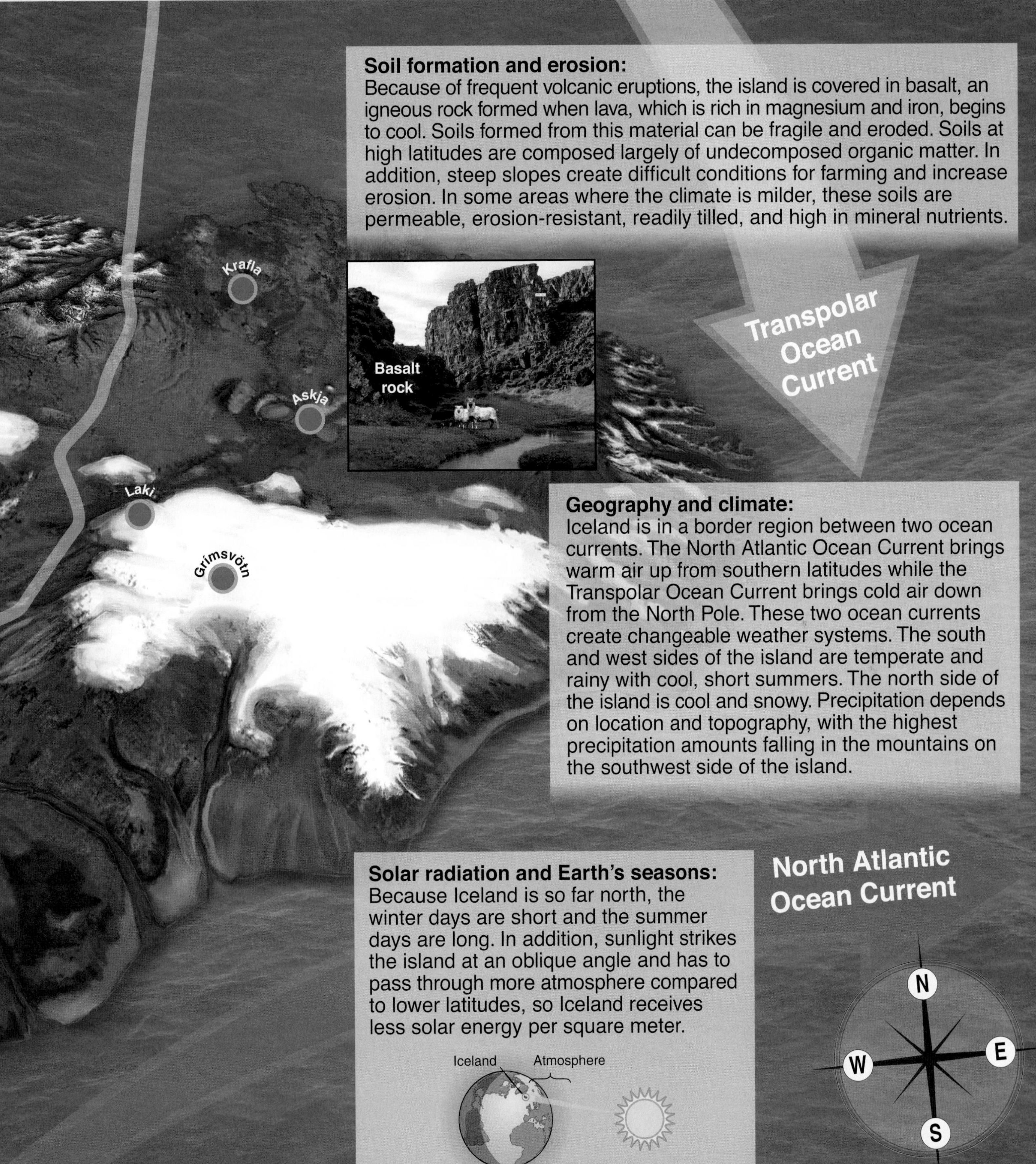

Soil formation and erosion:
Because of frequent volcanic eruptions, the island is covered in basalt, an igneous rock formed when lava, which is rich in magnesium and iron, begins to cool. Soils formed from this material can be fragile and eroded. Soils at high latitudes are composed largely of undecomposed organic matter. In addition, steep slopes create difficult conditions for farming and increase erosion. In some areas where the climate is milder, these soils are permeable, erosion-resistant, readily tilled, and high in mineral nutrients.
Krafla
Askja
Laki
Grímsvötn
Basalt rock
Transpolar Ocean Current
Geography and climate:
Iceland is in a border region between two ocean currents. The North Atlantic Ocean Current brings warm air up from southern latitudes while the Transpolar Ocean Current brings cold air down from the North Pole. These two ocean currents create changeable weather systems. The south and west sides of the island are temperate and rainy with cool, short summers. The north side of the island is cool and snowy. Precipitation depends on location and topography, with the highest precipitation amounts falling in the mountains on the southwest side of the island.
Solar radiation and Earth's seasons:
Because Iceland is so far north, the winter days are short and the summer days are long. In addition, sunlight strikes the island at an oblique angle and has to pass through more atmosphere compared to lower latitudes, so Iceland receives less solar energy per square meter.
Iceland
Atmosphere
North Atlantic Ocean Current
N
E
W
S

Module 23 AP® Review

Preparing for the AP® Exam

Learning Goals Revisited

23-1 How does ocean circulation affect weather and global climates?

Ocean circulation is driven by the warmer surface water at the equator expanding and flowing away from the equator. Winds and the Coriolis effect deflect the directions on the surface water flow and the presence of continents causes gyres to form, thereby distributing warm tropical water to regions far from the equator and altering the climates. Thermohaline circulation is a much slower movement of water that is driven by a combination of heat and salinity.

23-2 How do mountains interrupt ocean airflow to cause rain shadows?

Similar to the air circulation that occurs with Hadley cells, rain shadows occur when warm, moist air from the ocean encounters a mountain range. As the air rises up the mountain, the air cools and loses its moisture as precipitation. On the other side of the mountain, the dry air descends, causing dry climates on that side of the mountain.

23-3 What are the causes and consequence of El Niño and La Niña events?

El Niño events occur every 3 to 7 years when the trade winds that blow northwest from South America temporarily change direction. Because ocean circulation affects climates on continents, the change in ocean circulation dramatically alters climates around the world, leading to hot, dry conditions in some regions of the world and wet, cool conditions in other regions.

Practice Math and Graphing

Preparing for the AP® Exam

1. Practice Math

(a) Earth formed 4.6 billion years ago. Express this value in scientific notation.

(b) There is evidence of life on Earth in the fossil record as far back as approximately 3.8 billion years ago. How many years did the Earth exist before evidence of life appears in the fossil record? Express this in scientific notation.

2. Practice Graphing

Near Seattle, Washington, the predominant air flow over Mount Olympus is from the southwest to the northeast. Researchers have measured annual precipitation in different locations, along an approximate line from the southwest side of the mountain to the northeast side of the mountain.

(a) Create a line graph that plots the precipitation data versus the distance from the ocean.

Location	Distance from ocean from the west/ southwest (km)	Annual precipitation (cm)
Forks	19	302
Quinault	33	349
Mount Olympus	59	559
Elwha	69	143
Port Angeles	82	65
Sequim	100	42

(b) Based on your graph, which locations are in the rain shadow of Mount Olympus?

(c) What causes this rain shadow to occur?

AP® Practice Questions

Multiple-Choice Questions

Use the passage below to answer questions 1 & 2:

Ocean circulation is driven by the warmer surface water at the equator expanding and flowing away from the equator. Winds and the Coriolis effect deflect the direction of surface water flow, and the presence of continents causes gyres to form, thereby distributing warm tropical water to regions far from the equator and altering the climates. Warm surface currents such as the Gulf Stream contribute to a more temperate climate in areas that would otherwise be colder based on their latitude. For example, the United Kingdom is located at a latitude similar to that of Newfoundland in Canada, though its climate is more temperate.

1. Which of the following best describes a factor that drives ocean currents?
   (a) Salinity drives ocean currents, as saltier water rises to the surface.
   (b) Surface winds drive ocean currents, as all wind blows away from the equator.
   (c) Temperature drives ocean currents, as warm water expands in volume.
   (d) Precipitation drives ocean currents, as more water volume is added to the ocean each day.

2. Which of the following did the author use as evidence to support the claim that ocean currents affect global climate?
   (a) The Gulf Stream is a cold-water current that brings colder temperatures to western Europe.
   (b) Warm tropical waters are moved away from the equator, altering climates at higher latitudes.
   (c) The United Kingdom has colder temperatures than that of Newfoundland in Canada.
   (d) Cold, salty water sinks, which causes different climate conditions around the globe.

3. High productivity and nutrient availability in the ocean occur in areas with
   (a) gyres.
   (b) upwelling.
   (c) cold waters.
   (d) prevailing westerlies.

4. Which of the following statements correctly describes an El Niño–Southern Oscillation (ENSO) event?
   (a) It causes increased upwelling on the west coast of South America.
   (b) It causes wetter conditions in the southern United States.
   (c) It includes a reversal in polar ocean currents.
   (d) It causes wetter conditions in Australia.

5. Which of the following statements about gyres is correct?
   (a) Gyres redistribute heat in the ocean.
   (b) Gyres cause the Coriolis effect.
   (c) Gyres redistribute nutrients from the deep ocean.
   (d) Gyres decrease the temperatures of equatorial areas.

6. Which of the following plays a role in causing a rain shadow?
   (a) cold polar air
   (b) a flat grassland
   (c) the Coriolis effect
   (d) humid ocean air

Free-Response Question

As greenhouse gases continue to warm the planet slowly, the glaciers of Greenland are melting at a rapid rate. Scientists are concerned that this melting may dilute the salt water in that region of the ocean enough to shut down thermohaline circulation.

(a) **Identify** a mechanism that drives the thermohaline circulation. (1 pt.)
(b) **Describe** how the mechanism identified in part (a) affects the thermohaline circulation. (1 pt.)
(c) **Explain** how a shutdown of the thermohaline circulation would affect temperatures in western Europe. (1 pt.)
(d) A potential solution to combat the progression of global warming is the reduction of CO_2 emissions. **Justify** this solution by describing one potential advantage of lowering CO_2 emissions (other than a reduction in melting sea ice). (1 pt.)
(e) One of the strongest El Niño events took place in 1997–1998, where Peruvian fisheries' anchovy catches went from 6 million metric tons in 1997 to 1 million metric tons in 1998.
   (i) **Calculate** the percent decrease in anchovies caught in Peru from 1997 to 1998. Show all work. (2 pts.)
   (ii) In 2001, the Peruvian fishery exports were worth \$1.1 billion. If the anchovy catch was approximately 6.5 million metric tons, **calculate** the value per metric ton in 2001. Show all work. (2 pts.)
   (iii) In the same year, the value of fishery exports in the United States was \$3.3 billion, and estimated to be worth \$2,500 per metric ton. **Calculate** how many metric tons of fish were exported by the United States in 2001. Show all work. (2 pts.)

Preparing for the AP® Exam

UNIT 4
Practice Your Science Skills

Data Analysis

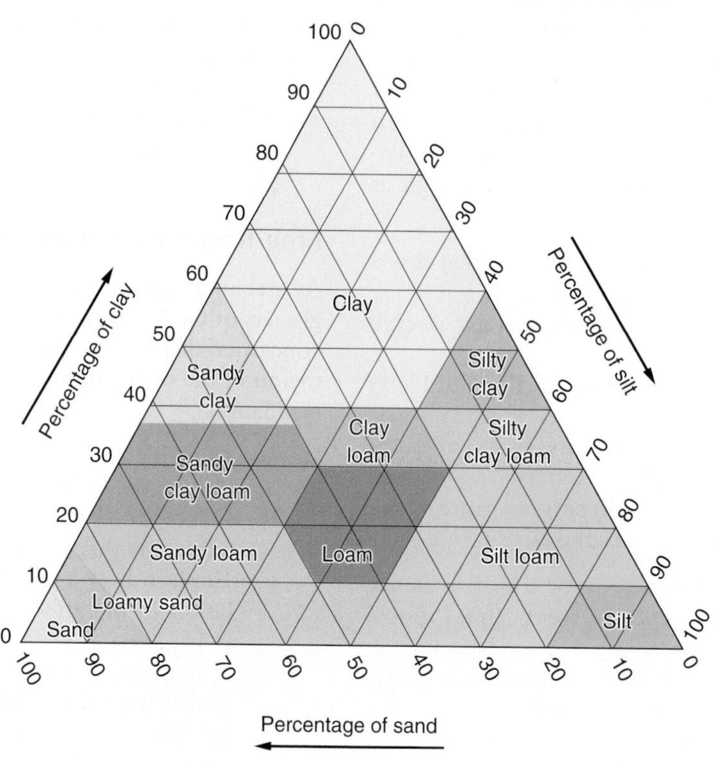

A student in the lab uses the mason jar method to determine the percentage of sand, silt, and clay in five soil samples that were provided by their teacher. The student records the data shown in the table below:

Sample #	% Sand	% Silt	% Clay
1	50	40	10
2	20	10	70
3	80	10	10
4	15	55	30
5	50	10	40

Questions

1. **Identify** the soil sample that is sandy clay loam.
2. **Identify** the soil sample that is silty clay loam.
3. **Identify** which soil sample is the best for growing crops.
4. **Identify** which soil sample is the best for making a bowl and **explain** why.
5. **Identify** which soil sample would drain the quickest and **explain** why.

Pursuing Environmental Solutions

Sequestering Carbon to Reduce Global Warming

As we have seen in the modules of this unit, the patterns of circulation in the atmosphere and the oceans are driven by the differential heating of Earth, with warmer temperatures near the tropics than the poles. As we have also learned, an important reason for this heating is the natural presence of greenhouse gases, which only comprise a small fraction of the gases in the atmosphere but play a key role in absorbing infrared energy and re-emitting it. Much of the re-emitted energy comes back to Earth, causing a much warmer planet than would exist without the natural levels of greenhouse gases. However, a problem arises when human activities produce additional amounts of greenhouse gases, which are causing the planet to warm to temperatures that are higher than existed before the Industrial Revolution. While we can work to emit fewer greenhouse gases, we can also think about how we might remove excess greenhouse gases that are already in the atmosphere.

Research for the capture and long-term storage of CO_2 from the atmosphere is a process known as **carbon sequestration**. The approaches to carbon sequestration include using plants to take up more CO_2 and depositing it in the soil, converting atmospheric CO_2 into a liquified form and pumping it deep underground, and converting captured CO_2 into products that people need.

The biological approach includes managing farms, grasslands, and forests to collect and store larger amounts of CO_2 from the atmosphere. In all of these cases, we have to think about the rates at which carbon is collected and then enters the soil in the form of organic matter, versus the rate at which the organic matter decomposes and releases CO_2 back to the atmosphere. To achieve a net storage of carbon, we need to collect and store carbon in the soil faster than it decomposes. For example, we can consider restoring forests to some areas since trees can store a large amount of carbon. In addition, most agricultural soils that have grown and removed crops from the fields for many years currently have a reduced amount of soil carbon compared to adjacent non-farmed areas. Thus, farmed soils have the potential to store much more carbon. Moreover, having more organic matter in the soil has the additional benefit of increasing the nutrients in the soil and improving plant productivity.

Winter cover crop. Planting a cover crop in the fall that can be cut down in the spring when crops are planted is an effective way to increase the amount of carbon in the soil. This farmer in Albion, Maine, is cutting down a cover crop of winter rye. *(Courtesy of Johnny's Selected Seeds, Johnnyseeds.com)*

We can achieve net carbon storage in agricultural soils through a number of management strategies. For example, we can leave more unused crop residues on the field, such as unused stems and leaves that are not consumed by humans. We can also add manure to field or plant cover crops, which are grown when the fields are not growing crops. For example, some farmers plant rye grass on their fields in the fall and the tall grass is plowed into the soil when the farmer is ready to plant a new crop in the spring. Farmers can also choose not to till the soil in the spring when planting a new crop, which slows the rates of decomposition of organic matter. We can also consider planting perennial crops instead of annual crops, which reduces the need to plow the soil and increases decomposition rates. As the idea of carbon sequestration has caught on, corporations such as Microsoft and General Mills, as well as governments and philanthropists, have invested in farms that manage for increases in soil carbon storage. This has also stimulated a great deal of soil research to determine how long the extra carbon is stored, the limit to how much carbon various soils can accumulate, and whether the extra soil carbon stored in the top meter of soil might be offset by an increased release

Carbon sequestration The capture and long-term storage of carbon dioxide from the atmosphere.

(a)

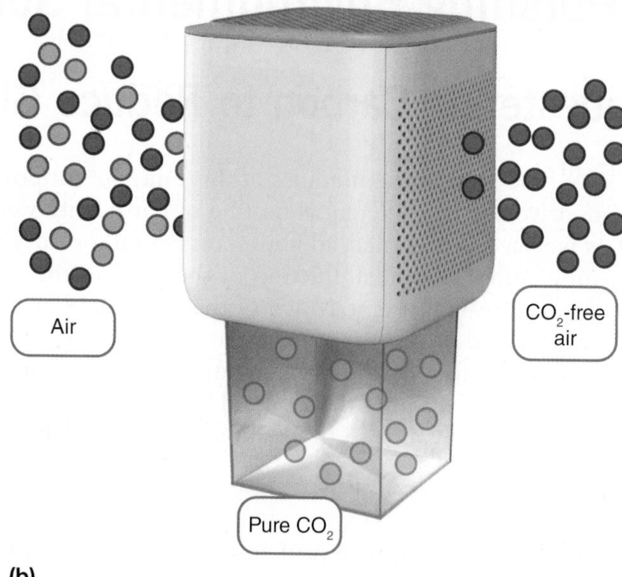

(b)

Removing CO_2 from the air. (a) New technologies are being developed to capture and store CO_2 from the atmosphere. (b) Using these technologies, air passes through a filter that captures CO_2 and concentrates it. The concentrated CO_2 can be sold for industrial applications of converted into a liquid form and injected deep into the ground. *(a: Meinrad Schade/laif/Redux)*

of carbon from the depths of 1 to 2 m. We will know much more about these research questions in the coming decade.

The second approach to reducing CO_2 in the atmosphere is by developing new technologies to pull CO_2 out of the air. They do this by passing the air through a chemical solution or binding it to a filter. This carbon can then be converted into a liquid form and pumped deep into the ground with the expectation that it will remain stored there for hundreds or thousands of years. While current technologies continue to improve, these systems are only able to remove a relatively small amount of CO_2 from the atmosphere. As more governments require reduced net CO_2 emissions, investment in these technologies should grow, as will their capabilities for removing CO_2.

The third approach to reducing CO_2 in the atmosphere is to collect carbon and use it in industrial and manufacturing processes. For example, carbon could be used to make building materials, fuels, and plastics. These processes are in the early stage and the likelihood of widespread use will depend on both the cost of production as well as the performance of the products made from CO_2 versus those made from conventional materials.

When it comes to reducing CO_2 in the atmosphere, experts agree that we need to embrace all the above approaches because no single approach will make a large enough impact by itself. The good news is that we have a diversity of options and some, such as improved soil management, can be readily used across the world on a very large scale.

Critical Thinking Questions

1. How might global warming affect the decomposition rate of organic matter in the soil and how might this cause a feedback loop on global warming?
2. In addition to soil decomposition releasing CO_2, what other farm activities related to crop production also contribute to atmospheric CO_2?

References

Hepburn, C., et al. 2019. The technological and economic prospects for CO_2 utilization and removal. *Nature* 575:87–97.

Paustian, K., E. Larson, J. Kent, E. Marx, and A. Swan. 2019. Soil C sequestration as a biological negative emission strategy. *Frontiers in Climate* 1:8 (doi: m10.3389/fclim.2019.00008)

Popkin, G. 2020. Can "carbon smart" farming play a key role in the climate fight? *Yale Environment* 360, March 31.

Science Applied 4: Concept Explanation

Can We Resolve the California Water Wars?

Pick up any tomato, broccoli spear, or carrot in the grocery store and chances are it came from California. California's $50 billion agricultural industry is more than twice the size of the agricultural industry of any other state, and it accounts for about 40 percent of all the fruits, nuts, and vegetables grown in the United States (**FIGURE SA4.1**). As a result of the combined effects of atmospheric circulation, ocean circulation, and rain shadow effects, California's temperatures are ideal for many crops, and its mild climate permits a long growing season. However, water is a scarce resource and most of California's growing regions lack the water that is necessary for farming. To remedy this situation, the state has developed an extensive irrigation system comprised of pipes and canals, collectively known as **aqueducts**, that move water from watersheds where it is abundant to areas where it is scarce, as you can see in **FIGURE SA4.2**.

FIGURE SA4.1 Water for agriculture. In California, large amounts of water are moved for the irrigation of agriculture to produce nearly one-half of all the fruits, nuts, and vegetables grown in the United States. *(ullstein bild/Getty Images)*

Aqueducts Pipes and canals that move water from where it is abundant to areas where it is scarce.

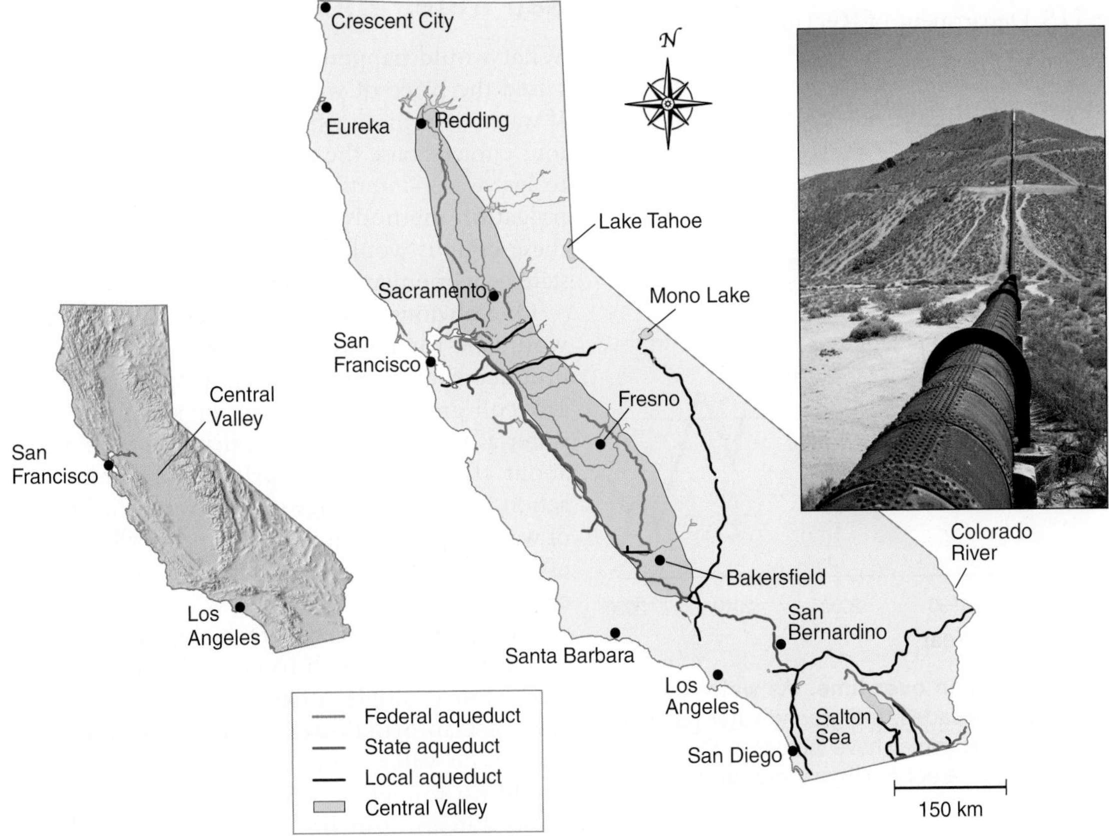

FIGURE SA4.2 Water distribution in California. To supply water to the agricultural areas in the Central Valley and to supply drinking water to municipalities, the state has developed an elaborate system of aqueducts, which consist of both open canals and pipes laid above or below the ground. *(Photo: Jim West/Alamy Stock Photo)*

Although agriculture in California has been economically successful, it has increased competition among the state's residents for water. Nearly two-thirds of California's population lives in the southern part of the state, where residential and business demand for water has long exceeded the locally available supply. Los Angeles, for example, has been importing water from other areas of the state for more than 100 years. Demand for water has continued to grow with the state's population, which has been increasing by about 1 percent per year.

Wildlife also depends on a readily available supply of fresh water. In many regions of California, diverting water away from streams and rivers threatens the habitats of fish and other aquatic organisms, including several endangered species. Salmon, for instance, require cold, clean water to lay their eggs. In the delta region where the San Joaquin and Sacramento rivers come together just east of San Francisco, so much water has been pumped out for use in the southern part of the state that the salmon in the region's rivers have experienced dramatic declines.

What historical factors have led to California's water wars?

At 144 trillion liters (38 trillion gallons) of fresh water per day, California's water use is more than twice that of any other state. The vast majority of that water is used in the Central Valley region, which is located between the Coast Range Mountains and the Sierra Nevada Mountains. The Central Valley produces more than 80 percent of California's farm income. In 1935, the U.S. Department of Reclamation began an immense endeavor, called the Central Valley Project, to divert 8.6 trillion liters (2.3 trillion gallons) of water per year from rivers and lakes throughout the state for both agricultural and municipal uses.

While the project made agriculture possible in many of southern California's desert regions, it also locked into place long-term water-use contracts that have subsidized several decades of cheap water for farmers, who pay rates that are only 5 to 10 percent of those paid by southern California towns and cities. Interestingly, northern California also has farms, but much of its abundant water is sent to the farms and municipalities of southern California. In addition to diverting stream and lake water, farmers also pump a good deal of groundwater. Over the years, this pumping of groundwater has exceeded the amount of recharged groundwater, particularly during droughts, as you can see in **FIGURE SA4.3**.

In 2005, the federal government renewed nearly 200 large water supply contracts with farmers. These contracts were for 25 years, with an option available to extend them another 25 years, at prices so low that the payments did not even cover the cost of the electricity used to transport the water. Organizations ranging from the Cato Institute to the Natural Resources Defense Council protested this decision. The controversy raised the question of whether farmers should continue to be given cheap access to water to protect the existing agricultural industry or be required to pay the same price for water as everyone else.

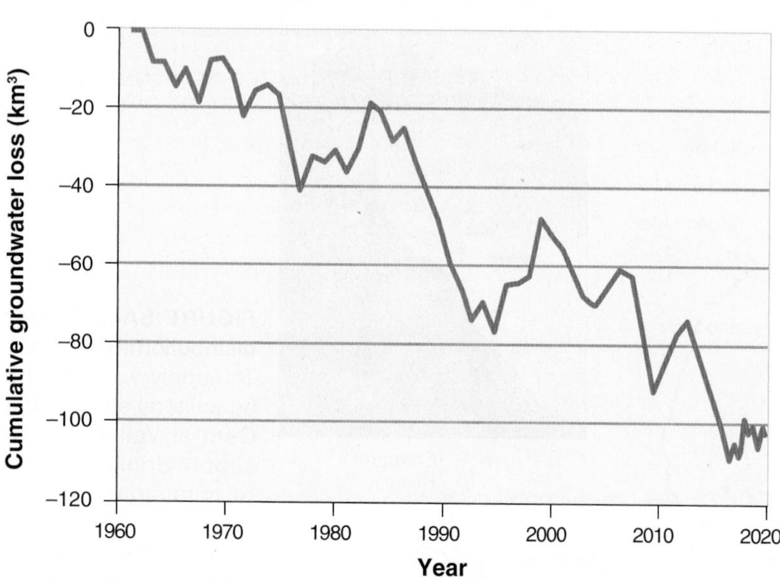

FIGURE SA4.3 Groundwater depletion over time. As water demand has grown in California over several decades, the groundwater has been depleted. Brief periods of above-average rainfall have increased the amount of groundwater, but not enough to counteract multi-year droughts. *(Data from USGS Professional Paper 1766.)*

How might farmers respond to increased water prices?

What would happen if the federal government raised the price of water for California farmers? If water were more expensive, farmers would face four choices: raise the prices of their crops, switch to less water-intensive crops, use more efficient irrigation methods, or sell their farms. Each of these options would have different effects on the state's environment and economy.

One obvious way to offset higher water prices would be for farmers to charge more for the produce they grow. For example, if California broccoli growers had to pay the same price for water as municipalities paid, that cost would add about 10 percent to the price of broccoli. This action would not change the environmental threat of water depletion but would have a small negative effect on the demand for broccoli and on the consumer's wallet.

Other farmers faced with higher water prices might decide to switch to crops that require less water. For example, alfalfa, which uses 25 percent of California's irrigation water and has a low resale value compared with other crops, would no longer be a profitable crop. A farm using 296 million liters (78 million gallons)

FIGURE SA4.4 Irrigating celery. Crops such as potatoes, tomatoes, and celery use less water and have a higher resale value than crops such as alfalfa. Growing such low-water crops would reduce the water needed for agriculture in California. *(Pgiam/Getty Images)*

of water can produce about $60,000 worth of alfalfa per year, according to the Natural Resources Defense Council. If farmers had to purchase water at the unsubsidized rate, that water would cost them $168,000; if the alfalfa continued to sell for only $60,000, the farmers would lose more than $100,000 a year growing alfalfa. Tomatoes, celery, and potatoes use about half as much water per acre as alfalfa and have a higher resale value (**FIGURE SA4.4**). Just by replacing alfalfa with these crops, California would reduce its statewide water use by 20 percent. This action would have a positive environmental impact, since less overall water would be needed for agriculture in the state.

Investment in new irrigation technology that uses less water would also reduce a farmer's costs. If farmers had to purchase water at the unsubsidized rate, converting to more-efficient irrigation techniques could rapidly pay for itself since the initial investment in equipment would be offset by savings on water. As long as the cost of water continues to be subsidized by the government, farmers have little incentive to upgrade their irrigation technology.

If water prices rise, not all farmers will be able to afford the new costs. This recognition brings us to the final option: Some farmers could be driven out of business and forced to sell their land. This outcome would probably reduce the amount of food grown in the region. Whether conversion from farming to other uses would reduce water demand would depend on how the land was used. For example, if it became a housing development in which residents practiced water conservation, then the residents would use much less water per acre than the amount used for agriculture. However, if the residents chose to plant and water large lawns and were not careful about their use of water indoors, they could use even more water than was used for agriculture.

Because charging more for water could force some farmers out of business, many politicians are reluctant to support this change. In response, some economists argue that farmers who possess long-term subsidized water contracts should be allowed to sell their water allotment to the highest bidder, something that is currently happening with some farms in California. In this scenario, a farmer can buy water from the government at the contracted low cost and then sell that water to a city at a substantial profit rather than using it on the farm. Many critics argue, however, that the government should not be giving preferential water rates to a few users who can then make a profit on the water.

The motivation provided by drought

The competition for water has become a higher priority among California residents due to a major drought. During the past two decades, California has experienced the worst drought ever recorded for the state, and the state has had no choice but to respond in substantial ways. In 2015, nearly 200,000 hectares (500,000 acres) of farm fields were allowed to go fallow (i.e., not planted with a crop) to reduce the amount of water used. A large portion of this fallow land were rice fields that need to be flooded with water. In addition, farmers spent nearly $300 million per year to pump water out of wells because the aqueducts were not delivering enough. Because the wells for large farms can be much deeper than wells for people's homes, pumping water for farms during a drought can draw down the water in the ground to a level that is below the homeowner's well, which leaves homeowners without water.

Many farmers also survived by planting crops that are better suited to the local climate and require less water. They also added water-saving devices where feasible. Farmers growing water-intensive crops such as hay and rice were hard hit. However, overall farm revenues increased during this time because the shortage of produce caused prices to rise. Those farmers who were able to continue crop production by pumping water to irrigate crops did well financially. Many scientists have expressed concern that the short-term gain from pumping groundwater might cause future problems since the groundwater may be slow to recharge.

By 2021, it became clear that the effort to offset the effects of repeated droughts could not be offset by pumping water from wells throughout the region. As well pumping increased, the amount of water in the ground continued to decline, causing more wells to go dry and leaving more farms and residents without water. Current estimates are that nearly 400,000 ha (1 million acres) of farmland in the Central Valley may have to stop growing food to make water use more sustainable. However, the land can be used for other purposes, including large solar farms to generate electricity for the region.

How have the water wars played out in the political arena?

The California water wars have developed because of the large number of competing interests involved, including northern California farmers and municipalities that are losing water, southern California farmers and municipalities that need more water, endangered species that are declining due to the diversion of water, and beautiful natural areas that have experienced reduced water capacity, seen increased salinity, or gone completely dry.

Competition for California's water has been marked by political maneuvering and numerous lawsuits. In the 1960s, for example, there was a large push to build dams that would create reservoirs to collect water. Today there are about 1,200 dams in the state. Because of the money and time needed to construct new dams and the environmental impact of dams, many Californians do not consider more dams to be a viable solution to today's water problems.

Over the past two decades there have been lawsuits that produced very interesting court decisions. For example, the Natural Resources Defense Council won a lawsuit regarding the 2005 water contracts for farmers in the eastern San Joaquin Valley, who use water diverted from the San Joaquin River. The judge ruled that the contracts were illegal because they did not consider how diverting the water might affect endangered species that rely on the river.

In 2007, a federal court was asked to consider how the large, powerful pumps used to divert water from the San Joaquin River and Sacramento delta region affected local fish populations—particularly the delta smelt (*Hypomesus transpacificus*), a federally protected fish whose population in the delta region was declining. The court ruled that pumping from the delta region—the source of much of the state's fresh water—must be reduced. Because this decision came during a multi-year drought, its impact on southern California was even larger than it otherwise would have been. A number of California congressmen accused the state and federal governments of caring more about fish than people.

When a drought hit California in 2008, the spring of 2008 was the driest in 88 years. The California legislature passed a bill to cut water use by 20 percent. But the legislature put most of the responsibility for reducing water use on residential users, despite the fact that agriculture uses a much larger portion of the state's water. Some municipalities started rationing water, and a number of municipalities began enforcing a 2001 state law that requires new building developments to show plans that ensure a 20-year water supply.

In 2009, the California legislature passed a series of bills to address the water crisis. The $40 billion package called for the restoration of the San Joaquin and Sacramento delta ecosystem as well as the building of new dams and a series of water conservation efforts to reduce water demand. A few days later, the governor signed the bill into law. However, these steps did not go far enough to resolve the problem.

In 2014, California passed the Sustainable Groundwater Management Act, which is intended to balance the amount of water pumped out of the ground with the amount of water entering the ground. It is the first statewide legislation to regulate how groundwater is being used. This allows local government agencies to develop their own plans for sustainable water use to reach a sustainable level of pumping water out of the ground by 2040. At the time, many legislators thought this timeline would be sufficient to meet the challenge of future droughts. However, major droughts continued to occur and by 2022 it became clear that immediate action needs to be taken to resolve the lack of water leaving watersheds into streams and lakes and the depleted water in the ground. The wide diversity of water users are coming together to solve a problem that is now dramatically affecting everyone's lives.

What can be done?

Given the scarcity of water in California, it stands to reason that its sustainable use is an important topic with many environmental, economic, and political implications. Economists generally argue that all users should be charged the same amount for a resource so that only those who value it most will pay to use it. Doing so would provide increased motivation for more efficient water use by farms, residences, and businesses. Increased water costs could be helpful in reducing the demand for water, but we cannot forget the simultaneous need for all users to better conserve water.

Scenarios similar to the California water wars are being played out in many other parts of the world. Competition for water exists not only among farmers and municipalities, but also among neighboring countries. Indeed, disagreements over water have turned into real wars between nations. If the competing interests in California can devise creative solutions to the state's water problems, those solutions can serve as a model for the rest of the world so that we can continue to have economic prosperity without severely harming the environment.

Questions

1. What are the challenges of balancing the need for water in agriculture with the need for water to protect endangered species?
2. How might a California farmer who receives subsidized costs of water argue that the current water distribution rules are beneficial not only to farmers but also to society?

Practice AP® Free-Response Question

The Los Angeles aqueduct spans 375 km (233 miles) from Mono Lake (see Science Applied 1) and through the Owens Valley, moving south through the Mojave Desert to the San Fernando Valley. The aqueduct collects snowmelt from multiple streams and rivers that drain from the mountains into the Owens Valley. This water then flows to Los Angeles by gravity. The initial aqueduct was completed in 1913 and subsequently extended farther north to obtain more water. Over the decades, the area in and around Owens Valley has become increasingly dry and farmers have suffered a tremendous loss of crop productivity.

(a) **Identify** a consequence of pumping deep water wells when aqueducts cannot provide enough water for farms and homeowners. (2 pts.)

(b) **Explain** why the assumption of consistent water availability from aqueducts has proven to be incorrect during the past two decades. (3 pts.)

(c) **Describe** two likely causes of increased water demand for the city of Los Angeles. (2 pts.)

(d) If farmers had to pay a higher cost for water, **explain** three ways in which farmers might respond. (3 pts.)

References

Charles, D. 2021. New protections for California's aquifers are reshaping the state's Central Valley. *NPR*, October 7. https://www.npr.org/2021/10/07/1037369959/new-protections-for-californias-aquifers-are-reshaping-the-states-central-valley.

Kasler, D. 2016. Drought costs California farms $600 million, but impact eases. *Sacramento Bee*, August 15.

Martin, G. 2005. The California water wars: Water flowing to farms, not fish. *San Francisco Chronicle*, October 23.

Obegi, D. 2011. California water wars: It isn't fish vs. farmers. *Los Angeles Times*, August 10.

Stokstad, E. 2020. Deep deficit. *Science*. April 16. https://www.science.org/content/article/droughts-exposed-california-s-thirst-groundwater-now-state-hopes-refill-its-aquifers.

UNIT 4 Review

Key Terms to Remember

Core (p. 230)
Mantle (p. 230)
Magma (p. 230)
Asthenosphere (p. 230)
Lithosphere (p. 230)
Plate tectonics (p. 231)
Earthquake (p. 231)
Hot spot (p. 231)
Volcano (p. 231)
Tsunami (p. 231)
Divergent boundary (p. 232)
Seafloor spreading (p. 232)
Convergent boundary (p. 232)
Subduction (p. 232)
Island arc (p. 232)
Collision zone (p. 232)
Transform boundary (p. 234)
Fault (p. 234)
Igneous rock (p. 239)
Sedimentary rock (p. 240)
Metamorphic rock (p. 240)
Rock cycle (p. 240)
Physical weathering (p. 241)
Chemical weathering (p. 242)
Acid precipitation (Acid rain) (p. 243)
Erosion (p. 243)
Parent material (p. 245)
Horizon (p. 246)
O horizon (p. 246)
Humus (p. 246)
A horizon (Topsoil) (p. 246)
E horizon (p. 246)
B horizon (p. 246)
C horizon (p. 246)
Porosity (p. 248)
Water holding capacity (p. 249)
Permeability (p. 249)
Cation exchange capacity (CEC) (p. 250)
Base saturation (p. 250)
Watershed (p. 253)
Insolation (p. 258)
Albedo (p. 260)
Troposphere (p. 262)
Stratosphere (p. 262)
Ozone (p. 262)
Mesosphere (p. 263)
Thermosphere (p. 263)
Exosphere (p. 263)
Saturation point (p. 263)
Adiabatic cooling (p. 264)
Adiabatic heating (p. 264)
Latent heat release (p. 264)
Atmospheric convection current (p. 264)
Hadley cell (p. 265)
Intertropical convergence zone (ITCZ) (p. 265)
Polar cell (p. 266)
Ferrell cell (p. 266)
Coriolis effect (p. 266)
Gyre (p. 272)
Upwelling (p. 272)
Thermohaline circulation (p. 272)
Rain shadow (p. 274)
El Niño–Southern Oscillation (ENSO) (p. 275)
La Niña (p. 275)
Carbon sequestration (p. 281)
Aqueducts (p. 283)

Unit 4 AP® Environmental Science Practice Exam — Preparing for the AP® Exam

Section 1: Multiple-Choice Questions

1. The correct vertical zonation of Earth above the core is
   (a) asthenosphere-mantle-soil-lithosphere.
   (b) asthenosphere-lithosphere-mantle-soil.
   (c) mantle-asthenosphere-lithosphere-soil.
   (d) mantle-lithosphere-soil-asthenosphere.

2. Evidence for the theory of plate tectonics includes
   (a) deposits of copper ore around the globe.
   (b) different rock formations on both sides of the Atlantic.
   (c) fossils of the same species on distant continents.
   (d) invasive species currently living on distant continents.

3. Using the figure below, which of the following best describes the islands in relation to the hot spot?

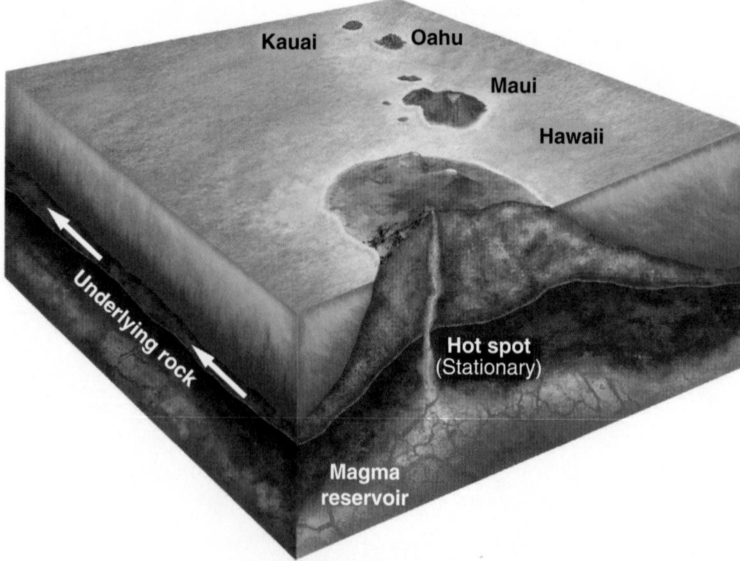

(a) Kauai is the oldest island.
(b) Maui is the youngest island.
(c) Oahu is the oldest island.
(d) Hawaii is the oldest island.

For questions 4–6, select from the following answer choices:

(a) divergent boundary
(b) convergent boundary
(c) transform boundary
(d) epicenter

4. At which type of boundary do tectonic plates move sideways past each other?

5. At which type of boundary does subduction occur?

6. At which type of boundary does seafloor spreading occur?

7. Which statement about earthquakes is true?
   (a) An earthquake of magnitude 8.0 on the Richter scale occurs about once every decade.
   (b) Earthquakes occur when rocks of the asthenosphere rupture along a fault.
   (c) The Ring of Fire describes a geological circle on Earth where earthquakes are likely to occur.
   (d) An earthquake of magnitude 4 on the Richter scale is 10,000 times greater than an earthquake of magnitude 2.

8. Which contributes to the physical weathering of rocks?
   (a) acidic rain
   (b) freeze-thaw cycles
   (c) volcanic activity
   (d) growth of plant roots

9. A field containing soil that is 60 percent clay can hold a tremendous amount of nutrients but often has poor crop growth. What is the most likely explanation for this phenomenon?
   (a) Excessive concentrations of nutrients in soil can become toxic to many types of plants.
   (b) Clay soils have a high cation exchange capacity that becomes toxic to plants.
   (c) Plant roots have difficulty obtaining water from soils high in clay.
   (d) The base saturation of clay soil is too high for most plants.

10. Which rock is formed by high temperature and pressure like that caused by continental collisions?
    (a) igneous
    (b) metamorphic
    (c) sedimentary
    (d) granite

11. New Zealand is a country with two main islands located on the boundary of two tectonic plates—the Australian and the Pacific—which are moving in different directions. The city of Christchurch, located on the South Island, is moving away from the city of Auckland, located on the North Island, at a rate of 40 mm per year. How many years will it take for Christchurch to be 500 km farther away from Auckland than it is today?
    (a) 12.5×10^5 years
    (b) 12.5×10^6 years
    (c) 20×10^5 years
    (d) 20×10^6 years

12. Which statement about oceanic plates is correct?
    (a) The crust of oceanic plates is mostly granitic rock.
    (b) The movement of oceanic plates toward each other results in seafloor spreading.
    (c) The dominant rock type in the crust of oceanic plates is basaltic rock.
    (d) Continental crust is subducted where oceanic and continental plate margins meet.

13. Research at Hubbard Brook showed that stream nitrate concentrations in two watersheds were ____________ before logging, and that after one watershed was logged, its stream nitrate concentration was ____________ immediately after the logging.
    (a) similar; decreased
    (b) similar; increased
    (c) different; increased
    (d) different; decreased

14. Which statement best explains why polar regions are colder than tropical regions?
    (a) Polar regions have lower albedo values.
    (b) Polar regions receive less solar energy per unit of surface area.
    (c) Tropical regions receive less direct sunlight throughout the year.
    (d) Sunlight travels through more atmosphere and loses more energy in tropical regions.

15. An increase in evaporation near the equator would most likely cause
    (a) increased precipitation at the ITCZ.
    (b) decreased precipitation at the ITCZ.
    (c) increased precipitation in Ferrell cells.
    (d) decreased precipitation in Ferrell cells.

16. The high heat capacity of water causes what effect when combined with ocean circulation?
    (a) the high salinity of deep polar water in the thermohaline cycle
    (b) warm temperatures in continental coastal areas
    (c) the suppression of upwelling during an ENSO event
    (d) the heat transfer between the ocean and atmosphere due to evaporation along the equator

17. Which statement about rain shadows is correct?
    (a) They occur on the eastern sides of mountain ranges in the Northern Hemisphere.
    (b) As air descends down a mountain range, water vapor condenses into precipitation.
    (c) They occur on the eastern sides of mountain ranges in the Southern Hemisphere.
    (d) The rain shadow side of a mountain range receives the most rain.

Section 2: Free-Response Questions

1. A soil study was conducted in a watershed to compare soil types and their effects on soil erosion into a river. The researchers took four soil samples from four different areas in the watershed, which are seen on the soil triangle below.

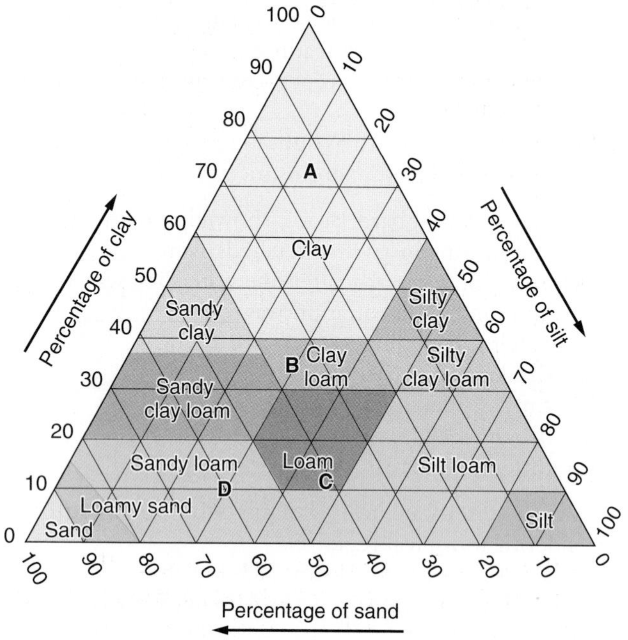

(a) **Identify** the sand content of Sample D. (1 pt.)
(b) **Describe** which soil sample would be most vulnerable to erosion. (1 pt.)
(c) **Explain** the relationship between pore space and permeability. (2 pts.)
(d) The researchers hypothesized that soil Sample B would provide the highest rate of infiltration. **Make a claim** using evidence that supports or refutes their hypothesis. (1 pt.)
(e) **Describe** an environmental problem that can be caused by soil erosion. (1 pt.)
(f) **Identify** a research question for the researchers' study. (1 pt.)
(g) **Identify** the independent variable for the study. (1 pt.)
(h) **Describe** TWO variables not mentioned in the study that could affect the results. (2 pts.)

2. The map below shows the ocean gyres and their component warm and cool surface currents in major ocean basins.

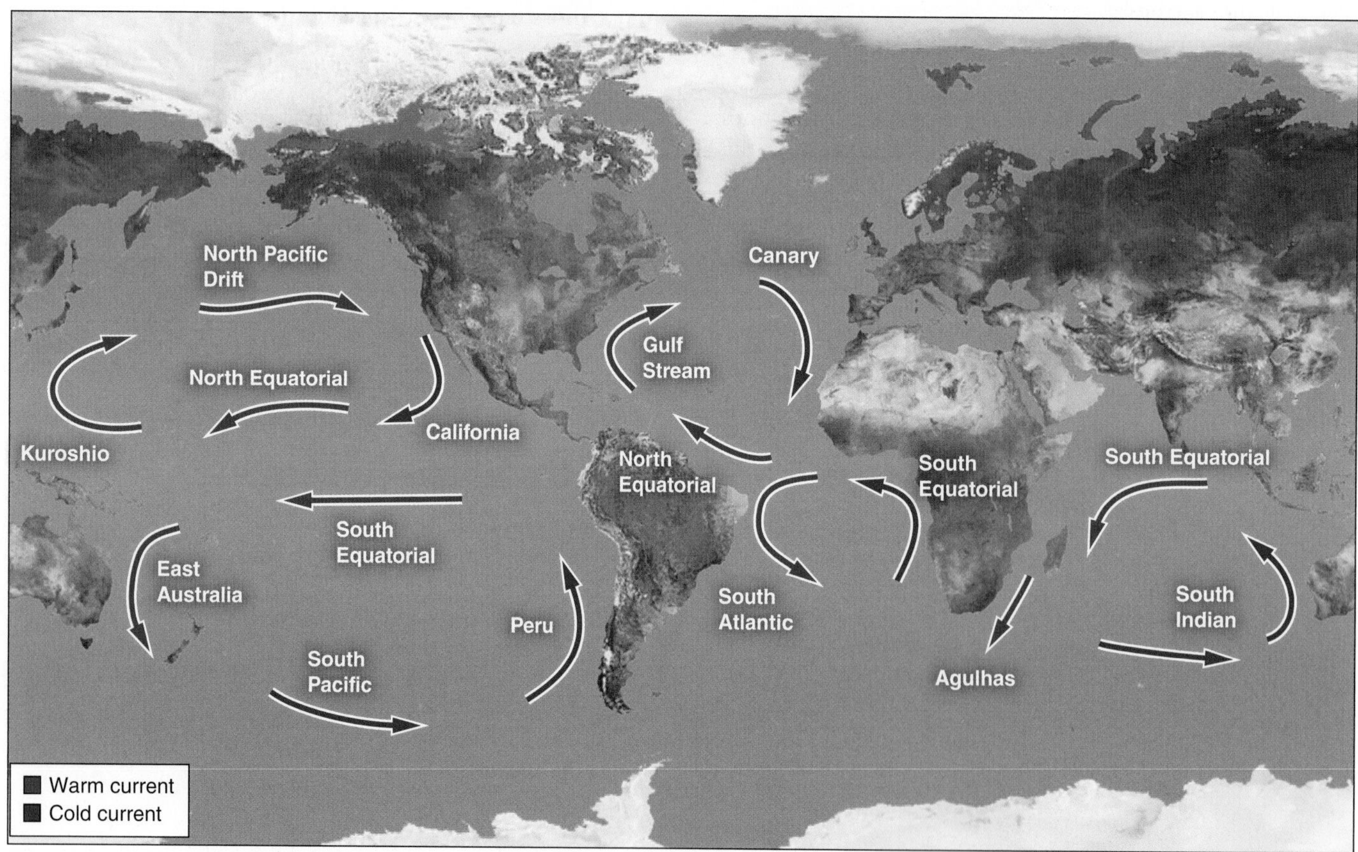

(a) **Identify** the directions gyres rotate in the Southern Hemisphere. (1 pt.)
(b) **Describe** how the California Current affects the climate in the Pacific Northwest region of the United States. (1 pt.)
(c) **Explain** why all gyres flow away from the equator. (1 pt.)
(d) **Identify** TWO factors that drive the movement of ocean gyres. (2 pts.)
(e) **Explain** how ocean gyres can create areas of high primary productivity along the west coasts of continents. (1 pt.)

El Niño is a phenomenon that occurs every 3 to 7 years as a result of changes in winds and surface currents in the tropical Pacific that can affect weather worldwide.

(f) **Describe** the change in winds that occurs during an El Niño event. (1 pt.)
(g) **Describe** one potential environmental problem that can be caused by an El Niño event. (1 pt.)
(h) A newspaper journalist makes a statement that the effects of El Niño events will worsen as an effect of climate change. **Make a claim** that supports or refutes the journalist's statement. (1 pt.)
(i) **Describe** a potential economic problem related to the effects of El Niño. (1 pt.)

3. The unequal heating of Earth's surface causes atmospheric circulation and global wind patterns.
(a) **Identify** the convection cell that occurs between the equator and 30° latitude. (1 pt.)
(b) **Describe** how the trade winds in the Northern Hemisphere are deflected to the west. (1 pt.)
(c) **Explain** why desert biomes exist at 30° latitude. (2 pts.)

Global climate change is caused in part by increasing carbon dioxide levels in the atmosphere, which are higher today than at any other point during at least the last 800,000 years.

(d) **Describe** an environmental problem that can be caused by increased carbon dioxide levels in the atmosphere. (1 pt.)
(e) Other than reducing CO_2 emissions, **make a claim** that proposes a solution that could mitigate the effects of climate change. (1 pt.)
(f) In 1965, the concentration of carbon dioxide in the atmosphere was 320 parts per million (ppm). By 2020, the concentration was 412 ppm. **Calculate** the percent difference in atmospheric carbon dioxide concentration from 1965 to 2020. Show all work. (2 pts.)
(g) It is estimated that as much as 23 percent of the world's natural habitats could be lost by the year 2100. If there are approximately 15.77 billion acres of inhabitable land on earth, **calculate** how much inhabitable land for habitats, in hectares, would remain in 2100 (there are 0.4 hectares in 1 acre). Show all work. (2 pts.)

UNIT 5

Land and Water Use

Environmentalists demonstrate in support of wolf protection in Olympia, Washington. *(AP Photo/Rachel La Corte, File)*

CASE STUDY

Who Can Use Our Public Lands? And for What Purpose?

There are many categories of public lands, a term for land that is held by a local or federal entity "in trust" for the citizens of a given country. Public lands, as the name suggests, are not privately held and are not owned by individuals. There have been many struggles over the best way to use and steward lands, both public and private. What are the rules regulating public lands? Who decides how they are used and how competing interests for the same land are evaluated? For over a century, the federal government of the United States has awarded the right to use federal lands to private individuals and corporations. The government awards or sells leases that permit a variety of activities including harvesting trees, mining copper, grazing cattle, and extracting oil and gas.

In 2008, a University of Utah student who objected to the designated use of federal lands in Utah attended a Bureau of Land Management oil and gas lease auction and was assigned bidder number 70. Tim DeChristopher, a 27-year-old undergraduate and former wilderness guide, successfully bid on thousands of acres of land. When federal officials realized that he had no intention of paying the $1.8 million lease fee, or extracting oil and gas from that land, he was arrested and charged with two felonies that carried a potential sentence of 10 years. He spent almost 2 years in federal prison for trying to obstruct a process that legally allocated federal land to be used for the extraction of fossil fuels that lie beneath that land.

The conflict over use of public lands and public waters raises many questions.

Some supporters argued that he was practicing a form of civil disobedience. Oil rigs, drilling platforms, and other evidence of resource extraction would have been visible from national parks in southern Utah. And certainly, the extraction of oil and gas would have consequences both locally and globally. Many environmentalists viewed the auction as a hasty attempt to develop as much land as possible for oil and gas exploration before environmental reviews could occur. Critics of DeChristopher maintained that the leasing of land for a variety of purposes has occurred for more than a century, and this auction was no different from any other. They stated that the government is obligated to allow certain lands to be used for a variety of purposes, most of which benefit some or many people in the United States. After his arrest and during his trial Tim DeChristopher continued to advocate for protecting public lands from drilling and expressed moral and legal objections to leasing public land.

Others have objected in less dramatic ways. In 2016, environmental writers and activists Terry Tempest and Brooke Williams tried to purchase oil and gas leases for federal land in Utah with the express intention of developing other energy resources from the land such as solar and wind. Their lease application was denied. They argued that they have been treated differently from other lease purchasers because they expressly stated they wanted to do something other than extract oil and gas. They sued the federal government for the right to lease the land and use it for purposes other than oil and gas exploration.

In 2021, the Department of Interior rejected their suit and affirmed that their lease application denial was allowable.

There have also been protests concerning public lands involving biological resources. Advocates for the protection of wolves called for the removal of cattle that had been allowed to graze on public lands in Colville National Forest, Washington. Cattle ranchers, who have grazed on public lands in northeastern Washington for over a hundred years, defended the right to graze their cattle. They also have supported actions by the Washington Department of Fish and Wildlife to protect their cattle, including shooting wolves when there was evidence that they had attacked or simply were in the vicinity of cattle. The state has authorized the killing of more than 34 wolves between 2013 and 2021. The wolves were believed to have killed and eaten cattle from nearby ranches. Strong sentiments, both pro and con, over wolf killing and cattle grazing on public lands has continued to increase. So far, the courts have allowed killings, as long as the agencies involved have studied the situation, collected and analyzed data, and have scientific or wildlife management support for their actions.

The question of who can use public water also generates controversy and conflict. There are numerous situations in the United States today where a government agency, company, or individual is being challenged in the courts for removing too much water from rivers and underground reservoirs. Lake Mead is an artificial lake straddling Nevada and Arizona created by the Hoover Dam along the Colorado River. Because of low water levels on Lake Mead in 2021 due to water removals for irrigation, cattle, and drinking water, exacerbated by drought and climate change, federal water authorities restricted withdrawals from the Colorado River. This affected millions of water users in California, Nevada, and Mexico. This action was preceded by a number of lawsuits questioning and challenging various parties' rights to use water from the Colorado River. It is very likely that more lawsuits will follow in the coming years.

The conflict over use of public lands and public waters raises many questions. What obligations does a government have when it holds land "in trust" for its citizens? Do citizens of the United States have the right to use public lands and water for purposes other than those designated by the government? Is Tim DeChristopher a hero or a lawbreaker? Are cattle ranchers entitled to use public lands and entitled to federal protection for their cattle? Can municipalities remove water from rivers or aquifers if that compromises other activities? These are challenging questions. From an environmental science perspective, we might begin to evaluate these situations by asking what land designations and uses can minimize harm to lands and increase sustainability.

Sources: B. Gage, *Bidder 70* (Film Documentary). First Run Features, 73 minutes, 2012; B. Loomis, DeChristopher Sentenced to Prison, 26 Protesters Arrested, *Salt Lake Tribune* (Salt Lake City, Utah), July 27, 2011; L. Mapes, With Cattle in Washington's Wolf Country, Ranchers Work and Worry, *Seattle Times*, July 15, 2017.

Practice your science skills

1. **Concept Explanation:** Describe the federal government's role in resource allocation on public lands.
2. **Environmental Solutions:** Describe why wolves might be necessary to the function of an ecosystem.
3. **Text Analysis:** Describe the opposing environmental viewpoints represented in the case study.

The use and misuse of land and water are prominent topics in environmental science. This unit-opening case study illustrates different land uses for mineral and biological resources and represents some of the topics we will cover in this unit. We will start with an examination of how public lands are used and misused, in part because they are public and with no specific individual owner, misuse is more likely. We will also cover impacts of clear-cutting forests, agricultural methods for raising crops and livestock, fishing, mining, urban impacts, and the concept of the ecological footprint. We will end the unit with definitions and explorations of sustainability for agriculture, aquaculture, and forestry.

Module 24

Unit 5 | 24 | 25 | 26 | 27 | 28 | 29 | 30 | 31 | 32 | 33 | 34

The Tragedy of the Commons; Clear-cutting

Humans value land for what it can provide: food, shelter, resources, and intrinsic beauty. However, human use of land for any purpose can create problems that may be addressed and possibly prevented by regulations. In this module we will examine the tragedy of the commons and the effects of clear-cutting on forests.

Learning Goals

After reading this module you should be able to

24-1 explain the tragedy of the commons.

24-2 describe rangeland and forest land management.

24-3 explain the environmental impacts of clear-cutting.

24-1 What is the tragedy of the commons?

The tragedy of the commons describes the harm that can occur to public and shared resources

Agriculture, forestry, housing, recreation, industry, mining, and waste disposal are all land uses that benefit humans. But, these activities also have negative consequences. Extensive clear-cutting may lead to erosion and water pollution, and deforestation of large areas contributes to climate change and many other environmental problems. Change to the landscape is the single largest cause of species extinctions today. Paving over land surfaces reroutes water runoff. These paved surfaces also absorb heat from the sun and reradiate it, creating urban "heat islands." Overuse of farmland can lead to soil degradation and water pollution and, in some cases, mudslides (**FIGURE 24.1**).

Every human use of land alters it in some way. Furthermore, individual activities on any parcel of land can have wide-ranging effects on other land. For this reason, communities around the world use a number of methods, including laws and informal norms, to influence or regulate both private and public land use.

As we saw in the unit-opening case study, people do not always agree on land use and management priorities. Do we protect a large tract of open land, or do we extract resources from it in order to gain benefits in the form of energy resources, jobs, profits, and economic development? To understand

FIGURE 24.1 Mudslide. This home in Silverado Canyon, California, was damaged by a mudslide in 2021. *(MediaNews Group/Orange County Register via Getty Images/Getty Images)*

Use of the commons is below the carrying capacity of the land. All users benefit.
If one or more users increase the use of the commons beyond its carrying capacity, the commons becomes degraded. The cost of the degradation is incurred by all users.
Unless environmental costs are accounted for and addressed in land use practices, eventually the land will be unable to support the activity.

FIGURE 24.2 The tragedy of the commons. If the use of common land is not regulated in some way — by the users or by a government agency — the land can easily be degraded to the point at which it can no longer support that use.

the issues involved in the use of land, we will examine two fundamental concepts essential to understanding land use and land conflicts: the tragedy of the commons and externalities.

The Tragedy of the Commons

In virtually every society throughout history, at least some land was viewed as a common resource: Anyone could use certain spaces for foraging, growing crops, felling trees, hunting, or mining. But as populations increased, common lands became degraded—overgrazed, overharvested, and deforested. An essay by the economist William Forster Lloyd in 1833 initially presented the idea of the village commons and overgrazing by livestock. The basic premise is that the village commons is a public area in the middle of the village that no single person owns. So everyone can use it, say to graze their animals, and if everyone overuses it, the land is harmed. In 1968, an ecologist brought the issue of overuse of common resources to the attention of the broader scientific community. Because people often act from self-interest for short-term gain, it was observed that when many individuals share a common resource without agreement on or regulation of its use, it is likely to become overused and depleted very quickly. The tendency for a shared, limited resource to become depleted if it is not regulated in some way is known as the **tragedy of the commons**. Regulation can be managed by the users or by a governmental agency.

Tragedy of the commons The tendency of a shared, limited resource to become depleted if it is not regulated in some way.

Let's explore the village commons example a little further. Imagine a communal pasture on which many farmers graze their sheep, like the one shown in FIGURE 24.2. At first, no single farmer appears to have too many sheep. But because an individual farmer gains from raising as many sheep as possible, each farmer may be tempted to add sheep to the pasture. However, if the total number of sheep owned by all the farmers continues to grow, the number of sheep will soon exceed the carrying capacity of the land. The sheep will overgraze the pasture to the point at which plants will not have a chance to recover. The common land will be degraded, through loss of vegetation and soil erosion, for example, and the sheep will no longer have an adequate source of nourishment. Over a longer period, the entire community will suffer. When the farmers make decisions that benefit only their own short-term gain and do not consider the common good, eventually, everybody loses. That is why it is called a "tragedy"—there are no winners. Note that this example describes a situation where the land is public (no individual owns it) and sustainability is not achieved (the land is degraded).

The tragedy of the commons applies not only to agricultural lands, but to any publicly available resource that is not regulated and not owned by an individual, including land, air, and water. For example, the use of global fisheries as commons has led to the overexploitation and rapid decline of many commercially harvested fish species and has upset the balance of entire marine ecosystems. The release of carbon dioxide to the atmosphere or sewage to an

FIGURE 24.3 Global fisheries. Global fisheries are one of the commons. These sardines were photographed in the North Atlantic Ocean. *(Jens Kuhfs)*

estuary are other examples of the tragedy of the commons (**FIGURE 24.3**).

There are a few important distinctions to make when referring to the tragedy of the commons. The land must be common—that is a public or shared resource. And in order for the tragedy of the commons to occur, the land must be degraded. If there is a public apple orchard in a town, and people in the town harvest apples without damaging the trees or the soil, that would not be a tragedy of the commons. The trees and the land will produce more apples the following year. The actions of the townspeople are sustainable, meaning that they are able to do something now without jeopardizing the ability of future generations to engage in similar activities. We first defined this term in Module 0 and continue to refer to it throughout the text as a central theme of this course.

Externalities

The tragedy of the commons is the result of an economic phenomenon called a negative externality. More generally, an **externality** is a cost or benefit of a good or service that is not included in the purchase price of that good or service, or otherwise not accounted for. For example, if a bakery moves into the building next to you and you wake up every morning to the delicious smell of freshly baked bread, you are benefiting from a positive externality. On the other hand, if the bakers arrive at three in the morning and make so much noise that they interrupt your sleep, making you less attentive at school later that day, you are suffering from a negative externality.

In environmental science, we are especially concerned with negative externalities because they so often lead to serious environmental damage for which no one is held legally, financially, or morally responsible. For example, if one farmer grazes more sheep than is beneficial for the community in a common pasture, their action will ultimately result in more total harm than total benefit. But, as long as the land continues to support grazing, the individual farmer will not have to pay for their overuse. Ultimately, though, other farmers will bear the burden of the overuse because the land will be degraded and they will have to find somewhere else to graze their sheep or they will fail to find enough feed for their sheep. But what if the farmer responsible for the extra sheep had to pay an extra fee for using more of the common resource? They probably would not graze the extra sheep on the commons because the cost of doing so would exceed the benefit. This concept of charging a fee for an externality is called internalizing the externality because in the field of economics, it makes what was external—using the common resource—part of the internal costs of the business or household. From this example, we can see that in order to calculate the true cost of using a resource, we must always include the cost of any negative externalities. In other words, we must account for any potential external harm or cost that comes from the use of that resource.

Some economists maintain that private ownership can prevent the tragedy of the commons. After all, a landowner is much less likely to overgraze and harm their own land than they would common land. Governmental regulation is another approach. For example, a local government could prevent overuse of a common pasture by passing an ordinance that permits only a certain number of sheep to graze there. In the air pollution area of environmental science, policy makers have utilized charging for permission to emit sulfur and carbon dioxide into the air, as a method of internalizing the negative externality. In other words, the individual who emits sulfur or carbon dioxide must pay for that privilege. If the individual is an automobile manufacturer, for example, they could pass along the cost of emitting pollutants on to the consumer, resulting in a higher price for the automobile. This would result in cars that generate higher pollution emissions costing more money to manufacture and that might result in people purchasing cars that pollute less when being manufactured because those cars would cost less.

Challenging the idea that government regulation is necessary to avoid the tragedy of the commons, the late Professor Elinor Ostrom (1933–2012) of Indiana University showed that many commonly held resources can be managed effectively at the community level or by user institutions. Ostrom's work, for which she was awarded the 2009 Nobel Prize in Economics, has shown that self-regulation by resource users can prevent the tragedy of the commons. Professor Ostrom also challenged the idea that private property rights can address all environmental problems.

Public lands are classified according to their use

All countries have public lands designated for a variety of purposes, including environmental protection. The use of

Externality The cost or benefit of a good or service that is not included in the purchase price of that good or service, or otherwise accounted for.

DO THE MATH Changes in the Amount of Protected Lands Over Time

Preparing for the AP® Exam

In recent years the United Nations has reported a significant increase in the total area protected worldwide. In 2004 the UN listed total area protected as 18,800,000 square kilometers. Let's convert this value in square kilometers to hectares, express it in scientific notation, and compare that amount to the current 2014 values indicated earlier in the module.

$$18{,}800{,}000\ \cancel{km^2} \times (100\ ha/1\ \cancel{km^2}) = 1{,}880{,}000{,}000\ ha$$

In scientific notation this value = 1.88×10^9 ha protected area in 2004

Protected area in 2014 = 3.2 billion ha = 3.2×10^9 ha

2004 protected area = 1.88×10^9 ha

2014 protected area = 3.20×10^9 ha

(3.20×10^9) ha − (1.88×10^9) ha = 1.32×10^9 ha = change in protected area 2004 to 2014

Change over original 2004 value = $(1.32 \times 10^9\ ha) \div (1.88 \times 10^9\ ha) = 0.70 \times 100\% = 70\%$

Protected areas increased 70 percent between 2004 and 2014.

YOUR TURN The most recent UN List of Protected Areas, released in 2018, shows that there are 238,563 protected areas across the world that cover a total area of 46,414,431 square kilometers.

1. Convert to hectares and express the total protected area in scientific notation.
2. Calculate the percent change in global protected area from 2014 to 2018.

protected areas may vary from excluding almost all human activities to permitting harvest of biological and other resources for human benefit. The 2014 United Nations List of Protected Areas—the most recent global study of protected areas—contains more than 209,000 protected areas around the world covering more than 3.2 billion hectares (7.9 billion acres). Roughly 14 percent of terrestrial areas and 3.4 percent of marine areas on the surface of Earth are protected in one way or another. "Do the Math: Changes in the Amount of Protected Lands over Time" gives you the opportunity to explore some of the data associated with protected lands.

Public Lands in the United States

All countries have public lands designated for a variety of purposes, including environmental protection (**FIGURE 24.4**). In the United States, publicly held land may be owned by federal, state, or local governments. Of the nation's

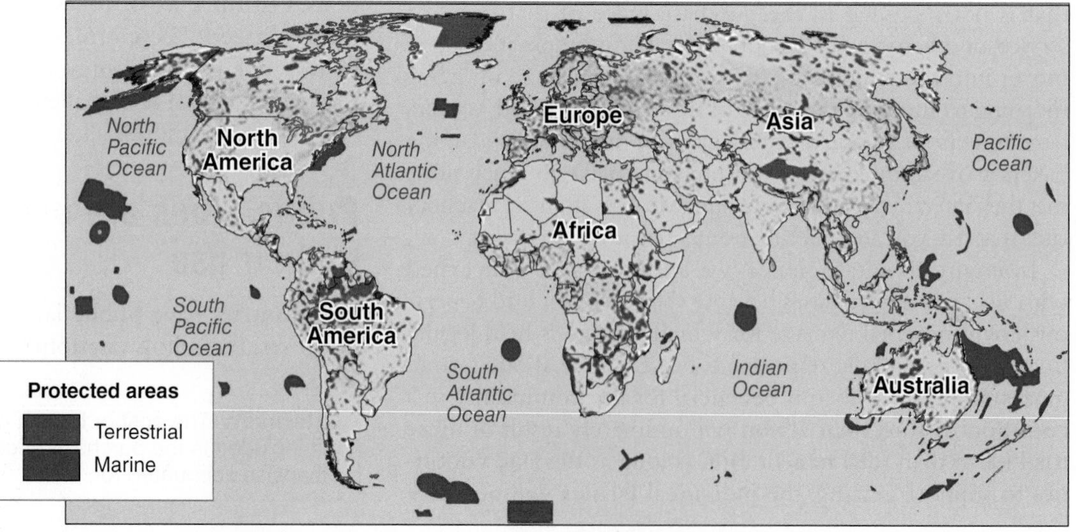

FIGURE 24.4 Protected land and marine areas of the world. Protected areas are distributed around the globe. *(Data from http://wdpa.s3.amazonaws.com/Files_pp_net/Global_map_template_Dec2016.png)*

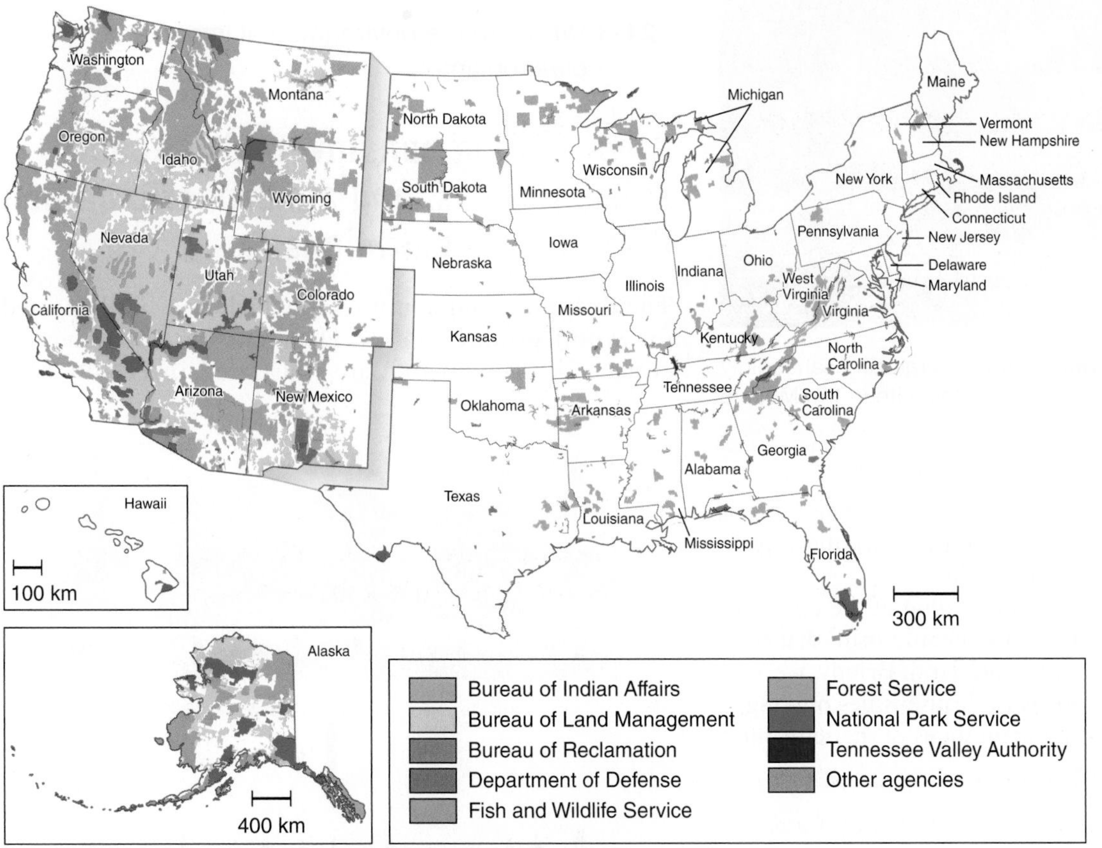

FIGURE 24.5 Federal lands in the United States. Approximately 42 percent of the land in the United States is publicly owned, with 25 percent of the nation's land owned by the federal government. *(Data from http://nationalatlas.gov)*

land area, 42 percent is publicly held—a larger percentage than in any other nation (**FIGURE 24.5**). As shown in the figure, land in the United States, both public and private, is used for many purposes. These uses can be divided into a number of categories. The probable use of public land determines how it is classified and which federal agency will manage it. More than 95 percent of all federal lands are managed by four federal agencies: the Bureau of Land Management (BoLM), the United States Forest Service (USFS), the National Park Service (NPS), and the Fish and Wildlife Service (FWS).

Although individual tracts may differ, the following are typical divisions of public land uses:

- BoLM lands: grazing, mining, timber harvesting, and recreation
- USFS lands: timber harvesting, grazing, and recreation
- NPS lands: recreation and conservation
- FWS lands: wildlife conservation, hunting, and recreation

AP® Exam Tip

As you learn about methods of timber harvest and timber management, try to identify the drawbacks and benefits of different techniques. Which drawbacks and benefits are environmental and which are economic? Remember, if it benefits humans it is considered an economic benefit.

We have seen that resources can be managed or they can be over-exploited, depending on a number of factors related to economics, human behavior, and specific management practices. Proper resource management also depends on specific aspects of the type of land involved as well as the agency and the guidelines under which it operates. Next, we will focus on tree and forest management practices.

24-2 What are rangeland and forest land management?

Land management differs in rangelands and forests

Some of the most challenging environmental science issues are related to land management theory and practice. Perhaps the most important work in the last century on this topic has come from Aldo Leopold, a University of Wisconsin professor, ecologist, writer, and environmentalist. His book *A Sand County Almanac*, published posthumously in 1949, characterizes land management, conservation, ecology, and interactions with people in a way that was revolutionary when published, and is still timely and important today. Leopold wrote about a "land ethic," which he described as a moral responsibility of humans to the natural world. The moral responsibility to protect land is not typically a topic in environmental science, but many contemporary discussions

FIGURE 24.6 Overgrazed rangeland. Overgrazing can rapidly strip an area of vegetation, as seen in this site in the Mojave Desert, California. *(Jeff Foott/Getty Images)*

about the conservation and protection of public and private lands inevitably reference Aldo Leopold.

Now that we have a basic understanding of the common and shared resources, let's turn to a specific issue that is of concern for both public and private land. Briefly, we will illustrate how grazing of animals can cause issues on rangelands. Then we will discuss the main focus of the remainder of this module: clear-cutting.

Rangelands

Rangelands are dry, open grasslands primarily used for grazing cattle, which is the most common use of land in the United States. Rangelands are semiarid ecosystems and are therefore particularly susceptible to fires and other environmental disturbances. When overused, rangelands can easily lose biodiversity. Grazing too many animals in a finite area can quickly denude a region of vegetation (**FIGURE 24.6**). Overgrazed land is exposed to wind erosion that makes it difficult for soils to absorb and retain water and nutrients when it rains.

Forests

Forests are land areas dominated by trees and other woody vegetation and sometimes used for commercial logging. Roughly three-quarters of the forests used for commercial timber operations in the United States are privately owned. Many national forests were originally established to ensure a steady and reliable source of timber, and commercial logging companies are currently allowed to harvest U.S. national forests, usually in exchange for a royalty that is a percentage of their revenues. The federal government manages the timber program and builds and maintains logging roads.

Rangelands Dry, open grasslands primarily used for grazing cattle.

Forest Land dominated by trees and other woody vegetation and sometimes used for commercial logging.

24-3 What are the environmental impacts of clear-cutting?

Clear-cutting can adversely impact soils, streams, and atmospheric carbon dioxide concentrations

The two most common ways in which trees are harvested for wood production are clear-cutting and selective cutting, both of which are illustrated in **FIGURE 24.7**.

Regrowth

(a) Clear-cutting

Regrowth

(b) Selective cutting

FIGURE 24.7 Timber harvest practices. (a) Clear-cutting removes most, if not all, trees from an area and is often coupled with replanting. The resulting trees are then all the same age. (b) In selective cutting, single trees or small numbers of trees are harvested. The resulting forest consists of trees of varying ages.

FIGURE 24.8 Clear-cut forest in western Washington state. A clear-cut on a steep slope increases the likelihood of erosion and delayed regeneration of vegetation. *(Phil Augustavo/Getty Images)*

Clear-cutting involves removing all, or almost all, of the trees within an area; it is the easiest harvesting method and, in most cases, the most economical. Sometimes the tree trunk is cut close to the ground surface. Sometimes the whole tree — roots included — are lifted out of the ground. As you can imagine, the latter method causes more disturbance and erosion. When a stand, or cluster, of trees has been clear-cut, foresters sometimes replant or reseed the area. Often the entire area will be replanted at the same time, so all the resulting trees will be the same age. Because they are exposed to full sunlight, clear-cut tracts of land are ideal for fast-growing tree species that achieve their maximum growth rates with large amounts of direct sunlight. Because often one species of tree is planted in the newly cleared area, and other species may not be so successful in these conditions, there can be an overall reduction in biodiversity. We will come back to this topic at the end of this section.

Clear-cutting

Clearcutting, especially on slopes, can increase wind and water erosion, which causes the loss of soil and nutrients (**FIGURE 24.8**). Erosion also adds silt and sediment to nearby streams, which harms aquatic populations (for example, sediment can clog the gills of fish, interfering with oxygen uptake). In addition, the denuded slopes are prone to dangerous mudslides, like the one shown in Figure 24.1 on page 295. Furthermore, clear-cutting with heavy machinery may lead to soil compaction. This means that there may be less water infiltration during rain events, which will lead to more erosion of soil and more runoff. This may lead to increased flooding. Clear-cutting also increases the amount of sunlight, and therefore heat, that reaches the soil. As you recall from Unit 4, in general, until moisture becomes limiting, increased temperatures lead to increased microbial decomposition. This leads to greater release of carbon dioxide from the soil, which reduces soil organic matter and increases atmospheric carbon dioxide concentrations. Sometimes clear-cutting is desirable when trying to achieve new stands of one or more tree species, such as when creating habitat for a specific bird or mammal population. There are also economic benefits to clear-cutting. However, in general, clear-cutting leads to increased atmospheric carbon dioxide concentrations, which contributes to global climate change.

Clear-cutting may also increase sunlight that reaches rivers and streams, which raises water temperatures. Since colder water allows greater amounts of dissolved oxygen, clear-cutting indirectly results in lower dissolved oxygen concentrations and may adversely reduce the viability of certain aquatic species. Even the replanting process can have negative environmental consequences. Logging companies often use fire or herbicides to remove bushy vegetation before a clear-cut is replanted. These practices reduce soil organic matter, and herbicides may contaminate water that runs off into streams and rivers. Many environmental scientists identify clear-cutting as a cause of habitat alteration and destruction, as well as a cause of breaking up large forests into smaller fragments. These effects decrease biodiversity and sometimes lower the aesthetic value of the affected forest. However, in heavily forested regions of the country, such as the northern New England states of Maine, New Hampshire, and Vermont, where forests can cover up to 85 percent of land area, a case can be made that clear-cutting increases habitat diversity and, as a result, species diversity. That's because there is relatively little open, cleared land—land that is not forested. So clear-cutting in certain heavily forested regions actually results in an increase in habitat diversity. And if the newly clear-cut area attracts different species, that too may contribute to an increase in

Clear-cutting A method of harvesting trees that involves removing all or almost all of the trees within an area.

TABLE 24.1 Tree harvesting benefits and consequences

Tree harvest technique	Pros	Cons
Clear-cutting	• Less expensive • Easiest method • Ideal for fast-growing tree species with high sunlight requirements • Can increase habitat diversity in heavily forested regions • Can create suitable habitat for certain desired bird and mammal species	• Facilitates erosion • Reduces biodiversity • Increased sunlight can raise the temperature of soils and of nearby rivers and streams • Releases carbon dioxide and contributes to climate change
Selective cutting	• Ideal for shade-tolerant tree species • Less extensive environmental impacts	• More expensive • More difficult • Still requires logging roads • May select against desirable individuals or species
Ecologically sustainable forestry	• Maintains forest in as natural a state as possible • Often done without using fossil fuels	• Costly • More difficult • Yields less timber

biodiversity. TABLE 24.1 highlights some of the effects of clear-cutting. See the "Do the Math: Expressing Protected Land Areas as a Percentage" to merge our discussion of protected land areas from earlier in the module, with our new understanding of clear-cutting.

Selective Cutting and Ecologically Sustainable Forestry

Selective cutting removes single trees or a relatively small number of trees from the larger forest. This method creates many small openings in a stand where trees can reseed or young trees can be planted, so the regenerated stand contains trees of different ages. Because seedlings and young trees must grow next to larger, older trees, selective cutting produces optimum growth only among shade-tolerant tree species.

The environmental impact of selective cutting is less extensive than that of clear-cutting. However, many of the negative environmental impacts associated with logging remain the same. For example, whether a company uses clear-cutting or selective cutting, it will need to construct logging roads to carry equipment and workers into the area that will be harvested. These roads fragment the forest habitat, which affects species diversity, and compact the soil, which causes harm to tree roots, nutrient loss, and reductions in water infiltration. In addition, selective cutting may result in removal of the most economically profitable trees, which may also be the largest or the most ecologically fit trees. Therefore, repeated selective cutting over multiple generations may be leaving future generations with less desirable trees in the forest.

A third approach to logging—**ecologically sustainable forestry**—removes trees from the forest in ways that do not unduly affect the viability of other, noncommercial tree species. This approach has a goal of maintaining both plants and animals in as close to a natural state as possible. Some loggers have even returned to using animals such as horses to pull logs in an attempt to reduce soil compaction. However, the costs of such methods make it difficult to compete economically with mechanized logging practices (FIGURE 24.9). We examine sustainable forestry further in Module 34.

Selective cutting The method of harvesting trees that involves the removal of single trees or a relatively small number of trees from the larger forest.

Ecologically sustainable forestry An approach to removing trees from forests in ways that do not unduly affect the viability of other noncommercial tree species.

FIGURE 24.9 **Sustainable forestry.** Logging without the use of fossil fuels, as is done in this forest in France, further enhances the sustainability of a forestry project. *(Andia/Getty Images)*

DO THE MATH Expressing Protected Land Areas as a Percentage

Preparing for the AP® Exam

Another initiative to track protected land cover is the Protected Planet Report, based on the World Database of Protected Areas. According to their 2020 Report, the total land cover of the United States is 9,490,391 square kilometers, of which 1,235,472 square kilometers are protected (excluding U.S. territories). We will convert these values to hectares, express them in scientific notation, and calculate what percentage of the United States is protected land.

$$9{,}490{,}391\ \cancel{km^2} \times (100\ ha/1\ \cancel{km^2}) = 949{,}029{,}100\ ha$$

In scientific notation this value is approximately = 9.49×10^8 ha

Protected area in the United States in 2020 = $1{,}235{,}472\ \cancel{km^2} \times (100\ ha/1\ \cancel{km^2}) = 123{,}547{,}200\ ha$

In scientific notation this value is approximately = 1.24×10^8 ha

The percentage of U.S. land that is protected is approximately $(1.24 \times 10^8\ ha) \div (9.49 \times 10^8\ ha) = 0.13 \times 100\% = 13\%$

YOUR TURN There are approximately 8,750,000 km^2 of forest in the tropics of Central and South America; 3,500,000 km^2 are in protected areas.

1. Convert to hectares and express in scientific notation for each.
2. Identify the percent of forest in protected areas.

Forest management influences biodiversity in a variety of ways

Almost 30 percent of all commercial timber in the world is produced in the United States and Canada. Compared with South America and Africa, deforestation and land loss and destruction in these two major timber-producing countries have been relatively small over the last several decades. Still, timber production presents important ecological challenges in the United States and Canada. For example, while timber production has always been a part of the mission of the United States Forest Service, maintaining biodiversity is an equally important goal.

All of the activities associated with logging disrupt habitat and usually have an effect, either negative or positive, on plant and animal species. One such species is the marbled murrelet (*Brachyramphus marmoratus*). This bird spends most of its life along the coastal waters of the Pacific Northwest, but it nests in coastal redwood forests. With the increase in logging of these forests, the marbled murrelet has experienced a reduction in the number of potential nesting sites it can use and as a result it has become scarce. It is now listed as a federally threatened species and as an endangered species by the state of California.

Reforestation

Commercial logging companies often replace complex forest ecosystems with **tree plantations**, which are large areas typically planted with a single fast-growing tree species. These same-aged stands can be easily clear-cut for commercial purposes, such as pulp to make paper, or wood for home building, and then replanted over and over. Because of this cycle of planting and harvesting, tree plantations never develop into mature, ecologically diverse forests. If too many planting and harvesting cycles occur, the soil may become depleted of important nutrients such as calcium.

Federal Regulation of Land Use

Since 1982, federal regulations have required the USFS to provide appropriate habitats for plant and animal communities while at the same time meeting multiple-use goals. However, because these regulations fail to specify how biodiversity protection should be achieved or how the results should be quantified, the USFS has had to choose its own approach to biodiversity management. Critics charge that the USFS is not adequately protecting biodiversity and forest ecosystems, but the USFS maintains that it is doing the best it can to meet many different objectives.

Government regulation can influence the use of private as well as public lands. The 1969 National Environmental Policy Act (NEPA) mandates an environmental assessment of all projects involving federal money or federal permits. Before a project can begin, NEPA rules require the project's developers to file an environmental impact statement. An environmental impact statement outlines the scope and purpose of the project, describes the environmental context, suggests alternative

Tree plantation A large area typically planted with a single fast-growing tree species.

approaches to the project, and analyzes the environmental impact of each alternative. Preparation of an environmental impact statement sometimes uncovers the presence of endangered species in the area under consideration. When this occurs, managers must apply the protection measures of the **Endangered Species Act**, a 1973 law designed to protect and restore plant and animal species that are threatened with extinction, and the habitats that support those species. The Endangered Species Act identifies and protects species that are both endangered, in danger of extinction through most of its range, and threatened, likely to become endangered in the foreseeable future.

Members of the public are entitled to comment on the environmental assessment and decision makers are required to respond. Although developers are not obligated to act in accordance with public wishes, in practice public concern often improves the project's outcome. For this reason, attending informational sessions and providing input is a good way for concerned citizens to learn more about local land use decisions and to help reduce the environmental impact of land development.

Endangered Species Act A 1973 U.S. law designed to protect plant and animal species that are threatened with extinction, and the habitats that support those species.

In this module we have seen that human use of public and private land has benefits and environmental consequences. The tragedy of the commons identifies the challenge of overusing public lands. Externality is the term for a cost not included in the purchase price of a good or service and is part of the tragedy of the commons story. There are different forest removal techniques, and each has its benefits and environmental consequences. In the following modules we explore the use of land, water, and other "commons" through agriculture, aquaculture, and raising meat and fish.

Module 24 AP® Review

Preparing for the AP® Exam

Learning Goals Revisited

24-1 What is the tragedy of the commons?

The tragedy of the commons describes the tendency of shared, common resources to be over-exploited. Externalities are costs or benefits not included in the price of a good or service, and the tragedy of the commons can be the result of negative externalities.

24-2 What are rangeland and forest land management?

Rangeland and forest land are two land types that are sometimes public lands and have the potential to be misused. Understanding these land types helps environmental scientists reduce environmental harm from their use.

24-3 What are the environmental impacts of clear-cutting?

Forests may be clear-cut or selectively cut. Clear-cutting may have greater environmental consequences than selective cutting. Tree plantations are sometimes established as replacement forest after clear-cutting.

Practice Math and Graphing

Preparing for the AP® Exam

Answer the following questions. Be sure to show all your work.

1. Practice Math

(a) If U.S. farmland is being converted to residential uses at a rate of 405,000 ha (1 million acres) per year, how much land will be converted in 10 years?

(b) Based on your answer in part (a), if there is roughly 369 million ha (911 million acres) of farmland in the United States, what percentage is lost each decade?

2. Practice Graphing

(a) Draw a forest clear-cutting graph based on the data shown in the table for the Brazilian Amazon. The x axis should show year and the y axis should show the hectares of trees cleared per year.

(b) Calculate the percent of intact forest that was clear-cut in 2019 if intact forest is $3,390,835\ km^2$.

Year	Deforestation (km^2)
2015	6,207
2016	7,893
2017	6,947
2018	7,900
2019	9,762

AP® Practice Questions

Preparing for the AP® Exam

Multiple-Choice Questions

1. The tragedy of the commons can be prevented by
   (a) externalities.
   (b) fisheries in international waters.
   (c) the use of harvest permits.
   (d) the use of public land for grazing.

2. Which is an example of an externality?
   (a) the cost to restore habitat in previously destroyed area
   (b) the use of the atmosphere for air pollution
   (c) tax breaks for use of high efficiency appliances
   (d) roads maintained with a tax on gasoline

3. Which is used to reduce the impact of timber harvesting?
   (a) tree plantations
   (b) selective cutting
   (c) prescribed burns
   (d) mechanized logging

4. Which of the following is a consequence of clear-cutting?
   (a) soil erosion
   (b) air pollution
   (c) stream cooling
   (d) groundwater contamination

5. Logging with horses is currently utilized with the goal of
   (a) giving exercise to horses.
   (b) reducing the use of fossil fuels.
   (c) increasing soil compaction.
   (d) increasing mechanization during forest removal.

6. Overgrazing can occur in common areas. Place the diagram below in the correct order.

   (a) A, B, C
   (b) C, B, A
   (c) A, C, B
   (d) B, A, C

Free-Response Question

Researchers set out to determine if there is a connection between clear-cutting and water quality levels for a nearby river. The researchers also measured the number of brown trout found in different sections of the river. The data collected are presented in the table below.

Clear-cutting area and dissolved oxygen levels

Clear-cutting percentage area	Dissolved oxygen levels	Brown trout
15%	6.2 mg/L	38
30%	5.3 mg/L	21
55%	3.1 mg/L	8

(a) Using the data in the table above, **identify** the percentage area of clearcutting with the lowest number of brown trout. (1 pt.)

(b) **Describe** the relationship between area clear-cut and dissolved oxygen levels indicated by the data in the table. (1 pt.)

(c) **Describe** how clear-cutting impacts species like the brown trout. (1 pt.)

(d) **Explain** the relationship between brown trout numbers and the percent of the area that was clear-cut. (2 pts.)

(e) For the experiment described above:
   (i) **Identify** the independent variable in the investigation. (1 pt.)
   (ii) **Describe** two possible research questions that may have spurred the investigation. (2 pts.)

(f) After the data from this research is published, the researchers recommend restoring areas by the river to increase numbers of brown trout. **Make a claim** to support or refute the researcher's prediction and **justify the claim** using evidence. (2 pts.)

Module 25

The Green Revolution

In the previous module we identified common resources such as air, water, and soil that are susceptible to harm by human activity as a consequence of the tragedy of the commons. In this module we examine agricultural methods that utilize the commons to produce a large amount of food at the lowest economic price, but not necessarily resulting in the least environmental impact. It is important to understand modern large-scale agricultural methods, their benefits, and their costs before examining alternatives.

Learning Goals

After reading this module you should be able to

25-1 describe changes in agricultural practices that have occurred throughout human civilization.

25-2 explain changes that developed as a result of the Green Revolution.

25-3 identify the benefits and consequences of Green Revolution innovations.

25-4 identify specific benefits and consequences of GMOs.

25-5 describe the resulting efficiency and fossil fuel impacts of mechanization.

25-1 What changes in agricultural practices have occurred throughout human civilization?

Agriculture has undergone a number of transformations over time

Agriculture is the planting and harvesting of domesticated plant species and the raising of domesticated animals for food. Domestication is the lengthy process by which humans selectively breed plants and animals that have been separated from wild populations to create distinct species. We call these plants and animals domesticated species. Early humans hunted their food and gathered wild plant species that they found and therefore have been named hunter-gatherers. Over thousands of years, toward the end of the Neolithic, or Stone, Age, early humans began to domesticate plants and animals. This occurred in multiple locations around the world and is typically called the First Agricultural Revolution, dating back to between 11,000 and 12,000 years ago. The Second Agricultural Revolution began in the late 1600s and continued through the 1930s. It involved mechanization of farming through the use of tools for plowing soil, planting seeds, and harvesting crops and toward the end of this period involved the use of the tractor, first powered by steam and then by gasoline and diesel. The Second Agricultural Revolution refers primarily to events happening in the United States and western Europe and is characterized by the move from **subsistence farming**, which means farming for consumption by the farming family and maybe a few neighbors, to farming with the purpose of selling the resulting food in village and city markets. It overlaps the timeframe of the Industrial Revolution and as a result, mechanization is part of the Second Agricultural Revolution. **FIGURE 25.1** shows a subsistence farm in Portugal and a larger farm in New York.

Just as a healthy ecosystem supports a wide range of species, a healthy soil supports abundant and continuous food production. Food grains such as wheat, corn, and rice provide more than half the calories and proteins that humans consume. However, the development of the Third Agricultural Revolution greatly increased food output but also began to strain ecosystems and soils to an extent that had not occurred before. This Third Agricultural Revolution is also called The Green Revolution, which we will explore in greater detail in the following section.

From the late 1800s through 1930, there were numerous famines worldwide that were caused by drought, other climatic conditions, and by government policies and wars. By the 1940s, the Green Revolution began and larger quantities of food became available than had ever been produced before. However, increases in the production of food initiated a higher level of environmental degradation than had ever been experienced prior to that time. In addition, the increased abundance of food contributed to the exponential growth of the human population. As we have seen, this growing human population has created even greater stresses on Earth's resources. An increase in food production leads to a positive feedback loop: With more food comes more people, who require even more food production.

Subsistence farming Farming for consumption by the farming family and maybe a few neighbors.

(a) (b)

FIGURE 25.1 **Subsistence versus larger-scale farming.** (a) A subsistence farm in Portugal. This farmer is watering and harvesting from her backyard garden that provides food for the farmer and the farmer's family, which leaves little, if any, surplus for sale or trade. (b) A larger farm in New York. The Soul Fire Farm in Petersburg, New York, grows food for distribution to hundreds of food-insecure people and for sale at its farmstand and mail order business. It also offers on-site training for hundreds of farmers of color each year. *(a: Westend61 GmbH/Alamy Stock Photo; b: Capers Rumph)*

In the twentieth century, farming became more mechanized, and the use of fossil fuel energy in food production increased. **Industrial agriculture**, or **agribusiness**, applies the techniques of mechanization and standardization to the production of food. Modern agribusiness today is quite different from the small family farms that dominated agriculture only a few decades ago.

25-2 What changes developed as a result of the Green Revolution?

The Green Revolution led to greater food production

How did we get to our current levels of energy use for food production? In the twentieth century, the agricultural system was dramatically transformed from a system of small farms relying mainly on human labor with relatively low fossil fuel inputs to a system of large industrial operations with fewer people and a lot more machinery. In the United States, there were roughly 6 million farms in 1900 and 2 million farms in 2020. In 1920, 32 million people or 32 percent of the U.S. population lived on farms and worked there, at least part-time. By 2020, there were roughly 2.6 million people or 1.3 percent of the U.S. workforce working on farms. This shift in farming practices, known as the **Green Revolution** (or Third Agricultural Revolution), involved new management techniques and mechanization as well as the triad of fertilization, irrigation, and improved crop varieties. These changes increased food production dramatically, where a smaller number of farmers were able to feed many more people. The changes also resulted in greater use of fossil fuels and greater environmental degradation.

The Green Revolution began with the work of crop scientists, particularly the American scientist Norman Borlaug (1914–2009), who won the Nobel Peace Prize for his contribution to increasing the world food supply. Through intensive breeding, Borlaug and other agricultural researchers developed strains of wheat that were disease resistant and produced higher yields. They also used fertilizers and irrigation to improve yields. Researchers brought these techniques to developing nations such as Mexico and the Philippines to help increase their agricultural output and feed their growing populations. From the 1950s through the 1970s, many countries, particularly those in the developing world, underwent similar shifts in the way they farmed. The upward trend in world grain production since 1950 (FIGURE 25.2) is a result of the Green Revolution. From the mid-1960s through the mid-1980s, world grain production doubled.

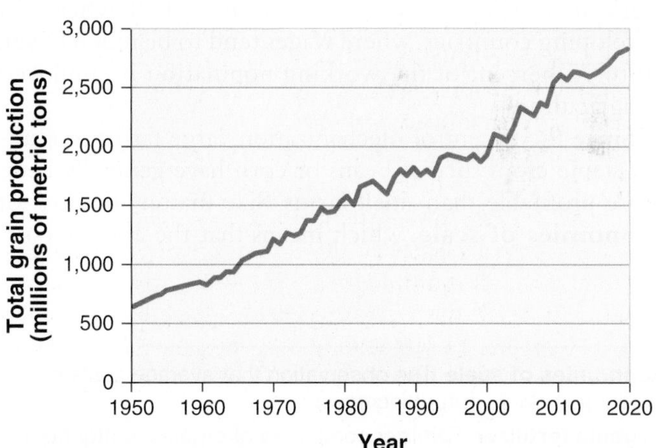

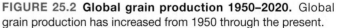

FIGURE 25.2 **Global grain production 1950–2020.** Global grain production has increased from 1950 through the present.

Industrial agriculture Agriculture that applies the techniques of mechanization and standardization to the production of food. *Also known as* **agribusiness.**

Green Revolution A shift in agricultural practices in the twentieth century that included new management techniques and mechanization, as well as the triad of fertilization, irrigation, and improved crop varieties, that resulted in increased food output.

25-3 What are the benefits and consequences of Green Revolution innovations?

The Green Revolution led to advances in mechanization, fertilization, irrigation, and the use of pesticides

By the 1990s, there were at least 18 organizations around the world promoting and developing Green Revolution techniques, including mechanization, use of fertilizers, irrigation, use of pesticides, and monocropping. However, the Green Revolution has also had negative environmental impacts. Let's examine some Green Revolution practices in more detail.

Mechanization

Farming involves many kinds of work. Fields must be plowed, planted, irrigated, weeded, protected from pests, harvested, and prepared for the next season. After harvesting, crops must be dried, sorted, cleaned, and prepared for market in almost as many different ways as there are different crops. Machines do not necessarily do this work better than humans or animals, but it can be economically advantageous to replace humans or animals with machinery, particularly if fossil fuels are abundant, fuel prices are relatively low, and labor prices are relatively high. In developed countries, where wages are relatively high, less than 5 percent of the workforce works in agriculture (both on the farm and elsewhere). In developing countries, where wages tend to be much lower, 40 to 75 percent of the working population is employed in agriculture.

Since the advent of mechanization, large farms producing staple crops such as beans or corn have generally been more profitable than small farms. Size matters because of **economies of scale**, which means that the average costs of production fall as output increases. For example, tractors and other large farm equipment can cost $150,000 or more. This large up-front expenditure is a good investment for a large farm, where the cost of the machine is justified by the profits on the increased production that comes from using it. A small farm, however, does not have enough land to be able to recoup the cost of such expensive equipment. Because of economies of scale, profits tend to increase with size, and large agricultural operations generally outcompete small ones. As a result, between 1950 and 2000, the average farm size in Iowa more than doubled, from about 70 ha (173 acres) to over 140 ha (346 acres).

Mechanization also means that single-crop farms are generally more efficient than farms that grow many crops. Because mechanized crop planting and picking require specialized, expensive equipment specific to each crop, planting and harvesting only a single type of crop reduces equipment costs.

Fertilizers

Growing crops and transporting them from the farm system to humans for consumption removes organic matter and nutrients from soil. If these materials are not replenished, they can be quickly depleted. Industrial agriculture, because it keeps soil in constant production, requires large amounts of fertilizers to replace lost organic matter and nutrients. Fertilizers contain essential nutrients for plants—primarily nitrogen, phosphorus, and potassium—and they foster plant growth where one or more of these nutrients is lacking. There are two types of fertilizers used in agriculture: organic and synthetic.

As their name suggests, **organic fertilizers** are composed of organic matter from plants and animals. They are typically made up of animal manure and crop residues that have been allowed to decompose. Traditional farmers often spread animal manure and crop wastes onto fields to return some of the nutrients that were removed from those fields when crops were harvested. **Synthetic fertilizers**, or **inorganic fertilizers**, are produced commercially, normally with the use of fossil fuels. Nitrogen fertilizers are often produced by combusting natural gas, which allows nitrogen from the atmosphere to be fixed and captured in fertilizer. Fertilizers produced in this way are highly concentrated, and their widespread use has increased crop yields tremendously since the Green Revolution began. Despite these advantages, synthetic fertilizers can have several adverse effects on the environment, such as their heavy reliance on fossil fuel energy. The United States uses somewhat less fertilizer and consequently experiences less nutrient runoff than other nations with similar agricultural output. Still, large amounts of nitrogen and other nutrients run into waterways in intensively farmed regions such as California's Central Valley, farming regions along the East Coast, and the Mississippi River watershed (**FIGURE 25.3**).

Economies of scale The observation that average costs of production fall as output increases.

Organic fertilizer Fertilizer composed of organic matter from plants and animals.

Synthetic fertilizer Fertilizer produced commercially, normally with the use of fossil fuels. *Also known as* **inorganic fertilizer**.

Waterlogging A form of soil degradation that occurs when soil remains under water for prolonged periods.

Salinization A form of soil degradation that occurs when the small amount of salts in irrigation water becomes highly concentrated on the soil surface through evaporation.

Pesticide A substance, either natural or synthetic, that kills or controls organisms that people consider pests.

FIGURE 25.3 **Fertilizer runoff after fertilization.** The proximity of this drainage ditch to an agricultural field may lead to runoff of fertilizer during a heavy rainstorm. *(Lynn Betts/USDA Natural Resources Conservation Service)*

Irrigation

Irrigation systems can increase crop growth rates or even enable crops to grow where they could not otherwise be grown (FIGURE 25.4). For example, irrigation has transformed approximately 400,000 ha (1 million acres) of former desert in the Imperial Valley of southeastern California into a major producer of fruits and vegetables. In other situations, irrigation can allow productive land to become extremely productive land. One estimate suggests that the 16 percent of the world's agricultural land that is irrigated produces 40 percent of the world's food.

While irrigation has many benefits, including more efficient use of water, it can also have a number of negative consequences over time. As we will explore in greater detail later in this unit, it can deplete groundwater, draw down aquifers, and cause saltwater intrusion into freshwater wells. It can also contribute to soil degradation through waterlogging and salinization, shown in FIGURE 25.5. **Waterlogging**, a form of soil degradation that occurs when soil remains under water for prolonged periods, impairs root growth because roots cannot obtain oxygen. **Salinization** occurs when the small amounts of salts in irrigation water become highly concentrated on the soil surface through evaporation. These salts can eventually reach toxic concentrations and impede plant growth.

Pesticides

Pesticides are substances, either natural or synthetic, that kill or control organisms that people consider pests. The use of pesticides has become routine and widespread in modern industrial agriculture. In the United States, the Environmental Protection Agency reports on the manufacture and use of

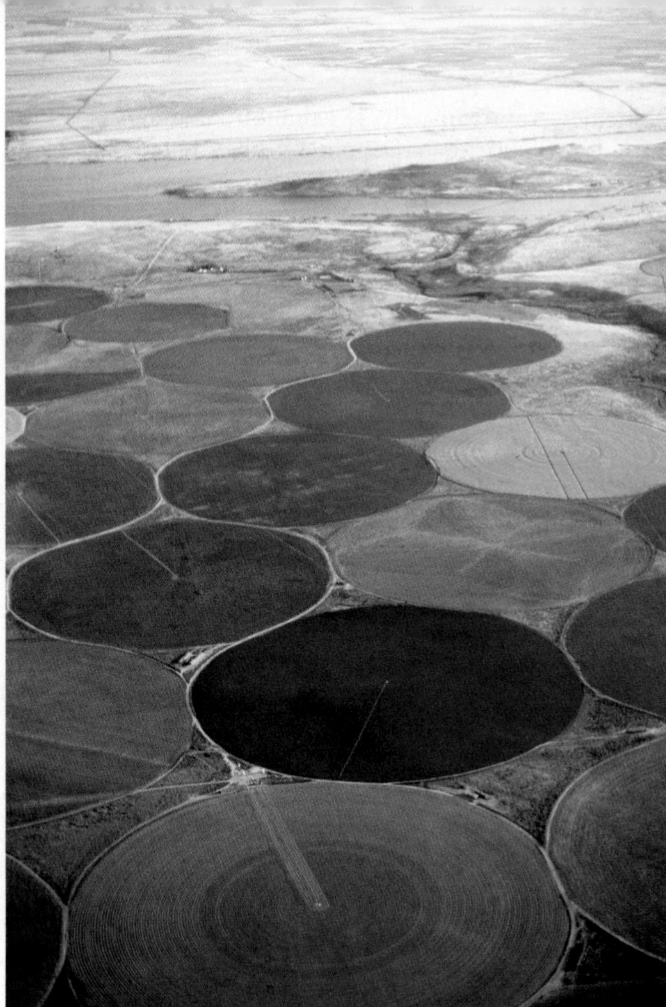

FIGURE 25.4 **Irrigation circles.** The green circles in this aerial photograph from Oregon are obvious evidence of irrigation. *(Doug Wilson/ARS/USDA)*

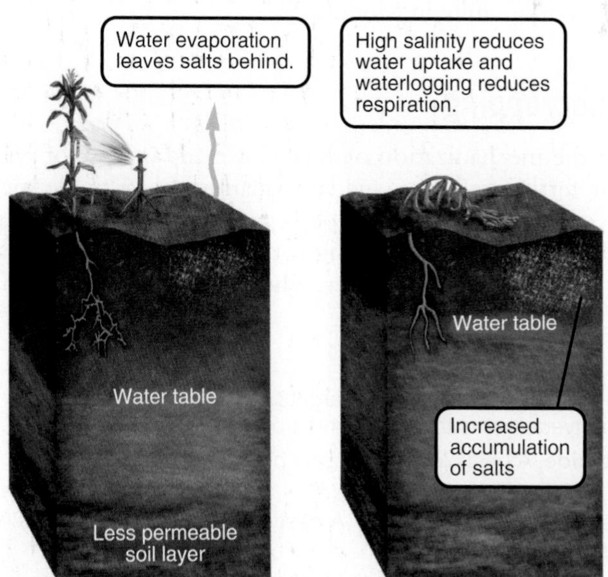

FIGURE 25.5 **Irrigation-induced salinization and water logging.** Over time, irrigation can degrade soil by leaving a layer of highly concentrated salts (such as sodium and calcium) at the soil surface and waterlogged soil below.

pesticides. Approximately 454 million kilograms (1 billion pounds) of pesticides are applied in the United States each year. Almost 90 percent are applied for agricultural purposes. The United States accounts for about one-fifth of worldwide pesticide use.

AP® Exam Tip

Many past free-response questions on the AP® Environmental Science Exam have asked about pesticide use. You should know the different categories of pesticides and understand the uses of each: insecticide, herbicide, broad-spectrum pesticide, and selective pesticide.

Insecticides target species of insects and other invertebrates that consume crops, and **herbicides** target plant species that compete with crops. Some pesticides are **broad-spectrum pesticides**, meaning that they kill many different types of pest, and some are **selective pesticides** or **narrow-spectrum pesticides** that target a narrow range of organisms. The broad-spectrum insecticide dimethoate, for example, kills almost any insect or mite—relatives of ticks and spiders—while the more selective narrow-spectrum acequinocyl kills only mites.

While pesticides can have the benefit of removing harmful organisms from crops and increasing yield, there are also many adverse consequences of applying them. They can harm beneficial organisms, leach into nearby waterways, and create possible health risks to the farm workers who apply the pesticides. We will explore these consequences of pesticide use more fully in Module 27.

Monocropping

Both the mechanization of agriculture and the use of synthetic fertilizers encourage large plantings of a single species or variety, a practice known as **monocropping** (**FIGURE 25.6**). Monocropping is the dominant agricultural practice in the United States, where corn, soybean, wheat, and cotton are frequently grown in monocrops of 405 ha (1,000 acres) or more.

Insecticide A pesticide that targets species of insects and other invertebrates that consume crops.

Herbicide A pesticide that targets plant species that compete with crops.

Broad-spectrum pesticide A pesticide that kills many different types of pest.

Selective pesticide A pesticide that targets a narrow range of organisms. *Also known as* **narrow-spectrum pesticide**.

Monocropping An agricultural method that utilizes large plantings of a single species or variety.

FIGURE 25.6 Monocropping. This large field in Iowa contains only one crop species, soybeans. There are both advantages and disadvantages to monocropping. *(Design Pics Inc/Alamy Stock Photo)*

Monocropping has greatly improved agricultural productivity. This technique allows large expanses of land to be planted at the same time and later harvested at the same time. With the use of large machinery, the harvest can be done easily and efficiently. If fertilizer or pesticide treatments are required, those treatments can also be applied uniformly over large fields, which, because they are planted with the same crop, have the same pesticide or nutritional needs.

Despite the benefits of increased efficiency and productivity, monocropping can lead to environmental degradation. First, soil erosion can become a problem. Because fields that are monocropped are readied for planting or harvesting all at once, soil will be exposed over many hectares at the same time. On a 405-ha (1,000-acre) field that has not yet been planted, wind can blow for over 1.6 km (1 mile) without encountering anything but bare soil. Under these circumstances, the wind can gain enough speed to carry dry soil away from the field. Certain farmland in the United States loses an average of 1 metric ton of topsoil per hectare (2.5 metric tons per acre) per year to wind erosion. As we saw in Unit 4, this topsoil contains important nutrients and its loss can reduce productivity.

Monocropping also makes crops more vulnerable to attack by pests. Large expanses of a single plant species represent a vast food supply for any pests that specialize on that particular plant. Such pests will establish themselves in the monocrop and reproduce rapidly. Their populations may experience exponential growth similar to what we saw in Georgy Gause's *Paramecium* populations that were supplied with unlimited food (see Figure 16.3 on page 184). Natural predators may not be able to respond rapidly to the exploding pest population. Many predators of crop pests, such as ladybugs and parasitic wasps, are attracted to the pests that feed on monocrops. But these predators also rely on non-crop plants for habitat, which monocropping removes. Therefore, predators that might otherwise control the pest population are largely absent.

25-4 What are the specific benefits and consequences of GMOs?

Genetic engineering and the introduction of genetically modified organisms are revolutionizing agriculture

Humans have modified plants and animals by artificial selection for thousands of years. The modern techniques of genetic engineering, however, go far beyond traditional practices. As we first discussed in Module 13, scientists today can isolate a specific gene from one organism and transfer it into the genetic material of another, often very different, organism to produce a genetically modified organism, or GMO. By manipulating specific genes, agricultural scientists can rapidly produce organisms with desirable traits that may be impossible to develop with traditional breeding techniques. Genetically modified organisms present both benefits and drawbacks.

The Benefits of Genetically Modified Organisms

Genetically modified crops and livestock offer the possibility of greater yield and food quality, reductions in pesticide use, and higher profits for the agribusinesses that use them. They are also seen as a way to help reduce world hunger by increasing food production and reducing losses to pests and varying environmental conditions.

Increased Crop Yield and Quantity

Genetic engineering can increase food production in several ways. It can create strains of organisms that are resistant to pests and harsh environmental conditions such as drought or high salinity. In addition, agricultural scientists have begun to engineer plants that produce essential nutrients for humans. For example, they have inserted a gene for the production of vitamin A into rice plants, creating new seeds known as golden rice (**FIGURE 25.7**). Although golden rice has not gained widespread acceptance and use, some scientists hope that it will help reduce the incidence of blindness resulting from vitamin A deficiency. Crop plants, animals, and bacteria have also been modified to produce pharmaceuticals and other compounds, a process that can make these products far less expensive to manufacture. A number of projects are under way to create genetically modified animals for food production, including salmon that grow to its full size of 3.6 kg (8 pounds) in 18 months—half the growing time of an unmodified fish.

FIGURE 25.7 White rice and golden rice. Crop scientists have inserted a gene that synthesizes a precursor to vitamin A in white rice. The resulting genetically modified rice is called golden rice. *(Reuters/Erik De Castro)*

Increased Profits

GMO seed crops can increase farm profits in two ways. Although GMO crop seeds do cost more to purchase than conventional crop seeds, they reduce pesticide use, which can be a significant savings. GMO crops can also raise revenues by producing greater yields. Both of these changes can lead to higher incomes for farmers, lower food prices for consumers, or both.

Concerns About Genetically Modified Organisms

Industrial agriculture relies more heavily on genetically modified crops each year. In 2020, 94 percent of the corn, 96 percent of the soybeans, and 95 percent of the cotton planted in the United States came from genetically modified seeds. However, many European countries, as well as a number of people in the United States, question the safety of GMOs. Genetically modified crops and livestock are the source of considerable controversy, and concerns have been raised about their safety for human consumption and their effects on biodiversity. Regulation of GMOs is also an issue, both in the United States and abroad.

Safety for Human Consumption

Some people are concerned that the ingestion of genetically modified foods may be harmful to humans, although so far there is little evidence to support these concerns. However, researchers are studying the possibility that GMOs may cause allergic reactions when people eat a food containing genes transferred from another food to which they are allergic.

Effects on Biodiversity

There is some concern that if genetically modified crop plants are able to breed with their wild relatives — as many domesticated crop plants are — the newly added genes will spread to the wild plants. The spread of such genes might then alter or eliminate natural plant varieties. Examples of GMOs crossing with wild relatives or living in the wild do exist. Because of these concerns, attempts have been made to introduce buffer zones around genetically modified crops.

The use of genetically modified seeds is contributing to a loss of genetic diversity among food crops. As with any reduction in biodiversity, we cannot know what beneficial genetic traits might be lost. For example, researchers recently discovered that one variety of sorghum, a cereal grass grown for its sweet juice extract, has a natural genetic variation that gives it resistance to a pest called the greenbug. This particular variety has since been used extensively to confer greenbug resistance on the U.S. sorghum crop. If growers had been growing only one or two genetically modified varieties of sorghum, this naturally resistant variety might have been eliminated before its beneficial trait was discovered.

Regulation of Genetically Modified Organisms

In 2016, the National Bioengineered Food Disclosure Standard, also known as the GMO labeling law, was passed. As of 2022, the United States Department of Agriculture (USDA), implemented a law that requires disclosure and labeling if a food contains a GMO product. Labeling is regarded as a measure of transparency, allowing a consumer to understand how a food has been grown or processed. The standard does not restrict the use of GMOs in plants. The European Union, in contrast, allows very few genetically modified foods. In France, Germany, and Italy, almost all GMOs are banned. Labeling opponents in the United States argue that labeling of foods containing GMOs might suggest to consumers that there is something wrong with GMOs. Those who want to avoid consuming GMOs can purchase organic food; the federal definition of "organic" excludes genetically modified foods.

As noted earlier, there are companies working on genetically modified animals including salmon. There are currently applications for a number of genetically modified animals approved and under appeal by the Food and Drug Administration. Genetically modified salmon were sold for the first time in the United States beginning in 2021 (**FIGURE 25.8**), although they are reportedly hard to find for sale in stores.

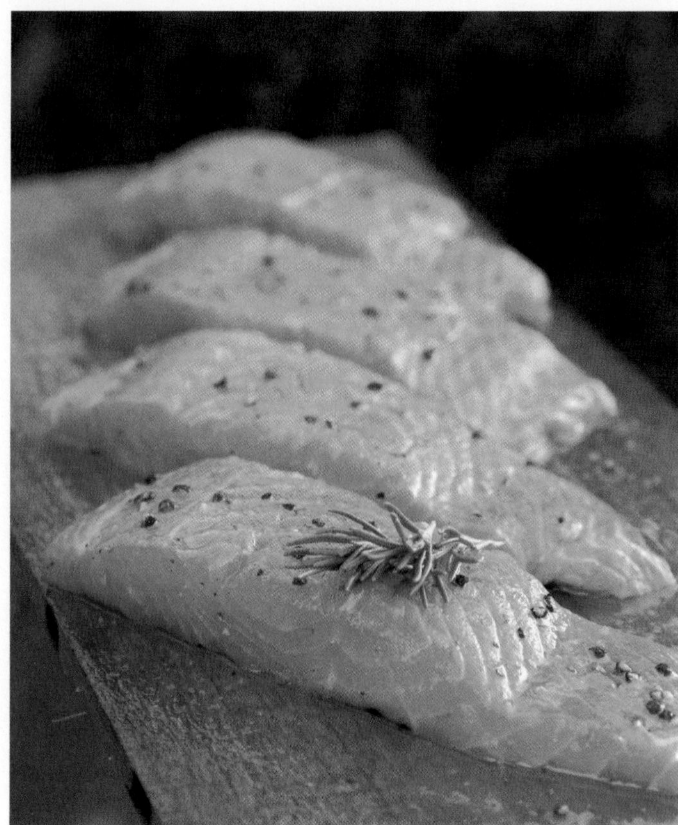

FIGURE 25.8 Genetically modified salmon ready to be cooked. Beginning in 2021, genetically modified salmon has been served in restaurants in the United States, although it is fairly uncommon to find it. *(AquaBounty/MCT/Sipa USA)*

25-5 What are the resulting efficiency and fossil fuel impacts of mechanization?

Mechanization results in the energy subsidy in agriculture

As we presented earlier in this module, mechanization has many benefits, especially in larger, commercial farms that can justify and earn profits to pay for expensive equipment. However, a farm that has a great deal of mechanization, whether it is a tractor to work the fields, a pump to irrigate water, or a truck to move around fertilizer, pesticides, or animal feed, is going to require fossil fuel energy or electrical energy to power the equipment.

In larger, mechanized farms, activities associated with planting, growing, fertilizing, irrigating, and harvesting food require a great deal of energy input beyond solar energy. Off the farm, processing and preparing food also require a great deal of energy input. The fossil fuel energy and human energy input per calorie of food produced — beyond that which is provided by the sun—is called the **energy subsidy**. In other words, if we use 5 calories of human and fossil fuel energy to produce food, and we receive 1 calorie of energy when we eat that food, then the food has an energy subsidy of 5. We can think of this in another way if we use a mass of inputs and outputs as a substitute measure for energy. If it takes 20 kg (44 pounds) of grain to feed cattle to produce 1 kg (2.2 pounds) of beef, the energy subsidy is 20. If it takes 2.8 kg (6 pounds) of grain to feed chickens to produce 1 kg (2.2 pounds) of chicken meat, the energy subsidy is 2.8, which is considerably smaller than the energy subsidy for beef. The comparison is not perfect, because the energy content of 1 kg of beef is greater than that of 1 kg of chicken. Nevertheless, it should help you to understand the relationship between energy inputs and outputs in producing food. And if it seems similar to the concept of the trophic pyramid that you learned about in Module 7, you are making a good connection: Both the trophic pyramid and the energy conversion losses explained here involve the second law of thermodynamics and inefficiencies during energy transformations. Whatever quantity of energy or food you put into the system (a cow or a farm), the resulting output is always going to be less.

As **FIGURE 25.9** shows, traditional small-scale agriculture and gathering of wild foods requires a relatively small energy subsidy beyond the energy provided by the sun: It uses a fraction of a calorie of energy input from the human collecting it per calorie of food produced. By contrast, food writer and journalism professor Michael Pollan estimates that if you eat the average modern U.S. diet, which contains foods mostly produced by modern agricultural methods, there is a 10-calorie energy input for every calorie you eat. Most of that energy input is fossil fuel energy. Therefore, food choices are also energy choices. And they are land choices as well. "Do the Math: Land Needed for Food" on page 314 shows you how to calculate the different land requirements for obtaining sufficient calories from corn and from beef.

Most of the energy subsidies in modern agriculture are in the form of fossil fuels, which are used to produce fertilizers and pesticides, to operate tractors, to pump water for irrigation, and to harvest food and prepare it for transport. Other energy subsidies take place off the farm. Although the average food item travels roughly 2,000 km (1,240 miles) from the farm to your plate, transportation accounts for only about 10 percent of the energy subsidy. The Department of Energy reports that in the United States, 17 percent of total commercial energy use goes into growing, processing, transporting, and cooking food. The largest fraction of the energy subsidy goes to food production at the farm and food processing in the factory and in the home. Those of us eating a supermarket diet in the developed world are highly dependent on fossil fuel for our food; the agricultural system we use in the developed world today could not exist without fossil fuels.

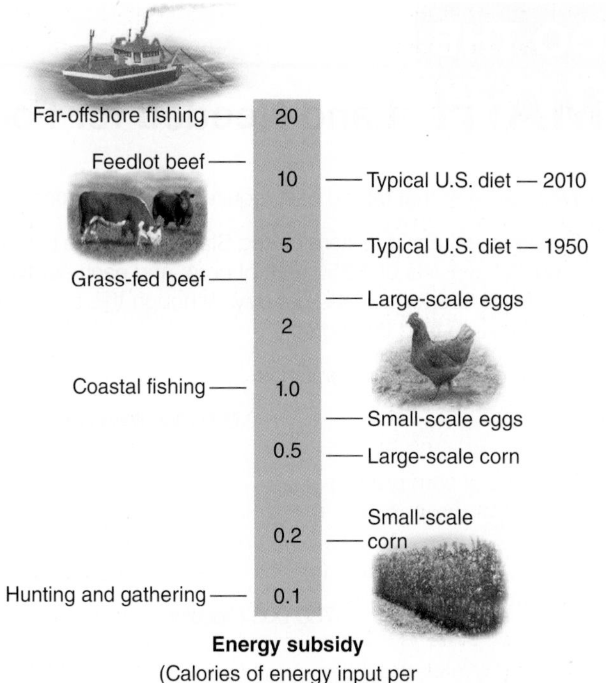

FIGURE 25.9 Energy subsidies for various methods of food production and diets. Additional energy (beyond that provided by the Sun) input per calorie of food obtained is greater for modern agricultural practices than for traditional agriculture. Energy inputs for hunting and gathering and for small-scale food production are mostly in the form of human energy, whereas fossil fuel energy is the primary energy subsidy for large-scale modern food production. All values are approximate, and for any given method there is a large range of values.

Earlier revolutions introduced efficiency improvements to agriculture. The Green Revolution introduced changes to agriculture in the 1950s and 1960s that continue to result in increased agricultural yield today. Mechanization, fertilization, irrigation, and the use of pesticides all improve crop yields but also have adverse environmental consequences. Genetic engineering has also changed agriculture, but it remains controversial. Large amounts of fossil fuel energy and fertilizer are common inputs to most agriculture in the developed world. In the following modules, we will explore some of the Green Revolution innovations in greater detail.

Energy subsidy The fossil fuel energy and human energy input per calorie of food produced.

DO THE MATH Land Needed for Food

Preparing for the AP® Exam

We have seen that raising beef requires more resources than growing grain. Let's look at some of the actual numbers.

On farms in the midwestern United States, a hectare of land yields roughly 370 bushels of corn (equivalent to 150 bushels per acre). A bushel consists of 1,250 ears of corn, and each ear typically contains 80 kilocalories. Assume that a person eats only corn and requires 2,000 kilocalories per day. Although this assumption is not very realistic, it allows an approximation of how much land it would take to feed that person.

The person's food requirement is

$$2{,}000 \text{ kilocalories/day} \times 365 \text{ days/year} = 730{,}000 \text{ kilocalories/year}$$

A hectare of corn produces

$$370 \text{ bushels/hectare} \times 1{,}250 \text{ ears/bushel} \times 80 \text{ kilocalories/ear}$$

$$= 37{,}000{,}000 \text{ kilocalories/hectare}$$

$$730{,}000 \text{ kilocalories/year} \div 37{,}000{,}000 \text{ kilocalories/hectare} = 0.02 \text{ ha } (0.05 \text{ acres})$$

of land to feed one person for a year

Thus, one person eating only corn can obtain sufficient calories in a year from 0.02 ha (0.05 acres) of land. What if that person ate only beef? We have seen that it takes 20 kg of grain to produce 1 kg of beef. So it would take 20 times as much land, or 0.4 ha (1 acre), to feed a person who ate only beef.

What happens if we extend this analysis to a global scale? If Earth has about 1.5 billion ha (3.7 billion acres) of land suitable for growing food, is there sufficient land to feed all 7.8 billion inhabitants of the planet if they all eat a diet of only beef?

$$7.8 \text{ billion people} \times 0.4 \text{ ha/person} = 3{,}120{,}000{,}000 \text{ ha}$$

So 3.1 billion ha (7.7 billion acres) would be needed, and the answer is no.

YOUR TURN How many people eating a beef-only diet can Earth support?

Module 25 AP® Review

Preparing for the AP® Exam

Learning Goals Revisited

25-1 What changes in agricultural practices have occurred throughout human civilization?

Early revolutions introduced efficiency improvements to agriculture. In the last century, mechanization resulted in drastic improvements in crop output.

25-2 What changes developed as a result of the Green Revolution?

Through intensive crop breeding, agricultural researchers developed strains of wheat that were disease resistant and produced higher yields. They also used mechanization, fertilizers, and irrigation to improve yields in a variety of crops and introduced these methods in the developed and developing worlds.

25-3 What are the benefits and consequences of Green Revolution innovations?

Green Revolution management techniques, mechanization, fertilization, irrigation, and improved crop varieties increased food production dramatically, and a smaller number of farmers were able to feed many more people.

While these innovations resulted in greater crop yields, there were also increased energy inputs and increases in fertilizer and pesticide runoff.

25-4 What are the specific benefits and consequences of GMOs?

By manipulating specific genes, agricultural scientists can rapidly produce organisms with desirable traits. This has changed agriculture, but it remains controversial. GMOs have yielded higher outputs but there are concerns about human safety and biodiversity.

25-5 What are the resulting efficiency and fossil fuel impacts of mechanization?

Large amounts of fossil fuel energy and fertilizer are common inputs to most agriculture in the developed world. Certain crops and certain agricultural activities are particularly energy intensive and therefore have higher environmental consequences.

Preparing for the AP® Exam

AP® Practice Questions

Multiple-Choice Questions

1. If 12,000,000 kilocalories of chicken can be produced per hectare per year, how much land is needed to provide a person with 2,000 kilocalories of chicken/day for a year?
   (a) 0.01 ha
   (b) 0.02 ha
   (c) 0.06 ha
   (d) 0.2 ha

2. The Green Revolution
   (a) discouraged the mechanization of agriculture.
   (b) decreased the energy subsidy of most food.
   (c) encouraged the use of monocropping.
   (d) pertained to leafy green plants only.

3. Which of the following is a typical change in genetically modified organisms?
   (a) increased reliance on pesticides
   (b) increased resistance to extreme weather
   (c) decreased crop yield
   (d) increased genetic diversity

4. The energy subsidy in modern agriculture
   (a) reduces financial cost for food.
   (b) identifies the additional energy beyond the sun added to food.
   (c) provides economic benefits for farmers as well as consumers.
   (d) can appear to be a violation of the Second Law of Thermodynamics.

Use the passage below to answer questions 5 & 6:

Two farmers growing corn received a cash loan from a federal stimulus plan. They used the loan to purchase updated equipment to harvest corn faster, and to purchase genetically modified seeds. Both farms increased their output by over 85 percent compared to the same month last year.

5. Which of the following correctly identifies the claim from the text?
   (a) Cash loans to farms has a negative impact on output.
   (b) Genetically modified seeds have adverse impacts on human health.
   (c) Updated mechanization can increase farm output.
   (d) The farmers harvested corn at the same rate as previous years.

6. Which of the following pieces of evidence supports the claim from the text?
   (a) Federal money was used to purchase GMO seeds.
   (b) The farms' output went up after purchasing GMO seeds.
   (c) The farms' output went down from year to year.
   (d) The farmers switched seeds from soy to corn.

Free-Response Question

In the twentieth century the Green Revolution transformed agricultural systems from small farms to large industrial operations. One method for improving crop yields is genetic modification.

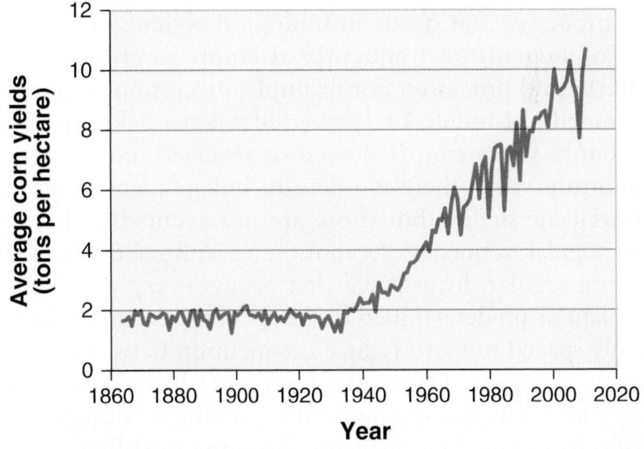

(a) **Identify** which year corn yields first reached 4 tons per hectare. (1 pt.)
(b) **Describe** two possible methods that were used to increase crop yields from the 1950s onward. (2 pts.)
(c) **Describe** one possible disadvantage to one of the methods you identified in part (b). (1 pt.)
(d) **Explain** how mechanization and the use of synthetic fertilizers led to monocropping. (2 pts.)
(e) **Describe** two sustainable alternatives to monocropping and their advantages. (2 pts.)
(f) **Describe** one possible disadvantage to an alternative you described in part (e). (2 pts.)

Module 26

Impacts of Agricultural Practices

In the previous module we introduced the history of agriculture and examined the Green Revolution. We also identified some of the impacts of large-scale modern agriculture. Here we examine in greater depth some of the agricultural practices that increase food output but also cause harm, specifically plowing, tilling, slash-and-burn agriculture, and the use of fertilizers.

Learning Goals

After reading this module, you should be able to

26-1 explain how plowing and tilling interfere with the natural progression of ecosystems.

26-2 identify the consequences of slash and burn farming.

26-3 describe the effects of fertilizer use.

26-1 How do plowing and tilling interfere with the natural progression of ecosystems?

Many agricultural practices work against succession and other natural processes

In Units 2 and 3, we explored a variety of natural patterns and processes that occur in biological systems. The practice of agriculture frequently attempts to change these patterns and processes. For example, succession, which we examined in Module 14, is reversed when a field is plowed. Certainly, succession is slowed or reversed under natural conditions when there is a density-independent event like a hurricane or fire, but those are not events that happen with regular frequency. Yet in most agricultural fields, plowing is a regular disturbance that occurs every year. Farmers plant at predetermined distances between each plant in evenly spaced rows, to reduce competition between a plant and its neighbor. That certainly doesn't happen in nature very often. Monocropping, or the planting of a single crop, results in minimal biodiversity, also something that is fairly rare in nature. And the existence of agricultural fields leads to fragmented habitats, which hinders biodiversity. So, while evolving agriculture practices have had remarkable impacts on food production, their unnatural human impacts can also harm Earth.

Plowing The process of digging deep into the soil and turning it over.

Tilling The preparation of soil through a variety of activities including plowing but also including stirring, digging, and cultivating.

Plowing and tilling reverse succession and mobilize organic matter

Plowing is the process of digging deep into the soil and turning it over. This action modifies the soil in a variety of ways including breaking up soil structure and exposing previously buried material to the air. It also brings nutrients from the B horizon to the surface and buries plants including undesired species (weeds) deeper into the soil where, deprived of sunlight, they will probably die. (You might want to turn back to Module 20 for a refresher on the different soil horizons.) Plowing is done on agricultural land currently in use and on land that has not previously been used for agriculture, like a field or pasture, at the beginning of a season before planting. Plow depths vary around the country and world based on soil and the desired crop, but 15 to 20 cm (6 to 8 inches) is a normal range for plowing depth. Typically, plowing turns over the A horizon of soil. Most commonly, it is done right before planting seeds at the beginning of the growing season. It is normally done with a tractor although it can be done with animals or by hand. Plowing has been done for thousands of years and there are many different kinds of tools used to achieve different kinds of plowing techniques and soil turnover. **Tilling** is the preparation of soil through a variety of activities including plowing but it also includes stirring, digging, and cultivating. This action modifies the soil but not usually to the degree that is achieved by plowing. It has many of the same effects on the soil as plowing does, but it is often used to bury undesired plants (weeds) so they do not interfere with the desired crop. Tilling can be done with a tractor and by hand

(a)

(b)

(c)

FIGURE 26.1 Plowing is usually done with a tractor or animal while tilling is done with a tractor or with a walk-behind power tool. (a) Oxen plowing a field; (b) tilling with tractor; (c) home gardener rototilling their garden. *(a: John Elk III/Getty Images; b: Graham Turner/Alamy Stock Photo; c: alvarez/Getty Images)*

and it is also done with a walk-behind power tool called a rototiller. Tilling can be done anytime during the season and is often done at the end of the growing season after the crop has been harvested. Some different plowing and tilling activities are shown in **FIGURE 26.1**.

Plowing and tilling both "turn over" the A or A and B soil horizons and aerate the soil as well as break up soil structure, mix soil horizons, and loosen the soil so that roots can more easily penetrate it. In addition to exposing nutrients that were previously deeper in the soil, plowing and tilling expose rock material in the soil to mechanical and chemical weathering, both of which also increase available nutrients in the soil. Plowing and tilling move organic matter from the surface to deeper in the soil. The net result of all these activities may lead to increased cation exchange capacity (CEC) and base saturation in the soil, concepts that you'll remember from Module 20. These actions typically benefit the soil and result in a higher crop yield. However, these same activities may also cause adverse consequences to the soil. For example, by loosening and disturbing the vegetation and A horizon, plowing and tilling may expose organic matter to erosion by wind and water. When organic matter is lost from the soil, it can reduce the CEC of the soil and reduce the nutrient content. Finally, erosion by wind may increase particulate matter in the air (an air pollutant we will discuss in Unit 7) (**FIGURE 26.2**) and it may increase sediments in nearby waterways, which increases the turbidity of the water (cloudiness) and may interfere with oxygen uptake by fish. In addition, organic matter from the A or B horizon being brought to the surface may readily decompose and contribute carbon dioxide, a greenhouse gas, to the atmosphere. Finally, the repeated plowing of a soil may

FIGURE 26.2 Plowing can lead to soil erosion by wind and water. This plow operation is overturning and breaking up the soil and exposing it to erosion by wind. *(Unknown/Alamy Stock Photo)*

ultimately lead to soil compaction at the bottom of the depth of disturbance. Soil compaction is a process where repeated trampling by humans, machinery, or animals causes a compaction of soil and a reduction in soil pore space. This can result in decreased soil water infiltration and root penetration. This compaction leads to a human-created horizon called plow layer. This layer is typically in the A horizon and is sometimes designated as an "Ap horizon," which means it is an A horizon that shows indications of being plowed, hence the lower case p. If an Ap horizon becomes compacted enough, it can interfere with root penetration and water infiltration.

26-2 What are the consequences of slash-and-burn farming?

Slash-and-burn farming causes short-term gains and long-term losses

In locations with a warm climate and relatively nutrient-poor soils, such as the rainforests of Central and South America, southeast Asia, and parts of Africa, most of the nutrients in a forest are contained within the vegetation. The A and B soil horizons contain smaller quantities of nutrients. In these locations, farmers sometimes use **slash-and-burn agriculture**, in which the land is cleared, the vegetation is burned, and nutrients are released to the soil. The land is farmed for only a few years until the soil is depleted of nutrients. This technique is sometimes called **shifting agriculture** (FIGURE 26.3). Swidden agriculture is sometimes used synonymously with slash-and-burn agriculture and sometimes it is used to refer to smaller-scale, short-term clearing done by indigenous communities. The resulting ash is rich in potassium, calcium, and magnesium, which increases the base saturation of the soil. As we know from Module 20, this improves the condition of the soil. However, because we are describing a situation that occurs mostly in tropical regions, where rain is plentiful, these nutrients are quickly leached from the soil. When the slash-and-burn agriculture occurs in areas of intense rainfall, both nutrients and soil may be washed away within a few years, which ultimately reduces the fertility of the soil. After a few years, the farmer usually moves on to another plot and repeats the process.

If a forest is cleared and burned and used for a few years and then abandoned for a number of decades, over time the soil may recover its organic content and nutrient supplies and the vegetation may have a chance to regrow. However, in recent decades, population pressure has caused the land to be overused without enough time to allow for its full recovery between uses. One additional consequence of slash-and-burn agriculture is soil compaction, which we have learned is also a consequence of tilling. When soil compaction happens, soil productivity can decrease rapidly, leaving the land suitable for animal grazing only, and not for farming crops. Furthermore, the disturbance of the soil during the clearing of vegetation and dragging it into piles creates a disturbance that is similar to plowing or tilling the soil. This contributes to soil compaction and exposes organic matter to erosion.

Slash-and-burn agriculture An agricultural method in which land is cleared and farmed for only a few years until the soil is depleted of nutrients. *Also known as* **shifting agriculture**.

FIGURE 26.3 **Shifting agriculture.** This forest in Thailand has been cleared for agriculture in a process called slash-and-burn farming or swidden agriculture. Clearing land in the tropics often involves burning it, which frees nutrients for uptake by crops, but also may make those nutrients susceptible to leaching and the soils vulnerable to erosion. *(UniversalImagesGroup/Getty Images)*

The burning of vegetation also has adverse consequences for air pollution and global climate change. Firstly, it oxidizes carbon, meaning that it converts it into the oxide compounds carbon monoxide (CO) and carbon dioxide (CO_2). The carbon monoxide and associated particulate matter contribute to local air pollution. And the carbon dioxide contributes to the atmospheric global carbon dioxide concentrations. In fact, clearing of land and burning the vegetation is the second largest human contributor to atmospheric carbon dioxide concentration increases after combustion of fossil fuels. This is often referred to as the net destruction or net clearing of forest and it merits a little more explanation: Each day, living trees remove CO_2 from the atmosphere during photosynthesis, and each night, respiring trees add CO_2 to the atmosphere. This part of the carbon cycle does not change the net atmospheric carbon because the inputs and outputs are approximately equal. However, when forests are destroyed by burning, as FIGURE 26.4 shows, all the carbon in those trees is released to the atmosphere and is not subsequently removed from the atmosphere by trees. In this way, the destruction of vegetation contributes to a net increase in atmospheric CO_2. And because of this process, slash-and-burn agriculture is a major source of both particulates and carbon dioxide.

Land clearing and subsequent environmental consequences can occur in regions other than the tropics. Large areas of brushland in California were burned in the 1950s to clear the land for agriculture. In semiarid environments, dry, nutrient-poor soils can be easily degraded after burning to the point where they are no longer viable for any production at all.

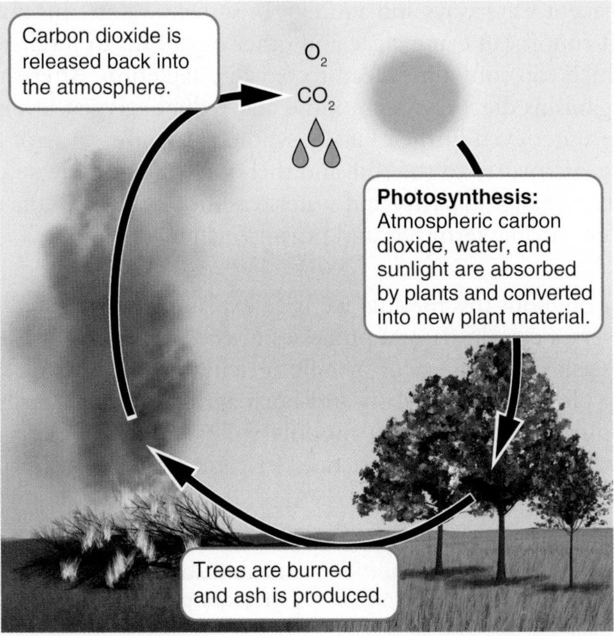

FIGURE 26.4 Clearing land leads to increased atmospheric carbon dioxide. When trees are cleared and burned and new trees are not allowed to replace them, there is a net increase in carbon dioxide.

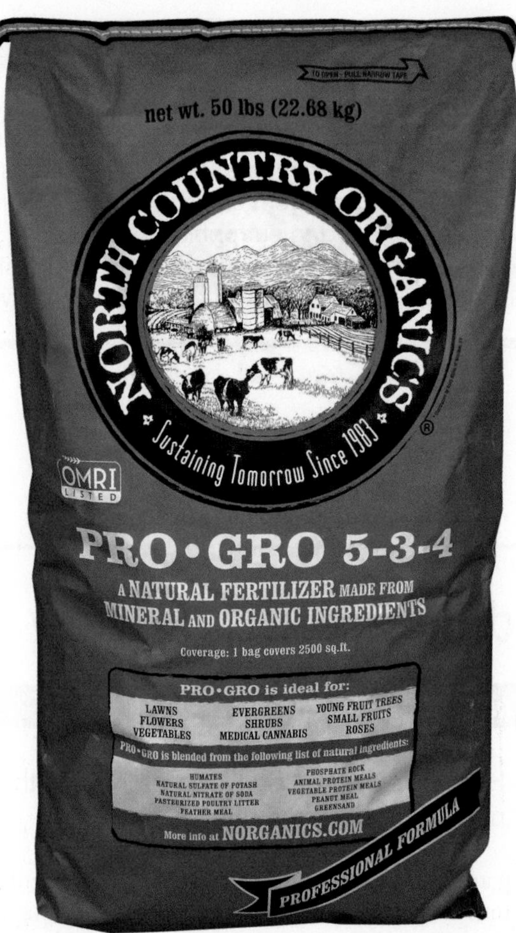

FIGURE 26.5 Most fertilizer bags contain a nutrient label. This bag of fertilizer shows that it contains 5 percent nitrogen, 3 percent phosphorus, and 4 percent potassium. *(North Country Organics)*

26-3 What are the effects of fertilizer use?

Fertilizer use affects soil and water

We briefly introduced fertilizers and their agricultural impacts in Module 25. Nitrogen (N), phosphorus (P), and potassium (K) are the most frequently added fertilizer nutrients. It is common to see fertilizer bags in garden centers and hardware stores with prominent N-P-K labels on them, such as 5-3-4, indicating 5 percent nitrogen, 3 percent phosphorus, and 4 percent potassium (**FIGURE 26.5**). Both organic and synthetic fertilizers can have any ratio of elements, although manufactured synthetic fertilizers are more likely to have higher numbers indicating more concentrated formulations. Because synthetic fertilizers are usually more concentrated and readily available to plants, they provide rapid fertilization, but they also allow more of the fertilizer to leave the soil and enter, and thus pollute, waterways. Organic fertilizers, because they are often made from decomposed plant material and animal manure, are generally in lower concentrations and more slowly available to plants. They also tend to be higher in organic matter, which means that organic fertilizers generally do more to increase the cation exchange capacity of the soil.

Synthetic calcium, magnesium, and potassium fertilizers are usually produced from crushed rock. Organic calcium, magnesium, and potassium fertilizers can be produced from crushed rock or organic material such as cow manure that is decomposed. Fossil fuel combustion during nitrogen production contributes carbon dioxide to the atmosphere and is a contributor to global climate change. In addition, release of nitrogen compounds during nitrogen fertilizer production and use contributes another greenhouse gas, nitrous oxide, to the atmosphere. As we will learn in Unit 9, nitrous oxide is a much more potent greenhouse gas than carbon dioxide.

Synthetic fertilizers have many advantages over organic fertilizers. They are designed for easy application, their nutrient content can be targeted to the needs of a particular crop or soil, and plants can easily absorb them, even in poor soils. Worldwide, synthetic fertilizer use increased from 20 million metric tons in 1960 to over 200 million metric tons in 2020. It can be argued that without synthetic fertilizers, we could not feed all the people in the world.

Despite these advantages, synthetic fertilizers can have several adverse effects on the environment. The process of manufacturing synthetic fertilizers uses large quantities of fossil fuel energy. Producing nitrogen fertilizer is an especially energy-intensive process. Whereas synthetic fertilizers are more readily available for plant uptake than are organic fertilizers, they are also more likely to be carried by runoff into

AP® Exam Tip

If the AP® Environmental Science Exam asks you to describe methods to reduce eutrophication of surface waters, make sure your answer keeps the nutrients used in agricultural methods from entering the water. Organic methods can still lead to eutrophication. Creating holding ponds, restoring shoreline vegetation, and restoring wetlands keep nutrients from entering the surface water.

adjacent waterways and aquifers. In surface waters, this nutrient runoff can cause algae and other organisms to proliferate, which can sometimes lead to oxygen depletion. After these organisms die, they decompose and reduce oxygen levels in the water even further. Finally, synthetic fertilizers do not add organic matter to the soil, and therefore do not increase cation exchange capacity and water retention abilities of the soil as do organic fertilizers and organic matter.

Throughout this module we have explained how agriculture reverses certain natural processes in ecosystems and we have described the effects of specific agricultural practices including plowing, tilling, slash-and-burn agriculture, and fertilizer application. In the next module we examine the effects of two other agricultural practices, irrigation and pest control.

Module 26 AP® Review

Preparing for the AP® Exam

Learning Goals Revisited

26-1 How do plowing and tilling interfere with the natural progression of ecosystems?

Plowing and tilling are not processes that occur regularly in the natural world, but they are practices that occur annually in agriculture. In this way, agriculture slows or reverses succession. The various types of plowing and tilling turn the A horizon soil upside down and expose roots, organic matter and minerals to the atmosphere while burying what had previously been the top of the A horizon. This is a disturbance not seen frequently in nature and exposes the soil to erosion by wind and water.

26-2 What are the consequences of slash-and-burn farming?

Cutting and burning trees creates ash that is rich in potassium, calcium, and magnesium, which increases the base saturation of the soil. However, when rainfall is plentiful, these nutrients are leached from the soil, ultimately reducing the overall availability of nutrients and causing degradation of the soil. Slash-and-burn farming also contributes to the release of carbon from vegetation and soil to the atmosphere as carbon dioxide.

26-3 What are the effects of fertilizer use?

Fertilizers contain essential nutrients for plants—primarily nitrogen, phosphorus, and potassium—and they promote plant growth. Synthetic fertilizers are designed for easy application, their nutrient content can be targeted to the needs of a particular crop, and plants can easily absorb them, even in poor soils. However, they are susceptible to leaching out of soils and contaminating waterways and they do not add organic matter to the soil, as do most organic fertilizers.

AP® Practice Questions

Preparing for the AP® Exam

Multiple-Choice Questions

1. Plowing and tilling allow nutrients from which horizon to rise to the surface?
   (a) O horizon
   (b) A horizon
   (c) B horizon
   (d) C horizon

2. Which of the following processes is known to reverse succession?
   (a) slash-and-burn agriculture
   (b) pesticide use
   (c) the use of GMOs
   (d) fertilization

3. Which of the following is an environmental impact caused by synthetic, inorganic nitrogen fertilizer?
   (a) Production relies on fossil fuel energy.
   (b) It reduces global concentrations of carbon dioxide.
   (c) It contains small concentrations of nutrients.
   (d) It is difficult for plants to absorb the nutrients contained within.

4. Ultimately, slash-and-burn agriculture leads to a net loss of nutrients due to
   (a) leaching from rainfall in tropical areas.
   (b) drought evaporating essential nutrients.
   (c) salinization from fertilizers.
   (d) reduction of soil horizons.

5. Slash-and-burn agriculture ultimately contributes to ___________ carbon dioxide concentrations.
   (a) a net decrease of
   (b) a net increase of
   (c) a maintenance of
   (d) no effect on

Free-Response Question

Certain agricultural practices, like fertilizer use, increase total crop yields while having negative impacts on ecosystems.

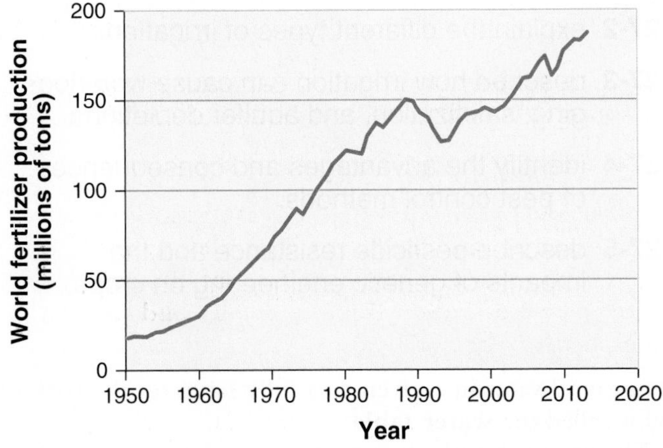

(a) **Identify** in which year fertilizer use first exceeded 140 million tons. (1 pt.)
(b) **Describe** one difference between organic and synthetic fertilizers. (1 pt.)
(c) **Explain** how production of synthetic fertilizers can lead to air pollution. (1 pt.)
(d) **Describe** one benefit to using synthetic fertilizers. (1 pt.)
(e) **Calculate** the percent change in world fertilizer consumption from 1970 to 2010. Show all work. (2 pts.)
(f) Additional data show that the cost of one ton of fertilizer is $660 dollars. **Calculate** the cost of fertilizers globally in 2013. Show all work. (2 pts.)
(g) Egypt currently uses the most fertilizer globally. If they purchased 28.4 percent of Earth's fertilizers, **calculate** the total amount Egypt spent on fertilizer in 2013. (2 pts.)

Module 27

Unit 5 | 24 | 25 | 26 | 27 | 28 | 29 | 30 | 31 | 32 | 33 | 34

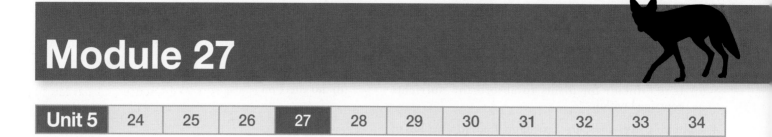

Irrigation and Pest Control Methods

In Module 25, we described the Green Revolution and its reliance on a variety of techniques to increase plant production including irrigation systems and pesticides. In this module we will explore irrigation systems and pesticides in greater detail.

Learning Goals

After reading this module, you should be able to

27-1 describe the sources and locations of water used for irrigation.

27-2 explain the different types of irrigation.

27-3 describe how irrigation can cause waterlogging, salinization, and aquifer depletion.

27-4 identify the advantages and consequences of pest control methods.

27-5 describe pesticide resistance and the impacts of genetic engineering on crops.

27-1 What are the sources and locations of water used for irrigation?

Groundwater and surface water are used for irrigation

When the agricultural crops grown in a particular location require more water than what is delivered to the soil naturally in rainfall, the farmer relies on irrigation to provide additional water to the crops. Agriculture is the largest use of water worldwide, and in the United States 70 percent of freshwater consumption is used for irrigation. The most common sources of irrigation water in the United States are groundwater and surface water (rivers and streams).

Because so much irrigation water is extracted from groundwater, and because over-irrigation can impact groundwater, in order to understand the environmental impacts of irrigation, we must learn more about groundwater.

Groundwater is stored in the pore spaces found within permeable layers of rock and sediment underneath the soil that are called **aquifers**. FIGURE 27.1 shows how different types of aquifers occur. Water can easily flow in and out of an **unconfined aquifer**, which is made of porous rock covered by soil. In contrast, **confined aquifers** are surrounded by a layer of impermeable rock or clay, which impedes water flow to or from the aquifer. The uppermost level at which the groundwater in a given area fully saturates the rock or soil is called the **water table**.

The process by which water from precipitation percolates through the soil and works its way into the groundwater is

FIGURE 27.1 Aquifers are sources of usable groundwater. Unconfined aquifers are rapidly recharged by water that percolates downward from the land surface. Confined aquifers are surrounded by an impermeable layer of rock or clay, which can cause water pressure to build up underground. Artesian wells are formed when a well is drilled into a confined aquifer and the natural pressure causes water to rise toward the ground surface.

Aquifer Pore spaces found within permeable layers of rock and sediment underneath the soil that store groundwater.

Unconfined aquifer Porous rock covered by soil.

Confined aquifer Surrounded by a layer of impermeable rock or clay, which impedes water flow to or from the aquifer.

Water table The uppermost level at which the groundwater in a given area fully saturates the rock or soil.

FIGURE 27.2 **Natural springs.** When an aquifer has an opening at the land surface, water can flow out to form a spring. Springs, such as this one in California, can be an important source of water for organisms and serves as the initial water source for many streams and rivers. *(Jerry Moorman/iStockphoto.com)*

known as **groundwater recharge**. If water falls on land that contains a confined aquifer, however, it cannot penetrate the impermeable layer of rock. A confined aquifer can only be recharged if the impermeable layer of rock has a surface opening that can serve as a recharge area.

Aquifers serve as important sources of fresh water for many organisms. Plant roots can access the groundwater in an aquifer and draw down the water table. Water from some aquifers naturally percolates up to the ground surface as **springs** (FIGURE 27.2). Springs serve as a natural source of water for freshwater aquatic biomes, and they can be used directly by humans as sources of drinking or irrigation water. Humans discovered centuries ago that water can also be obtained from aquifers by digging a well—essentially a hole in the ground. Most modern wells are very deep and water is pumped to the surface against the force of gravity. In some confined aquifers, however, the water is under tremendous pressure from the impermeable layer of rock that surrounds it. Drilling a hole into a confined aquifer releases the pressure, which allows the water to burst out of the aquifer and rise up in the well, as shown in Figure 27.1. A well created by drilling a hole into a confined aquifer is called an **artesian well**. If the pressure is sufficiently great, the water can rise all the way up to the ground surface, in which case no pump is required to extract the water from the ground. Irrigation water is often pumped from wells.

The age of water in aquifers varies, as does the rate at which aquifers are recharged. The water in an unconfined aquifer may originate from water that fell to the ground last year or even last week. This direct and rapid connection with the surface is one of the reasons water from unconfined aquifers is much more likely to be contaminated with chemicals released by human activities. Confined aquifers, on the other hand, are generally recharged very slowly, perhaps over 10,000 to 20,000 years. For this reason, water from a confined aquifer is usually much older and less likely to be contaminated by anthropogenic chemicals than is water from an unconfined aquifer. Large-scale use of water from a confined aquifer is unsustainable because the withdrawal of water is not balanced by recharge. The rapid removal of water from both confined and unconfined aquifers is an important topic in environmental science and has led to many studies suggesting that humans will not always have plentiful fresh, drinkable water where and when they need it. Some environmental scientists maintain that water, rather than oil, will be the resource most highly sought after and fought over by nations in the coming decades.

Major Uses of Water

If a supply of a resource is limited, it is common to look for more of that resource. However, many environmental scientists are fond of pointing out that if you lower demand for a resource, it is equivalent to increasing the supply of that resource. In order to determine how to reduce demand of a resource, it is necessary to understand how a resource is used. Because water is in high demand and because fresh water is relatively scarce, and becoming scarcer in certain regions of the world, it is important to learn how people use water, and how that differs in the United States compared to the rest of the world. With a better understanding of how we use water, we can better determine how water can be conserved, which will effectively increase the availability of water (by using less water for one purpose it will free up water to be used for another purpose). The per capita daily use of water varies dramatically among the nations of the world. FIGURE 27.3 shows total daily per capita use of fresh water for a number of countries, which is known as the **water footprint** of a nation. This reflects total water use by a country for agriculture, industry, and residences divided by the population of that country. This allows us to compare water use among nations. For example, a person living in the United States, Spain, or Canada uses about three times more water than a person living in Kenya or China. "Do the Math: Calculating Per Capita Water Use" on page 324 will help you understand how we obtain these estimates.

During the last 50 years, as agricultural output has grown along with the human population, the amount of water used for irrigation throughout the world has more than doubled. Indeed, producing a metric ton of grain (1,000 kg, or 2,200 pounds) requires more than 1 million liters of water (264,000 gallons). Together, India, China, the United States, and Pakistan account for more than half the irrigated land in the world. In the United States, approximately one-third of

Groundwater recharge The process by which water from precipitation percolates through the soil into groundwater.

Spring Water that naturally percolates up to the surface.

Artesian well A well created by drilling a hole into a confined aquifer.

Water footprint Total daily per capita use of fresh water for a country or the world.

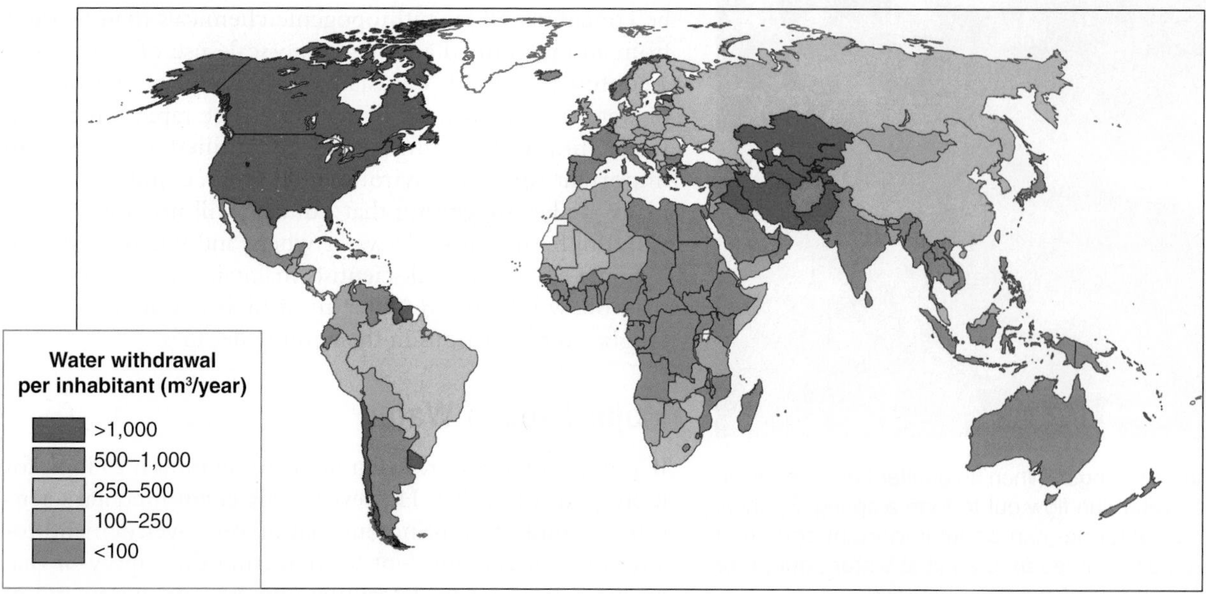

FIGURE 27.3 Total per capita water use per day around the world. The total water use per person per day for agriculture, industry, and households varies tremendously by country. *(Data from Food and Agriculture Association of the United Nations, http://www.fao.org/nr/water/aquastat/maps/WithT.Cap_eng.pdf)*

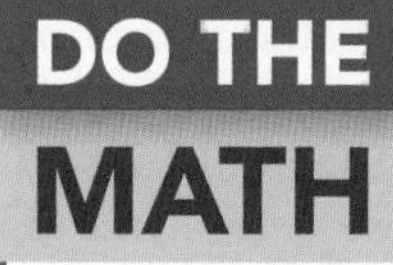

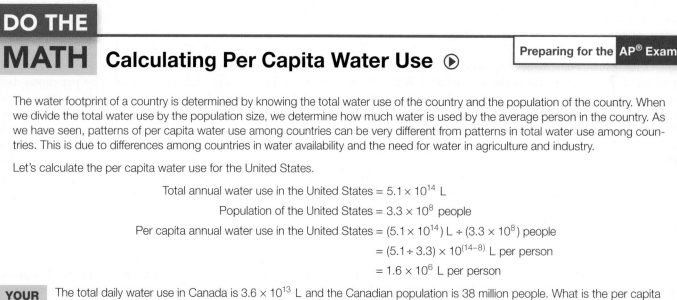

Calculating Per Capita Water Use

Preparing for the AP® Exam

The water footprint of a country is determined by knowing the total water use of the country and the population of the country. When we divide the total water use by the population size, we determine how much water is used by the average person in the country. As we have seen, patterns of per capita water use among countries can be very different from patterns in total water use among countries. This is due to differences among countries in water availability and the need for water in agriculture and industry.

Let's calculate the per capita water use for the United States.

$$\text{Total annual water use in the United States} = 5.1 \times 10^{14}\ \text{L}$$

$$\text{Population of the United States} = 3.3 \times 10^{8}\ \text{people}$$

$$\begin{aligned}\text{Per capita annual water use in the United States} &= (5.1 \times 10^{14})\ \text{L} \div (3.3 \times 10^{8})\ \text{people} \\ &= (5.1 \div 3.3) \times 10^{(14-8)}\ \text{L per person} \\ &= 1.6 \times 10^{6}\ \text{L per person}\end{aligned}$$

YOUR TURN The total daily water use in Canada is 3.6×10^{13} L and the Canadian population is 38 million people. What is the per capita daily water use for Canada?

all freshwater use is for irrigation. Raising livestock for meat also requires vast quantities of water. For example, producing 1 kg (2.2 pounds) of beef in the United States requires about 11 times more water than producing 1 kg of wheat.

There are many ways in which we use water that, when examined closely, will reveal opportunities for reductions in water usage. These will be considered when we examine meat production in Module 28, and in Unit 6 when we consider energy production. In this module we will consider some of the techniques used to irrigate crops and their relative water efficiencies.

27-2 What are the different types of irrigation?

Crop irrigation can be done by furrow, flood, spray, and drip irrigation

Since agriculture is the greatest consumer of fresh water throughout the world, agriculture poses a large potential for conserving water. One way to conserve water in agriculture is by changing irrigation practices. There are four major

FIGURE 27.4 Irrigation techniques. Several techniques are used for irrigating agricultural crops, each with its own set of costs and benefits. (a) Furrow irrigation is easy and inexpensive, but has a low water-use efficiency; only 67 percent of the water is accessible by the plants. (b) Flood irrigation is easy and inexpensive, with 70 to 80 percent water-use efficiency (c) Spray irrigation is more expensive, but water-use efficiency is 75 to 95 percent. (d) Drip irrigation is useful for perennial crops because the hoses do not have to be moved for plowing, and it has a water-use efficiency of over 95 percent. *(a: © Jenny E. Ross; b: Jeff Vanuga/NRCS/USDA; c: Deyan Georgiev/Shutterstock; d: Lynn Betts/NRCS/USDA)*

techniques for irrigating crops: furrow irrigation, flood irrigation, spray irrigation, and drip irrigation, and each presents benefits and drawbacks (**FIGURE 27.4**).

The oldest technique is **furrow irrigation** (Figure 27.4a), which is an easy and inexpensive method, where the farmer digs trenches, or furrows, along the crop rows and fills them with water, which seeps into the ground and provides moisture to plant roots. Furrow irrigation is about 67 percent efficient; 67 percent of the water is accessible by the plants and the other 33 percent either runs off the field or evaporates.

Flood irrigation (Figure 27.4b) involves flooding an entire field with water and letting the water soak in evenly. Flooding can be achieved by diverting a river or stream or by pumping water through a pipe to a field. This technique is generally more disruptive to plant growth than furrow irrigation, but is also slightly more efficient, ranging from 66 to 80 percent efficiency. Approximately 20 percent of the water used in flood irrigation is lost to runoff or evaporation. In addition, flood irrigation can lead to an increased risk of waterlogging of the soil, which we first discussed in Module 25. By depriving roots of oxygen, excess water applied during flood irrigation can reduce oxygen concentrations in soil pore space and impair root growth.

In **spray irrigation** (Figure 27.4c), water is pumped into an apparatus that contains a series of spray nozzles that spray

Furrow irrigation A form of irrigation where the farmer digs trenches, or furrows, along the crop rows and fills them with water.

Flood irrigation A form of irrigation where an entire field is flooded with water.

Spray irrigation A form of irrigation where water is pumped into an apparatus that contains a series of spray nozzles.

water across the field, like giant lawn sprinklers. The advantage of spray irrigation is that it is 75 to 95 percent efficient, so less than one-quarter of the water is lost to evaporation or runoff. However, it is more expensive than furrow or flood irrigation and uses a fair amount of energy to pump water and control the spray nozzle apparatus.

The most efficient method of irrigation, **drip irrigation** (Figure 27.4d), where a slowly dripping hose either on the ground or buried beneath the soil delivers water directly to the plant roots. Drip irrigation using buried hoses is over 95 percent efficient. It has the added benefit of reducing weed growth because the surface soil remains dry, which discourages weed germination. Drip irrigation systems are particularly useful in fields containing perennial crops such as orchard trees, where the hoses do not have to be moved each year in order to plow the field. However, the price of these systems limits the extent to which they are used.

The four irrigation methods described here each has its own benefits and disadvantages. Efficient irrigation technology benefits the environment by reducing both water consumption and the amount of energy needed to deliver the water. Many new technologies are being developed to more carefully control when plants are irrigated. As with all human activities, the costs and benefits of each irrigation technique need to be weighed to determine the best solution for each situation.

27-3 How does irrigation cause waterlogging, salinization, and aquifer depletion?

There are three major adverse consequences of irrigation

Waterlogging, salinization, and aquifer depletion are all related because each is caused by the overuse of water. While waterlogging and salinization can occur naturally, aquifer depletion can only occur by human activity.

Waterlogging

We've already discussed that waterlogging can result from flood irrigation. But what are the long-term impacts of this adverse consequence of irrigation? **FIGURE 27.5a** shows soil conditions in a normal situation where groundwater is below the A and B soil horizons. Plant roots extend into the A and B horizon and there is water in the pore space of the soil, but the pores are not saturated. Figure 27.5b shows soil conditions after a heavy rain or irrigation event. In this situation, the pores are completely saturated with water and there is very little oxygen available for plant roots. If this condition exists for a long period of time due to high clay content of the soil, heavy rains, or heavy irrigation, then the soil becomes waterlogged. And if a soil is waterlogged for a long time, the plant roots in that soil will not absorb sufficient oxygen to grow properly, thereby reducing crop yield and in some extreme cases killing the crop.

Drip irrigation A form of irrigation where a slowly dripping hose on the ground or buried beneath the soil delivers water directly to the plant roots.

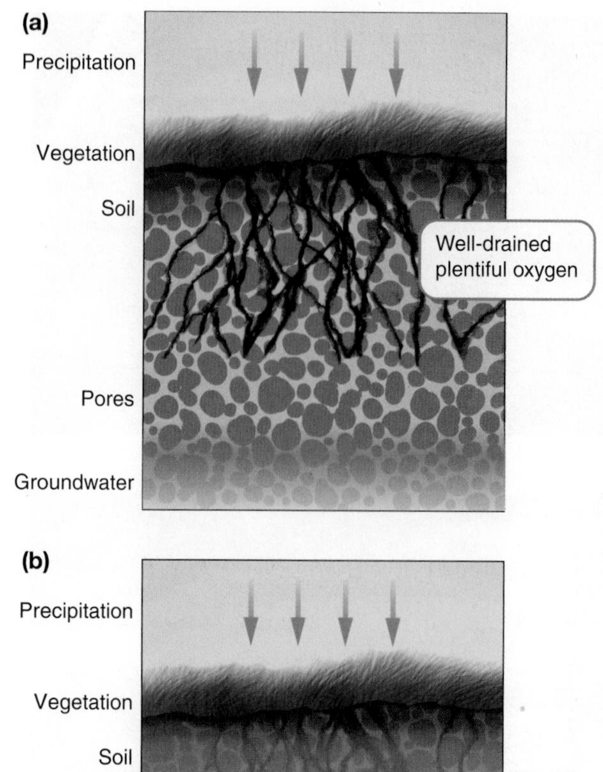

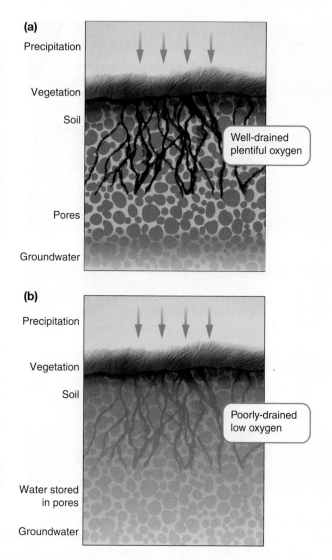

FIGURE 27.5 Irrigation-induced waterlogging of a soil. Over time, irrigation can lead to displacement of oxygen in soil pore space, leading to anaerobic conditions and lack of oxygen for plant roots. (a) An unsaturated soil with sufficient oxygen in pore spaces to allow healthy root growth. (b) A saturated soil with water filling pore spaces and little or no oxygen for plant roots.

Salinization

Salinization can occur when small amounts of salts that naturally occur in irrigated water (from surface or groundwater) become highly concentrated in soil after large quantities of water evaporate and leave the salt behind. Remember that when any water, including irrigation water

FIGURE 27.6 Irrigation-induced salinization of a soil. Over time, irrigation can degrade soil by leaving behind a layer of highly concentrated salts near the soil surface. *(Donald Suarez, USDA Salinity Laboratory)*

and seawater, evaporates, pure H_2O enters the atmosphere and the other contents of the water, such as salts, remain behind in the soil or the ocean. These salts can eventually reach toxic concentrations in some circumstances and impede plant growth. **FIGURE 27.6** shows irrigation-induced salinization of a soil.

Aquifer Depletion

The largest aquifer in the United States is the massive Ogallala aquifer in the Great Plains, which is located under portions of eight western states from Texas to South Dakota. Large amounts of water have been withdrawn from this aquifer for household, agricultural, and industrial uses. Unfortunately, the slow rate of recharge is not keeping pace with the fast rate of water withdrawal, as **FIGURE 27.7** shows. As a result, the depth of the water has declined by about 5 m (16 feet) from 1950 to 2015 and the Great Plains region could run out of water during this century.

FIGURE 27.8 shows what happens when more water is withdrawn from an aquifer than enters the aquifer. As the water table drops farther from the ground surface, springs that once bubbled up to the surface no longer emerge, and spring-fed streams dry up. In addition, some shallow wells no longer reach the water table. When water is rapidly withdrawn from a well, it can create an area that contains no groundwater around the well, which is known as a **cone of depression**. In other words, rapid pumping of a deep well can cause adjacent, shallower wells to go dry. The world's growing population and the associated expansion of irrigated agriculture have increased global water withdrawals more than fivefold in the last 100 years. Global water use is expected to continue to grow along with the human population through the early part of this century. However,

Cone of depression An area surrounding a well that does not contain groundwater.

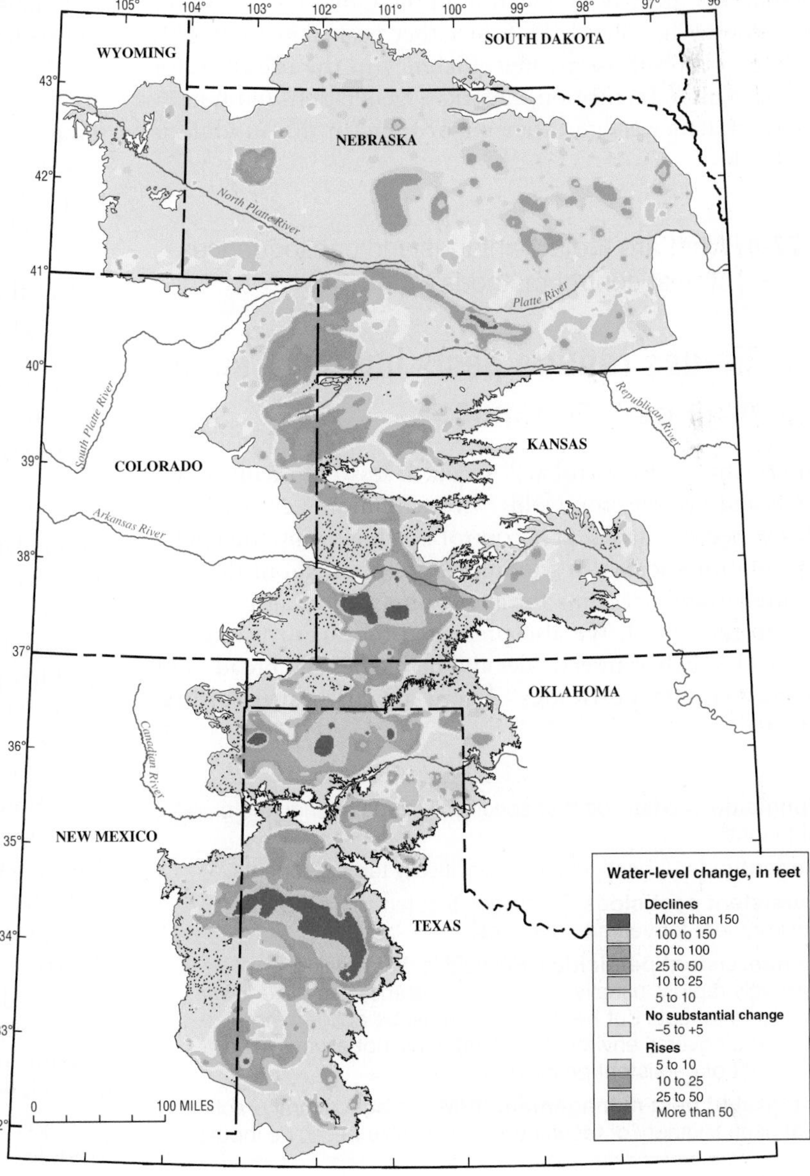

FIGURE 27.7 The Ogallala aquifer. The Ogallala aquifer, also called the High Plains aquifer, is the largest in the United States, with a surface area of about 450,000 km^2 (175,000 miles^2). The aquifer has declined by about 5 m (16 feet) from 1950 to 2015, mostly due to withdrawals for irrigation that have exceeded the aquifer's rate of recharge. *(Data from U.S. Geological Survey 2017)*

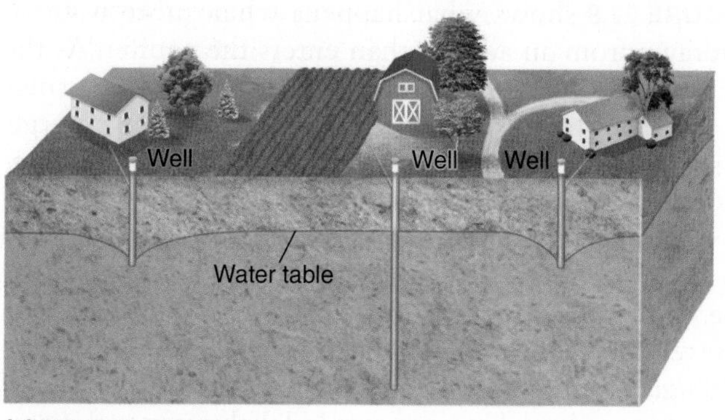

(a) Before heavy pumping

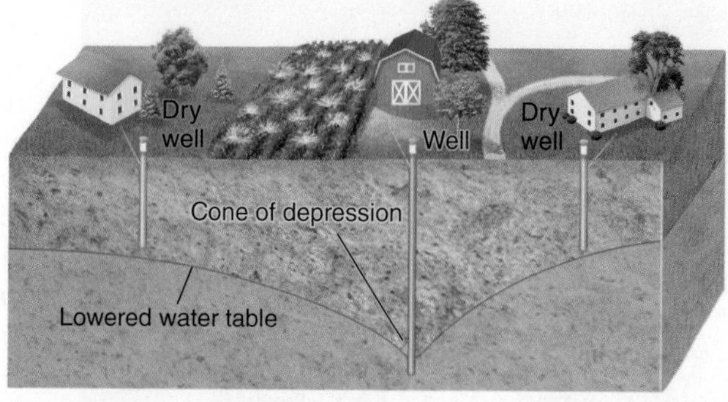

(b) After heavy pumping

FIGURE 27.8 Cone of depression. (a) When a deep well is not heavily pumped, the recharge of the water table keeps up with the pumping. (b) In contrast, when a deep well pumps water from an aquifer more rapidly than it can be recharged, it can form a cone of depression in the water table and cause nearby shallow wells to go dry.

despite the growing population in the United States, water withdrawals have leveled off since they peaked in 1980. This is largely a result of greater efficiency in the use of water for agricultural irrigation, electricity generation, and household appliances. The other major topic in this module is pesticides.

27-4 What are the advantages and consequences of pest control methods?

Pesticides reduce pests but have some adverse consequences

Successful pest control will result in less crop damage by pests and greater crop yields. Most commonly, especially in developed countries, pest control is achieved by the application of pesticides. In Module 25 we pointed out that the United States accounts for about one-fifth of all pesticide use in the world. We also presented the major groups of pesticides such as insecticides and herbicides and broad- and narrow-range pesticides. There are also **fungicides**, pesticides that specifically target fungi (the plural of fungus) and **rodenticides**, pesticides that specifically target rodents.

The application of pesticides allows a farmer to make a quick and relatively easy response to an infestation of pests on an agricultural crop. In many cases, a single application can significantly reduce a pest population. By preventing crop damage, a pesticide application can result in greater crop yields on less land, thereby reducing the area disturbed by agriculture and making agriculture more efficient.

But the application of pesticides, like many other industrial agricultural practices, presents some environmental problems. For example, pesticides often injure or kill more than their intended targets. Some pesticides, such as dichlorodiphenyltrichloroethane, also known as DDT, are **persistent pesticides**, meaning that they remain in the environment for years to decades. In 1972, DDT was banned in the United States, in part because it was found to accumulate in the fatty tissues of animals, such as eagles and pelicans. The increasing concentrations of DDT in these animals caused them to lay eggs with thin shells that easily cracked during incubation by the parents.

Other pesticides, such as the herbicide glyphosate, known by the trade name Roundup, are **nonpersistent pesticides**, meaning that they break down relatively rapidly, usually in weeks to months. Nonpersistent pesticides have fewer long-term effects, but because they must be applied more often, their overall environmental impact is not always lower than that of persistent pesticides.

There are other methods of pest control. An alternative agricultural practice, **integrated pest management (IPM)**, uses a variety of techniques designed to minimize pesticide inputs. These techniques include crop rotation and intercropping, the use of pest-resistant crop varieties, the creation of habitats for predators of pests, and limited use of pesticides. We will discuss IPM as well as some sustainable agricultural methods of pest control in more depth in Module 33.

Fungicide A pesticide that specifically targets fungi (the plural of fungus).

Rodenticide A pesticide that specifically targets rodents.

Persistent pesticides Pesticides that remain in the environment for years to decades.

Nonpersistent pesticides Pesticides that break down relatively rapidly, usually in weeks to months, and have fewer long-term effects but because they must be applied more often their overall environmental impact is not always lower than that of persistent pesticides.

Integrated pest management (IPM) An agricultural practice that uses a variety of techniques to minimize pesticide inputs.

AP® Exam Tip

In answering questions on the AP® Environmental Science Exam, make sure to treat "fertilizers" and "pesticides" as separate terms. Fertilizers and pesticides are different substances used for different purposes, so keep them separate in your responses.

27-5 What is pesticide resistance and what are the impacts of genetic engineering on crops?

Organisms can become pesticide resistant

One major disadvantage of pesticide use is that pest populations may evolve resistance to pesticides over time. Pest populations are usually large and thus contain significant genetic diversity. In those vast gene pools, there are usually a few individuals that are not as susceptible to a pesticide as others. In Module 25 we briefly introduced the concept that individuals that are exposed to a particular pesticide and survive are said to have **pesticide resistance** to that pesticide. If the pesticide is successful in reducing the non-resistant pest population, the next generation will contain a larger fraction of pesticide resistant individuals in the population. As time goes by, resistant individuals will make up a larger and larger portion of the population. Often the resistance becomes more effective, which makes the pesticide significantly less useful. At this point, crop scientists and farmers must search for a new pesticide. The cycle of pesticide development and pest resistance is a type of artificial selection, in that the pesticide is artificially selecting against individuals of the crop population that are susceptible to the pesticide. And at the same time, the pesticide is selecting for individuals that are pesticide resistant. This concept is illustrated in FIGURE 27.9. Because, over time, farmers have to use higher doses of pesticide to achieve the same results, and then eventually they have to develop new pesticides, again and again, this is also known as the pesticide "treadmill." The pesticide treadmill is an example of a positive feedback system.

Pesticides can cause even wider environmental effects. They may kill organisms that benefit farmers, such as predatory insects that eat crop pests, pollinator insects that pollinate crop plants, and plants that fix nitrogen and improve soil fertility. Furthermore, chemical pesticides, like fertilizers, can run off into surrounding surface waters and pollute groundwater, a problem we will examine in Unit 8. The toxicity of pesticides to farmworkers has been well documented. The risk to humans who ingest food that has been treated with pesticides is a subject of some debate.

Pesticide resistance A trait possessed by certain individuals that are exposed to a pesticide and survive.

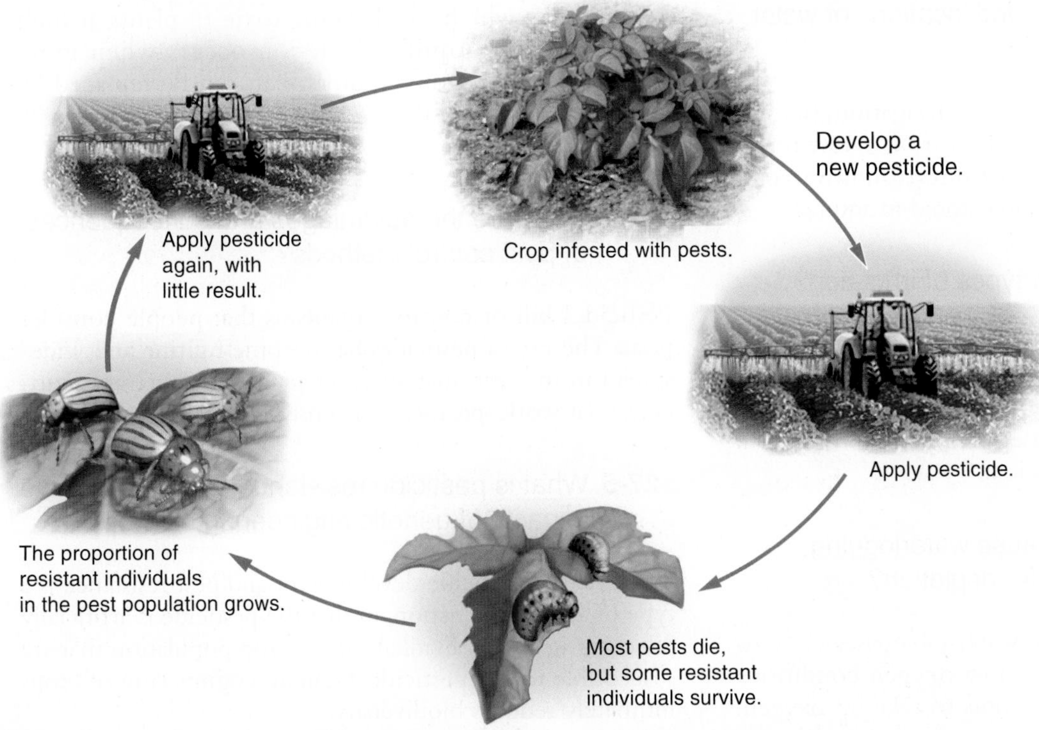
Crop infested with pests.
Develop a new pesticide.
Apply pesticide.
Most pests die, but some resistant individuals survive.
The proportion of resistant individuals in the pest population grows.
Apply pesticide again, with little result.

FIGURE 27.9 **Artificial selection also known as the pesticide treadmill.** Over time, pest populations evolve resistance to pesticides, which requires farmers to use higher doses or to develop new pesticides.

Potential Benefits of Genetic Engineering

Genetic engineering for resistance to pests could reduce the need for pesticides. Corn, for example, is subject to attacks from the bollworm, European corn borer (*Ostrinia nubilalis*), and lepidopteran (butterfly and moth) larvae. *Bacillus thuringiensis* is a natural soil bacterium that produces a toxin that can kill lepidopterans. The insecticidal gene of this bacterium, known as Bt, has been inserted into the genetic material of corn plants, resulting in a genetically modified plant that produces a natural insecticide in its leaves.

A similar technique has been used to create crop plants that are resistant to the herbicide Roundup. The "Roundup Ready" gene allows growers to spray the herbicide on their fields to control the growth of weeds without harming the crop plants. These genes are now widely used in corn, soybean, and cotton plants and are more generally referred to as herbicide tolerant (HT). The success of no-till agriculture, which we will discuss later in this unit, rests largely on the use of herbicides and the introduction of herbicide-resistant crops.

In 2020, roughly 82 percent of the land area planted with corn in the United States was planted with Bt and/or HT corn. Growers of Bt and HT corn have been able to reduce the amount of synthetic pesticide and herbicide used on their crops. However, because Bt and HT corn have become dominant in the United States, their use has greatly reduced the genetic diversity of the crop. This leads to less biodiversity in agricultural ecosystems and more susceptibility to a pathogen or other agent that could adversely affect a crop.

In this module we examined irrigation and the major source of irrigation water, aquifers and surface water. Different irrigation methods have varying efficiencies and effects on the environment. We also examined pesticides, their uses and consequences. One major effect of pesticide use is the development of pesticide resistance among certain pests. In the next module we will examine the impacts of obtaining two nutrient- and energy-rich foods, meat and fish.

Module 27 AP® Review

Preparing for the AP® Exam

Learning Goals Revisited

27-1 What are the sources and locations of water used for irrigation?

Because water use is so important in irrigation, first we must understand where water comes from. Most freshwater used for irrigation comes from groundwater or surface water. Groundwater is water stored in aquifers.

27-2 What are the different types of irrigation?

The four types of irrigation—furrow irrigation, flood irrigation, spray irrigation, and drip irrigation—each has advantages and disadvantages. Drip irrigation, for example, is the most expensive but it causes the least environmental harm because it is the most efficient.

27-3 How does irrigation cause waterlogging, salinization, and aquifer depletion?

Waterlogging occurs when the water table rises close to the ground surface. This causes low oxygen conditions that reduce the ability of plant roots to take up oxygen. Salinization occurs when water evaporates and leaves behind salts, which can become toxic to plants at high concentrations. Aquifer depletion occurs when more water is pumped from aquifers than can be replaced by recharge. All three of these adverse effects can be caused by irrigation.

27-4 What are the advantages and consequences of pest control methods?

Pesticides kill or control organisms that people consider pests. The use of pesticides has become routine and widespread in modern industrial agriculture. There are pesticides that work specifically against particular organisms.

27-5 What is pesticide resistance and what are the impacts of genetic engineering on crops?

The cycle of pesticide development and pest resistance is a type of artificial selection, in that the pesticide is artificially selecting against individuals of the crop population that are susceptible to the pesticide. Genetic engineering of crops ultimately reduces biodiversity.

AP® Practice Questions

Multiple-Choice Questions

1. In the United States, most water is used for
   (a) agriculture.
   (b) households.
   (c) industry.
   (d) hydroelectric plants.

2. Which irrigation method is the most efficient?
   (a) furrow irrigation
   (b) spray irrigation
   (c) flood irrigation
   (d) drip irrigation

3. When deep wells are heavily pumped, one result can be
   (a) decreased groundwater recharge.
   (b) spring formation.
   (c) a cone of depression.
   (d) increased groundwater recharge.

4. Groundwater recharge
   (a) is the result of precipitation.
   (b) sometimes occurs as a spring.
   (c) occurs rapidly in confined aquifers.
   (d) occurs after heavy irrigation.

5. Which statement best describes the results of repeated pesticide use over generations of pests?
   (a) Broad-spectrum pesticides degrade into selective pesticides, thereby killing a wide range of pests over a long period.
   (b) Pesticides accumulate in the fatty tissues of consumers and increase in concentration as they move up the food chain.
   (c) Certain individuals in pest populations evolve resistance to pesticides, which become less effective over time so that new pesticides must be developed.
   (d) Testing of the toxicity of pesticides to humans cannot keep pace with the discovery and production of new pesticides.

Use the diagram below to answer questions 6 & 7:

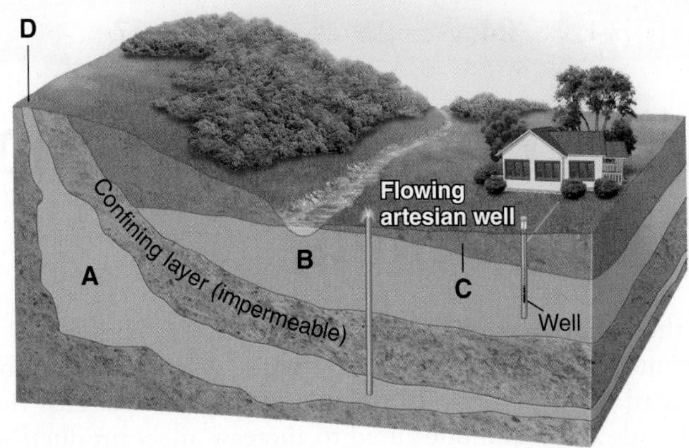

6. Which area in the diagram would most likely represent water from precipitation that has worked its way into the soil via percolation?
   (a) A
   (b) B
   (c) C
   (d) D

7. Overuse of a well can lead to a reduction in the area labeled as C, or the ____________.
   (a) confined aquifer
   (b) water table
   (c) recharge zone
   (d) artesian well

Free-Response Question

The photos below depict different methods of irrigation.

Lynn Betts/NRCS/USDA

(a)

Deyan Georgiev/Shutterstock

(b)

Jeff Vanuga/NRCS/USDA

(c)

(a) **Identify** which form of irrigation shown above would be the least efficient. (1 pt.)
(b) **Identify** which form of irrigation shown above would be the most efficient. (1 pt.)
(c) **Identify** another form of irrigation not depicted in the photos above. (1 pt.)
(d) **Explain** one advantage of drip irrigation. (1 pt.)
(e) **Explain** one disadvantage of drip irrigation. (1 pt.)
(f) **Describe** two disadvantages of modern irrigation methods. (2 pts.)
(g) **Describe** one type of pesticide and one disadvantage to using that pesticide. (2 pts.)
(h) **Propose a solution** (other than pesticides) for getting rid of unwanted pests. (1 pt.)

Module 28

Unit 5 | 24 | 25 | 26 | 27 | 28 | 29 | 30 | 31 | 32 | 33 | 34

Meat Production Methods and the Impacts of Overfishing

In the previous module we examined the benefits and disadvantages of irrigation and pesticides, two techniques for increasing crop yields. In this module we will discuss the methods and techniques used to increase meat production on land. We will also examine methods used to harvest fish from the oceans. Both meat production and fishing yield a more concentrated food product—more protein and energy per kg of food than vegetables—but also require greater land area and fossil fuel energy expenditure per kg of food obtained. Advances in recent decades that have allowed increased productivity and faster time to slaughter for meat and increased catch for fish have also led to greater environmental impact.

Learning Goals

After reading this module you should be able to

28-1 describe the benefits and consequences of different meat production methods.

28-2 describe the causes and consequences of overfishing.

28-1 What are the benefits and consequences of different meat production methods?

Concentrated animal feeding operations and free-range grazing each have their advantages and disadvantages

We saw in Module 25 that in order to remain economically viable and feed large numbers of people, modern agriculture in the United States has become larger and more mechanized. As the demand for grain and agricultural crops increased, so did the demand for meat. As agricultural methods changed to supply meat and poultry to larger numbers of people at a lower cost, the primary objective became implementing methods that promoted the faster growth of animals. If an animal grows larger in a shorter period of time, the profits to the farmer will be greater. But there are costs to both the animal and the environment from this accelerated growth program. Here we will examine some of the modern methods in large-scale production of farm animals.

Meat production is inherently less efficient than agricultural crop production. Efficiency can be measured in terms of land and in terms of energy. Land efficiency estimates vary widely based on the type of meat and the particular crop. **FIGURE 28.1**, for example, shows a variety of food products and the amount of land they require to produce 1,000 kilocalories of food. By this measure, beef requires more than 80 times as much land as wheat, and 20 to 30 times more land than poultry or farmed fish. As a general estimate, meat requires at least 20 times more land area than plants for the same caloric

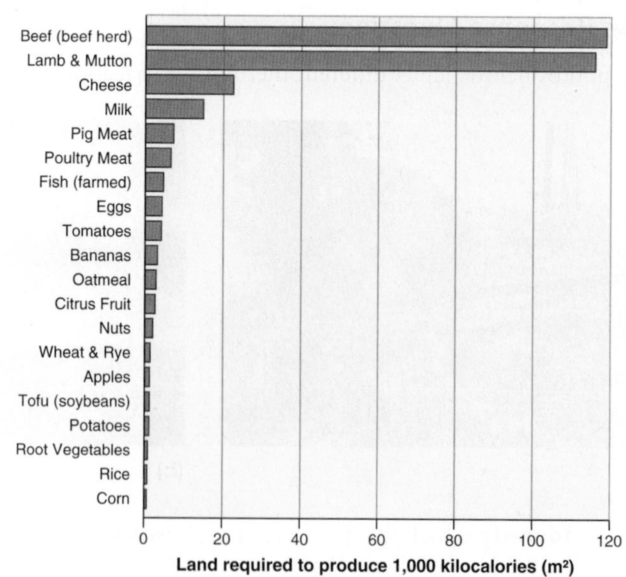

FIGURE 28.1 Amount of land needed to produce 1,000 kilocalories of food. The amount of land area required to produce the same amount of food energy varies widely depending on the food item. *(Data from J. Poore and T. Nemecek. Science 360 (2018): 987–992)*

output. The land requirements assigned to a particular food item, such as beef, include the land that the animal occupies while it is alive but also the land required to provide the food that it eats and for disposal of the wastes that it produces. So commercially raised beef cattle physically occupy very little land, as we will see in the next section, but the land area they require, or their "footprint," is much larger. We have already seen in Figure 25.9 that meat and wild-caught fish require more fossil fuel energy per calorie of food obtained than vegetables and grains.

High-Density Animal Farming

In 2020, according to the U.S. Department of Agriculture, roughly 167 million animals were slaughtered for beef, pork, and lamb, along with billions of chickens, turkeys, and ducks. Many of these animals were raised in feed-lots, or **concentrated animal feeding operations (CAFOs)**, which are large indoor or outdoor structures designed for maximum occupancy of animals and maximum output of meat (**FIGURE 28.2**). This type of high-density animal farming is used for beef cattle, dairy cows, hogs, and poultry, all of which are confined or allowed very little room for movement during all or the latter part of their life cycle. A CAFO may contain as many as 2,500 hogs or 50,000 turkeys in a single building. By keeping animals confined, farmers minimize land costs, improve feeding efficiency, and increase the fraction of food energy that goes into the production of animal body mass. By minimizing economic costs to the farmer or farming business operating the CAFO, food prices to consumers are much lower. Keeping animals confined to a small space, which is criticized by some on ethical grounds, ensures that less energy is expended by the animal in activities such as moving around and the increased respiration that will result. And this, in turn, results in animals that weigh more and yield more revenue for the farmer. Animals are fed grains or other feed rather than their natural foods such as grass. The animals are given antibiotics and nutrient supplements to reduce the risk of adverse health effects and diseases, which would normally be high in such concentrated animal populations. They are frequently given growth hormones to promote faster growth. CAFOs also require large amounts of clean water, which is ingested by the animals and also used to wash animal wastes from the buildings and concrete floors of the facility. All of these factors require money and energy to implement, and they produce heavier, fatter animals and have a variety of environmental impacts. High-density animal farming results in animal waste runoff and the need for waste disposal. There is also evidence that antibiotics given to confined animals are contributing to an increase in antibiotic-resistant strains of microorganisms that affect humans.

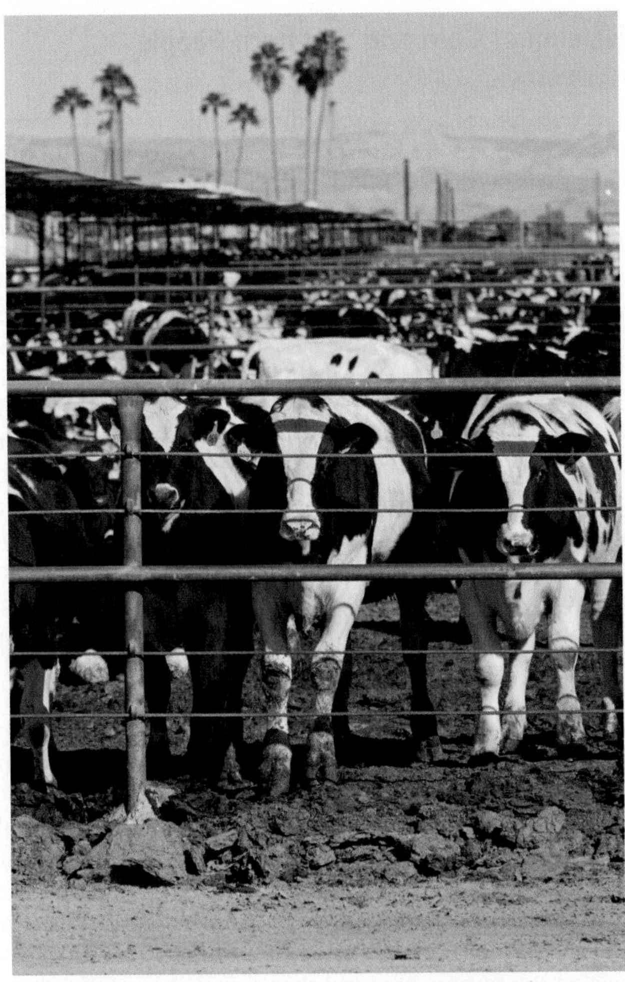

FIGURE 28.2 Cattle in a concentrated animal feeding operation in California. CAFOs are large indoor or outdoor structures that allocate a very small amount of space to each animal. *(Jim West/Alamy Stock Photo)*

AP® Exam Tip

There is no perfect dietary or agricultural solution. You should recognize that whatever protein is consumed results in certain drawbacks and consequences. Also note that some diets are not available in all regions or accessible to all people.

Waste Disposal and Waste Lagoons

A typical CAFO produces over 2,000 tons of manure annually, or about as much as a town of 5,000 people would produce. The waste is sometimes used to fertilize nearby agricultural fields, but if over-applied, it can cause the same nutrient runoff problems as the synthetic fertilizers that we discussed in Module 26. Sometimes animal wastes are stored adjacent to feedlots but, during heavy rainstorms, runoff from these locations can contaminate nearby waterways. Animal wastes have also been dumped, either inadvertently or intentionally, into natural waters. The U.S. Environmental Protection Agency has concluded that chicken, hog, and cattle waste has caused pollution along 56,000 km (35,000 miles) of rivers in 22 states and has caused some degree of groundwater contamination in 17 states.

Concentrated animal feeding operation (CAFO) A large indoor or outdoor structure designed for maximum occupancy of animals and maximum output of meat.

FIGURE 28.3 Feedlot and manure lagoon. Large-scale agricultural CAFOs, such as this one in Georgia, produce great amounts of waste that can be held in lagoons until pumped out and removed. *(Jeff Vanuga/USDA, Natural Resources Conservation Service)*

The largest CAFOs construct lagoons to dispose of the tremendous amount of manure produced (**FIGURE 28.3**). A **manure lagoon** is a large, human-made pond lined with rubber to prevent the manure from leaking into the groundwater. After bacteria has broken down the manure—the same process that occurs in sewage treatment plants—the manure can be spread onto farm fields to serve as a fertilizer. One major risk of manure lagoons is the possibility of developing a leak in the liner that would allow the waste to seep into and contaminate the underlying groundwater. Another risk is an overflow into adjacent water bodies. Like human wastewater, overflow of animal waste into rivers can lead to disease outbreaks in humans and wildlife. The application of manure as fertilizer can create runoff during heavy rains that moves the manure and associated bacteria and antibiotic residues into nearby water bodies. Manure lagoons are primarily anaerobic environments, so just as in the example of waterlogged soils presented in Module 27, there is very little oxygen present. In addition, there is a tremendous amount of organic matter contained in the manure, so anaerobic decomposition will lead to the release of carbon dioxide and methane, two important greenhouse gases. Nitrous oxide (N_2O), another greenhouse gas, is also released in wet, anaerobic environments when nitrate (NO_3^-) is available, and nitrate, a nitrogen compound, is plentiful in animal manure. So, as you can see, while manure lagoons might be an unpleasant visual image on the landscape, their intended purpose is to prevent further negative environmental impacts. However, these well-laid plans can have serious environmental consequences when even the smallest issue occurs.

Manure lagoon Human-made pond lined with rubber built to handle large quantities of manure produced by livestock.

Diversion of Corn and Soy from People to Livestock

Food researchers have observed that large amounts of agricultural resources are diverted to feed livestock and poultry rather than people. A Food and Agriculture Organization study in 2017 estimated that 13 percent of the grain grown in the world is used to feed livestock. Grass, leaves, and crop residue comprised 65 percent of livestock feed. However, in the United States, roughly half of the corn and over 70 percent of the soybeans are used for animal feed. When these foods are fed to livestock, the low efficiency of energy transfer causes much of the energy they contain to be lost from the system. Ultimately, perhaps only 10 to 15 percent of the calories in grain or soybeans fed to cattle are converted into calories in beef. If people ate producers, such as grains and soybeans, rather than primary consumers, such as cattle, more food would be available for human consumption (although presently much of the corn fed to livestock is not the variety of corn that humans desire to eat). This concept is referred to as "eating lower on the food chain." It was made popular in the United States in a book called *Diet for A Small Planet* by Frances Moore Lappé in 1971 and revised and reprinted many times, including most recently in 2021 (**FIGURE 28.4**). The main premise is that human beings can utilize fewer calories from agriculture and obtain sufficient nutrition (protein, vitamins, and calories) by eating primary producers (plants) rather than primary consumers (animals). This process is called "eating lower on the food chain" because producers are one level lower on the trophic pyramid than primary consumers such as chickens, pigs, and cows (see Figure 7.4 on page 76). The book also publicized the fact that with proper attention to food choices, one can obtain sufficient protein from plants.

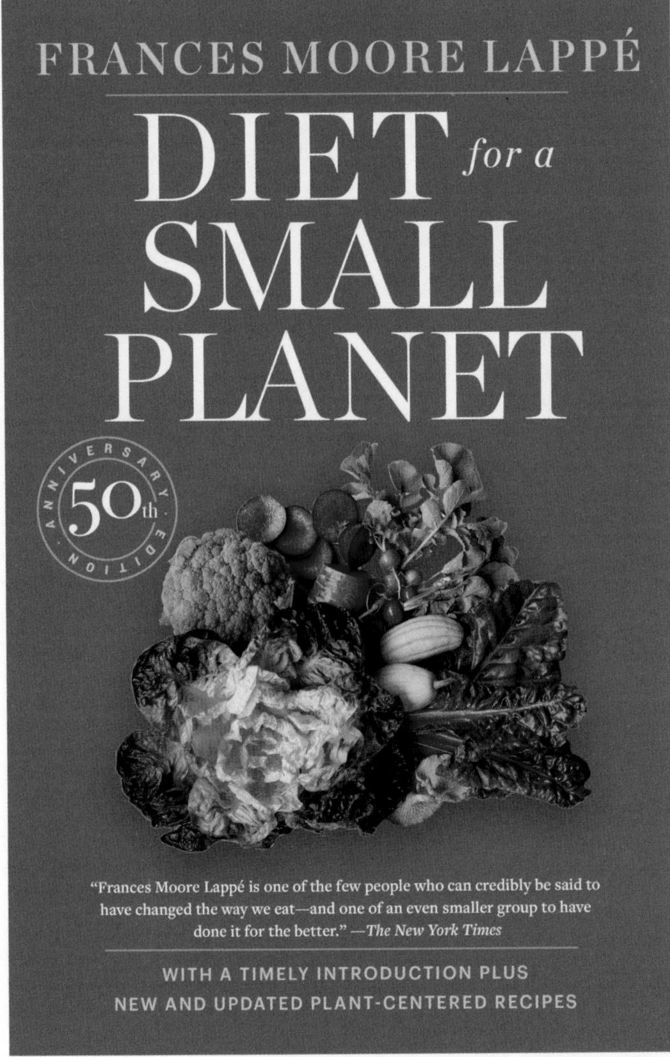

FIGURE 28.4 Book cover of *Diet for a Small Planet*. This book has been revised and reprinted many times and is still in print in 2022. *(Diet for a Small Planet by Frances Moore Lappé. Courtesy of Penguin Random House LLC. Photo by Paige Green; Styling by Alysia Andriola)*

Free-Range Grazing

Not all meat comes from CAFOs. Free-range chicken and beef are becoming increasingly popular in the United States. Some people find it more ethically acceptable to eat chicken or cattle that has wandered free than an animal that has spent its entire life confined in a small space. Currently, **free range grazing** is defined as allowing animals to graze outdoors on grass for most or all of their lifecycle. Free-range meat, if properly produced, is more likely to be sustainable than meat produced in CAFOs. Because these free-range animals are not as likely to spread disease as those that are kept in close quarters, the use of antibiotics and other medications can be reduced or eliminated. The animals graze or feed on the net primary productivity of the land, with little or no supplemental feeding, so less fossil fuel goes into the raising of free-range meat. Finally, manure and urine are dispersed over the range area and are naturally processed by detritivores and decomposers in the soil. As a result, there is no need to treat and dispose of massive quantities of manure. On the negative side, free-range operations use more land than CAFOs do, and the cost of meat produced using these techniques is usually significantly higher.

As we stated in Module 24, grazing livestock on open lands uses much less fossil fuel energy than raising them in feedlots. Furthermore, grazing domesticated animals for short, intense periods of time can mimic the pressure on grasslands that has been experienced over evolutionary time by migrating herds of bison and other herbivores. Therefore, the grazing of livestock can help maintain grasslands and prevent less desirable species from becoming dominant. In addition, grazing introduces animal waste and aerates the soil, both of which are beneficial to ecosystems. However, improperly managed livestock can damage stream banks and pollute surface waters.

To take the concept of free range even further, some farmers and indigenous groups mimic migrating herds of large herbivores. The most sustainable way for people to graze animals, and to do so when the available soils have only low or moderate productivity, is **nomadic grazing** in which they move herds of animals, often over very long distances, to seasonally productive feeding grounds. If grazing animals move from region to region without lingering in any one place for too long, the vegetation can usually regenerate. This grazing system attempts to mimic the aforementioned natural systems, where migrating herds of grazers such as the American bison and wildebeests in Africa travel long distances following lush grasses that sprout and thrive after rainfall. In a variety of countries including the United States, experiments that attempt to mimic nature are underway. Researchers and cattle farmers are experimenting with utilizing short-duration grazing in specific locations at relatively high stocking densities and then moving the animals to other locations for successive short duration grazing. Preliminary results suggest that the short periods of time of grass cutting followed by deposits of manure from the animals can increase soil carbon and nitrogen pools in certain grassland ecosystems.

Free range grazing Allowing animals to graze outdoors on grass for most or all of their lifecycle.

Nomadic grazing The feeding of herds of animals by moving them to seasonally productive feeding grounds, often over long distances.

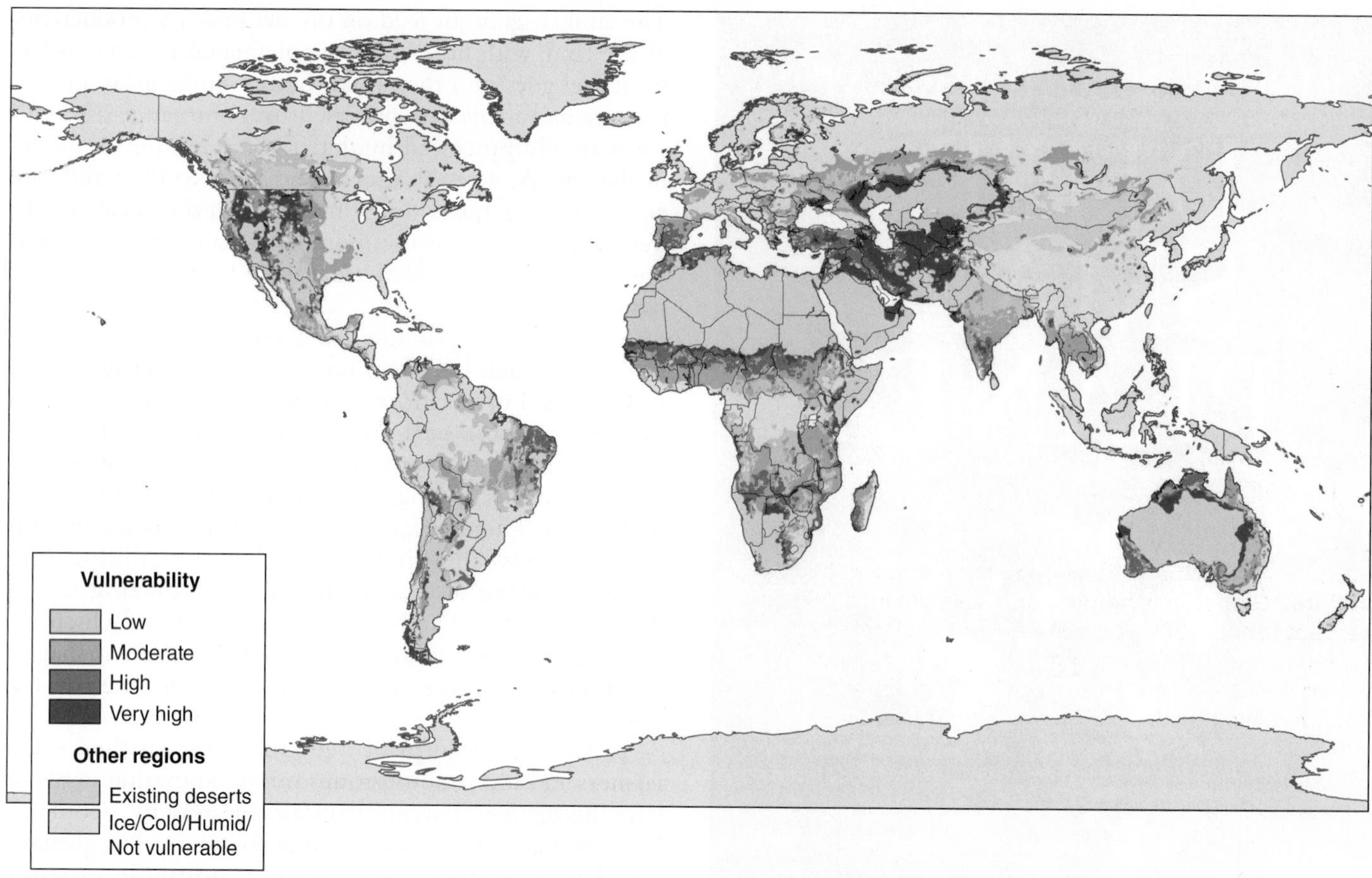

FIGURE 28.5 Global desertification vulnerability. Certain regions of the world are much more vulnerable to desertification than others.

Overgrazing and Desertification

When animals are allowed to graze in one location for too long, or when too many animals graze in a relatively small location, overgrazing occurs. **Overgrazing**, excessive grazing on a particular area of land that can reduce or remove vegetation and erode and compact the soil.

In semiarid environments, dry, nutrient-poor soils can be easily degraded by agriculture to the point at which they are no longer viable for any production at all. The transformation of arable, productive and relatively low-precipitation land to desert or unproductive land due to climate change or destructive land use such as overgrazing or logging is known as **desertification**. The world map in FIGURE 28.5 shows the locations of the world that are most vulnerable to desertification. Today, desertification is occurring most rapidly in Africa, where parts of the Sahara are expanding at a rate of up to 50 km (31 miles) per year. Unsustainable farming practices in northern China are also leading to rapid desertification.

If people chose to eat less meat, there would be less demand for it, and the environmental impacts of raising animals for consumption would decrease. The reduced impacts will be greatest in CAFOs. For example, fewer CAFOs could lead to a decrease in the production of greenhouse gases like carbon dioxide, methane, and nitrous oxide from the anaerobic processes taking place when animal manure accumulates in large piles or in lagoons. It could also lead to a reduction in greenhouse gases emitted during fossil fuel use from the meat production process. A reduction in the consumption of both CAFO and free-range meat would also result in less soil disturbance. Other aspects of meat production will be considered later in this unit. Next, we will consider another energy- and nutrient-dense food item, fish—specifically, the overfishing of wild species.

Overgrazing Excessive grazing that can reduce or remove vegetation and erode and compact the soil.

Desertification Transformation of arable, productive, low-precipitation land to desert or unproductive land due to climate change or destructive land use such as overgrazing and logging.

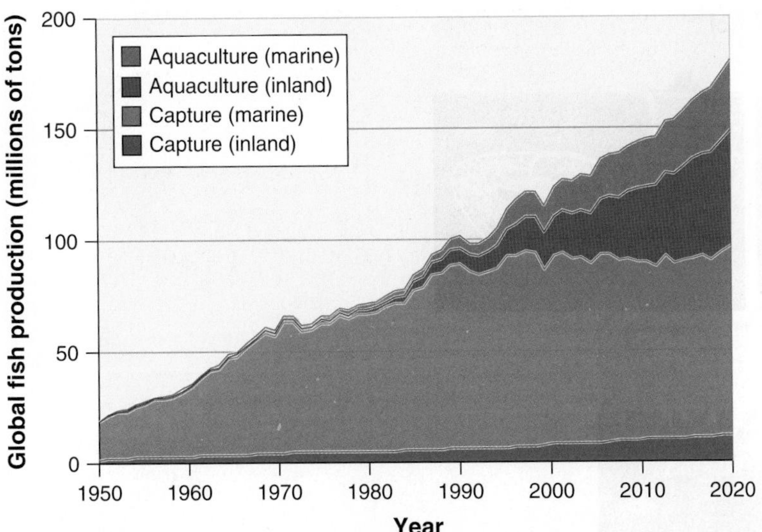

FIGURE 28.6 Global fish production from fisheries and aquaculture. Global fish production has increased by more than 100 percent since 1980, primarily as a result of the large increase in aquaculture. The graph shows data for capture (wild caught) fisheries-marine (orange), capture (wild caught) fisheries-inland (red), aquaculture-raised fish-marine (light blue), and aquaculture-raised fish-inland waters (dark blue). *(Data from The State of World Fisheries and Aquaculture, FAO. 2020, https://www.fao.org/3/ca9229en/ca9229en.pdf)*

28-2 What are the causes and consequences of overfishing?

Harvesting fish can result in fishery decline

Fish consumption provides over 3 billion people in the world with 20 percent of their animal protein. In many coastal areas of the world, particularly in Africa and Asia, fish accounts for nearly all animal protein that some people consume. As **FIGURE 28.6** shows, the global production of fish has doubled in the last 40 years. This increase masks two divergent trends: a decrease in wild fish caught in the world's oceans and a relatively rapid increase in aquaculture, which as we learned in Unit 2 is the farming of aquatic organisms such as fish, shellfish, and seaweeds. Here, we will consider wild-caught fish, primarily in marine systems although a small fraction of wild-caught fish is harvested in fresh-water systems.

A **fishery** is a commercially harvestable population of fish within a particular ecological region. The tragedy of the commons, which we learned about in Module 24, is particularly applicable to ocean fisheries. No individual country has an incentive to protect fish stocks or to attempt to replenish them because fish in the ocean do not belong to any one nation. Since most fish do not live their entire lives within one national border, one country working alone cannot solve the problem of fishery depletion. In fact, a voluntary limit by one or a few countries will give other countries more opportunities to overfish. In many parts of the oceans, the continuing competition for fish has led to a precipitous decline in fish populations.

In 2003, the journal *Nature* drew a great deal of attention by reporting a dramatic decline in the number of large predatory ocean fish caught over the previous 50 years, even though both the number of fishing vessels and the amount of time spent fishing had increased over that period. Fishers were working harder, but catching fewer fish. A study in 2006 found that 30 percent of fisheries worldwide had experienced a 90 percent decline in fish populations. The decline of a fish population by 90 percent or more is referred to as **fishery collapse**. Many studies on this subject focus on large predatory fish. However, a 2011 study in the *Proceedings of the National Academy of Sciences* noted an unexpected increase in fishery collapse among smaller species as well. A 2020 study by the Food and Agriculture Organization, based on long-term monitoring through 2017, concluded that marine fisheries have continued to decline. They stated that 34 percent of fisheries were overfished in 2017, meaning that they contained less than the number of fish than would allow the maximum harvest possible.

Whenever a fishery is overfished or in decline, it adversely impacts biodiversity of the aquatic system. It also adversely affects the people who make their livelihood either catching fish on the open water, or who participate in some aspect of processing, transporting, and selling the fish on land. So the collapse of a fishery adversely affects the biological aquatic system as well as the human system.

Ocean harvests used to be limited by the difficulty of finding fish in a vast ocean as well as by lack of capacity of small boats and nets. Neither limitation is an obstacle today. Current fishing methods make it easier to catch large numbers of fish. Factory ships can stay at sea for months at a time, processing and freezing their harvest without having to return to port.

Of the five major commercial fishing methods, four involve nets and the fifth uses a very long fishing lines bearing hundreds or even thousands of baited hooks. The four

Fishery A commercially harvestable population of fish within a particular ecological region.

Fishery collapse The decline of a fish population by 90 percent or more.

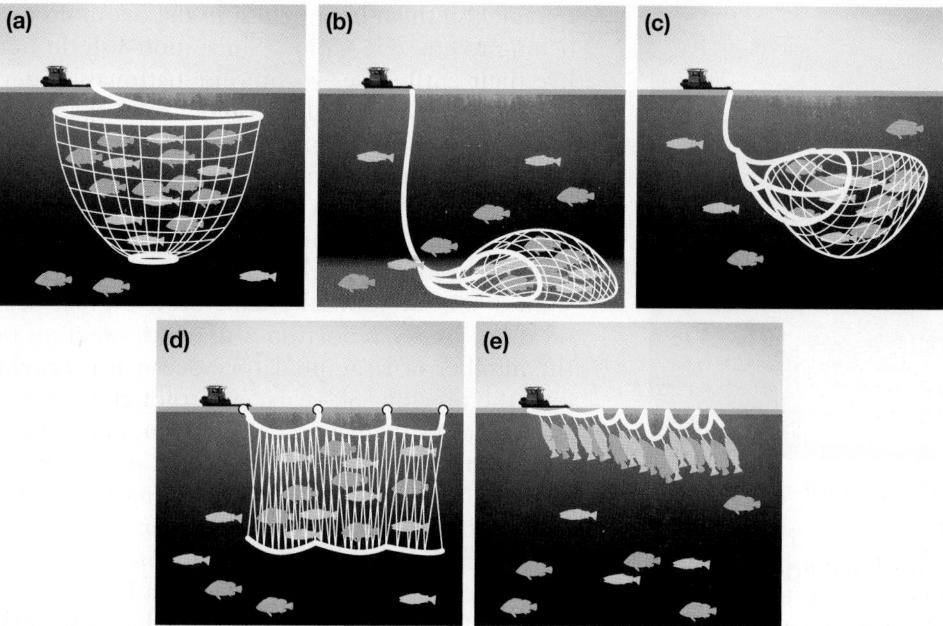

FIGURE 28.7 **The major commercial fishing methods.** (a) Purse seine nets surround a school of fish. (b) Bottom trawl nets run along the ocean bottom. (c) Midwater trawl nets are pulled through the water above the ocean bottom. (d) Gill nets have larger holes in them and are set up like a wall and fish swim into the holes, become caught, and can't escape. (e) Longlines can be more than a mile long and have baited hooks every meter (3.28 feet) or so.

net methods are shown in **FIGURE 28.7**, along with the fish-and-line method. Fishers in pursuit of high-value species such as tuna use spotter planes and sonar to locate schools. They encircle schools with nets that can capture up to 3,000 adult tuna at a time—weighing perhaps 450,000 kg (almost 1 million pounds). Fish species that live on or close to the ocean bottom, as well as many shellfish, are caught in drag-nets, which are weighted so that they can be pulled across the ocean floor.

Large-scale, high-tech fishing can adversely affect both target and nontarget species. Nets can damage ocean-bottom habitats by scouring them of coral, sea sponges, and plants. Many commercially important fish are keystone species, so a decline or loss of their populations can have cascading effects on other marine species. Intensive fishing also leads to the loss of juvenile fish of the target species as well as to the loss of noncommercial species that are accidentally caught by nets and lines. This unintentional catch of nontarget species, called **bycatch**, has significantly reduced populations of fish species such as sharks and has endangered other organisms such as sea turtles. Some countries now require the use of technology that minimizes bycatch or harm to endangered species.

Bycatch The unintentional catch of nontarget species while fishing.

More Sustainable Fishing

In the interest of creating and supporting sustainable fisheries, many countries around the world have developed fishery management plans, often in cooperation with one another. International cooperation is particularly important because fish migrate across national borders, some marine ecosystems span national borders, and many of the world's most important fisheries lie in international waters.

The Northwestern Atlantic fisheries, for example, comprise several continental shelf ecosystems that exist along the coast of the northeastern United States and eastern Canada. Historically, these fisheries were among the most productive in the world. However, overfishing by international fleets of factory ships led to a substantial and long-lasting depletion of fish stocks, particularly of cod and pollock, by the early 1990s, as **FIGURE 28.8** shows. The fisheries were forced to close because of the depleted stocks, and the Canadian and U.S. governments imposed a moratorium on bottom fishing in the area. For a period of time, a recovery occurred in some of the fisheries, seemingly in response to the closing of the fisheries. The majority of the Atlantic Canadian fisheries are still closed today. Some fisheries off the coast of the

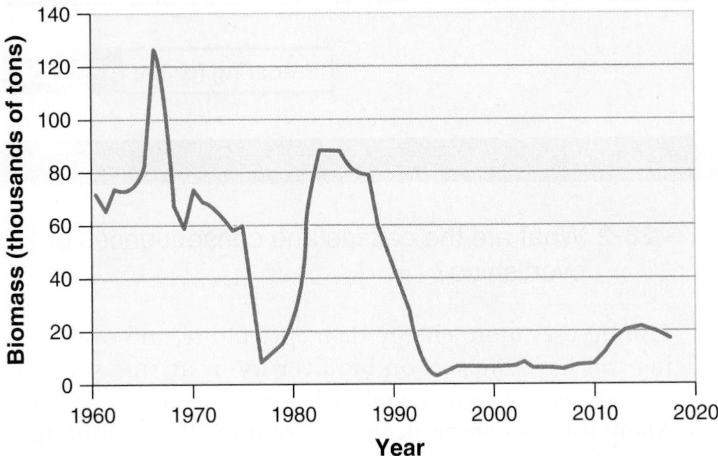

FIGURE 28.8 Fishery collapse in the Northwestern Atlantic Ocean. Cod biomass in the Grand Banks, Northwestern Atlantic Ocean. A decline beginning in 1985 persisted through 2010. Since then, these fisheries had begun to recover but then started declining again. *(Data through 2018 from FAO, Fisheries and Resources Monitoring System, http://firms.fao.org/firms/resource/10315/en)*

United States are now open, at least for part of the year, but with lower fish removals allowed.

In response to the fishery collapse, and in order to restore the depleted stocks and manage the ecosystem as a whole, the U.S. Congress passed the Sustainable Fisheries Act in 1996. This act shifted fisheries management from a focus on economic sustainability to an approach that increasingly stressed conservation and the sustainability of species. The act calls for the protection of critical marine habitat, which is important for both commercial fish species and nontarget species. For many commercial species considered to be in danger, such as cod, a sustainable fishery means that no fishing will be permitted until populations recover.

One successful fishery management plan was developed in Alaska, where the commercial salmon fishery declined rapidly between 1940 and 1970. Managers first tried to increase salmon populations by limiting the fishing season, but by 1970, when the season was restricted to less than a week, so many fishers participated that populations continued to drop.

In 1973, fishery managers introduced a system of quotas, a fishery management program in which individual fishers are given a total allowable catch of fish for a season that they can either catch themselves or sell to others. Before the start of each salmon season, fishery managers establish a total allowable catch and distribute or sell quotas to individual fishers or fishing companies, favoring those with long-term histories in the fishery. Fishers with quotas have a secure right to catch their quota so they have no need to spend money on bigger boats and better equipment in order to outcompete others. If fishers cannot catch enough salmon to remain economically viable, they can sell all or part of their quota to another fisher. **FIGURE 28.9** shows the results: Since the beginning of the quota program, the salmon population and harvest have increased—at times very rapidly—and the adverse effects on fishers and fishing companies have been reduced.

The Maine lobster fishery provides roughly 60 percent of the lobsters caught in the United States. Although it has had productivity declines, it is considered a relatively successful fishery. Rather than federal government regulation, it relies on local communities regulating catch size, which is the alternate approach to avoiding the tragedy of the commons as proposed by Elinor Ostrom, whom we introduced at the beginning of this unit.

Not all fisheries are declining, but it is often difficult for consumers to know which fish are being overharvested and which are not. To help consumers choose more sustainable fish, the Environmental Defense Fund and other organizations have compiled lists of popular food fish, dividing them into three categories, depending on how sustainable their stocks are. "Best" choices include wild Alaskan salmon and farmed rainbow trout. "Worst" choices include shark and Chilean sea bass. The Monterey Bay Aquarium has a seafood watch app that classifies seafood into four categories—"Best Choice," "Certified," "Good Alternative," or "Avoid"—depending on its status. Eating less fish will result in reduced impacts on fisheries and the ocean floor and it will result in lower fossil fuel use because wild-caught fish require more fossil fuel energy per calorie of food obtained than either vegetables or grains.

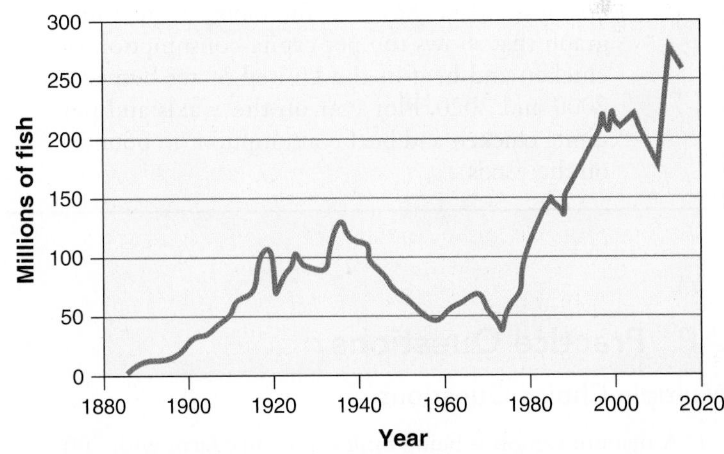

FIGURE 28.9 Commercial salmon harvest in the Alaska fishery. After a peak harvest in 1940, overfishing led to a decline in the number of fish caught. In 1973, fishery managers introduced a system of individual transferable quotas. By 1980, the fishery had rebounded. *(Data from http://www.adfg.alaska.gov/index.cfm?adfg=wildlifenews.view_article&articles_id=775)*

Module 28 AP® Review

Preparing for the AP® Exam

Learning Goals Revisited

28-1 What are the benefits and consequences of different meat production methods?

Meat is produced in concentrated animal feeding operations (CAFOs) and by various methods of free-range grazing. CAFOs produce meat more quickly and at lower cost but generally have much greater environmental impacts such as using more land and energy, impacting land more adversely, contaminating surface and groundwater, and producing more greenhouse gases.

28-2 What are the causes and consequences of overfishing?

Fishing uses more energy than agriculture, and overfishing can have impacts on biodiversity in marine systems and adversely impact people when fisheries are in decline. Many fisheries are in decline around the world although some have recovered, partially in response to regulation of catch size by communities or to national government intervention.

Practice Math and Graphing

Preparing for the AP® Exam

Answer the following questions. Be sure to show your work.

1. Practice Math

In 2020, the people in the United States consumed 115 pounds of chicken per capita, 84 pounds of beef per capita, and 67 pounds of pork per capita. Convert each value to kilograms and determine the total meat (chicken + beef + pork) consumption per capita in the United States in 2020 in kilograms.

2. Practice Graphing

Year	Beef, pounds per person	Chicken, pounds per person
2000	98	90
2005	97	100
2010	82	95
2015	82	102
2020	84	115

(a) Using the data shown in the table, create a line graph that shows the per capita consumption of chicken and beef in the United States between 2000 and 2020. Plot year on the x axis and per capita chicken and beef consumption (in pounds) on the y axis.

(b) Describe the trends in chicken and beef consumption in the United States between 2000 and 2020.

Data from https://farmdocdaily.illinois.edu/2021/05/an-overview-of-meat-consumption-in-the-united-states.html

AP® Practice Questions

Preparing for the AP® Exam

Multiple-Choice Questions

1. A manure lagoon is being built for a dairy farm with 700 cows, each of which produces 40 L of manure each day. How large must the lagoon be to hold 30 days' worth of manure?
   (a) 28,000 L
   (b) 120,000 L
   (c) 560,000 L
   (d) 840,000 L

2. If 12,000,000 kilocalories of chicken can be produced per hectare per year, how much land is needed to provide a person with 2,000 kilocalories of chicken/day for a year?
   (a) 0.01 ha
   (b) 0.02 ha
   (c) 0.06 ha
   (d) 0.2 ha

3. Which of the following describes bycatch?
   (a) It is a common problem with increased pesticide use.
   (b) It is a management technique in CAFOs.
   (c) It is a cause of fishery collapse.
   (d) It is common in large-scale fishing.

4. Concentrated animal feeding operations (CAFOs) can best be described as
   (a) facilities where a large number of animals are housed and fed in a confined space.
   (b) a method of producing more meat at a higher cost.
   (c) a means of producing great quantities of manure to fertilize fields organically.
   (d) the storing and compacting of grain for use as a nutrient supplement for cattle.

5. Which of the following is a human health problem associated with CAFOs?
   (a) the increase of antibiotic-resistant bacteria, which is potentially harmful to humans
   (b) the overgrazing of large tracts of land
   (c) lung cancer
   (d) the production of huge quantities of manure, creating a waste disposal problem

6. Which is an environmental effect of bycatch?
   (a) Juvenile fish are too small and slip through the nets, thereby ensuring the next generation of commercial fish.
   (b) Predators caught in the nets feed on the commercially important fish also caught, thereby reducing the total catch.
   (c) Nontarget fish populations have declined.
   (d) Cod populations have increased due to the removal of competitor species.

7. Suppose it takes 12 kg of grain to produce 1 kg of lamb and that a person eating only corn can obtain sufficient calories in a year from 0.02 ha of land. How much land would be required to support a person who only eats lamb?
   (a) 0.12 ha
   (b) 0.24 ha
   (c) 12.0 ha
   (d) 24 ha

Free-Response Question

Concentrated animal feeding operations (CAFOs) are indoor housing areas for raising cows and other animals for human consumption.

(a) **Identify** one organism, other than cows, that can be housed and grown in a CAFO. (1 pt.)
(b) **Describe** one economic benefit to utilizing CAFOs. (1 pt.)
(c) **Explain** one effect that CAFOs can have on water quality. (1 pt.)
(d) **Explain** one alternative to the use of CAFOs. (1 pt.)

Use the graph below to answer part (e).

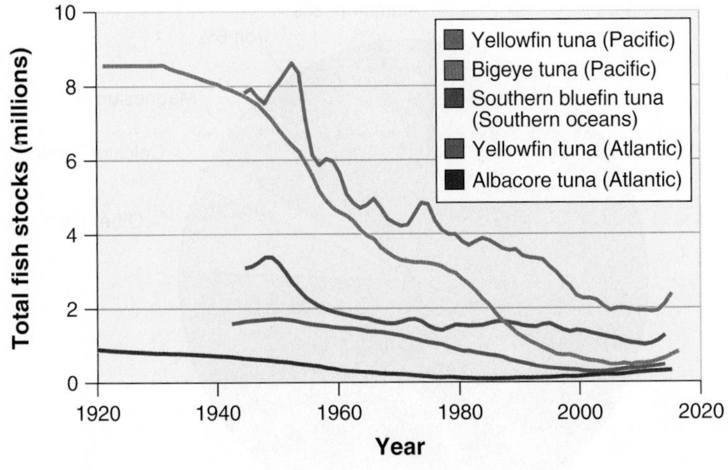

(e) Recently, consumption of fish and seafood products has increased. **Describe** the trend in fish stocks while consumption rises. (2 pts.)

Maintaining a dairy herd of cows in a CAFO requires large quantities of water and produces vast quantities of wastewater and manure. The increasing number of CAFO dairies in eastern New Mexico and western Texas is contributing to significant groundwater contamination and the depletion of the Ogallala aquifer. Consider a CAFO with 1,000 dairy cows when answering parts (f) and (g) below.

(f) According to the U.S. Department of Agriculture, the average dairy cow can consume up to 200 L of water daily. An additional 120 L per cow per day is required to wash the milking equipment and milking area. **Calculate** how many liters of water are required to operate this dairy daily. Show all work. (2 pts.)
(g) According to the U.S. Environmental Protection Agency, the average dairy cow produces 55 kg of wet manure daily. **Calculate** how many metric tons this dairy produces each day (1,000 kg = 1 metric ton). Show all work (2 pts.)

Module 29

Unit 5 | 24 | 25 | 26 | 27 | 28 | 29 | 30 | 31 | 32 | 33 | 34

Impacts of Mining

In the previous module we examined the impacts of raising meat and overfishing. Both activities provide nutrition for people but have impacts on terrestrial and marine systems. In this module we'll be exploring how mining provides benefits to human populations while impacting primarily terrestrial systems.

Learning Goals

After reading this module, you should be able to

29-1 describe how natural resources are extracted from Earth through mining.

29-2 identify the ecological and economic impacts of mining.

29-1 How are natural resources extracted from Earth through mining?

The distribution of mineral resources on Earth has social and environmental consequences

Plate tectonics, the rock cycle, soil formation, and erosion that were discussed in Unit 4 all influence the distribution of rocks and minerals on Earth. These resources, along with fossil fuels, exist in finite quantities, but are vital to modern human life. Some of these resources are abundant, whereas others are rare and extremely valuable.

Abundance of Ores and Metals

As we saw in Unit 4, early Earth cooled and differentiated into distinct vertical zones. Heavy elements sank toward the core, and lighter elements rose toward the crust. **Crustal abundance** is the average concentration of an element in Earth's crust. Looking at FIGURE 29.1, we can see that four elements—oxygen, silicon, aluminum, and iron—constitute over 88 percent of the crust. However, the chemical composition of the crust is highly variable from one location to another.

Crustal abundance The average concentration of an element in Earth's crust.

Ore A concentrated accumulation of minerals from which economically valuable materials can be extracted.

Metal An element with properties that allow it to conduct electricity and heat energy and to perform other important functions.

Environmental scientists and geologists study the distribution and types of mineral resources around the planet in order to locate them and to manage their extraction or conservation. **Ores** are concentrated accumulations of minerals from which economically valuable materials can be extracted. Ores are typically characterized by the presence of valuable metals, but accumulations of other valuable materials, such as salt or sand, can also be considered ores. **Metals** are elements with properties that allow them to conduct electricity and heat energy and to perform other important functions. Copper, nickel, and aluminum are

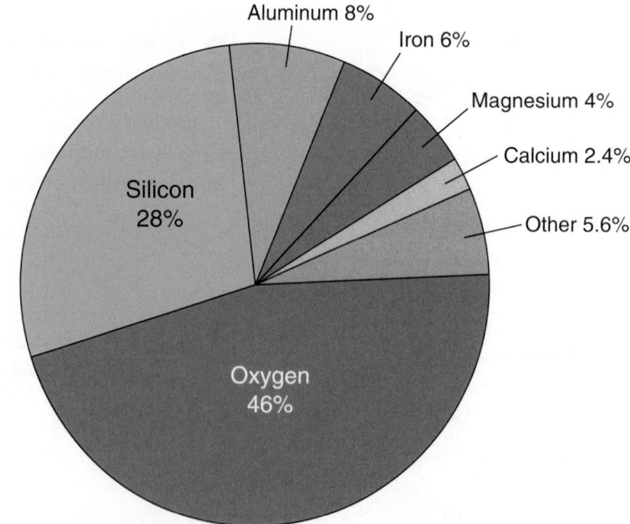

FIGURE 29.1 **Elemental composition of Earth's crust.** Oxygen is the most abundant element in the crust. Silicon, aluminum, and iron are the next three most abundant elements.

DO THE MATH What Fraction of Automobiles Are in the Developing World?

Preparing for the AP® Exam

Global mining of metals is influenced by the number of cars in the world. The following table shows the number of automobiles in the developed and developing worlds. Use scientific notation to calculate answers to the questions that follow.

Level of development	Number of automobiles
Developed countries	770,000,000
Developing countries	670,000,000

Express the number of automobiles in the developed and developing worlds in scientific notation. What is the total number of automobiles in the world? What percentage is in the developing world?

$$\text{Developed world} = 7.70 \times 10^8 \text{ automobiles}$$

$$\text{Developing world} = 6.70 \times 10^8 \text{ automobiles}$$

To find the total number of automobiles in the world, you need to add the developed and developing world numbers together:

$$7.70 \times 10^8 \text{ automobiles} + 6.70 \times 10^8 \text{ automobiles} = 1.44 \times 10^9 \text{ total automobiles in the world}$$

To determine the fraction of automobiles in the developing world, we divide the number of automobiles in the developing world by total automobiles in the world:

$$(6.70 \times 10^8 \text{ automobiles in the developing world}) \div (1.44 \times 10^9 \text{ total automobiles in the world}) = 0.47$$

To determine the value expressed as a percentage, multiply the decimal fraction by 100.

$$0.47 \times 100 = 47\% \text{ of all automobiles in the world are in the developing world.}$$

YOUR TURN If there are roughly 2.7×10^7 HEV, PHEV, and all-electric vehicles in the world, what percentage is this of all vehicles?

common examples of metals. They exist in varying concentrations in rock, usually in association with elements such as sulfur, oxygen, and silicon. Some metals, such as gold, exist naturally in a pure form. The manufacturing of hybrid electric vehicles (HEV), photovoltaic solar cells, and batteries use scarce metals, including neodymium, lithium, and lanthanum. As the number of vehicles that contain scarce metals increases, so will the demand for these metals. "Do the Math: What Fraction of Automobiles Are in the Developed World?" gives you an opportunity to use scientific notation to explore the number of automobiles in the developing and developed worlds and what percentage of those vehicles are HEV and all-electric.

Ores are formed by a variety of geologic processes. Some ores form when magma encounters water, heating the water and creating a solution from which metals precipitate, while others form after the deposition of igneous rock. Some ores occur in relatively small areas of high concentration, such as veins, and others, called disseminated deposits, occur in much larger areas of rock, although often in lower concentrations. Still other ores, such as copper, can be deposited both throughout a large area and in veins. Nonmetallic mineral resources, such as clay, sand, salt, limestone, and phosphate, typically occur in concentrated deposits. These deposits occur as a result of their chemical or physical separation from other materials by water, in conjunction with the tectonic and rock cycles. Some ores, such as bauxite—the ore in which aluminum is most commonly found—are formed by intense chemical weathering in tropical regions. The most accessible and high concentration ores are usually mined first. As high-concentration accessible ores are depleted, lower-concentration and less accessible ores are mined next. The lower the concentration and accessibility, the greater the energy expended and disturbance created, leading to more air pollution, waste, and soil disturbance and erosion.

The global supply of mineral resources is difficult to quantify. Because private companies hold the rights to extract certain mineral resources, information about the exact quantities of resources is not always available to the

TABLE 29.1	Approximate supplies of metal reserves remaining, assuming the same rate of metal use and same rate of recycling	
Metal	**Global reserves remaining (years)**	**U.S. reserves remaining (years)**
Aluminum (Al)	100	2
Copper (Cu)	45	20
Lead (Pb)	20	20
Zinc (Zn)	20	15
Gold (Au)	15	15
Nickel (Ni)	35	5
Cobalt (Co)	50	95
Manganese (Mn)	70	0
Chromium (Cr)	15	0

(Data from U.S. Geological Survey Mineral Commodity Summaries 2021, https://pubs.usgs.gov/periodicals/mcs2021/mcs2021.pdf; Bauxite and Alumina Report (USGS), https://pubs.usgs.gov/periodicals/mcs2021/mcs2021-bauxite-alumina.pdf)

public. The publicly known estimate of how much of a particular resource is available is based on its **reserve**: the known quantity of the resource that can be economically recovered. A resource is considered economically recoverable (or extractable) if the concentration in the host rock is high enough for it to be profitably mined. TABLE 29.1 lists the estimated number of years of remaining reserves for some of the most important metal resources commonly used in the United States, assuming that rates of use do not change, and that metal recycling continues at the current rates. The increased reliance on recycling of many metals will change the time estimates shown. So will innovations that allow substitutions and replacement materials. Some important metals, such as tantalum, have never been mined in the United States. The United States has consumed all of the economically extractable reserves of certain metals, such as manganese, which is used in steel production, and must now import those metals from other countries. While the estimates in this table are not certain, they do give a rough idea of the relative availability of different metals in the United States and around the world.

AP® Exam Tip

Recycling is generally a good solution for most challenges related to mining. Recycling increases the reserves of materials, reduces the amount of mining required, and reduces the impacts of mining. If you identify recycling as a solution, explain why.

Reserve In resource management, the known quantity of a resource that can be economically recovered.

Strip mining The removal of overlying vegetation and "strips" of soil and rock to expose underlying ore.

Mine tailings Unwanted waste material created during mining including mineral and other rock residues that are left behind after the desired metals are removed from the ore.

Mining Techniques

Mineral resources are extracted from Earth by mining the ore and separating any other minerals, elements, or residual rock away from the sought-after element or mineral. In general, when there is a relatively high concentration of the desired resource in the ore, there is less energy used and waste generated to obtain that resource. As the concentration of the desired resource decreases, energy use and waste generation increase. As illustrated in FIGURE 29.2, two kinds of mining take place on land: surface mining and subsurface mining. Each method has different benefits and costs in terms of environmental, human, and social perspectives.

Surface Mining

A variety of surface mining techniques can be used to remove a mineral or ore deposit that is close to the surface of Earth. **Strip mining**, or the removal of overlying vegetation and "strips" of soil and rock, called overburden, to expose the underlying ore, is used when the ore is relatively close to Earth's surface. Because it disturbs the soil, strip mining makes an area more susceptible to erosion. Sometimes the ore extracted from a strip mine runs parallel to the surface, which is often the case for deposits of sedimentary materials such as coal and sand. In these situations, miners remove a large volume of material, extract the resource, and return the unwanted waste material, called **mine tailings**, to the hole created during the mining. Mine tailings are the chemical compounds and rock residues that are left behind after the desired metal or metals are removed from the ore. They often contain trace quantities of the metals or compounds that are the desired resources. A variety of strategies can be used to restore the affected area to something close to its original condition.

Open-pit mining, a mining technique that creates a large visible pit or hole in the ground, is used when the resource is close to the surface but extends beneath the surface both horizontally and vertically. Copper mines are

FIGURE 29.2 **Surface and subsurface mining.** Surface mining methods include strip, open-pit, mountaintop removal, and placer mining.

usually open-pit mines. One of the largest open-pit mines in the world is the Kennecott Bingham Canyon mine near Salt Lake City, Utah (FIGURE 29.3). This copper mine is 4.4 km (2.7 miles) across and 1.1 km (0.7 miles) deep.

FIGURE 29.3 **Open-pit mining.** The Bingham Canyon Copper Mine, near Salt Lake City, Utah, is the largest open-pit mine in the world. *(Lee Prince/Shutterstock)*

In **mountaintop removal**, miners remove the entire top of a mountain with explosives. Large earth-moving equipment removes the resource and deposits the tailings in lower-elevation regions nearby, often in or near rivers and streams.

Placer mining is the process of looking for minerals, metals, and precious stones in river sediments. Miners use the river water to separate heavier items, such as diamonds, tantalum, and gold, from lighter items, such as sand and mud. Many of the prospectors in the California gold rush in the mid-1800s were placer miners, and the technique is still used today.

Subsurface Mining

When the desired resource, such as coal or a metal ore, is not easily accessible close to the surface of the Earth and is more than 100 m (328 feet) below the surface, miners must turn to **subsurface mining**. With few exceptions, subsurface

Open-pit mining A mining technique that creates a large visible pit or hole in the ground.

Mountaintop removal A mining technique in which the entire top of a mountain is removed with explosives.

Placer mining The process of looking for minerals, metals, and precious stones in river sediments.

Subsurface mining Mining techniques used when the desired resource is more than 100 m (328 feet) below the surface of Earth.

mining is almost always more economically expensive than surface mining. Typically, a subsurface mine begins with a horizontal tunnel dug into the side of a mountain or other feature containing the resource. From this horizontal tunnel, vertical shafts are drilled, and elevators are used to bring miners down to the resource and back to the surface. The deepest mines on Earth are up to 3.5 km (2.2 miles) deep. Coal, diamonds, and gold are some of the resources removed by subsurface mining.

29-2 What are the ecological and economic impacts of mining?

Mining impacts our ecosystems

While mining provides resources that are needed for many of the products we use on a daily basis at a relatively low cost, the extraction of mineral resources from Earth's crust has a variety of environmental impacts on water, soil, and biodiversity. In addition, mineral resource extraction can have economic and human health consequences that affect miners and others.

As you can see in **TABLE 29.2**, all forms of mining affect the environment. Mining almost always requires the construction of roads or railroad beds, which can result in soil erosion, damage to waterways, and cause habitat fragmentation. In addition, all types of mining produce mine tailings, which result in a wide variety of potential environmental and health consequences. Coal mining creates dust and methane, a potent greenhouse gas. Depending on the location and the amount of rainfall in a given area, mine tailings may contaminate land and water with acids and metals.

In mountaintop removal, the mine tailings are typically deposited in the adjacent valleys, sometimes obstructing or changing the flow of rivers. Mountaintop removal is used primarily in coal mining and is safer for workers than subsurface mining. In environmental terms, mining companies do sometimes make efforts to restore the mountain to its original shape. However, there is considerable disagreement about whether these reclamation efforts are effective. Damage to streams and nearby groundwater during mountaintop removal cannot be completely rectified by the reclamation process.

Placer mining can also contaminate large portions of rivers, and the areas adjacent to the rivers, with sediment and chemicals. In certain parts of the world, the toxic metal mercury is used in placer mining of gold and silver. Mercury is a highly volatile metal; that is, it moves easily among air, soil, and water. Mercury is harmful to plants and animals and can damage the central nervous system in humans; children are especially sensitive to its effects.

The environmental impacts of subsurface mining are often less apparent than the visible scars left behind by surface mining. One of these impacts is acid mine drainage. Acid mine drainage is the general term for water that passes through mine tailings. This water reacts chemically with mine tailings and becomes acidic and may contain metals that are leached from the tailings.

To keep underground mines from flooding, pumps must continually remove underground water from the mine. These pumps create a form of acid mine drainage. This drainage water lowers the pH of nearby soils and streams and can cause damage to ecosystems.

Mining affects human safety and the economy

Subsurface mining is a dangerous occupation. Hazards to miners include accidental burial, explosions, and fires. In addition, the inhalation of gases, dust, and particles over long periods can lead to a number of occupational respiratory

TABLE 29.2 Types of mining operations and their effects

Type of operation	Effects on air	Effects on water	Effects on soil	Effects on biodiversity	Effects on humans
Surface mining	Significant dust and particulate matter from earth-moving equipment	Contamination of water that percolates through tailings	Most soil removed from site; may be replaced if reclamation occurs	Habitat alteration and destruction over the surface areas that are mined	Minimal in the mining process, but air quality and water quality can be adversely affected near the mining operation
Subsurface mining	Minimal dust at the mining site, but emissions from fossil fuels used to power mining equipment can be significant	Acid mine drainage as well as contamination of water that percolates through tailings		Road construction to mines fragments habitat	Occupational hazards in mine; possibility of death or chronic respiratory diseases such as black lung disease

diseases, including black lung disease and asbestosis, a form of lung cancer. In the United States, in the twentieth century, more than 11,000 coal miners died in underground coal mine explosions and fires. A much larger number died from respiratory diseases. Today, there are relatively few deaths per year in coal mines in the United States, in part because of improved work safety standards and in part because there is much less subsurface mining. In other countries, especially China, mining accidents remain fairly common.

In the case of subsurface coal mining, there are usually deposits of methane interspersed with the coal beds. So coal mining leads to methane release which can be dangerous to miners because of the potential for explosions and suffocation due to lack of oxygen. It also emits methane to the atmosphere, where it becomes a potent greenhouse gas.

Over the past few hundred years, as human populations grew and developed nations industrialized, there was a great demand for mineral resources including coal. In developed nations, the demand for coal has slowed. As developing nations industrialize, the demand for certain mineral resources has been increasing. However, globally, the demand for coal is decreasing. In addition, as the most easily mined mineral resources are depleted, extraction efforts become more expensive and environmentally destructive. The ores that are easiest to reach and least expensive to remove are always recovered first. When these sources are exhausted, mining companies must turn to deposits that are more difficult to reach. These extraction efforts result in greater amounts of mining tailings and more of the environmental problems we have already noted. Designing systems that use and reuse limited mineral resources more efficiently will help protect the environment as well as human health and safety. It will also cost individuals and society less money because of lower health care costs.

Mining is responsible for a variety of environmental and human health problems. In the next module, we consider a very different subject that also is responsible for both environmental and human health issues: urbanization.

Module 29 AP® Review

Preparing for the AP® Exam

Learning Goals Revisited

29-1 How are natural resources extracted from Earth through mining?

Concentrated accumulations of elements and minerals in and below soils that are economically valuable are called ores. There are four surface mining techniques (strip mining, open-pit mining, mountaintop removal, and placer mining) plus subsurface mining that are commonly used to extract ores and coal from the ground.

29-2 What are the ecological and economic impacts of mining?

When ores are extracted, a variety of consequences affect humans and the environment. Surface mining usually has different environmental impacts than subsurface mining, which is more dangerous for workers.

AP® Practice Questions

Preparing for the AP® Exam

Multiple-Choice Questions

1. Tailings are
   (a) magma ore resulting from seafloor spreading.
   (b) the remaining supply of metals on Earth.
   (c) the nutrients that leach downward in soil.
   (d) the waste material from mining.

2. Which type of mining is usually most directly harmful to miners?
   (a) mountaintop removal
   (b) open-pit mining
   (c) placer mining
   (d) subsurface mining

3. Mining of this material is most likely going to release methane.
   (a) coal
   (b) aluminum
   (c) copper
   (d) lead

4. Which mining process is most likely to produce acid mine drainage?
   (a) mountaintop removal
   (b) open-pit mining
   (c) strip mining
   (d) placer mining

Use the table below to answer questions 5 & 6:

Metal	Global reserve remaining (years)	U.S. reserve remaining (years)
Aluminum	100	2
Copper	45	20
Gold	15	15
Cobalt	50	95

(Data from https://pubs.usgs.gov/periodicals/mcs2021/mcs2021.pdf; Bauxite and Alumina Report (USGS), https://pubs.usgs.gov/periodicals/mcs2021/mcs2021-bauxite-alumina.pdf)

5. According to the data, the resource that the United States has the most urgency to recycle and reuse is ____________.
   (a) aluminum
   (b) copper
   (c) gold
   (d) cobalt

6. Which metal does the United States have the most of compared to global resources?
   (a) aluminum
   (b) copper
   (c) gold
   (d) cobalt

Free-Response Question

The figure below shows several forms of mining. Use the figure to answer the questions below.

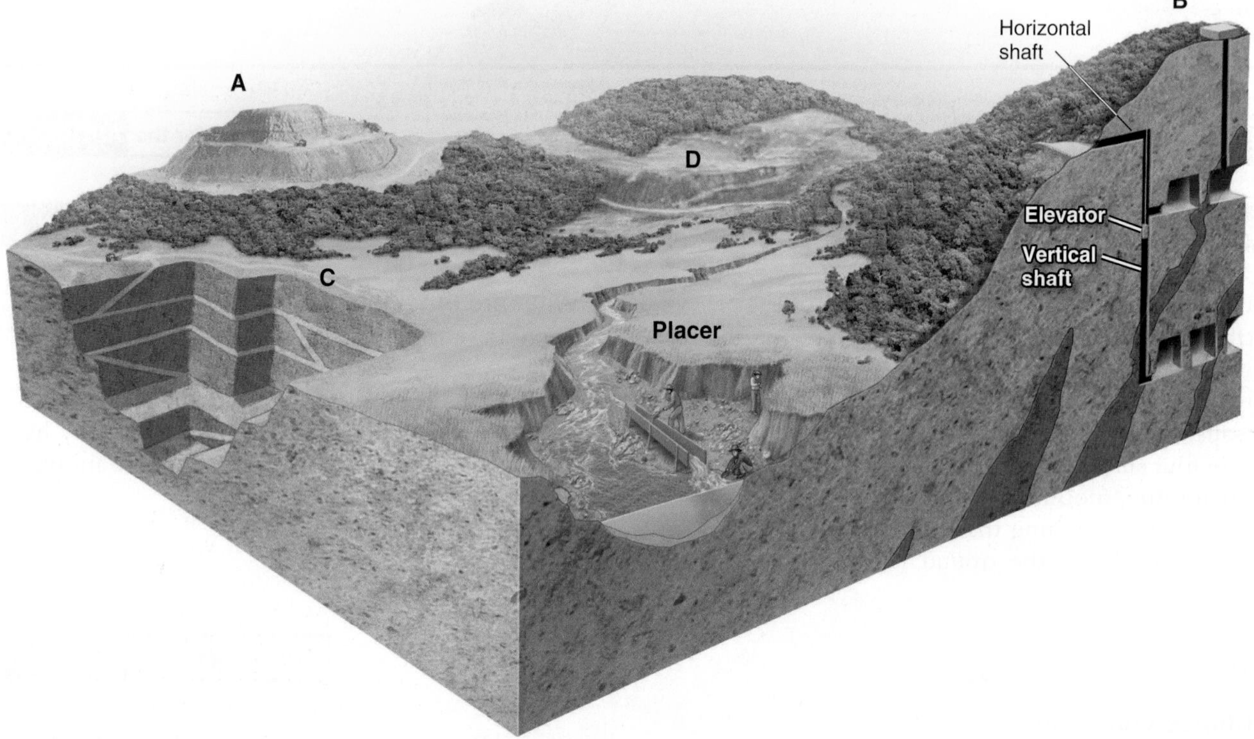

(a) **Identify** which type of mining removes plants, soils, and rocks to get to underlying resources, such as coal. (1 pt.)
(b) **Describe** the mining process used in part A in order to reach coal. (1 pt.)
(c) **Explain** what happens to the materials removed from part C. (2 pts.)
(d) **Describe** how mining operations can lead to acid deposition. (1 pt.)
(e) **Describe** the difference between tailings and overburden. (2 pts.)
(f) Mining operations have trended away from subsurface mining to other forms.
   (i) **Describe** one economic disadvantage to subsurface mining. (1 pt.)
   (ii) **Propose** a viable alternative to subsurface mining, that will still produce ore. (1 pt.)
   (iii) **Describe** one disadvantage to the proposed solution from part (ii). (1 pt.)

Module 30

Impacts of Urbanization and the Methods to Reduce Urban Runoff

In the previous module we examined the impacts of mining, which provides valuable resources to people while adversely affecting both the natural environment and humans. In this module we will explore how urbanization provides benefits to human populations while also impacting the natural environment and human health.

Learning Goals

After reading this module, you should be able to

30-1 explain urbanization and its effects on the environment.

30-2 describe the methods used to reduce urban runoff.

30-1 What is urbanization and what are its effects on the environment?

Increasing population density has many environmental implications

We just discussed how increasing human populations might impact the demand for different minerals extracted through mining. But there are many other ways that an increasing population impacts our environment. One of them is the move from less densely populated areas to urban areas as populations increase. Urban means relating to city or towns or densely populated areas. **Urbanization** is the process of an area becoming more urban, which means increasing the density of people per unit area of land. Impacts on the environment may occur locally—within the borders of a region, city, or country—or they may be global in scale. For example, a person may create a local impact by using products created within the city's borders. That same person may create a global impact by using materials that are imported from another region or another country.

Urban populations represent 55 percent of the human population and consume three-fourths of Earth's resources. The United Nations projects that by 2050, 68 percent of the human population will live in urban areas. While definitions vary by country, an **urban area**, according to the U.S. Census Bureau, contains more than 386 people per square kilometer (1,000 people per square mile). New York City is the most densely populated urban area in the United States, with 10,400 people per square kilometer (27,000 people per square mile). Mumbai, India, is the most densely populated urban area in the world, with 30,000 people per square kilometer (77,000 people per square mile).

Developed and Developing Country Differences

More than 75 percent of people in developed countries live in urban areas, as **FIGURE 30.1** shows, and that number is expected to increase slightly over the next 15 years. In developing countries, 48 percent of people live in urban areas. This number will probably increase to 56 percent by 2030. **TABLE 30.1** shows that, of the 20 largest cities in the world, 17 are in developing countries. Worldwide, almost 5 billion people are expected to live in urban areas by 2030.

Urban living in both developed and developing countries presents environmental challenges. Most developed countries employ city planning. As urban areas expand, experts design and install public transportation facilities, water and sewer lines, and other municipal services. All of these activities require resources and fossil fuels so one can be led to believe that urban areas deplete resources and increase the release of carbon dioxide to the atmosphere. However, while urban areas require more resources and produce greater amounts of solid waste, pollution, and carbon dioxide emissions than suburban or rural areas, they tend to have

Urbanization The process of making an area more urban, which means increasing the density of people per unit area of land.

Urban area An area that contains more than 386 people per square kilometer (1,000 people per square mile).

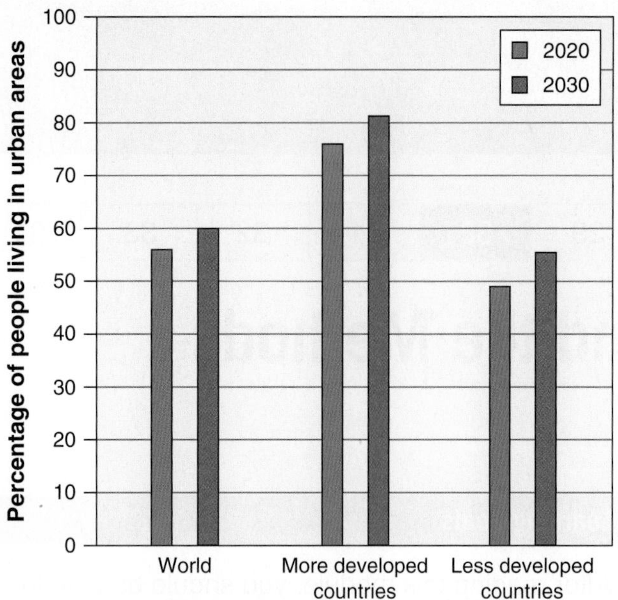

FIGURE 30.1 Urban growth. More than one-half of the world's population will live in urban settings by 2030. *(Data from U.N. POP Fund)*

a smaller impacts per person. There are many reasons for this difference including more dense housing (think large apartment buildings), easier access to public transportation and services that are nearby, such as shopping, which allow urban dwellers to be more efficient in their energy and material use and pollution emissions (think walking rather than driving to the store).

In developing countries, the relatively affluent sections of urban areas have safe drinking water, sewage treatment systems, and collection systems for the disposal of household solid waste. However, many less-affluent urban residents have limited access to these services. Rapid urbanization in the developing world often results in an influx of people without access to money or other resources. As a result, their shelters may be built of cardboard, plastic, or mud, rather than with appropriate building materials and infrastructure, and without adherence to local building codes. Whether they are squatter settlements, shantytowns, or slums, these overcrowded and underserved settlements are, unfortunately, a fact of life in the developing and, to some extent, in the developed world too. The United Nations organization UN-HABITAT estimates that 1 billion people live in "informal" settlements and other similar areas throughout the world. Most residents of informal settlements live in housing structures without flooring, safe walls and ceilings, or such basic amenities as water, sanitation, or health care (**FIGURE 30.2**).

Increasing Urbanization and Urban Sprawl in the United States

Roughly a century ago, many people in the United States began to move from rural areas to large cities. The last 50 to 60 years have seen an increased movement from those cities to the surrounding areas, which has resulted in the growth of two classes of communities: suburban and exurban. **Suburbs** are areas that surround metropolitan centers and have low population densities compared with urban areas. **Exurbs** are similar to suburbs, but are not connected to any central city or densely populated area. Since 1950, more than 90 percent of the population growth in metropolitan areas has occurred in suburbs, and two out of three people now live in suburban or exurban communities.

Nevertheless, the trend of more people in suburban and exurban settings does lead to a greater use of fossil fuels, which leads to greater carbon dioxide emissions. And the same trend leads to a greater collection of trash being disposed of in concentrated areas which leads to the possibility of anaerobic decomposition and the emission of methane. So even though there are some initial efficiency gains from urbanization, the overall trend of people moving toward urban and surrounding areas affects the carbon cycle negatively.

TABLE 30.1 The largest 20 urban areas in the world

Rank	City, Country	Population (millions)
1	Tokyo, Japan	37,500
2	Delhi, India	28,500
3	Shanghai, China	25,500
4	São Paulo, Brazil	21,600
5	Ciudad de México (Mexico City), Mexico	21,500
6	Al-Qahirah (Cairo), Egypt	20,000
7	Mumbai (Bombay)	19,980
8	Beijing, China	19,600
9	Dhaka, Bangladesh	19,600
10	Kinki M.M.A. (Osaka), Japan	19,300
11	New York-Newark, United States	18,800
12	Karachi, Pakistan	15,400
13	Buenos Aires, Argentina	15,000
14	Chongqing, China	14,800
15	Istanbul, Turkey	14,700
16	Kolkata (Calcutta), India	14,600
17	Manila, Philippines	13,500
18	Lagos, Nigeria	13,500
19	Rio de Janeiro, Brazil	13,300
20	Tianjin, China	13,200

Note: Data are from 2018 and contain the areas defined by the United Nations as "urban agglomerations." Other agencies agglomerate urban areas differently and obtain slightly different results. *(Data from United Nations Population Division)*

FIGURE 30.2 Informal settlements. Informal collections of dwellings such as this squatter settlement on the outskirts of Lima, Peru, expand outward from the urban center. *(Reuters/Mariana Bazo)*

(a)

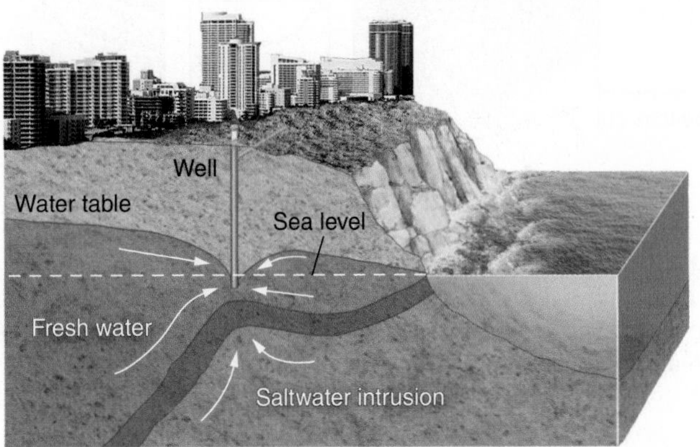

(b)

FIGURE 30.3 Saltwater intrusion. (a) When there are few wells along a coastline, the water table remains high and the resulting pressure prevents salt water from intruding. (b) After large population increases and more wells, rapid pumping of wells drilled in aquifers along a coastline can lower the water table. Lowering the water table reduces water pressure in the aquifer, allowing the nearby salt water to move into the aquifer and contaminate the well water with salt.

Greater populations in urban and suburban regions have led to greater use of water from aquifers and surface waters. We saw in Module 27 that greater use of water for agriculture led to a drawdown of the aquifer and a cone of depression (see Figure 27.8). A variation of this phenomena occurs when there is rapid development in urban and suburban areas close to the coastline. The drawdown of ground water from overpumping of wells in coastal aquifers creates a cone of depression but the water that moves in to replace the water that has been removed is partially seawater. In this case, the well eventually becomes contaminated with seawater as shown in **FIGURE 30.3**. This situation is known as **saltwater intrusion** and it is common in developed areas along coastlines such as Atlantic City, New Jersey. The increase in the number of houses, roads, and paved surfaces as a result of urbanization can cause other problems as well. Increased **impervious surfaces**, such as pavement or other surfaces that do not allow water penetration, causes a reduction in the infiltration of water to soil and aquifers, and can result in less aquifer recharge. Less infiltration means more surface runoff so it can also lead to more flooding and possibly erosion in unpaved areas. It can also overwhelm storm drainage systems that are designed to collect and treat water from urban surfaces. We will discuss urban runoff further in the next section of this module.

The rural population in the United States has been roughly the same since 1910 at a little more than 50 million people, and in 2022 it makes up less than one-sixth of the total U.S. population. Although this population shift from rural to urban areas has brought with it a new set of environmental problems associated with urban sprawl, attempts to find creative solutions have been increasingly successful.

Causes and Consequences of Urban Sprawl

If you have ever been to a strip mall, you are familiar with the phenomenon known as **urban sprawl**—urbanized areas that spread into rural areas and remove clear boundaries between the two. The landscape in these areas is characterized by clusters of housing, retail shops, and office parks, which are separated by miles of road. Large feeder roads and parking lots that separate "big box" and other retail stores from the road discourage pedestrian traffic.

Urban sprawl has had a dramatic environmental impact. Dependence on the automobile causes suburban residents to drive more than twice as much as people who live in cities. Because suburban house lots tend to be significantly

Suburbs Areas that surround metropolitan centers.

Exurbs Similar to suburbs, but are not connected to any central city.

Saltwater intrusion An infiltration of salt water in an area where groundwater pressure has been reduced as a result of a cone of depression from extensive pumping of wells.

Impervious surface Pavement or other surfaces that do not allow water penetration.

Urban sprawl Urbanized areas that spread into rural areas.

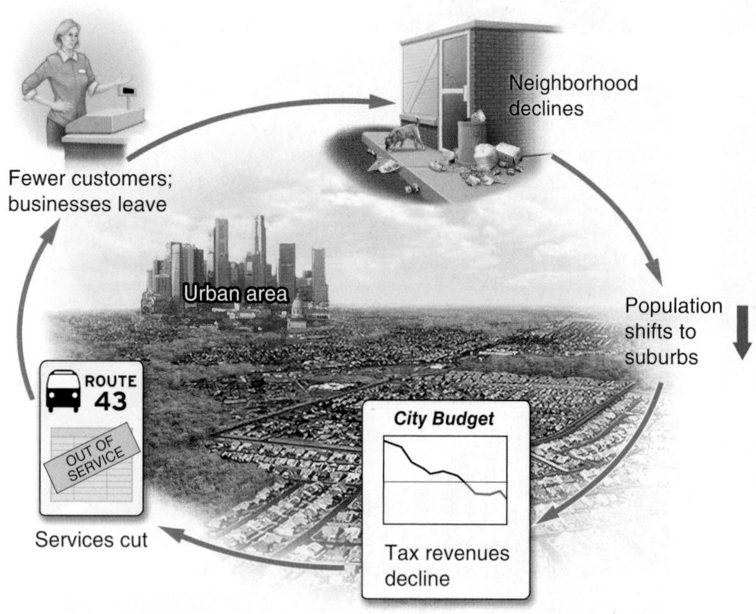

FIGURE 30.4 Urban blight. As people move away from a city to suburbs and exurbs, the city often deteriorates, which causes yet more people to leave. This cycle is an example of a positive feedback system. The green arrow indicates the starting point of the cycle.

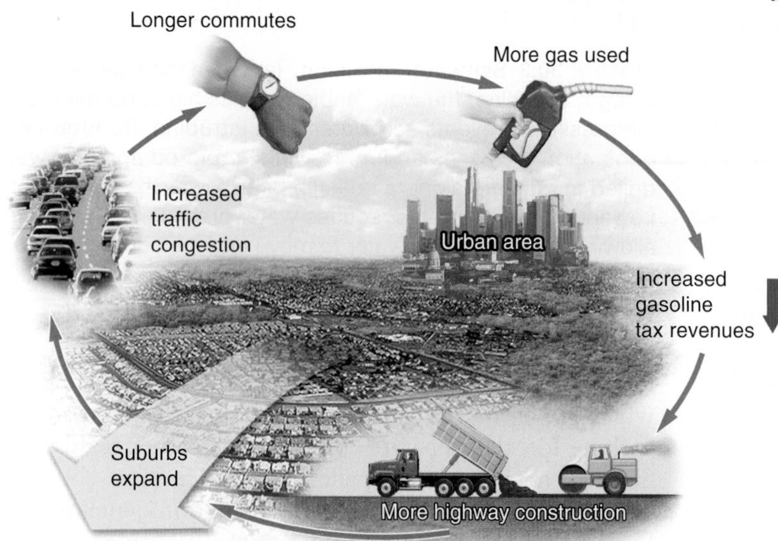

FIGURE 30.5 Induced demand as a cause of traffic congestion and urban sprawl. The use of gasoline tax money to build highways leads to the development of suburbs and traffic congestion, at which point yet more money is spent on highways to alleviate the congestion. The green arrow indicates the starting point of the cycle.

Urban blight A lack of support for and deterioration of urban communities.

Sense of place The feeling that an area has a distinct and meaningful character.

larger than their urban counterparts, suburban communities also use more than twice as much land per person as urban communities. Urban sprawl tends to occur at the edge of a city, often replacing farmland and increasing the distance between farms and consumers. There are four main causes of urban sprawl in the United States: increased availability of both automobiles and highways to allow access to the suburbs; reasonable living costs in the suburbs; lack of support for and deterioration of urban communities known as **urban blight**; and government policies that fund highways and subsidize mortgages for suburban areas. FIGURE 30.4 illustrates a positive feedback cycle for urban blight. FIGURE 30.5 illustrates a gasoline tax-highway construction positive feedback system.

Reducing Impacts of Urbanization and Urban Sprawl: Smart Growth

People are beginning to recognize and address the problems of urban sprawl. One approach, called smart growth, focuses on strategies that encourage the development of sustainable, healthy communities. Smart growth focuses on engaging residents and stakeholders—people with an interest in a particular place or issue—who need to work together to determine how their neighborhoods will appear and be structured. And it needs to foster distinctive communities that have a **sense of place**, the feeling that an area has a distinct and meaningful character. Many cities have such unique neighborhoods. For example, the French Quarter of New Orleans (FIGURE 30.6) has a sense of place that adds to the quality of life for people living there. Smart growth attempts to foster this sense of place through development that fits into the neighborhood. The Environmental Protection Agency lists basic principles of smart growth. These include creating walkable neighborhoods; mixing residential, retail, recreational, and business land uses; encouraging community and stakeholder collaborations in development decisions for the neighborhood, and preserving open space, farmland, and natural beauty—areas with a sense of place.

AP® Exam Tip

You might be asked to provide solutions to environmental problems that result from urbanization. Whether you offer an answer of increasing use of public transportation or increasing green space in the city, make sure to elaborate and explain why this helps improve negative impacts.

FIGURE 30.6 **The French Quarter of New Orleans.** One principle of smart growth is to foster communities with a strong sense of place. The French Quarter of New Orleans, Louisiana, is known for its architecture, food, and, especially, music. *(Design Pics Inc/Alamy)*

30-2 What are the methods used to reduce urban runoff?

A variety of methods can reduce urban runoff

Urban runoff is water in an urban area that does not evapotranspire or infiltrate the soil. It enters drainage systems such as sewers and urban rivers and streams. Reducing urban runoff entails reducing both the volume of water that is runoff and the amount of pollutants contained within the runoff water. There are a variety of approaches to reducing the amount of urban runoff (FIGURE 30.7). The first involves reducing the amount of water that runs off urban surfaces. This can be achieved in several ways. Firstly, the greater the area that is paved or covered with impervious surfaces, the less area there is for water to infiltrate the soil and recharge aquifers. Reductions in impervious surfaces can be achieved by replacing conventional pavement—that is highly impervious—with permeable pavement that allows water infiltration. It can also be achieved by reducing the overall amount of pavement in a city. Both increasing permeable pavement and reducing paved surfaces will decrease urban runoff. This includes building taller buildings so the "footprint" on the city surface stays the same while the capacity to provide living space or office space increases. Secondly, any urban systems that collect rainwater, such as rooftop rain collection systems that collect water to be used on a rooftop garden or for watering lawns or other needs, will decrease urban runoff because less water will reach the pavement surface. Thirdly, increased planting of trees, especially if there is

Urban runoff Runoff, water that does not evapotranspire or infiltrate the soil, that occurs in an urban area.

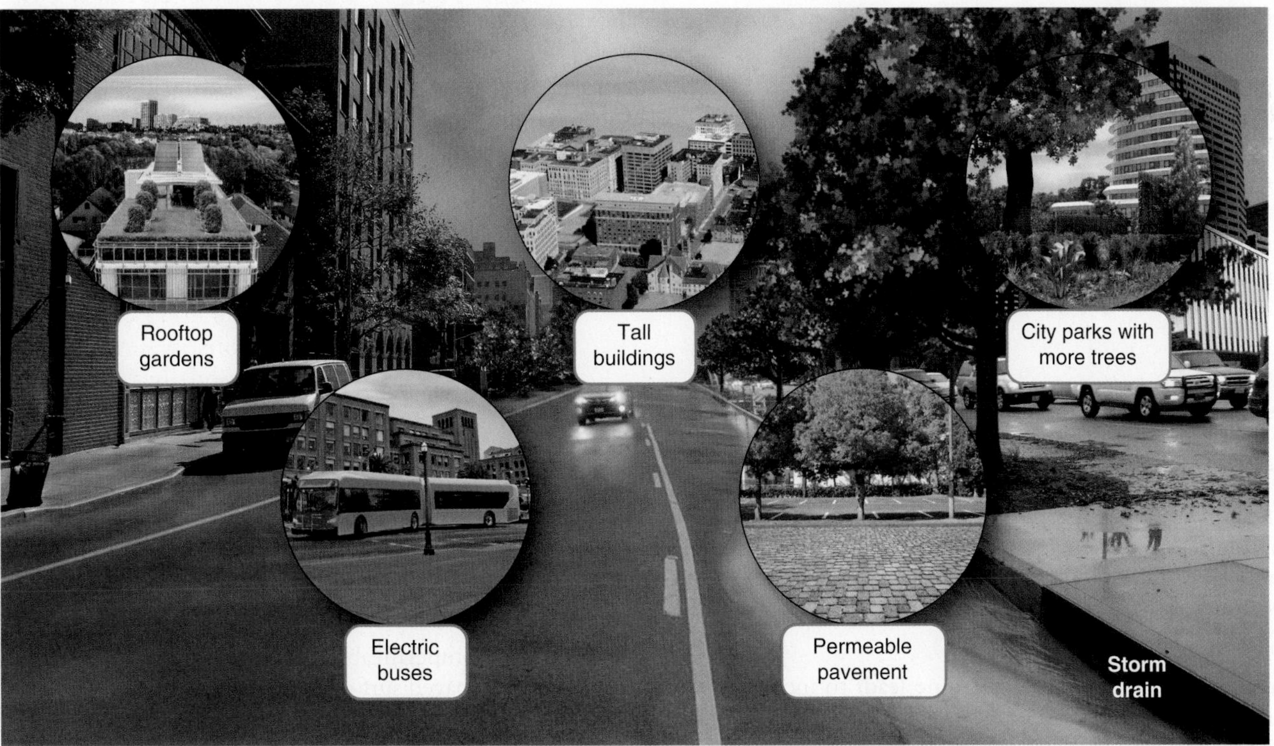

FIGURE 30.7 **Reducing Urban Runoff.** The reduction of urban runoff can be achieved by reducing the amount of water that runs off urban surfaces and by reducing the quantity of pollutants that are on the urban surfaces.

a large area of unpaved surface area surrounding the trees, can increase evapotranspiration and reduce runoff. These three approaches can be categorized as reducing the amount of water that will become runoff.

In addition, reducing the pollutants that are deposited on urban surfaces can reduce the amount of pollutants that wash away when water runs off the urban surface. When water does become runoff in an urban environment, it will usually flow to a body of water, or it will enter the drainage system for the city, usually called the storm water drainage system, and ultimately be diverted to a body of water. Depending on the city and the age of the storm sewer system, storm water may be treated or it may flow directly to a body of water without treatment. During heavy rain events, almost all cities lack the capacity to treat storm water, so, typically, heavy rain event water goes directly into a body of water without treatment. Usually, urban surfaces contain a variety of pollutants from vehicles that have leaked oil or other automotive fluids onto the road surface. When it rains, water that runs over the urban surface often picks up automotive and other pollutants. So one approach to reducing urban runoff pollutants is to reduce the amount of automotive and other waste that reaches the pavement. Methods for achieving this include increasing public transportation, particularly all-electric transportation which tends to have less or no motor oil so there's less likelihood for leaks onto the road surface. Also increased use of public transportation reduces the number of vehicles on the road. Although they might be larger vehicles, the greater efficiency should translate to fewer pollutants on the road surface.

When considering urban runoff, the most sound objectives are to reduce water flow and pollutants that can become urban runoff and eventually transfer to other environments. In the next module we consider a variety of environments as we integrate the environmental impact of a person or country in a metric called the ecological footprint.

Module 30 AP® Review

Preparing for the AP® Exam

Learning Goals Revisited

30-1 What is urbanization and what are its effects on the environment?

More than 75 percent of people in developed countries live in urban areas and that number will increase in the coming years. Urbanization and development of cities leads to consumption of resources and fossil fuels. However, urban dwellers tend to have smaller impacts per person due to efficiencies that occur with more people occupying smaller areas of land.

30-2 What are the methods used to reduce urban runoff?

Urban runoff can be reduced by decreasing the quantity of water that runs off urban surfaces and into storm drains and bodies of water. It can also be reduced by minimizing the amount of pollutants that are deposited on urban surfaces.

AP® Practice Questions

Preparing for the AP® Exam

Multiple-Choice Questions

1. The creation of walkable neighborhoods is
   (a) a result of urban sprawl.
   (b) one goal of the U.S. Geological Survey.
   (c) discouraged by the U.S. EPA.
   (d) a principal of smart growth.

2. When development along the coastline leads to increased withdrawals from coastal aquifers, which of the following is most likely to occur?
   (a) a decrease in impervious surfaces
   (b) saltwater intrusion
   (c) an accumulation of mine tailings at the surface
   (d) greater methane release

3. Which of the following typically contributes to reducing urban runoff?
   (a) increasing impermeable pavement
   (b) increasing paved surfaces
   (c) removing urban trees
   (d) installing rooftop rain collection systems

4. During heavy rain events, urban runoff usually
   (a) is diverted for use in local swimming pools.
   (b) undergoes intensive water treatment in a nearby water filtration plant.
   (c) is recirculated for car washes and urban irrigation systems.
   (d) goes directly into a nearby body of water without treatment.

5. The image below displays the phenomenon known as ____________.

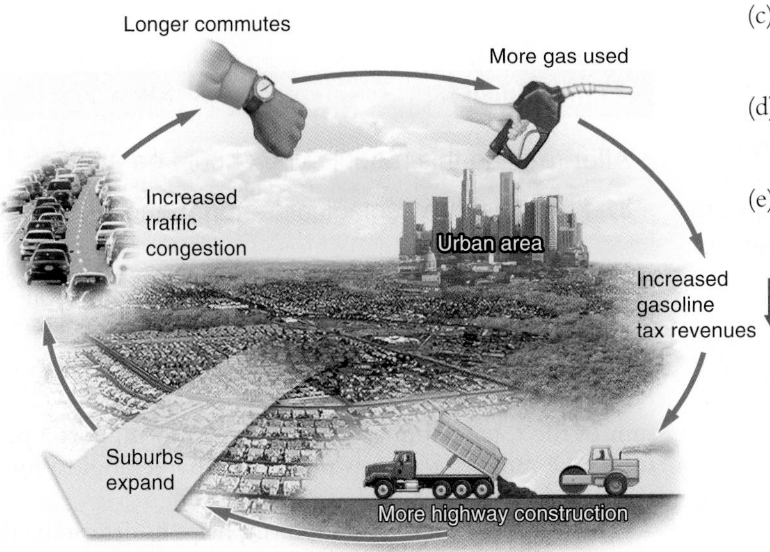

   (a) urban blight
   (b) urbanization
   (c) urban sprawl
   (d) suburbanization

Free-Response Question

Over the last 40 years, human populations have trended toward movement into urban and suburban areas. This has resulted in a variety of environmental issues.

(a) **Identify** one difference between urban areas in a developed nation versus urban areas in a developing nation. (1 pt.)

(b) **Describe** one health effect that a person in a developing urban area may have to deal with connected to your answer in part (a). (1 pt.)

(c) **Explain** how movement from urban areas to suburban areas may lead to an increase in atmospheric pollution. (1 pt.)

(d) Movement of people in concentrated areas such as urban and suburban locations can lead to saltwater intrusion. **Propose a solution** to this problem. (1 pt.)

(e) Populations have been increasing in urban areas over the last 40 years. The table below shows what percent of the global population is living in urban areas.

Year	% living in urban area
1980	39.4
1990	40.0
2000	46.7
2020	56.2

(Data from https://data.worldbank.org/indicator/SP.URB.TOTL.IN.ZS?end=2020&start=1980&view=chart)

(i) **Calculate** the percent increase in urban population from 1980 to 2020. Show all work. (2 pts.)

(ii) From 2000 to 2020, population moved at a steady rate. **Calculate** the percentage of people living in an urban area in 2010. Show all work. (2 pts.)

(iii) Assuming the same rate of urban population increase found part (ii), **calculate** the percentage of the population that will be living in urban areas in 2030. Show all work. (2 pts.)

Module 31

Unit 5 | 24 | 25 | 26 | 27 | 28 | 29 | 30 | 31 | 32 | 33 | 34

Ecological Footprints

Thus far in this book, we have identified ways in which human activities affect the environment. Now, in this module, we will describe a method for assessing environmental impact: the ecological footprint. We will also explore other more specialized indicators, including the carbon footprint. This module will allow us to explore the level of human impact on natural systems through indicators.

Learning Goals

After reading this module you should be able to

31-1 explain the ecological footprint and what it tells us.

31-2 describe the carbon footprint and how it differs from the ecological footprint.

31-1 What is the ecological footprint and what does it tell us?

The ecological footprint is an environmental assessment tool

As we study the way humans have altered the natural world, it is important to have methods for measuring and quantifying human impact. Environmental indicators allow us to assess the impact of humans on Earth. We have already used the term "footprint" to describe the water footprint in Module 27 and the area occupied or affected by a building in Module 30. And that is exactly how the term is used in one of the most common indicators in environmental science, the ecological footprint.

Ecological Footprint: Origins and Purpose

The tool many environmental scientists use to assess environmental impact, the ecological footprint, was developed in 1995 by Professor William E. Rees and his then-graduate student Mathis Wackernagel. The **ecological footprint** is a measure of the area of land and water an individual, population, or activity requires to produce all the resources it consumes and to process the waste it generates. It is typically expressed in biologically productive land in hectares (ha) (1 hectare = 2.47 acres). The sum of the total amount of land required to support a person's or country's activities represents the ecological footprint, as shown qualitatively in **FIGURE 31.1**.

The use of the ecological footprint helps us determine the current-day impact of humans on the environment. It also

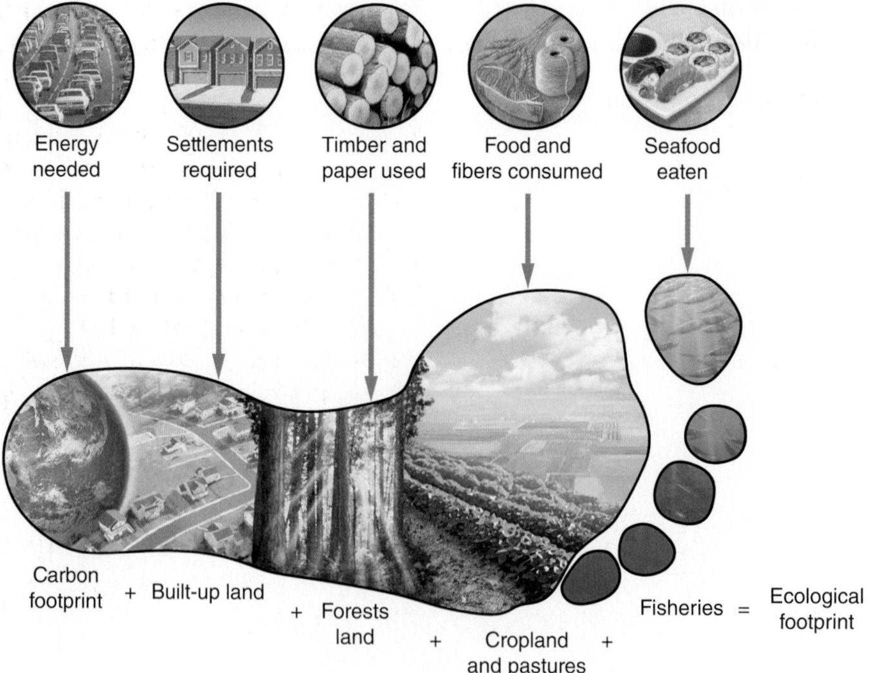

FIGURE 31.1 The ecological footprint. The ecological footprint is a measure of how much land is needed to supply the goods and services that individuals or countries use. Only some of the many factors that go into the calculation of the footprint are shown here. (The actual amount of land used for each resource is not drawn to scale.)

helps us determine whether impacts are increasing or decreasing over time. Ultimately, they inform us on whether we are achieving sustainability, which is the focus of the next module.

How do we determine what actions and what lifestyles have the greatest environmental impact? This is an important question for environmental scientists if we are to understand the effects of human activities on the planet and develop sustainable practices. Calculating sustainability, however, is more difficult than one might think because we must consider the impacts of our activities and lifestyles on different aspects of our environment. We use land to grow food and to build structures on, and for parks and recreation. We require water for drinking, for cleaning, and for manufacturing products such as paper, and we need clean air to breathe. Yet these goods and services are all interdependent: Using or protecting one has effects on the others. For example, using land for conventional agriculture may require water for irrigation, fertilizer to promote plant growth, and pesticides to reduce crop damage. This use of land to grow crops results in a reduction in the amount of water available for human use: Irrigation provides water that plants absorb and transpire. And fertilizers and pesticides may pollute water and introduce pollutants into the air. Finally, agriculture may generate waste products. Plant wastes take up land where they are composted. And waste products generated when manufacturing tractors and other farm equipment may end up in the landfill, which also takes up land. So determining the "footprint" of conventional agriculture requires measuring the land physically occupied by the crop, but also measuring an amount of land used by the aquifer (which is underground); measuring the amount of land used to manufacture fertilizer and pesticides to support the agricultural activity; and measuring the amount of land needed for landfills or compost piles to process the waste. With this brief example, you can see that the measurement is not simple!

Wackernagel and Rees maintained that if our lifestyle demands more land than is available, then we must be living unsustainably, that is, using up resources more quickly than they can be replenished, or producing wastes more quickly than they can be processed. For example, each person requires a certain number of food calories every day. We know the number of calories contained within a given amount of grain or meat. We also know how much farmland or rangeland is needed to grow the grain to feed people or to feed livestock such as sheep, chickens, or cows. If a person eats only grains or plants, the amount of land needed to provide that person with food is calculated as the amount of land needed to grow the plants they eat. If that person eats commercially grown meat and grains, however, the amount of land required to feed that person is greater because we must also consider the land required to raise and feed the livestock that ultimately become meat. Thus, one factor in the size of a person's ecological footprint is the amount of grain in their diet. And if they eat meat, their footprint will be larger.

Calculating the Ecological Footprint

We can calculate the total ecological footprint by summing the land requirements for the food we eat, the water and energy we use, the clothing we wear and the structures we occupy such as a house and car.

As you might remember, we discussed the environmental impacts of individuals and countries in Module 18, by presenting the relationship between wealth and consumption and the environmental impact equation, IPAT (impact = population × affluence × technology). The ecological footprint is another attempt at quantifying the same metric: environmental impact. Both consider the expenditure of resources and the impact of those resources. As countries prosper, their populations use more resources. Sometimes prosperity and economic development can improve environmental conditions and reduce environmental harm. For instance, wealthier countries may have the resources to implement pollution control measures and invest money to protect native species. So, for these two examples, people in developing countries may be less likely to have the financial resources to implement environmental protection measures. Nevertheless, in general, people in developing countries do not consume the same quantity of resources as those in developed countries, and therefore, usually have a smaller ecological footprint. For example, people in developing countries own fewer automobiles, refrigerators, and washing machines and therefore are responsible for less consumption of metals, plastics, and energy in the manufacturing process and in the use and disposal of those items when they have reached the end of their useful lives.

By estimating the ecological footprint for each person in each country and by determining the amount of biologically productive land on Earth, we can determine how much land each person is entitled to, and whether they are exceeding their ecological footprint or their "fair Earthshare."

AP® Exam Tip

The goal of calculating ecological footprints is to evaluate a person's or country's actions and how they impact the planet. They help individuals or groups reflect on where they should prioritize efforts to reduce negative impacts on the planet. See if you can describe one specific factor that could reduce the ecological footprint of an individual or group.

For example, this is a clear response:

The city of Fremont could provide access to light rail to reduce emissions of greenhouse gasses.

Compared to the following, which is less specific:

The city of Fremont should reduce its ecological footprint.

Ecological footprint A measure of the area of land and water an individual, population, or activity requires to produce all the resources it consumes and to process the waste it generates.

TABLE 31.1 Ecological footprints for selected countries and the world

Country	Hectares of biologically productive land per person
United States	8.0
Germany	4.7
China	3.7
India	1.2
World average	2.8
Fair Earthshare	1.6

There is roughly 12.2 billion hectares of biologically productive land on Earth. If we divide that by the number of people on Earth, 7.8 billion, we can calculate the "fair Earthshare" for each person:

$$(12.2 \text{ billion ha land}) \div (7.8 \text{ billion persons}) = 1.6 \text{ ha per person}$$

Therefore, the fair Earthshare, or area of biologically productive land on Earth divided by the number of people on Earth, is 1.6 hectares per person. An ecological footprint larger than 1.6 ha indicates that you or citizens of a given country use more than their "fair Earthshare" of resources. **TABLE 31.1** shows ecological footprints for a few countries and the world. Therefore, if the Ecological Footprint model is correct, citizens of the United States are exceeding the capacity of the Earth by a factor of five. Citizens of Germany are exceeding the capacity of Earth by a factor of three. And the entire world on average is exceeding the capacity of Earth by 75 percent. If this calculation is correct, it suggests that current human activities on Earth are not sustainable. Or as one thoughtful business executive stated: "If we grow by using more stuff [and increase our ecological footprint], I'm afraid we'd better start looking for a new planet." In more recent years, another metric that is more specialized than the ecological footprint has increased in popularity: the carbon footprint.

31-2 What is the carbon footprint and how does it differ from the ecological footprint?

The carbon footprint calculates carbon dioxide emitted into the atmosphere

The **carbon footprint** is a measure of the total carbon dioxide and other greenhouse gas emissions from the activities, both direct and indirect, of a person, country, or other entity. There are two aspects of this definition that must be clarified. First, carbon refers to the carbon released as carbon dioxide but also refers to the equivalent amount of carbon released in the other greenhouse gases, methane (CH_4) and nitrous oxide (N_2O). Since the other greenhouse gases have different potencies for warming, carbon footprints are often expressed in carbon dioxide equivalents. This means that if methane is 25 times more potent as a greenhouse gas, and you are responsible for releasing one unit of methane, this will be counted as 25 units of carbon dioxide equivalent. Second, being responsible for both direct and indirect emissions means that burning gasoline in a car and releasing carbon dioxide counts toward your carbon footprint. But so does purchasing a bicycle that required energy and resources for its manufacturing. Note that the carbon footprint does not take into account other types of environmental harm such as land used for disposal of solid waste, or alteration of habitat that leads to the extinction of a species of fish. The carbon footprint is more specific in what it quantifies whereas the ecological footprint is broader in its assessment of impact.

Carbon footprint A measure of the total carbon dioxide and other greenhouse gases emissions from the activities, both direct and indirect, of a person, country, or other entity.

There are many carbon calculators available on the web. They usually ask a few questions and make assumptions based on your answers to the questions. Direct expenditure is relatively easy to calculate. If you drive a known number of miles in a car and can calculate the quantity of gasoline that you combusted in the process, it is relatively easy to calculate how much carbon dioxide in kilograms or pounds per mile driven was released to the atmosphere during the combustion process. However, when you purchase an automobile, it is more difficult to assign an amount of energy and resulting carbon dioxide that was released during manufacture of the automobile. The energy required to manufacture a product is called the "embodied energy" of that product. Carbon footprint calculations require both the direct calculation of fuel combusted and the resulting carbon dioxide emitted, plus the carbon involved in all other aspects of the activity, including the embodied energy of the item used and the energy required to dispose of it.

When the carbon footprint is calculated for the activities of a country, all kinds of observations can be made. For example, transportation accounts for about 30 percent of the carbon footprint of the United States. **FIGURE 31.2** shows that cars, minivans, sports utility vehicles, and light-duty trucks, which are mostly used by individuals and small businesses, account for more than half of the carbon footprint associated with transportation in the United States. Heavier duty trucks make up 24 percent of the carbon footprint.

Food accounts for 15 to 30 percent of a typical household carbon footprint in the United States. **FIGURE 31.3** shows that meat accounts for 57 percent of the food carbon footprint and dairy accounts for 18 percent. By understanding the food portion of the household carbon footprint, an individual can gain an understanding of what activities contribute most to their carbon footprint. Based on a diagram

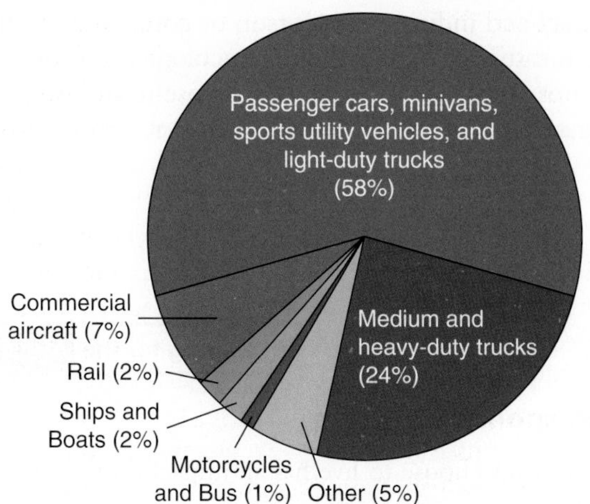

FIGURE 31.2 U.S. Transportation carbon footprint, 2019. Transportation is responsible for 30 percent of the carbon footprint of the United States. Cars, minivans, sports utility vehicles, and light trucks make up the largest part of the transportation carbon footprint. Numbers do not add to 100 percent due to rounding. *(Data from Center for Sustainable Systems, University of Michigan, and US EPA. https://css.umich.edu/factsheets/carbon-footprint-factsheet.)*

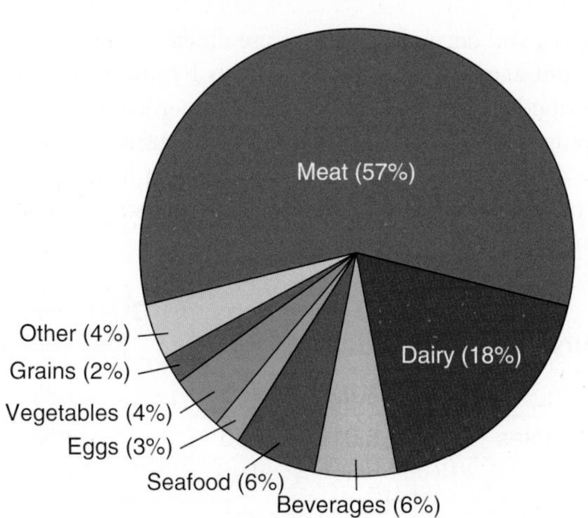

FIGURE 31.3 Food carbon footprint. A typical household carbon footprint in the United States illustrates that meat and dairy are the largest contributors. *(Data from Center for Sustainable Systems University of Michigan, https://css.umich.edu/factsheets/carbon-footprint-factsheet.)*

such as Figure 31.3, an individual might be able to conclude that if they wanted to reduce their food carbon footprint, they could reduce the amount of meat they eat each week. Or, after looking at Figure 31.3, they could draw the conclusion that if they substitute chicken for beef, they could also reduce their carbon footprint.

China is the largest contributor of carbon dioxide in the world followed by the United States, which emits about half the carbon dioxide of China. However, when you account for the populations of the two countries, the per person carbon footprint of the United States is much larger. According to the Rhodium Group, an international research group, China emitted 14 million metric tons of carbon dioxide equivalents from all activities in 2019. This was 27 percent of global emissions. With 1.4 billion people, China's carbon footprint was 10 metric tons per person. The United States emitted 5.7 million metric tons of carbon dioxide equivalent from all activities in 2019, which was 11 percent of global emissions. With 330 million people, the carbon footprint of the United States was 17.3 metric tons per person. Countries in the developing world, such as India, have carbon footprints as low as 2 metric tons per person. Because global climate change is such an important topic for the sustainability of humans on the planet, it is sometimes the only metric that people and organizations focus on. However, there are other environmental footprint metrics including the water footprint that we described in Module 27. The ecological footprint is the most comprehensive metric since it tries to measure the impact, in area of land utilized, of all aspects of the activities of a person or country. Because mitigating global climate change is so important, environmental scientists wanted a metric for assessing the climate change impact of an activity, or of a person or a country. Hence the carbon footprint was developed. As water became more and more scarce, the need for a footprint metric that specifically measured water was identified and developed.

While the ecological, carbon, and water footprints are useful tools in attempting to estimate the environmental impact of an activity, person, or country, ultimately there is one other assessment tool that is even more comprehensive: a measure of the overall sustainability of all the activities of a nation or the world. That is the focus of the next module.

Module 31 AP® Review

Preparing for the AP® Exam

Learning Goals Revisited

31-1 What is the ecological footprint and what does it tell us?

The ecological footprint quantifies the land area required to support a person's or a country's activities including the land and water needed to produce resources and to process waste generated. The fair Earthshare is the amount of land that each person is entitled to if all the biologically productive land on Earth is divided up equally among all the inhabitants of the world. Citizens of the United States exceed the fair Earthshare by a factor of 5.

31-2 What is the carbon footprint and how does it differ from the ecological footprint?

The carbon footprint is a measure of the total carbon dioxide and other greenhouse gas emissions from the activities, both direct and indirect, of a person or country. It is a more specific measure of impact than the ecological footprint and it does not assess factors other than those related to greenhouse gases. It is measured in equivalents of carbon dioxide.

AP® Practice Questions

Multiple-Choice Questions

1. A person's ecological footprint is
   (a) the land that a person lives on.
   (b) the amount of carbon dioxide a person contributes to climate change.
   (c) the land required to produce a person's food.
   (d) the land needed to support all of a person's activities.

2. The carbon footprint
   (a) includes carbon, nitrogen, phosphorus, and potassium.
   (b) is a broader measure of environmental impact than the ecological footprint.
   (c) measures other greenhouse gases including methane.
   (d) is another way to refer to the water footprint.

3. Which country has the largest ecological footprint?
   (a) China
   (b) the United States
   (c) India
   (d) Germany

4. The carbon footprint of a country
   (a) includes only the carbon dioxide emitted from vehicles.
   (b) includes carbon dioxide emitted both directly and indirectly.
   (c) includes the water footprint as well.
   (d) includes carbon dioxide and the land area for manufacturing and disposal.

5. The fair Earthshare is
   (a) the area of biologically productive land on Earth divided by the number of people on Earth.
   (b) the product of population × affluence × technology.
   (c) the sum of greenhouse gas emissions divided by the number of people on Earth.
   (d) the difference between the area of land on Earth used by people and the area unused.

6. Which of the following is likely to have the greatest impact on an ecological footprint?
   (a) use of land to raise pigs for slaughter
   (b) practicing subsistence fishing
   (c) use of compost fertilizers
   (d) living in a rural area

Preparing for the AP® Exam

Free-Response Question

Where humans choose to live has an impact on their ecological footprint.

(a) **Identify** one factor that goes into calculating an ecological footprint. (1 pt.)
(b) **Describe** the differences in ecological footprints between a person living in a developing nation and a person living in a developed nation who recycles and uses renewable energy. (2 pts.)
(c) **Explain** the difference between an ecological footprint and a carbon footprint. (1 pt.)

Developing and developed nations use different amounts of carbon dioxide per capita. A group of scientists hypothesize that people in developed nations release more CO_2 per capita than those in developing nations. Use the table below to answer parts d–f. Note that the data in this table are slightly different from the data presented on the previous page, due to different sources and year of reporting.

Country	CO_2 emissions (tons, 2016)	CO_2 emissions (per capita)
China	10,432,751,400	7.38
United States	5,011,686,600	15.52
India	2,533,638,100	1.91
Democratic Republic of Congo	6,564,773	0.08

(Data from https://www.worldometers.info/co2-emissions/co2-emissions-per-capita/)

(d) **Make a claim** about the scientists' hypothesis. (1 pt.)
(e) **Identify** an alternative hypothesis. (1 pt.)
(f) **Describe** one of the sources of carbon dioxide emissions that a developed nation would have, and one of the sources of carbon dioxide emissions that a developing nation would have. (2 pts.)
(g) **Identify** two other sources of greenhouse gases that can be emitted from both developed and developing nations. (2 pts.)

Module 32

Introduction to Sustainability

Throughout this book, we have identified ways in which human activities affect the environment. If you combine the various ways humans impact conditions on Earth, you are assessing sustainability. In this module we will explore in greater detail the topic of sustainability, which we first introduced in Module 0. Even though we are roughly halfway through the modules in this book, we have already examined a variety of ways that humans impact natural systems. While both the ecological footprint and the carbon footprint are useful tools in calculating the environmental impact of an activity, person, or country, ultimately there is one other assessment tool that is even more comprehensive: sustainability.

Learning Goals

After reading this module you should be able to

32-1 explain sustainability.

32-2 describe some of the environmental indicators of sustainability.

32-1 What is sustainability?

Sustainability is the most comprehensive assessment of environmental impact

As you may know from reading Module 0, **sustainability** means being able to use a resource or engage in an activity now without jeopardizing the ability of future generations to engage in similar activities later. If a resource is used too quickly, or if using that resource results in too many adverse environmental impacts, the status of the environment for future generations will be jeopardized—and that would mean that the activity is not sustainable. Sustainability is one of the Big Ideas or recurring themes in environmental science and many people maintain that achieving sustainability is the single most important goal for the human species. It is also one of the most challenging tasks we face. It is probable that achieving sustainability would be relatively easy with far fewer people on Earth. But as the population grows, the challenge of achieving sustainability grows with it. In addition, as with the ecological and carbon footprints, sustainability is challenging to measure.

The Impact of Consumption on the Environment

We have seen that people living in developed nations consume a far greater share of the world's resources than do people in developing countries. What effect does this consumption have on sustainability and our environment? It is easy to imagine a very small human population living on Earth without degrading the environment because there simply would not be enough people to do significant damage. Today, however, Earth's population is 7.8 billion people and still growing. And the environmental impact of all these people is substantial.

There are numerous examples of human beings overexploiting forests, fisheries, and other resources. We've already presented some of these examples in this book. When humans initially settle an area, they typically increase in number because their numbers are small relative to the carrying capacity of the local environment, something we discussed in Unit 3. They cut down trees to build homes and other structures. Eventually they overuse soil and water resources and possibly fish and wildlife. Without trees to hold the soil in place, erosion can occur, and the loss of soil causes food production to decrease.

Ultimately, consumption in the United States and in the world is related to the size of the human population. The human population did not reach a billion people until 1800 and has been growing rapidly since. Resource use has grown since then as well. **FIGURE 32.1** shows world population change since 1800. Each additional billion people on Earth are demanding and using more resources, which clearly has an impact on the environment. Many environmental scientists ask how we will be able to continue to produce sufficient food, build needed infrastructure, and process pollution and waste. As we have illustrated thus far, and will further in the following modules, our current actions to support the human population have already modified many environmental systems.

Sustainability Being able to use a resource or engage in an activity now without jeopardizing the ability of future generations to engage in similar activities later.

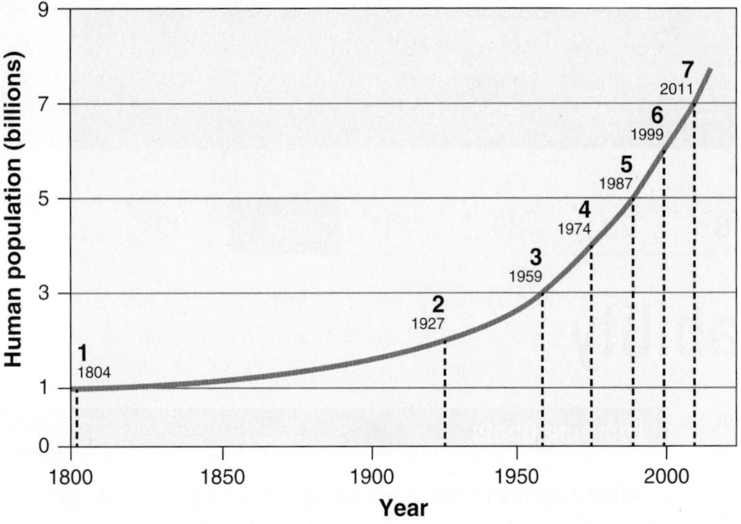

FIGURE 32.1 Adding 1 billion people. The number of years to add an additional billion people on Earth is shown since the first billion were added in 1800. It took 123 years to add 1 billion people from 1804 and only 12 years to add a billion people from 1999. *(Data from http://www.un.org/esa/population/publications/sixbillion/sixbilpart1.pdf, http://www.pragcap.com/seven-billion-people/)*

Can we continue our current level of resource consumption without jeopardizing the well-being of future generations? Most environmental scientists believe that there are limits to the supply of clean air and water, nutritious foods, and other life-sustaining resources our environment can provide. They also believe there is a point at which Earth will no longer be able to maintain a stable climate. We must meet several requirements in order to live sustainably:

- Environmental systems must not be damaged beyond their ability to recover.
- Renewable resources must not be depleted faster than they can regenerate.
- Nonrenewable resources must be used sparingly.

Sustainable development is development that balances current human well-being and economic advancement with resource management for the benefit of future generations. This is not as easy as it sounds. The issues involved in evaluating sustainability are complex, in part because sustainability depends not only on the number of people using a resource but also on how that resource is used. For example, eating chicken can be sustainable when people raise their own chickens and allow them to forage for food on the land. However, if all people, including city dwellers, wanted to eat chicken six times a week, the resources needed to raise that many chickens would probably make the practice of eating chicken unsustainable.

Living sustainably means acting in a way such that activities that are crucial to human society can continue. It includes practices such as conserving and finding alternatives to nonrenewable resources as well as protecting the capacity of the environment to continue to supply renewable resources (**FIGURE 32.2**).

Consider iron, a nonrenewable resource derived from metal ore removed from the ground. Iron is the major constituent of steel, which we use to make many things, including automobiles, bicycles, and strong frames for tall buildings. Historically, our ability to smelt iron for steel limited our use of that resource, but as we have improved steel manufacturing technology, steel has become more readily available and the demand for it has grown. Because of this increased demand, our current use of iron is unsustainable. What would happen if we ran out of iron? While not too long ago the depletion of iron ore might have been a catastrophe, today we have developed materials that can substitute for certain uses of steel—for example, carbon fiber—and we also know how to recycle steel. Developing substitutes and recycling materials are two ways to address the problem of resource depletion and to increase sustainability.

The example of iron leads us to a question that environmental scientists often ask: How do we determine the importance of a given resource? If we use up a resource such as iron for which substitutes exist, it is possible that the consequences will not be severe. However, if we are unable to find an alternative to the resource—for example, something to replace aviation fuel for flying airplanes across the Atlantic Ocean—people in the developed nations may have to make significant changes in their consumption habits. Finally, when harvesting or extracting resources, we must consider the rate at which they are being harvested or extracted.

Sustainable development Development that balances current human well-being and economic advancement with resource management for the benefit of future generations.

FIGURE 32.2 Living sustainably. Sustainable choices such as bicycling to work or school can help protect the environment and conserve resources for future generations. *(Jim West/Alamy Stock Photo)*

Maximum Sustainable Yield

When we want to obtain the maximum amount of a biological resource, we need to know how much of a given plant or animal population can be harvested without harming the resource as a whole. The **maximum sustainable yield (MSY)** is the largest quantity of a renewable resource that can be harvested indefinitely. Harvesting at the MSY keeps the resource population at about one-half the carrying capacity of the environment (**FIGURE 32.3**). If you're thinking this concept sounds familiar, that's because we have already explored a related concept in Module 15. There we described biotic potential as the maximum potential for growth of a population under ideal conditions with unlimited resources. We also addressed a related topic when we discussed fisheries in Module 28 and we asked how much fish can be removed without harming the fishery? Both of these subjects are related to MSY.

Imagine a situation in which deer hunting in a public forest is unregulated, with each hunter free to harvest as many deer as possible. As a result of unlimited hunting, the deer population could be depleted to the point of endangerment. This, in turn, would disrupt the functioning of the forest ecosystem. On the other hand, if hunting were prohibited entirely, in the absence of natural predators, the deer herd might grow so large that there would not be enough food in the forests and fields to support the herd. In extreme cases, such as that of the reindeer of St. Paul Island in Alaska (see Module 16), the population could grow unchecked until it crashed due to starvation.

Some intermediate amount of hunting will leave enough adult deer to reproduce at a rate that will maintain the population, but not leave so many that there is too much competition for food. This intermediate harvest is called the maximum sustainable yield. Specifically, the MSY of a renewable resource is the maximum amount that can be harvested without compromising the future availability of that resource—it is the maximum harvest that will be adequately replaced by population growth. In a sense, it is the sustainable harvest rate. MSY varies case by case. A reasonable starting point is to assume that population growth is the fastest at about one-half the carrying capacity of the environment, as shown on the S-shaped curve in Figure 32.3. Looking at the graph, we can see that at a small population size, the growth curve is shallow, and growth is relatively slow. As the population increases in size, the slope of the curve is steeper, indicating a faster growth rate. As the population size approaches the carrying capacity, the growth rate slows. The MSY is the amount of harvest that allows the resource population to be maintained at roughly one-half the carrying capacity of the environment.

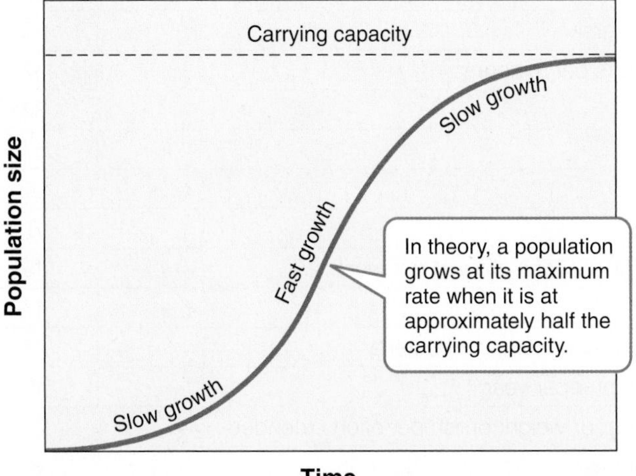

FIGURE 32.3 Maximum sustainable yield. Every population has a point at which a maximum number of individuals can be harvested sustainably. That point, the maximum sustainable yield, is often reached when the population size is about one-half the carrying capacity.

Forest trees, like animal populations, also have a maximum sustainable yield. Loggers may remove a particular percentage of the trees at a site to allow a certain amount of light to penetrate to the forest floor and reach younger trees. If they cut too many trees, however, an excess of sunlight will penetrate and dry the forest soil. This drying may create conditions inhospitable to tree germination and growth, thus inhibiting adequate regeneration of the forest. In theory, harvesting the maximum sustainable yield should permit an indefinite use without depletion of the resource. In reality, however, there are two problems with using MSY in environmental policy management.

First, it is very difficult to calculate MSY with certainty because in a natural ecosystem, it is difficult to obtain necessary information such as birth rates, death rates, and the carrying capacity of the system. Once an MSY calculation is made, we cannot know if a yield is truly sustainable until years later when we can evaluate the effect of the harvest on reproduction. By that time, if the harvest rate was too large, it is too late to prevent harm to the population.

In addition to the calculation difficulties, a sizeable group of environmental scientists argue that MSY is not an adequate guide to use for environmental protection. They believe that unmeasured externalities and harm to ecosystems can occur when using MSY as a harvest guideline, and that the best way to protect the environment is to harvest considerably less than the MSY. So MSY and sustainability are two goals, but there are definite challenges in measuring them.

AP® Exam Tip

Make sure that you understand the concept of maximum sustainable yield (MSY), as well as its application to sustainability and resources. MSY applies to reproducing biological populations including fish, timber, agricultural products, gathered plants, and hunted animals. Avoid using this term when referring to non-renewable and mineral resources. MSY often appears on the AP® Environmental Science Exam.

Maximum sustainable yield (MSY) The largest quantity of a renewable resource that can be harvested indefinitely.

32-2 What are some of the environmental indicators of sustainability?

While sustainability is the goal, it is often difficult to measure or establish with a high level of certainty. Environmental scientists have proposed many different metrics to use as indicators of environmental health and, ultimately, sustainability.

Scientists use environmental indicators to monitor the state of Earth

When examining sustainability, one critical question that environmental scientists investigate is whether the planet's natural life-support systems are being degraded by human-induced changes. Natural environments provide ecosystem services, which we defined in Unit 2, to provide life-supporting resources such as clean water, timber, fisheries, and agricultural crops. These are the resources necessary for organisms to survive. Ecosystem services may benefit people directly, for example by providing oxygen that we breathe. They may also benefit us indirectly, for example by providing a diversity of conditions, nutrients, and species, all of which make the ecosystem healthier for all organisms, including humans. Although we often take a healthy ecosystem for granted, we notice when an ecosystem is degraded or stressed because it is unable to provide the same services such as water purification or produce the same goods such as fruit or firewood.

Environmental indicators Describe the current state of an environmental system or the Earth.

To understand the extent of our effect on the environment, we need to be able to measure the health of Earth's ecosystems. Just as body temperature and heart rate can indicate whether a person is healthy or sick, **environmental indicators** describe the current state of an environmental system or the Earth. These indicators do not always tell us what is causing a change, but they do tell us when we might need to look more deeply into a particular issue. Environmental indicators provide valuable information about natural systems on both small and large scales. Some of these indicators and the modules in which they are covered are listed in **TABLE 32.1**.

In this book, we focus on five global-scale environmental indicators:

- Biodiversity
- Food production
- Average global surface temperature and carbon dioxide concentrations in the atmosphere
- Human population
- Resource depletion

TABLE 32.1 Some common environmental indicators

Environmental indicator	Unit of measure	Module
Human population	Individuals	17
Ecological footprint	Hectares of land	31
Total food production	Metric tons of grain	25
Food production per unit area	Kilograms of grain per hectare of land	32
Per capita food production	Kilograms of grain per person	25
Carbon dioxide	Concentration in air (parts per million)	32
Average global surface temperature	Degrees centigrade	32
Sea level change	Millimeters	57
Annual precipitation	Millimeters	2
Species diversity	Number of species	8
Fish consumption advisories	Present or absent; number of fish allowed per week	49
Water quality (toxic chemicals)	Concentration	47
Water quality (conventional pollutants)	Concentration; presence or absence of bacteria	47
Deposition rates of atmospheric compounds	Milligrams per square meter per year	42
Fish catch or harvest	Kilograms of fish per year or weight of fish per effort extended	28
Extinction rate	Number of species per year	8
Habitat loss rate	Hectares of land cleared or "lost" per year	32
Infant mortality rate	Number of deaths of infants under age 1 per 1,000 live births	17
Life expectancy	Average number of years an infant born today can be expected to live under current conditions	17

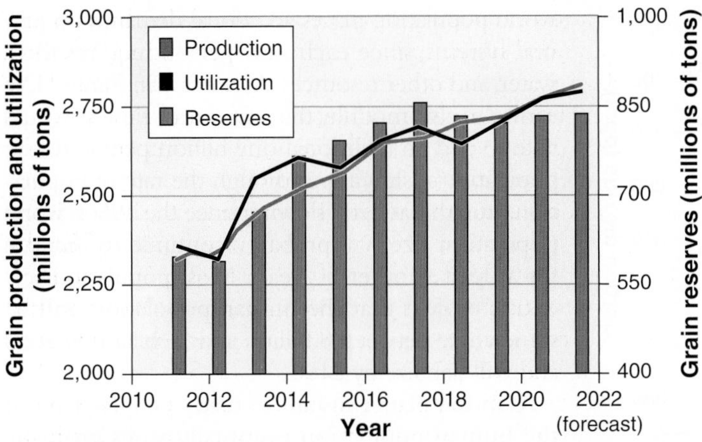

FIGURE 32.4 World grain production, utilization, and reserves. Grain production and utilization have roughly kept pace with one another and have slowly grown over time. There has been a slight drawdown in grain reserves over the past 5 years.

Biodiversity

Biodiversity was already covered in Module 8. We have seen that overall biodiversity is decreasing worldwide as genetic diversity, species diversity, and habitat diversity have all been decreasing in recent decades. This indicator shows that Earth systems are not healthy and that human beings are not living sustainably on the planet. Here we will briefly describe the other four indicators.

Food Production

The second of our five global indicators is food production: our ability to grow food to nourish the human population. Just as a healthy ecosystem supports a wide range of species, a healthy soil supports abundant and continuous food production. Food grains such as wheat, corn, and rice provide more than half the calories and protein humans consume and are dependent on a healthy soil and a stable climate, among other things.

In the past, we have used science and technology to increase the amount of food we can produce on a given area of land. As a result of the Green Revolution, world grain production increased fairly steadily between 1950 and 2014 due to expanded irrigation, fertilization, new crop varieties, and other innovations. Since 2015, grain production has kept pace with consumption. At the same time, there has been a slight drawdown of grain reserves. **FIGURE 32.4** shows current grain production and utilization through 2021.

In 2008, food shortages around the world led to higher food prices and even riots in some places. The amount of grain produced worldwide is influenced by many factors. These factors include climatic conditions, the amount and quality of land under cultivation, irrigation, and the human labor and energy required to plant, harvest, and bring the grain to market. In a given year or decade, grain production may not keep up with population growth because in some areas the productivity of agricultural ecosystems has declined as a result of soil degradation, crop diseases, and unfavorable weather conditions such as drought or flooding. In addition, demand can outpace supply. While the rate of human population growth has outpaced increases in food production, humans currently use more grain to feed livestock than they consume themselves. Finally, some government policies discourage food production by making it more profitable for land to remain uncultivated or by encouraging farmers to grow crops for fuels such as ethanol and biodiesel instead of food.

Will there be sufficient grain to feed the world's population in the future? In the past, whenever a shortage of food has loomed, humans have discovered and employed technological or biological innovations to increase production. However, these innovations often put a strain on the productivity of the soil. If we continue to overexploit the soil, its ability to sustain food production may decline. Therefore, although grain production should be a good indicator of environmental health and sustainability, it currently does not give a clear signal of whether or not there will be sufficient worldwide grain reserves in coming decades.

Average Global Surface Temperature and Carbon Dioxide Concentrations

We have seen that high biodiversity and abundant food production are necessary for life. A stable climate helps both of them to be possible. Earth's temperature has been relatively constant since the earliest forms of life began, about 3.5 billion years ago. The temperature of Earth allows the presence of liquid water, which is necessary for life.

What keeps Earth's temperature so constant? As **FIGURE 32.5** shows, our thick planetary atmosphere contains many gases. Some of these atmospheric gases are greenhouse gases that

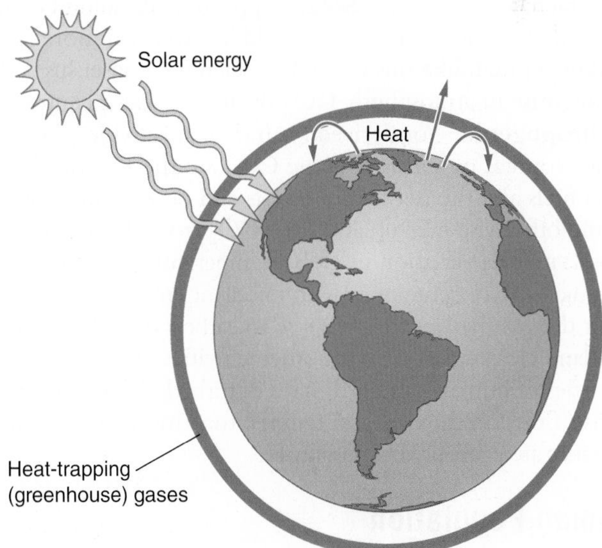

FIGURE 32.5 The Earth-surface energy balance. As Earth's surface is warmed by the Sun, it radiates heat outward. Heat-trapping gases absorb the outgoing heat and reradiate some of it back to Earth. Without these greenhouse gases, Earth would be much cooler.

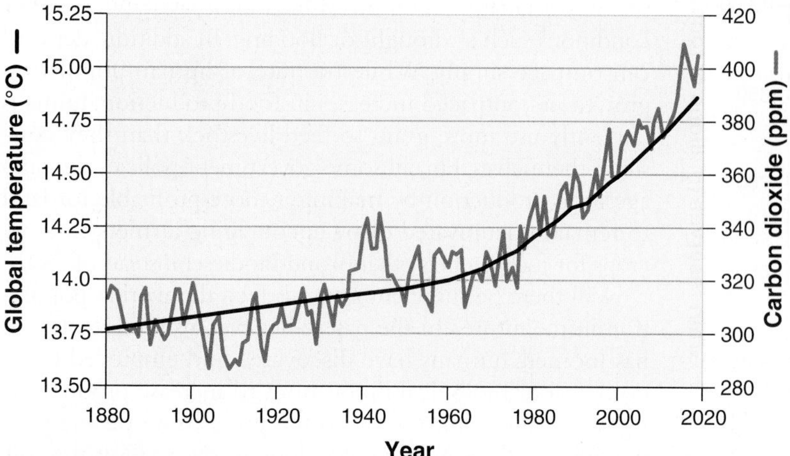

FIGURE 32.6 Changes in average global surface temperature and in atmospheric CO_2 concentrations. Earth's average global surface temperature has increased steadily for at least the past 100 years. Carbon dioxide concentrations in the atmosphere have varied over geologic time, but have risen rapidly since 1960.

trap heat near Earth's surface. The gas that contributes most to warming of the atmosphere is carbon dioxide (CO_2). During most of the history of life on Earth, greenhouse gases have been present in the atmosphere at fairly constant concentrations for relatively long periods. They help keep Earth's surface within the range of temperatures at which life can flourish.

In the past 2 centuries, however, the concentrations of CO_2 and other greenhouse gases in the atmosphere have risen. Today, atmospheric CO_2 concentrations are closer to 415 parts per million (ppm). During roughly the same period, as **FIGURE 32.6** shows, while global temperatures have fluctuated considerably, they have displayed an overall increase. (Note that this graph has two *y* axes. See the appendix "Reading Graphs" at the end of the book if you would like to learn more about reading a graph like this one.) The scientific consensus is that the increase in atmospheric CO_2 during the last 2 centuries is **anthropogenic**—or derived from human activities. The two major sources of anthropogenic CO_2 are the combustion of fossil fuels and the net loss of forests and other land areas that would otherwise take up and store CO_2 from the atmosphere. Observing the variation in global temperature and global carbon dioxide concentrations is an excellent environmental indicator that synthesizes the response to carbon dioxide emissions and land clearing, along with other activities. It shows that the activities of humans on Earth have disturbed global concentrations of carbon dioxide and temperature and therefore those activities are probably not sustainable.

Human Population

In addition to biodiversity, food production, and global surface temperature, the size of the human population can tell us a great deal about the health of our global environment. The human population is currently 7.8 billion and growing. The increasing world population places additional demands on natural systems, since each new person requires food, water, and other resources. As we saw in Figure 32.1 earlier in this module, the amount of time it would take to add an additional one billion people to the population is slowing. Although the rate of population growth has been slowing since the 1960s, world population size will probably continue to increase for at least another 50 years. Most population scientists project that the human population will be somewhere between 8 billion and 10 billion in 2050 and will stabilize by 2100.

Can the planet sustain so many people? Even if the human population eventually stops growing, the billions of additional people will create a greater demand on Earth's finite resources, including food, energy, and land (**FIGURE 32.7**). Unless humans work to reduce these pressures, the human population will put a rapidly growing strain on natural systems for at least the first half of this century. Therefore, monitoring the human population should be an excellent indicator of environmental health and yet it is difficult to make a definitive pronouncement of whether the current population is sustainable. It is probably not sustainable. As we know from the IPAT equation, population alone cannot be used to establish sustainability. We also need to consider resource use.

Resource Depletion

As we have seen from the IPAT equation, environmental impact is a function of human population and the resources used and the impact of those resources. So monitoring

Anthropogenic Derived from human activities.

FIGURE 32.7 The effect of crowded cities on natural and human systems. The human population will continue to grow for at least 50 years. Unless humans can devise ways to live more sustainably, these population increases will put additional strains on natural systems. Here, a street scene in Kolkata, India. *(Deshakalyan Chowdhury/AFP/Getty Images)*

resource availability and depletion is another good indicator of environmental health. We have already discussed biological resources such as fisheries and forests and natural resources such as coal and iron. As the human population grows, some resources necessary for our survival become increasingly depleted. In addition, as we have seen, extracting these resources can affect the health of our environment in many ways. Pollution and land degradation caused by mining, waste from discarded manufactured products, and air pollution from fossil fuel combustion are just a few of the negative environmental consequences of resource extraction and use.

Some natural resources, such as coal, oil, and uranium, are finite and cannot be renewed or reused. Others, such as iron or copper, also exist in finite quantities but can be used multiple times through reuse or recycling, as we discussed in Module 29. Renewable resources, such as timber, can potentially be grown and harvested indefinitely, but in some locations these resources are being used faster than they can be naturally replenished. "Do the Math: Rates of Forest Clearing" provides an opportunity to calculate rates of one type of resource depletion.

Sustaining the global human population requires vast quantities of resources. However, in addition to the total amounts of resources used by humans, we must consider per capita resource use.

Patterns of resource consumption vary enormously among nations depending on whether they are developing or developed. According to the United Nations Development Programme and many other sources, people in developed nations — including the United States, Canada, Australia, most European countries, and Japan — use a disproportionate share of the world's resources, or as we discussed in Module 31, their "fair Earthshare." **FIGURE 32.8**

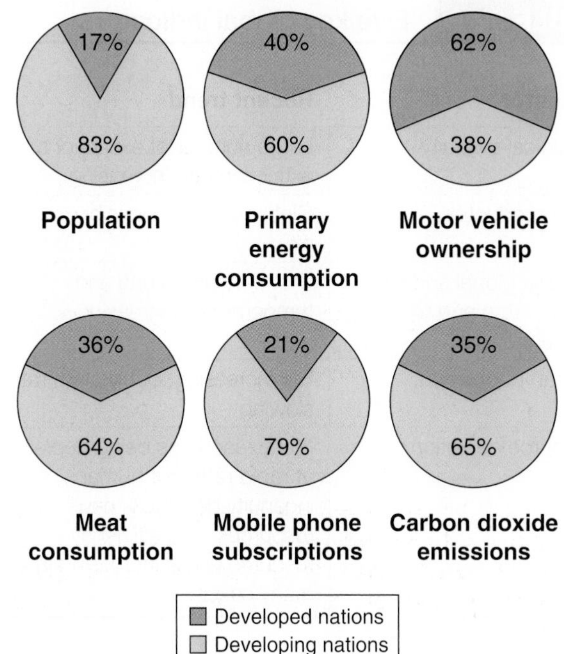

FIGURE 32.8 Proportion of resource use in developed and developing countries in 2020 for selected resources. Approximately 17 percent of the world's population lives in developed countries, but that 17 percent uses a disproportionate share of the world's resources. The remaining 83 percent of the population lives in developing countries and uses far fewer resources per capita. *(Data from https://ourworldindata.org/; energy data from https://www.bp.com/content/dam/bp/business-sites/en/global/corporate/pdfs/energy-economics/statistical-review/bp-stats-review-2021-full-report.pdf)*

DO THE MATH Rates of Forest Clearing

Preparing for the AP® Exam

A web search of environmental organizations yielded a range of estimates of the amount of forest clearing that is occurring worldwide:

Estimate 1: 1 acre per second

Estimate 2: 80,000 acres per day

Estimate 3: 32,000 ha per day

Convert Estimate 1 into hectares per year and Estimate 2 into hectares per day.

There are 2.47 acres per hectare. (See "Do the Math: Converting Between Hectares and Acres" on page 100.) Therefore, 1 acre = 0.40 ha.

Estimate 1: 1.0 acre/second × 0.40 ha/acre = 0.40 ha/second

0.40 ha/second × 60 seconds/minute × 60 minutes/hour × 24 hours/day × 365 days/year

= 12,614,400 ha cleared per year

Estimate 2: 80,000 acres/day × 0.40 ha/acre = 32,000 ha cleared per day

YOUR TURN Notice that Estimate 2, when converted to hectares, is identical to Estimate 3. Now convert the estimate of 32,000 ha/day into the amount cleared per year. How much larger is Estimate 1 than Estimate 2? Why might environmental organizations, or anyone else, choose to present similar information in different ways?

TABLE 32.2 Five key global indicators

Indicator	Recent trend	Outlook for the future	Overall impact on environmental quality
Biological diversity	Large number of extinctions, extinction rate increasing	Extinctions will continue	Negative
Food production	Global reserves possibly leveling off	Unclear	May affect the number of people Earth can support
Average global surface temperature and CO_2 concentration	CO_2 concentrations and temperatures increasing	Likely to continue to increase	Effects are varied and mostly detrimental
Human population	Still increasing, but growth rate slowing	Population leveling off; resource consumption rates also a factor	Negative
Resource depletion	Many resources being depleted at rapid rate, but human ingenuity develops "new" resources, and efficiency of resource use is increasing in many cases	Unknown	Increased use of most resources has negative effects; recycling of existing resources is mostly beneficial

shows that the 17 percent of the global population that lives in developed nations owns 62 percent of the world's automobiles, consumes 36 percent of all meat and fish, and releases 35 percent of carbon dioxide. Thus, even though the number of people in the developed countries is much smaller than the number in developing countries, their total consumption of resources is disproportionately larger.

So while it is generally true that a larger human population will have greater environmental impacts than a smaller population, a full evaluation requires that we look at economic development and consumption patterns as well. **TABLE 32.2** summarizes the five key global indicators that we have described in this module. In the next module we will examine sustainable approaches to agriculture by examining integrated pest management specifically and sustainable agriculture more broadly.

Module 32 AP® Review

Preparing for the AP® Exam

Learning Goals Revisited

32-1 What is sustainability?

Sustainability is the careful use of resources now so that future generations can also use those resources later. Sustainability is a Big Idea and overarching theme in the study of environmental science. A related concept, maximum sustainable yield (MSY), is the largest quantity of a renewable resource that can be harvested indefinitely. MSY is difficult to measure, and critics believe that unmeasured externalities and harm to ecosystems can occur when using MSY as a harvest guideline.

32-2 What are some of the environmental indicators of sustainability?

There are global-scale indicators of sustainability that allow us to monitor specific parameters over time. We previously identified that biodiversity is decreasing in Unit 2. Food production has leveled off. Atmospheric carbon dioxide concentrations are steadily increasing, and global temperatures fluctuate, although the overall change is increasing. The human population continues to increase in size but the rate of increase has been declining. Resource consumption is disproportionately larger in the developed world than the developing world.

AP® Practice Questions

Multiple-Choice Questions

1. Which of the following is considered a global scale environmental indicator?
   (a) atmospheric carbon dioxide concentrations
   (b) rural populations
   (c) available hectares for agriculture
   (d) pollution in a small stream.

2. How many hectares of land is a 500-acre park? (1 acre = 0.40 ha)
   (a) 200 ha
   (b) 250 ha
   (c) 500 ha
   (d) 750 ha

3. Refer to Figure 32.8 (on page 367). Relative to population size, how does meat consumption in developed and developing countries compare?
   (a) Developing countries consume more meat.
   (b) Developed countries consume more meat.
   (c) Developing and developed countries consume about the same amount of meat.
   (d) Developing and developed countries hardly consume any meat at all.

4. In 2020, 860,000 ha of the Amazon rainforest were cleared. Approximately how many hectares is that each hour?
   (a) 1.2 ha
   (b) 29 ha
   (c) 98 ha
   (d) 178 ha

Use the graph below to answer questions 5 & 6:

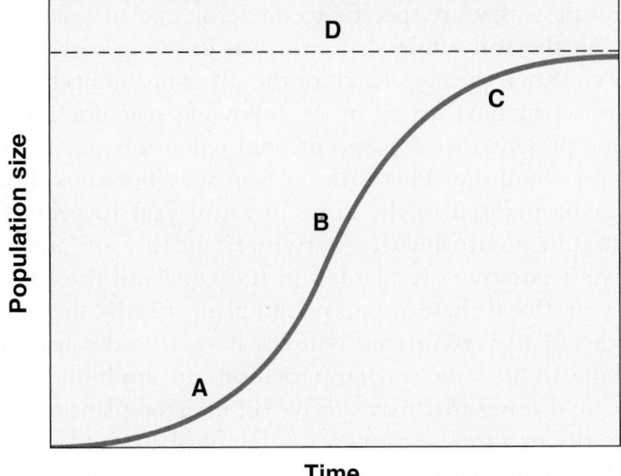

5. The maximum sustainable yield of a population occurs at
   (a) A.
   (b) B.
   (c) C.
   (d) D.

6. Fishers trying to make sure that a population of fish remains viable would most likely want the population to remain
   (a) between A and B.
   (b) between B and C.
   (c) anywhere below D.
   (d) above C.

Free-Response Question

The graph below shows the production of cereal grains from 2011 to the present.

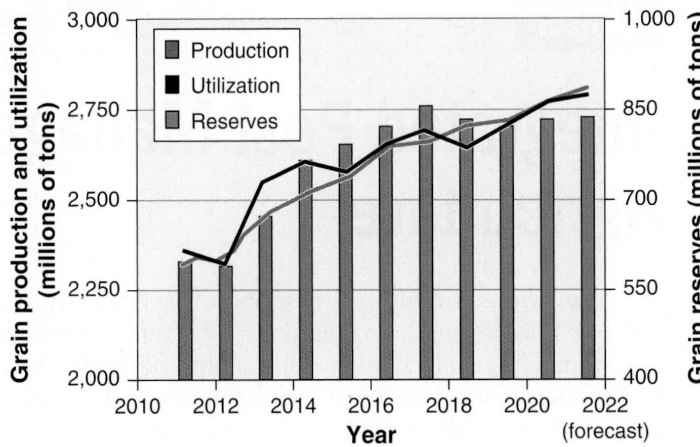

(a) **Identify** the first time that production went above the stock of cereal grains. (1 pt.)
(b) **Identify** the amount of cereal used and produced in 2014 to 2015. (2 pts.)
(c) **Explain** why utilization has increased over this time period. (1 pt.)
(d) **Explain** how production has been able to keep up with and exceed demand. (1 pt.)

Forestry involves the removal of trees for use as wood products and fuel.

(e) **Explain** how humans can maintain forests while still removing resources. (1 pt.)
(f) **Identify** one possible effect of removing too many trees from forests. (1 pt.)
(g) **Describe** a specific strategy to increase yield while mitigating the problem you identified in part (f). (1 pt.)
(h) **Identify** an environmental indicator used to tell if a forest ecosystem is healthy. (1 pt.)
(i) Other than biodiversity, **describe** one environmental indicator that can be measured. (1 pt.)

Module 33

Unit 5	24	25	26	27	28	29	30	31	32	33	34

Integrated Pest Management and Sustainable Agriculture

In the previous module we described sustainability and methods for assessing human impacts on natural systems. In this module we will examine sustainability of food systems by considering integrated pest management and the broader goals and methods of sustainable agriculture. Given the numerous adverse impacts of agriculture on the environment, it is important for students of environmental science to understand the potential for obtaining food with fewer harmful effects.

Learning Goals

After reading this module, you should be able to

33-1 describe the goals and techniques of integrated pest management (IPM).

33-2 explain the objectives of sustainable agriculture practices.

33-1 What are the goals and techniques of integrated pest management (IPM)?

Integrated pest management increases agricultural output while minimizing environmental harm

Integrated pest management (IPM) was introduced in Module 27 when we first discussed pest control methods. As a reminder, Integrated Pest Management uses a variety of techniques to control pest species while minimizing pesticide inputs into the environment and increasing agricultural output. These techniques include biological, physical, and chemical methods such as crop rotation and intercropping, the creation of habitats for predators of pests (also called biocontrol), and, under certain circumstances, the use of pesticides.

Crop Rotation and Intercropping

Crop rotation is a crop-planting strategy in which different types of crop species are planted from season to season or year to year on the same plot of land. For example, potatoes are planted one year and then cabbage is planted the following year. Crop rotation can reduce the success of insect pests that are specific to one crop and may have laid eggs in the soil while that crop was growing early in the season. When the eggs hatch or the surviving adults become active in the next season or the following year, if a different crop is present, the crop-specific pest will not have a suitable host and will die. This is the reason why potatoes should never be planted in the same location year to year: The Colorado potato beetle overwinters in the soil and will thrive if potatoes are planted in the same soil the following year. But if there are no potato plants nearby, the potato beetles that overwintered will not have a food source and will die. In the same way, crop rotation can also hinder crop-specific diseases that may survive on infected plant material from the previous season. If a different plant is growing in the following season, the crop-specific disease will not have a suitable host and will die. This relatively simple method of changing the physical location of which crops get planted where during which growing season in an agricultural field has led to great success in reducing pests and diseases.

Another related technique is **intercropping**, the physical spacing of different crops growing at the same time, in close proximity to one another, to promote biological interaction. Sometimes intercropping can literally be from one row to the next row. Intercropping of the appropriate crops makes it harder for specialized biological pests to locate or thrive than when only one crop is present. Sometimes, when paired properly, one crop may attract insect predators that

Crop rotation A crop-planting strategy in which different types of crop species are planted from season to season or year to year on the same plot of land.

Intercropping An agricultural technique that calls for physical spacing of different crops growing at the same time, in close proximity to one another, to promote biological interaction.

FIGURE 33.1 Intercropping vegetables and flowers. Farmers and backyard gardeners often interplant tomatoes with marigolds because of a compound that marigolds release that deters tomato whiteflies, a tomato plant pest. *(Botany vision/Alamy Stock Photo)*

predate on a pest that might feed on the nearby crop. When done properly this can improve plant yields and farmer income. For example, have you ever seen a backyard garden with tomatoes and marigold plants together? **FIGURE 33.1** shows a common technique in backyard gardens: planting tomatoes adjacent to marigolds. This is because the interplanting of this vegetable crop and flower crop combination provides benefits to the tomatoes. Although there are many suggested reasons for the benefits, one recent study demonstrated that marigolds release a chemical compound called limonene that repels tomato whiteflies, an insect that feeds on tomato leaves. Whatever the reason, this type of intercropping results in healthier tomatoes and a greater yield for your kitchen.

Biocontrol and Natural Predators

Biocontrol is a shortened way of saying "biological control," and it uses biological organisms to control agricultural pests. **Natural predators** are predators that occur naturally in the environment (as opposed to being introduced to the environment by humans). Farmers can provide a habitat that attracts certain natural predators that prey on crop pests. Caterpillars, for example, can be a pest to a farm because they eat the leaves of crops, harming the plant by reducing its ability to photosynthesize. Some wasps lay their eggs inside caterpillars and when the eggs hatch, the wasp larvae feed on the caterpillar. As a biocontrol method to mitigate caterpillar damage, farmers can provide an appropriate wasp habitat on their land, thus encouraging wasps to relocate to the farm where they will lay eggs in caterpillars, thereby reducing the numbers of caterpillars and the harm that caterpillars inflict on the crop. **FIGURE 33.2** shows a wasp that has paralyzed a caterpillar and is laying its eggs on the caterpillar. Another biocontrol method is intercropping trees and annual crops to encourage the presence of insect-eating birds, which will hopefully reduce harmful insect populations near the annual crops. However, sometimes an unintended consequence is that the birds may also eat or damage the crop!

FIGURE 33.2 Biocontrol with a natural predator. Practitioners of integrated pest management often provide habitat for insects that prey on crop pests. This wasp is laying eggs in a caterpillar, which it has paralyzed. *(USDA/Nature Source/Science Source)*

Although IPM practitioners do use pesticides, they limit applications through very careful observation. Farmers who practice IPM regularly examine their crops for the presence of insect pests and other potential crop hazards to catch them early so they can be treated using natural controls or smaller doses of pesticides than would be needed at later stages of an infestation. These more-targeted methods of pest control can result in significant cash savings on pesticides as well as improved yields. Although successful IPM requires training and skill, it can make a large difference in crop yields and reduced pesticide use. **FIGURE 33.3** shows a case study from IPM training in Indonesia. Farmers who learned how to determine whether a pesticide application was warranted were able to cut their pesticide applications and total quantity of pesticide used, and their exposure to and expenditures on pesticides, in half. Yields also improved after farmers learned and implemented IPM methods.

When farmers take time to inspect their fields carefully, as required by IPM methods, they often notice other crop needs, and this additional attention improves overall crop management

Biocontrol A shortened term for biological control, it uses biological organisms to control agricultural pests.

Natural predators Predators that occur naturally in the environment.

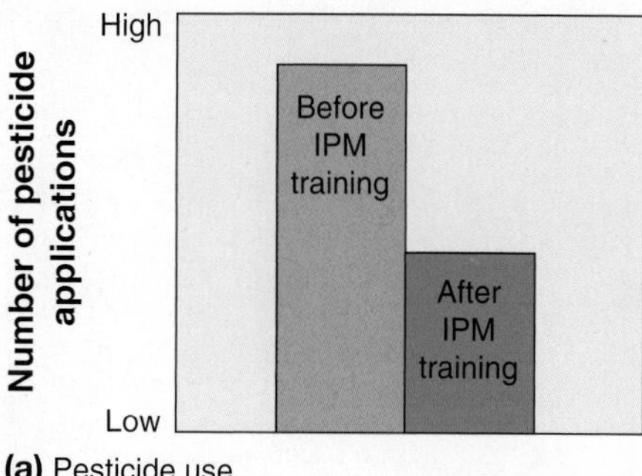

(a) Pesticide use

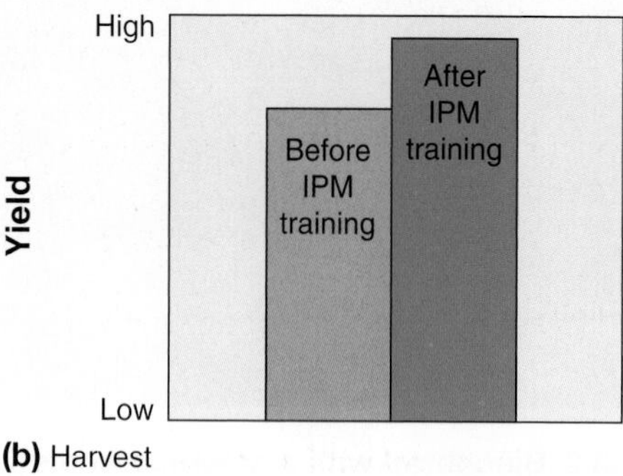

(b) Harvest

FIGURE 33.3 Effects of IPM training. (a) IPM training of farmers in Indonesia led to a significant reduction in pesticide applications. (b) Yield improvements also occurred after the training because of the additional attention the farmers gave to their crops. *(Data from FAO.)*

and may lower costs. The trade off for these benefits is that farmers must be trained in IPM methods and must spend more time inspecting their crops, both of which are costly. But once farmers are trained, the extra income and reduced economic costs associated with IPM sometimes outweigh the extra time they must spend in the field and the additional costs associated with that extra time. IPM has been especially successful in parts of the developing world where the high-input industrial farming model is not viable because labor costs are low and farmers lack financial resources to pay for costly pesticides. It also results in lower overall use of pesticides by farm workers, which means that farm workers, wildlife, and both surface and groundwater all receive less pesticide exposure.

Integrated pest management requires training and careful inspection of agricultural fields but it results in lower pesticide use and other benefits. It is a technique that uses fewer chemicals and less fossil fuel energy while improving environmental health, human health, and crop output. Reducing human impact on land and water is one of the benefits of IPM; reduced impact from humans also happens to be one of the goals of sustainable agriculture, which is the focus of the remainder of this module.

Sustainable agriculture Fulfills the need for food and fiber while enhancing the quality of the soil, minimizing the use of nonrenewable resources, and allowing economic viability for the farmer.

Soil conservation The prevention of soil erosion while simultaneously increasing soil depth and increasing the nutrient content and organic matter content of the soil.

Agroforestry An agricultural technique in which trees and vegetables are intercropped.

33-2 What are the objectives of sustainable agriculture practices?

Sustainable agriculture reduces environmental impacts

Is it possible to produce enough food to feed the world's population without destroying the land, polluting the environment, and reducing biodiversity? **Sustainable agriculture** fulfills the need for food and fiber while enhancing the quality of the soil, minimizing the use of nonrenewable resources, and allowing economic viability for the farmer. It emphasizes the ability to continue agriculture on a given piece of land indefinitely through **soil conservation**, the prevention of soil erosion while simultaneously increasing soil depth and increasing the nutrient content and organic matter content of the soil. Sustainable agriculture often requires more labor than industrial agriculture, which makes it more expensive in places where labor costs are high. But practitioners of sustainable agriculture consider the improved long-term productivity of the soil to be worth this extra cost.

Traditional Sustainable Farming Techniques

Many of the practices used in sustainable agriculture are traditional farming methods (**FIGURE 33.4**). Through soil conservation, these traditional farming methods almost always achieve the goals of reducing erosion, increasing nutrient and organic matter content, and increasing soil depth. Subsistence farmers in India, Kenya, Thailand, and other countries use animal and plant wastes as fertilizer because they cannot obtain or afford synthetic fertilizers. In developed countries such as the United States, we call these fertilizers compost or composted manure. Such traditional farmers may also practice intercropping (Figure 33.4a), which we described earlier in this module in relation to reducing pesticide use. There are additional benefits to intercropping. For example, corn, which requires a great deal of nitrogen, can be planted along with peas, a nitrogen-fixing plant. Nitrogen that the microorganisms associated with peas produce becomes available to the corn plants. Crop rotation, which we also discussed earlier in this module, achieves the same effect by rotating the crop species in a field from season to season. For example, peas can be planted in a field for 1 year, resulting in increased nitrogen concentrations in the soil at the end of the year. This increase in nitrogen will enhance the growth of the corn crop that is planted there in the following year.

Intercropping trees with vegetables — a practice that is sometimes called **agroforestry** — allows vegetation of

(a)

(c)

(b)

FIGURE 33.4 Sustainable farming methods. A variety of farming methods can be used to improve agricultural yield and retain soil and nutrients, including (a) intercropping such as this grain grown in a peach orchard in the state of Washington; (b) contour plowing, such as this farm in Iowa growing alfalfa and corn; and (c) agroforestry, such as this shade-grown coffee in Mexico. *(a: inga spence/Alamy Stock Photo; b: Tim McCabe/USDA National Resources Conservation Service; c: Chris R. Sharp/Science Source)*

different heights, including trees, to act as **windbreaks**, literally tall objects that "break" the wind and prevent soil erosion. Trees planted among annual vegetable crops will catch soil that might otherwise be blown away, greatly reducing erosion (Figure 33.4b). The trees not only protect the vegetable crops and the soil but also provide fruit and firewood.

Another variation on intercropping is **strip cropping**, a method of planting crops with different spacing and rooting characteristics in alternating sets of rows to prevent soil erosion. For example, corn, which is spaced widely apart, can be planted in groups of six rows and on each side of the six rows of corn, six rows of wheat can be planted with the rows closer together than the corn. Any soil that erodes from in between the corn plants will be captured by the surrounding rows of wheat.

Alternative methods of land preparation and land management can also help to conserve soil and prevent erosion. For instance, **contour plowing**—plowing and harvesting parallel to the topographic contours of the land—helps prevent erosion by water while still allowing for the practical advantages of plowing (Figure 33.4c). Some farmers plant an autumn crop, such as winter wheat, that will sprout before frost sets in, so that the land does not remain bare between regular plantings. This is called a cover crop, and by planting it when the soil might otherwise be unplanted, erosion is further reduced. **Terracing** is a technique of shaping sloping land into step-like terraces that are flat. This prevents erosion, or if there is erosion, the soil will move to the next lowest terrace, thereby avoiding loss from the agricultural

Windbreaks An agricultural technique that literally plants tall objects that "break" the wind and prevent soil erosion.

Strip cropping An agricultural method of planting crops with different spacing and rooting characteristics in alternating sets of rows to prevent soil erosion.

Contour plowing Plowing and harvesting parallel to the topographic contours of the land.

Terracing An agricultural technique where farms shape sloping land into step-like terraces that are flat.

area. Two other modern techniques are important in maintaining soil and reducing soil erosion. These are relatively recent and still considered experimental by some.

Modern Sustainable Farming Techniques

Most crops in farms and gardens are annuals. They die at the end of each growing season and need to be planted the following spring. Recall from Module 26 that conventional agriculture of annual plants therefore relies on plowing and tilling, processes that physically turn the soil upside down and push crop residues under the topsoil, thereby killing weeds and insect pupae. There are a few **perennial plants**, such as asparagus, that live for multiple years, and they do not need to be replanted at the beginning of each growing season. The advantage is that there is no need to plow, till, or disturb the soil so there is minimal erosion, oxidation of carbon, or other consequences of plowing that we discussed in Module 26. Proponents of the breeding and development of perennial crops maintain that the single most beneficial new development in agriculture would be the development of food crops that do not need to be replanted every year. Plant researchers are exploring a variety of ways to develop additional perennial crops. Using a combination of conventional selective breeding and technology, they are attempting to convert annual species such as wheat, sorghum, sunflowers, and corn into perennials. At the same time, they are collecting wild perennials, such as wheatgrass, then domesticating them and selecting for higher seed yield, size, and quality. Through these efforts, researchers hope to assemble communities of perennial plants, animals, fungi, and microorganisms that will be stable, productive, and resistant to insect pests and diseases. In the process, the need for plowing and tilling will be eliminated or at least greatly reduced.

No-till agriculture is an agricultural method used in fields of annual crops where farmers do not till or plow the soil between seasons. It is used as a means of reducing topsoil erosion and avoiding the soil degradation that comes with conventional agricultural techniques. Farmers using this method leave crop residues in the field between seasons (**FIGURE 33.5**). The intact roots hold the soil in place, reducing both wind and water erosion, and the undisturbed soil can regenerate natural soil horizons. No-till agriculture also reduces emissions of CO_2 because the intact soil undergoes less oxidation and carbon stored in the soil is more likely to remain there. "Do the Math: Comparing Carbon Accumulation in Tilled and No-Till Agricultural Soils" allows you to calculate the amounts of carbon in land that has been tilled versus land that has not been tilled to illustrate the carbon benefits of no-till agriculture.

Perennial plants Plants that live for multiple years and do not need to be replanted at the beginning of each growing season.

No-till agriculture An agricultural method used in fields of annual crops where farmers do not till or plow the soil between seasons.

Green manure Plant material deliberately grown in a field with the intention of plowing it under at the end of the season.

FIGURE 33.5 No-till agriculture. Rows of soybeans emerge between the residues of a corn crop left over from the previous season. *(Lynn Betts/USDA Natural Resources Conservation Service)*

In many cases, however, in order for no-till agriculture to be successful, farmers must apply herbicides to the fields before, and sometimes after, planting so that weeds do not outcompete with the crops. Therefore, the downside of no-till methods is an increase in the use of herbicides.

Soil Additives and Rotational Grazing

Conventional farmers add nutrients to the soil in the form of manufactured chemical fertilizers that require fossil fuel energy to manufacture and run the risk of leaching into waterways. Sustainable farmers often use composted animal manure. They also use **green manure**, which is plant material deliberately grown in a field with the intention of plowing it under at the end of the season. In this way, the green manure adds nutrients and organic matter to the soil, just as composted animal manure does. With green manure, which decomposes slowly after it has been plowed into the soil, there is no possibility that nutrients will run off the soil and contaminate waterways. However, it does require plowing to deliver it to the soil, so all the previously mentioned consequences of plowing would be possible. Typically, plants grown for green manure are legumes, so green manure is most likely going to deliver organic matter and nitrogen to the soil, along with smaller amounts of other nutrients.

As crops are removed from agricultural fields each year, base cations such as calcium and magnesium that were originally in the soil and have been taken up by the crop also leave the field. Over time, this base-cation removal leads

DO THE MATH

Comparing Carbon Accumulation in Tilled and No-Till Agricultural Soils

Preparing for the AP® Exam

Measuring soil carbon content is one way to determine the amount of organic matter in a soil. In theory, the practice of no-till agriculture on a farm should result in more soil carbon compared to a similar farm that practices conventional tilled agriculture. A native grassland that has not been farmed should be relatively undisturbed and therefore contain more organic matter and a greater carbon content than a plot of land tilled conventionally or a no-till plot.

A study in Nebraska measured the carbon content in soil on three adjacent areas of land: native grassland, tilled farmland, and no-till farmland. The investigators reported the findings for the first 5 cm of soil and then for the horizon from a 5 cm to a 20 cm depth. The study produced the following findings:

Treatment	Soil depth	Carbon content (g C m^{-2})	Nitrogen content (g N m^{-2})
Native grassland	0–5 cm	1,437	131
Native grassland	5–20 cm	2,653	265
Tilled farmland	0–5 cm	699	82
Tilled farmland	5–20 cm	2,208	230
No-till farmland	0–5 cm	1,129	108
No-till farmland	5–20 cm	2,299	366

Data from: J. Six et al. *Soil Sci. Soc. Am. J.* 62 (1998): 1367–1377.

1. Add the 0–5 cm and 5–20 cm horizon for each treatment and compare the results of the three treatments.

The carbon content in g C m^{-2} for the 0–20 cm depth can be calculated by adding the 0–5 cm and 5–20 cm horizons:

$$\text{Native grassland: } 1,437 \text{ g C m}^{-2} + 2,653 \text{ g C m}^{-2} = 4,090 \text{ g C m}^{-2}$$
$$\text{Tilled farmland: } 699 \text{ g C m}^{-2} + 2,208 \text{ g C m}^{-2} = 2,907 \text{ g C m}^{-2}$$
$$\text{No-till farmland: } 1,129 \text{ g C m}^{-2} + 2,299 \text{ g C m}^{-2} = 3,428 \text{ g C m}^{-2}$$

2. Describe the two treatments as a percentage of the native grassland carbon content.

Tilled farmland as a percentage of native grassland = $(2,907 \text{ g C m}^{-2}) \div (4,090 \text{ g C m}^{-2}) = 0.7108 \times 100\% = 71.08\%$

No-till farmland as a percentage of native grassland = $(3,428 \text{ g C m}^{-2}) \div (4,090 \text{ g C m}^{-2}) = 0.8381 \times 100\% = 83.81\%$

The no-till farmland contains 83.81 percent of the carbon contained in the native grassland while the tilled farmland contains only 71.08 percent of the native grassland carbon content.

3. Draw a conclusion about the effects of tilled agriculture versus no-till agriculture compared to native grassland.

This study supports the idea that there is more carbon retained in a no-till agricultural farm then in a conventionally tilled farm, relative to native grassland.

YOUR TURN Conduct a similar analysis for nitrogen.

1. Add the 0–5 cm and 5–20 cm horizon for each treatment and compare the results of the three treatments.
2. Describe the two treatments as a percentage of the native grassland nitrogen content.
3. Draw a conclusion about the effects of tilled agriculture versus no-till agriculture compared to native grassland.

many agricultural fields to become acidic (lower pH). If a farmer decides that the soil requires additional calcium to raise the pH, they will probably fertilize with **limestone**, which is calcium carbonate sedimentary rock that has been ground up or crushed for easy application as a fertilizer. If the soil needs both calcium and magnesium, they may use dolomitic limestone, which is calcium-magnesium carbonate sedimentary rock. Both limestone and dolomitic

Limestone A calcium carbonate sedimentary rock that has been ground up or crushed for easy application as a fertilizer.

limestone can be mined, transported, crushed, and spread on the surface of an agricultural field and tilled into the soil. While there is fossil fuel energy required for transport, processing, and application, the actual manufacturing of the fertilizer product does not require fossil fuels, which is true of most synthetic fertilizers. If a soil is too basic, then a farmer may need to acidify it, typically by adding elemental sulfur or another acidifying compound.

Farmers can maintain the nutrient status of fields and pastures by moving their animals to different locations. **Rotational grazing** is the rotation of farm animals to different pastures and fields to prevent overgrazing. We have already discussed the concept of using animals to help transfer nutrients from animal waste to the soil in Module 28, when we discussed free-range grazing of animals. Keep those ideas in mind as you consider rotational grazing as part of sustainable agriculture. As animals graze grass in a pasture or field, it is considered good farming practice to move animals to different pastures to prevent overgrazing.

There is one more aspect of sustainable agriculture that has become very well-known and prominent in discussion of agriculture: organic agriculture.

Organic Agriculture

Organic agriculture is the production of crops in a way that sustains or improves the soil, without the use of synthetic pesticides or fertilizers. Organic agriculture follows several basic principles, many of which have been discussed throughout this unit.

Organic agriculture:

- uses ecological principles and works with natural systems rather than dominating those systems.
- maintains the soil by increasing soil mass, organic matter, biological activity, and beneficial chemical properties.
- keeps as much organic matter and as many nutrients in the soil and on the farm as possible.
- avoids the use of synthetic fertilizers and synthetic pesticides.
- reduces the adverse environmental effects of agriculture.

Rotational grazing The rotation of farm animals to different pastures and fields to prevent overgrazing.

Organic agriculture The production of crops in a way that sustains or improves the soil, without the use of synthetic pesticides or fertilizers.

Delaney Clause A clause in the Food, Drug, and Cosmetic Act designed to prevent potentially harmful cancer-causing food ingredients.

FIGURE 33.6 Organic farming. Many customers are willing to pay higher prices for organically grown food, like the produce at this organic farm. *(silverkblackstock/Shutterstock)*

In the developed world, organic farming has increased in popularity over the past 3 decades. The U.S. Organic Foods Production Act (OFPA) was enacted as part of the 1990 farm bill to establish uniform national standards for the production and handling of foods labeled "organic." The **Delaney Clause** of the Food, Drug, and Cosmetic Act is another piece of legislation related to food. The goal of the Delaney Clause, passed in 1958, is to prevent any potentially harmful cancer-causing food ingredients. However, in the decades following passage of the bill, it has become clear that there are many naturally occurring compounds in food that both cause cancer and prevent cancer, so the utility of this clause has been questioned in recent years. "Science Applied 5: How Do We Define Organic Food?" on page 389 discusses organic food labeling in more detail.

Because most organic farmers plant diverse crops and engage in practices such as encouraging beneficial insects that are difficult to do at a large scale, it is common for organic farmers to keep their farms relatively small. Organic farmers also must manage the soil carefully because if they lose soil nutrients and see a decline in the health of their crops, they have fewer options than conventional farmers. These practices usually increase labor costs significantly. However, farmers can recoup extra labor costs by selling their harvest at a premium price to consumers who prefer to buy organic food and are sometimes willing to pay more for it (**FIGURE 33.6**).

Organic agriculture is not without adverse environmental consequences. Because organic farmers typically do not use herbicides, they are less likely than conventional farmers to be able to use no-till methods successfully. And alternative pest control methods are not always environmentally friendly. For example, to keep crops such as carrots free of weeds, organic farmers may treat the soil with a flame fueled by propane before planting. While this technique protects carrots without the use of herbicides, it does use propane, a fossil fuel, and causes the release of carbon dioxide to the atmosphere. In addition, because they are usually smaller, organic farms may not benefit from the economy of scale of a large farm. They may use a tractor more frequently because of different crops or irregular rows of crops, which contributes to additional fossil-fuel use. Finally, it is important to realize that just because a farm follows organic standards, that does not necessarily mean it is sustainable. There are organic farms that grow organic produce such as tomatoes in the middle of the winter in the northeast or north-central United States by using large amounts of fossil fuel to keep their greenhouses sufficiently warm. Such a practice may follow organic food standards, but because of the fossil fuel use, it may not be sustainable.

AP® Exam Tip

Draw a table in your notes listing all of the agricultural methods covered in your class. Include their benefits and drawbacks. Review this in the weeks before the exam.

Integrated pest management and sustainable agriculture are both practices that strive to increase agricultural yield while minimizing the harm to the environment that comes from growing food. Each has its economic and environmental costs and benefits. In the next module, we will consider growing food in water, aquaculture, and growing trees in forests. In both situations, many of the principles from IPM and sustainable agriculture that we examined in this module are also used.

Module 33 AP® Review

Preparing for the AP® Exam

Learning Goals Revisited

33-1 What are the goals and techniques of integrated pest management (IPM)?

IPM uses techniques to control pest species while minimizing pesticide inputs and increasing agricultural output. Techniques include crop rotation and intercropping, the creation of habitats for predators of pests, and the use of pesticides. IPM reduces pesticide exposure and other harms to natural systems and human health, but there are additional economic costs and added complexity for the farmer.

33-2 What are the objectives of sustainable agriculture practices?

Sustainable agriculture provides food while improving soil quality and minimizing use of nonrenewable resources. It emphasizes the ability to continue agriculture on a given piece of land indefinitely through conservation and soil improvement. It often uses traditional farming practices, many of which are sustainable.

AP® Practice Questions

Preparing for the AP® Exam

Multiple-Choice Questions

1. Which agricultural method is most similar to intercropping?
   (a) contour plowing
   (b) agroforestry
   (c) no-till agriculture
   (d) integrated pest management

2. Which agricultural method is used to prevent erosion?
   (a) intercropping
   (b) integrated pest management
   (c) contour plowing
   (d) biocontrol

3. Which is a key principle of organic farming?
   (a) avoiding synthetic fertilizers
   (b) high use of pesticides
   (c) increasing use of GMO crops
   (d) avoiding the use of fossil fuels

4. Which is a common practice of integrated pest management?
   (a) use of drip irrigation
   (b) elimination of pesticides
   (c) use of pest-resistant crops
   (d) use of only monoculture crops

5. Farmers who practice organic agriculture have less of an impact on the environment than farmers who practice industrial agriculture because they
   (a) use no-till agriculture exclusively.
   (b) import soil to maintain soil fertility.
   (c) maintain large farms with a single crop.
   (d) avoid pesticides and synthetic fertilizers.

Free-Response Question

Alternatives to industrial agriculture are now gaining more attention in some countries as people strive to practice agriculture more sustainably. The graph below shows how many people can be sustained with the use of synthetic fertilizers.

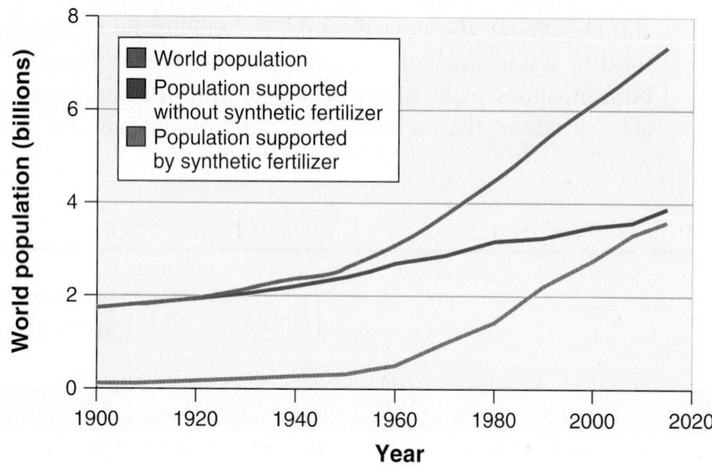

(a) **Identify** how many people were sustained without synthetic fertilizer in 1980. (1 pt.)
(b) **Identify** the world population sustained by synthetic fertilizer in the year 2000. (1 pt.)
(c) **Explain** one method that those who don't use synthetic fertilizers use to grow their food. (1 pt.)
(d) **Describe** an environmental issue that arises from increased use of synthetic fertilizers. (1 pt.)
(e) Biological controls are constantly released to control unwanted weeds that grow around crops. In 1930, an average of 50 new species were released every year, and in 1990, an average of 95 new species were released every year.
   (i) **Calculate** the percent increase of species from 1930 to 1990. Show all work. (2 pts.)
   (ii) **Calculate** the average number of new species released every year. Show all work. (2 pts.)
   (iii) **Calculate** the number of species added in new releases at the rate found in part (ii) from 1990 to 2020. Show all work. (2 pts.)

Module 34

Unit 5 | 24 | 25 | 26 | 27 | 28 | 29 | 30 | 31 | 32 | 33 | 34

Aquaculture and Sustainable Forestry

In the previous module we explored integrated pest management and sustainability of agriculture. Now, we'll explore two different types of farming: fish in water, and trees in forests. These two methods of extracting biological resources manipulate the natural environment in order to enhance the output of what would grow in the absence of human intervention. First, we will examine a method of obtaining fish for human consumption that has increased in importance in recent years and will continue to do so for many decades. Then we will learn about the optimal way to remove trees from forests while simultaneously attempting to reduce environmental harm.

Learning Goals

After reading this module you should be able to

34-1 explain the environmental benefits and consequences of aquaculture.

34-2 describe the methods for reducing human impacts on forest tree removal.

34-1 What are the environmental benefits and consequences of aquaculture?

Aquaculture is increasing in importance and its environmental impacts are less

In Module 28 we examined wild-caught fishing techniques and reviewed the definition of aquaculture, the farming of aquatic organisms such as fish, shellfish, and seaweeds. Here we will describe aquaculture in more detail and identify the environmental benefits and consequences relative to obtaining food on land. In Figure 28.6, we showed how global fish production has increased more than 100 percent, mostly because of the large increase in aquaculture. **FIGURE 34.1** shows that the increased importance of aquaculture is expected to continue beyond the year 2050.

You may recall from Module 28 that overfishing has caused fishery collapse in locations around the world at various times in the last 50 years. We defined a fishery as a commercially harvestable population of fish within a particular ecological region. Fishery collapse occurs when there is a decline of a fish population by 90 percent or more. The global demand for fish is increasing even as wild fish catches are decreasing. In response, many scientists, government officials, and entrepreneurs have developed ways to increase the production of seafood through aquaculture. Aquaculture requires constructing an entire aquatic ecosystem by stocking the organisms, feeding them, protecting them from diseases and predators and sometimes removing waste products. It usually requires keeping the organisms in constructed, netted enclosures (**FIGURE 34.2**). These netted enclosures can be floated in open ocean, near-coastal marine waters, coastal ponds (brackish water), inland ponds (fresh water), and in totally enclosed structures on land. Interestingly, most of the catfish, shrimp, and salmon eaten in the United States is produced by aquaculture. Norway and Chile are the two largest

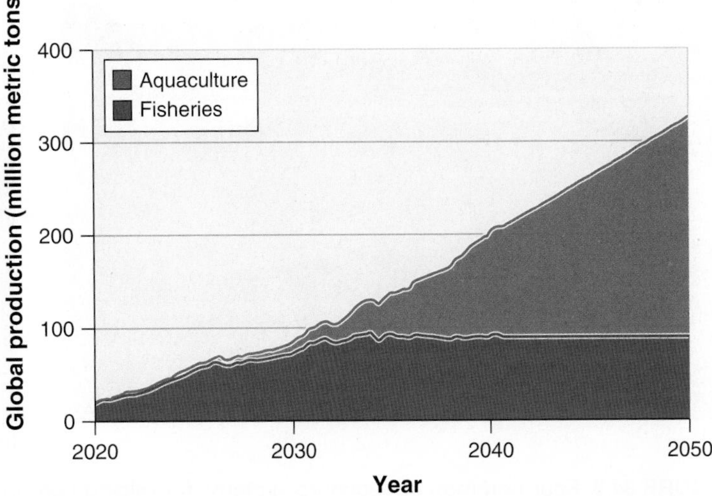

FIGURE 34.1 Projected global fish production. In the next 30 years, wild-caught fish production (fisheries) is expected to stay constant while aquaculture fish production is projected to be responsible for all of the increase in global fish production.

FIGURE 34.2 **A salmon farming operation in Chile.** Uneaten food and waste released from salmon farms can cause significant nutrient input into natural marine ecosystems. *(MARCELODLT/Shutterstock)*

aquaculture fish producers in the world. A typical salmon farm in one of those countries might contain netted enclosures that are equivalent to 3.8 million liters (1 million gallons) and might contain 75,000 fish.

Aquaculture Systems: Nets, Ponds, Recirculating Aboveground Tanks

There are many similar features in aquaculture fish enclosures, regardless of type and location. **FIGURE 34.3** shows the four most common types of aquaculture systems for raising fish.

The system that is probably most well-known is open-net pens that float in the water offshore or near-offshore. This system contains a frame and netting and entirely encloses the fish. Water flows spontaneously into the netted enclosure. Fish food, supplements, and medication delivery systems are contained within the netted enclosure as well. Depending on the density of fish enclosed in the netted area, antibiotics may be used to prevent infections among the fish. There are also a variety of waste removal and waste processing systems used to prevent

(a)

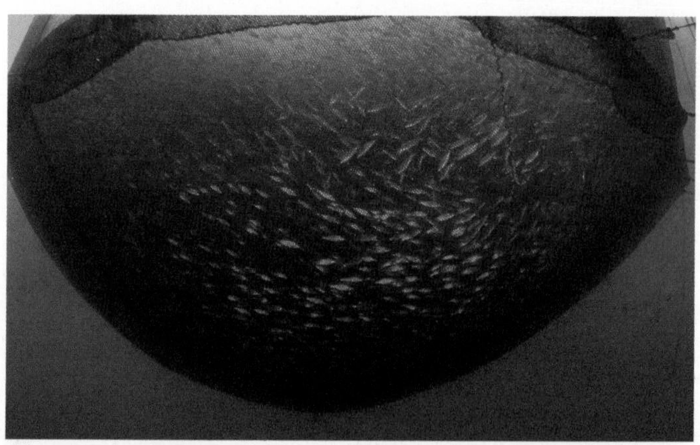

(b)

(c)

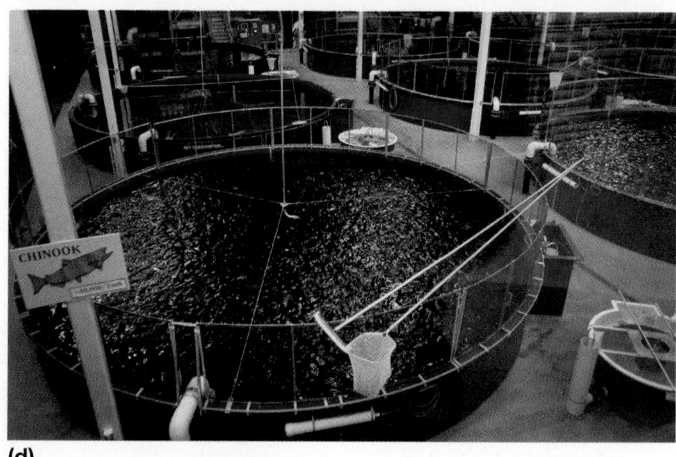

(d)

FIGURE 34.3 **Four common aquaculture systems for raising fish.** (a) An open pen floating system in the Faroe Islands, North Atlantic Ocean commonly used for salmon. (b) A submerged net system commonly used for fish like seabass that do not need to come up to the surface. (c) A pond farm system in natural or constructed freshwater ponds is used for catfish in the United States and Asia and can be 2 to 4 ha (5 to 10 acres) in size. (d) Recirculating aboveground tanks can be used for a variety of species. *(a: Smit/Shutterstock; b: Adnan Buyuk/Shutterstock; c: KobchaiMa/Shutterstock; d: Bloomberg/Getty Images)*

fish waste from building up to harmful concentrations within the enclosure. Submerged-net pens are similar to open-net pens except that the fish enclosure is completely submerged under water. Pond aquaculture uses open-net pens, but these pens are located either in constructed ponds or natural ponds on land. Recirculating above-ground tanks can be located inside buildings or outdoors in almost any location as long as there is a suitable source of water for use inside the tanks. Relative to wild-caught fish in the open ocean, aquaculture-grown fish—salmon is a good example—are efficient with respect to energy consumption and fuel use because fishing ships don't have to travel far out to sea to catch fish. Aquaculture systems cover a very small area compared to the large distances that wild-caught fishing requires. And they eliminate the demand for removing fish from wild populations. Above-ground tanks are also beneficial because there is a lesser chance of fish diseases contaminating wild populations than with open-net pens because the tanks are on land, at a distance from the wild populations. For the same reason, there is a lesser chance of farmed fish escaping to the wild and interbreeding with or contaminating the gene pool of wild fish.

Critics of aquaculture, though, point out that it can create many environmental problems. In a typical aquaculture facility, clean water is pumped in at one end of a pond or marine enclosure, and wastewater containing feces, uneaten food, and antibiotics is pumped back into the pond or ocean at the other end. The wastewater may also contain bacteria, viruses, and pests such as sea lice that thrive in the high-density habitat of aquaculture facilities and this can infect wild fish and shellfish populations outside the facility. In addition, fish that escape from aquaculture facilities may harm wild fish populations by competing with them, interbreeding with them and adversely affecting their gene pool, or spreading diseases and parasites.

If you think the tank in Figure 34.3d looks a lot like an above-ground swimming pool, you are right—it does! Above-ground tank aquaculture is a lot like raising fish in a swimming pool. And if you know that swimming pools utilize pumps and filters and sometimes heaters, then you probably realize that the same is true for aquaculture tanks. Filtration is required and possibly heating or cooling depending on the species of fish and the climate where the fish are being raised. Some water in the tanks will evaporate over time, so replacement water is needed. Accordingly, above-ground tank aquaculture systems require water and energy and possibly a means of disposing of fish waste. So a number of innovative methods have been developed to reduce environmental impact of above-ground tank aquaculture. One experimental, holistic food production system removes fish waste from above-ground aquaculture tanks to use as plant fertilizer in nearby greenhouses. Additional fish waste along with other nutrients are combined to fertilize algae, which is then fed back to the fish. Overall, aquaculture has many promising characteristics as a means of sustainable food production. Proponents of aquaculture believe it can alleviate some of the human-caused pressure on overexploited fisheries while providing much-needed protein for the more than 1 billion undernourished people in the world. Aquaculture also has been able to boost the economies of many developing countries. The final part of this module and this unit covers a very different kind of farming: the sustainable farming of trees. As in the case of aquaculture, sustainable tree farms attempt to reduce the environmental impact of removing a biological resource from ecosystems for human use.

AP® Exam Tip

Filters are not the solution for everything. Only mention using a filter to address wastewater for something like an aquaculture operation if you know that such a filter exists. There is no such thing as a "fertilizer filter" to filter runoff from fertilizer that has been spread on an agricultural field. Instead, describe the use of holding ponds and possibly discuss using less fertilizer, applying it at the appropriate time of the year, and other solutions.

34-2 What methods are used for reducing human impacts on forest tree removal?

Sustainable forestry uses principles found in agriculture to minimize the effects of logging

At the beginning of this unit in Module 24, we examined how forests may be clear-cut or selectively cut and identified that clear-cutting may have greater environmental consequences than selective cutting. We also explained that there may be natural or intentional (meaning "by people") restocking of trees after clear-cutting, a process called **reforestation**, that is done to repopulate a forest, reduce erosion, and begin the process of removing carbon dioxide from the atmosphere. Here we will describe the goals and methods for logging forests while maintaining the sustainability of those forests. **Sustainable forestry** is a methodology for managing forests so they provide wood while also providing clean water, maximum biodiversity, and maximum carbon sequestration in both trees and soil. Land managers strive to mimic nature whenever possible and preserve a balanced forested ecosystem. They try to remove trees with as little impact as possible on the forest, soil, and surrounding waterways. One technique for doing this is logging with horses, which was featured in Figure 24.9, or other animals,

Reforestation The natural or intentional restocking of trees after clear-cutting to repopulate the forest, reduce erosion, and begin the process of removing carbon dioxide from the atmosphere.

Sustainable forestry A methodology for managing forests so they provide wood while also providing clean water, maximum biodiversity, and maximum carbon sequestration in both trees and soil.

rather than fossil-fuel powered machinery. When choosing which trees to remove, sustainable foresters follow biodiversity goals such as maintaining species evenness and richness. Other practices include avoiding habitat fragmentation and conversion of land to other purposes and developing logging projects that will require the minimum of construction of new logging access roads. Sustainable forest removal of trees will encourage regrowth of natural forests, rather than plantations of non-native species. Sustainable forestry also includes human dimensions of logging such as respecting local communities and promoting activities that preserve or create jobs and keep the communities economically viable. Finally, sustainable forestry practices involve training forest workers properly and keeping them safe and healthy. This is especially important considering that the Occupational and Safety Administrations lists logging as the most dangerous occupation in the United States (**FIGURE 34.4**).

FIGURE 34.4 A logger removing a tree in sections. Logging is the most dangerous occupation in the United States. *(Steven Chadwick/Alamy Stock Photo)*

Other Sustainable Logging Techniques

Perhaps the most important first steps in sustainable forestry are to reduce the use of wood whenever possible and to reuse wood rather than create demand for new wood products. There are forest certification organizations that designate that wood for purchase has been sustainably harvested, similar to organizations that certify that food is organic. These organizations verify that certified wood was grown in a forest that was managed properly and that the maximum sustainable yield of the forest was not exceeded. They guarantee a "chain of custody" so that the consumer or builder purchasing sustainably grown lumber at a lumber yard or big box home supply store will have confidence that the wood they are buying truly is sustainably grown. Usually, certified lumber is sold at a higher cost than standard lumber, to help pay for the higher costs of the sustainable forestry program that was used to grow it. If a builder or homeowner chooses to purchase certified sustainably harvested wood rather than conventional wood, this would in theory create demand for products grown using sustainably harvested forestry methods and result in more forests being managed sustainably. However, there are many criticisms of sustainable forest certification programs. Some researchers have questioned the integrity of the chain of custody and have demonstrated that some unscrupulous vendors have called commercially grown, non-sustainably harvested wood as "certified sustainable" in order to charge higher prices.

Techniques similar to those used for integrated pest management in agriculture can be applied to forest management. However, the extent of the land area covered by forest is usually so much greater than in agricultural land that it is much more of a challenge to conduct any direct treatment of forests, such as spraying. Foresters can attempt to exclude the introduction of insect pests from their forests, although efforts of pest exclusion are usually done by country or for an entire state, rather than a plot of land. This entails not letting any firewood or untreated wood to be brought into an area. When a forest pest is already present, introduction of natural predators of the harmful pest is an option. This is similar to the biocontrol with a natural predator we showed being done for agriculture in Figure 33.2. Finally, when trees are infected with a pest, they can be removed and burned. However, it is important that they are kept close to the removal sites so that the infestation is not spread to neighboring counties or states if the wood were to be transported. Another important method to sustain forests uses fire.

Fire Management

In many ecosystems, fire is a natural process that is important for nutrient cycling and regeneration. As discussed briefly in Unit 1, when fires periodically move through ecosystems, they liberate nutrients tied up in dead biomass. In addition, areas where vegetation is killed by fire provide openings for early successional species.

Humans have used a variety of management policies with respect to fire and the sustainability of forests. For many years, managers of forest ecosystems, including the United States Forest Service, worked to suppress fires. This strategy led to the accumulation of large quantities of dead biomass on the forest floor, which built up until a large fire became inevitable. One method for reducing the accumulation of dead biomass is a **prescribed burn**, in which a fire is deliberately set under controlled conditions thereby decreasing the accumulation of dead biomass on the forest floor. In this way, prescribed burns help reduce the risk of uncontrolled natural fires and provide some of the other benefits of fire. More recently, there have been discussions about whether certain naturally occurring fires should be allowed to burn.

There are also many instances where human behavior and mismanagement can lead to large fires on public or private land. Both deliberate and accidental fire setting by individuals is certainly a cause of fires. But poor equipment management may also lead to fires. The northern California wildfires that have occurred almost every year since 2017 have received great attention because of the number, extent, and location of the fires in and near Sonoma County, famous for its wine making. Downed power lines are the cause of some of the California fires. A California electrical utility was deemed responsible for causing hundreds of wildfires in the past decade, burning land, destroying buildings, and causing deaths. One investigative study by a news organization reported that five of the ten most destructive fires in California since 2015 were linked to one utility's electrical network.

Until the aforementioned fires in California and similar fires in Oregon and Washington states, probably the best-known forest fires in the United States were those that occurred in Yellowstone National Park in the summer of 1988. That summer was the driest year on record at the park and a combination of human activity and lightning set off multiple fires. Over 25,000 people participated in fighting the fires. When the fires were out, more than one-third of Yellowstone National Park had burned, as shown in the map in **FIGURE 34.5a**. Initially, many people were upset that such a large fire had occurred within a national park. However, within a few years, it became clear that the fires had created new, nutrient-rich habitat for early successional plant species that attracted elk and other herbivores. Ultimately, researchers and forest managers concluded that the Yellowstone fires of 1988 provided many benefits to the Yellowstone ecosystem. Today, a typical visitor viewing a portion of the park that burned in 1988 probably wouldn't even know that there had been a major fire more than 30 years earlier (Figure 34.5b). Review the Unit 5 Visual Representation feature that shows many of the land and water use approaches we discussed throughout the unit. It also pairs the approaches with solutions to their environmental problems (pages 384–385).

Prescribed burn When a fire is deliberately set under controlled conditions, thereby decreasing the accumulation of dead biomass on the forest floor.

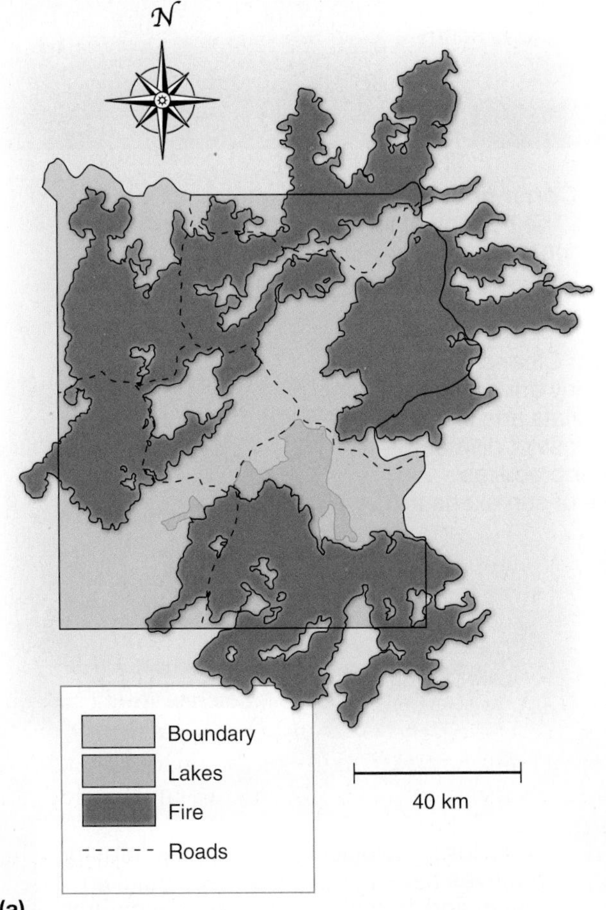

(a)

(b)

FIGURE 34.5 Yellowstone fires of 1988. (a) As can be seen from the map, extensive areas of the park were burned in this exceptionally hot and dry year. (b) Thirty years later, a casual observer might not know that the forest had burned, except for the sign in the foreground. *(b: Peter Haigh/Alamy)*

Preparing for the AP® Exam

Visual Representation 5 Land Use: Problems and Solutions: Human use of land and its resources has created detrimental environmental effects throughout history. Today people are working toward solutions that balance our need for land resources with sustainability and thoughtful stewardship.

Agriculture: Many agricultural practices stress ecosystems and soils. As modern agriculture has continued to develop, some farmers are starting to explore sustainable alternatives to techniques that have caused short-term gains and long-term problems.

Problem: In locations with a warm climate and relatively nutrient-poor soils, farmers sometimes use slash-and-burn agriculture. The land is farmed for only a few years until the soil is depleted of nutrients.

Solution: Rainforest farmers are developing techniques to reclaim depleted land and plant in sustainable ways with crops such as coffee and cocoa.

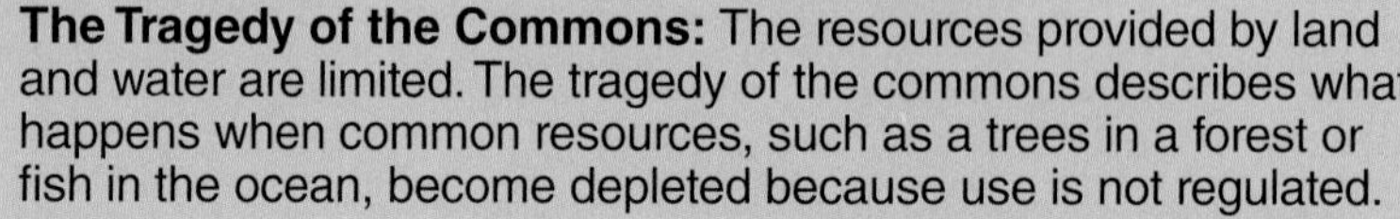

The Tragedy of the Commons: The resources provided by land and water are limited. The tragedy of the commons describes what happens when common resources, such as a trees in a forest or fish in the ocean, become depleted because use is not regulated.

Problem: In many parts of the oceans, the competition for fish and improved technology has led to a steep decline in fish populations and even fishery collapse.

Solution: Solutions to fishery collapse include quota systems, community self-regulation, and aquaculture.

Problem: Clear-cutting alters many animal and plant habitats and leads to erosion, loss of diversity, and other problems because of sun exposure.

Solution: Ecologically sustainable forestry removes trees in ways that help maintain plants and animals in as close to a natural state as possible. Using animals such as horses to remove logs reduces soil compaction from heavy equipment.

Problem: Pesticides can harm beneficial organisms, leach into nearby waterways, and create health risks to farm workers. Persistent pesticides can remain in the environment for years and harm wildlife. Organisms can become pesticide resistant, which leads to the pesticide treadmill.

Solution: Organic agriculture and integrated pest management techniques (IPM) both include crop rotation, intercropping, and encouragement of suitable conditions for predators of pests (biocontrol).

Problem: Irrigation agriculture is the largest use of water worldwide. Large-scale use of water from a confined aquifer is unsustainable because the withdrawal of water is not balanced by recharge. Some irrigation techniques can lead to inefficient use of water and waterlogging of soil.

Solution: Spray irrigation and drip irrigation are the most efficient methods of irrigation.

Urban Development and Runoff: Urban development increases impervious surfaces, such as pavement. Urban planners are developing new techniques to address the problems of runoff.

Problem: Runoff water often picks up contaminants such as roadway oil. Surface runoff leads to flooding, which can overwhelm storm drainage systems.

Solution: Planting trees in an unpaved urban area can increase evapotranspiration and reduce runoff.

Solution: Urban systems that collect rainwater decrease urban runoff. Some cities are experimenting with buildings that collect rainwater and use it to flush toilets or water rooftop gardens.

Solution: New, more pervious paving materials allow water to return to the aquifer.

Module 34 AP® Review

Preparing for the AP® Exam

Learning Goals Revisited

34-1 What are the environmental benefits and consequences of aquaculture?

Aquaculture is the fastest growing method for supplying fish to the human population. It reduces the need to remove fish from wild populations and requires less energy and fuel to obtain fish. It usually requires constructing an entire aquatic ecosystem in a body of water by stocking fish, feeding them, protecting them from diseases and predators, and sometimes removing waste products. The drawbacks of aquaculture can lead to contamination of wild populations from fish waste, and farmed fish may escape and harm wild fish populations by competing with them, interbreeding with them, and adversely affecting their gene pool.

34-2 What methods are used for reducing human impacts on forest tree removal?

Sustainable forestry will provide wood while also allowing the forest to provide clean water, biodiversity, and carbon sequestration. Integrated pest management (IPM), most commonly known for its use in agriculture, can be used in forestry as well. Practitioners remove trees with as little impact as possible and try to avoid habitat fragmentation and additional road building. Allowing fires to burn and deliberately setting controlled burns are methods used to reduce the impact of wildfires.

Practice Math and Graphing 

Preparing for the AP® Exam

1. Practice Math

(a) If U.S. farmland is being converted to residential uses at a rate of 405,000 ha (1 million acres) per year, how much land will be converted in 10 years?

(b) If there is roughly 369 million ha (911 million acres) of farmland in the United States, what percentage is lost each decade?

2. Practice Graphing

(a) Draw a maximum sustainable yield graph based on the data shown in the table. The *x* axis should show the fish population in metric tons. The *y* axis should show the annual removal of fish in tons of fish per year.

(b) Determine the MSY for a fishery based on this graph.

Fish population (metric tons)	Annual removal of fish (tons per year)
100	30
200	60
400	95
600	95
800	60
900	20

AP® Practice Questions

Preparing for the AP® Exam

Multiple-Choice Questions

1. Aquaculture is projected to be responsible for all of the increase in global fish production over the next 25 years. However, future production from aquaculture could become problematic because
   (a) concentrated waste from aquaculture facilities contaminates rivers and oceans.
   (b) raising fish in a protected environment could lead to a fish population overshoot.
   (c) the economies of developing countries would be negatively affected.
   (d) ocean fishing operations would go out of business.

2. Which of the following can be concluded from the information in the graph below?

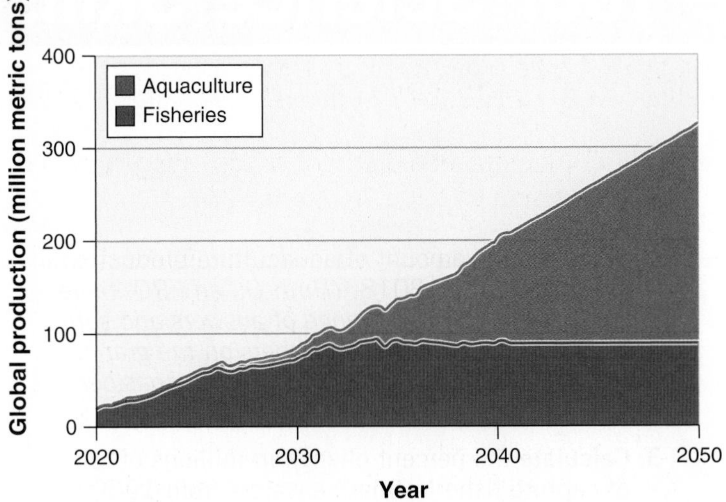

(a) Fish parasites may spread to wild populations.
(b) Fish production has remained stagnant.
(c) Aquaculture had allowed production to continue increasing.
(d) Fisheries may collapse without aquaculture.

3. For many years, forest fires were suppressed to protect lives and property. This policy has led to
(a) a buildup of dead biomass that can fuel larger fires.
(b) many forest species being able to live without having their habitats destroyed.
(c) increased solar radiation in most ecosystems.
(d) soil erosion on steep slopes.

4. Which of the following is a likely outcome of a prescribed forest burn?
(a) Forest floor biomass is replenished.
(b) Nutrients are stored in the soil.
(c) The risk of forest fires is increased.
(d) Habitat increases for early successional species.

5. Why are antibiotics added to aquaculture fish food?
(a) to provide protection to humans that eat the fish
(b) to protect fishers and those who handle the fish onshore from contracting diseases
(c) to prevent infection from spreading among the fish
(d) to promote a tastier fish

6. Recirculating aboveground tanks are beneficial because
(a) they don't require fossil fuel energy to maintain.
(b) there is only a relatively small chance that fish can escape to wild populations.
(c) there is no need to supplement the water with food.
(d) there is no need to supplement the food with nutrients.

Use the passage below to answer question 7:

Various public relations campaigns from the 1980s and 1990s have said, "Only you can prevent forest fires!" suggesting that people are the cause of forest fires because of their careless behavior. But it turns out that might not have been the case. A recent study shows that while many forest fires may have been ignited by individuals, pre-existing conditions may have played a bigger role than the careless behavior by a person. The study showed that management techniques from federal and state services were lacking and that improper preparation in advance of people setting wildfires may have had a bigger impact.

7. Which of the following are possible management steps that the author is alluding to when it comes to preventing wildfires?
(a) clear-cutting forests
(b) practicing prescribed burns
(c) practicing no-till agriculture
(d) prevention of slash-and-burn agriculture

Free-Response Question

The data below show three different species of marine aquaculture counts at the beginning of growing, to the final count at harvest. Species were held in open-ocean pens. Pens were counted at day 1 and day 21.

Holding tank	Individuals (initial count)	Individuals (final count)
Oysters	645	601
Clams	1344	1288
Shrimp	914	814

(a) **Identify** the trend across all three species. (1 pt.)
(b) **Describe** a likely reason for the trend identified in part (a). (1 pt.)
(c) **Describe** one way that the trend could be reversed. (1 pt.)
(d) Other than the problem identified in part (a), **identify** another issue that could arise from the change in counts of individuals. (1 pt.)
(e) **Calculate** which of the three groups had the largest percent change from the initial to the final count. Show all work. (2 pts.)
(f) **Calculate** the rate of shrimp lost per day. Show all work. (2 pts.)
(g) **Calculate** the rate of clams lost per day. Show all work (2 pts.)

UNIT 5 Practice Your Science Skills

Data Analysis

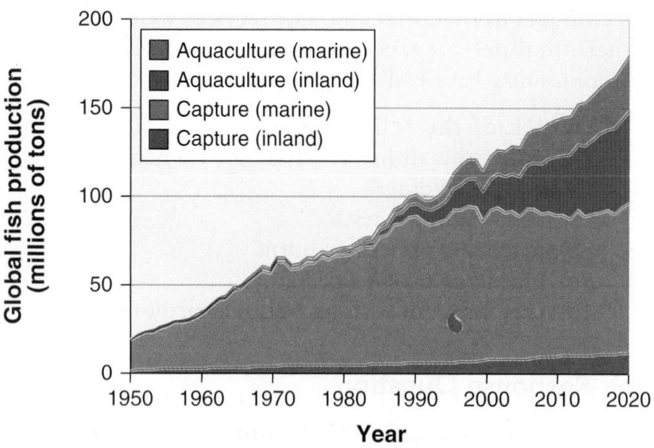

Questions

1. **Identify** the largest source of aquatic food production in 2018. *(Hint: Be sure to read the key at the top and notice that there is both marine and inland for each category of fish production.)*

2. **Calculate** the amount of aquaculture production in marine waters in 2018. *(Hint: On an FRQ, there will be an acceptable range of answers and setup for interpretation of the numbers on the graph. This is a stacked graph; make sure to consider more than just the top line.)*

3. **Calculate** the percent change in millions of tons of capture fisheries-marine waters from 1950 to 1970.

4. **Calculate** the percent change in millions of tons of capture fisheries-marine waters from 1970 to 1988. *(Hint: Note that this question asks you to find the year 1988, which is between labels on the x axis.)*

5. **Explain** why the percent change in capture from 1950 to 1970 (in question 3) and 1970 to 1988 (in question 4) declined.

Pursuing Environmental Solutions

Urban Agriculture

Urban farms are a small percentage of farmland by area, and they represent a small fraction of total food production, but they have the potential to introduce substantial benefits to humans and the natural environment. Remember from Module 30 that 55 percent of the people in the world live in cities. Around the United States and elsewhere in the world, an increasing number of urban gardeners are providing food for themselves and their neighbors. In doing so, they are enhancing ecosystem services by improving urban land that had been degraded, increasing habitat for plants and beneficial insects, and building up soil nutrients and storing atmospheric carbon. They are also providing knowledge about and access to nutritious, healthy food. Beyond environmental benefits, urban farms are a potential source of community revitalization, promoting jobs and helping redevelop neighborhoods that have been on the decline.

The USDA defines urban agriculture as "backyard, rooftop and balcony gardening" as well as farming and gardening in vacant lots and parks in urban and suburban areas. A group called the Detroit Black Community Food Security Network (DBCFSN) has taken advantage of unfortunate circumstances and urban blight to revitalize and energize residents in the city of Detroit, Michigan, with urban agriculture. In 2007, after a series of financial and political crises, one-third of the city's land area was designated as "vacant." It also didn't have any national food chain grocery stores until one relatively expensive food store arrived about 10 years ago. Forty percent of the population was living below the

A community garden. Local residents in a community garden in Detroit are growing both vegetables and flowers. Many gardens in Detroit are located in what used to be abandoned fields and lots. *(Jim West/Alamy Stock Photo)*

national poverty level, and although there is no subway system in Detroit and only a limited streetcar system, almost one-third of its residents did not own cars. In addition to unemployment, crime, and unpleasant living conditions, obtaining affordable, diverse, quality food was a challenge for many Detroit residents.

Created in 2006, DBCFSN started a 0.1 hectare (quarter-acre) organic farm called D-Town Farm that eventually grew to 3 hectares (7 acres). D-Town Farm was established as a model farm to demonstrate farming techniques to school groups and adults that can be used all over the city. The farm practices organic agriculture, so the thin, nutrient-depleted soil they started with has been improved over time, leading to storage of carbon and an increase in soil nutrients. The farm grows more than 30 crop species allowing for an increase in biodiversity of the area. There is greater water absorption since the farm was established, now that a rich soil and vegetation cover the land. They also keep bees at D-Town Farm that provide an ecosystem service to the area. From this model farm and others that started over the past 2 decades, there are now over 350 urban farms in Detroit. If you include community gardens in the count, a 2020 report estimated that there are 1900 farms and gardens. These farms grow produce that is sold at farmers markets, start seedlings that are distributed to family and school gardens, and educate children and adults about urban agriculture and environmental science. Some of the gardens supply food for a variety of food rescue organizations that gather food that is donated or might otherwise be wasted. They provide this food to people in need. Other gardens compost food that is no longer edible. Both of these actions reduce food waste and prevent spoiled food from ending up in the landfill, where it will decompose anaerobically and produce methane.

Urban farms and gardens in Detroit grew 220,000 kg (484,000 pounds) of produce in 2020. This is a small fraction of the produce consumed in Detroit, but it's a beginning to what could become a larger food contribution from urban areas. The Food and Agriculture Organization of the United Nations estimates that almost one billion people practice urban gardening and farming worldwide. In doing so, they are utilizing what could be called underutilized space such as backyards, abandoned lots, and rooftops and by these actions are increasing biodiversity, ecosystem services, and productivity of urban areas while also increasing carbon storage. There's another service provided by urban farming: People in urban areas are often unaware of their food sources. The writer Wendell Berry is famous for saying that "Eating is an agricultural act." Urban gardeners are able to do more than just eat their food—in some case, they are responsible for growing it. If that results in improvements in health, economic and environmental science knowledge, and reductions in food waste, there are many reasons to encourage more urban gardening.

Critical Thinking Questions

1. What are some of the benefits provided by urban agriculture compared to larger industrial agriculture?
2. How would you improve the USDA definition of urban agriculture?

References

Guzmán, M. 2016. Black farmers in Detroit are growing their own food. https://www.pri.org/stories/2016-03-30/black-farmers-detroit-are-growing-their-own-food-theyre-having-trouble-owning.

Keep Growing Detroit. 2021 Annual Report. Viewed on June 24, 2022. https://static1.squarespace.com/static/61ddad815f23d9286ca6ab1c/t/61f8501a9c1aeb0e3b188c39/1643663390602/2021_Annual-Report_1.18.22-compressed.pdf

Science Applied 5: Concept Explanation

How Do We Define Organic Food?

If you've ever spent time roaming through the aisles of a grocery store, you have undoubtedly seen food labeled "organic." You probably have purchased organic food, despite the fact that sometimes it is more expensive than conventionally grown food. Some people prefer organic food because they believe it is healthier for them and tastes better. Others buy it because they think organic food contains fewer pesticides and is safer to eat. Still others believe organic food is produced in a way that is healthier for the environment and safer for farmworkers. Are these beliefs accurate? What exactly does the organic food label mean?

How did the organic food movement begin?

The notion of organic food emerged in the 1940s in response to a time when farmers first started using synthetic pesticides in agriculture. Three decades later, interest in organic food increased as concerns about environmental contaminants became more widespread. Organic food enthusiasts wanted food that was free of chemicals, including pesticides. At the same time, the popularity of food grown locally on small farms was increasing. The push for organic produce in the 1970s was part of a counterculture movement against the farming practices of large-scale commercial agriculture. Organic food was perceived as the food of people rebelling against the mainstream culture.

FIGURE SA5.1 An organic farm. This strawberry farm in Florida is a certified organic producer. *(© Scott Keeler/Tampa Bay Times/ZUMAPRESS.com)*

As we saw in Unit 5, the organic farmer is fundamentally concerned with the health of the soil. They will work hard to ensure that the organic matter content, base saturation, and cation exchange capacity of the soil will increase each year. Actions that will degrade the soil or promote erosion are carefully avoided (**FIGURE SA5.1**).

Today, organic food is much more common and appeals to people from all walks of life. Indeed, organic farming has increased rapidly during the past 2 decades. According to the U.S. Department of Agriculture (USDA), the number of acres of organic farmland increased more than fivefold from 1992 to 2019 (the latest year from which USDA data are available). As you can see in **FIGURE SA5.2**, in 2016 there were about 2 million ha (5 million acres) of organic cropland and pastureland in the United States. The states with the largest area of organic certified land are California, Alaska, Montana, New York, Wisconsin, Texas, Nebraska, and Vermont. Although organic food has been the fastest-growing category of food in the nation—with a 2020 value of around $60 billion—it still represents only 6 percent of all food sales. The most popular products are fruits and vegetables (15 percent of all sales) and milk (8 percent of all sales).

What does it mean to be organic?

The rise in popularity of organic products has brought increased pressure to define what exactly it means to be organic. Without government guidelines, anyone could claim to be selling organic food. In 1990, the USDA began to develop guidelines for organic food. Its goal was to set national organic certification standards that would assure consumers that the food was produced and processed with minimal use of chemicals. Food that met these standards could carry the USDA Organic seal.

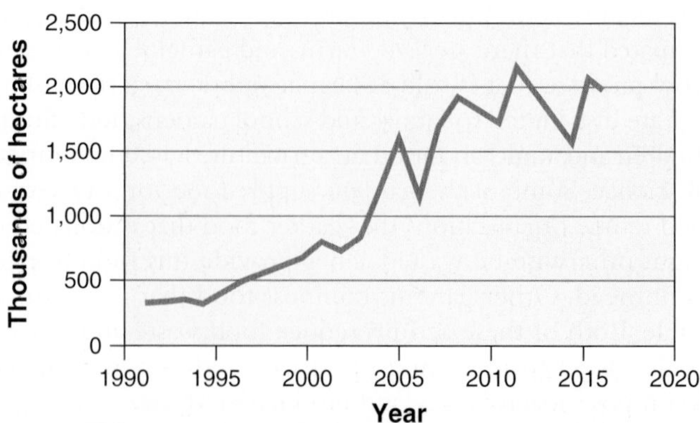

FIGURE SA5.2 Land devoted to organic farming. From 1992 to 2016, the amount of land dedicated to organic farms has increased more than fivefold. *(Data from USDA.)*

In 1995, the USDA determined that "Organic agriculture is an ecological production management system that promotes and enhances biodiversity, biological cycles, and soil biological activity. It is based on minimal use of off-farm inputs and on management practices that restore, maintain, and enhance ecological harmony." What does this statement mean in practice? First, it means that to be considered organic, food must be produced and processed with minimal use of pesticides. It also means that organic meat, eggs, and dairy products come from animals that have not been given antibiotics or growth hormones. Finally, it means that food labeled organic cannot have been fertilized with synthetic fertilizers or sewage sludge. Rather than relying on chemicals that come from outside the farm, organic farmers must use nonchemical methods of pest control and fertilization, as discussed in Module 34. It also means that the food cannot come from genetically modified organisms.

Pest control is perhaps one of the greatest challenges in organic farming. Agricultural pests include weeds that compete with the crops, insects that eat the crops, and fungal pathogens that can kill the crops. Organic farmers are required to try to control pests without synthetic pesticides. Such efforts can include mechanically destroying weeds and insect pests and using plant-derived insecticides or the release of insect enemies to combat the pests. If these tactics fail, organic farmers are permitted to use a synthetic pesticide from a relatively small list of approved chemicals. It came as a shock to consumers, however, when reports arose in 2017 that imported organic fruits and vegetables were being fumigated with pesticides when they entered the United States to ensure that they were not introducing any pests from other regions of the world.

Having standards for organic food is effective only if those standards are enforced. One of the important functions of the USDA's National Organic Program is to conduct inspections of organic food producers and processors to ensure that organic food meets the national standards. Since 1990, the government has allowed inspections to be conducted by independent companies that are hired by producers and processors to inspect organic farms and factories to ensure that the organic standards are being met.

As part of this effort, the certifying companies were required to conduct spot checks. In 2010, the USDA announced that it had failed to enforce the spot check requirement and that, as a result, many of the largest independent certifiers had been conducting spot checks on organic food only if they suspected a problem. When problems were not suspected, producers could operate for years without any spot checks. In fact, some growers had been selling nonorganic food under the USDA Organic seal and obtaining a higher price for their products. In response, the federal government increased the funding and staff of the National Organic Program to ensure that the certifying companies were conducting the required spot checks. Annual inspections are required, and unannounced inspections can occur. Organic farms that do not comply with guidelines can have their certification suspended or revoked.

FIGURE SA5.3 Large-scale organic farming. Although organic farms traditionally have been small-scale family farms, some organic farms, such as this organic dairy farm in California, operate at much larger scales. *(inga spence/Alamy)*

Does organic food mean family farms?

Most organic food production was originally conducted on small family farms. As we saw in Module 34, organic farming methods can require more labor, time, and money than conventional methods, so the price of organic food can be significantly higher. As farmers have learned, however, many consumers are willing to pay a premium for what they believe to be a premium product. This consumer behavior has attracted the attention of large factory farms owned by corporations, which have begun to conduct large-scale organic agriculture.

Although organic farms have a history of being small farms, there is no inherent reason why large farms cannot grow organic food (**FIGURE SA5.3**). An increasing number of corporations have decided to enter the business of organic food production. One of the most significant events to drive this movement occurred in 2006 when Wal-Mart, the largest grocery retailer in the United States, announced that it would begin selling a substantial number of organic foods in its stores at prices only slightly higher than those of nonorganic foods. The expansion into this market by Wal-Mart, Costco, Whole Foods, and Kroger has brought organic food to a much larger number of consumers. As a result, demand for organic food has increased, and more agribusinesses are producing it. Increased production of organic food by larger, more efficient farms has driven down the cost of organic food.

As more factory farms entered the organic food business, there was increasing pressure to permit the use of some nonorganic and synthetic chemicals during the production stage, when the food from the field is turned into a form that can be sold in the grocery store. Proponents of the rule change, including agribusinesses, argued that many processed foods, such as frozen meals, could not be made in an organic version unless the rules about additives were changed. Opponents argued that if processed foods could not be made without additives, then processed foods should not be certified as organic. In 2002, the USDA designated several nonorganic and synthetic additives as permissible in organic processed foods. Permitted nonorganic additives include citric acid, pectin, and cornstarch. Synthetic chemicals permitted in the processing of organic foods include alcohol (used for disinfecting machinery) and aspirin (used for medical treatment of animals).

What are the four different USDA organic designations?

There are four USDA labeling categories for organic food. "100% organic" means the products must be comprised of 100 percent certified organic ingredients. Foods can also be labeled "organic," which means the product and the ingredients it contains must be certified organic except for certain allowed nonorganic additives and synthetic chemicals that were described in the preceding paragraph. In addition, no more than 5 percent of the combined total ingredients may contain nonorganic items. The "made with organic ingredients" designation is for multi-ingredient food items where at least 70 percent of the food item contain certified organic ingredients. Lastly, if a multi-ingredient food item contains less than 70 percent certified organic ingredients, it can be labeled "contains specific organic ingredients."

What does the organic label not mean?

The USDA definition of "organic" does not match what many people think or hope "organic" should mean. For example, many people associate organic food with small family farms, but as we have seen, organic food can be grown on farms of all sizes. The USDA organic seal therefore tells us nothing about the size of the farm that grew the food. Nor does the seal tell us anything about the safety of the food, including whether it contains fewer dangerous food-related pathogens than nonorganic food. Nor does it indicate whether eggs, meat, and dairy products came from animals that were treated humanely or were allowed to roam freely outdoors (known as "free range"). It tells us nothing about where the food was grown, or about labor conditions, such as compensation for the farmworkers who produced it. Furthermore, there are currently no standards for certifying nonagricultural products, such as seafood or cotton, as organic. Although you may see these products labeled "organic," there is currently no agreed-upon definition for such products. Similarly, companies sometimes label their products "natural," a word that seems to imply "organic," but in fact has no standard definition.

In response to public demand, farmers sometimes use food labels that go beyond the federal requirements for organic certification. They may include information about humane treatment of animals, the conditions and wages of the farmworkers, and whether genetically modified organisms were used. Such detailed labels have seen growing popularity.

Although the USDA organic certification has a somewhat narrow definition, it does offer a number of important benefits. Researchers have examined hundreds of studies on organic versus traditional food and concluded that organic food in many cases is healthier, meaning it has a small to moderately higher nutrient content, and a lower pesticide content; organic meat and dairy contain more omega-3 fatty acids (which combat heart disease), and vegetables are higher in antioxidants (which provide a variety of health benefits). In addition, the certification program provides a standardized set of requirements so that informed consumers understand what they are buying. Given the reduction in synthetic pesticide use, the program also allows consumers to support farming practices that are much better for soils and ecosystems and better for farmworkers—who face decreased risks from applying pesticides—than conventional farming practices.

Questions

1. Why do many people think that organic food comes from small farms?
2. Why might large agricultural companies lobby the federal government to permit many more synthetic chemicals to be used in processing organic food?

Practice AP® Free-Response Question

The definition of organic agriculture involves more than the exclusion of pesticide and fertilizer use. Organic agriculture is also a method of enhancing the soil by building up organic matter content and nutrient status. These actions should sustain crop production for many years.

(a) Succession is the ecological process by which certain plants replace other plants in a community through competition. **Explain** how the principle of succession can be used to increase the sustainability of an organic farm. (3 pts.)

(b) **Describe** three methods of reducing soil erosion. (3 pts.)

(c) **Describe** integrated pest management and whether it would be used in organic farming. (2 pts.)

(d) Before a farm can become certified as organic, it must undergo a 3-year transition period. During this period, synthetic chemical fertilizers and pesticides cannot be used on the land. Due to the high cost of farming without the aid of synthetic chemicals, transitioning to organic farming represents a financial risk. **Describe** two ways that this risk can be reduced. (2 pts.)

References

Aubrey, A. 2016. Is organic more nutritious? A new study adds to the evidence. *Northeast Public Radio,* February 18. https://www.npr.org/sections/thesalt/2016/02/18/467136329/is-organic-more-nutritious-new-study-adds-to-the-evidence.

Chang, K. 2012. Organic food vs. conventional food. *New York Times,* September 4. http://well.blogs.nytimes.com/2012/09/04/organic-food-vs-conventional-food/.

Strom, S. 2012. Has organic been oversized? *New York Times,* July 7. http://www.nytimes.com/2012/07/08/business/organic-food-purists-worry-about-big-companies-influence.html.

U.S. Department of Agriculture. *Organic production/organic food: Information Access Tools.* October 2021. http://www.nal.usda.gov/afsic/pubs/ofp/ofp.shtml

UNIT 5 Review

Key Terms to Remember

Tragedy of the commons (p. 296)
Externality (p. 297)
Rangelands (p. 300)
Forest (p. 300)
Clear-cutting (p. 301)
Selective cutting (p. 302)
Ecologically sustainable forestry (p. 302)
Tree plantation (p. 303)
Endangered Species Act (p. 304)
Subsistence farming (p. 306)
Industrial agriculture (Agribusiness) (p. 307)
Green Revolution (p. 307)
Economies of scale (p. 308)
Organic fertilizer (p. 308)
Synthetic fertilizer (Inorganic fertilizer) (p. 308)
Waterlogging (p. 309)
Salinization (p. 309)
Pesticide (p. 309)
Insecticide (p. 310)
Herbicide (p. 310)
Broad-spectrum pesticide (p. 310)
Selective pesticide (Narrow-spectrum pesticide) (p. 310)
Monocropping (p. 310)
Energy subsidy (p. 313)
Plowing (p. 316)
Tilling (p. 316)
Slash-and-burn agriculture (Shifting agriculture) (p. 318)
Aquifer (p. 322)
Unconfined aquifer (p. 322)
Confined aquifer (p. 322)
Water table (p. 322)
Groundwater recharge (p. 323)
Spring (p. 323)
Artesian well (p. 323)
Water footprint (p. 323)
Furrow irrigation (p. 325)
Flood irrigation (p. 325)
Spray irrigation (p. 325)
Drip irrigation (p. 326)
Cone of depression (p. 327)
Fungicide (p. 328)
Rodenticide (p. 328)
Persistent pesticides (p. 328)
Nonpersistent pesticides (p. 328)
Integrated pest management (IPM) (p. 328)
Pesticide resistance (p. 329)
Concentrated animal feeding operation (CAFO) (p. 333)
Manure lagoon (p. 334)
Free range grazing (p. 335)
Nomadic grazing (p. 335)
Overgrazing (p. 336)
Desertification (p. 336)
Fishery (p. 337)
Fishery collapse (p. 337)
Bycatch (p. 338)
Crustal abundance (p. 342)
Ore (p. 342)
Metal (p. 342)
Reserve (p. 344)
Strip mining (p. 344)
Mine tailings (p. 344)
Open-pit mining (p. 344)
Mountaintop removal (p. 345)
Placer mining (p. 345)
Subsurface mining (p. 345)
Urbanization (p. 349)
Urban area (p. 349)
Suburbs (p. 350)
Exurbs (p. 350)
Saltwater intrusion (p. 351)
Impervious surface (p. 351)
Urban sprawl (p. 351)
Urban blight (p. 352)
Sense of place (p. 352)
Urban runoff (p. 353)
Ecological footprint (p. 356)
Carbon footprint (p. 358)
Sustainability (p. 361)
Sustainable development (p. 362)
Maximum sustainable yield (MSY) (p. 363)
Environmental indicators (p. 364)
Anthropogenic (p. 366)
Crop rotation (p. 370)
Intercropping (p. 370)
Biocontrol (p. 371)
Natural predators (p. 371)
Sustainable agriculture (p. 372)
Soil conservation (p. 372)
Agroforestry (p. 372)
Windbreaks (p. 373)
Strip cropping (p. 373)
Contour plowing (p. 373)
Terracing (p. 373)
Perennial plants (p. 374)
No-till agriculture (p. 374)
Green manure (p. 374)
Limestone (p. 375)
Rotational grazing (p. 376)
Organic agriculture (p. 376)
Delaney Clause (p. 376)
Reforestation (p. 381)
Sustainable forestry (p. 381)
Prescribed burn (p. 383)

Section 1: Multiple-Choice Questions

1. Which has contributed to urban sprawl over the past 50 years?
   (a) migration of people from suburban areas to rural areas
   (b) increased availability of public transportation
   (c) lower property taxes in urban areas
   (d) increased construction of highways

2. Which of the following would be an example of tragedy of the commons?
   (a) building roads for urban areas
   (b) farmers who siphon water from a county aqueduct for irrigation
   (c) slash-and-burn agriculture eroding nutrients from farms
   (d) a ranch that charges farmers to board their horses

3. Suppose it takes 12 kg of grain to produce 1 kg of lamb and that a person eating only corn can obtain sufficient calories in a year from 0.02 ha of land. How much land would be required to support a person who only eats lamb?
   (a) 0.12 ha
   (b) 0.24 ha
   (c) 12.0 ha
   (d) 24 ha

4. Sources of nitrogen pollution to the Chesapeake Bay were determined and results were plotted in the following pie diagram. If 300 million pounds of nitrogen entered the bay during the time period represented by the diagram, how much nitrogen came from agricultural activities during that time period?

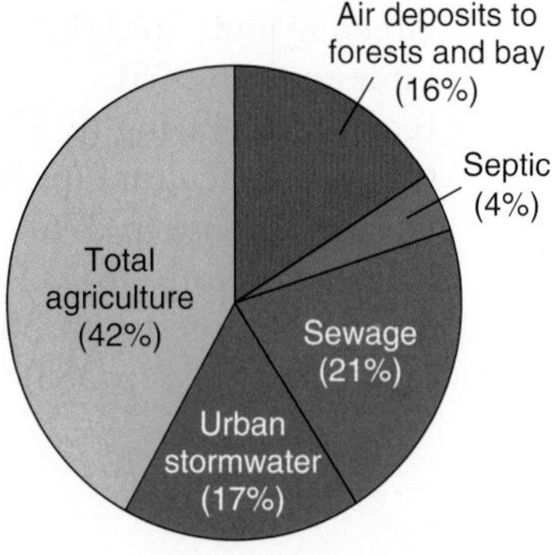

Nitrogen

   (a) 51 million pounds
   (b) 126 million pounds
   (c) 147 million pounds
   (d) 172 million pounds

5. Nitrogen and phosphorus are two elements seen in the waters of the Chesapeake Bay. What other substance might you expect to be found in higher quantities in the water with the increased use of CAFOs?
   (a) lead
   (b) growth hormones
   (c) methane
   (d) N_2O

6. Manure from feedlots (CAFOs) represent a large chunk of nitrogen/phosphorus as seen in the graphs. What is a viable method to reduce manure from CAFOs?
   (a) Integrated Pest Management
   (b) reduce antibiotic use
   (c) incorporate free-range grazing
   (d) reduce slash-and-burn farming

7. No-till agriculture and contour plowing are methods that typically result in a decrease in
   (a) pesticide use.
   (b) energy subsidies.
   (c) loss of topsoil.
   (d) pest outbreaks.

8. Which parameter is likely to contribute more to the ecological footprint of free-range cattle compared to cattle raised in a CAFO?
   (a) amount of land needed
   (b) amount of waste produced
   (c) amount of water required for the animals and to grow the feed
   (d) amount of antibiotics needed

9. Integrated pest management (IPM) is a combination of methods used to effectively control pest species while minimizing the disruption to the environment. Place the steps below in the proper order to follow IPM procedures.
   A. targeted pheromones
   B. physical trapping/removal of pests
   C. select pest resistant varieties
   D. spraying of pesticides

   (a) C, B, A, D
   (b) D, B, C, A
   (c) A, C, D, B
   (d) B, C, A, D

10. Which of the following strategies likely increased productivity of farmland around the world?
   (a) mechanization of farming
   (b) prescribed burns
   (c) increasing fines for illegally clearing land
   (d) intercropping

11. The table below shows water use for a small business in thousands of gallons for the first four months of 2016 and 2017. Which of the following methods would you use in order to determine the percent change in water use from the first four months of 2016 to the first four months of 2017?

Water use monthly (thousands of gallons)	2016	2017
January	38	26
February	15	32
March	22	17
April	24	14
Total	99	89

(a) $((89 - 99)/99) \times 100$
(b) $((26 - 38)/38) \times 100$
(c) $((99 - 89)/89) \times 100$
(d) $((32 - 15)/15) \times 100$

12. Soils filter water as it percolates down into the groundwater. Which method could be used by farmers to ensure soils are protected so they can effectively filter water?
(a) use of contour farming
(b) increase animal grazing
(c) use of monocropping each year
(d) use more pesticides

13. The diagram below shows the potential impacts of open-ocean aquaculture. Which of the outputs in the diagram would contribute to an increase in invasive species?

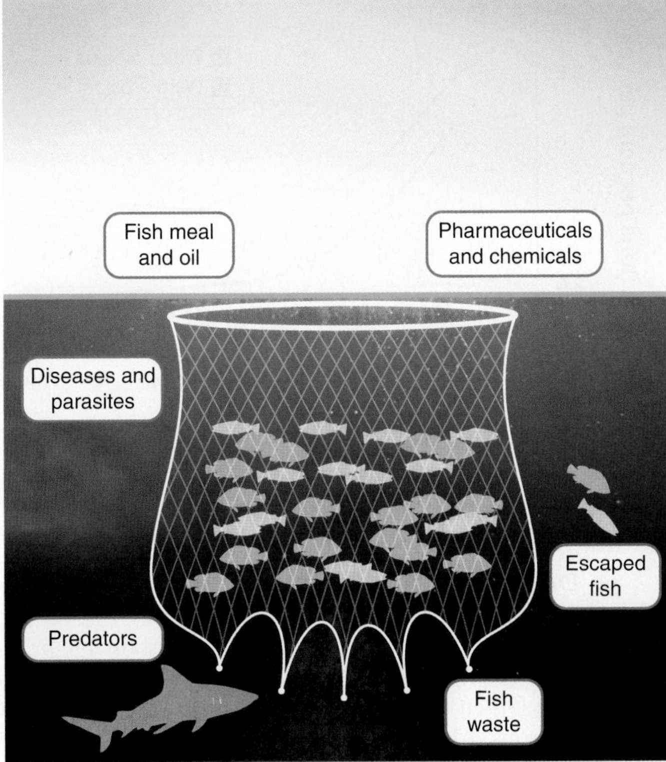

(a) escaped fish
(b) fish waste
(c) fish meal
(d) drugs and chemicals

14. Which method of forestry from the images below would lead to hypoxic/anoxic conditions in a river or stream running directly through the center of the diagrams?

(a)

(b)

(c)

(d)

(a) A
(b) B
(c) C
(d) D

15. Which of the following is a common benefit attributed to Integrated Pest Management?
(a) human health improvements
(b) water supply use increased
(c) native wildlife exposed to more pesticides
(d) expenses decrease

Use the graph of Canada's rural population below to answer questions 16–18:

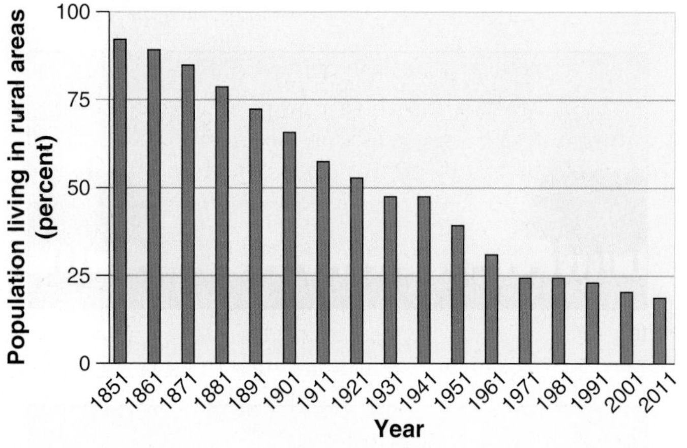

16. Which of the following would best describe the cause behind the trend in the graph?
    (a) A decrease in industrial and urban jobs led to a decrease in people living in rural areas.
    (b) An increase in agricultural and rural jobs led to an increase in people living in rural areas.
    (c) A decrease in industrial and urban jobs led to a decrease in people living in rural areas.
    (d) An increase in industrial and urban jobs led to a decrease in people living in rural areas.

17. Which of the following statements would best describe the small decrease in the percentage of the population living in rural areas from 1971 to 2011?
    (a) Populations living in rural areas experienced a population boom.
    (b) Populations living in urban areas experienced a population bust.
    (c) Factors causing the previous trend reversed and rural areas became popular again.
    (d) Factors causing the previous trend continued but became more gradual.

18. If the population of Canada was 24.8 million in 1981, what percent of the population was living in urban areas?
    (a) 6.2 million
    (b) 18.6 million
    (c) 25.1 million
    (d) 12.4 million

Section 2: Free-Response Questions

1. Industrial agriculture has contributed to an increase in food production, but it has also caused environmental problems.
    (a) **Describe** one effect that industrial agriculture has had on soils. (1 pt.)

Scientists devised a study to determine the effects of tilling on soils and plants. In the lab they planted seeds and monitored oxygen and nutrients in the soil, as well as plant growth. The same seeds were used in both treatments but one treatment was tilled and aerated frequently, while the other was left alone. They were watered the same amount daily.

(b) (i) **Identify** the likely scientific question being investigated. (1 pt.)
  (ii) **Identify** the dependent variable in the experiment. (1 pt.)
  (iii) **Explain** how the results might be changed if a high efficiency method of irrigation were used. (1 pt.)

In 2017, scientists gathered data on rates of water and wind erosion on cropland. The following data were found (use the graph to the right to answer parts c–e):

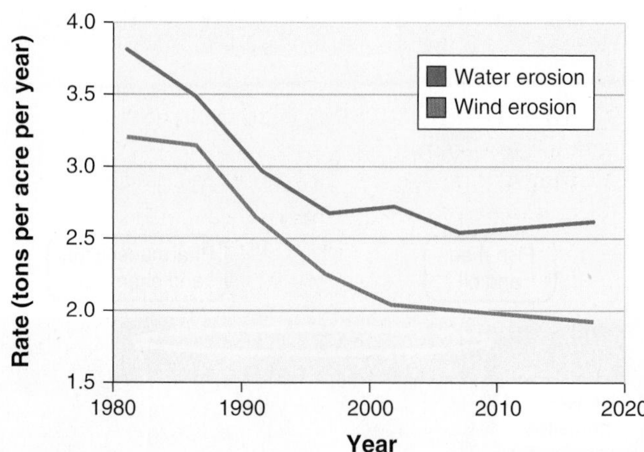

(c) **Describe** the trend in erosion rates from 1982 to 2017. (1 pt.)
(d) **Identify** the year with the highest rate of water erosion. (1 pt.)
(e) **Describe** how sustainable practices have led to the trends in the graph above. (1 pt.)
(f) **Identify** one disadvantage to using sustainable practices. (1 pt.)
(g) The Green Revolution led to increases in the output of crops. **Identify** two methods of the Green Revolution that aided in the increase. (1 pt.)
(h) **Describe** one aspect of integrated pest management that can also contribute to the sustainable growth of crops. (1 pt.)

2. Maintaining cattle in a Concentrated Animal Feeding Operation (CAFO) requires large quantities of water and produces vast quantities of wastewater and manure.
   (a) **Describe** how CAFOs can lead to the contamination of water. (1 pt.)
   (b) **Describe** one economic benefit to continuing the operation of CAFOs. (1 pt.)
   (c) **Identify** one alternative method to raising cattle that is more environmentally sustainable than a CAFO. (1 pt.)
   (d) **Describe** one negative aspect of the alternative method mentioned in part (c). (1 pt.)

As countries become more developed, meat consumption tends to rise. Use the graph below to answer parts e–g:

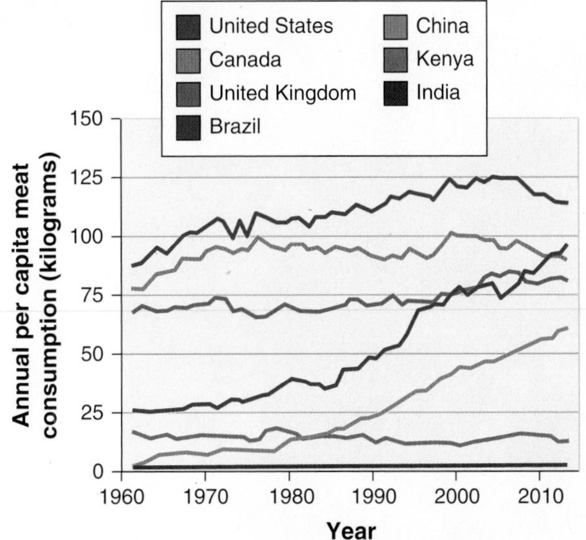

   (e) **Identify** the country with the slowest rate of increasing meat consumption. (1 pt.)
   (f) **Identify** the maximum average annual consumption that the United States has reached. (1 pt.)
   (g) **Explain** one social or economic development that has led to the trend in Brazil. (1 pt.)
   (h) With so many countries moving toward more meat consumption there are environmental issues that will arise.
      (i) **Identify** one environmental consequence, other than water contamination, that arises from consumption of meat. (1 pt.)
      (ii) **Make a claim** that proposes a long-term solution to the increase in meat consumption. (1 pt.)
      (iii) **Justify the claim** made in part (ii) with evidence that the strategy mentioned can work. (1 pt.)

3. Currents occur in the world's oceans. As the Earth rotates it forms circular patterns in the oceans called gyres. These are driven by the change in density of water due to differences in salt concentration and temperature. Plastic pollution has gathered in these gyres called, "garbage patches," the most well-known being the Great Pacific Garbage Patch.
   (a) **Identify** the socioeconomic concept that has led to the accumulation of plastic pollution in these gyres. (1 pt.)
   (b) **Explain** why the garbage patches continue to accumulate with no one cleaning them up. (1 pt.)
   (c) **Describe** another ocean resource that has also degraded under the concept from part (a). (1 pt.)
   (d) **Explain** one possible solution to the problems cause by plastic pollution in the ocean. (1 pt.)
   (e) In 2000, Atlantic bluefin tuna numbers were estimated to be around 310,000 breeding pairs, while in 1980, those numbers were estimated to be 615,000 breeding pairs.
      (i) **Calculate** based on the number of breeding pairs, how many individuals there were in 1980. Show all work. (2 pts.)
      (ii) **Calculate** the percent change in the bluefin tuna population from 1980 to 2000. Show all work. (2 pts.)
      (iii) **Calculate** the number of breeding pairs in 2020, if the percentage decrease from 1980 to 2000 is the same over the next 20-year period. Show all work. (2 pts.)

UNIT 6

Energy Resources and Consumption

In 2010, an explosion at the British Petroleum Deepwater Horizon oil extraction platform in the Gulf of Mexico killed 11 workers on the drilling platform, injured 17 others, and released more than 206 million gallons of oil into the ocean. *(U.S. Coast Guard/Getty Images)*

MODULE 35 Renewable and Nonrenewable Resources and Global Energy Consumption

MODULE 36 Fuel Types and Uses

MODULE 37 Distribution of Natural Energy Resources: Fossil Fuels

MODULE 38 Nuclear Power

MODULE 39 Biomass and Solar Energies and Hydroelectric Power

MODULE 40 Geothermal Energy and Hydrogen Fuel Cells

MODULE 41 Wind Energy and Energy Conservation

CASE STUDY

All Energy Use Has Consequences

The modern world is dependent on fossil and nuclear fuels for energy. Many aspects of our modern society — health care, comfortable living conditions, easy travel, abundant food — rely on readily accessible and relatively affordable fossil and nuclear fuels. But the costs to society are high.

Fossil fuels are named appropriately: They are fuels containing fossilized remains of carbon-based animals and plants. This means that the carbon released into the atmosphere when fossil fuels are burned is carbon that had been buried deep underground for millions of years. Therefore, fossil fuel combustion contributes additional fossil carbon into the atmosphere, contributing to climate change. Furthermore, fossil fuel combustion contributes other air pollutants that produce haze and smog in the atmosphere that harm human beings when inhaled and alters terrestrial and aquatic ecosystems.

In addition to the problems that occur while fossil fuel is used, extraction can cause significant ecosystem and property damage and even human injury and death. In October 2021, a pipeline running from an offshore oil platform to the California coast in Orange County ruptured, spilling 473,000 liters (125,000 gallons) of oil. Beaches were closed and numerous fish and birds were killed after being coated with oil. No humans were killed in this incident. However, a decade earlier, the worst oil spill in recent history occurred on April 20, 2010, when an explosion and fire occurred at the British Petroleum Deepwater Horizon oil extraction platform in the Gulf of Mexico. Oil gushed from the platform until it was capped 87 days later, on July 15. The accident killed 11 workers on the drilling platform, injured 17 others, and released more than 780 million liters (206 million gallons) of oil into the Gulf. Scientists from the Center for Biological Diversity later estimated that at least 6,000 sea turtles, 26,000 marine mammals, and more than 82,000 birds were killed as a result of the spill. The oil spread through the Gulf of Mexico and washed up on the shores of Louisiana, Mississippi, Alabama, and Florida. In 2018, scientists found evidence that oil released during the Deepwater Horizon accident was still becoming incorporated into the body fat of fish in the Gulf of Mexico.

The modern world is dependent on fossil and nuclear fuels for energy, but the costs to society are high.

Before the Deepwater Horizon accident, the largest spill in U.S. waters had been the March 1989 Exxon Valdez accident, when the Valdez, a supertanker carrying 200 million liters (53 million gallons) of oil, crashed into a reef in Prince William Sound, Alaska. Roughly 42 million liters (11 million gallons) of oil spilled into the sound, much of which washed up on shore, contaminating the coastline and killing perhaps half a million birds and thousands of marine mammals. The number of dead animals was so much greater in the Exxon Valdez accident compared to the Deepwater Horizon because the spill occurred in a relatively enclosed sound rather than in open waters.

Oil spills remain an ongoing environmental problem in the modern world. Some spills are caused by leaks or explosions at wells, while others, like the 2021 California spill, occur when oil is being transported by pipeline or tanker. Accidents can happen after oil is extracted and transported to a refinery. In 2005, 15 workers died in an explosion at a British Petroleum oil refinery in Texas.

Fossil fuels other than oil also pose similar risks. In the early 1900s, there were hundreds of accidental coal-mining deaths each year in the United States. While safety measures have improved since then, mining is still a dangerous business. In April 2010, the worst U.S. coal mine explosion in 40 years killed 29 miners in West Virginia. Coal mining accidents occur in many other countries as well. Fatal coal mine accidents occur almost yearly in China, with dozens of coal miners dead in 2021. In Siberia, a coal mine explosion killed over 50 people in November 2021. It was the worst accident in over a decade. But the major news stories of mine explosions aren't the only risks to miners. In fact, one of the leading causes of mining deaths occurs long after workers leave the mines. Tens of thousands of coal miners per decade have developed black lung disease and other respiratory ailments that lead to disability or death.

Natural gas is often considered to be "clean" because its combustion produces lower amounts of particulates, sulfur dioxide, and carbon dioxide than does oil or coal. However, combustion still results in emission of carbon dioxide, which is the major greenhouse gas produced by human activity. The production of natural gas also has negative consequences. "Thumper trucks," which generate seismic vibrations to identify natural gas deposits underground, can disturb soil and alter groundwater flow, causing some areas to flood and some wells to go dry. Drilling for natural gas can also contaminate drinking water. Construction of natural gas pipelines is disruptive to the environment and is often opposed by those who live in the affected communities. And a growing number of studies have demonstrated that an unknown but possibly substantial percentage of natural gas is lost to the atmosphere during extraction and transport.

Nuclear energy has contributed to a few energy catastrophes as well. A 2011 earthquake off the coast of Japan resulted in a tsunami that damaged the nuclear reactors at the Fukushima nuclear power plant, leading to the release of radioactive gases. In 1986, a fire and release of radioactivity at the Chernobyl nuclear power plant in Ukraine resulted in 31 deaths during the event, mostly due to acute radiation exposure. At least 4,000 deaths are estimated to have occurred in the following decades due to longer-term effects of radiation exposure, mostly radiation-induced cancer. Due to the substantial release of radioactive material, 116,000 people were evacuated from an area encompassing a 30 km (19 mile) radius from the plant. A larger area, and more people, were evacuated later and this area remains uninhabited by humans today more than 35 years after the accident. Even when operating properly, a nuclear power plant generates radioactive nuclear waste. Storage of this waste is currently an unresolved environmental challenge.

As you can see, there are consequences to all energy choices; every choice is problematic in one way or another. However, as we will see in the later modules of this unit, there are renewable energy choices that have many advantages and fewer disadvantages. The study of environmental science allows us to identify and articulate the advantages and disadvantages of each energy choice made by an individual or society and chart the path for a more environmentally sustainable energy future.

Sources: L. Margonelli, *Oil on the Brain: Petroleum's Long Strange Trip to Your Tank*, Broadway Books, 2008; B. Sovacool, J. Kim, and M. Yang. The hidden costs of energy and mobility: A global meta-analysis and research synthesis of electricity and transport externalities. *Energy Research & Social Science* 72 (2021), 101885.

Practice your science skills

1. **Environmental Solutions:** Describe an environmental problem associated with the combustion of fossil fuels.
2. **Text Analysis:** Evaluate the validity of the conclusion that there are consequences to all energy choices.
3. **Environmental Solutions:** Other than offering the suggestion "switch to renewables," make a claim that would lessen one environmental problem caused by the combustion of fossil fuels.

We use energy in all aspects of our daily lives: heating and cooling, cooking, lighting, communications, and travel. In these activities, humans convert renewable and nonrenewable energy resources such as the wind and the Sun, and natural gas and oil into useful forms of energy such as electricity, motion, and heat with varying degrees of efficiency and environmental impacts. Each energy choice we make has both positive and negative consequences. In a society like the United States, where each person averages 10,000 watts of energy use continuously—24 hours per day, 365 days per year—there are a lot of consequences to understand and evaluate. In this unit, we will examine the supplies and availability of fossil fuels, renewable resources and nuclear energy that we currently use, what we use them for and where in the world different energy sources are used. We will also examine the ways that conservation and efficiency can reduce our energy usage and reduce environmental impact and how our energy usage will likely change over the coming decades.

Module 35

Renewable and Nonrenewable Resources and Global Energy Consumption

In this module we will examine the renewable and nonrenewable energy sources that we use and describe patterns of energy use worldwide and in the United States. We will also introduce the concept of energy efficiency and conservation.

Learning Goals

After reading this module you should be able to

35-1 describe nonrenewable energy resource characteristics.

35-2 describe renewable energy resource characteristics.

35-3 describe trends of energy use worldwide and in the United States.

35-4 explain the importance of energy efficiency and conservation.

35-1 What are the characteristics of nonrenewable energy resources?

Nonrenewable energy resources are finite and important

Fossil fuels are fuels derived from biological material that became fossilized millions of years ago. Fuels from this source provide most of the energy used in both developed and developing countries. Most of the fossil fuels we use — coal, oil, and natural gas — come from deposits of organic matter that were formed 50 million to 350 million years ago. As we saw in Figure 4.1, when organisms die, decomposers break down most of the dead biomass aerobically, and it quickly reenters the food web. However, in an anaerobic environment — for example in places such as swamps, river deltas, and the ocean floor — a large amount of detritus may build up quickly. Under these conditions, decomposers cannot break down all of the detritus. As this material is buried under succeeding layers of sediment and exposed to heat and pressure, the organic compounds within it are chemically transformed into high-energy solid, liquid, and gaseous components that are easily combusted. When they are combusted, they release fossil carbon that has been stored for millions of years into the atmosphere. For this reason, fossil fuels contribute *additional* carbon dioxide to the atmosphere and contribute to global climate change.

Because a fossil fuel is an energy source with a finite supply, it is known as a **nonrenewable energy resource**. Nuclear fuel, derived from radioactive elements such as uranium ore that give off energy, is also an energy source with a finite supply, and another major source of nonrenewable energy on which we depend.

When we discuss quantities of energy used, it is helpful to use specific measures. The basic unit of energy is the joule (J), which is 1 W-s (Watt-second). One gigajoule (GJ) is 1 billion (1×10^9) joules, or about as much energy as is contained in 30 L (8 gallons) of gasoline. One exajoule (EJ) is 1 billion (1×10^9) gigajoules. In some of the figures in this unit, we also present the quad, a unit of energy used by the U.S. government to report energy consumption. The quad is 1 quadrillion, or 1×10^{15}, British thermal units, or Btu. One quad is equal to 1.055 EJ. We will also use the kilowatt-hour, a measurement for quantities of electrical energy. Additional energy conversions are shown in TABLE 35.1. See "Do the Math: Comparing Energy Content in Fuels" on page 402 to gain familiarity with some of these energy units.

Fossil fuels Fuels derived from biological material that became fossilized millions of years ago.

Nonrenewable energy resource An energy source with a finite supply, primarily fossil fuels and nuclear fuels.

TABLE 35.1 Common units of energy and their conversion into joules			
Unit	**Definition**	**Relationship to joules**	**Common uses**
calorie	Amount of energy it takes to heat 1 gram of water 1°C	1 calorie = 4.184 J	Energy expenditure and transfer in ecosystems; human food consumption
Calorie (with a capital "C")	Calories in food	1 Calorie = 1,000 calories = 1 kilocalorie (kcal) = 4,184 J	Food labels; human food consumption
British thermal unit (Btu)	Amount of energy it takes to heat 1 pound of water 1°F	1 Btu = 1,055 J	Energy transfer in air conditioners and home water heaters
Kilowatt-hour (kWh)	Amount of energy expended by using 1 kilowatt of electricity for 1 hour	1 kWh = 3,600,000 J = 3.6 megajoules (MJ)	Energy use by electrical appliances; often given in kWh per year

DO THE MATH Comparing Energy Content in Fuels

Preparing for the AP® Exam

Which contains more energy: 400 gallons of gasoline or 8,000 kilowatt hours of electricity?

To compare different quantities of energy in different units, in this case gallons and kWh, we convert both to joules.

400 gallons of gasoline

1 gallon gasoline = 132 megajoules energy content

400 ~~gallons~~ × 132 megajoules/~~gallon~~ = 52,800 megajoules of energy content in the gasoline

1 kilowatt-hour = 3.6 megajoules energy content

8,000 ~~kWh~~/year × 3.6 megajoules/~~kWh~~ = 28,800 megajoules in the electricity

Answer: The energy content of the gasoline is greater than the energy content of the electricity.

YOUR TURN Which is greater: the energy contained in 4 metric tons of coal or the energy contained in 1,000 liters of diesel oil?

Note that 1 metric ton of coal = 29,300 megajoules energy equivalent and 1 liter of diesel fuel = 36 megajoules

AP® Exam Tip

Remember that a prefix in front of a unit of measure indicates a multiple or fraction of that unit. For example, the prefix *kilo* (k) means 1,000, or 10^3, so a kilogram (kg) is equal to 1,000 g. The prefix *mega* (M) means 1,000,000 or 10^6, and the prefix *giga* (G) means one billion or 10^9. You will encounter many prefixes for units of energy in this chapter and on the AP® Environmental Science Exam. Note that you are expected to know prefixes such as kilo, mega, and giga, but you are not expected to know the energy content of fuels such as a gallon of gasoline or a kWh. Those will always be given to you on an exam.

There are two aspects of nonrenewable energy sources that make them so important in the world today. Fossil fuels contain a relatively large amount of energy in a relatively small volume and mass, although this varies depending on the fuel. And they can be converted to useful heat energy at a relatively rapid rate, although this also varies depending on the fuel. For these two reasons, nonrenewable energy resources play a vital role in almost all aspects of society in much of the world and finding replacement fuels has posed many challenges. The best candidates for replacing nonrenewable fossil fuels are renewable energy resources such as solar and wind.

35-2 What are the characteristics of renewable energy resources?

Renewable energy resources are infinite and becoming more important

In the preceding section, we learned that conventional energy resources, such as petroleum, natural gas, coal, and uranium ore, are nonrenewable. From a systems analysis perspective, fossil fuels constitute an energy reservoir that we will deplete in a few hundred years. It takes tens of millions to hundreds of millions of years for new fossil fuels to form, so we are depleting them much faster than they can be replenished. Therefore, we call them finite and nonrenewable because in the timespan of human civilization, they cannot be replenished. We also have a finite amount of uranium ore available to use as fuel in nuclear reactors. There is no additional uranium being formed so when we use it up, there will be no more.

In contrast, certain other sources of energy that are infinite are called **renewable energy resources**. There are two categories of renewable resources. Biomass energy resources are **potentially renewable** because those resources can be regenerated indefinitely as long as we do not consume them more quickly than they can be replenished. For example, if all the trees in a forest are removed, then there would be no more renewable energy available in that forest at that time. Similarly, if the water behind a hydroelectric dam dries up, then there would not be any hydropower energy available, at least until additional rainfall occurs. There are still other energy resources that cannot be depleted no matter how much we use them. Solar, wind, geothermal, and tidal energy are essentially **nondepletable** in the span of human time; no matter how much we use, they cannot be used up. The amount of a nondepletable resource available tomorrow does not depend on how much we use today. In this book we refer to potentially renewable and nondepletable energy resources together as renewable energy resources. **FIGURE 35.1** illustrates the categories of renewable and nonrenewable energy resources.

Many renewable energy resources have been used by humans for thousands of years. In fact, before humans began using fossil fuels, the only available energy sources were wood, plants, animal manure, and fish or animal oils. Today, in parts of the developing world where there is limited access to fossil fuels, people still rely on local biomass energy sources such as manure and wood for cooking and heating—sometimes to such an extent that they overuse the resource. For example, according to the International Energy Agency, biomass is currently the source of 65 percent of the energy consumed in sub-Saharan Africa (excluding South Africa) and much of it is not harvested sustainably. This means that, over time, it is being depleted.

FIGURE 35.1 Renewable and nonrenewable energy resources. Fossil fuels and nuclear fuels are nonrenewable energy resources. Renewable energy resources include potentially renewable energy sources such as biomass, which is renewable as long as humans do not use it faster than it can be replenished, and nondepletable energy sources, such as solar radiation and wind. Globally, hydroelectric is nondepletable but it can be depleted in localized areas when water is not available, such as during a drought. *(Ian Hamilton/iStockphoto.com; babyblueut/Getty Images; MichaelUtech/Getty Images; RelaxFoto.de/istockphoto; BerndLang/istockphoto; Inga Spence/Science Source; acilo/istockphoto; Blend Images - Don Mason/Getty Images; Kyodo News/Getty Images; Rhoberazzi/Getty Images)*

35-3 What are the trends of energy use worldwide and in the United States?

Trends in energy use are changing around the world and in the United States

Energy resources are not evenly distributed around the world and neither is energy use. This section will examine patterns and trends in energy usage globally and in the United States.

Renewable energy resources Sources of energy that are infinite.

Potentially renewable An energy source that can be regenerated indefinitely as long as it is not overharvested.

Nondepletable An energy source that cannot be used up.

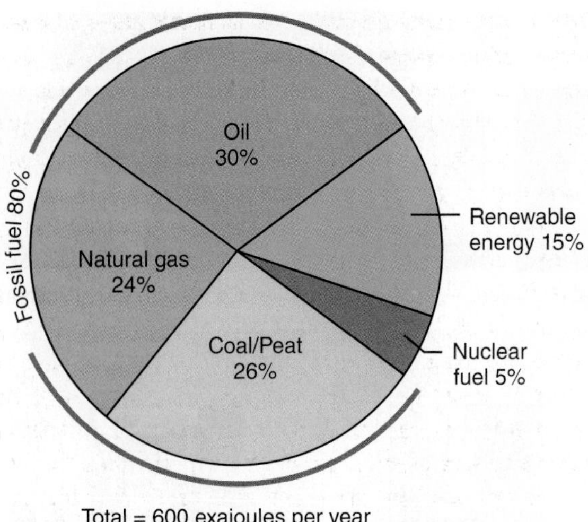

FIGURE 35.2 Worldwide annual energy consumption, by resource, in 2020. Oil, coal, and natural gas are the major sources of energy for the world. *(Data from U.S. DOE Energy Information Administration, 2021)*

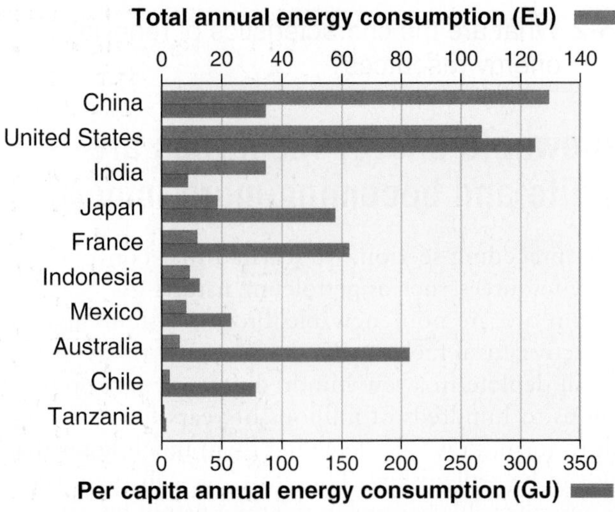

FIGURE 35.3 Global variation in total annual energy consumption and per capita energy consumption. The 10 countries shown are among the largest and the smallest energy users in the world. *(Data from U.S. DOE Energy Information Administration)*

Worldwide Patterns of Energy Use

As **FIGURE 35.2** shows, in 2020 total world energy consumption was approximately 600 EJ per year. This number amounts to roughly 77 GJ per person per year. Oil, coal, and natural gas were the three largest energy sources and comprise 80 percent of total energy use. Renewable energy comprises 15 percent of global energy use. Hydroelectricity, such as electricity that can be generated from water flowing over a waterfall, is the largest source of renewable energy. Wind and solar are the second and third largest sources and are increasing rapidly each year. We will discuss renewable energy in greater detail in Modules 39 to 41.

Energy use is not evenly distributed throughout the world. **FIGURE 35.3** shows that energy consumption in the United States was 320 GJ per person per year, almost five times greater than the world average. In fact, although only 20 percent of the world's population lives in developed countries, people in those countries use more than 40 percent of the world's energy each year. Note that of the countries shown in Figure 35.3, China has the greatest total energy consumption, whereas the United States has the greatest per capita energy consumption. At 0.17 EJ per year, Tanzania has the lowest annual energy consumption of the countries shown; annual per capita energy consumption in Tanzania is approximately 3 GJ per person per year.

There are a variety of reasons for the patterns we see in Figure 35.3. In developed countries and in the urban areas of some developing countries, individuals are likely to use fossil fuels such as coal, oil, and natural gas—either directly, or indirectly through the use of electricity that is generated by burning those fuels. However, people living in rural areas of developing countries still utilize fuels such as wood, charcoal, or animal waste. These differences lead us to distinguish between commercial and subsistence energy sources. **Commercial energy sources** are those that are bought and sold, such as coal, oil, and natural gas, although sometimes wood, charcoal, and animal waste are also sold commercially. **Subsistence energy sources** are those gathered by individuals for their own immediate needs and include straw, sticks, and animal dung. There is much greater use of subsistence energy sources in the developing world, especially in rural areas.

Changes in energy demand generally reflect the level of industrialization in a country or region. As energy demand increases, societies change the types of fuels they use. Today, we see the same patterns of changing energy use in developing countries that have been observed historically in the United States. For example, as more people own automobiles, demand for gasoline and diesel fuel increases, and as more people own electric vehicles, demand for electricity increases. As industries develop and factories are built, and as households acquire more refrigerators and washing machines, demand for electricity also increases. Although worldwide energy use varies considerably, the United States is a particularly large energy consumer.

Commercial energy sources Energy sources that are bought and sold, such as coal, oil, and natural gas.

Subsistence energy sources Energy sources gathered by individuals for their own immediate needs including straw, sticks, and animal dung.

Patterns of Energy Use in the United States

Wood was the predominant energy source in the United States until about 1875, when coal came into wider use. Starting in the early 1900s, oil and natural gas joined coal as the primary sources of energy. **FIGURE 35.4** shows the history of energy use in the United States from 1950 through the present. By 1960, electricity generated by nuclear energy became part of the mix, and hydroelectricity became more prominent. In the 1970s, there was a decline of oil and a resurgence of coal. These changes were the result of political, economic, and environmental factors that will continue to shape energy use into the future. Today, the three resources that supply the majority of the energy used in the United States—in order of importance—are oil, natural gas, and coal. The recent increase in the consumption of natural gas and reduction in coal consumption shown in Figure 35.4 is a result of the increase in availability of natural gas, largely because of hydraulic fracturing (fracking), which has reduced the cost. The recent rise in natural gas consumption has resulted in oil and natural gas use becoming almost equal. And the decrease in coal usage has resulted in the sum of all renewable energy sources surpassing coal. Also note the sharp decrease in total energy usage in 2020 due to the pandemic, which, as we noted in Module 0, resulted in less fuel consumption due to less travel and commuting to work.

Energy use in the United States today is the result of the inputs and outputs of an enormous system. The boundaries of the system are political and technological as well as physical. For example, oil inputs enter the U.S. energy system from both domestic production and imports from other countries. Hydroelectric energy comes from water that flows within the physical boundaries of the country, as well as from neighboring Canada, but it is not an energy input until we move it into a technological system, such as a hydroelectric dam. One major output from the system is work—the end use of the energy, such as in transportation, residential, commercial, and industrial. The other major output is waste: heat, CO_2, and other pollutants that are released as energy is converted and entropy increases.

As you can see in **FIGURE 35.5**, U.S. energy consumption is around 100 EJ (1.0×10^{18} J) per year (95 quads per year). The energy mix of U.S. consumption is 79 percent fossil fuel, 9 percent nuclear fuel, and 12 percent renewable energy resources. Also note that all renewable energies combined contribute more to the total than does coal (10 percent). This is a new phenomenon. Only 5 years ago, coal was 15 percent of the energy mix in the United States and renewables were 10 percent. The United States produces 85 percent of the energy it needs, while the remainder—roughly 15 percent—comes from other countries, primarily in the form of petroleum imports. The percentage of energy imports has been decreasing as domestic production of natural gas and oil increases; these trends are expected to continue. Industry uses the most energy, followed closely by the transportation sector.

Energy use varies regionally and seasonally. In the midwestern and southeastern states, coal is the primary fuel burned for electricity generation. The western and northeastern states generate electricity using a mix of nuclear fuels, natural gas, and hydroelectric dams. Highly populated areas tend to use less coal, which creates more air pollution than any other fuel. Regional use of fuel varies according to the climate and the season. Northern areas consume more oil and natural gas during the winter months to meet the demand for heating while southern areas consume more electricity in the summer months to meet the demand for air conditioning.

Expanded domestic production of natural gas and oil has led to a rapidly changing energy portfolio in the United States and elsewhere. However, the type of energy used for a particular application is a function of many factors, including its characteristics. Factors to be considered include the ease with which the fuel can be transported, and the amount of energy a given mass of the fuel contains.

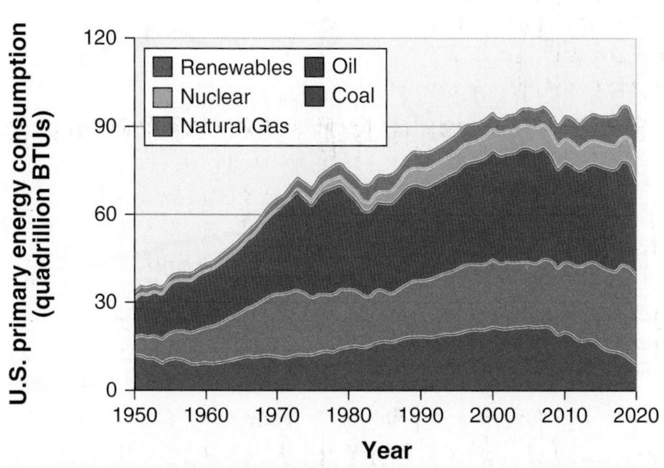

FIGURE 35.4 United States energy consumption by fuel source in quadrillion Btu (quads) from 1950 through 2020. In recent years, the importance of coal has decreased and the importance of natural gas has increased. *(Data from U.S. DOE Energy Information Administration, https://www.eia.gov/energyexplained/us-energy-facts/)*

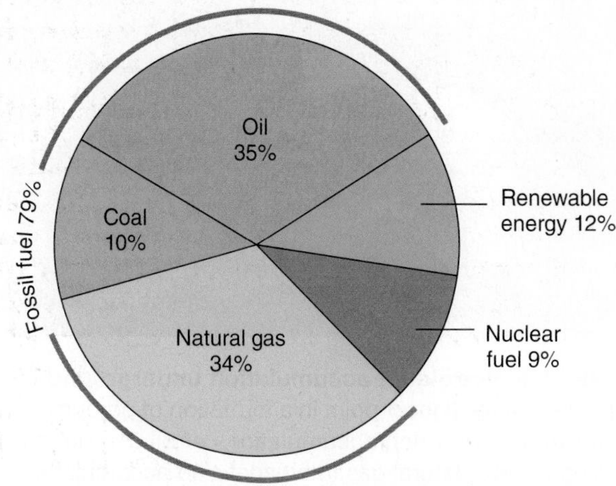

Total = 100 exajoules (95 quads) per year

FIGURE 35.5 United States energy consumption by resource. This graph shows energy consumption in the United States in 2020. *(Data from U.S. DOE Energy Information Administration)*

Quantities of Fossil Fuels in the United States and Worldwide

Because fossil fuels are a finite resource and of such great importance to the functioning of the economies of virtually all countries, many organizations in the United States and worldwide have spent decades and billions of dollars assessing the worldwide reserves of different fuels. The U.S. Geological Survey (USGS) assesses geologic formations for undiscovered oil and gas—such as that depicted in **FIGURE 35.6**—and coal resources in the United States and worldwide. Other agencies and fossil fuel exploration companies also assess formations. The U.S. Department of Energy, Energy Information Agency (DOE EIA) and the International Energy Agency (IEA) use the estimated reserves and make projections on how much of each fuel is in the ground and potentially recoverable and how usage rates will change over time. Based on estimated reserves and projected consumption rates, they estimate the number of years remaining of a particular fuel. **TABLE 35.2** shows the estimated number of years remaining for natural gas, oil, and coal for the United States and the world. The table also shows the estimates for uranium ore that can be used to supply nuclear power plants. Since these estimates are based on many, many assumptions, they should be considered highly uncertain. Furthermore, remember that all three of these fossil fuels contribute fossil carbon dioxide to the atmosphere and most countries of the world have agreed to goals for reducing carbon dioxide emissions. This introduces even more uncertainty to these estimates.

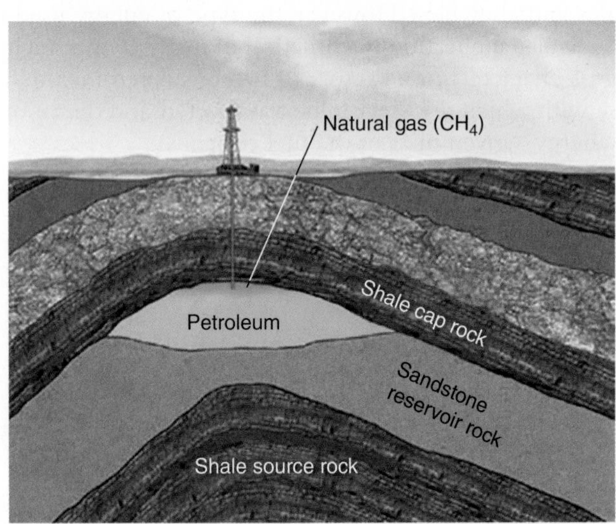

FIGURE 35.6 Petroleum accumulation underground. Crude oil migrates to the highest point in a formation of porous rock and accumulates there. Such accumulations of oil can be removed by drilling a well. Natural gas (methane) also accumulates at the highest point.

Energy intensity The energy use per unit of gross domestic product (GDP).

TABLE 35.2 Years supply remaining for natural gas, oil, coal, and uranium ore in the United States and the world

	Years Supply Remaining (Approximate)	
	U.S.	World
Natural gas	85	50
Oil	5	50
Coal	470	130
Uranium ore	20	130–200

Based on current consumption levels and excluding projected but unproven reserves. Data from U.S. DOE, Statistical Review of World Energy and International Energy Agency.

Although we know that the supply of fossil fuels is finite, there is some discussion within the environmental science community on whether it matters. Recall our discussion in Module 17 about how human creativity and innovation can lead to greater efficiency of fuel use or to development of a new method for obtaining energy. Many people believe that we will apply human ingenuity to develop new energy sources and reduce our demand for current resources. So far, this has proven to be true with the various energy conservation and efficiency improvements that we will discuss later in this module and the tremendous expansion of renewable energy sources that we have just discussed and will discuss further in Modules 39–41. For now, total energy use in the United States has leveled off, with small variations from year to year, and energy use per person has been decreasing slightly (**FIGURE 35.7**). An important measure of energy use and productivity is **energy intensity**, which is the energy use per unit of gross

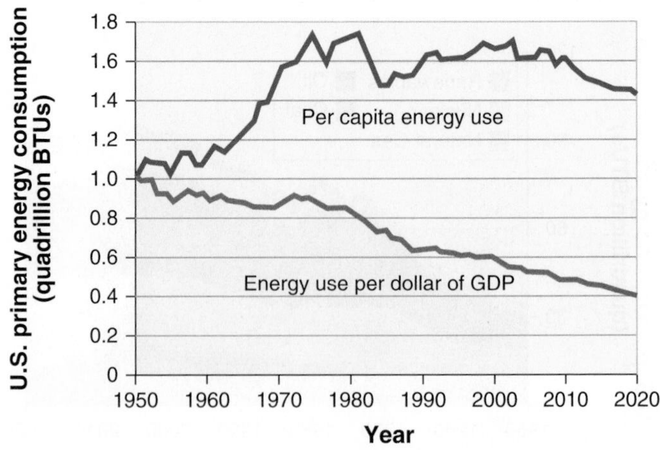

FIGURE 35.7 U.S. energy use per capita and energy intensity. Energy use per capita in the United States was roughly level between 1990 and 2005 and has been dropping slowly in recent years. Our energy intensity, or energy use per dollar of GDP, has been decreasing steadily since 1980. *(Data from the U.S. Department of Energy, Energy Information Administration, 2021)*

domestic product (GDP). For the last 5 decades, energy intensity in the United States has been steadily decreasing, as Figure 35.7 shows. In other words, we are using energy more efficiently relative to each dollar of GDP. But because the U.S. population has grown, and we are doing more things that use energy, our overall energy use has leveled off but it has not decreased, even though energy intensity has. Prior to that, it had been growing steadily for the previous 2 centuries. You can partially understand why there isn't a decrease if you think about how many electronic devices you have, then ask your parents or grandparents how many electronic devices they had when they were high school students. There is one more concept we need to examine to help us understand future patterns of fossil fuel energy and whether energy supply is the most important factor: the Hubbert curve.

The Hubbert Curve

We have seen that fossil fuels, our primary source of energy, are finite. Because fossil fuels take millions of years to form, they are nonrenewable resources, at least on a human timescale. By definition, then, the use of fossil fuels is not sustainable because there is no way to limit our consumption to the rate at which these fuels are being formed. Energy scientists and others have debated whether our economy will at some point be limited by the availability of this energy resource. In recent years, concerns have shifted away from the actual supply of fossil fuels to the consequences of **fossil fuel combustion**, which is the chemical reaction between any fossil fuel and oxygen. The resulting product is carbon dioxide, water and the release of energy. In recent decades, the release of CO_2 and other greenhouse gases and their contribution to global climate change have been given more and more attention. Many environmental scientists believe that these consequences will manifest themselves in adverse ways long before we run out of fossil fuels. However, there is still a group of energy analysts deeply concerned about "the end of oil."

In 1969, M. King Hubbert, a geophysicist and oil company employee, published a graph showing a bell-shaped curve representing oil use. Shown in **FIGURE 35.8**, the **Hubbert curve** represents oil use and projects both when world oil production will reach a maximum and when world oil will finally be depleted. Hubbert used two estimates of total world petroleum reserves: an upper estimate and a lower estimate. He found that the total reserves did not greatly influence the time it would take to use up all of the oil in known reserves. He predicted that oil extraction and use would increase steadily until roughly half the supply had been used up, a point known as **peak oil**. Hubbert predicted that at peak oil, extraction and use would begin to decline. Some oil experts believe we have already reached peak oil, while others maintain that we may reach it very soon. Back in 1969, Hubbert predicted that 80 percent of the world's total oil supply would be used up in roughly 60 years.

Although there have been discoveries of large oil fields since Hubbert did his work, the conclusion he drew from his model still holds. Regardless of the exact amount of the total reserves, the total number of years we use petroleum

Fossil fuel combustion The chemical reaction between any fossil fuel and oxygen resulting in the production of carbon dioxide, water, and the release of energy.

Hubbert curve A graph that represents oil use and projects both when world oil production will reach a maximum and when world oil will be depleted.

Peak oil The point at which oil extraction and use would increase steadily until roughly half the supply had been used up.

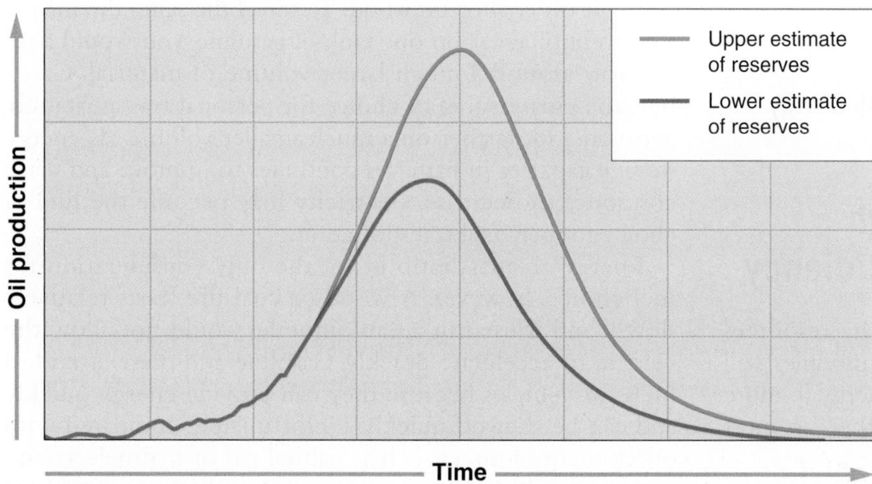

FIGURE 35.8 A generalized version of the Hubbert curve. Regardless of whether an upper estimate or a lower estimate of total petroleum reserves is used, the date by which petroleum reserves will be depleted does not change substantially. A growing body of environmental scientists believes that the consequences of using fossil fuels will require a switch to other fuels long before supplies become limited.

will fall within a relatively narrow window of time, perhaps 200 years. When we identify a fuel source, we tend to use it until we come upon a better fuel source. As a number of energy experts are fond of saying, "We did not move on from the Stone Age because we ran out of stones." In a similar vein, many people believe that ingenuity and technological advances in the renewable energy sector will one day render oil, and other fossil fuels, much less desirable. A growing number of environmental scientists maintain that wondering about how much oil remains and whether we have reached peak oil are no longer meaningful questions to ask. As fossil fuels become less desirable, many believe a more important question is, "What fuel or energy sources will we use next?" That is the subject of upcoming Modules 39–41.

The Future of Fossil Fuel Use

We have already seen in Table 35.2 that the remaining fossil fuel availability projections are for years to hundreds of years. If current global use patterns continue, and no significant additional petroleum supplies are discovered, we may run out of conventional oil and gas supplies in 50 years. Coal supplies will potentially last for hundreds of years. While these projections assume that we will continue our current use patterns, we have already observed that technological advances, a shift to renewable fuels, and changes in social choices and population patterns are decreasing these usage patterns and rates. As more people and nations have come to accept the concept that anthropogenic increases in atmospheric greenhouse gas concentrations are causing global climate change, a large number of researchers have suggested that we should turn our attention to how we might transition from fossil fuels before their use causes further environmental problems. For this reason, some researchers focus less on the projected reserves of fossil fuels and more on the transition to other sources of energy.

35-4 What is the importance of energy efficiency and conservation?

We can use less energy through conservation and increased efficiency

This module has provided an overview of the resources that provide energy to society. The following modules will examine both nonrenewable and renewable energy in more detail. However, before we explore energy further, we must consider sustainability and energy use. A truly sustainable approach to energy use must incorporate both energy conservation and energy efficiency. **Energy conservation** means finding and implementing ways to use less energy. **Energy efficiency** is the ratio of the amount of energy expended in the form you want to the total amount of energy that is introduced into the system. We will explore both of these concepts below.

Why are we presenting these two topics in a module on energy resources? Because conservation and efficiency efforts save energy that can then be used later, just as you might save money in a bank account to use later when the need arises. In this sense, conservation and efficiency are sustainable energy "sources."

Energy conservation and energy efficiency are the least expensive and most environmentally sound options for maximizing our energy resources. In many cases, they are also the easiest approaches to implement because they often require fairly simple changes to existing systems rather than a switch to a completely new technology. Increasing energy efficiency means obtaining the same work from a smaller amount of energy. Energy conservation and energy efficiency are closely linked. One can conserve energy by not using an electrical appliance because doing so results in less energy consumption. But one can also conserve energy by using a more efficient appliance—one that does the same work but uses less energy.

Different Forms of Energy

The best form of energy to use depends on the particular purpose for which it is needed. For example, for transportation, we usually prefer gasoline or diesel fuel—liquid energy sources that are relatively compact, meaning that they have a high energy-to-mass ratio. Imagine running your car on coal or firewood: To travel the same distance as a conventional car on one tank of gasoline, you would have to carry around a much larger volume of material. Gasoline, the current fuel of choice for personal transportation, gets you a lot farther on a much smaller volume. As energy storage capacity in batteries continues to improve and price continues to decrease, electricity may become the fuel of choice for personal transportation.

Energy-to-mass ratio is not the only consideration for fuel choice, however. A wood or coal fire starts relatively slowly and if used in an automobile, would not allow the vehicle to accelerate quickly. Gasoline and diesel are ideal fuels for vehicles because they can provide energy quickly and can be shut off quickly. Unfortunately, compared with other energy sources such as natural gas or hydroelectricity, gasoline produces large amounts of air pollution per joule of energy released. Furthermore, unlike coal or wood, gasoline requires a good deal of refining—chemical processing—to produce. So there are many factors to consider when determining the most suitable energy source for a particular application.

Energy conservation Methods for finding and implementing ways to use less energy.

Energy efficiency The ratio of the amount of energy expended in the form you want to the total amount of energy that is introduced into the system.

Quantifying Energy Efficiency

Although all conventional nonrenewable energy sources have environmental impacts, energy efficiency is an important factor in making energy-use decisions. Energy efficiency refers to both the efficiency of the process we use to obtain the fuel, and the efficiency of the process that converts the fuel into the work that is needed.

In Module 0 and on the previous page we discussed energy efficiency as well as energy quality, a measure of the ease with which stored energy can be converted into useful work. The second law of thermodynamics dictates that when energy is transformed, its ability to do work diminishes because some energy is lost during each conversion. In addition to these losses, there is an expenditure of energy involved in obtaining almost every fuel that we use.

FIGURE 35.9 outlines the process of energy use from extraction of a resource to electricity generation and disposal of waste products from the power plant. The red arrows indicate that there are many opportunities for energy loss, each of which reduces energy efficiency. Due to the second law of thermodynamics, about two-thirds of the energy that enters a coal-burning electricity generation plant ends up as waste heat or other undesired outputs. If we included the energy used to extract the coal as well as the energy used to build the coal extraction machinery, to construct the power plant, and to remove and dispose of the waste material from the power plant, the efficiency of the process would be even lower. These other energy inputs will be discussed in greater detail in Module 37.

Every energy source, from coal to oil to wind, requires an expenditure of energy to obtain. The most direct way to account for the energy required to produce a fuel, or energy source, is by calculating the energy return on energy investment. The **energy return on energy investment (EROEI)** is the amount of energy we get out of an energy source for every unit of energy expended on its production. EROEI is calculated as follows:

$$\text{EROEI} = \frac{\text{Energy obtained from the fuel}}{\text{Energy invested to obtain the fuel}}$$

For example, in order to obtain 100 J of coal from a surface coal mine, 5 J of energy is expended. Therefore,

$$100\text{ J EROEI} = \frac{100\text{ J}}{5\text{ J}} = 20$$

As you might expect, a larger value for EROEI suggests a more efficient and more desirable process. "Science Applied 6: Should Corn Become Fuel?" at the end of this unit discusses the EROEI for ethanol, a fuel made from corn.

In this first module in Unit 6, we examined the characteristics of nonrenewable and renewable energy sources and we identified patterns of energy use around the world and in the United States. We also introduced the topics of energy conservation and efficiency. Next, we will learn about the different fuel types and their distribution around the world.

Energy return on energy investment (EROEI) The amount of energy we get out of an energy source for every unit of energy expended on its production.

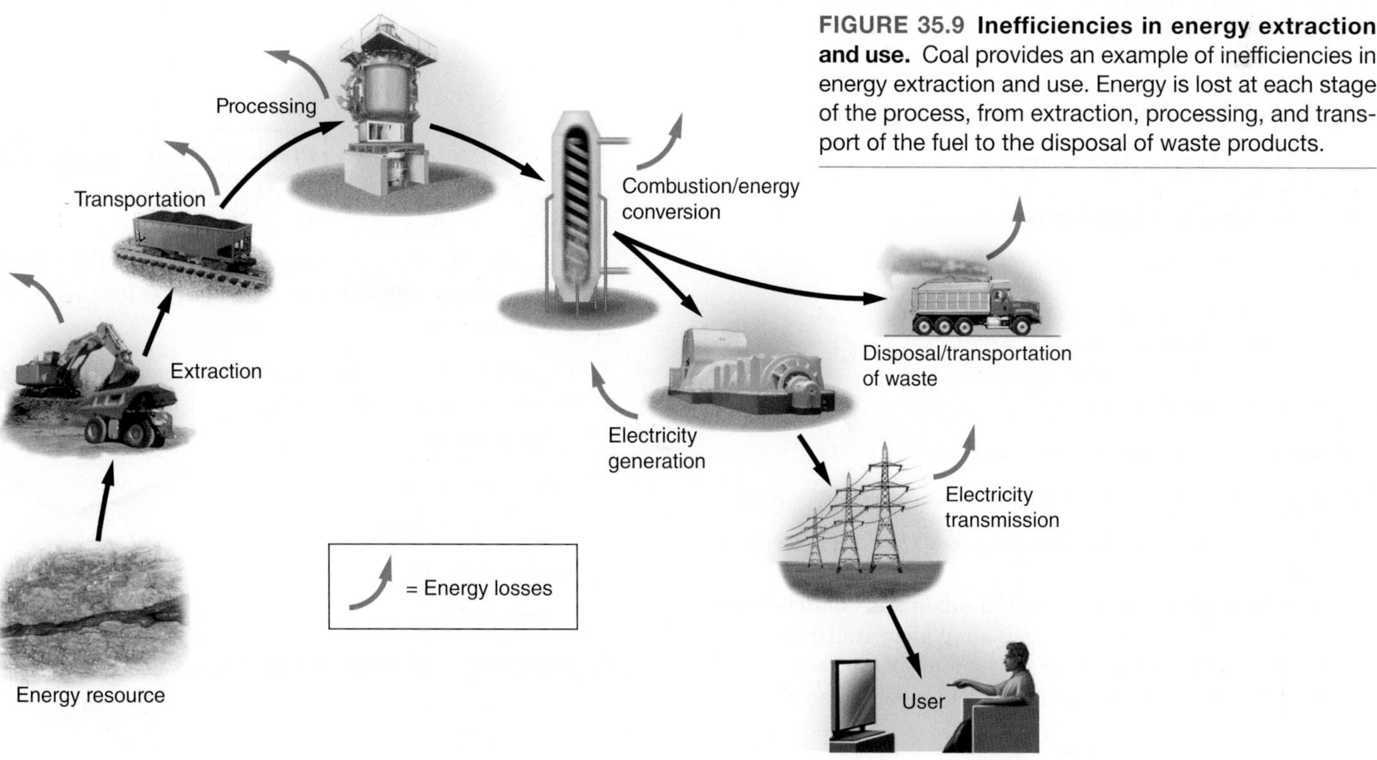

FIGURE 35.9 Inefficiencies in energy extraction and use. Coal provides an example of inefficiencies in energy extraction and use. Energy is lost at each stage of the process, from extraction, processing, and transport of the fuel to the disposal of waste products.

Module 35 AP® Review

Preparing for the AP® Exam

Learning Goals Revisited

35-1 What are the characteristics of nonrenewable energy resources?

Coal, oil, and natural gas are fossil fuels derived from biological material that became fossilized millions of years ago. When they are combusted, they release fossil carbon that has been stored for millions of years into the atmosphere. Because a fossil fuel cannot be replenished once it is used up, it is known as a nonrenewable energy resource.

35-2 What are the characteristics of renewable energy resources?

Other sources of energy that are infinite are called renewable. There are two categories of renewable resources. Biomass energy resources are potentially renewable because those resources can be regenerated indefinitely if they are not consumed too quickly. Solar, wind, geothermal, hydroelectric, and tidal energy are nondepletable.

35-3 What are the trends of energy use worldwide and in the United States?

Worldwide, oil, coal, and natural gas are the three largest energy sources and comprise 80 percent of total energy use. Renewable energy comprises 15 percent of global energy use. Hydroelectricity is the largest source of renewable energy. Energy use is highly variable in countries around the world. In the United States, total energy use is 79 percent fossil fuel, 9 percent nuclear fuel, and 12 percent renewable energy resources. Based on reserves and projected consumption rates, estimates can be prepared for the number of years remaining for a particular fossil fuel. These estimates are uncertain because of new energy sources and reduced demand for current resources. As anthropogenic increases in atmospheric greenhouse gas concentrations are causing global climate change, the transition from fossil fuels to other energy sources has become even more important.

35-4 What is the importance of energy efficiency and conservation?

Energy conservation is finding and implementing ways to use less energy while energy efficiency is the ratio of the amount of energy expended in the form you want to the total amount of energy that is introduced into the system. By avoiding the use of energy resources, conservation and efficiency are actually sustainable energy "sources."

AP® Practice Questions

Preparing for the AP® Exam

Multiple-Choice Questions

Use the choices below to answer questions 1 & 2:

(a) nonrenewable resource
(b) nondepletable resource
(c) potentially renewable resource
(d) renewable resource

1. Which type of resource represents nuclear energy?

2. Which type of resource represents hydroelectric power?

3. Which of the following best describes a subsistence energy source?
   (a) These are commonly used in urban areas of developed countries, where fossil fuels are widely available.
   (b) Examples include hydropower and solar power.
   (c) Examples include natural gas and nuclear power.
   (d) These are commonly used in rural areas of developing countries, where humans gather fuel for their own immediate use.

4. Which of the following best represents the correct method to determine the per capita energy use of a country?
   (a) $\frac{\text{population}}{\text{total energy use}}$
   (b) population × total energy use
   (c) $\frac{\text{total energy use}}{\text{population}}$
   (d) population × personal energy use

5. In 2021, the United States generated 4,119.85 billion kilowatt hours (kWh) of energy. If there are 3,600 kilojoules (kJ) in a kilowatt hour, how much total energy, in exajoules (EJ), was generated by the United States in 2021?
   (a) 14.8 EJ
   (b) 1.48×10^3 EJ
   (c) 1.48×10^{10} EJ
   (d) 1.48×10^{16} EJ

Use the following graph to answer questions 6 & 7:

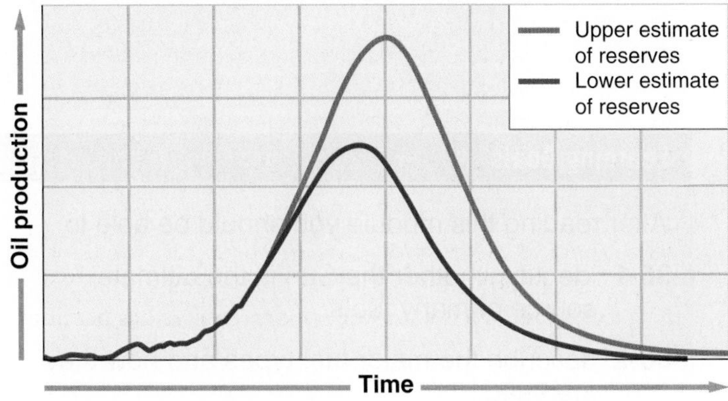

6. Which of the following best represents the information in the Hubbert Curve diagram?
   (a) The Hubbert Curve is a bell-shaped curve representing the amount of oil traded between nations over time.
   (b) The Hubbert Curve is a bell-shaped curve representing how much oil has existed on Earth over time.
   (c) The Hubbert Curve is a bell-shaped curve representing the amount of oil being refined into gasoline over time.
   (d) The Hubbert Curve is a bell-shaped curve representing oil production in the world over time.

7. Which of the following best describes what can be predicted using the Hubbert curve?
   (a) Peak oil will occur once half of the supply is used up.
   (b) Finding additional reserves could greatly increase the time to peak oil.
   (c) Natural gas will be cheaper than oil by the early twenty-first century.
   (d) The cost of oil is independent of the world's oil supply.

Free-Response Question

In 2020, the United States generated approximately 4 trillion kilowatt hours (4×10^{12} kWh) of electricity. The share of this total by energy source is shown in the table below.

(Some sources may not add up due to rounding.)

Energy Source	Share of Total Electricity Generated
Fossil Fuels	**60%**
Coal	19.3%
Natural gas	41.0%
Petroleum & other gas	0.7%
Nuclear	**20%**
Renewables	**20%**
Biomass	1.4%
Geothermal	0.4%
Hydroelectric	7.2%
Solar	2.3%
Wind	8.4%

(a) **Identify** the potentially renewable energy source used by the United States for electricity generation in 2020. (1 pt.)

(b) **Describe** why fossil fuels are considered nonrenewable energy sources. (1 pt.)

(c) **Identify** an environmental problem that can be caused by fossil fuel use. (1 pt.)

(d) **Make a claim** that proposes a potential solution to the problem described in part (c). (1 pt.)

(e) **Justify** the solution proposed in part (d) by describing an economic advantage of the solution. (1 pt.)

(f) In 2020, the United States produced a total of 100 exajoules (EJ) of energy.
   (i) If the U.S. population in 2020 was 330 million people, **calculate** the per capita energy use for the United States in gigajoules (GJ). Show all work. (2 pts.)
   (ii) Using the information in the table above, **calculate** the per capita share of fossil fuels in the United States in kilowatt hours (kWh) if there are 3,600 kJ per kWh. Show all work. (2 pts.)
   (iii) Before extraction, the energy available for oil is 100 megajoules (MJ). After extraction, transportation, and refinement, the total energy remaining is 58 MJ. **Calculate** the energy return on energy investment (EROEI) using the equation below:

$$\text{EROEI} = \frac{\text{Energy obtained from the fuel}}{\text{Energy invested to obtain the fuel}}$$

(1 pt.)

Module 36

Unit 6	35	36	37	38	39	40	41

Fuel Types and Uses

We use a wide variety of fuels for different purposes in our daily lives as individuals and as a society. In this module we will describe the fuels we use and the advantages of each. We will also explain specialized uses of certain fuels for transportation and electricity generation.

Learning Goals

After reading this module you should be able to

36-1 identify whether the Sun is the ultimate source of many fuels.

36-2 describe the major fuel types and how they are used.

36-3 identify other uses of fossil fuel.

36-4 describe electricity generation and cogeneration.

36-1 Is the Sun the ultimate source of many fuels?

The Sun is the ultimate source of many of the fuels we use

As **FIGURE 36.1** shows, all fossil fuel and most renewable energy sources ultimately come from the Sun. Biomass energy resources encompass a large class of fuel types that include wood and charcoal, animal products and manure, plant remains, and solid waste collected from households and small businesses, as well as liquid fuels such as ethanol and biodiesel. Many forms of biomass used directly as fuel, such as wood and manure, are readily available all over the world. Because these materials are inexpensive and abundant, they account for about 12 percent of world energy consumption, with a much higher percentage in many developing countries. Biomass can also be processed or refined into liquid fuels such as ethanol and biodiesel, known collectively as **biofuels**. These fuels are used in more limited quantities due to the technological demands associated with their use. For example, it is easier to burn a log in a fire than it is to develop the technology to produce a compound such as ethanol.

Biomass—including ethanol and biodiesel—accounts for roughly 40 percent of the renewable energy and approximately 5 percent of all the energy consumed in the United States today. However, the mix of biomass used in the United States differs from that found in the developing world. A little less than half of the biomass energy used in the United States comes from wood, and a similar amount comes from biofuels.

Biofuel A liquid fuel such as ethanol or biodiesel created from processed or refined biomass.

Modern carbon Carbon in biomass that was recently in the atmosphere.

Fossil carbon Old carbon contained in fossil fuels.

Modern Carbon versus Fossil Carbon

Like fossil fuels, biomass contains a great deal of carbon, and burning it releases that carbon into the atmosphere. Given the fact that both fossil fuels and biomass raise atmospheric carbon concentrations, is it really better for the environment to replace fossil fuels with biomass? The answer depends on how the material is harvested and processed and on how the land is treated during and after harvest. It also depends on how long the carbon has been stored. The carbon found in plants growing today was in the atmosphere in the form of carbon dioxide until recently when it was incorporated into the bodies of the plants through photosynthesis. Depending on the type of plant it comes from, the carbon in biomass fuels may have been captured through photosynthesis as recently as a few months ago, as in the case of a corn plant, or perhaps up to several hundred years ago, as in the case of wood from a large tree. We call the carbon in biomass that was recently in the atmosphere **modern carbon**, in contrast to the old carbon in fossil fuels, which we call **fossil carbon**.

Unlike modern carbon, fossil carbon has been buried for millions of years. Fossil carbon is carbon that was out of circulation until humans discovered it and began to use it. The burning of fossil fuels results in a rapid increase in

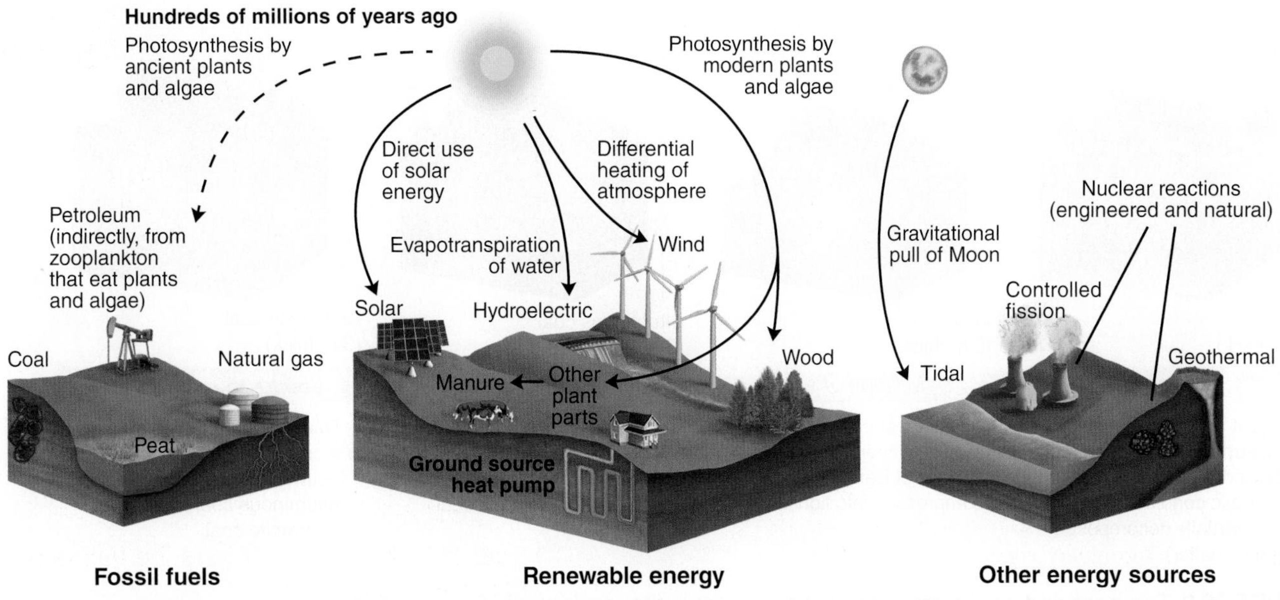

FIGURE 36.1 Energy from the Sun. The Sun is the ultimate source of almost all types of energy.

atmospheric CO_2 concentrations because we are unlocking or releasing stored carbon that was last in the atmosphere millions of years ago. In theory, the burning of biomass (modern carbon) should not result in a net increase in atmospheric CO_2 concentrations because we are returning the carbon to the atmosphere, where it had been until recently. And, if we allow vegetation to grow back in areas where biomass was recently harvested, that new vegetation will take up an amount of CO_2 more or less equal to the amount we released earlier by burning the biomass. Over a long period of time, the net change in atmospheric CO_2 concentrations should be zero. An activity that does not change atmospheric CO_2 concentrations is referred to as **carbon neutral**.

Whether using biomass is truly carbon neutral, however, is an important question that is currently being discussed by scientists and policy makers around the world. Sometimes the use of modern carbon involves cutting down trees and releasing CO_2 into the atmosphere that would otherwise have remained in the soil, so in actuality, it is not carbon neutral.

36-2 What are the major fuel types and how are they used?

Each fuel has specific optimal applications

At the end of Module 35, we examined how different forms of energy are best suited for specific purposes. We gave the example of trying to power a car with wood and how large the fuel tank would have to be to hold enough wood. This illustrates how gasoline contains much more energy for a given volume of fuel than does wood. And it demonstrates that each activity that we engage in usually has one fuel that is optimal for that application.

Wood

Throughout the world, 2 billion to 3 billion people rely on wood for heating or cooking. In the United States, approximately 3 million homes use wood as the primary heating fuel, and more than 20 million homes use wood for energy at least some of the time. In addition, the pulp and paper industries, power plants, and other industries use wood waste and by-products for energy. In theory, cutting trees for fuel is sustainable if forest growth keeps up with forest removal. Unfortunately, many forests, like those in Indonesia and parts of Africa and South America, are cut intensively, allowing little chance for regrowth. Wood is also used to make charcoal, which is lighter than wood and contains approximately twice the energy content of wood per unit mass.

Coal and Peat

Coal is a solid fuel formed primarily from the remains of trees, ferns, and other plant materials that were preserved

Carbon neutral An activity that does not change atmospheric CO_2 concentrations.

Coal A solid fuel formed primarily from the remains of trees, ferns, and other plant materials that were preserved 280 million to 360 million years ago.

Peat
(50 m thick)
Lignite
(10 m thick)
Bituminous coal
(5 m thick)
Anthracite coal
(3 m thick)
0 years
Millions of years
Hundreds of millions of years
280 to 360 million years
Ancient forests cover much of land surface. The vegetation dies and is buried under anaerobic conditions, forming peat (partially decomposed organic matter).
As layers of peat become buried deeper below the surface of Earth, they are compressed into lignite.
Lignite layers are buried even deeper. The increased pressure compresses lignite into soft, bituminous coal.
Deeper burial, years of increased pressure, tectonic activity, and heat transform bituminous coal into harder anthracite coal.

FIGURE 36.2 The peat and coal formation process. Peat is the raw material from which coal is formed. Over millions of years and under increasing heat and pressure due to greater depth of burial under more and more layers of rock and sediment, various types of coal are formed.

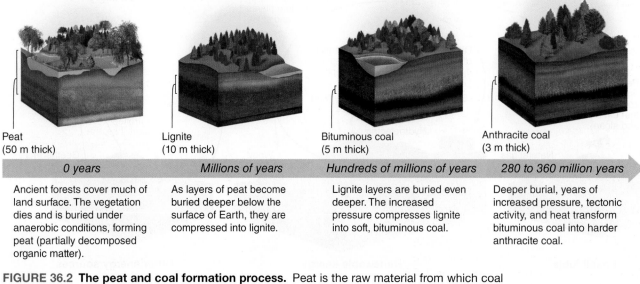

280 million to 360 million years ago. It is available in many areas of the world and often is relatively easy to extract, handle, and process. It is used for electricity generation and industrial processes in many parts of the world. A precursor to coal, called **peat**, is made up of partly decomposed organic material, including mosses, that can be burned for fuel. There are three types of coal, ranked from lesser to greater age, exposure to heat and pressure, depth of burial, and energy content; they are: lignite, bituminous, and anthracite. **Lignite** is sometimes called brown coal and is a soft sedimentary rock that sometimes shows traces of plant structure from the plants that it is derived from. It typically contains 60 to 70 percent carbon. **Bituminous coal** is a black or dark brown coal that contains bitumen, also known as **asphalt**. It typically contains up to 80 percent carbon and is hard but can be broken by hand. **Anthracite** is also known as **hard coal** and usually contains greater than 90 percent carbon. It has the highest quantity of energy per volume of coal and the fewest impurities. The formation of coal takes hundreds of millions of years. **FIGURE 36.2** represents factors involved in the formation of peat and the different types of coal. Starting with an organic material such as peat, increasing time and pressure, greater heat, and greater depth of burial produce successively denser coal with more carbon molecules, and more potential energy, per kilogram.

Peat A precursor to coal, made up of partly decomposed organic material, including mosses.

Lignite A brown coal that is a soft sedimentary rock that sometimes shows traces of plant structure; it typically contains 60 to 70 percent carbon.

Bituminous coal A black or dark brown coal that contains bitumen, *also known as* **asphalt**. It typically contains up to 80 percent carbon.

Anthracite *Also known as* **hard coal**, it contains greater than 90 percent carbon. It has the highest quantity of energy per volume of coal and the fewest impurities.

Natural gas A relatively clean fossil fuel containing 80 to 95 percent methane (CH_4) and 5 to 20 percent ethane, propane, and butane.

Natural Gas

In the unit opener, we observed that natural gas is often considered to be "clean" because its combustion produces smaller amounts of particulates, sulfur dioxide, and carbon dioxide than does oil or coal. However, combustion still results in emission of carbon dioxide, which is the major greenhouse gas produced by human activity. **Natural gas** is 80 to 95 percent methane (CH_4) and 5 to 20 percent ethane, propane, and butane. Remember that methane is released during anaerobic respiration (decomposition in the absence of oxygen). There are very few impurities in natural gas and that is why it burns cleanly, except for carbon dioxide emissions.

The two largest uses of natural gas in the United States are for electricity generation and industrial processes. Natural gas is also used to manufacture nitrogen fertilizer, and it is used in some homes as a fuel for cooking, heating, and operating clothes dryers and water heaters. Compressed natural gas can be used as a fuel for vehicles, but because it must be transported by pipeline, it is not accessible in all parts of the United States and is therefore unlikely to become an important fuel for cars. Liquefied petroleum gas (LPG)—which is similar to natural gas, but in a liquid form—is a slightly less energy-dense substitute. LPG can be transported via train or truck and stored at the point of use in tanks. This fuel is available practically everywhere in the United States and is used in place of natural gas and for portable barbecue grills and outdoor restaurant heaters. Overall, natural gas and LPG supply 34 percent of the energy used in the United States. Globally, natural gas contributes 24 percent of total energy consumption, after oil and coal (Figure 35.2). It is used for industrial processes, and heating and electricity generation. The United States is the largest user of natural gas in the world. Russia, China, Iran, Japan, and countries in Europe are some of the other major users of natural gas.

Crude Oil

Crude oil is a mixture of hydrocarbons such as oil, gasoline, and kerosene as well as water and sulfur that exists in a liquid state underground and when brought to the surface. There is often methane (natural gas) dissolved in the oil. Although there are slight differences, the U.S. Department of Energy refers to oil, crude oil, and petroleum as equivalent substances, and we will do the same in this book. While coal is ideal for stationary combustion applications such as those in power plants and industry, the fact that they are liquid fuels make oil and gasoline more suitable for mobile combustion applications, such as in vehicles.

Crude oil can be refined into a variety of compounds. These compounds, including tar and asphalt, gasoline, diesel, and kerosene, are distinguished by the temperature at which they boil and can therefore be separated by heating the crude oil. This process takes place in an oil refinery, a large factory-like structure (**FIGURE 36.3**). The refining process is complex and dangerous and requires a major financial investment. There are roughly 130 oil refineries in the United States; some of the larger ones can refine over 80 million liters (21 million gallons) per day. Many of them are old and reaching the end of their useful lives. Oil production and sales are measured in barrels of oil; one barrel equals 160 L (42 gallons).

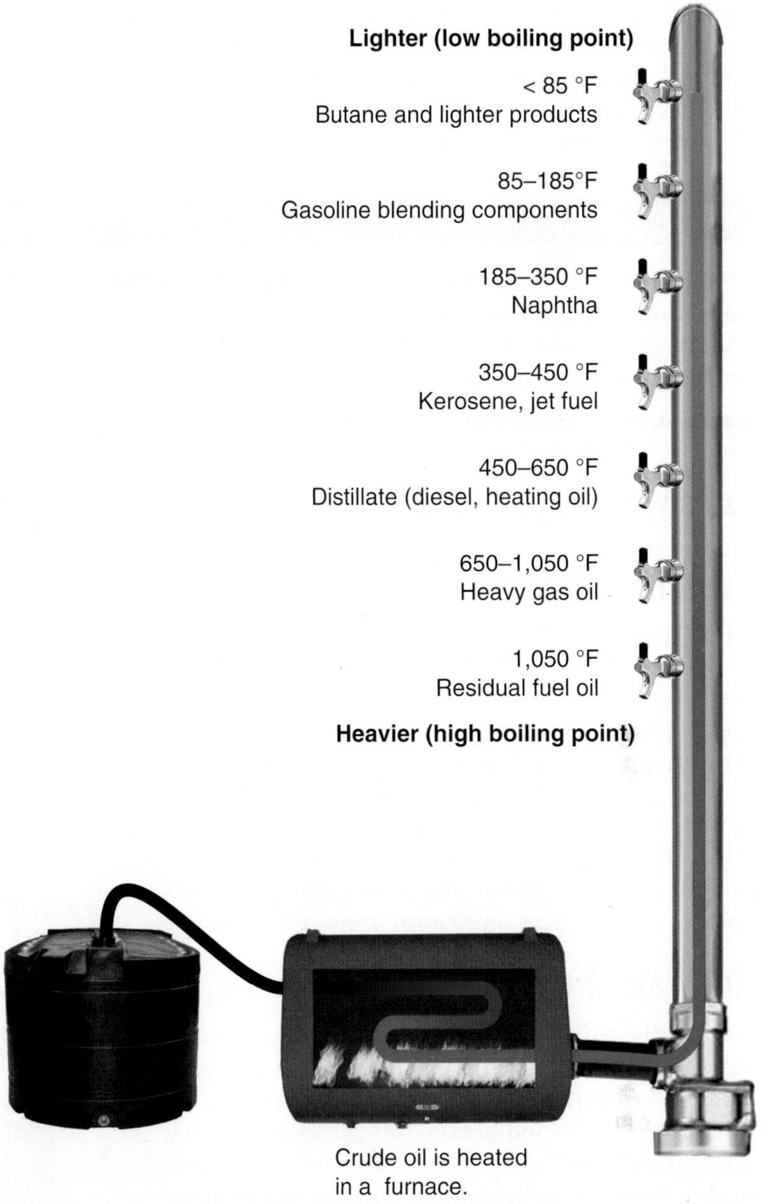

FIGURE 36.3 A crude oil distillation unit. Distillation separates crude oil into different fractions. Crude oil is heated and transferred to a distillation column, where different products boil off based on their boiling points and are then recovered separately. *(Information from Energy Information Administration, U.S. Department of Energy)*

Tar Sands

Tar sands, also known as **oil sands**, are slow-flowing, viscous deposits of bitumen or asphalt—described in the coal section above—mixed with sand, water, and clay. Bitumen is a degraded type of crude oil that forms in certain oil deposits. The oil migrates close to the surface, where bacteria metabolize some of the light hydrocarbons while other hydrocarbons evaporate. The remaining mix no longer flows at ambient temperatures and pressures, which is why it is called "tar" sands. Tar is a thick, oily substance that does not flow at room temperature, but it can be extracted by surface mining. The end product after tar sands extraction is crude oil.

Although tar sands exploration and extraction appear likely to increase the supply of available crude oil, they could have serious negative environmental impacts. The mining of tar sands is much more energy-intensive than conventional drilling for crude oil. As we saw in Module 29, surface mining creates large open pits. Extraction of the bitumen from the surrounding material contaminates roughly 2 to 3 L of water for every liter of bitumen obtained, and many

Crude oil A mixture of hydrocarbons such as oil, gasoline, kerosene as well as water and sulfur that exists in a liquid state underground, and when brought to the surface.

Tar sands Slow-flowing, viscous deposits of bitumen or asphalt, mixed with sand, water, and clay; *also known as* **oil sands**.

tar sands are located in areas where water is not an abundant resource. In addition, because tar sands require so much energy before they arrive at the refinery, the overall system efficiency is lower, and the resulting CO_2 release greater, than for conventional oil production.

AP® Exam Tip

Past AP® Environmental Science Exams have asked about tar sands. Make sure that you know how tar sands are mined and processed, and that you understand the environmental consequences of this process.

36-3 What are other uses of fossil fuel?

Fossil fuels have specialized uses for motor vehicles and electricity

We have already discussed in Module 35 how different fossil fuels are each best suited for specific purposes. When deciding between two energy sources for a given job, it is essential to consider the best fuel for the work needed as well as the overall system efficiency. Sometimes the trade offs are not immediately apparent. The home hot water heater, also called a domestic hot water heater, is an excellent illustration of this principle.

Hot Water Heaters

Electric hot water heaters are often described as being highly efficient. Even though it is not possible to convert an energy supply entirely to its intended purpose and obtain 100 percent efficiency, converting electricity into hot water in a water heater comes very close. Electric hot water heaters contain an electric heating element (a resistance coil) that generates heat inside the tank of water (**FIGURE 36.4a**). The heat is transferred to the surrounding water, which increases in temperature. Any heat energy that is "lost" to the surrounding environment is actually captured by the water in the tank that surrounds the heating element. Even though nothing is 100 percent efficient (remember the Second Law of Thermodynamics), the electric hot water comes close; the efficiency of this process is 99 percent. In contrast, a typical natural gas water heater transfers energy to water with a flame burning natural gas below the tank (Figure 36.4b). Wasted heat and by-products of combustion warm the surrounding environment, as well as the tank of water, and the by-products are vented to the outside. This type of water heater is about 60 percent efficient.

However, that's not the entire story. In order to compare the *total* system efficiency of each water heater we need to consider the fuel each uses. The energy expended to extract, process, and deliver natural gas to the home can be considerable. Think of Figure 35.6 in the previous module that showed qualitatively the energy loses in the coal-extraction process. Similar losses occur in the natural gas production process. And there are similar losses for electricity used in an electric powered hot water heater. If a coal-fired power plant is the source of the electricity that fuels the electric water heater, we have to consider that conversion of coal into electricity is only about 36 percent efficient. This means that even though an electric water heater has a higher direct efficiency than a natural gas water heater, the overall efficiency of the electric water heating system is lower—(0.36 efficiency × 0.99 efficiency = 0.35)—35 percent for the electric water heater compared with roughly 60 percent for the gas water heater. There may be many situations in which it is a better choice to heat water with electricity rather than with natural gas. If the source of the electricity is renewable—wind or solar for example—it might make sense to use electricity. Also, a natural gas water heater requires a combustion source inside the house. In a well-insulated home, there may be air pollution introduced

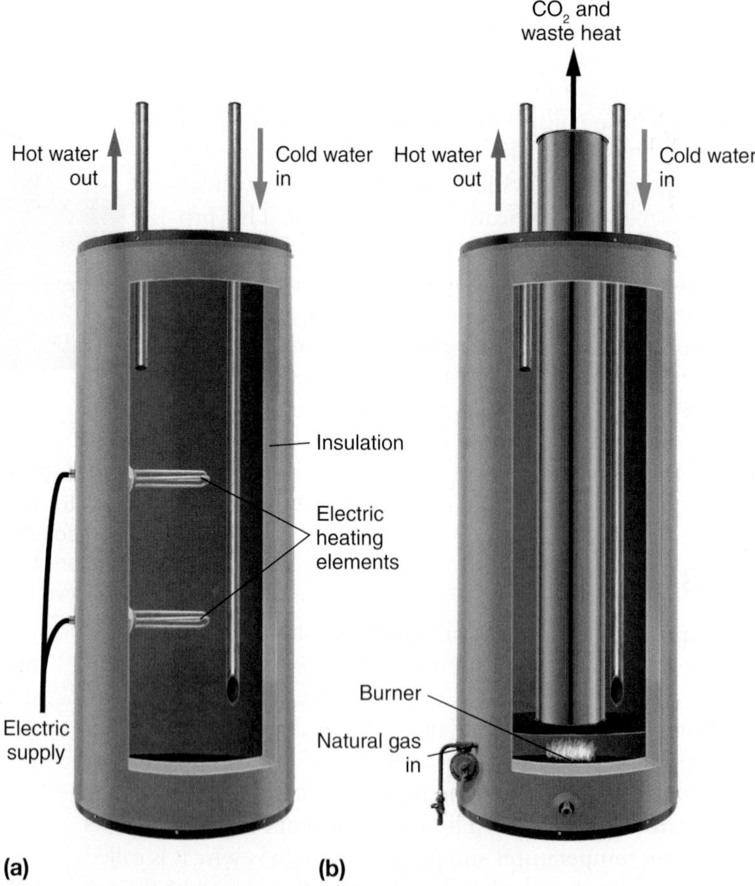

FIGURE 36.4 Hot water heaters for the home. (a) In an electric hot water heater, heat is transferred from an electric element to the surrounding water. The efficiency of this process is 99 percent. (b) In contrast, a typical natural gas water heater transfers energy to water with a flame below the tank and the efficiency is about 60 percent.

to the living environment from a gas hot water heater. There is also the option of using an electric hot water heat pump, a relatively new and even more efficient home hot water heater that is increasingly common in homes and apartment buildings. We will examine the hot water heat pump in more detail in Module 40, but from an environmental science perspective, you can see that it is important to examine the fuel, its appropriateness for a given task, and to look at the overall system efficiency when considering the pros and cons of a device and an associated energy choice. Similar fuel decisions are also made with transportation.

Fossil Fuel Choices and Transportation

In 2020, 35 percent of energy use in the United States was for transportation. Therefore, making the most energy-efficient choice in transportation and the type of fuel to use are particularly important. The movement of people and goods occurs primarily by means of vehicles that are fueled with petroleum products, such as gasoline and diesel fuel, and by electricity. These vehicles contribute to air pollution and greenhouse gas emissions. However, specific modes of transportation use certain fuels, and some are more efficient than others.

As you might expect, public transportation—train or bus travel—is much more efficient than traveling by car, especially when there is only one person in the car. And public ground transportation is usually more efficient than air travel. **TABLE 36.1** shows the fuels used and the efficiencies of different modes of transportation. Note that the energy values report only energy consumed, in megajoules (MJ, 10^6 J), per passenger-kilometer traveled and do not include the embodied energy used to build the different vehicles.

TABLE 36.1 Energy expended for different modes of transportation in the United States

Mode (fuel)	MJ per passenger-kilometer
Air (jet fuel)	2.1
Passenger car (gasoline) (driver alone; no passengers)	3.6
Motorcycle (gasoline)	1.1
Train (Amtrak – electricity)	1.1
Bus (diesel)	1.7

Trains, primarily using electricity but sometimes diesel fuel, and motorcycles using gasoline are the most energy-efficient modes of transportation shown. Most cars in the United States use gasoline. If a car contained four passengers, we would divide the value in Table 36.1 for a lone driver by four, and thus the car would then be the most efficient means of transportation. However, cars traveling with four riders are relatively rare in the United States; single-occupant vehicles are the most common means of transportation. "Do the Math: Efficiency of Travel" shows you how to calculate the efficiencies of different modes of transportation.

Transportation efficiency calculations do not consider convenience, comfort, or style. Many people in the developed world are quite particular about how they get from place to place and want the independence of a personal vehicle. They also tend to have strong feelings about what

DO THE MATH Efficiency of Travel

Preparing for the AP® Exam

Imagine that you had unlimited time and you needed to get from Ann Arbor, Michigan, to St. Louis, Missouri. The distance is roughly 800 km (500 miles). For each mode of transportation, calculate how many megajoules of energy you would use.

Using the data in Table 36.1, we can determine the following:

Air 2.1 MJ/passenger-kilometer × 800 km/trip = 1,680 MJ/passenger-trip

Car 3.6 MJ/passenger-kilometer × 800 km/trip = 2,880 MJ/passenger-trip

Train 1.1 MJ/passenger-kilometer × 800 km/trip = 880 MJ/passenger-trip

Bus 1.7 MJ/passenger-kilometer × 800 km/trip = 1,496 MJ/passenger-trip

If a gallon of gasoline contains 120 MJ, how many gallons of gasoline does it take to make the trip by car?

2,880 MJ/passenger-trip/120 MJ/gallon = 24 gallons/passenger-trip

Examining total joules expended makes it clear that the train is the most energy-efficient means of travel. Driving alone in a car is the least energy-efficient.

YOUR TURN If you could carpool with three other people who needed to make the same trip, what would the energy expenditure be for each person?

type of personal vehicle is most desirable. In the United States in 2020, light trucks—a category that includes sport-utility vehicles (SUVs), crossover vehicles, and pickup trucks—accounted for roughly 75 percent of total automobile sales. Hybrid-electric vehicles accounted for 3.5 percent and electric vehicles accounted for less than 2 percent of total sales in the United States. In China, electric cars comprised 5.7% of new car sales in 2020 and in Europe electric cars accounted for 10% of new car sales. To help you choose the best mode of transportation, and recognize that all transportation decisions involve trade offs, review the Unit 6 Visual Representation feature (pages 420-421).

Light trucks are comparatively heavy vehicles and so generally have mileage ratings of less than 8.5 km per liter, or 20 miles per gallon (mpg). Because they are exempt from certain vehicle emission standards, they emit more of certain air pollutants per liter of fuel combusted than passenger cars. Smaller cars with standard internal combustion engines can travel up to 19 km per liter (45 mpg) on the highway. Some hybrid passenger cars, which use a gasoline engine, electric motors, and special braking systems, obtain closer to 21 km per liter (50 mpg). Electric cars and plug-in hybrid electric cars, which are becoming more common in the United States, obtain even better fuel efficiency. Unfortunately, the technology that powers self-driving cars uses a fair amount of energy and may decrease the fuel efficiency of a given vehicle by 5 to 10 percent. However, a self-driving car can be programmed to drive as fuel efficiently as possible and use less fuel than the most careful and energy-mindful human driver. And presumably, self-driving cars will eventually make fewer navigation and other errors. So over time, self-driving cars may allow fuel efficiency gains to continue.

Energy efficiency is an important consideration when making fuel and technology choices, but it is not the only factor we must consider. Fuel choice is also important. Determining the best fuel for a particular activity is not always easy, and it involves trade offs among convenience, ease of use, safety, cost, and pollution.

36-4 What are electricity generation and cogeneration?

Generation and cogeneration convert fuels to electricity

Because electricity can be generated from many different sources, including fossil fuels, wind, water, and the Sun, electricity is a form of energy in its own category. Coal, oil, and natural gas are primary sources of energy. Electricity is a secondary source of energy, meaning that we obtain it from the conversion of a primary source. As a secondary source, electricity is an **energy carrier**—something that can move and deliver energy in a convenient, usable form to end users. End users are residents of homes, students at a school—anyone who uses electricity such as turning on a light or charging a phone at the end point of use.

Energy carrier An energy source such as electricity that can move and deliver energy in a convenient, usable form to end users.

Approximately 40 percent of the energy consumed in the United States is used to generate electricity. But because of conversion losses during the electricity generation process, of that 40 percent, only 13 percent of total energy consumed in the United States is available electricity for end uses. In this section we will look at some of the basic concepts and issues related to generating electricity from fossil fuels.

The Process of Electricity Generation

Electricity is produced by conversion of primary sources of energy such as coal, natural gas, or wind. Electricity is clean at the point of use; no pollutants are emitted in your home when you use a light bulb or computer. When electricity is produced by combustion of fossil fuels, however, pollutants are released at the location of its production. And, as we have seen, the transfer of energy from a fuel to electricity is only about 36 percent efficient. Therefore, although electricity is highly convenient, from the standpoint of efficiency of the overall system and the total amount of pollution released, it is usually more desirable to transfer heat directly to a home with wood or oil combustion, for example, rather than via electricity generated from the same materials. The energy source that entails the fewest conversions from its original form to the end use is likely to be the most efficient. Many types of fossil fuels, as well as nuclear fuels, can be used to generate electricity. Regardless of which fuel is used, all thermal power plants work in the same basic way—they convert the potential energy of a fuel into electricity. We will show the details of how a thermal power plant converts the potential energy in coal or natural gas to electricity in the next module. Here, we will discuss the efficiency of electricity generation and cogeneration and the capacity of an electricity generation plant.

AP® Exam Tip

Recall from Module 0 that potential energy is energy that has been stored and not yet released, while kinetic energy is the energy of motion. Also make sure you know the first law of thermodynamics, which states that energy cannot be created or destroyed, it can only change forms. These concepts are critical for the AP® Environmental Science Exam and will help you understand what happens in a coal-fired electricity generation plant.

Efficiency of Electricity Generation

Whereas a typical coal-burning power plant has an efficiency of about 36 percent, newer coal-burning power plants may have slightly higher efficiencies of approximately 40 percent. Power plants using other fossil fuels can be even more efficient. An improvement in gas combustion technology has led to the **combined cycle** natural gas–fired power plant, which uses both a steam turbine to generate electricity and a separate turbine which is powered by the exhaust gases from natural gas combustion. For this reason, a combined cycle plant can achieve efficiencies of up to 60 percent.

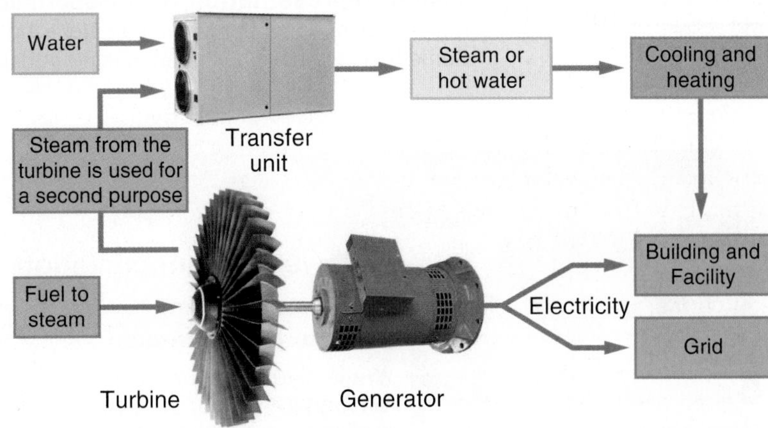

FIGURE 36.5 A cogeneration power plant. Steam is generated and used to turn a turbine, as in a coal-burning power plant, but the steam is also used to heat a building. Greater efficiency is achieved compared to operating two separate plants. Also known as combined heat and power. *(Modified after U.S. EPA)*

Capacity

The **capacity** of a power plant is its maximum electrical output. A typical power plant in the United States might have a capacity of 500 megawatts (MW). Recall that a watt is a joule per second, so a power plant with a 500 MW capacity generates 500 MW continuously. If the plant operated for one day, it would generate 500 MW × 24 hours = 12,000 megawatt-hours (MWh). Most home electricity is measured in kilowatt-hours (kWh). Since 1 MWh equals 1,000 kWh, to convert MWh per month into kWh per month, we multiply by 1,000.

The typical power plant we've just described would generate 12,000 MWh × 1,000 kWh/MWh = 12,000,000 kWh in a day. If it operated for 365 days per year, it would generate 365 times that daily amount.

Most power plants, however, do not operate every day of the year. They must be shut down for maintenance, refueling, or repairs. Other factors may also influence whether a plant is shut down for a time, including the demands of users on the grid and electricity production by other power plants on the grid. Therefore, to determine a plant's output, it is useful to report the amount of time it actually operates in a year. This number—the fraction of time a plant is operating in a year—is known as its **capacity factor**. Most thermal power plants have capacity factors of 0.9 or greater, except when they are voluntarily shut down by the operators due to low demand for electricity from that plant or that type of fuel. Nuclear fueled power plants have a capacity factor of 0.9 or greater as do coal-burning power plants in parts of the United States where coal is an abundant and preferred fuel. As we will see in Modules 39 and 41, power plants using some forms of renewable energy, such as wind and solar, may have a capacity factor of only about 0.25. "Do the Math: Calculating Energy Supply" on page 422 shows you how to calculate the amount of energy a power plant can supply.

A power plant, both nuclear and coal-fired plants, may take hours, or even a day, to come up to full generating capacity. Because of the time it takes for them to become operational, electric companies tend to keep nuclear, natural gas, and some coal-fired plants running at all times. As demand for electricity changes during the day or week, plants that are more easily powered up, such as those that use water (hydro), are used.

Cogeneration

The use of a fuel to both generate electricity and deliver heat to a building or industrial process is known as **cogeneration**, also called **combined heat and power** (**FIGURE 36.5**). Don't confuse this with combined cycle electricity generation described earlier on this page. They are similar in some ways, but not the same. Cogeneration is a method employed by certain power plants for obtaining greater efficiencies. When steam is used to generate electricity, there is a good amount of waste heat that is not used for any purpose. So the power plant design allows for the waste heat to be used for industrial purposes or to heat buildings, and in this way, the power plant will achieve greater overall efficiency than by generating electricity and heat separately in two different power plants. Cogeneration efficiencies can be as high as 90 percent, whereas steam heating alone might be 75 percent

Combined cycle A feature in some natural gas–fired power plants that uses both a steam turbine to generate electricity and a separate turbine that is powered by the exhaust gases from natural gas combustion to turn another turbine to generate electricity.

Capacity The maximum electrical output of something such as a power plant.

Capacity factor The fraction of time a power plant operates during a year.

Cogeneration The use of a fuel to both generate electricity and deliver heat to a building or industrial process. *Also known as* ***combined heat and power***.

Preparing for the AP® Exam

Visual Representation 6 **Transportation Decisions** While buses and trains are almost always the most efficient option for individuals traveling any distance, all transportation decisions involve trade offs. Some things to consider when making transportation decisions are fuel efficiency, carbon emissions, the

Human-Powered Transportation

Bike: Riding a bicycle is an energy-efficient option. An average person cycling at a moderate pace will burn 50 calories per mile.

Walk: If you live close enough to work or school, walking is a good option. Depending on your weight it may take between 100–300 calories to walk half an hour.

E-bike: An electric bicycle requires less work from the rider and may be useful for commuting longer distances, carrying a trailer, or negotiating hilly terrain. Electric bicycles use a rechargeable battery and may offer pedal assist or power-on-demand. E-bikes are more expensive to purchase and maintain, and they are slow to charge.

Manufacturing the batteries used in electric-powered vehicles requires the use of nonrenewable minerals. Vehicles charged with electricity generated from renewable energy sources are more fuel-efficient and reduce carbon emissions more than vehicles charged with electricity generated by fossil fuels.

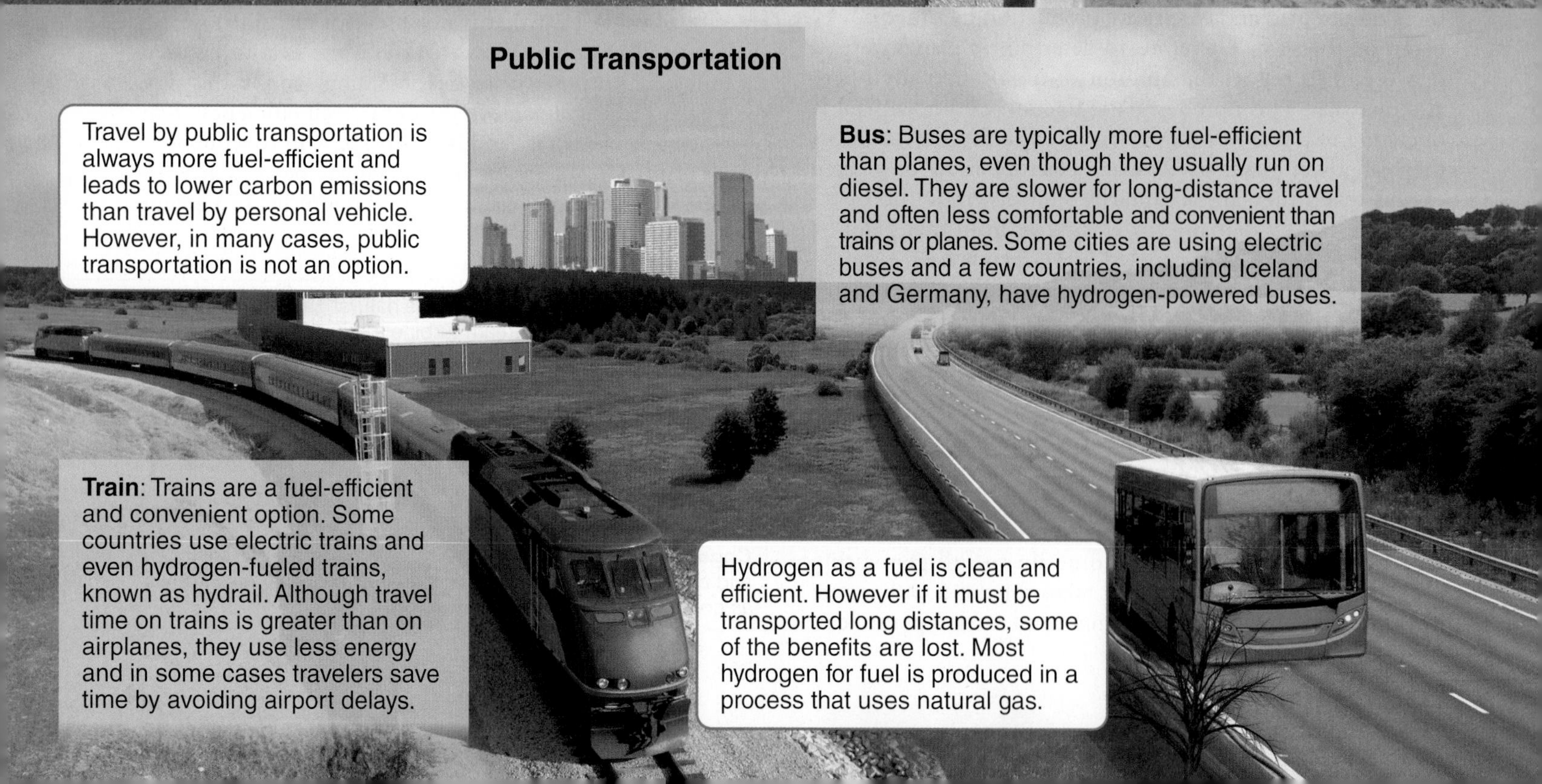

Public Transportation

Travel by public transportation is always more fuel-efficient and leads to lower carbon emissions than travel by personal vehicle. However, in many cases, public transportation is not an option.

Bus: Buses are typically more fuel-efficient than planes, even though they usually run on diesel. They are slower for long-distance travel and often less comfortable and convenient than trains or planes. Some cities are using electric buses and a few countries, including Iceland and Germany, have hydrogen-powered buses.

Train: Trains are a fuel-efficient and convenient option. Some countries use electric trains and even hydrogen-fueled trains, known as hydrail. Although travel time on trains is greater than on airplanes, they use less energy and in some cases travelers save time by avoiding airport delays.

Hydrogen as a fuel is clean and efficient. However if it must be transported long distances, some of the benefits are lost. Most hydrogen for fuel is produced in a process that uses natural gas.

amount of luggage needed, time available, and distance to be covered. This figure examines pros and cons of some of the common options individuals have when selecting transportation, with the addition of some up and coming technology that may be more widely available in the future.

Personal Vehicles
Hybrid electric vehicle: HEVs have both an internal combustion engine and an electric motor that uses energy stored in batteries. The battery is charged by the internal combustion system and regenerative braking that captures kinetic energy of the vehicle's motion. HEVs do not travel on the battery alone.
Electric vehicles: EVs run exclusively on an electric motor powered by a battery pack. Electric vehicles emit no exhaust during operation and can travel between 250–300 miles on a single charge. Charging stations are becoming more common. Manufacturers are developing bi-directional charging which allows an electric vehicle to serve as a backup source of electricity to a residence during power outages.
$O_2 + 2H_2$
$2H_2O$
Hydrogen fuel cell electric vehicle: FCEVs run on an electric motor powered by electricity generated by hydrogen. A vehicle can recharge in just a few minutes, has no emissions, and does not use electricity from the grid. One tank can run for 350 miles. Cost and lack of recharging stations currently limit their use.
Plug-in hybrid electric vehicles: PHEVs have an internal combustion engine and a battery-powered electric motor. The battery can be charged with electricity and through regenerative braking. The car will use electric power until the battery is almost depleted and then switch to the internal combustion engine. Gasoline usage is decreased by as much as 60 percent, and the vehicle can travel approximately 20 miles on the battery.
Combustion engine: A traditional combustion engine uses gasoline or diesel fuel. Fossil fuel use and emissions are highest with this option. These vehicles are the least expensive to purchase.

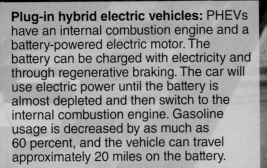

DO THE MATH Calculating Energy Supply

Preparing for the AP® Exam

According to the U.S. Department of Energy, an average electricity customer in the United States uses approximately 900 kWh of electricity per month. On an annual basis, this is

$$900 \text{ kWh/month} \times 12 \text{ months/year} = 10,800 \text{ kWh/year}$$

How many homes can a 500 MW power plant with a 0.9 capacity factor support?

Begin by determining how much electricity the plant can provide per month:

$$500 \text{ MW} \times 24 \text{ hours/day} \times 30 \text{ days/month} \times 0.9 = 324,000 \text{ MWh/month}$$

1 MWh equals 1,000 kWh, so to convert MWh per month into kWh per month, we multiply by 1,000:

$$324,000 \text{ MWh/month} \times 1,000 \text{ kWh/MWh} = 324,000,000 \text{ kWh/month}$$

So,

$$\frac{324,000,000 \text{ kWh/month}}{900 \text{ kWh/month/home}} = 360,000 \text{ homes}$$

On average, a 500 MW power plant can supply roughly 360,000 homes with electricity in a given year.

YOUR TURN During summer months, in hot regions of the United States, some homes run air conditioners continuously. How many homes can the same power plant support if average electricity usage increases to 1,200 kWh/month during summer months?

efficient, and electricity generation alone might be 35 percent efficient.

There are roughly 11,000 power plants in the United States with generation capacity of 1 MW or greater. In 2020, they generated approximately 4 billion MWh (4 trillion kWh). **FIGURE 36.6** shows the fuels that were used to generate this electricity. Natural gas–fired power plants are the largest component of electricity generation in the United States, responsible for 40 percent of all electricity produced. Three other sources account for the remaining 60 percent: the combination of all renewables (wind, hydroelectric, solar, and others) at 21 percent, nuclear energy at 20 percent, and coal at 19 percent. Expanded domestic production of natural gas has increased the fraction of electricity generated from natural gas while the fraction of electricity generated from coal has decreased. In 2005, coal accounted for 50 percent of electricity generation. That is a large decrease in coal-produced electricity in only 15 years. And the fact that renewables combined account for more electricity generation than either coal alone or nuclear alone is an important indicator of the direction energy use and electricity generation are headed in the coming years.

In this module we examined fossil fuels and described the role of the Sun in fossil fuels and many renewable fuels. We described the major fuel types and how they are used. And we examined specialized uses of fossil fuels and started to examine electricity generation and cogeneration. In the next module we will examine fossil fuel distribution and its advantages and disadvantages and we will explore electricity generation in greater detail.

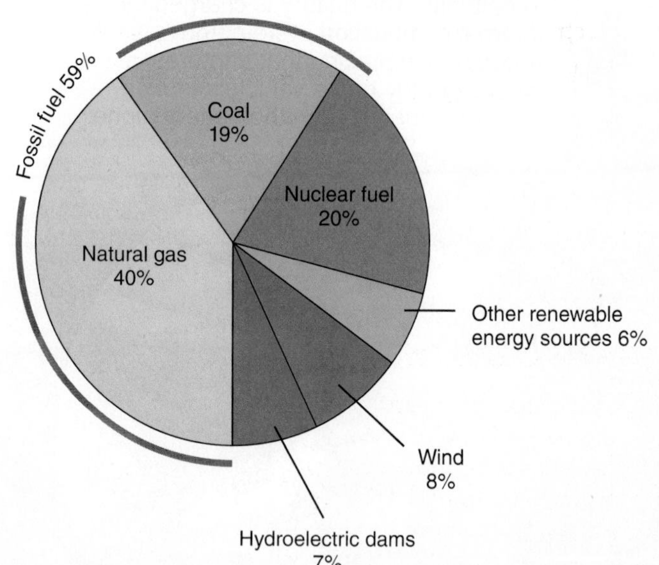

FIGURE 36.6 Fuels used for electricity generation in the United States in 2020. Natural gas is the fuel most used for electricity generation. The electricity fuel mix has changed considerably in the last 20 years due to the increased availability of and decreasing price of natural gas and renewables. Note that all renewables combined account for more electricity generation than either coal alone or nuclear alone. *(Data from U.S. Department of Energy, Energy Information Administration, 2021)*

Module 36 AP® Review

Preparing for the AP® Exam

Learning Goals Revisited

36-1 Is the Sun the ultimate source of many fuels?

The Sun is the source of energy contained within fossil fuels and is also the energy source for many renewable fuels including biomass, solar, wind and hydro. We call the carbon in biomass modern carbon, in contrast to the carbon in fossil fuels, which we call fossil carbon. Unlike modern carbon, which was recently in the atmosphere, fossil carbon has been buried for millions of years.

36-2 What are the major fuel types and how are they used?

Different fuel types are best suited for specific purposes. Wood is used in much of the developing world as firewood and charcoal. Peat and three different types of coal are derived from fossil plant material and are used in electricity generation, industrial processes, and heating. Natural gas is comprised of mostly methane and is used for electricity generation, industrial processes, and home use such as heating interior living spaces and making hot water. Crude oil can be extracted from the ground as oil or as tar sands. Oil can be distilled to a variety of products such as kerosene, aviation fuel, and gasoline.

36-3 What are other uses of fossil fuel?

Electric hot water heaters contain a resistance coil that generates heat inside the tank of water and are very efficient. But generating electricity can have a variety of efficiencies so the overall process may be less efficient. Natural gas heaters are less efficient, but their overall efficiency may be greater. Different modes of transportation use different fuels and have different efficiencies. In general, one person alone in a car is more energy intensive than more people in a car and in public transportation.

36-4 What are electricity generation and cogeneration?

Natural gas and coal are two fuels used to generate electricity. Cogeneration is a process where the heat from fuel combustion is used to generate electricity and heat buildings. It is a more efficient process than conducting these activities separately.

AP® Practice Questions

Preparing for the AP® Exam

Multiple-Choice Questions

1. Which pair of energy sources originate from the Sun?
   (a) coal and nuclear
   (b) solar and geothermal
   (c) oil and hydropower
   (d) nuclear and tidal power

2. Traveling alone in a car uses 3.6 megajoules (MJ) of energy per kilometer. If four people go on a trip of 400 miles, how many MJ are used per person? (Hint: 1.609 kilometers = 1 mile)
   (a) 200 MJ
   (b) 320 MJ
   (c) 580 MJ
   (d) 860 MJ

3. The largest source of energy for electricity in the United States is
   (a) natural gas.
   (b) coal.
   (c) oil.
   (d) nuclear.

4. Which of the following best describes cogeneration?
   (a) the use of two or more energy sources to generate electricity
   (b) the use of two separate turbines to generate electricity
   (c) a method of electricity generation that includes renewable energy
   (d) a method to generate electricity and heat a building

Use the diagram below to answers questions 5 & 6:

The diagram below depicts a fossil fuel-fired power plant. Four components of the power plant are labeled A through D.

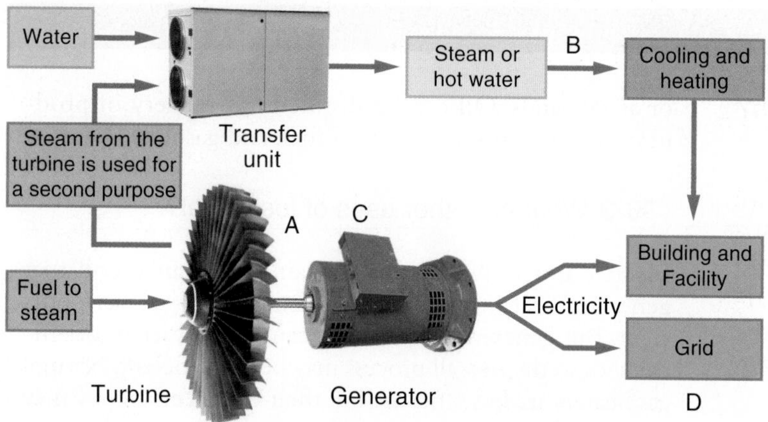

5. Which component of the power plant represents kinetic energy moving a turbine?
   (a) A
   (b) B
   (c) C
   (d) D

6. Which component of the power plant represents a secondary source of energy?
   (a) A
   (b) B
   (c) C
   (d) D

Free-Response Question

Much of the discussion about climate change focuses on the extraction and use of nonrenewable resources such as fossil fuels.

(a) **Identify** one nonrenewable energy source that derives its energy from the Sun. (1 pt.)

(b) **Describe** why fossil fuels are not considered to be modern carbon sources. (1 pt.)

(c) **Explain** why fossil fuel-burning power plants have a low overall efficiency. (1 pt.)

(d) **Describe** how the process of cogeneration can allow a power plant to increase its total efficiency. (1 pt.)

A natural-gas fired power plant has a maximum output of 655 megawatts (MW) of power and a capacity factor of 0.9.

(e) If the power plant was operational for one year, **calculate** how many kilowatt hours (kWh) of electricity would be generated by the power plant. Show all work. (2 pts.)

(f) The average home in the United States uses approximately 900 kWh of electricity each month. **Calculate** how many homes the power plant can support in one year. Show all work. (2 pts.)

(g) If the natural–gas power plant is 60 percent efficient, **calculate** how much natural gas, in cubic feet (ft^3), would need to be burned in order to operate at full capacity if 3.3 cubic feet of natural gas produces 1 kWh of electricity. Show all work. (2 pts.)

Module 37

Unit 6 | 35 | 36 | 37 | 38 | 39 | 40 | 41

Distribution of Natural Energy Resources: Fossil Fuels

In Module 36 we began our examination of the fuels we use and some of the advantages of each. We also described specialized uses of fuels for transportation and began to describe electricity generation and cogeneration. In this module we will explain why particular fossil fuels occur in specific locations and how many years use of each resource are available. We will also describe fossil fuel advantages and disadvantages, oil extraction and fracking, and explore electricity generation in more detail.

Learning Goals

After reading this module you should be able to

37-1 explain why fossil fuels and ores are found only in certain locations.

37-2 describe the advantages and disadvantages of fossil fuels including oil extraction and fracking.

37-3 describe how fossil fuels are used for electricity generation.

37-1 Why are fossil fuels and ores found only in certain locations?

Fossil fuel and ore distribution around the globe depends on the geology of the region

In Module 36 we stated that coal is formed from the remains of trees and ferns that were preserved 280 to 360 million years ago. Figure 36.2 illustrated the process with different kinds of peat, lignite, and coal. The organic matter that is eventually going to become coal must experience the unique circumstances that prevent it from decomposing, which is what it normally does as part of the carbon cycle. The organic matter must become buried quickly without being exposed to air. This typically happens in tropical locations, perhaps in wetlands and river deltas, where material is quickly deposited over ferns and leaves. Locations where we find coal today experienced those circumstances roughly 300 million years ago. The largest coal reserves today are found in the United States, Russia, China, Australia, and India. The countries that are currently producing the greatest amounts of coal are China, the United States, India, and Australia.

Crude oil, or petroleum as it is also known, is formed from the remains of ocean-dwelling phytoplankton (microscopic algae) that died 50 million to 150 million years ago. As with coal, these deposits of organic matter did not decompose as they normally do as part of the carbon cycle. Deposits of phytoplankton are found in locations where porous sedimentary rocks, such as sandstone, are capped by nonporous rocks. Crude oil forms over millions of years and fills the pore spaces in the rock. Geologic events related to the tectonic cycle we discussed in Module 19 may deform the rock layers so that they form a dome, as we saw in Figure 35.6. The fossilized organic matter that is becoming crude oil is less dense than the rock, so over time it migrates upward toward the highest point in the porous rock, where it is trapped by the nonporous rock. The natural gas often associated with crude oil is even less dense, so it migrates to the absolute highest point in the dome, above the oil. In certain locations, crude oil flows out under pressure the way water flows from an artesian well, as described in Module 27. But usually, petroleum producers must drill wells into a deposit and extract the oil with pumps. After extraction, the oil must be transported to a petroleum refinery, by pipeline if the well is on land, or by supertanker if the well is underwater.

Crude oil contains natural gas. As we saw in Figure 35.6, some of this gas separates out naturally and floats to the top of the dome. The large flame that is sometimes associated with oil wells is a gas flare that is created when oil workers flare, or burn off, the natural gas, which is done under

FIGURE 37.1 Flaring natural gas at an oil well. Natural gas (methane) is frequently associated with oil wells and workers commonly burn the natural gas (called flaring) as a way to remove the gas and reduce the likelihood of an explosion. *(Terrance Klassen/AGE Fotostock)*

controlled conditions to prevent an explosion (**FIGURE 37.1**). This gas can also be captured for use as fuel. As we saw in Figure 35.4, the United States uses just slightly more oil than natural gas. In 2020, this was roughly 3.2 billion liters (850 million gallons) of petroleum products per day. Gasoline accounts for roughly one-half of that amount. While the primary use of petroleum products is for transportation, petroleum is also the raw material for petrochemicals, such as plastics, and for lubricants, pharmaceuticals, and cleaning solvents. Worldwide petroleum consumption is almost 15 billion liters (4 billion gallons) per day, of which the United States is responsible for about 21 percent. The top petroleum-producing countries in 2020 were the United States, Saudi Arabia, Russia, Canada, and China in roughly that order. These five countries account for approximately one-half of worldwide oil production. The United States and China are responsible for roughly 35 percent of world oil consumption.

As we discussed in Module 29, ores are concentrated accumulations of minerals and are formed by a variety of geologic processes. The distribution of certain ores, such as uranium, has importance in relation to energy resources because certain uranium deposits can be mined and refined and used to fuel nuclear power plants. We will discuss this further in the next module.

37-2 What are the advantages and disadvantages of fossil fuels including oil extraction and fracking?

Fossil fuels have many advantages and disadvantages

There are many advantages and disadvantages to coal, oil, and natural gas and we have identified a few of these in Modules 35 and 36. Here we will outline a few more, and also describe oil extraction and fracking. We will discuss the advantages and disadvantages of uranium ore in the next module when we discuss nuclear-powered electricity generation plants.

Advantages of Coal

Because it is energy-dense and plentiful, coal is used to provide heat energy for industrial processes such as making steel and for electricity generation. In many parts of the world, coal reserves are relatively easy to exploit by surface mining. The technological demands of surface mining are relatively small, and the economic costs are low. Once coal is extracted from the ground, it is relatively easy to move and needs little refining before it is burned. It can be transported to power plants and factories by train, barge, or truck. All of these factors make coal a relatively easy fuel to use, regardless of technological development or infrastructure.

Disadvantages of Coal

Although coal is a relatively inexpensive fossil fuel, its use does have several disadvantages. As we saw in Module 29, the environmental consequences of the tailings from surface mining are significant. As surface coal is used up and becomes harder to find, subsurface mining becomes necessary. With subsurface mining, the technological demands and costs increase, as do the consequences for human health.

Coal contains a number of impurities, including sulfur, that are released into the atmosphere when the coal is burned. The sulfur content of coal typically ranges from 0.4 to 4 percent by weight. Lignite and anthracite have relatively low sulfur contents, whereas the sulfur content for bituminous coal is often much higher. Trace metals such as mercury, lead, and arsenic are also found in coal. Combustion of coal results in the release of these elements, which leads to an increase of sulfur dioxide and other air pollutants, such as particulates, in the atmosphere, as we will see in Module 43. Compounds that are not released into the atmosphere remain behind in the resulting ash.

In order to reduce the chemical compounds released into the air, coal companies wash their coal in a variety of organic compounds. In some cases, these compounds have not been widely tested for toxicity or their effect on humans and ecosystems. In West Virginia in 2014, a chemical storage tank containing one such coal cleaning compound leaked onto the surrounding ground surfaces and eventually the underlying aquifer. Residents nearby immediately noticed a sweet odor and ultimately over 300,000 residents were advised not to drink their water for a number of days because the compound is poisonous and potentially cancer causing. Unfortunately, chemical spills are not the only accidents related to coal combustion. The residual ash from coal combustion can also be problematic.

According to the U.S. Department of Energy, there were approximately 550 coal-producing mines in the United States and they produced roughly 480 million metric tons of coal in 2020, although this is many fewer mines and much less coal production than 5 or 10 years earlier. Most of this coal is burned in the United States, with anywhere from 3 to 20 percent of the coal remaining behind as ash. Large deposits of this ash are often stored near coal-burning power plants. One such ash deposit—a mixture of ash and water—was kept in a holding pond at a power plant near Knoxville, Tennessee. In December 2008, the retaining wall that contained the ash gave way and spilled 4.2 billion liters (1.1 billion gallons) of ash onto the surrounding ground surfaces (**FIGURE 37.2**). Three houses were destroyed by the flow of muddy ash, which covered over 121 ha (300 acres) of land. The event was the largest of its kind in U.S. history. It took 5 years and over $1 billion to clean up the bulk of the ash.

Finally, coal is a significant source of air pollution. Coal is 60 to 80 percent carbon. When it is burned, most of that carbon is converted into CO_2 and energy is released in the process. Coal produces far more CO_2 per unit of energy released than either oil or natural gas and it contributes to the increasing atmospheric concentrations of CO_2.

FIGURE 37.2 Results of a coal ash spill. This Tennessee home was buried in ash when an ash holding pond at a nearby power plant gave way. *(J. Miles Cary/Knoxville News Sentinel/AP Images)*

Advantages of Oil

Because oil is a liquid, it is extremely convenient to transport and use. It is relatively energy-dense and is cleaner-burning than coal. For these reasons, it is an ideal fuel for mobile combustion engines such as those found in automobiles, trucks, and airplanes. Because it is a fossil fuel, it releases CO_2 when burned, although for every joule of energy released, oil produces only about 85 percent as much CO_2 as coal.

Disadvantages of Oil

Oil, like coal, contains sulfur and trace metals such as mercury, lead, and arsenic, which are released into the atmosphere when it is burned. Some sulfur can be removed during the refining process, so it is possible, though more expensive, to obtain low-sulfur oil. Since 2010, all diesel fuel for trucks sold in the United States is ultra-low sulfur diesel, meaning it contains less than 15 mg L^{-1} (parts per million) sulfur.

As we have seen, oil must be extracted from under the ground or beneath the ocean. Whenever oil is extracted and transported, there is the potential for oil to leak from the wellhead or to be spilled from a pipeline or tanker, as we discussed in the opener to this unit. Some oil naturally escapes from the rock in which it was stored and seeps into water or out onto land. However, commercial oil extraction has greatly increased the number of leakage and spillage events and the amount of oil that has been lost to land and water around the world. As noted at the beginning of this unit, the largest oil spill in the United States until 2010 was the *Exxon Valdez* oil tanker accident in 1989. The blowout of the BP Deepwater Horizon oil well, drilled 81 km (50 miles) off the coast of Louisiana in 1,524 m (5,000 feet) of water, led to a spill of over 780 million liters (206 million gallons) of oil.

Larger oil spills have occurred elsewhere in the world. For example, during the 1991 Persian Gulf War, approximately 912 million liters (240 million gallons) of oil were spilled when wellheads were deliberately sabotaged or destroyed by the Iraqi army in Kuwait.

Oil is spilled into the natural environment in various ways. A 2003 National Academy of Sciences study found that oil extraction and transportation were responsible for a relatively small fraction of the oil spilled into marine waters worldwide. It found that roughly 85 percent of the oil entering marine waterways came from runoff from land, airplanes, small boats, and personal watercraft, including both deliberate and accidental releases of waste oil.

In the United States, debates continue over the trade off between domestic oil extraction and the consequences for habitat and species living near oil wells or pipelines. For example, when a 1,300-km (800-mile) pipeline was constructed to transport oil overland from the North Slope of Alaska to tankers that would carry it south to the contiguous United States, wildlife biologists predicted that the pipeline might melt permafrost and interfere with the calving grounds of caribou. Scientists continue to monitor the pipeline, but so far have not reached definitive conclusions about its environmental impact. The transportation of oil by means other than pipeline can have significant consequences as well. There have been a number of serious railway

FIGURE 37.3 The Lac-Mégantic accident. In 2013, a railroad train carrying oil derailed and exploded in Quebec, Canada, near the Maine border. *(Paul Chiasson/AP Images)*

accidents in recent years as a result of increased domestic drilling for oil in the United States. In a 2013 accident in Quebec, Canada, near the Maine border, a railroad train carrying oil derailed and exploded in a small town in the early morning hours, killing 47 people and destroying many buildings (**FIGURE 37.3**).

The debate about the environmental effects of land-based oil extraction continues with periodic proposals to allow oil exploration in the Arctic National Wildlife Refuge (ANWR), a 7.7 million hectare (19 million acre) tract of land in northeastern Alaska, shown in **FIGURE 37.4**. Proponents of exploration suggest that ANWR might yield 95 billion liters (25 billion gallons) to 1.4 trillion liters (378 billion gallons) of oil and substantial quantities of natural gas. Opponents maintain that opening ANWR to exploration and petroleum extraction will harm pristine habitat for many species as well as adversely affect people living in the area. Congress opened parts of ANWR in 2017 and the first oil and gas leases were issued in early 2021. The next day, they were temporality halted and remain halted as this book went to press.

Humans, as well as wildlife, have been harmed by oil extraction. In Nigeria and many other developing countries, oil fields are adjacent to villages. Thick, gelatinous crude oil covers the ground where people walk, sometimes in bare feet. Oil flaring—the burning off of excess natural gas—takes place close to homes and causes local air pollution problems. Concerns about the effects of oil extraction on health, human rights, and environmental justice have led to violent political protests against oil companies in Nigeria and elsewhere.

Advantages of Natural Gas

Because of the extensive natural gas pipeline system in many parts of the United States, roughly one-half of homes use natural gas for heating. Compared with coal and oil, natural gas contains fewer impurities and therefore emits almost no sulfur dioxide or particulates during combustion. And for every joule of energy released during combustion, natural gas emits only 60 percent as much CO_2 as coal. So natural gas is the cleanest of the fossil fuels and as long as it can be supplied by pipeline, it is a very convenient and desirable fossil fuel. In some locations where natural gas pipelines are not present, liquified petroleum gas (LPG) is used although it is slightly less convenient because it has to be delivered to businesses and homes by a fuel truck.

Disadvantages of Natural Gas

While natural gas when combusted releases the least carbon dioxide of all the fossil fuels, unburned natural gas—methane—that escapes into the atmosphere is itself a potent greenhouse gas that is 25 times more efficient at absorbing heat from the Earth than CO_2.

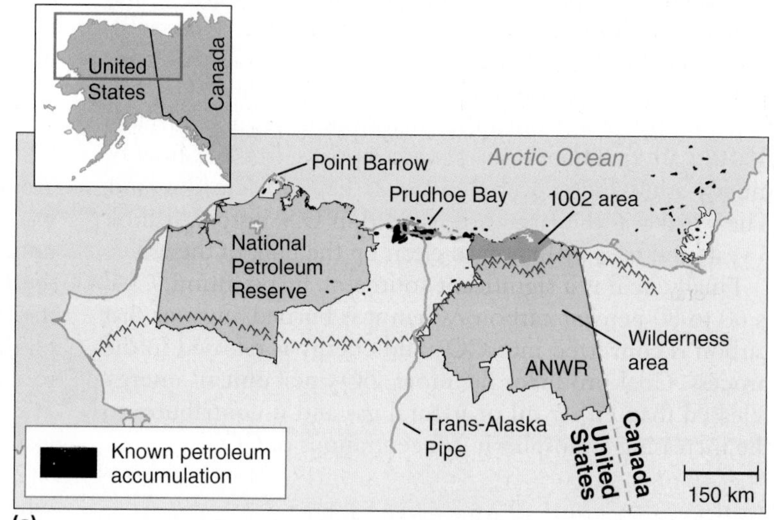

FIGURE 37.4 The Arctic National Wildlife Refuge (ANWR). (a) This map shows ANWR and adjacent areas on the North Slope of Alaska. The area in orange is the coastal plain, also called the 1002 area (pronounced as "ten-o-two"), which is part of the area where leases were issued and then halted for petroleum exploration and extraction. (b) Caribou are among the many species that live in ANWR. *(b: Jim Goldstein/DanitaDelimont.com)*

Natural gas that leaks after extraction is a suspected contributor to the steep rise in atmospheric methane concentrations that was observed in the 1990s. While natural gas is referred to as the "clean" fossil fuel, extraction and use still lead to environmental problems such as the broad extent of gas fields shown in **FIGURE 37.5**. A great deal of attention has been paid to one fairly recent method of extracting natural gas: fracking.

Fracking

The process of exploring for natural gas involves the "thumper trucks" that have already been mentioned in the opener to this unit. Because of major advances in extraction technology, in the last 2 decades oil and mining companies have increased their reliance on one specific method of recovering natural gas and sometimes oil: fracking. **Fracking**, short for hydraulic fracturing, is a method of oil and gas extraction that uses high-pressure fluids to force open existing cracks in rocks deep underground. This technique allows extraction of natural gas from locations that were previously so difficult to reach that extraction was economically unfeasible. As a result, large quantities of natural gas are now available in the United States at a lower cost than before. Fracking has increased our reliance on a domestic energy source and created jobs. Roughly 38 percent of energy consumed in the United States is used to generate electricity and from 1980 through 2000, 50 percent of that energy came from coal and only 15 percent came from natural gas. As stated in Module 36, as a result of greater natural gas availability and lower price due to fracking, in 2020, 40 percent of electricity generated came from natural gas and only 19 percent from coal. Since coal emits more air pollutants—including carbon dioxide—than does natural gas, fracking and the increase in natural gas use for electricity generation initially appeared to be beneficial to the environment.

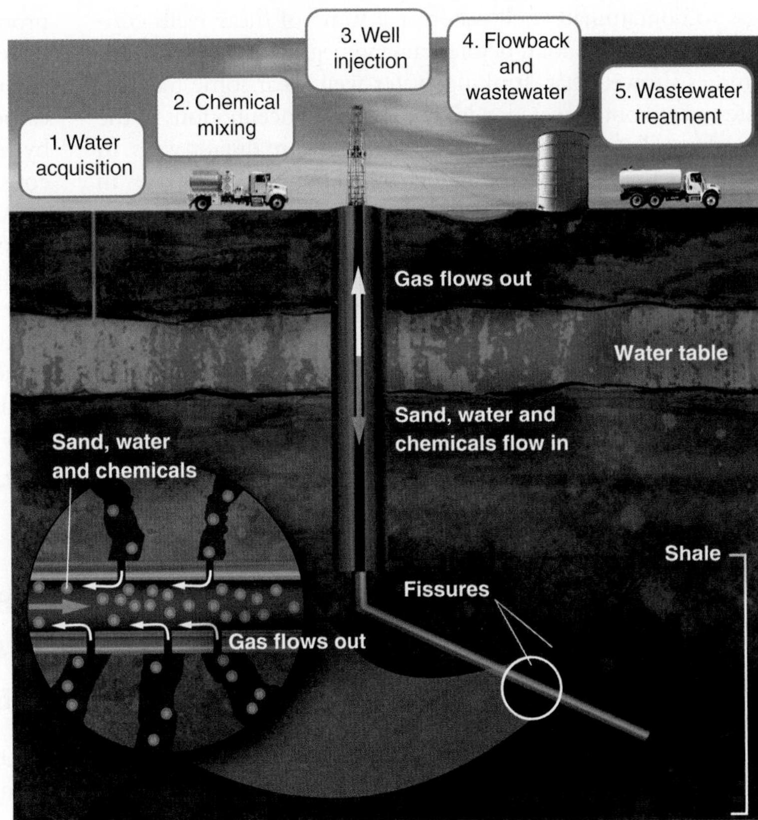

FIGURE 37.6 **Basic aspects of natural gas fracking.** A well is drilled vertically then horizontally. Water, sand, and chemical fracking fluid are injected into the well. Pressure from the fluid and sometimes explosives are used to expand existing fissures and gas flows out. *(Modified after a diagram from the U.S. Environmental Protection Agency)*

FIGURE 37.5 **Natural gas field in Wyoming.** Even though natural gas is relatively clean compared with other fossil fuels, its extraction impacts large amounts of land. *(Jim Havey/Alamy stock photo)*

However, reports gained attention both in the popular press and in scientific journals about the negative consequences of fracking. Large amounts of water are used in the fracking process, with millions of gallons of water taken out of local streams and rivers and pumped down or injected into each gas well. A portion of this water is later removed from the well and must be properly treated after use to avoid contaminating local water bodies.

A variety of chemicals are added to the fracking fluid to facilitate the release of natural gas (**FIGURE 37.6**). Mining companies are not required to publicly identify these chemicals. Environmental scientists and concerned citizens began to wonder if fracking was responsible for chemical contamination of nearby aquifers. Some drinking-water wells near fracking sites became contaminated with natural gas, and homeowners and public health officials asked if fracking was the culprit. Water with high concentrations of natural gas can be flammable, and footage of flames shooting from kitchen faucets after someone ignited the water became popular on YouTube, in documentaries, and in feature films. However, it wasn't clear if fracking caused natural

Fracking Short for hydraulic fracturing, a method of oil and gas extraction that uses high-pressure fluids to force open existing cracks in rocks deep underground.

gas to contaminate well water or if some of these wells contained natural gas long before fracking began. Several reputable studies showed that drinking-water wells near some fracking sites were contaminated, with natural gas concentrations in the nearby wells being much higher than in more distant wells. In addition, there has been a higher frequency of earthquakes in central and southwestern United States and the U.S. Geological Survey, the primary government agency responsible for understanding and assessing earthquakes and other geological hazards, reported in 2016 that the injection of wastewater from fracking is believed to be the primary cause. These issues are receiving further study.

Fracking also releases **volatile organic compounds (VOCs)**, a category of organic compound air pollutants that evaporate at typical atmospheric temperatures, from both the fracking fluid and from the vehicles and machinery that are used in the fracking operation. VOCs are a precursor to other types of air pollution, and some can harm human health. We will cover VOCs in greater detail in Module 43.

In a direct connection to global climate change, more and more studies have begun to suggest that significant quantities of natural gas escape as "leakage" during the fracking and gas extraction process. Estimating the amount of escaped natural gas, also called fugitive gas, has become a controversial subject due to the large uncertainties in the natural gas extraction process, measurement difficulties, and industry secrecy. Estimates range from 2 to 9 percent. As we have discussed, methane is 25 times more efficient at trapping heat from Earth than carbon dioxide, the greenhouse gas most commonly produced by human activity. Due to its high heat trapping ability, the consequences of methane leakage are potentially substantial.

Almost certainly, using natural gas is better for the environment than using coal, though using less fossil fuel—or using no fossil fuel at all—would be even better. However, at present it is difficult to know whether the benefits of using natural gas outweigh the problems that extraction causes. Many years may pass before the extent and nature of harm from fracking is known.

37-3 How are fossil fuels used for electricity generation?

Fuel is converted to electricity and releases carbon dioxide and heat energy

FIGURE 37.7 illustrates the major features of a typical coal-burning power plant. Fuel—in this case, coal—is delivered

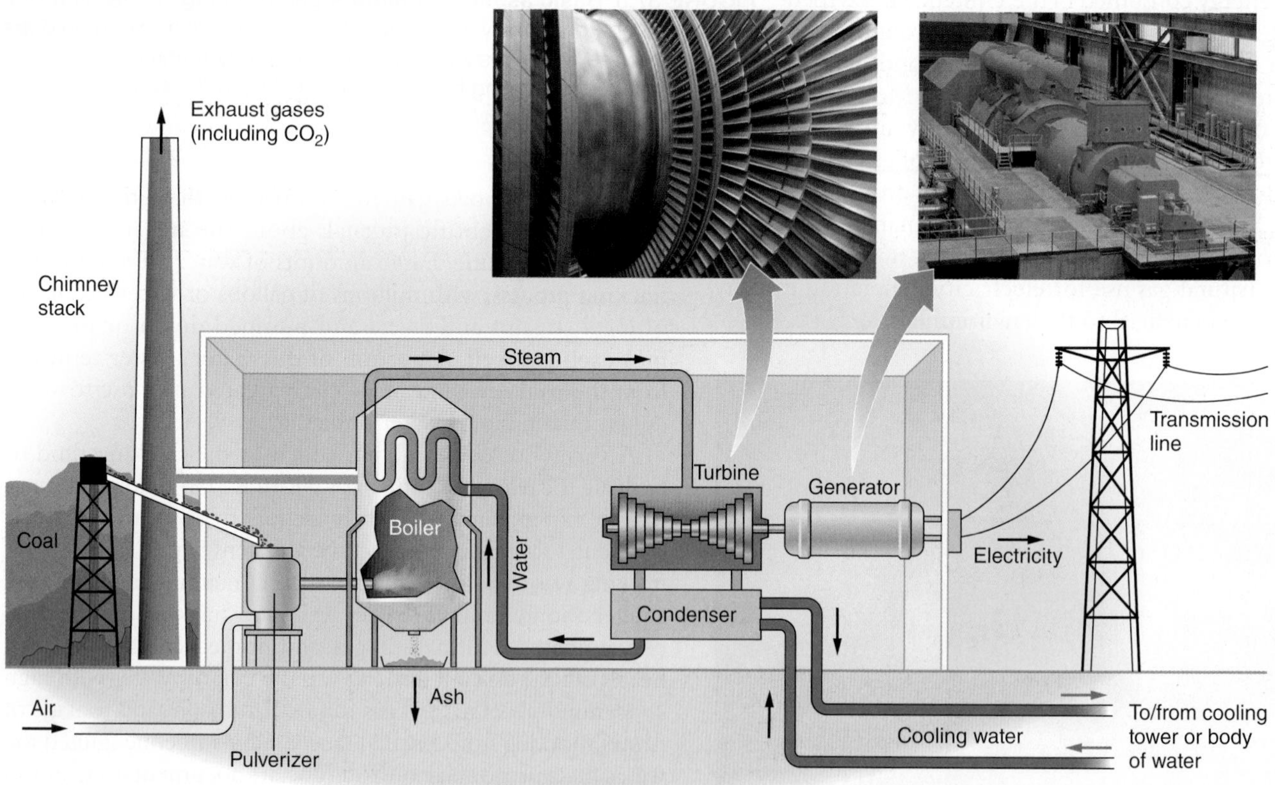

FIGURE 37.7 Coal is delivered to a boiler, where it is burned. The energy released from the combusted fuel is transferred to water, which becomes steam. The steam is transferred to the blades of a turbine, which turns the generator, which generates electricity, which is distributed to the grid. *(Photos: CHRISTIAN LUNIG/Science Source; Peter Bowater/Science Source)*

to a boiler, where it is burned. The energy contained within the combusted fuel is transferred to water, which becomes steam. The kinetic energy contained within the steam is transferred to the blades of a **turbine**, a device that can be turned by water, steam, or wind to produce power such as electricity. As the energy in the steam turns the turbine, the shaft in the center of the turbine turns the generator, which generates electricity. The electricity that is generated is then transported along a network of interconnected transmission lines known as the **electrical grid**, which connects electricity generation sources and links them with end users of electricity. Once the electricity is on the grid, it is distributed to homes, businesses, factories, and other consumers of electricity, where it may be converted into heat energy for cooking, kinetic energy in motors, or radiant energy in lights or used to operate electronic and electrical devices. After the steam passes through the turbine, it is condensed back into water. Sometimes the water is cooled in a cooling tower or discharged into a nearby body of water. Once-through use of water for thermal electricity generation is responsible for a substantial amount of water consumption in the United States.

AP® Exam Tip

You may be asked to describe how electricity is produced by a generator. For most forms of electricity it will involve a turbine: coal, natural gas, hydroelectric, wind, and nuclear power. Your answer should include three details:

- The fluid (steam, water, wind) rotates a turbine.
- The turbine drives or is attached to a generator.
- The generator produces electricity.

In Module 35 we described efficiency and here we introduced the coal-burning power plant, a system that converts potential energy in coal to electricity, heat, and pollution, including carbon dioxide. **FIGURE 37.8** illustrates the broader process, beyond the power plant process. A modern coal-burning power plant can convert 1 metric ton of coal, containing 24,000 megajoules of chemical energy, into about 8,400 MJ of electricity. Since 8,400 is 35 percent of 24,000, this means that the process of turning coal into electricity is about 35 percent efficient. The rest of the energy from the coal—65 percent—is lost as waste heat.

In the electrical transmission lines between the power plant and the house, 10 percent of the electrical energy from the plant is lost as heat and sound, so the transport of energy away from the plant is about 90 percent efficient. We know that the conversion of electrical energy into light in an incandescent bulb is 5 percent efficient; again, the rest of the energy is lost as heat. (Incandescent light bulbs are excellent heaters—that's why they are sometimes used as incubator heating sources when raising chickens from eggs in your home or science class!). From beginning to end, we can calculate the energy efficiency of converting coal into incandescent lighting by multiplying all the individual efficiencies:

Volatile organic compounds (VOCs) A type of organic compound air pollutants that evaporate at typical atmospheric temperatures.

Turbine A device that can be turned by water, steam, or wind to produce power such as electricity.

Electrical grid A network of interconnected transmission lines.

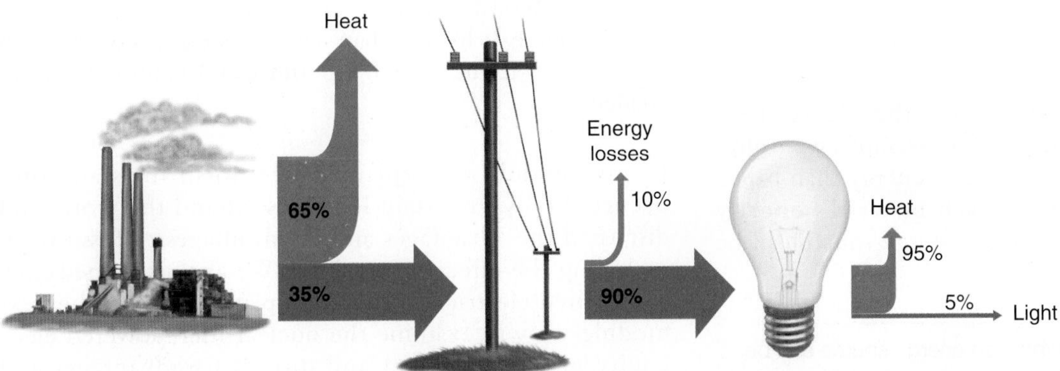

Calculation: (35%) × (90%) × (5%) = 1.6% efficiency

FIGURE 37.8 The second law of thermodynamics. Whenever one form of energy is transformed into another, some of that energy is converted into a less usable form of energy, such as heat. In this example, we see that the conversion of coal into the light of an incandescent bulb is only 1.6 percent efficient.

DO THE MATH Comparing the Efficiencies of Two Systems

Preparing for the AP® Exam

Figure 37.8 illustrates an overall efficiency of 1.6 percent in a system that converts coal to electricity that lights an incandescent bulb. Calculate the overall efficiencies of the two systems shown in the table.

Power Plant	Transmission	Light bulb
(a) Coal to electricity 35%	90%	Compact fluorescent 20%
(b) Natural gas to electricity 50%	90%	LED (Light emitting diode) 25%

Efficiencies are multiplicative. Rounding to two significant figures:

(a) Coal to electricity: $0.35 \times 0.90 \times 0.20 = 0.06 \times 100\% = 6\%$ efficient

(b) Natural gas to electricity: $0.50 \times 0.90 \times 0.25 = 0.11 \times 100\% = 11\%$ efficient

YOUR TURN The most efficient natural gas to electricity generation plants are 60 percent efficient. The most efficient electric cars are about 60 percent efficient. What is the efficiency of this natural gas to electricity to electric car system?

Calculating energy efficiency:

$$\text{Coal to electricity} \times \text{transport of electricity} \times \text{light bulb efficiency} = \text{overall efficiency}$$

$$0.35 \times 0.90 \times 0.05 = 0.016$$

(1.6% efficiency)

"Do the Math: Comparing the Efficiencies of Two Systems" gives you an opportunity to explore this topic further.

Energy Quality

Most of us have an intuitive sense about the relative effectiveness of various energy sources. For example, we realize that gasoline is a more useful source of energy than paper. This difference is a function of each material's **energy quality**, or the ease with which an energy source can be used to do work. A high-quality energy source has a convenient, concentrated form so that it does not require too much energy to move it from one place to another. Energy quality is one important factor humans must consider when they make energy choices.

Gasoline, for example, is a high-quality energy source because its chemical energy is concentrated (about 44 MJ/kg), and because we have an infrastructure that can conveniently transport it from one location to another. In addition, it is relatively easy to convert gasoline energy into work and heat. Wood, on the other hand, is a lower-quality energy source. It has less than half the energy concentration of gasoline (about 20 MJ/kg) and using it presents more challenges.

Energy quality The ease with which an energy source can be used to do work.

In this module we explained why fossil fuels and ores occurred only in certain locations around the world and discussed the advantages and disadvantages of fossil fuels, including the effects of fracking. We also examined how a coal-fired electricity generation plant works. In the next module we will examine the nuclear fuel–powered electricity generation plant and discuss its advantages and disadvantages.

Module 37 AP® Review

Preparing for the AP® Exam

Learning Goals Revisited

37-1 Why are fossil fuels and ores found only in certain locations?

Fossil fuel distribution around the world depends on the geology of the region. Organic matter that will eventually become coal or oil must become buried quickly without being exposed to air. This typically happens in tropical locations. Oil and natural gas are also dependent on geologic events related to tectonics that create geologic domes underground. Over time, oil and gas migrate to the top of these structures.

37-2 What are the advantages and disadvantages of fossil fuels including oil extraction and fracking?

Coal, oil, and natural gas each have their advantages and disadvantages. All three release heat energy and carbon dioxide with differing amounts of pollutant released from each. Fracking has increased the availability of natural gas in the United States and has led to groundwater contamination and a suspected increase in earthquakes. It also causes the release of a category of air pollutant—volatile organic compounds—from both the fracking fluid and from the machinery used at the fracking site. VOCs are a precursor to other types of air pollution, and some can harm human health.

37-3 How are fossil fuels used for electricity generation?

A modern coal-burning power plant converts the potential energy of coal into electricity at about 35 percent efficiency. The remaining 65 percent of the energy from the coal is lost as waste heat. In the electrical transmission lines between the power plant and the house, and in the conversion of electricity to the end use such as lighting or computing, there are other efficiency losses.

Practice Math and Graphing

Preparing for the AP® Exam

Answer the following questions. Be sure to show all your work.

1. Practice Math

Coal contains a number of impurities, including sulfur, that are released into the atmosphere when the coal is burned. The sulfur content of coal typically ranges from 0.4 to 4 percent by weight. Assume that a power plant burns one metric ton (1,000 kg) of coal with a 1 percent sulfur content to generate electricity. A ton of coal contains the energy equivalent of 7,700 kWh of electricity.

(a) Assuming typical efficiency, how much electricity will be generated?

(b) How much sulfur will be released?

2. Practice Graphing

The daily maximum 1-hour average concentration of sulfur dioxide (measured in parts per billion or ppb) in the United States has been decreasing over the past few decades.

(a) Use the data in the table to graph sulfur dioxide concentrations from 2000 to 2016.

(b) State the percentage decrease in 2016 relative to the 2000 concentration.

Year	SO_2 concentration
2000	79.5
2001	79.7
2002	71.4
2003	73.2
2004	69.9
2005	70.5
2006	66.8
2007	63.7
2008	57.6
2009	49.1
2010	45.4
2011	38.3
2012	36.8
2013	30.7
2014	29.7
2015	25.7
2016	22.4

AP® Practice Questions

Multiple-Choice Questions

Use the passage below to answer questions 1 & 2:

Fossil fuel distribution around the world depends on the geology of the region. Organic matter that will eventually become coal or oil must become buried quickly without being exposed to air, which typically occurs in tropical locations. Millions of years ago, however, much of Earth had a more tropical climate, which led to coal and oil deposits being formed in areas that are currently temperate or polar.

Oil and natural gas are also dependent on geologic events related to tectonic activity that create geologic domes underground. Over time, oil and gas migrate to the top of these structures and are stored within permeable rock, which is typically why oil and natural gas can both be recovered from the same extraction sites over time.

1. According to the passage, why are oil and natural gas often recovered together?
   (a) Oil and natural gas are less dense than surrounding rock, so they migrate to the top of the spaces in permeable rock.
   (b) Oil and natural gas are more dense than surrounding rock, and get trapped in impermeable rock layers underground.
   (c) Oil and natural gas are found as shallow deposits in rock layers that have not undergone any tectonic activity.
   (d) Oil and natural gas deposits are only found in polar regions where tectonic activity is present.

2. Which of the following best describes a location in which coal deposits would be found?
   (a) under a mountain formed during the last ice age
   (b) under an area that was once a prehistoric wetland
   (c) under a mountain that was once a polar desert
   (d) under an area that was once a prehistoric boreal forest

3. Bitumen is
   (a) a form of liquid coal.
   (b) a degraded type of petroleum.
   (c) a by-product of natural gas extraction.
   (d) a fast-forming fossil fuel.

4. One major problem associated with natural gas is
   (a) sulfur.
   (b) radiation.
   (c) leakage.
   (d) ash spills.

Use the table below to answer questions 5–7:

Fuel source	Description	Energy quality (MJ/kg)
A	Solid	20
B	Solid	24
C	Liquid	44
D	Gas	55

5. Which fuel source is most likely a low-quality type of coal?
   (a) A
   (b) B
   (c) C
   (d) D

6. Which fuel source is most likely used for transportation?
   (a) A
   (b) B
   (c) C
   (d) D

7. Which fuel source is most likely to be recovered from the fracking process?
   (a) A
   (b) B
   (c) C
   (d) D

Free-Response Question

Energy sources used in the United States have changed over time, as shown in the graph below.

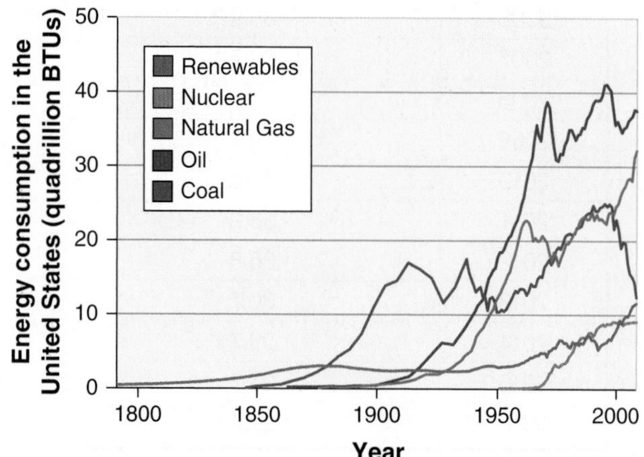

(a) Using the graph above, **identify** an energy source that has increased since approximately 2000. (1 pt.)

(b) Using the graph above, **describe** the overall trend of the use of coal as an energy source since 1850. (1 pt.)

(c) According to the EPA, total emissions of sulfur compounds into the atmosphere have decreased significantly since 2000. Using the graph above, **explain** why emissions of sulfur compounds into the atmosphere have been decreasing. (2 pts.)

(d) **Describe** one reason why petroleum in the United States has been the most-used energy source since 1950. (1 pt.)

(e) **Explain** how electricity is produced from fossil fuels in a power plant. (1 pt.)

Natural gas is extracted using a technique called hydraulic fracturing, which uses a combination of water and chemicals that is injected into the ground to break up rock and release the trapped gas.

(f) **Identify** one reason the use of natural gas has increased significantly over the last 20 years. (1 pt.)

(g) **Describe** an environmental problem associated with hydraulic fracturing. (1 pt.)

(h) The use of natural gas has been touted as an environmentally friendly alternative to coal-fired power plants.

(i) Other than those related to fracking, **describe** one environmental advantage to using natural gas for electricity production rather than coal. (1 pt.)

(ii) **Describe** one economic advantage to using natural gas for electricity production rather than coal. (1 pt.)

Module 38

Unit 6	35	36	37	38	39	40	41

Nuclear Power

It is very likely that you've seen items about electricity generated from nuclear energy in the news lately. Electricity generation from the nuclear energy contained in nuclear fuel is called **nuclear power**. Concerns about nuclear power include radioactivity, the proliferation of radioactive fuels that could be used in weapons, and the potential for accidents. Recently, however, nuclear power has received positive attention, even from environmentalists, because of its relatively low emissions of CO_2. In this module we will examine how nuclear fuels are used to generate electricity and the advantages and disadvantages of using nuclear power.

Learning Goals

After reading this module you should be able to

38-1 explain how nuclear energy is used to generate electricity.

38-2 describe advantages and disadvantages of nuclear power.

38-3 explain radioactivity and radioactive waste.

38-4 identify the three major nuclear accidents.

38-1 How is nuclear energy used to generate electricity?

Nuclear reactors use fission to generate electricity

Generating electricity from nuclear fuel uses the same basic process as electricity generation from fossil fuels: The fuel releases heat that converts water to steam that turns a turbine that turns a generator that generates electricity. The difference is that a nuclear power plant uses a radioactive isotope, uranium-235 (^{235}U), as its fuel.

The nuclei of atoms can be stable or unstable, depending on the mass number of the isotope and the number of neutrons it contains. Unstable isotopes are radioactive. **Radioactivity** is the emission of ionizing radiation or particles caused by the spontaneous disintegration of atomic nuclei. Radiation contains energy that is transferred to the surrounding environment, which becomes hotter. This heat is used to create steam to generate electricity in the same way that coal and natural gas are burned to release heat to create steam to generate electricity.

The naturally occurring isotope ^{235}U, as well as other radioactive isotopes, undergoes a process called fission. **Fission**, depicted in **FIGURE 38.1**, is a nuclear reaction in which a neutron strikes a relatively large atomic nucleus, which then splits into two or more parts, releasing additional neutrons and energy in the form of heat. The additional neutrons can, in turn, promote additional fission reactions, which leads to a chain reaction of nuclear fission that gives

Nuclear power Electricity generated from the nuclear energy contained in nuclear fuel.

Radioactivity The emission of ionizing radiation or particles caused by the spontaneous disintegration of atomic nuclei.

Fission A nuclear reaction in which a neutron strikes a relatively large atomic nucleus, which then splits into two or more parts, releasing additional neutrons and energy in the form of heat.

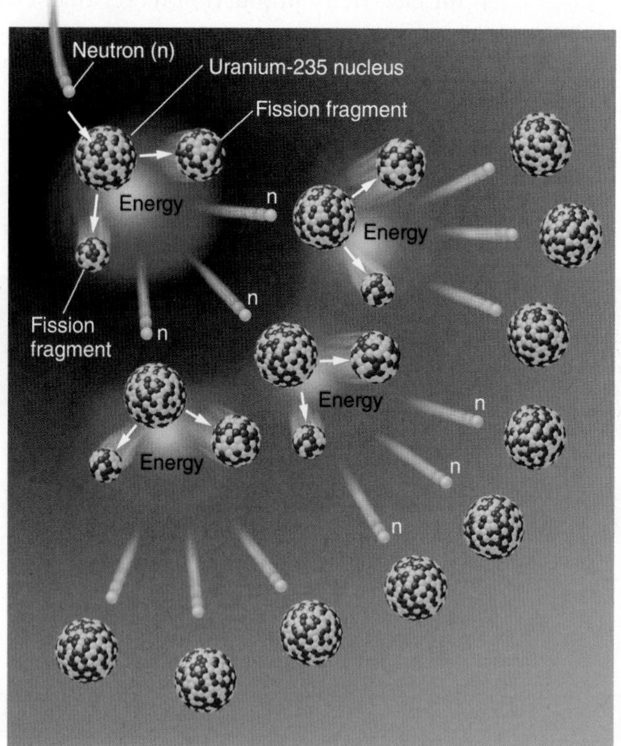

FIGURE 38.1 **Nuclear fission.** Energy is released when a neutron strikes a large atomic nucleus, which then splits into two or more parts.

off an immense amount of heat energy. In a nuclear power plant, that heat energy is used to produce steam, just as in any other thermal power plant. However, 1 g of ^{235}U contains 2 million to 3 million times the energy of 1 g of coal.

A properly designed nuclear reactor will harness the kinetic energy from the additional neutrons in motion to produce a self-sustaining chain reaction of nuclear fission. The by-products of the nuclear reaction include radioactive waste that remains hazardous for many half-lives—that is, sometimes for millions of years.

FIGURE 38.2 shows how a nuclear reactor works. It is very similar to the electricity generation power plants we saw in Module 36, with a few differences. The containment structure encloses the nuclear fuel—which is contained within cylindrical tubes called **fuel rods**—and the steam generator. Uranium fuel is processed into pellets, which are then put into the fuel rods. A typical nuclear reactor might contain hundreds of bundles of fuel rods in the reactor core within the containment structure.

Heat from nuclear fission is transferred to water within the containment structure, which circulates in a loop. This loop passes close to another loop of water, and heat is transferred from one loop to the other. In the process, steam is produced, which turns a turbine, which turns a generator, just as in most other thermal power plants. The nuclear power plant shown in Figure 38.2 is a light-water reactor, the only type of reactor used in the United States and the most common type used around the world.

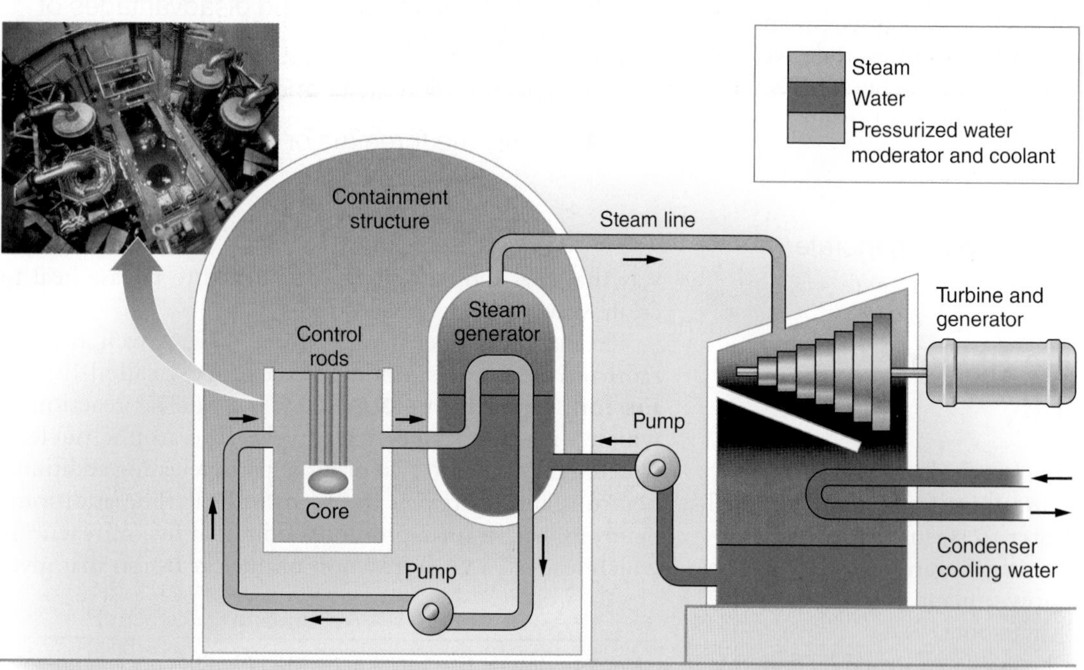

FIGURE 38.2 **A nuclear reactor.** This schematic shows the basic features of the light-water reactor, the type of reactor found in the United States. *(US Department of Energy/Science Source)*

A nuclear power plant is designed to harness heat energy from fission to make steam. But the plant must be able to slow the fission reaction to allow collisions to take place at the appropriate speed. To do this, nuclear reactors contain a moderator, such as water, to slow down the neutrons so that they can effectively trigger the next chain reaction. Because there is also a risk that the reaction will run out of control, nuclear reactors contain **control rods**, cylindrical devices that can be inserted between the fuel rods in a nuclear reactor to absorb excess neutrons, thus slowing or stopping the fission reaction. This is done routinely during the operation of the plant because nuclear fuel rods left uncontrolled will quickly become too hot and melt—an event called a meltdown—or cause a fire, either of which could lead to a catastrophic nuclear accident. Control rods are also inserted when the plant is being shut down during an emergency or for maintenance and repairs.

Concentrating the Uranium Ore

Depending on the ore, it may take up to 900 kg (2,000 pounds) of uranium ore to produce 3 kg (6.6 pounds) of nuclear fuel. In order to obtain uranium, miners remove large amounts of the host rock, extract and concentrate the uranium, and leave the remaining material in tailings piles. Australia, the western United States, and parts of Canada have large commercial uranium mining operations for nuclear fuel. As we saw in Module 29, the mining of any material requires fossil fuel energy and results in mine tailings. This is also true for uranium, although, as we have noted, a much smaller volume and mass of uranium is needed to generate a similar quantity of electricity than would be the case with coal.

Nuclear power plants rely on ^{235}U as their fuel. However, most uranium ores contain as much as 99 percent ^{238}U, another isotope of uranium that occurs with ^{235}U but does not fission as easily. Therefore, when uranium ore is mined, it must be chemically enriched—a process to increase its concentration of ^{235}U to be useful as a fuel. Typically, suitable nuclear fuel contains 3 percent or greater ^{235}U.

38-2 What are the advantages and disadvantages of nuclear power?

Nuclear power has advantages and disadvantages

Nuclear power plants do not produce air pollution during their operation, so proponents of nuclear energy consider it "clean" energy. In countries with limited fossil fuel resources, nuclear energy is one way to achieve independence from imported oil. Nuclear power generates 70 percent of electricity in France, and it is widely used in China, Russia, South Korea, Canada, and Ukraine, as well as in other countries.

As we saw in Figure 36.6, 20 percent of the electricity generated in the United States comes from nuclear energy. Early proponents of nuclear energy in the 1950s and 1960s claimed that it would be "too cheap to meter," meaning that it would be so inexpensive that there would be no point in trying to figure out how much each customer used. However, construction of new nuclear power plants became more and more expensive in the United States, in part because public protests, legal battles, and other delays increased the cost of construction. Public protests arose because of concerns that a nuclear accident would release radioactivity into the surrounding air and water. Other concerns included uncertainty about appropriate locations for radioactive waste disposal and fear that radioactive waste could fall into the hands of individuals seeking to make a nuclear weapon. These concerns are well founded since it takes 704 million years for the radioactivity of ^{235}U to decrease to half its original level. By the 1980s, it had become prohibitively expensive—both monetarily and politically—to attempt the construction of new nuclear power plants.

There are 93 nuclear reactors operating in the United States today—lower than the peak number of 110 that were operating in 1990. However, the relatively low CO_2 emissions associated with nuclear energy has caused a resurgence of interest in constructing additional nuclear power plants. There are certainly CO_2 emissions related to mining, processing, and transporting nuclear fuel as well as constructing the actual nuclear power plant. These emissions are perhaps 10 percent or less of those related to generating an equivalent amount of electricity from coal. Two major issues of environmental concern remain: the disposal of radioactive waste and the possibility of accidents; those are the subjects of the next two sections of this module.

38-3 What are radioactivity and radioactive waste?

Nuclear power depends on radioactivity but as a result, it generates radioactive waste

Because of the large amount of energy released from a small amount of radioactive fuel undergoing fission, the fuel required to operate a nuclear powerplant is so much less than the fuel required for a coal-fired powerplant. However, the uranium fuel is radioactive and it remains radioactive after it is no longer useful for generating electricity. Radioactivity refers to the emission of radiation or particles from the spontaneous decay of atomic nuclei.

Fuel rod A cylindrical tube that encloses nuclear fuel within a nuclear reactor.

Control rod A cylindrical device inserted between the fuel rods in a nuclear reactor to absorb excess neutrons and slow or stop the fission reaction.

Radioactive Isotopes Undergo Radioactive Decay

Radioactive isotopes undergo **radioactive decay**, the spontaneous release of material from the nucleus, when a parent radioactive isotope emits alpha or beta particles or gamma rays. We measure radioactive decay by recording the average rate of decay of a quantity of a radioactive element. This measurement is commonly stated in terms of the **half-life** of the element, which is the time it takes for one-half of the original radioactive parent atoms to decay. An element's half-life is a useful parameter to know because some elements that undergo radioactive decay emit harmful radiation. Knowledge of the half-life allows scientists to determine the length of time that a particular radioactive element may be dangerous. For example, using the half-life allows scientists to calculate the period of time that people and the environment must be protected from depleted nuclear fuel, like that generated by a nuclear power plant. As it turns out, many of the elements produced during the decay of ^{235}U have half-lives of tens of thousands of years and more. From this we can see why long-term storage of radioactive nuclear waste is so important.

Radioactive Waste: The By-Product of Electricity Generated from Nuclear Power

When nuclear fuel is no longer useful in a power plant, yet continues to emit radioactivity, it is considered **radioactive waste**. Long after nuclear fuel can produce enough heat to be useful in a power plant, it continues to emit radioactivity. Because radioactivity can be extremely damaging to living organisms, radioactive materials—the nuclear waste—must be stored in special, highly secure locations.

The use of nuclear fuels produces three kinds of radioactive waste: high-level waste in the form of used fuel rods; low-level waste in the form of contaminated protective clothing, tools, rags, and other items used in routine plant maintenance; and uranium mine tailings, the residue left after uranium ore is mined and enriched. High-level radioactive waste releases ionizing radiation in doses large enough to potentially cause immediate damage to the human body including organ failure, radiation sickness, and death. At lower exposure to high-level waste, and exposure to low-level radioactive waste, humans can experience damage to DNA, which may ultimately lead to cancer or tumors. There are also many other sublethal harmful effects to the eyes, brain, and immune and reproductive systems. Fetuses, children, and adolescents are most vulnerable to the harmful effects of radiation and radioactive waste.

Radioactive decay When a parent radioactive isotope emits alpha or beta particles or gamma rays.

Half-life The time it takes for one-half of the original radioactive parent atoms to decay.

Radioactive waste Nuclear fuel that can no longer produce enough heat to be useful in a power plant but continues to emit radioactivity.

Becquerel (Bq) A measurement of the rate at which a sample of radioactive material decays; 1 Bq is equal to the decay of one atom per second.

Curie A unit of measure for radiation, a curie is 37 billion decays per second.

AP® Exam Tip

The AP® Environmental Science Exam might ask for drawbacks of using nuclear power during the normal operations of a nuclear power plant. Accidents in nuclear power plants are not part of their normal operation. Instead describe the challenge of storing the high-level radioactive waste or thermal pollution from their cooling towers or discharging into nearby bodies of water.

Measuring Half-Lives

After a time, nuclear fuel rods become spent, meaning they are not sufficiently radioactive to generate electricity efficiently. Each radioactive isotope has a specific half-life: the time it takes for one-half of its radioactive atoms to decay. Uranium-235 has a half-life of 704,000,000 years, which means that 704 million years from today, a sample of ^{235}U will be half as radioactive as it is today. In another 704 million years it will lose another half of its radioactivity, so it will be one-fourth as radioactive as it is today. Because of this very long half-life, the safe and proper disposal of nuclear waste from electricity generation is one of the most important topics in discussions about nuclear power.

Radiation can be measured with a variety of units. A **becquerel (Bq)** measures the rate at which a sample of radioactive material decays; 1 Bq is equal to the decay of one atom per second. A **curie**, another unit of measure for radiation, is 37 billion decays per second. If a material has a radioactivity level of 100 curies and has a half-life of 50 years, the radioactivity level in 200 years will be

$$\frac{200 \text{ years}}{50 \text{ years/half-life}} = 4 \text{ half-lives}$$

$$100 \text{ curies} \rightarrow 50 \text{ curies (one half-life)} \rightarrow 25 \text{ curies (two half-lives)} \rightarrow 12.5 \text{ curies (three half-lives)} \rightarrow 6.3 \text{ curies (four half-lives)}$$

You can gain more experience in working with these calculations in "Do the Math: Calculating Half-Lives."

Radioactive Waste Disposal

Because spent fuel rods remain a threat to human health for 10 or more half-lives, they must be stored until they are no longer dangerous. At present, nuclear power plants

DO THE MATH Calculating Half-Lives

Preparing for the AP® Exam

Strontium-90 is a radioactive waste product from nuclear reactors. It has a half-life of 29 years. How many years will it take for a quantity of strontium-90 to decay to $\frac{1}{16}$ of its original mass?

It will take 29 years to decay to $\frac{1}{2}$ its original mass; another 29 years to $\frac{1}{4}$; another 29 years to $\frac{1}{8}$; and another 29 years to $\frac{1}{16}$:

$$29 + 29 + 29 + 29 = 116 \text{ years}$$

YOUR TURN You have 180 g of a radioactive substance. It has a half-life of 265 years. After 1,325 years, what mass remains of the radioactive substance?

are required to store spent fuel rods at the plant itself. Initially, all fuel rods were stored in pools of water at least 6 m (20 feet) deep. The water acts as a shield from radiation emitted by the rods. Currently, more than 100 sites around the country are temporarily storing spent fuel rods. However, some facilities have run out of pool storage space and are now storing them in lead-lined dry containers on land (**FIGURE 38.3**). Eventually, all of this material will need to be moved to a permanent radioactive waste disposal facility.

FIGURE 38.3 Storage of nuclear waste. Nuclear waste is usually stored in metal containers at the power plant where it is produced. *(Toby Talbot/AP Images)*

Disposing of radioactive waste is a challenge. This waste cannot be incinerated, safely destroyed using chemicals, shot into space, dumped on the ocean floor, or buried in an ocean trench because all of these options involve the potential for large amounts of radioactivity to enter the oceans or atmosphere. Therefore, at present, the only solution is to store it safely somewhere on Earth indefinitely. The physical nature of the storage site must ensure that the waste will not leach into the groundwater or otherwise escape into the environment. It must be far from human habitation in case of any accidents and be secure against terrorist attacks. In addition, the waste, if it is eventually moved to a permanent storage site, will need to be transported in a way that minimizes the risk of accidents or theft by terrorists.

Long-Term Storage

In 1978, the U.S. Department of Energy (DOE) began examining a site at Yucca Mountain, Nevada, 145 km (90 miles) northwest of Las Vegas, as a possible long-term repository for the country's spent nuclear fuel. The Yucca Mountain proposal has generated enormous protest and controversy. In 2006, the Department of Energy released a report confirming the soundness of the research supporting the Yucca Mountain site. However, a few years later, in 2010, the DOE abandoned its efforts to license a waste repository at Yucca Mountain and so, effectively, the project was canceled. Every few years, interest in the project is generated in Congress or elsewhere but as of 2022, there are no specific plans to proceed with Yucca Mountain, or any other location, as a national nuclear waste repository.

Nuclear energy raises a unique question of sustainability. It is a source of electricity that releases much less CO_2 than coal, and none during the actual generation of electricity. But some argue that the use of a fuel that consistently generates large quantities of high-level radioactive waste is not a sustainable practice. Many critics of nuclear energy maintain that there will never be a place safe enough to store high-level radioactive waste.

38-4 What are the three major nuclear accidents?

Three Mile Island, Chernobyl, and Fukushima are the three nuclear accidents

Two nuclear accidents occurred in the 1980s and 1990s and contributed to the global protests against nuclear energy at that time. On March 28, 1979, at the Three Mile Island nuclear power plant in Pennsylvania, operators did not notice that a cooling water valve had been closed the previous day. This oversight led to a lack of cooling water around the reactor core, which overheated and suffered a partial meltdown. The reactor core was severely damaged, and a large part of the containment structure became highly radioactive with an unknown amount of radiation released to the outside environment. A few thousand school children and pregnant women were evacuated from the area surrounding the plant by order of the governor of Pennsylvania, although an estimated 200,000 other people chose to evacuate as well.

In the days following the accident, residents of the area experienced a great deal of anxiety and fear, especially as reports were released to the press of a potentially explosive gas bubble in the containment structure. Although there has been no documented increase in local health problems as a result of this accident, concerns remain, and several investigators maintain that infant mortality rates and cancer rates increased in the following years. The nuclear reactor in which the incident occurred, one of two reactors at the Three Mile Island nuclear facility, has not been used since the accident.

A more serious accident occurred on April 26, 1986, at a nuclear power plant in Chernobyl, Ukraine. The accident occurred during a test of the plant when, in violation of safety regulations, operators deliberately disconnected emergency cooling systems and removed the control rods. With no control rods and no coolant, the nuclear reactions continued without control and the plant overheated. These "runaway" reactions led to an explosion and fire that damaged the plant beyond use (**FIGURE 38.4**). At the time of the accident, 31 plant workers and firefighters died from acute radiation exposure and burns, and many more died later of causes related to exposure.

After the accident, winds blew radiation from the plant across much of Europe, where it contaminated crops and milk from cows grazing on contaminated grass. More than a hundred thousand people were evacuated from the area around Chernobyl. Estimates of health effects vary widely, in part because of the paucity of information provided by the Soviet government, but a U.S. National Academy of Sciences panel estimated that 4,000 additional cancer deaths (over and above the average number of expected deaths) would occur over the next 50 years among people who lived near the plant or worked on the cleanup. There have been approximately 5,000 cases of thyroid cancer, most of them nonfatal, among children who were younger than 18 at the time of the accident and lived near the Chernobyl plant. Thyroid cancer may be caused by the absorption of radioactive iodine, one of the radioactive elements emitted during the accident.

FIGURE 38.4 **The Chernobyl Power Plant accident.** Reactor #4 at the power plant in Chernobyl, Ukraine, experienced an explosion and fire on April 26, 1986. *(Wojtek Laski/Getty Images)*

The third and most recent nuclear power plant accident occurred in Japan on March 11, 2011. A magnitude 9.0 earthquake off the country's coast generated a tsunami that was 15 m (49 feet) high. This caused flooding in northeastern Japan, where the Fukushima nuclear power plant is located. The resulting structural damage to the plant and interruption of the regional electrical supply led to fires, hydrogen gas explosions, and the release of radioactive gases from nuclear reactors within the containment structure. Ultimately, four out of a total of six reactors were damaged beyond repair, radioactive gases were released into the surrounding environment, and over 100,000 people were evacuated from their homes. Approximately 20,000 people, including three workers at the Fukushima plant, were killed by the earthquake and tsunami. However, at present, there have been no deaths attributed to the release of radioactivity from the reactors. After the Fukushima accident, Japan shut down all 54 of its nuclear reactors. As of mid-2022, only 10 have been restarted but the Japanese government is tentatively planning to restart more reactors in the coming years.

Nuclear Power Compared with Other Fuels

At present, our reliance on nuclear energy in the United States is subject to speculation and projections, but it is hard to know whether the positive aspects of nuclear energy will come to outweigh the risks of accidents and containment of radioactive waste. Indeed, in 2006, applications for new nuclear power plants in the United States were filed with the Nuclear Regulatory Commission for the first time in more than 2 decades. In 2012, licenses were approved for two nuclear power plants. In 2018, there were two plants under construction, but then construction was halted. In

2022, there are two plants under construction and two plants being planned. It is possible—but not a certainty—that up to five new nuclear power plants may come online in the United States in the next decade. There are currently 14 nuclear power plants under construction in China. **TABLE 38.1** summarizes the major benefits and consequences of nuclear power plants in comparison to fossil fuel power plants that we have discussed in the preceding modules.

TABLE 38.1 Comparison of nuclear with other nonrenewable energy fuels

Energy Type	Advantages	Disadvantages	Pollutant and greenhouse gas emissions	Electricity (cents/kWh)	Energy return on energy investment*
Oil/gasoline	• Ideal for mobile combustion (high energy/mass ratio) • Quick ignition/turn-off capability • Cleaner burning than coal	• Significant refining required • Oil spill potential effect on habitats near drilling sites • Significant dust and emissions from fossil fuels used to power earthmoving equipment • Human rights/ environmental justice issues in developing countries that export oil • Will probably be much less available in the next 40 years or so	• Second highest emitter of CO_2 among fossil fuels • Hydrocarbons • Hydrogen sulfide	• Relatively little electricity is generated from oil 20 cents/kWh	4.0 (gasoline) 5.7 (diesel)
Coal	• Energy-dense and abundant – U.S. resources will last at least 200 years • No refining necessary • Easy, safe to transport • Economic backbone of some small towns	• Mining practices frequently risk human lives and dramatically alter natural landscapes • Coal power plants are slow to reach full operating capacity • A large contributing factor to acid rain in the United States	• Highest emitter of CO_2 among energy sources • Sulfur • Trace amounts of toxic metals such as mercury	5 cents/kWh	14
Natural Gas	• Combined cycle power plants can have efficiencies up to 60% • Efficient for cooking, home heating, etc. • Fewer impurities than coal or oil	• Risk of leaks/explosions • Twenty-five times more effective as a greenhouse gas than CO_2 • Not available everywhere because it is transported by pipeline	• Methane • Hydrocarbons • Hydrogen sulfide • Less CO_2 than both coal and oil	3–4 cents/kWh	8
Nuclear Energy	• Emits no CO_2 once plant is operational • Offers independence from imported oil • High energy density, ample supply	• Very unpopular; generates protests • Plants are very expensive to build in part because of legal challenges • Meltdown could be catastrophic • Possible target for terrorist attacks	• Radioactive waste is dangerous for hundreds of thousands of years • No long-term plan currently in place to manage radioactive waste • No air pollution during production • Thermal pollution warms nearby bodies of water	12–15 cents/ kWh	8

*Estimates vary widely.

In this module we described that nuclear powered electricity generation is a nonrenewable energy source. It is considered by many to be a relatively clean energy source because it does not produce air pollutants during operation of the nuclear power plant. However, as with other thermal power plants, it does release thermal pollution, that is, it warms waterways such as bays and estuaries if the plant is located near a body of water. And it generates radioactive, hazardous solid waste and risks the possibility of a meltdown of the radioactive core. In comparison, many of the renewable energy sources have consequences that are less severe. In the next module we will begin to consider renewable fuels in more detail with a focus on biomass and solar energy and hydroelectric power.

Module 38 AP® Review

Preparing for the AP® Exam

Learning Goals Revisited

38-1 How is nuclear energy used to generate electricity?

Electricity generation from nuclear energy uses the same basic process as electricity generation from fossil fuels: Steam turns a turbine that turns a generator that generates electricity. The difference is that a nuclear power plant uses a radioactive isotope, uranium-235 (^{235}U), as its fuel source. ^{235}U emits a tremendous amount of heat as it undergoes fission and decays. The nuclear fuel is contained within cylindrical tubes called fuel rods. A typical nuclear reactor might contain hundreds of bundles of fuel rods and heat from nuclear fission is transferred to water that turns the turbine.

38-2 What are the advantages and disadvantages of nuclear power?

Nuclear energy is a relatively clean means of electricity generation with respect to CO_2 and other air pollutants, although fossil fuels are used in constructing nuclear power plants and mining and processing the uranium fuels. The possibility of accidents during plant operation and the difficulty of radioactive nuclear waste disposal are major environmental hazards of nuclear energy used for the generation of electricity. Neither of these issues has been satisfactorily resolved at present. Throughout the nuclear electricity generation cycle, carbon dioxide emissions are much less than during electricity generation from fossil fuels.

38-3 What are radioactivity and radioactive waste?

The use of nuclear fuels produces three kinds of radioactive waste: high-level waste in the form of used fuel rods; low-level waste in the form of contaminated protective clothing and tools; and uranium mine tailings, the residue left after uranium ore is mined and enriched. Each radioactive isotope has a specific half-life: the time it takes for one-half of its radioactive atoms to decay. ^{235}U has a half-life of 704,000,000 years. Because of the long half-life of radioactive waste, the proper and safe disposal of nuclear waste is a major issue. At present, there is no acceptable long-term storage facility for nuclear waste.

38-4 What are the three major nuclear accidents?

In 1979, Three Mile Island nuclear power plant in Pennsylvania experienced a lack of cooling water around the reactor core, which overheated and suffered a partial meltdown. The reactor core was severely damaged, and an unknown amount of radiation was released to the outside environment. In 1986, the Chernobyl, Ukraine, nuclear power plant experienced "runaway" reactions that led to an explosion and fire that damaged the plant beyond use. Thirty-one plant workers and firefighters died from acute radiation exposure and burns, and many more died later of causes related to exposure. In Fukushima, Japan, in 2011, a magnitude 9.0 earthquake off the coast generated a tsunami. This caused flooding and structural damage to the nuclear power plant leading to fires, hydrogen gas explosions, and the release of radioactive gases from the nuclear reactors. Radioactive gases were released into the surrounding environment and over 100,000 people were evacuated from their homes.

AP® Practice Questions

Multiple-Choice Questions

1. Control rods slow nuclear reactions by
   (a) reducing the amount of fuel available.
   (b) absorbing the heat produced.
   (c) increasing the rate heat transfers to the water.
   (d) absorbing excess neutrons.

2. Interest in nuclear power has recently increased because of
   (a) low energy costs.
   (b) lack of significant accidents.
   (c) low carbon dioxide emissions.
   (d) new solutions for waste disposal.

3. For a sample of Ti-44 with a half-life of 63 years, how long would it take until $\frac{1}{16}$ of the original sample is left?
   (a) 126 years
   (b) 189 years
   (c) 252 years
   (d) 315 years

Use the graph below to answer questions 4–6:

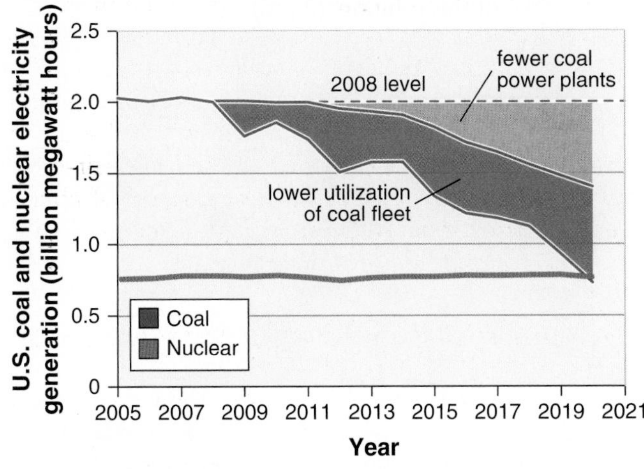

4. Which of the following best describes one aspect of the graph?
   (a) Coal generation has increased since 2005.
   (b) Nuclear generation has decreased since 2005.
   (c) Nuclear generation has increased since 2005.
   (d) Coal generation is equal to nuclear generation.

5. Which of the following best explains the trend in coal use in the United States since 2005?
   (a) Coal use has increased because more power plants are burning more coal.
   (b) Coal use has decreased because more nuclear power plants are being built.
   (c) Coal use has increased because nuclear energy is being used less.
   (d) Coal use has decreased because fewer coal plants are being used.

6. Which of the following best explains a potential outcome related to the energy use shown in the graph?
   (a) The United States will need to obtain more uranium to produce electricity from nuclear energy.
   (b) Lower sulfur emissions will be recorded due to the decrease in coal power plants.
   (c) More nuclear waste will be produced, requiring the need to store it.
   (d) More coal power plants will need to be built in response to decreased demand for nuclear power.

Use the passage below to answer questions 7 & 8:

In the United States, most uranium used for nuclear power is mined from other countries and imported to the United States. According to the Energy Information Administration (EIA), 90 percent of the uranium purchased by the United States was imported from other counties, including Canada, Kazakhstan, Australia, and Russia. Imports from Canada have been the largest source of uranium for the United States for the last 4 years, because Canada is home to large uranium reserves. Only 10 percent of all uranium used by nuclear reactors in the United States is produced domestically, and this amount has been decreasing over the past several years. According to the EIA, U.S. uranium production in 2018 was 1.6 million pounds, the lowest since 1950.

7. According to the passage, how much of the uranium used for nuclear power came from sources in the United States in 2018?
   (a) 10 percent
   (b) 24 percent
   (c) 83 percent
   (d) 90 percent

8. Based on the passage, which of the following best describes evidence used by the author to support the claim that uranium production in the United States has fallen?
   (a) U.S. uranium production increased to 1.6 million pounds in 2018.
   (b) The United States has the highest concentrations of uranium reserves in the world.
   (c) In 2018, the uranium production in the United States was the lowest since 1950.
   (d) U.S. uranium purchases accounted for 3.9 million pounds in 2018.

Free-Response Question

Several U.S. states use nuclear power for their electricity production. Their relative production amounts are shown in the map below.

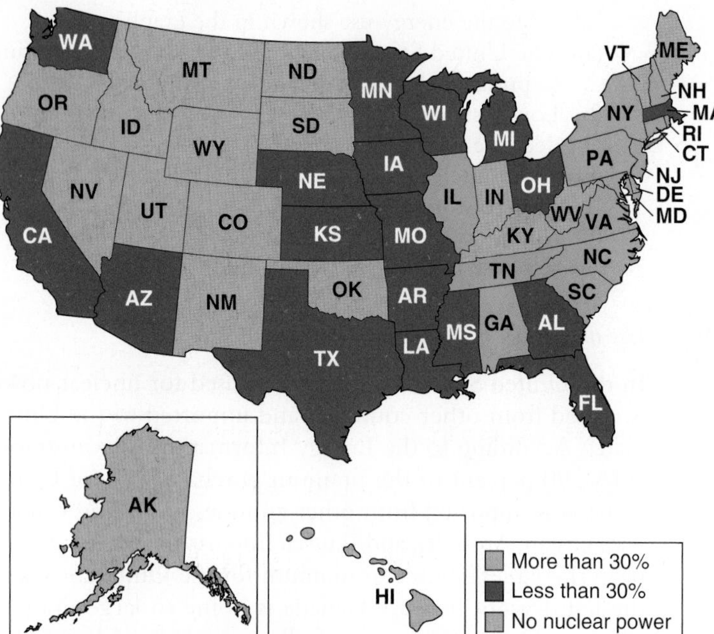

(a) **Identify** a state that produces more than 30 percent of its electricity from nuclear power. (1 pt.)

(b) **Identify** one state that does not use nuclear power to produce electricity. (1 pt.)

(c) **Describe** one reason a state may decide to reduce its use of nuclear energy for electricity generation. (1 pt.)

(d) **Explain** the process by which electricity is generated by a nuclear power plant. (1 pt.)

A number of U.S. electric companies have filed applications with the Nuclear Regulatory Commission for permits to build new nuclear power plants to meet future electricity demands.

(e) **Describe** one environmental benefit of generating electricity from nuclear energy rather than coal or natural gas. (1 pt.)

(f) One concern of using nuclear energy for electricity production is the risk of a nuclear accident.

(i) **Describe** one nuclear accident that has occurred that contributes to widespread concern about the safety of nuclear power plants. (1 pt.)

(ii) Other than the risk of a nuclear accident, **describe** one environmental disadvantage of using nuclear energy for electricity generation. (1 pt.)

(g) **Describe** one way humans store radioactive waste so radiation is not released into the environment. (1 pt.)

(h) One proposed solution to nuclear waste storage was the Yucca Mountain storage plan. **Identify** one advantage and one disadvantage of this solution. (2 pts.)

Module 39

Unit 6	35	36	37	38	39	40	41

Biomass and Solar Energies and Hydroelectric Power

In the previous module we examined electricity generation from nuclear fuel, which releases little carbon dioxide into the atmosphere relative to fossil fuels, and none during plant operation. Here, we will examine biomass and solar energy sources and hydroelectric power. All of them have advantages over fossil fuels and nuclear power but there are also disadvantages to each.

Learning Goals

After reading this module you should be able to

39-1 explain positive and negative consequences of biomass energy resources.

39-2 describe the various types of solar energy systems.

39-3 explain how hydroelectric power is generated and describe its environmental effects.

39-1 What are the positive and negative consequences of biomass energy resources?

Biomass energy resources are derived from biological material and can displace fossil fuels

Biomass is literally what the name suggests: biological material that has mass. The common biomass energy resources that are most frequently referred to are solid materials such as wood, charcoal, and animal manure. However, there are also liquid biomass fuels such as ethanol, an alcohol, and a diesel oil-like fuel from biological material called biodiesel. First, we will examine the solid biomass fuels.

Solid Biomass: Wood, Charcoal, and Manure

Three of the most commonly used solid biomass resources are wood, charcoal, and manure. Wood is derived from woody material such as trees. **Charcoal** is wood that has been heated in the absence of oxygen so that water and some volatile compounds are driven off. Manure is the waste products from animals. They are used much more extensively in the developing world than in the United States. While each of these three types of energy can been seen as having advantages over fossil fuels or nuclear power, they do have some environmental consequences.

Wood and Charcoal

As you know from Module 36, there are between 2 billion and 3 billion people in the world that rely on wood for a relatively inexpensive form of heating or cooking. In the United States, more than 20 million homes use wood for home heating at least some of the time (including the homes of the two authors of this textbook!). Remember also from Figure 36.1 that most fuels ultimately come from the Sun. The energy contained within biomass is captured by leaves from sunlight via photosynthesis and stored in plant structures such as leaves and wood. Biomass as an energy resource is potentially renewable if trees are not overharvested.

Tree removal can be sustainable if we allow time for forests to regrow. In addition, in some heavily forested areas, extracting individual trees of abundant species and opening up spaces in the canopy will allow other plants to grow and increase habitat diversity, which may even increase total photosynthesis. More often, however, tree removal for fuel has the potential to cause deforestation and soil erosion, to increase soil temperatures and water temperatures in nearby rivers and streams, and to fragment forest habitats when logging roads and cleared areas divide them. Tree removal may also harm species that are dependent on old-growth forest habitat.

Biomass Biological material that has mass.

Charcoal Woody material that has been heated in the absence of oxygen so that water and some volatile compounds are driven off.

Many people in the developing world use wood to make charcoal, which is a superior fuel for many reasons. Charcoal is made from wood by using heat to burn off or release water and organic compounds in a low oxygen environment. It is lighter than wood and contains approximately twice as much energy per unit of weight. A charcoal fire produces much less smoke than wood and does not need to be tended constantly, as does a wood fire. But both wood and charcoal produce certain air pollutants in greater quantities than are created during combustion of fossil fuels. These pollutants are also more problematic to human health because wood and charcoal are often burned indoors such as for cooking. A few of the major air pollutants associated with wood and charcoal are:

- **Particulates** or **particulate matter**: Solid or liquid particles suspended in the air (also known as **soot**)
- **Carbon monoxide**: Colorless, odorless gas that is formed during incomplete combustion of most materials
- **Nitrogen oxides**: A by-product of combustion of any fuel in the atmosphere (which contains 78 percent nitrogen)
- **Volatile organic compounds**: Organic compounds that evaporate easily
- **Carbon dioxide**: A by-product of all combustion, carbon dioxide from biofuels contains modern carbon from woody material, rather than fossil carbon from fossil fuels

It should be noted that these pollutants are released during the combustion of most fossil fuels as well. Although it is more expensive than wood, because it is lighter, cleaner-to-burn, and more convenient, charcoal is a fuel of choice in urban areas of the developing world and for families who can afford it. However, harvesters who clear an area of land for charcoal production often leave it almost completely devoid of trees (**FIGURE 39.1**). Air pollutants and indoor air pollutants will be addressed more fully in Unit 7.

Particulates (Particulate matter) Solid or liquid particles suspended in the air. *Also known as* **soot**.

Carbon monoxide A colorless, odorless gas that is formed during incomplete combustion of most materials.

Nitrogen oxides A by-product of combustion of any fuel in the atmosphere (which contains 78 percent nitrogen).

Carbon dioxide A by-product of all combustion, carbon dioxide from biofuels contains modern carbon from woody material, rather than fossil carbon from fossil fuels.

Manure

In regions where wood is scarce, such as parts of Africa and India, people often use dried animal manure as a fuel for indoor heating and cooking. Since manure is usually a free waste product from owning livestock, it is an ideal and cost-effective energy source. However, burning manure releases particulate matter, carbon monoxide, and other pollutants that cause a variety of respiratory illnesses, from emphysema to cancer. The problem is exacerbated when the manure is burned indoors in poorly ventilated rooms, a common occurrence in many developing countries. The World Health Organization estimates that indoor air pollution is responsible for more than 3 million deaths annually. Module 44 covers indoor air pollution and human health in more detail. Whether indoors or out, burning biomass fuels produces a

(a)

(b)

FIGURE 39.1 Charcoal as fuel in the developing world. (a) Many people in developing countries rely on charcoal for cooking and heating. This photo shows a charcoal market in the Philippines. (b) Charcoal production can strip the land of all trees. *(a: JAY DIRECTO/Getty Images; b: Eduardo Martino/Panos Pictures)*

FIGURE 39.2 Particulate emissions from burning biomass fuels. Burning biomass fuels contributes to air pollution. This photo shows smog and decreased visibility in Montreal, Canada, caused by emissions of particulate matter due to the extensive use of woodstoves as either a primary or secondary heating source. *(alank/Getty Images)*

variety of air pollutants, including particulate matter, carbon monoxide, nitrogen oxides, volatile organic compounds, and carbon dioxide, all of which are important components of air pollution (**FIGURE 39.2**).

Liquid Biofuels: Ethanol and Biodiesel

Biofuels are liquid fuels created from processed or refined biomass. The liquid biofuels—ethanol and biodiesel—can be used as substitutes for gasoline and diesel, respectively. **Ethanol** is an alcohol made by converting starches and sugars from plant material into alcohol and CO_2. More than 90 percent of the ethanol produced in the United States comes from corn and corn by-products, although ethanol can also be produced from sugarcane, wood chips, crop waste, or switchgrass. **Biodiesel**, a diesel substitute produced by extracting and chemically altering oil from plants, is a substitute for regular petroleum diesel. It is usually produced by extracting oil from algae and such plants as soybean and palm. In the United States, policies have encouraged the production of ethanol and biodiesel to reduce the need to import foreign oil while also supporting U.S. farmers and declining rural economies.

Ethanol

The United States is the world leader in ethanol production, manufacturing roughly 53 billion liters (14 billion gallons) in 2020. Brazil, the world's second largest ethanol producer, is making biofuels a major part of its sustainable energy strategy. Brazil manufactures ethanol from sugarcane, which is easily grown in its tropical climate. Unlike corn, which must be replanted every year, sugarcane is replanted every 6 years and is sometimes harvested by hand, factors that reduce the amount of fossil fuel energy needed to grow sugarcane.

Ethanol is usually mixed with gasoline, most commonly at a ratio of one part ethanol to nine parts gasoline. The result is gasohol, a fuel that is 10 percent ethanol. (If you ever fill a car with gasoline at a service station, look for a sign identifying the percent ethanol in the gasoline posted by the pump.) Gasohol has a higher oxygen content than gasoline alone and produces less of some air pollutants when combusted.

Proponents of ethanol claim that it is a more environmentally friendly fuel than gasoline since it does not introduce fossil carbon into the atmosphere (although that claim is disputed by some). Nonetheless, ethanol does have environmental disadvantages. The carbon bonds in alcohol have a lower energy content than those in gasoline, which means that a 90 percent gasoline/10 percent ethanol mix reduces gas mileage by 2 to 3 percent when compared with 100 percent gasoline fuel. As a result, a vehicle needs more gasohol to go the same distance it could go on gasoline alone. Furthermore, growing corn to produce ethanol uses a significant amount of fossil fuel energy, as well as land that can otherwise be devoted to growing food. Ethanol production has led to concern among economic analysts that

Biofuel Liquid fuel created from processed or refined biomass.

Ethanol Alcohol made by converting starches and sugars from plant material into alcohol and CO_2.

Biodiesel A diesel substitute produced by extracting and chemically altering oil from plants.

this production periodically contributes to short-term food shortages. Furthermore, some scientists argue that using ethanol actually creates a net increase in atmospheric CO_2 concentrations. The benefits and drawbacks of using ethanol as a fuel are discussed in more detail in "Science Applied 6: Should Corn Become Fuel?" that follows this unit.

Biodiesel

Biodiesel is a direct substitute for petroleum-based diesel fuel. It is usually more expensive than petroleum diesel, although the difference varies depending on market conditions and the price of petroleum. Biodiesel is typically diluted to "B-20," a mixture of 80 percent petroleum diesel and 20 percent biodiesel and is available at some gas stations around the United States. It can be used in any diesel engine without modification.

In the United States, most biodiesel comes from soybean oil or processed vegetable oil. In addition, some species of algae appear to have great potential for producing biodiesel. Algae are photosynthetic microorganisms that can be grown almost anywhere and, of all biodiesel options, produce the greatest yield of fuel per hectare of land area per year and utilize the least amount of energy and fertilizer per quantity of fuel. One study reported that algae produce 15 to 300 times more fuel per area used than did conventional crops. Algae can be grown on marginal lands, in brackish water, on rooftops, and in other places that are not traditionally thought of as agricultural space.

Emissions of carbon monoxide from combustion of biodiesel are lower than those from petroleum diesel. Since it contains modern carbon rather than fossil carbon, biodiesel should be carbon neutral, although, as with ethanol, some critics question whether biodiesel is truly carbon neutral. For instance, producing biodiesel from soybeans requires less fossil fuel input per liter of fuel than producing ethanol from corn, but soybeans require more cropland, and so they may actually transfer more carbon from the soil to the atmosphere. In contrast, producing biodiesel from wood waste or algae may require very little or no cropland and a minimal amount of other land.

Any diesel vehicle can be modified to operate on 100 percent straight vegetable oil (SVO), typically obtained as a waste product from restaurants and filtered for use as fuel. Because a major source of SVO is fast-food restaurants, some people call it French fry oil. Groups of college students in the United States have driven buses around the country almost exclusively on SVO. Some municipalities even have community-based SVO recycling facilities (**FIGURE 39.3**).

AP® Exam Tip

Make sure that you can differentiate between the impact of ethanol and fossil fuels on atmospheric CO_2 concentrations.

Passive solar A use of energy from the Sun that takes advantage of solar radiation without active technology.

FIGURE 39.3 A cooking oil recycling program. Used cooking oil collected from restaurants in San Francisco, California, is transferred to dumpsters where it will be recycled into biodiesel, reducing demand for fossil fuels and keeping the oil from entering the sewer system. Sometimes the oil is called French fry oil because it is often a waste product of making French fries at fast-food restaurants. *(Justin Sullivan/Getty Images)*

39-2 What are the various types of solar energy systems?

The energy of the Sun can be captured passively and actively

In addition to driving the natural cycles of water and air movement that we can harness as energy resources, the Sun also provides energy directly. Every day, Earth is bathed in solar radiation, an almost limitless source of energy. The amount of solar energy available in a particular place varies with the amount of cloudiness, time of day, and season. The average amount of solar energy available varies geographically. As **FIGURE 39.4** shows, average daily solar radiation in the continental United States ranges from 3 kWh of energy per square meter in the Pacific Northwest to almost 7 kWh per square meter in parts of the Southwest.

Passive Solar Heating

Passive solar is a use of energy from the Sun that takes advantage of solar radiation without active technology. Passive solar cannot be stored; it gets used when it is received. You may have observed applications of passive solar heating in daily life, including the positioning of windows in the Northern Hemisphere on south-facing walls to admit solar radiation in winter, and covering buildings or swimming pools with dark roofing and covers in order to absorb the maximum amount of heat. None of these strategies relies on intermediate pumps or electrical equipment to supply heat and therefore these strategies are passive. Solar ovens are another practical application of passive solar heating. For

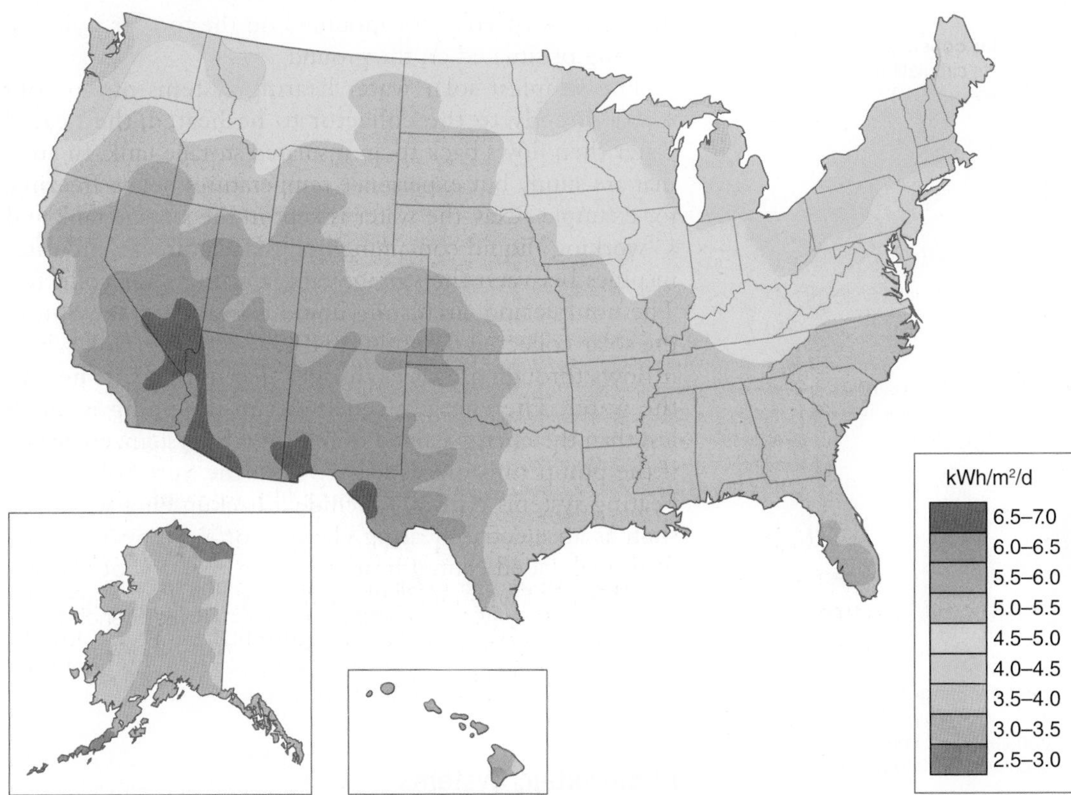

FIGURE 39.4 Geographic variation in solar radiation in the United States. This map shows the amount of solar energy available to a flat photovoltaic solar panel in kilowatt-hours per square meter per day, averaged over a year. *(Data from National Renewable Energy Laboratory, U.S. Department of Energy)*

FIGURE 39.5 Solar cookers. This woman in a Mali village in West Africa is preparing food with a solar cooker. *(Joerg Boethling/Alamy)*

instance, a simple "box cooker" concentrates sunlight as it strikes a reflector on the top of the oven. Inside the box, the solar energy is absorbed by a dark base and a cooking pot and is converted into heat energy. The heat is distributed throughout the box by reflective material lining the interior walls and is kept from escaping by a glass top. On sunny days, such box cookers can maintain temperatures of 175°C (347°F), can heat several liters of water to boiling in under an hour, or cook dishes of rice, beans, or chicken in 2 to 5 hours.

Solar ovens have both environmental and social benefits. The use of solar ovens in place of firewood reduces deforestation and, in areas unsafe for travel, having a solar oven means not having to leave the relative safety of home to seek firewood. For example, over 10,000 solar ovens have been distributed in refugee camps in the Darfur region of western Sudan in Africa, where leaving the camps to find cooking fuel would put people at risk of attack (**FIGURE 39.5**).

Active Solar Energy Technologies

In contrast to passive solar design, **active solar energy** technologies capture and store the energy of sunlight with

Active solar energy A use of technology that captures and stores the energy of sunlight with electrical equipment and devices.

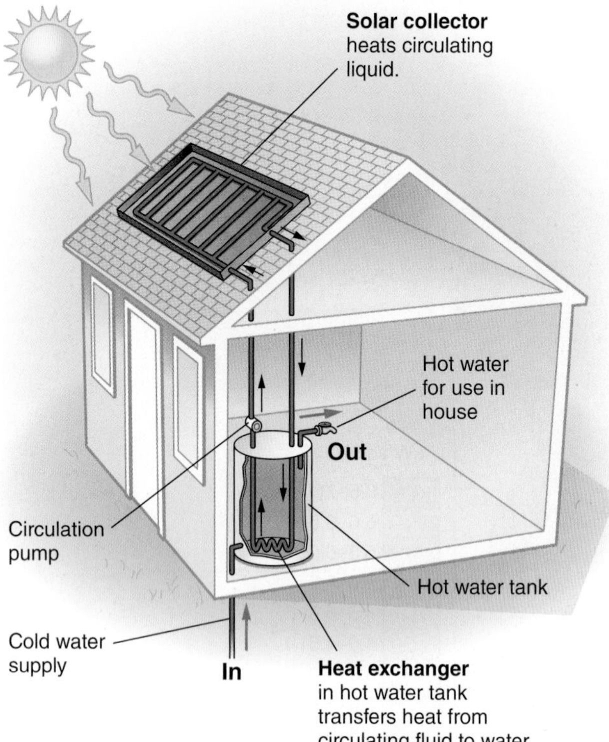

FIGURE 39.6 A solar domestic hot water system. When a solar hot water system is used to heat water, a nonfreezing liquid is circulated by an electric pump through a closed loop of pipes. This circulating liquid moves from a water storage tank to a solar collector on the roof, where it is heated, and then sent back to the tank, where a heat exchanger transfers the heat to water that can be used for showers and sinks in the house.

the use of electric equipment and technologies. Active solar energies include small-scale solar water heating systems, photovoltaic solar cells, and large-scale concentrating solar thermal systems for electricity generation.

Solar Water Heating Systems

Solar water heating applications range from providing domestic hot water and heating swimming pools to a variety of heating purposes for businesses and homes. In the United States, heating swimming pools is the most common application of solar water heating, and it is also the one that pays for itself the most quickly.

A household solar water heating system, like the domestic hot water system shown in **FIGURE 39.6**, allows heat energy from the Sun to be transferred directly to water or another liquid, which is then circulated to a hot water heating system. The circulation of the liquid is driven either by a pump (in active systems) or by natural convection (in passive systems). In both cases, cold liquid is heated as it moves through a solar collector mounted on the roof or wall of a building or situated on the ground.

The simplest solar water heating systems pump cold water directly to the collector to be heated; the heated water then flows back to an insulated storage tank. In areas that are sunny but experience temperatures below freezing, for example Texas, the water is kept in the storage tank and a "working" liquid containing nontoxic antifreeze circulates in pipes between the storage tank and the solar collector. The nonfreezing circulating liquid is heated by the Sun in the solar collector, then returned to the storage tank where it flows through a heat exchanger that transfers its heat to the water. The energy needed to run the pump is much less than the energy gained from using the system, especially if the pump runs on electricity from the Sun. Solar water heating systems typically include a backup energy source, such as an electric heating element or a connection to a fossil fuel–based central heating system, so that hot water is available even when it is cloudy or very cold. Although solar water heating systems seemed promising 15 or 20 years ago, they have been surpassed by the most common type of solar energy, converting sunlight to electricity.

Photovoltaic solar cells A use of energy from the Sun as light, not heat, and converting it directly into electricity.

Photovoltaic Systems

In contrast to solar water heating systems, **photovoltaic solar cells** capture energy from the Sun as light, not heat, and convert it directly into electricity. **FIGURE 39.7a** shows how a photovoltaic system, also referred to as PV, delivers electricity to a house. A photovoltaic solar cell makes use of the fact that certain semiconductors—very thin, ultraclean layers of material—generate a low-voltage electric current when they are exposed to direct sunlight (Figure 39.7b). The low-voltage direct current is usually converted into higher-voltage alternating current for use in homes or businesses. Most solar photovoltaic cells generate electricity well beyond their warranty period of 20 to 25 years, although their efficiency does decline slowly over time. The photovoltaic solar cell efficiency of newly manufactured panels has steadily increased over time. In fact, commercial, widely available photovoltaic solar cells are now almost 25 percent efficient in converting the energy of sunlight into electricity. "Do the Math: Calculating Home PV Generation and Capacity Factor" shows you how to calculate electricity production and dollar savings from a home photovoltaic solar system.

Electricity produced by photovoltaic systems can be used in several ways. Solar panels—arrays of photovoltaic solar cells—on a roof can be used to supply electricity to appliances or lights directly, or they can be used to charge batteries. Most photovoltaic systems in the United States are tied to the electrical grid, meaning that any extra electricity generated and not needed is sent to the electric utility, which buys it or gives the customer credit toward the cost of future electricity use. Homes that are "off the grid" may rely on photovoltaic solar cells as their only source of electricity, using batteries to store the electricity until it is needed. Photovoltaic solar cells have

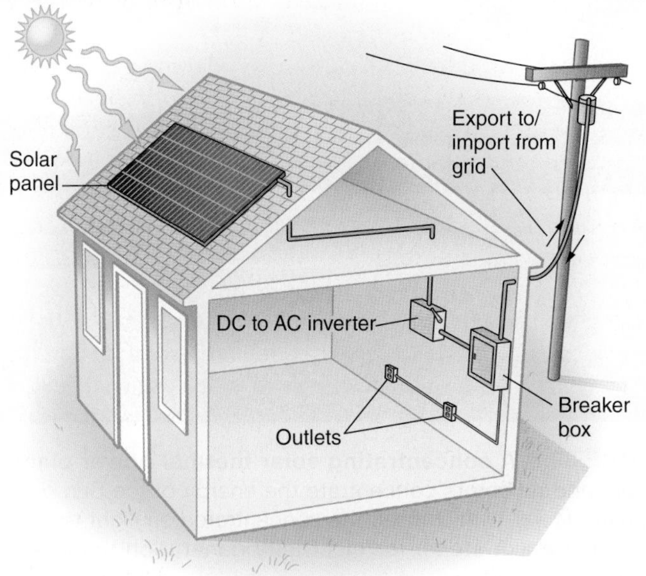

(a)

(b)

FIGURE 39.7 **Photovoltaic solar energy.** (a) In this domestic photovoltaic system, photovoltaic solar panels convert sunlight into direct current (DC). An inverter converts DC into alternating current (AC), which supplies electricity to the house. Any electricity not used in the house is exported to the electrical grid. (b) Photovoltaic panels on the roof of this house provide nearly all of the electricity this family uses. *(b: altrendo images/Getty Images)*

DO THE MATH Calculating Home PV Generation and Capacity Factor

Preparing for the AP® Exam

One of this book's authors, Andy Friedland, has a 4,000 watt photovoltaic solar array along the side of the driveway of his house in Vermont. It consists of 16 250-watt solar panels. In 2020, this array generated 5,300 kWh of electricity. Electricity in Vermont costs \$0.15/kWh. What is the capacity factor of this system and how much did Andy offset in electricity costs?

If 4,000 watts were generating continuously, it would generate this much electricity in a year:

$$4{,}000 \text{ watts} \times 1 \text{ kW}/1{,}000 \text{ watts} = 4 \text{ kW}$$

4 kW × 24 hours per day × 365 days per year = 35,000 kWh per year (after rounding to two significant figures) if the solar panels were generating 24 hours per day.

In actuality, if the system generated 5,300 kWh of electricity, the capacity factor is:

$$5{,}300 \text{ kWh/year} \div 35{,}000 \text{ kWh/year} = 0.15 \times 100 = 15\% \text{ capacity factor}$$

The capacity factor or percentage of time that the photovoltaic cells were generating electricity at full capacity was 15 percent.

To determine the dollar amount of the offset with this system, multiply the amount generated by the cost of electricity:

5,300 kWh × \$0.15/kWh = \$795. Professor Friedland offset, or avoided paying, \$795 to the electrical utility in 2020.

YOUR TURN A 390 MW concentrated solar thermal energy system was installed in a location where the capacity factor was measured to be 24 percent.

1. How much electricity will this solar plant generate in a year?
2. How much revenue would a utility receive if they were paid \$0.05/kWh for this electricity?

other uses in locations far from the grid where a small amount of electricity is needed on a regular basis. For example, small photovoltaic solar cells charge the batteries that keep highway signs and emergency telephones working. In many U.S. cities, photovoltaic solar cells provide electricity for new "smart" parking meter systems that have replaced aging coin-operated parking meters. While photovoltaic systems are increasing in popularity throughout the United States, their usefulness is limited by the availability of the Sun. In areas that receive limited sunlight for longer periods of time, coastal Washington State for example, modifications to standard systems must be made. For example, the necessary amount of battery storage will be designed into a system in a location where the number of cloudy days is relatively large. Photovoltaic solar cells usually generate the largest quantity of electricity per day when the air is cooler, such as in the spring and fall seasons. In the hottest part of the summer, photovoltaic cells are less efficient at higher temperatures, and the air usually contains more moisture and haze and therefore solar cells typically generate less electricity.

Concentrating Solar Thermal Electricity Generation

Concentrating solar thermal (CST) systems are a large-scale application of solar energy to electricity generation. CST systems use lenses or mirrors and tracking systems to focus the sunlight falling on a large area into a small beam, in the same way you might use a magnifying glass to focus energy from the Sun and perhaps burn a hole in a piece of paper. In this case, however, the heat of the concentrated beam is used to evaporate water and produce steam that turns a turbine to generate electricity. CST power plants operate much like conventional thermal power plants; the only difference is that the energy to produce the steam comes directly from the Sun, rather than from fossil fuels. The arrays of lenses and mirrors required are large, so CST power plants are best constructed in desert areas where there is consistent sunshine and plenty of open space (**FIGURE 39.8**). However, the coverage of land by large solar panels or mirrors can reduce solar radiation reaching the desert floor, which will reduce net primary productivity—a term that we introduced in Module 6. This will lead to changes in the temperature and humidity of the desert floor which can impact both plant and animal desert species.

Although CST systems have been available for more than 20 years, they are now becoming more common. In the United States, several plants generate electricity in California and in the Southwest. For example, the Ivanpah plant in California contains over 390 MW of capacity on 1,420 hectares (3,500 acres). These plants, though, have drawbacks that include the large amount of land required, especially in arid and desert ecosystems, and their inability to generate electricity at night.

AP® Exam Tip

Be sure that you understand the different types of solar energy systems, their respective benefits and costs, and their use and design. Be able to identify what type of energy is produced with each system: heat or electricity.

FIGURE 39.8 A concentrating solar thermal power plant. Mirrors and reflectors concentrate the energy of the Sun onto a "power tower," which uses the concentrated sunlight to heat water and make steam for electricity generation. *(craig coker/Shutterstock)*

Benefits and Drawbacks of Active Solar Energy Systems

Active solar energy systems offer many benefits such as generating hot water or electricity without producing CO_2, or polluting the air or water during operation. In addition, photovoltaic solar cells and CST power plants can produce electricity when it is needed most: on hot, sunny days when demand for electricity is high, primarily for air conditioning. By producing electricity during peak demand hours, these systems can help reduce the need to build new fossil fuel–power plants.

In many areas, small-scale solar energy systems are economically feasible. For a new home located miles away from the grid, installing a photovoltaic system may be much less expensive than running electrical transmission lines to the home site. When a house is near the grid, the initial cost of a photovoltaic system may be expensive, taking 5 to 20 years for payback. Once the initial cost is paid back, however, the electricity it generates is almost free.

Despite these advantages, a number of drawbacks have inhibited the growth of solar energy use in the United States. Photovoltaic solar panels are expensive to manufacture and install. Although the technology is changing rapidly as industrial engineers and scientists seek more efficient, cheaper photovoltaic materials and systems, the initial cost to install a photovoltaic system can be daunting and the payback period is a long one. In parts of the country where the average daily solar radiation is low, the payback period can be even longer. Some countries, such as Germany, have made solar energy a part of their sustainable energy agenda by subsidizing their solar industry. In the United States, tax breaks, rebates, and funding packages instituted by various states and the federal government have made solar electricity and water heating more affordable for consumers and businesses.

The use of photovoltaic solar cells has environmental as well as financial costs. Manufacturing photovoltaic solar cells requires a great deal of energy and water. It also involves a variety of toxic metals that need to be mined and industrial chemicals that can be released into the environment during the manufacturing process, although newer types of solar cells may reduce reliance on toxic materials. For systems that use batteries for energy storage, there are environmental costs associated with manufacturing, disposing of, or recycling the batteries, as well as energy losses during charging, storage, and recovery of electricity in batteries. The end-of-life reclamation and recycling of photovoltaic solar cells is another potential source of environmental contamination, particularly if the cells are not recycled properly. However, the energy expended to manufacture photovoltaic solar cells is usually recovered within a year or two of their operation. As the life span of photovoltaic solar cells and their efficiency continues to increase, they are a very promising source of renewable energy.

39-3 How is hydroelectric power generated and what are its environmental effects?

The kinetic energy of water can generate electricity but there are consequences

Hydroelectricity is electricity generated by the kinetic energy of moving water. It is the second most used form of renewable energy in the United States and in the world after biomass, and it is the form most widely used for electricity generation. As we saw in Figure 36.6, hydroelectricity provides 7 percent of the electricity generated in the United States. More than one-half of that hydroelectricity is generated in five states: Washington, Oregon, New York, California, and Alabama. Worldwide, 17 percent of all electricity comes from hydroelectric power plants, with China being the leading producer, followed by Canada, Brazil, the United States, India, and Russia. We will examine the ways in which hydroelectricity is generated, and we will consider whether hydroelectricity is sustainable.

Methods of Generating Hydroelectricity

Moving water, whether flowing with a river or tide, or falling over a distance vertically, such as in a waterfall, contains kinetic energy. A hydroelectric power plant captures this kinetic energy and uses it to turn a turbine in the same way that the kinetic energy of steam turns a turbine in a coal-fired power plant. The turbine transforms the kinetic energy of water into electricity, which is then exported to the electrical grid via transmission lines.

The amount of electricity that can be generated at any particular hydroelectric power plant depends on the flow rate of water, the vertical distance the water falls, or both. Where falling water is the source of the energy, the amount of electricity that can be generated depends on the vertical distance the water falls; the greater the distance, the more potential energy the water has, and the more electricity it can generate. In Module 0 we observed that water behind a dam is not in motion and doesn't seem to contain a great deal of energy. But water behind a dam does have potential energy. When you release water from the dam by opening a gate, the water rushes out and moves to a lower elevation due to the gravitational pull of Earth. The potential energy of the impounded water becomes kinetic energy of moving water. The amount of electricity a hydroelectric power plant can generate also depends on the flow rate: the amount of water that flows past a certain point per unit of time. The higher the flow rate, the more kinetic energy is present, and the more electricity can be generated. These basic principles apply to two different kinds of hydro dams as well as to tidal electricity generating systems.

Water Impoundment Systems

Storing water in a reservoir behind a dam is known as **water impoundment**. FIGURE 39.9 illustrates the various features of a water impoundment system. By managing the opening and closing of the gates, the dam operators control the flow rate of the water that turns the turbine—and in turn the generator—and thereby influence the amount of electricity produced.

Water impoundment is the most common method of hydroelectricity generation because it usually allows for the generation of electricity on demand. The largest hydroelectric water impoundment dam in the United States is the Grand Coulee Dam in Washington State, which generates 6,800 MW at peak capacity. The Three Gorges Dam on the Yangtze River in China (see Figure 48.2) is the largest dam in the world. It has a capacity of 22,500 MW and can generate almost 50 billion kilowatt-hours per year, approximately 7 percent of China's total electricity demand.

Run-of-the-River Systems

In **run-of-the-river** hydroelectricity generation, water is retained behind a low, small dam, or no dam, and passes through a channel that has a submerged turbine before returning to the river. Run-of-the-river systems do not store water in a reservoir. These systems have several advantages that reduce their environmental impact: Relatively little flooding occurs upstream, and seasonal changes in river flow are not disrupted. However, run-of-the-river systems are generally small and, because they rely on natural water flows, electricity generation can be intermittent. Heavy spring runoff from rains or snowmelt cannot be stored, and the system cannot generate any electricity in hot, dry periods when the flow of water is low.

Hydroelectricity Electricity generated by the kinetic energy of moving water.

Water impoundment The storage of water in a reservoir behind a dam.

Run-of-the-river Hydroelectricity generation in which water is retained behind a low, small dam or no dam.

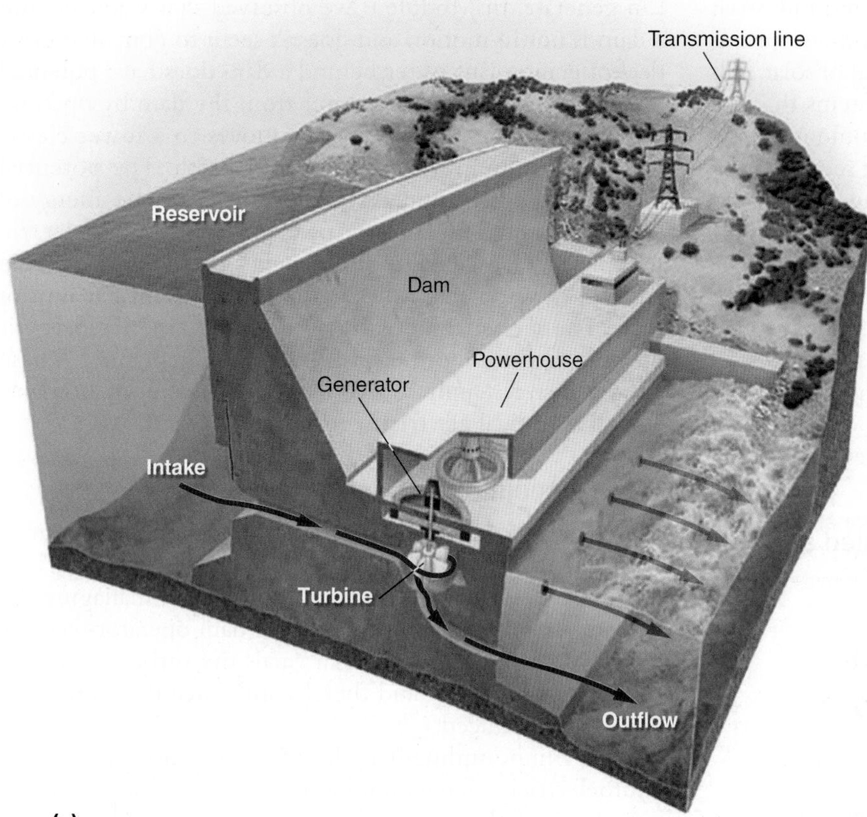

(a)

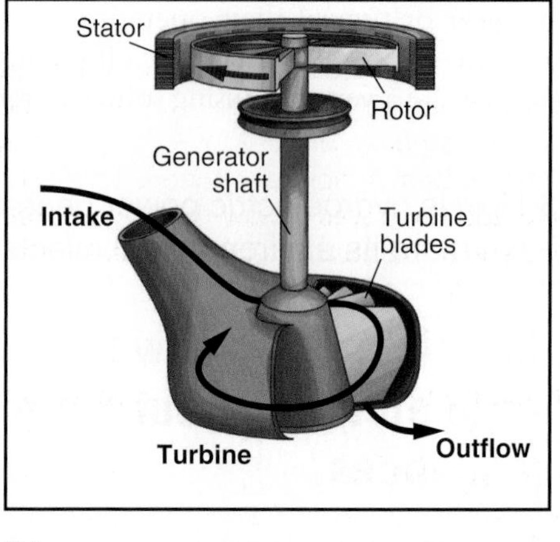

(b)

FIGURE 39.9 A water impoundment hydroelectric dam. (a) Water impoundment, a common method of hydroelectricity generation, allows for electricity generation on demand. Dam operators control the rate of water flow by opening and closing the gates. Increased water flow results in increased rotation of the turbines (b) resulting in a greater amount of electricity generated. The arrows indicate the path of water flow.

Tidal Systems

Tidal energy also comes from the movement of water, although the movement in this case is driven by the gravitational pull of the Moon. Tidal energy systems use gates and turbines similar to those used in run-of-the-river and water impoundment systems to capture the kinetic energy of water flowing through estuaries, rivers, and bays and convert this energy into electricity.

Although tidal power plants are operating in many parts of the world, including France, Korea, and Canada, tidal energy does not have the potential to become a major energy source. In many locations around the world, the difference in water level between high and low tides is not great enough to provide sufficient kinetic energy to generate a large amount of electricity. In addition, in order to transfer the energy generated, transmission lines must be constructed on or near a coastline or estuary. This infrastructure may have a disruptive effect on coastal, shoreline, and marine ecology as well as on tourism that relies on the aesthetics of a coastal region.

Hydroelectricity and Sustainability

Major hydroelectric dam projects have brought renewable energy to large numbers of rural residents in many countries, including the United States, Canada, India, China, Brazil, and Egypt. Although hydroelectric dams are expensive to build, once built, they require a minimal amount of fossil fuel for operation. In general, the benefits of water impoundment hydroelectric systems are large: They generate large quantities of electricity without creating air pollution, waste products, or CO_2 emissions. Electricity from hydroelectric power plants is usually less expensive for the consumer than electricity generated using nuclear fuels or natural gas. In the United States, the price of hydroelectricity ranges from 5 cents to 11 cents per kilowatt-hour.

In addition, the reservoir behind a hydroelectric dam can provide recreational and economic opportunities as well as downstream flood control for flood-prone areas. For example, Lake Powell, the reservoir impounded by the hydroelectric Glen Canyon Dam, draws more than 3 million visitors to the Glen Canyon National Recreation Area each year and generates more than $400 million annually for the local and regional economies in Arizona and Utah (**FIGURE 39.10**). However, droughts in recent years have reduced water flow into Lake Powell, and other lakes behind hydro dams. This has led to lower lake levels and what some call a "bathtub ring" around the sides of steep canyon walls surrounding the lake. In China, the Three Gorges Dam provides flood control and protection to many millions of people.

Water impoundment, however, does have negative environmental consequences. In order to form an impoundment, a free-flowing river must be held back. The resulting reservoir

Tidal energy Energy that comes from the movement of water driven by the gravitational pull of the Moon.

FIGURE 39.10 A recreation area created by water impoundment. Lake Powell at the Glen Canyon National Recreation Area was created by water impoundment at the hydroelectric Glen Canyon Dam in northern Arizona. In 2021, water flow from the Colorado River was so low that water levels in Lake Powell dropped and a "bathtub ring" was evident. *(Justin Sullivan/Getty Images)*

may flood hundreds or thousands of hectares of prime agricultural land or canyons with great aesthetic or archeological value. It may also force people to relocate. As we will see in Module 48, the construction of the Three Gorges Dam displaced more than 1.3 million people and submerged ancient cultural and archaeological sites as well as large areas of farmland. Impounding a river in this way may also make it unsuitable for organisms or recreational activities that depend on a free-flowing river. Large reservoirs of standing water hold more heat and contain less oxygen than do free-flowing rivers, thereby affecting which species can survive in the waters. Certain human parasites also become more abundant in impounded waters in tropical regions.

By regulating water flow and flooding, dams also alter the dynamics of the river ecosystem downstream. Some rivers, for example, have sandbars created during periods of low flow that follow periods of flooding. Some plant species, such as cottonwood trees, cannot reproduce in the absence of these sandbars. The life cycles of certain aquatic species, such as salmon, certain trout species, and freshwater clams and mussels, also depend on seasonal variations in water flow. Impoundment systems disrupt these life cycles by controlling the flow of water so it is consistently plentiful for hydroelectricity generation.

It is possible to address some of these problems. For example, as we will see in greater detail in Module 48, the installation of a "fish ladder" may allow fish to travel upstream around a dam. Such solutions are not always optimal, however; some fish species fail to utilize them and some predators learn to monitor the fish ladders for their prey.

Other environmental consequences of water impoundment systems include the release of greenhouse gases to the atmosphere, both during dam construction and after filling the reservoir. Production of cement—a major component of dams—is responsible for approximately 5 percent of global anthropogenic CO_2 emissions to the atmosphere. Once the dam is completed, the impounded water usually covers forests or grasslands. The dead plants and organic materials in the flooded soils decompose anaerobically and release methane, a potent greenhouse gas. Some researchers assert that in tropical regions, the methane released from a hydroelectric water impoundment contributes more to climate warming than a coal-fired power plant with about the same electricity generation output.

The accumulation of sediments in reservoirs has negative consequences not only for the environment but also for the electricity-generating capacity of hydroelectric dams. A fast-moving river carries sediments that settle out when the river feeds into the reservoir. The accumulation of these sediments on the bottom of the reservoir is known as **siltation**. Over time, as the reservoir fills up with sediments, the amount of water that can be impounded, and thus the generating capacity and life span of the dam, is reduced. This process may take hundreds of years or only decades, depending on the geology of the area. The only way to reverse the siltation process is by dredging, which is removal of the sediment, usually with machinery that runs on fossil fuels.

Because of either environmental concerns or heavy siltation, a number of hydroelectric dams are being dismantled. In 1999, the Edwards Dam was removed from the Kennebec River in Maine. More than a decade later, native fishes such as bass and alewives have returned to the waters and are flourishing. In 2007, the Marmot Dam on Oregon's Sandy River was removed using explosives. The restored river now hosts migrating salmon and steelhead trout (*Oncorhynchus mykiss*) for the first time since 1912 (**FIGURE 39.11**). In 2022, work was in progress to remove four large dams on the Klamath River in Oregon and northern California. Some are calling this the largest dam removal project in American history.

AP® Exam Tip

You should be able to explain the processes by which alternative sources of energy are converted to useable forms of energy. On an FRQ make certain that your answer is directly related to the prompt. For example, if the prompt asks for ways to heat your home using solar energy, make certain your answer pertains to heating the home. *For example: Passive solar installation of windows will trap heat inside the home, heating the home.*

In this module we examined biomass, solar energy, and hydroelectric power. Solid biomass, such as wood and charcoal, is a source of heat energy at relatively low cost and is most commonly used in the developing world. It produces pollutants such as carbon monoxide and particulates, which are especially problematic when biomass is burned indoors. Liquid biomass includes ethanol and biodiesel

Siltation Sediments from moving water that accumulate on the bottom of a reservoir.

(a)

(b)

FIGURE 39.11 Dam removal and environmental restoration. In recent years, a number of dams have been dismantled due to environmental concerns or heavy siltation. (a) This photo shows Oregon's Marmot Dam, on the Sandy River, before removal. (b) The same stretch of the Sandy River after dam removal in 2007 shows how the natural landscape has been restored. *(a and b: Portland General Electric)*

that do not release fossil carbon into the atmosphere. Solar energy is harnessed actively, using technology, and passively, using direct absorption of heat. Active solar can be stored in batteries and transferred elsewhere via the electrical grid. Passive solar can only be used at the location where it is absorbed. Water that is dammed or flowing and water that flows in and out with the tides can be harnessed to generate electricity. During operation, hydroelectricity does not generate pollutants but there are adverse consequences of damming a river and impounding water. In the next module we will examine geothermal energy and hydrogen fuel cell electricity generation.

Module 39 AP® Review

Preparing for the AP® Exam

Learning Goals Revisited

39-1 What are the positive and negative consequences of biomass energy resources?

Biomass is a modern source of carbon, which was formed between a few years ago and hundreds of years ago, as opposed to fossil fuels, which contain carbon formed millions of years ago. Solid biomass includes wood, charcoal, and animal manure, all of which are relatively low-quality sources of energy and release particulates, carbon monoxide, and other pollutants when burned. The use of biomass can also lead to overharvesting of trees. Liquid fuels include ethanol, an alcohol derived from plant material such as corn, and biodiesel, produced from vegetable oils such as soybean. Algae is another source of oil for biodiesel.

39-2 What are the various types of solar energy systems?

Passive solar energy takes advantage of relatively inexpensive strategies such as the direction windows are facing in a building. Active solar technologies use mechanical and electrical equipment to obtain heat or electrical energy from the Sun and have high initial costs but can potentially supply relatively large amounts of energy. Active solar applications can be small, such as those that fit on a rooftop or in a field, or they can be extremely large, on an industrial scale. Large solar systems can negatively impact land, including in desert ecosystems.

39-3 How is hydroelectric power generated and what are its environmental effects?

Hydroelectricity is generated from the energy in water. The largest hydroelectric projects come from impounding water behind a large dam and releasing it periodically when electricity is needed. The impounded water behind a dam can promote recreational and economic opportunities but can also have numerous impacts on the environment, especially upstream of the dam. Hydroelectricity can also be generated from run-of-the-river systems and intertidal systems.

AP® Practice Questions

Multiple-Choice Questions

Use the choices below to answer questions 1–3:

(a) biomass
(b) passive solar
(c) photovoltaic cells
(d) impoundment system

1. Energy source that can result in millions of deaths from indoor air pollution annually
2. Energy source that includes south-facing windows on a home
3. Energy source that uses kinetic energy of water to turn a turbine
4. A hydroelectric power plant's rate of electricity generation depends on which of the following pairs?
   (a) the flow rate of the water and the amount of water behind a dam
   (b) the amount of water behind a dam and the vertical distance the water falls
   (c) the flow rate of the water and the length of the river
   (d) the vertical distance the water falls and the flow rate of the water

Use the following table to answer questions 5 & 6:

U.S. Energy Consumption by Source, 2015–2018
Partial list of sources

Source	2015 consumption (quadrillion BTU)	2018 consumption (quadrillion BTU)
Hydroelectric	2.3	2.7
Solar	0.4	0.9
Wood biomass	2.3	2.4
Biofuels	2.2	2.3
Waste biomass	0.5	0.5
Petroleum	35.2	36.5
Coal	15.6	13.2
Natural gas	28.3	31.4

5. Which renewable energy source grew the most from 2015 to 2018?
   (a) hydroelectric
   (b) solar
   (c) wood biomass
   (d) biofuels
6. Which of the following best describes the relationship between renewable energy sources and fossil fuel sources from 2015 to 2018?
   (a) Energy consumption of renewable sources increased, while all fossil fuel consumption decreased.
   (b) Consumption of coal decreased, while all other sources increased.
   (c) Total energy consumption increased, though consumption of energy from coal decreased.
   (d) Consumption of renewable sources decreased, while consumption from petroleum sources increased.

Use the following passage to answer questions 7 & 8:

One proposed method for reducing carbon emissions from the transportation sector is to use biofuels made from staple crops such as corn. In 2020, the United States produced 14 billion gallons of corn-based ethanol for fuel, making it the largest producer in the world. Benefits of using corn-based ethanol instead of gasoline may in some cases include fewer greenhouse gas emissions and a decrease in the emission of various air pollutants. However, it requires a large expanse of land to grow the crops used for ethanol production, and the corn needs to be processed into fuel before it can be used, which can require fossil fuel use.

Recent increases in corn-based ethanol production have required large amounts of land to grow the corn for fuel, replacing land used to grow other crops or corn for food. Experts estimate that even if all of the cropland in the United States were converted to land to grow corn for ethanol, we could not produce enough ethanol to supply our gasoline use, and our food supply would need to be imported from other countries. Furthermore, while fuel prices may decrease in this scenario, it is likely that food costs would increase greatly.

7. Which of the following best identifies the author's claim in the article?
   (a) Growing corn for biofuel is a carbon-neutral practice.
   (b) Increasing demand for ethanol has led to an increase in land used to grow corn.
   (c) Corn-based ethanol will replace gasoline in vehicles within 30 years.
   (d) Increasing demand for ethanol will decrease the price of corn and other grains.
8. According to the article, which of the following best describes the author's perspective on the relationship between ethanol production and food production?
   (a) Ethanol production has increased in the United States and growing and processing corn for ethanol releases greenhouse gas emissions.
   (b) The ten bushels of corn it takes to produce ethanol for 25 gallons of fuel is enough calories to feed one person for 1 year.
   (c) It is not a practical solution to reducing gasoline use to convert all of the cropland in the United States for ethanol production, because all food crops would need to be imported.
   (d) Increasing the amount of land used for fuel production increases erosion.

Free-Response Question

In 2019, renewable energy sources accounted for approximately 11 percent of all energy use in the United States, and nearly 20 percent of electricity production.

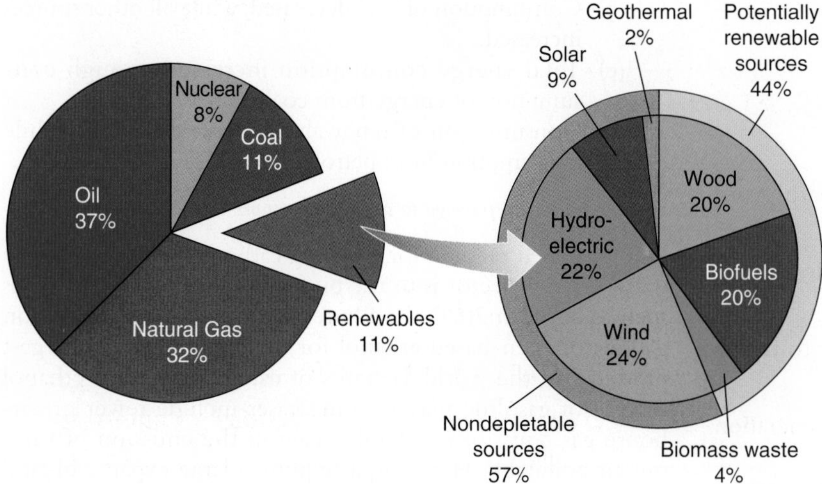

(a) **Calculate** the total amount of renewable energy consumed in the United States that comes from biomass. (1 pt.)

(b) **Identify** the nondepletable source of energy that provides the greatest amount of energy in the United States. (1 pt.)

(c) **Explain** why biomass is considered to be a potentially renewable resource rather than a nondepletable resource. (1 pt.)

(d) Biomass energy requires combustion of biomass and biofuels for electricity generation and transportation, which releases carbon into the atmosphere.

(i) Other than carbon emissions, **describe** an environmental problem related to using biomass for energy. (1 pt.)

(ii) **Explain** why biomass is widely considered to be a modern carbon source rather than a fossil carbon source. (2 pts.)

(e) A research team studying solar energy sets up an experiment to determine how the surface area of a solar panel containing photovoltaic cells affects the amount of electricity it can produce. They set up three solar arrays consisting of five panels, each with a different surface area.

(i) **Identify** the dependent variable in the experiment. (1 pt.)

(ii) **Identify** a research question for this experiment. (1 pt.)

(iii) **Identify** a variable not mentioned that should be kept constant in the experiment. (1 pt.)

(iv) **Describe** one modification the researchers can use to increase the reliability of their study. (1 pt.)

Module 40

Unit 6 | 35 | 36 | 37 | 38 | 39 | 40 | 41

Geothermal Energy and Hydrogen Fuel Cells

In the previous module we examined biomass and solar energies and hydroelectric power sources. All three renewable fuels contribute considerable amounts of energy to people around the world. In this module we examine geothermal energy and hydrogen fuel cells, both of which have potential for producing clean energy but also have limitations.

Learning Goals

After reading this module you should be able to

40-1 explain how humans harness geothermal energy.

40-2 describe the hydrogen fuel cell and its potential for providing electricity.

40-1 How do humans harness geothermal energy?

Earth's internal heat is transferred to water that we use for heating and electricity generation

Geothermal energy Heat energy that comes from the natural radioactive decay of elements deep within Earth.

Unlike most forms of energy discussed in Modules 39–41, geothermal energy does not come from the Sun. **Geothermal energy** is heat energy that comes from the natural radioactive decay of elements deep within Earth. As we saw in Module 22, convection currents in Earth's mantle bring hot magma toward the surface of Earth. Wherever magma comes close enough to groundwater, that groundwater is heated. The pressure of the hot groundwater sometimes drives it to the surface where it visibly manifests itself as geysers and hot springs, like those at Old Faithful in Yellowstone National Park (**FIGURE 40.1**). Where hot groundwater does not naturally rise to the surface, humans may be able to reach it by drilling.

Many countries obtain clean, renewable energy from geothermal resources. The United States, Indonesia, Philippines, and Turkey, all of which have substantial geothermal resources, are the largest geothermal energy producers. Nevertheless, the global geothermal power generation capacity was only around 15 GW at end of 2020. By comparison, wind generation capacity was 743 GW at the end of 2020.

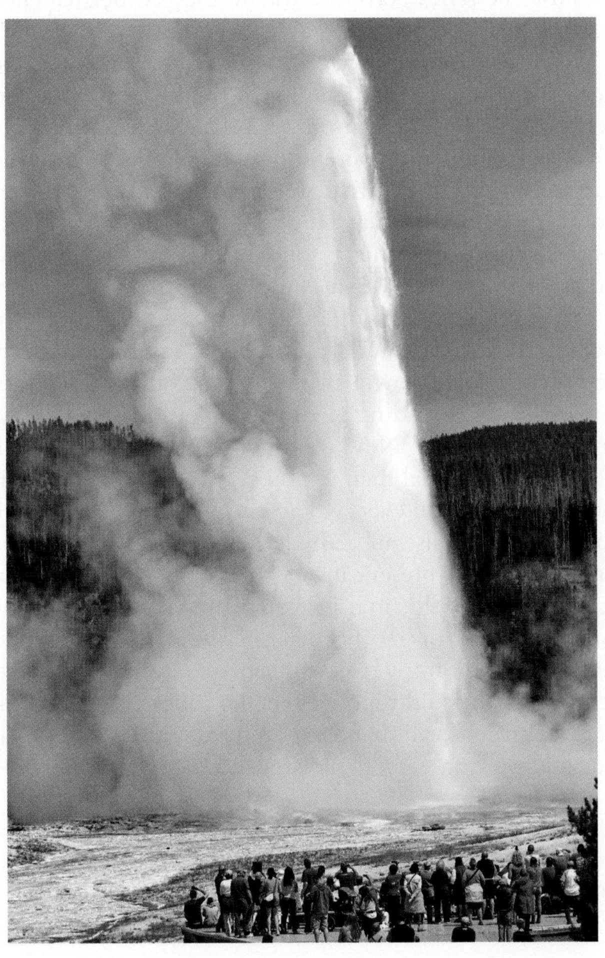

FIGURE 40.1 Geysers at Yellowstone National Park. Old Faithful, the well-known tourist attraction at Yellowstone, is an example of geothermal energy. The average temperature of the water is 96°C (204°F) and the steam is over 177°C (350°F). *(Marco Brivio/Getty Images)*

Harvesting Geothermal Energy

Geothermal energy can be used directly as a source of heat. Hot groundwater can be piped directly into household radiators to heat a home. In other cases, heat exchangers can collect heat by circulating cool liquid underground, where heat from the ground flows to the cool circulating liquid, and then returns to the surface. Iceland, a small nation with vast geothermal resources, heats 90 percent of its homes and businesses this way.

Geothermal energy can also be used to generate electricity. The electricity-generating process is much the same as that in a conventional thermal power plant although, in this case, the steam to run the turbine comes from water evaporated by Earth's internal heat instead of by burning fossil fuels or harnessing a nuclear reaction.

The heat released by decaying radioactive elements deep within Earth is essentially nondepletable in the span of human time. However, the groundwater that so often carries that heat to Earth's surface can be depleted. As we learned in Module 27, groundwater, if used sustainably, is a renewable resource. Unfortunately, long periods of harvesting groundwater from a site may deplete it to the point at which the site is no longer a viable source of geothermal energy. Returning the water to the ground to be reheated is one way to use geothermal energy sustainably.

Iceland currently produces about 25 percent of its electricity using geothermal energy. In the United States, geothermal energy accounts for less than 1 percent of the renewable energy used but because the United States uses so much electricity, the United States is the largest producer of geothermal electricity in the world. Within the continental United States, geothermal resources are primarily in the West and Southwest. **FIGURE 40.2** shows geothermal resources in the United States and the temperatures commonly encountered just below the ground surface at geothermal sites. Geothermal power plants are currently in operation in many states including California, Nevada, Utah, Hawaii, Oregon, Idaho, and New Mexico. Geothermal energy has less growth potential than wind or solar energy because it is not easily accessible everywhere and the expense of drilling multiple deep wells can be high. Hazardous gases and steam that are naturally present in the areas being drilled may also escape from geothermal power plants, which is an additional drawback of geothermal energy. Hydrogen sulfide is the most common gas released from geothermal wells and is the cause of the sulfurous "rotten egg" odor that is common near hot springs. Methane, a potent greenhouse gas, is also released from some geothermal plants.

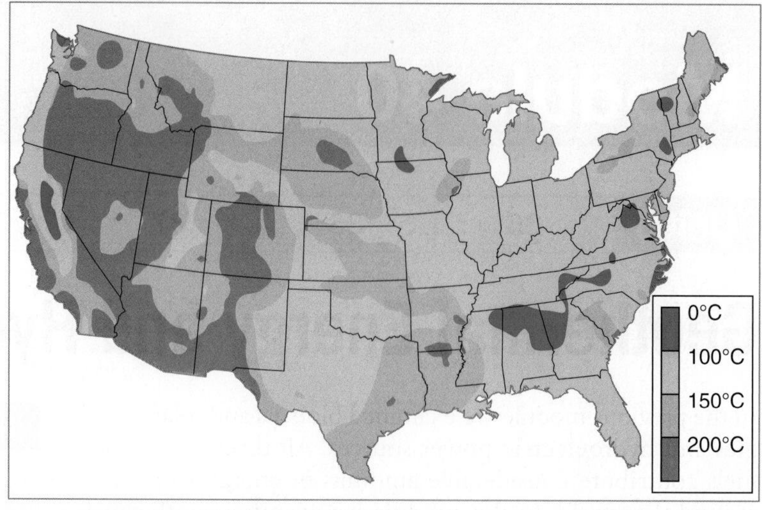

FIGURE 40.2 Geothermal resources in the United States. Although there are geothermal resources throughout the continental United States and Hawaii, the greatest amounts are in the West and Southwest. Temperatures shown are those commonly found just below the ground surface. *(Information from U.S. D.O.E. Energy Information Administration)*

Ground Source Heat Pumps

Another approach to tapping Earth's thermal resources is the use of **ground source heat pumps**, a technology that transfers heat from the ground to a building. Ground source heat pumps take advantage of the high thermal mass of the ground. Earth's temperature about 3 m (10 feet) underground remains fairly constant year-round, at 10°C to 15°C (50°F–60°F) because the ground retains the Sun's heat more effectively than does the ambient air. We can take advantage of this fact to heat and cool residential and commercial buildings. Although the heat tapped by ground source heat pumps is often referred to informally as geothermal, it comes not from geothermal energy but from solar energy. So in a sense, it is actually more of a nondepletable energy source than geothermal energy that comes from the radioactivity of elements deep within the Earth.

FIGURE 40.3 shows how a ground source heat pump transfers heat from the ground to a house. In contrast to the geothermal systems just described, ground source heat pumps do not remove steam or hot water from the ground. In much the same way that a solar water heating system works, a ground source heat pump cycles fluid through pipes buried underground. In winter, this fluid absorbs heat from underground. The slightly warmed fluid is compressed in the heat pump to increase its temperature even more, and the heat is distributed throughout the house. The fluid is then allowed to expand, which causes it to cool and run through the cycle again, picking up more heat from the ground. In summer, when the underground temperature is lower than the ambient air temperature, the fluid is cooled underground and then pulls heat from the house as it circulates, resulting in a cooler house as heat is transferred underground.

Ground source heat pump A technology that transfers heat from the ground to a building.

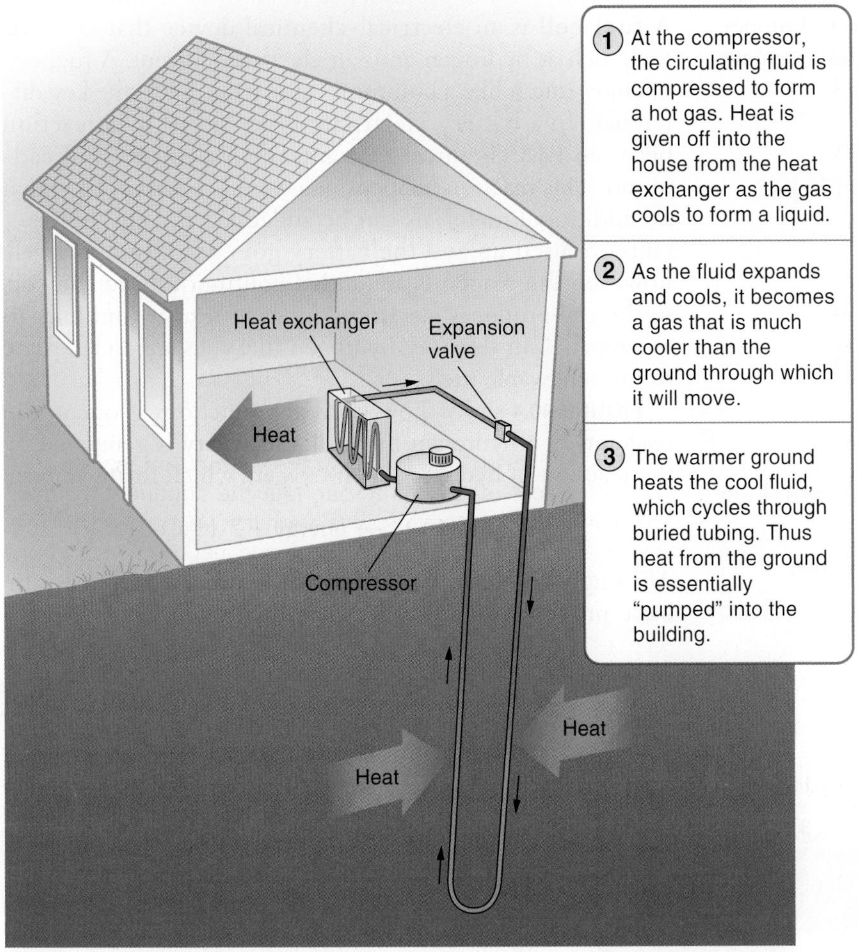

(1) At the compressor, the circulating fluid is compressed to form a hot gas. Heat is given off into the house from the heat exchanger as the gas cools to form a liquid.

(2) As the fluid expands and cools, it becomes a gas that is much cooler than the ground through which it will move.

(3) The warmer ground heats the cool fluid, which cycles through buried tubing. Thus heat from the ground is essentially "pumped" into the building.

FIGURE 40.3 Heating and cooling with a ground source heat pump. By exchanging heat with the ground, a ground source heat pump can heat and cool a building using 30 to 70 percent less energy than traditional furnaces and air conditioners.

Ground source heat pumps can be installed anywhere in the world, regardless of whether there is geothermal energy accessible in the vicinity. The operation of the pump requires some energy, but in most cases the system uses 30 to 70 percent less energy to heat and cool a building than a standard furnace or air conditioner.

Hot Water Heat Pumps

The hot water heat pump, a variation of the ground source heat pump, extracts heat from the air in a garage or basement and transfers it to water in a domestic hot water tank. This water is then used for household activities such as washing dishes and taking showers. Hot water heat pump systems are similar to those found in air conditioners and refrigerators as well as the ground source heat pump we just described. A refrigerator extracts heat from the inside of an insulated box—the refrigerator—thereby lowering the temperature of the food and air inside, while discharging heat to the outside—the kitchen. A hot water heat pump works in reverse: It extracts heat from the surrounding air in the basement or garage and pumps it into a tank filled with water destined for the kitchen sink or bathroom shower. In Module 36 we stated that a resistance coil water heater is 99 percent efficient, meaning that 99 percent of the energy in the electricity is transferred to the water and we explained that nothing is 100 percent efficient but an electric hot water heater comes close. In a hot water heat pump, between 200 and 250 percent of the amount of energy in the electricity used to run the hot water heat pump is transferred to the water in the tank. You might be wondering if this violates the first law of thermodynamics. How can we obtain more energy from the system than the amount of energy put into the system? Remember our efficiency equation introduced in Module 35? Energy efficiency is the ratio of the amount of energy obtained in the desired form to the total amount of energy introduced into the system. The heat pump technology utilizes the energy in the electricity and also extracts heat energy from the surrounding air. The amount of electrical energy required to run the heat pump plus the energy extracted from the air in the room is greater than the electrical energy used to run the heat pump. So this yields a number greater than 100 percent of the electricity used to run the pump. For this reason, hot water heat pumps are becoming more and more popular in homes and rental properties in the United States and can be found in all plumbing supply and big box home improvement stores. Many states give financial incentives for homeowners and apartment building managers to choose a hot water heat pump when replacing a conventional water heater. One of the authors of this textbook (Andy Friedland) has had a hot water heat pump in his house since 2017. He has saved fuel, carbon dioxide emissions, and money by using electricity to extract heat from the air in his basement and transferring it to the water in his hot water tank.

40-2 What is a hydrogen fuel cell and what is its potential for providing electricity?

Hydrogen fuel cells use hydrogen as an energy source and are almost pollution free

We need to consider one other energy resource that uses an energy technology and an energy carrier. This technology, hydrogen fuel cells, has received a great deal of attention for many years as a potential environmental savior but it has never achieved the prominence its enthusiasts said it would. A **fuel cell** is an electrical-chemical device that converts fuel, such as hydrogen, into an electrical current. A fuel cell operates much like a common battery, but with one key difference. In a battery, electricity is generated by a reaction between two chemical reactants, such as nickel and cadmium. This reaction happens in a closed container to which no additional materials can be added; eventually the reactants are used up and the battery goes dead. In a fuel cell, however, the reactants are added continuously to the cell, so the cell produces electricity for as long as it continues to receive fuel. In this way, hydrogen fuel cells are an alternate to nonrenewable fuels.

Fuel cell An electrical-chemical device that converts fuel, such as hydrogen, into an electrical current.

FIGURE 40.4 shows how hydrogen functions as one of the reactants in a hydrogen fuel cell. Electricity is generated by the reaction of hydrogen with oxygen, which forms water:

$$2\ H_2 + O_2 \rightarrow \text{energy} + 2\ H_2O$$

Although there are many types of hydrogen fuel cells, the basic process forces protons from hydrogen gas through a

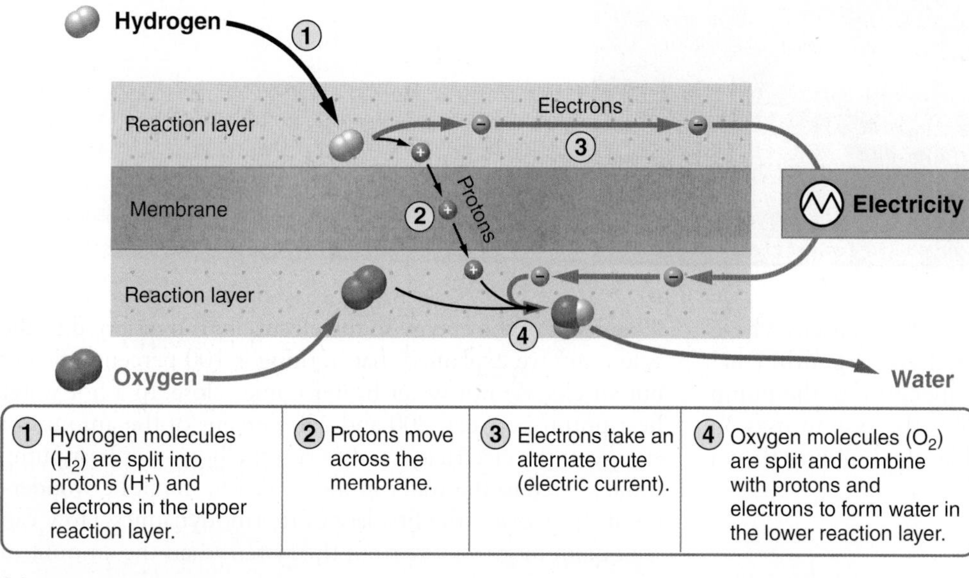
Hydrogen
1
Reaction layer
Electrons
3
Membrane
2
Protons
Electricity
Reaction layer
4
Oxygen
Water
1 Hydrogen molecules (H_2) are split into protons (H^+) and electrons in the upper reaction layer.
2 Protons move across the membrane.
3 Electrons take an alternate route (electric current).
4 Oxygen molecules (O_2) are split and combine with protons and electrons to form water in the lower reaction layer.

(a) One common fuel cell design

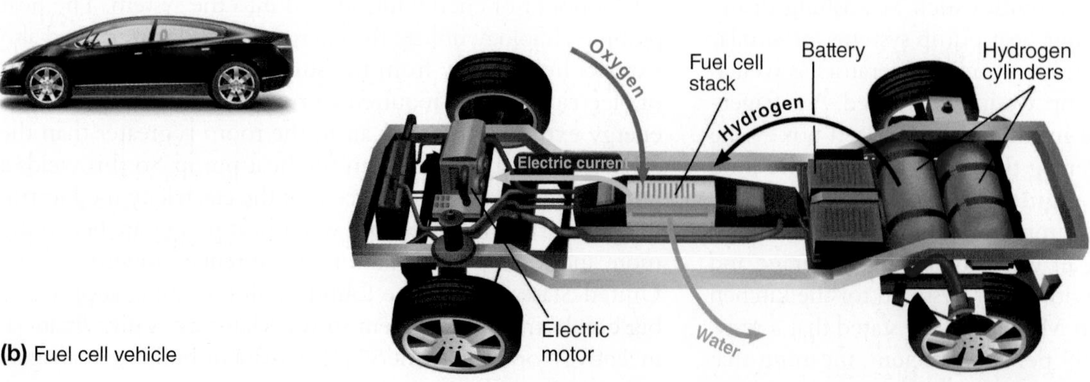
Oxygen
Fuel cell stack
Battery
Hydrogen cylinders
Hydrogen
Electric curren
Electric motor
Water

(b) Fuel cell vehicle

FIGURE 40.4 **Power from a hydrogen fuel cell.** (a) Hydrogen gas enters the cell from an external source. Protons from the hydrogen molecules pass through a membrane, while electrons flow around it, producing an electric current. Water is the only waste product of the reaction. (b) In a fuel cell vehicle, hydrogen is the fuel that reacts with oxygen to provide electricity to run the motor.

membrane, while the electrons take a different pathway. The movement of protons in one direction and electrons in another direction generates an electric current. The most important take-home message is this: The only waste products are energy and water. During the fuel cell process, there are no other waste products and, therefore, there are no pollutants released.

Using a hydrogen fuel cell to generate electricity requires a supply of hydrogen. Supplying and storing hydrogen is a challenge, however, because free hydrogen gas is relatively rare in nature and because the gas is explosive. Hydrogen tends to bond with other molecules, forming compounds such as water (H_2O) or natural gas (CH_4). Producing hydrogen gas requires separating it from these compounds using either heat or electricity. Currently, most commercially available hydrogen is produced by an energy-intensive process of using natural gas in order to extract its hydrogen; carbon dioxide and other pollutants are waste products of this process. In an alternative process, known as **electrolysis**, an electric current is applied to water to split it into hydrogen and oxygen and if the electricity comes from a renewable source, the process would not require fossil fuel nor would it generate carbon dioxide and air pollutants. Energy scientists are looking for other ways to obtain hydrogen; for example, under certain conditions, some photosynthetic algae and bacteria, using sunlight as their energy source, can give off hydrogen gas.

Although it may seem counterintuitive to use electricity to create electricity, the advantage of hydrogen is that it can act as an energy carrier. Renewable energy sources such as wind and the Sun cannot produce electricity constantly, but the electricity they produce can be used to generate hydrogen, which can be stored until it is needed. Thus, if we could generate electricity for electrolysis using a clean, nondepletable energy resource such as wind or solar energy, hydrogen could potentially be a sustainable energy carrier. A utility or network of energy producers could generate electricity from the Sun when it is sunny and from the wind when it is windy and when it is neither sunny nor windy, it could generate electricity from stored hydrogen gas. This electricity could be generated from a powerplant fueled by hydrogen gas. That is one of the appeals of the hydrogen fuel cell as part of an energy mix.

AP® Exam Tip

Be prepared to describe hurdles to the use of alternative technologies for transportation.

The Viability of Hydrogen

Some energy scientists consider hydrogen fuel cells to be the future of providing energy for humans and the solution to many of the world's energy problems. As of 2021, there were 160 fuel cells operating at more than 100 power plants in the United States with roughly 250 megawatts (MW) of electric generation capacity, which is less than 0.2 percent of total electricity generation capacity. So although possible, it does not appear that hydrogen will be used for generating electricity on a large scale anytime soon in the United States. Some energy analysts and transportation scientists still promote hydrogen fuel cells for use in automobiles. Even in 2022, when it appears that all-electric vehicles are possibly on the way to becoming the dominant automobile technology within a decade, an important role for hydrogen fuel cell cars is still considered possible. Hydrogen fuel cells are 80 percent efficient in converting the potential energy of hydrogen and oxygen into electricity. And most importantly, water is the only by-product. During operation of a hydrogen powered automobile, while converting hydrogen to electricity, there is no carbon dioxide released and no air pollutants produced. In contrast, thermal fossil fuel power plants are only 35 to 50 percent efficient, and they produce a wide range of pollutants as by-products. However, there are many detractors who believe that hydrogen fuel cells will not provide a solution to our energy problems.

Despite the advantages of hydrogen as a fuel, it also has a number of disadvantages. First, scientists must develop ways to obtain hydrogen without expending more fossil fuel energy than its use would save. This means that the energy for the hydrogen generation process must come from a renewable resource such as wind or solar energy rather than fossil fuels. Second, suppliers will need a distribution network to safely deliver hydrogen to consumers—something similar to our current system of gasoline delivery trucks and gasoline stations. Hydrogen can be stored as a liquid or as a gas, although each storage medium has its limitations. In a fuel cell vehicle, hydrogen would probably be stored in the form of a gas in a large tank under very high pressure. Vehicles would have to be redesigned with fuel tanks much larger than current gasoline tanks to achieve an equivalent travel distance per tank. There is also the risk of a tank rupture, in which case the hydrogen might catch fire or explode.

Given these obstacles, why is hydrogen even considered a viable energy alternative? Ultimately, hydrogen-fueled vehicles could be a sustainable means of transportation because a hydrogen-fueled car would use an electric motor, as do current electric vehicles like the Chevy Bolt and the Tesla. Electric motors are more efficient than internal combustion engines: While an internal combustion engine converts about 20 percent of the fuel's energy into the motion of the drive train, an electric motor can convert 60 percent of its energy into motion. So if we generated electricity from hydrogen at 80 percent efficiency, and used an electric motor to convert that electricity into vehicular motion at 60 percent efficiency, we would have a vehicle that is much more efficient than one with an internal combustion engine. Thus, even if we obtained hydrogen by burning natural gas, the total amount of energy used to move an electric vehicle

Electrolysis The application of an electric current to water molecules to split them into hydrogen and oxygen.

FIGURE 40.5 Hydrogen fuel cell–powered automobiles. Toyota offered one hydrogen fuel cell vehicle for sale in 2022. Access to hydrogen fueling stations in the United States is limited to about 40 locations, most of which are in California. *(Philip Cheung/The New York Times/Redux)*

using hydrogen might still be substantially less than the total amount needed to move a car fueled by gasoline. Using solar or wind energy to produce the hydrogen would lower the environmental cost even more, and the energy supply would be renewable. In those circumstances, an automobile could be fueled by a truly renewable source of energy that is both carbon neutral and pollution free.

Toyota, one of the leading car manufacturers in the world, offers a hydrogen fuel cell vehicle (**FIGURE 40.5**). In the United States, access to hydrogen fueling stations is confined to about 40 locations, mostly in California. Nevertheless, it appears that this manufacturer believes there is merit in developing hydrogen fuel cell cars. Both geothermal and hydrogen fuel cells are relatively small contributors to total energy use, both globally and in the United States. However, they both have the potential to contribute to the energy mix in the United States with minimal release of air pollutants and carbon dioxide from fossil fuels. In the next module we consider wind energy, which is the largest and fastest growing source of renewable energy in the United States.

In this module we examined geothermal energy and hydrogen fuel cells. We described how both energy sources have potential for producing clean energy but also have limitations. Geothermal energy contributes minimally to pollution but its use is limited by availability only in specific locations around the globe. Hydrogen fuel cells are a great source of clean electricity but currently require the use of natural gas to produce hydrogen. In the next module we will examine wind energy and energy conservation.

Module 40 AP® Review

Preparing for the AP® Exam

Learning Goals Revisited

40-1 How do humans harness geothermal energy?

Geothermal energy is renewable and comes from internal energy of the Earth caused by radioactive decay. Internal heat from the Earth is transferred to water and delivered to buildings for heating or to electricity generation power plants. It is ideal because its use does not produce pollution, although it does contribute to the release of gases such as hydrogen sulfide and methane. Its access is restricted to certain locations around the globe.

40-2 What is a hydrogen fuel cell and what is its potential for providing electricity?

Using a hydrogen fuel cell to generate electricity requires a supply of hydrogen. The only waste product from the reaction is water. Most commercially available hydrogen is presently generated by utilizing natural gas. If produced by a renewable energy source, hydrogen has the potential to be a pollution-free source of electricity.

AP® Practice Questions

Preparing for the AP® Exam

Multiple-Choice Questions

Use the following passage to answer questions 1 & 2:

Geothermal energy originates from the heat produced from radioactive decay in the core of the Earth. Geothermal reservoirs are areas of heated groundwater that are found along major tectonic plate boundaries around the world, such as those around the Ring of Fire in the Pacific Ocean. Most geothermal reservoirs are found close to Earth's surface, heated from magma in the mantle. In the United States, most of this geothermal energy is found in western states and Hawaii, where there is more tectonic activity and hotspots. California produces the most electricity from geothermal energy in the country.

1. According to the passage, which is true about geothermal energy?
   (a) Geothermal energy relies on heated surface water.
   (b) Most geothermal energy sources occur in areas where groundwater is heated by magma.
   (c) One of the most active geothermal areas is along the Ring of Fire, which encircles the Atlantic Ocean.
   (d) Most of the heat used for geothermal energy is located deep within Earth, far from the surface.

2. According to the passage, geothermal resources in the United States are most commonly found in
   (a) the northeast, where the Ring of Fire is located.
   (b) the mid-Atlantic states, where there is a large amount of tectonic activity.
   (c) Florida, where the low-lying land gets heated by the Sun.
   (d) the West, where hotspots are located.

Use the following diagram to answer questions 3–5:

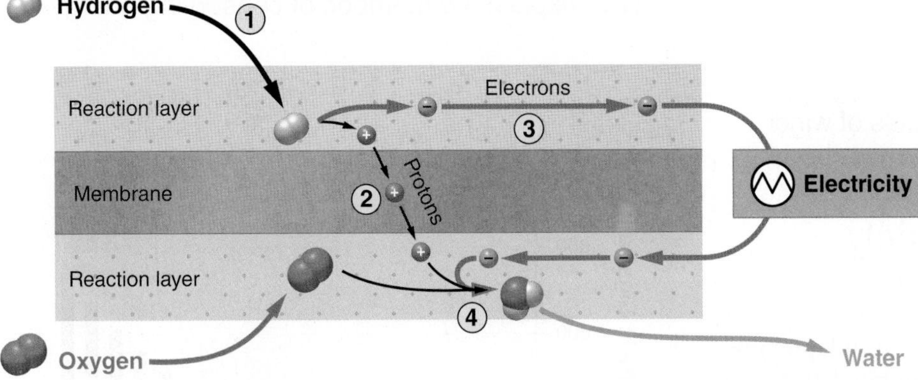

3. According to the diagram, electricity in a fuel cell is generated by the reaction of hydrogen with which element?
   (a) oxygen (c) helium
   (b) carbon (d) nitrogen

4. According to the diagram, using a fuel cell for electricity generation is desirable because the only waste product is
   (a) carbon dioxide. (c) air.
   (b) methane. (d) water.

5. Which two aspects of the diagram illustrate the breakdown of hydrogen molecules to produce electricity?
   (a) 1 and 2 (c) 1 and 3
   (b) 2 and 3 (d) 2 and 4

6. A hydrogen fuel cell is most similar to
   (a) an engine. (c) a source of coal.
   (b) a photovoltaic cell. (d) a battery.

Free-Response Question

The map below shows where the most concentrated areas of geothermal energy are located in the United States.

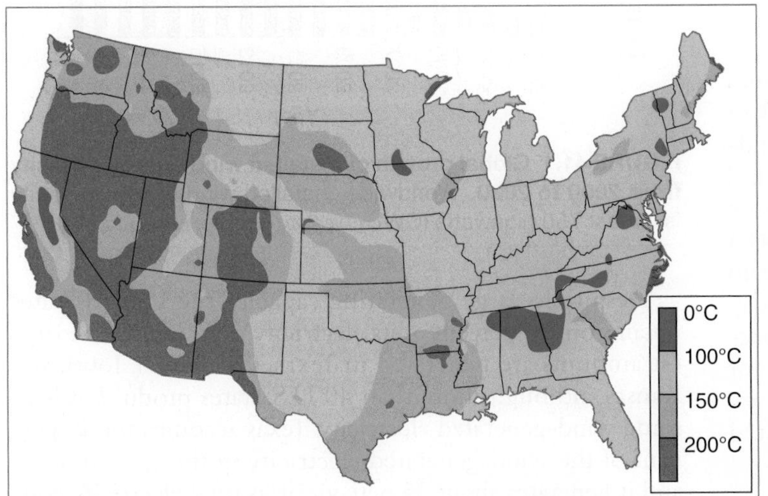

(a) **Identify** a state that has a range of 0°C to 250°C of geothermal energy available. (1 pt.)

(b) **Describe** why the western states have more geothermal energy available than eastern states. (1 pt.)

(c) **Explain** how geothermal energy is harnessed to produce electricity. (1 pt.)

(d) Other than issues related to location, **describe** two drawbacks related to using geothermal energy for heating and/or electricity. (2 pts.)

(e) Iceland produces 25 percent of its electricity using geothermal power, and heats 90 percent of its homes and businesses directly using groundwater heated by geothermal energy. In 2019, Iceland produced 19,489 GWh of electricity.
   (i) **Calculate** how much electricity, in GWh, are produced by geothermal energy in Iceland. Show all work. (1 pt.)
   (ii) In 2019, the population of Iceland was 360,563 people. **Calculate** the total electricity, in MWh, consumed per capita in 2019. Show all work. (2 pts.)
   (iii) **Calculate** the daily rate of electricity use, in kWh, of the citizens of Iceland in 2019. Show all work. (2 pts.)

Module 41

Unit 6 | 35 | 36 | 37 | 38 | 39 | 40 | 41

Wind Energy and Energy Conservation

In the previous module we examined geothermal energy sources and hydrogen fuel cells, both of which are minor contributors to the energy mix in the United States. In this module we examine wind energy, which is the fastest growing component of renewable energy in the United States and many other parts of the world. We also close this unit with a consideration of energy conservation and energy efficiency.

Learning Goals

After reading this module you should be able to

41-1 describe the benefits and impacts of wind energy.

41-2 explain the methods of conserving energy.

41-1 What are the benefits and impacts of wind energy?

Wind energy is the most rapidly growing source of electricity

Wind is currently the largest source of nondepletable, renewable energy. **Wind energy** is generated from the kinetic energy of moving air. As discussed in Module 22, winds are the result of the unequal heating of the surface of Earth by the Sun. Warmer air rises and cooler, denser air sinks, creating circulation patterns similar to those in a pot of boiling water. Ultimately, the Sun is the source of all winds—it is both solar radiation and ground surface heating that drive air circulation.

Before the electrical grid reached rural areas of the United States in the 1920s, windmills dotted the landscape. Most of those windmills pumped water or were used for grinding grain and they have since been removed. Today, wind energy from modern wind turbines is the fastest-growing major source of electricity in the world. As **FIGURE 41.1** shows, global installed wind energy capacity has risen from less than 24 gigawatts in 2001 to almost 750 gigawatts today. **FIGURE 41.2** shows installed wind energy generating capacity and the percentage of electricity generated by wind for a number of countries. China has the largest wind energy generating capacity in the world, followed by the United States, Germany, India, and Spain (Figure 41.2a).

Wind energy Energy generated from the kinetic energy of moving air.

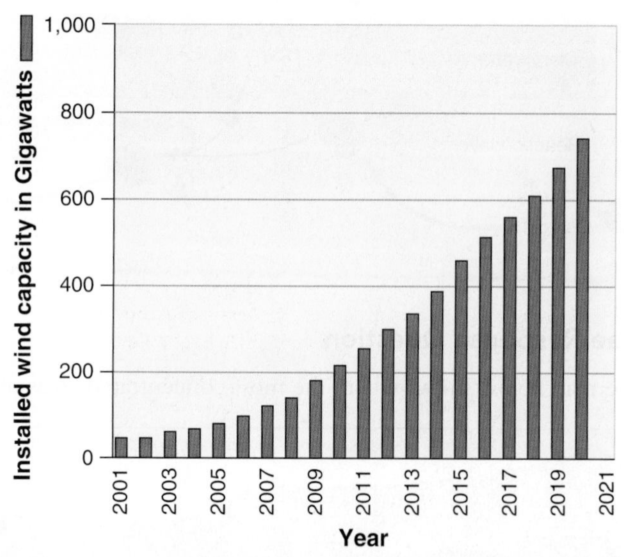

FIGURE 41.1 Global growth of installed wind energy capacity from 2000 to 2020. Worldwide, installed wind energy capacity is almost 750 gigawatts (GW). *(Data from Global Wind Energy Council)*

Despite its large generating capacity, the United States obtains only 8 percent of its electricity from wind. The largest amounts are generated in Texas, Oklahoma, Iowa, and Kansas, although more than 40 U.S. states produce at least some wind-generated electricity. Texas accounts for 40 percent of the wind-generated electricity in the United States and it generates about 25 percent of its total electricity consumption from wind. Denmark, a country of 5.8 million people, generates almost 47 percent of its electricity from wind (Figure 41.2b). Although the United States currently obtains only a small percentage of its electricity from wind, it is the fastest growing source of electricity in the country.

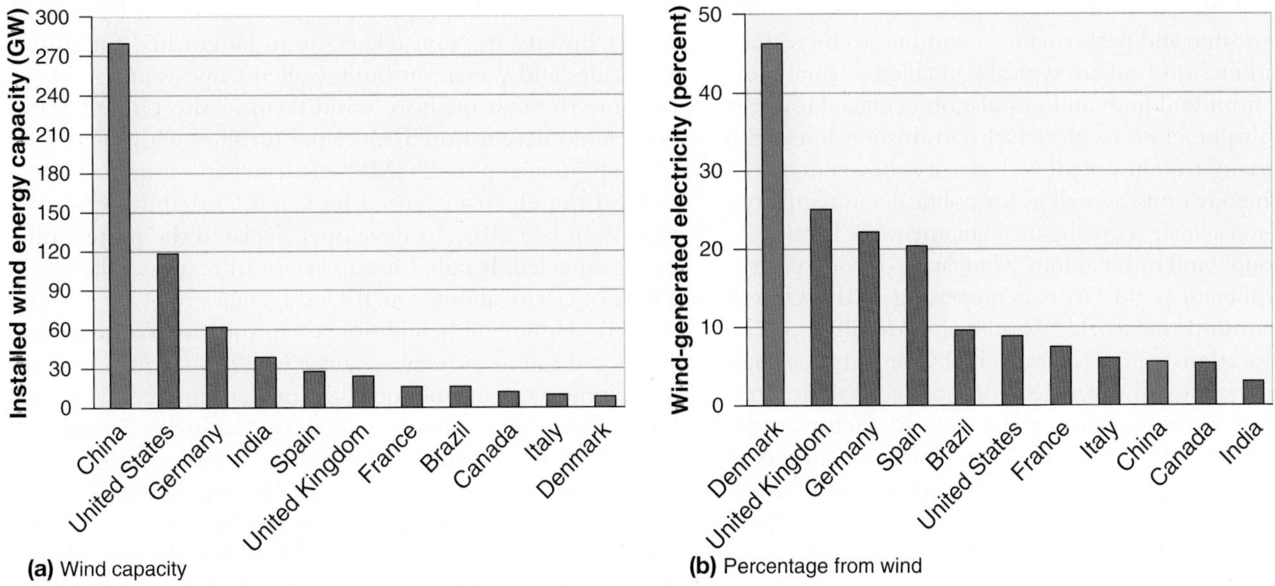

FIGURE 41.2 **Installed wind energy capacity and percentage of electricity from wind, by country.** (a) China generates more electricity from wind energy than any other country. (b) However, some relatively small countries, such as Denmark, generate a much higher percentage of their electricity from wind. *(Data from Global Wind Energy Council)*

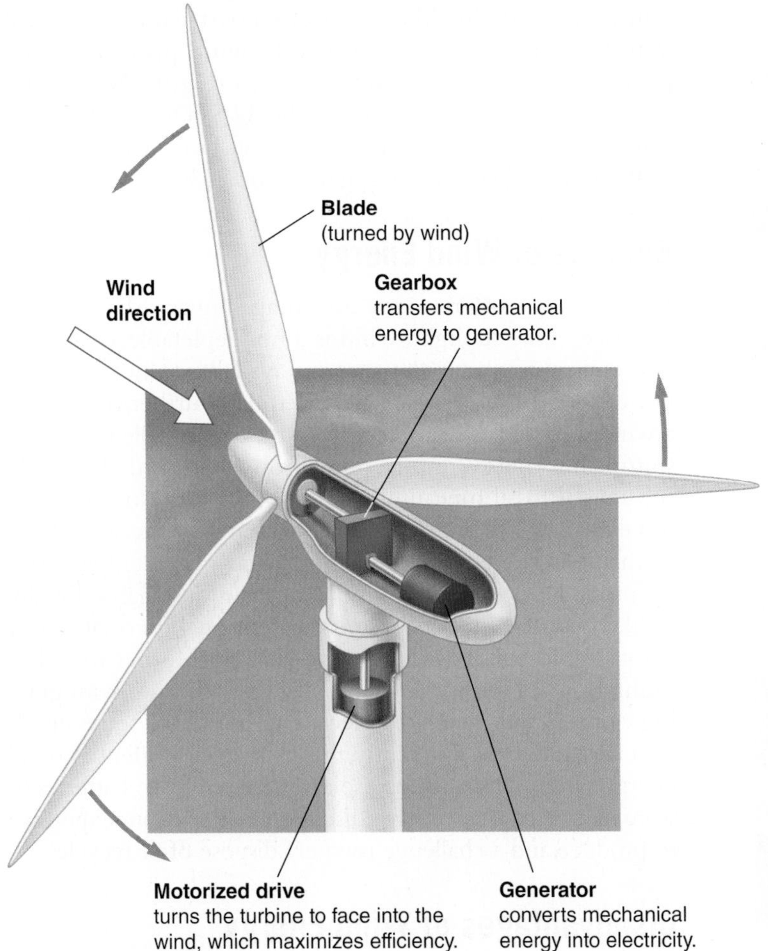

FIGURE 41.3 **Generating electricity with a wind turbine.** The wind turns the blade, which is connected to the generator, which generates electricity.

Generating Electricity from Wind

A **wind turbine** converts the kinetic energy of moving air into electricity in much the same way that a hydroelectric turbine harnesses the kinetic energy of moving water. As you can see in **FIGURE 41.3**, wind turns the blades of the wind turbine and the blades transfer energy to the gear box that in turn transfers energy to the generator that generates electricity. A modern wind turbine, like the one shown, may sit on a tower 100 m (330 feet) high and have blades 40 to 75 m (130–250 feet) long. There has been a gradual move to larger and taller wind turbines, so wind turbines installed since 2020 are even larger. The larger turbine blades increase turbine capacity. Taller towers increase electricity generation substantially because wind speeds are faster at higher altitudes and wind power is proportional to the cube of wind speed. Under average wind conditions, a wind turbine on land might have a capacity factor of 25 percent, but the newest installations are achieving up to a 40 percent capacity factor. While it is spinning, a wind turbine might generate 2,500 kW (2.5 MW), and in a year it might produce more than 6 million kilowatt-hours of electricity, enough to supply more than 600 homes. Offshore wind conditions are even more desirable for electricity generation with capacity factors of 40 percent to 50 percent, and turbines can be made even larger in an offshore environment. Every year for the

Wind turbine A turbine that converts the kinetic energy of moving air into electricity.

past decade, turbines continue to be built larger, capacity factors are greater, and performance continues to increase.

Wind turbines on land are typically installed in rural locations, away from buildings and population centers. However, they must also be close to electrical transmission lines with enough capacity to transport the electricity they generate to users. For these reasons, as well as for political and regulatory reasons and to facilitate servicing the equipment, the usual practice is to group wind turbines into wind farms or wind parks.

The number of wind farms is increasing in the United States and around the world. Wind farms are often placed on land in locations where the capacity factor can be as high as 35 to 40 percent. However, near-offshore coastal locations are even more desirable because the capacity factor can be up to 50 percent. Offshore wind parks, which are clusters of wind turbines, are often located in the ocean within a few miles of the coastline (**FIGURE 41.4**). Such parks are operating in Denmark, the Netherlands, the United Kingdom, Sweden, and elsewhere.

Until recently, the United States has lagged behind many other countries in the development of near-offshore wind. An infamous proposed project located off Cape Cod, Massachusetts, in Nantucket Sound languished for almost 2 decades and was never built. Called Cape Wind, it was to become the first offshore wind farm in the United States and would have featured 130 wind turbines with the potential to produce up to 420 MW of electricity, or up to 75 percent of the electricity used by Cape Cod and the nearby islands. In late 2016, its developer declared the project officially canceled. It failed for a variety of reasons. There was a lack of clarity about which U.S. agencies regulate offshore wind development. In addition, the project was close to sensitive and scenic waterways near Cape Cod. And perhaps most importantly, some local landowners and legislators did not want to see a wind farm close to Nantucket Sound.

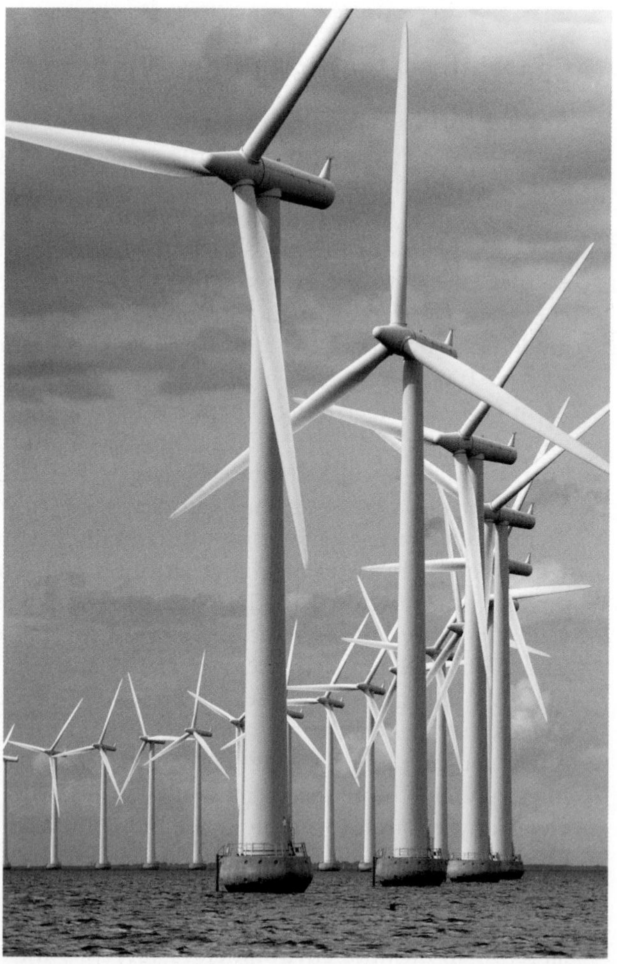

FIGURE 41.4 Offshore wind parks. Capacity factors at near-offshore locations like this one in Denmark approach 50 percent. *(Max Mudie/Alamy)*

In 2016, the Block Island Wind Farm project went online in the near-offshore environment of Rhode Island. Officially the first near-offshore wind project to be completed in the United States, it contains five turbines that are much larger than commonly found on land, each 6 MW, for a total capacity of 30 MW. In 2020, the project generated 120 million kWh with a capacity factor of 46 percent and provided electricity to around 12,000 homes. There is currently a second, smaller wind project operating off the coast of Virginia and in 2021, construction began on an 800 MW wind farm off the Massachusetts coast called Vineyard Wind. There are 15 near-offshore wind projects in the permitting process in New Jersey, Oregon, New York, Massachusetts, and Virginia. The U.S. Departments of Energy anticipates the expansion of offshore wind capacity in the United States to 30 gigawatts by 2030.

Benefits of Wind Energy

Wind energy offers many advantages over other energy resources. Like sunlight, wind is a nondepletable, clean, and free energy resource; the amount available tomorrow does not depend on how much we use today. Furthermore, once a wind turbine has been manufactured and installed, the only significant energy input comes from the wind. The only substantial fossil fuel input required, once the turbines are installed, is the fuel workers need to travel to the wind farm to maintain the equipment. Thus, wind-generated electricity produces no pollution and no greenhouse gases. Finally, unlike hydroelectric, CST, and conventional thermal power plants, wind farms can share the land with other uses. For example, wind turbines on land may share the area with grazing cattle. Wind-generated electricity does have some disadvantages, however. Currently, most off-grid residential wind energy systems, like off-grid solar systems, rely on batteries to store electricity. As we have discussed, batteries are expensive to produce and a challenge to safely dispose of or recycle.

Disadvantages of Wind Energy

Birds and bats are killed by collisions with wind turbine blades. According to a variety of studies that have occurred in the last decade, 200,000 to as many as 500,000 birds may be killed by collisions with wind turbine blades in the United States each year. Songbirds and raptors are the two groups of birds most affected. For comparison, it is estimated that songbird deaths from domestic house cats are 1 million to 5 million per year. Bird collisions with and electrocutions from electrical transmission lines, and from all electricity generation and transmission, not just wind turbines, is in the millions per year. Raptors such as bald eagles are known to nest in the tops of electrical transmission towers and electrocutions have been documented. Wind energy companies have instituted various measures that appear to be reducing bird deaths. Known bird migratory pathways are avoided when installing wind turbines, and some wind turbines are turned off during known peak migration time periods. Painting one blade a different color than the other blades has also been shown to help birds notice and avoid turbines blades. Bat deaths from wind turbines are not as well quantified, although new turbine designs and location of wind farms away from bat migration paths have reduced these deaths to some extent.

There are other problems with wind energy as well. New turbine designs have reduced some of the noise and aesthetic problems associated with wind farms. There is a small but vocal minority of people in various regions of the country who find a wind farm visually objectionable. Some people also find the sound of a wind turbine bothersome or intrusive, especially when it can be heard in their homes. There are people who experience anxiety or irritability from the background noise of a wind turbine. For these and other reasons, there has been resistance to wind farms in some regions of the United States. For example, the Cape Wind project that we described earlier experienced numerous hearings, protests, and court decisions that slowed development and eventually led to its failure. In Vermont, a state often considered to be very environmentally friendly, more and more towns and individuals have argued against the installation of commercial wind projects on or near ridgelines, citing habitat fragmentation and alteration, noise, and aesthetics, among other reasons. But there are also many locations in the United States where wind farms are welcomed and there are few or no protests to their construction. While wind may be the most desirable and least environmentally harmful source of electricity, there is one more way to effectively obtain energy—by reducing the current demand for energy. That is achieved by conservation, including increasing energy efficiency.

41-2 What are the methods of conserving energy?

We can use less, and use different technologies to conserve energy

In Module 35 we defined energy conservation as finding and implementing ways to use less energy while conducting the same activities, which is also known as doing more with less. **TABLE 41.1** lists some of the ways that an individual might conserve energy, including adjusting the household thermostat to use less heat during cold months and less air conditioning in hot months, consolidating errands in order to drive fewer miles, conserving water, or turning off a computer when it is not being used. A person or household could also use energy-efficient refrigerators and air conditioners, and plant appropriate trees around a house to reduce the need for air conditioning, all of which result in less energy use. One can also reduce **phantom loads**, electrical demand by a device that draws electrical current, even when it is turned off. Cable boxes and gaming consoles are often a source of phantom loads. On a larger scale, a government might implement energy conservation measures that encourage or even require individuals to adopt strategies or habits that use less energy. Some of these methods are also referred to as increasing the efficiency of appliances and vehicles. One such top-down approach is to improve the availability of public transportation (**FIGURE 41.5**). Governments can also facilitate energy conservation by increasing taxes on electricity, oil, and natural gas, since higher taxes discourage use. Alternatively, governments offer rebates or tax credits for retrofitting a home or business so it will operate using less energy; they can also use similar approaches to encourage use of more efficient automobiles such as hybrid-electric vehicles (HEVs), plug-in hybrid-electric vehicles (PHEVs), and battery electric vehicles (BEVs). Hybrid-electric vehicles are more efficient than conventional

Phantom loads Electrical demand by a device that draws electrical current, even when it is turned off.

TABLE 41.1 Ways to conserve energy

Home	Transportation	Electrical and electronic devices
• Weatherize; insulate, seal air gaps • Turn thermostat down in winter, up in summer • Reduce use of hot water; do laundry in cold water; take shorter showers • Plant landscaping that shades a house from hot sun in summer	• Walk or ride a bike • Use public transportation • Carpool • Consolidate trips • Choose a vehicle with higher fuel economy • Choose a hybrid or battery-electric vehicle	• Buy Energy Star/energy-efficient devices and appliances • Reduce phantom loads or use a power strip • Use a laptop rather than a desktop computer

FIGURE 41.5 Reducing energy use. There are many ways individuals can reduce their energy use in and outside the home. Taking public transportation rather than traveling by personal vehicle is one method of energy conservation. These students are taking the Washington, D.C., Metro. *(Chicago Tribune/Tribune News Service/Getty Images)*

internal-combustion engine vehicles because they capture some of the kinetic energy of motion lost when applying the brakes of the car to slow it. This captured energy is converted to electricity, stored in batteries, and used to power an electric motor to propel the car, thereby using less gasoline. PHEVs are similar to HEVs, but they can also run entirely on electricity for a limited distance (27 to 71 km or 17 to 44 miles), relying on stored electricity from being plugged into the grid prior to use. BEVs are entirely reliant on the grid for the electricity in their batteries and run entirely on electricity. We described in Module 40 how a car that runs on electricity, some of the time or all of the time, is more efficient than an internal combustion engine car.

The demand for energy varies with time of day, season, and weather. When electricity-generating plants are unable to handle the electrical demand during high-use periods, the utility sometimes reduces the amount of electricity it sends to a neighborhood (a brownout) or they cut electricity completely to that neighborhood (a blackout). To avoid this problem, electric companies must be able to provide enough electricity to satisfy **peak demand**, the greatest quantity of energy used at any one time. Peak demand may be several times the overall average demand, which means that substantially more electricity must be available than is needed under average conditions. To meet peak demand for electricity, electric companies often keep backup generators of electricity available—typically fossil fuel–fired generators, and in some cases, batteries.

Therefore, an important aspect of energy conservation is the reduction of peak demand, which would make it less necessary for electric companies to build excess generating capacity that is used only sporadically. One way of reducing peak demand is to establish a variable price structure under which customers pay less to use electricity when demand is lowest (typically in the middle of the night and on weekends) and more when demand is highest. This approach helps even out the use of electricity, which both reduces the burden on the generating capacity of the utility and rewards the electric consumer at the same time.

The second law of thermodynamics tells us that whenever energy is converted from one form into another, some energy is lost as unusable heat. In a typical thermal fossil fuel or nuclear power plant, only about one-third of the energy expended goes to its intended purpose; the rest is lost during energy conversions. We need to consider these losses in order to fully account for all energy conservation savings. So, the amount of energy we save is the sum of both the energy we did not use together with the energy that would have been lost in converting that energy into the form in which we would have used it. For example, if we can reduce our electricity use by 100 kWh, we may actually be conserving 300 kWh of an energy resource such as coal, since we save both the 100 kWh that we decide not to use and the 200 kWh that would have been lost during the conversion process to make the 100 kWh available to us.

Peak demand The greatest quantity of energy used at any one time.

Efficiency

Modern changes in electric lighting are a good example of how steadily increasing energy efficiency results in overall energy conservation. Compact fluorescent light bulbs use one-fourth as much energy to provide the same amount of light as incandescent bulbs. LED (light-emitting diode) light bulbs are even more efficient; they use one-sixth as much energy as incandescent bulbs. Over time, the widespread adoption of these efficient bulbs has resulted in substantially less energy used to provide the same amount of lighting.

Another way in which consumers can increase energy efficiency is by switching to products that meet the efficiency standards of the Energy Star program set by the U.S. Environmental Protection Agency. For example, an Energy Star air conditioner may use 0.2 kWh (200 watt-hours) less electricity per hour than a non–Energy Star unit. In terms of cost, a single consumer may save only 2 to 5 cents per hour by switching to an Energy Star unit. However, if 100,000 households in a city switched to Energy Star air conditioners, the city would reduce its energy use by 20 MW, or 4 percent of the output of a typical power plant. "Do the Math: Energy Star" on page 472 shows you how to calculate Energy Star savings.

AP® Exam Tip

One of the FRQs on the AP® Environmental Science Exam will require you to do calculations. Sometimes, there will be one point awarded for showing your work. This is called the "set up." *You must show units in both the set up and your final answer.*

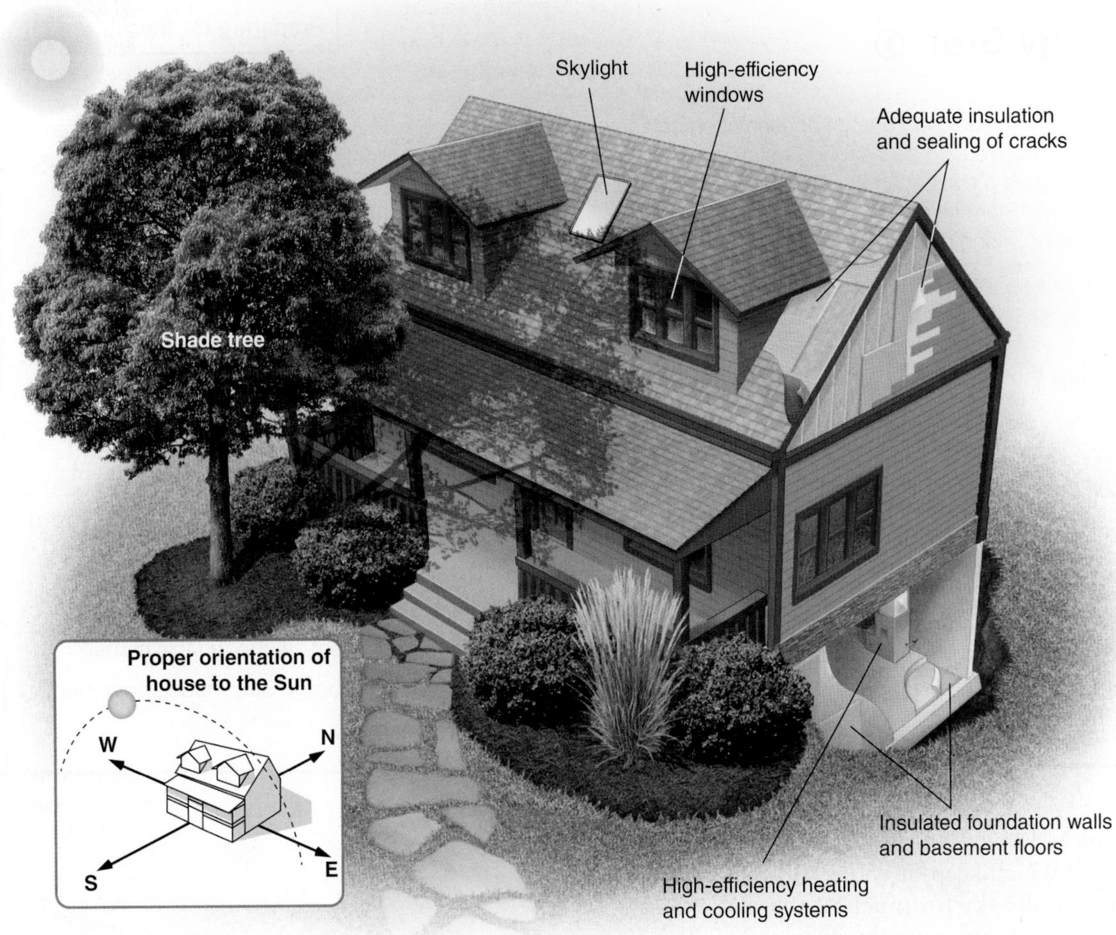

FIGURE 41.6 Green building design features in an energy-efficient home. A sustainable green building design incorporates proper solar orientation and landscaping as well as insulated windows, walls, and floors. In the Northern Hemisphere, a southern exposure allows the house to receive more direct rays from the Sun in winter when the path of the Sun is in the southern sky.

Sustainable Design

Sustainable design can improve the efficiency of the buildings and communities in which we live and work. **FIGURE 41.6** shows some key features of green building design as they are applied to a single-family dwelling. Insulating foundation walls and basement floors, orienting a house properly in relation to the Sun, and planting shade trees in warm climates are all appropriate design features. As we saw in Module 30, good community planning also conserves energy. Building houses close to where residents work reduces reliance on fossil fuels used for transportation, which in turn reduces the amount of pollution and carbon dioxide released into the atmosphere.

Buildings consume a great deal of energy for cooling, heating, and lighting. Many sustainable building strategies rely on **passive solar design**, a construction technique designed to take advantage of solar radiation without the use of active technology. **FIGURE 41.7** illustrates key features of passive solar design along with other features. Passive solar design stabilizes indoor temperatures without the need for pumps or other mechanical devices. For example, in the Northern Hemisphere, constructing a house with windows along a south-facing wall allows the Sun's rays to penetrate and warm the house, especially in winter when the Sun is more prominent in the southern sky. Double-paned windows insulate while still allowing incoming solar radiation to warm the house. Carefully placed windows also allow the maximum amount of light into a building and reduce the need for artificial lighting. Dark materials on the roof or exterior walls of a building absorb more solar energy than light-colored materials, further warming the structure.

Passive solar design Construction technique designed to take advantage of solar radiation without active technology.

DO THE MATH Energy Star

Preparing for the AP® Exam

Thinking clearly about energy efficiency and energy conservation can save you a lot of money in some surprising ways. If you are saving money on your electric bill, you are also saving energy and reducing the emission of pollutants. Consider the purchase of an air conditioner. Suppose you have a choice: an Energy Star unit for $300 or a standard unit for $200. The two units have the same cooling capacity but the Energy Star unit costs 5 cents per hour less to run. If you buy the Energy Star unit and run it 12 hours per day for 6 months of the year, how long does it take to recover the $100 extra cost?

You would save

$$\$0.05/\text{hour} \times 12 \text{ hours/day} = \$0.60/\text{day}$$

Six months is about 180 days, so in the first year you would save

$$\$0.60/\text{day} \times 180 \text{ days} = \$108$$

Spending the extra $100 for the Energy Star unit actually saves you $8 in just 1 year of use. In 3 years of use, the savings will more than pay for the entire initial cost of the unit (3 × $108 = $324), and after that you pay only for the operating costs.

YOUR TURN You are about to invest in a 66-inch flat screen TV. These TVs come in both Energy Star and non–Energy Star models. The cost of electricity is $0.15 per kilowatt-hour, and you expect to watch TV an average of 4 hours per day.

1. The non–Energy Star model uses 0.5 kW (half a kilowatt). How much will the electricity cost you per year to run this model?
2. If the Energy Star model uses only 40 percent of the amount of electricity used by the non–Energy Star model, how much money would you save on your electric bill over 5 years by buying the efficient model?

Conversely, using light-colored materials on a roof reflects heat away from the building, which keeps it cooler. In summer, when the Sun is high in the sky for much of the day, an overhanging roof helps block out sunlight during the hottest period, which makes the indoor temperature cooler and reduces the need for ventilation fans or air conditioning. Window shades can also reduce solar energy entering the house.

To reduce demand for heating at night and for cooling during the day, builders can use construction materials that have high thermal mass. **Thermal mass** is a property of a building material that allows it to retain heat or cold. Materials with high thermal mass stay hot once they have been heated and cool once they have been cooled. Stone and concrete have high thermal mass, whereas wood and glass do not; think of how a cement sidewalk stays warm longer than a wooden boardwalk after a hot day. A south-facing room with stone walls and a stone floor will heat up on sunny winter days and retain that heat long after the Sun has set.

Although building a house into the side of a hill or covering the roof of a building with soil and plants are less-common approaches, these measures also provide insulation and reduce the need for both heating and cooling. While

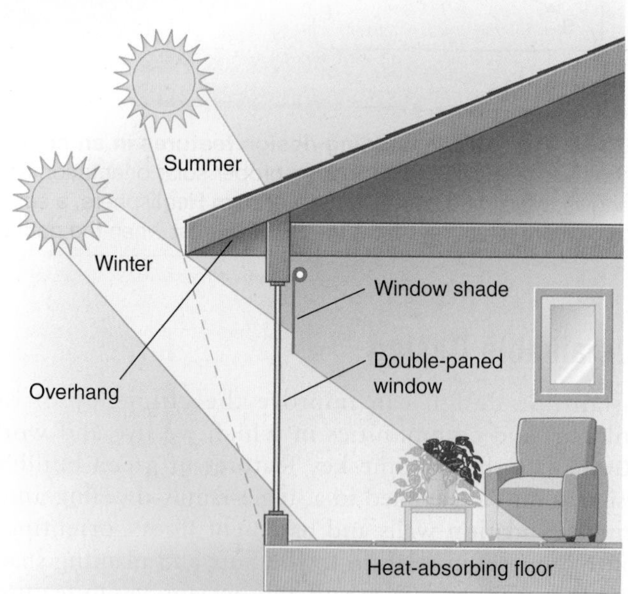

FIGURE 41.7 Passive solar design. Passive solar design uses solar radiation to maintain indoor temperature. Roof overhangs make use of seasonal changes in the Sun's position to reduce energy demand for heating and cooling. In winter, when the Sun is low in the sky, it shines directly into the window and heats the house. In summer, when the Sun is higher in the sky, the overhang blocks incoming sunlight and the room stays cool. High-efficiency windows and building materials with high thermal inertia are also components of passive solar design.

Thermal mass A property of a building material that allows it to maintain heat or cold.

"green roofs"—roofs with soil and growing plants—are somewhat unusual in the United States, many European cities, such as Berlin, Germany, have them on new or rebuilt structures. They are especially common on high-rise buildings in downtown areas that have little natural plant cover. These green roofs cool and shade the buildings and the surrounding environment. And the addition of plants to an urban environment also improves overall air quality.

The use of recycled building materials is another method of energy conservation. Recycling reduces the need for new construction materials, which reduces the amount of energy required to produce the components of the building. For example, many buildings now use recycled denim insulation in the walls and ceilings, and fly ash (a by-product recovered from coal-fired power plants) in the foundation.

Homes constructed today may incorporate some or all of these sustainable design strategies, but it is possible to achieve energy efficiency even in very large buildings. The building that houses the California Academy of Sciences in Golden Gate Park in San Francisco is a showcase for several of these sustainable design techniques (**FIGURE 41.8**). This structure, which incorporates a combination of passive solar design, radiant heating, solar panels, and skylights, actually uses 30 percent less energy than the amount permitted under national building energy requirements. Natural light fills 90 percent of the office space and many of the public areas. Windows, blinds, and skylights open as needed to allow air to circulate, capturing the ocean breezes and ventilating the building. Recycled denim insulation in the walls and a soil-covered rooftop garden provide insulation that reduces heating and cooling costs. As an added benefit, the living green roof grows native plants and captures 13.6 million liters (3.6 million gallons) of rainwater per year, which is then used to recharge groundwater stores.

In addition to these passive techniques, the designers of this building incorporated active technologies that further reduce its use of energy. An efficient radiant heating system carries warm water through tubes embedded in the concrete floor, using a fraction of the energy required by a standard forced-air heating system. To produce some of the electricity used in the building directly, the designers added 60,000 photovoltaic solar cells to the roof. These solar panels convert energy from the Sun into 213,000 kWh of electricity per year and reduce greenhouse gas emissions by about 200 metric tons per year. The California Academy of Sciences took an innovative approach to meeting its energy needs through a combination of energy efficiency and use of renewable energy resources. However, many, if not all, of these approaches will have to become commonplace if we are going to use energy in a sustainable way.

FIGURE 41.8 The California Academy of Sciences. The sustainable design of this San Francisco research institution maximizes the use of natural light and ventilation. The building generates much of its own electricity with solar panels on its roof and captures water in its rooftop garden. *(Nancy Hoyt Belcher/Alamy)*

Energy Summary and Synthesis

Although renewable energy is a more sustainable energy choice than nonrenewable energy, using any form of energy has an impact on the environment. Biomass, for instance, is a renewable resource only if it is used sustainably. Overharvesting wood leads to deforestation and degradation of the land, as we discussed in Module 39. Wind turbines can kill birds and bats, and hydroelectric turbines kill millions of fish. Manufacturing photovoltaic solar panels requires heavy metals and a great deal of water. Because all energy choices have environmental consequences, minimizing energy use through conservation and efficiency should be the first approach and is usually the best approach. After we achieve that, we must make energy choices wisely, depending on a variety of environmental, economic, and convenience factors.

Efficiency, Conservation, and the Development of Renewable and Nonrenewable Energy Resources

Each of the renewable energy resources we have discussed in this unit has unique advantages. None of these resources, however, is a perfect solution to our energy needs. **TABLE 41.2** lists some of the advantages and limitations of each of these resources. In short, no single energy resource that we are currently aware of can replace nonrenewable energy resources in a way that is completely renewable, nonpolluting, and free of impacts on the environment. A sustainable energy strategy, therefore, must combine energy efficiency, energy conservation, and the development of renewable and nonrenewable energy resources, taking into account the costs, benefits, and limitations of each. Convenience and reliability are also important factors. Finally, logistical considerations, such as where an energy source is located and how we transport the energy from that source to users, are also important. This is particularly important with the generation of electricity from renewable sources in remote regions, which requires an electrical transmission grid to get it to users.

A Renewable Energy Strategy

Energy expert Amory Lovins suggests that innovation and technological advances, not the depletion of a resource, have provided the driving force for moving from one energy technology to the next. Extending this concept to the present, one can argue that we will develop new energy technologies before we run out of the fuels on which we currently depend.

An increased reliance on renewable energy means that energy will be obtained in many locations and will need to be delivered to other locations. Delivery can be particularly problematic when electricity for an urban area is generated at a remote location. The electrical distribution system—the electrical grid that we described in Module 37—was not originally designed for this purpose. So in addition to investing in new energy sources, the United States will have

TABLE 41.2 Comparison of renewable energy resources

Energy resource	Advantages
Liquid biofuels	• Potentially renewable • Can reduce our dependence on fossil fuels • Reduce trade deficit • Possibly more environmentally friendly than fossil fuels
Solid biomass	• Potentially renewable • Eliminates waste from environment • Available to everyone • Minimal technology required
Photovoltaic solar cells	• Nondepletable resource • After initial investment, no cost to harvest energy
Solar water heating systems	• Nondepletable resource • After initial investment, no cost to harvest energy
Hydroelectricity	• Nondepletable resource • Low cost to run • Flood control • Recreation
Tidal energy	• Nondepletable resource • After initial investment, no cost to harvest energy
Geothermal energy	• Nondepletable resource • After initial investment, no cost to harvest energy • Can be installed anywhere (ground source heat pump)
Wind energy	• Nondepletable resource • After initial investment, no cost to harvest energy • Low up-front cost
Hydrogen fuel cell	• Efficient • Zero Pollution

Disadvantages	Emissions (pollutants and greenhouse gases)	Electricity cost ($/kWh)	Energy return on energy investment
• Loss of agricultural land • Higher food costs • Lower gas mileage • Possible net increase in greenhouse gas emissions	• CO_2 and methane		1.3 (from corn) 8 (from sugar cane)
• Deforestation • Erosion • Indoor and outdoor air pollution • Possible net increase in greenhouse gas emissions	• Carbon monoxide • Particulate matter • Nitrogen oxides • Possible toxic metals from MSW • Danger of indoor air pollutants		
• Manufacturing materials requires high input of metals and water • No plan in place to recycle solar panels • Geographically limited • High initial costs • Storage batteries required for off-grid systems	• None during operation • Some pollution generated during manufacturing of panels	0.1	8
• Manufacturing materials requires high input of metals and water • After initial investment, no cost to harvest energy and water • No plan in place to recycle solar panels • Geographically limited • High initial costs	• None during operation • Some pollution generated during manufacturing of panels		
• Limited amount can be installed in any given area • High construction costs • Threats to river ecosystems • Loss of habitat, agricultural land, and cultural heritage; displacement of people • Siltation	• Methane from decaying flooded vegetation	.05–.11	12
• Potential disruptive effect on some marine organisms • Geographically limited	• None during operation		15
• Emits hazardous gases and steam • Geographically limited	• None during operation	.05–.30	8
• Turbine noise • Deaths of birds and bats • Geographically limited to windy areas near transmission lines • Aesthetically displeasing to some • Storage batteries required for off-grid systems	• None during operation	.04–.06	18
• Producing hydrogen is an energy-intensive process • Lack of distribution network • Hydrogen storage challenges	• None during operation		8

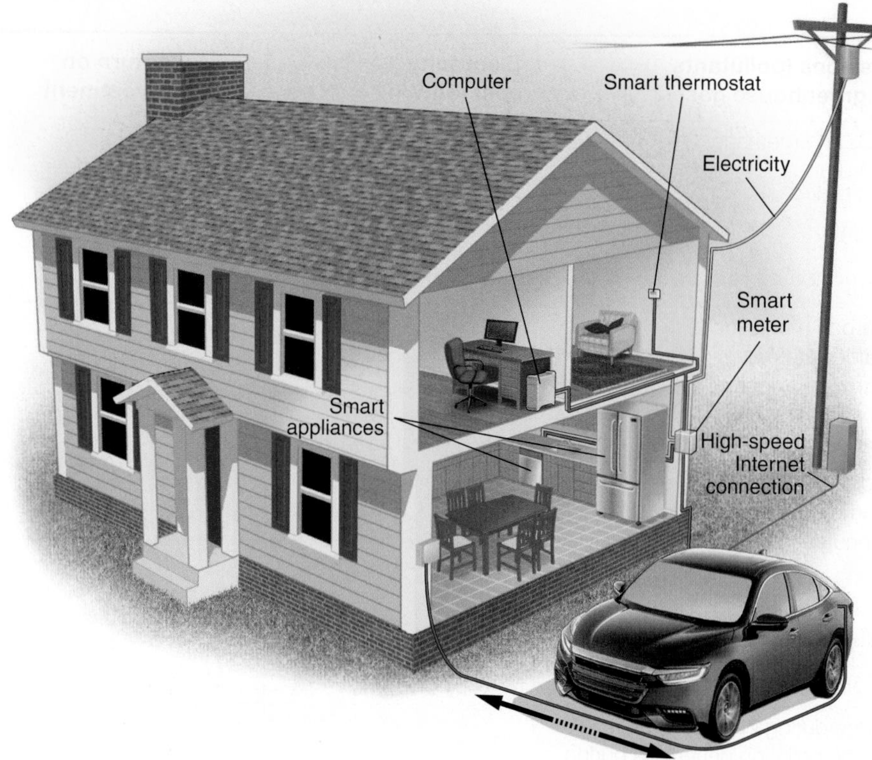

FIGURE 41.9 Using a smart grid. A smart grid optimizes the use of energy in a home by continuously coordinating energy use with energy generation.

to upgrade its existing electrical infrastructure—its power plants, storage capacity, and distribution networks. Approximately 40 percent of the energy used in the United States is used to generate electricity. Approximately 5 to 10 percent of the electricity generated is lost as it is transported along electrical transmission lines, and the greater the distance, the more that is lost. While the storage capacity of batteries improves each year, batteries are probably not a sustainable solution for this problem of energy loss. Many people are focusing their attention on improving the electrical grid to make it as efficient as possible at moving electricity from one location to another, thereby reducing the need for storage capacity.

One solution currently being implemented is the **smart grid**, an efficient, self-regulating electricity distribution network that accepts any source of electricity and distributes it automatically to end users. A smart grid uses computer programs and the Internet to tell electricity generators when electricity is needed and electricity users when there is excess capacity on the grid. In this way, it coordinates electricity use with electricity availability. Since 2008, government and industry contributions have brought total investment in the smart grid to almost $150 billion, which has been used to fund smart grid projects around the country.

How does a smart grid work? **FIGURE 41.9** shows one example. With "smart" appliances plugged into a smart grid, at bedtime a consumer could set an appliance such as a dishwasher to operate overnight. A computer in the dishwasher would be programmed to run the appliance anytime between midnight and 5:00 AM, depending on when there is a surplus of electricity. The dishwasher's computer would query the smart grid and determine the optimal time, in terms of electricity availability, to turn on the appliance. Conservation and efficiency need to be considered along with all the renewable and nonrenewable energy resources discussed in this unit. Taken together, they potentially chart a sustainable energy future for the United States and the world.

Smart grid An efficient, self-regulating electricity distribution network that accepts any source of electricity and distributes it automatically to end users.

In this module we examined wind energy and conservation. We described how a wind turbine operates and identified that wind energy is the fastest growing component of renewable energy in the United States. We described benefits and impacts of wind energy. We also examined methods of conserving energy and identified that if conserving energy saves energy that would have been used, it is equivalent to generating energy from another source.

Module 41 AP® Review

Preparing for the AP® Exam

Learning Goals Revisited

41-1 What are the benefits and impacts of wind energy?

Wind energy is harnessed through the use of a wind turbine, which converts the kinetic energy of moving air into electricity. Wind is the fastest growing form of new electricity generation in the world. Among renewable energy resources, it has many advantages, but it is responsible for killing bats and birds and some people object to it for concerns of noise and aesthetics.

41-2 What are the methods of conserving energy?

Minimizing energy use through conservation and efficiency should be considered first. Conservation can be summarized as doing the same activities with less energy, such as getting to school or work with public transportation rather than a personal vehicle. One aspect of conservation includes increasing efficiency, which refers to obtaining the same amount of usable work from a device with less energy input, such as traveling in a hybrid electric vehicle the same distance as a gasoline-powered car but with less gasoline input.

Practice Math and Graphing

Preparing for the AP® Exam

Answer the following questions. Be sure to show all your work.

1. Practice Math

(a) Wind capacity in the United States has undergone a rapid increase over the past 2 decades. Total capacity of wind turbines in the United States was 4,000 MW in 2000 and 118,000 MW in 2020. What was the percentage increase per year over this 20-year period? Round to the appropriate number of significant figures.

(b) Geothermal energy has decreased much less in the past 20 years. There was approximately 2.7 GW in 2000 and 3.7 GW in 2020 of installed geothermal capacity in the United States. What was the percentage increase over this 20-year period?

2. Practice Graphing

Smart meters are electric meters installed in houses and businesses that have two-way communication between a home or business and an electrical utility. They range from recording hourly electricity usage to lowering electricity demand at specific times of peak demand on the grid. In 2007, there were approximately 1 million smart meters installed in the United States. By 2021, there were more than 115 million. Create a graph with year on the x axis and number of smart meters in millions on the y axis.

Show the pattern of smart meter growth in the United States, assuming the growth was roughly steady from 2007 to 2021.

AP® Practice Questions

Preparing for the AP® Exam

Multiple-Choice Questions

1. Which of the following is an energy efficiency improvement?
   (a) adjusting the thermostat in a building
   (b) using a power strip
   (c) using cold water instead of hot water
   (d) replacing compact fluorescent bulbs with LEDs

2. Peak demand
   (a) decreases electricity-generating capacity.
   (b) is caused by higher electricity prices.
   (c) occurs equally in all seasons.
   (d) can be managed using a variable price structure.

3. If an Energy Star refrigerator costs 2 cents less per hour to run, and if it runs for 16 hours a day, how much will it save in a year?
   (a) \$44
   (b) \$82
   (c) \$117
   (d) \$174

Use the following graphs to answer questions 4 & 5:

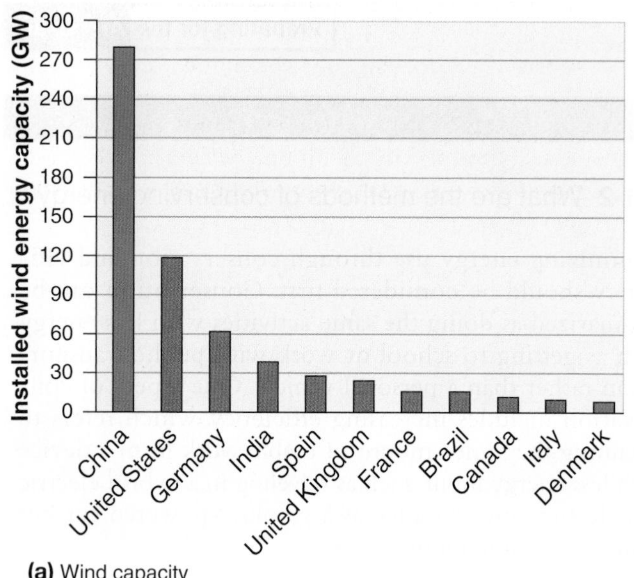

(a) Wind capacity

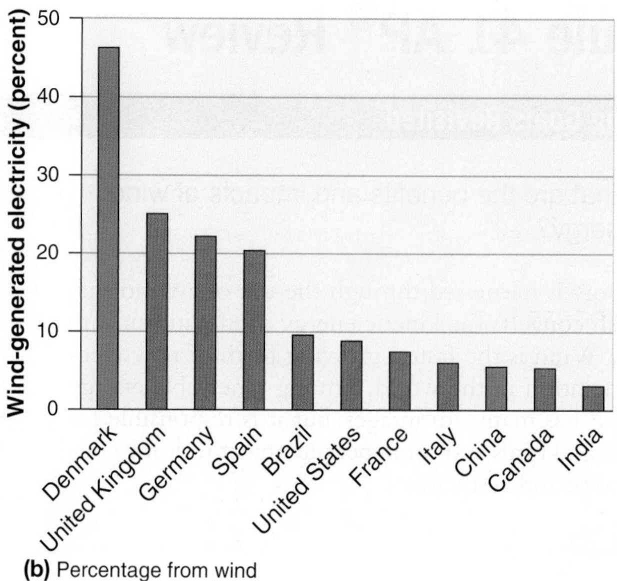

(b) Percentage from wind

4. Which country uses approximately 45 percent wind-driven electricity?
   (a) China
   (b) Denmark
   (c) Germany
   (d) France

5. Which of the following best explains why China and the United States have high amounts of installed wind capacity and low percentages of wind-generated electricity?
   (a) China and the United States have large amounts of electricity, and rely more heavily on electricity sources other than wind.
   (b) China and the United States don't have the space available to produce electricity from wind, and therefore have low rates of installation.
   (c) China and the United States each produce more than half of their electricity from wind power, which requires a large amount of installed wind energy capacity.
   (d) China and the United States each produce less than 10 percent of their total electricity from wind power, and therefore have less than 60 GWh produced from installed wind power.

Use the following diagram to answer questions 6–8:

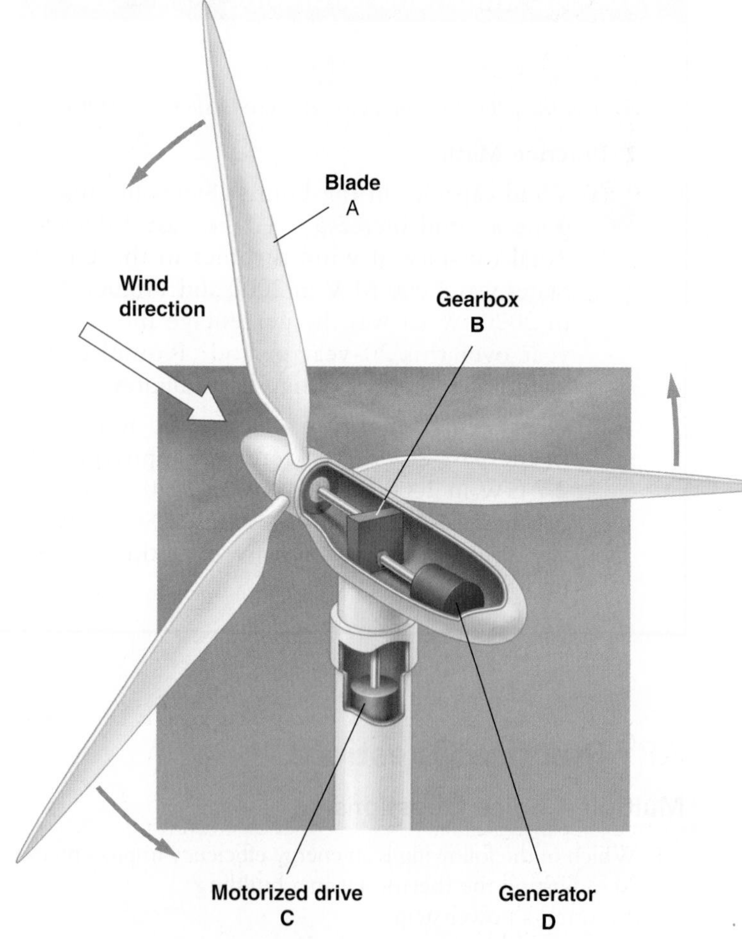

6. This part of a wind turbine transfers mechanical energy directly to the generator.
   (a) A
   (b) B
   (c) C
   (d) D

7. The size of this part of a wind turbine is directly related to the overall capacity of the turbine.
   (a) A
   (b) B
   (c) C
   (d) D

8. It is estimated that 200,000–500,000 birds may be killed by collisions with wind turbine blades each year. Which of the following best explains a realistic solution to this problem?
   (a) More birds are killed each year by house cats and electrical transmission lines than by wind turbine blades.
   (b) Install shorter wind turbines with smaller blades, to reduce the probability of a collision with a flying bird.
   (c) Relocate all wind turbines offshore, rather than installing them inland on ridgelines.
   (d) Install wind turbines with one blade painted in a different color, to increase the ability of birds to notice and avoid the blades.

Free-Response Question

One way to conserve energy is to increase energy efficiency using smart grid technology.

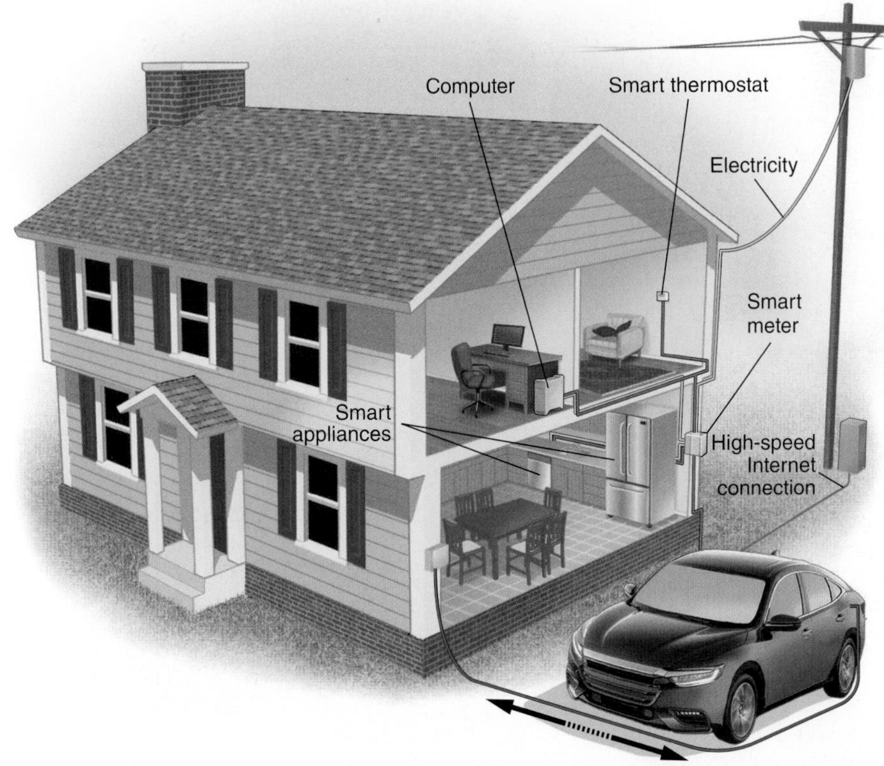

(a) **Identify** an aspect of the home in the diagram that can be programmed to operate during times of energy surplus. (1 pt.)

(b) **Describe** how a smart thermostat can be used to increase energy efficiency at home. (1 pt.)

(c) **Explain** a potential environmental drawback to owning a plug-in electric vehicle. (1 pt.)

Many communities are switching to wind power to supplement or replace their electricity production from fossil fuels.

(d) **Describe** two environmental benefits of switching to wind power. (2 pts.)

(e) A community plans to replace their electricity source from a natural-gas power plant to a new, local offshore wind farm. They would need to purchase and install several turbines, but first they must ask a team of researchers to determine if there is enough wind available in their area to meet demand. The researchers set up an experiment using five small turbines of the same height, each with three blades, but with varying blade length to see which produces the most electrical energy.
   (i) **Identify** the independent variable of this experiment. (1 pt.)
   (ii) **Identify** a possible hypothesis for this experiment. (1 pt.)
   (iii) **Identify** a variable not mentioned that should be kept constant in this experiment. (1 pt.)
   (iv) **Describe** a modification the researchers can make that will increase the reliability of their study. (1 pt.)
   (v) **Describe** one potential environmental problem that could result if this community switches to wind power. (1 pt.)

UNIT 6
Practice Your Science Skills

Data Analysis

This map shows the amount of solar energy available to a flat photovoltaic solar panel in kWh per square meter per day, averaged over a year.

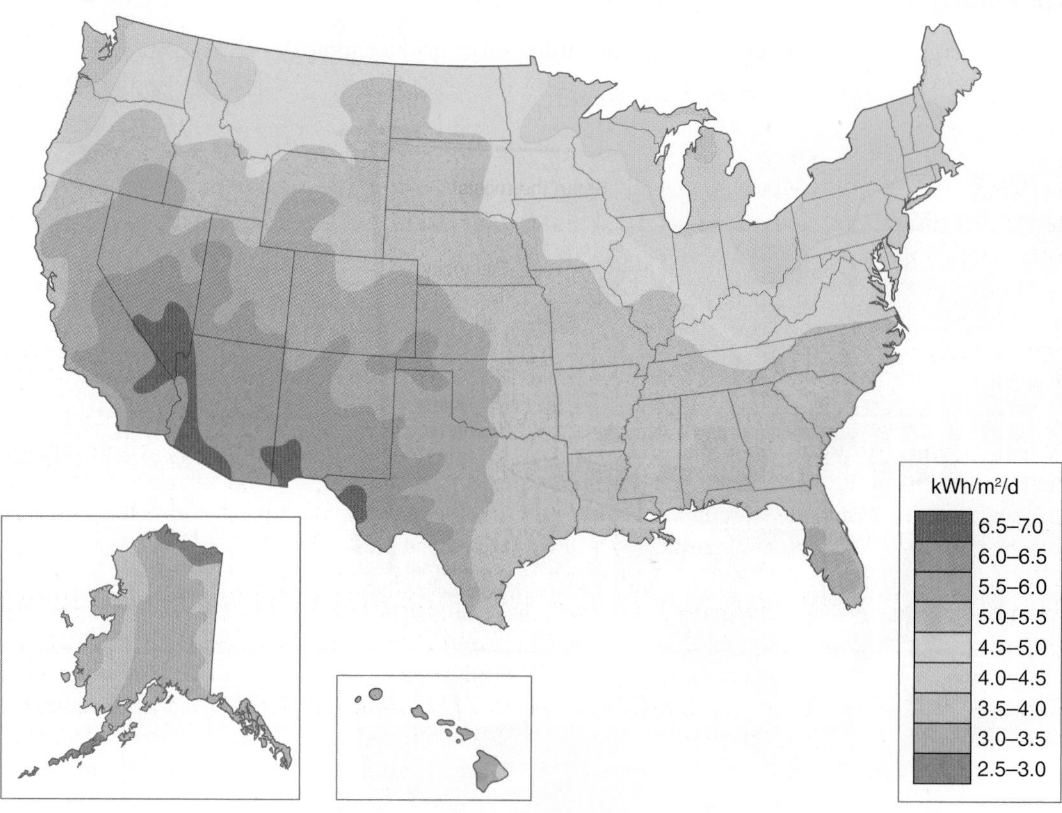

Questions

1. Using the map above, **identify** the region where solar power would be the most effective in the contiguous 48 states. *(Hint: In a question like this, you must be specific. Southern, middle, or western United States is too vague.)*
2. **Describe** solar energy available for flat PV cells in relation to latitude. *(Hint: In an FRQ, you must use specific words such as "increasing" and "decreasing" to describe a relationship like this. Make sure you know your lines of latitude with the equator at 0, the North Pole at 90°N, and the South Pole at 90°S.)*
3. A community in Hawaii wants to replace their petroleum powerplant with a PV solar plant. They hypothesize that the average solar radiation in the area is enough to pay back the cost of the PV solar plant in 15 years. Using the data in the map, **make a claim** as to whether solar PV is a viable solution for electricity in Hawaii. *(Hint: When asked to "make a claim," you must include evidence. If the prompt and figure (or graph or chart) has numbers, use numbers in your answer.)*
4. **Justify** your claim with an environmental benefit of switching to solar power. *(Hint: On an FRQ, try to answer with a specific example instead of just responding with "pollution.")*

Pursuing Environmental Solutions

Energy from Innovation and Renewables

In a small village in the African nation of Malawi, a 14-year-old boy named William Kamkwamba and his family did not have enough to eat because of a famine. His family could not afford the required school tax; in many parts of Africa a child whose parents cannot pay the school tax cannot attend school. So instead of attending school, William spent his days in a public library funded by the U.S. government, trying to teach himself. In the library, he studied one book over and over—a textbook titled *Using Energy*. The cover of the book featured a series of wind turbines. Although William had never seen a wind turbine, within months he was building his own from abandoned bicycles and old parts he found in scrap heaps. William used his knowledge of the first and second laws of thermodynamics that he learned from the book and his inherent skills at tinkering and fixing things. He did not have any teachers or mentors, but he did rely on assistance from some of his friends. He worked hard and made many attempts to construct something that in his world was seemingly impossible. At first his neighbors thought he was mentally ill or was practicing magic. But when he was able to illuminate a small light bulb at the top of what they had called his "junk" tower, people rushed from great distances to see it, and he became a local hero. William had generated electricity without any conventional fuel and far from the nearest power plant or electrical grid. Because there were no visible inputs like fuel and no waste piles or pollution outputs, in many ways it did seem like magic. William used the electricity he generated from wind to light his house and charge cell phones, and eventually to irrigate his family's crops.

As this unit has illustrated, the use of wind turbines to generate electricity has been rapidly increasing in both the developing and developed worlds. The mechanics of how to build a windmill are widely discussed in online sources including YouTube, Wikipedia, and "how to build it" instructional videos. However, William had only one book and no access to the Internet. A few years after he built his first windmill, William exclaimed to Jon Stewart on *The Daily Show*, "Where was this Internet when I needed it?" Today, companies with relatively high technological abilities are installing windmills around the world. Combining wind turbines with photovoltaic panels, they are providing electricity in locations so remote there is no grid.

In Kigutu, Burundi, 1,500 kilometers (930 miles) to the north of William's home and another area without an electrical grid, a hospital and women's health pavilion is providing high-quality health care. Operated by an organization called Village Health Works, this hospital currently has about 40,000 patient-visits per year in an area where health care is limited for a variety of reasons, including the lack of an electrical grid. But the use of renewable energy sources has allowed this hospital to achieve high-quality health care. One of the sources of electricity at this hospital is a nearby small-scale hydroelectric plant. Two new buildings are under construction with its rooftops covered with photovoltaic panels from end to end. Combined, these two

William Kamkwamba's cousin climbs one of the wind turbines that William built. *(Lucas Oleniuk/Getty Images)*

The rooftop of the Village Health Works hospital and women's health pavilion in this architectural photograph are covered from end to end with photovoltaic panels. *(© Bergen Street Studio and Stantec)*

sources of electricity have become a "micro-grid" providing electricity to the hospital 24 hours per day, allowing even higher quality health care and more patient-visits per year.

Innovators for projects like these utilize renewable energy and benefit from local and imported knowledge. In the case of William Kamkwamba, after he built his wind turbines and electrified his home and his family's irrigation system, he attended and graduated from Dartmouth College, where he majored in environmental studies. He is coauthor of the book *The Boy Who Harnessed the Wind*, that has sold thousands of copies and has been adopted as summer reading in high schools and colleges around the United States and elsewhere in the world. The book was also the basis of a BBC feature film that was released in 2018 starring Chiwetel Ejiofor. Since graduating from Dartmouth, William has worked for a variety of a nonprofit organization in the United States and Africa that innovate in a number of areas including delivering renewable energy to countries such as Malawi.

Critical Thinking Questions

1. What are some of the challenges of developing renewable energy projects such as wind in developing countries?
2. Why is combining wind turbines with a hydroelectric power plant more desirable than installing wind turbines alone?

References

Kamkwamba, W., and Mealer, B. 2009. *The Boy Who Harnessed the Wind*. Harper.

Moving Windmills Website: https://movingwindmills.org/.

Village Health Works Website. https://www.villagehealthworks.org/.

Wolfson, R. 2017. *Energy, Environment and Climate*. 3rd ed. Norton.

Science Applied 6: Concept Explanation

Should Corn Become Fuel?

Corn-based ethanol is big business—so big, in fact, that in an effort to offset demand for petroleum, U.S. policy has resulted in the production of 53 billion liters (14 billion gallons) in 2020. The United States is the largest producer of corn ethanol in the world. Ethanol proponents maintain that substituting ethanol for gasoline decreases air pollution, greenhouse gas emissions, and our dependence on foreign oil. Opponents counter that when we consider all the inputs used to grow and process corn into ethanol, it increases air pollution and greenhouse gas emissions. Moreover, opponents maintain that growing corn and converting it into ethanol uses more energy than we obtain when we burn ethanol for fuel and that the impact of ethanol on reducing our import of foreign oil is very small. What does the science tell us?

Does ethanol reduce air pollution?

Ethanol (C_2H_6O) and gasoline (a mixture of several compounds, including heptane, C_7H_{16}) are both hydrocarbons. Under ideal conditions, in the presence of enough oxygen, burning hydrocarbons produces only water and carbon dioxide. In reality, however, gasoline-only vehicles always produce some carbon monoxide (CO) because, even with modern engines, there can be insufficient oxygen during combustion. Carbon monoxide has direct effects on human health and also contributes to the formation of photochemical smog.

Oxygenated fuel A fuel with oxygen as part of the molecule.

Because ethanol is an **oxygenated fuel**—a fuel with oxygen as part of the molecule—adding ethanol to the fuel mix of a car should ensure that more oxygen is present and that combustion is more complete, which would reduce the production of CO. When it comes to combustion in vehicles, it is true that ethanol produces lower amounts of air pollutants than gasoline. However, when we compare the entire life cycle of growing, harvesting, and processing the corn for ethanol, it turns out that ethanol produces more of many air pollutants than gasoline.

Is ethanol neutral in the production of greenhouse gases?

Biofuels are modern carbon, not fossil carbon. As a result, burning ethanol should not introduce additional carbon into the atmospheric reservoir because the carbon captured in growing the corn kernels and the carbon released in burning the ethanol should cancel each other out. That is, ethanol should be neutral in terms of the amount of carbon produced. However, when we once again consider the entire life cycle of ethanol, we are reminded that corn production requires fossil fuels for driving a tractor, fertilizer production, and processing the corn kernels to make ethanol. These are all sources of fossil carbon, which means that ethanol production causes a net increase in the amount of greenhouse gases being produced.

Various aspects of corn production, such as plowing and tilling, may release additional CO_2 into the atmosphere from

FIGURE SA6.1 An ethanol-producing factory. Ethanol-producing factories, such as this one in South Dakota, convert corn into ethanol to produce a fuel that can be mixed with gasoline to power automobiles. *(Jim Parkin/Alamy)*

organic matter that otherwise would have remained undisturbed in the A and B horizons of the soil. Furthermore, greater demand for corn will increase pressure to convert land that is forest, grassland, or pasture into cropland. There is increasing evidence that these conversions result in a net transfer of carbon from the soil to the atmosphere and lead to additional increases in atmospheric CO_2 concentrations. Moreover, in the United States, the ethanol production process currently uses natural gas or coal. Quite possibly, producing ethanol with coal releases as much carbon into the atmosphere as simply burning gasoline in the first place. This means that when we consider the entire life cycle of ethanol production, it is not carbon neutral (**FIGURE SA6.1**).

Does ethanol provide a substantial return on energy investment?

We can also ask to what extent does ethanol production provide a return on the investment, which is how much energy we get out of ethanol for every unit of energy we put in. Scientists at the U.S. Department of Agriculture have analyzed this problem, examining the energy it takes to grow corn and convert it into ethanol (the inputs) and the return on this energy investment (the outputs). The energy inputs include the energy to run farm machinery, to produce chemicals (especially nitrogen fertilizer), dry the corn, transport the corn, convert it into ethanol, and ship it. The primary output is the ethanol, although several by-products are produced, including distiller's grains, corn gluten, and corn oil. Each by-product would have required energy to produce if it had been manufactured independently of the ethanol manufacturing process, and so energy "credit" is assigned to these by-products. As **FIGURE SA6.2** shows, there is a slight gain of usable energy when corn is converted into ethanol: For every unit of energy we put in, we obtain about 1.3 units of output. For comparison, for every unit of fossil fuel we invest in producing gasoline, we obtain about 15 units of output.

Does ethanol reduce our dependence on gasoline?

If there is a positive energy return on energy investment, then using ethanol should reduce the amount of gasoline we use and therefore the amount of oil we must produce and refine, or import. Our 2020 production of 53 billion

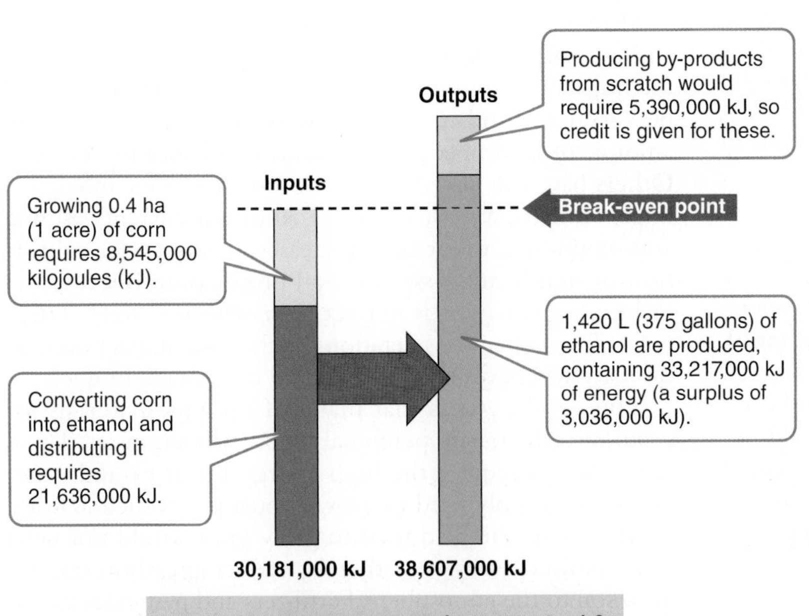

Outputs
Producing by-products from scratch would require 5,390,000 kJ, so credit is given for these.
Inputs
Growing 0.4 ha (1 acre) of corn requires 8,545,000 kilojoules (kJ).
Break-even point
1,420 L (375 gallons) of ethanol are produced, containing 33,217,000 kJ of energy (a surplus of 3,036,000 kJ).
Converting corn into ethanol and distributing it requires 21,636,000 kJ.
30,181,000 kJ
38,607,000 kJ
Net energy return on energy investment = 1.3
$\left(\frac{38,607,000}{30,181,000}\right)$

FIGURE SA6.2 Energy required to produce ethanol. An analysis of the energy costs of growing and converting 0.4 ha (1 acre) of corn into ethanol shows a slight gain of usable energy when corn is converted into ethanol. For every 1 unit of energy input, we obtain 1.3 units of energy output.

liters (14 billion gallons) of ethanol equaled about 10 percent of gasoline consumption in the United States. If our goal is to reduce gasoline consumption, much larger gains could be accomplished by establishing higher fuel mileage standards for cars and trucks or encouraging greater adoption of electric cars.

What are the unintended impacts of ethanol production?

If we create greater demand for a crop that until now has been primarily used as a food source, there are many unintended impacts. For example, increased ethanol production has led to the large-scale conversion of cropland from food to fuel production. Even if we converted every acre of potential cropland to ethanol production, we could not produce enough ethanol to substantially reduce U.S. annual gasoline consumption. Furthermore, if all cropland in the United States were devoted to ethanol production, all agricultural products destined for human consumption would have to be imported from other countries. Clearly this is not a practical solution.

What seems more likely is that we will be able to replace some smaller fraction of gasoline consumption with biofuels. The Earth Policy Institute points out, however, that the 10 bushels of corn that it takes to produce enough ethanol to fill a 95-L (25-gallon) fuel tank in a vehicle contain the number of calories needed to feed a person for about a year. Higher ethanol demand would increase corn and other grain prices and would thus make it harder for lower-income people around the world to afford food.

Indeed, in the summer of 2007, corn prices in the United States rose to $4 per bushel, roughly double the price in prior years, primarily because of the increased demand for ethanol. In more recent years, prices have stayed above $3 per bushel and at times have gone above $8 per bushel. People in numerous countries have had difficulty obtaining food because of these higher grain prices. In a number of years, most notably 2008 and 2016, there were food riots in certain countries around the globe.

Are there alternatives to corn ethanol?

Stimulating demand for corn ethanol may spur the development of another ethanol technology—**cellulosic ethanol**, an ethanol derived from cellulose. Cellulose is the material that makes up the cell walls of plants: Grasses, trees, and plant stalks are made primarily of cellulose. If we were able to produce large quantities of ethanol from cellulose, we could replace fossil fuels with fuel made from a number of sources. Ethanol could be manufactured from fast-growing grasses such as switchgrass (*Panicum virgatum*) or Miscanthus grass (**FIGURE SA6.3**), tree species that require minimal energy input, discarded paper and agricultural waste, including corn stalks that are left behind when corn kernels are harvested for conventional ethanol production. It is also possible that algae could be used as the primary material for ethanol.

Cellulosic ethanol An ethanol derived from cellulose, the cell wall material in plants.

FIGURE SA6.3 A potential source of cellulosic ethanol. Miscanthus, a fast-growing tall grass, may be a source of cellulosic ethanol in the future. *(Frank Dohleman)*

Producing cellulosic ethanol requires breaking cellulose into its component sugars before distillation. This is a difficult and expensive task because the bonds between the sugar molecules are very strong. One method of breaking down cellulose is to mix it with enzymes that sever these bonds. In 2007, the first commercial cellulosic ethanol plant was built in Iowa. At the moment, however, cellulosic ethanol is more expensive to produce than corn ethanol and the future of these plants around the world is uncertain.

How much land would it take to produce significant amounts of cellulosic ethanol? Some scientists suggest that the impact of extensive cellulosic ethanol production would be very large and it would require large increases in the amount of land needed for cellulosic ethanol production. Others have calculated that, with foreseeable technological improvements, we could replace all of our current gasoline consumption without large increases in land under cultivation or significant losses in food production. Because the technology is new, it is not yet clear who is correct. There will still be other considerations, such as the impact on biodiversity whenever land is dedicated to growing biofuels.

The good news is that many of the raw materials for cellulosic ethanol are perennial crops such as grasses. These crops do not require the high energy, fertilizer, and water inputs commonly used to grow annual plants such as corn. Furthermore, the land used to grow grass would not need to be plowed every year, thereby reducing carbon transfer from soil to the atmosphere. Fertilizers and pesticides would also be unnecessary, eliminating the large inputs of energy needed to produce and apply them. Algae may be an even more attractive raw material for cellulosic ethanol because its production would not need to use land that could otherwise be used for growing food crops.

What's the bottom line?

When we compare the full life cycle of ethanol production, it is clear that ethanol results in a greater production of many air pollutants and it is not neutral in terms of greenhouse production. However, it produces fewer greenhouse gases than gasoline does. The return on the energy invested to create ethanol is quite low compared to producing gasoline. Moreover, corn-based ethanol has the potential to replace only a small amount of total U.S. gasoline consumption. Increasing corn ethanol consumption to the levels suggested by some policy makers may require troublesome trade offs between driving vehicles and feeding people. Cellulosic ethanol shows the potential to have a significant effect on fossil fuel use, at least in part because of lower-energy inputs required to obtain the raw material and convert it into a fuel.

Questions

1. Legislation requiring that ethanol be added to gasoline was passed to improve air quality and reduce dependence on foreign oil. Do you think the latter goal is achieved by adding ethanol to gasoline? Explain.
2. What are some of the other environmental impacts of using corn-based ethanol that are not illustrated by the energy calculations shown here?

Practice AP® Free-Response Question

Write your answer to each part clearly. Support your answers with relevant information and examples. Where calculations are required, show your work.

The production of ethanol from corn is much like the production of any alcoholic beverage. Corn is harvested, ground, and cooked to form a "mash." Yeast that is added to the mash use the sugar to grow and reproduce. As they grow, yeast respire CO_2 and produce ethanol as a waste product through fermentation. Since ethanol has a lower boiling point than water, the fermented product can be boiled to evaporate the ethanol, which is a process known as distillation. Evaporated ethanol is then collected through condensation.

(a) **Identify** SIX energy inputs in the process of producing corn ethanol. (2 pts.)

(b) Using the answers provided in question (a), write an equation that would **calculate** the energy return on energy investment (EROEI) for ethanol production. (2 pts.)

(c) The leftover mash consists of water and corn. Ethanol producers will often dry this mash, collect the corn, and sell it as livestock feed. **Explain** why this livestock feed might be less beneficial for livestock than unfermented corn. (2 pts.)

(d) If the price of food increases around the world as a result of corn ethanol production, poorer communities may revert to subsistence energy production. **Explain** how this reversion could impact the total environmental benefit of ethanol production. (2 pts.)

(e) Cellulosic ethanol production is similar to corn ethanol production, except producers make use of enzymes that increase the sugar content of the mash, which in turn increases the energy available for yeast to grow. **Describe** two benefits and two disadvantages of cellulosic ethanol production relative to corn ethanol production. (2 pts.)

References

Biofuels Fact Sheet. Center for Sustainable Systems. University of Michigan. https://css.umich.edu/factsheets/biofuels-factsheet.

Conca, J. 2014. It's final: Corn ethanol is of no use. *Forbes*, April 20, 2014. https://www.forbes.com/sites/jamesconca/2014/04/20/its-final-corn-ethanol-is-of-no-use/#6385d53867d3.

Runge, C. F. 2016. The case against more ethanol: It's simply bad for the environment. *Yale Environment 360*. http://e360.yale.edu/features/the_case_against_ethanol_bad_for_environment.

Searchinger, T. D. et al. 2009. Fixing a critical climate accounting error. *Science* 326:527–528.

Service, R. F. 2013. Battle for the barrel. *Science* 339:1374–1379.

UNIT 6 Review

Key Terms to Remember

Fossil fuels (p. 401)
Nonrenewable energy resource (p. 401)
Renewable energy resources (p. 403)
Potentially renewable (p. 403)
Nondepletable (p. 403)
Commercial energy sources (p. 404)
Subsistence energy sources (p. 404)
Energy intensity (p. 406)
Fossil fuel combustion (p. 407)
Hubbert curve (p. 407)
Peak oil (p. 407)
Energy conservation (p. 408)
Energy efficiency (p. 408)
Energy return on energy investment (EROEI) (p. 409)
Biofuel (p. 412)
Modern carbon (p. 412)
Fossil carbon (p. 412)
Carbon neutral (p. 413)
Coal (p. 413)
Peat (p. 414)
Lignite (p. 414)
Bituminous coal (Asphalt) (p. 414)
Anthracite (Hard coal) (p. 414)
Natural gas (p. 414)
Crude oil (p. 415)
Tar sands (Oil sands) (p. 415)
Energy carrier (p. 418)
Combined cycle (p. 419)
Capacity (p. 419)
Capacity factor (p. 419)
Cogeneration (Combined heat and power) (p. 419)
Fracking (p. 429)
Volatile organic compounds (VOCs) (p. 430)
Turbine (p. 431)
Electrical grid (p. 431)
Energy quality (p. 432)
Nuclear power (p. 435)
Radioactivity (p. 435)
Fission (p. 435)
Fuel rod (p. 436)
Control rod (p. 437)
Radioactive decay (p. 438)
Half-life (p. 438)
Radioactive waste (p. 438)
Becquerel (Bq) (p. 438)
Curie (p. 438)
Biomass (p. 445)
Charcoal (p. 445)
Particulates (Particulate matter; Soot) (p. 446)
Carbon monoxide (p. 446)
Nitrogen oxides (p. 446)
Carbon dioxide (p. 446)
Biofuel (p. 447)
Ethanol (p. 447)
Biodiesel (p. 447)
Passive solar (p. 448)
Active solar energy (p. 449)
Photovoltaic solar cells (p. 450)
Hydroelectricity (p. 453)
Water impoundment (p. 453)
Run-of-the-river (p. 453)
Tidal energy (p. 454)
Siltation (p. 455)
Geothermal energy (p. 459)
Ground source heat pump (p. 460)
Fuel cell (p. 462)
Electrolysis (p. 463)
Wind energy (p. 466)
Wind turbine (p. 467)
Phantom loads (p. 469)
Peak demand (p. 470)
Passive solar design (p. 471)
Thermal mass (p. 472)
Smart grid (p. 476)
Oxygenated fuel (p. 482)
Cellulosic ethanol (p. 484)

Section 1: Multiple-Choice Questions

1. Within a developing nation, an increase in the use of subsistence energy sources would most likely be caused by
   (a) a decrease in the availability of straw, sticks, animal dung, and other local sources of fuel.
   (b) an increase in the availability of oil.
   (c) an increase in the cost of oil.
   (d) the loss of forested land.

Use the following table to answer questions 2 & 3:

Energy type	Energy return on energy investment	MJ per kilogram of fuel
Biodiesel	1	40
Coal	80	24
Ethanol from corn	1	30
Ethanol from sugarcane	5	30

2. Which fuel is most likely the least expensive fuel to produce per MJ of fuel?
   (a) biodiesel
   (b) coal
   (c) ethanol from corn
   (d) ethanol from sugarcane

3. How many kilograms of fuel will be consumed if someone travels 200 km in a car that uses 3 MJ of biodiesel per kilometer?
   (a) 7.5 kg
   (b) 15 kg
   (c) 22 kg
   (d) 66 kg

4. Which is likely to reduce the efficiency of a power plant?
   (a) increasing the capacity factor of the plant
   (b) using anthracite coal instead of bituminous coal
   (c) shutting off a turbine driven by exhaust gases
   (d) pumping cold water from a nearby stream into the condenser

5. Which is true regarding natural gas?
   (a) Extraction and combustion of coal have less effect on global warming than extraction and combustion of natural gas.
   (b) Contamination of water during the extraction process is of little concern.
   (c) Liquefied petroleum gas is slightly more energy-dense than natural gas.
   (d) Pipelines are the primary means of transporting natural gas.

6. In a nuclear power plant, control rods are used to
   (a) control the placement of fuel rods.
   (b) increase the efficiency of nuclear reactions.
   (c) transfer heat energy from the fuel rods into water.
   (d) absorb excess neutrons emitted by fuel rods.

7. One particularly large nuclear power plant produces about 400 kilocuries of krypton per year. Krypton has a half-life of 10 years. After 30 years, what will be the radioactivity of the krypton waste generated in a single year?
   (a) 50 kilocuries
   (b) 100 kilocuries
   (c) 200 kilocuries
   (d) 800 kilocuries

8. Which of the following best describes something that would reduce the capacity factor of nuclear power plants?
   (a) the need for long-term storage of radioactive waste
   (b) government regulation of low-level radioactive waste
   (c) competition with plants that use coal to produce energy
   (d) the need to periodically replace fuel rods

9. Which of the following is a renewable energy source that does not originate from solar radiation?
   (a) biomass
   (b) geothermal
   (c) CST
   (d) wind

10. A homeowner in the Northern Hemisphere wants to save money and energy by using a passive solar design for heating. Which of the following best describes a strategy the homeowner should consider?
    (a) opening window blinds on south-facing windows on sunny winter days
    (b) using only Energy Star appliances
    (c) placing photovoltaic cells on the roof
    (d) installing a ground source heat pump

11. Suppose you own a diesel car and the current price of diesel gasoline is $3.00 per gallon. You pay $750 for parts that allow the engine to be converted so it can operate on straight vegetable oil (SVO), which costs $1.50 per gallon. The car gets 50 miles per gallon when running on either type of fuel. After how many miles will you recoup the cost of engine conversion?
    (a) 1,500
    (b) 5,000
    (c) 10,000
    (d) 25,000

Use the following diagram to answer questions 12–14:

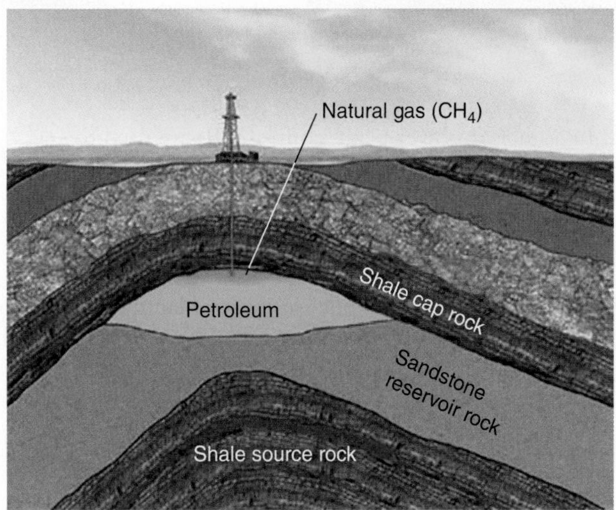

12. According to the diagram, where is petroleum found?
   (a) trapped in shale cap rock
   (b) underneath shale source rock
   (c) above natural gas deposits
   (d) trapped in sandstone reservoir rock

13. Based on the diagram, which of the following best describes how petroleum and gas deposits are extracted from rock layers?
   (a) Wells are drilled into the rock layers, where natural pressure releases the petroleum and gas deposits.
   (b) Petroleum and gas are extracted using surface mining techniques such as mountaintop removal.
   (c) Rock layers are dredged with large machinery until the petroleum and gas deposits are reached.
   (d) Rock layers are eventually eroded by wind and water, which exposes the petroleum and gas deposits underneath.

14. Based on the diagram, which of the following best explains why natural gas is found above petroleum deposits?
   (a) Natural gas is trapped in shale source rock, which is denser than petroleum.
   (b) Natural gas is less dense than liquid petroleum, so it rises above it and is trapped by cap rock.
   (c) Petroleum is formed from ancient marine organisms such as plankton, which rise to the top of sediment deposits.
   (d) Petroleum is denser than natural gas, so it sinks into impermeable cap rock layers such as shale.

15. Relative to burning fossil carbon, the use of modern carbon for energy
   (a) rarely contributes to the removal of vegetation.
   (b) does not use energy that originates from the Sun.
   (c) is cheaper and more efficient.
   (d) is more likely to be carbon neutral.

16. Which consequence of water impoundment for hydroelectric energy production is most likely to release greenhouse gases?
   (a) flooding of forests and grasslands
   (b) siltation
   (c) the generation of electricity by use of turbines
   (d) transfer of energy to power lines

17. The combined use of concentrated solar thermal (CST) and fossil fuel energy generation can provide for greatest grid reliability when
   (a) CST and fossil fuel plants are built near each other.
   (b) fossil fuels are able to be used when CST is unavailable.
   (c) CST and fossil fuel energy are equally distributed to the power grid.
   (d) fossil fuel plants are built in desert areas where there is consistent sunshine and plenty of open space.

18. Which methods of electricity production employ the use of turbines?
   (a) photovoltaic cells, wind, run-of-the-river hydroelectric plants
   (b) photovoltaic cells, wind, water impoundment hydroelectric dams
   (c) concentrated solar thermal plants, nuclear plants, wind
   (d) concentrated solar thermal plants, water impoundment hydroelectric dams, fuel cells

19. Which is most likely to increase the efficiency of energy use in the United States?
   (a) government subsidies for active solar construction designs
   (b) decreasing the cost of fossil fuels
   (c) replacement of large power plants with several smaller plants
   (d) replacement of incandescent lighting with LEDs

Use the following choices to answer questions 20 & 21:

(a) bitumen
(b) ethanol
(c) wind
(d) wood

20. Which is a subsistence energy source?

21. Which energy source is found in Canada and requires millions of gallons of hot water to be extracted from sand?

22. At a particular wind farm, 3 J of energy is expended in order to obtain 45 J of energy, and its capacity factor is 0.25. What is the EROEI of this wind farm?
   (a) 3.75
   (b) 15
   (c) 34
   (d) 60

Use the passage below to answer questions 23 & 24:

In a small village in the African nation of Malawi, a 14-year-old boy named William Kamkwamba and his family did not have enough to eat because of a famine. His family could not afford the required school tax to attend school. So instead of attending school, he spent his days in a public library funded by the U.S. government, trying to teach himself. In the library, he studied one book over and over—a textbook titled *Using Energy*. The cover of the book featured a series of windmills. Although William had never seen a windmill, within months he was building his own from abandoned bicycles and old parts he found in scrap heaps.

William used the fundamentals of physics he learned from the book and his inherent skills at tinkering and fixing things. He did not have any teachers or mentors but he did rely on assistance from some of his friends. At first his neighbors thought he was practicing magic, but when he was able to illuminate a small light bulb at the top of what they had called his "junk" tower, people rushed from great distances to see it, and he became a local hero.

William had generated electricity without any conventional fuel, far from the nearest power plant. Because there were no visible inputs like fuel and no waste piles or pollution outputs, in many ways it did seem like magic. William used the electricity he generated from wind to light his house and eventually to irrigate his family's crops.

23. According to the author, which of the following best describes William's motivation?
    (a) To produce electricity without the need for a power plant or conventional fuel sources.
    (b) To produce enough food to feed his village, as a famine was making food scarce.
    (c) To show his friends and neighbors that he did not need to go to school.
    (d) To produce enough energy to charge his cell phone without the need for conventional fuel.

24. Which of the following was necessary to support William's eventual achievements?
    (a) Raising enough money to pay the school tax and attend school in Malawi.
    (b) Deciding not to attend school in order to stay home and plant crops for his family.
    (c) The conversion of wind energy to electricity via the use of a generator.
    (d) The ability to travel to school in a village farther from his home.

Section 2: Free-Response Questions

1. Hydroelectricity provides approximately 7 percent of the electricity generated in the United States in 2020. The graph below shows the hydroelectricity generation in the United States from 1950 to 2020.

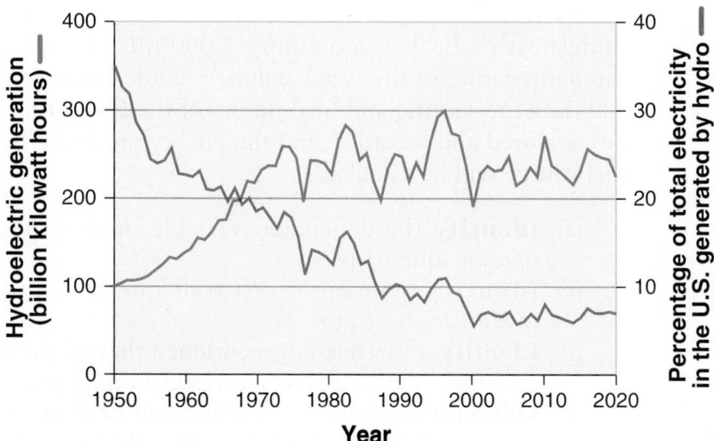

(a) Using the graph, **identify** the approximate year in which the total kilowatt hours produced by hydropower in the United States reaches its long-term average. (1 pt.)

(b) Using the graph, **describe** the overall trend of hydroelectricity generation in the United States from 1950 to 2020. (1 pt.)

(c) Using the graph, **explain** one reason why the share of total U.S. electricity for hydro power has been decreasing since 1950. (1 pt.)

(d) **Explain** how a hydroelectric power plant converts energy stored in water into electricity. (1 pt.)

(e) **Identify** TWO factors that determine the amount of electricity that can be generated by an individual hydroelectric power plant. (2 pts.)

(f) Hydroelectric power is commonly produced by constructing a dam in a river as a way to produce sustainable energy that is not reliant on fossil fuels.
   (i) **Describe** TWO economic advantages of damming a river. (2 pts.)
   (ii) Other than those related to construction, **describe** one environmental disadvantage of dams. (1 pt.)
   (iii) **Propose a solution** to the disadvantage described in part (ii). (1 pt.)

2. Electricity accounts for approximately 40 percent of overall energy use in the United States. It is a secondary source of energy since we obtain it through the conversion of a primary source of energy, like coal or wind. The diagram below depicts a typical coal-fired power plant.

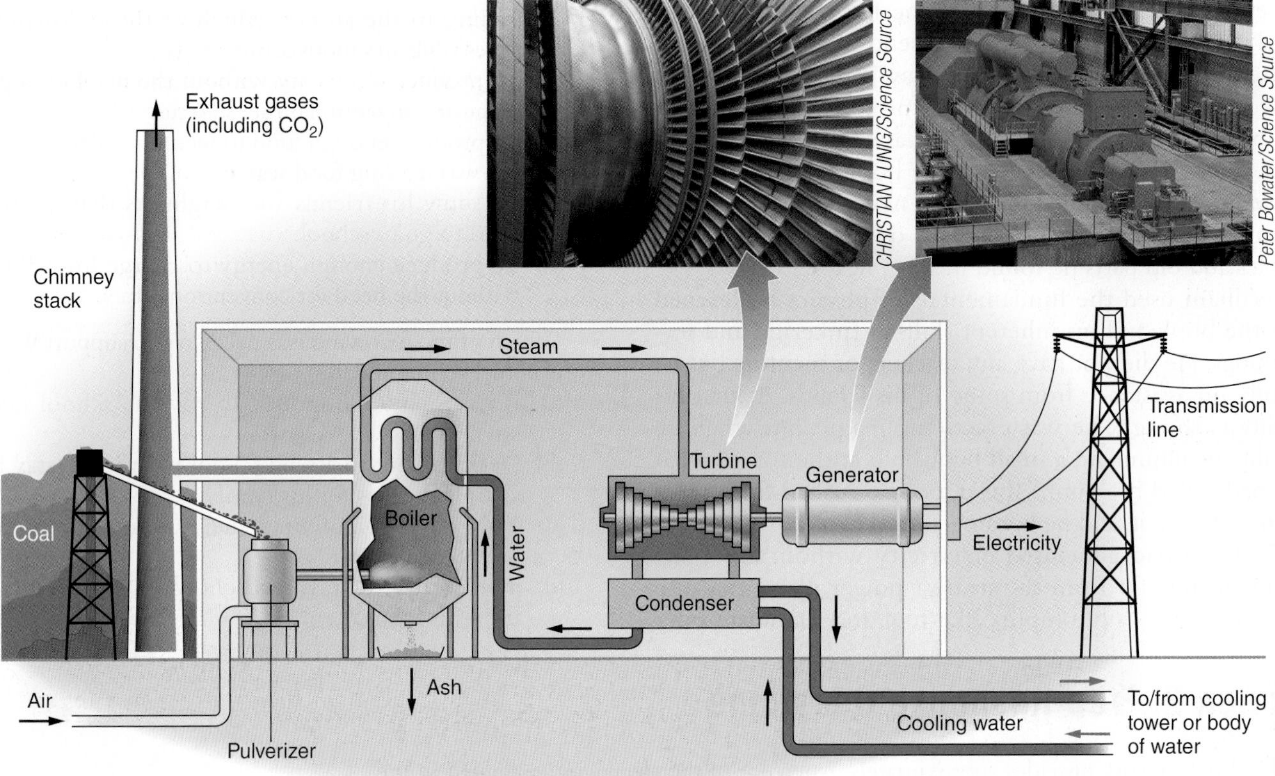

(a) **Identify** a part of the diagram that uses water. (1 pt.)

(b) Using the diagram, **describe** how electricity is produced in a typical coal-fired power plant. (1 pt.)

(c) **Explain** why the overall efficiency of a coal-fired power plant is relatively low. (1 pt.)

(d) Energy demands can vary depending on season, weather, and time of day.

  (i) **Describe** one reason why power companies must be able to produce more electricity than consumers may need on average. (1 pt.)

  (ii) **Identify** two environmental problems related to your reason in (i). (2 pts.)

An AP® Environmental Science class is investigating factors that affect electricity production. They set up an experiment to test the energy-generating capacity of various fuel types: solid, liquid, and gas. Each fuel type will be burned underneath a beaker containing 1,000 mL of water, and the temperature of the water will be recorded over time. In addition, the starting and final masses of the fuel source will be measured and recorded, and the energy produced by the fuel source will be calculated.

(e) **Identify** the dependent variable in the students' investigation. (1 pt.)

(f) **Identify** a possible research question for the investigation. (1 pt.)

(g) **Identify** a variable not mentioned that might affect the results of the students' investigation. (1 pt.)

(h) **Describe** a modification that could be made to the investigation that would increase the reliability of the students' results. (1 pt.)

3. Although hybrid and electric vehicles release less carbon dioxide per distance traveled than internal combustion engine (gasoline-powered) vehicles, there remains some debate over whether hybrid vehicles are better for the environment and whether their overall efficiency is greater.
   (a) **Identify** TWO reasons why hybrid and electric vehicles might be less energy-efficient than gas-powered cars. (2 pts.)
   (b) **Describe** how the average lifetime of a hybrid or electric vehicle affects the total ecological footprint of the vehicle. (1 pt.)
   (c) **Explain** how the use of "smart grid" technology could change the ecological footprint of an electric vehicle. (1 pt.)

A new all-electric vehicle costs $40,000 and requires 1.5 MJ per passenger-mile. You trade in your old gasoline-powered vehicle, which required 4 MJ per passenger-mile, for $2,000.

   (d) If a gallon of gasoline contains 20 MJ and costs $4.00, **calculate** how many miles you must drive before your purchase becomes cost effective. Show all work. (2 pts.)
   (e) Burning a gallon of gasoline released 19 pounds of carbon dioxide into the atmosphere. In 2020, Americans burned approximately 123 billion gallons of gasoline in vehicles, or about 1.022 gallons per day per person.
      (i) Using the information provided, **calculate** the approximate total carbon dioxide emissions of the United States in 2020 from vehicle gasoline combustion. Show all work. (2 pts.)
      (ii) Using the information provided, **calculate** how many pounds of carbon dioxide were produced by each American in 2020. Show all work. (2 pts.)

UNIT 7

Atmospheric Pollution

In this recent aerial view of Chattanooga, Tennessee, Lookout Mountain is clearly visible in the background. Fifty years ago, this view was often obscured by air pollutant haze. *(Rock Creek Aviation)*

CASE STUDY

Cleaning Up in Chattanooga

The fall of 2016 was not kind to Chattanooga, Tennessee. Nearby wildfires covered downtown Chattanooga with a smoky haze resulting in a number of poor air quality days in a city that is generally known for very good air quality. A few years earlier, dust storms in the West had also reduced the generally good air quality of the city. Protecting air quality in cities in the United States is something of a challenge. Regulators and government officials can impose standards, but a vast array of natural and human-caused regional and continental pollutants can conspire to cause local air pollution problems for any number of reasons. Local geography, direction of prevailing winds, the activities of other cities and sometimes other countries, and many other factors can affect air quality. Unfortunately, most of these factors are beyond the control of one city. The story of Chattanooga is representative of many cities in the United States and around the world.

Chattanooga sits along the Tennessee River in a natural basin formed by the Appalachian Mountains, one of which — Lookout Mountain — rises 600 m (1,970 feet) over the city. After the Civil War, foundries, textile mills, and other industrial plants were quickly built, and Chattanooga became one of the leading manufacturing centers in the nation.

The economic boom in Chattanooga had an environmental cost. Because it is surrounded by mountains, it traps pollutants that hover above the city. By 1957, Chattanooga had the third-worst particulate pollution in the country and rates of respiratory diseases were well above the national average. Over the next decade conditions worsened and by the 1960s, people were often unable to see Lookout Mountain even from a short distance of a quarter mile. In 1969, a U.S. survey of the nation's air quality confirmed what many Chattanooga residents suspected: Their city topped the list of the worst cities in the United States for particulate air pollution.

By 1957, Chattanooga had the third-worst particulate pollution in the country and rates of respiratory diseases were well above the national average.

Obviously, the poor quality of the air needed to be addressed. In 1969, Chattanooga and the county in which it is located created its own air pollution legislation by enacting the Air Pollution Control Ordinance. It controlled the emissions of sulfur oxides, allowed open burning by permit only, placed regulations on odors and dust, outlawed visible automobile emissions, capped the sulfur content of fuel at 4 percent, and limited visible emissions from industry. At the same time, the city and county governments implemented new pollution monitoring techniques to ensure that the ordinance was being followed.

The city and county governments, along with private industry, spent approximately $40 million on the cleanup effort. Actions to improve air quality did not hinder business, as some people feared, but created new industrial opportunities related to the cleanup effort, such as the establishment of a local manufacturer of smokestack scrubbers. As a result of all these measures, in 1972 — just 3 years after passage of the city ordinance and 2 years after the passage of the Federal Clean Air Act — Chattanooga achieved compliance with Clean Air Act air-quality standards.

The people of Chattanooga and the local governments recognized that keeping their air clean and maintaining economic sustainability would be an ongoing effort. To maintain their newly improved air quality, the city government and local businesses began several programs. One such program was a comprehensive recycling program, chosen as an alternative to a waste incinerator that would have added more particulates to the air. Public and private sectors successfully partnered to achieve both environmental and economic sustainability in creating the largest municipal fleet of electric buses in the United States, manufactured by another local business.

Unfortunately, while Chattanooga's efforts dramatically reduced the levels of particulate pollutants, the concentration of ozone, mostly from automotive pollutant precursors within and beyond the city limits, continued to climb. Ozone concentrations exceeded the 1997 air quality standard of 0.08 parts per million by volume set by the Environmental Protection Agency. Chattanooga has responded to the new air pollution problem in the same way it faced the particulate pollutant problem of the 1960s — through a combined effort of government, public, and local industries. The city and county governments formed an Early Action Compact with the EPA, agreeing to improve ozone concentrations ahead of EPA requirements in return for not being labeled a "non-attainment area," a designation that can result in the loss of federal highway funds and create a negative image that makes industrial recruitment and economic development more difficult. Like the 1969 Air Pollution Control Ordinance, the new Early Action Compact called for a concerted effort by private and public sectors and included educating people on how they can take action to limit ozone production on high-ozone days.

Chattanooga attained the 0.08 parts per million standard in 2007, 2 years ahead of schedule. Then, national legislation again lowered the ozone standard, this time to 0.075 parts per million. By 2011, before the dust storms from the West and the local forest fires, Chattanooga had met the new, lower ozone standard. In 2017, the American Lung Association rated Chattanooga one of the cleanest cities in the United States and in 2020 it received its best rankings ever for ozone and particulate pollution. In 2022, massive fires in the western United States did not adversely impact Chattanooga. However, residents know that to achieve their goals of an economically vibrant city with clean air, they must continue to encourage cooperation among government, people, and business. They also need fewer dust storms from the West and fewer distant and nearby forest fires, which are, unfortunately, not in their control.

Sources: Chattanooga Area Chamber of Commerce, *Summary of the Chattanooga Area Chamber of Commerce's Position on Strengthening the National Ambient Air Quality Standard for Ozone*, 2010, www.chattanoogachamber.com; S. Johnson, Report: Chattanooga air quality among best in U.S., *Chattanooga Times Free Press*, April 19, 2017.

Practice your science skills

1. **Concept Explanation:** By 1957, Chattanooga's pollution levels were some of the highest in the country and rates of respiratory diseases were well above the national average. The foundries, textile mills, and other industrial plants built after the Civil War released large amounts of air pollution. Describe the category of air pollutants that cause respiratory diseases in Chattanooga.
2. **Concept Explanation:** Chattanooga's efforts dramatically reduced the levels of particulate pollutants; but the concentration of ozone, mostly from automotive pollutant precursors within and beyond the city limits, continued to climb. Describe the environmental process of ozone formation.
3. **Concept Explanation:** The mountains surrounding Chattanooga trap hovering air pollutants produced by industry. Describe the atmospheric and geological conditions that would trap large amounts of pollution in the city.

Air pollution occurs over terrestrial, aquatic, and marine natural systems and it also occurs in human-made indoor systems. To understand air pollution and its effects, we need to examine the wide variety of air pollutants, where they come from, and the harm they cause after they are released into the air. In this unit we will identify the major air pollutants found around the globe and discuss the specific air pollution conditions that promote photochemical smog and thermal inversions. We will describe natural sources of carbon dioxide and particulates. We will identify indoor air pollutants and consider air pollution control measures currently used. And finally, we will conclude the unit with a discussion of acid rain and noise pollution. After reading the modules within this unit, you will be able to recognize the physical, chemical, and biological consequences of human activities on our atmosphere, and you will recognize that with appropriate measures, in many instances, improved air quality is achievable.

Module 42

Introduction to Air Pollution

Air pollution is defined as the introduction of chemicals, particulate matter, or microorganisms into the atmosphere at concentrations high enough to harm plants, animals, and materials such as buildings, or to alter ecosystems. Generally, the term air pollution refers to pollution in the troposphere, the first 16 km (10 miles) of the atmosphere above the surface of Earth. Tropospheric pollution is also sometimes called ground-level pollution. In this module we will examine the major air pollutants that occur in the troposphere and identify where they come from and what harm they cause. We will also distinguish between primary and secondary pollutants.

Learning Goals

After reading this module you should be able to

42-1 identify the sources and effects of major air pollutants.

42-2 describe primary and secondary pollutants and how they change over time.

42-1 What are the sources and effects of major air pollutants?

Air pollution sources are widespread and their effects vary

Since one of the major repositories for air pollutants is the atmosphere, which envelops the entire globe, we must think of the air pollution system as a global system. In fact, air pollution can travel from one side of the globe to the other. For example, in recent years, sulfur emissions in Asia have been responsible for acidic rainfall on the West Coast of the United States (**FIGURE 42.1**).

The air pollution system has many inputs, which are the sources of pollution. It also has many outputs, which are components of the atmosphere and biosphere that remove air pollutants. Air pollution inputs can come from automobiles on the ground, airplanes in the sky, or tree leaves that emit volatile gases. Similarly, air pollution can be removed or altered by plant surfaces, soil, and components of the atmosphere such as clouds, particles, or gases. As a starting point for understanding the global air pollution system, we will identify the major pollutants and where they come from.

Classifying Pollutants

Even though air pollution has existed for millennia, both the specific definition of pollution and the classification of a substance as a pollutant have changed considerably in the

FIGURE 42.1 Air pollution and visibility. Air pollutants can reduce visibility in cities, such as this location in China. Some of those same air pollutants can also be transported long distances and cause problems far from the source. Pollutants such as sulfur dioxide from China have been detected on the West Coast of the United States as acid deposition. *(Natalie Behring/Panos Pictures)*

Air pollution The introduction of chemicals, particulate matter, or microorganisms into the atmosphere at concentrations high enough to harm plants, animals, and materials such as buildings, or to alter ecosystems.

last few decades. The atmosphere is a public resource—in effect, a global commons—and consequently the science of air pollution is closely intertwined with political and social perspectives. In formulating the U.S. Clean Air Act of 1970 and subsequent amendments, legislators used information from environmental scientists and human health scientists on the most important air pollutants to monitor and control. The original act identified six pollutants that significantly threaten human well-being, ecosystems, and structures: sulfur dioxide, nitrogen oxides, carbon monoxide, particulate matter, tropospheric ozone, and lead. These were called "criteria air pollutants" because under the Clean Air Act, the U.S. Environmental Protection Agency (EPA) uses allowable concentrations of each pollutant to set criteria for whether the air in a particular location should be considered "clean."

Although carbon dioxide was not included among the major air pollutants identified in the 1970s, today it is widely accepted that carbon dioxide is altering ecosystems in a substantial way. In 2007, the U.S. Supreme Court ruled that carbon dioxide should be considered an air pollutant under the Clean Air Act, and in 2012, a federal appeals court agreed that the EPA is required to impose limits on harmful greenhouse gas emissions, including carbon dioxide. In addition, volatile organic compounds and mercury, though not officially listed in the Clean Air Act, are commonly measured air pollutants that have the potential to be harmful. Traditionally, the six criteria air pollutants originally identified in the Clean Air Act of 1970 are the most commonly identified air pollutants. Carbon dioxide may or may not be considered an air pollutant, depending on the context of the statement or question.

Coal and Oil

Different fuels are the sources of different types and amounts of air pollutants. In general, coal combustion releases the greatest amount of air pollutants per unit of energy obtained, including carbon dioxide, sulfur dioxide, toxic metals, and particulates. Oil releases somewhat fewer pollutants than coal. It will release less carbon dioxide per unit of energy released and less sulfur dioxide. Oil also releases fewer particulates and toxic metals than coal. Natural gas, as you might have guessed, is even cleaner. It releases half the carbon dioxide of coal per unit of energy obtained and relatively small amounts of sulfur dioxide, metals, and particulates. The combustion of all fuels in the atmosphere results in the release of nitrogen oxides, which contribute to ozone formation and acid rain. Combustion of fuels also results in release of carbon monoxide and hydrocarbons.

The sources and effects of the major air pollutants, including the six criteria air pollutants, are summarized in **TABLE 42.1**. Let's take a closer look at each of these major air pollutants: sulfur dioxide, nitrogen oxides, carbon monoxide, carbon dioxide, particulate matter, photochemical oxidants, lead and other metals, and volatile organic compounds.

Sulfur Dioxide

Sulfur dioxide (SO_2) is a corrosive gas that comes primarily from combustion of fuels such as coal and oil, including diesel fuel from trucks. Since 2010, diesel sold for on-road vehicles has been ultra-low sulfur diesel, so it contributes to less sulfur pollution than it used to. Sulfur dioxide emissions affect air quality in many ways: It is a respiratory irritant and can adversely affect plant tissue, and it also is a precursor to one component of acid rain. Because all plants and animals contain sulfur in varying amounts, the fossil fuels derived from their remains contain sulfur. When these fuels are combusted, the sulfur combines with oxygen to form sulfur dioxide. Sulfur dioxide is also released in large quantities from volcanoes during volcanic eruptions and can be released, though in much smaller quantities, during forest fires.

AP® Exam Tip

Always be as specific as you can when answering questions about air pollution. Give the specific pollutant name or compound rather than referring to "air pollution" broadly.

Nitrogen Oxides

Nitrogen oxides are generically designated NO_x, with the X indicating that there may be either one or two oxygen atoms per nitrogen atom: nitrogen oxide (NO), a colorless, odorless gas; and nitrogen dioxide (NO_2), a pungent, reddish-brown gas, respectively. When we use the term nitrogen oxides in our discussion, we will be referring to either nitrogen oxide or nitrogen dioxide since they easily transform from one to the other in the atmosphere. The atmosphere is 78 percent nitrogen gas (N_2), and all combustion in the atmosphere leads to the formation of nitrogen oxides. Motor vehicles and stationary fossil fuel combustion are the primary anthropogenic sources of nitrogen oxides, which can transform to nitric acid and eventually acid rain. Natural sources of nitrogen oxides include forest fires, lightning, and microbial activity in soils. Atmospheric nitrogen oxides play a role in forming ozone and other components of smog. We will take a closer look at this process in Module 43.

Sulfur dioxide (SO_2) A corrosive gas that comes primarily from combustion of fuels such as coal and oil, including diesel fuel from trucks.

TABLE 42.1 Major air pollutants

Compound	Symbol	Human-derived sources	Effects/Impacts
Criteria air pollutants			
Sulfur dioxide	SO_2	• Combustion of fuels that contain sulfur, including coal, oil, gasoline	• Respiratory irritant, can exacerbate asthma and other respiratory ailments • SO_2 gas can harm stomata and other plant tissue • Converts to sulfuric acid in atmosphere, which is harmful to aquatic life and some vegetation
Nitrogen oxides	NO_x	• All combustion in the atmosphere including fossil fuel combustion, wood, and other biomass burning	• Respiratory irritant, increases susceptibility to respiratory infection • An ozone precursor, leads to formation of photochemical smog • Converts to nitric acid in atmosphere, which is harmful to aquatic life and some vegetation • Contributes to over-fertilization of terrestrial and aquatic systems
Carbon monoxide	CO	• Incomplete combustion of any kind • Malfunctioning exhaust systems and poorly ventilated cooking fires	• Bonds to hemoglobin, thereby interfering with oxygen transport in the bloodstream • Causes headaches at low concentrations • Can cause death with prolonged exposure at high concentrations
Particulate matter	PM_{10} (smaller than 10 micrometers) $PM_{2.5}$ (2.5 micrometers and smaller)	• Combustion of coal, oil, and diesel, and of biofuels such as manure and wood • Agriculture, road construction, and other activities that mobilize soil, soot, and dust	• Can exacerbate respiratory and cardiovascular disease and reduce lung function • May lead to premature death • Reduces visibility and contributes to haze and smog; correlated with heart disease and higher incidence of lung cancer
Lead	Pb	• Gasoline additive, oil and gasoline, coal, old paint	• Impairs central nervous system • At low concentrations, can have measurable effects on learning and ability to concentrate
Ozone	O_3	• A secondary pollutant formed by the combination of sunlight, water, oxygen, VOCs, and NO_x	• Reduces lung function and exacerbates respiratory symptoms • Degrades plant surfaces • Damages materials such as rubber and plastic
Other air pollutants			
Volatile organic compounds	VOC	• Evaporation of fuels, solvents, paints • Improper combustion of fuels such as gasoline	• A precursor to ozone formation
Mercury	Hg	• Coal, oil, gold mining	• Impairs central nervous system • Bioaccumulates in the food chain
Carbon dioxide	CO_2	• Combustion of fossil fuels and clearing of land	• Affects climate and alters ecosystems by increasing greenhouse gas concentrations

Carbon Monoxide and Carbon Dioxide

As you first learned in Unit 6, Carbon monoxide (CO) is a colorless, odorless gas that is formed during incomplete combustion of all fuels including fossil fuels, and therefore is a common pollutant in vehicle exhaust and most other combustion processes. Carbon monoxide can be a significant component of air pollution in urban areas. It also can be a dangerous indoor air pollutant in developed countries when exhaust systems from natural gas heaters malfunction. Carbon monoxide is a particular problem in developing countries, where people may cook with manure, charcoal, or kerosene indoors.

Carbon dioxide (CO_2), as you learned in Unit 6, is a colorless, odorless gas that is formed during the complete combustion of matter, including fossil fuels and biomass. While it has not historically been considered an air pollutant, its classification as such is undergoing a transition. It is absorbed by plants during photosynthesis and is released during respiration.

In general, the complete combustion of matter that produces carbon dioxide is more desirable than the incomplete combustion that produces carbon monoxide and other pollutants. Burning fossil fuels has contributed additional carbon dioxide to the atmosphere and has led to it being considered a major pollutant. Carbon dioxide concentrations have been steadily increasing in conjunction with the increase in the use of fossil fuels, and in 2022, concentrations exceeded 415 parts per million. This topic will be covered in more detail later on in the textbook when we discuss greenhouse gases and climate change.

Particulate Matter

Particulate matter (PM), also called particulates or particles, is solid or liquid particles suspended in air. It may also be referred to as "soot particles." Particulate matter comes from the combustion of wood, animal manure and other biofuels, coal, oil, and gasoline. In the developed world, it is most commonly known as a class of pollutants released from the combustion of fuels such as coal and oil. Diesel-powered vehicles give off more particulate matter, in the form of black soot particles, than do gasoline-powered vehicles. Particulate matter can also come from road dust and rock-crushing operations. Volcanoes, forest fires, and dust storms are important natural sources of particulate matter. Particulate matter contributes to **haze**, which is reduced visibility. We will discuss particulate matter in greater detail in Module 43.

Photochemical Oxidants

Oxides are reactive compounds that remove electrons from other substances. **Photochemical oxidants** are a class of air pollutants formed as a result of sunlight (*photo*) acting on chemical compounds such as nitrogen oxides and sulfur dioxide. There are many photochemical oxidants, and they degrade plant tissue, human respiratory tissue, and construction materials such as brick and concrete. However, environmental scientists frequently focus on one particular photochemical oxidant: ozone. Ozone (O_3), as we first learned in Unit 4, is a secondary pollutant made up of three oxygen atoms bound together. It is harmful to both plants and animals and impairs respiratory function.

In the presence of sulfur and nitrogen oxides, photochemical oxidants can enhance the formation of certain particulate matter, which contributes to scattering light. The resulting mixture is called **smog**, a mixture of oxidants and particulate matter. We will examine smog in greater detail in the next module.

Lead and Other Metals

Lead (Pb) is a trace metal that occurs naturally in rocks and soils, is present in small concentrations in fuels including oil and coal, and is a neurotoxin. Supplemental lead compounds were used as a gasoline additive for many years to improve vehicle performance. During that time, lead compounds released into the air traveled with the prevailing winds and were deposited on the ground by rain or snow. They became pervasive around the globe, including in polar regions far from combustion sources.

One provision of the Clean Air Act required that lead be phased out as a gasoline additive in the United States between 1975 and 1996, and since then its concentration in the air has dropped considerably. A campaign to phase out lead use in gasoline globally finally succeeded. In July 2021, the United Nations Environment Program announced that lead is no longer being added to gasoline anywhere in the world. Another persistent source of lead is lead-based paint in buildings constructed in the 1960s or earlier. When the paint peels off, the resulting dust or chips can be toxic to the central nervous system and can affect learning and intelligence, particularly for young children who may ingest the dust or chips.

Mercury (Hg), another trace metal, is also found in coal and oil and, like lead, is toxic to the central nervous system of humans and other organisms. Mercury emissions into the atmosphere in the United States is primarily from the combustion of fossil fuels, especially coal, and waste incineration. However, due to reductions in both coal combustion and waste incineration, the concentrations of mercury in both air and water have decreased in recent years. "Do the Math: Unleaded versus Leaded Gasoline" shows you how to calculate lead emissions from leaded and unleaded gasoline.

Volatile Organic Compounds

Organic compounds that evaporate at typical atmospheric temperatures are called volatile organic compounds (VOCs). Many VOCs are **hydrocarbons**—pollutant compounds that contain carbon-hydrogen bonds—such as gasoline and other fossil fuels, lighter fluid, dry-cleaning fluid, oil-based paints, and perfumes. Compounds that give off a strong aroma are often VOCs since the fact that you can smell them indicates that the chemicals are easily released into the air. VOCs play an important role in the formation of photochemical oxidants such as ozone. VOCs are not necessarily hazardous; many, such as VOCs given off by conifer trees, cause no direct harm. VOCs are not currently

Haze Reduced visibility.

Photochemical oxidant A class of air pollutants formed as a result of sunlight acting on chemical compounds such as nitrogen oxides and sulfur dioxide.

Smog A type of air pollution that is a mixture of oxidants and particulate matter.

Lead (Pb) A trace metal that occurs naturally in rocks and soils, is present in small concentrations in coal and oil and is a neurotoxin.

Hydrocarbons Pollutant compounds that contain carbon-hydrogen bonds, such as gasoline and other fossil fuels, lighter fluid, dry-cleaning fluid, oil-based paints, and perfumes.

DO THE MATH Unleaded versus Leaded Gasoline

Preparing for the AP® Exam

In 1970, when gasoline in the United States contained the additive tetra-ethyl lead, there was roughly 0.5 g of lead added to each liter of gasoline. How much lead was emitted to the air per year by one car if that car burned 2,000 liters of gasoline per year?

0.5 g lead/liter × 2,000 liters/year = 1,000 g of lead released in a year from one car in 1970

YOUR TURN Although lead is not added to gasoline today, gasoline typically contains about 0.01 g of lead per liter from lead that is in the petroleum before the gasoline is refined. How much lead is released from one car that burns 2,000 liters of unleaded gasoline per year?

considered a criteria air pollutant, but because they can lead to the formation of photochemical oxidants, they have the potential to be harmful and are therefore of concern to air pollution scientists. It is important to be aware that hydrocarbons such as gasoline become air pollutants when they evaporate and enter the air from usage and spillage. They also become air pollutants when they pass through the internal combustion engine and are emitted in a raw or partially combusted form. Small engines like lawn mowers are notorious for releasing through their exhaust system a large fraction of their fuel as raw, uncombusted hydrocarbons. Understanding what comes directly from an emission source and what undergoes transformations in the atmosphere after emission is the subject of the next section.

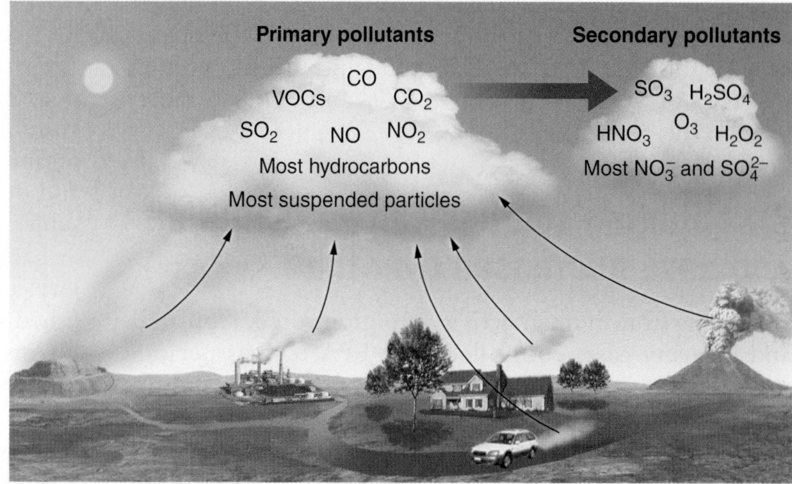

FIGURE 42.2 **Primary and secondary air pollutants.** The transformation from primary to secondary pollutants requires a number of factors including sunlight, water (clouds), and the appropriate temperature.

42-2 What are primary and secondary pollutants and how do they change over time?

Primary and secondary pollutants have decreased over time

When trying to understand pollution sources and effects, and when attempting to reduce pollution emissions, it is important to know if a particular pollutant is coming directly from an emission source such as a smokestack or tailpipe, or if it has undergone transformations in the atmosphere after emission. Depending on the source, pollutants in the air can be categorized as primary or secondary.

AP® Exam Tip

Make sure that you know the difference between primary and secondary pollutants and can describe how each type of pollutant is formed.

Primary Pollutants

Primary pollutants are polluting compounds that come directly out of a smokestack, exhaust pipe, or natural emission source. As you can see in **FIGURE 42.2**, they include CO, CO_2, SO_2, NO_x, and most suspended particulate matter. Many VOCs are also primary pollutants. For example, as gasoline is burned in a car, it volatilizes from a liquid to a vapor, some of which is emitted from the exhaust pipe in an uncombusted form. The effect is more pronounced if the car is not operating efficiently. The resulting VOC becomes a primary air pollutant.

Secondary Pollutants

Secondary pollutants are primary pollutants that have undergone transformation in the presence of sunlight, water,

Primary pollutant A polluting compound that comes directly out of a smokestack, exhaust pipe, or natural emission source.

Secondary pollutant A primary pollutant that has undergone transformation in the presence of sunlight, water, oxygen, or other compounds.

oxygen, or other compounds. Because solar radiation provides energy for many of these transformations, and because water is usually involved, the conversion to secondary pollutants occurs more rapidly during the day and in wet environments.

Ozone is an example of a secondary pollutant. Ozone is formed in the atmosphere as a result of the emission of the primary air pollutants NO_x and VOCs in the presence of sunlight. The main components of acid deposition—sulfate (SO_4^{2-}) and nitrate (NO_3^-)—are also secondary pollutants. These secondary pollutants will be discussed further in Module 43.

When trying to control secondary pollutants, it is necessary to consider the primary pollutants that create them, as well as factors that may lead to the formation, breakdown, or reduction in the secondary pollutants themselves. For example, when municipalities such as Chattanooga try to reduce ozone concentrations in the air, as described at the beginning of this unit, they focus on reducing the compounds that lead to ozone formation—NO_x and VOCs—rather than on the ozone itself.

Air pollution sources can be identified and have decreased over time

Emissions from human activity are monitored, regulated, and in many cases controlled. The EPA reports periodically on the emission sources of the criteria air pollutants for the entire United States, listing pollution sources in a variety of categories, such as on-road vehicles, power plants, industrial processes, and waste disposal (**FIGURE 42.3**). Highway vehicles are the largest sources of carbon monoxide and nitrogen oxides. Nonroad mobile sources such as farm equipment, airplanes, and boats are sources of carbon monoxide and nitrogen oxides as well. Electricity generation and other stationary sources of fuel combustion are the major sources of anthropogenic sulfur dioxide. A substantial amount of the sulfur dioxide emissions come from the 19 percent of electricity that is fueled by coal. Particulate matter comes from a variety of sources including stationary fuel combustion of coal and oil.

Through the National Ambient Air Quality Standards (NAAQS), the EPA specifies concentration limits for each air pollutant. The NAAQS notes a concentration for each pollutant that should not be exceeded over a specified time period. Our discussion of air pollution in Chattanooga at the beginning of this unit partially described the standard for ozone: For each locality in the United States, the average ozone concentration for any 8-hour period should not exceed 0.075 parts of ozone per million parts of air by volume more than 4 days per year, averaged over a 3-year period. If a locality violates the ozone air-quality standard and does not make an attempt to improve air quality, it is subject to penalties.

Each year, the EPA issues a report that shows the national level of the six criteria air pollutants relative to the published standards. **FIGURE 42.4** shows that all criteria air pollutants have decreased considerably in the United States in recent decades to well below published standards. Only ozone concentrations have been close to the NAAQS in the last few years. Lead has decreased most significantly because it is no longer added to gasoline.

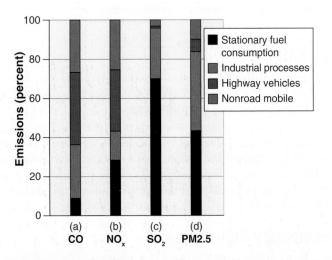

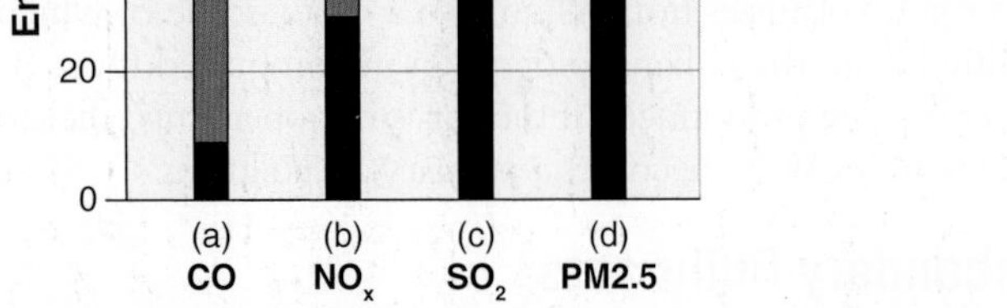

Emissions (percent)
100
80
60
40
20
0
Stationary fuel consumption
Industrial processes
Highway vehicles
Nonroad mobile
(a) CO
(b) NO_x
(c) SO_2
(d) PM2.5

FIGURE 42.3 Emission sources of criteria air pollutants for the United States in 2020. Recent EPA data show that highway vehicles are the largest source of (a) carbon monoxide and (b) nitrogen oxides. The major sources of (c) anthropogenic sulfur dioxide and (d) particulate matter are stationary fuel combustion from industrial plants and power plants that generate electricity with coal. *(Data from Our Nation's Air USEPA 2020, https://gispub.epa.gov/air/trendsreport/2020)*

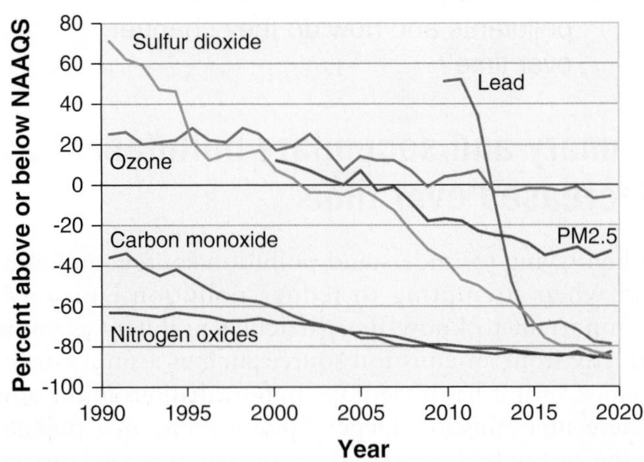

FIGURE 42.4 Criteria air pollutant trends for the past 30 years. Trends in the criteria air pollutants in the United States between 1990 through 2020 are shown. All criteria air pollutants have decreased during this time period. Note that data are shown as a percentage above or below the national air quality standard. Lead has shown the most substantial decrease of any pollutant in recent years. *(Data from EPA Our Nation's Air, Trends Through 2020, https://gispub.epa.gov/air/trendsreport/2021/#introduction)*

The situation is less positive in other parts of the world. Large areas in Germany, Poland, and the Czech Republic contain a great deal of "brown" coal or lignite that provides fuel for nearby coal-fired power plants and other industries. Emissions from combustion of this high-sulfur-content coal once caused this region to become one of the most polluted areas in the world. In addition to human health problems such as respiratory illnesses, forest ecosystems in this region have also been damaged in the last 40 years. In many parts of Asia, air quality has been so severely impaired by particulate matter and sulfates that visibility has been reduced, in some cases by more than 20 percent. Many of the most polluted cities in the world are in China and India.

In this module we identified the major air pollutants and described their sources and effects. The Clean Air Act identified six "criteria" air pollutants. Coal combustion releases the greatest amount of air pollutants per unit of energy obtained, including carbon dioxide, sulfur dioxide, toxic metals, and particulates. Oil releases fewer pollutants than coal. Air pollutants can be both primary, meaning directly emitted from pollution sources, or secondary, meaning they undergo transformations in the atmosphere. In the next module we will examine photochemical smog, inversions, particulates, and carbon dioxide more closely.

Module 42 AP® Review

Preparing for the AP® Exam

Learning Goals Revisited

42-1 What are the sources and effects of major air pollutants?

Air pollution occurs across the troposphere, the portion of the atmosphere closest to Earth's surface. Coal is the dirtiest fossil fuel and releases the greatest amount of carbon dioxide, sulfur dioxide, and particulates per unit of energy obtained. Oil emits fewer pollutants than coal and natural gas releases fewer pollutants than oil. Nitrogen oxides lead to ozone formation, smog, and acid rain. Lead concentrations in air have decreased considerably due to EPA regulations banning it as a gasoline additive.

42-2 What are primary and secondary pollutants and how do they change over time?

Primary pollutants are released directly from the emission source while secondary pollutants undergo transformations in the atmosphere in the presence of sunlight and water at the appropriate temperatures. Implementation of air quality standards has led to reductions in most air pollutants, especially lead.

AP® Practice Questions

Preparing for the AP® Exam

Multiple-Choice Questions

1. Which form of legislation is responsible for regulating lead in fuels, which ultimately ends up in the atmosphere?
   (a) Safe Drinking Water Act
   (b) Clean Air Act
   (c) Montreal Protocol
   (d) Clean Water Act

2. Secondary pollutants tend to be the result of a combination of a pollutant and ____________.
   (a) water
   (b) soil
   (c) photochemical smog
   (d) ozone

3. Which is a major source of sulfur dioxide found in nature?
   (a) ruminant animals
   (b) lightning strikes
   (c) plant emissions
   (d) volcanoes

4. Carbon monoxide
   (a) increases lung cancer rates.
   (b) leads to the formation of photochemical smog.
   (c) is most problematic in rural areas.
   (d) is produced by incomplete combustion.

5. Which criteria air pollutant has seen the most substantial decrease in recent years?
   (a) carbon monoxide
   (b) nitrogen oxides
   (c) ozone
   (d) lead

Use the passage below to answer questions 6 & 7:

Air quality data were gathered near a downtown area at 5:00 PM, when most businesses close, and at 8:00 PM. Readings show a high concentration of volatile organic compounds (VOCs) at 5:00 PM, and a reduction of them at 8:00 PM. The same data also show low concentrations of ozone (O_3) at 5:00 PM, and an increase at 8:00 PM.

6. Which of the following explains the difference in concentrations of VOCs and Ozone at the two reading times?
   (a) The VOCs are a primary pollutant that react and are reduced in quantity.
   (b) Ozone comes from vehicle emissions that are used more in evenings.
   (c) VOCs are a secondary pollutant that peak in the afternoon.
   (d) Ozone is a primary pollutant that peaks in the evenings.

7. Which of the following claims is supported by the evidence in the paragraph?
   (a) Burning fossil fuels increases ozone levels.
   (b) Vehicle emissions cause VOCs to decrease.
   (c) Ozone forms as a secondary pollutant from VOCs.
   (d) Industrial processes are responsible for a drop in VOCs.

Free-Response Question

There are a number of fossil fuel sources that provide energy.

	Tons of pollutant per 1,000 tons of fuel		
	NO_2	SO_2	Particulates
Natural gas	3.5	0.0	0.0
Bituminous Coal	63.6	9.2	65.3
Fuel oil	5.2	35.3	0.3

(Data from https://www.researchgate.net/figure/Levels-of-emissions-of-pollutants-into-the-atmosphere-when-burning-various-types-of-fuel_tbl2_351818804)

(a) **Identify** the fuel that releases the least amount of overall pollutants into the atmosphere. (1 pt.)
(b) **Identify** the fuel source that releases the highest amount of overall pollutants into the atmosphere. (1 pt.)
(c) **Describe** why the fuel source in part (b) typically has more pollutants. (1 pt.)
(d) **Describe** which fuel source is used primarily for transportation purposes. (1 pt.)
(e) **Explain** which fuel source is more prevalent in powering homes. (1 pt.)
(f) **Describe** one environmental drawback, other than air pollution, of using bituminous coal as a fuel source. (1 pt.)
(g) A group of researchers gathered data on particulate matter at four different times of the day:

12:00 AM – 113 ppm
6:00 AM – 105 ppm
12:00 PM – 91 ppm
6:00 PM – 93 ppm

(i) **Identify** two possible hypotheses for this investigation. (2 pts.)
(ii) **Identify** the dependent variable for this investigation. (1 pt.)
(iii) **Describe** a modification the researchers could do to improve the accuracy of their investigation. (1 pt.)

Module 43

Unit 7 | 42 | 43 | 44 | 45 | 46

Photochemical Smog, Thermal Inversions, Atmospheric CO_2, and Particulates

As we saw in the last module air quality in the United States has greatly improved in recent decades. However, some cities and regions of the United States continue to experience high air pollution events, often related to smog formation. In this module we will examine photochemical smog, which is a problem in the United States and elsewhere in the world, and thermal inversions that can lead to serious air pollution episodes and human health problems. We will also describe natural sources of CO_2 and particulates, the latter of which can contribute to photochemical smog.

Learning Goals

After reading this module you should be able to

43-1 explain photochemical smog and how to reduce it.

43-2 describe thermal inversions and how they relate to air pollution.

43-3 describe the natural sources of particulates and CO_2.

43-1 What is photochemical smog and how can we reduce it?

Photochemical smog is a complex combination of compounds and it can be reduced by decreasing emissions of its precursors

A recent news article stated that "More than 4 in 10 Americans live in counties where the air is unhealthy to breathe." You might think this was from a newspaper in the 1970s, before the Clean Air Act was fully in effect. But in *State of the Air 2021*, the American Lung Association reported that 40 percent of people in the United States—over 135 million people—were exposed to air that did not comply with the maximum allowable ozone concentration. An allowable ozone concentration is more than 75 parts of ozone per billion parts of air averaged over an 8-hour period. Although sulfur, nitrogen, and carbon monoxide pollution have been reduced well below the specified standards since the Clean Air was implemented, controlling the secondary pollutants photochemical smog and ozone have proven to be especially challenging. As we saw in Figure 42.4 in the previous module average ozone concentrations in 2020 were just below the Clean Air Act standards. For ozone, the average is taken over 8 hours during a day. If the average concentration is just below the standard, that means that shorter-term ozone concentrations, such as those measured hourly, are sometimes higher than the CAA standards. For example, in September 2020, Los Angeles, California, experienced its worst smog in 30 years. **FIGURE 43.1** shows the hourly ozone and nitrogen oxide concentrations during one day, September 4. As you can see, the ozone concentration

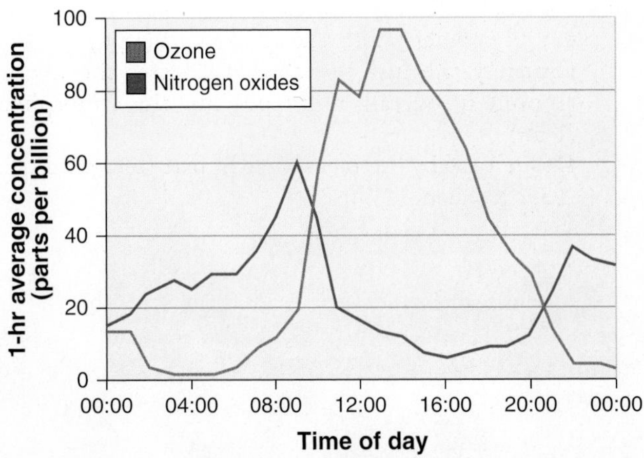

FIGURE 43.1 Tropospheric ozone concentrations measured each hour in Los Angeles, California, on September 4, 2020. In most cities, mid- to late-afternoon ozone concentrations are higher than the daily average as a result of an increase in nitrogen oxides in the early morning hours. *(Data from South Coast Air Quality Management District, http://www.aqmd.gov/)*

at 6 AM is less than 5 parts per billion (ppb) and at 2 PM it was more than 95 ppb. The reason for this large daytime increase lies in the chemistry of smog formation and the behavior of the atmosphere during changing weather conditions. These various environmental factors make smog formation complex and somewhat difficult to predict and even more difficult to reduce.

Chemistry of Ozone and Photochemical Smog Formation

We defined smog in the last module as a combination of oxidants and particulate matter. The word smog was derived by combining smoke and fog. Smog is partly responsible for the hazy view and reduced sunlight observed in many cities. Smog can be divided into two categories: photochemical smog and sulfurous smog. **Photochemical smog** is dominated by oxidants such as ozone and is sometimes called **Los Angeles–type smog** or **brown smog**. FIGURE 43.2a shows Los Angeles during the September, 2020 smog event represented by the data in Figure 43.1. Nearby forest fires also contributed to the poor air quality during this time period. **Sulfurous smog** is dominated by sulfur dioxide, sulfate compounds, and particulate matter and is also called **London-type smog**, **gray smog**, or **industrial smog**. Figure 43.2b shows the city of London, England during a gray smog event.

Photochemical smog Smog that is dominated by oxidants such as ozone. *Also known as* **Los Angeles–type smog; brown smog.**

Sulfurous smog Smog dominated by sulfur dioxide, sulfate compounds, and particulate matter. *Also known as* **London-type smog; gray smog; industrial smog.**

AP® Exam Tip

Remember that the prefix "photo" means light. You should be able to explain that photochemical smog is caused in part by the presence of sunlight. And on days when there are more hours of sunlight, if all other factors remain equal, there will be more photochemical smog.

In addition to human health problems, particulate matter and photochemical oxidants also cause economic harm, since poor visibility in popular vacation destinations can reduce tourism revenues for recreation areas, including lower incomes for hotels and restaurants in these areas.

Today, Los Angeles–type brown photochemical smog is still a problem in many U.S. cities. The formation of this photochemical smog is complex and still not well understood. A number of pollutants are involved, and they undergo a series of complex transformations in the atmosphere that involve both intensity and number of hours of sunlight as well as water and the presence of volatile organic compounds (VOCs).

FIGURE 43.3 shows a portion of the chemical process that creates photochemical smog. The first part of the process, shown in Figure 43.3a, takes place during the day, in the presence of sunlight, including ultraviolet energy from the Sun, which warms the atmosphere. When an abundance of nitrogen oxides is present in the atmosphere, with very few VOCs present, nitrogen dioxide (NO_2) splits to form nitrogen oxide (NO) and a free oxygen atom (O). In the presence of energy inputs from sunlight, this free oxygen atom combines with diatomic oxygen (O_2) to form ozone (O_3). When there is abundant nitrogen dioxide in the atmosphere and the Sun provides light and warms the atmosphere, ozone accumulates. Because of the dependence on sunlight and temperature, ozone concentrations are normally greatest in the early to middle

(a)

(b)

FIGURE 43.2 **There are two different types of smog.** (a) Los Angeles–type brown smog. The smog seen here, photographed in Los Angeles, September 2020, is also a result of nearby wildfires. (b) London-type smog, gray smog, also known as industrial smog. The smog seen here was photographed in 2014. *(a: Citizen of the Planet/Alamy Stock Photo; b: Dinendra Haria/Alamy Stock Photo)*

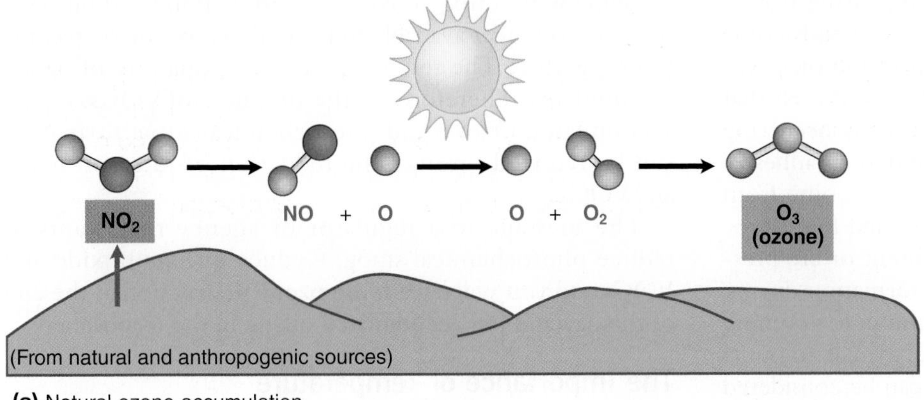

(a) Natural ozone accumulation

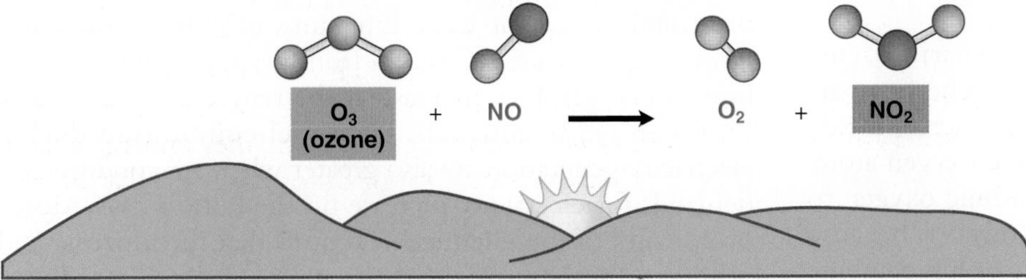

(b) Natural ozone destruction

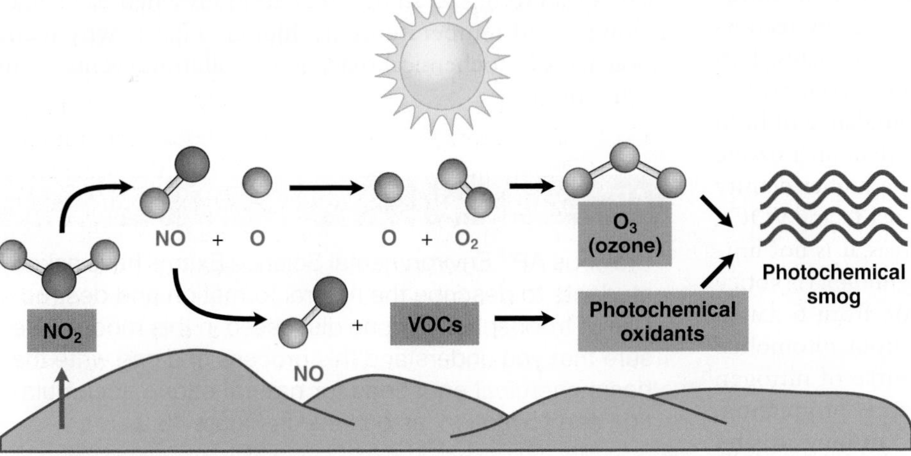

(c) Buildup of photochemical smog

FIGURE 43.3 Tropospheric ozone and photochemical smog formation. (a) In the absence of VOCs, ozone will form during the daylight hours. (b) After sunset, the ozone will break down. (c) In the presence of VOCs, ozone will form during the daylight hours. The VOCs combine with nitrogen oxides to form photochemical oxidants, which reduce the amount of ozone that will break down later and contribute to prolonged periods of photochemical smog.

afternoon and they are also highest in the summer months, when there is the greatest number of hours of sunlight.

Figure 43.3b shows that a few hours later, when sunlight intensity decreases and nitrogen oxide is still present in the atmosphere, the ozone combines with nitrogen oxide (NO), and re-forms into $O_2 + NO_2$. This is referred to as ozone destruction and it is a natural process that happens in the latter part of every day and evening. But sometimes, there are compounds in the air that interfere with the natural ozone destruction that normally occurs in the latter part of the day as the Sun goes down.

In order to explore this further, we first need to discuss volatile organic compounds (VOCs). We described VOCs as having a strong aroma, stating that if you can smell them, this indicates that the chemicals within the VOC are easily released into the air. This means that the VOCs either **evaporate**, the process of converting from a liquid to a gas

Evaporate The process of converting from a liquid to a gas or vapor.

or vapor, or **sublimate**, the process of converting from a solid to a gas or vapor. Volatile organic compounds come from human activity such as spilling of gasoline on pavement. They also come from combustion of fossil fuels that release VOCs such as **formaldehyde**, a naturally occurring compound that is used as a preservative and as an adhesive in plywood and carpeting. In addition, VOCs come from natural sources such as tree leaves in a forest and forest fires. When volatile organic compounds are absent or are present in small amounts, the cycle of ozone formation during the day (Figure 43.3a) and ozone destruction in the evening (Figure 43.3b) occurs. As a result, relatively small amounts of photochemical smog accumulate. This can be considered the natural cycle of ozone formation and ozone destruction that occurs around the world on a daily basis.

As shown in Figure 43.3c, a different scenario occurs when VOCs are present in larger quantities. The first part of the process is the same: Sunlight causes nitrogen dioxide to break apart into nitrogen oxide and a free oxygen atom. The free oxygen atom combines with diatomic oxygen to form ozone. And ozone accumulates. However, because VOCs are present in larger quantities, more VOCs combine with nitrogen oxide in a strong bond and nitrogen oxide is no longer available to combine with ozone. Since the nitrogen oxide is not available to break down ozone by recombining with it, a larger amount of ozone accumulates. This explains, in part, the daytime accumulation of ozone particularly in urban areas that experience an abundance of both VOCs and nitrogen dioxide. Part of the urban area ozone accumulation problem occurs because of the higher density of automobiles that release both nitrogen oxide and VOCs. Although smog is associated with urban areas, it is not limited to such areas. Take another look at Figure 43.1. Notice that there is an increase in nitrogen oxide from 6 AM to 9 AM. The nitrogen oxide comes primarily from automobile exhaust during "rush hour," and is the source of nitrogen oxide needed in Figure 43.3a to cause ozone production. Rush hour is so named because people are "rushing" to and from work, so although rush hour time varies by city and by country, the approximate times of rush hour are from 7 AM to 9 AM and from 5 PM to 8 PM. To summarize this complex series of reactions in the atmosphere: Sunlight, including ultraviolet radiation from the Sun, causes nitrogen dioxide to break apart into nitrogen oxide and a free oxygen atom which combines with diatomic oxygen to form ozone. Ozone accumulates. In the absence of VOCs, the nitrogen oxide accumulates in the atmosphere and later in the day it contributes to ozone destruction. However, when air pollutant VOCs are present in large quantities, VOCs combine with nitrogen oxide in a strong bond and nitrogen oxide is no longer available to break down ozone by recombining with it. The result is that a large quantity of ozone accumulates. Therefore, in the presence of VOCs, ozone accumulates. Ozone and photochemical smog formation can be reduced through the reduction of nitrogen oxide and VOCs.

The message to a regulator or agency that wants to reduce photochemical smog: Reduce nitrogen oxide and VOCs and you will have more ozone destruction at the end of the day and less accumulated ozone in the troposphere!

The Importance of Temperature

Atmospheric temperature influences the formation of smog in several important ways. Emissions of VOCs from vegetation such as trees, as well as from evaporation of volatile liquids like gasoline, increase as the temperature increases. Nitrogen oxide emissions from fuel combustion during electricity generation are also greater when air-conditioning demands for electricity increase on the hottest days. Moreover, many of the chemical reactions that form ozone and other photochemical oxidants proceed more rapidly in sunny conditions at higher temperatures. These and other factors lead to higher smog concentrations when day length is longer and temperatures are higher. That is why many smog and photochemical oxidant air pollution events occur in the summer.

AP® Exam Tip

Previous AP® Environmental Science Exams have asked students to describe the natural formation and destruction of tropospheric ozone discussed in this module. Be sure that you understand this process and can write the basic chemical equations for natural ozone accumulation and formation, as outlined in Figure 43.3.

Smog and Human Health

Secondary pollutants such as smog and photochemical oxidants, including ozone, can affect human health in a number of ways. They are responsible for burning and itchy eyes and can irritate the nose and throat, causing coughing and a scratchy, sore throat. They can make it harder to breathe deeply and cause pain when taking a deep breath. They can also inflame the air passageways and lungs, making them more susceptible to infection. And they aggravate existing lung conditions such as asthma (constriction and inflammation of the airway to the lungs), emphysema (a disease affecting the air sacs or alveoli in the lungs), and bronchitis (an inflammation of the bronchial tubes that transport air to the lungs). The U.S. Environmental Protection Agency reports that long-term exposure to ozone is possibly one of many factors that causes asthma to develop. The U.S.

Sublimate The process of converting from a solid to a gas or vapor.

Formaldehyde A naturally occurring compound that is used as a preservative and as an adhesive in plywood and carpeting.

Centers for Disease Control reports that smog is responsible for both decreased lung capacity and increased hospital admissions for asthma and causes some number of premature deaths each year. When smog levels are above a certain level, many public health officials recommend that people reduce or eliminate outdoor activity like cycling or going for a run. Most smart phone weather apps give an air quality index for cities susceptible to high concentrations of smog and inform residents when it is advisable to limit time outside. However, people in lower income groups who may be more dependent on public transportation may have no choice but to be outside, for example, while waiting for a bus, during poor air quality events.

AP® Exam Tip

Remember to be specific when answering questions about air pollution effects on human health. You should provide more details than simply stating that air pollution causes respiratory problems. Identify specific pollutants such as sulfur dioxide and ozone and specific conditions such as asthma, bronchitis, and emphysema.

There is one more factor, a combination of air temperature and topography of an urban location and its surroundings, that can influence smog accumulation. This factor is called a thermal inversion.

43-2 What are thermal inversions and how do they relate to air pollution?

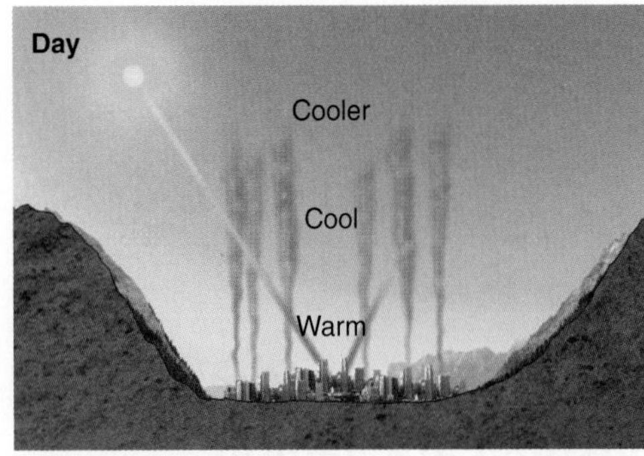

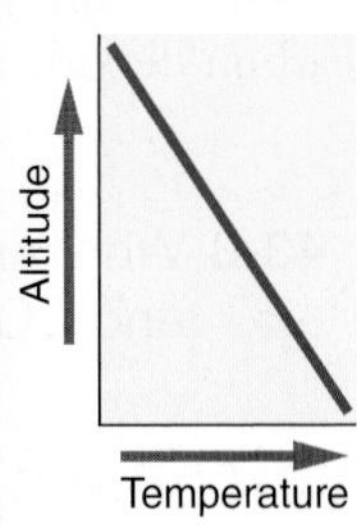

(a) Normal conditions

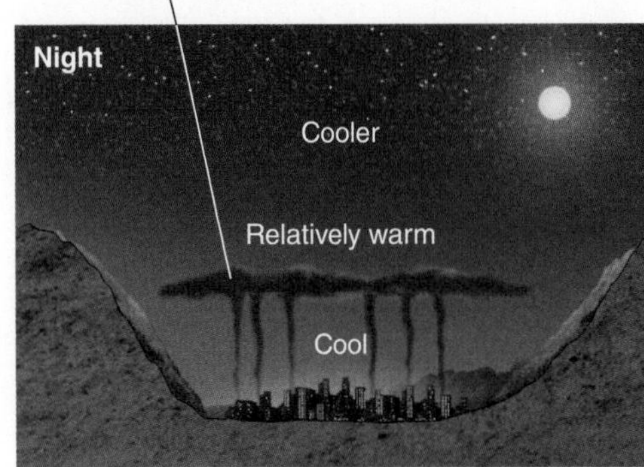

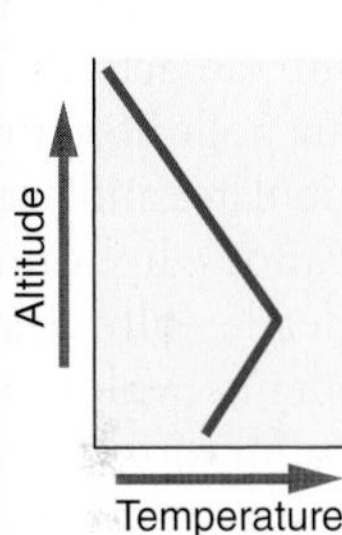

(b) Thermal inversion

FIGURE 43.4 A thermal inversion. (a) Under normal conditions, where temperatures decrease with increasing altitude, emissions rise into the atmosphere. (b) When a mid-altitude, relatively warm inversion layer blankets a cooler layer, emissions are trapped and accumulate.

Thermal inversions trap air pollutants close to Earth's surface

Temperature also influences air pollution conditions in more complex ways. Normally, temperature decreases as altitude increases. As shown in **FIGURE 43.4a**, the warmest air is closest to Earth. This warm air, which is less dense than the colder air above it, can easily rise, dispersing pollutants into the upper atmosphere. This allows pollutants from the surface to be reduced or diluted by all of the atmosphere above. However, during a **thermal inversion**—shown in Figure 43.4b—the typical temperature gradient shown in Figure 43.4a is flipped: a relatively warm layer of air at mid-altitude covers a layer of cold, dense air below. This layer of warm air that traps emissions is known as an **inversion layer**. Because in a thermal inversion the air closest to the surface of Earth is denser than the air above it, the cool air and the pollutants within it do not rise. Thus, the inversion layer traps pollutants that accumulate beneath it close to the ground. These trapped emissions, mostly smog and particulates, can cause a severe pollution event. Thermal inversions that create pollution events are particularly common in certain cities, where high concentrations of vehicle exhaust and industrial emissions are easily trapped by the inversion layer.

Thermal inversions can also lead to problems resulting from other forms of pollution. A striking example occurred in spring 1998 in the northern Chinese city of Tianjin. A cold spell that occurred after the city had shut off its district heating system for the season led many households to use individual coal-burning stoves for heat. A temperature inversion trapped the carbon monoxide and particulate matter from the coal used in these stoves and caused over 1,000 people to suffer carbon monoxide poisoning or respiratory ailments from the polluted air. Eleven people died. Inversions are a natural process although some researchers have begun to investigate whether climate change has led to an increase in the number of inversions that occur each year around the world. Here and in the previous module we

Thermal inversion An atmospheric condition in which a relatively warm layer of air at mid-altitude covers a layer of cold, dense air below.

Inversion layer The layer of warm air that traps emissions in a thermal inversion.

have examined anthropogenic pollutants. It is also important to understand the natural sources of pollutants, especially carbon dioxide and particulates.

43-3 What are the natural sources of particulates and CO_2?

Volcanoes, forest fires, respiration, and decomposition are natural processes responsible for particulate emissions and carbon dioxide

Environmental scientists and concerned citizens direct much of their attention toward air pollution that comes from human activity. But human activity is not the only source of air pollution. Processes in nature are also the sources of particulates and carbon dioxide among other air pollutants. Volcanoes, lightning, forest fires, and plants—both living and dead—all release compounds that can be classified as pollutants. Volcanoes release carbon dioxide, particulate matter, sulfur dioxide, carbon monoxide, and nitrogen oxides. Lightning strikes create nitrogen oxides from atmospheric nitrogen. Forest fires release particulate matter, nitrogen oxides, carbon dioxide, and carbon monoxide (**FIGURE 43.5a**). Living plants release a variety of VOCs, including terpenes. The fragrant smell from conifer trees such as pine and fir and the smell from citrus fruits are mostly from terpenes; though we enjoy their fragrance, they can be precursors to photochemical smog. Long before anthropogenic pollution was common, the natural VOCs from plants gave rise to smog and photochemical oxidant pollution—hence the names of the forested mountain ranges in the southeastern United States, the Blue Ridge and the Smoky Mountains, which were given their names by their indigenous residents who predated the use of modern combustion fuels (Figure 43.5b).

Natural Sources of Particulates

Large nonindustrial areas such as agricultural fields can give rise to particulate matter when they are plowed, as happened in the Dust Bowl of the 1930s. Averaged globally, sulfur dioxide emissions are 30 percent natural, nitrogen oxide emissions are 44 percent natural, and volatile organic compound emissions are 89 percent natural. However, in certain locations, such as North America, the anthropogenic contribution is much greater, perhaps as much as 95 percent for nitrogen oxides and sulfur dioxide.

The effects of these various compounds, especially when major natural events occur, depend in part on natural conditions such as wind direction. In May 1980, the prevailing westerly winds carried volcanic emissions from a massive eruption of Mount St. Helens in Washington State and distributed particulate matter and sulfur oxides from the volcano across the United States. The rainfall pH was noticeably lower in the eastern United States that summer due to the sulfur oxides emitted by Mount St. Helens.

Particulate matter is a frequent pollutant from natural sources. Volcanoes, forest fires, and dust storms are all events that result in the release of particulate matter into the atmosphere. **FIGURE 43.6** shows the sources of particulate matter and its effects. Particulate matter ranges in size from 0.01 micrometer (μm) to 100 μm (1 micrometer = 0.000001 m). For comparison, a human hair has a diameter of roughly

(a)

(b)

FIGURE 43.5 Natural sources of air pollution. There are many natural sources of air pollution, including volcanoes, lightning strikes, forest fires, and plants. (a) A forest fire in Ojai, California, produces air pollution. (b) The Great Smoky Mountains were named for the natural air pollutants that reduce visibility and give the landscape a smoky appearance. *(a: Bill Boch/Getty Images; b: Bill Lea/Dembinsky Photo Associates/Alamy)*

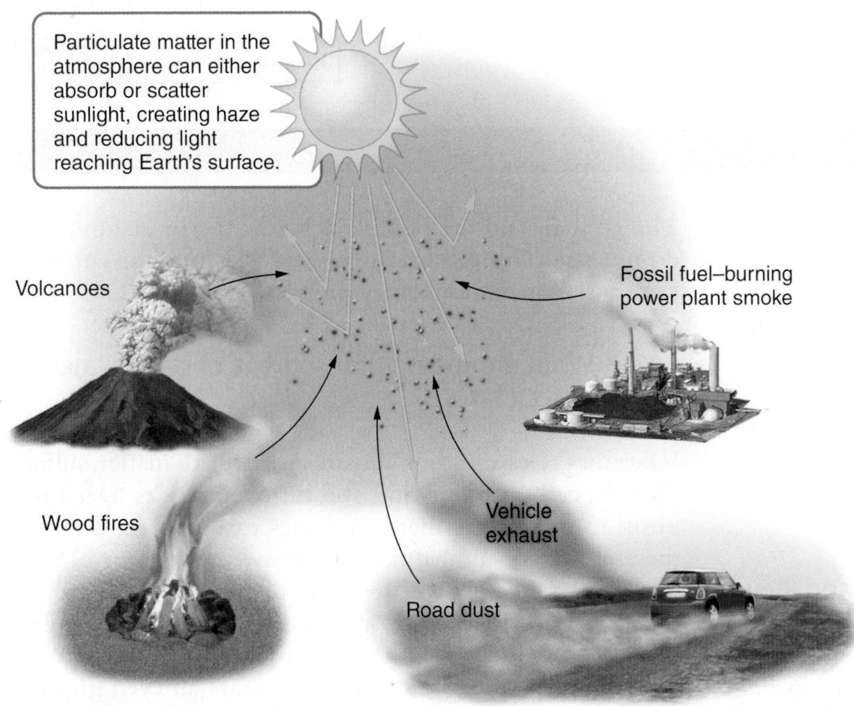
Particulate matter in the atmosphere can either absorb or scatter sunlight, creating haze and reducing light reaching Earth's surface.
Volcanoes
Fossil fuel–burning power plant smoke
Wood fires
Vehicle exhaust
Road dust

FIGURE 43.6 The sources of particulate matter and its effects. Particulate matter can be natural as well as anthropogenic. Particulate matter in the atmosphere ranges considerably in size and can absorb or scatter light, which creates a haze and reduces the light that reaches the surface of Earth.

50 to 100 μm. Particulate matter larger than 10 μm is usually filtered out by the nose and throat; particulate matter of this size is not regulated by the EPA. Particles smaller than 10 μm are called Particulate Matter-10, written as $\mathbf{PM_{10}}$, and are of concern to air pollution scientists because they are not filtered out by the nose and throat and can be deposited deep within the respiratory tract. Particles of 2.5 μm and smaller, called $\mathbf{PM_{2.5}}$, are an even greater health concern because they can travel further within the respiratory tract and they tend to be composed of more toxic substances than particles in larger size ranges. They are also associated with a higher incidence of lung cancer.

Particulate matter also scatters and absorbs sunlight. If the atmospheric concentration of particulate matter is high enough, as it would be immediately following a large forest fire or a volcanic eruption, incoming solar radiation in the region will be reduced enough to affect photosynthesis. This happened in 1816, a year after a large volcanic eruption in Java released more than 150 million metric tons of particles that slowly spread around the globe, reducing and scattering sunlight and creating haze. That year is commonly referred to as the year without a summer. Haze, which we described as reduced visibility in Module 42, occurs when particulate matter from air pollution scatters light. As we saw earlier in this module, ozone and photochemical oxidants can also play an important indirect role in the formation of haze.

Natural Sources of Carbon Dioxide

We have referred to photosynthesis and respiration several times throughout this text. Carbon dioxide is a by-product of respiration, decomposition, and combustion. All three processes are similar in that they convert an energy source such as biomass or sugar into energy, water, and carbon dioxide. As we saw in Figure 6.1, respiration is the opposite of photosynthesis: Cells convert sugar and oxygen into energy, carbon dioxide, and water. So respiration is one natural source of carbon dioxide. We also learned in Module 6 that there are two types of respiration: aerobic, or with oxygen, and anaerobic, in the absence of oxygen. Both types of respiration result in producing carbon dioxide. Decomposition also occurs in the presence or absence of oxygen, and globally respiration and decomposition are important natural sources of carbon dioxide. Finally, combustion of biomass in fires also results in the release of carbon dioxide. The decomposition reactions can be represented by

Aerobic decomposition (oxygen present):
$$C_6H_{12}O_6 + 6\,O_2 \rightarrow 6\,CO_2 + 6\,H_2O + \text{energy}$$

For decomposition of organic matter in the absence of oxygen, we use the aerobic respiration formula that we presented in Module 6

Anaerobic decomposition (oxygen not present):
$$C_6H_{12}O_6 \rightarrow 3\,CO_2 + 3\,CH_4 + \text{less energy}$$

In this module we examined the conditions under which photochemical smog forms and is destroyed. We described the connection between nitrogen oxide, volatile organic compounds, and ozone formation, accumulation, and destruction. Thermal inversions occur when the air temperature at the surface of Earth is cooler than higher up. This will trap air pollutants close to Earth's surface, giving rise to harmful air pollution events. Carbon dioxide and particulate matter both have natural and human sources. In the next module we will examine indoor air pollutants, some of which are derived from outdoor pollutants.

$\mathbf{PM_{10}}$ Particles smaller than 10 μm are called Particulate Matter-10 and are not filtered out by the nose and throat and can be deposited deep within the respiratory tract.

$\mathbf{PM_{2.5}}$ Particles of size 2.5 μm and smaller can travel further within the respiratory tract and are of even greater health concern.

Module 43 AP® Review

Preparing for the AP® Exam

Learning Goals Revisited

43-1 What is photochemical smog and how can we reduce it?

Photochemical smog accumulates in the troposphere in the presence of sunlight when nitrogen oxides and volatile organic compounds are present. Despite many improvements in air quality in the United States, photochemical smog is still a problem in a number of urban and some relatively rural areas. Different types of smog are dominated by different compounds.

43-2 What are thermal inversions and how do they relate to air pollution?

During a thermal inversion, a warm layer of air covers a colder layer below it. The warm layer that is trapped between two cooler layers is known as an inversion layer. The cool air and the pollutants below the inversion layer do not rise. This can lead to a severe pollution event that can cause human respiratory problems.

43-3 What are the natural sources of particulates and CO_2?

Volcanoes release carbon dioxide, particulate matter, sulfur dioxide, carbon monoxide, and nitrogen oxides. Decomposition of organic matter releases carbon dioxide into the atmosphere and natural fires as well as dust storms can release particulates into the atmosphere. Particles smaller than 10 μm, PM_{10}, are of concern to air pollution scientists because they are not filtered out by the nose and throat. Particles of 2.5 μm and smaller, $PM_{2.5}$, are an even greater health concern because they can travel further within the respiratory tract and cause even greater harm.

AP® Practice Questions

Preparing for the AP® Exam

Multiple-Choice Questions

1. The data below show sulfur dioxide and nitrogen dioxide levels within an urban metro area. Which time of day would have the highest likelihood of photochemical smog formation?

Time of day	SO_2 (PPM)	NO_2 (PPM)
8:00 AM	0.1	0.2
11:00 AM	0.2	0.2
2:00 PM	0.3	0.4
5:00 PM	0.6	0.8

(a) 8:00 AM
(b) 11:00 AM
(c) 2:00 PM
(d) 5:00 PM

2. Thermal inversions
(a) increase the rate of smog formation.
(b) trap high concentrations of pollution at ground level.
(c) result in lower levels of particulate matter.
(d) are caused by high levels of precipitation.

3. Which size of particulate matter causes the greatest health concern?
(a) $PM_{2.5}$
(b) PM_{10}
(c) PM_{100}
(d) PM_{30}

4. Which of the following is considered a natural source of carbon dioxide in the atmosphere?
(a) decomposition
(b) transportation
(c) agricultural practices
(d) energy generation

5. Which of the following best represents the atmospheric changes during a thermal inversion?

(a)
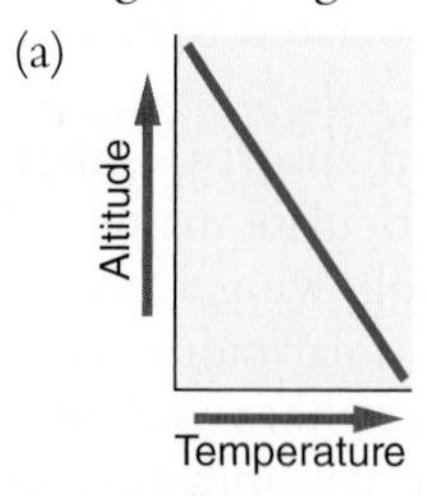

(c)
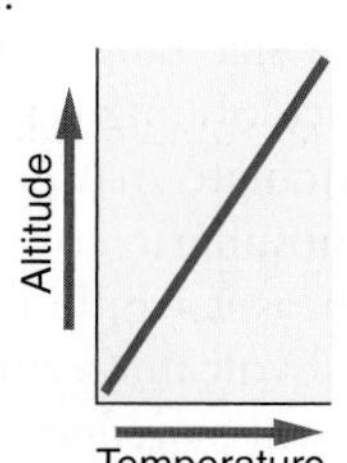

(b)
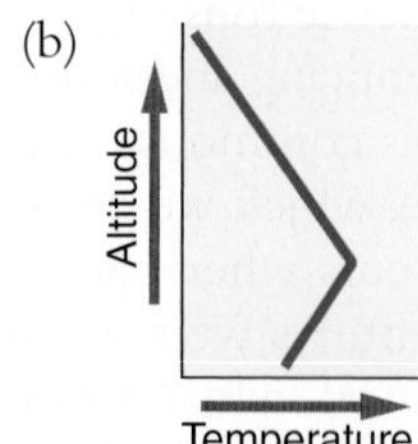

(d)
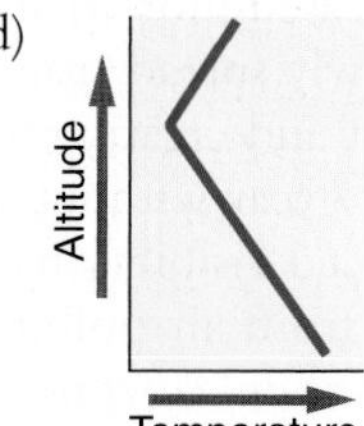

Free-Response Question

The figure below shows a volcano eruption.

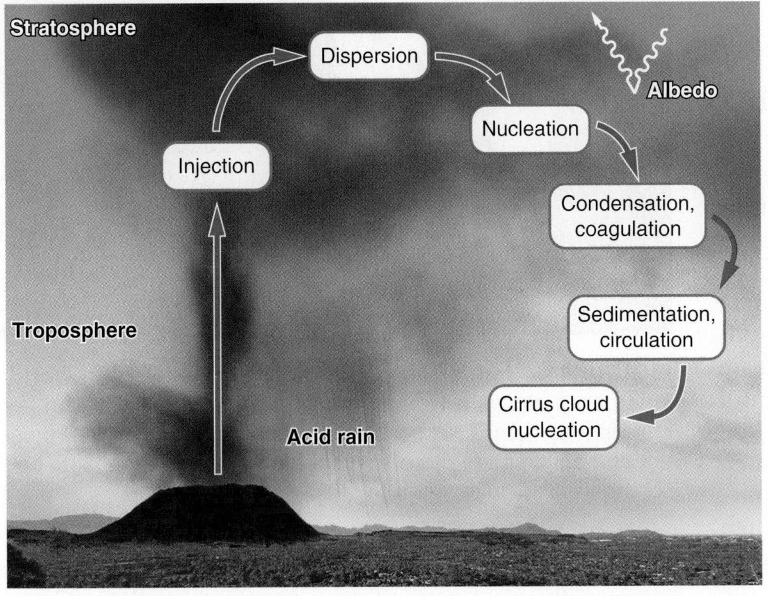

(a) **Identify** one pollutant that comes out of a volcano at an eruption. (1 pt.)

(b) **Describe** the class of pollutants identified in part (a). (1 pt.)

(c) Once the pollutants come out of the volcano, **explain** one way those pollutants change into secondary pollutants. (1 pt.)

(d) **Identify** one way that humans add more of the pollutant mentioned in part (a) into the environment. (1 pt.)

(e) **Explain** how the pollutant mentioned in part (a) can impact ecosystems. (1 pt.)

(f) While volcanoes are natural emitters of pollution, there are also natural ways to mitigate particulates and carbon dioxide emissions.

 (i) **Describe** a natural method to reduce carbon dioxide emissions. (1 pt.)

 (ii) **Identify** a potential drawback to the method described in part (i). (1 pt.)

 (iii) **Describe** an additional benefit of the solution proposed in part (i). (1 pt.)

(g) **Make a claim** as to whether natural emissions or anthropogenic emissions have a greater impact on ecosystems. (1 pt.)

(h) **Justify** the claim in part (g), using evidence. (1 pt.)

Module 44

Unit 7	42	43	44	45	46

Indoor Air Pollutants

When we think of air pollution, we usually don't associate it with air inside our buildings, but indoor air pollution causes more deaths each year than does outdoor air pollution. Sadly, most of these deaths occur in the developing world, but indoor air pollution is a significant hazard in both developing and developed countries. The quality of indoor air is highly variable and when polluting activities take place indoors, and time spent indoors is high, exposure to pollutants can be a significant health risk. In this module we will learn about the various types of indoor air pollutants, where they come from, what effects they have on human health, and how they differ around the world.

Learning Goals

After reading this module you should be able to

44-1 identify the major indoor air pollutants and where they come from.

44-2 describe indoor air pollution differences in the developing and developed world.

44-1 What are the major indoor air pollutants and where do they come from?

Indoor air pollutants come from natural, human-made, and combustion sources

Indoor air pollutants are compounds that adversely affect the quality of air in buildings and structures. Air quality indoors influences both health and comfort of the occupants. Because a home or other building is a closed system with an abundance of manufactured materials, there is ample opportunity for indoor air pollutants to accumulate and for the occupants of that space to encounter harmful substances. Some indoor air pollutants are the exact same compounds of concern found in outdoor air pollution, such as carbon monoxide and particulates, while other compounds are only of concern when they are indoors, such as radon and asbestos. Sources of indoor air pollution are activities such as combustion within the building and the release of gases and compounds from materials within the building; other indoor air pollutants occur after transfer of gases and compounds from outside into the building. We will review the major indoor air pollutants and their effects.

Carbon Monoxide

Carbon monoxide is perhaps the most dangerous indoor air pollutant because when concentrations are high enough, it can cause asphyxiation (oxygen deprivation) and death within minutes. We have already described carbon monoxide as an outdoor air pollutant, but it can be even more dangerous as an indoor air pollutant. In the human body, carbon monoxide binds with hemoglobin more efficiently than oxygen, essentially displacing oxygen from hemoglobin, thereby interfering with oxygen transport in the blood. Extended exposure to high concentrations of carbon monoxide in air can lead to oxygen deprivation in the brain and, ultimately, death. Lower exposures can lead to headache, dizziness, nausea, and loss of consciousness.

Carbon monoxide can come from urban outdoor air pollution that infiltrates a building or it can come from within the building. In the developed world, carbon monoxide exposure indoors occurs most often as a result of malfunctioning exhaust systems on household furnace heating systems, most commonly natural gas heaters. When the exhaust system in a natural gas heater malfunctions, exhaust air escapes into the living space of the house instead of being vented outside. Because natural gas burns relatively cleanly with little odor, a malfunctioning natural gas burner can cause the colorless, odorless carbon monoxide to

Indoor air pollutants Compounds that adversely affect the quality of air in buildings and structures.

FIGURE 44.1 Indoor air pollution in the developing world. This photo shows a woman cooking on an open fire with her child in their home in Zimbabwe. Carbon monoxide is likely present and children are more susceptible to carbon monoxide poisoning than adults. *(Crispin Hughes/Panos Pictures)*

accumulate in a house without the occupants noticing. This is especially problematic if the occupants are asleep. That is why carbon monoxide detectors are so important in homes and apartments and why they should be maintained and tested periodically.

In the developing world, exhaust fans are not that common in homes, so cooking is done with little or no mechanical ventilation. As we discussed in Module 39, biomass is burned in open-pit fires that lack the proper mix of fuel and air to allow complete combustion. Usually, there is no exhaust system and little or no ventilation available in the home, which makes indoor air pollution from carbon monoxide, and also particulates, a hazard in developing countries (**FIGURE 44.1**). Children are at an even higher risk for carbon monoxide poisoning because their respiration rate is higher than adults.

Particulates

As we know from Module 42, particulates are solid particles or particulate matter suspended in the air. Among indoor air pollutants, the most important particles are smoke, dust, and mold.

Smoke

Smoke and soot from fire, either from outside or from within the building, and from burning tobacco are sources of particulate matter in buildings. We described particulate matter in Modules 42 and 43 and the same principles and health concerns apply to indoor particulate matter. Respiratory infections, pneumonia, bronchitis, and even cancer can develop from prolonged exposure to indoor particulate matter. Obviously, a person smoking tobacco is going to be exposed to high concentrations of particulate matter but other people in the same building will likely be exposed to second-hand smoke, which is a source of particulate matter that can increase the likelihood of developing respiratory ailments and cancer.

FIGURE 44.2 Dust mites in a pillow. An entire ecosystem lives in a pillow. The energy source is dead human skin cells. These dust mites are approximately 250 microns in size and can cause respiratory and skin allergic reactions. *(selvanegra/Getty Images)*

Dust and Mold

Both dust and molds are natural sources of indoor air pollutants. Dust is a combination of particulate matter, pollen, bacteria, and the waste product of an intricate ecosystem within all households. Humans shed millions of skin cells each day. Within each household, there is an entire ecosystem supported by the food energy of dead human skin cells including dust mites that live in our bedding, pillows, flooring, and carpeting (**FIGURE 44.2**). Dead human skin cells, the droppings from dust mites, and dust mite body fragments all become part of household dust. While household dust might seem like it is simply an inconvenience, it actually contributes to a variety of conditions including asthma and skin and respiratory allergic reactions. Particulate matter enters buildings from airborne particulates and from soil that is carried into the building on shoes and clothing. Pollen can enter buildings through open windows and clothing. Bacteria is introduced into buildings through just about anything that is brought inside. Mold, also known as mildew, is a fungal growth that can rapidly spread on damp organic matter. It occurs in homes, sometimes behind walls where it cannot be easily detected or seen. Molds can be introduced from outside and thrive in moist environments such as bathrooms, kitchens, and basements. Allergies, lung inflammation, and asthma are three conditions that can occur after exposure to damp conditions and mold.

Asbestos

Asbestos is a long, thin, fibrous silicate mineral with insulating properties that can cause cancer when inhaled (**FIGURE 44.3a**). There are actually six different minerals that are classified as asbestos and all are harmful to the respiratory tract. It is found in metamorphic rocks and mined in many locations around the world. For many years it was used as an insulating material on steam and hot water pipes (Figure 44.3b) and in shingles for the siding of buildings. The greatest health risks from asbestos have been respiratory diseases such as asbestosis, a chronic lung condition, and mesothelioma, a form of lung cancer caused from exposure to asbestos. Both of these diseases occur at very high rates among workers who have mined asbestos. In manufactured form, asbestos is relatively stable and not dangerous until it is disturbed. When insulating materials become old or are damaged or disrupted, however, the fine fibers can become airborne within a building and can

Asbestos A long, thin, fibrous silicate mineral with insulating properties, which can cause cancer when inhaled.

(a)

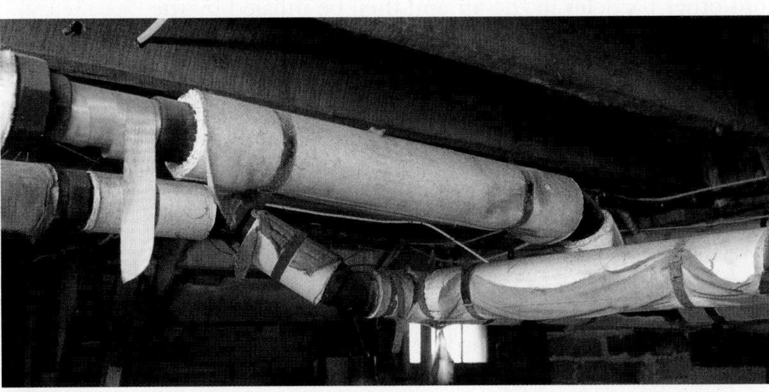

(b)

FIGURE 44.3 Asbestos insulation in a school Asbestos mineral (a) is a good insulator and used to be installed on steam pipes in buildings. (b) It still remains in some older buildings. *(a: Education Images/Getty Images; b: Joe_Potato/Getty Images)*

enter the respiratory tract. In the United States, asbestos is no longer used as an insulating material, but it can still be found in older buildings, including schools. Removal of asbestos must be done under tightly controlled conditions so that the fibers, typically less than 10 microns in diameter, cannot contaminate air inside the building. Some studies have shown that when asbestos removal is complete, the concentration of asbestos fibers in the air of the remediated building is greater in the year after removal than during the year before removal. For this reason, it is absolutely necessary that asbestos removal be carefully done by qualified asbestos abatement personnel.

AP® Exam Tip

The AP® Environmental Science Exam may ask about solutions to challenges such as asbestos in a residence. Make certain you describe "professional removal" or "professional abatement." Merely stating "remove the asbestos" suggests removing it on your own, and improper removal can be more dangerous for the occupants of the building and those doing the removal than leaving it in place.

Radon

Radon-222, a radioactive gas that occurs naturally from the decay of uranium, exists in granitic and some other rocks and soils in many parts of the world, and is an indoor air pollutant. The map in **FIGURE 44.4** shows areas for potential radon exposure in the United States. Humans can receive significant exposure to radon if it enters a building through cracks in the foundation from underlying rock or soil, or enters the house via groundwater that came from a well. Radon-222 decays within 4 days to a radioactive daughter product, polonium-210. Either the radon or the polonium can attach to dust and other particles in the air and then be inhaled by the inhabitants of the home. Exposure to radon can lead to lung cancer.

The EPA, the federal agency most responsible for identifying, measuring, and addressing environmental risks, estimates that about 21,000 people die each year from radon-induced lung cancer. This is 15 percent of yearly lung cancer deaths, and makes radon the second leading cause of lung cancer, after smoking. The EPA suggests that people test their homes for airborne radon. If radon air concentrations are determined to be high, it is important to increase ventilation in the home and ventilate the underground space below the house. These radon mitigation systems are highly successful at reducing indoor air radon concentrations. Other relatively inexpensive actions, such as sealing cracks in the basement, can be beneficial if radon is coming from underlying soil and bedrock. "Do the Math: Averaging Radon over Time" asks you to calculate radon concentrations in a home based on the readings of two radon collection devices.

Radon-222 A radioactive gas that occurs naturally from the decay of uranium and is an indoor air pollutant.

AP® Exam Tip

Remember that it is not advisable to write a phrase such as "air pollution causes cancer." Be specific: radon, particulate matter (specifically $PM_{2.5}$ from diesel exhaust), and airborne asbestos are associated with lung cancer. But many air pollutants are not directly associated with lung cancer.

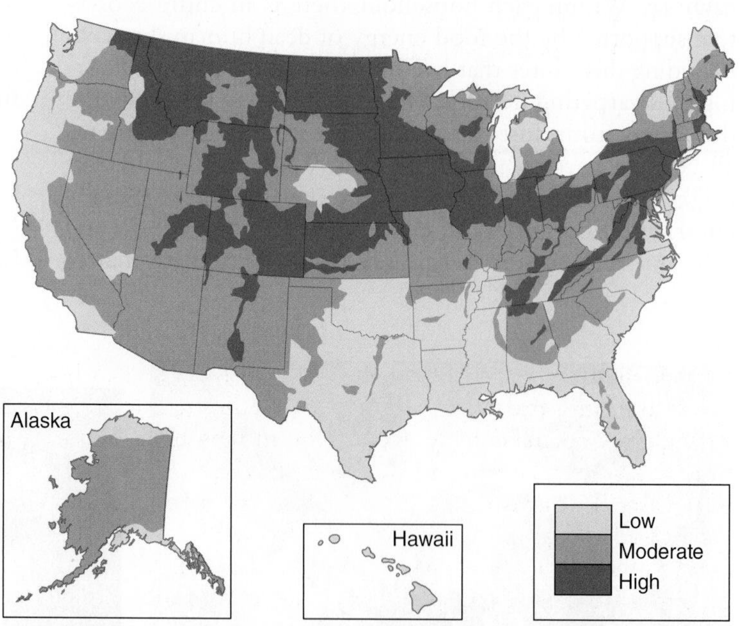

FIGURE 44.4 Potential radon exposure in the United States. Depending on the underlying bedrock and soils, the potential for exposure to radon exists in houses in certain parts of the United States. *(Data from U.S. Geological Survey and http://www.epa.gove/radon/zonemap.html)*

DO THE MATH Averaging Radon over Time

Preparing for the AP® Exam

Since radon in homes is believed to cause approximately 21,000 lung cancer deaths in the United States each year, homeowners are encouraged to measure radon concentrations in their homes at least once. There are short-term (e.g., 4 day or 8 day measurements) and long-term (up to 1 year) radon collection devices available. Longer measurement times are likely to produce more accurate readings. After the designated collection time period, the device is mailed to a laboratory and the air concentration of radon is estimated from the collection device. The EPA has established an action level for home radon: If concentrations of 4.0 pCi/L (picoCuries per liter) or greater are found, the homeowner should install abatement equipment to lower radon concentrations in the home.

For one brand of a short-term collection device, the homeowner should place two devices within 1 meter (3 feet) of each other and leave them undisturbed for 4 days. Another brand requires that the device be left undisturbed for 8 days. These devices are monitoring the same space at the same time for the same length of time. The concentration of radon in the home is determined by the average of the two devices.

As an example, imagine two devices with the following readings:

Device 1 reading after 8 days: 5.0 pCi/L

Device 2 reading after 4 days: 1.2 pCi/L

What is the concentration of radon in the home?

We will take the average of 5 pCi/L and 1.2 pCi/L, weighted by their respective measurement times by splitting the 8-day measurement to two 4-day measurements:

$$(5.0 \text{ pCi/L [for 4 days]} + 5.0 \text{ pCi/L [for 4 days]} + 1.2 \text{ pCi/L [for 4 days]}) \div 3 = 3.73 \text{ pCi/L}$$

There are two significant figures in the measurements, so we round the answer to 3.7 pCi/L.

3.7 pCi/L is less than 4.0 pCi/L so radon abatement is not recommended by the EPA.

YOUR TURN A homeowner uses two different short-term radon collection devices to measure the same space. One device measured the air for 4 days and the result was 1 pCi/L. The other device measured the air for 8 days and the result was 6 pCi/L. Calculate an estimate of the radon concentrations in the home, taking into consideration the different measurement periods.

Volatile Organic Compounds

Many volatile organic compounds are used in building materials, furniture, and other home products such as glues and paints. One of the most toxic of these compounds is formaldehyde, which is used widely to manufacture a variety of building products such as particle board and carpeting glue. Formaldehyde is common in new homes and new products made from pressed wood, such as certain bookcases. Bookcases made from actual wood and most hardwood flooring do not contain formaldehyde. The pungent smell that you may have noticed in a new home or one with new carpeting comes from formaldehyde, which is volatile and emits gases over time. A high enough concentration in a confined space can cause a burning sensation in the eyes and throat, and breathing difficulties and asthma in some people. There is evidence that people develop a sensitivity to formaldehyde over time; though they may not be very sensitive at first, with continued exposure they can experience irritation from increasingly smaller exposures. Formaldehyde has been shown to cause cancer in laboratory animals and has recently been suspected of being a human carcinogen.

Many other consumer products such as detergents, dry-cleaning fluids, deodorizers, and solvents may contain VOCs and can be harmful if inhaled. Plastics, fabrics, paints, construction materials, and synthetic carpets may also release VOCs over time. In response to concerns over manufactured products in the home, some homeowners choose to utilize wood flooring rather than carpeting, or natural fiber carpeting rather than synthetics. While such actions will not ensure zero exposure to VOCs, it will decrease the likelihood of exposure. VOC exposure can also come from outside if the building is in a neighborhood with a source of high VOC emissions such as a chemical plant or waste facility that releases VOCs into the air.

Lead

We have already discussed lead as an outdoor air pollutant in Module 42. It is also an indoor air pollutant. The greatest exposure to lead in buildings comes from leaded paint, which was last used in the United States in the 1960s. Certain older buildings have lead paint on the walls and if that paint is sanded or scraped, there may be some lead mobilized and dispersed into the air. This mobilized lead can be inhaled by human occupants. Lead from outside can be introduced to a building in air pollution or contaminated soil brought inside. However, the greatest potential lead exposure comes from lead in paint chips that can be ingested by young children, as we discussed in Module 42.

Other Pollutants

Other pollutants that occur outside such as nitrogen oxide and sulfur dioxide can also be present in indoor air pollution. In buildings that are not well insulated and tightly air sealed, which includes most older buildings, outdoor air pollution can be a source of indoor air pollutants such as nitrogen oxide and sulfur dioxide. Nitrogen oxide can also be present inside buildings as a result of combustion from cooking. But proximity to outdoor sources of air pollution is an important driver of indoor air pollution, not just for nitrogen and sulfur oxides but for many other pollutants as well.

44-2 How does indoor air pollution differ in the developing and developed world?

Indoor air pollution in the developing world comes mostly from indoor combustion while in the developed world tightly sealed buildings and VOCs are a bigger problem

Although it generally receives less attention than outdoor air pollution, indoor air pollution is a hazard all over the world. The World Health Organization estimates that indoor air pollution is responsible for approximately 3.8 million deaths annually worldwide. Pneumonia, heart disease, and pulmonary disease are the top three causes of death. Most of the deaths attributable to indoor air pollution occur in lower- and middle-income countries, though the reasons for indoor air pollution differ between the developing world and the developed world.

Sick building syndrome A buildup of toxic pollutants in weatherized spaces, such as newer buildings in the developed world.

Developing Countries

We have discussed the use of wood, animal manure, or coal indoors for heat and cooking in the developing world in Module 39 and earlier in this module. Because the source of these air pollutants is a combustion source from inside the home and usually in close proximity to the occupants, the World Health Organization estimates that indoor particulate matter concentrations can be 100 times higher than acceptable levels. Because many buildings in developing countries may not be well insulated and sealed from the outside, there may be some ventilation that occurs, allowing indoor air pollutants to escape. However, if outdoor air quality is also poor, ventilating with outside air might only improve the situation slightly.

Developed Countries

There are a number of factors that have caused the quality of air in homes in developed countries to take on greater importance in recent decades. First of all, people in much of the developed world have begun to spend more and more time indoors. Secondly, improved insulation and tightly sealed building envelopes have been constructed or retrofitted to reduce energy consumption. This is beneficial for energy savings due to the increased efficiency of the building but it can have adverse health effects as well. The tightly sealed buildings also keep existing air in contact with the inhabitants of homes, schools, and offices for greater amounts of time, exposing them for longer periods to any indoor air pollutants that may be present. Finally, an increasing number of materials in the home and office are made from volatile organic compounds including formaldehyde, as we discussed earlier in this module. In addition, plastics and other petroleum-based materials can give off chemical vapors harmful to our health. As **FIGURE 44.5** illustrates, all of these factors combine to allow many possible sources of indoor air pollution to impact occupants. In fact, a phenomenon has been observed often enough in new or renovated buildings for it to be given a name: **sick building syndrome**, which describes a buildup of toxic pollutants in weatherized spaces such as newer buildings in the developed world. Because new buildings contain many products made with synthetic

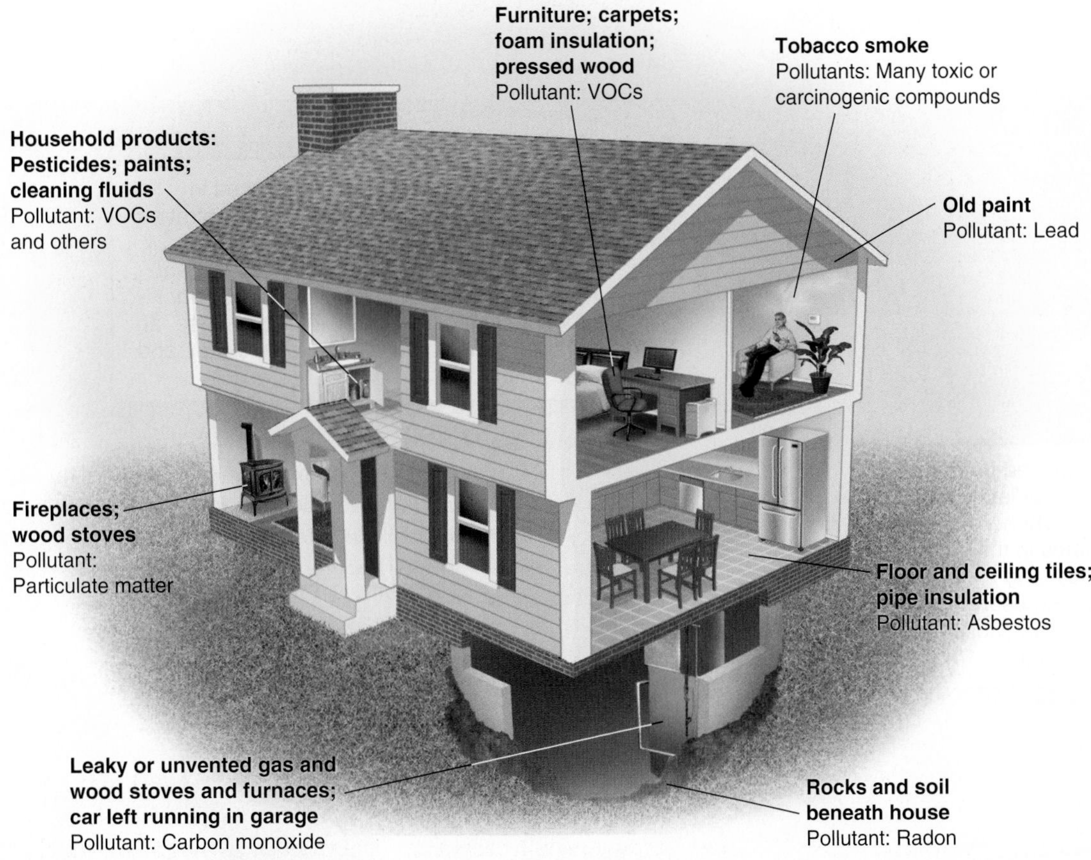

FIGURE 44.5 Some sources of indoor air pollution in the developed world. A typical home in the United States may contain a variety of chemical compounds that could, under certain circumstances, be considered indoor air pollutants.

materials and glues that may not have fully dried out, a significant amount of off-gassing occurs, which usually means that the indoor levels of VOCs, hydrocarbons, and other potentially toxic materials are quite high. Sick building syndrome has been observed particularly in office buildings, where large numbers of workers have reported a variety of maladies such as headaches, nausea, throat or eye irritations, and fatigue. The EPA has identified four specific reasons for sick building syndrome: inadequate or faulty ventilation; chemical contamination from indoor sources such as glues, carpeting, furniture, cleaning agents, and copy machines; chemical contamination in the building from outdoor sources such as vehicle exhaust transferred through building air intakes; and biological contamination from inside or outside, such as molds and pollen.

In this module we examined indoor air pollutants and observed that some, such as carbon monoxide and particulates, are the same pollutants that we considered in outdoor air pollution. Others, such as radon and asbestos, are specifically pollutants of concern indoors. Both radon and asbestos cause serious health problems including cancer. Sources of indoor air pollution include combustion within the building, and the release of gases and compounds from materials within the building; other indoor air pollutants occur after transfer of gases and compounds from outside into the building or structure. These can be both natural and anthropogenic compounds. Review the Unit 7 Visual Representation feature to continue your study of the specific human health effects of atmospheric pollutants. (pages 518-519).

Preparing for the AP® Exam

Visual Representation 7 Sources of Atmospheric Pollution According to the World Health Organization, 91 percent of the world's population lives in locations where atmospheric pollution exceeds

Outdoor Air Pollution

Nitrogen Oxides (NO_x) are produced in combustion processes. They can penetrate deep in the lung and cause respiratory problems, requiring emergency department visits and hospital admissions. Long-term exposure can cause chronic lung disease, an impaired sense of smell, and eye, throat, and nose irritation.

Lead (Pb) in the air is emitted from some engines, batteries, radiators, and waste incinerators. When inhaled, lead accumulates in the blood, brain, liver, bones, and cardiovascular system. Exposure in a fetus or young child can cause serious neurological damage.

Coal-fired power plants and industrial machinery emit NO_x and SO_2. Workers, nearby residents, and those who live downwind are all affected.

Volcanic eruptions and forest fires emit VOCs, NO_x, CO, SO_2, and PM. Smoke from fires can travel for hundreds of miles.

Combustion engines in gasoline and diesel-powered vehicles emit NO_x, SO_2, CO, VOCs, and Pb. Commuters are especially vulnerable to exposure.

Sulfur dioxide (SO_2) is emitted mainly from fossil fuel consumption and industrial activities. It penetrates deep into the lungs. Health consequences include respiratory irritation, bronchitis, mucus production, and breathing difficulties. Exposure also causes eye damage, damage to mucus membranes, and worsens pre-existing cardiovascular disease. Smoke is associated with an increase in deaths.

Ground level ozone (O_3) is generated in a chemical reaction between NO_x and VOCs from natural and anthropogenic sources. It is associated with skin cell damage, and increased deaths from respiratory illness and cardiovascular illness.

guidance limits. Atmospheric pollution can come from natural sources such as volcanoes or from anthropogenic sources. It can occur both outdoors and indoors. The effects of air pollution on human health can be severe and long-lasting.

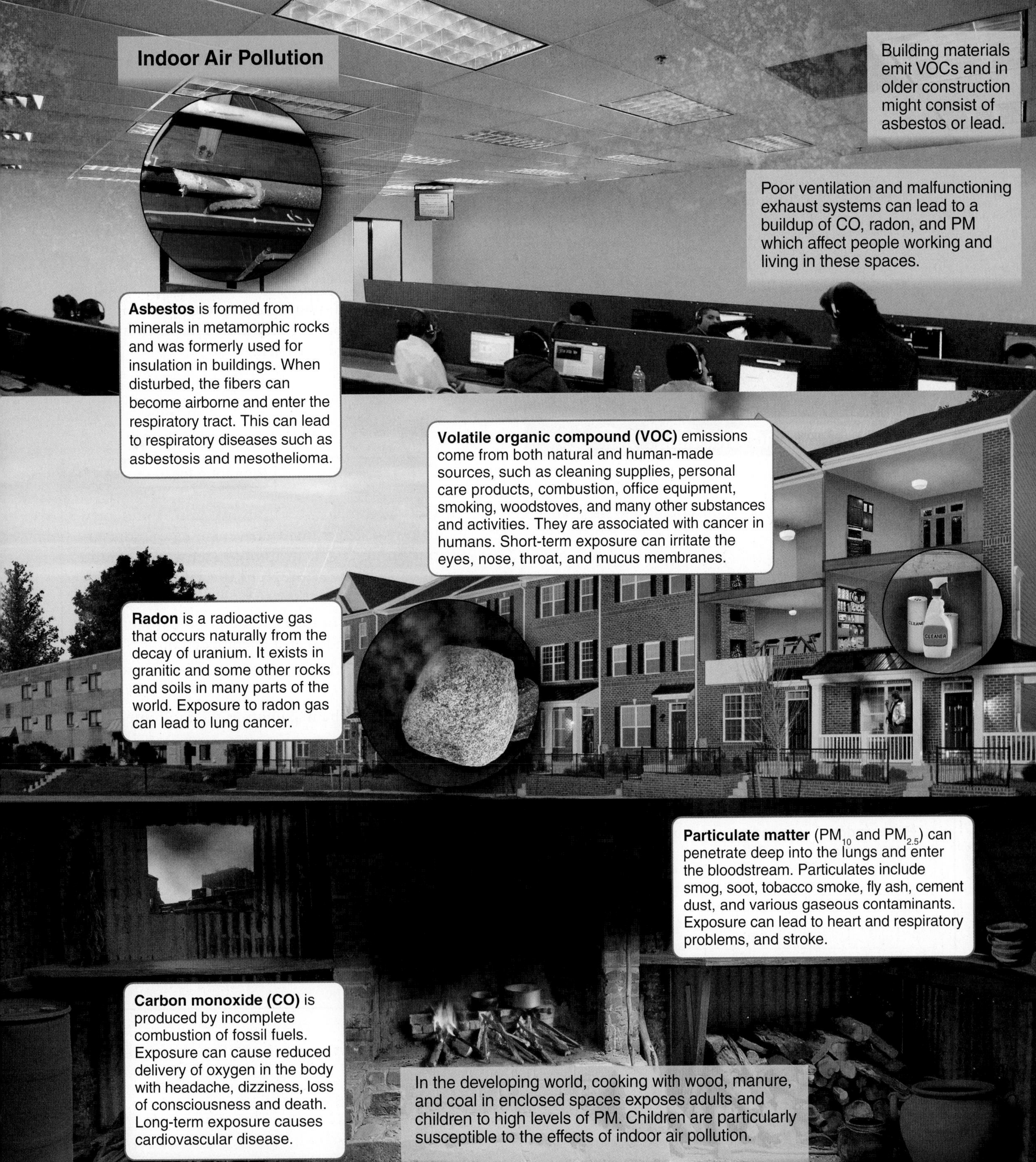

Indoor Air Pollution
Building materials emit VOCs and in older construction might consist of asbestos or lead.
Poor ventilation and malfunctioning exhaust systems can lead to a buildup of CO, radon, and PM which affect people working and living in these spaces.
Asbestos is formed from minerals in metamorphic rocks and was formerly used for insulation in buildings. When disturbed, the fibers can become airborne and enter the respiratory tract. This can lead to respiratory diseases such as asbestosis and mesothelioma.
Volatile organic compound (VOC) emissions come from both natural and human-made sources, such as cleaning supplies, personal care products, combustion, office equipment, smoking, woodstoves, and many other substances and activities. They are associated with cancer in humans. Short-term exposure can irritate the eyes, nose, throat, and mucus membranes.
Radon is a radioactive gas that occurs naturally from the decay of uranium. It exists in granitic and some other rocks and soils in many parts of the world. Exposure to radon gas can lead to lung cancer.
CLEANER
Particulate matter (PM_{10} and $PM_{2.5}$) can penetrate deep into the lungs and enter the bloodstream. Particulates include smog, soot, tobacco smoke, fly ash, cement dust, and various gaseous contaminants. Exposure can lead to heart and respiratory problems, and stroke.
Carbon monoxide (CO) is produced by incomplete combustion of fossil fuels. Exposure can cause reduced delivery of oxygen in the body with headache, dizziness, loss of consciousness and death. Long-term exposure causes cardiovascular disease.
In the developing world, cooking with wood, manure, and coal in enclosed spaces exposes adults and children to high levels of PM. Children are particularly susceptible to the effects of indoor air pollution.

Module 44 AP® Review

Preparing for the AP® Exam

Learning Goals Revisited

44-1 What are the major indoor air pollutants and where do they come from?

Indoor air pollutants are compounds that adversely affect the quality of air in buildings and structures. Indoor air pollutants can come from natural, human-made, and combustion sources. Carbon monoxide is perhaps the most dangerous indoor air pollutant because when concentrations are high enough, it can cause asphyxiation and death. Smoke from fire and from burning tobacco are sources of particulate matter in buildings. Dust contributes to a variety of conditions including asthma and skin and respiratory allergic reactions. The greatest health risks from asbestos have been respiratory diseases such as asbestosis, a chronic lung condition, and lung cancer. Humans can receive significant exposure to radon, a naturally occurring gas, in buildings if sources of radon below the building are present.

44-2 How does indoor air pollution differ in the developing and developed world?

Indoor air pollution is a problem around the world, but its presence in the developing and the developed worlds differ. In the developed world, houses sealed up for greater energy efficiency are often the causes of exposure to radon, carbon monoxide, and volatile organic compounds for people who spend substantial amounts of time indoors. In the developing world, combustion sources from inside the house are the largest problem.

Practice Math and Graphing

Preparing for the AP® Exam

Answer the following questions. Be sure to show all your work.

1. Practice Math

Radon in homes is believed to cause approximately 21,000 cancer deaths in the United States each year. The EPA has established an action level for home radon: If concentrations of 4.0 pCi/L (picoCuries per liter) or greater are found, the homeowner should install abatement equipment to lower radon concentrations.

(a) Use the information from five homes presented below to determine the average radon concentration and a maximum concentration found in each home. Complete the table with your calculations.

(b) For each row of the table, determine if the house should have radon abatement based on

(i) whether the average is the most important indicator.

(ii) whether the maximum is the most important indicator.

Home radon concentration	First reading pCi/L	Second reading pCi/L	Third reading pCi/L	Average	Max	Abatement yes or no if average most important	Abatement yes or no if maximum most important
1	3.2	3.5	3.9				
2	3.8	3.5	4.9				
3	4.5	6.5	12.7				
4	5.5	3.3	1.9				
5	9.6	4.4	8.3				

2. Practice Graphing

The table lists $PM_{2.5}$ air concentrations in Beijing, China, from 2011 through 2021. Plot the data on a graph with year on the x axis and $PM_{2.5}$ concentrations on the y axis. Then describe the trend of $PM_{2.5}$ over time.

Year	$PM_{2.5}$ air concentration $\mu g/m^3$
2011	99
2012	91
2013	100
2014	98
2015	83
2016	73
2017	59
2018	51
2019	43
2020	39
2021	41

AP® Practice Questions

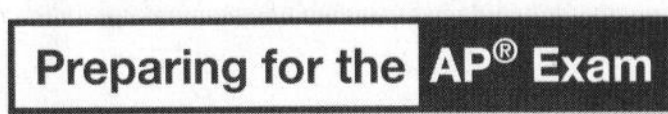

Multiple-Choice Questions

1. Carbon monoxide is an important indoor air pollutant because
   (a) it binds with hemoglobin causing asphyxiation.
   (b) it comes from complete combustion.
   (c) it is related to carbon dioxide.
   (d) it doesn't occur in outdoor air pollution.

2. New furniture and carpeting may release which indoor air pollutant?
   (a) sulfur dioxide
   (b) nitrogen oxides
   (c) carbon monoxide
   (d) formaldehyde

3. The primary source of radon is
   (a) electronics.
   (b) indoor fires.
   (c) household chemical fumes.
   (d) rocks and soils.

4. Which of the following would be a viable solution for indoor air pollution of radon?
   (a) reduction of asbestos in homes
   (b) use of a wet scrubber on residential chimneys
   (c) addition of an electrostatic precipitator
   (d) increasing ventilation

5. Asbestos
   (a) causes respiratory ailments and cancer.
   (b) can be easily removed and treated.
   (c) can be a problem in new construction.
   (d) causes skin irritation, nausea, and fatigue.

Use the graph below to answer questions 6 & 7:

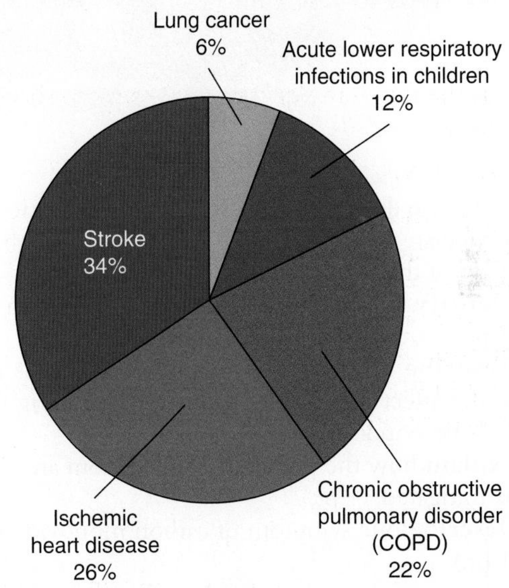

6. In 2020, indoor air pollution led to approximately 4 million deaths worldwide. The data above show the causes of death. How many people died of lung cancer due to indoor air pollution?
   (a) 240,000 (4 million × 0.06)
   (b) 840,000
   (c) 2,380,000
   (d) 1,820,000

7. What percent of indoor air pollution deaths were caused by direct effects on the lungs?
   (a) 60 percent
   (b) 40 percent
   (c) 18 percent
   (d) 48 percent

Free-Response Question

Indoor air pollutants come from many different sources. The diagram below shows some of these sources.

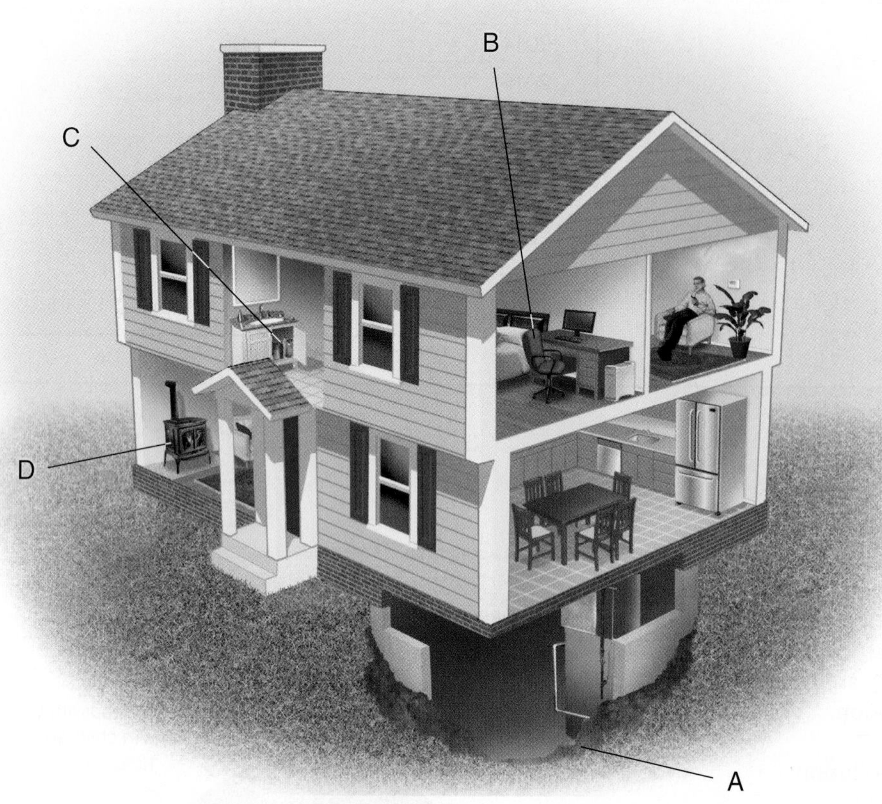

(a) Using the diagram above, **identify** the location where particulate matter comes from and **identify** the health effect of this indoor air pollutant. (2 pts.)

(b) **Identify** the pollutant likely coming from area A. (1 pt.)

(c) **Describe** how volatile organic compounds (VOCs) would affect the health of someone spending much of their time indoors. (1 pt.)

(d) **Explain** how the pollutant coming from area D could be mitigated. (1 pt.)

(e) **Describe** one symptom of carbon monoxide indoors. (1 pt.)

(f) Individuals in developed and developing nations tend to deal with different forms of indoor air pollution. **Identify** one indoor air pollutant that is more prevalent in a developing nation. **Explain** why this is the case. (2 pts.)

(g) Individuals in developed nations tend to set up their homes for greater energy efficiency. **Make a claim** to support or refute how this affects indoor air pollution, and **justify** the claim using evidence. (2 pts.)

Module 45

Unit 7	42	43	44	45	46

Reduction of Air Pollutants

The previous modules in this unit have identified the sources of both outdoor and indoor pollutants. Here, we'll examine methods of reducing outdoor air pollution. For air pollution reduction, the most sustainable solutions utilize conservation and efficiency first. It requires less energy and resources to clean up pollution if we seek ways to avoid creating it in the first place. Preventing pollution is usually much less expensive and energy intensive then controlling it. Unfortunately, pollution prevention is not always possible. In this module we will examine a variety of methods for reducing and controlling outdoor air pollution.

Learning Goals

After reading this module you should be able to

45-1 explain how air pollutants can be reduced.

45-2 describe the various air pollution control devices and how they are used.

45-1 How can air pollutants be reduced?

Air pollution can be controlled by prevention, conservation, fuel switching, and regulatory practices

As with other types of pollution, the best way to decrease air pollution emissions is to avoid them in the first place. Pollution prevention can be achieved in a number of ways and the first one is by using less fuel. So, any technique that utilizes conservation and increasing efficiency, topics we discussed in Module 41, will result in less air pollution. Switching to an alternative fuel is also a method for reducing air pollution. In general, oil is cleaner than coal. The sulfur concentrations of both coal and oil vary with type of fuel and location in the world. In addition, during refining and processing, the concentration of sulfur can be lowered in both fuels.

Choosing an alternative fuel can also involve switching from bituminous coal to anthracite coal, or from coal from one region to another, or from one type of oil to another type of oil.

Use of a low-sulfur coal or oil is certainly one of the best means of controlling sulfur dioxide pollution, although typically low-sulfur coal or oil is more expensive to purchase than coal or oil containing higher sulfur concentrations. While these measures reduce emissions by a certain amount, wherever fuel is combusted, pollutants such as nitrogen oxides, carbon monoxide, and carbon dioxide will be emitted. Depending on the fuel, particulates and sulfur dioxide may be emitted as well. So ultimately, no matter what measures are implemented to reduce air pollution, there will ultimately be a need to control the exhaust stream after combustion has occurred. That is the focus of the next section of this module. But first, let's consider regulatory practices that can be implemented.

Regulatory Practices

Regulatory practices at the city, state, and national levels have reduced air pollution. Municipalities and counties have passed measures, for example, to reduce the amount of gasoline spilled at gasoline stations by requiring a **vapor recovery nozzle** on all gasoline pumps at filling stations. This device prevents VOCs from escaping into the atmosphere while a person is fueling their vehicle (**FIGURE 45.1**). Other measures include regulations to restrict the evaporation of dry-cleaning fluids (a VOC) or restrict the use of lighter fluid (a VOC) for starting charcoal barbecues. All of these regulations have resulted in less VOC emissions into the atmosphere and less photochemical smog formation. Both urban and suburban areas around the United States have taken additional actions such as calling for a reduction in the use of wood-burning stoves or fireplaces that would reduce emissions of not only nitrogen oxide but also particulate matter, VOCs, and carbon monoxide. A number of California municipalities have even discussed reducing the number of bakeries within certain areas, as the emissions from rising bread contain VOCs. This proposal was not very

Vapor recovery nozzle A device that prevents VOCs from escaping into the atmosphere while a person is fueling their vehicle.

FIGURE 45.1 Vapor recovery nozzle at a gasoline filling station. This control device on the nozzle of a gasoline dispenser reduces VOC fumes that escape to the atmosphere at a gasoline fueling station. *(Radharc Images/Alamy Stock Photo)*

(a)

(b)

FIGURE 45.2 Reducing photochemical smog. The photos show the same locations in Beijing, China, (a) during the Beijing Summer Olympics in 2008 and (b) after the pollution restrictions were removed. *(PETER PARKS/Getty Images)*

popular, as you can imagine, but emissions from bakeries along with many other businesses are sometimes regulated by local air-quality ordinances.

Since cars are responsible for large emissions of nitrogen oxides and VOCs in urban areas, and these two compounds are the major contributors to smog formation, some municipalities have tried to achieve lower smog concentrations by restricting automobile use. A number of cities, including Mexico City, have instituted plans permitting automobiles to be driven only every other day—for example, those with license plates ending in odd numbers may be used on one day and those with even-numbered license plates on alternate days. In Mexico City, as of 2020, the plan had not had much of an impact. In China, during the 2008 Beijing Summer Olympics, the government successfully expanded public transportation networks, imposed motor vehicle restrictions, and temporarily shut down a number of industries as a way to reduce photochemical smog and improve visibility. For a short period, the air pollution control measures were successful. However, long-term improvements in air quality in China have been harder to achieve. Just prior to the 2022 Beijing winter Olympics, levels of smog and particulate matter reached unhealthy levels (**FIGURE 45.2**).

In 1990 and again in 1995, scientists, policy makers, and academics collaborated on amendments to the Clean Air Act that would allow the free market to determine the least expensive ways to reduce emissions of sulfur dioxide. The free-market program was implemented in two phases between 1995 and 2000, and each phase has led to significant reductions in sulfur emissions. One of the most innovative aspects of the Clean Air Act amendments is "Cap and Trade," which we discuss in "Science Applied 7: Can Cap-and-Trade Reduce Pollution Emissions Equitably?" This is a program that institutes the buying and selling of allowances that authorize the owner of power plants and factories to release a certain quantity of sulfur dioxide (SO_2). Each allowance authorizes a power plant or industrial source to emit one ton of SO_2 during a given year. Sulfur allowances are awarded annually to existing sulfur emitters proportional to the amounts of sulfur they were emitting before 1990, and the emitters are not allowed to emit more sulfur than the amount for which they have permits. At the end of a given year, the emitter must possess a number of allowances at least equal to its annual emissions. In other words, a facility that emits 1,000 tons of SO_2 must possess at least 1,000 allowances that are usable in that year. Facilities that emit quantities of SO_2 above their allowances must pay a financial penalty.

Sulfur allowances can be bought and sold on the open market by anyone. If emitters wanted to exceed their allowance level—say, because they intended to increase their industrial output—they would be required to purchase more allowances from another source. If, on the other hand, a company decreased its sulfur emissions more than it needed to in order to comply with its allowance amount, it could sell any unused sulfur emission allowances. Over time, the number of allowances distributed each year were gradually reduced so, ultimately, sulfur dioxide emissions have gone down considerably. "Do the Math: Calculating Annual Sulfur Reductions" shows you how to calculate these decreases as percentages of the original value.

DO THE MATH Calculating Annual Sulfur Reductions

Preparing for the AP® Exam

SO_2 emissions from all sources in the United States declined from 23.5 million metric tons (26 million U.S. tons) in 1982 to 10.3 million metric tons (11.4 million U.S. tons) in 2008. Calculate the total percentage reduction and the annual percentage reduction of SO_2 emissions.

23.5 million metric tons – 10.3 million metric tons = 13.2 million metric tons total reduction

Divide the reduction by the original amount and multiply by 100 to obtain a percent reduction:

13.2 million ÷ 23.5 million metric tons × 100% = 56%

The total reduction was 56 percent between 1982 and 2008.

To calculate the reduction per year, divide 56 percent by the number of years from beginning to end:

2008 – 1982 = 26 years

56% ÷ 26 years = 2.2%/year

YOUR TURN In the United States, SO_2 emissions continued to decrease from 10.3 million metric tons (11.4 million U.S. tons) in 2008 to 1.8 million metric tons (2 million U.S. tons) in 2020. Calculate the total percentage reduction and the annual percentage reduction for SO_2 from 2008 to 2020.

Even with regulations, there is still going to be pollution emissions. Therefore, air pollution control devices are necessary to further reduce air pollution.

45-2 What are the various air pollution control devices and how are they used?

Control devices convert pollutants to less harmful compounds or remove them from the exhaust stream

Once combustion has occurred, pollution control devices function in two basic ways: by converting pollutant compounds to other, less harmful compounds, or by removing the pollutant compounds from the waste stream. Let's examine the first method.

Converting Pollutants to Less Harmful Compounds

One of the most common air pollution control devices is the **catalytic converter**, a device that uses chemicals to convert pollutants such as nitrogen oxide and carbon monoxide to nitrogen gas and carbon dioxide. They have been in the news in recent years mostly because they are expensive and relatively easy for thieves to steal by crawling under an unattended car or truck and cutting the converter out. An internal combustion engine emits combustion products, including pollutants, into the exhaust system that is mounted underneath the chassis of the car. Until 1975, the exhaust system consisted of a pipe with a muffler, which is designed to reduce the noise of the engine. Beginning in 1975, as part of the Clean Air Act, a catalytic converter was required to be added before the muffler in all new cars. It contains a metal such as platinum that serves as a catalyst; it increases the rate of a chemical reaction without itself undergoing a change (**FIGURE 45.3**). The catalytic

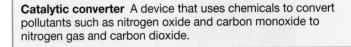

Catalytic converter A device that uses chemicals to convert pollutants such as nitrogen oxide and carbon monoxide to nitrogen gas and carbon dioxide.

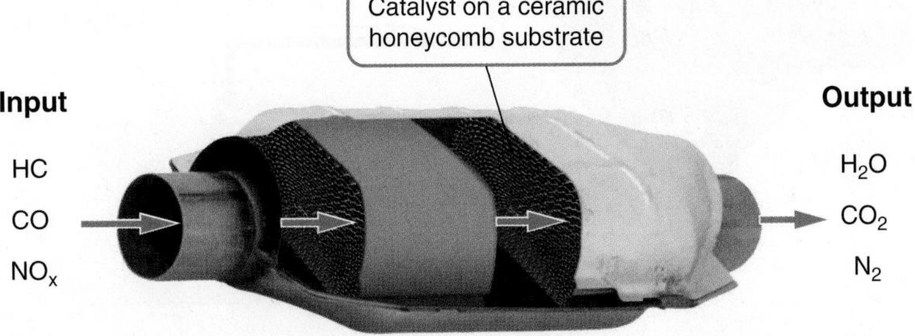

FIGURE 45.3 Catalytic converter. This honeycomb device coated with a metal such as platinum, treats pollutant gases from the engine such as nitrogen oxides, carbon monoxide, and hydrocarbons and converts them into nitrogen gas, carbon dioxide, and water. Because the device can be damaged by the metal lead, Clean Air Act legislation that required the catalytic converter also mandated that lead no longer be added to gasoline.

converter in cars and trucks removes oxygen from nitrogen oxides resulting in the formation of nitrogen gas and oxygen gas. It also adds oxygen to carbon monoxide, resulting in the formation of carbon dioxide. Catalytic converters also treat hydrocarbons. The net result of all three reactions is a reduction in nitrogen oxides, carbon monoxide, and hydrocarbons and an increase in nitrogen gas, carbon dioxide, oxygen, and water. This is an example of converting pollutants to less harmful compounds. Catalytic converters are effective. In fact, nitrogen oxide emissions from automobiles have decreased significantly in the United States over the last 45 years. Figure 42.4 shows the decrease from 1990 to 2020. During this same time period, improvements in the combustion technology of power plants and factories also contributed to reduced emissions of nitrogen oxides.

Scrubber A device that uses a combination of lime and or water to separate and remove particles from industrial exhaust streams.

Removing Pollutants from Waste Streams

A substantial number of air pollution control measures have been directed toward removal of particulates, sulfur oxides, and nitrogen oxides in stationary sources of air pollution such as coal-burning power plants and factories. A **scrubber** is an air pollution control device that uses a combination of air and lime (dry scrubber) or air and water (wet scrubber) to separate and remove particles and sulfur dioxide from industrial exhaust streams (**FIGURE 45.4**). Particles and gases coalesce in the scrubber in a liquid or sludge form and relatively cleaner gas exits through the smokestack. Scrubbers are used on industrial plants and coal-burning power plants. In a scrubbing process known as fluidized bed combustion, granulated coal is burned in close proximity to calcium carbonate. The heated calcium carbonate absorbs sulfur dioxide and produces calcium sulfate, which is removed from the exhaust stream and can be used in the production of gypsum wallboard, also known as sheetrock, for houses.

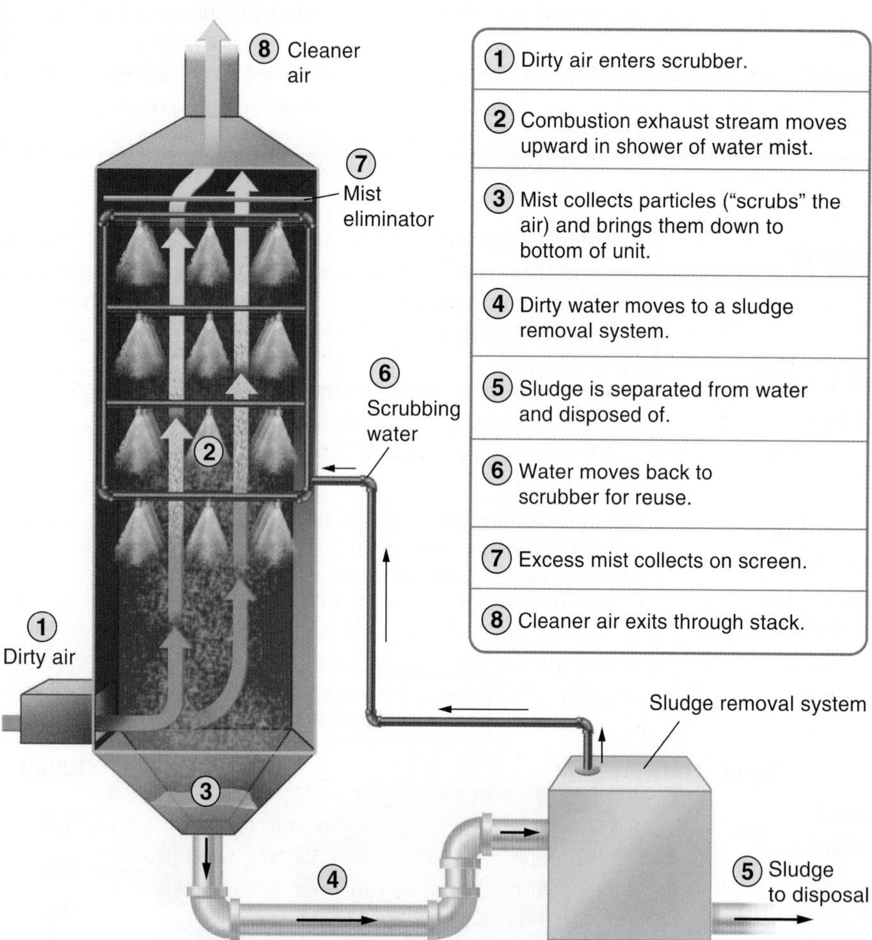

8 Cleaner air
7 Mist eliminator
6 Scrubbing water
2
1 Dirty air
3
4
Sludge removal system
5 Sludge to disposal
1 Dirty air enters scrubber.
2 Combustion exhaust stream moves upward in shower of water mist.
3 Mist collects particles ("scrubs" the air) and brings them down to bottom of unit.
4 Dirty water moves to a sludge removal system.
5 Sludge is separated from water and disposed of.
6 Water moves back to scrubber for reuse.
7 Excess mist collects on screen.
8 Cleaner air exits through stack.

FIGURE 45.4 The scrubber, wet variety. In this air pollution control device, particles and sulfur dioxide are "scrubbed" from the exhaust stream by water droplets and air. A water-particle "sludge" is collected and processed for disposal. Another variation, the dry scrubber, uses compounds such as lime and air to remove particles.

There are a variety of methods used to remove particulate matter. Sometimes the process of removing particulate matter incidentally also removes sulfur. The simplest method is gravitational settling, which relies on gravity as the exhaust travels through the smokestack. The particles simply settle out to the bottom. The ash residue that accumulates must be disposed of in a landfill. Depending on the fuel that was burned, the ash may contain sufficiently high concentrations of metals that require special disposal.

Fabric filters are a type of filtration device that allow gases to pass through them but remove particulate matter. Often called baghouse filters, certain fabric filters can remove almost 100 percent of the particulate matter emissions. **Electrostatic precipitators** remove particulate matter by using an electrical charge to make particles coalesce so they can be removed from the exhaust stream (**FIGURE 45.5**). Polluted air enters the precipitator and the electrically charged particles within are attracted to negative or positive charges on the sides of the precipitator. The particles collect on the collection electrodes and relatively clean gas exits the precipitator. Electrostatic precipitators are common in industrial plants and coal-burning power plants.

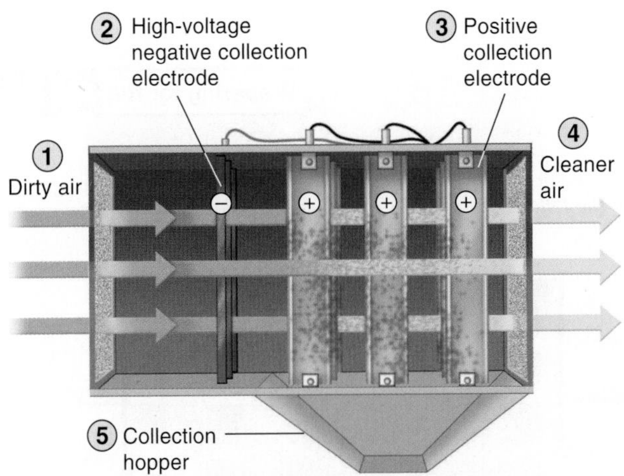

(1) Dirty air enters precipitator unit.

(2) Particles in combustion exhaust stream pass by negatively charged plates, which gives them a negative charge.

(3) The negatively charged particles are attracted to positively charged collection plates.

(4) Cleaner air moves out of the unit.

(5) The positive collection plates are periodically discharged, which causes the particles to fall off so that they can be removed from the system.

FIGURE 45.5 The electrostatic precipitator. In this air pollution control device, particles are given a negative charge. This causes them to be attracted to a positively charged plate, where they are held. Periodically, the particles are removed from the plate and collected for disposal.

Nitrogen can also be removed from combustion exhaust in stationary sources. As you know, the atmosphere of Earth is 78 percent nitrogen gas and, as a result, nitrogen oxides are produced in virtually all combustion processes. Hotter burning conditions and the presence of oxygen allow proportionally more nitrogen oxide to be generated per unit of fuel burned. In order to reduce nitrogen oxide emissions, burn temperatures must be reduced and the amount of oxygen must be controlled—a procedure that is sometimes utilized in factories and power plants by means of certain air pollution control technologies. However, lowering temperatures and oxygen supply can result in less-complete combustion, which reduces the efficiency of the process and increases the number of particulates and carbon monoxide. Finding the exact mix of air, temperature, oxygen, and other factors is a challenge being met by the introduction of computer-based controls that regulate these factors.

All pollution control devices, because they use additional energy and increase resistance to air flow in the factory or power plant, require the use of more fuel and result in a slight increase in carbon dioxide emissions. And it is much harder—if not impossible—to remove pollutants from the environment after they have been dispersed over a wide area. That is why it is preferable to conserve, increase efficiency, and switch to fuels that contain fewer contaminants.

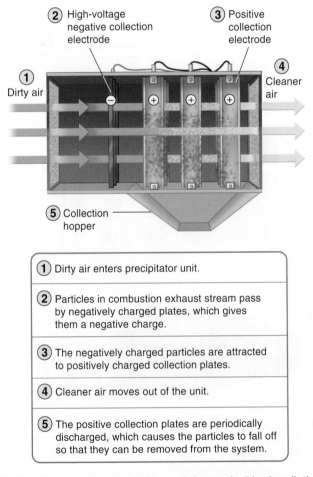

AP® Exam Tip

When describing solutions to reducing emissions from combustion sources, don't use the word "filter." Describe a specific technology such as catalytic converters for cars or scrubbers and electrostatic precipitators on smokestacks. Be familiar with what these technologies remove.

In this module we have seen that the best approach for decreasing air pollution emissions is by avoiding them in the first place. This can be done by conservation, efficiency, and switching fuels. It can also be achieved by regulatory practices. When those methods have been utilized, additional air pollution reduction can be achieved by treating the pollutants immediately after combustion. Air pollution control devices such as the catalytic convertor, scrubber, and electrostatic precipitator use a variety of methods to remove sulfur dioxide and particulates from the waste stream.

Electrostatic precipitator A device that removes particulate matter by using an electrical charge to make particles coalesce so they can be removed from the exhaust stream.

Module 45 AP® Review

Preparing for the AP® Exam

Learning Goals Revisited

45-1 How can air pollutants be reduced?

Air pollution is best treated by preventing it or reducing it before combustion occurs. This can be achieved by conservation, increasing efficiency, and choosing fuels that contain fewer impurities. Regulatory practices can reduce air pollution by requiring implementation of devices, such as the vapor recovery nozzle at gasoline filling stations or restricting vehicle use on certain days for certain drivers.

45-2 What are the various air pollution control devices and how are they used?

Once combustion occurs, pollution control devices are used to prevent pollutants from entering the atmosphere. The catalytic converter in cars and trucks removes oxygen from nitrogen oxides resulting in the formation of nitrogen gas and oxygen gas. It also adds oxygen to carbon monoxide, resulting in the formation of carbon dioxide. The scrubber literally scrubs the exhaust stream, creating a sludge that can be collected and removed. The electrostatic precipitator uses an electrical charge to coalesce particles so they can be removed.

AP® Practice Questions

Preparing for the AP® Exam

Multiple-Choice Questions

1. Which method is used to reduce nitrogen oxide emissions?
   (a) installation of catalytic converters
   (b) increased temperature of combustion
   (c) the addition of oxygen to combustion processes
   (d) the use of fluidized bed combustion

2. Which is used to prevent the emission of particulate matter?
   (a) automotive mufflers
   (b) increased ventilation
   (c) electrostatic precipitators
   (d) catalytic converters

3. A vapor recovery nozzle is commonly found at
   (a) coal-burning power plants.
   (b) factories.
   (c) gasoline stations.
   (d) natural-gas power plants.

4. If carbon monoxide emissions decreased from 145 million tons annually to 80 million tons annually, by what percentage have emissions been reduced?
   (a) 41 percent
   (b) 45 percent
   (c) 47 percent
   (d) 52 percent

5. The main purpose of an electrostatic precipitator is to remove
   (a) carbon monoxide.
   (b) nitrogen oxide.
   (c) particulates.
   (d) sulfur dioxide.

Use the following diagrams to answer questions 6 & 7:

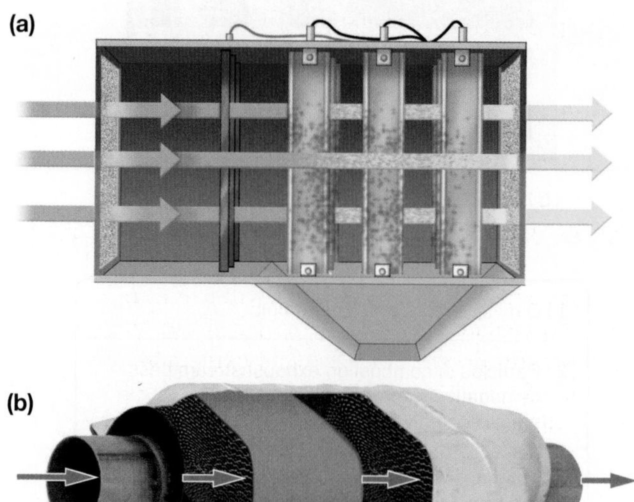

6. Figure A represents an electrostatic precipitator and Figure B represents a catalytic converter. Which of the following is the product of a catalytic converter?
   (a) carbon monoxide
   (b) nitrogen
   (c) particulate matter
   (d) nitrogen oxides

7. Which statement is true about the two figures?
   (a) Figure A is found on vehicles to alter pollutants before they reach the atmosphere.
   (b) Figure B removes particulate matter from the air.
   (c) Figure A uses electrodes to remove particulates.
   (d) Figure B is placed on a smokestack of coal-burning factories.

Free-Response Question

Industrial processes and fossil fuel-burning tend to lead to different air pollutants being released into the atmosphere. One of these pollutants is sulfur dioxide (SO_2).

(a) **Identify** a method or device that removes air pollution from the exhaust stream of an industrial process. (1 pt.)

(b) **Identify** a piece of legislation that regulates emissions of sulfur dioxide. (1 pt.)

(c) **Describe** one by-product of removing air pollution through the method identified in part (a). (1 pt.)

(d) **Explain** one reason why air pollution removal strategies are not typically used unless there is legislation requiring them. (1 pt.)

(e) A smokestack emits 31 micrograms of particulate matter each hour. After a wet scrubber is installed this number drops to 2 micrograms of particulate matter each hour.

   (i) **Calculate** how much particulate matter is released in 24 hours before the scrubber is added. Show all work. (2 pts.)

   (ii) **Calculate** how much particulate matter is released in 24 hours after the scrubber is added. Show all work. (2 pts.)

   (iii) **Calculate** the percent decrease in particulate matter released in the two 24 hour periods. Show all work. (2 pts.)

Module 46

Unit 7 | 42 | 43 | 44 | 45 | **46**

Acid Rain and Noise Pollution

As we saw in the last module, there are a variety of methods for controlling air pollution and many methods have been successful. In this module we will examine acid rain, also known as acidic deposition, which is caused by the air pollutants sulfur dioxide and nitrogen oxide, and describe its causes and effects. We will also describe noise pollution and its effects on humans and marine life.

Learning Goals

After reading this module you should be able to

46-1 describe acid rain and its effects.

46-2 explain the sources and effects of noise pollution.

46-1 What is acid rain and what are its effects?

Acid rain is acidified precipitation derived from air pollutants, which adversely affects terrestrial and aquatic systems and buildings

You may recall that acidified rainfall and snow, or acid rain, was first defined in Module 20 as rain or snow containing sulfuric and nitric acid. In order to discuss acid rain further, we must first examine the pH scale. The **pH** indicates the relative strength of acids and bases in a substance. A water molecule, H_2O, is made up of two hydrogens (2 H+) and one oxygen (O). Many substances dissolve easily into water, which can be represented in solution as one hydrogen ion (H+) and one hydroxide ion (OH–). A pH value of 7—the pH of pure water—is neutral, meaning that the number of hydrogen ions is equal to the number of hydroxide ions.

pH The relative strength of acids and bases in a substance. It is a logarithmic scale, meaning that each number on the scale represents a change by a factor of 10.

Anything above 7 is basic, or alkaline, and anything below 7 is acidic. Another way to define acids and bases is to state that an **acid** is a substance that contributes hydrogen ions to a solution. A **base** is a substance that contributes hydroxide ions to a solution. Both acids and bases dissolve in water. The pH scale tells us that the lower the number, the stronger the acid, and the higher the number, the stronger the base. The pH scale is logarithmic, meaning that each number on the scale represents a change by a factor of 10. For example, a substance with a pH of 5 has 10 times the acidity or 10 times the hydrogen ion concentration of a substance with a pH of 6—or in simpler terms, it is 10 times more acidic. Water in equilibrium with Earth's atmosphere typically has a pH of 5.65 because carbon dioxide from the atmosphere dissolves in it, which makes it weakly acidic. **FIGURE 46.1** lists the pH of many familiar substances on the pH scale, which ranges from 0 to 14.

As you can see from Figure 46.1, normal rainwater is naturally somewhat acidic; the reaction between water and atmospheric carbon dioxide lowers the pH of precipitation on the pH scale from a neutral 7.0 to 5.6. We'll explain this idea in further detail when we discuss ocean acidification in Module 58. What's important to know now, is that the increase in carbon dioxide in the atmosphere has led to more carbon dioxide in ocean water, which has led to acidification of ocean water. When an acid is dissolved in water, it dissociates or breaks apart into positively charged hydrogen ions (H^+) and negatively charged ions. When air pollution mixes with rainwater, it creates additional acidity. The two acids that comprise **acid rain**, also known as **acid deposition**, are nitric acid (HNO_3) and sulfuric acid (H_2SO_4), both of which dissociate to contribute hydrogen ions to water and soil. Acid deposition is largely the result of human activity, although natural processes, such as volcanoes, may also contribute to its formation. Let's examine how acid deposition forms and how it travels in the atmosphere.

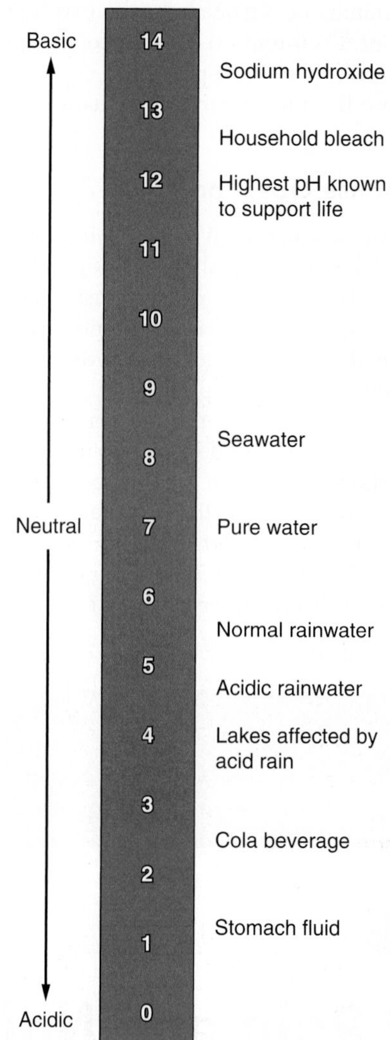

FIGURE 46.1 Acid rain and the pH scale. The pH scale indicates how acidic or how basic a solution is. The pH scale is logarithmic, meaning that each number on the scale represents a change by a factor of 10. Acid rain is more acidic than normal rainwater.

AP® Exam Tip

You are expected to understand pH on the AP® Environmental Science Exam and to know the pH of water — rainwater, acid rain, and ocean water — as shown in Figure 46.1. You should also be able to recognize some common acids and bases.

Acid A substance that contributes hydrogen ions to a solution.

Base A substance that contributes hydroxide ions to a solution.

Acid rain Precipitation high in sulfuric acid and nitric acid. *Also known as* **acid deposition**.

How Acid Deposition Forms and Travels

FIGURE 46.2 shows how acid deposition forms. Nitrogen oxides (NO and NO_2) and sulfur oxides (primarily SO_2) are released into the atmosphere by natural and anthropogenic combustion processes. Volcanoes emit sulfur oxides and nitrogen oxides and are a natural source of acid rain. Automobile and truck exhaust from gasoline and diesel fuels as well as stationary combustion of coal and oil are important human sources of nitrogen oxides and sulfur dioxide. Through a series of reactions with atmospheric oxygen and water, these primary pollutants are transformed into the secondary pollutants nitric acid (HNO_3) and sulfuric acid (H_2SO_4). Sulfuric acid breaks down into 2 H^+ and SO_4^{2-} (sulfate) and nitric acid breaks down into H^+ and NO_3^- (nitrate). The sulfate and nitrate are inorganic

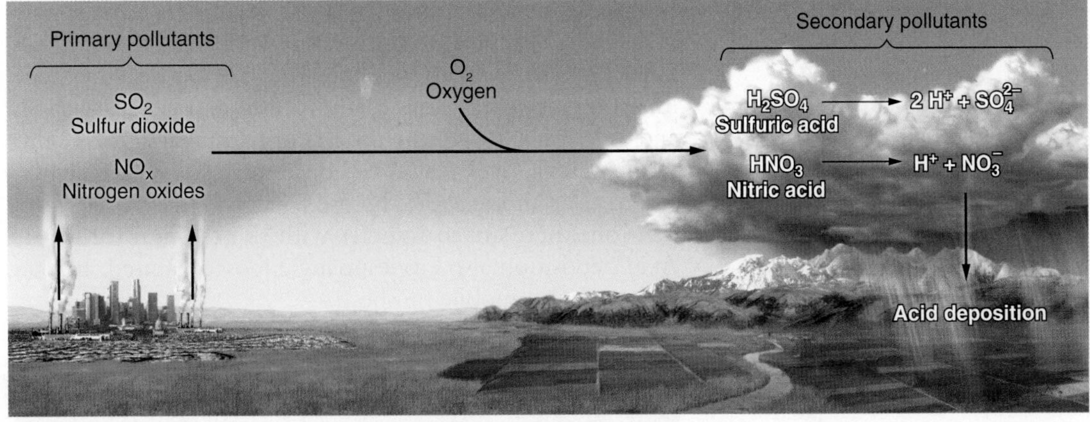

FIGURE 46.2 Formation of acid deposition. The primary pollutants sulfur dioxide and nitrogen oxides are precursors to acid deposition. After transformation to the secondary pollutants — sulfur acid and nitric acid — dissociation occurs in the presence of water. The resulting ions — hydrogen sulfate and nitrate — cause the adverse ecosystem effects of acid deposition.

compounds and the hydrogen ions (H^+) generate the acidity in acid deposition. These transformations occur over several days, and during this time, the pollutants may travel a thousand kilometers (600 miles) or more. Eventually, these secondary acidifying pollutants are washed out of the air and deposited either as precipitation or in dry form on vegetation, soil, or water.

Acid deposition has been reduced in the United States as a result of lower sulfur dioxide and nitrogen oxide emissions, as we saw in Figure 42.4. As we discussed in Module 42, much of this improvement is a result of the Clean Air Act Amendments that were passed in 1990 and implemented in 1990 and 1995. Prior to the implementation of the Clean Air Act Amendments, acid rain from midwestern states was carried by the prevailing westerlies to New York and New England where it was deposited in lakes and forests.

In addition to locations in the United States, studies have documented regional acid deposition in West Africa, South America, Japan, China, and many areas in eastern and central Europe. Acid deposition crosses international borders between the United States and Canada and is carried from England, Germany, and the Netherlands to Scandinavia. Because of this mobility, the precursors to acid deposition emitted in one region may have a significant impact on another region or another country. For example, over the years, there have been legislative and legal attempts to restrict emissions from coal-burning power plants in the midwestern United States that fall as acid deposition in Canada. Many of the regions that once received high amounts of acid deposition are receiving less deposition today. However, recently observed acid deposition documented along the West Coast of the United States is believed to be the result of coal combustion in China; sulfur and nitrogen oxides are released in China and elsewhere in Asia and are carried by the prevailing westerlies from one continent to another, across the Pacific Ocean. When they reach the western United States, they are deposited on the ground in rain and snow and intercepted by vegetation in California, Oregon, and Washington State.

Effects of Acid Deposition on Terrestrial and Aquatic Ecosystems

Acid deposition in the United States increased substantially from the 1940s through the 1990s due to the combustion of coal and oil and other industrial activities. It leads to acidification and degradation of building materials and automobiles, on agricultural lands, and on both aquatic and terrestrial natural habitats. News reports from the United States and Europe in the 1980s and 1990s contained frequent reports about adverse effects of acid deposition on forests, lakes, and streams.

Effects of acid deposition may be direct, such as a decrease in the pH of lake water, or indirect, such as when acid deposition acidifies a soil (direct) and mobilizes a toxic metal (indirect) that adversely affects fish. When soils are high in calcium and magnesium, usually because of limestone bedrock underlying the soil, acid rain can be partially neutralized or "buffered." This will reduce the impact of acid rain on plants growing in that soil, and on aquatic systems that are adjacent to the limestone bedrock. Regional differences in vegetation and bedrock can mitigate acidification but they can also make the situation worse.

In locations without underlying limestone bedrock, the lower pH from acid deposition can lead to mobilization of metals, an indirect effect. When this happens, metals bound in organic or inorganic compounds in soils and sediments are released into surface water. Because metals such as aluminum and mercury can impair the physiological functioning of aquatic organisms, exposure can lead to species loss. Decreased pH can also affect the food sources of aquatic organisms, creating indirect effects at several trophic levels. On land, at least one species of tree, the red spruce (*Picea rubens*), at high elevations of the northeastern United States, was shown to have been harmed by acid deposition. It is likely that these trees

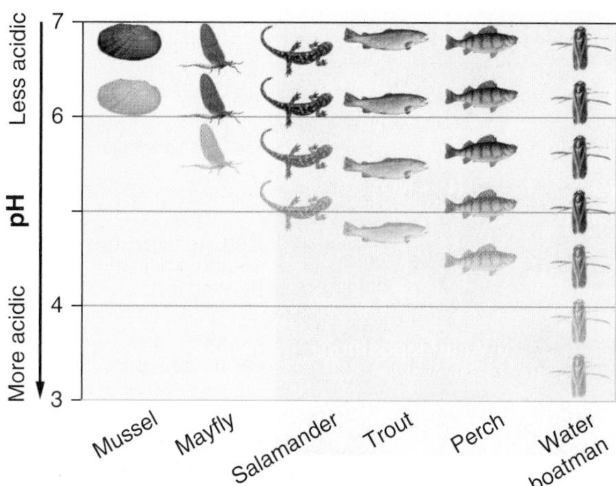

FIGURE 46.3 Aquatic organisms and the pH ranges in which they survive and reproduce. For each organism, the solid color represents the pH range in which they thrive, meaning they survive and reproduce. The lighter colors for each organism represents the pH range in which they survive but cannot reproduce. When the organism is absent below a certain pH, it indicates they cannot survive.

have been harmed by a combination of the acidity of the deposition as well as by the nitrate and sulfate ions.

The greatest effects of acid deposition have been on aquatic ecosystems. Lower pH of lakes and streams in areas of northeastern North America, Scandinavia, and the United Kingdom has caused decreased species diversity of aquatic organisms. As we saw in Unit 2, many species are able to survive and reproduce only within a narrow range of environmental conditions, called range of tolerance. Many amphibians, including salamanders, for instance, will survive when the pH of a lake is 6.5, but when the lake acidifies to pH 6.0 or 5.5, the same organism will begin to have developmental or reproductive problems. In water below pH 5.0, most salamander species cannot survive. **FIGURE 46.3** shows a variety of aquatic organisms and the pH range in which they survive and reproduce, the range in which they survive only but cannot reproduce, and the pH below which they cannot survive. As you can see, the pH range in which each organism thrives is different.

Effects of Acid Deposition on Humans and Infrastructure

People are not harmed by direct contact with acid rain at the pH levels of acid rain commonly experienced in the United States and elsewhere in the world because human skin is a sufficiently robust barrier. Acid deposition mainly affects ecosystems and materials that are downwind from sources of acids such as coal-burning power plants and factories. Acid deposition can harm human-built structures such as statues, monuments, and buildings. For example, buildings from ancient Greece such as those on and near the Acropolis in Athens, many of which have stood for approximately 2,000 years, have been seriously eroded over the last half century by acid deposition (**FIGURE 46.4**). The damage happens because acid deposition reacts with building materials. When the hydrogen ion in acid deposition interacts with limestone or marble, the calcium carbonate reacts with H^+ and gives off Ca^{2+}. In the process of neutralizing the acidity in the rainfall, the calcium carbonate material is partially dissolved. The more acidic the precipitation, the more hydrogen ions there are to interact with the calcium carbonate. Acid deposition also erodes many exposed painted surfaces, including automobile finishes.

FIGURE 46.4 Material damage from acid deposition. Hadrian's Arch, near the Acropolis in Athens, Greece, has been damaged by acid deposition. It is made of marble, which contains calcium carbonate, and is susceptible to deterioration from acid deposition and acids in the air. *(Heritage Image Partnership Ltd/Alamy Stock Photo)*

Acid rain comes from sulfur dioxide and nitrogen oxide pollutant emissions. A portion of the legislation that addresses acid rain is contained within the Clean Air Act Amendments of 1990, Title IV. When those amendments were passed, Congress did not repeal the previously existing Title IV, which covered noise pollution. So although seemingly disconnected, it is appropriate to cover these concepts together since they are both regulated by Title IV of the Clean Air Act.

46-2 What are the sources and effects of noise pollution?

Noise pollution in the air can affect human health and behavior; noise pollution in water may interfere with animal communication

Noise pollution is unwanted sound that interferes with normal activities such as sleeping or conversation, or sound that is loud enough to cause physiological stress and other health issues including hearing loss. You might be aware that noise pollution affects humans, although disproportionately more in low-income communities, but did you know it also affects wildlife? We'll explore how it affects all life below.

Noise Pollution and Humans

Noise pollution is measured in two ways: loudness and frequency, which is also known as pitch. The standard unit for measuring loudness is the decibel (db), which is one-tenth (0.1 or deci) of a bel. Like the pH scale, it is logarithmic. Therefore, a sound that is 10 db higher than another is perceived by the average listener as twice as loud. A sound that is 20 db higher is perceived as four times as loud. Frequency or pitch is measured in hertz (Hz). Humans can hear in the frequency range of 20 Hz to 20,000 Hz. People are more sensitive to some frequencies than others, and as people age, some lose hearing in the higher frequency ranges. The lowest frequency range that people can hear is called low-frequency noise (20 Hz to 200 Hz) and is a sound range sometimes emitted by wind turbines. There is more than one decibel scale, but the most commonly reported decibel scale is the **decibel A scale (db(A))**, a logarithmic scale that measures both the loudness of sound, and the frequency combined with the way that the human ear perceives that sound. **FIGURE 46.5** shows the db(A) scale with some common sounds and their db(A) ratings. This rating incorporates both the loudness and the frequency of the sound. If you have ever heard the sound of a leaf blower used by gardeners and cleanup crews, or the whine of a garbage truck compressing garbage, you may be able to appreciate how the frequency as well as the loudness affects perception of sound.

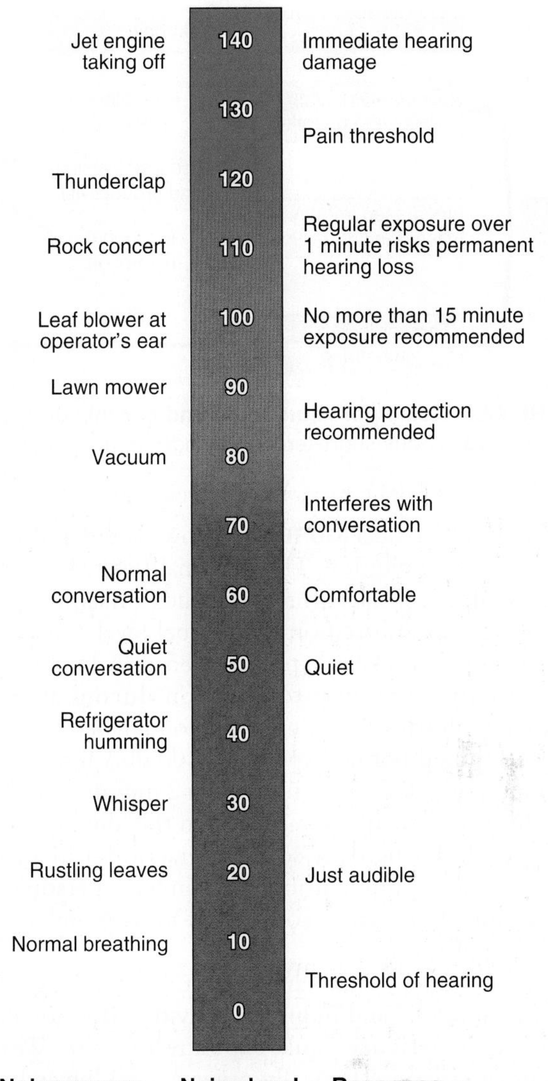

FIGURE 46.5 The decibel scale. Like the pH scale, the decibel scale is logarithmic. So a normal conversation at 60 db is perceived as twice as loud as a quiet conversation at 50 db.

In humans, unwanted noise and long-term exposure to noise can lead to hearing loss but can also lead to stress-related illnesses, sleep disruption, high blood pressure, and difficulty with concentration. Noise Induced Hearing Loss is probably the most common and commonly recognized effect from noise pollution, but other conditions can be

Noise pollution Unwanted sound that interferes with normal activities that is loud enough to cause health issues including hearing loss.

decibel A scale (db(A)) A logarithmic scale that measure both the loudness of sound and the frequency.

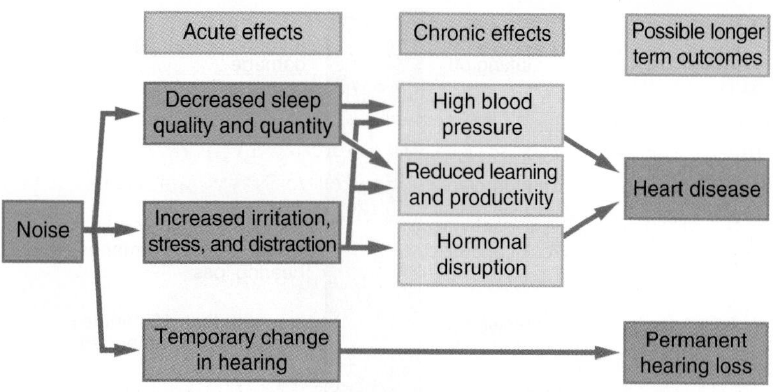

FIGURE 46.6 Noise pollution acute and chronic effects. Noise pollution can have serious short-term and longer-term effects.

equally problematic. **FIGURE 46.6** shows acute and chronic effects of noise pollution. The National Institute for Occupational Safety and Health recommends a maximum exposure time to noise based on the decibel level. For example, a person should not be exposed to 85 dbA for more than 8 hours (corresponding to a person during an 8-hour work day). Shorter limits exist for louder decibel ratings. A person should not be exposed to 100 dbA for more than 14 minutes or 110 dbA for more than 2 minutes. In addition to health issues, small impairments in the ability to understand speech during daily social interactions can adversely affect speech comprehension and hinder a person's social functioning. This can become even more problematic in elderly populations.

Noise pollution occurs in cities due to traffic, construction, and domestic and industrial activities. But not all parts of the city experience equal exposure to noise. Wealthier sections of cities are typically quieter. And lower-income areas are generally noisier. Within a given city, housing adjacent to elevated rail lines, for example, experiences noise that is much louder and the cost of the housing is less expensive than housing that is closer to a park (**FIGURE 46.7**). Suburban and rural areas are also subject to the same noise pollution sources but perhaps traffic noise is not as loud as in more densely populated, congested areas. Industries such as factories and waste treatment facilities such as landfills, incinerators, and recycling centers can all be sources of noise pollution.

FIGURE 46.7 Noise pollution in urban areas. Transportation infrastructure, such as this city street with an elevated train structure above it, can be a major source of noise pollution in urban areas. *(Franck Michel, CC)*

Noise Pollution and Environmental Justice

People of color, lower income communities, and recent immigrants are more likely than Caucasians to live in proximity to solid waste incinerators, chemical production plants, petroleum refineries, toxic waste plants, and other so-called loud and dirty industries. In a number of now-famous studies in the 1980s and 1990s, investigators used the distribution of minority residents by postal zip code to relate race and class to the location of hazardous chemical and waste facilities and sites. In Atlanta, 83 percent of the Black population lived within the same zip code as the 94 uncontrolled toxic waste sites, while only 60 percent of Caucasians lived in those areas. In Los Angeles, roughly 60 percent of Hispanics lived in the same areas as the toxic waste sites, while only 35 percent of the Caucasian population lived in those areas. One study in five southern states compared the size of specific landfill facilities with the percentage of minorities in the zip code in which the landfill was located. The study concluded that the largest landfills are located in areas that have the greatest percentage of minorities. Not only are people who work or live near these facilities subjected to unwanted noise, but they may also be subjected to outdoor or indoor air pollution as well.

Noise Pollution and Wildlife

Noise pollution can adversely affect animal species in multiple ways. Amphibians, birds, mammals, fish, and reptiles are all affected. As it does with humans, excess anthropogenic noise can cause stress in animal populations and it can damage hearing. Traffic noise has been shown to reduce the hunting efficiency of bats. Noise can interfere with warning signals and mating calls being heard by birds and amphibians. Noise has also been shown to alter migratory routes of certain animals. A 2019 comprehensive review of noise pollution and wildlife suggested that rather than just a few sensitive species being affected, it is more likely that most species are affected by anthropogenic noise.

Noise pollution may also occur in the oceans. Sounds emitted by propellers and mechanical systems on ships and submarines can interfere with underwater animal communication. Sonar, which is underwater sound propagation for use in navigation and map making, can negatively affect species such as whales that rely on low-frequency, long-distance

FIGURE 46.8 Aquatic noise pollution. Noise from navy sonar and oil exploration may interfere with the normal behavior of whales. *(Lazareva/Getty Images)*

communication (**FIGURE 46.8**). Several instances of beached whales in the Bahamas, the Canary Islands, and the Gulf of California have been connected to the use of military sonar and the use of loud, underwater air guns by energy companies searching for oil deposits under the oceans.

In 2003, a federal judge rejected the U.S. Navy's request to install a network of long-range sonar systems across the ocean floor to detect incoming submarines because of suspected negative impacts on endangered whales and other species of marine animals. In 2008, however, the U.S. Supreme Court ruled that the president of the United States could exempt the Navy from environmental laws that were related to potential sonar effects on ocean life. To better understand noise pollution in the ocean, the U.S. National Oceanic and Atmospheric Administration (NOAA) announced in 2012 that it had completed the first step in mapping the noises across different regions of the ocean. By 2014, NOAA concluded that the Navy's activities would have a negligible effect on fish and whales in the ocean. At the same time, an increased awareness of noise pollution in the ocean has inspired some ship builders to design ships equipped with quieter propellers.

In this module we described acid rain and its formation from nitrogen oxides and sulfur dioxide. We identified that the major sources of acid rain are vehicle exhaust and stationary combustion. Acid rain can potentially affect forests, soils, and aquatic communities downwind from the location where acid rain precursors are released. Aquatic ecosystems are the most adversely affected. If a soil is underlain by limestone bedrock, the impact of acid rain on terrestrial and aquatic systems can be partially neutralized. Noise pollution is unwanted noise at levels high enough to cause hearing damage and other health effects. Transportation, construction, and industry are sources of noise pollution. People of color, lower income communities, and recent immigrants are more likely to experience anthropogenic noise pollution than Caucasians. Many different animal populations are also subject to noise pollution including marine mammals that rely on sound for long-distance communication. Outdoor air pollution, indoor air pollution, acid deposition and noise are all components of atmospheric pollution that require assessment and mitigation for the safety of humans, animals, and ecosystems.

Module 46 AP® Review

Preparing for the AP® Exam

Learning Goals Revisited

46-1 What is acid rain and what are its effects?

The pH indicates the relative strength of acids and bases in a substance and is a logarithmic unit of measure. Acid rain is rain or snow containing nitric and sulfuric acids that come from the release of the primary pollutants nitrogen oxides and sulfur dioxide. Motor vehicles and stationary combustion such as from coal-burning power plants are two important sources of acid rain. Volcanoes are a natural source of acid rain. Vegetation, soils, and aquatic communities are potentially affected by acid rain. Watersheds underlain by limestone bedrock may have the capacity to partially neutralize acid rain. Structures such as buildings and automobiles can be degraded by acid rain. The acidity of rainfall in the United States has decreased due to legislation passed with the Clean Air Act and subsequent amendments.

46-2 What are the sources and effects of noise pollution?

Noise pollution is unwanted sound that causes hearing loss, stress, and other health issues in humans. Amphibians, birds, mammals, fish, and reptiles are also affected by unwanted noise. Noise pollution in air is measured in decibels, which is a logarithmic unit of measure. Transportation, construction, and industrial activity generate noise pollution. Urban areas can be major sources of anthropogenic noise. People of color, lower income people, and recent immigrants are more likely to experience anthropogenic noise pollution than Caucasians.

AP® Practice Questions

Multiple-Choice Questions

1. Acid deposition
   (a) is caused by combustion of fossil fuels.
   (b) is made worse by limestone.
   (c) occurs as a result of mining.
   (d) leads to photochemical smog.

2. The pH scale and the decibel scale both
   (a) are logarithmic.
   (b) range from 0 to 100.
   (c) increase with increasing harmful impact.
   (d) are measured in air and water.

3. Lakes affected by acid rain could have a pH of approximately
   (a) 2.
   (b) 4.
   (c) 10.
   (d) 12.

4. Noise pollution can cause which of the following health effects?
   (a) heart disease
   (b) hair loss
   (c) diabetes
   (d) cancer

5. Which of the following is an acute effect of noise pollution?
   (a) high blood pressure
   (b) sleep loss
   (c) reduced productivity
   (d) hormonal disruptions

Free-Response Question

The diagram below shows the formation of acid rain with anthropogenic sources.

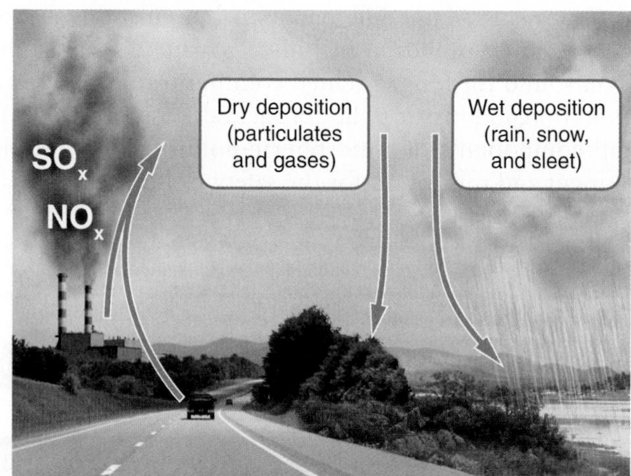

(a) **Identify** one type of acid that forms in the atmosphere. (1 pt.)
(b) **Describe** one effect of acid rain on human-made structures. (1 pt.)
(c) **Describe** one effect of acid rain on ecological systems. (1 pt.)
(d) **Explain** what occurs to the pH of aquatic ecosystems when acid rain falls. (1 pt.)
(e) Local scientists suggest that when acid rain falls, and enters a lake, one method for remediation is to add limestone to the lake. **Justify** whether you support or refute this method. (2 pts.)
(f) The data table below shows the pH preference of salamanders in a freshwater ecosystem. Scientists measured the abundance of salamanders at different pH levels in the ecosystem.

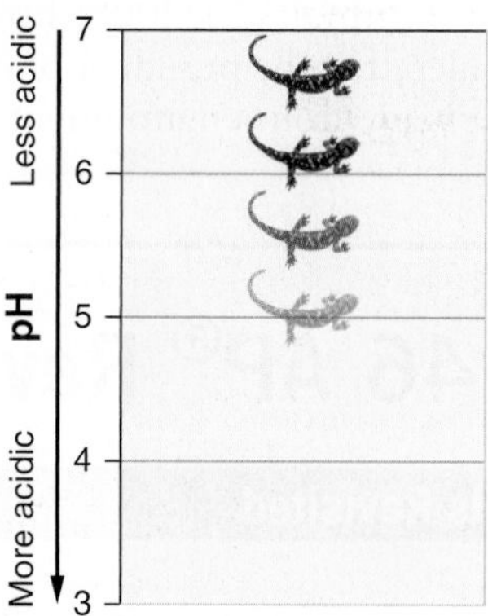

(i) **Identify** the independent variable for this data collection. (1 pt.)
(ii) **Identify** a likely hypothesis for this data collection. (1 pt.)
(iii) **Identify** the dependent variable for this data collection. (1 pt.)
(iv) **Describe** a modification to this data collection that would clarify the data. (1 pt.)

UNIT 7
Practice Your Science Skills

Preparing for the AP® Exam

Data Analysis

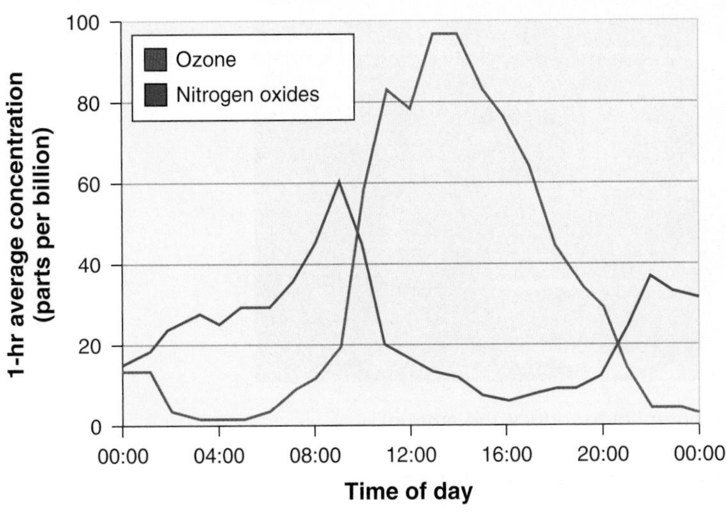

Questions

1. **Describe** the pattern of ozone over the course of the day.
2. **Explain** why this pattern of ozone occurs.
3. **Identify** one source of NO_2 emissions that most likely contributes to the pattern of NO_2 seen in the graph.
4. **Explain** how this graph would change on a typical day in winter.

Pursuing Environmental Solutions

Better Indoor Cook Stoves

In China, India, and sub-Saharan Africa, people in 80 to 90 percent of households cook food using wood, animal manure, and crop residues as their fuel. Since women do most of the cooking, and young children are with the women of the household for much of the time, it is the women and young children who receive the greatest exposure to carbon monoxide and particulate matter. When biomass is used for cooking, concentrations of particulate matter in the home can be 200 times higher than the exposure limits recommended by the U.S. Environmental Protection Agency. A wide range of diseases has been associated with exposure to smoke from cooking. Earlier in this unit, we described that indoor air pollution is responsible for 3.8 million deaths annually around the world, and indoor cooking is a major source of indoor air pollution.

There are hundreds of projects and business ventures underway around the world to enable women to use more efficient cooking stoves, ventilate cooking areas, cook outside whenever possible, and change customs and practices that will reduce their exposure to indoor air pollution. The use of an efficient cook stove will have the added benefit of consuming less fuel. This improves air quality and reduces the amount of fuel needed, which reduces the amount of nearby vegetation that must be collected and also reduces the amount of time that a woman must spend searching for fuel.

Increasing the efficiency of the combustion process requires the proper mix of fuel and oxygen. One effective method of ensuring a cleaner burn is the use of a small fan to facilitate greater oxygen delivery. However, because most homes in developing countries with significant indoor air pollution problems do not have access to electricity, some sort of internal source of energy for the fan is needed.

Two innovators from the United States developed a cook stove for backpackers and other outdoor enthusiasts who needed to cook a hot meal with little impact on the environment. The BioLite CampStove, which works by burning

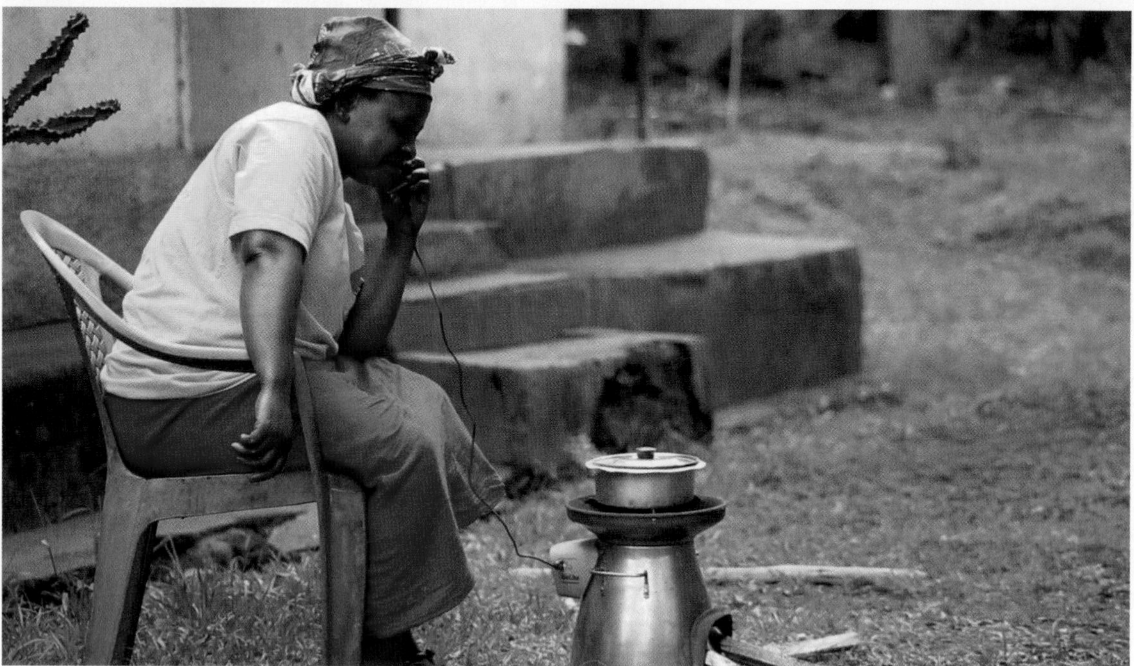

BioLite HomeStove. This small stove and others like it have the potential to reduce the amount of firewood needed to cook a meal, as well as lower the amount of indoor air pollution emitted. This woman in a village in Kenya is also using the stove to charge her cell phone. *(Courtesy BioLite Inc.)*

solid biomass and weighs only 935 grams (2 pounds), needs no gasoline or batteries—a desirable feature for people carrying all their belongings on their backs. The stove utilizes two small wires made from different metals to generate electricity from the heat of the stove. The electricity can be used to charge a battery or phone.

The inventors of the CampStove soon realized that the same technology could make an important contribution to improving indoor air quality in the developing world. For home use in the developing world, they created the BioLite HomeStove, which weighs 8 kg (17.6 lbs). Like the CampStove, it physically separates the solid fuel from the gases that form when the fuel is burned, which allows the stove to burn the gases. In addition, a small electric fan, located inside the stove, harnesses energy from the heat of the fire and moves air through the stove at a rate that ensures complete combustion. The result is a more efficient burn, less fuel use, and less release of carbon monoxide and particulate matter. BioLite states that its HomeStove uses "less than half the fuel and produces 90% fewer emissions." The stove generates enough electricity to power the fan and has additional capacity to power a small USB port suitable for charging a cell phone or flashlight.

The BioLite CampStove stove is readily available on consumer goods websites; the BioLite HomeStove is sold in Kenya and Uganda. The manufacturer estimates that three-quarters of a million people have benefitted from the use of their stove. BioLite stoves won awards in 2009, 2012, 2014, and 2021 for low emissions, innovative design, and promoting social good. There are many other small stoves under development and in use around the world that increase combustion efficiency and reduce carbon monoxide and particulate emissions. The challenge is to make them affordable, available, and locally manufactured so that large numbers of people in the developing world will have access to them and use them. They also need to be easily and simply repairable. Other promising ways to reduce fuel use and improve indoor air quality include the solar cooker shown in Figure 39.5 on page 449.

Critical Thinking Questions

1. Why are women and children exposed more often to indoor air pollution than men in developing countries?
2. How can technology offer solutions to cooking over open fires?

References

Bilger, B. 2009. Annals of Invention, Hearth Surgery, *The New Yorker*, December 21, p. 84; https://www.newyorker.com/magazine/2009/12/21/hearth-surgery.

https://www.bioliteenergy.com/, homepage of BioLite stove.

https://youtu.be/mV1MRlhLnWQ Biolite Mission, Everyone Plays A Part. Posted on YouTube on February 10, 2016.

Science Applied 7: Concept Explanation

Can Cap-and-Trade Reduce Pollution Emissions Equitably?

Throughout this unit we have identified the need for reducing pollutant emissions. Pollutant reductions have been achieved in the United States for just about every major pollutant except carbon dioxide. As we identified in Module 42 and will further explore in Unit 9, a current challenge for human civilization is the need to reduce carbon dioxide concentrations in the atmosphere. There are even international protocols agreed to by almost every nation calling for specific carbon dioxide reductions through 2050. While there is something of a consensus on the need for further carbon reduction, there has been a great deal of debate regarding the most effective and fair way to achieve these reductions.

Can successes controlling sulfur translate to carbon emission reductions?

As we described in Module 45, the cap-and-trade approach has been used for the control of sulfur dioxide emissions from coal-burning power plants since enactment of the Clean Air Act Amendments (CAAA) of 1990 to reduce acid precipitation (**FIGURE SA7.1**). In effect, the CAAA was a nation-wide experiment to see if controls of sulfur dioxide emissions at power plants could result in reductions in the acidity of rainfall. Motivated by the newly instituted cost of buying

(a)

(b)

(c)

(d)

FIGURE SA7.1 Carbon cap-and-trade approach. A diversity of industries emits large amounts of CO_2, including (a) coal-fired electricity-generating power plants, (b) automobiles, (c) airplanes, and (d) steel manufacturers. Under cap-and-trade, all of these industries could be involved in buying and selling permits for CO_2 emissions. *(a: RGB Ventures/Alamy Stock Photo; b: NadyGinzburg/Shutterstock; c: Dmitry Marchenko/Alamy Stock Photo; d: Andrey Radchenko/Alamy Stock Photo)*

pollution permits, companies that owned coal-fired electricity generation plants were required to retrofit existing facilities, construct new facilities that polluted less, or buy pollution permits from other companies that had already reduced their emissions. Even though some older power plants were exempt from certain provisions of the act, the CAAA had achieved a 50 percent reduction in sulfur emissions by 2021. The cap-and-trade system encouraged investment in new technologies, which is one of the underlying reasons for the success of this program. For example, an air pollution control system from coal-burning power plants was developed that recovers sulfur from sulfur dioxide emissions and converts it to synthetic gypsum (calcium sulfate that is used to make sheetrock, a building material). Because gypsum is a valuable product used in construction, the economic incentive to reduce sulfur emissions motivated new research and development that produced a profitable commercial product as part of the solution. Furthermore, the reductions in sulfur dioxide emissions were accompanied by reductions in particulate matter and carbon dioxide as additional benefits.

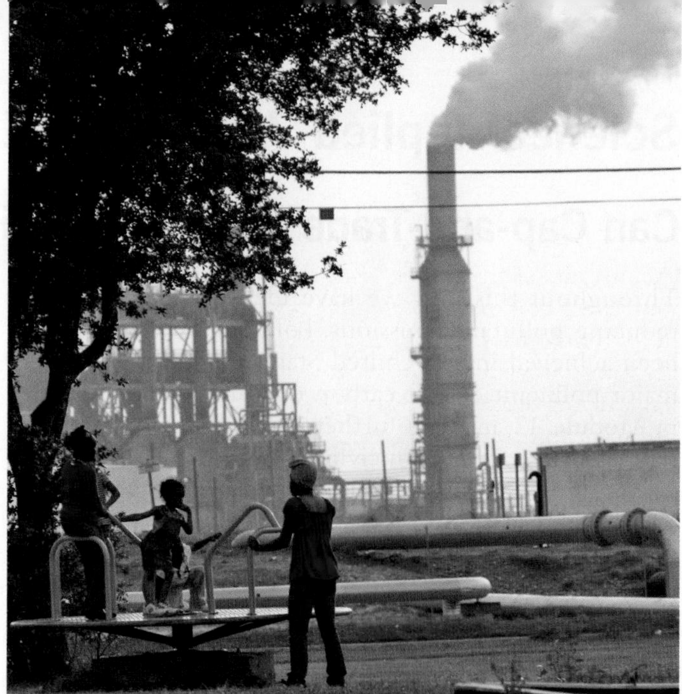

FIGURE SA7.2 An oil refinery located near a community of color in Port Arthur, Texas. Residents are exposed to a variety of pollutants including sulfur dioxide, volatile compounds, particulates, and noise. *(AP Photo/LM Otero)*

What were the implications of sulfur cap-and-trade for the environmental justice movement?

The substantial decreases in emissions that were achieved did not benefit everyone equally. While the total amount of sulfur dioxide emitted by coal-burning power plants was reduced, certain communities were subjected to even higher rates of pollution. By design, the CAAA created a market for emissions credits that allows heavily polluting plants to continue polluting or even increase their annual emissions, as long as they purchase the right to do so. Through this system, plants can pay to pollute and the nearby communities that are downwind of the plants experience the adverse consequences—they experience increased air pollution. A number of studies have shown that the negative impacts of pollution disproportionately fall on lower-income communities and people of color. Critics of the sulfur dioxide cap-and-trade regulations argue that the program has exacerbated these inequalities, by leading to higher emissions downwind of certain plants and exposing communities of color to more pollution, not less.

A core concern of environmental justice is the importance of finding solutions to environmental challenges that distribute the benefits and costs equitably. Environmental justice examines whether there is equal enforcement of environmental laws and elimination of disparities—intended or unintended—in the exposure to pollutants and other environmental harms affecting different ethnic and socioeconomic groups within a society (**FIGURE SA7.2**). Some environmental injustices occur because wealthier communities sometimes have residents with time and financial resources to object to a proposed landfill or incinerator through protests, letter writing campaigns, social media posts, and legal action. When a landfill or incinerator is proposed in a lower-income community, the local residents may not possess the same resources to successfully prevent the proposed entity from being permitted and constructed. Delegates to the First National People of Color Environmental Leadership Summit in 1991 established a total of 17 principles of environmental justice. Professor Robert Bullard of Texas Southern University is probably best known for his 1990 book *Dumping in Dixie*, which demonstrated that minority and lower socioeconomic groups were often the recipients of pollution from dumping of municipal solid waste and hazardous waste. Newer studies have reassessed the unequal distribution of hazardous waste dumping in the United States and found that the situation is worse than previously reported.

Sulfur dioxide and other pollutants from coal-burning plants are of special concern to environmental justice activists because of their direct implications for human health. A 2012 study determined that 78 percent of Black Americans lived within 30 miles of a coal-fired power plant and, as a result, Black children were significantly more likely to have asthma and require emergency department visits for an asthma attack than white children. The concentration of coal-burning power plants in lower-income communities has also been implicated in higher rates of lung disease and stroke. As more and more coal-burning power plants have been shutting down, the situation has improved but there is still a disproportionate exposure of communities of color to coal-burning power plant emissions. A 2017 study based on EPA data found that almost 8 million Black Americans live within a half-mile of oil and gas facilities including refineries where they are exposed to air pollution, noise, and groundwater contamination. Overall, they found that Black Americans were 75 percent more likely than white Americans to live in communities that were next to commercial facilities that had harmful levels of emissions, odors, noise, and/or vehicular traffic.

Could cap-and-trade be used to reduce carbon dioxide emissions in the United States?

The inequitable distribution of air pollution did not improve from sulfur cap-and-trade regulations, but cap-and-trade has the potential to yield better results in other situations. In fact, many environmental scientists and economists argue that carbon dioxide is an ideal pollutant to be managed through cap-and-trade. One of the reasons for this is the fact that the need to reduce carbon dioxide emissions is due to its greenhouse warming potential rather than its health effects. Carbon dioxide does express indirect implications for human health—climate change increases the vulnerability of at-risk populations to natural disasters, can exacerbate the spread of infectious diseases, and may reduce the availability of fresh produce in some areas—but living near a large carbon dioxide emitter does not directly threaten health in the same way that living near sulfur dioxide or particulate polluters does. The cap-and-trade approach gives companies the freedom to choose the carbon-reduction approach that is most affordable while also offering an economic incentive to reduce pollution. The rate at which the carbon dioxide reduction occurs would depend on how quickly one government or an international treaty reduced the number of pollution permits. In 2005, a carbon cap-and-trade system was instituted in the European Union (EU). The system started slowly and was criticized for not lowering the number of permits quickly enough. In 2010, the European Union started to lower the cap and more companies, including electric generating plants and airlines began paying for the permits. A study in the Proceedings of the National Academy of Science estimated that the EU carbon cap-and-trade program had avoided more than 1 billion tons of carbon dioxide emissions between 2008 and 2016. This amounted to an almost 4 percent reduction of what would have been emitted in the EU had the program not been enacted. However, there are differences between controlling sulfur and carbon dioxide. As we discussed in Modules 37 and 42, sulfur is an impurity found in coal and oil. Sulfur can be removed before or after burning and certain fuels can be selected that have a lower sulfur content. Chemical bonds from carbon, on the other hand, are the source of the energy in fossil fuels and carbon dioxide is the actual end product of combusting that carbon, so it is much more difficult to control.

If carbon cap-and-trade is used in the United States, energy companies will probably be targeted including those that produce petroleum products and those that generate electricity. A recent study by Point Carbon, a market research company, found that the major oil companies including ExxonMobil and Chevron would likely pay hundreds of millions of dollars in CO_2 permits while electric generating companies that have diversified their electricity generation portfolios to include hydroelectric and nuclear energy will pay substantially less. Ultimately, if cap-and-trade is successful in the European Union, it could serve as an effective model for other locations in the world and help reduce the growing problem of excessive CO_2 emissions and the resulting global change.

Questions

1. In general, is cap-and-trade an acceptable pollution control approach? When is it not such a desirable approach?
2. What is a difference between controlling emissions of sulfur dioxide versus carbon dioxide?

Practice AP® Free-Response Question

Cap-and-trade controls on air pollution can be general, or they can be very specifically designed to control one specific pollutant. Cap-and-trade measures can also result in certain pollutant emitters purchasing pollution permits and emitting large amounts of pollution in one specific location.

(a) **Describe** one method of reducing air pollution that is general, meaning it will control virtually all air pollutants at the same time. (2 pts.)

(b) There are also very specific air pollution control methods that target one or a few air pollutants. **Explain** how one air pollution control technology will reduce or remove only specific air pollutants, but not usually all air pollutants. (2 pts.)

(c) Fuel switching is a term for achieving the same work or activity in an environmentally less harmful way by switching from one fuel to another.

(i) **Describe** an example of fuel switching that will result in lower sulfur emissions. (2pts.)

(ii) **Describe** an example of fuel switching that will result in lower carbon dioxide emissions. (2 pts.)

(d) When considering human health effects, critics of cap-and-trade express concern about localized areas of high pollutant emissions that will occur during a properly designed cap-and-trade pollution reduction system. **Explain** how a cap-and-trade program may result in an increase in adverse human health effects in specific geographic locations. (2 pts.)

References

Becker, R. 2020. Legacy of a clean-air czar: Clearer skies, bold alliances and bitter controversy. *CalMatters*, December 23. https://calmatters.org/environment/2020/12/mary-nichols-czar-legacy-skies-controversy/.

Food and Water Watch. 2019. Cap and Trade: More Pollution for the Poor and People of Color. https://foodandwaterwatch.org/wp-content/uploads/2021/04/ib_1911_rggicaptradeej-web.pdf.

Toomey, D. 2013. Coal Pollution and the Fight for Environmental Justice. *Yale Environment* 360, June 19. https://e360.yale.edu/features/naacp_jacqueline_patterson_coal_pollution_and_fight_for_environmental_justice.

UNIT 7 Review

Key Terms to Remember

Air pollution (p. 495)
Sulfur dioxide (SO_2) (p. 496)
Haze (p. 498)
Photochemical oxidant (p. 498)
Smog (p. 498)
Lead (Pb) (p. 498)
Hydrocarbons (p. 498)
Primary pollutant (p. 499)
Secondary pollutant (p. 499)
Photochemical smog (Los Angeles–type smog; Brown smog) (p. 504)
Sulfurous smog (London-type smog; Gray smog; Industrial smog) (p. 504)
Evaporate (p. 505)
Sublimate (p. 506)
Formaldehyde (p. 506)
Thermal inversion (p. 507)
Inversion layer (p. 507)
PM_{10} (p. 509)
$PM_{2.5}$ (p. 509)
Indoor air pollutants (p. 512)
Asbestos (p. 513)
Radon-222 (p. 514)
Sick building syndrome (p. 516)
Vapor recovery nozzle (p. 523)
Catalytic converter (p. 525)
Scrubber (p. 526)
Electrostatic precipitator (p. 527)
pH (p. 529)
Acid (p. 530)
Base (p. 530)
Acid rain (Acid deposition) (p. 530)
Noise pollution (p. 533)
decibel A scale (dbA) (p. 533)

Unit 7 **AP® Environmental Science Practice Exam** Preparing for the AP® Exam

Section 1: Multiple-Choice Questions

1. Which of the following is an example of a secondary pollutant?
   (a) sulfur dioxide
   (b) carbon dioxide
   (c) tropospheric ozone
   (d) nitrogen oxide

2. The accumulation of tropospheric ozone during the middle of the day depends mainly upon the atmospheric concentration of nitrogen oxides and
   (a) carbon dioxide.
   (b) volatile organic compounds.
   (c) chlorofluorocarbons.
   (d) sulfates and nitrates.

3. Air pollution generally refers to pollution in the
   (a) stratosphere.
   (b) troposphere.
   (c) lithosphere.
   (d) mesosphere.

4. Which of the following is an effect of acid deposition?
   (a) mobilization of metal ions from the soil into surface water
   (b) increased numbers of salamanders in ponds and streams
   (c) increased food sources for aquatic organisms
   (d) increased incidence of photochemical smog

5. Two major factors involved in the conversion of primary pollutants into secondary pollutants are the presence of
   (a) sunlight and water.
   (b) sulfates and sunlight.
   (c) water and volatile organics.
   (d) nitrogen oxides and sulfates.

6. The pollutant least likely to be emitted from a smokestack would be
   (a) carbon monoxide.
   (b) carbon dioxide.
   (c) ozone.
   (d) sulfur dioxide.

7. Which of the following is a source of particulate matter in the atmosphere?
   (a) the chemical by-products of acid rain formation
   (b) incomplete natural gas combustion
   (c) runoff from irrigation
   (d) volcanoes

8. Which air pollutant is sourced from bedrock typically in foundations of homes?
   (a) ozone
   (b) radon-222
   (c) volatile organic compounds
   (d) particulate matter

9. Pollutants can be categorized as primary or secondary. Which pairing is correct?
   (a) primary: ozone
   (b) primary: sulfate
   (c) secondary: nitrate
   (d) secondary: carbon monoxide

10. A thermal inversion
   (a) occurs when cool and warm air are intermixed throughout the year.
   (b) occurs when air pollution from China reaches the West Coast of the United States.
   (c) occurs when cool air is present through the troposphere.
   (d) occurs when a warm air layer overlies a cooler layer.

11. The juvenile Western Bluebird (*Sialia mexicana*) is a small thrush that has been found to be affected by noise pollution. A study showed that a small increase of 10-decibels can reduce the juveniles ability to hear background noise by as much as 90 percent. Birds can't hear predators or mating calls, and have drastically different migratory patterns due to background noises. One solution to reducing noise pollution would include
   (a) increased transportation.
   (b) decreased use of pesticides on crops.
   (c) increase in protected habitats.
   (d) decreased use of fossil fuels.

12. In the troposphere, O_3 is formed by the combination of ____________ and ____________ in the presence of sunlight due to fossil fuel combustion.
   (a) carbon dioxide, particulate matter (PM_x)
   (b) sulfur oxides, carbon dioxide
   (c) photochemical smog, acid rain
   (d) volatile organic compounds (VOCs), nitrogen oxides (NO_x)

Use the graph below to answer questions 13–15:

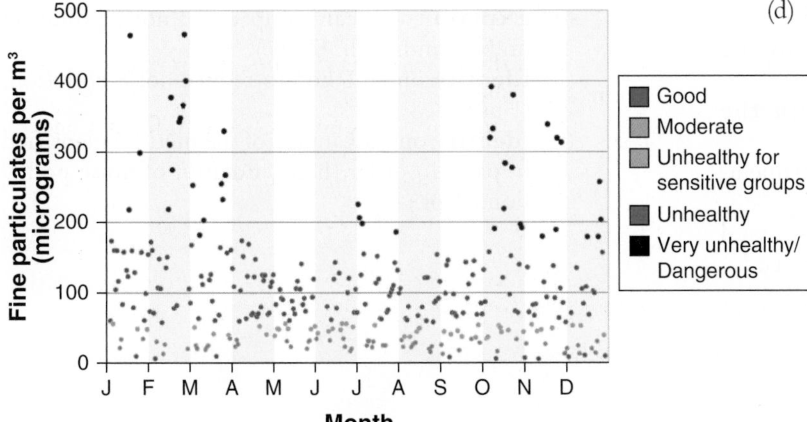

13. Using the graph, which month had the widest range of particulate matter around the U.S. Embassy in Beijing?
   (a) November
   (b) June
   (c) February
   (d) August

14. During the months of April through August, there are 0 reported "good days," (0–12 micrograms). What other air pollutant can be expected to rise during these months of the year?
   (a) ozone
   (b) carbon dioxide
   (c) asbestos
   (d) radon

15. January through March saw more days above 250 micrograms than the rest of the year. Which of the following would be a good explanation for this finding?
   (a) Respiration and decomposition increase, causing an increase in particulate matter.
   (b) Particulate matter increases due to the increase of acid deposition in winter months.
   (c) Coal is burned during winter months, increasing particulate matter.
   (d) During the winter months, trees release more particulate matter.

16. Migrant farm workers picking fruit on a farm would not have to worry as much about which of the following pollutants?
   (a) asbestos
   (b) photochemical smog
   (c) particulate matter
   (d) sulfur dioxide

17. A city has been experiencing rain with a pH of 5.5. Given this information, which is the likely source of energy used in this city?
   (a) tidal energy
   (b) wind turbines
   (c) coal power plant
   (d) natural gas power plant

18. The graph below shows carbon monoxide levels in the period of 1979 to 2011 in the city of Boston and surrounding areas. Which of the following is most likely responsible for the trend in carbon monoxide in this area?

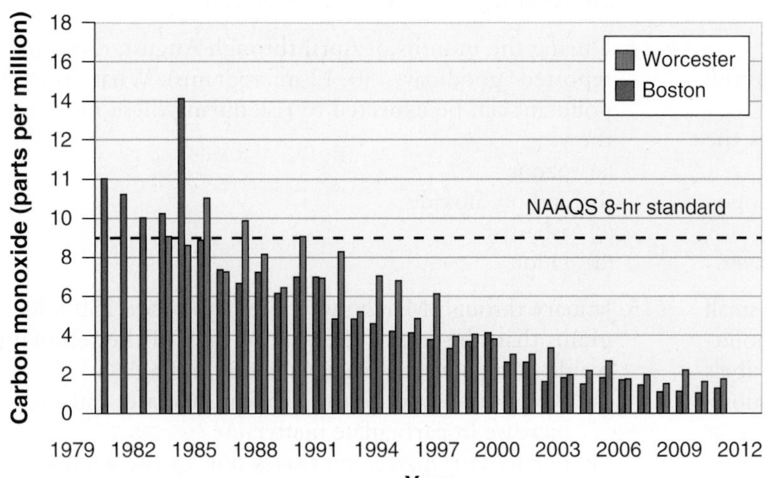

(a) catalytic converters in vehicles
(b) wet scrubbers in factories
(c) vapor recovery nozzles on fuel nozzles
(d) electrostatic precipitators in factories

Use the following passage to answer questions 19 & 20:

The Clean Air Act established criteria for air pollution and limits how much of certain air pollutants can be released. Further amendments also created programs to reduce acid rain, creating cap-and-trade programs for SO_2. The CAAA has been found to reduce emission by 77 percent since 1970, and prevented an estimated 2.3 million premature deaths, 17 million lost workdays, and 5.4 million lost school days.

19. Which of the following best describes the author's claim?
    (a) The Clean Air Act has had a positive effect on ecosystems.
    (b) the Clean Air Act has had a negative impact on ecosystems.
    (c) The Clean Air Act has had a positive impact on health and safety in the United States.
    (d) The clean Air Act has had no impact on health and safety in the United States.

20. Which of the following is a method to determine the accuracy of the author's claim?
    (a) analyzing data to see if deaths from respiratory and cardiovascular causes have decreased since the implementation of the law
    (b) measuring amounts of nitrates and phosphates in water systems
    (c) testing soils for sand, silt, and clay particles
    (d) obtaining data to measure the amount of solid waste in ponds and lakes

Section 2: Free-Response Questions

1. A study counted the number of bird nests found around a small city. The number of nests was counted around three different areas: a park without a forest, an area of land near a small airport, and an area of land around a small nature center at the edge of the city.
    (a) **Identify** the independent variable in the experiment. (1 pt.)
    (b) **Identify** the dependent variable in the experiment. (1 pt.)
    (c) **Identify** a likely hypothesis for the experiment. (1 pt.)
    (d) **Describe** one variable that isn't discussed that might affect the results of the study. (1 pt.)
    (e) **Describe** an environmental effect of noise pollution on terrestrial species. (1 pt.)
    (f) **Describe** an environmental effect of noise pollution on aquatic species. (1 pt.)
    (g) **Explain** two health impacts of noise pollution on human health. (2 pts.)
    (h) **Identify** one method to reduce noise pollution. (1 pt.)
    (i) **Identify** one advantage of the method identified in part (h) other than reduction of noise pollution. (1 pt.)

2. The table below shows the ambient air data collected for Pittsburgh, Pennsylvania, in 2008. Examine the data and answer the following questions.

2008 Monthly ambient air monitoring report, Pittsburgh

Month	Monthly maximum ozone concentration (ppb)	Monthly average ozone concentration (ppb)	Monthly average solar radiation
January	37	14	65
February	49	15	63
March	56	23	86
April	76	31	81
May	75	27	152
June	77	32	208
July	95	31	215
August	92	27	204
September	89	20	153
October	48	14	109
November	57	12	64
December	30	14	45

Data from http://www.ahs.dep.pa.gov/aq_apps/aadata/

(a) **Identify** the month that had the highest average ozone concentration. (1 pt.)

(b) **Describe** why the month in part (a) was likely the highest. (1 pt.)

(c) Based on the National Ambient Air Quality Standards (NAAQS), average ozone concentrations are not to exceed 0.075 ppm (75 ppb) in any 8-hour period. **Describe** how this NAAQS may not truly reflect the overall air quality. (1 pt.)

(d) Ozone is classified as a secondary pollutant. **Identify** one of the primary pollutants necessary for its formation and **describe** how tropospheric ozone is formed. (2 pts.)

(e) **Explain** one environmental effect ozone can have on ecosystems. (1 pt.)

(f) **Describe** one method to reduce tropospheric ozone. (1 pt.)

(g) **Explain** how ozone can contribute to other secondary pollutant formation. (1 pt.)
The same sources that contribute to ozone formation can also contribute to other secondary pollution.

(h) **Identify** another secondary pollutant formed by the same sources that form ozone. (1 pt.)

(i) **Propose a solution** to the source mentioned in part (f). (1 pt.)

3. The diagram below shows the typical formation and buildup of photochemical smog.

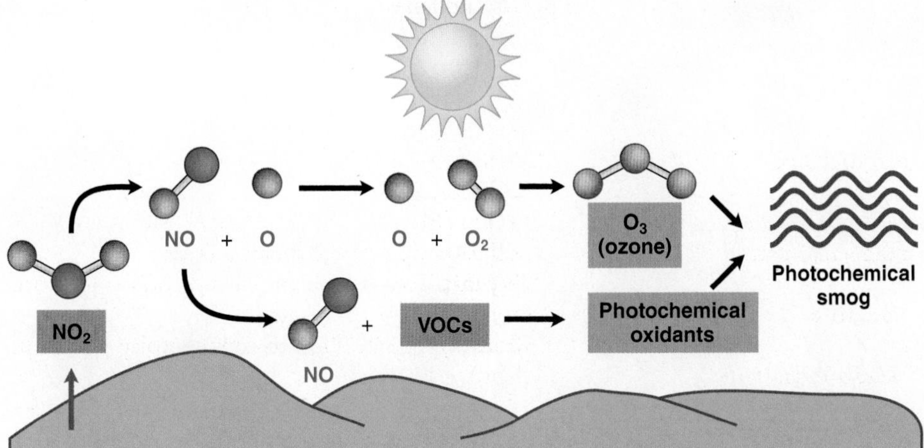

(a) **Identify** one primary pollutant in the diagram. (1 pt.)

(b) **Identify** one secondary pollutant formed in the accumulation of photochemical smog. (1 pt.)

(c) Volatile Organic Compounds (VOCs) have many anthropogenic as well as natural sources.
  (i) **Describe** one natural source of VOCs. (1 pt.)
  (ii) **Describe** one anthropogenic source of VOCs. (1 pt.)

(d) Nitrogen dioxide is a gas that occurs naturally and can be released anthropogenically into the atmosphere. In 2000, the national average air concentration of nitrogen dioxide was 98 ppb, and in 2015, the average concentration had changed to 76 ppb.
  (i) **Calculate** the average decrease of nitrogen dioxide per year. (2 pts.)
  (ii) **Calculate** the percent change in nitrogen concentrations from 2000 to 2015. (2 pts.)
  (iii) **Calculate** the estimated concentration of nitrogen in 2030 if the rate remains the same as in part (i). (2 pts.)

UNIT 8

Aquatic and Terrestrial Pollution

A diversity of alternatives for household products. Single-use and multiple-use products differ in their environmental costs and benefits, including the amount of energy, water, and global warming impact needed to manufacture the items, the water and energy used to wash the items, and the environmental impact of disposal. *(ARISTIDIS VAFEIADAKIS/Alamy Stock Photo)*

CASE STUDY

Paper, Plastic, or Reusable?

As the human population on Earth has grown, our activities have resulted in increasing amounts of pollution, including a diversity of chemicals that are released into the environment and solid waste that is brought to landfills for disposal. As we think about ways of reducing our impact on the planet, we often consider alternative approaches to living. One example of this is the question of whether to use paper, plastic, or a reusable option for items such as grocery bags, packaging, and the cups that hold our favorite beverages.

Polystyrene is a plastic polymer that has high insulation value and cushions items that are shipped in packaging. More commonly known by its trade name, Styrofoam, it is particularly useful for food packaging because it minimizes temperature changes in food and beverages. Polystyrene is lighter, insulates better, and is less expensive than the alternatives. However, a number of years ago, polystyrene was deemed harmful to the environment because, like all plastics, it is made from petroleum and it does not decompose in landfills. In response to public sentiment, most food businesses greatly reduced or eliminated their use of polystyrene. All over the country, schools, businesses, and public institutions began purging their cafeterias of polystyrene cups and most have replaced them with disposable paper cups. For example, Dunkin Donuts in the United States completed its phase out of foam cups in 2020. This resulted in 1 billion fewer foam cups going into landfills each year. In addition, six states have banned Styrofoam containers as of 2022.

All over the country, schools, businesses, and public institutions began purging their cafeterias of polystyrene cups and most have replaced them with disposable paper cups.

But is the elimination of polystyrene actually an environmental victory? It is difficult to quantify the environmental benefits and costs of using plastic products versus the alternatives. For example, because a paper cup does not insulate as well as a Styrofoam cup, paper cups filled with hot drinks are usually too hot to hold and vendors often wrap them in a cardboard band, or use two cups, both of which add to waste. To fully quantify the environmental costs and benefits of each alternative, one must create a list of inputs and outputs related to their manufacture, use, and disposal. When we make a list of all the materials and all the energy required to produce and then dispose of each type of cup, we find that it is not easy to determine which choice is better for the environment. For example, one study found that making a paper cup requires approximately 2 grams of petroleum and 33 grams of wood and bark, which are renewable materials. A polystyrene cup requires 3 grams of petroleum, a nonrenewable material, but no wood or bark. Manufacturing the paper cup requires about twice as much energy, and much more water. More energy is needed to make and transport a paper cup. A paper cup is normally used once or at most a few times while the polystyrene cup can, at least in theory, be reused many times. However, both types of cups are usually thrown away after a single use. In a landfill, the paper cup will degrade and eventually produce methane gas, while the polystyrene cup, because it is made of an inert material, will remain undegraded for a very long time.

In 2021, researchers examined a wide variety of single-use and multiple-use household items and asked whether the multiple-use items, which are washed between uses, are a better option. They considered the amount of energy, water, and global warming impact caused by manufacturing the items, the water and energy used in washing the items, and the environmental impact of disposal. The researchers examined four categories of kitchen items: coffee cups, sandwich wrappers, drinking straws, and forks. They then calculated how many times a reusable item would have to be reused before the environmental impact was the same as a single-use plastic item. For example, they compared single-use coffee cups made of Styrofoam or paper to reusable coffee cups made of ceramic, plastic, or metal.

The results of this study were quite surprising. For all of the items, the break-even time was much longer if a person washed the item by hand rather than in a dishwasher, because dishwashers clean dishes more efficiently. Of the twelve reusable items examined, three of them never broke even in terms of their environmental impact compared to comparable single-use items. For example, beeswax paper and silicone bags, both used to wrap sandwiches, have to be washed by hand with warm water. The environmental impact of this washing exceeds the impact of producing a single-use sandwich bag or plastic wrap each time a sandwich is wrapped. In contrast, reusable forks made of metal, plastic, or bamboo had a break-even point after just 12 uses. Among the different types of coffee cups, ceramic cups had the lowest environmental impact.

These types of studies illustrate that analyzing the environmental effects of the products we use is complex because it involves synthesizing many aspects of environmental studies. Not only does it include science, ethics, and social judgments, it also necessitates a systems-based understanding of waste generation, waste reduction, and waste disposal.

Sources: H. Fetner and S. A. Miller, Environmental payback periods of reusable alternatives to single-use plastic kitchenware products, *The International Journal of Life Cycle Assessment* 26 (2021): 1521–1537; M. B. Hocking, Paper versus polystyrene: A complex choice, *Science* 251 (1991): 504–505; E. van der Harst, J. Potting, and C. Kroeze, Comparison of different methods to include recycling in LCAs of aluminum cans and disposable polystyrene cups, *Waste Management* 48 (2016): 565–583.

Practice your science skills

1. **Mathematical Routines:** As mentioned, Dunkin' Donuts phased out foam cups and this resulted in 1 billion fewer foam cups going into landfills each year. Scientists have estimated that as of 2021, people in the United States dispose of 25 billion foam cups annually. Which mathematical formula should be used to determine the change between the numbers of foam cups going into U.S. landfills in 2019 versus the number in 2021?
2. **Mathematical Routines:** Using the information from question 1, calculate the percent change in the number of foam cups found in landfills before and after Dunkin' Donuts phased out their foam cups in 2020. Show your work.
3. **Environmental Solutions:** Make a claim that provides a solution to the environmental problem of single-use compostable forks found in landfills.

In the previous unit we discussed the various forms of air pollution, including photochemical smog, atmospheric CO_2, and indoor air pollutants. In this unit we consider the various pollutants that occur on land and in the water. We begin by considering the sources of the pollutants and how different types of pollutants impact human health in unique ways. Next, we explore the effects of excess nutrients and excess heat on aquatic ecosystems. We also examine the materials that we dispose as trash and ways to reduce the amount of trash that we generate. We wrap up the unit by discussing how to safely treat sewage, followed by an explanation of how we determine the impacts of pollutants on human health.

Module 47

Unit 8	47	48	49	50	51	52	53	54

Sources of Pollution, Human Impacts on Ecosystems, and Endocrine Disruptors

As we saw in the unit opener, the issue of pollution on land and water can be complex and often involves trade offs in terms of the best solution. Pollution in aquatic and terrestrial ecosystems is generally defined as the contamination of land, water bodies, or groundwater with substances produced through human activities. As we will see, the broad range of pollutants that can be found on land and in water includes discarded items such as cups, chemical pollutants, and even oil pollution. The prevalence and impact of pollutants also varies. Given the many ecological connections between aquatic and terrestrial ecosystems, we will consider pollution in both types of ecosystems and, in many cases, how pollution moves between the ecosystems. In this module we will consider the origin of each category of pollutant, its negative effects on humans and the environment, and what can be done to reduce each pollutant.

Learning Goals

After reading this module you should be able to

47-1 identify the difference between point and nonpoint sources of pollution.

47-2 explain why species differ in their tolerance to pollutants.

47-3 identify the major groups of chemical pollutants and where they come from.

47-4 describe the impacts of chemical pollutants.

47-5 identify the sources of oil pollution.

47-1 What is the difference between point and nonpoint sources of pollution?

Point sources of pollution have single locations while nonpoint sources have diffuse locations

Regardless of the specific contaminant, aquatic and terrestrial pollution can come from either point sources or nonpoint sources. In this section we will look briefly at each. A **point source** is a distinct location from which pollution is directly produced. Examples of point sources include a particular factory that buries its waste underground or a sewage treatment plant that discharges its wastewater from a pipe into the ocean, as shown in **FIGURE 47.1a**. For atmospheric pollution, which we discussed in Module 42, a point source would be a smokestack that releases harmful chemicals into the air.

In contrast, a **nonpoint source** is a more diffuse area that produces pollution, such as an entire farming region, a suburban community with many lawns and septic systems, or storm runoff from a large number of parking lots. In each of these cases, there is a large number of relatively small contributions to pollution that are spread over a large area. For example, we can consider the thousands of cattle that live on hundreds of farms near a stream or river, as shown in Figure 47.1b. While each cow makes a small contribution of manure that can release excess nutrients that are carried to the stream, collectively the thousands of cattle spread across a large area represent a nonpoint source of pollution.

The distinction between the two sources can help us control pollutants on land and in water. If a city can determine one or two point sources for most of its pollution, it can target those specific sources to reduce pollution outputs. An interesting example of a point source of pollutants is the cruise ship industry. In recent years, it has been reported that many cruise ships have been dumping their sewage and other waste overboard into the oceans. In fact, the larger cruise ships can produce 1 metric ton of garbage per day. Public outcry over this behavior has led many cruise ship companies to improve their programs for recycling, incinerating, and treating their waste rather

Point source A distinct location from which pollution is directly produced.

Nonpoint source A diffuse area that produced pollution.

(a)

(b)

FIGURE 47.1 Two types of pollution sources. Pollution can enter water bodies in two ways. (a) Point sources, such as a sewage pipe, are distinct locations where water pollution is produced. (b) Nonpoint sources are more diffuse areas that produce pollution. For example, rainwater that runs off hundreds of square kilometers of agricultural fields and into streams can carry pollutants with it. *(a: age fotostock/Superstock; b: The Irish Image Collection/Superstock)*

than dumping all of it into the oceans. Controlling pollution from nonpoint sources is much more challenging because the sources can span a large area. For example, in Module 5 we saw that the enormous Mississippi River watershed is subject to pollution from tens of thousands of farms because of fertilizer and pesticide runoff.

47-2 Why do species differ in their tolerance to pollutants?

Organisms differ in their tolerance to various pollutants

As you likely recall from Module 11, every species has a niche that includes a range of conditions that it can tolerate. Living under this ideal range of conditions allows organisms to maintain **homeostasis**, which is the ability to experience relatively stable internal conditions in their bodies. Within this range of conditions, the organism can survive, grow, and reproduce. Outside this range of conditions, organisms can no longer maintain homeostasis, so they experience impaired survival, growth, and reproduction. The range of ideal conditions can differ a great deal among species that have evolved to live in different biomes with different biotic and abiotic conditions. In a similar way, species commonly differ in how they are affected by different concentrations of pollutants in the environment.

Homeostasis The ability to experience relatively stable internal conditions in their bodies.

Like other abiotic conditions, such as temperature and precipitation, the performance of individuals begins to decline as the concentration of a pollutant increases. Initial increases in pollutant concentration can impact the physiology and behavior of an individual, but higher concentrations can impede its growth and survival. For example, low concentrations of a pesticide that impact an animal's nervous system may cause their movement to be slower and less coordinated. At moderate concentrations, the pesticide more strongly impairs nerve transmission and muscle movement to the point where it is difficult to find food and grow. At even higher concentrations, the pesticide can cause the nervous system to stop stimulating heart contractions and the animal dies. As a result, it is not merely the presence of a pollutant that harms a species, but rather the concentration of the pollutant and the ability of the species to tolerate it.

Because different species possess very different genetics, each species typically has a different level of tolerance to a pollutant. For example, you may recall from Module 14 our discussion of scientists using the presence of certain mayfly species in streams as bioindicators of stream health. Given that some species of mayflies are very sensitive to a variety of pollutants, their absence in a stream is a rapid indicator that the stream may be polluted. In contrast, most species of fish are much more tolerant to pollutants, so a stream that contains fish but no longer contains mayflies suggests a moderate concentration of pollutants, while a stream that no longer contains mayflies or fish suggests a much higher concentration of pollutants.

In some cases, species are impacted by combinations of pollutants and other stressors caused by human activities. For example, the corals that slowly build coral reefs over thousands of years serve as important biomes for a large diversity of ocean life, as we discussed in Module 3. Today, coral reefs are being harmed by several simultaneous challenges, including pesticides that impact the homeostasis of the corals combined with being smothered by increased sediments that run off the land from human activities. At the same time, the increase in CO_2 in the atmosphere has created warmer ocean temperatures. When the higher atmospheric CO_2 dissolves into the water, the ocean experiences a lower pH that causes further stress on the corals. Together, these environmental changes can make the corals more susceptible to various diseases, which can lead to reduced growth or, ultimately, death. When these harmful impacts are combined with destructive fishing practices, including tourists removing corals for souvenirs, the biodiversity and abundance of coral reefs leads to their decline in abundance in many regions of the world.

47-3 What are the major groups of chemical pollutants and where do they come from?

Chemical pollutants include heavy metals and synthetic compounds that are produced naturally and by human activities

In Unit 1 we saw that some compounds such as nitrogen and phosphorus cause environmental problems by overfertilizing the land and water. Other inorganic compounds, including heavy metals (lead, arsenic, and mercury), acids, and synthetic compounds (pesticides, pharmaceuticals, and hormones), can directly harm humans and other organisms because species have a limited range of tolerance to such compounds. In this section, we will examine how each of these chemicals enters ecosystems and the effects that they can have on organisms including humans.

Heavy Metals

Heavy metals are a group of chemicals that can pose serious health threats to humans and other organisms. Three heavy metals are of particular concern: lead, arsenic, and mercury.

Lead

Lead is rarely found in natural sources of drinking water, but it contaminates water that passes through lead-lined pipes and other materials that contain lead such as brass fittings and solder used to fasten pipes together. Fetuses and infants are the most sensitive to lead—exposure can damage the brain, nervous system, and kidneys. Since the federal government required a gradual elimination of lead in gasoline and paint in the 1970s, lead exposure in the United States has declined by about 90 percent. However, lead contamination in children remains a serious problem in low-income neighborhoods due to the presence of old lead paint in buildings, which over time can peel off walls and be inadvertently eaten by small children.

A series of federal guidelines for building construction implemented during the past 3 decades now requires the installation of lead-free pipes, pipe fittings, and pipe solders. These changes and the increased use of water filtration systems in many homes are gradually reducing the problem of lead in drinking water, although it remains a concern in older houses and apartments. A high-profile example of the problem of lead water pipes happened in Flint, Michigan, in 2014. For many years, the city had been buying its water from Detroit, which supplied water that contained an added chemical that coated pipes and prevented lead from leaching into household water. However, Flint was struggling financially and determined it would save $5 million over two years if it did not buy the Detroit water but obtained water from the local Flint River. The Flint River has higher chloride concentrations, which can be corrosive to lead pipes, and the city of Flint did not add the chemical to coat and protect the pipes. As you can see in **FIGURE 47.2**, lead pipes in the Flint water system became

(a)

(b)

FIGURE 47.2 Lead contaminated water. (a) When the city of Flint, Michigan, switched its water source from Detroit to the Flint River it failed to add a chemical that prevents pipe corrosion. The old lead pipes in the water distribution system began to leak lead and poison residents of the city. (b) As a result of this corrosion, tap water in many homes and businesses often was contaminated by lead from the corroding pipes. *(a & b: Courtesy Virginia Tech College of Engineering)*

corroded, the water turned brown, and lead began to leach from the pipes into household water supplies.

The EPA guidelines state that there are health concerns for lead in water at 5 parts per billion (ppb). Water tests in 2015 revealed that concentrations of lead in Flint's water supply were much higher—the highest sample concentration measuring 13,000 ppb. Late in 2015, the Flint water system was reconnected to the Detroit water system, but the corroded pipes will likely continue to leach lead.

Exposure to lead attacks the nervous system and causes children to experience reduced IQ, attention problems, and even kidney damage. It is estimated that 8,000 children were exposed to high lead concentrations in the city of Flint. In an attempt to save $5 million in water costs, the city incurred an estimated $400 million in current and future costs associated with lead poisoning. This is in addition to the $1.5 billion it will cost to replace the old, now corroded pipes.

Arsenic

Arsenic is a compound that occurs naturally in Earth's crust and can dissolve into groundwater. As a result, naturally occurring arsenic in rocks can lead to high concentrations of arsenic in groundwater and drinking water. Human activity also contributes to higher arsenic concentrations in groundwater. For example, mining breaks up rocks deep underground, and industrial uses of arsenic for items such as wood preservatives can add to the amount of arsenic found in drinking water. Fortunately, arsenic can be removed from water using water filtration systems.

FIGURE 47.3 indicates where high concentrations of arsenic have been found in well water throughout the United States. As you can see, concentrations are highest in the Midwest and West. Arsenic in drinking water is associated with cancers of the skin, lungs, kidneys, and bladder. These illnesses can take 10 years or more after exposure to develop. Even very low concentrations of arsenic, such as those found in many wells around the world and measured in ppb, can cause severe health problems. In 1999, the EPA set the upper "safe" limit to 10 ppb, a compromise reached after much debate between environmental groups pushing for 5 ppb and water, mining, and wood preservative industries arguing that even the 10-ppb limit would be too expensive to implement. As a result, after substantial investment required to improve many water treatment facilities, people in the United States will now have lower amounts of cancer-causing arsenic in their drinking water.

The problem of arsenic in drinking water is a worldwide issue. During the 1980s and 1990s, for example, water engineers in regions of Bangladesh and eastern India drilled millions of deep wells in an attempt to avoid surface waters that contaminated by local pollution. While this had the desired effect of obtaining less-polluted water, officials were not aware that this deeper groundwater was contaminated by naturally occurring arsenic from the rocks deep underground. Currently, 140 million people in this region drink arsenic-contaminated water and thousands of individuals have been diagnosed with arsenic poisoning. Current efforts to solve this problem include plans to collect uncontaminated rainwater as a source of drinking water and research to develop inexpensive filters that can remove arsenic from the well water. It is a challenge in many parts of the world for people to know that they should have their water tested and, more importantly, having facilities available to do this testing.

Mercury

Mercury is another naturally occurring heavy metal found in increased concentrations in water as a result of human

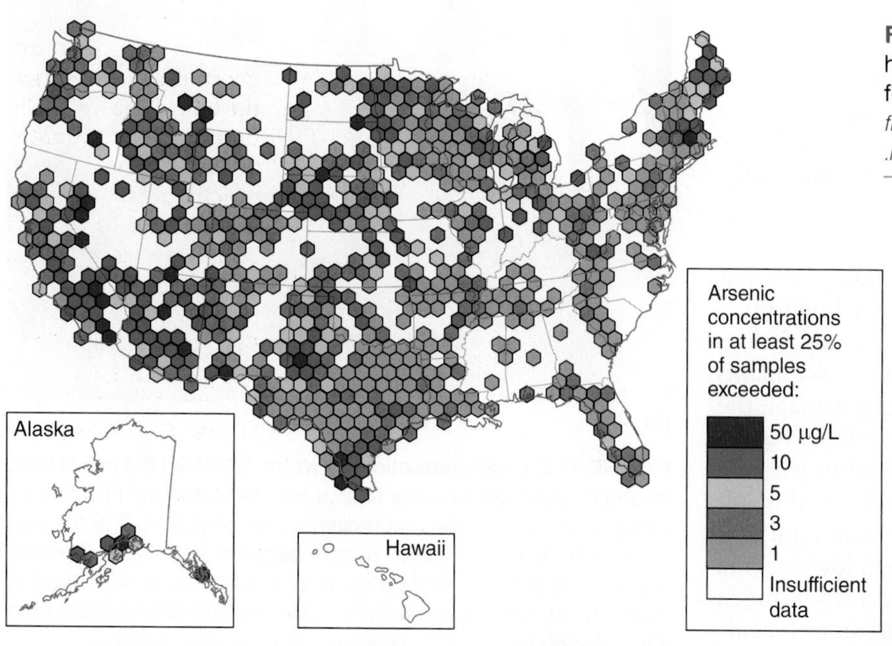

FIGURE 47.3 Arsenic in U.S. well water. The highest concentrations of arsenic are generally found in the upper Midwest and the West. *(Data from http://water.usgs.gov/nawqa/trace/pubs/geo_v46n11/fig3.html)*

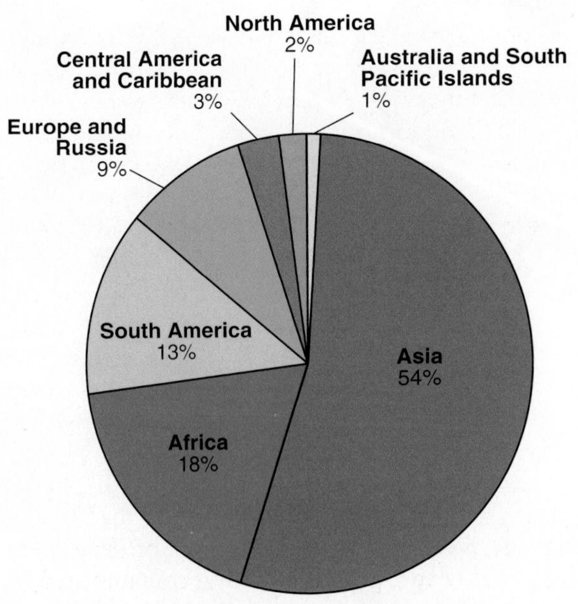

FIGURE 47.4 World mercury emissions. Mercury emissions from human activities vary greatly among regions of the world. *(Data from the 2018 Global Mercury Assessment)*

activities. **FIGURE 47.4** shows mercury releases from different regions of the world, as estimated by the 2018 Global Mercury Assessment. Among regions of the world, 2 percent of human-produced mercury comes from North America, where mercury emissions have continued to decline, while more than half comes from Asia.

Approximately two-thirds of all mercury produced by human activities comes from the burning of fossil fuels, especially coal. Other important sources of mercury include the incineration of garbage, hazardous waste, and medical and dental supplies. One of the lesser-known sources of mercury comes from the raw materials that go into the manufacturing of cement for construction. The limestone used to make cement can contain mercury, and this mercury is released during a heating process that is used when making cement. Moreover, the source of heat for cement manufacturing is often coal that also releases mercury when burned.

Petroleum exploration is a source of both mercury and lead pollution. Each petroleum well produces roughly 681,374 L (180,000 gallons) of contaminated wastewater and mud over its lifetime. This water is usually dumped at the drilling site and, depending on the soils and topography, can either run into nearby waterways or infiltrate the soil and contaminate the underlying groundwater.

The mercury emitted by these activities eventually finds its way into water. Inorganic mercury (Hg) is not particularly harmful, but its release into the environment can be hazardous because of a chemical transformation it undergoes. In wetlands and lakes, bacteria convert inorganic mercury into methylmercury, which is highly toxic to humans. Methylmercury damages the central nervous system, particularly in young children and in the developing embryos of pregnant women. The result impairs coordination and the senses of touch, taste, and sight.

Human exposure to methylmercury occurs mostly through eating fish and shellfish. Methylmercury can move up the food chain in aquatic ecosystems, which results in the top consumers containing the highest concentrations of mercury in their bodies. Given that oceans are contaminated with mercury and that tuna are top predators, it is not surprising that these fish contain high concentrations of mercury. In 2017, researchers reported that mercury concentrations in two species of tuna were increasing by 4 to 5 percent annually, which translates to a 40 to 50 percent increase over a decade. When researchers examined meat from the yellowfin tuna (*Thunnus albacares*) that were caught in different regions of the world's oceans, they found that some regions contained individuals that frequently exceeded the U.S. Environmental Protection Agency guideline of 0.3 μg methylmercury per gram of tuna, as you can see in **FIGURE 47.5**.

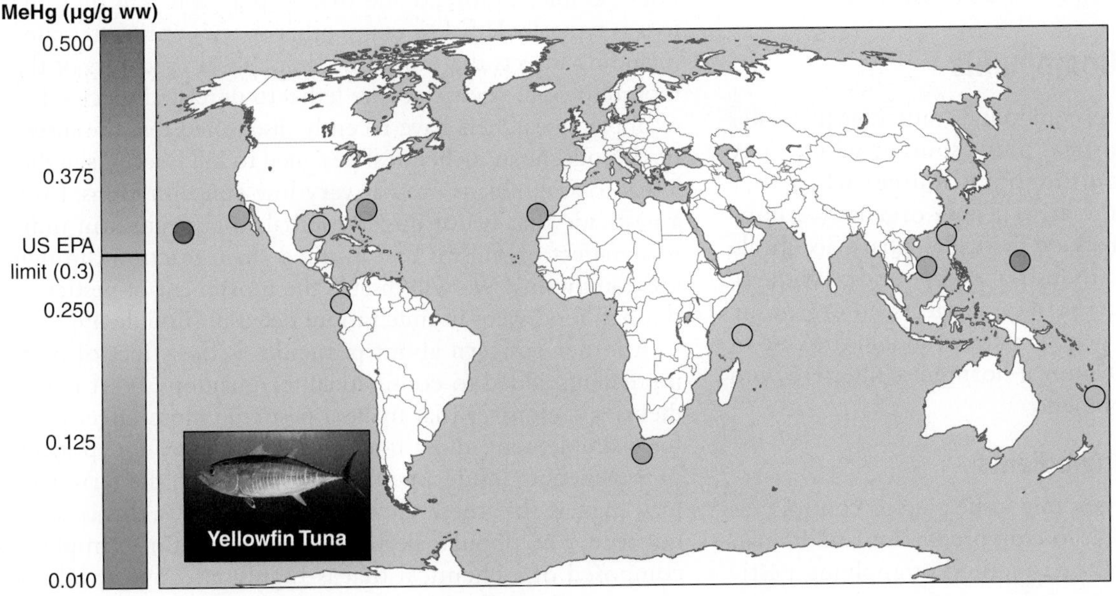

FIGURE 47.5 Mercury concentration in yellowfin tuna. The yellowfin tuna lives in different regions of the oceans, which contain different concentrations of mercury in the water. Tuna in the North Pacific contain the highest concentrations of methylmercury, at a level that is above the U.S. EPA limit of 0.3 μg/g of tuna. *(Data from Nicklish et al. 2017. Environmental Pollution 229:87–93)*

FIGURE 47.6 Applying pesticides. Pesticides provide benefits to humans, but they also can have unexpected effects on humans and other nonpest organisms that are not fully understood and have not been adequately investigated. This airplane is spraying pesticides over crops in New South Wales, Australia. *(incposterco/Getty Images)*

What can be done about mercury pollution? In 2013, the United States announced a new agreement with more than 140 other countries to reduce global mercury pollution. In a step toward this goal, the EPA has proposed that cement-manufacturing plants reduce mercury emissions by 92 percent. New EPA rules are also being written to reduce mercury emissions from other major sources, including coal-burning power plants. These changes have allowed a reduction in mercury emissions in North America from 6 percent of the global contribution in 2013 to only 2 percent in 2018.

Synthetic Organic Compounds

Synthetic, or human-made, compounds can enter the water supply either from industrial point sources where they are manufactured or from nonpoint sources when they are applied over very large areas. These organic (carbon-containing) compounds include pesticides, pharmaceuticals, military compounds, and industrial compounds. Synthetic organic compounds have a variety of effects on organisms. They can be toxic, cause genetic defects, and, in the case of compounds that resemble animal hormones, interfere with growth and sexual development.

Pesticides and Inert Ingredients

Pesticides serve an important role in helping to control pest organisms that pose a threat to crop production and human health. **FIGURE 47.6** shows one method of applying pesticides on crops. Although natural pesticides such as arsenic have been used for centuries, the first generation of synthetic pesticides was developed during World War II. These chemicals are very effective in killing a variety of undesired plants (herbicides), fungi (fungicides), and insects (insecticides). In the decades that have followed, however, environmental scientists have identified a number of concerns about the unintended effects of pesticides.

Most pesticides do not target particular species of organisms, but generally kill a wide variety of related organisms. For example, an insecticide that is sprayed to kill mosquitoes is typically lethal to many other species of invertebrates, including insects that might be desirable as predators of the mosquito. Some pesticides are lethal to unrelated species. For example, researchers have recently discovered that the insecticide endosulfan, a chemical designed to kill insects, is highly lethal to amphibians even at very low concentrations. Even a pesticide that is not directly lethal to a species can indirectly affect organisms by causing a chain reaction through the community. We will discuss the movement of pesticides through food webs in much more detail in Module 49.

Another concern about pesticides is the effect of inert ingredients added to commercial formulations. Inert ingredients are additives that make a pesticide more effective—for example, they allow it to dissolve in water for spraying or to penetrate inside a pest species. Although the term *inert* may suggest that these chemicals are harmless, this is often not true. The popular herbicide Roundup, for example, is composed of a chemical that is highly effective at killing plants but has difficulty getting past the waxy outer layer of leaves without the help of an added inert ingredient. Since inert ingredients are legally classified as trade secrets and most are not required to be tested for safety, their effects are not always known before a product comes to market. In the case of Roundup, recent research has discovered that the herbicide is highly toxic to amphibians, not because of the plant-killing chemical but because of the inert ingredients. It appears that the same properties that allow the penetration of leaves also allow the penetration of tadpole gill cells. The gills burst and the tadpoles suffocate. Unintended consequences such as these have roused interest in both Europe and North America to require that inert ingredients be tested for potentially harmful effects.

Pharmaceuticals and Hormones

While most people know that pesticides are commonly found in the environment, they often are surprised to learn that pharmaceutical drugs are also common. For example, the U.S. Geological Survey tested 139 streams across the United States for a variety of chemical contaminants. **FIGURE 47.7** shows data for the frequency of detection. Among the different types of chemicals that were present at detectable levels, approximately 50 percent of all streams contained antibiotics and reproductive hormones, 80 percent contained nonprescription drugs, and 90 percent contained steroids. In a follow-up study in the American Midwest, the U.S. Geological Survey sampled 100 streams and reported in 2017 that the average stream contained 52 pesticides. In most cases the concentrations of these chemicals are quite low and currently are not thought to pose a risk to environmental or human health. Some chemicals such as hormones, however, operate at very low concentrations inside the tissues of organisms and, as we will see later in this module, we are just beginning to understand their effects.

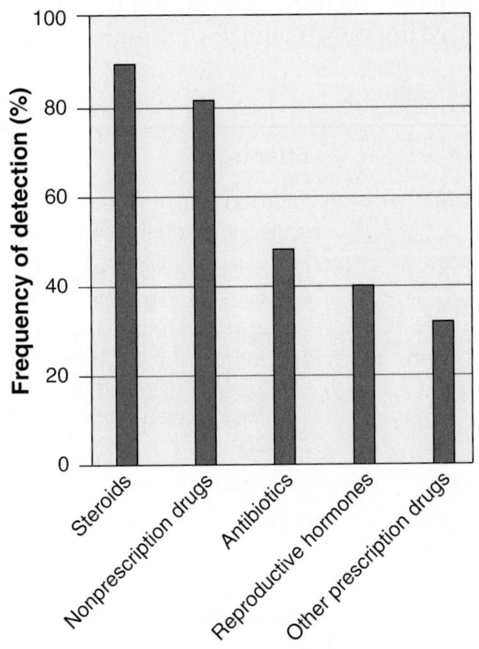

FIGURE 47.7 Contaminants in streams. Streams contain a wide variety of chemicals, including pharmaceutical drugs and hormones. These come from a combination of wastewater inputs, agriculture, forestry, and industry. *(Data from D.W. Kolpin et al. 2002. Pharmaceuticals, hormones, and other organic wastewater contaminants in U.S. streams, 1999–2000: A national reconnaissance.* Environmental Science and Technology *36: 1202–1211)*

Military Compounds

Perchlorates, a group of harmful chemicals used for rocket fuel, sometimes contaminate the soil in regions of the world where military rockets are manufactured, tested, or dismantled. Perchlorates come in many forms. The U.S. space shuttle, for example, used a booster rocket that contained 70 percent ammonium perchlorate. Perchlorates easily leach from contaminated soil into the groundwater, where they can persist for many years. Human exposure to perchlorates comes primarily through consumption of contaminated food and water. In the human body, perchlorates can affect the thyroid gland and reduce the production of hormones necessary for proper functioning of the human body.

Industrial Compounds

Industrial compounds are chemicals used in manufacturing. Unfortunately, it was once common for manufacturers in the United States to dispose of industrial compounds directly into bodies of water. One of the most widely publicized consequences of this practice occurred in the Cuyahoga River of Ohio. For more than 100 years, industries along the river dumped industrial wastes that formed a slick of pollution along the surface, killing virtually all animal life. The problem became so bad that the river actually caught fire and burned several times over the decades (**FIGURE 47.8**). The fire on the river in 1969 garnered national attention and led to a movement to clean up America's waterways.

FIGURE 47.8 A river on fire. In 1952, the polluted Cuyahoga River caught fire after a spark ignited the film of industrial pollution that was floating on the surface of the water. *(Bettmann/Getty Images).*

Today, the Cuyahoga River and most other rivers in the United States are much cleaner because of legislation that substantially reduced the amount of industrial and other waste that can be legally dumped into waterways.

There are many other industrial compounds that can harm people and organisms in nature. For example, **polychlorinated biphenyls (PCBs)** are a group of industrial compounds that were once used to manufacture plastics and insulate electrical transformers. Ingested PCBs are lethal and carcinogenic. While PCBs have long been a concern, there is a growing concern over compounds known as PBDEs (polybrominated diphenyl ethers). PBDEs are most commonly known as flame retardants added to a wide variety of items that include construction materials, furniture, electrical components, and clothing. They make buildings and their contents considerably less flammable than they would be otherwise. Since the 1990s, however, scientists have been detecting PBDEs in some unexpected places, including fish, aquatic birds, and human breast milk. Exposure to some types of PBDEs can lead to brain damage, especially in children. As a result, the European Union and several states, including Washington and California, have banned the manufacture of several types of PBDEs.

There is also a growing concern over another group of chemicals that have been used for decades in fire retardants, nonstick pans, food packaging, stain-resistant carpets and water-repellant clothing. Known as PFAS (per- and polyfluoroalkyl substances), these chemicals are being increasingly discovered in groundwater and public drinking water, often as a result of companies dumping the waste chemicals in waterways. Because the products containing these chemicals are so pervasive and can persist for thousands of years, PFAS are now detected in the blood of nearly everyone in the world. These chemicals can have a variety of impacts on humans including causing cancer and harming the development of fetuses.

Polychlorinated biphenyls (PCBs) A group of industrial compounds that were once used to manufacture plastics and insulate electrical transformers.

Neurotoxin A chemical that disrupts the nervous systems of animals.

47-4 What are the impacts of chemical pollutants?

Chemical pollutant impacts can be categorized as neurotoxins, carcinogens, teratogens, allergens, or endocrine disruptors

Chemicals can have many different effects on organisms, and some of the most harmful are common in our environment; **TABLE 47.1** lists those of current concern. They can be grouped into five categories: neurotoxins, carcinogens, teratogens, allergens, and endocrine disruptors.

Neurotoxins

Neurotoxins are chemicals that disrupt the nervous systems of animals. Many insecticides, for example, are neurotoxins

TABLE 47.1 Some chemicals of major concern

Chemical	Sources	Type	Effects
Lead	Paint, gasoline	Neurotoxin	Impaired learning, nervous system disorders, death
Mercury	Coal burning, fish consumption	Neurotoxin	Damaged brain, kidneys, liver, and immune system
Arsenic	Mining, groundwater	Carcinogen	Cancer
Asbestos	Building materials	Carcinogen	Impaired breathing, lung cancer
Polychlorinated biphenyls (PCBs)	Industry	Carcinogen	Cancer, impaired learning, liver damage
Radon	Soil, water	Carcinogen	Lung cancer
Vinyl chloride	Industry, water from vinyl chloride pipes	Carcinogen	Cancer
Alcohol	Alcoholic beverages	Teratogen	Reduced fetal growth, brain and nervous system damage
Atrazine	Herbicide	Endocrine disruptor	Feminization of males, low sperm counts
DDT	Insecticide	Endocrine disruptor	Feminization of males, thin eggshells of birds
Phthalates	Plastics, cosmetics	Endocrine disruptor	Feminization of males

that interfere with an insect's ability to control its nerve transmissions. Insects and other invertebrates are highly sensitive to neurotoxin insecticides. These animals can become completely paralyzed, cannot obtain oxygen, and quickly die. Other important neurotoxins include lead and mercury, which are heavy metals that can damage the human kidneys, brain, and nervous system.

Carcinogens

Carcinogens are chemicals that cause cancer. Carcinogens cause cell damage and lead to uncontrolled growth of these cells either by interfering with the normal metabolic processes of the cell or by damaging the genetic material of the cell. Carcinogens that cause damage to the genetic material of a cell are called **mutagens** (although not all mutagens are carcinogens). Some of the most well-known carcinogens include asbestos, radon, formaldehyde, and the chemicals found in tobacco. Asbestos was once commonly used as a fireproof insulation material, radon is a naturally occurring gas that can enter through building foundations, and formaldehyde has been used in a wide range of preservatives including in many building materials.

Teratogens

Teratogens are chemicals that interfere with the normal development of embryos or fetuses. One of the most infamous teratogens was the drug thalidomide, prescribed to pregnant women during the late 1950s and early 1960s to combat morning sickness. Sadly, tens of thousands of these mothers around the world gave birth to children with defects before the drug was taken off the market in 1961. One of the most common modern teratogens is alcohol. Excessive alcohol consumption reduces the growth of the fetus and damages the brain and nervous system of the fetus, a condition known as fetal alcohol syndrome. This is why physicians recommend that women not consume alcoholic beverages while they are pregnant.

> **AP® Exam Tip**
>
> You should be able to identify specific examples of chemical toxins, their source, and their human health impacts.

Allergens

Allergens are chemicals that cause allergic reactions. Although allergens are not pathogens, allergens are capable of causing an abnormally strong response from the immune system. In some cases, this response can cause breathing difficulties and even death. Typically, a given allergen only causes allergic reactions in a small fraction of people. Some common chemicals that cause allergic reactions include the chemicals naturally found in peanuts and milk, and several drugs including penicillin and codeine.

Endocrine Disruptors

Endocrine disruptors are chemicals that interfere with the normal functioning of hormones in an animal's body. Hormones are normally manufactured in the endocrine system and released into the bloodstream in very low concentrations. As the hormones move through the body, they bind to specific cells. Binding stimulates the cell to respond in a way that regulates the functioning of the body including growth, metabolism, and the development of reproductive organs. As **FIGURE 47.9** shows, an endocrine disruptor can bind to receptive cells and cause the cell to respond in ways that are not beneficial to the organism.

One high-profile example of endocrine disruptors in our environment is the group of reproductive hormones that can be found in the **wastewater**, which is the water produced by livestock operations and human activities, including human sewage from toilets and gray water from bathing and washing clothes and dishes. In waterways exposed to hormones through wastewater, scientists are increasingly finding that male fish, reptiles, and amphibians are becoming feminized; males possess testes that have low sperm counts and, in some cases, testes that produce both eggs and sperm. Males normally convert the female hormone estrogen into the male chemical testosterone. Reproductive hormones in wastewater can interfere with the production of testosterone, which causes males to have higher concentrations of estrogen and lower concentrations of testosterone in their bodies.

In the Chesapeake Bay, the U.S. Fish and Wildlife Service announced that estrogen from sewage in the water was the most likely reason that 23 percent of male largemouth bass and 82 percent of male smallmouth bass developed into hermaphrodites; their male sex organs grow eggs, a function that normally only occurs in the sex organs of female fish. This impact not only concerns fish, but also is a concern for humans who share many similarities in their reproductive systems and may consume water contaminated with estrogens.

In addition to real animal hormones, some pesticides can mimic animal hormones and also serve as endocrine disruptors. Such discoveries raise serious concerns about whether

Carcinogen A chemical that causes cancer.

Mutagen A type of carcinogen that causes damage to the genetic material of a cell.

Teratogen A chemical that interferes with the normal development of embryos or fetuses.

Allergen A chemical that causes allergic reactions.

Endocrine disruptor A chemical that interferes with the normal functioning of hormones in an animal's body.

Wastewater The water produced by livestock operations and human activities, including human sewage from toilets and gray water from bathing and washing clothes and dishes.

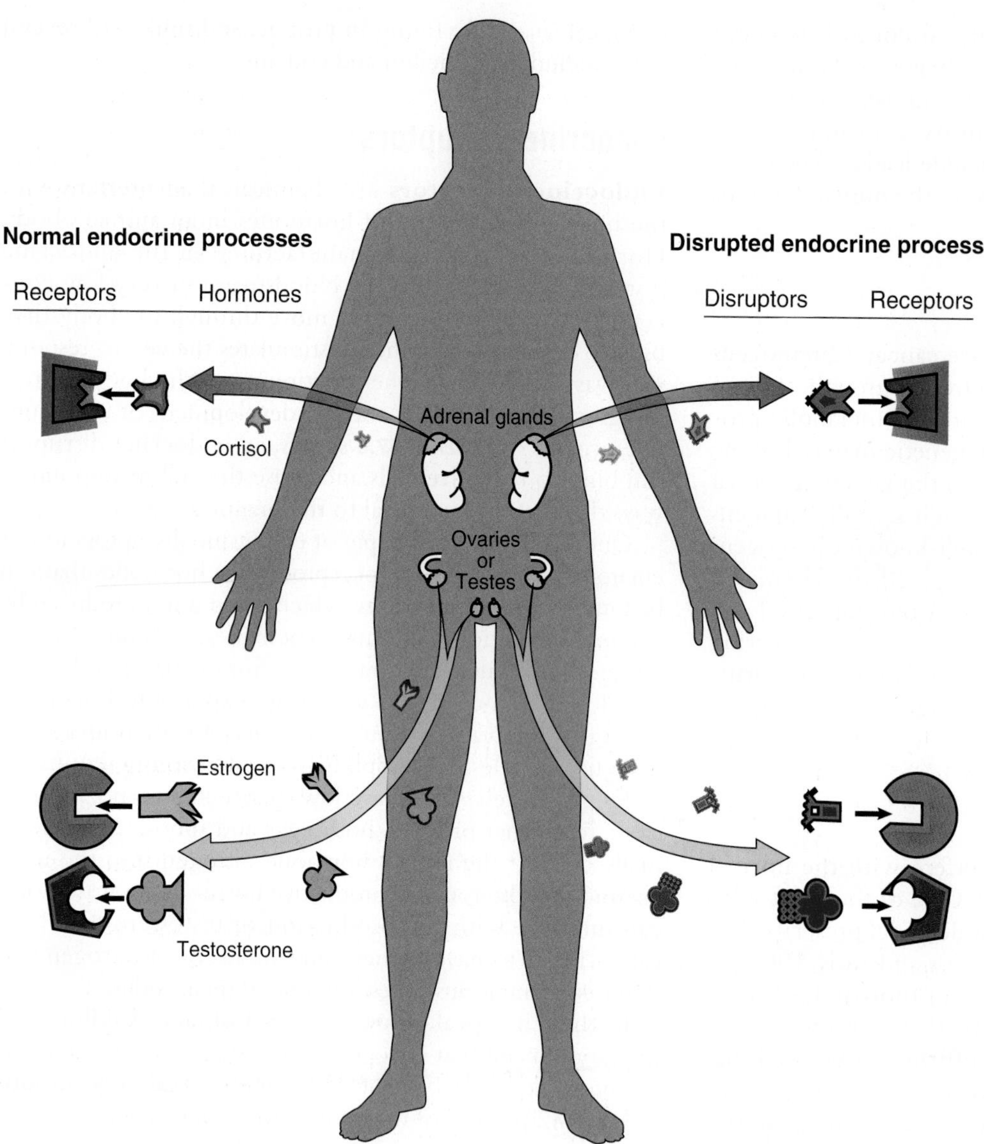

FIGURE 47.9 Endocrine disruption. In normal endocrine processes, hormones bind with receptors on cells to regulate the functioning of the body including growth, metabolism, and the development of reproductive organs. Hormone-disrupting chemicals mimic the hormones in the body and also bind to receptive cells and cause the cell to respond in ways that are not beneficial to the organism.

endocrine disruptors might affect the normal functioning of human hormones. These effects include developmental disorders, low sperm counts in men, and an increased risk of breast cancer in women.

47-5 What are the sources of oil pollution?

Oil pollution comes from multiple sources

The pollution of Earth's oceans and shorelines from crude oil and other petroleum products is an ongoing problem that can have large economic consequences on recreation, tourism, and the fishing industry. Oil is a persistent substance that can spread below and across the surface of the water for hundreds of kilometers and leave shorelines with a thick, viscous covering that is extremely difficult to remove. It is highly toxic to marine organisms, including birds, mammals, and fish, as well as to algae and microorganisms that form the base of the aquatic food chain. For example, exposure to oil can cover the feathers of birds and the fur of mammals, such as seals and otters. This sticky outer layer of oil is difficult to remove and is not only toxic, but also reduces the animals' layer of insulation so they cannot stay warm. It also impairs their ability to swim and fly. Whales and dolphins can inadvertently inhale the oil, causing it to get into their lungs, or inadvertently consume the oil as they capture their prey in the water. It can also sink to the ocean bottom where it smothers and kills bottom-dwelling organisms. In this section, we will examine the many sources of oil pollution and then talk about some of the ways currently used to clean up oil and to reduce its harmful effects.

Sources of Oil Pollution

There are many different sources of oil pollution in water. Oil and other petroleum products can enter the oceans as spills from oil tankers or pipelines. One of the best-known

FIGURE 47.10 The *Exxon Valdez* oil spill. In 1989, the oil tanker ran aground and spilled millions of liters of crude oil onto the shores of Alaska, where the oil killed thousands of animals and harmed many others such as this red-necked grebe (*Podiceps grisegena*) on Knight's Island, located about 35 miles from the spill. *(AP Images)*

spills involved the tanker *Exxon Valdez* that ran aground off the coast of Alaska in 1989 (**FIGURE 47.10**). The ship spilled 42 million liters (11 million gallons) of crude oil that spread across the surface of several kilometers of ocean and along hundreds of kilometers of coastline. The accident killed 250,000 seabirds, 2,800 sea otters, 300 harbor seals, and 22 killer whales. Cleanup efforts went on for 2 decades.

Twenty years after the *Exxon Valdez* spill, scientists evaluated the state of the contaminated Alaskan ecosystem. They concluded that the harmed populations of many species have rebounded, including bald eagles and salmon. However, several species have not yet rebounded, including killer whales (*Orcinus orca*) and sea otters (*Enhydra lutris*). Nor has the oil been completely removed from the environment. Pits dug into the shoreline suggest that approximately 55,000 L (14,500 gallons) of oil remain. It is estimated that this oil will take more than 100 years to break down and the long-lasting effects will only become apparent over the coming decades.

For its part, Exxon has paid $1 billion for the cleanup and $500 million in damages. The company also changed the ship's name, although the ship has been banned from carrying oil in North America. The *Valdez* accident sparked new rules for oil tankers in North America. The *Exxon Valdez* had a single-hull design, but tankers must now have a double-hull design with two steel walls to contain leaking oil. As a result, the number of large oil spills from oil tankers has dramatically declined in recent decades.

Offshore drilling is another source of oil pollution. There are approximately 5,000 offshore oil platforms in North America and another 3,000 worldwide. Drilling platforms often experience leaks. The best estimate for the amount of petroleum leaking into North American waters is 146,000 kg (322,000 pounds) per year. In other parts of the world, antipollution regulations are often less stringent. Estimates of the amount of petroleum leaking into the ocean annually from foreign oil platforms range from 0.3 million to 1.4 million kilograms (0.6 million to 3.1 million pounds).

One of the most famous oil leaks from an offshore platform occurred in 2010 on a BP operation in the Gulf of Mexico. In this case, an explosion on the *Deepwater Horizon* platform caused a pipe to break on the ocean floor nearly 1.6 km (1 mile) below the surface of the ocean. From the time of the explosion in April until the well was sealed in August 2010, the broken pipe released an estimated 780 million liters (206 million gallons) of crude oil into the Gulf of Mexico. This spill contaminated beaches, wildlife, and the estuaries that serve as habitats for the reproduction of commercially important fish and shellfish. The magnitude of the oil spill was nearly 20 times larger than that of the *Exxon Valdez*. However, because much of the oil spilled into the ocean, scientists may not be able to assess the full impact of the oil spill for several decades. The accident has the potential to become one of the largest environmental disasters in history.

In addition to oil spills, oil pollution in the ocean occurs naturally. In fact, the U.S. National Academy of Sciences recently estimated that natural releases of oil from seeps in the bottom of the ocean account for 60 percent of all oil in the waters surrounding North America and 45 percent of all oil in water worldwide. **FIGURE 47.11** shows the proportion

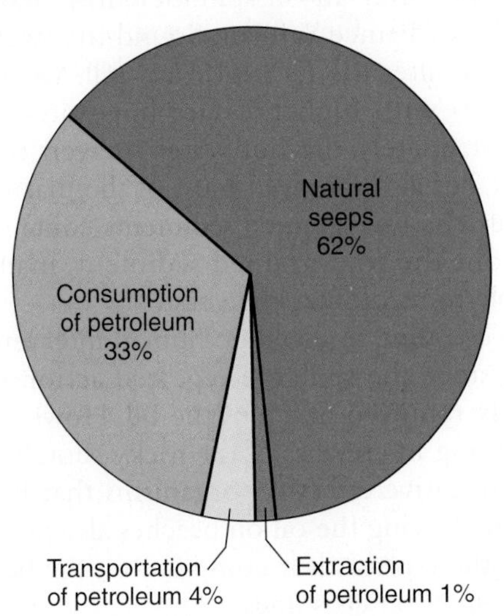

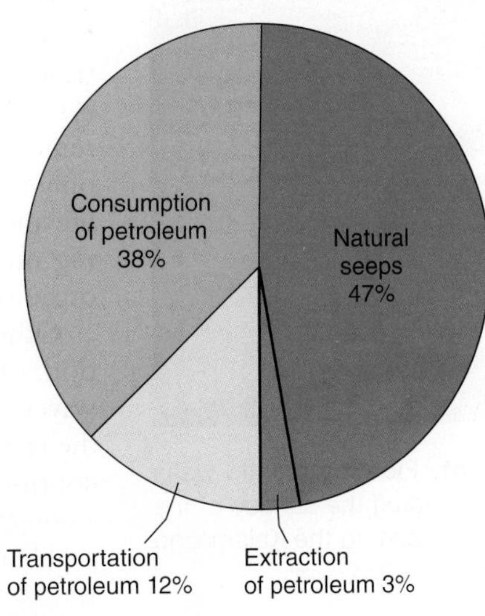

FIGURE 47.11 Sources of oil in the ocean. Oil contamination in the ocean, both (a) in North America and (b) worldwide, comes from a variety of sources including natural seeps, extraction of oil from underneath the ocean, transport of oil by tanker or pipeline, and consumption of petroleum-based products. *(Data from http://oceanworld.tamu.edu/resources/oceanography-book/contents.htm)*

of different sources of oil in water for both North American and worldwide marine waters. In the waters controlled by the United States, the ocean seeps more than 270,000 L (70,000 gallons) of oil every day. This means that when we assess the environmental impact of oil in our oceans, we must consider the combination of both natural and anthropogenic releases of oil. "Do the Math: Calculating the Magnitude of Oil Pollution" provides an opportunity to work with converting units of measurement to determine the amount of oil seepage.

Remediating Oil Pollution

Since the 1989 *Exxon Valdez* oil spill, researchers have been investigating how best to remediate oil spills. Contaminated mammals and waterfowl must be cleaned by hand. Bird feathers that are covered with oil, for example, become heavy and lose their ability to insulate. The best approach to cleaning up the spilled oil, however, is not always clear.

Oil spilled in the ocean can either float on the surface or remain far below in the form of underwater plumes. For oil floating on the surface of the open ocean, a common approach is to contain the oil within an area and then suck it off the surface of the water. Containment occurs by laying out oil containment booms that consist of plastic barriers floating on the surface of the water and extending down into the water for several meters. These plastic walls keep the floating oil from spreading further. Once the oil is contained, boats equipped with giant vacuums suck up as much oil as possible (**FIGURE 47.12**). In shallow areas and along the coastline, absorbent materials are used to suck up the spilled oil.

FIGURE 47.12 Oil-spill containment. Floating plastic walls can contain oil spills while the oil is sucked off the surface of the water. This worker is using a vacuum to clean up the Kalamazoo River in Battle Creek, Michigan. *(Jim West/Alamy Stock Photo)*

A second approach to treating oil floating on the surface is to apply chemicals that help break up and disperse the oil before it hits the shoreline and causes damage to the coastal ecosystems. Although the dispersants can be effective, they can also be toxic to marine life. Current research is examining ways to make chemical dispersants more environmentally friendly.

A third approach is to burn the oil slicks. This was done during the *Deepwater Horizon* oil spill to rapidly remove floating crude oil. However, the burning causes a great deal of air pollution and leaves behind a thick by-product that can be harmful to wildlife.

A fourth approach to cleaning up oil uses genetically engineered bacteria. Several years ago, scientists discovered a naturally occurring bacterium that obtained its energy by consuming oil emerging from natural seeps. These bacteria were typically rare in the ocean but were very abundant in areas where oil spills or seeps occurred. Scientists are currently trying to determine the genes that confer the bacteria's ability to consume oil and hope to insert copies of these genes into genetically modified bacteria to consume oil spills even faster.

Research on oil spill cleanups continues today. For example, researchers discovered that specially designed sponges could absorb oil while repelling seawater. These sponges can be used repeatedly and absorb 50 times their own weight. Researchers also have discovered a new method to burn oil slicks by heating the oil to allow more complete combustion with cleaner emissions from the fires.

In contrast to oil floating on the surface of the water, oil in underwater plumes persists as a mixture of water and oil, similar to the mixture of vinegar and oil in a salad dressing. In the case of the BP platform explosion in the Gulf of Mexico, scientists reported observing an oil plume moving approximately 1,000 m (3,000 feet) below the surface of the ocean. The plume was approximately 24 to 32 km (15–20 miles) long, 8 km (5 miles) wide, and hundreds of meters thick. There is currently no agreed-upon method of removing underwater plumes from the water.

When spilled oil comes to shore, the best solution in not always clear. For example, there is some debate over how to treat rocky coastlines after an oil spill. Scientists have been monitoring parts of Prince William Sound that were treated in different ways after the *Exxon Valdez* spill. Workers cleaned some areas with high-pressure hot water to remove the oil. Unfortunately, the hot water sprayers not only removed the oil, but also removed most of the marine life and, in some cases, the fine-grained sediments containing nutrients. Without the fine-grained sediment, many organisms were unable to recolonize the coast.

Other parts of the coastline received no human intervention. Over the years since the spill, the repeated action of waves and tides slowly removed much of the oil. However, the remaining oil existing in crevices of the rocky shoreline continues to have a negative effect on organisms that live among the rocks. Thus, leaving the oil on beaches also poses problems. At present, there is no clear consensus on the best way to respond to oil spills on coastlines.

DO THE MATH Calculating the Magnitude of Oil Pollution

Preparing for the AP® Exam

As we have discussed, there are many sources of oil pollution including from natural seeps, oil consumption, extraction, and transportation. Oil pollution is typically estimated in terms of metric tons of crude oil, but how does this translate into gallons of oil?

1 metric ton of crude oil = 320 gallons of crude oil

Given that 160,000 metric tons of oil seeps from the ocean floor surrounding North America, how many gallons of crude oil are seeping into the ocean in this region?

$$160{,}000 \text{ metric tons} \times 320 \text{ gallons/ton} = 5.12 \times 10^7 \text{ gallons}$$

If we assume that 1 gallon of crude oil can be used to make 1 gallon of heating oil (which is a simplifying assumption), and the average house in the northeastern United States uses 800 gallons of heating oil each year, how many northeastern homes could be heated by the oil that is naturally seeping into the ocean around North America?

$$(5.12 \times 10^7) \text{ gallons} \div (8 \times 10^2) \text{ gallons/home} = (5.12 \div 8) \times (10^{7-2}) = 0.64 \times 10^5 = 6.4 \times 10^4 \text{ homes} = 64{,}000 \text{ homes}$$

YOUR TURN

1. The amount of pollution from oil extraction activities is 3,000 metric tons. How many gallons does this represent?
2. Using the assumptions above, how many homes in the northeastern United States could be heated by this amount of oil?

As we have seen in this module, humans are impacting ecosystems by adding a wide variety of chemical pollutants that impact the health and performance of animals and humans. In the next module we continue with this theme by examining how human activities contribute excess nutrients to ecosystems and how these excess nutrients have harmful impacts on ecosystems.

Module 47 AP® Review

Preparing for the AP® Exam

Learning Goals Revisited

47-1 What is the difference between point and nonpoint sources of pollution?

A point source is a distinct location from which pollution is directly produced while a nonpoint source is a more diffuse area that produces pollution.

47-2 Why do species differ in their tolerance to pollutants?

Every species has a niche that includes that range of conditions it can tolerate, including pollutants to which an individual might be exposed. The performance of individuals begins to decline as the concentration of a pollutant increases. Initial increases in pollutant concentration can impact the physiology and behavior of an individual, but higher concentration can impede the growth and survival of an individual. Because different species possess very different genetics, each species typically has a different level of tolerance to a pollutant.

47-3 What are the major groups of chemical pollutants and where do they come from?

Several inorganic compounds are common pollutants, including heavy metals (lead, arsenic, and mercury), acids, and synthetic compounds (pesticides, pharmaceuticals, and hormones). These chemicals can directly harm humans and other organisms.

47-4 What are the impacts of chemical pollutants?

Chemicals can have many different effects on organisms, and they can be grouped into five categories: neurotoxins, carcinogens, teratogens, allergens, and endocrine disruptors.

47-5 What are the sources of oil pollution?

The sources of oil pollution include oil spills from tanks and pipelines, offshore drilling, and natural releases from oil seeps in the bottom of the ocean. Oil spilled in the ocean can either float on the surface or remain far below in the form of underwater plumes. For oil floating on the surface of the open ocean, a common approach is to contain the oil within an area and then suck it off the surface of the water.

AP® Practice Questions

Preparing for the AP® Exam

Multiple-Choice Questions

Use the following diagram to answers questions 1 & 2:

1. Which location in the diagram best represents a nonpoint source of water pollution?
   (a) location A, sewage from a town
   (b) location B, wastewater from a city
   (c) location C, algae growth in a stream
   (d) location D, agricultural runoff

2. Which of the following is the best description of a pollutant originating from location A?
   (a) industrial waste released into the stream
   (b) animal waste from a dairy farm
   (c) pesticide runoff from lawns and parks
   (d) mercury pollution from storm water

3. Which of the following is a point source of water pollution?
   (a) agricultural runoff
   (b) discharge from a sewage treatment plant
   (c) stormwater runoff
   (d) a suburban community

4. Mercury
   (a) is harmless once converted into methylmercury.
   (b) exposure often occurs through shellfish.
   (c) is most concentrated in herbivores.
   (d) can be safely trapped during the production of concrete.

Use the answer choices below to answer questions 5–7:

(a) arsenic
(b) lead
(c) PCBs
(d) pesticides

5. This can run off from agricultural land; many are endocrine disruptors.

6. This is a drinking water contaminant from plumbing; it can cause nervous system damage.

7. These are industrial compounds once used to manufacture plastics; they are persistent in the environment as a known human carcinogen.

Free-Response Question

Mercury is both an air pollutant and a common source of water pollution. The graph below depicts data collected by the United States Environmental Protection Agency for the release of mercury into the atmosphere from 2011 to 2020.

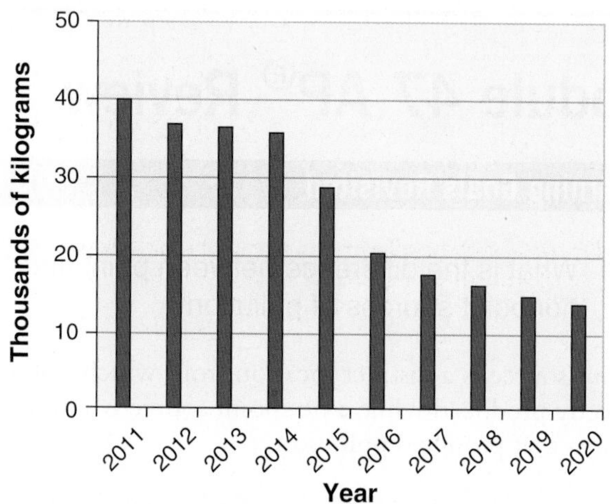

(a) **Identify** the approximate amount of mercury released into the atmosphere in 2014. (1 pt.)
(b) **Describe** the overall trend in mercury emissions from 2011 to 2020. (1 pt.)
(c) **Describe** one probable reason for the trend described in part (b). (1 pt.)

(d) Other than mercury, **describe** one point source of water pollution. (1 pt.)

(e) **Explain** how aquatic organisms, such as shellfish, can become contaminated with mercury released into the atmosphere. (1 pt.)

(f) **Calculate** the approximate percent change in mercury emissions from 2011 to 2020. Show all work. (1 pt.)

(g) According to the EPA, iron and steel manufacturing accounted for 37 percent of all mercury emissions in 2020. **Calculate** the total mercury emissions, in kilograms, of the mercury produced by the iron and steel manufacturing sector in 2020. Show all work. (2 pts.)

(h) In 2020, the U.S. population was 329.5 million people. **Calculate** the per capita mercury emissions, in grams, for the United States in 2020. Show all work. (2 pts.)

Module 48

Human Impacts on Wetlands and Mangroves, Eutrophication, and Thermal Pollution

Water is a critical resource for all organisms but it can be challenging to obtain. As we saw in Module 2, climate variation causes some regions of the world to possess abundant supplies of water, whereas other regions have very little. This not only affects the plants and animals that live in different regions of the world, but it also affects growing human populations that have limited water availability. In this module we will examine how humans have developed ways of altering the availability of water on Earth. With this foundational knowledge, we will also consider how humans are impacting freshwater biomes, with a particular focus on marine and freshwater wetlands, as well as the impact of excess nutrients being released in many different types of aquatic biomes.

Learning Goals

After reading this module you should be able to

48-1 describe the human impacts on water availability.

48-2 explain the human impacts on wetlands and mangroves.

48-3 explain the causes and consequences of eutrophication and sediments.

48-4 describe the sources of thermal and noise pollution.

48-1 What are the human impacts on water availability?

Humans are altering the availability of water by controlling its movement

To understand how human activities are polluting aquatic ecosystems, we first need to understand the many ways in which humans have altered the availability of water. As we discussed in Module 3, only 3 percent of all water on Earth is present as fresh water, the type of water that humans consume. This small percentage of fresh water exists as ice, groundwater, and the variety of aquatic biomes including streams, rivers, wetlands, and lakes. In Module 27 we discussed how humans affect the availability of groundwater. In this section, we will examine how humans have devised creative ways to move water from streams, rivers, wetlands, and lakes. We can channel the flow of flood waters, block the flow of rivers to store water, divert water from rivers and lakes and transport it to distant locations, and even obtain fresh water by removing the salt from salt water. In this section, we will look at each of these water distribution methods and examine their costs and benefits.

Levees and Dikes

Before the intervention of humans, most rivers periodically overflowed their banks. These occasional overflows of nutrient-rich water made their floodplains particularly fertile. While throughout history, humans have taken advantage of these fertile floodplains for agriculture, in more recent times, they have sought ways to prevent flooding so floodplain land could be developed for residential and commercial use. One way to prevent flooding is by constructing a **levee**, an enlarged bank built up on each side of the river. The Mississippi River has the largest system of levees in the world. This river is enclosed by more than 2,400 km (1,500 miles) of levees that offer flood protection to more than 6 million hectares (15 million acres) of floodplains.

The use of levees has produced several major challenges. First, the fertility of the surrounding lands is reduced because natural floodwaters no longer add fertility to floodplains by depositing sediments there. Second, because the sediments do not leave the river, they are carried farther downstream and settle out where the river enters the ocean. In the case of the Mississippi River, massive amounts of sediments enter the Gulf of Mexico, forming a fan-shaped delta of river sediments where the fast-flowing river water enters the slow-moving Gulf. Third, levees may prevent flooding at one location, but by doing so they force floodwater farther downstream where it can cause even worse flooding. Finally, the building of levees encourages development in floodplains, although these areas will still occasionally flood. Thus, the practice of developing land in flood-prone areas raises questions about human safety and economic risk.

When floodwaters become too high, levees can either collapse due to the tremendous pressure of the water, or water can come over the top and quickly erode a large hole in the levee. Both events result in massive flooding (**FIGURE 48.1**). One of the most famous levee failures occurred in New Orleans in 2005 as the result of Hurricane Katrina. The levees could not contain the storm surge and heavy rainfall associated with the hurricane. The water overtopped nearly 50 levees, quickly washing out the banks of dirt and flooding many of the neighborhoods that the levees were built to protect. The hurricane and the devastating floodwaters caused more than 1,800 deaths and over $80 billion in damage to homes and businesses.

Dikes are structures built to prevent ocean waters from flooding adjacent land. Functionally, dikes are similar to levees in that they are built to keep water from flooding onto the land. Dikes are common in northern Europe, where large areas of farmland lie below sea level. Perhaps the best-known dikes are those of the Netherlands, where dikes have been used for nearly 2,000 years. Approximately 27 percent of the land in the Netherlands is below sea level. The dikes, combined with pumps that move any intruding water back out to the ocean, have allowed the country to inhabit and farm areas that would otherwise be under water. The original pumps were powered by windmills, which is why we often associate the Netherlands with wind energy. Today, however, the water is pumped with electric and diesel pumps.

FIGURE 48.1 Levees. Levees are built to prevent rivers from flowing over their banks and onto the floodplain. In 2008, this levee, just north of St. Louis, Missouri, collapsed and allowed the floodwaters to spread over the surrounding fields. *(Anthony Souffle/MCT/Newscom)*

Levee An enlarged bank built up on each side of the river.

Dikes Structures built to prevent ocean waters from flooding adjacent land.

Dam A barrier that runs across a river or stream to control the flow of water.

Reservoir The water body created by damming a river or stream.

Dams

A **dam** is a barrier that runs across a river or stream to control the flow of water. The water body created by damming a river or stream is called a **reservoir**. Dams hold water for a wide variety of purposes, including human consumption, generation of electricity, flood control, and recreation. There are more than 800,000 dams in the world and more than 84,000 dams in the United States. The largest reservoir system in the United States is along the Missouri River. With six dams and reservoirs in Montana, North Dakota, South Dakota, and Nebraska, this system stores nearly 90 trillion liters (24 trillion gallons) of water.

For centuries, humans have used dams to do work, from turning waterwheels that operated grain mills to powering modern turbines that generate electricity in hydroelectric plants. Although hydroelectric dams (discussed in Unit 6) are some of the largest dams in the world, they represent only 3 percent of all dams in the United States. In contrast, 18 percent of U.S. dams were built with the primary purpose of flood control—to reduce or prevent flooding farther down the river. Another 38 percent of U.S. dams were constructed primarily for recreation. Dams have also been built to create scenic lakes for housing developments.

Dams provide substantial benefits to humans, but they come with financial, societal, and environmental costs. The world's largest dam, the Three Gorges Dam built across the

Yangtze River in China (**FIGURE 48.2**), illustrates these costs and benefits well. This massive structure is 2 km (1.3 miles) wide, 185 m (610 feet) high, and has created a reservoir 660 km (410 miles) long behind it. The Three Gorges Dam, completed in 2006, took 13 years to construct. The reservoir required another 2 years to fill with water.

The benefits of the Three Gorges Dam include the large amount of hydroelectric power it generates, which reduces the extraction and use of fossil fuels. In addition, the dam helps prevent the seasonal flooding that damaged downstream cities and villages and killed more than 1 million people over the past 100 years.

The costs of building dams, both to people and the environment, are substantial. Building a dam, like other large-scale construction, uses large amounts of energy and materials. Projects on such a large scale may displace many people. For example, the Three Gorges Dam and the reservoir it created flooded 13 cities, 140 towns, and 1,350 villages; more than 1.3 million people were forced to relocate. Today, nearly two-thirds of the world's rivers are no longer free flowing due to dams and other human-caused disruptions. Many of those that continue to flow freely are in remote areas, including the Amazon, the Congo, and the Arctic.

One of the most common environmental problems associated with dams is the interruption of the natural flow of water to which many organisms are adapted. For migrating fish such as salmon, dams represent an insurmountable obstacle to breeding. The loss of these fish can have a cascading effect on other organisms that depend on them, such as bears that feed on migrating salmon. To alleviate this problem, **fish ladders**, a stair-like structure with water flowing over them allowing migrating fish to get around a dam have been added to some dams. Migrating fish can swim up the fish ladders and reach their traditional breeding grounds (**FIGURE 48.3**). Fish are also impacted when water is released from a dam to generate electricity. Passing millions of gallons of water through a spinning turbine kills millions of fish each year.

FIGURE 48.3 Fish ladder. Because dams are an impediment to fish such as salmon that migrate upstream to breed, fish ladders like this one on the Snake River at Little Goose Lock and Dam in Washington State have been designed to allow the fish to get around the dam and continue on their traditional path of migration. *(Theodore Clutter/Science Source)*

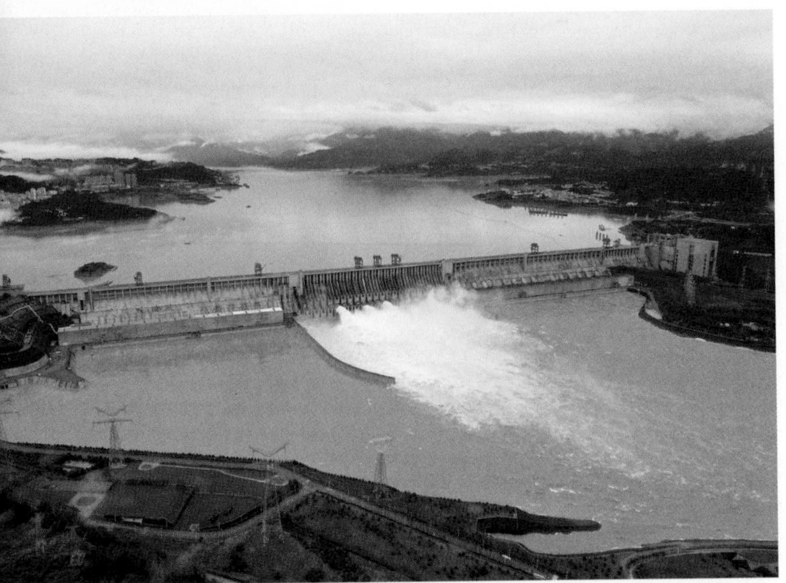

FIGURE 48.2 Dams. The photo shows the world's largest dam, the Three Gorges Dam, on the Yangtze River in China. Dams serve a wide variety of purposes, including flood control and electricity generation. *(Imagine China/Newscom)*

Using dams to prevent seasonal flooding has other consequences for the ecology of an area. Seasonal flooding represents a natural disturbance that scours out pools and shorelines, and this disturbance favors colonization by certain plants and animals. Some of the species that are highly dependent on such disturbances are very rare. In recent years, managers have begun to experiment with releasing large amounts of water from reservoirs to simulate seasonal flooding; the results have been very promising.

In some areas, dams have been removed where they are no longer needed. For example, the Klamath River is a beautiful stretch of water that runs for 400 km (250 miles) from southern Oregon into northern California, where it empties into the ocean. The Klamath River was once a spectacular habitat for salmon and was the third largest salmon fishery on the West Coast. An estimated 800,000 mature Chinook salmon (*Oncorhynchus tshawytscha*)—which can grow as large as 45 kg (100 pounds)—migrated up the river each year. During the past 100 years, however, the Klamath River has seen many changes. It has been dammed for electricity, and farmers have settled the land and diverted

Fish ladder A stair-like structure with water flowing over them, which allows migrating fish to get around a dam.

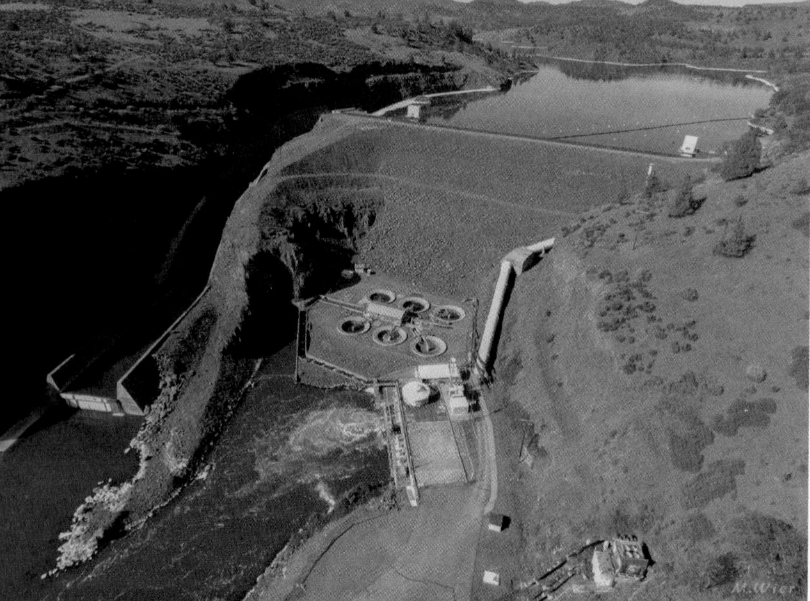

FIGURE 48.4 The Iron Gate Dam on the Klamath River. The Iron Gate Dam is one of four dams that are scheduled to be removed by 2024 to return a more natural flow to the Klamath River. Doing so should help salmon migration and improve the abundance of plants along the river that rely on the disturbances caused by seasonal scouring by rushing waters. *(Michael Wier/Cal Trout)*

water to irrigate their crops (**FIGURE 48.4**). The salmon population has been greatly reduced, and local Native American tribes and the commercial fishing industry have suffered as a result.

In 2009, the various parties reached a decision. The hydroelectric company, which realized it would soon have to spend \$460 million on dam renovations, agreed to remove the four dams. The farmers agreed to conserve water by planting crops that required less water and by updating their irrigation technology. These changes are expected to improve river flow, thereby improving conditions for the salmon and those who depend on them. In 2016, a formal plan was signed by the states of California and Oregon, the federal government, the hydroelectric company, and several Native American tribes. At a cost of \$450 million (\$250 million from the state of California and \$200 from the utility's customers), this endeavor represents the largest dam removal project in history. Final approval for dam removal by the Federal Energy Regulatory Commission is expected in 2022, with actual dam removal occurring by 2024. When completed, salmon will have 676 km (420 miles) of uninterrupted river for the first time in a century. The hope is that this change comes in time for the salmon and all of the other animals and plants of the river to recover.

To date, nearly 1,300 dams have been removed in the United States. In such cases, the natural flow of water has been restored and much of the ecology has quickly rebounded to its natural state, including dramatic increases in plant species that live along the riverbanks and critically rely on the scouring disturbances of free-flowing rivers.

Aqueducts

While dams are designed to hold water back and store it, we often also need to move water from one place to another. Aqueducts, which we first learned about in Unit 4, are canals, ditches, or pipes used to carry water from one location to another. Typically, aqueducts remove water from a lake or river to a place where it is needed. Although aqueducts are famously associated with the Roman Empire, their use dates back to at least the seventh century BCE in Greece. The earliest aqueducts were made of limestone, but modern aqueducts include concrete canals and pressurized steel pipes laid above or under the ground. These structures are more efficient water carriers than the ditches and stone-lined canals of historic times. Older aqueducts, many of which are still in use, can lose as much as 55 percent of the water they carry through leakage or evaporation. These losses can be particularly troublesome in such arid regions as Israel and Jordan where water is already scarce.

In the United States, two of the country's largest cities, New York and Los Angeles, depend on aqueducts to meet their daily water needs. The Catskill Aqueduct brings clean, fresh water over 200 km (120 miles) from the streams and lakes of the Catskill Mountains to New York City. The Colorado River Aqueduct is a canal that carries water 400 km (250 miles) from the Colorado River to Los Angeles (**FIGURE 48.5**).

Using aqueducts to transport water supplies comes with costs and benefits. Although bringing water to cities from pristine areas may ensure a clean supply of water, the construction of aqueducts is expensive and disturbs natural habitats. Aboveground aqueducts can fragment an environment. Even if an aqueduct is buried as an underground pipeline, a great deal of

FIGURE 48.5 Aqueducts. Aqueducts are canals, ditches, or pipes that deliver water from places where it is abundant to places where it is needed. This section of the Colorado River Aqueduct, which diverts water from the Colorado River to Los Angeles, passes through the Mojave Desert. *(iofoto/Shutterstock)*

FIGURE 48.6 Aqueduct pipes. Massive water pipes, such as this one for an aqueduct from the Catskill Mountain reservoirs to New York City, are buried deep underground. *(Allyse Pulliam)*

disruption occurs during construction. In **FIGURE 48.6**, you can see an example of the massive underground pipes that are typically used. Often water is diverted from a natural river where it has flowed for millennia. Some major rivers in the United States, including the Colorado River and the Rio Grande, now lose so much water in multiple locations that during certain periods the rivers go dry before they reach the ocean, thereby destroying the habitat for species living in those locations.

Some water diversion projects have international impacts. India and Bangladesh, for example, share nearly 250 rivers that originate in the Himalayas of India and flow into Bangladesh. In 2016, India initiated a large-scale water diversion project, involving more than 50 rivers, to move water from the north, where it is abundant, to other parts of the country where it is scarce. By diverting water from these rivers, India stands to gain 170 billion liters (55 billion gallons) of water each year for agricultural and household use. However, Bangladesh is situated downstream of many of these river diversion projects and Bangladeshi officials worry that the project will end up dramatically reducing the flow of river water into their country and affect local fish populations and river navigability for commerce. Even farther downstream, the project may reduce the flow of fresh water into estuaries, allowing salty ocean water to move farther up into the river. This influx of salt water raises the salinity of the estuary and harms aquatic life that is not adapted to such salty water.

At the same time that India was diverting water that would naturally flow into Bangladesh, neighboring China began the construction of dams on the Yarlung-Zangbo River, which flows from China into India. The first dam was completed in 2015 and additional downstream dams are scheduled to be built. In fact, the Chinese government announced in 2020 that it was planning to build a "super dam" in the lower part of the Yarlung-Zangbo River near India; this dam would be three times larger than the massive Three Gorges Dam. The Chinese dams are primarily used to generate electricity, but there is concern that the water behind the dams could later be diverted to agricultural use in China, which could result in substantially less water flowing through India and Bangladesh. These conflicts over water rights have led many people to call for treaties among the countries in the region to reach a mutually agreeable decision regarding water use.

The most infamous river diversion project happened in the 1950s, when the Soviet Union diverted two rivers that fed the Aral Sea in Central Asia. Diversion of the rivers dramatically decreased freshwater input into the Aral Sea, increased the salinity of the remaining lake water, and destroyed the local fish populations. Although the Aral Sea was originally the fourth largest lake in the world in terms of surface area, the diversion project reduced its surface area by more than 60 percent, as shown in **FIGURE 48.7**. The withdrawals of water caused the Aral Sea to split into two parts: the North Aral Sea (in modern-day Kazakhstan) and the South Aral Sea (spanning Kazakhstan and Uzbekistan). Dust storms have eroded soil, salt, and pesticide residues from the dry lake bed and carried particles of these substances into the atmosphere. The reduction in the size of the

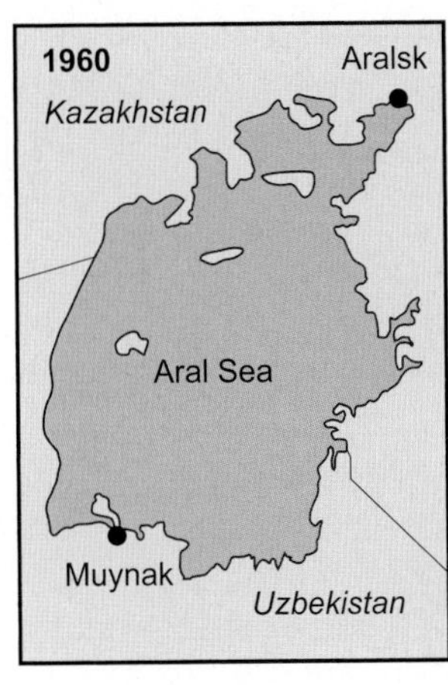

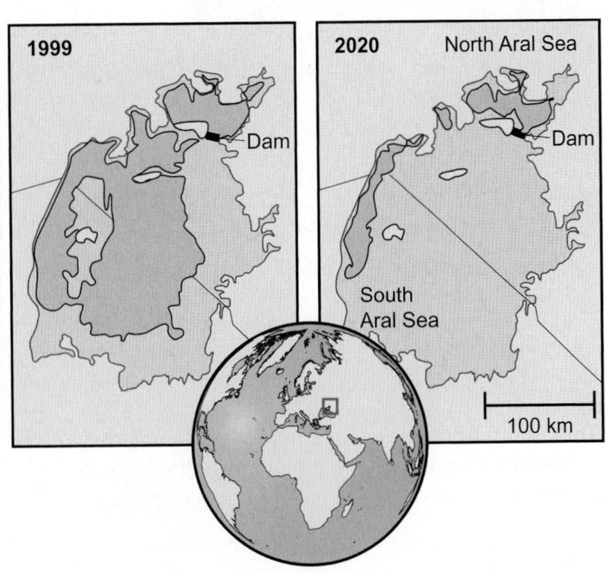

FIGURE 48.7 Consequences of river diversion. Diverting river water can have devastating impacts downstream. The Aral Sea, on the border of Kazakhstan and Uzbekistan, was once the world's fourth largest lake. Since the two rivers that fed the lake were diverted, its surface area has declined by 60 percent and the lake has split into two parts: the North and South Aral seas.

lake has also affected the local climate. Without the moderating effect of a large body of water on climate, summers in the region are now hotter and winters are colder.

In 2005, the World Bank worked with the Kazakhstan government to fund dam construction to save the North Aral Sea by reducing the amount of water flowing into the South Aral Sea. Efforts were also made to increase river flows into the North Aral Sea. Within a year, the water depth of the North Aral Sea rapidly increased by 3.3 m (11 feet). Within a decade, the commercial fishing industry has dramatically increased, creating an economic boom for the region. Unfortunately, saving the South Aral Sea has been deemed too expensive and the water continues to be diverted to support farms growing cotton. As a result, the drying of the South Aral Sea continues; in 2009, the shallower eastern half of the South Aral Sea completely dried for the first time in more than 600 years.

Humans are converting salt water into fresh water by desalination

Today, some water-poor countries are able to obtain fresh water by removing the salt from salt water, a process called **desalination**, or **desalinization**. The salt water usually comes from the ocean, but it can also come from salty inland lakes. Currently, the countries of the Middle East produce 50 percent of the world's desalinated water. During the past decade, a number of technological advances have been made and this has greatly reduced the cost of desalination.

Desalination A process for obtaining fresh water by removing the salt from salt water. *Also known as* **desalinization.**

Distillation A process of desalination in which water is boiled and the resulting steam is captured and condensed to yield pure water.

Reverse osmosis A process of desalination in which water is forced through a thin semipermeable membrane at high pressure.

The two most common desalination technologies are *distillation* and *reverse osmosis*. **Distillation** is a process of desalination in which water is boiled and the resulting steam is captured and condensed to yield pure water. As the water is converted into steam, salt is left behind. A great deal of energy is required to boil the water and then condense it, so distillation can be a monetarily and environmentally expensive process.

Reverse osmosis is a process of desalination in which water is forced through a thin semipermeable membrane at high pressure. Water can pass through the membrane, but salt cannot. Reverse osmosis is a newer technology that is more efficient and often less costly than distillation. However, the liquid that remains, called brine, has a very high salt concentration. It cannot be deposited on land because its high salt content would contaminate the soil, potentially harming plant and animal life. If dumped into a bay or coastal area, it can harm fish and other aquatic life. Typically, brine is returned to the open ocean, though its high salt concentrations can still cause harm to ocean life in areas where it is dumped.

Ultimately, all water management systems require a large investment to build, maintain, and repair. This means that many water-poor countries are not able to implement large-scale methods of water management to provide for their needs. Water availability varies greatly around the world, as **FIGURE 48.8** shows. Middle Eastern and North African

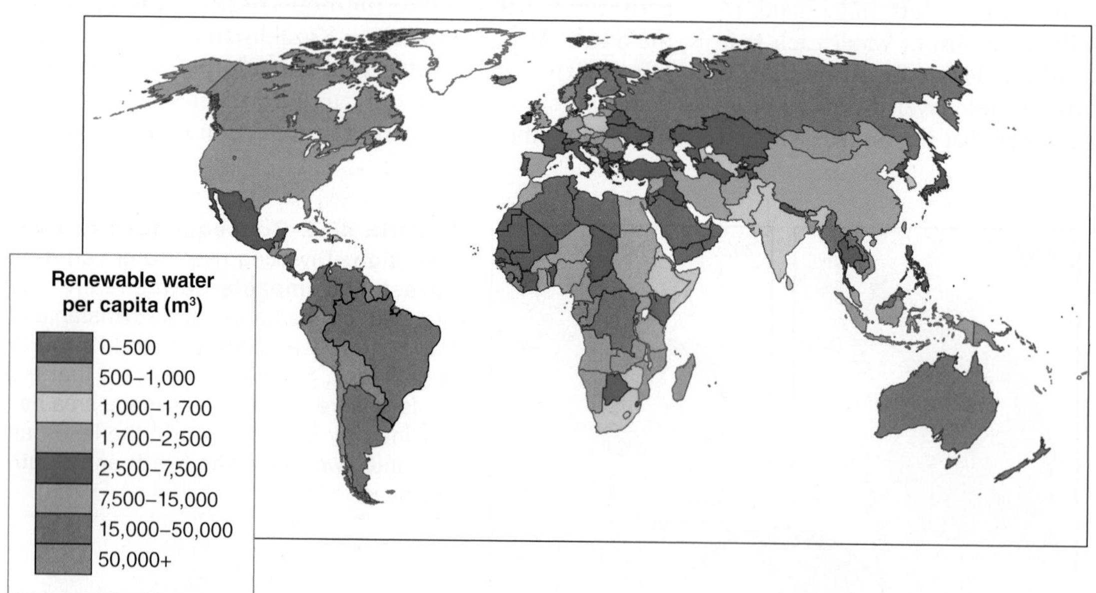

FIGURE 48.8 Water availability per capita. The amount of water available per person varies tremendously around the world. North Africa and the Middle East are the regions with the lowest amounts of available fresh water. *(Data from United Nations World Water Development Report 2015, http://www.unesco.org/fileadmin/MULTIMEDIA/HQ/SC/images/WWDR2015_03.pdf)*

countries represent about 5 percent of the world's population, yet they have less than 1 percent of the fresh water available for drinking. The United Nations estimates that nearly 1.2 billion people live in regions that have a scarcity of water.

48-2 What are the human impacts on wetlands and mangroves?

Humans are impacting wetlands and mangroves through development, dams, overfishing, and pollutants

In addition to controlling the movement of water, humans have also impacted waterbodies through a variety of activities and nowhere is this more apparent than in wetlands. You may recall our discussion of different types of freshwater and marine wetlands in Module 3, including freshwater wetlands—such as bogs, marshes, swamps—and coastal wetlands—such as estuaries and mangrove swamps. We categorize these biomes as wetlands because they have standing water or water-saturated soil during some part of the year.

Wetlands provide a number of important ecosystem services. For example, plants in wetlands can purify water by removing pollutants and excess nutrients. Wetlands also have the capacity to absorb floodwaters and buffer shorelines from the damaging effects of major storms, including hurricanes. Mangrove swamps, for example, have roots that hold onto the shoreline soil and this helps to prevent the soil from being eroded by large waves (Figure 3.7). Wetlands provide critical habitat for fish and wildlife. For instance, migrating birds such as ducks and geese that stop to rest and feed in wetlands along their long trips headed north in the spring and headed south in the fall. The wetland vegetation provides protective habitat for young fish and shellfish, as we discussed for salt marshes and mangrove swamps in Module 3. Finally, wetlands can serve as major carbon sinks, which is important for slowing the rate of global warming.

Given these important ecosystem services, it is easy to understand why environmental scientists are concerned about the impacts of human activities on wetlands around the world. Historically, many freshwater wetlands have been drained so that the land can be used for agriculture and, in many cases, to reduce populations of disease-carrying mosquitoes (**FIGURE 48.9**). For example, the Montezuma National Wildlife Refuge in New York State represents the remnants of a much larger wetland that was drained after the Erie Canal was built and a series of connecting canals drained the wetland. During the most recent decades, the number of wetlands being converted to agriculture has declined dramatically, but there continues to be a substantial conversion of wetlands to development, including the construction of homes and businesses. A smaller number of wetlands are being converted into lakes when dams are built to protect against flooding or to provide recreational areas. Wetlands not destroyed by conversion to agriculture or development still face threats from pollutants, as discussed in the previous module, and from overfishing. For example, in the massive wetlands of the Amazon and Pantanal in South America, intense commercial fishing that began in the 1970s has caused some fish species to decline by 90 percent. These declines are altering the food webs of the wetlands. This includes impacts on the dispersal of plant seeds, because several species of wetland fish eat the fruits of plants and later defecate the seeds elsewhere in the wetland. Sharp declines in the fish populations will make it difficult for the plants to disperse and sustain their populations.

FIGURE 48.9 Draining wetlands for agriculture. During the past two centuries, many wetlands have been drained by digging canals to provide land for pastures and growing crops, including this site in Vlist, Netherlands. *(André Meyer-Vitali/EyeEm/Getty Images)*

Today, the amount of wetland habitat is less than half of what existed in the United States during the 1600s, with some states like California having lost 90 percent of their original wetlands. A recent study found that the combination of hurricanes, rising sea levels, and human development has caused the loss of more than 145,000 hectares (360,000 acres) of freshwater and saltwater wetlands. In fact, the United Nations recently concluded that the world's wetlands are disappearing three times faster than the world's forests. Losing such a large percentage of this habitat means that we have far fewer places for those plants and animals that critically rely on wetlands. It also means that we have far less of the ecosystem functions provided by wetlands, including reduced water purification and buffers against flooding.

48-3 What are the causes and consequences of eutrophication and sediments?

Aquatic ecosystems are being harmed by eutrophication and sediment inputs

When excess nutrients from human activities make their way into waterbodies, it causes nutrient pollution that alters food webs and harms water quality in a process known as **eutrophication**.

Algal Blooms

As we discussed in Unit 1, phosphorus and nitrogen are the most important nutrients to impact aquatic biomes including streams, rivers, lakes, ponds, wetlands, and oceans. In oceans, algal blooms are also known as red tides, because the blooming algae cause the water to become red. The major sources of these nutrients are improperly treated sewage (which we will discuss later in this unit), water running off fertilized lawns and farm fields, and storm water running off buildings and streets, carrying organic matter into water bodies.

When these nutrients enter the water, the water has unnaturally high fertility and this causes a rapid increase in the abundance of algae in the water, which we refer to as an algal bloom. As noted in Module 5, algal blooms can cause clear water to become pea-soup green in a matter of days or weeks. While the green water is less attractive for swimming and recreating, a much larger concern is the fact that some species of blooming algae start producing toxins when they bloom, including nerve and liver toxins that can harm people, pets, farm animals, and wildlife (**FIGURE 48.10**).

When waterbodies experience algal blooms, the large population eventually dies and decomposes. The decomposition of the algae by microbes requires a large amount of oxygen, so waterbodies experiencing large amounts of algal decomposition can experience low concentrations of oxygen in the water. When this happens, we refer to the water as being hypoxic. Hypoxic water is not suitable to most species of fish and shellfish, given that they require oxygen, so they must either leave the hypoxic region of water or die from a lack of sufficient oxygen. As noted in Module 5, we call these areas dead zones. The number of dead zones around the world has been rapidly increasing. A century ago, there were four dead zones identified along the coasts of the world's continents. This number increased to 87 dead zones by the 1980s and more than 500 dead zones today (**FIGURE 48.11**). Clearly, human impacts on eutrophication continue to increase.

Eutrophication Excess nutrients from human activities that make their way into waterbodies; it causes nutrient pollution that alters food webs and harms water quality.

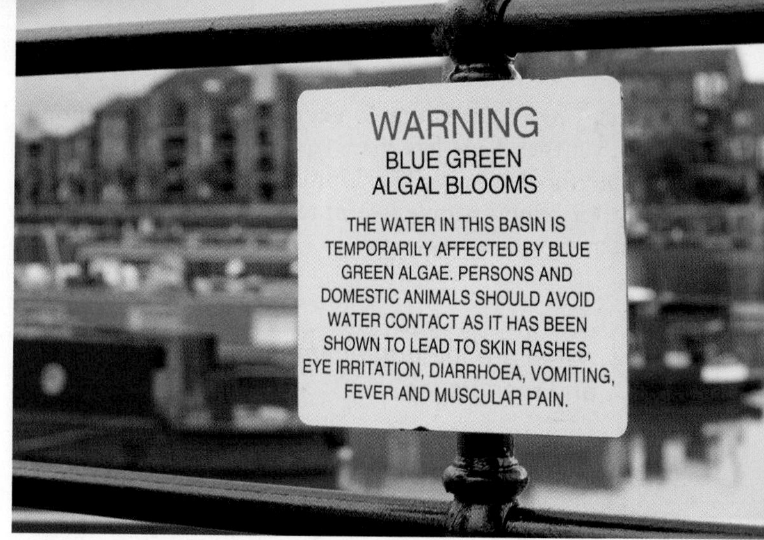

FIGURE 48.10 Harmful algal blooms. When waterbodies experience eutrophication, some of the blooming species of algae can produce toxins that can harm people, pets, farm animals, and wildlife. This algal bloom site is at the *Preston Marina in Lancashire, UK.* *(Ashley Cooper/Getty Images)*

In the past decade, there has been increased concern about freshwater lakes experiencing harmful algal blooms. The risk of a lake having algal blooms is typically affected by their natural fertility. As noted in Module 3, lakes range from being naturally low in nutrients, known as oligotrophic, to being naturally high in nutrients, known as eutrophic. Given that oligotrophic lakes have naturally low nutrient levels, they also have low algal populations, very clear water, and high oxygen concentrations. They are also less susceptible to having harmful algal blooms. In contrast, eutrophic lakes may naturally experience algal blooms, but inputs of additional nutrients from human activities can cause these algal blooms and hypoxic events to be more intense and longer lasting.

Oxygen Sag Curves

A similar effect on oxygen concentrations can be observed in water bodies that experience inputs of sewage or other organic pollutants that rapidly decay. Researchers can

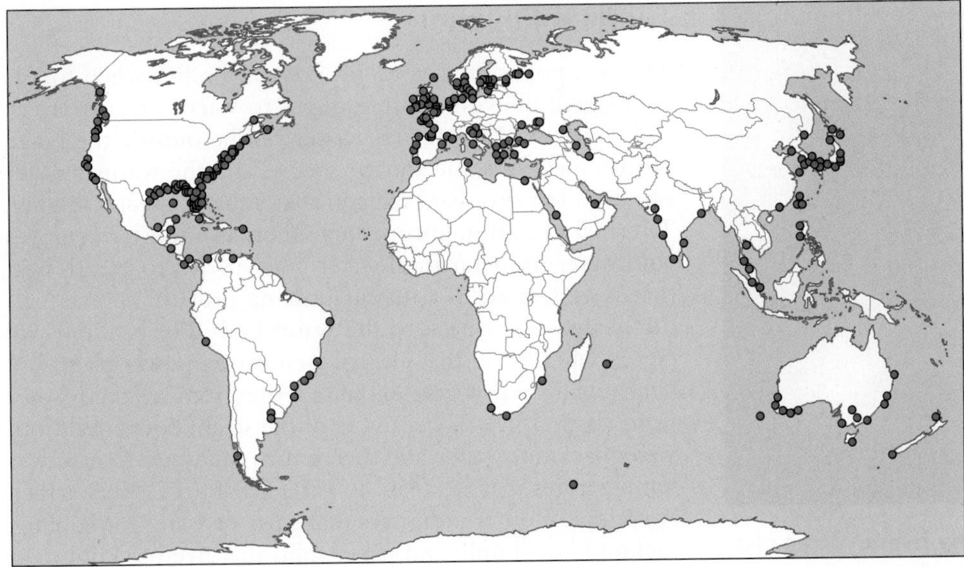

FIGURE 48.11 **Dead zones around the world.** In coastal waters, researchers identified more than 500 dead zones. *(Data from Breitburg, D., et al. 2018. Declining oxygen in the global ocean and coastal waters.* Science *359: 6371)*

measure oxygen concentrations in the water at different locations to determine the location that has a sudden drop in oxygen due to a point-source input. For example, if a city releases sewage into a river at a single point source, researchers will measure a sharp drop in oxygen in the river next to the sewage pipe because bacteria are consuming much more oxygen as they break down the sewage. Farther downstream from the pollution source, there is less sewage to decompose, so the oxygen rises again in the river while plants and algae continue to produce oxygen as part of their normal photosynthesis. The relationship of oxygen concentration to the distance from a point source of decomposing sewage or other pollutants in known as an **oxygen sag curve**, which you can see illustrated in FIGURE 48.12. By detecting oxygen sag curves, researchers can discover previously unknown point sources of pollutants in water bodies.

Sediments

The inputs of nutrients to aquatic ecosystems are often associated with inputs of sediments that erode from terrestrial landscapes and get carried long distances with flowing streams and rivers. Earlier in this module we mentioned the collection of sediments where the flowing Mississippi River enters the slow-moving waters of the Gulf of Mexico. We can see the same effects in many lakes around the world, where fast-moving streams dump water into lakes. These accumulations of sediment can make it difficult to navigate boats because the water depths become more and more

Oxygen sag curve The relationship of oxygen concentrations to the distance from a point source of decomposing sewage or other pollutants.

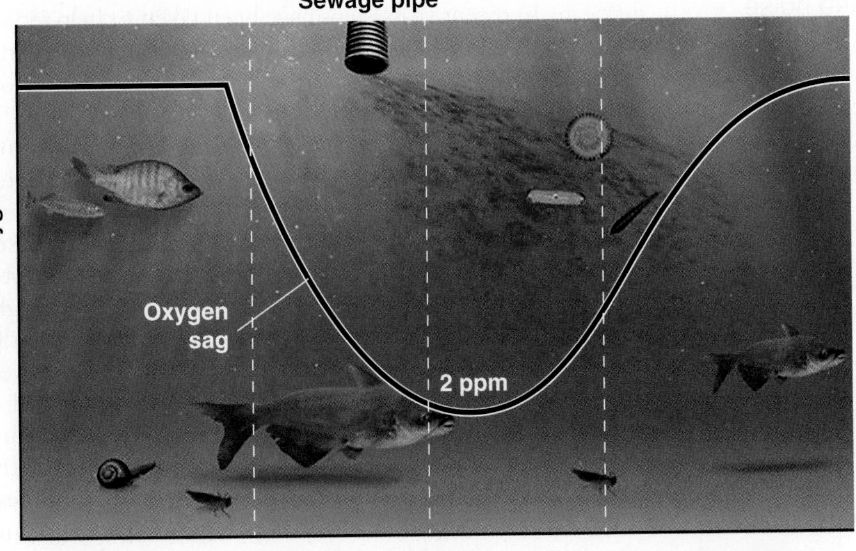

FIGURE 48.12 **Oxygen sag curve.** When there is a point source of decomposing sewage or other pollutants that is released into a waterbody.

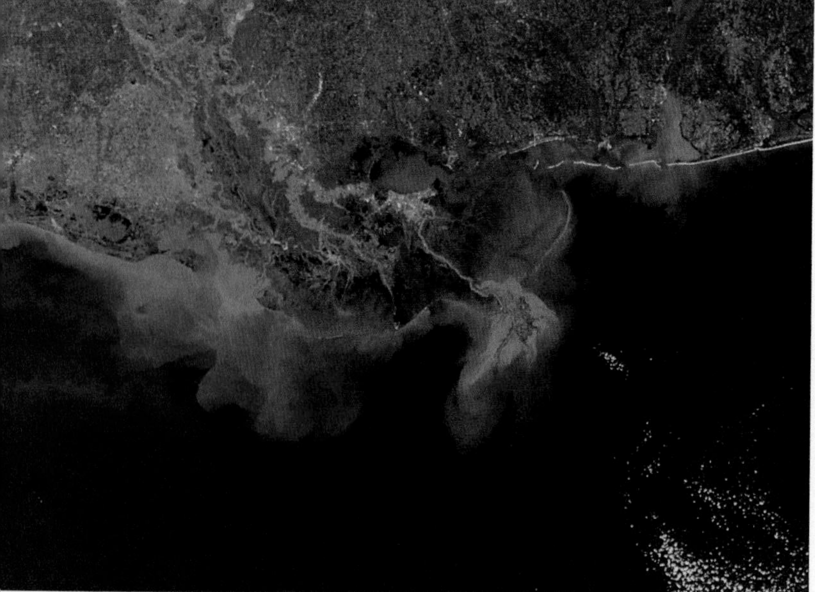

FIGURE 48.13 Sediments deposited by rivers. When fast-moving rivers carrying sediment reach the slow-moving waters of lakes or oceans, the sediments settle out and accumulate over time. This image shows sediments from the Mississippi River emptying into the Gulf of Mexico. *(Science History Images/Alamy Stock Photo)*

shallow over time (**FIGURE 48.13**). Often these sediments have to be dug up, in a process known as dredging, to allow boats to move through the area.

Sediment inputs also have important impacts on organisms. For aquatic plants, an increase in sediments in the water can substantially reduce the penetration of sunlight in the water, which leads to the plants growing more slowly or even dying. Sediments can also make it hard for visual predators to find prey and harm many aquatic animals by clogging their gills, making it difficult to obtain sufficient oxygen from the water. Finally, sediments can bury organisms that live stationary lives at the bottom of waterbodies, such as plants, clams, and mussels.

48-4 What are the sources of thermal and noise pollution?

Thermal pollution comes from warming water bodies while noise pollution comes from human sounds

In addition to a variety of chemical pollutants, humans also cause **thermal pollution**, which occurs when humans cause a substantial change in the temperature of a water body, and noise pollution, which occurs when humans cause elevated sounds that harm organisms. In this section, we will explore both of these phenomena.

Thermal pollution Occurs when humans cause a substantial change in the temperature of a water body.

Thermal shock A dramatic change in temperature that can kill many species.

Thermal Pollution

Thermal pollution can occur in water bodies when forests are logged, allowing more sunlight to warm streams, rivers, and wetlands. This warmer water can be outside the range of tolerance for some aquatic species and the warmer water can hold less dissolved oxygen that aquatic species require. Thermal pollution also occurs when an industry removes cold water from a natural water body, uses it to absorb heat that is generated in a manufacturing process, and returns the heated water back to the water body. For example, we saw in Module 39 that electric generating power plants use nearly half of all water extracted; they remove cold water from rivers, lakes, or oceans, cool the steam converted from water back into water, and then return the water to nature at temperatures that are 10°C to 15°C (18°F to 27°F) warmer. A variety of other industries make use of water for cooling, including steel mills and paper mills that need to cool their machines.

AP® Exam Tip

Remember that thermal pollution is caused not by global warming but by industrial activities that return heated water back to the natural water body. It is also caused by land conversion practices including urban development and deforestation.

Thermal pollution is an environmental problem because species are generally adapted to a particular natural range of temperatures. Therefore, a dramatic change in temperature can kill many species, a phenomenon called **thermal shock**. High temperatures also cause organisms to increase their respiration rate. But because warmer water does not contain as much dissolved oxygen as cold water, an increase in respiration and a decrease in oxygen will cause many animals to suffocate. In recent years, steps have been taken to help reduce thermal pollution, including pumping the heated water into outdoor holding ponds where it can cool further before being pumped back into natural water bodies.

In the United States, the EPA regulates how much heated water can be returned to natural water bodies. Compliance can become a real challenge in the summer when high demand for electricity causes an increased demand for cooling water even though the water in rivers and lakes tends to be at its lowest volume and at its highest temperature in the summer. One common solution to this problem has been the construction of cooling towers that release the excess heat into the atmosphere instead of into a water body. A cooling tower relies on the cooling power of evaporation to reduce the temperature of the water, much as we depend on our own sweat and a breeze to cool ourselves on a hot day. Some industries have built closed systems in which they cool the hot water in a cooling tower and then recycle the water to be heated again. In this way, the industries neither extract water from natural

FIGURE 48.14 Thermal pollution. Nuclear reactors, such as the Three Mile Island nuclear plant in Middletown, Pennsylvania, use water to generate steam. To cool this steam, they either use cooling towers, as shown here, that cool through evaporation, or dump the heated water into holding ponds where it is cooled before being returned to the natural source of water, such as a lake or river. *(Bloomberg/Getty Images)*

water bodies, nor do they release any heated water back into nature (**FIGURE 48.14**).

Noise Pollution

While we have discussed noise pollution in the air in Module 46, it may strike you as odd to think of noise as a type of water pollution. Indeed, noise pollution has received the least amount of attention from environmental scientists and, as a result, we know the least about it. When we think of noise pollution, what often comes to mind is the sound of city traffic. However, noise pollution also occurs in the water. For example, sounds emitted by ships and submarines that interfere with animal communication are the major concern. Especially loud sonar could negatively affect species such as whales that rely on low-frequency, long-distance communication (**FIGURE 48.15**). Several instances of beached whales in the Bahamas, the Canary Islands, and the Gulf of California have been connected to the use of military sonar and the use of loud, underwater air guns by energy companies searching for oil deposits under the oceans.

Naval operations have also raised concerns about noise pollution in the ocean. For example, the U.S. Navy conducts training exercises that include using ship and submarine sonar, firing torpedoes, and detonating bombs underwater. In 2014, the National Oceanic and Atmospheric Administration (NOAA) concluded that the Navy's activities would have a negligible effect on fish and whales in the ocean, although NOAA researchers continue to assess impacts of noise pollution on aquatic species. By 2020, however, public concern had grown regarding naval activities off the West Coast of the United States, when the U.S. Navy estimated their training exercises over 7 years will negatively impact 29 species of marine mammals, including several species of whales, dolphins, and seals. This increased awareness of noise pollution in the ocean has inspired some ship builders to design ships equipped with quieter propellers to reduce their contributions to noise pollution.

In this module we have seen that humans have developed a number of creative ways to control and move water away from where it is abundant to areas where it is needed to meet our needs. We have also learned that our activities can have harmful effects on water bodies, with a prime example being the impacts of development, dam construction, overfishing, and pollutants on wetlands. We have seen how inputs of excess nutrients and sediments are causing algal blooms, dead zones, and a variety of harmful impacts of aquatic organisms. Finally, we have learned that thermal and noise pollution can also harm aquatic organisms. See the "Unit 8 Visual Representation" feature to explore how many of the factors discussed in this module have contributed to the significant water pollution problems in the Florida Everglades (pages 642–643). In the next module, we will examine the impacts of persistent chemicals in the environment and how they can move up through a food chain in an ecosystem.

FIGURE 48.15 Noise pollution. Noise from navy sonar and oil exploration may interfere with the normal behavior of whales. *(Steve Woods Photography/Getty Images)*

Module 48 AP® Review

Preparing for the AP® Exam

Learning Goals Revisited

48-1 What are the human impacts on water availability?

Humans channel the flow of flood waters with levees and dikes, block the flow of rivers with dams to store water, divert water from rivers and lakes and transport it to distant locations, and even obtain fresh water by removing the salt from salt water.

48-2 What are the human impacts on wetlands and mangroves?

Wetlands provide a number of important ecosystem services including water purification, flood protection, and critical habitat for plants and animals. Historically, many freshwater wetlands have been drained so that the land can be used for agriculture and, in many cases, to reduce populations of disease-carrying mosquitoes. During the most recent decades, there has been a substantial conversion of wetlands to development. Wetlands are also experiencing impacts form pollutants and overfishing.

48-3 What are the causes and consequences of eutrophication and sediments?

Excess nutrients cause water bodies to have unnaturally high fertility, leading to rapid increases in the abundance of algae, which can lead to low oxygen conditions and dead zones. The input of sewage and other organic pollutants can also cause declines in oxygen, which can be observed as oxygen sag curves that are used to identify point sources of pollution. Sediments in rivers and streams can cause large deposits where they reach slow-moving lakes and oceans. Sediments can also bury stationary organisms, reduce sunlight for plants in the water, and clog animal gills.

48-4 What are the sources of thermal and noise pollution?

Thermal pollution most commonly occurs when water is removed from a water body to absorb heat from electric-generating facilities or factories and then the warmed water is returned to the water body, where it can kill aquatic organisms by thermal shock. Noise pollution occurs when humans cause elevated sounds that harm organisms, such as by interfering with their communication abilities.

AP® Practice Questions

Preparing for the AP® Exam

Multiple-Choice Questions

1. Which statement about dams is correct?
   (a) Dams are used to flood areas for agriculture.
   (b) Dams can cause decreased water temperatures.
   (c) Dams are used to stop the flow of rivers for migrating fish.
   (d) Most dams are built to generate electricity.
2. Which statement about human impacts on wetlands is correct?
   (a) Wetlands contain contaminated water.
   (b) Draining for agriculture has been a major cause of wetland loss .
   (c) Wetlands are no longer lost due to development.
   (d) Some wetlands are not susceptible to overfishing due to their high biodiversity.
3. Eutrophication is a problem because
   (a) it reduces the abundance of algae in the water, which increases dissolved oxygen.
   (b) it helps prevent dead zones by an influx of nutrients.
   (c) it is more common in oligotrophic lakes, due to their high BOD.
   (d) it causes algae blooms, which can produce toxins.
4. Thermal pollution
   (a) is primarily a problem in the winter.
   (b) is rarely lethal.
   (c) is not regulated in the United States.
   (d) has been reduced by the use of cooling towers.

Use the diagram below to answer questions 5–7:

5. Which is the zone with the highest BOD?
   (a) Zone A
   (b) Zone B
   (c) Zone C
   (d) Zone D

6. Which zone is most likely to be considered a "dead zone"?
   (a) Zone A
   (b) Zone B
   (c) Zone C
   (d) Zone D

7. In which zone has the water recovered from oxygen-demanding organic waste?
   (a) Zone A
   (b) Zone B
   (c) Zone C
   (d) Zone D

Free-Response Question

The data table below shows the data recorded at four sampling sites along a stream, where Site A is the farthest upstream.

Site	pH	Temperature (°C)	BOD (mg/L)	Turbidity (NTU)
A	6.5	10	10	15
B	6.6	13	12	7
C	6.6	19	12	100
D	6.5	15	11	28

(a) Based on the data, **identify** the sampling site with the lowest turbidity. (1 pt.)
(b) Based on the data, **identify** the relationship between turbidity and temperature. (1 pt.)
(c) Based on the data, **describe** one likely source of water pollution entering the stream near site C. (1 pt.)
(d) **Explain** how the pollutant described in part (c) could lead to a dead zone if it accumulated in a small pond. (2 pts.)
(e) Humans use pesticides for many different purposes, which is why they are a common source of surface water pollution. Other than pesticides, **identify** an anthropogenic water pollutant that is likely to be present in an agricultural area containing dairy farms. (1 pt.)
(f) **Describe** how the pollutant identified in part (e) can enter aquatic ecosystems. (1 pt.)
(g) **Explain** the sequence of ecological changes that could result from an accumulation of the water pollutant identified in part (e). (1 pt.)
(h) **Propose a solution** to prevent the environmental effects explained in part (g). (1 pt.)
(i) **Describe** one additional advantage of the solution proposed in part (h). (1 pt.)

Module 49

Unit 8 | 47 | 48 | 49 | 50 | 51 | 52 | 53 | 54

Persistent Organic Pollutants (POPs), Bioaccumulation, and Biomagnification

In this unit we have been exploring aquatic and terrestrial pollution, including excess nutrients and chemical pollutants. In this module we begin by examining chemical pollutants that persist in the environment for very long periods of time, from years to centuries. We then consider the importance of how much of a given pollutant an organism experiences, since it is the concentration of the pollutant the determines the impact the pollutant has on an individual. This turns out to be a complex question, because the amount of a pollutant in the environment can be very different from the amount of a pollutant inside the body of an individual in that environment. The amount of a pollutant inside the body of an individual depends on how the pollutant gets inside the individual and whether the pollutant is more soluble in water or fat. Finally, we will discover that some pollutants can become more accumulated inside an individual, thereby achieving much higher concentrations than what exists in the environment and causing much greater harm.

Learning Goals

After reading this module you should be able to

49-1 describe how persistent organic pollutants affect ecosystems.

49-2 explain how routes of exposure and solubility determine the concentrations of chemicals that organisms experience.

49-3 explain how bioaccumulation and biomagnification can increase the concentrations of chemicals in organisms.

49-1 How do persistent organic pollutants affect ecosystems?

Some chemicals can persist in nature and harm ecosystems for years

Earlier in this unit we discussed a wide variety of chemical pollutants that were categorized by their impacts on organisms, including neurotoxins, carcinogens, teratogens, allergens, and endocrine disruptors. While each of these chemicals cause harm, a key factor in how much harm they can do is related to how long they persist in nature. In this section, we discuss chemicals that can persist for long periods of time.

Chemical Persistence

The **persistence** of a chemical refers to how long the chemical remains in the environment. Persistence depends on a number of factors including temperature, pH, whether the chemical is in water or soil, and whether it can be degraded by sunlight or broken down by microbes. Scientists often measure persistence by observing the time needed for a chemical to degrade to half its original concentration, known as the half-life of the chemical, which we discussed in Module 38. **TABLE 49.1** lists the persistence of various chemicals in the environment measured according to half-life.

TABLE 49.1 The persistence of various chemicals in the environment

Chemical	Source	Half-Life
Malathion	Insecticide	1 day
Radon	Rocks and soil	4 days in air
Vinyl chloride	Industry, water from vinyl chloride pipes	4.5 days in air
Phthalates	Plastics, cosmetics	2.5 days in water
Roundup	Herbicide	7 to 70 days in water
Atrazine	Herbicide	224 days in wetland soils
Polychlorinated biphenyls (PCBs)	Industry	8 to 15 years in water
DDT	Insecticide	30 years in soil

DDT, for example, has a half-life in soil of up to 30 years. Thus, even after DDT is no longer sprayed in an area, half of the chemical that was absorbed in the soil would still be present after 30 years, and one-fourth would be present after 60 years. Such synthetic, carbon-based molecules that break down very slowly in the environment are known as **persistent organic pollutants (POPs)**. Because they take so long to break down, there is the potential for POPs to travel long distances by wind and water and therefore to contaminate large areas of the world. Moreover, POPs can also be stored in the fat tissues of animals, a problem that we will explore later in the module.

Chemicals that cause harmful effects on humans and other organisms may become even larger risks when they persist for many years. For this reason, many modern chemicals are designed to break down rapidly so that any unintended effects will be short-lived. "Do the Math: Estimating Half-Lives of Toxic Chemicals" shows you how half-lives of chemicals are calculated.

Persistent Organic Chemicals of High Concern

As we have seen, chemical pollutants can vary a great deal in their persistence. In Module 47 we introduced a group of chemicals known as PCBs, which were once used to manufacture plastics and insulate electrical transformers. As scientists began to study PCBs, they discovered they were not only lethal and carcinogenic, but they were also highly persistent in the environment. Although the manufacture of PCBs ended in 1979 and they are no longer used in the United States, they are still present in the environment because of their long-term persistence.

One particularly high-profile case involves two General Electric manufacturing plants in New York State that dumped 590,000 kg (1.3 million pounds) of PCBs into the Hudson River from 1947 to 1977. In 2002, the EPA ruled that General Electric must pay for the dredging and removal of approximately 2.03 million cubic meters (2.65 million cubic yards) of PCB-contaminated sediment from a 64-km (40-mile) stretch of the upper Hudson River in New York

Persistence The length of time a chemical remains in the environment.

Persistent organic pollutants (POPs) Synthetic, carbon-based molecules that break down very slowly in the environment.

DO THE MATH Estimating Half-Lives of Toxic Chemicals

Preparing for the AP® Exam

As we have discussed, it is important that we know about the persistence of toxic chemicals in the environment to evaluate how much of that toxin a person, plant, or animal might experience. Using the table below, quantify how long it will take the insecticide malathion to break down to one-eighth of its initial concentration.

Chemical	Source	Half-life
Malathion	Insecticide	1 day
Radon	Rocks and soil	4 days
Roundup	Herbicide	40 days
Atrazine	Herbicide	224 days
PCBs	Industry	10 years

We know that the amount of any chemical is reduced by half after the duration of a half-life (by definition), so we will have one-half as much malathion after one day, one-quarter as much malathion after 2 days, and one-eighth as much malathion after 3 days. In other words, we will have one-eighth as much of any chemical after the time required for three half-lives has passed. In the case of malathion, three half-lives require 3 days.

$$3 \text{ half-lives} \times 1 \text{ day/half-life} = 3 \text{ days}$$

As another example, how much time will it take until we have one-eighth as much radon present?

$$3 \text{ half-lives} \times 4 \text{ days/half-life} = 12 \text{ days}$$

YOUR TURN Based on the half-lives provided in the table, how long will it take to have only one-eighth as much Roundup, atrazine, and PCBs?

FIGURE 49.1 Dredging PCBs from the Hudson River. Following decades of dumping PCBs into the Hudson River, the EPA ruled that General Electric must pay for the dredging and removal of approximately 2 million cubic meters of PCB-contaminated sediment from a 64-km stretch of the upper Hudson River in New York State. *(Nathaniel Brooks/ The New York Times/Redux)*

State. General Electric argued that the dredging should not occur because it would stir up the sediments and re-suspend the PCBs in the river water, causing further problems. The courts and EPA scientists disagreed and in 2009 the dredging of the PCBs finally began (**FIGURE 49.1**). The dredging was completed in 2015 and today researchers are continuing to monitor PCB levels in the water, soil, and fish. It is expected that the Hudson River will still require 15 to 30 years before PCB levels are low enough in fish to allow their consumption by humans. As this case demonstrates, though it can take decades of scientific research and debate, in the end we can partially reverse the contamination of pollutants in our environment.

In Module 47 we also discussed a group of industrial chemicals known as PFAS, which are synthetic organic chemicals that are used in nonstick pans, stain-resistant carpets, water-repellant clothing, food packaging, and fire retardants. These chemicals are known carcinogens, but they have also been linked to weakened immune systems and reproductive problems. Given their widespread use, these chemicals are pervasive in the environment. Equally alarming is the fact that they persist so long in the environment that they are known as "forever chemicals," with some formulations currently expected to last thousands of years. Scientists are currently discovering more and more locations in the United States where these chemicals have been dumped onto land and in the water, resulting in entire towns now experiencing contaminated groundwater, which provides water to town wells. When this happens, there are only two solutions and both are expensive: installing water filtration systems in every home in the town or installing public water lines to every home to provide each home with uncontaminated water.

AP® Exam Tip

For prompts that ask you to propose methods to reduce water pollution do not use the word "filter" on its own. Only mention use of filters if the proposed filter exists and removes the targeted contaminant. Clearly describe this in your answer.

49-2 How do routes of exposure and solubility determine the concentrations of chemicals that organisms experience?

Concentrations of chemicals experienced by organisms depends on routes of exposure and solubility

Knowing the concentrations of chemicals that can harm humans or other animals is important, but it is only useful when combined with information about the concentrations that an individual might actually experience in the environment. If a chemical is quite harmful at some moderate concentration but individuals only experience lower concentrations of that chemical, we might not be particularly concerned. Therefore, to identify and understand the effects of chemical concentrations that organisms experience, we need to know something about how the chemicals behave in the environment.

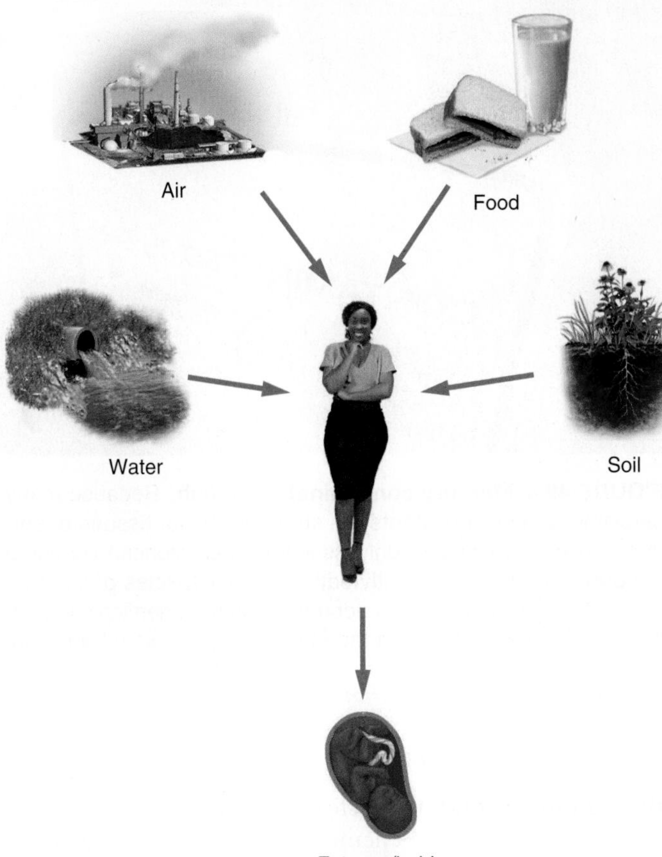

FIGURE 49.2 Routes of exposure. Despite a multitude of potential routes of exposure to chemicals, most chemicals have a limited number of major routes.

Routes of Exposure

The ways in which an individual might come into contact with an environmental hazard, such as a chemical, are known as **routes of exposure**. As FIGURE 49.2 illustrates, the full range of possibilities is complex because it includes potential exposures from the air, from water used for drinking, bathing, or swimming, from food, and from the environments of places where people live, work, or visit. For any particular chemical, however, the major routes of exposure are usually limited to just a few of the many possible routes. For example, bisphenol A is a chemical used in manufacturing hard plastic items such as toys, food containers, and baby bottles. Recent research has raised concerns that bisphenol A may be responsible for early puberty and increased rates of cancer. While these effects are being debated and investigated, it is clear that a child's routes of exposure to bisphenol A are limited to toys, food containers, and baby bottles.

Routes of exposure are also important for animals. For example, pesticides can enter the bodies of many organisms when they ingest food. However, pesticides can also enter animal bodies through the skin when pesticides are sprayed into the air and the droplets land on the animals. Some animals, such as amphibians, have a moist skin that is more permeable to pesticides compared to the feathers of birds or the hair of mammals, which can block pesticides from reaching the skin's surface. For some species, such as snakes, exposure can also occur as they slither over the surface of the ground that has been sprayed with chemicals. For aquatic species, some chemicals can be absorbed by their gills from the water. As you can see, understanding the various routes of exposure is important to understanding how much chemical an animal experiences in its body.

Solubility of Chemicals

Once we know the potential routes of exposure, scientists can then determine how well the chemical can move in the environment as a function of its **solubility**, which is how well a chemical can dissolve in a liquid. For example, some chemicals such as herbicides are readily soluble in water whereas others such as insecticides are much more soluble in fats and oils. When a chemical is highly soluble in water, it can be washed off surfaces, percolate into groundwater, and run off into surface waters, including rivers and lakes. In contrast, chemicals that are soluble in fats and oils are not very soluble in water so they tend not to be found percolating into the groundwater or running off into surface waters. Instead, they can be found in higher concentrations bound to soils, including the benthic soils that underlie bodies of water.

49-3 How can bioaccumulation and biomagnification increase the concentrations of chemicals in organisms?

Bioaccumulation and biomagnification can dramatically increase the concentration of a chemical in organisms by storing them in fat

Now that we understand the different pathways in which chemicals can enter the body and the importance of chemical solubility, we can begin to understand how chemicals can accumulate in an individual's body and how this process can accumulate even more as chemicals are moved up a food chain from one trophic group to the next.

Route of exposure The way in which an individual might come into contact with an environmental hazard, such as a chemical.

Solubility How well a chemical dissolves in a liquid.

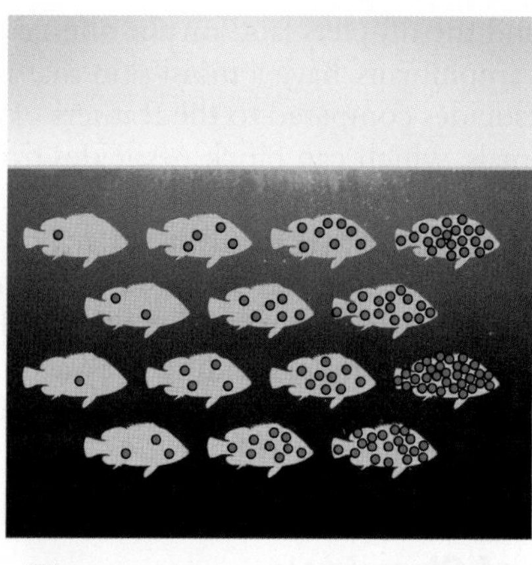

FIGURE 49.3 Bioaccumulation. When chemicals in the environment are fat soluble and they are consumed by organisms such as fish, the chemical gets stored in the individual's fat. Over time, the amount of chemical, represented by red dots in the figure, can continue to accumulate, resulting in a much higher concentration in the individual organism compared to the environment.

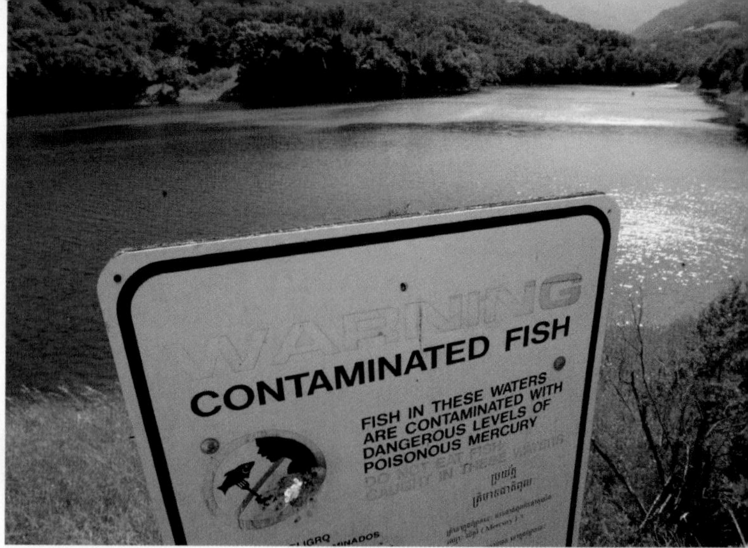

FIGURE 49.4 Mercury contamination of fish. Because many persistent organic pollutants are stored in the fat tissues of animals, governments commonly issue health advisories to educate the public to avoid frequently eating certain species of fish that are particularly vulnerable to accumulating the chemicals in their bodies. This warning is from the Almaden Reservoir in San Jose, CA. *(AP Photo/Paul Sakuma)*

Bioaccumulation

Chemicals that are soluble in fats and oils are more likely to become stored in the fatty tissues of animals. For example, in Unit 5 we mentioned that DDT accumulates in the fatty tissues of birds, such as pelicans and eagles, that feed on aquatic animals. This process, known as **bioaccumulation**, is the selective absorption and concentration of a chemical within an organism over time. As shown in **FIGURE 49.3**, the process of bioaccumulation begins when an individual is exposed to small amounts of a chemical from the environment and incorporates the chemical into its tissues, typically its fat tissues. Fish, for example, are exposed to low concentrations of methylmercury when they drink water, pass water over their gills to breathe, and consume food that contains mercury. A fish stores mercury in its fat tissues and, over time, the mercury accumulates. This accumulation of methylmercury is why many governments issue recommendation that people not eat meals of fish, such as tuna, more than once per month (**FIGURE 49.4**). You might recall that we discussed the impact of accumulating mercury of several species of tuna in Module 47.

Other chemicals that bioaccumulate include many of the pesticides that were developed after World War II. Given the dangers of bioaccumulation, most modern pesticides are designed to have chemical properties that reduce the likelihood of being stored in fat tissues of animals in the environment. The rate of accumulation for any animal depends on the concentration of the chemical in the environment, the rate at which the animal takes up each source of the chemical, the rate at which the chemical breaks down inside the animal, and the rate at which the chemical is excreted by the animal.

Bioaccumulation The selective absorption and concentration of a chemical within an organism over time.

Biomagnification The increase in chemical concentration in animal tissues as the chemical moves up the food chain.

Biomagnification

Biomagnification is the increase in chemical concentration in animal tissues as the chemical moves up the food chain. In this way, the original concentration in the environment is magnified to occur at a much higher concentration in the top predator of the community. The classic example of biomagnification is the case of DDT, an insecticide that has been widely used to kill insect pests in agriculture and to kill the mosquitoes that carry malaria and other diseases. When DDT was initially invented, it was viewed as a miracle insecticide that would kill pests such as mosquitoes without having any harmful effects on people or the environment (**FIGURE 49.5**).

An important trait of DDT is that it is not water soluble. When sprayed over water it quickly binds to particulates in the water, such as algae, and the underlying soil. It is then quickly taken up by the tiny zooplankton that act as primary consumers on algae. As we see in **FIGURE 49.6**, the very low concentration of DDT in the water bioaccumulates in the bodies of the zooplankton where it becomes approximately 1,000 times more concentrated. Small fish eat the zooplankton for many weeks or months and the DDT is further concentrated approximately sixfold. Large fish spend their lives eating the contaminated smaller fish and the DDT in the large fish is further concentrated approximately fivefold. Finally, fish-eating birds such as pelicans and eagles spend years eating the large fish and further magnify the DDT in their own bodies.

Because of biomagnification along the food chain, the concentration of DDT in the birds is nearly 276,000 times higher than the concentration of DDT in the water. The concentrated DDT in the fish-eating birds causes them to produce thin-shelled eggs that often break when the parent birds try to incubate the eggs. This was a primary cause in the decline of these birds in the 1960s. Since DDT was banned in the United States in 1972, the populations of fish-eating birds

FIGURE 49.5 **Historic advertising for DDT.** In the 1940s and 1950s, DDT was viewed as a very effective insecticide that posed no harm to people or wildlife. Subsequent research over the decades found a number of unexpected impacts. *(Courtesy of Science History Institute)*

have dramatically increased. While banned in the United States, DDT is still used in other parts of the world, including in Africa where there is a priority on killing mosquitoes that can carry pathogens that make people sick, such as malaria.

While the combined effects of bioaccumulation and biomagnification result in increased concentrations of fat-soluble pollutants in animals, it can have similar effects in humans. This is one of the reasons that fish-consumption advisories exist for areas contaminated by mercury, PCBs, and other persistent organic pollutants. Humans can readily store these persistent chemicals in their fat tissues and suffer substantial health consequences including cancer, reproductive problems, and nervous system issues. For example, the methylmercury in fish such as tuna can accumulate in the human body and harm the central nervous system. In addition, pregnant women who consume too much methylmercury can have fetuses with developmental abnormalities and long-term learning and motor skill problems throughout childhood.

In this module we learned that some chemical pollutants can persist for years or even centuries and pose a long-term risk to humans and ecosystems. This persistence becomes even more problematic for fat-soluble pollutants that can bioaccumulate and biomagnify up the food chain. In the next module, we will change our focus from chemical pollutants in the environment to the issue of disposing solid waste.

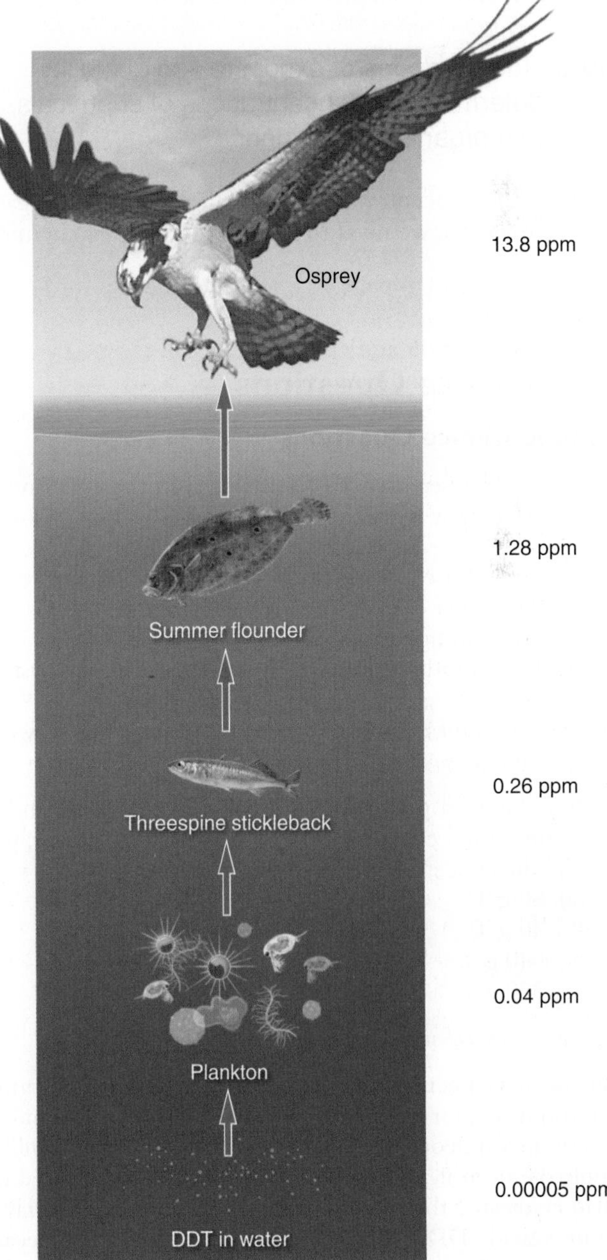

FIGURE 49.6 **The biomagnification of DDT.** The initial exposure is primarily in a low trophic group such as the plankton in a lake. Consumption causes the upward movement of the chemical where it is accumulated in the bodies at each trophic level. The combination of bioaccumulation at each trophic level and upward movement by consumption allows the concentration to magnify to the point where it can be substantially more concentrated in the top predator than it was in the water. *(Data from G. M. Woodwell, C. F. Wurster, Jr., and Peter A. Isaacson, DDT Residues in an East Coast Estuary: A Case of Biological Concentration of a Persistent Insecticide,* Science, *New Series, 156 (3776) (May 12, 1967): 821–824, http://www.jstor.org/stable/1722018)*

Module 49 AP® Review

Preparing for the AP® Exam

Learning Goals Revisited

49-1 How do persistent organic pollutants affect ecosystems?

While chemical pollutants can be categorized as neurotoxins, carcinogens, teratogens, allergens, and endocrine disruptors, a key factor in how much harm they can do is related to how long they persist in nature. Persistent organic pollutants (POPs) are synthetic, carbon-based molecules that require years, decades, or centuries to break down in the environment. As a result, their harmful effects on humans and ecosystems can last for a very long time.

49-2 How do routes of exposure and solubility determine the concentrations of chemicals that organisms experience?

To know the concentrations of chemicals to which organisms are exposed, we need to know the routes of chemical exposure from the land, air, water, and ingested foods. We also need to know the solubility of the chemical in water and in fats.

49-3 How can bioaccumulation and biomagnification increase the concentrations of chemicals in organisms?

For chemicals that are fat-soluble, they can bioaccumulate in an individual by being stored in the fat tissues. They can also biomagnify up the food chain as each trophic level consumes contaminated prey and then stores the chemical pollutant in its fat tissues.

AP® Practice Questions

Preparing for the AP® Exam

Multiple-Choice Questions

1. Why is the persistence of a chemical in the environment an important factor in considering the impact of the chemical?
   (a) Persistent chemicals are always more soluble in water.
   (b) The route of exposure for persistent chemicals is typically through the skin.
   (c) Pollution by persistent chemicals can pose a risk for decades.
   (d) Persistent chemicals can be rapidly broken down by sunlight.

2. If a chemical pollutant has a half-life of 2 years and we begin with a concentration of 1,280 g/L in a river, what will the concentration be in the river after 10 years?
   (a) 80 g/L
   (b) 40 g/L
   (c) 640 g/L
   (d) 320 g/L

Use the passage below to answer questions 3 & 4:

Humans have been affected by malaria, which is transmitted by mosquitoes, for thousands of years. Each year, 350 million to 500 million people are infected with malaria, and 1 million people die from it. In 1951, the United States announced that it had eradicated the malaria pathogen by extensively spraying the insecticide DDT. However, the spraying of DDT became controversial in the 1960s when its effects on other species were investigated. The insecticide was causing the eggshells of large predatory bird species, such as bald eagles, to become thin and fragile. It was discovered that DDT chemicals bioaccumulate in species as they are eaten by their predators. It is widely accepted that the recovery of the bald eagle population in the United States was caused by the banning of DDT in 1972.

3. According to the passage, why was DDT so widely used before it was banned?
   (a) It was used to prevent infection by the malaria pathogen.
   (b) It was sprayed on crops to kill pest insect species.
   (c) It was used to increase the growth of native wildflower species.
   (d) It was used to control invasive species.

4. According to the passage, what caused the eggshells of bald eagles to become thin and break?
   (a) Mercury concentrations were ingested by their prey species.
   (b) An inert ingredient in herbicides was ingested by their prey species.
   (c) DDT was ingested by their prey species.
   (d) A pesticide was used to control pest birds such as crows.

Use the passage below to answer questions 5 & 6:

A research team is investigating the biomagnification and bioaccumulation of persistent organic pollutants in food webs by testing blood levels of DDT, which was banned in 1972, in the tissues of primary, secondary, and tertiary consumers in an ecosystem.

5. Which of the following best identifies a testable hypothesis for the researchers' investigation?
   (a) Blood levels of DDT in consumers will increase as the trophic level decreases.
   (b) Blood levels of DDT in consumers will increase as the trophic level increases.
   (c) Blood levels of DDT in producers will increase as the trophic level increases.
   (d) There is no relationship between blood levels of DDT in producers and the trophic level.

6. Which of the following best identifies the dependent variable in the researchers' investigation?
   (a) time
   (b) number of organisms sampled
   (c) DDT concentration
   (d) trophic level

Free-Response Question

Below is a model of a marine food chain, where a persistent organic pollutant (POP) is present in the water at a concentration of 0.00005 parts per million (ppm).

(a) Using the diagram, **identify** the trophic level of the summer flounder. (1 pt.)

(b) Using the diagram, **describe** how the concentration of the persistent organic pollutant changes within the food chain. (1 pt.)

(c) Using the diagram, **explain** which organism would have the highest concentration of the persistent organic pollutant in their body tissues. (1 pt.)

(d) **Describe** one way in which humans can be affected by the persistent organic pollutant present in this food chain. (1 pt.)

(e) A medical research team is studying the long-term exposure of PFAS, a group of persistent organic chemicals that was manufactured beginning in the 1940s and widely used as a component in nonstick pans and water-repellent clothing. PFAS has been found in drinking water sources and in trace amounts in the blood of a large proportion of the human population around the world. The research team has been testing the blood levels of PFAS chemicals in a group of adults over the last 20 years to determine how that concentration has changed over time.
   (i) **Identify** a possible research question for this study. (1 pt.)
   (ii) **Identify** the independent variable for this study. (1 pt.)
   (iii) **Identify** two variables not mentioned that should be kept constant in this study. (2 pts.)
   (iv) **Describe** one modification the research team could make in order to increase the reliability of their study. (1 pt.)

(f) **Explain** why it would be difficult for the research team to have a valid control group for this study. (1 pt.)

Module 50

Unit 8	47	48	49	50	51	52	53	54

Solid Waste Disposal

When we think about terrestrial and aquatic pollution, we commonly envision scenes of dirty water contaminated with toxic chemicals or massive oil spills along shorelines. However, another major pollutant is **solid waste**, which is waste produced by humans as discarded materials that are not in liquid or gas form and do not pose a toxic hazard to humans and other organisms. Much solid waste is what we call garbage, as well as the sludge produced by sewage treatment plants, which we will describe in greater detail in the next module. The challenge is how to dispose of solid waste as the human population continues to grow and produces more waste. In this module we will explore how the production of solid waste has increased over the decades and the major sources of this waste. We will then explore different ways to properly dispose of this waste and the trade offs involved in each disposal option. In the next module, we will examine the ways in which we can work to reduce the amount of waste we produce, including the special category of hazardous waste.

Learning Goals

After reading this module you should be able to

50-1 describe how solid waste production has changed over time and identify the sources.

50-2 explain how landfills are used to dispose of municipal solid waste.

50-3 describe how incineration is used to dispose of municipal solid waste.

50-4 explain why some municipal solid waste does not go to landfills or incinerators.

50-5 identify how hazardous waste is safely disposed.

50-1 How has solid waste production changed over time and what are the sources?

Solid waste pollution has increased over time from residences, businesses, industries, and agricultural activities

As life in many countries has become increasingly dependent on disposable items, the generation of solid waste has become more of a problem for both natural and human environments. In 1900 in the United States, for example, virtually all metal, wood, and glass materials were recycled. If a wooden bookcase broke and could not be repaired, the pieces might be used to make a step stool. When the step stool broke, the wood was burned in a wood stove to heat the house. After World War II, consumption patterns changed. The increasing industrialization and wealth of the United States, as well as cultural changes, made it possible for people to purchase household conveniences that could be used and then thrown away. Families were large, and people were urged to buy labor-saving household appliances and to dispose of them as soon as a new model was available. Disposable plate "TV dinners," throw-away napkins, and disposable plates and forks also became common. In the 1960s disposable diapers became widely available and largely replaced reusable cloth diapers, though some families choose to use cloth diapers today for a variety of reasons. The components of household materials also changed. Objects were more likely to contain mixtures of different materials, which makes them harder to use for another purpose and difficult to recycle. The United States became the leader of what came to be known as the "throw-away society." In this section, we examine how solid waste production has increased over the years and where it comes from.

Solid waste The waste produced by humans as discarded materials that is not in liquid or gas form and do not pose a toxic hazard to humans and other organisms.

Municipal Solid Waste

Solid waste collected by municipalities from households, small businesses, and institutions such as schools, prisons, municipal buildings, and hospitals is known as

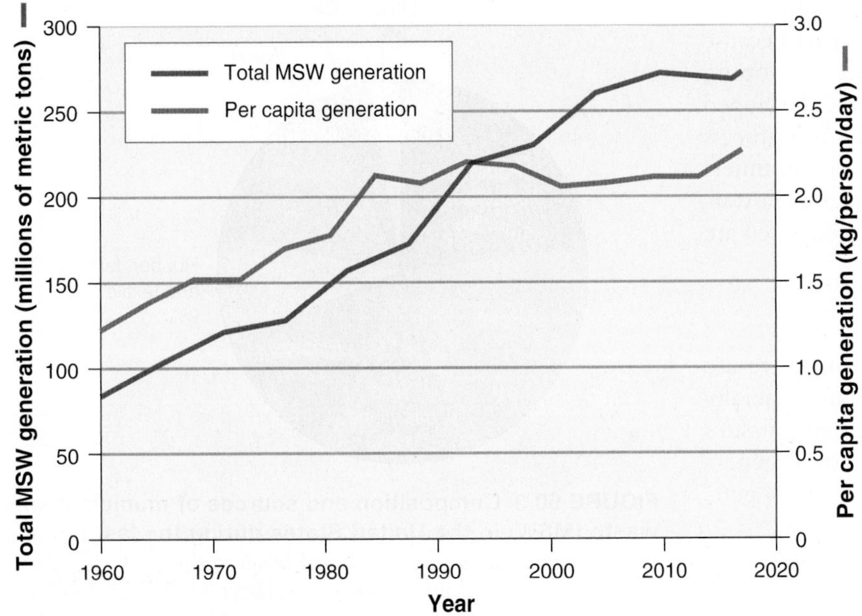

FIGURE 50.1 Municipal solid waste generation in the United States, 1960–2018. Total MSW generation and per capita MSW generation have been increasing since 1960, although per capita MSW generation has levelled off since 1990. *(Data from Advancing Sustainable Materials Management: 2018 Fact Sheet, U.S. EPA, 2020)*

municipal solid waste (MSW). The Environmental Protection Agency (EPA) estimates that approximately 60 percent of MSW comes from residences and 40 percent from commercial and institutional facilities. Additional kinds of waste generated in the United States include agricultural waste, mining waste, and industrial waste. Waste other than MSW is typically deposited and processed on-site rather than transferred to a different location for disposal. Although some of these other categories generate a much greater percentage of yearly total solid waste, this module focuses on MSW.

Each year since 1960, the U.S. Environmental Protection Agency has monitored the amount of MSW produced in the United States, as shown in **FIGURE 50.1**. Over the six decades, the total amount of MSW generated in the United States increased. However, the per capita MSW, which is the amount of waste produced per person, has levelled off since 1990. Thus, the increase in total MSW is mostly the result of a growing population of people in the United States. During this time, average MSW generation was approximately 2.0 kg (4.4 pounds) per person per day. Waste generation varies by season of the year, socioeconomic status of individuals, and geographic location.

Compared to people in the United States, per capita waste generation is much lower in many other countries. The UN-HABITAT estimates for the developed world ranges from 0.8 to 2.2 kg (1.8–4.8 pounds) per person per day while the estimate for the developing world is 0.55 kg (1.2 pounds) per person per day. Some indigenous people create virtually no waste per day, with as much as 98 percent of MSW being reused for something by someone. The remaining 2 percent ends up in a landfill. Even there, impoverished people scavenge and reuse some of the discarded material (**FIGURE 50.2**).

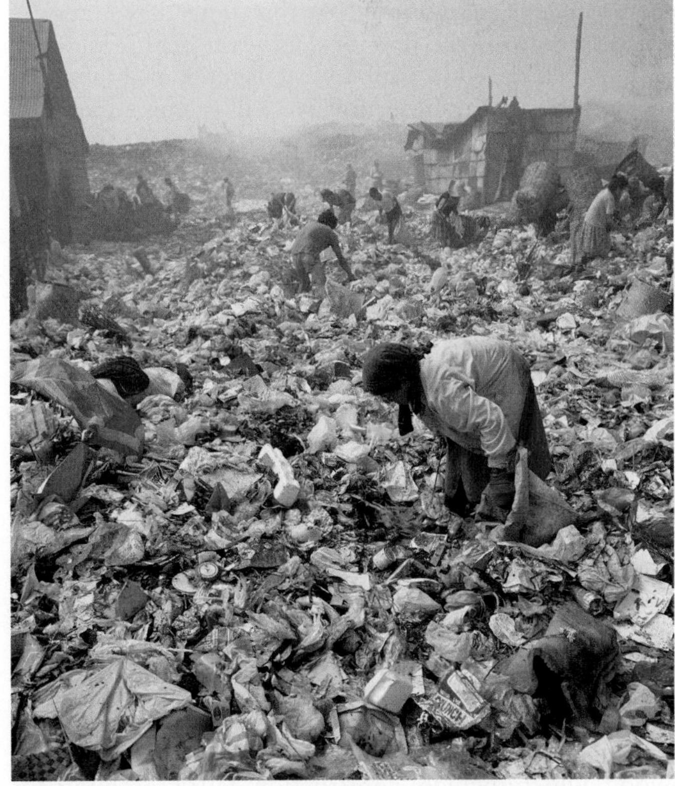

FIGURE 50.2 A large dump in Manila, Philippines. Throughout the world, impoverished people scavenge dumps. *(Stockbyte/Getty Images)*

Municipal solid waste (MSW) Solid waste collected by municipalities from households, small businesses, and institutions such as schools, prisons, municipal buildings, and hospitals.

Developing countries have become responsible for a greater portion of total MSW in the world, in part because of their growing populations. In addition, as developing countries produce more of the goods used in the developed world, they generate more waste in the production process for these goods. For example, computers sold to consumers in the United States are assembled in such places as Taiwan, Singapore, and China, and the waste products generated are disposed of at the manufacturing location.

Sources of Municipal Solid Waste

We have seen that MSW is made up of the things we use and then throw away. The goods that we use are generally a combination of organic items, fibers, metals, and plastics made from petroleum. A certain amount of waste is generated during any manufacturing process. Waste is also generated from the packaging and transporting of goods.

Depending on the particular materials, products, and goods that consumers use, such items can remain in the consumer-use system for a long time. For example, a ceramic plate or drinking mug might last for 10 to 20 years. In most cases, a disposable paper cup is disposed within minutes or hours after it is used. Ultimately, all products wear out, lose their value, or are discarded. At this point they enter the **waste stream**—the flow of solid waste that is recycled, incinerated, placed in a solid waste landfill, or disposed of in another way. In this section, we will examine the composition of MSW and then put a special focus on electronic waste.

Composition of Municipal Solid Waste

Each year the U.S. EPA estimates the total MSW for the country and breaks it up into the different categories of waste, as shown in **FIGURE 50.3**. In this figure, which estimates the amount of waste produced prior to any recycling, we can see that the category "paper and paperboard" makes up 23 percent of the 265 million metric tons (292 million U.S. tons) of waste generated in the United States. This category includes newsprint, office paper, cardboard, and boxboard such as cereal and food boxes. Interestingly, the fraction of paper in the solid waste stream has been decreasing; only a decade ago it was 40 percent of MSW. This decline in paper waste has been aided by emails replacing paper letters and websites replacing paper magazines, newspapers, and catalogs. Organic materials other than paper products make up another large category, with yard waste and food scraps together making up 34 percent of MSW. Wood, which includes construction debris, accounts for another 6 percent. The combination of all plastics makes up approximately 12 percent of MSW.

Waste stream The flow of solid waste that is recycled, incinerated, placed in a solid waste landfill, or disposed of in another way.

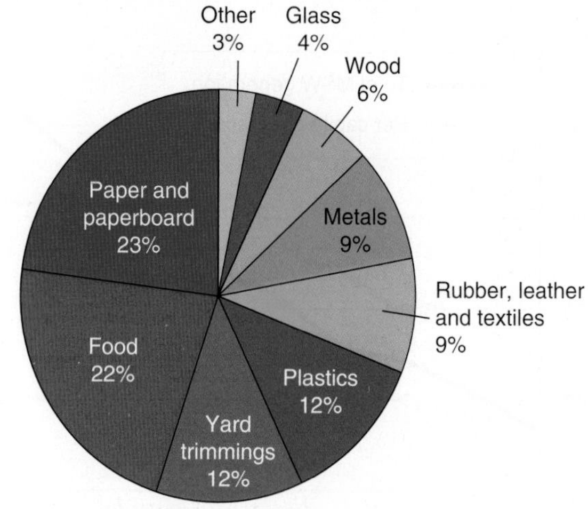

FIGURE 50.3 Composition and sources of municipal solid waste (MSW) in the United States during the last decade. (a) The composition, by weight, of MSW in the United States in 2018 before any of the material was recycled. Paper, food, and yard waste make up more than half of the MSW by weight. *(Source: Advancing Sustainable Materials Management: 2018 Fact Sheet. U.S. EPA, 2020)*

Electronic Waste

Electronic waste, or e-waste, encompasses consumer electronics, including televisions, computers, portable music players, and cell phones. E-waste accounts for only 2 percent of the waste stream, but the environmental effect of these discarded objects is far greater than their weight. The older-style cathode-ray-tube (CRT) television or computer monitor contains 1 to 2 kg (2.2–4.4 pounds) of the heavy metal lead as well as other toxic metals, such as mercury and cadmium. These metals may eventually leach out of the bottom of the landfill into groundwater or surface water. The toxic metals and other components can be extracted, but at present there is little formalized infrastructure or incentive to recycle them. However, many communities have begun voluntary programs to divert e-waste from landfills.

In the United States, most electronic devices are not designed to be easily dismantled after they are discarded. As a result, it generally costs more to recycle a computer than to put it in a landfill. In 2020, the U.S. EPA estimated that approximately 38 percent of discarded electronics were sent to recycling facilities. Unfortunately, much e-waste from the United States is exported to Asia, where adults as well as some children separate valuable metals from other materials using fire and acids in open spaces with no protective clothing and no respiratory gear (**FIGURE 50.4**). So even when consumers in the United States do send electronic products to be recycled, there is a good chance that the recycling will not be done in a way that protects worker health and the environment.

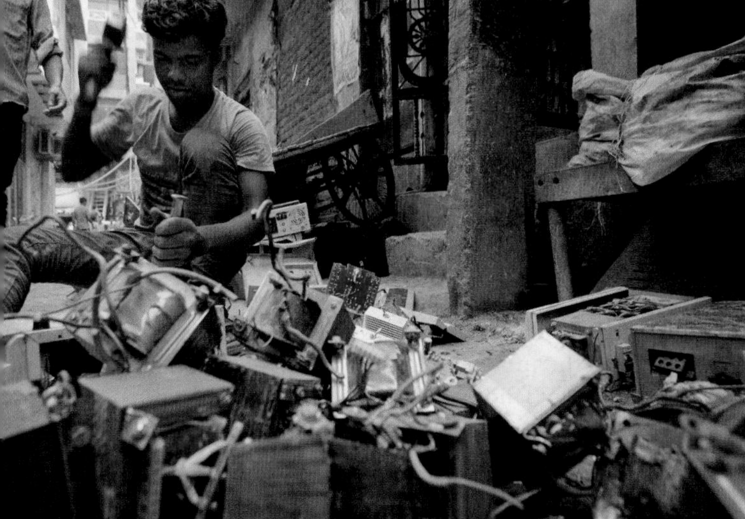

FIGURE 50.4 Electronic waste recycling in New Delhi, India. Much of the recycling is done without the protective gear and respirators that would typically be used in the United States. *(Bloomberg/Getty Images)*

50-2 How are landfills used to dispose of municipal solid waste?

Landfills are the primary destination for municipal solid waste

Now that we understand the sources of solid waste, we next need to think about where it all goes. As **FIGURE 50.5** shows, in the United States 50 percent of our municipal waste is discarded in landfills, 38 percent is recovered through recycling and composting, and 12 percent is incinerated. In this section, we explore the history of landfills and then examine some landfill basics. Then we consider how to choose a site for a landfill and the issues in site selection. Finally, we examine the environmental consequences of landfills. In the next section, we will explore the process of waste incineration and in the next module we will examine recycling and composting.

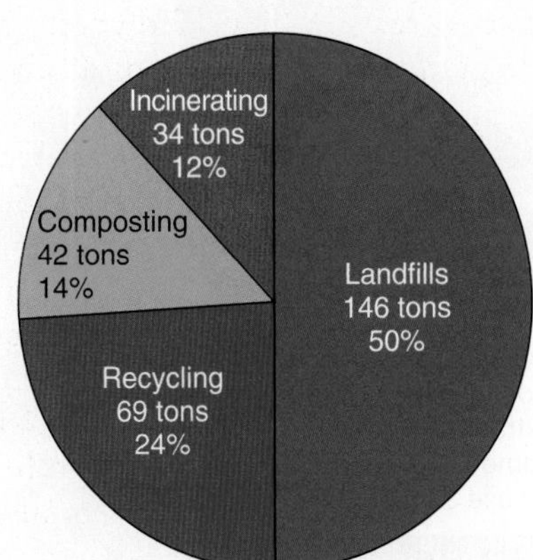

FIGURE 50.5 The fate of municipal solid waste in the United States. The majority of MSW is disposed of in landfills. *(Data from Advancing Sustainable Materials Management: 2018 Fact Sheet, U.S. EPA, 2020)*

A Brief History of Landfills

More than a century ago, municipalities deposited their solid waste into open landfills, which were just large holes in the ground where the waste could be deposited away from people who didn't want to experience the sights and foul smells of the garbage. Such open landfills were not regularly covered with soil and anything deposited into the hole could eventually leach out through the bottom and potentially contaminate the groundwater of the community. These open landfills were also places where scavenging animals could find food and occasionally fires could erupt from the gases produced by the decomposing waste. In short, the open dumps were dangerous to humans and the environment

Beginning in the 1930s, however, public opposition to open landfills in the United States began to increase. Municipalities started to dispose of MSW in ways that were more protective of human health and the surrounding environment. Though open landfills are now rare in developed countries, they still exist in the developing world, where they pose a considerable health hazard. In the United States, we have modified the design of landfills to minimize the hazards they present to the surrounding environment.

Landfill Design

To understand how to design landfills, we need to be aware of a few processes that occur in landfills. One important process is the decomposition of waste by microbes. In many ecosystems, microbial decomposition occurs under aerobic conditions and produces CO_2. However, the burial of waste in a landfill creates an anaerobic environment. In an anaerobic environment, microbial decomposition produces methane (CH_4), which you may recall is a potent greenhouse gas. A second important process is that groundwater and precipitation can pass through a landfill and leach chemicals from the landfill. Known as **leachate**, this is a liquid that can contain elevated levels of pollutants as a result of passing through the solid waste of a landfill. Understanding these two processes is important as we design landfills to minimize the release of methane and prevent the leachate from escaping the landfill.

Leachate Liquid that can contain elevated levels of pollutants as a result of having passed through the solid waste of a landfill.

In the United States today, repositories for municipal solid waste, known as **sanitary landfills**, are engineered to hold solid waste with as little contamination of the surrounding environment as possible, as you can see illustrated in **FIGURE 50.6**. Sanitary landfills are constructed with a clay or plastic lining at the bottom. Clay is often used because it impedes water flow and retains positively charged ions, such as metals. A system of pipes is constructed below the landfill to collect leachate, which is the excess water and chemicals accumulating at the bottom of the landfill. The leachate is sometimes recycled back into the landfill. Finally, a top layer of soil, called a cap, is added when the landfill reaches capacity.

The input of rainfall and other water sources are kept to a minimum because excess water in the landfill increases the rate of anaerobic decomposition and subsequent methane release. Also, with a large amount of water entering the landfill from both the solid waste and rainfall, there is a greater likelihood that some of that water will leave the landfill as leachate. Leachate that is not captured by the collection system may leach into nearby soils and groundwater. Leachate is tested regularly for its toxicity and if it exceeds certain toxicity standards, the landfill operators could be required to collect it and treat it as a toxic waste.

Sanitary landfill An engineered ground facility designed to hold municipal solid waste (MSW) with as little contamination of the surrounding environment as possible.

Once the landfill is constructed, it is ready to accept MSW. Perhaps the most important component of operating a safe modern-day landfill is controlling inputs. The materials that are suitable for a landfill are those least likely to cause environmental damage through leachate or by generating methane. Composite materials made of plastic and paper, such as juice boxes for children, are good candidates for a landfill because they are difficult to recycle, while aluminum and other metals such as copper may contribute to leaching and they are valuable as recyclables. Therefore, aluminum and copper should never go into a landfill. Glass and plastics are both chemically inert, making them suitable for a landfill when reuse or recycling is not possible. Toxic materials (such as household cleaners, oil-based paints,

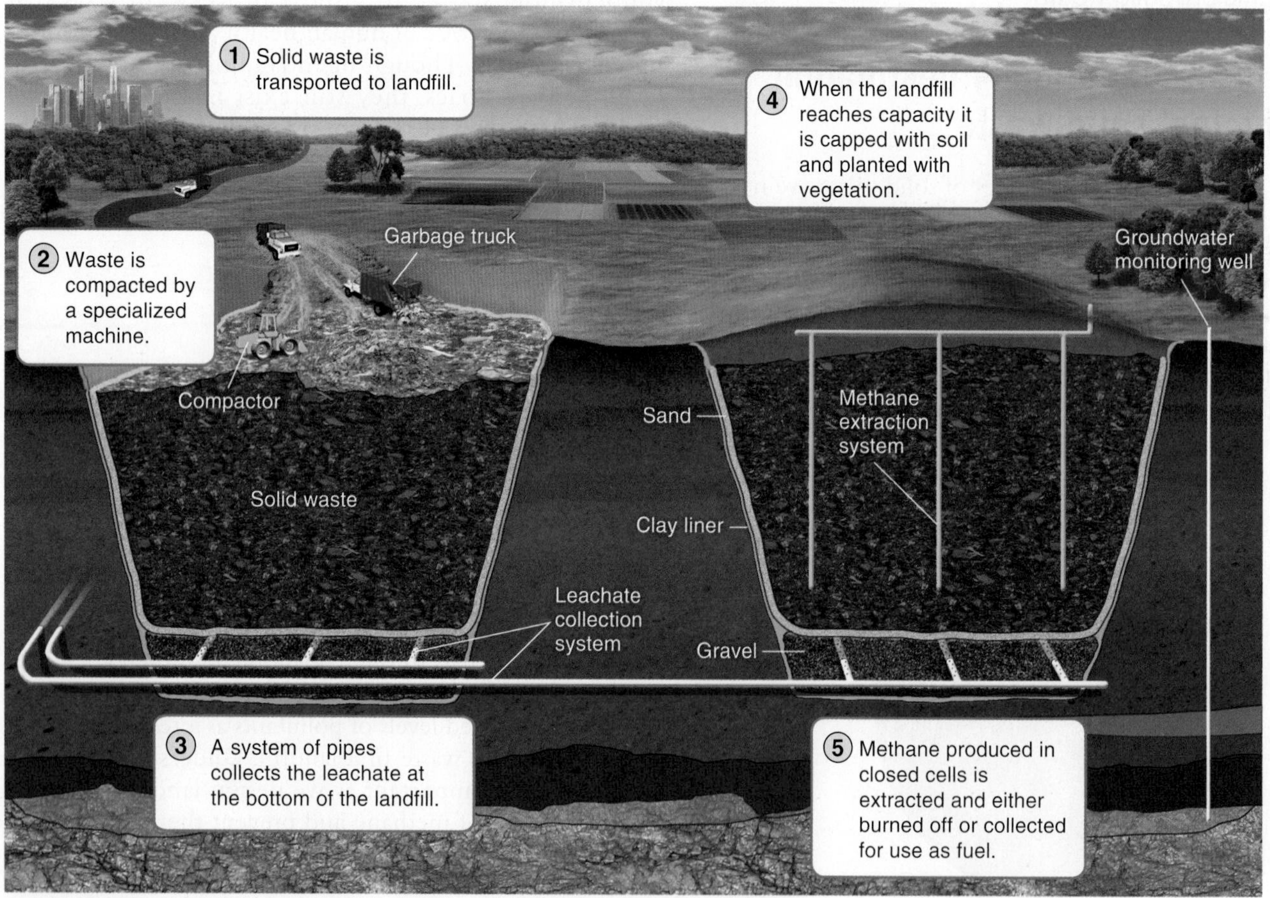

FIGURE 50.6 A modern sanitary landfill. A landfill constructed today has many features to keep components of the solid waste from entering the soil, water table, or nearby streams. Some of the most important environmental features are the clay liner, the leachate collection system, the cap — which prevents additional water from entering the landfill — and, if present, the methane extraction system.

and automotive additives like motor oil and antifreeze), consumer electronics, appliances, batteries, and anything that contains substantial quantities of metals should not be deposited in landfills. All organic materials, such as food and garden scraps and yard waste, are potential sources of methane and should not be placed in landfills.

AP® Exam Tip

Many past AP® Environmental Science Exams have asked students to describe and explain landfill design and function. Be sure you know how a sanitary landfill works.

The solid waste added to a landfill is periodically compacted into compartments or "cells," which reduce the volume of solid waste, thereby increasing the capacity of the landfill. Filled cells are then covered with soil. When a landfill is full, it must be closed off from the surrounding environment so that the input and output of water are reduced or eliminated. Once a landfill is closed and capped, the waste within it is more or less sealed off. Some water may enter from the outside environment, but this is minimal if the landfill is well-designed and properly sealed. The design and topography of the landfill cap, which is a combination of clay soil and sometimes plastic, encourage water to flow off to the sides, rather than moving down into the landfill. Closed landfills can be reclaimed, meaning that shallow-rooted vegetation can be planted on the landfill cap to reduce soil erosion and appear more aesthetically pleasing. Construction on the surface of a closed landfill is normally restricted for many years, although parks, playgrounds, and even golf courses have been built on reclaimed landfills.

A great example of converting a landfill can be found at the Freshkills Landfill on Staten Island, New York, which is named after the nearby Freshkills Creek. This landfill received solid waste for nearly 50 years and was the world's largest landfill. Once it reached full capacity in 2001, it closed and over the past 2 decades it has been transformed into an 890-ha (2,200-acre) park for people to enjoy (**FIGURE 50.7**). Pipes have been installed to capture the methane gas produced by anaerobic decomposition and the leachate is being collected and treated to remove any pollutants. The new Freshkills Park is the largest park built in New York City in the past century; when fully opened in 2037, it will be three times larger than Central Park in Manhattan.

A municipality or private enterprise constructs a landfill at a tremendous cost. These costs are recovered by charging a fee for waste delivered to the landfill, called a **tipping fee** because each truckload is put on a scale and, after the solid waste is weighed, it is tipped into the landfill. Tipping fees at solid waste landfills average $55 per ton in the United States, although these fees can vary in different regions of the country. These fees create an economic incentive to reduce the amount of waste that goes to the landfill. Many localities

Tipping fee A fee charged for trucks that deliver and tip solid waste into a landfill or incinerator.

(a)

(b)

FIGURE 50.7 Reclamation of a landfill. The Fresh Kills landfill in Staten Island, New York, was open from 1947 to 2001. It is currently in the process of being converted into a large park and is scheduled to be completed in 2037. The mounds of the landfill have been capped and planted with vegetation. *(a & b: Richard Levine/Alamy Stock Photo)*

accept recyclables at no cost but charge for disposal of material destined for a landfill. This practice encourages individuals to separate recyclables from their solid waste. Some localities mandate that recyclable material be removed from the waste stream and disposed of separately. However, if tipping fees become too high, and regulations too stringent, a locality may inadvertently encourage illegal dumping of waste materials.

Choosing a Landfill Site

Many considerations go into designating a landfill location. A landfill should be located in a loam or clay loam soil lined with clay or plastic to reduce the leaking of contaminants. It should be located away from rivers, streams, and other bodies of water and, especially, drinking-water supplies. A landfill should also be sufficiently far from population centers so that trucks transporting the waste and animal scavengers such as seagulls and rats present minimal risks to people. However, the energy needed to transport the municipal solid waste must also be considered in siting; as distance from a population center increases, so does the amount of energy required to move the waste to the landfill. Regional landfills, though, are becoming more common because sending all waste to a single location often offers the greatest economic advantage.

A landfill location is always highly controversial and sometimes politically charged. Because landfills are unsightly and smell bad, they are not considered desirable neighbors. Landfill locations have been the source of considerable environmental injustice. People with financial resources or political influence often adopt what has been popularly called a "not-in-my-backyard," or NIMBY, attitude about landfill sites. Because of this, a site may be chosen not because it meets the safety criteria better than other options but because its neighbors lack the resources to mount an effective opposition.

In Fort Wayne, Indiana, for example, the Adams Center Landfill was located in a densely populated, low-income, and predominantly minority neighborhood. A University of Michigan environmental justice study quotes Darrell Leap, a hydrogeologist and professor at Purdue University, as saying that on a scale of 1 to 10, with 10 being a geologically ideal site, the Adams Center Landfill would rate a "3, possibly 4" because the site held a substantial risk of water contamination. When the communities surrounding the Adams Center Landfill learned of the report and this danger, they protested the renewal of the federal permit and fought expansion of the landfill at both the state and local levels. Ultimately, the Indiana Department of Environmental Management closed the landfill.

Environmental Consequences of Landfills

Though sanitary landfills are an improvement over open dumps, they present many problems. Locating landfills near populations that do not have the resources to object is a global problem. No matter how carefully the landfill has been designed, there is always the possibility that leachate will contaminate underlying and adjacent waterways. The EPA estimates that some leaching has occurred at all landfills in the United States. Even after a landfill is closed, the potential to harm adjacent waterways remains. The amount of leaching, the substances that have leached out, and how far they will travel are impossible to know in advance. To get an idea of how much leachate is generated from a landfill and how much might be collected, see "Do the Math: How Much Leachate Might Be Collected?"

The risk to humans and ecosystems from leachate is uncertain. Public perception is that landfill contaminants

DO THE MATH How Much Leachate Might Be Collected?

Preparing for the AP® Exam

Annual precipitation at a landfill in the town of Fremont is 100 mm per year, and 50 percent of this water runs off the landfill without infiltrating the surface. The landfill has a surface area of 5,000 m^2. Underneath the landfill, the town installed a leachate collection system that is 80 percent effective; the remaining 20 percent of the leachate enters the surrounding soil and groundwater. This leachate contains cadmium and other toxic metals.

Calculate the volume of water in cubic meters (m^3) that infiltrates the landfill per year.

$$100 \text{ mm/year} = 0.1 \text{ m/year}$$

$$0.1 \text{ m/year} \times 5{,}000 \text{ m}^2 \times 50\% = 250 \text{ m}^3$$

So, the volume of leachate in m^3 that is collected per year is: $250 \text{ m}^3 \times 80\% = 200 \text{ m}^3\text{/year}$

YOUR TURN In a neighboring landfill with the same surface area of 5,000 m^2, 70 percent of annual precipitation runs off without infiltrating the surface and the leachate collection system is 90 percent effective. Calculate in cubic meters (m^3) the volume of leachate that is collected each year.

pose a great threat to human health. However, the EPA has ranked this risk as fairly low compared with other risks such as global climate change and air pollution. But methane and other organic gases generated from decomposing organic material in landfills do release greenhouse gases, which contributes to global climate change.

50-3 How is incineration used to dispose of municipal solid waste?

Incinerators reduce solid waste by burning it

Given all the problems of landfills, people have turned to a number of other means of solid waste disposal, including incineration. **Incineration** is the process of burning waste materials to reduce their volume and mass and, sometimes, to generate electricity or heat. More than three-quarters of the material that constitutes municipal solid waste is easily combustible. Because paper, plastic, and food and yard waste are composed largely of carbon, hydrogen, and oxygen, they are excellent candidates for incineration. An efficient incinerator operating under ideal conditions may reduce the volume of solid waste by up to 90 percent and the weight of the waste by 75 percent, although the reductions vary greatly depending on the incinerator and the composition of the waste. Because incineration often does not operate under ideal conditions, the volume of ash produced is typically about one-quarter of the volume of the original solid waste. In this section we will look at incineration basics and then explore the environmental consequences of incineration.

Incineration Basics

FIGURE 50.8 shows a mass-burn municipal solid waste incinerator. At a typical incinerator, the solid waste is sorted and certain recyclables are diverted to recycling centers. The remaining material is dumped onto a platform where certain materials such as metals are identified and removed. A moving grate or other delivery system transfers the waste to a furnace. Combustion rapidly converts much of the waste into carbon dioxide and water, which are released into the atmosphere along with heat after being passed through a filter to remove harmful particles in the combustion gases. The residual nonorganic material that does not combust during incineration, commonly known as **ash** is an end product of combustion.

Disposal of ash from an incinerator is determined by its concentration of toxic metals. If the ash is relatively low in concentration of contaminants such as lead and cadmium, the ash can be disposed of in a conventional landfill. It can also be used for other purposes such as fill in road construction or as an ingredient in cement blocks and cement flooring. If deemed toxic, the ash goes to a special ash landfill that is designed specifically for toxic substances.

Incineration The process of burning waste materials to reduce volume and mass, and sometimes to generate electricity or heat.

Ash The residual nonorganic material that does not combust during incineration.

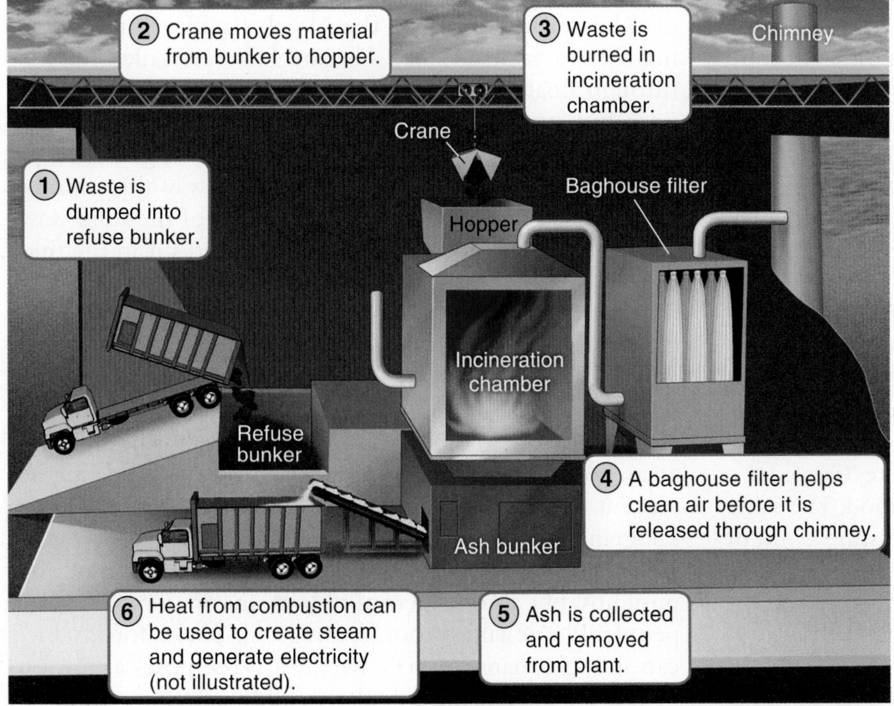

(2) Crane moves material from bunker to hopper.
(3) Waste is burned in incineration chamber.
Chimney
Crane
(1) Waste is dumped into refuse bunker.
Baghouse filter
Hopper
Incineration chamber
Refuse bunker
(4) A baghouse filter helps clean air before it is released through chimney.
Ash bunker
(6) Heat from combustion can be used to create steam and generate electricity (not illustrated).
(5) Ash is collected and removed from plant.

FIGURE 50.8 A municipal mass-burn waste-to-energy incinerator. In this plant, MSW is combusted and the exhaust is filtered. Remaining ash is disposed in a landfill. The resulting heat energy is used to make steam, which turns a generator that generates electricity in the same manner as was illustrated in Figure 37.7 in Unit 6.

The exhaust gases from an incinerator may also contain metals and other toxins, such as organic compounds from the incomplete combustion of plastics and metals contained in the solid waste that was burned. Some compounds, such as sulfur dioxide and nitrogen oxides, move through collectors and other devices that reduce their emission to the atmosphere. These collectors are similar in design to those described in Unit 7 on air pollution. Acidic gases such as hydrogen chloride (HCl), which results from the incineration of certain materials including plastic, are recovered in a scrubber, neutralized, and sometimes treated further before disposal in a conventional landfill or ash landfill.

Incineration also releases a great deal of heat energy, which is often used to heat the incinerator building or to generate electricity, using a process similar to that of a coal, natural gas, or nuclear power plant. When heat generated by incineration is used as an energy source rather than released into the surrounding environment, it is known as a **waste-to-energy** system. Although energy generation is a positive benefit of incineration, as we shall see, there are a number of environmental problems with incineration as a method of waste disposal.

Environmental Consequences of Incineration

Though incinerators address some of the problems of landfills, they also have shortcomings. To cover the costs of construction and operation, incineration facilities also charge tipping fees. Tipping fees are generally higher at incinerators than at landfills; national averages are around $70 per U.S. ton. As we have seen, an incinerator may release air pollutants. In addition, the ash from incineration is more concentrated and thus more toxic than the original solid waste. As we have already mentioned, incinerator ash that is deemed toxic must be disposed of in a special landfill for toxic materials.

Because incinerators are typically large and expensive to build and operate, they require large quantities of solid waste on a daily basis to burn efficiently and to be profitable. As a means of supporting these costs, communities that use incinerators may be less likely to encourage recycling. One solution is to use rate structures and other programs to encourage waste reduction and diversion, with the goal of using incineration only as a last resort. However, to be successful, this usually must be a community effort.

Incinerators may not completely burn all the waste deposited in them. Plant operators can monitor and modify the oxygen content and temperature of the burn, but because the contents of the waste stream are extremely variable and lumped all together, it is difficult to have a uniformly hot burn to incinerate all of the waste. Consider a truckload of solid waste from your neighborhood. The same load may contain food waste with high moisture content and, right next to that, packaging and other dry, easily burnable material. It is difficult for any incinerator—even a state-of-the-art modern facility—to burn all of these materials uniformly. In addition to the problems of air pollution, ash, and uneven burning, the location of an incinerator raises NIMBY and environmental justice issues similar to those of landfills.

Waste-to-energy A system in which heat generated by incineration is used as an energy source rather than released into the surrounding environment.

50-4 Why does some municipal solid waste not go to landfills or incinerators?

A lot of solid waste ends up in the ocean

While depositing municipal solid waste into landfills and incinerators is the most common approach for proper waste disposal, not all solid waste gets properly disposed. In some cases, solid waste is illegally dumped on land or in the water to avoid the cost of tipping fees. For example, you may have noticed tires from cars and trucks that are dumped along the sides of roads because most landfills charge a fee for each tire; these tires not only add to roadside litter, but they also hold rainwater and serve as breeding grounds for mosquitoes that can carry disease-causing pathogens.

In other cases, solid waste is illegally dumped because landfills and incinerators will not accept the waste due to the toxic material it contains. In developing countries that lack the resources to build enough sanitary landfills and incinerators or to collect solid waste, it accumulates and can be carried away during rainstorms that transport the waste into streams and rivers. One outcome is massive quantities of municipal solid waste in the oceans.

In 1997, Captain Charles Moore discovered a large area of solid waste floating in the North Pacific while returning from a yachting race. He and his crew sailed for days through floating, discarded plastics, which do not decompose but simply break into tinier pieces of plastic. This area, which he named the Great Pacific Garbage Patch, was assumed to be pristine since it was far away from human populations, but it actually collects much of the solid waste that is dumped into waters and concentrates it in the middle of the rotating currents in an area the size of Texas. Today, we know that the Pacific Ocean contains multiple massive garbage patches, as you can see in **FIGURE 50.9**. Because other oceans also circulate, scientists have found massive garbage patches in the circulating North Atlantic that are hundreds of kilometers across. With the vastness of these areas, no one is certain how much solid waste is floating but current estimates are in the range of hundreds of millions of kilograms.

Garbage on beaches and in the ocean is not only unsightly, but also dangerous to both marine organisms and people. Plastic rings from beverage six-packs, for instance, can strangle many animals. Medical waste poses a threat to people on the beach, particularly children. During the 1970s

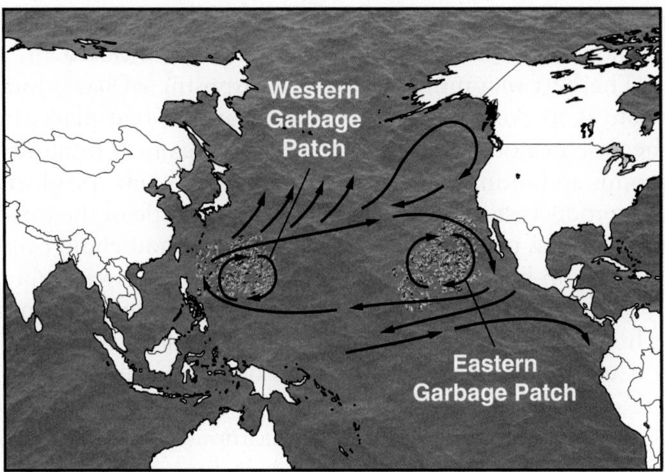

FIGURE 50.9 Garbage patches in the Pacific. As humans continue to dump solid waste into streams, rivers, lakes, and oceans, much of the waste is transported to the open ocean where ocean circulation patterns concentrate the waste into several massive patches. Similar patches exist in the circulating ocean waters of the North Atlantic.

FIGURE 50.10 A river of garbage. Environmental regulations have greatly reduced the amount of solid waste that is dumped in U.S. waters, but other parts of the world, such as the Citarum River in Indonesia, still face a major environmental challenge. *(REUTERS/Alamy Stock Photo)*

and 1980s, the public turned its attention to garbage dumping off the coasts of the United States by both municipalities and cruise ships. The outcry grew when garbage was found to include medical waste such as used hypodermic needles. As a result, the practice of dumping garbage in the ocean was curtailed in the early 1980s. The problem remains in many developing countries where there is often no political mechanism to prevent dumping or it is economically difficult to manage proper garbage disposal (**FIGURE 50.10**).

50-5 How is hazardous waste safely disposed?

Hazardous waste requires proper handling and disposal

When solid waste material is deemed toxic or otherwise harmful to people or natural ecosystems, it is not categorized as municipal solid waste and it cannot be placed in a conventional landfill or incinerated because special means of handling and disposal are required. In this section, we will discuss the proper treatment of hazardous waste and some of the legislation regulating its disposal.

Producing Hazardous Wastes

Hazardous waste is liquid, solid, gaseous, or sludge waste material that is harmful to humans, ecosystems, or materials. The U.S. EPA identifies four characteristics of hazardous waste: ignitability—it can catch on fire; corrosivity—it can cause materials to corrode or degrade; reactivity—it is not stable under normal conditions; toxicity—its chemical components are harmful or fatal when ingested or absorbed through the skin. Although the terms toxic and hazardous are often used interchangeably, toxic waste is a category of hazardous waste. While many substances are hazardous and can cause harm, substances that are toxic cause harm through chemical means when ingested or absorbed through the skin.

AP® Exam Tip

You should be able to differentiate between toxic waste and hazardous waste. You should clearly describe the danger of the chemical to organisms, humans, or the ecosystem. Don't use the words "toxic" or "hazardous" without identifying the specific risk.

According to the EPA, during the past decade the United States produced about 25 million metric tons (28 million U.S. tons) of hazardous waste each year. Only about 4 percent of that waste was recycled. The majority of hazardous waste is the by-product of industrial processes such as textile production, cleaning machinery, and manufacturing computer equipment, but it is also generated by small businesses such as dry cleaners, automobile service stations, and small farms. Even households generate hazardous waste, including oven cleaners, batteries, and lawn fertilizers. All of these materials have a likelihood of causing harm to humans and ecosystems and should not be disposed of in regular landfills.

Hazardous waste Liquid, solid, gaseous, or sludge waste material that is harmful to humans, ecosystems, or materials.

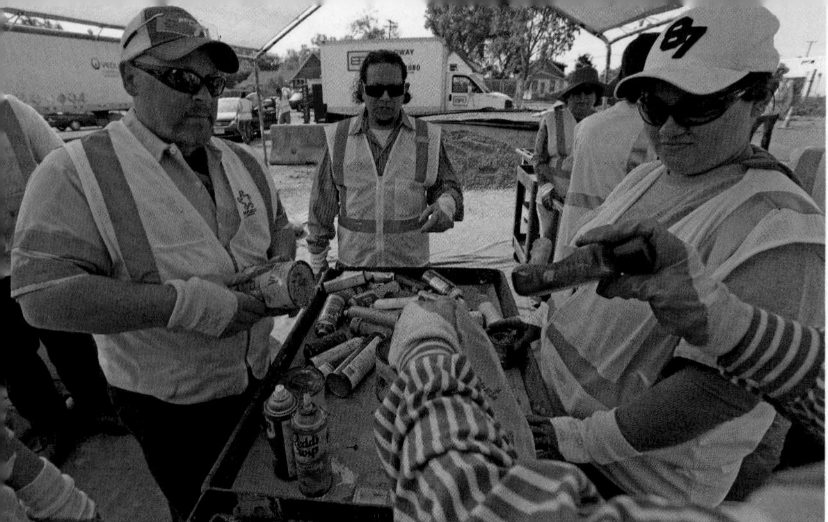

FIGURE 50.11 A typical household hazardous waste collection site in Minneapolis, Minnesota. Residents are encouraged to keep hazardous household waste separate from their regular household waste. Collections are held periodically. *(ZUMA Press Inc/Alamy)*

Most municipalities do not have regular collection sites for hazardous waste. Rather, homeowners and small businesses are asked to keep their hazardous waste in a safe location until periodic collections are held (**FIGURE 50.11**).

Disposing of Hazardous Waste

Every aspect of the treatment and disposal of hazardous waste is more expensive and more difficult than the disposal of ordinary MSW. Households use numerous substances, such as oil-based paints, motor oil, or chemical cleaners, that are easy to purchase but become regulated hazardous waste when the municipality collects them. Hazardous waste must be treated before disposal. Treatment, according to the EPA, means making it less environmentally harmful. To accomplish this, the waste must usually be altered through a series of chemical procedures.

Collection sites designated as hazardous waste collection facilities must be staffed with specially trained personnel. Sometimes the materials brought to a collection facility are unlabeled and unknown and must be treated with extreme caution. Ultimately, the wastes may be sorted into a number of categories, such as fuels, solvents, and lubricants. Some items, such as paint, can be reused, while others are sent to a special facility for treatment. There are no good options for disposing of hazardous waste. Hazardous waste landfills are more expensive to construct and require much more monitoring than a conventional MSW landfill. They must be monitored for at least 30 years after they are closed.

Superfund Act The common name for the Comprehensive Environmental Response, Compensation, and Liability Act (CERCLA); a 1980 U.S. federal act that imposes a tax on the chemical and petroleum industries, uses those funds for the cleanup of abandoned and nonoperating hazardous waste sites, and authorizes the federal government to respond directly to the release or threatened release of substances that may pose a threat to human health or the environment.

The best recommendation for disposing of hazardous waste is to avoid creating the waste in the first place. In the case of household hazardous waste, many community groups and municipalities encourage consumers to substitute products that are less toxic or to use as little of the toxic substances as possible. For example, a can of household oven cleaner can be purchased at a grocery store, hardware store, or home center for roughly $5. Once the consumer brings it home, it is potentially household hazardous waste, and if it were to be disposed of before being completely used, the disposal cost might be $25 to $50. A consumer who brings a product to a household hazardous waste collection site typically does not pay for product disposal directly; the municipality or some other entity bears the cost. Therefore, product substitution or product avoidance is often the best solution to the household hazardous waste problem.

Another major source of hazardous waste is the coal ash and coal tailings that remain behind when coal is burned. As we saw in Module 37, such waste contains a number of harmful chemicals including mercury, arsenic, and lead. The solid waste products from burning coal are typically dumped into landfills, ponds, or abandoned mines where it can contaminate groundwater. In the United States, the Environmental Protection Agency considers the waste from burning coal and other fossil fuels to be "special waste" that is exempt from federal regulations for the disposal of hazardous waste.

Regulation and Oversight of Handling Hazardous Waste

The dangers that hazardous waste present have brought disposal to the attention of the general public. In the United States there are several laws and acts that specifically cover hazardous waste. Regulation and handling of hazardous waste in the United States falls under two pieces of federal legislation, the U.S. Resource Conservation and Recovery Act (RCRA) and the Comprehensive Environmental Response, Compensation, and Liability Act (CERCLA).

In 1976, the RCRA expanded previous solid waste laws. Its main goal was to protect human health and the natural environment by reducing or eliminating the generation of hazardous waste. Under RCRA's provision for "cradle-to-grave" tracking, the EPA maintains lists of hazardous wastes and works with businesses and state and local authorities both to minimize hazardous waste generation and to make sure that the waste is tracked until proper disposal. In 1984, RCRA was modified with the federal Hazardous and Solid Waste Amendments (HSWA), which encouraged waste minimization and phased out the disposal of hazardous wastes on land. The amendments also increased law enforcement authority to punish violators.

The CERCLA, usually referred to as the **Superfund Act**, is a 1980 U.S. federal act that imposes a tax on the chemical and petroleum industries, uses those funds for the

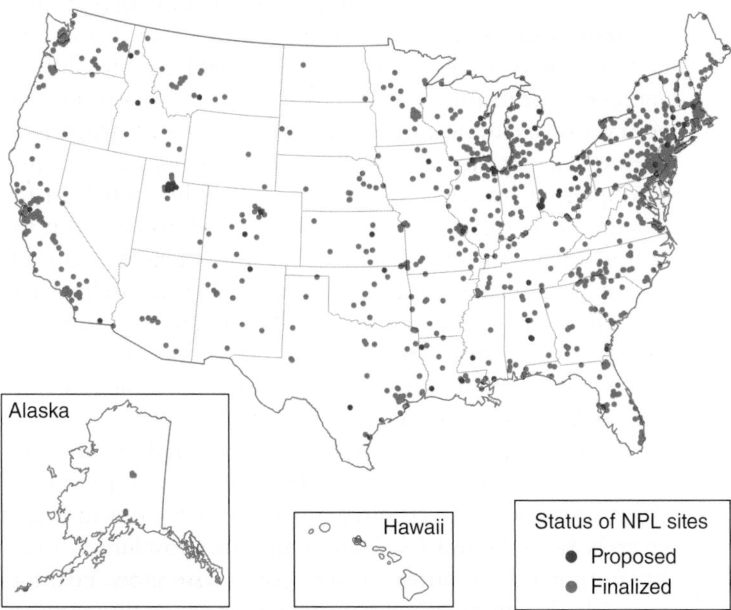

FIGURE 50.12 Distribution of Superfund sites in the United States. Under the Superfund Act, the EPA maintains the National Priorities List of contaminated sites that are eligible for cleanup funds. *(U.S. EPA, 2018)*

cleanup of abandoned and nonoperating hazardous waste sites, and authorizes the federal government to respond directly to the release or threatened release of substances that may pose a threat to human health or the environment. The Superfund Act is well known because of a number of high-profile cases that have fallen under its jurisdiction. Originally passed in 1980 and amended in 1986, the tax revenue is used to fund the cleanup of abandoned and nonoperating hazardous waste sites where a responsible party cannot be established. This tax fund came to be known as the "Superfund" for cleaning up hazardous waste sites.

AP® Exam Tip

Make sure you know the provisions of the Superfund Act (CERCLA) and the U.S. Resource Conservation and Recovery Act (RCRA). These laws are frequently tested in multiple-choice questions on the AP® Environmental Science Exam.

Under Superfund, the EPA maintains the National Priorities List of contaminated sites that are eligible for cleanup funds. The map in **FIGURE 50.12** shows the locations of the sites. As of 2021, there were 1,332 Superfund sites, with at least one in every state. The states with the most sites are New Jersey, California, Pennsylvania, and New York. For a long time, very little Superfund money was disbursed and few sites underwent remediation. By 2021, 430 sites had been appropriately remediated and removed from the Superfund list.

Perhaps the best-known Superfund site is Love Canal, a neighborhood within Niagara Falls, N.Y. (**FIGURE 50.13**). Originally a hazardous waste landfill, Love Canal was covered with fill and topsoil and used as a site for a school and a housing development. In 1978 and 1980, known cancer-causing wastes were found in the basements of homes in the area. These substances included the solvents benzene, dioxin, and a degreasing agent known as trichloroethylene. When it became clear that a disproportionately large number of illnesses, possibly connected to the chemical waste, had been diagnosed in the local population, the situation attracted national attention. Much of the attention that Love Canal received was due to the mother of a then 5-year-old boy in the school near Love Canal. Lois Gibbs, who had no previous experience in community activism, is credited with drawing attention to

FIGURE 50.13 Lois Gibbs and Love Canal, New York. Love Canal became a symbol of hazardous chemical pollution in the United States in the 1970s. Lois Gibbs became a spokesperson for the neighborhood of Love Canal and is credited with bringing national attention to her community. *(Harry Scull Jr./Getty Images)*

the problems created by Love Canal and other locations. The contamination was so bad that in 1983 Love Canal was listed as a Superfund site and the inhabitants of the area were evacuated. In 1994, the EPA removed Love Canal from the National Priorities List because the physical cleanup had been completed and the site was no longer deemed a threat to human health. However, the site remains a symbol of the dangers presented by improper disposal of hazardous waste in a residential neighborhood. It also shows that a private citizen with no previous experience in community activism can greatly improve the safety of a neighborhood.

Since its inception, the CERCLA has not had enough funding to clean up the numerous hazardous waste sites around the country. The listing of NPL sites in a state does not necessarily include all contaminated sites within that state. For example, the Department of Environmental Protection in New Jersey, a state that has 114 listed Superfund sites, believes that there are more than 9,000 sites where soil or groundwater is contaminated with hazardous chemicals.

Brownfields

The Superfund designation is reserved strictly for those locations with the highest risk to public health, and Superfund sites are managed solely by the federal government. In 1995, the EPA created the Brownfields Program. **Brownfields**, like Superfund sites, are contaminated industrial or commercial sites that may require environmental cleanup before they can be redeveloped or expanded. The Brownfields Program assists state and local governments in cleaning up contaminated industrial and commercial land that did not achieve conditions necessary to be in the Superfund category. Old factories, industrial areas and waterfronts, dry cleaners, gas stations, landfills, and rail yards are common examples of brownfield sites. Brownfields legislation has prompted the revitalization of several sites throughout the country. One notable instance is Seattle's Gas Works Park. In 1962, the city of Seattle purchased the land, which had housed a coal gasification plant, with a plan to rehabilitate the site into a park. After undergoing chemical abatement and environmental cleanup, the park has become a distinctive landmark for the city.

The Brownfields Program has been criticized as an inadequate solution to the estimated 450,000 contaminated locations throughout the country. Since the cleanup is managed entirely by state and local governments, brownfields management can vary widely from region to region. Furthermore, the brownfields legislation lacks legal liability controls to compel polluters to rehabilitate their properties. Without legal recourse, many brownfield sites remain unused and contaminated, posing a continued risk to public health.

Brownfields Contaminated industrial or commercial sites that may require environmental cleanup before they can be redeveloped or expanded.

As we have seen in this module the amount of municipal solid waste produced in the United States continues to increase over time and there is no ideal choice for waste disposal. Sometimes the decision between whether to construct a landfill or build an incinerator is in part a decision about the kind of pollution a community prefers. The best choice is that we strive to simply produce less waste material. We also learned that some solid waste is categorized as hazardous and must be handled very carefully to prevent contamination of the environment. In the next module, we will explore a number of options for reducing municipal solid waste.

Module 50 AP® Review

Preparing for the AP® Exam

Learning Goals Revisited

50-1 How has solid waste production changed over time and what are the sources?

Modern societies produced refuse that we call municipal solid waste (MSW). In the United States, total MSW has increased steadily over the past 6 decades while the per capita MSW has leveled off. Of the total MSW produced, 50 percent is discarded in landfills, 38 percent is recovered through recycling and composting, and 12 percent is incinerated. E-waste is a relatively small component of MSW but because of the toxicity of the metals it often contains, its disposal is a significant problem. Of the total MSW, 23 percent is paper and paperboard, 34 percent is food and yard waste, and 6 percent is wood. The remaining 37 percent is plastic, glass, metals, rubber, leather, textiles, and other items. E-waste is a small percentage of solid waste, but it contains highly toxic metals that need to be specially handled.

50-2 How are landfills used to dispose of municipal solid waste?

Sanitary landfills that are designed to contain the solid waste and prevent dangerous leachate and gases from leaving the landfill. Sanitary landfills are designed to be sealed at the bottom and capped with soils at the top, while collecting and monitoring the pollutants that emerge as leachate or gases so that they do not pollute the surrounding environment. The high cost of creating and running landfills is offset by tipping fees charged to trucks that bring MSW to the landfill. Landfill locations have also been the source of considerable environmental injustice.

50-3 How is incineration used to dispose of municipal solid waste?

Incinerators burn waste materials to reduce their volume and mass and, sometimes, to generate electricity or heat. Combustion rapidly converts the waste into carbon dioxide, water, and ash. Ash containing low levels of toxins is deposited in landfills while ash containing high levels of toxins is deposited in special landfills that are designed specifically for toxic substances. The exhaust gases from an incinerator may also contain metals and other toxins, which need to be collected whenever possible.

50-4 Why does some municipal solid waste not go to landfills or incinerators?

When landfills and incinerators are not available or they are avoided to save on the cost of tipping fees, solid waste gets dumped on land or in the water. This waste can be transported by streams and rivers to the ocean, where we now have massive areas of ocean with floating waste as a result of ocean circulation patterns.

50-5 How is hazardous waste safely disposed?

Hazardous wastes are a special category of waste material that is harmful to humans or ecosystems. Hazardous wastes must be handled and treated separately from the MSW stream. There are a number of legislative acts that specifically address hazardous waste and the cleanup of sites that have been abandoned.

Practice Math and Graphing 

Preparing for the AP® Exam

1. Practice Math

The table below shows the amounts of waste a factory generated in the first 5 years of operation. If by year 5 the factory was able to implement a source reduction of 25 percent, and assuming waste generation remained proportionately the same, how much waste would the factory generate in year 5?

	Waste generated (tons)
Year 1	1,472
Year 2	1,566
Year 3	1,900
Year 4	1,950
Year 5	2,000

2. Practice Graphing

Using the table from "Practice Math," plot the amount of waste generation from the factory in years 1 through 4 with a vertical bar graph. Plot year 5 using your calculation for a 25 percent waste reduction. Show the year on the *x* axis and the amount of waste in tons on the *y* axis.

AP® Practice Questions

Multiple-Choice Questions

1. The material that makes up the highest proportion of MSW is
   (a) plastic.
   (b) rubber, leather, and textiles.
   (c) paper and paperboard.
   (d) food.

2. Which of the following best describes electronic waste?
   (a) It accounts for over 10 percent of the waste stream.
   (b) It is almost always recycled.
   (c) It contains few toxic components.
   (d) It is more expensive to recycle than to put in a landfill.

3. Which material, when placed in a landfill, is most likely to cause problems as a result of leaching into the surrounding soil?
   (a) paper
   (b) food matter
   (c) glass
   (d) aluminum

4. A 500 m^2 landfill experiences 150 mm of rain each year and 60 percent of the rain is runoff. If the landfill has a 90 percent effective leachate collection system, how much leachate escapes each year?
   (a) 3 m^3
   (b) 5 m^3
   (c) 16 m^3
   (d) 27 m^3

5. Household hazardous waste
   (a) costs much less to recycle than regular waste.
   (b) is primarily generated by individual households.
   (c) should be disposed of in landfills intermixed with MSW.
   (d) can include items such as paints and oil.

6. Which legislation calls for listing hazardous waste to use in cradle-to-grave tracking?
   (a) Resource Conservation and Recovery Act
   (b) Comprehensive Environmental Response, Compensation, and Liability Act
   (c) Hazardous and Solid Waste Amendments
   (d) Superfund Act

7. Which best describes why hazardous waste disposal in the United States could become an international issue?
   (a) Lower disposal costs elsewhere encourage U.S. industries to export waste.
   (b) Air pollution from waste treatment often crosses international boundaries.
   (c) The United States continues to dump its hazardous waste into oceans.
   (d) The United States produces the majority of the world's hazardous waste.

Free-Response Question

The graph below illustrates the different categories of Municipal Solid Waste (MSW) generated in the United States in 2018.

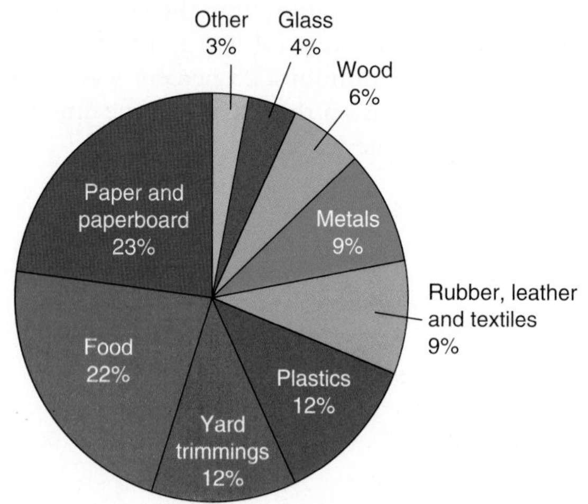

(a) **Identify** a material that made up nearly one-quarter of MSW in 2018. (1 pt.)

(b) **Describe** a component of MSW in 2018 that would be suitable for disposal in an incinerator. (1 pt.)

(c) **Explain** a component of MSW in 2018 that should not be disposed of in a sanitary landfill. (1 pt.)

(d) Sanitary landfills are often the location of small biogas power plants, which burn collected methane for electricity production. **Justify** the use of biogas plants by explaining one potential advantage of this practice. (1 pt.)

(e) For the following calculations, assume the following:
   - The U.S. population in 2018 was 327 million.
   - There are 2,000 pounds in a U.S. ton.
   - There are 2.2 pounds in one kilogram.

   (i) **Calculate** the MSW generated in the United States in 2018 in kilograms per person per day. Show all work. (2 pts.)

   (ii) Composting is a common method of reducing the amount of MSW that ends up in landfills or incinerators, and can include food scraps and yard trimmings. **Calculate** how many tons of MSW generated in 2018 could potentially be composted. Show all work. (2 pts.)

   (iii) In 2018, only 14 percent of all MSW was actually composted. **Calculate** the percent difference between the number of tons actually composted and the number of tons potentially composted from part (ii). Show all work. (2 pts.)

Module 51

Unit 8	47	48	49	50	51	52	53	54

Waste Reduction Methods

In the previous module, we explored the challenges of disposing municipal solid waste using landfills and incinerators. Given the trade offs of these waste disposal methods, we also need to think about how we can reduce the amount of waste we produce so that it never has to enter a landfill or incinerator. As we discussed in the unit-opening story, recycling and reusing materials are two important options that we can use. However, we can also think about how to reduce the amount of waste that is created when we manufacture items. We can also consider composting items that rapidly decay, such as food scraps. The most effective way to assess the best approach for any community is to consider combinations of these efforts, including the limitations and trade offs of each approach. In this module we focus on how we can reduce municipal solid waste when we reduce, reuse, recycle, and compost.

Learning Goals

After reading this module you should be able to

51-1 describe the three Rs that divert materials from the waste stream.

51-2 explain how composting reduces materials entering the waste stream.

51-3 explain how life-cycle analysis and integrated waste management reduce municipal solid waste.

51-1 What are the three Rs that divert materials from the waste stream?

The three Rs: reduce, reuse, recycle

Starting in the 1990s, people in the United States began to promote the idea of diverting materials from the waste stream with a popular phrase "**Reduce, Reuse, Recycle**," also known as **the three Rs** (FIGURE 51.1). The phrase incorporates a practical approach to solid waste management, with the techniques presented from the most environmentally beneficial to the least beneficial. In this section, we explore each of these approaches. Note that each action uses less energy than the next action.

FIGURE 51.1 **Reduce, Reuse, Recycle.** This popular slogan emphasizes the actions to take in the proper order, from the most environmentally effective to the least. *(Becky Stares/Shutterstock)*

Reduce

"Reduce" is the most energy efficient among the three Rs, which makes reducing inputs the optimal way to achieve a decline in solid waste generation. This strategy is also known as waste minimization and waste prevention. When less material is used, there will be less material to discard. One approach, known as **source reduction**, seeks to cut waste by reducing the use of potential waste materials in the early stages of design and manufacture. In many cases, source reduction also increases energy efficiency because the manufacturing produces less waste to begin with and can

Reduce, Reuse, Recycle A popular phrase promoting the idea of diverting materials from the waste stream. *Also known as* **the three Rs**.

Source reduction An approach to waste management that seeks to cut waste by reducing the use of potential waste materials in the early stages of design and manufacture.

minimize disposal processes. Since fewer resources are being expended, source reduction also provides economic benefits.

Source reduction can be implemented by individuals, companies, and institutions. For example, if teachers have a two-page assignment to hand out for a class, they could reduce paper use by 50 percent if they provide students with double-sided photocopies. Further source reduction could be achieved by using electronic copies of assignments, where possible, presuming that students do not print the documents.

Companies that manufacture products can work on source reduction in several ways. One way is to create new packaging that provides the same amount of protection to the product with less material. For example, consider how purchasing music has changed over the past two decades. People used to buy music as compact discs (CDs) that were packaged in large plastic sleeves that were three times the size of the CD. Years later, CDs were wrapped with a small amount of plastic that just covered the CD case. Today, most people download or listen to music from the Web. Decreased wrapping of CDs is an example of source reduction by companies. Downloading music instead of purchasing CDs is an example of source reduction by individuals.

Source reduction can also be achieved by substituting materials. In an office where people drink from paper cups, providing each person with a reusable mug will reduce MSW. This change could be considered reuse rather than source reduction. Nevertheless, cleaning the mugs will require water, energy to heat the water, soap, and processing of wastewater. The break-even point, beyond which there are gains achieved by using a ceramic mug (though this depends on a variety of factors), might be approximately 50 uses. The break-even point is shorter for a reusable plastic mug, in part because less energy is required to manufacture the plastic mug and, because it is lighter, less energy is used to transport it from the manufacturing facility to the point of use. Source reduction may also involve substituting less toxic materials. For example, switching from an oil-based paint that contains toxic petroleum derivatives to a relatively nontoxic latex paint is a form of source reduction.

All of these examples achieve a reduction in material use and, ultimately, a reduction in solid waste, without an additional expenditure of materials or energy. For that reason, reduction is the first "R"—and it is the most environmentally beneficial.

FIGURE 51.2 Reusable bottles. Some beverage containers, such as these glass milk bottles, can be used by customers and then brought back to the store. The returned bottles can then be washed and refilled with milk for another customer. *(Homestead Creamery)*

AP® Exam Tip

You should be able to describe ways to divert waste from the waste stream and be ready to discuss which action is the most environmentally beneficial and why. Remember that fences—which are above ground and useful for stopping paper and plastic from being blown away from a landfill site—do not prevent runoff.

Reuse

The second "R" is the **reuse** of a product or material that would otherwise be discarded, allowing an item to be used longer before it becomes waste. Ideally, no additional energy or resources are needed for the object to be reused. For example, a mailing envelope can be reused by covering the first address with a label and writing the new address over it. Or we could reuse a disposable polystyrene cup more than once, though reuse might involve cleaning the cup, which would add some energy cost and generate some wastewater. Sometimes reuse may involve repairing an item, which costs time, labor, energy, and materials but produces less waste than discarding the object.

Sometimes energy is required to prepare or transport an object for reuse by someone other than the original user. For example, some companies reuse containers such as milk bottles or soda bottles by shipping them to the bottling factory where they are washed, sterilized, and refilled (**FIGURE 51.2**). Although energy is involved in the transport and preparation of the containers, it is still less than the energy that would be required for recycling or disposal.

Reuse is common in many countries and it was once common practice in the United States. However, it is still practiced in many ways that we might not consider as reuse. For example, people often reuse newspapers for animal bedding

Reuse Using a product or material that would otherwise be discarded.

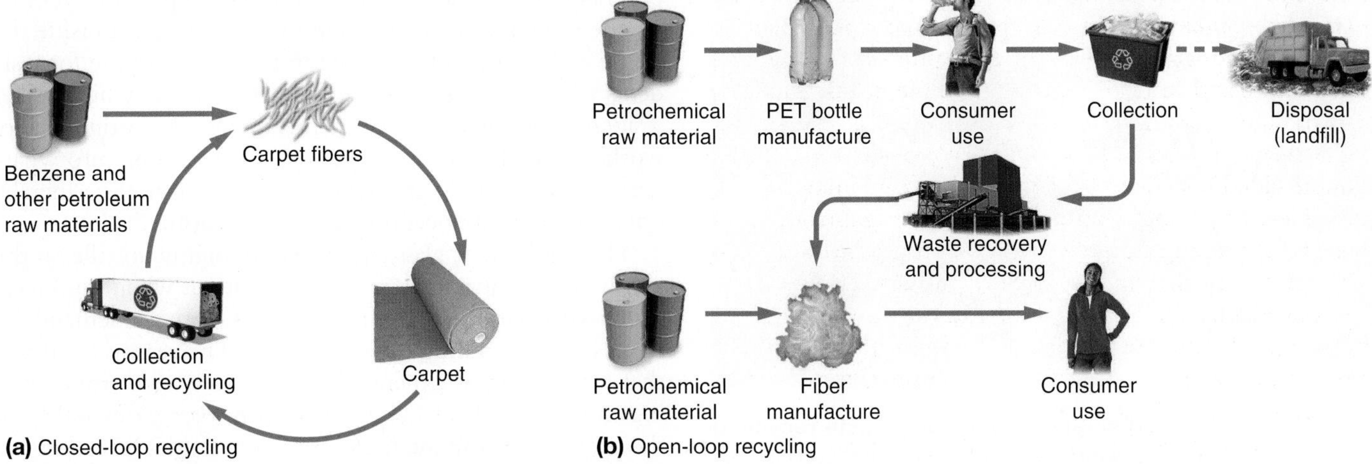

FIGURE 51.3 Closed- and open-loop recycling. (a) In closed-loop recycling, a discarded carpet can be recycled into a new carpet, although some additional energy and raw material are needed. (b) In open-loop recycling, a material such as a beverage container is used once and then recycled into something else, such as a fleece jacket.

or art projects. Many businesses and universities have surplus equipment agents who help find a home for items no longer needed. Flea markets, swap meets, and even popular websites such as eBay, craigslist, and The Freecycle Network are all helping to increase the reuse of items. Reuse often involves the expenditure of energy and generates waste. For example, to transport, wash, and sterilize the glass milk bottles in our earlier example requires energy and generates wastewater. Therefore, reuse involves environmental costs. However, it is typically more energy efficient than manufacturing and distributing new material that will later be discarded.

Recycle

The third "R" is **recycling**, the process by which materials destined to become MSW are collected and converted into raw materials that are used to produce new objects. We divide recycling into two categories: closed-loop and open-loop. **FIGURE 51.3** shows the process for each. **Closed-loop recycling**, shown in Figure 51.3a, is the recycling of a product into the same product. Aluminum cans are a familiar example; they are collected, brought to an aluminum plant, melted down, and made into new aluminum cans. This process is called closed-loop recycling because it is possible to keep making aluminum cans from only old aluminum cans almost indefinitely. In **open-loop recycling**, shown in Figure 51.3b, a product such as a plastic soda bottle is recycled into a different product, such as polar fleece jackets. Although recycling plastic bottles into other materials avoids sending the plastic bottles to a landfill, it does not reduce demand for the raw material—in this case petroleum—to make plastic for new bottles.

Recycling is not new in the United States, but over the past 25 years it has been embraced enthusiastically by both individuals and municipalities in the belief that it measurably improves environmental quality. **FIGURE 51.4** shows the

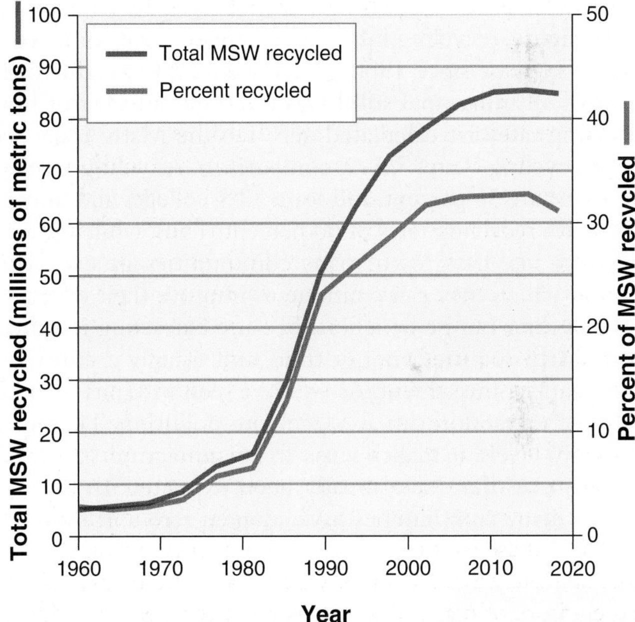

FIGURE 51.4 The amount of municipal solid waste recycled in the United States since 1960, in terms of MSW and the percent of total MSW in the waste stream. Both the total weight of MSW that is recycled and the percentage of MSW that is recycled have increased over time. *(Data from Advancing Sustainable Materials Management: 2018 Fact Sheet, U.S. EPA, 2020)*

Recycling The process by which materials destined to become municipal solid waste (MSW) are collected and converted into raw materials that are then used to produce new objects.

Closed-loop recycling Recycling a product into the same product.

Open-loop recycling Recycling one product into a different product.

FIGURE 51.5 A mixed single-stream solid waste recycling facility in Elkridge, Maryland. With single-stream recycling, also called no-sort or zero-sort recycling, consumers no longer have to worry about separating materials. This facility takes in and sends out a thousand tons of material daily. *(The Washington Post/Getty Images)*

increase in the weight of MSW in the United States from 1960 to 2018 and the increase in the percent of waste that was recycled over the same period of time. As you can see in the figure, recycling rates have increased over time, with a sharp increase since 1985. Today, we recycle roughly one-third of all municipal solid waste. To get an idea of how recycling rates are calculated, see "Do the Math: Calculating Recycling Rates." As a comparison, recycling rates in Germany are 56 percent, and some U.S. colleges and universities have recycling rates of 60 percent. Thus, while the U.S. recycling rate has grown, many communities are exploring ways in which they can continue to improve their efforts.

Recycling can be beneficial because extracting resources from Earth requires energy, time, and usually a considerable financial investment. As we have seen in Units 5 and 6, resource extraction can also generate pollution. Therefore, on many levels, it makes sense for manufacturers to recycle resources that have already been extracted. In the past decade, many communities have adopted zero-sort recycling programs, which allow residents to mix all types of recyclables in one container that they deposit on the curb outside the home or bring to a transfer station. This saves time for residents who were once required to sort materials, which makes residents more likely to recycle. At the MSW sorting facility, workers sort the materials destined for recycling into whatever categories are in greatest demand at a given time, thereby offering the greatest economic return (**FIGURE 51.5**).

The markets for glass and paper are highly volatile, so the sorting facility may only recycle these items when it is financially worthwhile. In contrast, there is always a demand for metals such as aluminum and copper. The reason for this is that it is much less expensive to make new cans from existing metal cans than making cans from ores extracted from the ground. In contrast, the cost of new bottles from existing plastic bottles is much higher than making new bottles from petroleum. In addition, recycling metal cans does not result in a degradation of the metal while each recycling of plastic bottles results in degraded plastic. You can learn much more about the costs and benefits of recycling in "Science Applied 8: Is Recycling Always Good for the Environment?".

Given that recycling requires time, processing, cleaning, transporting, and possible modification before the solid waste is usable as raw material, recycling materials usually requires more energy than reducing or reusing materials. For example, high recycling costs caused New York City to make a controversial decision about 20 years ago to suspend glass and plastic recycling. This was a major shift in policy for the city, which had been encouraging as much recycling as possible, including collection of mixed recyclables—glass, plastic, newspapers, magazines, and boxboard—at the same time as waste materials. After collection, the recyclables were sent to a facility where they were sorted. According to city officials, the entire process of sorting glass and plastic from other recyclables and selling the material was not cost effective.

As people have come to increasingly embrace the idea of recycling over recent decades, much of the recycled material was being shipped to China to be processed into new products. However, in 2018 China announced it was going

DO THE MATH Calculating Recycling Rates

Preparing for the AP® Exam

Imagine that it is the year 2030. People in the United States are able to recycle 40 million metric tons of material that was destined to be MSW. The total amount of material destined to be MSW before recycling is 80 million metric tons. What would be the percent recycling rate in 2030?

Divide the amount recycled by the total amount that would have gone to the landfill without recycling:

$$40 \text{ million metric tons} \div 80 \text{ million metric tons} = 0.5$$

To convert to percentage, multiply by 100%:

$$0.5 \times 100\% = 50\%$$

YOUR TURN Using Figure 51.4, determine the total MSW generated in 1960.

to stop accepting recycled paper and plastic products. As a result, there is now a much smaller market for purchasing recycled items, so waste-management companies now have to charge more for picking up recycled items to make their efforts financially viable. Many towns and cities cannot afford these rising costs, so they are reducing the items they recycle or completely ending their recycling programs. If this situation persists, we can expect a larger portion of solid waste to be going to landfills and incinerators in the coming years.

Given these challenges, we can see why recycling is the last choice among the three Rs. This doesn't necessarily suggest that we should abandon recycling programs, but it also encourages people to be more aware of the consequences of their consumption patterns. Nevertheless, in terms of the impact on the environment, efforts to reduce and reuse materials are preferable to recycling materials.

FIGURE 51.6 Composting. Good compost has a pleasant smell and will enhance soil quality by adding nutrients to the soil and by improving moisture and nutrient retention. In a compost pile that is turned frequently, compost can be ready to use in a few months. *(Deborah Vernon/Alamy Stock Photo)*

51-2 How does composting reduce materials entering the waste stream?

Composting reduces food and yard trimmings in the waste stream and enhances soil quality

While we often hear an emphasis on the three Rs for reducing materials in the waste stream, a fourth approach to assist in this goal is focused on organic materials such as food and yard waste. When these items end up in landfills it causes two problems. Like any material, they take up space, but unlike glass and plastic that are chemically inert, organic materials can break down. However, the absence of oxygen in landfills causes the organic material to decompose anaerobically, which produces methane gas, a much more potent greenhouse gas than carbon dioxide.

One way to avoid the problems of organic waste disposal in landfills and incinerators is through **composting**, which is the breakdown of organic materials into organic matter (humus). Vegetables and vegetable by-products, such as cornstalks, grass, animal manure, yard wastes (like leaves and branches), and paper fiber not destined for recycling are suitable for composting (**FIGURE 51.6**). Normally, meat and dairy products are not composted because they do not decompose as easily, they produce foul odors, and are more likely to attract unwanted visitors such as rats, skunks, and raccoons.

Outdoor compost systems can be as simple as a pile of food and yard waste in the corner of a yard, or as sophisticated as compost boxes and drums that can be rotated to ensure mixing and aeration. To encourage rapid decomposition, it is important to have the ratio of carbon to nitrogen (C:N) that will best support microbial activity—about 30:1. While it is possible to calculate the carbon and nitrogen content of each material that is put into a compost pile, most compost experts recommend layering dry material such as leaves or dried cut grass—known as "brown materials"—with wet material such as kitchen vegetables—known as "green materials." This will provide the correct carbon to nitrogen ratio for optimal composting. Frequent turning or agitation is usually necessary to ensure that decomposition processes are aerobic and to maintain appropriate moisture levels—otherwise the compost pile will produce methane and emit a foul odor. If the pile becomes particularly dry, water needs to be added. Although many people assume that a compost pile must smell bad, with proper aeration and moisture, the only odor will be the pleasant smell of fresh compost in 2 to 3 months. The resulting humus is an organic-rich material that enhances soil structure, cation exchange capacity, and fertility for lawns and gardens.

To help divert large amounts of organic materials from the waste streams of cities and towns, large-scale composting facilities exist in many municipalities in the United States. Some facilities are indoors, but most employ the same basic process we have described, although on a much larger scale. As shown in **FIGURE 51.7**, once compostable waste is separated from noncompostable waste, the compostable waste is piled up in long, narrow rows of compost. The material is mixed frequently to expose it to a combination of air and water that will speed aerobic decomposition. As with household composting, the organic material must include the correct combination of green (fresh) and brown (dried) organic material so that the ratios of carbon and nitrogen are optimal for bacteria. Various techniques are used to turn the organic material over periodically, including the use of rotating blades that move through the piles of organic material or a tractor with a large bucket that turns over the piles. The respiration activity of the microbes generates enough heat to kill any pathogenic bacteria that may be contained in food scraps, which is typically a concern only in large municipal composting systems. If the pile becomes too hot,

Composting The breakdown of organic materials into organic matter (humus).

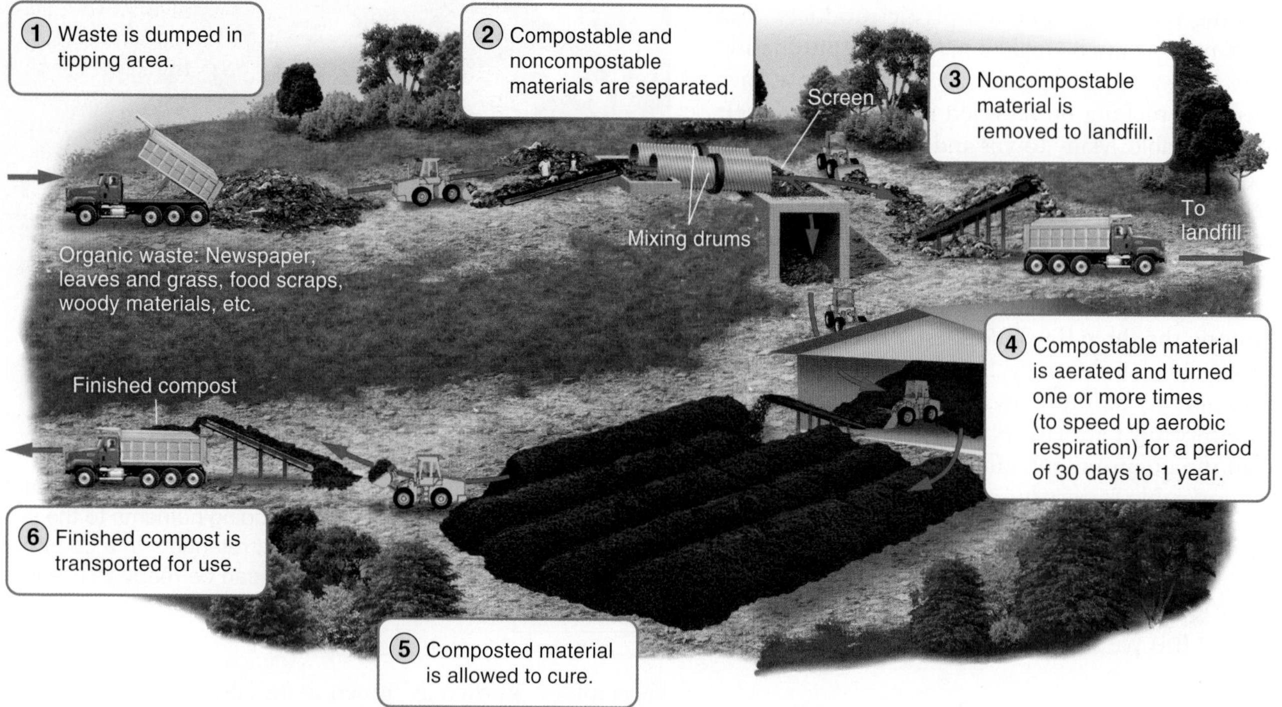

1 Waste is dumped in tipping area.
2 Compostable and noncompostable materials are separated.
Screen
3 Noncompostable material is removed to landfill.
To landfill
Mixing drums
Organic waste: Newspaper, leaves and grass, food scraps, woody materials, etc.
4 Compostable material is aerated and turned one or more times (to speed up aerobic respiration) for a period of 30 days to 1 year.
Finished compost
6 Finished compost is transported for use.
5 Composted material is allowed to cure.

FIGURE 51.7 A municipal composting facility. A typical facility collects almost 100,000 metric tons of food scraps and paper per year and turns it into usable compost. Most facilities have some kind of mechanized system to allow mixing and aeration of the organic material, which speeds conversion to compost.

it is turned more frequently. Within a matter of weeks or months, the organic waste is converted into compost.

While an outdoor space is convenient for composting household waste, composting is possible even in a city apartment or a dorm room. It is even possible to set up a composting system in a kitchen or basement. The very popular book *Worms Eat My Garbage: How to Set Up and Maintain a Worm Composting System* by Mary Appelhof has encouraged thousands of individuals across the country to compost kitchen waste using red wiggler worms. A small household recycling bin is large enough to serve as a worm box. As with an outdoor compost pile, a properly maintained worm box does not give off bad odors.

The composting process does take time and space. Separating out organic materials can be an inconvenience or, in some situations, not possible. Also, in certain environments, storing materials before they are added to the compost pile can attract flies or vermin. Finally, the compost pile itself can attract unwanted animals such as rats, skunks, raccoons, and even bears. But because compost is high in organic matter, which has a high cation exchange capacity and contains nutrients, it enhances soil quality when added to agricultural fields, gardens, and lawns.

Life-cycle analysis A systems tool that examines the materials used and released throughout the lifetime of a product — from the product design and procurement of raw materials through their manufacture, use, and disposal. *Also known as* **cradle-to-grave analysis.**

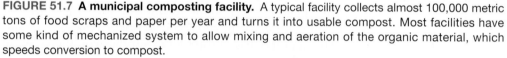

51-3 How do life-cycle analysis and integrated waste management reduce municipal solid waste?

New ways to think about municipal solid waste

Throughout this unit we have described a variety of ways of managing solid waste, from creating less of it to burying it in a landfill to burning it. Each method has both benefits and drawbacks. Because there is no obvious best method, the problem of waste disposal seems overwhelming. How can an individual, a small business, an institution, or a municipality best address solid waste? The answer is highly specific to each situation and varies by region. In addition, every method of waste disposal will have adverse environmental effects; the challenge is to find the least detrimental option. How do we decide which choices are best? Life-cycle analysis and a holistic approach are two useful strategies to help determine what we should do with our solid waste.

Life-Cycle Analysis

Recall the beginning of the unit where we attempted to make an objective assessment of solid waste disposal options by comparing different types of drinking cups. This process, known as **life-cycle analysis**, also known as **cradle-to-grave analysis**, is an important tool that examines the

FIGURE 51.8 A life-cycle analysis. To fully understand our options for solid waste disposal, we need to start with thinking about how we might modify the design and resource needs of a manufactured item. We can then think about ways to improve the manufacturing, transport, and sale of the item. Finally, we need to consider the options for disposal, including recycling and reusing the item versus placing it in a landfill or incinerator.

materials used and released throughout the lifetime of a product—from product design and procurement of raw materials through their manufacture, use, and disposal. For instance, in **FIGURE 51.8** you can see the many steps in conducting a life-cycle analysis. In this figure, you can see how we need to think about every step in an item's life cycle, from how it is designed and manufactured, to how it is transported and sold, to how it gets disposed. As we saw when we compared the various types of cups, the full inventory of a life-cycle analysis sometimes yields surprising results.

In theory, conducting a life-cycle analysis should help a community determine whether incineration is more or less desirable than using a landfill. However, such an analysis has limitations. For example, it can be difficult to determine the overall environmental impact of a specific material. It is not possible to know whether the particulates and nitrogen oxides released from incinerating food waste are better or worse for the environment than the amount of methane that might be released if the same food waste were placed in a landfill. So, in the case of waste that contains food matter, it is not possible to directly compare the full environmental impact of disposal in a landfill versus incineration. Similarly, it is a challenge to compare the sulfur dioxide released when making a paper cup with the volatile organic compounds released when making a plastic cup. While a manufacturer can conduct a life-cycle analysis from cradle-to-grave, a municipality typically begins a life-cycle analysis from the point of receiving the solid waste to the point of its final disposal. Although life-cycle analysis may not be able to determine absolute environmental impact, it can be very helpful in assessing economics and energy use, which are usually relatively easy to quantify.

In terms of economics, a municipality might use a life-cycle analysis to compare the costs of different disposal methods. For example, a glass manufacturing plant might pay the municipality $5 per ton for green glass that it will recycle into new glass. Economically, a municipality might do better if it received $5 for a ton of glass from a bottle manufacturing plant than if it paid a $50 per ton tipping fee to deposit the material into a landfill. But the municipality must also consider the lower cost of transporting the glass to a relatively close landfill rather than to a distant glass plant.

Whether direct or indirect, there is always a cost to waste disposal. In some parts of the country, the cost of waste disposal is covered by local taxes; in other locations, municipalities, businesses, or households may have to pay directly for disposal of their solid waste. Normally, disposal of recyclables costs less than disposal of material into a landfill because the landfill always involves a tipping fee while recyclables have a lower tipping fee or generate revenue. However, as we have seen, costs change depending on many factors, including the fluctuating prices paid for recyclables. For example, in a particular year, there may not be a market for recycled newspapers in the United States, but in the following year a large increase in overseas demand for newsprint may cause the price to go up in the United States. It is therefore essential for municipalities to be able to modify their choices as market prices for recyclables fluctuate.

From the perspective of energy use, a life-cycle analysis should also consider the energy content of gasoline or diesel fuel used and the pollution generated in trucking material to each destination, as well as the monetary, energy, and pollution savings achieved if the new glass is made from old glass rather than from raw materials (sand, potash, lime). Reconciling all these competing factors is very challenging and the decisions based on such analyses are often debatable. As a result, some municipalities are taking a more holistic approach, known as integrated waste management.

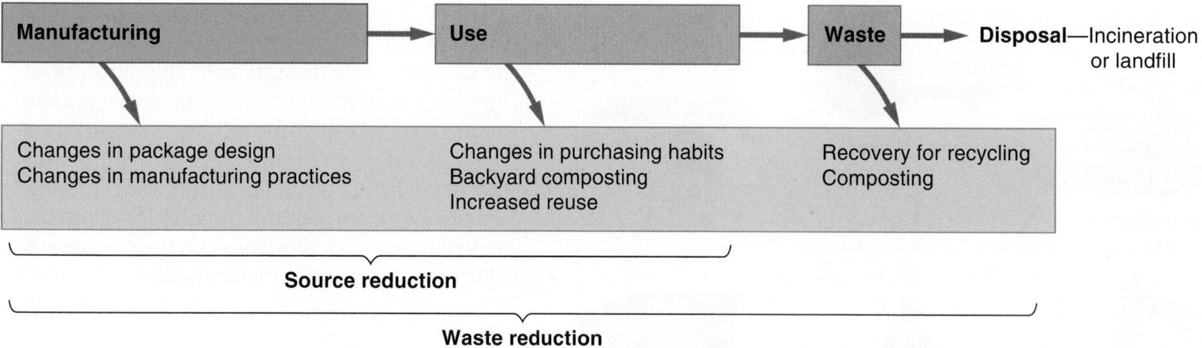

FIGURE 51.9 A holistic approach to waste management. Depending on the kind of waste and the geographic location, reducing waste can take much less time and money than disposing of it. Horizontal arrows indicate the waste stream from manufacture to disposal and curved arrows indicate ways in which waste can either be reduced or removed from the stream, thereby reducing the amount of waste incinerated or placed in landfills.

Integrated Waste Management

Integrated waste management employs several waste reduction, management, and disposal strategies to reduce their costs and reduce the environmental impact of MSW. Such options include a major emphasis on source reduction and include any combination of recycling, composting, use of landfills, incineration, and whatever additional methods are appropriate to the particular situation. **FIGURE 51.9** shows how a community could consider a series of steps, starting with source reduction during manufacturing and procurement of items. After that, behavior related to use and disposal can be considered and possibly altered to obtain the desired outcome of reduced waste. According to this approach, no community should be forced into any one method of waste disposal. If a region makes a large investment in an incinerator, for example, there is a risk that it would then need to attract large quantities of waste to pay for that incinerator, thereby reducing the incentive to recycle or use a landfill. Landfill space may be abundant or scarce, depending on the location of the community. If the municipality is free to consider all options, it can make the choice or choices that are efficient, cost effective, and least harmful to the environment. In short, every community can select the best strategy for their particular situation

The architect, innovator, and former University of Virginia dean William McDonough looks at holistic waste management from a more far-reaching perspective. In the book *Cradle to Cradle*, McDonough and coauthor Michael Braungart propose a novel approach to the manufacturing process. They argue in favor of minimizing waste generation before, during, and after manufacturing. Beyond that, manufacturers of goods such as automobiles, computers, appliances, and furniture strive to design products so that when the products are no longer useful, components can be recycled with as little of the material as possible becoming part of the waste stream.

Some industries have designed future recycling into their products. Most automobile manufacturers, for example, design and build their cars so that they can be easily taken apart and materials of different composition easily separated to allow recycling. The U.S. Environmental Protection Agency estimates that 75 percent by weight of automobiles were recycled and that most of the recycled materials were metals. Certain carpet manufacturers design their carpets so that when worn out they can be easily recycled into new carpeting (**FIGURE 51.10**). This is typically done by making a base that is extremely durable with a top portion

FIGURE 51.10 A recyclable carpet. FLOR carpet tiles are designed to be easily replaced and easily recycled when the carpet wears out. *(Glenn Asakawa/Getty Images)*

Integrated waste management An approach to waste disposal that employs several waste reduction, management, and disposal strategies to reduce their costs and reduce the environmental impact of MSW.

of the carpet that can be changed when the color fades, is worn out, or is no longer desired. Finally, McDonough and Braungart point out that many organisms in the natural world, such as the turtle, produce very hard, impact-resistant materials, such as a shell, without producing any toxic waste. They suggest that humans should examine how a turtle creates such a hard shell without the production of toxic wastes. Humans can use this example as a goal for other kinds of production where no toxic wastes would be produced. More recently, McDonough and Braungart have introduced the term "upcycle" to describe the conversion of a waste material to something of higher quality and greater value than the original product.

In this module we have seen that reduce, reuse, recycle, and compost can help decrease municipal solid waste headed to landfills and incinerators. Although communities continue to look for ways to improve handling of municipal solid waste, the nation has taken source reduction seriously. An effective way to help guide this effort is through the use of life-cycle analysis and integrated waste management, which provides the most effective waste-reduction strategy for each community's unique situation. In the next module, we will focus on sewage waste and the ways that it is treated.

Module 51 AP® Review

Preparing for the AP® Exam

Learning Goals Revisited

51-1 What are the three Rs that divert materials from the waste stream?

The three Rs for reducing municipal solid waste are reduce, reuse, and recycle. This is a practical approach to solid waste management, with the techniques presented from the most environmentally beneficial to the least beneficial.

51-2 How does composting reduce materials entering the waste stream?

Composting is a fourth approach to reducing solid waste that is focused on organic materials such as food and yard waste. Composting breaks down organic materials aerobically into humus, which is highly beneficial to the plants and soils of lawns, gardens, and agricultural fields.

51-3 How do life-cycle analysis and integrated waste management reduce municipal solid waste?

To evaluate the best strategies for reducing municipal solid waste, there is increasing use of life-cycle analyses to assess the cradle-to-cradle impact of manufactured products. A more holistic approach is integrated waste management, which allows communities to weigh all of their waste-reduction options and find the best strategy for their particular situation that has the lowest cost and the lowest environmental impact.

AP® Practice Questions

Preparing for the AP® Exam

Multiple-Choice Questions

1. Which of the following provides the best description of an example of source reduction?
   (a) printing pages double-sided instead of single-sided
   (b) purchasing disposable alkaline batteries in bulk
   (c) replacing plastic water bottles with disposable paper cups
   (d) substituting a nontoxic material for something toxic

2. Which material usually uses closed-loop recycling?
   (a) aluminum
   (b) glass
   (c) paper
   (d) cardboard

3. Organic matter in landfills is a problem primarily because
   (a) it contains bacteria.
   (b) as it breaks down it can dissolve the containment barrier.
   (c) it creates excessive heat.
   (d) it produces methane gas.

Use the passage below to answer questions 4 & 5:

In 2021, researchers examined a wide variety of single-use and multiple-use household items and asked whether the multiple-use items, which are washed between uses, are a better option. They considered the amount of energy, water, and global warming impact caused by manufacturing the items, the water and energy used in washing the items, and the environmental impact of disposal. The researchers examined four categories of kitchen items: coffee cups, sandwich wrappers, drinking straws, and forks. They then calculated how many times a reusable item would have to be reused before the environmental impact was the same as a single-use plastic item. For example, they compared single-use coffee cups made of Styrofoam or paper to reusable coffee cups made of ceramic, plastic, or metal.

The results of this study were quite surprising. For all of the items, the break-even time was much longer if a person washed the item by hand rather than in a dishwasher, because dishwashers clean dishes more efficiently. Of the twelve reusable items examined, three of them never broke even in terms of their environmental impact compared to comparable single-use items. Among the different types of coffee cups, ceramic cups had the lowest environmental impact. For example, beeswax paper and silicone bags, both used to wrap sandwiches, have to be washed by hand with warm water. The environmental impact of this washing exceeds the impact of producing a single-use sandwich bag or plastic wrap each time a sandwich is wrapped. In contrast, reusable forks made of metal, plastic, or bamboo had a break-even point after just 12 uses.

4. According to the article, which type of coffee cup has the least impact on the environment?
   (a) metal
   (b) paper
   (c) Styrofoam
   (d) ceramic

5. According to the article, why is it important to gather more information about the products we use?
   (a) Our environmental impacts can be greater in making new products than reusing those that already exist.
   (b) Reusable products are always better.
   (c) The need to wash reusable products makes them all poorer options for the environment.
   (d) Overall, single-use items and reusable items have the same ecological footprint.

Free-Response Question

The data table below shows the amount of municipal solid waste (MSW) generated in the United States that was recycled and composted from 1960 to 2018 in four categories.

Recycling and composting as a percentage of MSW generation

	1960	1970	1980	1990	2000	2005	2010	2015	2017	2018
Paper and paperboard	17%	15%	21%	28%	43%	50%	63%	67%	66%	68%
Glass	2%	1%	5%	20%	23%	21%	27%	28%	25%	25%
Plastics	<0.05%	<0.05%	<1%	2%	6%	6%	8%	9%	9%	9%
Yard trimmings	<0.05%	<0.05%	<0.05%	12%	52%	62%	58%	61%	69%	63%

(Data from https://www.epa.gov/facts-and-figures-about-materials-waste-and-recycling/national-overview-facts-and-figures-materials)

(a) **Identify** the component of MSW that was recycled the most in 2015. (1 pt.)

(b) **Describe** how one component of MSW could undergo closed-loop recycling. (1 pt.)

(c) **Explain** one possible reason why the least-recycled component of MSW in 2018 was not recycled. (1 pt.)

(d) **Propose a solution** that could reduce the amount of the MSW component from part (c) in landfills. (1 pt.)

(e) In 2018, the United States population was 327 million people, and produced approximately 2.23 kilograms of MSW per person per day.

(i) **Calculate** how many tons of glass were recycled in 2018, if 1 U.S. ton is equal to 907 kilograms. Show all work. (2 pts.)

(ii) **Calculate** how much paper and paperboard was recycled in the United States in 2018 in kilograms per person per year. Show all work. (2 pts.)

(iii) **Calculate** how much more glass, in kg, was recycled than plastic in the United States in 2018. Show all work. (2 pts.)

Module 52

Unit 8	47	48	49	50	51	52	53	54

Sewage Treatment

As we mentioned in Module 47, a major type of water pollution is wastewater, which is the water produced by livestock operations and human activities, including sewage from toilets and gray water from bathing and washing clothes and dishes. We begin the module by looking at the effects that wastewater pollution can have on water bodies. We will then discuss the technologies that have been developed to treat wastewater and minimize its effects.

Learning Goals

After reading this module you should be able to

52-1 describe the three major problems caused by wastewater pollution.

52-2 explain how we treat wastewater to prevent pollution.

52-1 What three major problems are caused by wastewater pollution?

Wastewater pollution is caused by oxygen demand, nutrient release, and disease-causing organisms

Environmental scientists are concerned about wastewater as a pollutant for three major reasons. First, wastewater dumped into bodies of water naturally undergoes decomposition by bacteria, which creates a large demand for oxygen in the water. Second, the nutrients that are released from wastewater decomposition can make the water more fertile. Third, wastewater can carry a wide variety of disease-causing organisms.

Oxygen Demand

When organic matter enters a body of water, microbes feed on it. Because these microbes require oxygen to decompose the waste, the more waste that enters the water, the more the microbes grow and the more oxygen they demand. The amount of oxygen a quantity of water uses over a period of time at a specific temperature is its **biochemical oxygen demand (BOD)**. For example, a researcher typically collects a sample of water and measures the amount of dissolved oxygen in units of mg/L (**FIGURE 52.1**). This water is then incubated for 5 days in a laboratory at 20°C and then a second measurement of dissolved oxygen is taken. By subtracting the second measurement from the first, we can determine how much oxygen was consumed by the biological and chemical

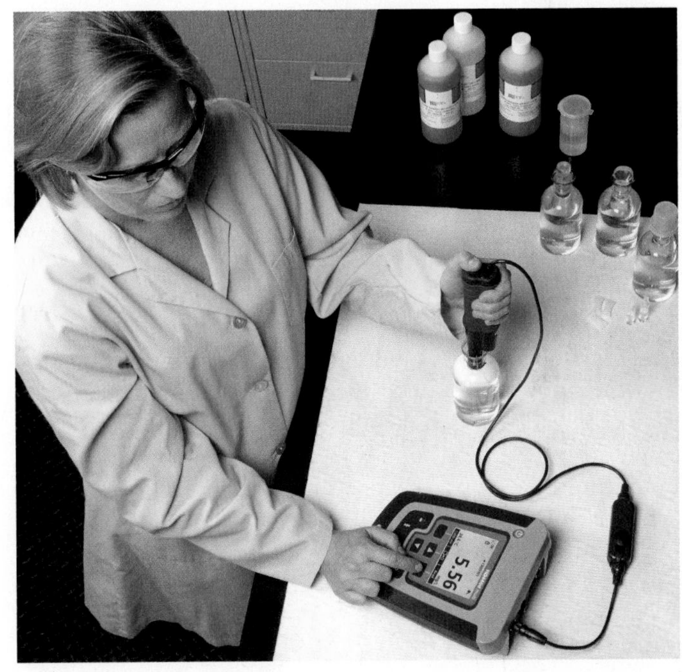

FIGURE 52.1 Measuring biochemical oxygen demand. To measure the biochemical oxygen demand of water collected from a water body, researchers measure the initial concentration of oxygen dissolved in the water and five days later they measure it again. The decline in oxygen over the five days provides an estimate of the biochemical oxygen demand due to biological and chemical reactions occurring in the water. *(Courtesy Hach® Company)*

Biochemical oxygen demand (BOD) The amount of oxygen a quantity of water uses over a period of time at a specific temperature.

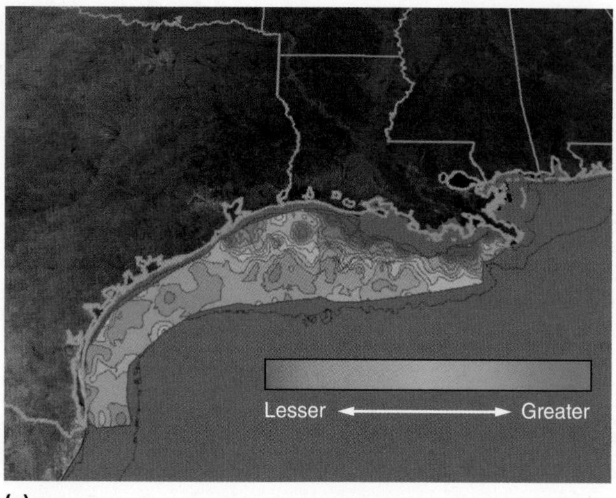

(a)

(b)

FIGURE 52.2 Dead zone. When raw sewage is dumped directly into bodies of water, subsequent decomposition by microbes can consume nearly all of the oxygen in the water and cause a dead zone to develop. Shown here are (a) oxygen concentrations in Gulf Coast waters and (b) a massive fish die-off that occurred as a result of low oxygen conditions in Lake Trafford, Florida.
(a: NASA/Goddard Space Flight Center Scientific Visualization Studio; b: Steven David Miller/Nature Picture Library)

processes occurring in the water sample. Lower BOD values indicate that a water body is less polluted by wastewater while higher BOD values indicate that a water body is more polluted by wastewater. If we were to test the BOD of natural waters over a 5-day period in a liter of water, we might find a BOD of 5 to 20 mg/L of oxygen coming from the decomposition of leaves, twigs, and perhaps a few dead organisms. In contrast, wastewater might have a BOD of 200 mg/L of oxygen.

When microbial decomposition uses a large amount of oxygen in a body of water, the amount of oxygen remaining for other organisms can be very low and we refer to them as dead zones, as we discussed in Module 48. Low oxygen concentrations are lethal to many organisms, including fish. Low oxygen can also be lethal to organisms that cannot move, such as many plants and shellfish. Dead zones can be self-perpetuating; organisms that die from lack of oxygen decompose and cause microbes to use even more oxygen.

Nutrient Release

When wastewater decomposes it releases nitrogen and phosphorus. As you may recall from Unit 1, nitrogen and phosphorus are generally the two most important elements for limiting the abundance of producers in aquatic ecosystems. The decomposition of wastewater releases nitrogen and phosphorus, which in turn provide an abundance of fertility to a water body, a phenomenon known as eutrophication that we first mentioned in Module 48. For example, the Chesapeake Bay experiences eutrophication from wastewater decomposition as well as from nutrients leached from fertilized agricultural lands during precipitation. When a body of water experiences an increase in fertility due to anthropogenic inputs of nutrients, it is called **cultural eutrophication**.

Cultural eutrophication An increase in fertility in a body of water, the result of anthropogenic inputs of nutrients.

Eutrophication initially causes a rapid growth of algae, which we call an algal bloom. As we discussed in Module 5, this enormous amount of algae eventually dies and microbes then rapidly begin digesting the dead algae, which causes additional biochemical oxygen demand. The increase in microbes consumes most of the oxygen in the water (see Figure 5.2). In short, the release of nutrients from wastewater initiates a chain of events that eventually leads to a lack of oxygen and the creation of dead zones once again. One of the largest dead zones in the world occurs where the Mississippi River flows into the Gulf of Mexico, shown in **FIGURE 52.2a**. The Mississippi River receives water from 41 percent of the land of the continental United States. Each summer there is an influx of wastewater and fertilizer that causes large algal blooms followed by substantial decreases in oxygenated water and massive die-offs of fish (Figure 52.2b). While some dead zones arise naturally, most are caused by human activities. In 1910, 4 dead zones were known around the world. This number increased to 87 dead zones in the 1980s and more than 500 dead zones by 2018, which is the most recent estimate.

FIGURE 52.3 Washing clothes in the Tuo River of China. Using water for bathing, washing, and disposing of sewage without contaminating sources of drinking water is a long-standing challenge. *(Bamboosil/AGE Fotostock)*

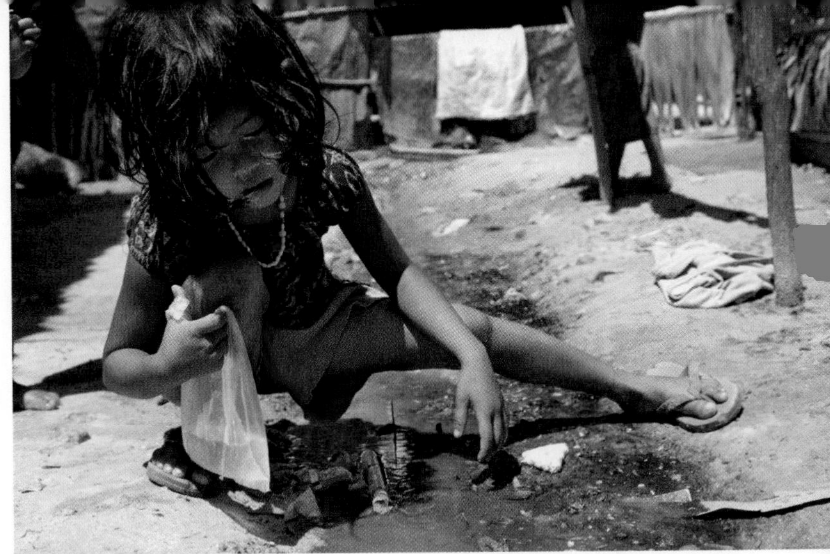

FIGURE 52.4 Cholera is prevalent in raw sewage. Children who play in water contaminated by raw sewage, such as this girl in Cambodia, face a high risk of contracting the cholera pathogen. *(imageBROKER/Alamy)*

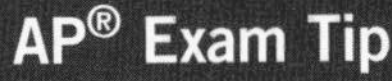

AP® Exam Tip

Make sure you can explain that fertilizers are transported from their application site into the environment through runoff or groundwater infiltration.

Disease-Causing Organisms

For centuries, humans have faced the challenge of keeping wastewater from contaminating drinking water. This can be difficult because throughout the world many people routinely use the same water source for drinking, bathing, washing, and disposing of sewage (**FIGURE 52.3**). Wastewater can carry a variety of pathogens including viruses, bacteria, and protists. Pathogens in wastewater are responsible for a number of diseases that can be contracted by humans or other organisms coming in contact with or ingesting the water.

Worldwide, the major waterborne diseases from pathogens are cholera, typhoid fever, hepatitis, and gastrointestinal illness (commonly known as dysentery). Cholera, which claims thousands of lives annually in developing countries (**FIGURE 52.4**), is not common in the United States. However, hepatitis A is appearing more frequently in the United States, usually originating in restaurants that lack adequate sanitation practices. According to the Centers for Disease Control and Prevention, there were nearly 20,000 cases of hepatitis A in the United States annually in the mid-1990s. From 2010 to 2016, that number declined to approximately 1,000 cases. In 2019, however, that number jumped to more than 18,000 cases as outbreaks of the disease occurred in 31 states. The bacterium *Cryptosporidium* has caused a number of outbreaks of gastrointestinal illness in this country. Large-scale disease outbreaks from municipal water systems are relatively rare in the United States, but they do occasionally occur. They are relatively common in many regions of the developing world.

Pathogens often infect people due to unsafe drinking water. The World Health Organization estimates that 2.1 billion people—nearly one-fourth of the world's population—do not have access to sufficient supplies of safe drinking water. Approximately 2.3 billion people lack access to proper sanitation and over half of these people live in China and India. In sub-Saharan Africa, only 40 percent of the population has access to proper sanitation. It is estimated that 361,000 children around the world die from diarrhea each year because they do not have safe drinking water, proper sanitation, and proper hygiene.

Monitoring for Wastewater Contamination

You may recall that in Module 14, we discussed the use of indicator species to determine when an ecosystem is affected by human activities. For wastewater pollution, scientists use the presence of a species of **fecal coliform bacteria**, known as *E. coli*, which lives in the intestines of humans, other mammals, and birds. While most varieties of *E. coli* are not harmful to humans—except for those who are very old, very young, or have compromised immune systems—they are common in wastewater. Therefore, the presence of *E. coli* is a reliable indicator that a body of water is being contaminated by inputs of wastewater from human sewage or other animal waste.

Fecal coliform bacteria A group of microorganisms that live in the intestines of humans, other mammals, and birds that serve as an indicator species for potentially harmful microorganisms associated with contaminated sewage.

FIGURE 52.5 Monitoring for *E.coli* to ensure safe water. Health departments routinely monitor beaches for the presence of *E. coli* bacteria in the water. The presence of *E. coli* can indicate inputs of human sewage or animal waste. *(AP Photo/Charlie Neibergall)*

Public water supplies, such as drinking water sources and water used for swimming pools, are routinely tested for *E. coli*. Homeowners with a single-family well might test their water less frequently or not at all. Public health authorities recommend declaring water unsuitable for human consumption if any *E. coli* are present. For safe swimming and fishing, the acceptable level of *E. coli* is higher; for example, swimming at a public beach or in a river is considered safe as long as the fecal coliform bacteria levels are less than 500 to 10,000 colonies per 100 megaliters of water. A pool, beach, or campground with contaminated water likely would be posted with a sign from the local health department (**FIGURE 52.5**). In some cases, beaches are closed not due to inputs of human sewage, but from inputs of animal waste, such as when hundreds of Canada geese spend time on the beach and a rainstorm washes goose feces accumulated over weeks into the water.

52-2 How do we treat wastewater to prevent pollution?

We have modern technologies used to treat wastewater

Because proper sanitation is so important, we need ways of treating wastewater to reduce concentration of nutrients, the biochemical oxygen demand, and the risk of waterborne pathogens. Humans have devised a number of ways to handle wastewater. The various solutions all have the same basic approach: Bacteria are used to break down the organic matter into carbon dioxide and inorganic compounds, which include nitrogen and phosphorus, and the harmful pathogens are outcompeted by nonharmful pathogens. In this section we will look at the two most widespread systems for treating human sewage: septic systems and sewage treatment plants. We will also describe the treatment of waste from large livestock operations.

Septic system A relatively small and simple sewage treatment system, made up of a septic tank and a leach field, often used for homes in rural areas.

Septic tank A large container that receives wastewater from a house as part of a septic system.

Sludge Solid waste material from wastewater.

Septage A layer of fairly clear water found in the middle of a septic tank.

Septic Systems

In rural areas, each house commonly treats its own sewage with a **septic system**, which is a relatively small and simple sewage treatment system that is made up of a septic tank and a leach field. As shown in **FIGURE 52.6**, the **septic tank** is a large container that receives wastewater from the house. Having a capacity of 1,900 to 4,700 L (500–1,250 gallons), the septic tank is buried underground adjacent to the house. Wastewater from the house flows into the tank at one end and leaves the tank at the other end. After the tank has been operating for some time, three layers develop. Anything that floats will rise to the top of the tank and form a scum layer. Anything heavier than water, including many pathogens, will sink to the bottom of the tank and form the **sludge** layer, which we define as the solid waste material from wastewater. In the middle is a fairly clear water layer called **septage**. The septage contains large quantities of bacteria and may also contain some pathogens and inorganic nutrients such as nitrogen and phosphorus.

Through gravity, septage moves out of the septic tank into several underground pipes laid out across a lawn below the surface. The combination of pipes and lawn makes up the

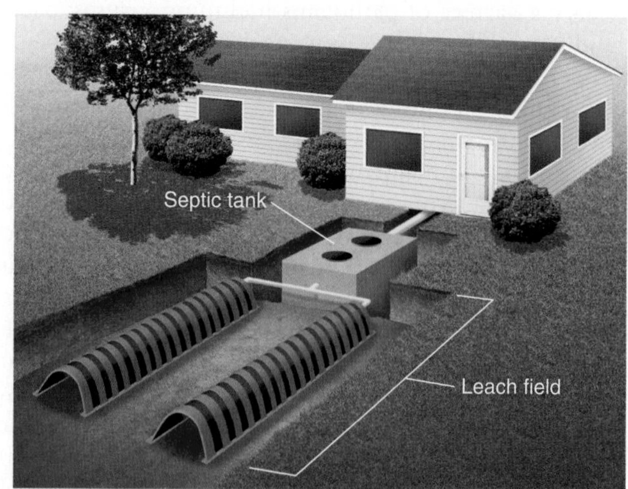

FIGURE 52.6 A septic system. Wastewater from a house is held in a large septic tank where solids settle to the bottom and bacteria break down the sewage. The liquid moves through a pipe at the top of the tank and passes through perforated pipes that distribute the water through a leach field.

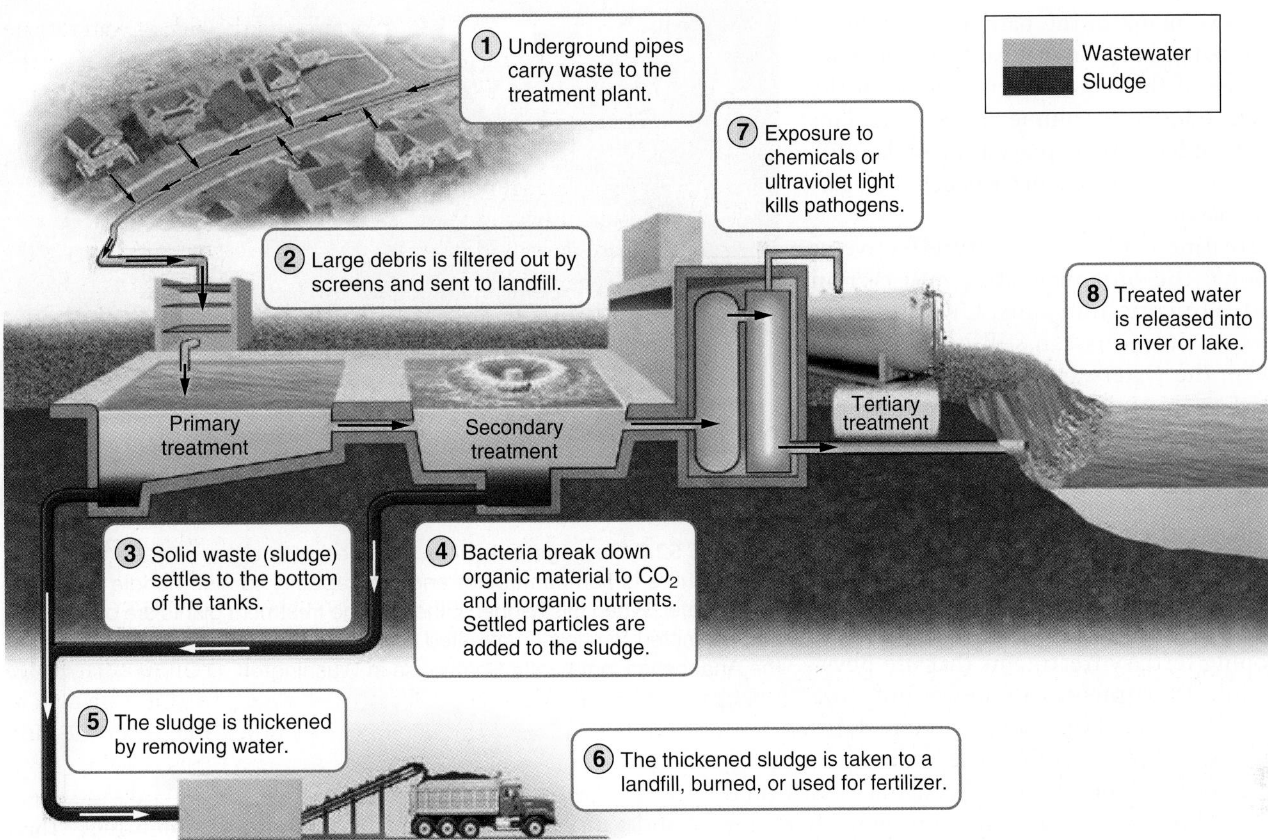

FIGURE 52.7 A sewage treatment plant. In large municipalities, great volumes of wastewater are handled by separating the sludge from the water and then using bacteria to break down both components.

leach field. The pipes contain small perforations so the water can slowly seep out and spread across the leach field. The septage that seeps out of the pipes is slowly absorbed and filtered by the surrounding soil. The harmful pathogens in the septage can be outcompeted by other microorganisms in the septic tank and therefore diminish in abundance, or they can be degraded by soil microorganisms after the septage seeps out into the leach field. The organic matter found in the septage is broken down into carbon dioxide and inorganic nutrients. Eventually, the soil-filtered water and nutrients are taken up by plants or enter a nearby stream or aquifer.

There are significant environmental advantages to septic systems. Because most septic systems rely on gravity—water from the house flows downhill to the septic tank, and water from the septic tank flows downhill to the leach field—no electricity is needed to run a septic system. However, sludge from the septic tank must be pumped out periodically (every 5 years) and taken to a sewage treatment plant.

Sewage Treatment Plants

Although household septic systems work well for rural areas in which each house has sufficient land for a leach field, they are not feasible for more developed areas with greater population densities and little open land. In developed countries, municipalities use centralized sewage treatment plants that receive the wastewater from hundreds or thousands of households via a network of underground pipes. In a traditional sewage treatment plant, wastewater is handled using a primary treatment followed by a secondary treatment.

As shown in **FIGURE 52.7**, the primary treatment involves the physical removal of any large objects by passing the sewage through grates, and then allowing the solid waste material to settle into a sludge layer. The remaining wastewater undergoes a secondary treatment that uses bacteria to break down 85 to 90 percent of the organic matter and convert it into carbon dioxide and inorganic nutrients such as nitrogen and phosphorus. The secondary treatment typically aerates the water to add oxygen; this promotes the growth of aerobic bacteria, which emit less offensive odors than anaerobic bacteria. The treated water sits for several days to allow particles to settle out. Settled particles are added to the sludge from the primary treatment.

The combined sludge from the primary and secondary treatments must be taken away from the sewage treatment plant. To reduce the volume of material prior to transport

Leach field A component of a septic system, made up of underground pipes laid out below the surface of the ground.

and help remove many of the pathogens, sludge is typically exposed to bacteria that can digest it. After digestion, most of the water is removed from the sludge, which reduces its volume and weight. This final form of sludge can be placed into a landfill, burned, or converted into fertilizer pellets for agricultural fields, lawns, and gardens.

Sewage treatment plants are very effective at breaking down the organic matter into carbon dioxide and inorganic nutrients. Unfortunately, these inorganic nutrients can still have undesirable effects on the waterways into which they are released. Although nitrogen and phosphorus are important nutrients for increasing primary productivity, when high concentrations of these nutrients are released into bodies of water from sewage treatment plants, they fertilize the water, which can lead to unnaturally large increases in the abundance of algae and aquatic plants. In response to this problem, large sewage treatment plants are now developing tertiary treatments that use physical and chemical treatments to remove inorganic compounds (e.g., phosphorus and heavy metals), bacteria, viruses, and parasites from the wastewater. Tertiary treatment commonly involves chemicals that bind to inorganic compounds to cause them to precipitate out of the water. It also involves using fine mesh filters to capture microorganisms followed by a disinfectant step such as chlorine, ozone, or ultraviolet light to kill any remaining microorganisms. Unfortunately, these treatments do not remove pharmaceuticals from the wastewater. The ultimate goal is to release wastewater that is similar in quality to the water body that is receiving it.

FIGURE 52.8 Sewage overflow. In cities with older sewage treatment plants, there is sometimes not enough capacity to also handle incoming storm water. When this happens, the sewage treatment plants are overloaded and permitted to release untreated sewage into water bodies. This site is at the Anacostia Community Boathouse in Washington, D.C. *(The Washington Post/Getty Images)*

Legal Sewage Dumping

Sewage treatment plants are critical to human health because they remove a great deal of harmful organic matter and associated pathogens that cause human illness. It might surprise you to know that even in developed countries, raw sewage can sometimes be directly pumped into rivers, lakes, and the ocean (**FIGURE 52.8**). Sewage treatment plants are typically built to handle wastewater from local households and industries. However, many older sewage treatment plants also receive water from stormwater drainage systems. During periods of heavy rain, the combined volume of storm water and wastewater overwhelms the capacity of the sewage treatment plants. When this happens, the treatment plants are allowed to bypass their normal treatment protocol and pump vast amounts of water directly into an adjacent body of water.

How big a problem is this? According to the U.S. Environmental Protection Agency, overflows of raw sewage occur 23,000 to 75,000 times per year in the United States. This problem is concentrated in the older cities of the Northeast and Midwest—where some sewage pipes date back to the 1870s—although it also occurs in cities such as Atlanta and San Francisco. In Indianapolis, for example, more than 3.8 billion liters (1 billion gallons) of raw sewage have been dumped into surrounding water bodies each year during periods of heavy rain. Around the country, 11 to 38 billion liters (3–10 billion gallons) are released and such incidents result in the contamination of drinking water, beaches, fish, and shellfish, and can lead to human illness. Between 1.8 million and 3.5 million illnesses are associated with swimming in sewage-contaminated water in the United States each year, and 500,000 illnesses are linked to drinking sewage-contaminated water. Illnesses caused by eating contaminated shellfish are estimated to cost \$2.5 million to \$22 million annually.

The solution to this problem is straightforward though expensive. Municipalities facing sewage overflow need to modernize their sewage treatment systems to prevent the influx of storm water from overwhelming their capacity to treat human wastewater. Unfortunately, the cost of such work can be considerable. In the case of Indianapolis, for example, a 4-km (2.8-mile) long tunnel is being constructed to hold the excess storm water and raw sewage during rainstorms until it can later be treated at nearby sewage treatment plants; this construction project is part of a \$2 billion upgrade. Similarly, three tunnels are being dug under Washington, D.C. at a cost of \$2.7 billion. Once completed, these tunnels will hold up to 720 billion liters (190 million gallons) of raw sewage and storm water, so that the city can stop dumping the sewage into the Anacostia River during large rainstorms. Given that global climate change is predicted to cause more intense rainstorms in the coming decades, solutions to this problem are being viewed as high priorities for many cities.

Animal Feed Lots and Manure Lagoons

Although the impact of human waste on the environment and human health is generally understood, people are less familiar with the problem of animal waste. On a small scale, animal manure can contaminate waters when farm animals are allowed access to streams for food and water because, inevitably, they will defecate in the stream. As you may recall from Module 28, large-scale manure issues arise with concentrated animal feeding operations that raise thousands of cattle, hogs, and poultry. The manure from these operations contains digested food and often also contains a variety of hormones and antibiotics that farmers use to improve the growth and health of their animals.

Farms that raise thousands of animals in a single location often use a manure lagoon to dispose of the tremendous amount of manure produced (**FIGURE 52.9**). A manure lagoon, which we first introduced in Unit 5, is a large, human-made pond lined with rubber to prevent the manure from leaking into the groundwater. After bacteria have broken down the manure—the same process that occurs in septic tanks and sewage treatment plants—the manure can be spread onto farm fields to serve as a fertilizer. One major risk of manure lagoons is the possibility of developing a leak in the liner that would allow the waste to seep into and contaminate the underlying groundwater. Another risk is a possible overflow into adjacent water bodies. Like human wastewater, overflow of animal waste into rivers can lead to disease outbreaks in humans and wildlife. Finally, the application of manure as fertilizer can create runoff that moves into nearby water bodies. "Do the Math: Building a Manure Lagoon" shows you how to calculate the appropriate size for a manure lagoon.

FIGURE 52.9 **Feedlot and manure lagoon.** Large-scale agricultural operations, such as this one in Georgia, produce great amounts of waste that can be held in lagoons until pumped out and removed. *(Jeff Vanuga/USDA, Natural Resources Conservation Service)*

In this module we learned that wastewater from livestock and humans can cause harmful effects on the water bodies that receive it. One effect is high oxygen demand associated with the decomposition of the organic matter which can cause dead zones. Other effects include the production of inorganic nutrients such as nitrogen and phosphorus that can cause algal blooms, and the introduction of disease-causing organisms. Fortunately, humans have invented several technologies to handle wastewater including septic tanks, sewage treatment plants, and manure lagoons. In the next module, we will explore how we assess the concentrations of pollutants that cause harm to ecosystems and people.

DO THE MATH Building a Manure Lagoon

Preparing for the AP® Exam

Concentrated animal feeding operations typically use manure lagoons to hold the manure produced by cattle. If an individual animal produces 50 L of manure each day and the average concentrated animal feeding operation holds 800 cattle on any given day, how much manure is produced each day?

$$\text{Daily manure production} = 50\text{ L/animal} \times 800\text{ animals} = 40{,}000\text{ L}$$

YOUR TURN

1. If a manure lagoon needs to hold 30 days' worth of manure production for 900 cattle, what is the minimum capacity of the lagoon a farmer would need?
2. After the manure has broken down, the manure must be spread onto farm fields. A modern manure spreader can hold 40,000 L of liquid manure. How many trips will it take for the manure spreader to remove the 30 days' worth of manure that is held in the manure lagoon?

Module 52 AP® Review

Preparing for the AP® Exam

Learning Goals Revisited

52-1 What three major problems are caused by wastewater pollution?

Wastewater dumped into water bodies undergoes decomposition and creates a large demand for oxygen in the water. The nutrients that are released from wastewater decomposition also can make the water more fertile. As algal blooms occur in response to the high fertility, they eventually die and their decomposition can cause dead zones in the water. Finally, wastewater can carry a wide variety of disease-causing organisms.

52-2 How do we treat wastewater to prevent pollution?

Humans have devised a number of ways to handle wastewater including septic systems for individual houses and sewage treatment plants that receive the sewage from large populations via underground pipes. In some large and older American cities, sewage treatment plants also receive storm water during large rain events. The combined volume overwhelms the plants, leading to a decision to release raw sewage into water bodies.

AP® Practice Questions

Multiple-Choice Questions

1. Fecal coliform bacteria
   (a) are used as an indicator of water quality.
   (b) are the cause of many waterborne diseases.
   (c) usually cause diarrhea.
   (d) are rarely found in septage.

2. A manure lagoon is being built for a dairy farm with 700 cows, each of which produces 40 L of manure each day. How large must the lagoon be to hold 30 days' worth of manure?
   (a) 28,000 L
   (b) 120,000 L
   (c) 560,000 L
   (d) 840,000 L

Use the following choices to answer questions 3–5. Choices may be used once, more than once, or not at all.

(a) primary treatment
(b) secondary treatment
(c) tertiary treatment
(d) disinfection

3. Which step of sewage treatment requires the use of bacteria?

4. Which step of sewage treatment uses filters, screens, and sieves?

5. Which step of sewage treatment can produce nitrates and phosphates?

6. Which location is most likely to have experienced an algae bloom based on oxygen concentrations shown on the map below?

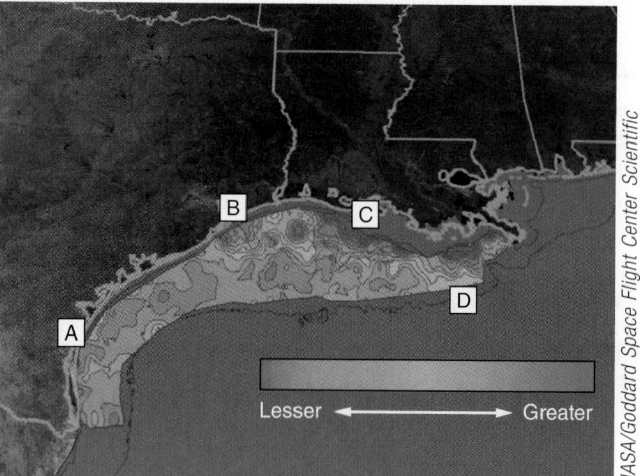

NASA/Goddard Space Flight Center Scientific Visualization Studio

   (a) A
   (b) B
   (c) C
   (d) D

Free-Response Question

The diagram below illustrates the process used in a wastewater treatment plant.

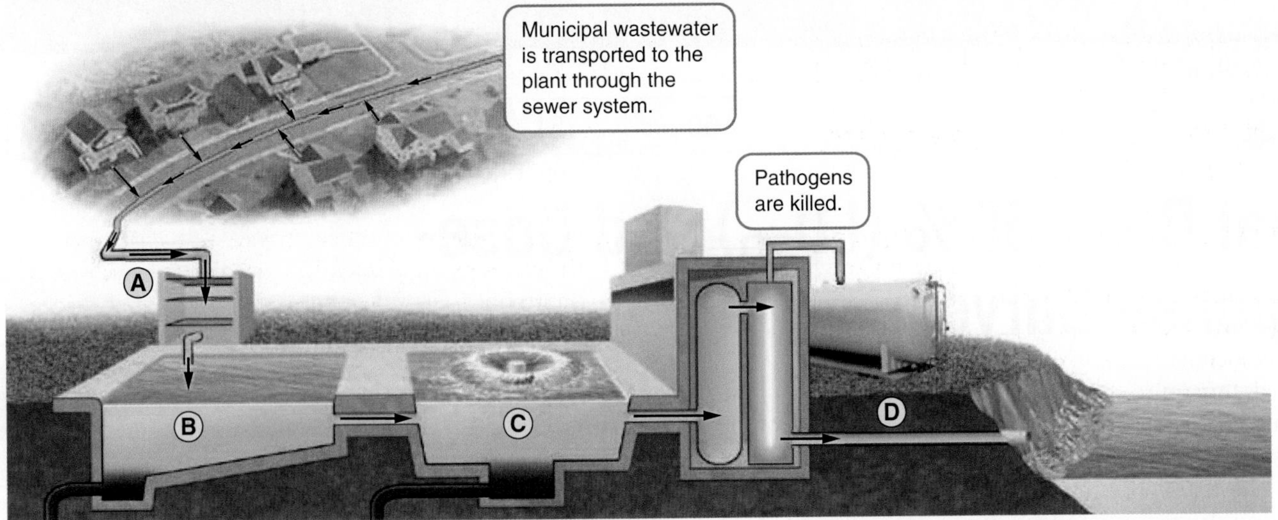

Municipal wastewater is transported to the plant through the sewer system.
Pathogens are killed.
A
B
C
D

Wastewater treatment plant

(a) **Identify** the part of the diagram where large debris and solid waste is removed. (1 pt.)

(b) **Identify** one example of large debris or solid waste removed in primary wastewater treatment. (1 pt.)

(c) **Describe** how solid waste is produced during wastewater treatment. (1 pt.)

(d) **Explain** the process of wastewater treatment labeled "C" in the diagram. (1 pt.)

(e) Many older sewage treatment plants also receive water from stormwater drainage systems. During periods of heavy rain, the combined volume of storm water and wastewater overwhelms the capacity of the sewage treatment plants. When this happens, the treatment plants are allowed to bypass their normal treatment protocol and pump vast amounts of water directly into an adjacent body of water.

(i) **Identify** one potentially dangerous component of human sewage. (1 pt.)

(ii) **Describe** how the component identified in part (i) could be dangerous if released directly into surface water. (1 pt.)

(iii) **Describe** one way the component identified in part (ii) is removed during the wastewater treatment process. (1 pt.)

(f) **Identify** one environmental problem that can be caused by the release of treated waste from a wastewater treatment plant. (1 pt.)

(g) **Describe** one way to reduce the problem identified in part (f). (1 pt.)

(h) **Describe** one additional environmental advantage of using the solution proposed in part (g). (1 pt.)

Module 53

Lethal Dose 50% (LD_{50}) and Dose-Response Curves

Throughout this unit we have examined the various types of pollutants and the impacts that they can have on people and ecosystems. A key detail, however, is that pollutants only impact species and ecosystems once they reach a particular concentration. Increasing concentrations of pollutants initially cause changes in behavior, physiology, and growth. It is only as the concentrations exceed a higher level that they can cause death. We need to know which concentrations of different pollutants impact humans and a wide variety of other species in nature. If we know this, we can set guidelines for each pollutant to ensure that we never exceed harmful concentrations of each pollutant. In this module we focus on the approaches that scientists take to determine how much of a given chemical it takes to cause harm to different species. We start by examining how scientists determine the concentrations of a chemical that can harm a species and, using this information, estimate the doses that would prove lethal to various species, including humans. With an understanding of pollutant impacts, we can then assess the risk of a pollutant and how to best manage that risk. Finally, we will consider two different philosophies for determining risk that are used in different regions of the world.

Learning Goals

After reading this module you should be able to

53-1 explain how dose-response curves are used to estimate lethal doses of chemicals.

53-2 identify how we estimate potential harm of chemicals in the environment.

53-3 describe the major philosophies of regulating chemicals in the environment.

53-1 How are dose-response curves used to estimate lethal doses of chemicals?

Scientists can determine the concentrations of chemicals that harm organisms

Given that we are often concerned with the harm that various pollutants can have on humans and other species, we first need to know the concentrations that cause harm. Scientists have three techniques to determine harmful concentrations: dose-response studies, prospective studies, and retrospective studies.

Dose-response study A study that exposes animals or plants to different amounts of a chemical and then looks for a variety of possible responses, including mortality or changes in behavior or reproduction.

Acute study An experiment that exposes organisms to an environmental hazard for a short duration.

Chronic study An experiment that exposes organisms to an environmental hazard for a long duration.

Dose-Response Studies

Dose-response studies expose animals or plants to different amounts of a chemical and then look for a variety of possible responses including mortality or changes in behavior or reproduction. For example, dose-response studies of aquatic animals such as tadpoles are used to determine the concentrations of various pesticides that cause 50 percent of the animals to die (**FIGURE 53.1**). The concentration of the chemicals being considered can be measured in air, water, or food. They can also be measured as the dose of a chemical, which is the amount that an organism absorbs or consumes. For reasons of efficiency, most dose-response studies only last for 1 to 4 days. Experiments that expose organisms to an environmental hazard for a short duration are called **acute studies**. Studies that are conducted for longer periods of time are called **chronic studies**.

Dose-response studies most commonly measure mortality as a response. At the end of a dose-response experiment, scientists count how many individuals die after exposure to each concentration. When the data are graphed, they generally follow an S-shaped curve, like the one in **FIGURE 53.2a**. If you examine the purple curve, you will see that at the lowest dose no individuals die. At slightly higher doses, a

(a)

(b)

FIGURE 53.1 Conducting dose-response experiments. (a) Researchers determine how chemicals affect the mortality of animals using dose-response experiments in the laboratory. (b) In the experiment shown, researchers are examining the effects of different insecticide concentrations on the survival of tadpoles. *(a: Jessica Hua, Binghamton University; b: Courtesy of Jason Hoverman)*

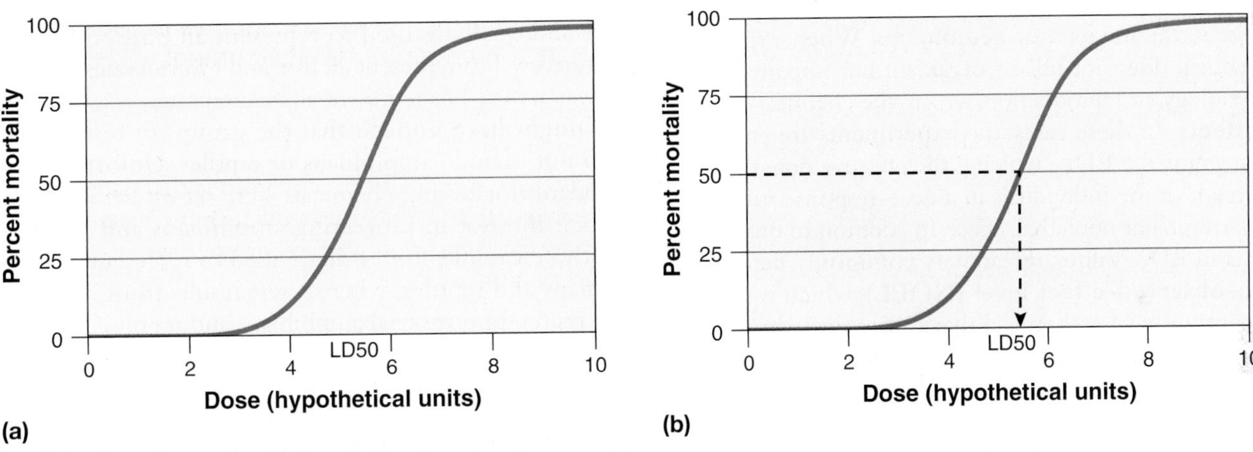

FIGURE 53.2 Dose-response curves. To determine which chemical concentrations cause harm, scientists expose a species to different concentrations of a chemical. (a) At low doses, no individuals experience mortality. As the concentration increases, more individuals die. At even higher concentrations, all individuals die. Such an experiment typically produces an S-shaped curve, which is known as a dose-response curve. (b) To determine the dose of a chemical that causes 50 percent mortality, scientists draw a horizontal line at 50 percent mortality on the *y* axis until it intersects the dose-response curve. At this intersection, we draw a vertical line to the *x* axis, which tells us the concentration of the chemical that causes 50 percent mortality.

few individuals die. The dose at which an effect can be detected is called the threshold. These individuals generally are in poorer health or genetically are not very tolerant of the chemical. As the dose is further increased, many more individuals begin to die. At the highest concentrations all individuals die.

To compare the harmful effects of different chemicals, scientists measure the **LD_{50}**, which is an abbreviation for the lethal dose that kills 50 percent of the individuals in a dose-response study. To estimate the LD_{50} value, we can draw a horizontal line from the 50 percent point on the *y* axis until it intersects the purple dose-response curve, as illustrated in Figure 53.2b. At this intersection, we then draw a vertical line down to the *x* axis. This is the

LD_{50} The lethal dose of a chemical that kills 50 percent of the individuals in a dose-response study.

concentration that causes 50 percent mortality in a population of organisms.

The LD_{50} value helps assess the relative toxicity of a chemical to a particular species. For example, scientists can compare the LD_{50} value of a new chemical with the LD_{50} values from thousands of previously tested chemicals to determine whether the new chemical is more or less lethal to a given organism than other chemicals. Scientists can also compare the LD_{50} value for a single chemical across different species of plants of animals to determine which types of organisms are the most sensitive. For example, as a result of conducting hundreds of experiments on different animal species over the decades, we now know that insecticides are lethal to insects and other invertebrates at very low concentrations but they are only lethal to fish, birds, and mammals at high concentrations.

Although most toxicology studies are only conducted for a few days, chronic studies will often last from the time an organism is very young to when it is old enough to reproduce. For some species such as fish, chronic experiments can take several months. The goal of chronic studies is to examine the long-term effects of chemicals, including how they affect survival and reproduction.

As we noted, not all dose-response experiments measure death as a response to chemicals. In many cases, scientists are interested in other harmful effects, including chemicals acting as teratogens, carcinogens, or neurotoxins. When exposure to a chemical does not kill an organism but impairs its behavior, physiology, or reproduction, we say the chemical has **sublethal effects**. In these cases, the experiments are conducted to determine the **ED_{50}**, which is the effective dose that causes 50 percent of the individuals in a dose-response study to display a harmful, but nonlethal, effect. In addition to quantifying LD_{50} and ED_{50} values, researchers commonly determine the **No-observed-effect level (NOEL)**, which is the highest concentration of a chemical that causes no lethal or sublethal effects. The NOEL indicates how much of a chemical an organism can experience with no harmful effect. For example, in the hypothetical example shown in Figure 53.2, the dose of 3 units is the highest concentration that causes no harmful effect. Above this concentration, individuals begin to experience mortality. Thus, the NOEL value is 3.

AP® Exam Tip

You should be able to read and interpret an LD_{50} graph for the AP® Environmental Science Exam.

Sublethal effect The effect of an environmental hazard that does not kill an organism but which may impair an organism's behavior, physiology, or reproduction.

ED_{50} The effective dose of a chemical that causes 50 percent of the individuals in a dose-response study to display a harmful, but nonlethal, effect.

No-observed-effect level (NOEL) The highest concentration of a chemical that causes no lethal or sublethal effects.

LD_{50} Studies

In the United States, chemicals that affect humans and other species are regulated by the EPA. The Toxic Substances Control Act of 1976 gives the EPA the authority to regulate many chemicals, but does not include food, cosmetics, and pesticides. Pesticides are regulated under a separate law—the Federal Insecticide, Fungicide, and Rodenticide Act of 1996. Under this act, a manufacturer must demonstrate that a pesticide "will not generally cause unreasonable adverse effects on the environment."

Because no chemical can be tested on every one of the approximately 10 million species of organisms on Earth, scientists have devised a system of testing a few species—a bird, mammal, fish, and invertebrate—that are thought to be among the most sensitive in the world. The particular species tested from each of the four animal groups can vary, depending on which species is thought to be the most sensitive to a particular chemical. The reasoning for this is that regulations devised to protect the most sensitive species in a group will automatically protect all other species in that group. Since conducting LD_{50} studies on humans would be unethical, results from studies conducted on mice and rats are extrapolated to humans. For nonhuman animals, test results from mice and rats are used to represent all mammals, birds such as pigeons and quail are used to represent all birds, fish such as trout are used to represent all fish, and invertebrates such as water fleas are used to represent all invertebrates.

You might have noticed that the groups of tested animals do not include amphibians or reptiles. Unfortunately, the standards for testing chemicals were set up before there was much interest in protecting amphibians and reptiles. Currently, test results from fish are used to represent aquatic amphibians and reptiles, whereas test results from birds are used to represent terrestrial amphibians and reptiles. Because amphibians and reptiles are now experiencing population declines throughout the world, there is increased interest in requiring tests on species from these two groups as well.

Using the LD_{50} and ED_{50} values from dose-response experiments, regulatory agencies such as the EPA can determine the concentrations in the environment that should cause no harm. For most animals, a safe concentration is obtained by taking the LD_{50} value and dividing it by 10. The logic is that if the LD_{50} value causes 50 percent of the animals to die, then 10 percent of the LD_{50} value should cause few or no individuals to die.

The regulatory agencies are much more conservative in setting concentrations for humans. Scientists determine the LD_{50} or ED_{50} values for rats or mice and then divide by 10 to determine a safe concentration for rats and mice. This value is divided by 10 again to reflect that rats and mice may be less sensitive to a chemical than humans. Finally, this value is often divided by 10 again to ensure an extra level of caution. In short, the LD_{50} and ED_{50} values obtained from rats and mice are divided by 1,000 to set the safe values for humans. "Do the Math: Estimating LD_{50} Values and Safe Exposures" shows you how to make this calculation.

DO THE MATH

Estimating LD_{50} Values and Safe Exposures

Preparing for the AP® Exam

Using our knowledge of how scientists conduct LD_{50} studies, we can consider an example. Let's imagine that you are a scientist charged with determining the safe levels for mammals of a pesticide in the environment. Using lab rats, you feed them a diet that contains different amounts of the pesticide, ranging from 0 to 4 mg of pesticide per kg of the rat's mass. After feeding them these diets for 4 days, you count how many rats are still alive. When you plot the data, you obtain the graph shown.

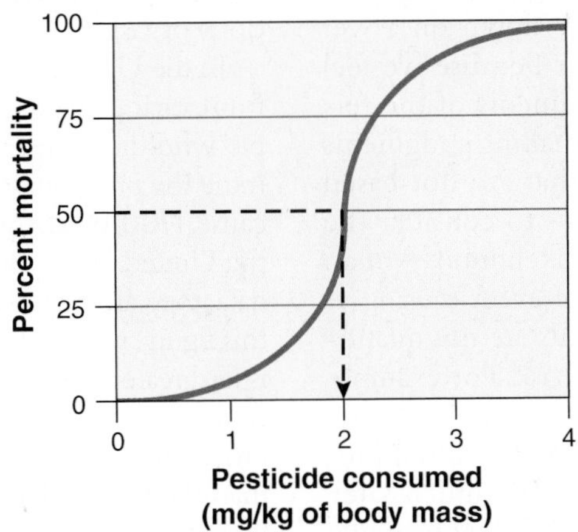

What is the LD_{50} value for lab rats? To determine this, we can draw a horizontal line at the point of 50 percent mortality on the *y* axis. Where this line intersects the purple line, we can draw another line straight down to the *x* axis. This second line crosses the *x* axis at 2 mg/kg of mass.

Based on this LD_{50} study, what amount of pesticide would be considered safe for mammals to ingest? Recall that we can calculate this number by taking the LD_{50} value and dividing it by 10. Thus, the safe amount of pesticide for a rat is:

$$\frac{2 \text{ mg/kg of mass}}{10} = 0.2 \text{ mg/kg of mass}$$

YOUR TURN Using the same LD_{50} study, what amount of pesticide would be considered safe for a human to ingest?

53-2 How do we estimate potential harm of chemicals in the environment?

We can estimate potential harm using risk assessment, risk acceptance, and risk management

Most people face some kind of environmental hazard every day. The hazards we face may be voluntary, as when we decide to smoke tobacco, or they may be involuntary, as when we are exposed to air pollution. When assessing the risk of different environmental hazards, regulatory agencies, environmental scientists, and policy makers usually follow three steps for risk analysis: risk assessment, risk acceptance, and risk management. In this module we will examine each of the three steps.

Risk Assessment

As illustrated in **FIGURE 53.3**, risk assessment is the first of the three steps involved in risk analysis. Risk analysis seeks to identify a potential hazard and determine the magnitude of the potential harm. There are two types of risk assessment—qualitative and quantitative. Each of us has some idea of the risk associated with different environmental hazards. For our purposes, an **environmental hazard** is anything in our environment that can potentially cause harm. Environmental hazards include substances such as pollutants or other chemical contaminants, human activities such as driving cars or flying in airplanes, or natural catastrophes such as volcanoes and earthquakes.

Environmental hazard Anything in the environment that can potentially cause harm.

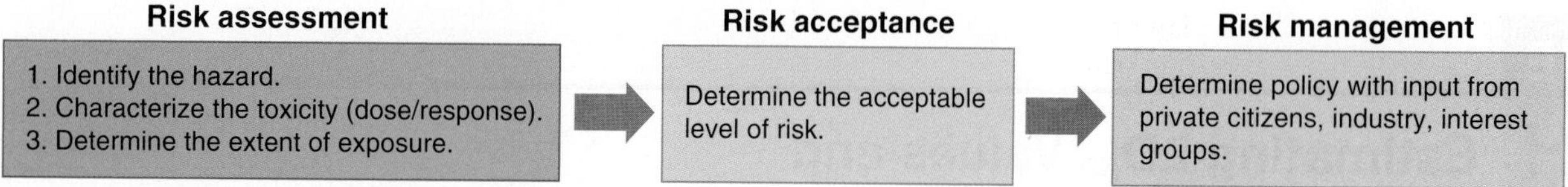

FIGURE 53.3 The process of risk analysis. Risk analysis involves risk assessment, risk acceptance, and risk management.

We generally make qualitative judgments in which we might categorize our decisions as having low, medium, or high risks. When we choose to slow down on a wet highway or to buy a more expensive car because we feel it is safer, we are making qualitative judgments of the relative risks of various decisions. We are making judgments that are based on our perceptions but that are not based on actual data. It would be unusual for us to consider the actual probability—that is, the statistical likelihood—of an event occurring and the probability of that event causing us harm. Because our personal risk assessments are not quantitative, they often do not match the actual risk. For example, some people find air travel very stressful because they are afraid the plane might crash. These same people often prefer riding in a car, which they perceive to be much safer. As another example, a person may be very cautious while walking in an area with heavy traffic but never consider the health dangers of smoking or a lack of exercise. To manage our risk effectively, we need to ask how closely our perceptions of risk match the reality of actual risk.

In the United States, we have data on the probability of death from various hazards. By looking at the total number of people who die in a year and their causes of death, we can determine the probability that an individual will die from a particular cause. **FIGURE 53.4** provides current data on causes of death in the United States. Because these risk estimates are based on real data, they are quantitative rather than qualitative. If we examine this figure, we see that the probability of dying in an automobile is far greater than the probability of dying in an airplane, which is contrary to the perception of many people. Similarly, the probability of dying from heart disease is monumentally greater than the risk of dying in a pedestrian accident. These numbers underscore the fact that our perceptions of risk can often be very different from the actual risk. Because a catastrophic event,

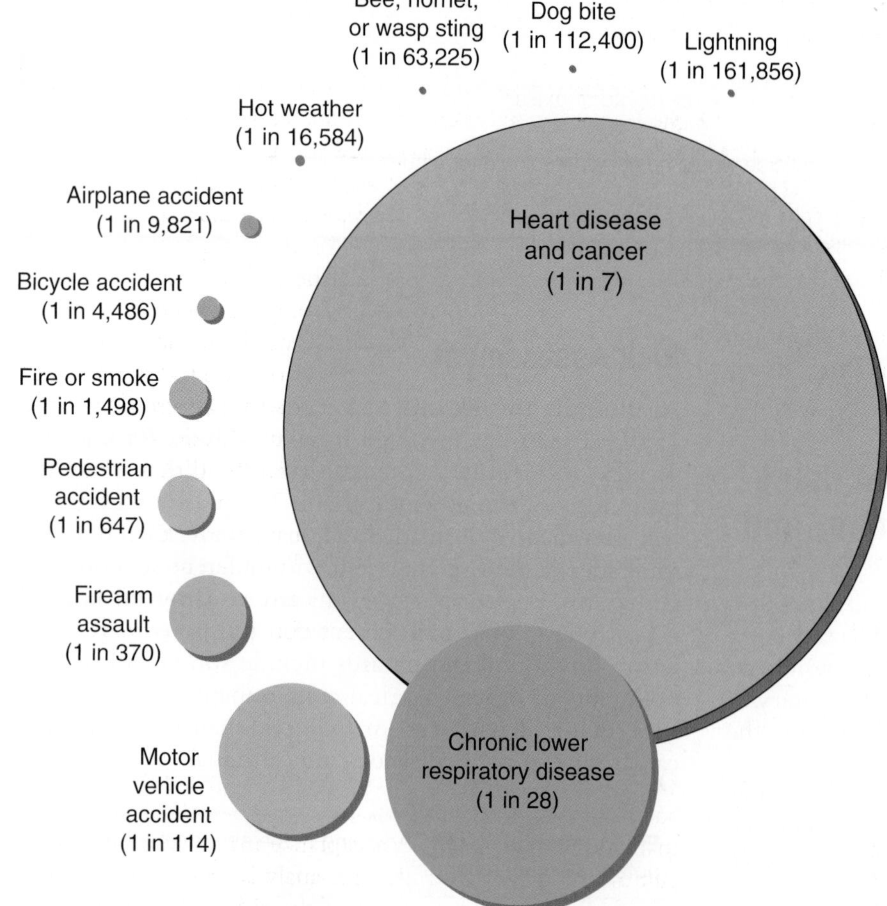

FIGURE 53.4 The probabilities of death in the United States. Some causes of death that people perceive as having a high probability of occurring, such as dying in an airplane crash, actually have a low probability of occurring. In contrast, some causes of death that people rate as having a low probability of occurring, such as dying from heart disease, actually have a very high probability of occurring. *(Data from http://www.nsc.org/learn/safety-knowledge/Pages/injury-facts-chart.aspx.)*

such as a nuclear plant meltdown or a plane crash, can do a great deal of harm and receives great media attention, people believe that it is very risky to use nuclear reactors or to fly in airplanes. However, these events rarely occur so the risk of harm is low. In contrast, we tend to downplay the risk of activities that provide us with cultural, political, or economic advantages such as drinking alcohol or working in a coal mine.

Quantitative Risk Assessment

The most common approach to conducting a quantitative risk assessment can be expressed with a simple equation:

$$\text{Risk} = \text{probability of being exposed to a hazard} \times \text{probability of being harmed if exposed}$$

Using this equation, we could ask whether it is riskier in a year to fly on commercial airlines for 1,609 km (1,000 miles) per year or to eat 40 tablespoons of peanut butter, which contains tiny amounts of a carcinogenic chemical produced naturally by a fungus that sometimes occurs in peanuts. The risk of dying in a plane crash depends on the probability of experiencing a plane crash, which is very low, multiplied by the probability of dying if the plane does crash, which approaches 100 percent. The risk of dying of cancer from consuming peanut butter depends on your probability of eating peanut butter, which may be near 100 percent, multiplied by the probability that consuming peanut butter will cause you to develop lethal cancer, which is very low. It turns out that both behaviors produce a 1 in 1 million chance of dying. This example demonstrates a fundamental rule of risk assessment: The risk of a rare event that has a high likelihood of causing harm can be equal to the risk of a common event that has a low likelihood of causing harm.

A Case Study in Risk Assessment

As we saw in our discussion of water pollution in Module 49, from the 1940s to the 1970s some companies manufacturing electrical components dumped PCBs (polychlorinated biphenyls) into rivers. Beginning in the 1960s, there was increasing evidence that PCBs might have harmful health effects on organisms that come into contact with them, including liver damage in animals and impaired learning in human infants.

Once the EPA identified PCBs as a potential hazard, it began a risk assessment. The agency brought together a range of data. First, scientists had to determine which concentrations of PCBs might cause cancer. To accomplish this objective, they examined dose-response studies on laboratory rats exposed to different concentrations of PCBs. They also examined studies of cancer cases in workers employed by industries that used PCBs. Next, they had to determine what concentrations people might experience. To accomplish this, scientists examined data on current concentrations in the air, soil, and water and considered the half-life of the chemical. Because PCBs were found throughout the environment and because they are very persistent, the probability of coming into contact with PCBs was considered relatively high. They also considered the potential routes of exposure: eating contaminated fish, drinking contaminated water, and breathing contaminated air.

FIGURE 53.5 The outcome of a risk assessment of PCBs. Based on a risk assessment of humans consuming fish, the EPA determined that the fish living in the Hudson River in New York State and in Silver Lake in Massachusetts had unacceptably high concentrations of PCBs due to illegal dumping of PCBs by General Electric. As a result, anglers were not allowed to keep and consume the fish that they caught from those water bodies. *(Suzanne DeChillo/New York Times/Redux)*

The final result of the risk assessment on PCBs showed that the risk from eating contaminated fish is higher than the risk from drinking contaminated water and much higher than breathing contaminated air. As a result, signs were posted on the Hudson River and at Silver Lake in Massachusetts instructing anglers not to consume any fish they caught (**FIGURE 53.5**). With limited fish consumption, the EPA concluded that the absolute risk of an individual developing cancer from PCB exposure was low. However, the risk was high enough to cause the EPA to recommend a dredging of the Hudson River to remove a large fraction of the PCBs that had settled at the bottom of the river. As we discussed in Module 49, this decision led to a long court battle between the EPA and General Electric, the company that had been dumping PCBs into the Hudson River. The dredging was finally initiated in 2009 and completed in 2015. While the dredging has removed a large amount of the contaminated river mud, scientists estimate that it may be another 15 to 30 years before people can safely consume any fish from the river.

Risk Acceptance

Once the risk assessment is completed, the second step in risk analysis is to determine risk acceptance—the level of risk that can be tolerated. Risk acceptance may be the most difficult of the three steps in the risk-analysis process. No amount of information on the extent of the risk will overcome the conflict between those who are willing to live with some amount of risk and those who desire zero risk, which is often not realistic. Among those people who are willing to accept some risk, the precise amount of acceptable risk is often hotly debated. For example, according to the EPA, a risk of 1 in 1 million is acceptable for most environmental hazards. Some people believe this is too high. Others feel that a risk such as a 1 in 1 million chance of death from radiation leaks is a small price to pay for the electricity generated by nuclear energy.

Risk Management

The third step of the risk-analysis process, risk management, integrates the scientific data on risk assessment and the analysis of acceptable levels of risk with a number of additional factors including economic, social, ethical, and political issues. Risk management highlights the fact that there are trade offs in deciding the amount of a given chemical that should be permitted in the environment. Whereas risk assessment is the job of environmental scientists, risk management is a regulatory activity that is typically carried out by local, national, or international government agencies.

The regulation of arsenic in drinking water provides an excellent example of the difference between risk assessment and risk management. Despite the fact that scientists knew that 50 ppb of arsenic could cause cancer in people, from 1942 to 1999 the federal government set the acceptable concentration of arsenic at 50 ppb. In 1999, the EPA announced it was lowering the maximum concentration of arsenic in drinking water to 10 ppb, which matched the standards set by the European Union and the World Health Organization. This regulation threatened to place a large financial burden on mining companies that produced arsenic as a by-product of mining, and an economic burden on several municipalities in western states with naturally high concentrations of arsenic in their drinking water. Both groups lobbied hard against the lower arsenic limits. In 2001, weeks before the new lower limits were to go into effect, the EPA announced that it would return to the 50 ppb. The agency argued that further risk assessments needed to be conducted and any risk assessment had to be balanced by economic interests. Later in 2001, the National Academy of Sciences concluded that the acceptable amount of arsenic was a mere 5 ppb, which was lower than some previous estimates. This new risk assessment played a key role in striking a balance between the scientific data and economic interests and the EPA revised its ruling, ultimately setting the acceptable arsenic concentration at 10 ppb.

53-3 What are the major philosophies of regulating chemicals in the environment?

Worldwide standards of risk are guided by two different philosophies

There are currently about 80,000 regulated chemicals in the world but they are not regulated the same way around the globe. A key factor determining the type of chemical regulation is whether the regulations are guided by the innocent-until-proven-guilty principle or the precautionary principle, both illustrated in **FIGURE 53.6**.

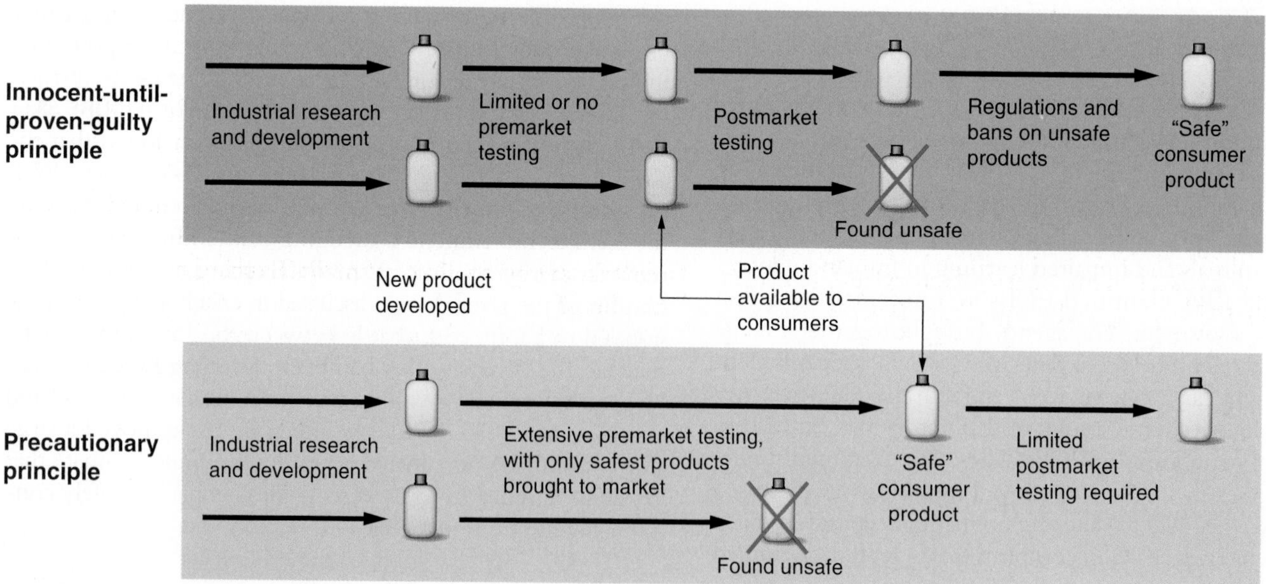

FIGURE 53.6 The two different approaches to managing risk. The innocent-until-proven-guilty principle requires that researchers prove harm before the chemical is restricted or banned. The precautionary principle requires that when there is scientific evidence that demonstrates a plausible risk, the chemical must then be further tested to demonstrate it is safe before it can continue to be used.

The **innocent-until-proven-guilty principle** is based on the belief that a potential hazard should not be considered an actual hazard until the scientific data definitively demonstrate that it actually causes harm. This strategy allows beneficial chemicals to be sold sooner. The downside is that harmful chemicals can affect humans or wildlife for decades before sufficient scientific evidence accumulates to confirm that they are harmful.

In contrast, the **precautionary principle** is based on the belief that when a hazard is plausible but not yet certain, we should take actions to reduce or remove the hazard. The plausibility of the risk cannot be speculation; it must have a scientific basis. In addition, the intervention should be in proportion to the potential harm that might be caused by the hazard. This approach allows fewer harmful chemicals to enter the environment. However, if the initial assessment indicates a plausible risk and the chemical ultimately proves harmless but beneficial, its introduction and use can be delayed for many years. Moreover, the slower pace of approval can reduce the financial motivation of manufacturers to invest in research for new chemicals. In short, there is a trade off between greater safety with slower introduction of beneficial chemicals versus greater potential risk with a greater rate of discovery of helpful chemicals. Use of the precautionary principle has been growing throughout many parts of the world and was instituted by the European Union in 2000. The United States, however, continues to use the innocent-until-proven-guilty principle.

The benefit of the precautionary principle can be illustrated using the case of asbestos. Asbestos is a white, fibrous mineral that is very resistant to burning. This made asbestos a popular building material throughout much of the twentieth century. It is now widely accepted that dust from asbestos can cause a number of deadly diseases including asbestosis (a painful inflammation of the lungs) and several types of cancer including mesothelioma, which causes extreme chest pain and shortness of breath. When asbestos was first mined in 1879, there was no evidence that it harmed humans. The first report of deaths in humans was in 1906 and the first experiment showing harmful effects in rats was conducted in 1911. In 1930, it was reported that 66 percent of workers in an asbestos factory suffered from asbestosis. In 1955, researchers found that asbestos workers had a higher risk of lung cancer than other groups. In 1965, a study linked a rare form of cancer with workers who were exposed to asbestos dust.

Despite all of the growing scientific evidence that asbestos was harming human health, little was done to reduce the exposure of workers and the public. Indeed, it was not until 1998 that the European Union banned asbestos. Today, workers go to great lengths not to be exposed to asbestos dust such as by wearing protective suits and using respirators (**FIGURE 53.7**). A study in the Netherlands estimated that had asbestos been banned in 1965 when the harm to health became clear, the country would have had 34,000 fewer deaths from asbestos and would have saved approximately $25 billion in cleanup and compensation costs. Because the effects of asbestos can take several decades to harm a person's health, the European Union estimates that from 2005 to 2040 it will have 250,000 to 400,000 additional people die as a result of past exposures to asbestos. Had the European Union been using the precautionary principle decades earlier, the number of deaths would have been considerably less.

FIGURE 53.7 The risks of asbestos dust. Despite nearly a century of studies on the risks of asbestos dust to human health, only recently have workers been required to go to great lengths to prevent exposure. Today, they dress in chemical suits and respirators when removing asbestos from a building. Applying the precautionary principle would have required protection of workers many decades earlier and saved hundreds of thousands of lives. *(Phanie/Alamy)*

International Agreements on Hazardous Chemicals

In 2001, a group of 127 nations gathered in Stockholm, Sweden, to reach an agreement on restricting the global use of some chemicals. The agreement, known as the **Stockholm Convention**, produced a list of 12 chemicals to be banned, phased out, or reduced. These 12 chemicals came to be known as the "dirty dozen" and included pesticides such as DDT, industrial chemicals such as PCBs, and certain chemicals that are by-products of manufacturing processes. All of the chemicals were known to be endocrine disruptors, and a number of them had already been banned or were experiencing declining use in many countries. However, bringing

Innocent-until-proven-guilty principle A principle based on the belief that a potential hazard should not be considered an actual hazard until the scientific data definitively demonstrate that it actually causes harm.

Precautionary principle A principle based on the belief that when a hazard is plausible but not yet certain, we should take actions to reduce or remove the hazard.

Stockholm Convention A 2001 agreement among 127 nations concerning 12 chemicals to be banned, phased out, or reduced.

countries together in a forum to discuss controlling the most harmful chemicals was the great achievement of the Stockholm Convention. By 2017, the Convention listed 32 chemicals that countries agreed should be eliminated, restricted in their use, or reduced in their release into the environment.

In 2007, the 27 nations of the European Union put into effect an agreement on how chemicals should be regulated within the European Union. Known as **REACH**, an acronym for registration, evaluation, authorization, and restriction of chemicals, the agreement embraces the precautionary principle by putting more responsibility on chemical manufacturers to confirm that chemicals used in the environment pose no risk to people or the environment. This regulation was enacted because many chemicals used for decades in the European Union had not been subjected to rigorous risk analyses. The new regulations were phased in through 2018 to permit sufficient time for chemical manufacturers to complete the required testing.

REACH A 2007 agreement among the nations of the European Union about regulation of chemicals; the acronym stands for registration, evaluation, authorization, and restriction of chemicals.

In this module we learned how scientists determine the concentrations of chemicals that can cause harm to humans and other species. We also learned about how we can assess risk and how international agreements provide guidance for regulating chemicals. In the next module, we will focus on how pollutants can affect human health.

Module 53 AP® Review

Preparing for the AP® Exam

Learning Goals Revisited

53-1 How are dose-response curves used to estimate lethal doses of chemicals?

Scientists determine the concentrations of chemicals that will harm organisms using LD_{50} experiments to assess lethal effects and ED_{50} studies to assess sublethal effects.

53-2 How do we estimate potential harm of chemicals in the environment?

We can estimate potential harm of chemicals in the environment by using risk assessment, risk acceptance, and risk management.

53-3 What are the major philosophies of regulating chemicals in the environment?

A key factor determining the type of chemical regulation is whether the regulations are guided by the innocent-until-proven-guilty principle or the precautionary principle. Based on these philosophies, the Stockholm Convention produced a list of 12 chemicals to be banned, phased out, or reduced, which have come to be known as the "dirty dozen."

Practice Math and Graphing

Preparing for the AP® Exam

Answer the following questions. Be sure to show all your work.

1. Practice Math

To determine the LD_{50} of a new pesticide, a scientist exposes shrimp to a range of concentrations. The scientist places the shrimp in tubs of water, with 12 shrimp in each tub. The table shows how many shrimp are dead after 24 hours of being exposed to the various concentrations. Based on these data, calculate the percentages of dead shrimp in each concentration.

Concentration (mg/L)	Number dead	Percent dead
1	0	
2	1	
3	3	
4	6	
5	9	
6	11	
7	12	

2. Practice Graphing

(a) Using your percentage data in the table from "Practice Math", create a graph that shows the relationship between chemical concentration and the percentage of dead shrimp. Then draw a vertical and horizontal line to indicate the LD_{50} value of the chemical.

(b) Based on this graph, what is the no-observed-effect level for shrimp exposed to this chemical?

AP® Practice Questions

Multiple-Choice Questions

Use the following graph to answer questions 1 & 2:

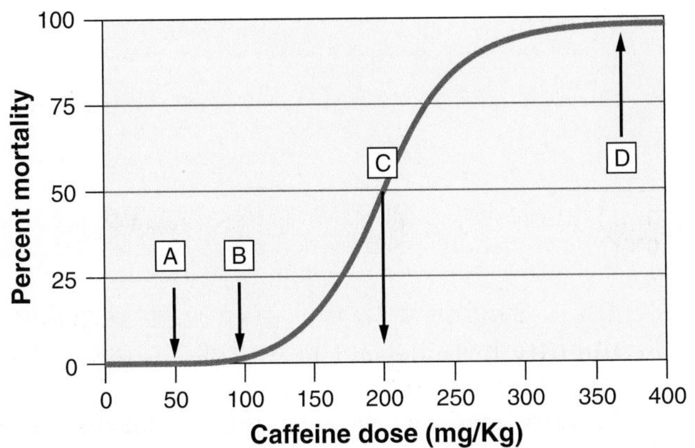

1. Which of the points on the graph above best represents the LD_{50} of caffeine?
   (a) A
   (b) B
   (c) C
   (d) D

2. According to the graph, which dose of caffeine is most likely to cause effects in the most sensitive individuals?
   (a) 50 mg/kg
   (b) 100 mg/kg
   (c) 200 mg/kg
   (d) 400 mg/kg

3. Which risk has the highest probability of death in the United States?
   (a) motor vehicle accident
   (b) lightning
   (c) fire
   (d) firearm assault

4. Which of the following best describes a benefit of utilizing the precautionary principle?
   (a) It decreases financial incentives for chemical development.
   (b) It was used in considering the use of asbestos.
   (c) It is primarily used in the United States.
   (d) It increases the risk of harmful chemicals being used.

5. Which of the following could be a set of treatment and control groups for a retrospective study on the effects of endocrine disruptors on humans?
   (a) a group of construction workers that have been exposed to asbestos and a group of construction workers that have been exposed to PCBs
   (b) a group of children with mothers exposed to PCBs during pregnancy and a group of children who were exposed to DDT after they were born
   (c) residents of a town that was downwind from a commercial farm that sprayed pesticides and residents of a town that was upwind of the farm
   (d) humans raised when DDT was used and humans raised after DDT was no longer used

Free-Response Question

AP® Environmental Science students conducted an investigation on the effects of soil salinization on seed germination and mortality. Seven treatments were set up with five dishes, each containing 10 seeds. The seeds were exposed to different saltwater concentrations. The number of seeds that died was recorded after 15 minutes, and their percent mortality is shown on the graph below.

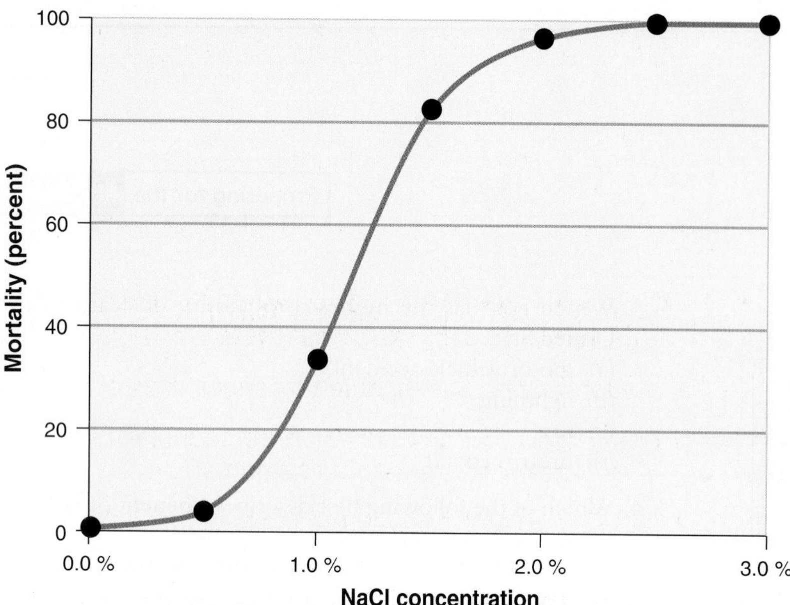

(a) According to the graph, **identify** the NaCl concentration that caused an approximate mortality of 30 percent. (1 pt.)

(b) According to the graph, **describe** the lethal-dose 50 (LD_{50}) of salt for the seeds. (1 pt.)

(c) According to the graph, **explain** the threshold level of salt toxicity for the seeds. (1 pt.)

(d) **Explain** how scientists could determine whether this species of plant is more sensitive to salt than other species of plants. (2 pts.)

(e) **Identify** the independent variable for the students' investigation. (1 pt.)

(f) **Identify** a testable hypothesis for the students' investigation. (1 pt.)

(g) **Identify** a variable not mentioned that could affect the results of the investigation. (1 pt.)

(h) **Identify** the control group for the students' investigation. (1 pt.)

(i) **Describe** a modification the students could make that would improve the reliability of their results. (1 pt.)

Module 54

Unit 8 | 47 | 48 | 49 | 50 | 51 | 52 | 53 | 54

Pollution, Human Health, Pathogens, and Infectious Diseases

Throughout the modules in this unit on terrestrial and aquatic pollution, we have learned about the many types of pollutants and their effects on humans and other organisms. These various types of pollutants have included chemical pollutants, thermal pollution, noise pollution, persistent organic pollutants, solid waste, and sewage. We have also learned that some forms of pollution—particularly from improperly treated sewage—can cause people to be exposed to pathogens, such as those that cause cholera and dysentery. In this module we begin by examining approaches to track human populations to monitor how pollution exposures during people's lives alter their health. We then examine the full suite of infectious and noninfectious diseases that pose a risk to humans. With this understanding, we discuss a number of infectious diseases that have been important in human history and new infectious diseases that have become prevalent in recent decades. Finally, we discuss the major legislation that helps protect people from pollutants and pathogens. The Covid-19 pandemic taught us that it is more important than ever to understand that infectious diseases continue to emerge with potentially devastating consequences.

Learning Goal

After reading this module you should be able to

54-1 describe how we establish cause and effect between pollutants and human health.

54-2 describe the different types of human diseases.

54-3 identify historic human pathogens that have cycled through the environment.

54-4 identify the major emergent infectious diseases in humans.

54-5 identify the laws that protect human health from pollutants and pathogens.

54-1 How can we establish cause and effect between pollutants and human health?

We can establish cause and effect between pollutants and human health using retrospective and prospective studies

Estimating the effects of chemicals on humans is a major challenge. We have seen that one approach is to conduct dose-response experiments on rats and mice and extrapolate the results to humans. An alternative approach is to examine large populations of humans or animals that are exposed to chemicals in their everyday lives and then determine whether these exposures are associated with any health problems. Such investigations fall within the study of epidemiology, a field of science that strives to understand the causes of illness and disease in human and wildlife populations. There are two ways of conducting this type of research: retrospective studies and prospective studies.

Retrospective Studies

Retrospective studies monitor people who have been exposed to an environmental hazard, such as a harmful chemical, at some time in the past. In such studies, scientists identify a group of people who have been exposed to a potentially harmful chemical and a second group of people who have not been exposed to the chemical. Both groups are then monitored for many years to see if the exposed group experiences more health problems than the unexposed group. In 1984, for example, there was an accidental release of methyl isocyanate gas from a Union Carbide pesticide

Retrospective study A study that monitors people who have been exposed to an environmental hazard, such as a harmful chemical, at some time in the past.

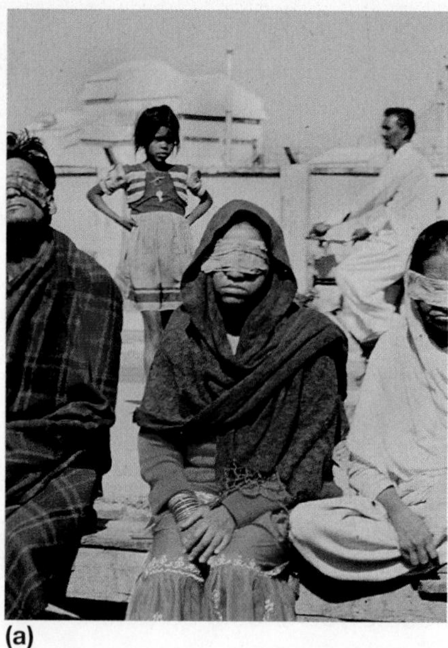

(a)

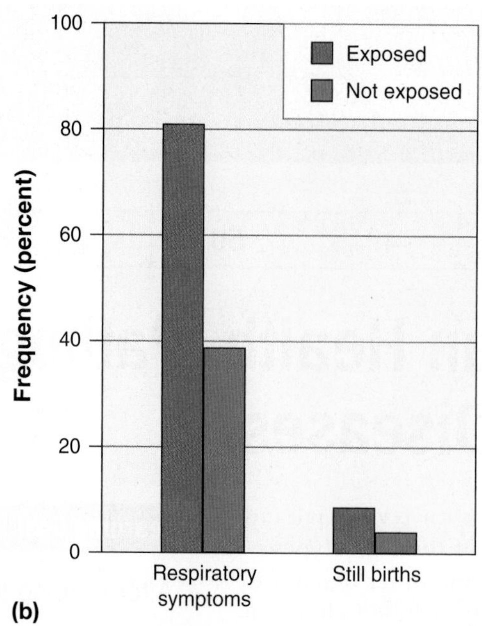

(b)

FIGURE 54.1 The chemical disaster in Bhopal, India. (a) In 1984, a massive release of methyl isocyanate gas killed and injured thousands of people, with many of them suffering blindness. (b) Retrospective studies that followed the survivors of the accident have identified a large number of longer-term health effects from the accident. *(Data from P. Cullinan, S.D. Acquilla, and V.R. Dhara, Long-term morbidity in survivors of the 1984 Bhopal gas leak, National Medical Journal of India, Jan.–Feb. (1996) 9(1): 5–10. Photo by STR/Getty Images)*

factory in Bhopal, India (**FIGURE 54.1a**). This turned out to be one of the largest industrial accidents ever; more than 36,000 kg (80,000 pounds) of hazardous gas spread through the city of 600,000 inhabitants. An estimated 2,000 people died that night and another 14,000 died later from effects related to the exposure. For more than 3 decades, scientists monitored many citizens of Bhopal to determine if survivors of the accident have developed any additional health problems. The retrospective studies have found that approximately 100,000 people are still suffering illnesses from the accidental exposure to the gas. The survivors have higher rates of genetic abnormalities, infant mortality, kidney failure, and learning disabilities. As shown in Figure 54.1b, they also have higher rates of respiratory problems and stillbirths.

Retrospective studies have been helpful in identifying the harmful impacts of several common pollutants. For example, in Module 52 we discussed the human health risks of untreated sewage. By examining groups of people who were exposed versus not exposed to raw sewage, we have learned that exposed people can be exposed to a variety of harmful chemicals and pathogens that have led to an increased emphasis on sewage treatment throughout the world. Similarly, we explored the history of people being exposed to asbestos in Module 53 and learned that studies of people exposed to asbestos were developing a number of deadly diseases including asbestosis (a painful inflammation of the lungs) and several types of cancer including mesothelioma. Finally, in Module 43 we learned about the harmful effects of tropospheric ozone and how individuals exposed to this pollutant are more likely to experience increased lung damage—including emphysema, bronchitis, and asthma—compared to people not exposed to this ozone.

Prospective study A study that monitors people who might become exposed to an environmental hazard, such as a harmful chemical, at some time in the future.

Synergistic interaction A situation in which two risks together cause more harm than expected based on the separate effects of each risk alone.

Prospective Studies

In contrast to retrospective studies, **prospective studies** monitor people who might become exposed to an environmental hazard, such as a harmful chemical, at some time in the future. In this case, scientists might select a group of 1,000 participants and ask them to keep track of the food they eat, the tobacco they use, and the alcohol they drink over a period of several decades. As time passes, the researchers can determine if the habits of the participants are associated with any future health problems. Prospective studies can be quite challenging because a participant's habits, such as tobacco use, can also be associated with many other risk factors, such as socioeconomic status. Of particular concern is when multiple risks cause **synergistic interactions**, in which two risks together cause more harm than expected based on the separate effects of each risk alone. For example, the health impact of a carcinogen such as asbestos can be much higher if an individual also smokes tobacco.

Studies of lead poisoning in children are often prospective. In one study, researchers at Harvard University looked at the effects of lead on children's intelligence by following 276 children in Rochester, New York, from 6 months to 5 years of age. IQ tests are reliable at the age of 5. In addition to lead exposure, the researchers also accounted for other factors that might affect childhood IQ including the mother's IQ, exposure to tobacco, and the intellectual environment of their homes. After controlling for these other factors, the researchers found that among children who had been exposed to lead in the environment—primarily from breathing lead dust and consuming lead paint chips—those with higher lead exposures scored lower on subsequent IQ tests. Such prospective studies can help regulators determine acceptable levels of chemical exposure.

54-2 What are the different types of human diseases?

Human diseases can be infectious or noninfectious

Human causes of death are tracked by countries around the world and brought together by the World Health Organization (WHO). As you can see in **FIGURE 54.2**, approximately three-quarters of all deaths are caused by diseases. A **disease** is any impaired function of the body with a characteristic set of symptoms. **Infectious diseases** are diseases caused by infectious agents, known as pathogens, which we discussed in Module 1. Infectious diseases cause about one-quarter of worldwide deaths; examples include pneumonia and sexually transmitted diseases. Noninfectious diseases are not caused by pathogens; these include most cardiovascular diseases, respiratory and digestive diseases, and most cancers.

The pathogens that cause most infectious diseases include viruses, bacteria, fungi, protists, and a group of parasitic worms called helminths. However, only four types of infectious disease account for nearly three-quarters of all deaths from infectious diseases: respiratory infections, HIV/AIDS, tuberculosis, and diarrheal diseases. We will discuss many of these infectious diseases later in the module.

All diseases can also be categorized as either acute or chronic. **Acute diseases** rapidly impair the functioning of a person's body. In some cases, death can come in a matter of days or weeks. In contrast, **chronic diseases** slowly impair the functioning of a person's body. Heart disease and most cancers, for example, are chronic diseases that develop over several decades.

Risk Factors for Chronic Disease in Humans

Numerous factors cause people to be at a greater risk for chronic diseases such as cancer, cardiovascular diseases, diabetes, and chronic infectious diseases. The WHO has found that these risk factors differ substantially between low- and

Disease Any impaired function of the body with a characteristic set of symptoms.

Infectious disease A disease caused by a pathogen.

Acute disease A disease that rapidly impairs the functioning of a person's body.

Chronic disease A disease that slowly impairs the functioning of a person's body.

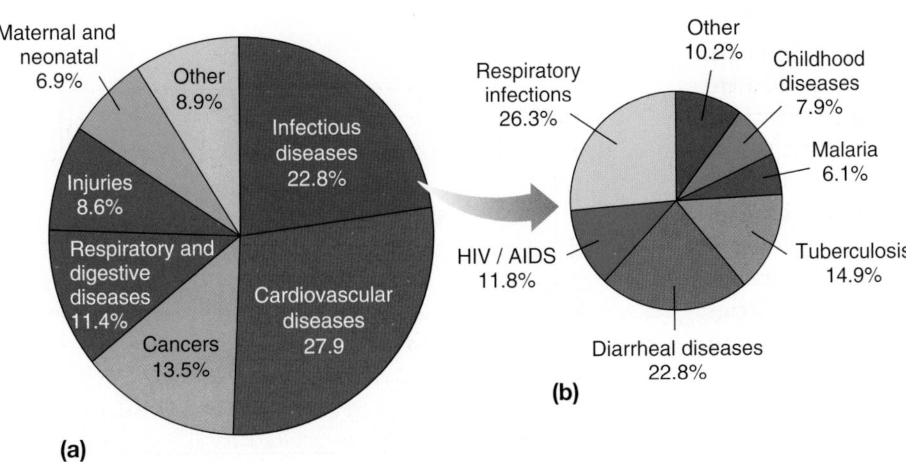

FIGURE 54.2 Leading causes of death in the world. (a) About three-quarters of all world deaths are caused by diseases, including respiratory and digestive diseases, various cancers, cardiovascular diseases, and infectious diseases. (b) Among the world's deaths caused by infectious diseases, 90 percent are caused by only six types of diseases. *(Data from Global Health Estimates 2020. Geneva, World Health Organization, 2020)*

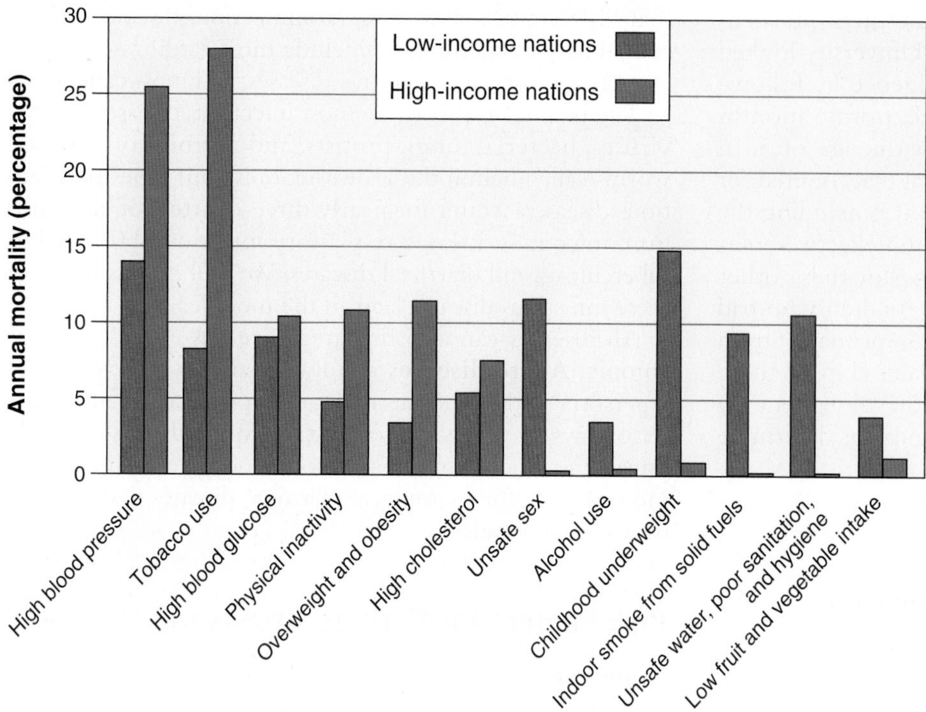

FIGURE 54.3 Leading health risks in the world. If we consider all deaths that occur and separate them into different causes, we can examine which categories cause the highest percentage of all deaths. The leading health risks for low-income countries include issues related to poor nutrition and poor sanitation. The leading risks for high-income countries include issues related to tobacco use, inactivity, obesity, and urban air pollution. *(Data from World Health Organization, 2009)*

high-income countries. **FIGURE 54.3** shows the WHO data for a variety of leading health risks.

As you can see in the figure, in low-income countries, the top risk factors leading to chronic disease are associated with poverty and include underweight children, unsafe drinking water, poor sanitation, and malnutrition. As an example of poverty leading to chronic disease, nearly half of the children under the age of 5 who die from pneumonia succumb to the disease because they suffer from poor nutrition. Similarly, nearly three-quarters of children who die from diarrhea are malnourished. Children with good nutrition are better able to fight infectious diseases and to survive.

Affluence changes the major health risk factors for chronic disease. Because people in high-income countries can afford better nutrition and proper sanitation, fewer die at a young age from diseases such as pneumonia and diarrhea. Risk factors for chronic disease in high-income countries include increased availability of tobacco, and a combination of less-active lifestyles, poor nutrition, and overeating that leads to high blood pressure and obesity.

The change in risk factors between low- and high-income countries occurs over time as a given country becomes more affluent. The graph in **FIGURE 54.4** illustrates how this transition in economic development affects health risk. People living in low-income countries face the challenge of fulfilling basic needs such as food and proper sanitation. As a result, their greatest causes of mortality are things such as children being underweight and unsafe water. As a country begins to accumulate wealth, the health risks to its citizens will change in a predictable fashion. In high-income countries, people are more able to have sufficient food and high-quality sanitation. However, their major causes of mortality are related to smoking, high blood pressure, obesity, and inactivity.

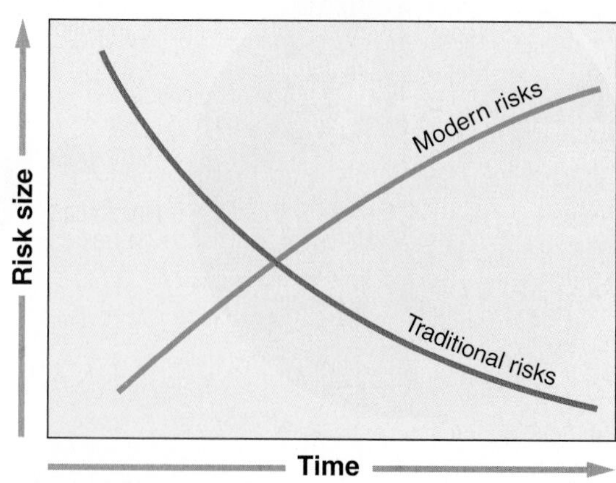

FIGURE 54.4 The transition of risk. As a nation becomes more developed over time and attains higher income levels, the risks of inadequate nutrition and sanitation decline while the risks of tobacco, obesity, and poor urban air quality rise.

54-3 What historic human pathogens have cycled through the environment?

Many pathogens have been historically important

Diseases can have both genetic and environmental causes and these two factors often interact, especially for diseases caused by pathogens such as fungi, bacteria, and viruses. In addition, pathogens have evolved a wide variety of pathways for infecting humans, including transmission through food, water, other humans, and other animals. **FIGURE 54.5** illustrates some of these pathways.

Historically, disease-causing pathogens have taken a large toll on human health and mortality. When a pathogen causes a rapid increase in disease, we call it an **epidemic**. When an epidemic occurs over a large geographic region such as an entire continent, we call it a **pandemic**. Among the diversity of human diseases that have caused epidemics and pandemics, we will consider both those that have been historically important and those that have emerged recently. In Module 52 we looked at several diseases historically associated with poor sanitation and unsafe drinking water including cholera and hepatitis. Cholera is caused by a bacteria (*Vibrio cholerae*) that is a leading cause of **dysentery**, which is an infection of the intestines that causes diarrhea. The infection results in dehydration and can cause death. Cholera claims thousands of lives annually in developing countries. Hepatitis is an inflammation of the liver caused by a virus. There are different forms of the hepatitis virus; some types are transmitted by eating contaminated food while other types are transmitted directly between individuals. Below, we consider some historically important diseases that are cycled in the environment and among hosts, including plague, malaria, and tuberculosis.

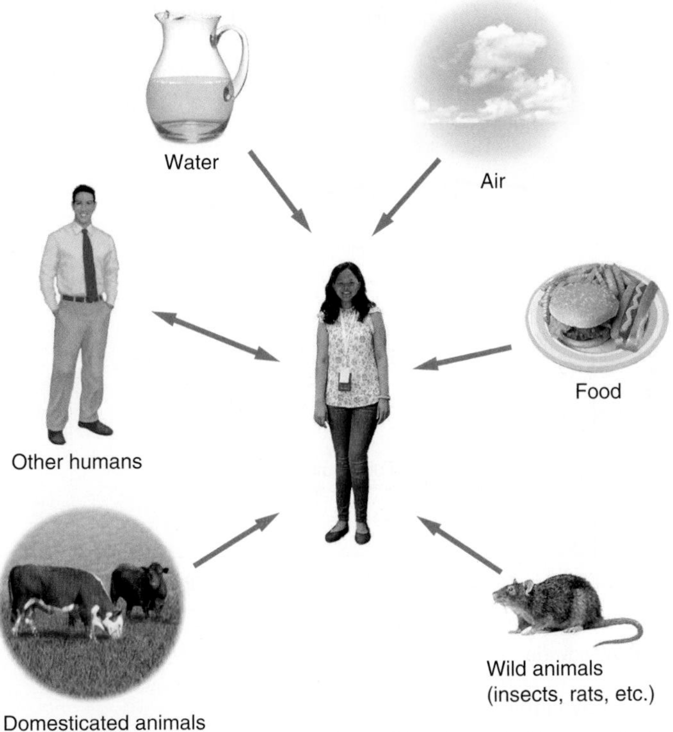

FIGURE 54.5 Pathways of transmitting pathogens. Pathogens have evolved a wide variety of ways to infect humans.

FIGURE 54.6 The Black Death in Europe. As depicted in Carlo Coppola's *The Marketplace in Naples During the Plague of 1656*, plague pandemics repeatedly swept through Europe from the 1300s through the 1800s and killed millions of people. Because the disease caused black sores on people's bodies, it also had the name *Black Death*. *(The Picture Art Collection/Alamy Stock Photo)*

Plague

Plague is caused by an infection from a bacterium (*Yersinia pestis*) that is carried by fleas. Fleas attach to rodents such as mice and rats, which gives the fleas tremendous mobility. One of the most well-known diseases of human history, plague also carries several historical names including *bubonic plague* and *Black Death*. When humans live in close contact with mice and rats, the bacterium can be transmitted either by flea bites or by handling the rodents. Individuals who become infected often experience swollen glands, black spots on their skin, and extreme pain. Plague is estimated to have killed hundreds of millions of people throughout history, including nearly one-fourth of the European population in the 1300s (**FIGURE 54.6**). The last major pandemic

Epidemic A situation in which a pathogen causes a rapid increase in disease.

Pandemic An epidemic that occurs over a large geographic region, such as an entire continent.

Dysentery An infection of the intestines that causes diarrhea, which results in dehydration and can cause death.

Plague An infectious disease caused by a bacterium (*Yersinia pestis*) that is carried by fleas.

of plague occurred in Asia in the early 1900s. Today there are still occasional small outbreaks of plague around the world. For example, on the island of Madagascar off the eastern coast of Africa, there was an outbreak of plague in 2017 that infected more than 2,000 people and killed more than 200 of them. The plague has also continued to infect small numbers of people each year in Arizona, California, Colorado, and New Mexico; these infections occur because a small number of rodents in the American Southwest continue to carry the bacteria. Fortunately, modern antibiotics are highly effective at killing the bacterium and preventing human death.

Malaria

Malaria, caused by an infection from several species of protists in the genus *Plasmodium*, is another widespread disease that has killed millions of people over the centuries. The malaria parasite spends one stage of its life inside a mosquito and another stage of its life inside a human. Infections cause recurrent flulike symptoms. Each year, 350 to 500 million people in the world contract the disease and 1 million people, mostly children under 5 years of age, die from it. The regions hardest hit include sub-Saharan Africa, Asia, the Middle East, and Central and South America. Since 1951, the malaria parasite has been eliminated from the United States by mosquito eradication programs. Although more than 1,000 cases of malaria are diagnosed in the United States each year, these are people who have returned from regions of the world where the malaria parasite lives. The traditional approach to combating malaria was widespread spraying of insecticides such as DDT to eradicate the mosquitoes. Eradication efforts have proven to be ineffective in many parts of the world.

Malaria An infectious disease caused by one of several species of protists in the genus *Plasmodium*.

Tuberculosis A highly contagious disease caused by the bacterium *Mycobacterium tuberculosis* that primarily infects the lungs.

Tuberculosis

Tuberculosis is a highly contagious disease caused by a bacterium (*Mycobacterium tuberculosis*) that primarily infects the lungs. Tuberculosis is spread when a person coughs and expels the bacteria into the air. The bacteria can persist in the air for several hours and infect a person who inhales them. Symptoms of an infection include feeling weak, sweating at night, and coughing up blood. As is the case with many pathogens, a person can be infected but not develop the tuberculosis disease. Indeed, it is estimated that one-third of the world's population is infected with tuberculosis. Each year 9 million people develop the disease and 2 million die.

A year-long course of antibiotics can treat most tuberculosis infections. In countries such as the United States, where the appropriate antibiotics are readily available, there has been a dramatic drop in both the number of new cases and the number of deaths from tuberculosis since the mid-1950s. As shown in **FIGURE 54.7**, the decline in tuberculosis

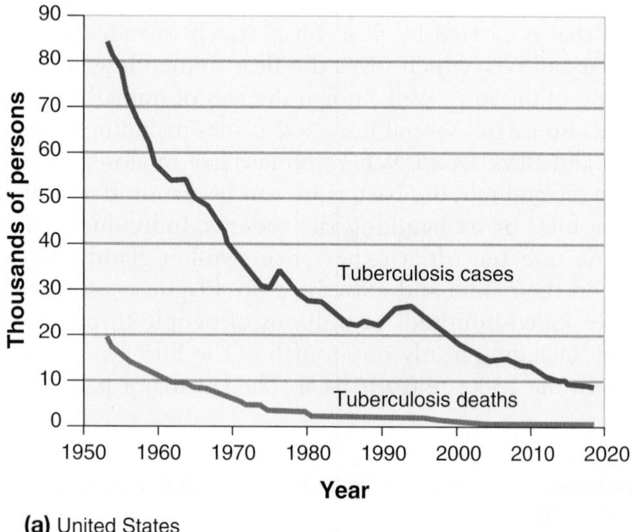

(a) United States

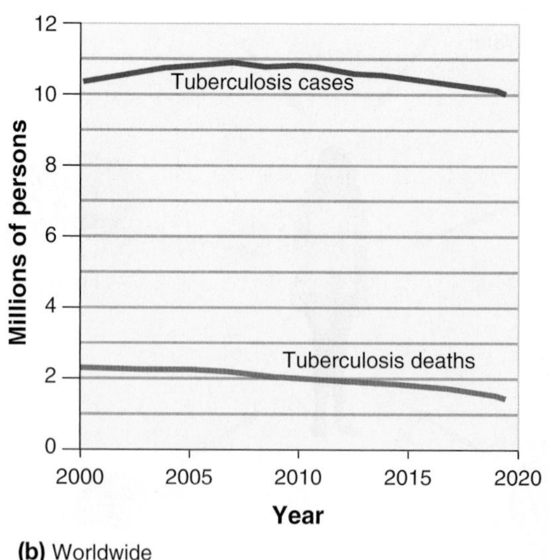

(b) Worldwide

FIGURE 54.7 **Tuberculosis cases and deaths.** (a) Due to effective and available medicines, tuberculosis has gone from being one of the deadliest diseases in the United States to a disease that rarely kills. (b) Worldwide, tuberculosis has continued to infect and kill millions of people, especially in low- and middle-income countries. *(U.S. data from https://www.cdc.gov/tb/statistics/tbcases.htm; global data from World Health Organization. Global tuberculosis report 2019. Geneva, Switzerland: World Health Organization; 2019)*

elsewhere in the world has been much slower. Today, tuberculosis remains the leading cause of death by disease in the developing world. In these countries, the medicines are not as available or affordable and those who receive the medicine sometimes do not take the full course of medicine at the prescribed dose. When patients stop taking antibiotics before the last few bacteria in their bodies have been killed, there are two consequences. First, the pathogen can quickly rebuild its population inside the person's body. Second, because the last few bacteria are generally the most drug-resistant, stopping the antibiotics before the bacteria are eradicated selects for drug-resistant strains. Drug-resistant strains of tuberculosis are becoming a major concern, particularly in parts of Africa and in Russia, where up to 20 percent of the people infected with tuberculosis carry a drug-resistant strain. Such strains are much harder to kill and therefore require newer antibiotics that can cost 100 times more than the traditional drugs.

54-4 What are the major emergent infectious diseases in humans?

Emergent infectious diseases pose new risks to humans

In recent decades, we have witnessed the appearance of many **emergent infectious diseases**, which are defined as infectious diseases that were previously not described or have not been common for at least the prior 20 years. **FIGURE 54.8** locates some of the best-known emergent infectious diseases. Since the 1970s, the world has observed an average of one emergent disease each year. Many of these new diseases have come from pathogens that normally infect animal hosts but then unexpectedly jump to human hosts. This typically occurs because the diseases can mutate rapidly and eventually produce a genotype that can infect humans. Such infections initially emerge in one locality and spread locally. Over time, the pathogens can spread over large areas of the world. Some of the most high-profile diseases that have jumped from animals to humans include HIV/AIDS, Ebola, mad cow disease, bird flu, SARS, and West Nile virus. These emerging diseases are of increasing concern because of the increased movement of people and cargo throughout the world during the past century. In fact, currently diseases can spread to nearly any place on Earth within 24 hours. Moreover, these emerging diseases can spread through developed nations that we commonly think of as having very sanitary conditions. An excellent example of such global spread is that of the SARS-CoV-2 virus (which causes the Covid-19 disease) that was first discovered in China in 2019 and had spread throughout the world by 2020.

HIV/AIDS

In the late 1970s, rare types of pneumonia and cancer began appearing in individuals with weak immune systems.

Emergent infectious disease An infectious disease that has not been previously described or has not been common for at least the prior 20 years.

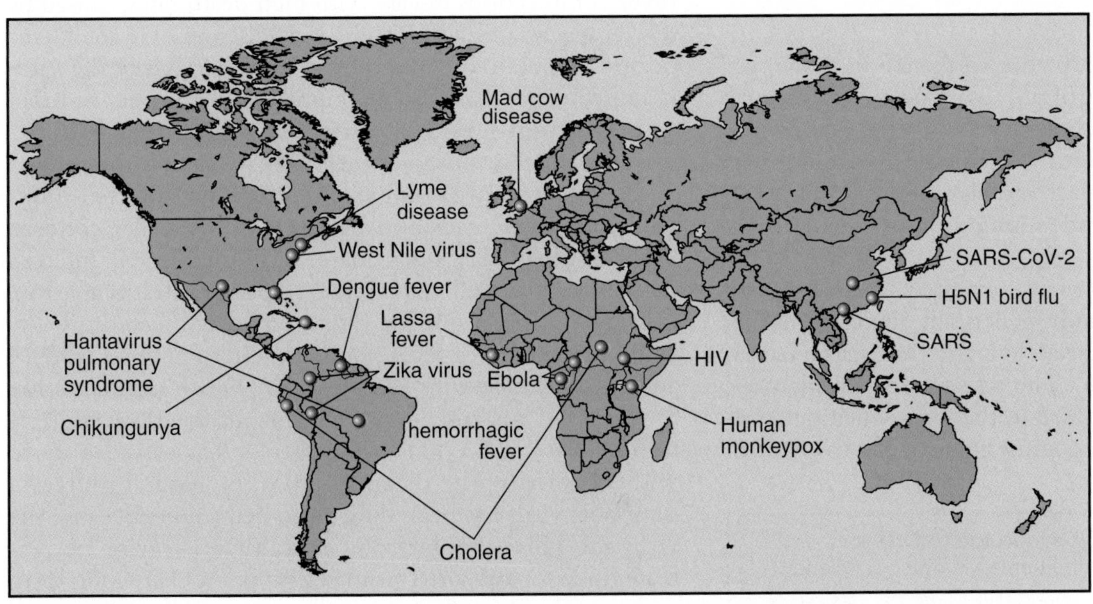

FIGURE 54.8 The emergence of new diseases. Since the 1970s, new diseases, or diseases that have been rare for more than 20 years, have been appearing throughout the world at a rate of approximately one per year. *(Data from https://www.niaid.nih.gov/sites/default/files/public%3A//images/news/main%20map.jpg)*

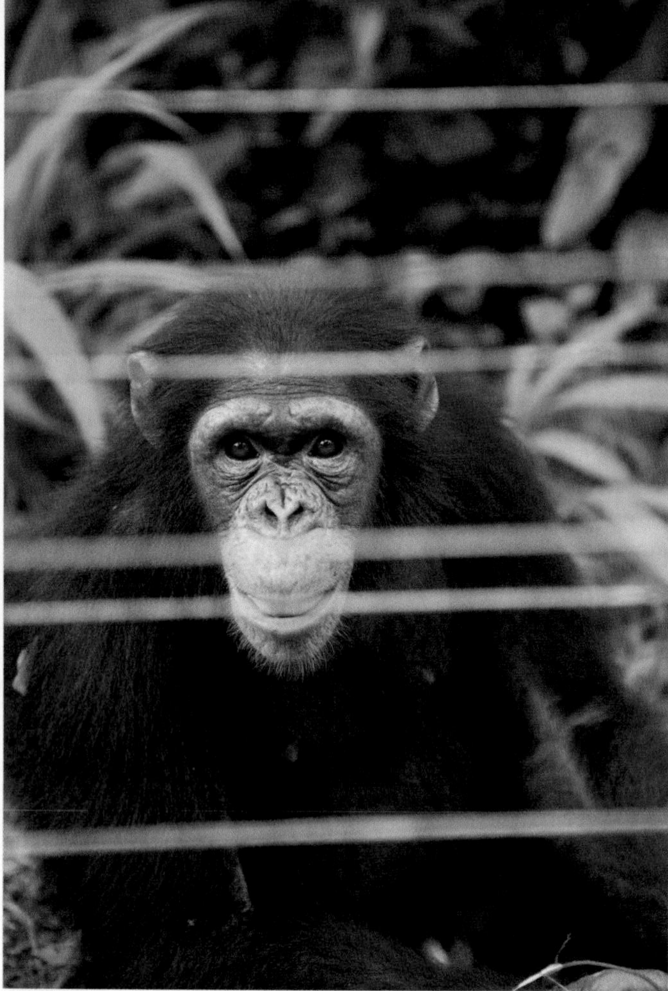

FIGURE 54.9 The source of HIV. In 2006, researchers found that chimpanzees in Cameroon carried a virus that was genetically very similar to HIV. Thus, these chimps are the most likely source of this emerging human disease. *(Tim E White/Alamy)*

The condition responsible for the weakened immune systems was the disease **Acquired Immune Deficiency Syndrome (AIDS)**, which is caused by a virus known as **Human Immunodeficiency Virus (HIV)**. This virus spreads through sexual contact and blood transfusions, from mothers to their fetuses, and among drug users who share unsanitized needles.

The origin of this new virus remained a mystery until 2006 when researchers found a genetically similar virus in a wild population of chimpanzees living in the African nation of Cameroon (**FIGURE 54.9**). The researchers hypothesized that local hunters were exposed to the virus when butchering or eating the chimps (a common practice in that part of the world). With this exposure, the virus was able to infect a new host, humans. According to the World Health Organization, more than 70 million people in the world have been infected with HIV and about 36 million people have died from HIV-related illnesses.

Fortunately, new antiviral drugs are able to maintain low viral populations inside the human body and thereby substantially extend life for those who are infected with the virus. In 2020, 680,000 people died from AIDS; thanks to the new treatments, this represented a 47 percent decline in deaths compared to a decade earlier. From the lessons learned in combating other diseases such as tuberculosis, combinations of antiviral drugs are being used to reduce the risk that the virus will evolve resistance to any single drug. Unfortunately, many of these drugs are expensive and most people living in low-income countries cannot afford them, although this is changing with wider availability and improved distribution of these drugs to the poor.

Acquired Immune Deficiency Syndrome (AIDS) An infectious disease caused by the human immunodeficiency virus (HIV).

Human Immunodeficiency Virus (HIV) A type of virus that causes Acquired Immune Deficiency Syndrome (AIDS).

Ebola hemorrhagic fever An infectious disease with high death rates, caused by several species of Ebola viruses.

AP® Exam Tip

Be cautious not to use HIV and AIDS interchangeably. HIV is the virus that can lead to AIDS. AIDS is the medical condition that results from HIV attacking the immune system. One is the virus. The other is the condition.

Ebola Hemorrhagic Diseases

In 1976, researchers first discovered **Ebola hemorrhagic fever**, an infectious disease with high death rates, caused by several species of Ebola viruses. First discovered in the Democratic Republic of Congo near the Ebola River, the virus has infected several hundred humans and a variety of other primates from several countries in central Africa. Infections have been sporadic since Ebola was first discovered, but there was a large outbreak in 2014 that infected thousands of people. The Ebola virus is of particular concern because it kills a large percentage of those infected. Infected individuals have suffered a 50 to 90 percent death rate from different outbreaks of the disease. Those infected quickly begin to experience fever, vomiting, and sometimes internal and external bleeding (**FIGURE 54.10**). Death occurs within 2 weeks, and currently only experimental drugs are available to fight the virus. Unlike the progress that has been made with identifying the origin of HIV, the natural source of the Ebola virus has been difficult to determine. Because the virus also kills other primates at high rates, leaving no primate hosts for the virus, primates are not a likely long-term source of the virus. In 2013, however, researchers discovered that fruit bats carried the Ebola virus and were likely the reservoir species that spread the virus to primates. In 2019, the U.S. FDA approved the first vaccine to protect people against the Ebola virus.

FIGURE 54.10 Ebola hemorrhagic fever. The Ebola virus is highly lethal to humans and there is now a vaccine. When treating a person infected with the virus, such as this patient who escaped hospital quarantine from Elwa hospital in Monrovia, Liberia, researchers and medical workers have to exercise extreme caution to avoid getting infected. *(REUTERS/Reuters TV)*

Mad Cow Disease

In the 1980s, scientists first described a neurological disease, later known as **mad cow disease**, in which prions mutate into deadly pathogens and slowly damage a cow's nervous system. The cow loses coordination of its body (a condition compared with a person going mad), and then dies (**FIGURE 54.11**). Scientists now know that **prions**, which are small, beneficial proteins in brains of cattle, occasionally mutate into deadly proteins that act as pathogens and subsequently cause mad cow disease. Prions are not well understood and represent a new category of pathogen.

In 1996, scientists in Great Britain announced that mad cow disease, also known as bovine spongiform encephalopathy (BSE), could be transmitted to humans who ate meat from infected cattle. Unlike harmful bacteria that can be killed with proper cooking, prions are difficult to destroy by cooking. Infected humans developed variant Creutzfeldt-Jakob Disease (CJD) and suffered a fate similar to the infected cattle.

FIGURE 54.11 Mad cow disease. Cattle that have been fed the remains of dead cows and sheep can become infected with harmful prions. These prions damage the nervous system and cause the cows to develop glazed eyes, body tremors, and a loss of coordination, eventually leading to death. Humans who consume beef from infected cows can become infected and suffer a similar fate. *(C. E. V./Science Source)*

Mutant prions cannot be transmitted among cattle that live together and eat grass. Transmission requires an uninfected cow to consume the nervous system of an infected cow. As a result, when cattle feed on grass together in a pasture, a rare mutation in a prion would be restricted to a single cow and not spread to other cattle. In the 1980s, however, the diets of European cattle commonly included the ground-up remains of dead cattle, which farmers included as a source of additional protein. When these dead cattle happened to contain a mutant prion, the prions spread rapidly through the entire cattle population and, in turn, infected humans who ate the beef. In Britain, more than 180,000 cattle have become infected and 178 people have died. It is estimated that several thousand people are currently infected, but the prions can exist in the human body for many years before they begin to cause symptoms of the disease. The European Union temporarily banned British beef imports in 1996 and the British government destroyed millions of cattle. Since that time, the disease has been found in several other countries including Canada and the United States, but only in a few cattle and no humans. Today, new rules exist that forbid the feeding of animal remains to cattle. As a result, the current risk of mad cow disease to humans has been greatly reduced; only six cases were detected in the United States from 2003 to 2021.

Swine Flu and Bird Flu

Humans commonly contract many types of flu viruses and some can be quite lethal. For example, the Spanish flu of 1918 killed up to 100 million people. Spanish flu was a type of influenza, known as **swine flu**, caused by the H1N1 virus. This virus is similar to a flu virus that humans normally contract, but H1N1 normally infects only pigs. Occasionally, however, the flu jumps from pigs to humans. Another pandemic of swine flu occurred around the world in 2009 and 2010 and it caused more than 18,000 deaths. The most recent outbreaks have occurred in Venezuela in 2013 and in India in 2017. In India, more than 20,000 people became ill and at least 1,000 of them died. Fortunately, there are vaccines available to help prevent swine flu infections and there are drugs available to combat the swine flu virus if a person becomes infected.

Mad cow disease A disease in which prions mutate into deadly pathogens and slowly damage a cow's nervous system.

Prion A small, beneficial protein that occasionally mutates into a pathogen.

Swine flu A type of flu caused by the H1N1 virus.

FIGURE 54.12 Bird flu. A farm worker feeds a large number of chickens on a farm in the Republic of Niger in western Africa. The virus that causes bird flu normally infects only birds. In 2006, however, the virus began jumping to human hosts where people and birds were in close contact. *(PIUS UTOMI EKPEI/Getty Images)*

In 2006, reports emerged from Asia that a related virus, known as H5N1, or **bird flu**, had jumped from birds to people, primarily to people who were in close contact with birds (**FIGURE 54.12**). Infections are rarely deadly to wild birds but can frequently cause domesticated birds such as ducks, chickens, and turkeys to become very sick and die. Humans often contract a variety of flu viruses. Because humans have no evolutionary history with the H5N1 virus, they have few defenses against it. As of 2021, more than 800 people had become infected by H5N1 and over half of them died. Governments responded to this risk by destroying large numbers of infected birds. Currently the H5N1 virus is not easily passed among people and cases have sharply declined since 2017, but if a future mutation makes transmission easier, scientists estimate that H5N1 has the potential to kill 150 million people.

SARS, MERS-CoV, and SARS-CoV-2

In 2003, an unusual form of pneumonia was spreading through human populations in Southeast Asia that was eventually named **severe acute respiratory syndrome (SARS)**. While some of the respiratory symptoms are similar to bird flu and swine flu, SARS is a disease caused by a different type of virus known as a coronavirus. The virus is believed to have come from a wild animal infecting a person and was then able to spread from one person to another. During the 2002–2003 outbreak, there were more than 8,000 people infected and nearly 10 percent of them died.

Another coronavirus, Middle Eastern Respiratory Syndrome (**MERS-CoV**), appeared on the Arabian Peninsula in 2012. Researchers are investigating the possibility that this disease originated from a bat and may currently also infect camels, which can then transmit the virus to people. To date, approximately 900 people have died from MERS-CoV.

In 2019, a new SARS virus was discovered, known as **SARS-CoV-2**, which causes the disease known as Covid-19. The virus rapidly traveled around the world and quickly became a pandemic. As of May 2022, the World Health Organization estimated that more than 500 million people had been infected and more than 6 million people died from the virus. Fortunately, multiple vaccines were rapidly developed and they provide very good protection against the virus, particularly when combined with social distancing and wearing protective masks to reduce the transmission of the virus between people (**FIGURE 54.13**).

West Nile Virus

The **West Nile virus** lives in hundreds of species of birds and is transmitted among birds by mosquitoes. Although the virus can be highly lethal to some species of birds, including blue jays (*Cyanocitta cristata*), American crows (*Corvus brachyrhynchos*), and American robins (*Turdus migratorius*), most species of birds survive the infection. During the latter half

FIGURE 54.13 The Covid-19 pandemic. The pandemic that began in 2019 spread throughout the world, resulting in mass vaccinations and the wearing of masks to help reduce the spread of the virus. *(INA FASSBENDER/Getty Images)*

Bird flu A type of flu caused by the H5N1 virus.

Severe acute respiratory syndrome (SARS) A type of flu caused by a coronavirus.

MERS-CoV A coronavirus that causes the disease known as Middle Eastern Respiratory Syndrome.

SARS-CoV-2 A coronavirus that causes the disease known as Covid-19.

West Nile virus A virus that lives in hundreds of species of birds and is transmitted among birds by mosquitoes.

of the twentieth century there were increasing reports that the virus could sometimes infect horses and humans who had been bitten by mosquitoes. The first human case was identified in 1937 in the West Nile region of Uganda, thus giving the virus its name. In humans, the virus causes an inflammation of the brain leading to illness and sometimes death. In 1999, the virus appeared in New York and quickly spread throughout much of the United States. **FIGURE 54.14** shows the history of the West Nile virus in the United States. The highest numbers of infections and deaths from the virus occurred in 2002 and 2003. Increased efforts to combat mosquito populations and protect against mosquito bites are causing a decline in the disease, although there was a brief spike in West Nile virus cases in 2012, which was related to warm, rainy conditions that favored larger mosquito populations.

Lyme disease

Lyme disease is a disease caused by a bacterium (*Borrelia burgdorferi*) that is transmitted by ticks. The primary vector is the black-legged tick (*Ixodes scapularis*), which is also known as the deer tick (**FIGURE 54.15**). After hatching from eggs on the forest floor, the tick spends its first stage of life attached to birds and small rodents such as mice and chipmunks. If a bird or rodent is infected with the bacterium, it can be transferred to the tick feeding on the host's blood. After dropping off these hosts and spending the winter in the leaves of the forest, they attach themselves to larger mammals, including deer and people. It is when they attach their mouthparts to people that they transfer the disease-causing bacteria.

The CDC currently estimates that there are about 30,000 cases of Lyme disease in the United States annually, with most infections happening in the northeastern United States. Infected people often experience a red bullseye at the site of tick attachment. This is followed by symptoms that resemble the flu, arthritis, and various neurological disorders. Most victims can be cured using modern antibiotics, although some people have persistent health problems years after becoming infected.

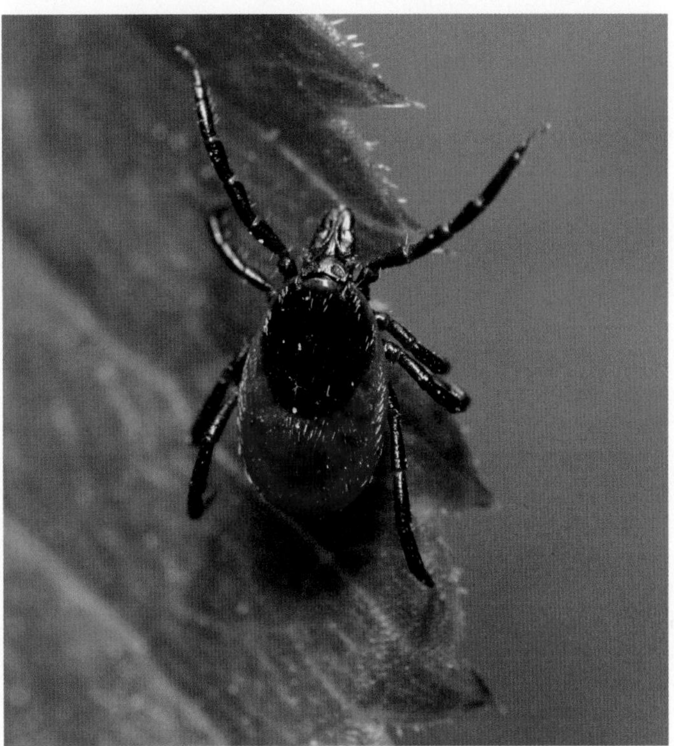

FIGURE 54.15 Deer tick. When deer ticks attach to birds and rodents infected by the Lyme bacteria, the pathogen can be transferred to the tick. After the tick drops off the bird or rodent and attaches to a person, the pathogen can be transferred to the person. *(Jiri Prochazka/Shutterstock.com)*

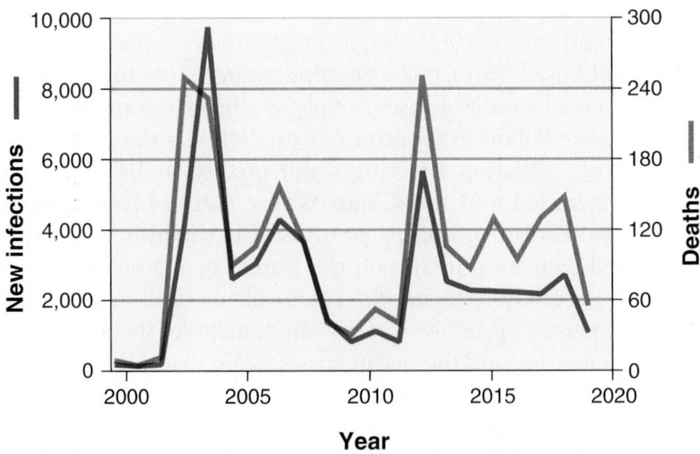

FIGURE 54.14 West Nile virus in the United States. Following the first appearance of West Nile virus in the United States in 1999, the number of human infections and deaths rapidly increased. Efforts to control populations of mosquitoes that carry the virus are helping to reduce the prevalence of the disease. *(Data from https://www.cdc.gov/westnile/statsmaps/cumMapsData.html)*

Lyme disease was first discovered in school children living in Lyme, Connecticut, in the 1970s, which is how the disease got its name. However, researchers have recently discovered the bacteria in a frozen mummy found in the European Alps, suggesting that the disease may have been infecting humans for more than 5,000 years.

Zika Virus Disease

The **Zika virus disease** is caused by a pathogen that is carried by mosquitoes and transmitted to humans when bitten by the mosquito or through sexual contact with an infected person. Most infected people only experience mild effects for a few days or weeks after being infected,

Lyme disease A disease caused by a bacterium (*Borrelia burgdorferi*) that is transmitted by ticks.

Zika virus disease A disease caused by a pathogen that causes fetuses to be born with unusually small heads and damaged brains.

such as rashes, fevers, and headaches. However, the major risk is that the virus can be passed between pregnant mothers and their fetuses. When this happens, the virus causes babies to be born with unusually small heads and damaged brains.

Zika is a very recent emerging infectious disease. Although it was first discovered in a monkey in Uganda in 1947, it was only found in 14 people around the world prior to 2007. In that year, more than 10,000 people became infected on Yap Island in Southeast Asia. In 2013, more than 28,000 people became infected in Tahiti. Today, it is present in Central America, South America, central Africa, and Southeast Asia. In 2015, the virus appeared in Brazil and more than 1 million people became rapidly infected. There is currently no known treatment for the Zika virus and attempts to control its spread are focused on reducing mosquito populations.

Future Challenges to Human Health

While humans face a large number of health risks, we have an excellent understanding of important risk factors and the ways to combat many historical and emerging infectious diseases. Combating diseases in low-income countries requires improvements in nutrition, wider availability of clean drinking water, and proper sanitation. In high-income countries, we need to promote healthier lifestyle choices such as increased physical activity, a balanced diet, and limiting excess food consumption and tobacco use. In all countries, continued education is needed to reduce the spread of diseases such as AIDS, tuberculosis, and Covid-19.

As we combat many diseases, an issue of growing concern is the ability of pathogens to evolve resistance. As we noted in the case of tuberculosis, patients often feel much better long before they complete the full year of prescribed medicine. Because they feel better or because they cannot afford a full year of drugs, they stop the drug treatment early. This allows a small fraction of highly resistant bacteria to survive in the body, reproduce, and then potentially spread to other people. When a pathogen evolves resistance to one drug, physicians often prescribe a different drug to combat the pathogen. Over time, however, some pathogens such as tuberculosis have evolved multiple drug resistance. Without new drugs, little can be done for patients with pathogen strains that possess multiple drug resistance.

Though many historical diseases are either currently under control or likely to be soon if the financial resources become available, changing climates offer the potential for some equatorial diseases—and the animals that carry them—to expand their range into more temperate regions. Of particular concern is the potential for multiple species of mosquitoes to expand their ranges toward the poles as the planet continues to warm. As we have discussed, mosquitoes act as disease vectors and carry a considerable number of diseases, including Zika virus, West Nile virus, and malaria. With expanded ranges, we will likely see regions of the world that have never experienced these diseases before now exposed to an expanded risk from these diseases.

Emerging infectious diseases present a major challenge. New diseases often arise from new pathogens with which we have no experience. Since we cannot predict which diseases will emerge next, public health officials throughout the world must develop rapid response plans when a particular disease does appear. As we all witnessed with Covid-19, this includes rapid worldwide notification of newly identified diseases and strategies to isolate infected persons, which will slow the spread of the disease and provide time for researchers to develop appropriate tactics to combat the threat.

Clean Water Act Legislation that supports the "protection and propagation of fish, shellfish, and wildlife and recreation in and on the water" by maintaining and, when necessary, restoring the chemical, physical, and biological properties of surface waters.

54-5 What laws protect human health from pollutants and pathogens?

Laws exist to protect human health from pollutants and pathogens

To protect human health from aquatic pollution and pathogen, the United States has two important laws: The Clean Water Act and the Safe Drinking Water Act. Other developed countries have similar laws. Many developing countries are rapidly growing and experiencing substantial water pollution, but legislation to improve the water will likely come as these countries become more affluent.

The Clean Water Act

As recently as the 1960s, water quality was very poor in much of the United States, but a growing awareness of the problem encouraged a series of laws to fight water pollution. The Federal Water Pollution Control Act of 1948 was the first major piece of legislation affecting water quality. In 1972, the act was expanded into the **Clean Water Act**, which supports the "protection and propagation of fish, shellfish, and wildlife and recreation in and on the water" by maintaining and, when necessary, restoring the chemical, physical, and biological properties of surface waters, which includes streams, rivers, lakes, estuaries, and the ocean coasts. Note that this legislation does not include the protection of groundwater.

The Clean Water Act originally focused mostly on the chemical properties of surface waters. More recently, there has been an increased focus on ensuring that the biological properties of the waters also receive attention, including the abundance and diversity of various species. Most importantly, the Clean Water Act issued water-quality standards that defined acceptable limits of various pollutants in U.S. waterways. To help enforce these limits, the act allowed the

EPA and state governments to issue permits to control how much pollution municipalities and industries can discharge into the water. Over time, more and more categories of pollutants have been brought under the jurisdiction of the Clean Water Act, including animal feedlots and storm runoff from municipal sewer systems.

The Safe Drinking Water Act

In addition to the Clean Water Act, other legislation has been passed to regulate water pollution, including the **Safe Drinking Water Act** (1974, 1986, 1996), which sets the national standards for safe drinking water. Under the Safe Drinking Water Act, the EPA is responsible for establishing **maximum contaminant levels (MCL)** for 77 different elements or substances in both surface water and groundwater. This list includes some well-known microorganisms, disinfectants, organic chemicals, and inorganic chemicals (**TABLE 54.1**). These maximum concentrations consider both the concentration of each compound that can cause harm as well as the feasibility and cost of reducing the compound to such a concentration.

MCLs are somewhat subjective and are subject to political and economic pressures. As we saw in the case of arsenic, the maximum amount allowed in drinking water used to be 50 ppb. With more studies completed, scientists pushed to reduce this level to 5 ppb while industries pushed to retain the 50 ppb level. As a compromise, the EPA set the new limit at 10 ppb.

What has been the impact of these water pollution laws? In general, they have been very successful. The EPA defines bodies of water in terms of their designated uses, including aesthetics, recreation, protection of fish, and as a source of safe drinking water. The EPA then determines if a particular waterway fully supports all of the designated uses. According to the EPA's most recent report, 47 percent of wetlands, 46 percent of all rivers and streams, 30 percent of lakes and ponds, and 22 percent of bays and estuaries in the United States now fully support their designated uses. This is a large improvement from decades past but, as **TABLE 54.2** shows, we still have a lot of work to do to improve the remaining waterways. Today, the water in municipal water systems in the United States is generally safe. Water regulations have greatly reduced contamination of waters and nearly eliminated major point sources of water pollution. But nonpoint sources such as oil from parking lots and nutrients and pesticides from suburban lawns are not covered under existing regulations. In addition, the U.S. government has exempted fracking for natural gas (Module 37) from the Safe Drinking Water Act, despite the fact that fracking injects a suite of harmful chemicals deep into the ground.

AP® Exam Tip

You should be able to differentiate between the Clean Water Act and the Safe Drinking Water Act, but because these acts get modified every few years, the AP® Environmental Science Exam isn't likely to ask for specifics.

TABLE 54.1 The maximum contaminant levels (MCL) for a variety of contaminants in drinking water as determined by the U.S. Environmental Protection Agency, in parts per billion (ppb)

Contaminant category	Contaminant	Maximum contaminant level (ppb)
Microorganism	*Giardia*	0
Microorganism	Fecal coliform	0
Inorganic chemical	Arsenic	10
Inorganic chemical	Mercury	2
Organic chemical	Benzene	5
Organic chemical	Atrazine	3

(Data from U.S. Environmental Protection Agency, http://www.epa.gov/safewater/contaminants/index.html)

TABLE 54.2 The current leading causes and sources of impaired waterways in the United States

	Causes of impairment	Sources of impairment
Wetlands	Organic enrichment, mercury, arsenic, selenium	Agriculture, atmospheric deposition, petroleum, and natural gas production
Streams and rivers	Bacterial pathogens, sediments, excess nutrients	Agriculture, atmospheric deposition, water diversions, dam construction
Lakes, ponds, and reservoirs	Mercury, PCBs, nutrients	Atmospheric deposition, agriculture
Bays and estuaries	Bacterial pathogens, oxygen depletion, mercury	Atmospheric deposition, municipal discharges including sewage

(Data from U.S. Environmental Protection Agency. 2017. National Water Quality Inventory: Report to Congress)

Safe Drinking Water Act Legislation that sets the national standards for safe drinking water.

Maximum contaminant level (MCL) The standard for safe drinking water established by the EPA under the Safe Drinking Water Act.

Preparing for the AP® Exam

Visual Representation 8 **Water pollution in the Florida Everglades.** The Florida Everglades includes 18,000 square miles of rivers, lakes, wooded uplands, swamps, estuaries, sawgrass prairie, and coral reefs. It provides drinking water for 9 million people. It is one of the most biodiverse locations in

Runoff and Sedimentation
Historical water flow
Current water flow
Large building projects in Florida rerouted the water flow and drained wetlands. The reduction in water flow across the Everglades coupled with development and agriculture led to sedimentation buildup, agricultural runoff of phosphate fertilizers, and trash accumulation. As a result, lake Okeechobee, which supplies the Everglades with water, has experienced eutrophication and massive algal blooms.
Ongoing efforts to reduce the problems of runoff and sedimentation include creation of stormwater treatment and trash collection areas, working with farmers to curb phosphorus use, and removal of excess sedimentation. Long-term goals include reversing changes in water flow to increase the amount of water reaching the Everglades.
In mangroves and estuaries on the coast, the flow of nutrients from agricultural runoff also leads to harmful algal blooms. The subsequent reduction in oxygen availability leads to loss of wildlife in these breeding grounds that many birds and fish depend upon.

the world, and home to many endangered and rare species. Agriculture, oil exploration and drilling, land reclamation for development, wastewater, and air pollution have all contributed to significant water pollution problems in the Everglades. Intense remediation efforts are restoring the historical water flow and cleaning up waterways.

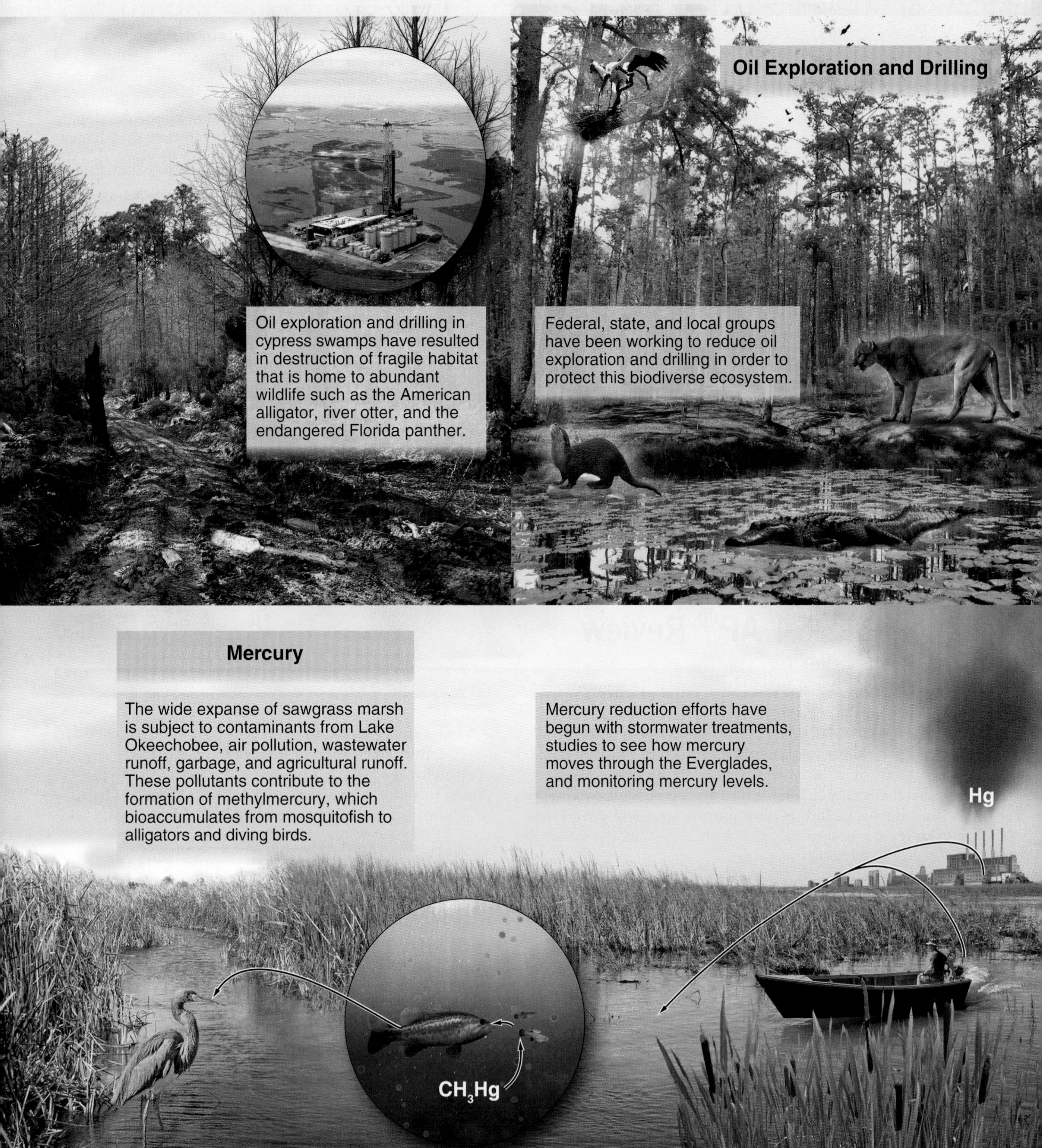

Oil Exploration and Drilling
Oil exploration and drilling in cypress swamps have resulted in destruction of fragile habitat that is home to abundant wildlife such as the American alligator, river otter, and the endangered Florida panther.
Federal, state, and local groups have been working to reduce oil exploration and drilling in order to protect this biodiverse ecosystem.
Mercury
The wide expanse of sawgrass marsh is subject to contaminants from Lake Okeechobee, air pollution, wastewater runoff, garbage, and agricultural runoff. These pollutants contribute to the formation of methylmercury, which bioaccumulates from mosquitofish to alligators and diving birds.
Mercury reduction efforts have begun with stormwater treatments, studies to see how mercury moves through the Everglades, and monitoring mercury levels.
Hg
CH3Hg

FIGURE 54.16 The Tietê River in Brazil. The Tietê River, which passes through the large city of São Paolo, was badly polluted in the 1950s but is much cleaner today. *(Pulsar Imagens/Alamy)*

Water Pollution Legislation in the Developing World

If we look at water pollution legislation around the world, there is a clear difference between developed and developing countries. Developed countries, including those in North America and Europe, experienced tremendous industrialization many decades ago and widely polluted their air and water at that time. More recently, they have addressed the problems of pollution by cleaning up polluted areas and passing legislation to prevent pollution in the future. Developing countries are still in the process of industrializing. They are less able to afford water-quality improvements such as wastewater treatment plants or the costs associated with restrictive legislation. Moreover, political instability and corruption often make enforcement of legislation difficult. In some cases, polluting industries move from developed countries to developing countries. Although the developing countries suffer from the additional pollution, they benefit economically from the additional jobs and spending that the new industries bring with them.

Water pollution problems are prevalent in many of the developing nations of Africa, Asia, Latin America, and eastern Europe. China and India, for example, have undergone rapid industrialization and have many areas of dense human population—a recipe for major water pollution problems. However, as a nation becomes more affluent, people often show more interest in the environment and resources available to address environmental issues. In Brazil, for example, industrialization began to take off in the 1950s. By the 1990s, the Tietê River, which passes through the large city of São Paulo, was badly polluted. More than a million Brazilians signed a petition in 1992 requesting that the government regulate the industrial and municipal pollution being dumped into the river. Today the Tietê River is much cleaner (**FIGURE 54.16**).

Module 54 AP® Review

Preparing for the AP® Exam

Learning Goals Revisited

54-1 How can we establish cause and effect between pollutants and human health?

Scientists examine long-term effects of chemical exposure on humans including both retrospective and prospective studies.

54-2 What are the different types of human diseases?

Human diseases are either noninfectious or infectious. Diseases can also be categorized as either acute or chronic and the risk factors for chronic diseases differ between low- and high-income countries.

54-3 What historic human pathogens have cycled through the environment?

Some infectious diseases that have a long history of harming humans include plague, malaria, and tuberculosis.

54-4 What are the major emergent infectious diseases in humans?

Some infectious diseases have emerged more recently and include HIV/AIDS, mad cow disease, swine flu, bird flu, SARS, SARS-CoV-2, MERS-CoV, West Nile virus, Zika virus disease, and Lyme disease.

54-5 What laws protect human health from pollutants and pathogens?

The Clean Water Act is designed to protect surface water from pollution, but it does not protect groundwater that serves as the source of drinking water for many people. In contrast, the Safe Water Drinking Act sets maximum levels for microorganisms, disinfectants, organic chemicals, and inorganic chemicals in drinking water. Developing countries are currently experiencing industrialization and widespread water pollution. As these countries become more affluent, it is expected that they will also turn their attention to improving water quality.

AP® Practice Questions

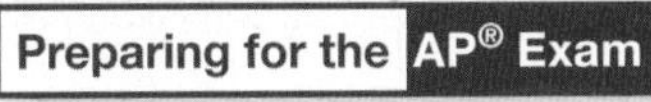

Multiple-Choice Questions

1. An infectious disease is always
   (a) caused by a virus.
   (b) transmitted between humans and animals.
   (c) treatable with antibiotics.
   (d) caused by a pathogen.

2. Tuberculosis
   (a) is transmitted by mosquitoes.
   (b) has been almost entirely eliminated in the world.
   (c) is caused by a virus.
   (d) has strains that have developed resistance to antibiotics.

3. The major legislation that protects drinking water is
   (a) the Federal Water Pollution Control Act.
   (b) the Clean Water Act.
   (c) the Safe Drinking Water Act.
   (d) the Resource Conservation and Recovery Act.

4. Maximum contaminant levels for groundwater were set in
   (a) the Clean Water Act.
   (b) the Safe Drinking Water Act.
   (c) the Resource Conservation and Recovery Act.
   (d) the Water Quality Act.

5. Which water quality issue is not covered in existing water quality legislation?
   (a) organic chemicals
   (b) offshore drilling
   (c) nonpoint sources
   (d) groundwater

6. Which of the following can contribute to poor water quality in developing countries?
   (a) sparse population density
   (b) rapid industrialization
   (c) high unemployment
   (d) relocation of industry to developed nations

Free-Response Question

The maps below illustrate the differences in the total population of the mosquito species *Aedes aegypti* in 1950 and 2000, with projections for 2050. Net decreases in population are shown in blue and green, and net increases in population are shown in orange and red.

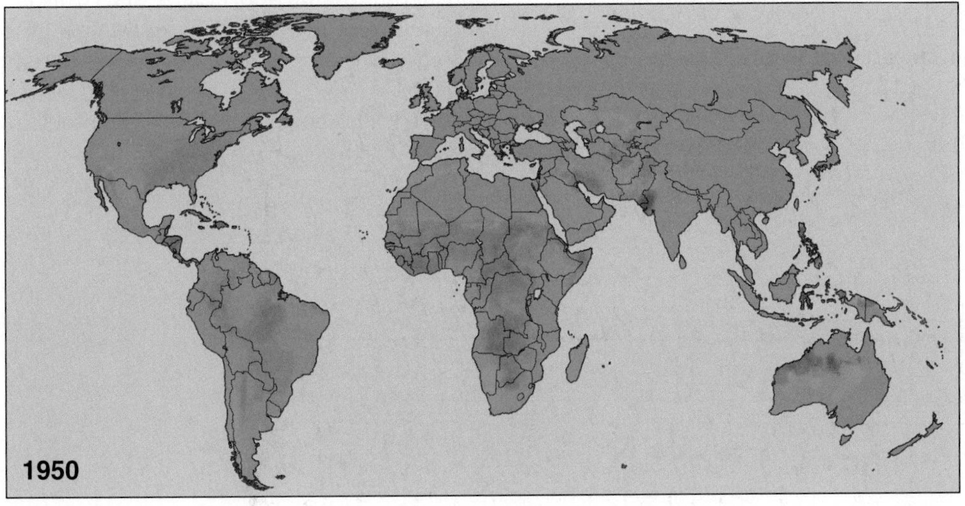

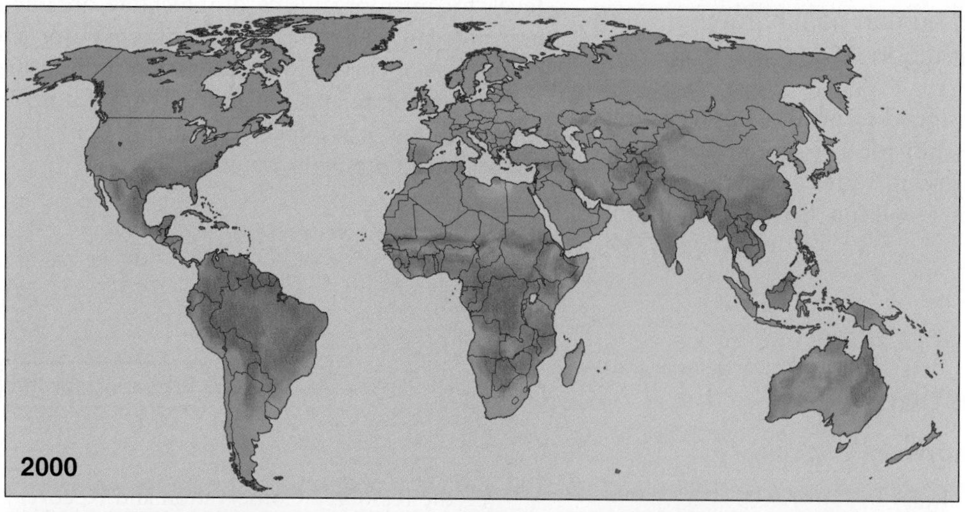

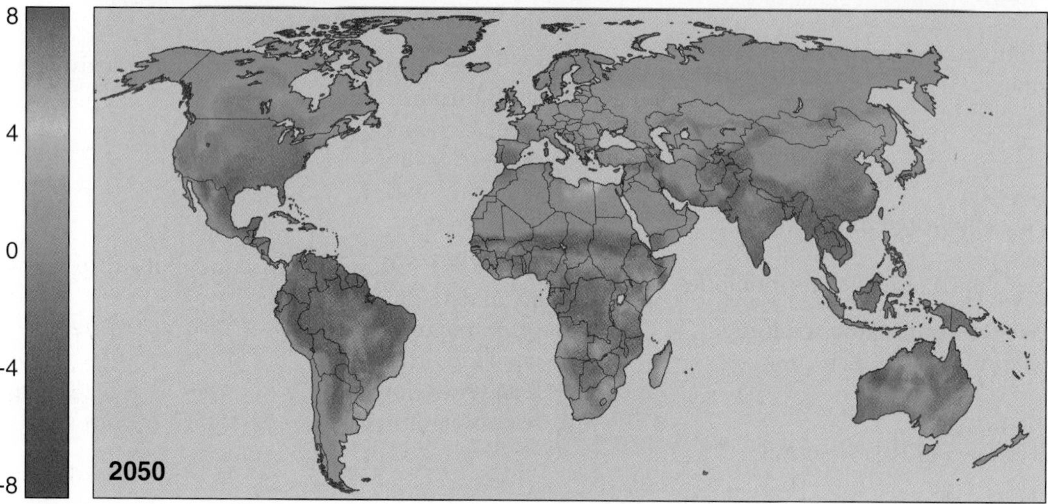

(a) **Identify** the change in the population of mosquitoes in the United States from 1950 to 2050. (1 pt.)

(b) **Describe** the overall trend in the population shift of the mosquitoes presented in these maps. (1 pt.)

(c) **Explain** one possible reason for the population shift described in part (b). (1 pt.)

(d) Global average temperatures have increased by about 1.1°C since the 1880s. Two-thirds of that change has occurred since 1975.

  (i) **Identify** a human disease that is spread by mosquitoes. (1 pt.)

  (ii) **Explain** how increasing global average temperatures can affect the prevalence of the disease identified in part (i). (2 pts.)

(e) A solution to the spread of mosquito-borne illnesses has been the extensive use of pesticides such as DDT, which was banned in 1972. **Describe** an advantage and a disadvantage to using pesticides to control the spread of mosquito-borne illnesses. (2 pts.)

(f) **Describe** an additional environmental problem associated with increasing global average temperatures. (1 pt.)

(g) **Identify** a potential solution to the problem described in part (f). (1 pt.)

UNIT 8

Practice Your Science Skills

Preparing for the AP® Exam

Data Analysis

A local stream flows from site 1 to site 4. Students tested these four sites to determine its water quality. They collected and counted macroinvertebrates and determined the diversity index and measured pH, nitrates, turbidity, and dissolved oxygen and grew bacteria from samples back in the lab. Students hypothesized that a local farm between sites 1 and 2 may be harming water quality. The results are in the following table.

Site	Macroinvertebrate Diversity Index	pH	Nitrates (ppm)	Turbidity (NTU)	Dissolved Oxygen (mg/L)	Coliform Bacteria (# colonies/100mL)
1	0.945	7.3	8.0	15.0	8.4	48
2	0.334	7.1	45.0	45.0	1.5	450
3	0.450	7.2	20.0	35.0	4.0	34
4	0.441	5.1	7.0	80.0	8.0	25

Questions

1. **Identify** which site would be the best for a control. *(Hint: Note where the farm is located and how that might impact where a control should be taken.)*
2. **Make a claim** about whether the students' hypothesis was correct and explain why it was correct. *(Hint: A claim question must have evidence to back it up.)*
3. Students found some unexpected results at site 4. **Describe** what could have caused the drop in pH or the increase in turbidity. Explain your answer. *(Hint: This question integrates information about impacts of mining on water quality from Unit 5 with water quality in Unit 8.)*
4. **Describe** how you would modify the experiment to strengthen the results.

Pursuing Environmental Solutions

Purifying Water for Pennies

Nearly 1 billion people drink unsafe water that contains soil sediments, pesticides, heavy metals, and disease-causing organisms. The problem of unsafe drinking water is largely a problem of developing countries that lack the financial resources to build proper sewage and water treatment facilities. Any solution to this challenge would need to be both low-cost and effective.

More than a decade ago, the Procter & Gamble Company teamed up with the Centers for Disease Control and Prevention to come up with a solution. Together they developed a powder with two components that purify water. The first component is known as a flocculant; it attaches to soil sediments, heavy metals, and pesticides that are in the water and forces them to settle out of the water. The second component is a chlorine compound that kills 99.99 percent of all harmful bacteria and viruses. To use the powder, you mix a small amount into 10 L of water and allow the flocculant to work for 5 minutes. Once the flocculant settles to the bottom, it is filtered out through a piece of cotton fabric. Waiting an additional 20 minutes allows the chlorine to kill any bacteria and viruses. In short, for a small amount of effort, a person can create 10 L of safe drinking water in just 30 minutes.

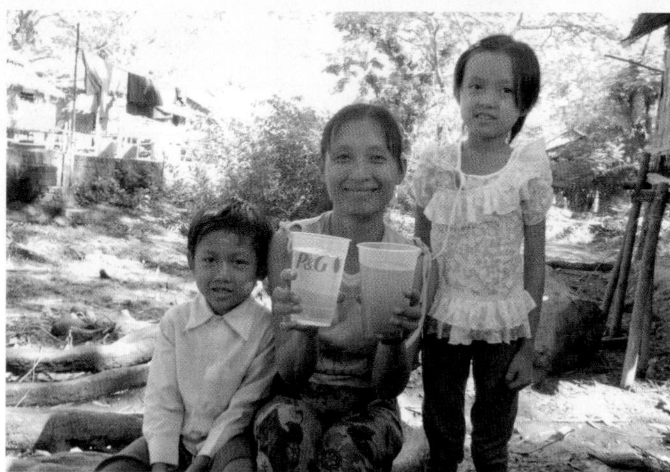

Making safe drinking water. Nearly 1 billion people do not have access to safe drinking water. Procter & Gamble partnered with many other organizations to distribute its water purifying powder technology to people in developing countries. The powder technology kills bacteria and viruses and removes parasites and solid materials. In this photo, a family poses with a cup of typical dirty water in one hand and the same water after adding the power in the other hand. *(Courtesy P&G)*

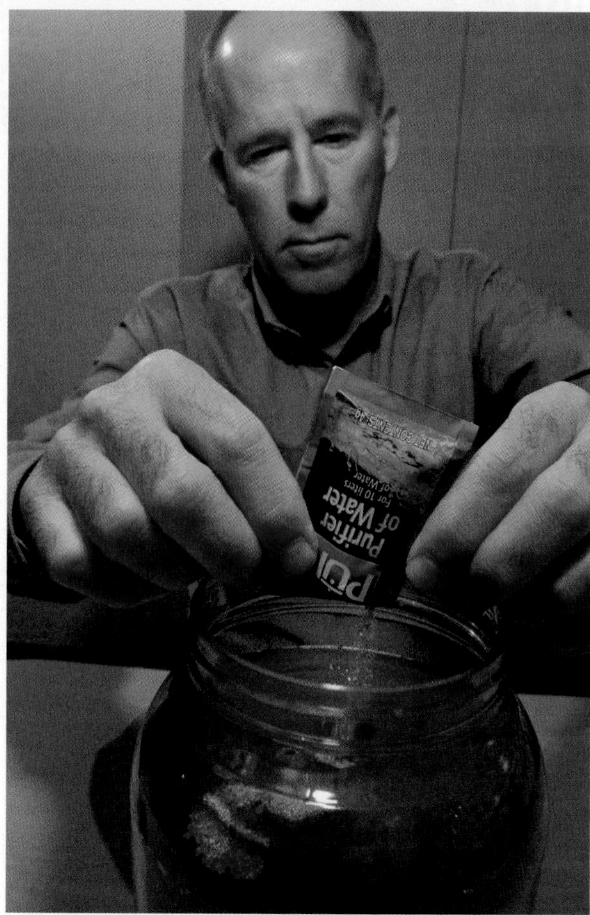

Water purifying packet. The contents of each small packet can remove soil and kill harmful organisms in 10 L of water. *(TOM UHLMAN/AP Images)*

Although Procter & Gamble typically develops products designed to make a profit, this project was different. Purified powder is needed in developing countries with populations that cannot afford to pay much even for something as essential as clean drinking water. Procter & Gamble decided to create a nonprofit organization, called the Children's Safe Drinking Water Program, that would supply the powder to people in developing countries at no profit. They created small powder packets that could clean 10 L of contaminated water but were not much bigger than a fast-food ketchup packet. The powder is inexpensive for Procter & Gamble to manufacture and the company is able to sell the packets at 3.5 cents each. Partnering with more than 150 governments and humanitarian groups including UNICEF, the company distributes the packets throughout the world. In 2021, Procter & Gamble announced that they had distributed enough packets to filter 19 billion liters of water across 93 countries and estimated that this saved tens of thousands of lives. Their current goal is to provide 25 billion liters of clean drinking water by 2025.

As the water purification program became known around the world, Procter & Gamble realized that the product they created with no expectation for profit could be sold at a profit to wealthier people who need to purify water that they collect from streams and lakes when they go hiking and camping. In 2007, the company announced that it would begin selling the packets to consumers under the brand name PUR, for $2.50 each. This price yields a considerable profit, part of which supports the goals of the nonprofit program.

The Children's Safe Drinking Water Program is an excellent example of new technologies that can be developed to be effective, inexpensive solutions to major environmental problems. It also demonstrates that such solutions can ultimately be profitable while providing a positive corporate image.

Critical Thinking Questions

1. Why is it important for a water-purifying system to contain both chlorine and a flocculant?
2. How might a company benefit when it produces a product that purifies water at no profit?

References

Deutsch, C. H. 2007. Procter & Gamble to benefit in all but name from water purifier. *New York Times*, July 23.

P&G Children's Safe Drinking Water. http://www.csdw.org/csdw/index.shtml.

P&G sets new goal to deliver 25 billion liters of clean drinking water to families in need worldwide. *Business Wire*. 2019. https://www.businesswire.com/news/home/20190318005078/en/PG-Sets-New-Goal-to-Deliver-25-Billion-Liters-of-Clean-Drinking-Water-to-Families-in-Need-Worldwide.

Water-purification plant the size of a fast-food ketchup packet saves lives. 2013. ScienceDaily, September 9. http://www.sciencedaily.com/releases/2013/09/130909092343.htm.

Science Applied 8: Concept Explanation

Is Recycling Always Good for the Environment?

As we have discussed in this unit, one of the three ways to reduce solid waste is to recycle. When we recycle items such as paper, plastic, bottles, and cans, less material ends up in landfills and fewer natural resources need to be extracted to produce these items in the future. The EPA estimates that Americans recycle 81 million metric tons (90 million U.S. tons) of trash. This represents about 35 percent of all the trash that we generate. As impressive as that sounds, it is estimated that the average household generates more than twice that amount of recyclables, so there is more work to be done.

At first glance, recycling appears to make a lot of sense both economically and environmentally. Indeed, many state and local governments have encouraged or required recycling programs and the public generally associates recycling with being good for the environment (**FIGURE SA8.1**). But what do the data tell us? When we decide to recycle, what are the measurable benefits for the environment? How do these benefits compare with benefits from other decisions we make, such as the type of car we drive? The answers to these questions may surprise you.

How do we begin to assess the benefits of recycling?

To determine the overall effect of recycling any type of waste, we need to consider the full range of costs and benefits of recycling and then compare these with the costs and benefits of manufacturing the same item from raw materials. For example, to assess the benefits of recycling paper, we need to compare the cost of recycling old paper into new paper products versus the cost of manufacturing new paper products from trees.

As we saw in Module 51, the best way to answer these questions is to complete a life-cycle analysis. Let's look at two examples: aluminum cans and plastic containers. To compare the environmental and economic costs of transporting and manufacturing these items from recycled materials versus raw materials, we begin at the manufacturing facility.

The recycling of aluminum, primarily from aluminum cans, is widespread in the United States. According to the Aluminum Association, more than 47 billion cans were recycled in the United States in 2020, which represents nearly 60 percent of all aluminum cans that were manufactured in that year (**FIGURE SA8.2**). To manufacture aluminum cans

FIGURE SA8.1 Recycling. There is increasing interest in recycling many materials. From a perspective of energy savings, some items are more important to recycle than others. *(sunsetman/Shutterstock)*

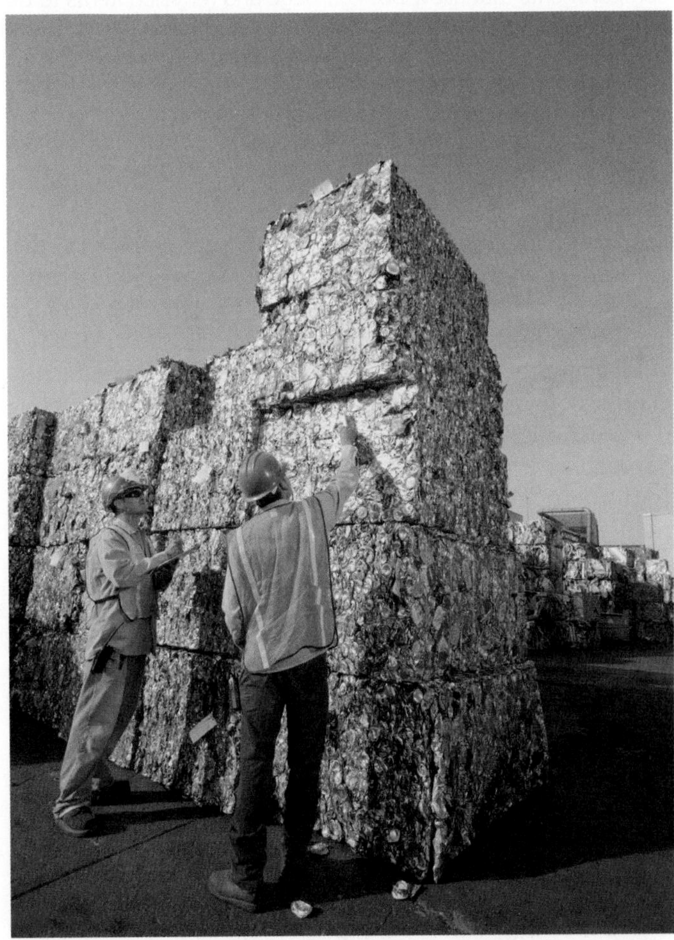

FIGURE SA8.2 Recycling aluminum cans. Converting old aluminum cans into new aluminum cans requires only 5 percent of the energy used to convert aluminum ore from a mine into new aluminum cans. *(Erik Isakson/AGE Fotostock)*

FIGURE SA8.3 Transportation costs. Although early recycling programs had garbage trucks make separate trips to pick up trash and recycling materials, modern trucks have separate compartments that allow both garbage and recycled items to be picked up at the curb in a single trip. A single trip saves time and money and reduces consumption of fossil fuels as well as the production of air pollutants. *(japatino/Getty Images)*

from raw materials, aluminum ore or bauxite must be mined and processed into pure aluminum. Not only does mining have environmental impacts as discussed in Unit 5, but this processing of aluminum from ore also takes a substantial amount of energy. In contrast, manufacturing aluminum cans from recycled cans requires only 5 percent as much energy. In short, making new cans from recycled cans saves a large amount of energy and therefore saves manufacturers a lot of money. In fact, according to the EPA, recycling 0.9 metric ton (1 U.S. ton) of aluminum cans saves about 32 barrels of oil. When this energy comes from burning fossil fuels, it also means that manufacturing recycled cans reduces the amount of carbon dioxide and other pollutants that are released into the atmosphere. Moreover, aluminum can be recycled over and over again without any loss of quality. In fact, 75 percent of all aluminum ever produced is still in circulation today thanks to recycling efforts. However, Americans still throw nearly $1 billion worth of aluminum cans into landfills each year, so there is room for improvement.

The recycling of plastic containers is also widely practiced. According to the American Chemistry Council and the Association of Postconsumer Plastic Recyclers, the recycling of plastic continues to grow with 1.3 million kilograms (2.9 million pounds) of bottles recycled in 2018, representing 29 percent of all plastic bottles that are manufactured. The cost of energy required to make new plastic bottles from raw material—in this case petroleum—is substantially less than the cost of recycling plastic bottles. As a result, manufacturing plastic bottles from recycled plastic bottles results in much smaller energy savings; washing and reprocessing plastic bottles requires nearly 50 percent of the energy required in manufacturing plastic bottles from oil. In addition, recycled plastic typically degrades in quality, so plastic from recycled bottles is typically used for products such as carpets and insulation for jackets and sleeping bags rather than new plastic bottles. This means that the economic and environmental benefits of using recycled plastic are much smaller than the economic benefits of using recycled aluminum.

What other costs of recycling do we need to consider?

Regardless of the type of material that is being recycled, we have to remember that there are several additional costs of recycling beyond the cost of energy used in manufacturing. To understand these costs, let's start at your house. If you rinse out your cans and bottles before recycling, energy is needed to get the water to your sink, particularly if you use hot water. If you use hot water to rinse out the peanut butter from a plastic peanut butter jar, for example, you are likely using more energy to clean the jar than is saved when you recycle the jar.

After they are cleaned, the materials to be recycled must be transported to a central recycling facility. Depending on location, the homeowner must either set out recycled items on the curb for pickup by a collection truck (**FIGURE SA8.3**) or bring them to a central facility. Both scenarios require burning fossil fuels for transportation. Additional fossil fuels must be consumed to transport the recycled items from the collection facility to the manufacturing facility. Although transportation costs will vary among different towns and cities, in terms of energy consumed and pollutants produced, they reduce the benefits of recycling. However, we can easily compare the cost of transporting recycled materials to manufacturers against the cost of transporting raw materials from their source, such as an aluminum mine. In addition, when homeowners pay for transporting the items to the collection center through taxes or trash collection fees, they will avoid the costs of putting the waste in a landfill.

What other benefits of recycling do we need to consider?

The primary argument for recycling is that it reduces the need for raw materials and keeps solid waste out of landfills. During the 1990s, there was a growing concern that the United States was running out of landfill space and that recycling was critical to extending the life of existing landfills. While it is true that many landfills are nearing capacity, particularly in the northeastern United States, there is still a large amount of land throughout the country that could serve as landfill space if people in those areas agreed to the construction of new landfills.

Reducing the amount of solid waste going into landfills allows existing landfills to operate longer. This, in turn, reduces the costs of closing and monitoring existing landfills. It also reduces the costs of building more landfills in the future as well as the costs of trucking the waste to new landfills likely to be farther away. Increased trucking raises both the economic cost and the environmental impact.

When we consider how we can improve the environment it is often helpful to gather the scientific data to make objective comparisons rather than simply make decisions based on perceptions. In the case of recycling, the analysis of the data makes it clear that recycling certain materials will have much greater environmental and economic benefit than recycling other materials. This helps us understand why manufacturers might be much more inclined to promote the recycling of certain items such as aluminum cans. Identifying the full range of costs and benefits also helps us identify the complexity of the question. The energy costs of recycling, for example, are wide-ranging and include homeowner costs, transportation costs, manufacturing costs, and landfill costs. By identifying all of the costs and benefits, we can strive to design more efficient recycling programs.

Questions

1. Why is it much more beneficial to recycle aluminum than plastic?
2. Why is it important to take a life-cycle approach when considering the benefits of recycling?

Practice AP® Free-Response Question

Suppose you are planning a party and want to determine the most environmentally friendly way to serve drinks to your friends. You can choose either one-use, recyclable plastic cups, or glass cups that can be kept and reused.

(a) **Identify** two factors that are likely to increase the environmental benefits of using recyclable plastic cups. (2 pts.)

(b) **Identify** two factors that are likely to decrease the environmental benefits of using recyclable plastic cups. (2 pts.)

(c) **Describe** two reasons why using plastic cups at a party may be less environmentally costly than using glass cups. (2 pts.)

(d) To reduce the amount of waste that ends up in landfills, engineers have developed plastic cups made with biodegradable plastic that are designed to be composted, where oxygen can help break the plastics down into very small pieces that can then be further degraded by microbes. However, many of these cups end up in landfills because people do not have compost piles or do not understand that they can be composted. Despite their biodegradability, these cups persist for many years in landfills. **Explain** why biodegradable plastic might persist in a landfill for a long time. (2 pts.)

(e) **Propose a solution** for one method to reduce, one method to reuse, and one method to recycle waste at your party. (2 pts.)

References

The Aluminum Association. http://www.aluminum.org/industries/production/recycling.

The Association of Plastic Recyclers. 2019. *2018 United States National Postconsumer Plastic Bottle Recycling Report*. https://plasticsrecycling.org/images/pdf/resources/reports/Rate-Reports/National-Postconsumer-Plastics-Bottle-Recycling-Rate-Reports/2018_UNITED_STATES_NATIONAL_POSTCONSUMER_PLASTIC_BOTTLE_RECYCLING_REPORT.pdf.

United States EPA. 2020. *Advancing Sustainable Materials Management: 2018 Fact Sheet*. https://www.epa.gov/sites/default/files/2021-01/documents/2018_ff_fact_sheet_dec_2020_fnl_508.pdf.

UNIT 8 Review

Key Terms to Remember

Point source (p. 549)
Nonpoint source (p. 549)
Homeostasis (p. 550)
Polychlorinated biphenyls (PCBs) (p. 556)
Neurotoxin (p. 556)
Carcinogen (p. 557)
Mutagen (p. 557)
Teratogen (p. 557)
Allergen (p. 557)
Endocrine disruptor (p. 557)
Wastewater (p. 557)
Levee (p. 564)
Dikes (p. 564)
Dam (p. 564)
Reservoir (p. 564)
Fish ladder (p. 565)
Desalination (Desalinization) (p. 568)
Distillation (p. 568)
Reverse osmosis (p. 568)
Eutrophication (p. 570)
Oxygen sag curve (p. 571)
Thermal pollution (p. 572)
Thermal shock (p. 572)
Persistence (p. 576)
Persistent organic pollutants (POPs) (p. 577)
Route of exposure (p. 579)
Solubility (p. 579)
Bioaccumulation (p. 580)
Biomagnification (p. 580)
Solid waste (p. 584)
Municipal solid waste (MSW) (p. 585)
Waste stream (p. 586)
Leachate (p. 587)
Sanitary landfill (p. 588)
Tipping fee (p. 589)
Incineration (p. 591)
Ash (p. 591)
Waste-to-energy (p. 592)
Hazardous waste (p. 593)
Superfund Act (p. 594)
Brownfields (p. 596)
Reduce, Reuse, Recycle (the three Rs) (p. 599)
Source reduction (p. 599)
Reuse (p. 600)
Recycling (p. 601)
Closed-loop recycling (p. 601)
Open-loop recycling (p. 601)
Composting (p. 603)
Life-cycle analysis (Cradle-to-grave analysis) (p. 604)
Integrated waste management (p. 606)
Biochemical oxygen demand (BOD) (p. 609)
Cultural eutrophication (p. 610)
Fecal coliform bacteria (p. 611)
Septic system (p. 612)
Septic tank (p. 612)
Sludge (p. 612)
Septage (p. 612)
Leach field (p. 613)
Dose-response study (p. 618)
Acute study (p. 618)
Chronic study (p. 618)
LD_{50} (p. 619)
Sublethal effect (p. 620)
ED_{50} (p. 620)
No-observed-effect level (NOEL) (p. 620)
Environmental hazard (p. 621)
Innocent-until-proven-guilty principle (p. 625)
Precautionary principle (p. 625)
Stockholm Convention (p. 625)
REACH (p. 626)
Retrospective study (p. 629)
Prospective study (p. 630)
Synergistic interaction (p. 630)
Disease (p. 631)
Infectious disease (p. 631)
Acute disease (p. 631)
Chronic disease (p. 631)
Epidemic (p. 633)
Pandemic (p. 633)
Dysentery (p. 633)
Plague (p. 633)
Malaria (p. 634)
Tuberculosis (p. 634)
Emergent infectious disease (p. 635)
Acquired Immune Deficiency Syndrome (AIDS) (p. 636)
Human Immunodeficiency Virus (HIV) (p. 636)
Ebola hemorrhagic fever (p. 636)
Mad cow disease (p. 637)
Prion (p. 637)
Swine flu (p. 637)
Bird flu (p. 638)
Severe acute respiratory syndrome (SARS) (p. 638)
MERS-CoV (p. 638)
SARS-CoV-2 (p. 638)
West Nile virus (p. 638)
Lyme disease (p. 639)
Zika virus disease (p. 639)
Clean Water Act (p. 640)
Safe Drinking Water Act (p. 641)
Maximum contaminant level (MCL) (p. 641)

Section 1: Multiple-Choice Questions

1. Which statement best describes nonpoint source (NPS) pollution?
   (a) NPS can only affect groundwater.
   (b) NPS is easier to control, measure, and regulate than point source pollution.
   (c) NPS includes sediment from improperly managed construction sites as a pollutant.
   (d) NPS is water pollution that originates from a distinct source such as a pipe or tank.

2. Human wastewater results in which water-pollution problem?
   (a) reduced concentrations of dissolved oxygen
   (b) reduced nutrient abundance
   (c) reduced in *E. coli* abundance
   (d) reduced algal abundance

3. Which indicates that a body of water is contaminated by human wastewater?
   (a) low BOD and a fecal coliform bacteria count of zero
   (b) high levels of nutrients, such as nitrogen and phosphorus, and high BOD
   (c) low BOD and low levels of nutrients, such as nitrogen and phosphorus
   (d) low levels of nutrients, such as nitrogen and phosphorus, and a fecal coliform bacteria count of zero

4. Sewage treatment plants use bacteria to break down organic matter. Where in this process does this process occur?
   (a) primary treatment and secondary treatment
   (b) primary treatment and chlorination
   (c) treatment only
   (d) secondary treatment only

5. Under which circumstance is a sewage treatment plant legally permitted to bypass normal treatment protocol and discharge large amounts of sewage directly into a lake or river?
   (a) when the population of the surrounding community surpasses the plant's capacity
   (b) when combined volumes of storm water and wastewater exceed the capacity of an older plant
   (c) when a permit to modernize the plant is denied by the Environmental Protection Agency
   (d) when an extended period of drought restricts water flow in a lake or river

6. Tertiary treatment of wastewater
   (a) removes pathogens.
   (b) reduces sediment.
   (c) reduces eutrophication.
   (d) reduces the amount of sludge.

7. Which inorganic substance is naturally occurring in rocks, soluble in groundwater, and toxic at low concentrations?
   (a) mercury
   (b) lead
   (c) PCBs
   (d) arsenic

8. A forest surrounding a stream is cleared to create pasture for cattle grazing. Which is likely to occur in the stream as a result?
   (a) decreased water clarity due to increased erosion and sedimentation
   (b) decreased water temperature due to reduced tree cover
   (c) eutrophication caused by increased availability of sunlight
   (d) dead zones caused by increasing heavy metal concentrations

Use the following graph to answer questions 9 & 10:

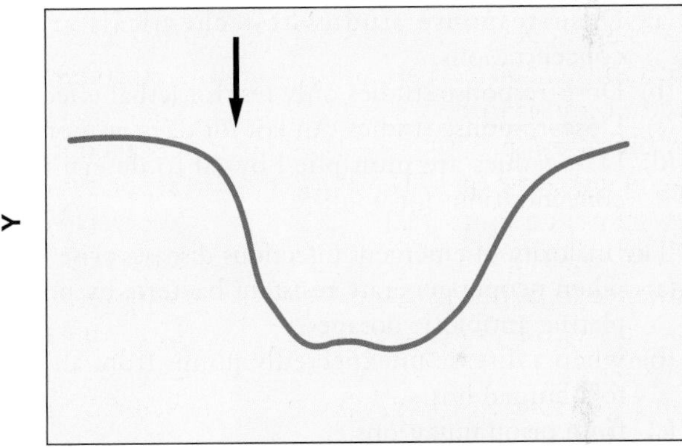

9. The graph plots variable Y as it changes over the course of a stream. The arrow indicates an area where wastewater from a sewage treatment plant is released into the stream. Which does Y most likely represent?
   (a) BOD
   (b) pH
   (c) dissolved oxygen
   (d) fecal coliform bacteria

10. Which of the following statements best describes indicators that the contamination of the stream in the graph (indicated by the arrow) is from human wastewater?
    (a) low BOD and a fecal coliform bacteria count of zero
    (b) high levels of nutrients, such as nitrogen and phosphorus, and high BOD
    (c) low BOD and low levels of nutrients, such as nitrogen and phosphorus
    (d) low levels of nutrients, such as nitrogen and phosphorus, and a fecal coliform bacteria count of zero

11. Which of the following statements about oil pollution is correct?
    (a) oil breaks down rapidly in water
    (b) oil pollution only exists on the ocean's surface
    (c) nearly half of global oil pollution comes from natural seeps
    (d) oil tankers in North America use a single-hull design

12. Which statement about the relationship between health risks and income is correct?
    (a) A major risk in high-income countries is poor sanitation.
    (b) A major risk in low-income countries is obesity.
    (c) A major risk in low-income countries is a lack of food.
    (d) The major risks in high- and low-income countries are similar.

13. Which is an example of an infectious disease transmitted by mosquitoes?
    (a) AIDS
    (b) pneumonia
    (c) malaria
    (d) leukemia

14. Which statement best describes the process of dose-response studies?
    (a) Dose-response studies test chemicals at single concentrations.
    (b) Dose-response studies only test for lethal effects.
    (c) Dose-response studies can last for days or months.
    (d) LD_{50} values are multiplied by 10 to determine safe concentrations for wildlife.

15. The majority of emergent infectious diseases arise
    (a) when people generate resistant bacteria by not completing antibiotic dosages.
    (b) when a disease unexpectedly jumps from an animal to a human host.
    (c) from prion mutations.
    (d) from sexual transmission.

16. Which host/disease pairing is correct?
    (a) Lyme disease; birds
    (b) Zika; mosquitoes
    (c) HIV/AIDS; fruit bats
    (d) Ebola; deer

17. Although thermal and sediment pollution have different sources, they both can
    (a) lead to respiratory problems in aquatic animals.
    (b) contaminate waterways with heavy metals.
    (c) lower the temperature of water.
    (d) harm fish in the open ocean.

18. Which is regulated by the EPA Clean Water Act or Safe Drinking Water Act?
    (a) stormwater runoff
    (b) sludge disposal from sewage treatment plants
    (c) nonpoint sources of oil pollution
    (d) inorganic chemicals in groundwater

19. Which factor is most likely to reduce the total amount of material that ends up as municipal solid waste in a landfill?
    (a) utilizing packing materials that are recyclable
    (b) increasing production of one-time-use materials
    (c) lowering the cost of disposable products
    (d) manufacturing products that are less durable

20. Open-loop recycling refers to
    (a) the recycling of a product into a product that will enter a different waste stream.
    (b) the recycling of a product into compost.
    (c) a process where only part of the waste is ultimately recycled.
    (d) a process where one product is recycled into a different product.

21. Which factor is least likely to contribute to a decision on where to locate a landfill?
    (a) proximity of landfill to residential neighborhoods
    (b) location of underlying aquifers
    (c) underlying soil texture
    (d) net primary productivity of habitat

Use the following table on the costs associated with landfills and incinerators to answer questions 22 & 23:

Landfill		**Incinerator**	
Transporting material to landfill	$10/ton	Transporting material to incinerator	$10/ton
Tipping fee to dump waste in landfill	$60/ton	Tipping fee to incinerate waste in incinerator and remove ash	$90/ton

22. Assume an incinerator can produce 500 kWh of energy by incinerating 1 ton of waste and can sell electricity for $0.06/kwh. What is the overall dollar cost of incineration relative to a landfill?
    (a) Incinerator is $10 more per ton.
    (b) Incinerator is $20 more per ton.
    (c) They are the same.
    (d) Incinerator is $10 less per ton.

23. Which of the following statements best describes why an incinerator charges more per ton in tipping fees?
(a) Waste in an incinerator must be transported to the site, and these are often farther away from population centers.
(b) Incinerators produce solid waste, which must be removed periodically and transported offsite.
(c) Landfills accept waste that is not accepted in an incinerator, leading to higher tipping fees.
(d) Incinerators don't produce any methane, which makes tipping fees more affordable.

24. The Comprehensive Environmental Response, Compensation, and Liability Act is referred to as "Superfund" because it
(a) imposes taxes collected from chemical and petroleum industries to fund the cleanup of abandoned hazardous waste sites.
(b) provides a fund to compensate workers exposed to hazardous situations during national emergencies such as hurricanes and floods.
(c) provides a fund to help other countries design environmental responses to emergencies involving the accidental release of hazardous waste.
(d) distributes tax revenues from various government agencies to chemical and petroleum industries.

25. Given the mortality curve in the graph for a rat, what would be a safe dosage of caffeine for a rat?

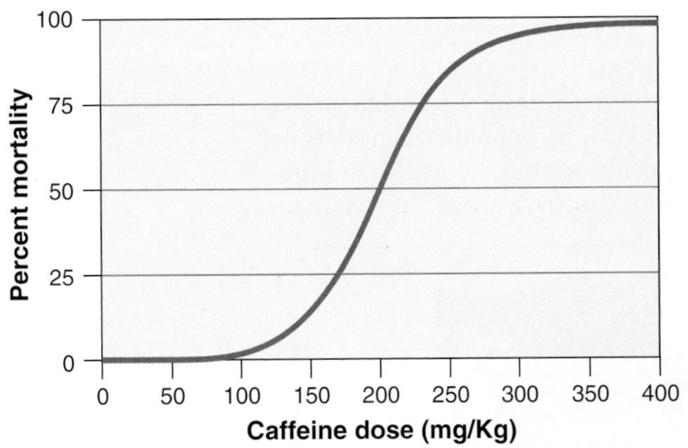

(a) 10 mg
(b) 20 mg
(c) 40 mg
(d) 100 mg

26. Under the Safe Drinking Water Act, the EPA is responsible for
(a) regulating pollution caused by fracking for natural gas.
(b) setting maximum contaminant levels.
(c) mitigating the effects of cultural eutrophication.
(d) eradicating fecal coliform from all surface waters.

27. What is usually the best solution to household production of hazardous waste?
(a) hazardous waste landfills
(b) product substitution
(c) source separation
(d) reuse and recycling

Use the following passage to answer questions 28 & 29:

Nearly 1 billion people in the world drink unsafe water that contains soil sediments, pesticides, heavy metals, and disease-causing organisms. The problem of unsafe drinking water is largely a problem of developing countries that lack the financial resources to build proper sewage and water treatment facilities. Any solution to this challenge would need to be both low-cost and effective.

More than a decade ago, the Procter & Gamble Company teamed up with the Centers for Disease Control and Prevention to come up with a solution. Together they developed a powder with two components that purify water. The first component is known as a flocculant; it attaches to soil sediments, heavy metals, and pesticides that are in the water and forces them to settle out of the water. The second component is a chlorine compound that kills 99.99 percent of all harmful bacteria and viruses. To use the powder, you mix a small amount into 10 L of water and allow the flocculant to work for 5 minutes. Once the flocculant settles to the bottom, it is filtered out through a piece of cotton fabric. Waiting an additional 20 minutes allows the chlorine to kill any bacteria and viruses. In short, for a small amount of effort, a person can create 10 L of safe drinking water in just 30 minutes.

28. According to the passage, how long will it take for Procter & Gamble's water purification powder to filter and disinfect contaminated water?
(a) 5 minutes
(b) 10 minutes
(c) 20 minutes
(d) 25 minutes

29. According to the passage, which is the best description of the purpose of Procter & Gamble's powder to purify drinking water?
(a) Their purpose was to make as much profit as possible in the developing world.
(b) Their purpose was to show the Centers for Disease Control that they were unable to help solve this global issue.
(c) Their purpose was to help provide clean drinking water to those without access to it.
(d) Their purpose was to show developed nations how to disinfect drinking water without the need for water treatment plants.

Section 2: Free-Response Questions

1. A research team collected data from two polluted streams, measuring pH, biological oxygen demand (BOD), fecal coliform bacteria, and aluminum. Their results are presented in the table below.

	Stream A	Stream B
pH	7.5	3.0
BOD	230 mg	15 mg
Aluminum	0.001 mg/L	1.0 mg/L
Fecal coliform	100 colonies/mL	10 colonies/mL

(a) **Identify** the stream with the most acidic water. (1 pt.)

(b) **Describe** a pollutant that is likely affecting Stream A. (1 pt.)

(c) Stream A flows into a local pond. **Explain** the likely change in dissolved oxygen concentration in that pond as a result of the pollution described in part (b). (2 pts.)

(d) **Identify** a likely cause of the pollution affecting Stream B. (1 pt.)

(e) **Describe** a potential environmental problem that may be caused if Stream B overflowed beyond its banks into the surrounding soil. (1 pt.)

(f) Researchers have monitored the population of northern leopard frogs in and around Stream A since 1990. In 2004, a farm near the stream changed their pesticide use practices. The data of the researchers' study is shown in the table below.

Northern Leopard frog population (frogs/square mile)	Year
2,000	1990
2,050	1995
2,030	2000
1,700	2005
1,500	2010
900	2015
600	2020

(i) **Identify** a possible research question for the study in Stream A. (1 pt.)

(ii) **Identify** the dependent variable or the study in Stream A. (1 pt.)

(iii) **Identify** a variable not mentioned that could affect the results of the study in Stream A. (1 pt.)

(iv) **Make a claim** using evidence to explain the decline in frog population beginning in 2005. (1 pt.)

2. The Earth is covered with water, but only about one percent of that water is readily available surface fresh water. On the surface, the availability of water around the world can be unpredictable, yet water is critical for the survival of living things. Water availability varies greatly around the world, and the United Nations estimates that nearly 1.2 billion people live in regions that have a scarcity of water.

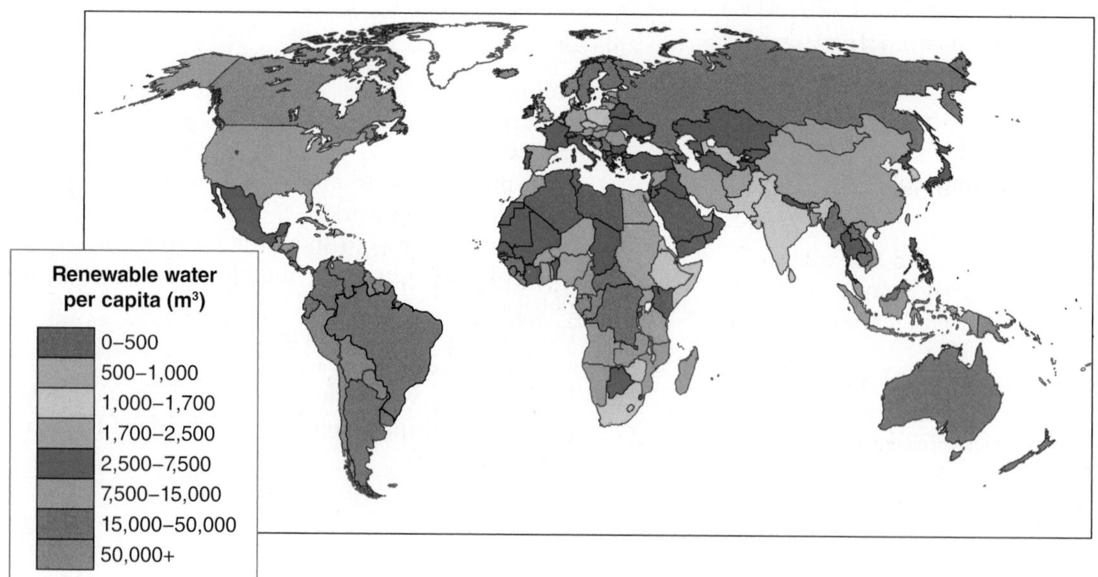

(a) **Identify** the amount of renewable water per capita in the United States. (1 pt.)

(b) **Describe** why the amount of water available in the United States per capita varies widely. (1 pt.)

(c) **Explain** the difference in renewable water per capita in China as compared to that of the United States, even though they have similar total amounts of renewable freshwater sources available. (1 pt.)

(d) To help combat water scarcity, humans have learned to divert surface water to fit our needs. **Describe** one specific way in which humans divert fresh surface water to suit our needs. (1 pt.)

(e) For the water diversion method described in part (d), **describe** one potential benefit and one potential consequence of this method for diverting surface water. (2 pts.)

Recycling water has been recently considered for use in the United States as a solution to a diminishing surface water supply. After being used by humans in their homes, this water is filtered and disinfected at a treatment plant. Flowing through purple pipes so plumbers can distinguish it from other utility lines, states have begun to install systems to use recycled water in a variety of ways. Municipalities use it to water golf courses, parks, school yards and road medians, fill lakes and enhance natural wetlands, and to fight fires. Energy producers can substitute it for fresh water in cooling towers and manufacturers can use recycled water throughout the production process. In the Northeast, reclaimed water is sometimes used to make snow at ski resorts.

(f) **Describe** one advantage and one disadvantage to using recycled/reclaimed water as an alternative to surface water. (2 pts.)

(g) **Identify** one pollutant that can still be present in disinfected water treated at a sewage treatment plant that can cause eutrophication if released into surface water. (1 pt.)

(h) **Propose a solution** to reduce the pollutant identified in part (g). (1 pt.)

3. The data below was collected by the Maryland Department of Natural Resources and the Chesapeake Bay Program, showing the trends in the population of blue crabs from 1990 to 2017.

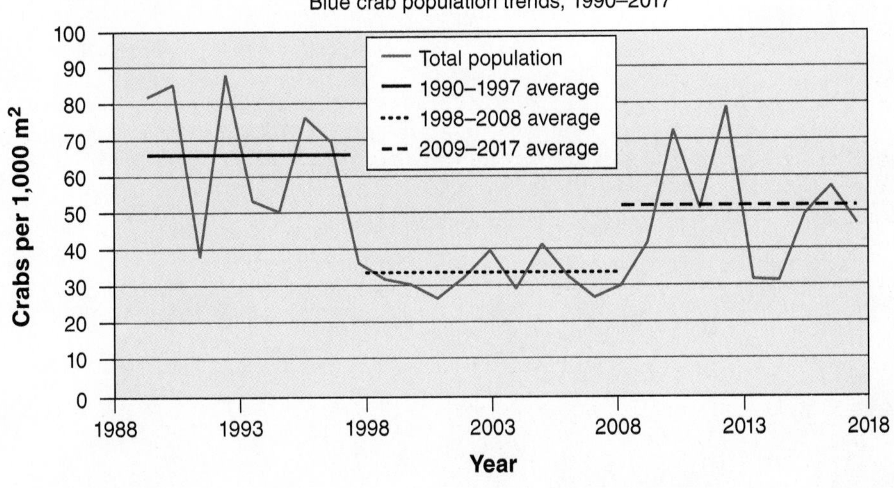

(a) **Identify** the year in which Maryland blue crabs were approximately 30 per 1,000 square meters. (1 pt.)

(b) **Describe** the overall trend in blue crab population from 1990 to 2017. (1 pt.)

(c) **Explain** one potential reason for the change in blue crab population from 1997 to 2007. (1 pt.)

(d) **Describe** one possible way to reduce the effects of the pollutant identified in part (d). (1 pt.)

(e) **Calculate** the percent change in the average crab population between 1990 and 1997 and 2010 and 2017. Show all work. (2 pts.)

(f) The Chesapeake Bay has an area of 4,479 square miles. **Calculate** the area of the Bay in hectares, assuming the following. Show all work. (2 pts.)
- 1 square mile is equal to 2.6 square kilometers
- 1 square kilometer is equal to 100 hectares

(g) If 1 hectare is equal to 10,000 square meters, **calculate** approximately how many blue crabs were present in the Chesapeake Bay, on average, in 2005. Show all work. (2 pts.)

UNIT 9

Global Change

Global warming in the Great Barrier Reef of Australia has caused a dramatic increase in the proportion of female green sea turtles. *(Georgette Douwma/Getty Images)*

CASE STUDY

Sea Turtle Responses to a Warming World

When we think about global warming, we typically think about small increases in the average temperature of Earth over decades, causing glaciers to melt and sea levels to rise. Given that the temperature increases are small, we might expect plants and animals to be minimally affected. It was therefore shocking when researchers discovered that warming temperatures were causing green sea turtles (*Chelonia mydas*) to change their sex from male to female.

In 2018, researchers reported that they had examined the sex of green sea turtles that were feeding around the Great Barrier Reef of Australia. They used genetic markers that told them whether the turtles had originally hatched from populations living in the cooler southern beaches or in the warmer northern beaches around the reef. In this species, turtles from the northern and southern populations can be found feeding together, but when the turtles are ready to breed, they go back to the place where they hatched. While most animal populations are comprised of about 50 percent females and 50 percent males, the green sea turtles from southern beaches had a modest bias, with 65 to 69 percent females. However, those from the northern beaches had an extreme bias; juvenile turtles were 99.1 percent female, subadult turtles were 99.8 percent female, and adults were 86.8 percent female. Given that these turtles take 25 years to reach adulthood, these data suggest that the cause of the most extreme bias, found in the juveniles and subadults, has occurred in the past 25 years.

An increase in global temperatures of just a few degrees can have major effects on the species experiencing these temperatures.

How did these northern turtles come to produce so many females? It turns out that in some species of reptiles and fish, an individual's sex is not determined by the genes, such as the X and Y chromosomes that determine sex in humans. Instead, sex is determined by temperature. In the case of green sea turtles, if the hatchlings develop in sandy nests on beaches with warmer temperatures, they become female. If they experience cooler nest temperatures, they develop into males. When the researchers examined weather records for the southern and northern nesting regions of Australia, they found that the threshold temperature for creating females was consistently exceeded in the northern populations starting in 1990. Such biased sex ratios are of great concern to researchers; as the 99 percent female juveniles and subadults become breeding adults over the next 2 decades, they will likely have a difficult time finding males with which they can breed in the northern nesting sites.

The problem of warming global temperatures altering sex ratios is not limited to the green sea turtles. In New Zealand, researchers have been examining the sex ratios of a lizard-like reptile known as the tuatara (*Sphenodon punctatus*). The sex ratios of tuataras are also temperature dependent, but in this species warmer nest temperature cause more male offspring to be produced. During surveys of the tuataras from 1988 to 1998, researchers found that the populations were modestly biased with 62 percent males. Subsequent surveys from 2005 to 2012 found that the bias had increased to 70 percent male.

Such a biased sex ratio not only reduces the number of females that can lay eggs in the future, but it also causes the males to compete more with each other and to harass females more as they try to court them. During a 24-year period, researchers found that increased competition for females caused a decline in body condition among both males and females. Based on these results, the researchers examined worst-case scenarios for global warming in New Zealand, which is a 3.3°C to 4°C temperature increase. Under this scenario, they predict that the tuatara hatchlings will become 100 percent male and the population will go extinct. As in the case of the green sea turtles, an increase in global temperatures of just a few degrees can have major effects on the species experiencing these temperatures.

Sources: K.L. Grayson et al., Sex ratio bias and extinction risk in an isolated population of tuatara (*Sphenodon punctatus*), *PLOS ONE* 9 (2014): e94214; M.P. Jensen et al., Environmental warming and feminization of one of the largest sea turtle populations in the world, *Current Biology* 28 (2018):154–159.

Practice your science skills

1. **Concept Explanation:** Describe how global temperature change is altering the sex ratio of certain species' populations.
2. **Data Analysis:** The case study states most animal populations are about 50 percent females and 50 percent males. Researchers discovered green sea turtles from southern beaches contained 65 to 69 percent females, but northern beaches contained adult green sea turtles that were 86.8 percent female. What do the data imply about global warming?
3. **Environmental Solutions:** Propose a solution to avoid the extinction of green sea turtle populations from unbalanced sex ratios due to global climate change.

The temperature-induced change in reptile sex ratios is just one example of many changes taking place on Earth over the past few decades. In this unit, we will examine how humans have altered the world's atmosphere and climate as well as the underlying causes of these changes. We will also investigate the consequences of these changes for humans, and other species, and consider predictions of future consequences. This unit ties together many of the themes we have developed throughout the book: the interconnectedness of systems on Earth, the indicators that enable us to measure and evaluate the environmental status of Earth, and the interaction of environmental science and policy.

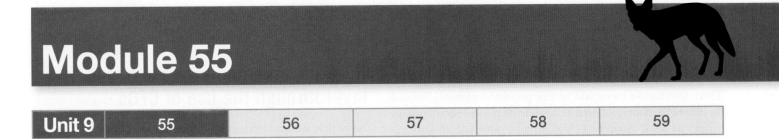

Module 55

Stratospheric Ozone Depletion and Its Reduction

In Unit 7, we saw that ground-level pollution contributes to a number of problems in the natural world, exacerbates asthma and breathing difficulties in humans, and contributes to the incidence of cancer. Here, we turn to the effects of pollutants in the stratosphere that have a substantial impact on the health of humans and ecosystems. As you may recall from Unit 7, ozone (O_3) in the troposphere—the layer of atmosphere closest to the ground—acts as an oxidant that can harm respiratory systems in animals and damage a number of structures in plants. However, ozone in the stratosphere forms a necessary, protective shield against radiation from the Sun; it absorbs ultraviolet light and prevents harmful ultraviolet radiation from reaching Earth (**FIGURE 55.1**).

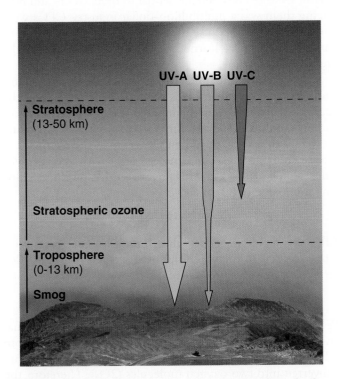

FIGURE 55.1 Stratospheric ozone concentration. Surface-level ozone in the troposphere – a product of phytochemical oxidants such as nitrogen oxides and sulfur dioxide – can cause human health problems. However, natural ozone in the stratosphere is highly beneficial because it filters out harmful ultraviolet radiation, especially UV-B and UV-C radiation.

Learning Goals

After reading this module you should be able to

55-1 explain how stratospheric ozone forms and the benefits it provides.

55-2 identify the cause of depleted stratospheric ozone.

55-3 describe efforts made to reduce ozone depletion.

55-1 How does stratospheric ozone form and what benefits does it provide?

Stratospheric ozone is beneficial to the health and survival of life on Earth

To understand the importance of stratospheric ozone to life on Earth, we need to first understand how it forms. With this background, we can then examine the benefits it provides and how it is being depleted as a result of human activities.

Formation of Stratospheric Ozone

How does ozone form in the stratosphere? When solar radiation strikes O_2 in the stratosphere, 13 to 50 km (8–31 miles) above Earth's surface, a series of chemical reactions produces ozone. In the first step, UV-C radiation breaks the molecular bond holding an oxygen molecule together:

$$O_2 + \text{UV-C} \rightarrow O + O$$

This happens to only a small percentage of oxygen molecules at any given time. The vast majority of the oxygen in the atmosphere remains in the form O_2.

In the second step, a free oxygen atom (O) produced by the first reaction encounters an oxygen molecule, and they combine to form ozone:

$$O + O_2 \rightarrow O_3$$

Both UV-B and UV-C radiation can break a bond in this new ozone molecule, forming molecular oxygen and a free oxygen atom once again:

$$O_3 + \text{UV-B or UV-C} \rightarrow O_2 + O$$

Thus, formation of ozone in the presence of sunlight and its subsequent breakdown is a cycle that can occur indefinitely as long as there is UV energy entering the atmosphere. Under normal conditions, the amount of ozone in the stratosphere remains relatively constant.

Benefits of Stratospheric Ozone

We care about ozone in the stratosphere because it filters out harmful ultraviolet (UV) radiation. As you may know, the Sun radiates energy at many different wavelengths, including wavelengths in the ultraviolet range (see Figure 6.4). The UV wavelengths are further classified into three groups: UV-A, or low-energy ultraviolet radiation, and the shorter, higher-energy UV-B and UV-C wavelengths. All three types of UV radiation can damage the tissues and DNA of living organisms. Exposure to UV-B radiation increases the risk of skin cancer and cataracts; it also suppresses the human immune system. In addition, exposure to UV-B is harmful to plant cells and reduces their ability to convert sunlight into usable energy by photosynthesis. UV-B exposure can therefore harm entire biological communities. Should high levels of UV-B cause a decline of phytoplankton—the microscopic algae that form the base of many marine food chains—it will subsequently cause the depletion of fisheries that are further up the food chain. As you can see, having a layer of ozone in the stratosphere that can filter out these harmful rays plays an important role in the health and survival of species on Earth.

It is easy to confuse stratospheric ozone with tropospheric, or ground-level, ozone that we discussed earlier in Unit 7 because it *is* the same gas, O_3. However, stratospheric ozone occurs higher in the atmosphere where its ability to absorb ultraviolet radiation and thereby shield the surface below makes stratospheric ozone critically important to life on Earth.

AP® Exam Tip

The AP® Environmental Science Exam regularly asks questions about the difference between stratospheric ozone and ground level (tropospheric) ozone. Remember that tropospheric ozone is a pollutant, and stratospheric ozone is beneficial.

Chlorofluorocarbons (CFCs) Chemical that can be used for cooling refrigerators and air conditioners.

55-2 What has caused the depletion of stratospheric ozone?

Humans have depleted the ozone layer through the use of CFCs

Now that we understand how stratospheric ozone is formed and its benefits, we can explore how ozone can be destroyed chemically and the role that humans have played in destroying ozone in the stratosphere.

The Chemical Reaction Between CFCs and Ozone

To understand how ozone can be destroyed in the stratosphere, we need to examine a group of chemicals known as **chlorofluorocarbons**, or **CFCs**, which are chemicals that can be used for cooling refrigerators and air conditioners. As you likely appreciate, we rely on refrigeration to keep our foods safe and edible; we depend on air conditioning to keep us comfortable in hot weather. Once we started using CFCs as refrigerants starting in the 1920s, we rapidly increased their use over the subsequent decades. CFCs were also used in a host of other consumer items, including aerosol spray cans and products such as Styrofoam. These chemicals were considered essential to modern life and producing them was a multibillion-dollar industry. CFCs were considered "safe" because they are both nontoxic and nonflammable. Later we learned that these chemicals cause the breakdown of ozone.

The widespread manufacturing of CFCs for refrigerants and aerosols meant that CFCs were getting released into the air, both when they were being used and when refrigerators and air conditioners were thrown away without removing the CFCs they contain. When CFCs are in the air, they can release chlorine (Cl). Chlorine can react with ozone molecules, which causes a chemical reaction that produces chlorine monoxide (ClO) and O_2:

$$O_3 + Cl \rightarrow ClO + O_2$$

After this reaction occurs, chlorine monoxide reacts with a free oxygen atom, which pulls the oxygen from the ClO, producing free chlorine again:

$$ClO + O \rightarrow Cl + O_2$$

When we consider these two reactions together, we see that chlorine starts as a free atom (Cl) and ends as a free atom (Cl). In contrast, ozone (O_3) and a free oxygen atom (O) are converted into two oxygen molecules ($2O_2$). Therefore, the chlorine atom does not get used up. In fact, a single chlorine atom can break down as many as 100,000 ozone molecules until finally one chlorine atom finds another chlorine atom and the process is stopped.

> **AP® Exam Tip**
>
> Make sure you understand the source of ozone depletion and can outline the chemical reactions involved in the formation and destruction of stratospheric ozone.

CFC Destruction of the Ozone Layer

Now that we understand how CFCs react with ozone, we can examine whether this reaction causes a depletion of ozone in the stratosphere, thereby allowing more UV-B radiation to reach Earth's surface and cause harm to organisms. In the mid-1980s, atmospheric researchers noticed that stratospheric ozone in Antarctica was decreasing each year, beginning in about 1979. Since the late 1970s, global ozone concentrations had decreased by more than 25 percent. You can see an example of this in the data collected by NASA from multiple locations in the Southern Hemisphere, which show a decreasing trend of ozone concentrations from 1980 through 2020 (**FIGURE 55.2**).

Researchers determined that ozone depletion in the stratosphere was occurring worldwide, but it was greatest at the poles. In the Antarctic, ozone depletion is seasonal; each year the declines occur from August through November (late winter through early spring in the Southern Hemisphere). This caused an area of severely reduced ozone concentrations over most of Antarctica, creating what has come to be called the "ozone hole."

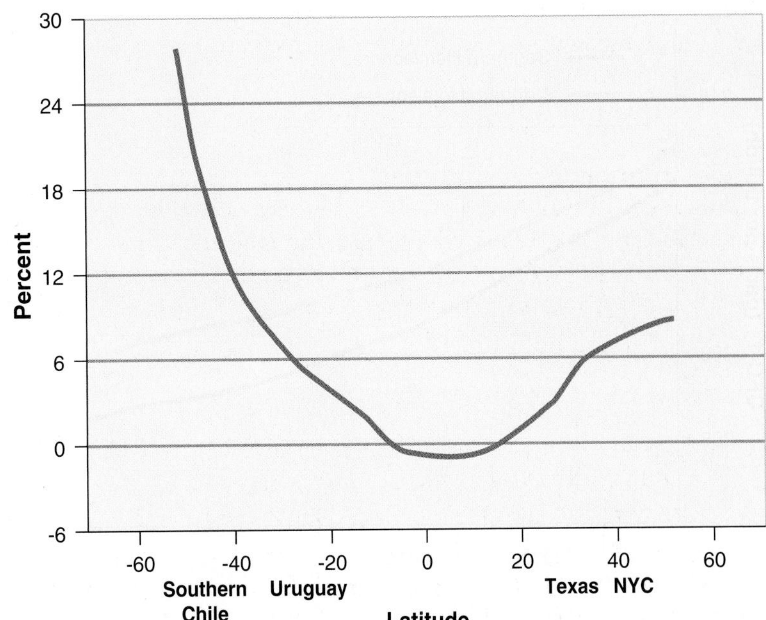

FIGURE 55.3 Increases in UV radiation at different latitudes. Using data collected from 1979 to 2009, researchers discovered that the largest percent increases in UV radiation were happening near Antarctica. *(Data from NASA's Goddard Space Flight Center/Jay Herman NASA's Goddard Space Flight Center/Jay Herman.)*

A depletion of ozone also occurs over the Arctic in January through April, but it is not as severe.

The cause of the ozone hole near Antarctica has been studied intensively and is quite complex. A key cause of this ozone depletion is the production of CFCs. When the CFCs are exposed to ultraviolet radiation from the Sun, the molecules are broken up and Cl atoms are released. These Cl atoms then react with ozone molecules (O_3), converting them into O_2 molecules, in a reaction that occurs under springtime weather conditions. Thus, the increased manufacturing of CFCs causes an increase in chlorine atoms in the stratosphere, which then causes the destruction of ozone in the stratosphere each spring.

Given what we now know about the protective layer in the ozone in the stratosphere, it is perhaps not surprising that decreased stratospheric ozone has led to a rise in the amount of UV-B radiation that reaches the surface of Earth. A study by NASA researchers showed the tropics have only experienced a small increase in UV-B radiation. In mid-latitudes of North and South America—the latitudes of Texas and Uruguay—UV radiation at the surface of Earth increased about 6 percent between 1979 and 2008 (**FIGURE 55.3**). Since 2008, this radiation in North America has been relatively constant. Closer to Antarctica, however, there has been a 25 percent increase in UV-B radiation. As a result, countries near the Antarctic ozone hole such as Chile and Australia have observed significant increases in people diagnosed with skin cancers.

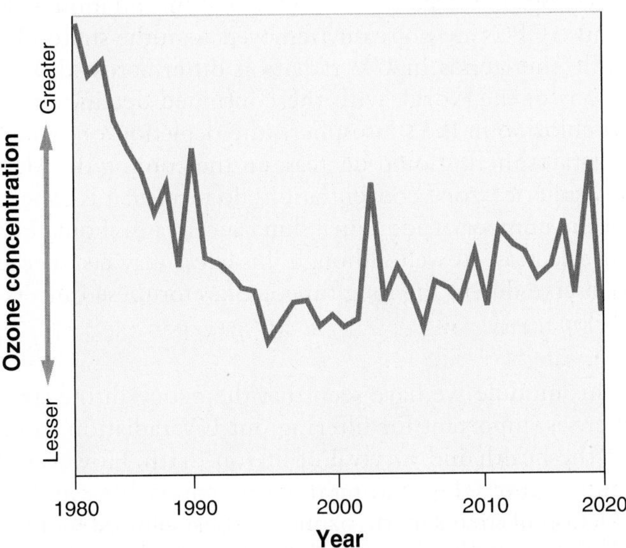

FIGURE 55.2 Declining stratospheric ozone concentration. These data are from the Southern Hemisphere and show a decreasing trend from 1970 to 2020. *(Data from NASA Ozone Watch, https://ozonewatch.gsfc.nasa.gov/)*

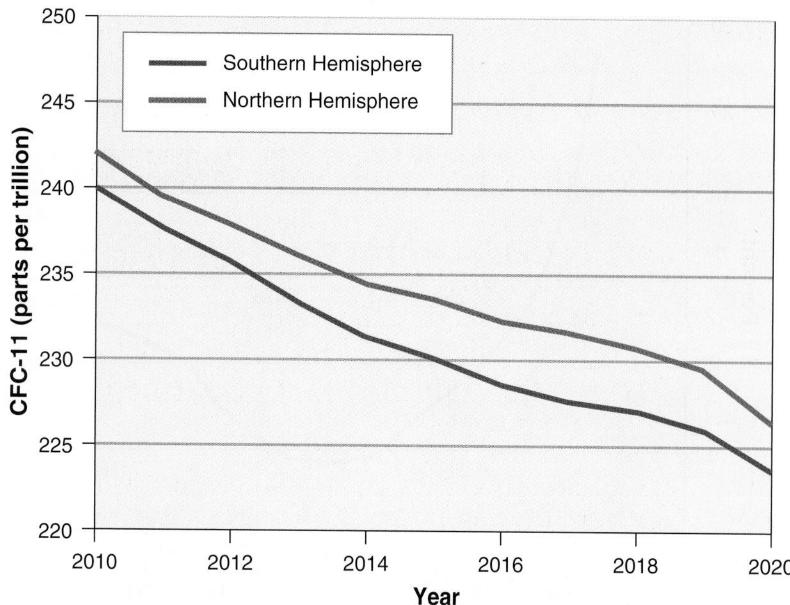

FIGURE 55.4 Global declines of CFCs. From 2010 to 2020, the world has witnessed a decline in CFCs in the atmosphere, both in the Northern and Southern Hemispheres. *(Data from Montzka, S.A., Dutton, G.S., Portmann, R.W., et al. A decline in global CFC-11 emissions during 2018–2019.* Nature *590 (2021): 428–432.)*

55-3 What efforts have been made to reduce ozone depletion?

Nations around the world have agreed to reduce CFC production to reduce ozone depletion

In response to the decrease in stratospheric ozone, 24 nations in 1987 signed the **Montreal Protocol** on Substances That Deplete the Ozone Layer. This agreement was a commitment to reduce CFC production by 50 percent by the year 2000. It was the most far-reaching environmental treaty to date, in which global CFC exporters like the United States prioritized the protection of the global biosphere over their short-term economic self-interest. More than 180 countries eventually signed a series of increasingly stringent amendments that required the elimination of CFC production and use in the developed world by 1996. In total, the protocol addressed 96 ozone-depleting compounds. Since that time, industries have developed new types of refrigerants for refrigerators and air conditioners and new propellants for aerosol cans that do not cause the destruction of stratospheric ozone. One group of alternative chemicals that do not deplete the ozone layer are known as hydrofluorocarbons (HFCs). Unfortunately, these chemicals act as potent greenhouse gases. Researchers are currently working to identify chemicals that can work effectively as refrigerants and not act as greenhouse gases. For example, ammonia gas is a good alternative gas for some types of refrigerators.

Because of these efforts, the concentration of CFCs in the stratosphere has been declining (**FIGURE 55.4**). As you can see in the figure, both the Northern and Southern Hemispheres have experienced a 7 percent decline from 2010 to 2020. In addition, the concentration of chlorine in the stratosphere, which peaked at about 4 ppb, is now decreasing. The reduction in chlorine concentration is slow because CFCs are not easily removed from the stratosphere. While the trends in UV radiation differ across different regions of the world, with the continued decline of CFS and chlorine in the stratosphere, the depletion of ozone in the stratosphere should decrease in the coming decades. If stratospheric ozone concentrations do recover, it is expected that the number of additional skin cancer cases should eventually decrease as well, although this effect may take decades to observe due to the long time it takes for these cancers to develop.

Montreal Protocol A commitment by 24 nations to reduce CFC production by 50 percent by the year 2000.

In this module we have seen that the ozone in the stratosphere is important for filtering out UV radiation to protect the health and survival of life on Earth. However, the production of CFCs during the past century has caused the depletion of stratospheric ozone. This has allowed more UV radiation to strike the planet's surface, which can lead to a variety of impacts on humans and ecosystems. In the next module we will examine how other human activities have led to global climate change.

Module 55 AP® Review

Preparing for the AP® Exam

Learning Goals Revisited

55-1 How does stratospheric ozone form and what benefits does it provide?

Stratospheric ozone forms when UV-C radiation breaks apart an oxygen molecule (O_2) into two free oxygen atoms (O). A free oxygen atom (O) can then react with an oxygen molecule (O_2) to produce ozone (O_3). These ozone molecules absorb harmful ultraviolet radiation, thereby preventing much of the UV radiation from reaching ground level where it could be harmful to plants and animals, including humans.

55-2 What has caused the depletion of stratospheric ozone?

The introduction of human-synthesized chlorofluorocarbons led to increased amounts of chlorine in the stratosphere, leading to destruction of stratospheric ozone. This caused an increase in UV-B radiation in certain locations on Earth.

55-3 What efforts have been made to reduce ozone depletion?

Since more than 100 nations signed the Montreal Protocol on Substances That Deplete the Ozone Layer, CFC concentrations have declined, and stratospheric ozone depletion has begun to slow. Some alternative gases are being used that do not destroy ozone, but they can act as greenhouse gases. The current challenge is to find gases that do not cause either of these harmful impacts.

AP® Practice Questions

Preparing for the AP® Exam

Multiple-Choice Questions

Use the image below to answer questions 1 & 2:

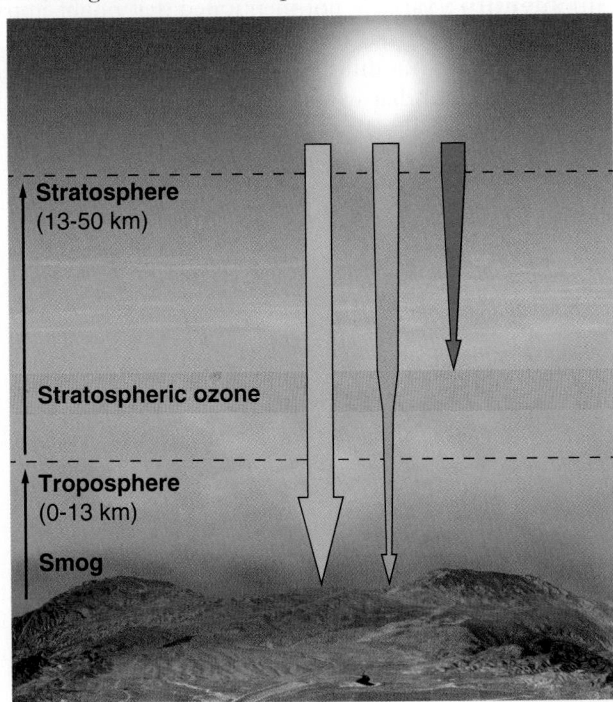

1. The ozone layer blocks various types of UV radiation. Which UV is completely blocked by the ozone layer, as shown above?
   (a) UV-A
   (b) UV-B
   (c) UV-C
   (d) UV-D

2. Which type of UV is allowed to partially pass through the ozone layer, typically causing sunburn?
   (a) UV-A
   (b) UV-B
   (c) UV-C
   (d) UV-D

3. The formation of ozone begins when an O_2 molecule is split by
   (a) UV-A radiation.
   (b) UV-B radiation.
   (c) UV-C radiation.
   (d) UV-A or UV-B radiation.

4. How many ozone molecules can a single chlorine atom break down?
   (a) 5,000
   (b) 10,000
   (c) 50,000
   (d) 100,000

5. Which of the following is a result of reduced stratospheric ozone?
   (a) increased photosynthetic activity
   (b) increased chances of skin cancer
   (c) decreased chance of cataracts
   (d) increased birth defects

6. Through international cooperation, the concentration of chlorine in the atmosphere
   (a) will never change.
   (b) has continued to increase.
   (c) has peaked and is now decreasing.
   (d) goes through a cyclical change each year depending on industrial activity.

7. In what season in the Antarctic is the ozone hole largest?
   (a) early spring
   (b) late summer
   (c) mid-summer
   (d) late fall

Free-Response Question

The ozone layer protects the surface of Earth from harmful UV radiation. Ozone-depleting substances break down the ozone layer. The diagram below depicts the typical process of how those ozone depleting substances break down ozone.

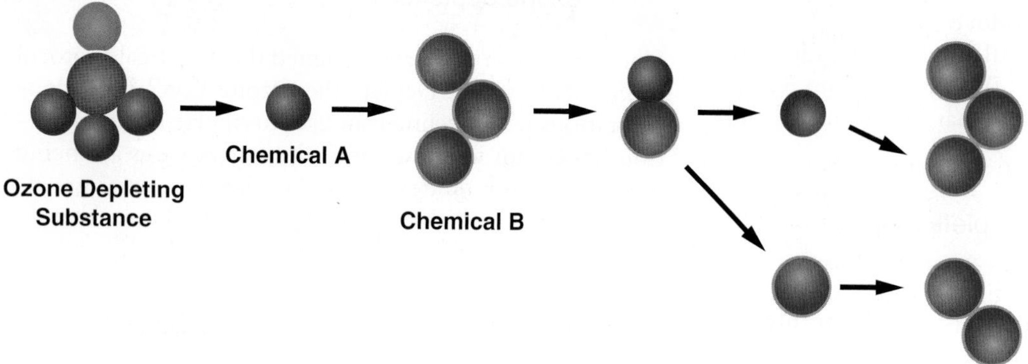

(a) The ozone depleting substance in the image above is the chemical that breaks down ozone. **Identify** the chemical that is primarily responsible for this breakdown. (1 pt.)
(b) **Describe** how the chemical named in part (a) breaks down ozone. (2 pts.)
(c) **Explain** the source of ozone depleting substances. (1 pt.)
(d) **Describe** the form of legislation responsible for reduction of ozone-depleting substances. (1 pt.)
(e) **Identify** the layer of the atmosphere where ozone blocks UV rays. (1 pt.)

Students in an AP® Environmental class measure UV-B radiation at the surface of Earth. Students used different forms of sunblock SPF to determine which prevented more UV radiation from passing through. The following data was collected.

Sunblock SPF	UV intensity (no sunblock)	UV intensity (sunblock)
4	559	425
15	597	182
50	402	74
70	549	38
100	622	53

(f) **Identify** the dependent variable in the students' investigation. (1 pt.)
(g) **Identify** a possible research question for the investigation. (1 pt.)
(h) **Identify** a variable not mentioned that might impact the results of the students' investigation. (1 pt.)
(i) **Describe** a modification that could be made to the investigation that would increase the reliability of the students' results. (1 pt.)

Module 56

Unit 9 | 55 | 56 | 57 | 58 | 59

The Greenhouse Effect

In the previous module we examined how human activities have caused a depletion in the planet's stratospheric ozone and the steps taken to reverse this impact. In this module we will continue our exploration of how humans are altering the planet by considering the greenhouse effect. In previous units we have mentioned how greenhouse gases cause the planet to be warmer. Here, we are going into much greater detail to understand how this process works. In later modules we will explore how humans are increasing the abundance of greenhouse gases and the consequences of these changes on humans and ecosystems.

Learning Goals

After reading this module you should be able to

56-1 describe how we can distinguish between global change, global climate change, and global warming.

56-2 explain the process underlying the greenhouse effect.

56-3 identify the sources of greenhouse gases.

56-1 How do we distinguish between global change, global climate change, and global warming?

Global change includes global climate change and global warming

Throughout this book, we have highlighted a wide variety of ways in which the world has changed as a result of a rapidly growing human population. Human activity has placed increasing demands on natural resources such as water, trees, minerals, and fossil fuels. We have also emitted growing amounts of carbon dioxide, nitrogen compounds, and sulfur compounds into the atmosphere. We learned that our agricultural methods depend on chemicals, including fertilizers and pesticides. Finally, a growing population faces challenges of waste disposal, sanitation, and the spread of human diseases.

Change that occurs in the chemical, biological, and physical properties of the planet is referred to as **global change**. As you can see in **FIGURE 56.1**, some types of global change are natural and have been occurring for millions of years. Global temperatures, for example, have fluctuated over millions of years. During periods of cold temperatures, Earth has experienced ice ages. In modern times, however, the rates of change have often been much higher than those that occurred historically. Many of these changes are the result of human activities, and they can have significant, sometimes cascading, effects. For example, as we

Global change Change that occurs in the chemical, biological, and physical properties of the planet.

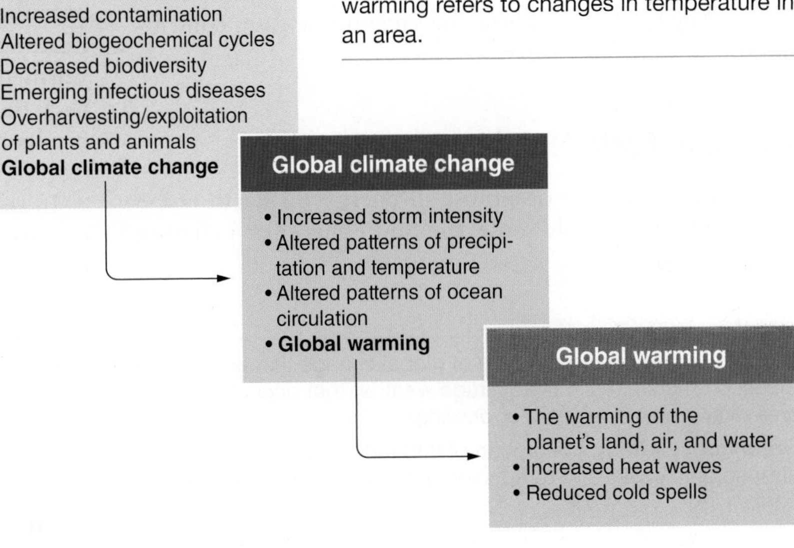

Global change
• Rising sea levels
• Increased extraction of fossil fuels
• Increased contamination
• Altered biogeochemical cycles
• Decreased biodiversity
• Emerging infectious diseases
• Overharvesting/exploitation of plants and animals
• Global climate change
Global climate change
• Increased storm intensity
• Altered patterns of precipitation and temperature
• Altered patterns of ocean circulation
• Global warming
Global warming
• The warming of the planet's land, air, and water
• Increased heat waves
• Reduced cold spells

FIGURE 56.1 Global change. Global change includes a wide variety of factors that are changing over time. Global climate change refers to those factors that affect the average weather in an area of Earth. Global warming refers to changes in temperature in an area.

saw in Module 42, emissions from coal-fired power plants and waste incinerators have increased the amount of mercury in the air and water, with concentrations roughly triple those of preindustrial levels. This mercury bioaccumulates in fish caught thousands of kilometers away from the sources of pollution, as we discussed in Module 49. Because mercury has harmful effects on the nervous system of children, women who might become pregnant and children are advised to avoid eating top-predator fish such as swordfish and tuna. Far-reaching effects on this scale were unimaginable just 50 years ago.

One type of global change of particular concern to scientists is **global climate change**, which refers to changes in the average weather that occurs in an area over a period of years or decades. Changes in climate can be categorized as either natural or anthropogenic. For example, you might recall from Module 23 that El Niño events, which occur every 3 to 7 years, alter global patterns of temperature and precipitation (see Figure 23.5 on page 275). Anthropogenic activities such as fossil fuel combustion and deforestation also have major effects on global climates. Of the many ways that we can experience global climate change, global warming (which you were first introduced to in Module 4) refers to a specific aspect of climate change: the increase in global temperatures due to humans producing more greenhouse gases.

56-2 What are the processes underlying the greenhouse effect?

Solar radiation and greenhouse gases make our planet warm

The physical and biogeochemical systems that regulate temperature at the surface of Earth—the concentrations of gases, distribution of clouds, atmospheric currents, and ocean currents—are essential to life on our planet. It is therefore critical that we understand how the planet is warmed by the Sun and how the greenhouse gases contribute to the warming of Earth.

The Sun–Earth Heating System

The ultimate source of almost all energy on Earth is the Sun. In the most basic sense, the Sun emits solar radiation that strikes Earth. As the planet warms, it emits radiation back toward the atmosphere. However, the types of energy radiated from the Sun and Earth are different. Because the Sun is very hot, most of its radiated energy is in the form of high-energy visible radiation and ultraviolet radiation—which we generally refer to as visible light and ultraviolet light. When this radiation strikes Earth, the planet warms and radiates energy. Earth is not nearly as hot as the Sun, so it emits most of its energy as infrared radiation—also known as infrared light. We cannot see infrared radiation, but we can feel it being emitted from warm surfaces like the heat that radiates from an asphalt road on a hot day.

Differences in the types of radiation emitted by the Sun and Earth, in combination with processes that occur in the atmosphere, cause the planet to warm. Using **FIGURE 56.2**, we can walk through each step of this process. As radiation from the Sun travels toward Earth, about one-third of the radiation is reflected back into space. Although much of the ultraviolet radiation is absorbed by the ozone layer in the stratosphere, as noted in the previous module, the remaining ultraviolet radiation, as well as visible light, passes through the atmosphere. Once it has passed through the atmosphere, this solar radiation strikes clouds and the surface of Earth. Some of this radiation is reflected from the surface of the planet back into space. The remaining radiation is absorbed by clouds and the surface of Earth, which become warmer and begin to emit lower-energy infrared radiation back toward the atmosphere. Unlike ultraviolet and visible radiation, infrared radiation does not easily pass through the atmosphere. Instead, it is absorbed by gases, which causes these gases to become warm. The warmed gases emit infrared radiation out into space and back toward the surface of Earth. The infrared radiation that is emitted toward Earth causes Earth's surface to become even warmer. This absorption of infrared radiation by atmospheric gases and reradiation of the energy back toward Earth is the **greenhouse effect**.

The greenhouse effect gets its name from the idea that solar radiation causes a gardener's greenhouse to become very warm. However, the process by which a gardener's greenhouse is warmed by the Sun involves glass windows holding in heat whereas the process by which Earth is warmed involves greenhouse gases radiating infrared energy back toward the planet's surface. This warming of the planet results in surface temperatures that allow life on Earth to exist.

In the Sun–Earth heating system, the net flux of energy is zero; the inputs of energy to Earth equal the outputs from Earth. Over the long term—thousands or millions of years—the system has been in a steady state. However, in the shorter term—over years or decades—inputs can be slightly higher or lower than outputs. Factors that influence short-term fluctuations include changes in incoming solar radiation from increased solar activity and changes in outgoing radiation from an increase in atmospheric gases that absorb infrared radiation. If incoming solar energy is greater than the sum of reflected solar energy and radiated infrared energy

Global climate change A type of global change that is focused on changes in the average weather that occurs in an area over a period of years or decades.

Greenhouse effect Absorption of infrared radiation by atmospheric gases and reradiation of the energy back toward Earth.

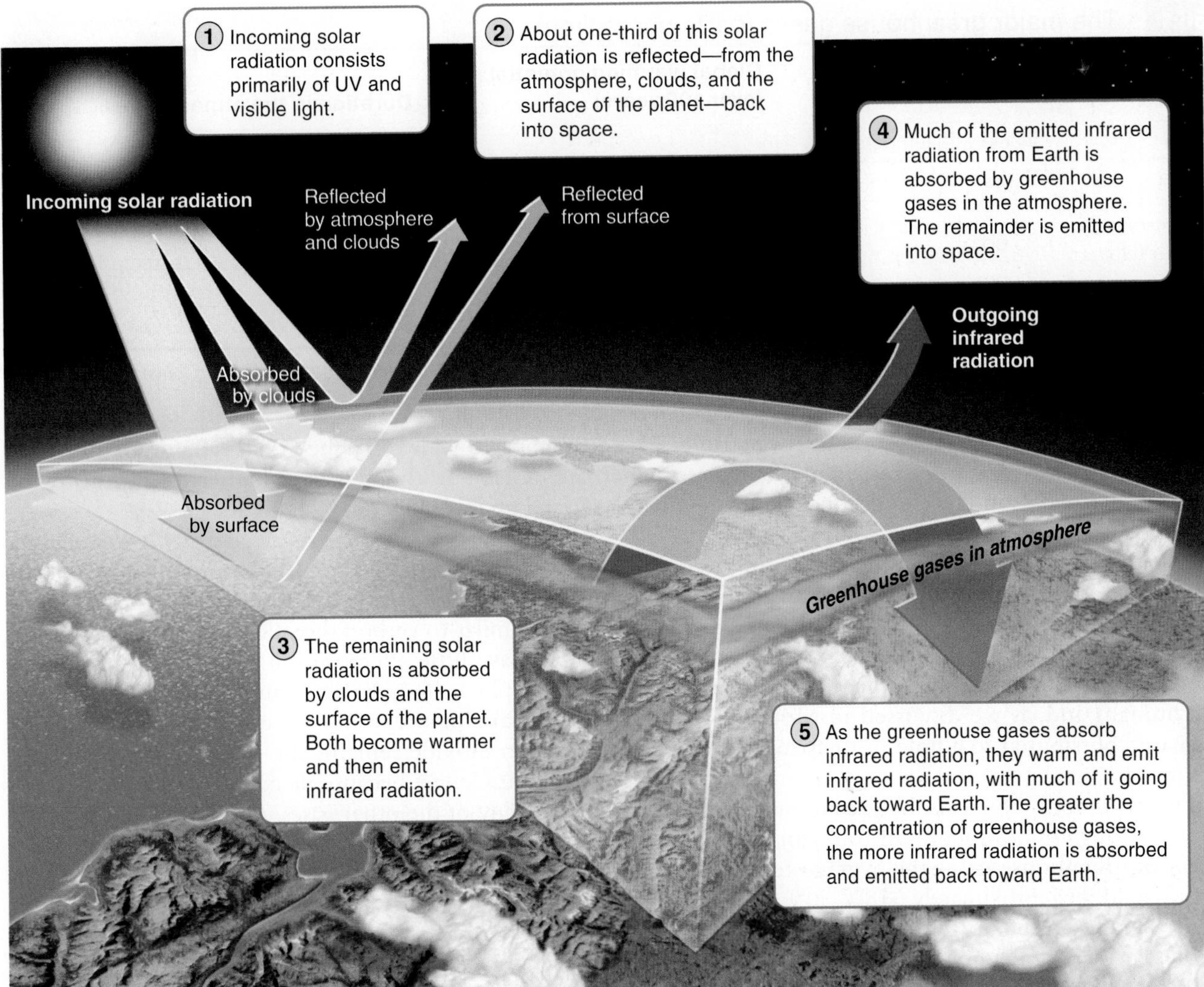
1 Incoming solar radiation consists primarily of UV and visible light.
2 About one-third of this solar radiation is reflected—from the atmosphere, clouds, and the surface of the planet—back into space.
4 Much of the emitted infrared radiation from Earth is absorbed by greenhouse gases in the atmosphere. The remainder is emitted into space.
Incoming solar radiation
Reflected by atmosphere and clouds
Reflected from surface
Outgoing infrared radiation
Absorbed by clouds
Absorbed by surface
Greenhouse gases in atmosphere
3 The remaining solar radiation is absorbed by clouds and the surface of the planet. Both become warmer and then emit infrared radiation.
5 As the greenhouse gases absorb infrared radiation, they warm and emit infrared radiation, with much of it going back toward Earth. The greater the concentration of greenhouse gases, the more infrared radiation is absorbed and emitted back toward Earth.

FIGURE 56.2 **The greenhouse effect.** When the high-energy radiation from the Sun strikes the atmosphere, about one-third is reflected from the atmosphere, clouds, and the surface of the planet. Much of the high-energy ultraviolet radiation is absorbed by the ozone layer where it is converted to low-energy infrared radiation. Some of the ultraviolet radiation and much of the visible light strikes the land and water of Earth where it is also converted into low-energy infrared radiation. The infrared radiation radiates back toward the atmosphere where it is absorbed by greenhouse gases that radiate much of it back toward the surface of Earth. Collectively, these processes cause warming of the planet.

from Earth, then the energy accumulates faster than it is dispersed and the planet becomes warmer. If incoming solar energy is less than the sum of the two outputs, the planet becomes cooler. Such natural changes in inputs and outputs cause natural changes in the temperature of Earth over time.

Gases That Cause the Greenhouse Effect

Throughout this book we have seen that certain gases in the atmosphere can absorb infrared radiation emitted by the surface of the planet and radiate much of it back toward the surface. These are known as greenhouse gases.

The two most common gases in the atmosphere, N_2 and O_2, compose 99 percent of the atmosphere. Because these two gases do not absorb infrared radiation, they are not greenhouse gases and do not contribute to the warming of Earth. This means that greenhouse gases make up a tiny fraction of the atmosphere. The most common greenhouse gas is water vapor (H_2O). Water vapor absorbs more infrared radiation from Earth than any other compound, although a molecule of water vapor does not persist nearly as long as other greenhouse gases, and therefore isn't a significant contributor to the global climate change caused by humans. Other important greenhouse gases include carbon dioxide (CO_2), methane (CH_4), nitrous oxide (N_2O), and ozone (O_3). All of these gases have been a part of the atmosphere for millions of years and have kept Earth warm enough to be habitable. In the case of ozone, we have seen

TABLE 56.1 The major greenhouse gases (parts per million)

Greenhouse gas	Concentration in 2017	Global warming potential (over 100 years)	Duration in the atmosphere
Water vapor	Variable with temperature	<1	9 days
Carbon dioxide	407 ppm	1	Highly variable (ranging from years to hundreds of years)
Methane	1.85 ppm	25	12 years
Nitrous oxide	.33 ppm	300	114 years
Chlorofluorocarbons	0.0007 ppm	1,600 to 13,000	55 to >500 years

(Data from The National Oceanic and Atmospheric Administration, www.esrl.noaa.gov/gmd/aggi, and The United Nations Framework Convention on Climate Change.)

that its effects on Earth are diverse. Ozone in the stratosphere is beneficial because it filters out harmful ultraviolet radiation. In contrast, ozone in the lower troposphere acts as a greenhouse gas and can cause increased warming of Earth. It is also an air pollutant in the lower troposphere because it can cause damage to plants and human respiratory systems. There is one other type of greenhouse gas, chlorofluorocarbons (CFCs), which does not exist naturally. It occurs in the atmosphere exclusively due to production of CFCs by humans and, as we discussed in Module 55, these CFCs have contributed to a hole in the ozone layer over Antarctica.

Although we commonly think of the greenhouse effect as detrimental to our environment, without our natural concentration of greenhouse gases the average temperature on Earth would be approximately −18°C (0°F) instead of its current average temperature of 14°C (57°F). Concern about the danger of greenhouse gases is based on our understanding that an increase in the concentration of these gases—caused by human activities—can cause the planet to warm even more than usual.

The contribution of each gas to global warming depends in part on its greenhouse warming potential. The **greenhouse warming potential (GWP)** of a gas estimates how much a molecule of any compound can contribute to global warming over a period of 100 years relative to one molecule of CO_2, which is defined as having a GWP = 1. In calculating the GWP, scientists consider the amount of infrared energy that a given gas can absorb and how long a molecule of the gas persists in the atmosphere. Because greenhouse gases can differ a great deal in these two factors, GWPs span a wide range of values. For example, water vapor has a lower potential compared with carbon dioxide. The remaining greenhouse gases have much higher values, either because they absorb more infrared radiation than a molecule of CO_2 or because they persist much longer in the atmosphere than a molecule of CO_2. **TABLE 56.1** shows the GWP for five common greenhouse gases. Compared with CO_2, the GWP is 25 times higher for methane (CH_4), nearly 300 times higher for nitrous oxide (N_2O), and up to 13,000 times higher for CFCs.

Although carbon dioxide has a relatively low warming potential, it is much more abundant than most other greenhouse gases, except for water vapor, which can have a concentration similar to carbon dioxide. While human activity appears to have little effect on the amount of water vapor in the atmosphere, humans have caused substantial increases in the amount of the other greenhouse gases. Among these, carbon dioxide remains the greatest contributor to the greenhouse effect because its concentration is so much higher than any of the other gases. As a result, scientists and policy makers focus their efforts on ways to reduce carbon dioxide in the atmosphere.

Given what we now know about how greenhouse gases work, the concentrations of each gas, and how much infrared energy each gas absorbs, we can understand how changes in the concentrations of greenhouse gases can contribute to global warming. Increasing the concentration of any historically present greenhouse gas should cause more infrared radiation to be absorbed in the atmosphere, which will then radiate more energy back toward the surface of the planet and cause the planet to warm. Likewise, producing new greenhouse gases that can make their way into the atmosphere, such as CFCs, should also cause increased absorption of infrared radiation in the atmosphere and further cause the planet to warm.

Greenhouse warming potential (GWP) An estimate of how much a molecule of any compound can contribute to global warming over a period of 100 years relative to one molecule of CO_2.

56-3 What are the sources of greenhouse gases?

Sources of greenhouse gases are both natural and anthropogenic

As we have seen, greenhouse gases include a variety of compounds such as water vapor, carbon dioxide, methane, nitrous oxide, ozone, and CFCs. These gases have natural and anthropogenic sources. After reviewing the different sources of greenhouse gases, we will discuss the relative ranks of the different anthropogenic sources.

Natural Sources of Greenhouse Gases

Natural sources of greenhouse gases include volcanic eruptions, decomposition, digestion, denitrification, evaporation, and evapotranspiration.

Volcanic Eruptions

Over the scale of geologic time, volcanic eruptions can add a significant amount of carbon dioxide to the atmosphere. Other gases and the large quantities of ash released during volcanic eruptions can also have important, short-term climatic effects. A volcanic eruption emits a large quantity of ash into the atmosphere. The ash reflects incoming solar radiation back out into space, which has a cooling effect on Earth. In 2009, for example, Mount Pinatubo in the Philippines was a major volcanic eruption that spewed millions of tons of ash into the atmosphere, as far as 20 km (12 miles) high (**FIGURE 56.3**). The large amount of ash in the atmosphere reduced the amount of radiation striking Earth, which caused a 0.5°C (0.9°F) decline in the temperature on the planet's surface. Because the ash and small particles eventually settle out of the atmosphere, such effects usually last only a few years.

Decomposition and Digestion

When decomposition occurs under high-oxygen conditions, the dead organic matter is ultimately converted into carbon dioxide. As we saw in our discussion of landfills in Module 50, methane is created when there is not enough oxygen available to produce carbon dioxide. This is also a common occurrence at the bottom of wetlands where plants and animals decompose and oxygen is in low supply. Wetlands are the largest natural source of methane.

FIGURE 56.3 Ash from volcanic eruptions. Volcanic eruptions, such as this eruption of Mount Pinatubo in the Philippines in 1991, send millions of tons of ash into the atmosphere where it can reduce the amount of radiation striking Earth, and cause Earth to cool. *(Exactostock/SuperStock)*

A similar situation occurs when certain animals digest plant matter. Animals that consume significant quantities of wood or grass, including termites and grazing antelopes, require gut bacteria to digest the plant material. Because the digestion occurs in the animal's gut, the bacteria do not have access to oxygen and methane is produced as a by-product. For example, a single termite colony can contain more than a million termites that have gut bacteria to help them digest the wood that they consume (**FIGURE 56.4**). Termites are abundant throughout the world—especially in the tropics—and represent the second largest natural source of methane.

Denitrification

As we learned in Module 4, nitrous oxide (N_2O) is a natural component of the nitrogen cycle that is produced through the process of denitrification. Denitrification occurs in the

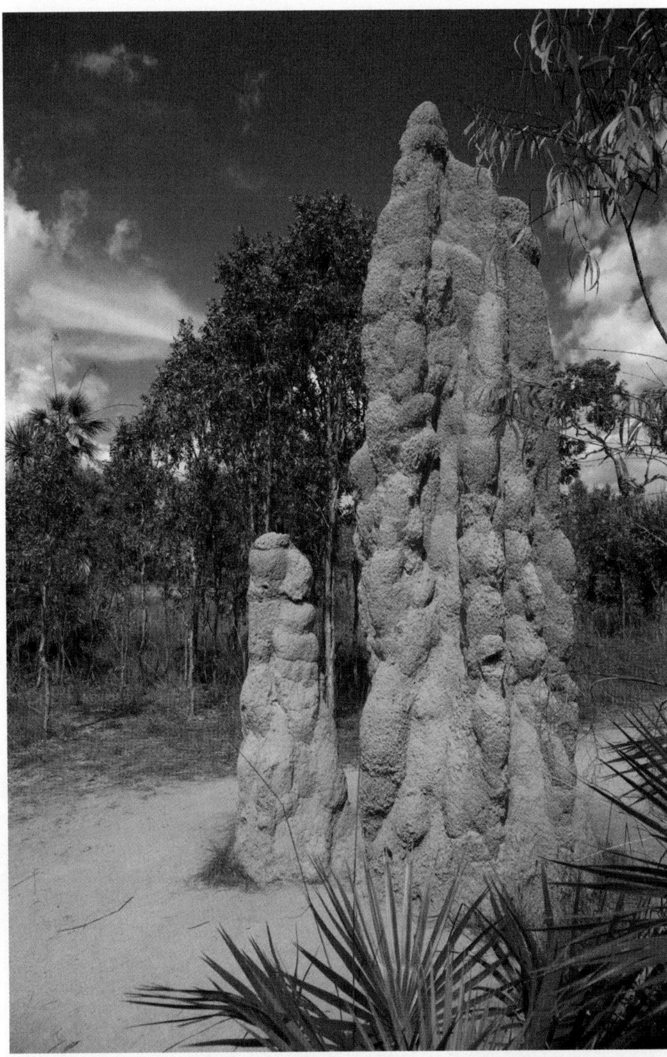

FIGURE 56.4 Termites and methane. The bacteria that live in the anaerobic gut environment of herbivores such as termites produce methane as a by-product of their digestive activities. Because termite colonies, such as this one in Australia, can achieve population sizes of more than one million, collectively they can produce large amounts of methane. *(Anton Harder/Shutterstock)*

low-oxygen environments of wet soils and at the bottoms of wetlands, lakes, and oceans. (Figure 4.4 on page 56 shows the nitrogen cycle.) In these environments, nitrate is converted to nitrous oxide gas, which then enters the atmosphere as a powerful greenhouse gas.

Evaporation and Evapotranspiration

As we stated earlier, water vapor is the most abundant greenhouse gas in the atmosphere and the greatest natural contributor to global warming. In Module 5 we examined the role of water vapor in the hydrologic cycle. (Figure 5.4 on page 63 shows the hydrologic cycle.) Water vapor is produced when liquid water from land and water bodies evaporates and by the evapotranspiration process of plants. Because the amount of evaporation into water vapor varies with climate, the amount of water vapor in the atmosphere can vary regionally.

Anthropogenic Sources of Greenhouse Gases

As shown in **FIGURE 56.5**, there are many anthropogenic sources of greenhouse gases. The most significant of these are the burning of fossil fuels, agricultural practices, deforestation, landfills, and industrial production of new greenhouse chemicals.

AP® Exam Tip

Be careful not to relate anthropogenic climate change to stratospheric ozone loss. The loss of stratospheric ozone does not contribute to climate change.

Burning Fossil Fuels

Tens to hundreds of millions of years ago, organisms were sometimes buried without first decomposing into carbon dioxide. In Figure 4.1 on page 53, we outlined the process by which the carbon contained in these organisms, called fossil carbon, is slowly converted to fossil fuels deep underground. When humans burn these fossil fuels, we produce CO_2 that goes into the atmosphere. Because of the long time required to convert carbon into fossil fuels, the rate of putting carbon into the atmosphere by burning fossil fuels is much greater than the rate at which producers take CO_2 out of the air and both the producers and consumers contribute to the pool of buried fossil carbon.

Because fossil fuels differ in how they store energy, each type of fossil fuel produces different amounts of carbon dioxide. For a given amount of energy, burning coal produces the most CO_2. In comparison, burning oil produces 85 percent as much CO_2 as coal, and burning natural gas produces 56 percent as much. As we saw in Unit 7, from the perspective of CO_2 emissions, natural gas is considered better for the environment than coal. The production of fossil fuels, such as the mining of coal, and the combustion of fossil fuels can also release methane and, in some cases, nitrous oxide.

Particulate matter may also play an important role in global warming. Although particulate matter, also known as black soot, may reflect solar radiation under some conditions, recent findings suggest that it may be responsible for up to one-quarter of observed global warming during the past century. Particulates that fall on ice and snow in the higher latitudes absorb more energy of the Sun by lowering the albedo. As the snow and ice begin to melt, the particulates become more concentrated on the surface. The increased concentration raises the amount of solar radiation absorbed, which increases melting. This positive feedback system might help explain warming that occurred early in the last century, when atmospheric concentrations of greenhouse gases had not yet increased much but soot from the burning of coal was widespread.

Agricultural Practices

Agricultural practices can produce a variety of greenhouse gases. Agricultural fields that are overirrigated, or those that are deliberately flooded for cultivating crops such as rice, create low-oxygen environments similar to wetlands and therefore can produce methane and nitrous oxide. Synthetic fertilizers, manures, and crops that naturally fix atmospheric nitrogen—for example, alfalfa—can create an excess of nitrates in the soil that are converted to nitrous oxide by the process of denitrification.

Raising livestock can also produce large quantities of methane. Many livestock such as cattle and sheep consume large quantities of plant matter and rely on gut bacteria to

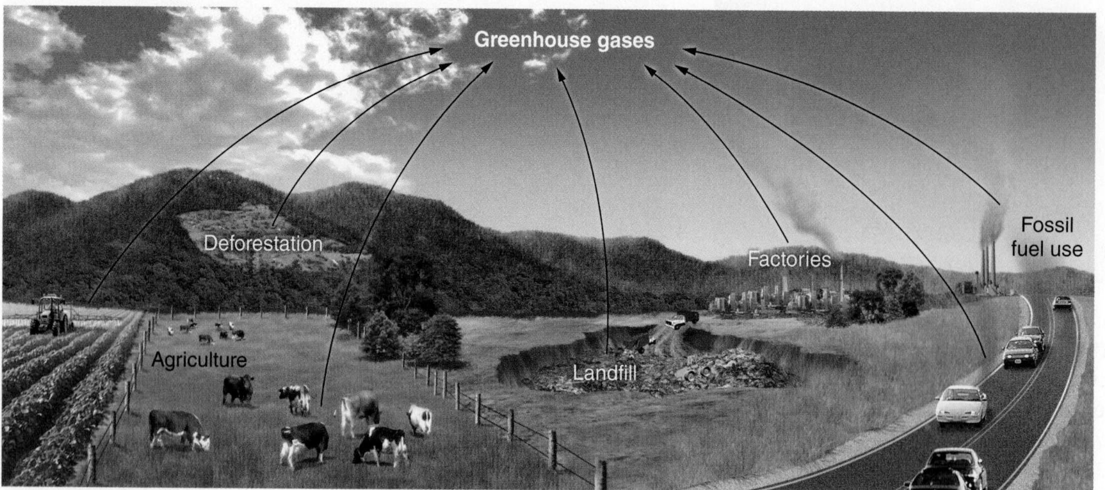

FIGURE 56.5 Anthropogenic sources of greenhouse gases. Human activities are a major contributor of greenhouse gases, including carbon dioxide (CO_2), methane (CH_4), and nitrous oxide (N_2O). These activities include the use of fossil fuels, agricultural practices, the creation of landfills, and the industrial production of new greenhouse gases.

digest this cellulose. As we saw in the case of termites, gut bacteria live in a low-oxygen environment and digestion in this environment produces methane as a by-product. In addition, manure from livestock operations will decompose to CO_2 under high-oxygen conditions, but in low-oxygen conditions, for example in manure lagoons that are not aerated, it will decompose to methane.

Deforestation

Each day, living trees remove CO_2 from the atmosphere during photosynthesis, and decomposing trees add CO_2 to the atmosphere. This part of the carbon cycle does not change the net atmospheric carbon because the inputs and outputs are approximately equal. However, when forests are destroyed by burning or decomposition and not replaced, as can happen during deforestation, the destruction of vegetation will contribute to a net increase in atmospheric CO_2. This is because the mass of carbon that made up the trees is added to the atmosphere by combustion or decomposition. The shifting agriculture described in Module 26, which involves clearing forests and burning the vegetation to make room for crops, is a major source of both particulates and a number of greenhouse gases, including carbon dioxide, methane, and nitrous oxide.

Landfills

As we saw in Module 50, landfills receive a great deal of household waste that slowly decomposes under layers of soil. When the landfills are not aerated properly, they create a low-oxygen environment, like wetlands, in which decomposition causes the production of methane as a by-product.

Industrial Production of New Greenhouse Chemicals

The creation of new industrial chemicals often has unintended effects on the atmosphere. In the previous module we looked at CFCs, the family of chemicals that serves as refrigerants used in air conditioners, freezers, and refrigerators. CFCs were widely used for decades until scientists discovered that they were damaging the protective ozone layer in the stratosphere. As we discussed, the nations of the world joined together to sign the Montreal Protocol on Substances That Deplete the Ozone Layer, which phased out the production and use of CFCs by 1996. Unfortunately, many of the alternative refrigerants that are less harmful to the ozone layer, including a group of gases known as hydrofluorocarbons (HFCs) and hydrochlorofluorocarbons (HCFCs), still have very high greenhouse warming potentials. As a result, the U.S. EPA decided to phase out the use of HFCs by 2020 and phase out HCFCs by 2030.

Ranking the Anthropogenic Sources of Greenhouse Gases

We have seen that there are multiple anthropogenic sources of greenhouse gases. What is the relative contribution of each source? **FIGURE 56.6** shows the major anthropogenic

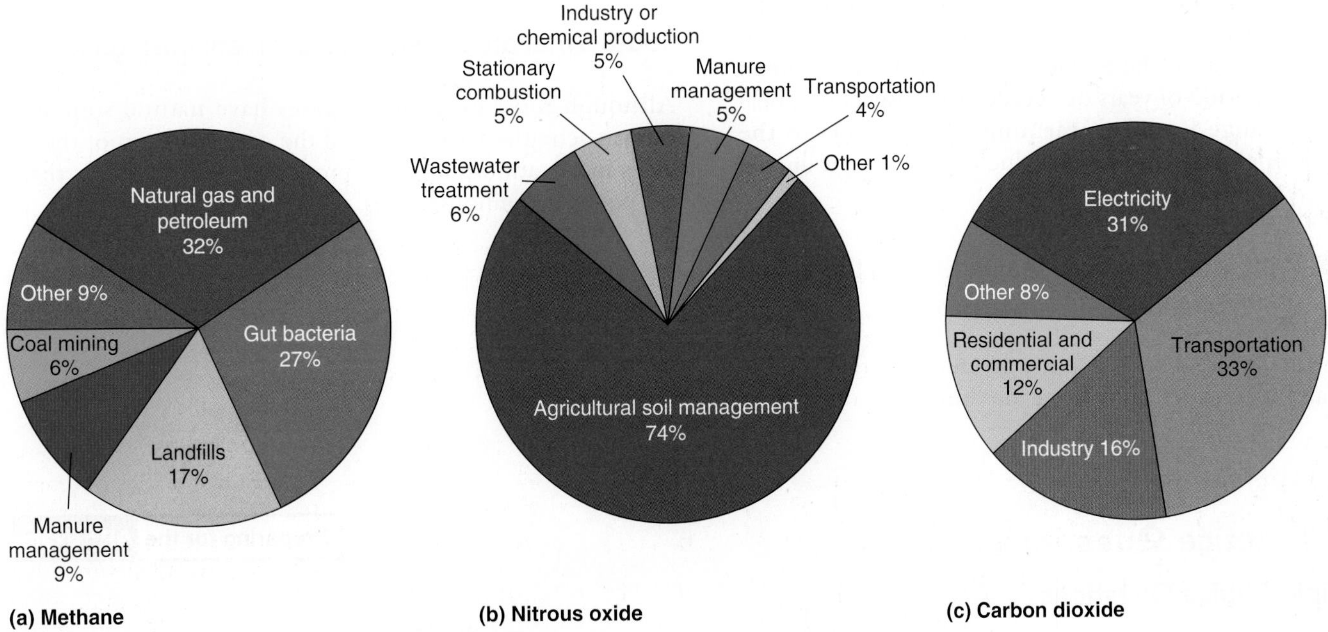

FIGURE 56.6 Anthropogenic sources of greenhouse gases in the United States. (a) The largest contributions of methane in the atmosphere arise from gut bacteria that help many livestock species digest plant matter, landfills that experience decomposition in low-oxygen environments, and the production, storage, and transport of natural gas and petroleum products from which methane escapes. (b) The largest contributions of nitrous oxide in the atmosphere arise from the agricultural soils that obtain nitrogen from applied fertilizers, combustion, and industrial production of fertilizers and other products. (c) Nearly all anthropogenic CO_2 emissions come from the burning of fossil fuels. *(Data from https://www.epa.gov/ghgemissions/overview-greenhouse-gases–carbon-dioxide; https://www.epa.gov/ghgemissions/overview-greenhouse-gases–methane; https://www.epa.gov/ghgemissions/overview-greenhouse-gases–nitrous-oxide.)*

sources of greenhouse gases in the United States. For methane, Figure 56.6a shows that the three major contributors in the atmosphere are the digestive processes of livestock, landfills, and the production of natural gas and petroleum products. For nitrous oxide, Figure 56.6b shows the major contributor is agricultural soil because it receives nitrogen from synthetic fertilizers, combustion, and industrial production of fertilizers and other products. Finally, for carbon dioxide, shown in Figure 56.6c, we see that approximately 93 percent of all CO_2 emissions come from industrial processes and the burning of fossil fuels for transportation, residents, businesses, and generating electricity.

AP® Exam Tip

You should be able to identify the various greenhouse gases and discuss how humans have influenced increased concentrations of these gases.

In this module we have learned the concepts of global change, global climate change, and global warming. We learned in detail the processes underlying the greenhouse effect and the most important greenhouse gases that contribute to the greenhouse effect. In the next module we will consider the evidence that human activities have been causing increases in these greenhouse gases.

Module 56 AP® Review

Preparing for the AP® Exam

Learning Goals Revisited

56-1 How do we distinguish between global change, global climate change, and global warming?

Global change represents change that occurs in the chemical, biological, and physical properties of the planet. One type of global change is global climate change, which refers to changes in the average weather that occurs in an area over a period of years or decades. One aspect of global climate change is global warming, which refers to the increase in global temperatures due to humans producing more greenhouse gases.

56-2 What are the processes underlying the greenhouse effect?

In a natural process known as the greenhouse effect, visible light and ultraviolet light from the Sun strike our planet and this energy is converted to infrared radiation that is emitted back to the atmosphere. A tiny percentage of gases in the atmosphere, known as greenhouse gases, absorb this infrared radiation and emit a portion of it back to Earth and this causes the planet to warm even more.

56-3 What are the sources of greenhouse gases?

Although most greenhouse gases have natural sources, human activities have increased the concentration of these gases in the atmosphere and produced new chemicals that are potent greenhouse gases.

AP® Practice Questions

Preparing for the AP® Exam

Multiple-Choice Questions

1. The greenhouse gas with the highest greenhouse warming potential is
   (a) carbon dioxide.
   (b) methane.
   (c) chlorofluorocarbon.
   (d) nitrous oxide.

2. Methane is naturally produced by
   (a) anaerobic decomposition.
   (b) volcanic eruptions.
   (c) denitrification.
   (d) forest fires.

3. The greenhouse effect is due to
   (a) the absorption and reradiation of infrared radiation by the atmosphere.
   (b) the reflection of ultraviolet radiation by the atmosphere.
   (c) the absorption of ultraviolet radiation by the atmosphere.
   (d) the reflection of infrared radiation from Earth's surface.

4. Most nitrous oxide emissions are from
   (a) fossil fuel-combustion.
   (b) agricultural practices.
   (c) refrigerants.
   (d) industrial processes.

5. Particulate matter can increase global warming by
   (a) reacting with chlorofluorocarbons.
   (b) reducing the surface absorption of ultraviolet radiation.
   (c) reflecting radiation.
   (d) lowering surface albedo.

Free-Response Question

The amount of greenhouse gases produced from different sources is monitored each year by the EPA. The percentages for three major greenhouse gases are illustrated below.

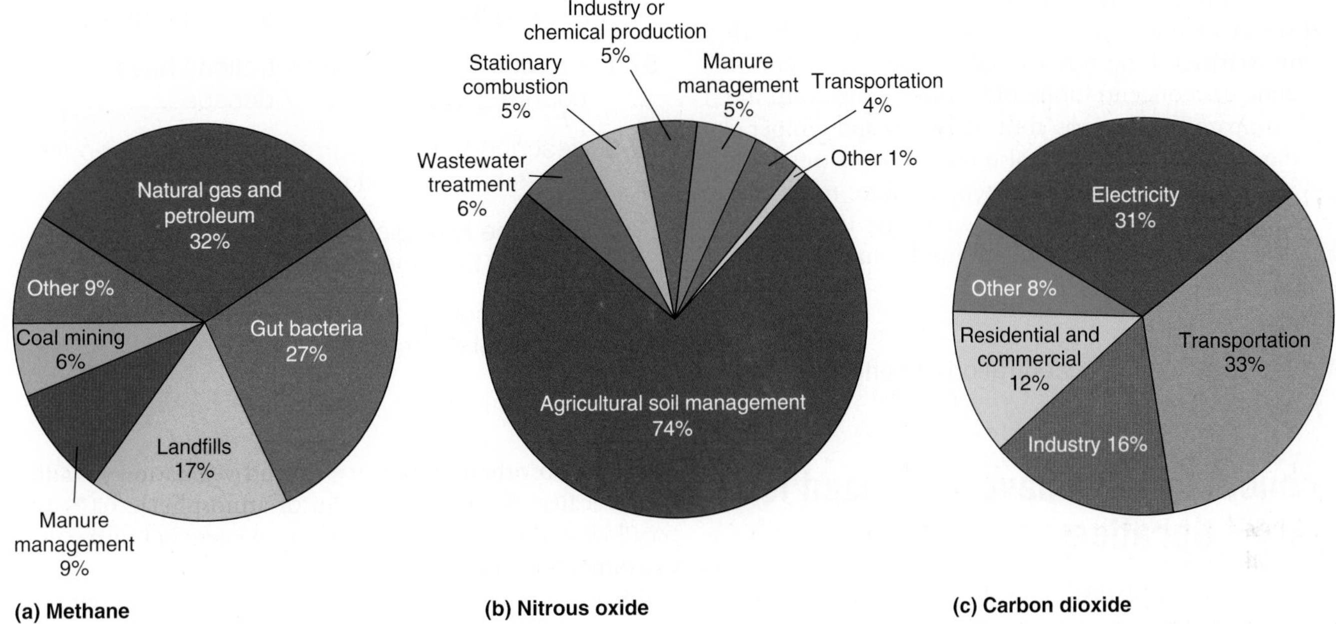

(a) **Identify** the largest source of methane. (1 pt.)
(b) **Identify** the largest source of nitrous oxide. (1 pt.)
(c) **Identify** the largest source of carbon dioxide. (1 pt.)
(d) **Explain** why landfills are a large source of methane. (1 pt.)
(e) **Explain** why generating electricity is an important source of carbon dioxide. (1 pt.)
(f) **Describe** TWO sources of methane that we can reduce. (2 pts.)
(g) **Describe** one way we could reduce nitrous oxide production in agriculture. (1 pt.)
(h) **Describe** how we can reduce two sources of carbon dioxide. (2 pts.)

Module 57

Unit 9 | 55 | 56 | 57 | 58 | 59

Increases in the Greenhouse Gases and Global Climate Change

In the previous module we reviewed the role of greenhouse gases in global warming. Here, we will examine the evidence that human activities have caused increased concentrations of greenhouse gases, which are causing Earth to become warmer. One way to make these assessments is to determine gas concentrations and temperatures from the past and compare them to gas concentrations and temperatures in the present day. We can also use information about changes in gas concentrations and temperatures to predict future climate conditions. With these changes in climates, we can also examine how this is altering biomes and how populations of organisms are responding.

Learning Goals

After reading this module you should be able to

57-1 explain how CO_2 concentrations have changed over the past 7 decades.

57-2 describe how temperatures and greenhouse gases have varied historically.

57-3 describe how global climate change has affected the environment.

57-4 explain how global climate change is affecting populations.

57-1 How have CO_2 concentrations changed over the past 7 decades?

CO_2 concentrations have increased for the past 7 decades

In 1988, the United Nations and the World Meteorological Organization created the Intergovernmental Panel on Climate Change (IPCC), a group of more than 3,000 scientists from around the world working together to assess climate change. Their mission is to understand the details of the global warming system, the effects of climate change on biodiversity and energy fluxes in ecosystems, and the economic and social effects of climate change. The IPCC enables scientists to assess and communicate the state of our knowledge and to suggest research directions that would improve our understanding in the future. This effort has produced an excellent understanding of how greenhouse gases and temperatures are linked. Through the work of the IPCC, we now understand that CO_2 is an important greenhouse gas that can contribute to global warming, but we didn't always realize this.

Measuring CO_2 Concentrations

In the first half of the twentieth century, most scientists believed that if any excess CO_2 were being produced, it would be absorbed by the oceans and vegetation. In addition, because the concentration of atmospheric CO_2 was very low compared to gases such as oxygen and nitrogen, it was difficult to measure accurately.

Charles Keeling was the first scientist to overcome the technical difficulties in measuring CO_2. When Keeling set out to measure the precise level of CO_2 in the atmosphere, most atmospheric scientists believed that two measurements several years apart would be sufficient to answer the question of whether human activities were causing increased concentrations of CO_2 in the atmosphere. Keeling did not agree and in 1958 he began collecting data throughout the year at the Mauna Loa Observatory in Hawaii. After just 1 year of work, Keeling found that CO_2 levels varied seasonally and that the concentration of CO_2 increased from year to year. His results prompted him to take measurements for several more years,

AP® Exam Tip

Graphs that show changes over time, like Figure 57.1, frequently appear on the AP® Environmental Science Exam. You should be able to describe the general trends shown in such graphs and explain fluctuations.

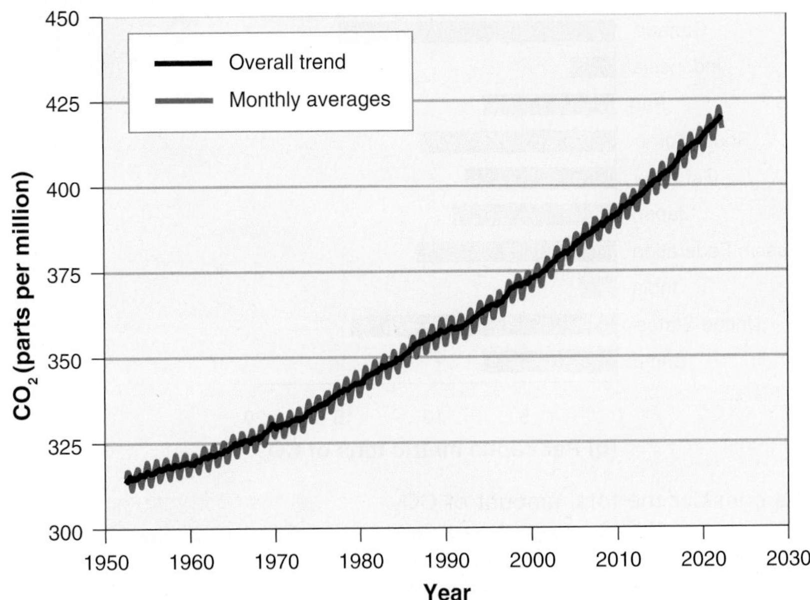

FIGURE 57.1 Changes in atmospheric CO_2 over time. Carbon dioxide levels have risen steadily since measurement began in 1958. *(Data from https://www.esrl.noaa.gov/gmd/ccgg/trends/full.html)*

and he and his students have continued this work into the twenty-first century. The results, shown in **FIGURE 57.1**, confirm Keeling's early findings; although CO_2 concentrations vary between seasons, there is a clear trend of rising CO_2 concentrations across the years. This increase over time is correlated to increased human emissions of carbon from the combustion of fossil fuels and net destruction of vegetation. "Do the Math: Projecting Future Increases in CO_2" gives you an opportunity to estimate the increase in CO_2 concentrations by the end of the century.

What causes the seasonal variation in CO_2 that Keeling observed? Each spring, as deciduous trees, grasslands, and farmlands in the Northern Hemisphere turn green, they increase their absorption of CO_2 as they carry out photosynthesis. At the same time, bodies of water begin to warm and the algae and plants also begin to photosynthesize. In doing so, these producers take up some of the CO_2 in the atmosphere. Conversely, in the fall, as leaves drop, crops are harvested, and bodies of water cool, the uptake of atmospheric CO_2 by algae and plants declines and the amount of CO_2 in the atmosphere increases.

AP® Exam Tip

Do the Math, "Projecting Future Increases in CO_2," contains two problems that have appeared on past exams. Make sure you understand how to calculate the average annual increase in CO_2 and how to predict the concentration of CO_2.

DO THE MATH Projecting Future Increases in CO_2

Preparing for the AP® Exam

Because Charles Keeling and his colleagues began measuring CO_2 in 1958, we have an excellent record of how CO_2 concentrations have changed in the atmosphere over time. From 1959 to 2021, the concentration of CO_2 in the atmosphere increased from 316 to 415 ppm (parts per million).

Based on these two points in time, what has been the average annual increase of CO_2 in the atmosphere?

$$\text{Time} = 2021 - 1959 = 62 \text{ years}$$

$$\text{Increase in } CO_2 = 415 \text{ ppm} - 316 \text{ ppm} = 99 \text{ ppm}$$

$$\text{Average annual increase in } CO_2 = 99 \text{ ppm} \div 62 \text{ years} = 1.60 \text{ ppm/year}$$

YOUR TURN

1. If the annual rate of CO_2 increase is 1.60 ppm, what concentration of CO_2 do you predict for the year 2100? Round to the nearest whole number.
2. From 2011 to 2021, the rate of increase grew to 2.5 ppm per year. Based on this faster rate, what concentration of CO_2 do you predict for the year 2100? Round to the nearest whole number.

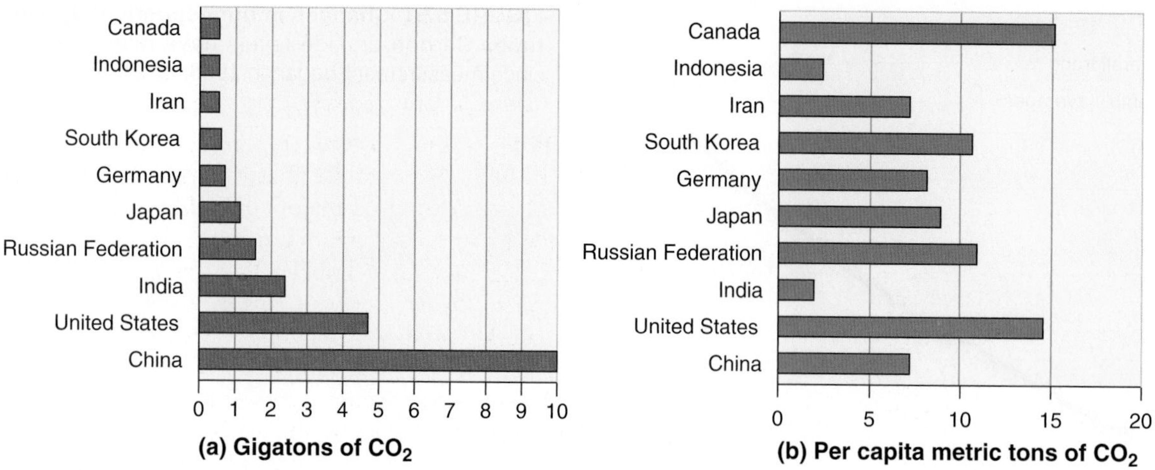

FIGURE 57.2 CO_2 emissions by country. (a) When we consider the total amount of CO_2 produced by a country, we see that the largest contributors are the developed and rapidly developing countries of the world. (b) Some major CO_2 emitters, such as China and India, have relatively low per capita CO_2 emissions. *(Data from IEA Atlas of Energy, https://www.ucsusa.org/resources/each-countrys-share-co2-emissions)*

CO_2 Emissions Among Nations

Throughout this book we have seen that the consumption of fossil fuel and materials is greatest in developed countries. It is not surprising, then, that the production of carbon dioxide has also been greatest in the developed world. For many decades, the 20 percent of the population living in the developed world—roughly 1 billion people—produced about 75 percent of the carbon dioxide. However, these percentages are changing as some developing nations industrialize and acquire more vehicles that burn fossil fuels. In 2009, developing countries surpassed developed countries in the production of CO_2. Today, developing countries are responsible for about two-thirds of the world's CO_2 production while developed nations only produce one third.

Development has been especially rapid in China and India, which together contain one-third of the world's people. For example, from 2000 to 2009, China more than doubled its emissions of carbon dioxide as the country built many new coal-powered electrical plants that increased its ability to burn coal. As shown in **FIGURE 57.2a** China is the leading emitter of CO_2. China emitted nearly 10 gigatons of CO_2, which represents 29 percent of all global CO_2 emissions. The United States was in second place, emitting nearly 5 gigatons, which represents 14 percent of all global CO_2 emissions. However, the United States contains only 5 percent of the world's population. The other nations in the top five emitters are India, the Russian Federation, and Japan.

If we consider the amount of per capita CO_2 emissions, shown in Figure 57.2b, we obtain a very different picture of which countries produce the most CO_2. Of the top 10 emitting countries, South Korea, Russia, the United States, and Canada are the leading per capita emitters of CO_2. Despite the fact that China and India rank among the top producers of CO_2, their per capita production ranks much lower. This reflects the fact that these two countries both have very large populations.

57-2 How have temperatures and greenhouse gases varied historically?

Global temperatures and greenhouse gases have dramatically increased over the past two centuries

Before we can determine if global increases in temperature and greenhouse gases are a recent phenomenon and if these increases are unusual, we must establish how temperatures and greenhouse gases have changed in the past. Since about 1880, there have been enough direct measurements of land and ocean temperatures that NASA Goddard Institute for Space Studies has been able to generate a graph of global temperature change over time. This graph is shown in **FIGURE 57.3**. Comprising thousands of measurements from around the world, the graph shows global temperatures have increased 1.1°C (2.0°F) from 1880 through 2020. In fact, of the 20 warmest years since 1880, all of them have occurred between 1998 and 2020. Moreover, the seven warmest years since 1880 have been the past 7 years (2014–2020).

While an increase in average global temperature of 1.1°C (2.0°F) may not sound very substantial, it is not evenly distributed around the globe. As the map in **FIGURE 57.4** shows, some regions, including parts of Antarctica, have experienced cooler temperatures. Other regions, including certain ocean areas, have experienced no change in temperature. However, regions such as those in the extreme northern latitudes have experienced increases of 1°C to 4°C (1.8°F to 7.2°F). The substantial increases in temperatures in the

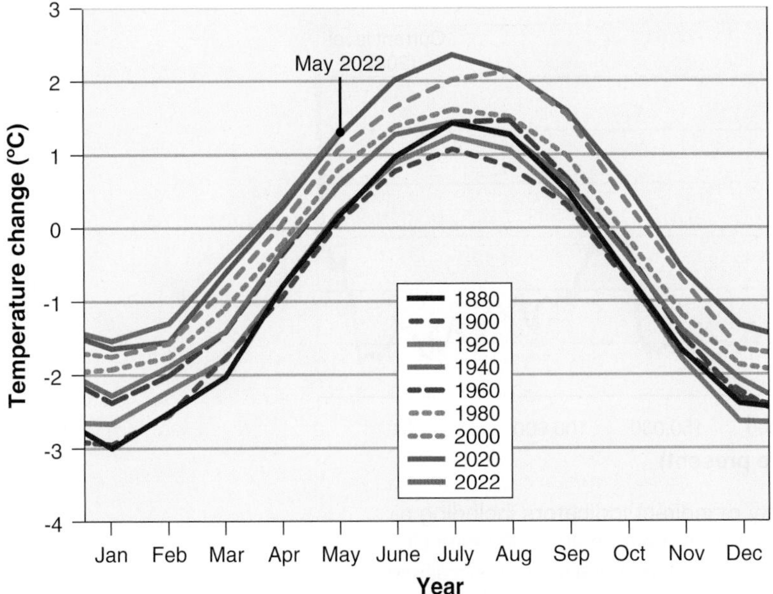

FIGURE 57.3 Changes in mean global temperatures over time. Although annual mean temperatures can vary from year to year, temperatures have slowly increased from 1880 to 2022. This pattern becomes much clearer when scientists compute the average temperature for 20-year periods. Note that the zero value is set at the mean temperature from 1980 to 2015. *(Data from https://data.giss.nasa.gov/gistemp/news/20170714/cycle_201706_1600.tif)*

northern latitudes have caused—among other problems—nearly 45 percent of the northern ice cap to melt.

The data collected by NASA demonstrate that the globe has been slowly warming during the past 120 years. However, it is possible that such changes in temperature are a natural phenomenon. If we want to know whether these changes are unusual, we must examine a much longer span of time. As you may recall from Module 12, we discussed how scientists have documented changes in greenhouse gases and global temperatures over the past 500,000 years. Scientists have estimated temperatures based on the changing species composition of preserved protists, known as foraminifera, because different species prefer different temperatures (Figure 12.4 on page 131). These data demonstrate that the planet's temperature has experienced major shifts during periods of warming and cooling. Scientists have also discovered that ancient air bubbles have been preserved in ice for hundreds of thousands of years. From these bubbles, they can determine historic concentrations of CO_2. In addition, the isotopes of oxygen in these air bubbles indicate the air temperature of the atmosphere when the bubbles were captured in the ice (Figure 12.5 on page 132). Based on these indirect estimates of CO_2 in the atmosphere and global temperatures, scientists have observed a pattern of increasing and decreasing concentrations of CO_2 being closely associated with increasing and decreasing temperatures (Figure 12.6 on page 133), which is exactly what we expect based on our understanding of how greenhouse gases work.

We can also look at changes in CO_2 concentrations over the past 400,000 years (**FIGURE 57.5**). Notice that for over 400,000 years, the atmosphere never contained more than 300 ppm of CO_2. In contrast, from 1958 to 2021 the concentration of CO_2 in the atmosphere rapidly climbed from 310 to 415 ppm. Thus, the recent rise of CO_2 in the atmosphere is unprecedented during the past 400,000 years.

CO_2 is not the only greenhouse gas concentration that has increased. As you can see in **FIGURE 57.6**, methane and nitrous oxide show a pattern of increase that is similar to the pattern we saw for CO_2. For all three gases, the concentrations have exhibited small changes during the past 10,000 years. After 1800, however, concentrations of the three gases began to rise dramatically and are now far above what

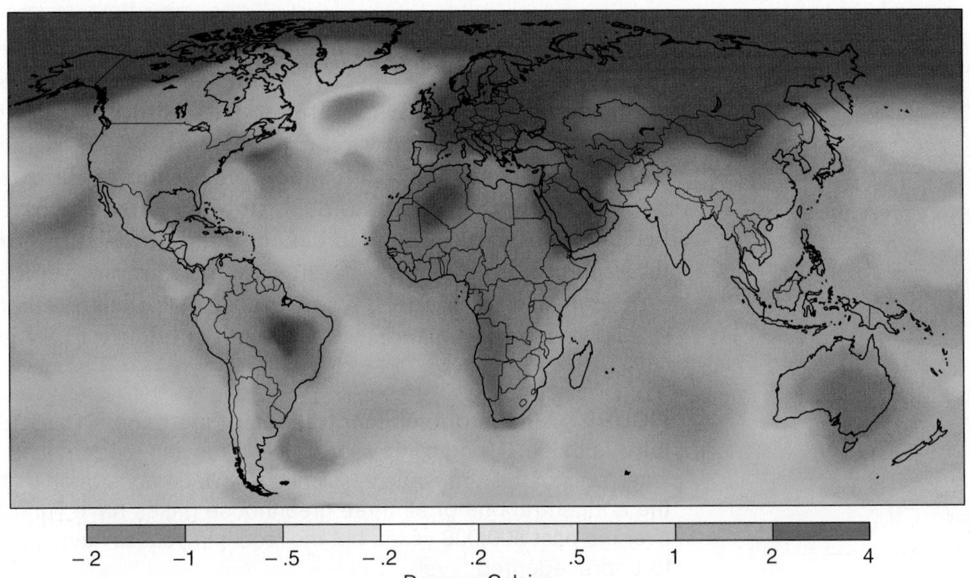

FIGURE 57.4 Changes in mean annual temperature in different regions of the world. During the period 2016 to 2020, some regions became cooler, some regions had no temperature change, and the northern latitudes became substantially warmer than the long-term average temperature. The surface temperatures plotted on the map represent differences relative to the average temperature from 1951 to 1980. *(Data from https://svs.gsfc.nasa.gov/4882)*

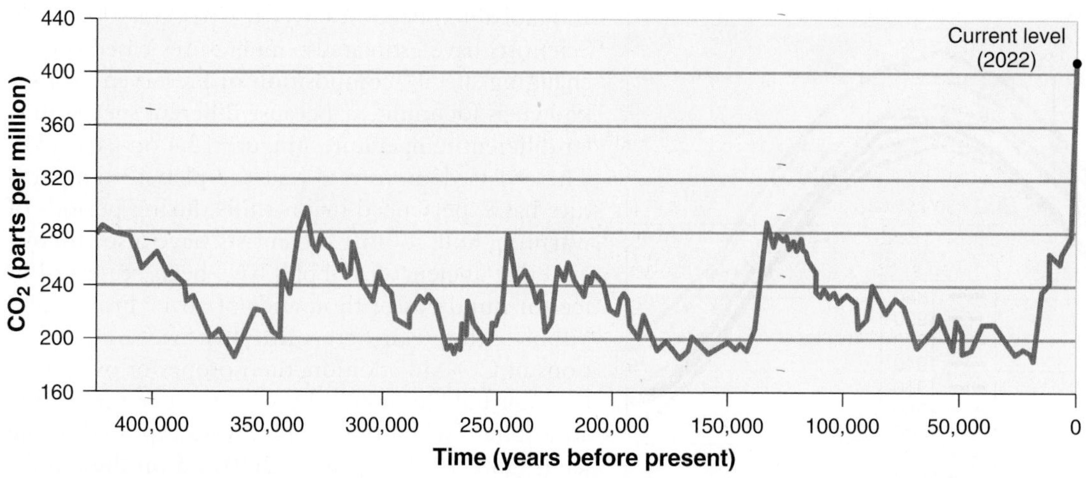

FIGURE 57.5 Historic CO_2 concentrations. Using a variety of indirect indicators including air bubbles trapped in ancient ice cores, scientists have found that for more than 400,000 years CO_2 concentrations never exceeded 300 ppm. After 1950, CO_2 concentrations have sharply increased to their current level of more than 415 ppm. *(Data from https://data.giss.nasa.gov/gistemp/graphs_v3/)*

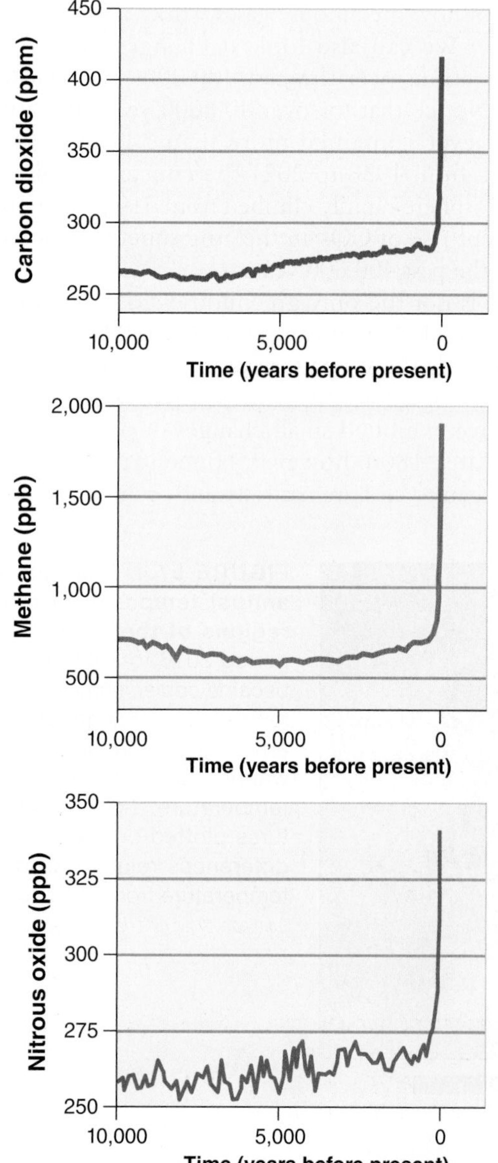

existed during the previous 10,000 years. Given what we now know about the anthropogenic sources of greenhouse gases, this increase in greenhouse gases occurred because this time period marks the start of the Industrial Revolution when humans began burning large amounts of fossil fuel that produced a variety of greenhouse gases.

Climate Models Predicting Future Global Temperatures

Just as indirect indicators can help us get a picture of what the temperature has been in the past, computer models can help us predict global temperatures in the future. Researchers can determine how well a model approximates real-world processes by applying it to a time in the past for which we have accurate data on conditions such as air and ocean temperatures, CO_2 concentration, extent of vegetation, and sea ice coverage at the poles. Modern models accurately reproduce the temperature fluctuations that we measure over large spatial scales. From this work modelers are fairly confident that climate models capture the most significant features of today's climate.

Although climate models cannot forecast future climates with total accuracy, as the models improve scientists have been able to place more confidence in their predictions of future temperature changes. Because assumptions vary among different climate models, when multiple models predict similar changes, we can have increased confidence that the

FIGURE 57.6 Historic concentrations of methane, CO_2, and nitrous oxide. Using samples from ice cores and modern measurements of the atmosphere, scientists have demonstrated that the concentrations of all three greenhouse gases have varied over the past 800,000 years, but the recent increases have risen to unprecedented levels. *(Data from https://www.epa.gov/climate-indicators/climate-change-indicators-atmospheric-concentrations-greenhouse-gases)*

predictions are robust. **FIGURE 57.7** shows recent predictions from climate models. Scientists generally agree that average global temperatures will rise by 1.8°C to 4°C (3.2°F–7.2°F) by the year 2100, depending on whether CO_2 emissions experience slow, moderate, or high growth rates over time. As you can see in the figure, northern latitudes will continue to experience temperatures well above these averages.

As temperatures increase, scientists also predict that we will experience changes in our current climates. For example, we expect to experience more extreme weather events, including more heat waves, more intense snowstorms and rainstorms, and more intense hurricanes. In addition, we may experience changing patterns of precipitation around the planet, in part because Hadley cell air circulation, which we discussed in Module 22, may change with global warming. At the same time, we may experience changes in ocean currents, which we will explore further later in this module. With regional climate changes, regions receiving increased precipitation

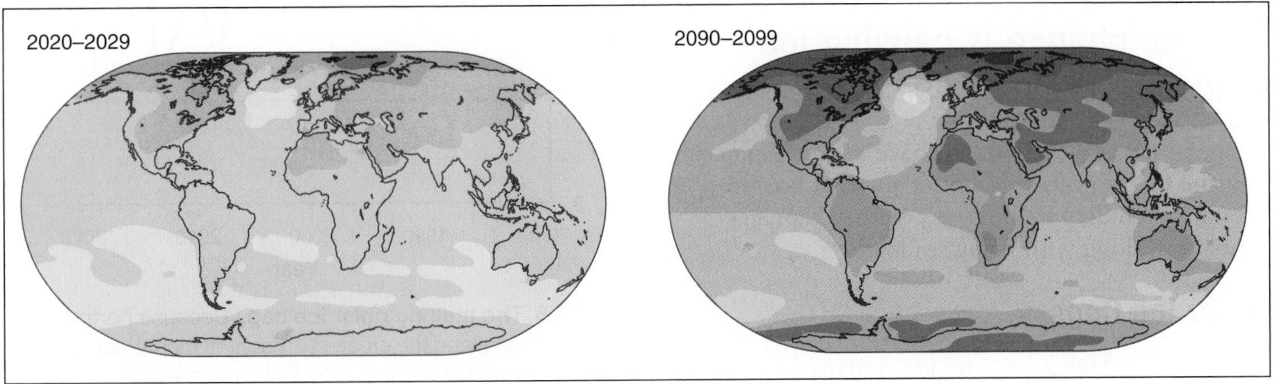

(a) Low increase in CO_2

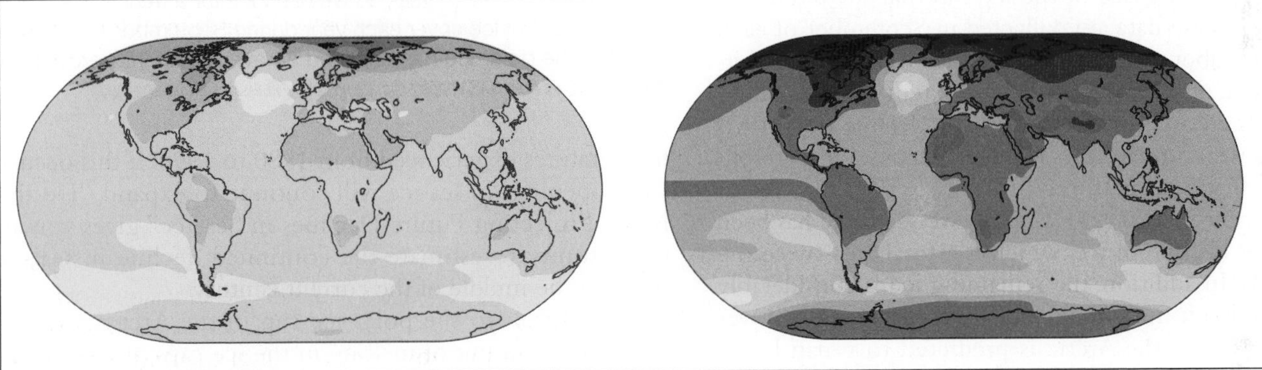

(b) Moderate increase in CO_2

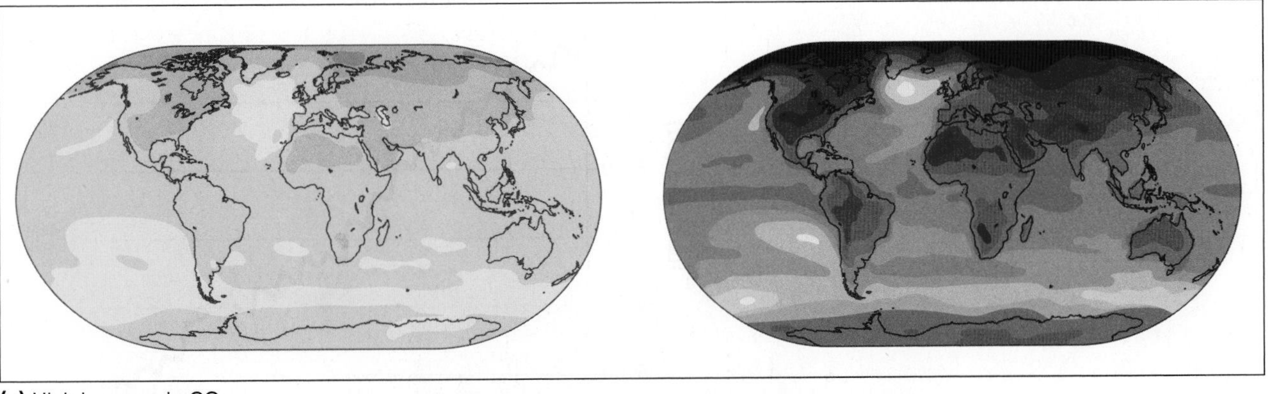

(c) High increase in CO_2

Surface temperature change (°C)

0 1 2 3 4 5 6 7 8

FIGURE 57.7 Predicted increase in global temperatures by 2100. The predictions depend on whether we expect (a) low, (b) moderate, or (c) high increases in how much CO_2 the world emits during the current century. These changes in temperature are relative to the mean temperatures from 1961 to 1990. *(Data from Moberg, A., et al. Highly variable Northern Hemisphere temperatures reconstructed from low- and high-resolution proxy data.* Nature *433 (2005): 613–617.)*

would benefit from an increased recharge to aquifers and higher crop yields, but they could also experience more flooding, landslides, soil leaching of nutrients, and soil erosion. In contrast, other regions of the world are predicted to receive less precipitation, making it more difficult to grow crops and requiring greater efforts to supply water.

57-3 How has global climate change affected the environment?

Global climate change is causing ice to melt and sea levels to rise

Warming temperatures are expected to have a wide range of impacts on the environment. Many of these effects are already happening, including melting of polar ice caps, glaciers, and permafrost, along with rising sea levels.

Melting Polar Ice Caps

As we have seen, the Arctic has already warmed by as much as 4°C (7.2°F). **FIGURE 57.8** illustrates the extent of the reduction in the size of the ice cap that surrounds the North Pole. These data are collected in September of each year, which is about the time that the sea ice has reached its minimum extent. As you can see, the extent of sea ice fluctuates, but there is an overall trend of a decline from 1979 to 2021. If we compare the average extent of ice present during the first 5 years of monitoring (1979–1983) versus the most recent 5 years (2017–2021), we see there has been a 38 percent decline in ice, which is a total area twice the size of Texas. In addition, the remaining ice is considerably thinner, making it more vulnerable to future melting. Over the next 70 years, the Arctic is predicted to warm by an additional 4°C to 7°C (7°F–13°F) compared to the mean temperatures experienced from 1980 to 1990. If this occurs, large openings in sea ice will continue to expand. "Do the Math: Projecting Future Declines in Sea Ice" gives you an opportunity to estimate the continued decline in sea ice through the middle of the current century.

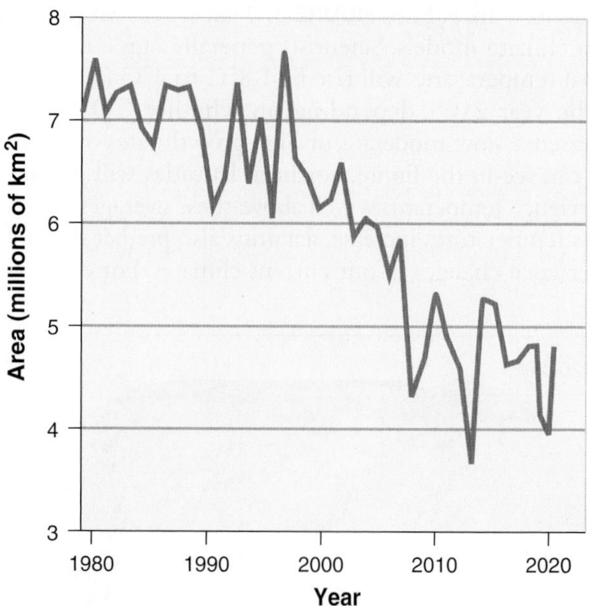

FIGURE 57.8 The melting polar ice cap. Because northern latitudes have experienced the greatest amount of global warming, the extent of the ice cap near the North Pole has been declining over the past 4 decades. The polar ice cap reaches its minimum late in the summer of each year, so we can look for a trend by examining the extent of ice cover each year during September. From 1979 to 2021, the polar ice has declined an average of 13 percent per decade. *(Data from NASA https://climate.nasa.gov/vital-signs/arctic-sea-ice/)*

In addition to the polar ice cap in the Arctic, we also see a large amount of melting in the ice caps of Greenland and Antarctica. As you can see in **FIGURE 57.9**, sea ice mass

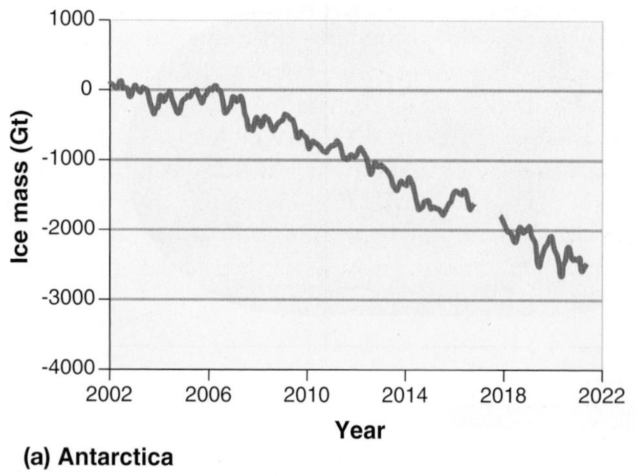

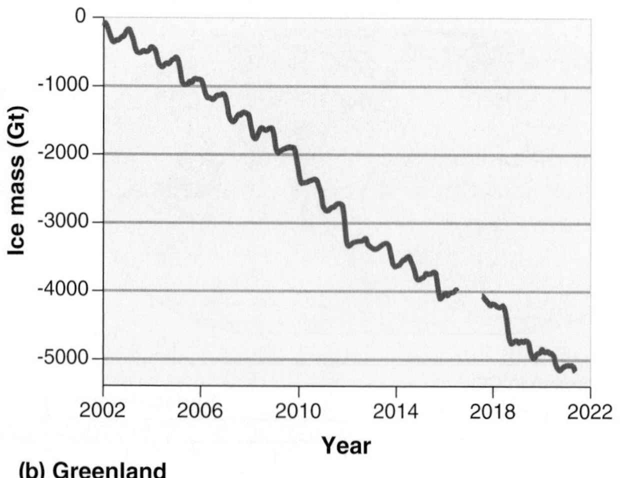

FIGURE 57.9 Declining ice in Antarctica and Greenland. Measurements of lost ice mass from 2002 to 2021 have detected declines in both (a) Antarctica and (b) Greenland. In both graphs, the *y* axes zero is set at 2002 to show the decline in ice compared to 2002. The gap in the data represents data lost between NASA missions. *(Data from NASA https://climate.nasa.gov/vital-signs/ice-sheets/)*

DO THE MATH Projecting Future Declines in Sea Ice

Preparing for the AP® Exam

Polar sea ice has been declining at a rate of 13 percent per decade since 1979, which means that 87 percent remains after each decade. Given that the extent of sea ice was approximately 8 million km^2 in 1979, how much sea ice was present in 1989, 1999, 2009, and 2019? To do this calculation, we have to calculate the amount of sea ice decline in each of the past 3 decades (rounding to the nearest hundredth):

$$\begin{aligned}\text{Sea ice in 1989} &= \text{Sea ice in 1979} \times 87\% \\ &= 8 \text{ million km}^2 \times 0.87 \\ &= 6.96 \text{ million km}^2\end{aligned}$$

$$\begin{aligned}\text{Sea ice in 1999} &= \text{Sea ice in 1999} \times 87\% \\ &= 6.96 \text{ million km}^2 \times 0.87 \\ &= 6.06 \text{ million km}^2\end{aligned}$$

$$\begin{aligned}\text{Sea ice in 2009} &= \text{Sea ice in 1999} \times 87\% \\ &= 6.06 \text{ million km}^2 \times 0.87 \\ &= 5.27 \text{ million km}^2\end{aligned}$$

$$\begin{aligned}\text{Sea ice in 2019} &= \text{Sea ice in 2009} \times 87\% \\ &= 5.27 \text{ million km}^2 \times 0.87 \\ &= 4.58 \text{ million km}^2\end{aligned}$$

YOUR TURN

1. If the annual rate of sea ice melting were to continue declining at this same rate of 13 percent per decade, how much ice should be left by 2049?
2. What percentage of the original 1979 sea ice will remain in 2049?

was measured in Antarctica and Greenland from 2000 to 2021. During this time, Antarctica lost nearly 2,500 gigatons (about 2.8 billion tons) of ice while Greenland lost nearly 5,000 gigatons (5.5 billion tons) of ice. Of course, the mass of the ice caps includes both area and thickness. This makes the change in the polar ice cap of Antarctica quite interesting. It has shown a small increase in total area over the past 3 decades, but the ice cap is losing thickness due to melting, so its overall mass has been reduced. Current evidence shows that the melting rate of these ice-covered regions is continuing to increase. As we will see, such large amounts of melted ice have caused sea levels to rise.

Melting Glaciers

Global warming has caused the melting of many glaciers around the world. Glacier National Park in northwest Montana, for example, had 150 individual glaciers in 1850 but has only 25 glaciers today. It is estimated that by 2030 Glacier National Park will no longer have any glaciers. In fact, we can see the impact by looking at photos that were taken many decades ago and comparing them to photos from today at the same location. As you can see in **FIGURE 57.10**, the melting of glaciers has been substantial.

The loss of glaciers is not simply a loss of an aesthetic natural wonder. In many parts of the world the melting of glaciers starting each spring provides a critical source of water for many communities. Historically, these glaciers partially melted during the spring and summer and then grew back to their full size during the winter. However, as summers become warmer, glaciers are melting faster than they can grow back in the winter, which leaves some people without a reliable water supply. In addition, when the melting exposes underlying soil, this soil will absorb more solar radiation than the highly reflective ice caps of the glaciers, which increases the albedo effect that we discussed in Module 22. The ability of the underlying soil to absorb more solar radiation favors additional warming of the air, so it provides a positive feedback loop.

Ocean Currents

Global ocean currents may shift as a result of more fresh water being released from melting ice. In Module 23 we saw that ocean currents have major effects on the climate of nearby continents. If the currents change, the distribution of heat on the planet could be disrupted. Scientists are particularly concerned about the thermohaline circulation, which, as we saw in Module 23, is a deep ocean circulation

(a)

(b)

FIGURE 57.10 Melting glaciers. (a) In Glacier National Park, photos from a century ago show the many glaciers that were in the park. (b) Photos taken from the same location in modern times show that these glaciers have substantially receded or disappeared due to global warming.
(a & b: Science History Images/Alamy Stock Photo)

driven by water that comes out of the Gulf of Mexico and moves up to Greenland where it becomes colder and saltier and sinks to the ocean floor. This sinking water mixes with the deep waters of the ocean basin, resurfaces near the equator, and eventually makes its way back to the Gulf of Mexico (see Figure 23.2 on page 273). This circulating water moves the warm water from the Gulf of Mexico up toward Europe and moves cold water from the North Atlantic down to the equator. However, increased melting from Greenland and the northern polar ice cap could dilute the salty ocean water sufficiently enough to stop the water from sinking near Greenland and thereby shutting off the thermohaline circulation. If this occurs, much of Europe would experience significantly colder temperatures, especially along the coastlines.

Warming Soils and Permafrost

Global soils contain more than twice as much carbon as the atmosphere. Higher temperatures are expected to increase the biological activity of decomposers in these soils, which should have cascading effects on the rest of the food web that lives in the soil. Given that this decomposition will cause the release of additional CO_2 from the soil into the atmosphere, the temperature change will be amplified even more.

A similar, but more troubling, scenario is expected in tundra biomes containing permafrost. You may recall from the discussion on biomes in Module 2 that permafrost is permanently frozen ground that exists in the cold regions of high altitudes and high latitudes, which include the tundra and boreal forest biomes. About 20 percent of land on Earth contains permafrost; in some places it can be as much as 1,600 m (1 mile) thick. Melting of the permafrost causes overlying lakes to become smaller as the lake water drains deeper down into the ground. Melting can also cause substantial problems with human-built structures that are anchored into the permafrost, including houses and oil pipelines. As the frozen ground melts, it can subside and slide away, causing these structures to collapse.

Melting permafrost also means that the massive amounts of organic matter contained in the tundra will begin to decompose. Because this decomposition would be occurring in wet soils under low-oxygen conditions, it would release substantial amounts of methane, increasing the concentration of this potent greenhouse gas. This chain of events could produce a positive feedback in which the warming of Earth melts the permafrost, releasing more methane that causes further global warming.

Rising Sea Levels

The rise in global temperatures affects sea levels in two ways. First, the water from melting glaciers and ice sheets on land adds to the total volume of ocean water. Second, as the water of the oceans becomes warmer, it expands. As a result of both processes, sea levels have risen 210 mm (8.3 inches) since 1900, as illustrated in **FIGURE 57.11a**. Over the past 30 years, the sea level has risen by 3.4 mm per year. At this rate, scientists predict that by the end of the twenty-first century, sea levels could rise an additional 130 to 540 mm (5–21 inches).

Rising sea levels could endanger coastal cities and low-lying island nations by making them more vulnerable to flooding, especially during storms, with more saltwater intrusion into aquifers and increased soil erosion (Figure 57.11b). Currently, 100 million people live within 1 m (3 feet) of sea level. The actual impact on these areas of the world will depend on the steps taken to mitigate these effects. For example, as we saw in Module 48, some countries may be able to build up their shorelines with dikes to prevent inundation from rising sea levels. Countries possessing less wealth are not expected to be able to respond as effectively to coastal flooding and their residents will likely be displaced by rising sea levels. At the same time, rising sea levels will create new habitats for aquatic organisms along the newly flooded shorelines while causing some deep-water species to no longer be in the photic zone that experiences sunlight.

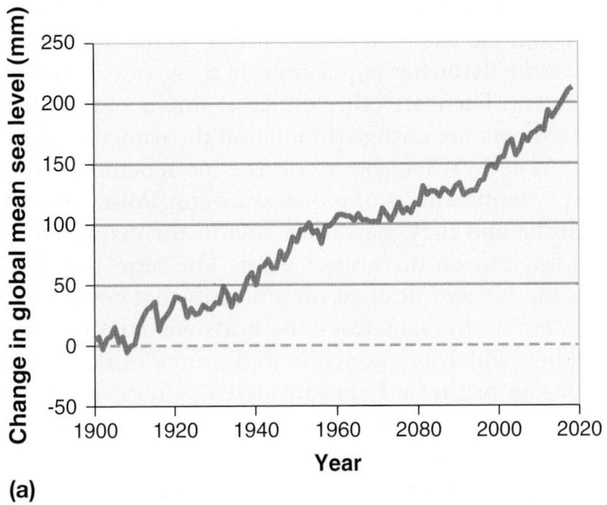

FIGURE 57.11 Rising sea levels. (a) Since 1900, sea levels have risen by 210 mm (8.3 inches). Future sea level increases are predicted to be 180 to 590 mm (7 to 23 inches) above 1999 levels by the end of this century. Note that sea level is graphed by setting the sea level in 1900 equal to zero. (b) Nearly 100 million people live within 1 m (3 feet) of sea level, such as on this island in the Maldives in the Indian Ocean. *(a: Data from NASA, https://climate.nasa.gov/vital-signs/sea-level/; b: Mohamed Abdulla Shafeeg/Getty Images)*

57-4 How is global climate change affecting populations?

Global climate change is affecting the timing and performance of plants and animals

The warming of the planet is not only affecting polar ice caps and sea levels but it is also affecting populations of organisms. These effects range from temperature-induced changes in the timing of plant flowering and animal behavior to the ability of plants and animals to disperse to more hospitable habitats.

During the last decade, the IPCC reviewed approximately 2,500 scientific papers that reported the effects of warmer temperatures on plants and animals. The panel concluded that over the preceding 40 years, the growing season for plants had lengthened by 4 to 16 days in the Northern Hemisphere, with the greatest increases occurring in higher latitudes where the temperature increase has been the largest. Indeed, scientists are finding that in the Northern Hemisphere many species of plants now flower earlier, amphibians start breeding earlier, birds arrive at their breeding grounds earlier, and insects emerge earlier. At the same time, the geographic ranges occupied by different species of plants, birds, insects, and fish have shifted toward both poles. These changes not only affect wild plants and animals, but also domesticated plants and animals that are bred to grow well in particular climates. For example, you may recall from Unit 1 that grape growers in France are finding it harder to grow some varieties of wine grapes due to rising air temperatures, rising soil temperatures, and changes in precipitation patterns. At the same time, grape growers in England, who have historically experienced a cooler climate, are now able to successfully grow French varieties of grapes.

Climate change has the potential to affect human health and agriculture. Continued warming of the planet could affect the geographic range of temperature-limited disease vectors, allowing them to shift from the tropics toward the poles. For example, the mosquitoes that carry West Nile virus and malaria could spread beyond their current geographic range and bring health threats to regions that were once relatively untouched. Infectious diseases and bacterial and fungal illnesses might extend over a wider range than they do at present. Some of these diseases directly impact human health, while others are crop pests that cause a decline in the amount of crops that can be produced, which lessens the amount of food we have to feed the world. Such changes in human health and agriculture has the potential to motivate large numbers of people to move to more desirable locations.

Rapid temperature changes have the potential to cause harm if organisms do not have the option of moving to more hospitable climates and do not have sufficient time to evolve adaptations. Historically, organisms have migrated to different latitudes and elevations in response to changing climates. This ability to migrate is one reason that temperature shifts have not been catastrophic over the past few million years. Today, however, fragmentation of certain habitats by roads, farms, and cities has made movement to cooler latitudes and cooler, higher elevations more difficult. This fragmentation may be the primary factor that allows a warming climate to cause the extinction of species.

The pied flycatcher (*Ficedula hypoleuca*) is a bird that provides an interesting example. In the Netherlands, the pied flycatcher has evolved to synchronize the time that its chicks hatch with the time of peak abundance of caterpillars, a major

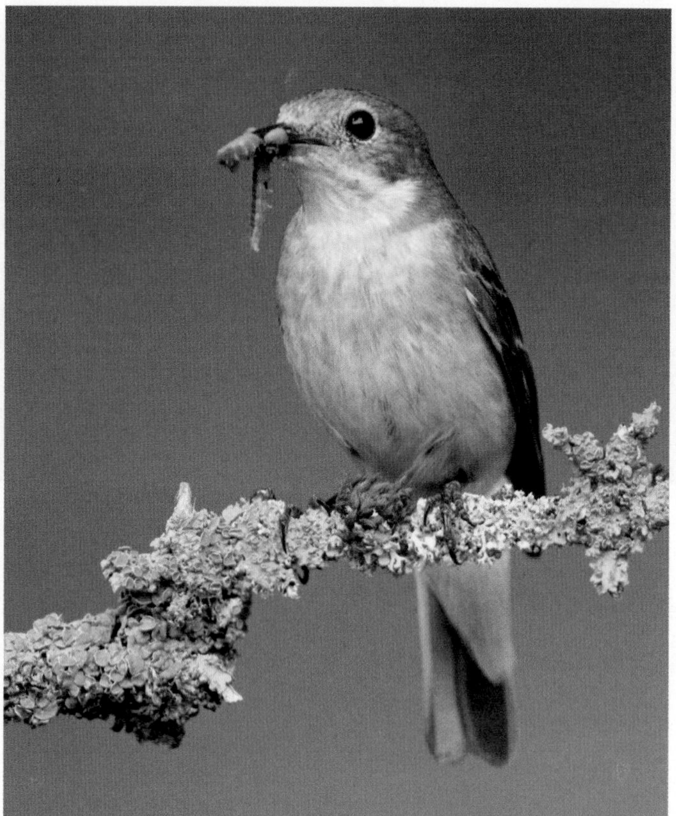

FIGURE 57.12 Effects of global warming on the pied flycatcher. Due to global warming in the Netherlands, the bird's main food source, a caterpillar, now becomes abundant 2 weeks earlier. However, the time when eggs hatch for the flycatcher has not changed. As a result, the birds hatch after the caterpillar population has begun to decline. *(Kats Edwin/AGE Fotostock)*

source of food for the newly hatched birds (**FIGURE 57.12**). In 1980, when the birds completed their spring migration from Africa to the Netherlands, the date of their chicks hatching preceded the peak in caterpillar abundance by a few days, so there was plenty of food for the new chicks. Twenty years later, however, warmer spring temperatures caused trees to produce leaves earlier. Because the caterpillars feed on tree leaves, the peak abundance of caterpillars now occurs about 2 weeks earlier than it did in 1980. However, the pied flycatchers still migrate from Africa at the same time, so the hatching date of the pied flycatcher has not changed. Thus, by the time the chicks hatch, the caterpillars are no longer abundant and the hatchlings lack a major source of food; this has caused the flycatcher populations in these areas to decline by 90 percent. There are other similar examples of the effects of rapid temperature change throughout the natural world.

As we noted earlier, the Arctic is experiencing some of the most extreme effects of global warming. Polar bears live in the Arctic and they play a key role in the ecosystem by hunting for seals on the polar ice cap. The bears search for holes in the ice and pounce on any seals that come up to the holes for air. In many cases, the bears will consume only seal blubber, which is a concentrated source of the energy critical for an organism living in such a cold environment. The portion of the seal carcass that remains is a significant food source for other animals, including the Arctic fox (*Vulpes lagopus*). As the polar ice cap retreats far away from the land each summer, there are fewer weeks in which polar bears can stand on the ice to hunt. In some regions, the sea ice melts 3 weeks earlier than it did 30 years ago. Because of the shorter time they are able to hunt for seals, male polar bears near Hudson Bay in Canada currently weigh 68 kg (150 pounds) less than they weighed 30 years ago. With less seal predation, there will be fewer seal carcasses for other animals like the Arctic fox to consume, so there is a cascade of effects when polar bears are affected by global warming. While the populations of polar bears near Hudson Bay are declining, in other regions of the Arctic their populations are stable or increasing.

Large openings in the Arctic sea ice will also affect humans by creating new shipping lanes that will reduce the distance some ships have to travel by thousands of kilometers. Also, it is estimated that nearly one-fourth of all undiscovered oil and natural gas lies under the polar ice cap and a melted polar ice cap might make these fossil fuels more easily obtainable. However, the combustion of these fossil fuels would further facilitate global warming.

In this module we learned that greenhouse gases have been increasing during the past 2 centuries. Both the rate of increase and the concentrations of greenhouse gases currently in the atmosphere are unprecedented. Associated with these increases are higher temperatures, especially at northern latitudes. We also examined the many consequences of global warming including melting ice, rising sea levels, thawing permafrost, and changing populations of organisms.

Module 57 AP® Review

Preparing for the AP® Exam

Learning Goals Revisited

57-1 How have CO_2 concentrations changed over the past 7 decades?

The concentration of CO_2 in the atmosphere has been steadily increasing. The production of CO_2 differs among nations, both in terms of total production and per capita production.

57-2 How have temperatures and greenhouse gases varied historically?

Global temperatures have increased 1.1°C (2.0°F) from 1880 through 2020. Of the 20 warmest years since 1880, all have occurred between 1998 and 2020. While the average global temperature has increased by 1.1°C (2.0°F), far northern latitudes have experienced increases up to 4°C (7.2°F). Three major greenhouse gases (CO_2, methane, and nitrous oxide) have all dramatically increased their concentrations in the atmosphere during the past 2 centuries. Climate models predict continued warming through the year 2100, especially at far northern latitudes.

57-3 How has global climate change affected the environment?

Global warming has caused declines in the polar ice cap, declines in the ice masses of Greenland and Antarctica, melting of glaciers, thawing of permafrost, and increases in sea level.

57-4 How is global climate change affecting populations?

Global warming has caused changes in the dates of plant flowering, bird migrations, amphibian breeding, insect emergence, and the length of growing seasons.

Practice Math and Graphing

Preparing for the AP® Exam

Answer the following questions. Be sure to show all your work.

1. Practice Math

Figure 57.2 shows the current amount of CO_2 emissions for the top 10 emitting countries. In the table shown, you can see the CO_2 emissions in 2019, with a comparison to emissions for each country in 1990.

(a) Based on these data, calculate the percent change in CO_2 emissions from 1990 to 2019. Round to the nearest whole number.

(b) Which country has increased its emissions by the greatest percentage over the 29-year period and which country has decreased its emissions by the greatest percentage?

Country	CO_2 emissions in 2019 (billions of metric tons)	CO_2 emissions in 1990 (billions of metric tons)	Percent change
China	9.9	2.1	
United States	4.7	4.8	
India	2.3	0.5	
Russian Federation	1.6	2.2	
Japan	1.1	1.0	
Germany	0.6	0.9	
South Korea	0.6	0.2	
Iran	0.6	0.2	
Canada	0.6	0.4	
Saudi Arabia	0.5	0.2	

(Data from International Energy Agency. 2021. CO_2 emissions from fuel combustion. Union of Concerned Scientists, 2021. IEA Atlas of Energy.)

2. Practice Graphing

Using the percentage data that you calculated in "Practice Math," create a bar graph that shows how each country has altered its CO_2 emissions in 2019 as a percentage of its emissions in 1990.

AP® Practice Questions

Multiple-Choice Questions

1. What is the evidence that solar radiation is less important to global warming than the increase in greenhouse gases?
   (a) Temperatures have increased more in the summer than the winter.
   (b) Temperatures have increased more in the winter than the summer.
   (c) There is a historic correlation of CO_2 and temperature.
   (d) There is a lack of temperature change in some areas.

2. If the annual rate of CO_2 increase is 2.5 ppm and the concentration in 2021 is 415 ppm, what concentration would you expect in 2051?
   (a) 420 ppm
   (b) 490 ppm
   (c) 505 ppm
   (d) 525 ppm

3. How much has the average global temperature increased from 1880 to 2020?
   (a) 0.5°C
   (b) 1.1°C
   (c) 2.0°C
   (d) 3.1°C

4. Which data are used to estimate historic temperatures and carbon dioxide concentrations?
   (a) marine mammals
   (b) air trapped in ice cores
   (c) glacier depth
   (d) fossilized birds

5. How might climate change increase the range of pests?
   (a) higher intensity weather events
   (b) changing precipitation patterns
   (c) increasing number of heat waves
   (d) decreased duration of cold spells

6. Temperatures have been rising since the proliferation of industry. Which is the correct way to calculate the percent change in temperature, based on the data shown below?

Year	Average temperature
1920	13.83°C
2020	14.90°C

   (a) $\frac{14.90 - 13.83}{13.83} \times 100$
   (b) $\frac{13.83 - 14.90}{14.90} \times 100$
   (c) $\frac{1920 - 2020}{1920} \times 100$
   (d) $14.90 - 13.83$

Free-Response Question

Loggerhead Sea Turtles, which can be found in the Atlantic, Indian and Pacific oceans, come ashore to lay eggs in the sand, laying clutches ranging from 110 to 130 eggs at a time. These loggerhead sea turtles have temperature-dependent sex determination, meaning that the ratio of males to females is determined by the temperature of the sand the clutch has been laid in, with warmer temperatures producing more females. A study conducted in 2008 off the coast of Japan found that the temperature of the oceans where loggerheads laid their eggs had climbed 0.5°C.

(a) **Explain** the possible effects on the loggerhead population from the warming oceans and possibly warming beaches. (1 pt.)
(b) **Describe** the effect a warming beach might have on a sea turtle's biotic potential. (1 pt.)
(c) **Propose a solution** to maintain the population of loggerhead sea turtles to mitigate the impacts of a warmer ocean and beach. (1 pt.)

Fluctuations in climate have frequently occurred in Earth's history. Ice cores can be analyzed by using air trapped in the cores. Carbon dioxide levels and temperatures have also been measured. It was found that in the 800,000 year history of the planet, carbon dioxide levels have fluctuated between 170 and 300 parts per million. At the same time, temperatures have also fluctuated between 9°C below modern surface temperatures to 2°C above.

(d) **Describe** one way species have responded to changes in global temperature over the last 800,000 years. (1 pt.)
(e) **Calculate** the range of total carbon dioxide change in those last 800,000 years. Show your work. (2 pts.)
(f) **Calculate** the range of total temperature change in those last 800,000 years. Show your work. (2 pts.)
(g) **Calculate** the percent change in carbon dioxide levels from the maximum during those 800,000 years and the current total of 419 ppm. Show your work. (2 pts.)

Module 58

Unit 9 | 55 | 56 | 57 | 58 | 59

Ocean Warming and Ocean Acidification

In the previous module we explored how human activities have caused a rise in greenhouse gases and global temperatures during the past 200 years. We then explored how these changes were causing ice caps and glaciers to melt, sea levels to rise, and climate and weather patterns to change, all of which are impacting populations and ecosystems. In this module we will examine how global climate change is altering the temperature and acidity of ocean water. We will also finish our coverage of global climate change by discussing two major international agreements that are focused on slowing and eventually reversing global climate change.

Learning Goals

After reading this module you should be able to

58-1 describe how ocean warming is altering ocean ecosystems.

58-2 explain how climate change affects ocean pH.

58-3 explain the international agreements on global climate change.

58-1 How is ocean warming altering ocean ecosystems?

Ocean warming is affecting a variety of marine species

Given that water bodies exchange heat with the atmosphere, the warming of Earth's air due to increased greenhouse gases has also been warming small and large water bodies on our planet. For example, many lakes around the world have warmed by approximately 2°C, over just the past 4 decades, which means these lakes are warming at a faster rate than the air. Similarly, the oceans have been warming by 1.5°C to 2.0°C. Oceans cover about 70 percent of the planet and their massive volume can absorb a great deal of heat. It is estimated that the oceans have absorbed about 90 percent of the excess heat that humans have produced through global warming. In fact, the top 700 m (2,300 feet) of ocean waters have increased by 1.5°C (2.7°F) during the past century; without this absorption of heat, temperatures would have increased much more on land. This ocean warming has had a number of impacts on marine species.

Impacts on Marine Species

The warming of the oceans can alter the distribution of marine species because each species has a preferred range of temperature where they can optimize their physiology, growth, and reproduction. As the ocean warms, a given habitat experiences a change in temperature, so marine species may try to move to new areas that remain in their preferred temperature range. Some species will be able to locate such areas, while others—such as those in the highest latitudes—may not have cooler options as their habitat warms.

A recent example of marine species shifting their distribution in response to global warming can be found in the North Sea, which is located between England, Norway, and the Netherlands. In the North Sea, the bottom temperature has increased by 2°C over 3 decades. Researchers wondered whether this increase in temperature would be enough to motivate more southern species to move up into the North Sea while existing species in the North Sea might move farther north to colder waters. Over 2 decades, researchers pulled large nets (called "trawls") across the bottom on the North Sea to determine how many fish species were present. At the beginning of the study, the North Sea contained 60 fish species. With 2°C warmer water 2 decades later, there were 90 fish species, as you can see in **FIGURE 58.1**. Most of the new species were fish that formerly lived south of the North Sea but had spread northward as the temperature of the ocean bottom warmed. At the same time, many species that had formerly lived in the North Sea migrated farther north to cooler waters. Such shifts in the distributions of fish not only alter the local ecosystem but can also have large economic impacts on commercial fishing in the North Sea; the incoming species are of low commercial value whereas many of the departing species have high commercial value as seafood sold to consumers.

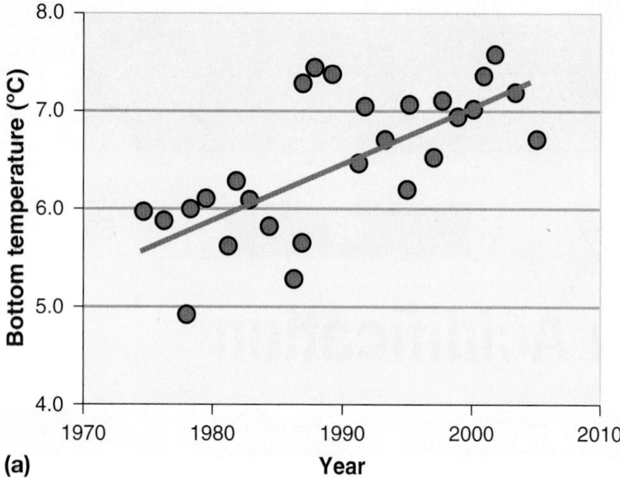

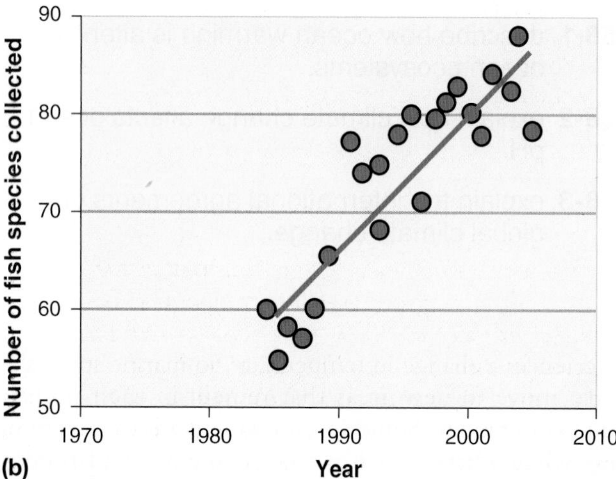

FIGURE 58.1 Effects of ocean warming on fish distributions. (a) The bottom waters of the North Sea have warmed over several decades. (b) As the North Sea warmed, the number of fish species increased from 60 to 90, with many of the new species arriving from more southern waters.

Corals are another group of organisms that are particularly sensitive to global warming because their range of temperature tolerance is quite small. With an increase in peak summer ocean temperatures of just 1°C, corals can experience "bleaching." Coral bleaching occurs when stressed corals eject their mutualistic algae, which provide corals with energy. This loss of algae causes the coral to turn white. The underlying causes of coral bleaching appear to be a combination of warming oceanwater, pollution, decreasing pH, diseases, and sedimentation. While bleaching can be temporary, if it lasts for more than a few weeks, the corals die. While new corals should colonize regions at higher latitudes, it will take centuries before major new reef systems can be built. More coral bleaching is expected from additional global warming in the future, even if climate changes are kept relatively small.

Ocean acidification A process in which an increase in ocean CO_2 causes more CO_2 to be converted to carbonic acid, which lowers the pH of the water.

58-2 How does climate change affect ocean pH?

Increasing CO_2 concentrations are causing ocean acidification

As CO_2 concentrations increase in the atmosphere due to burning fossil fuels, vehicle emissions, and deforestation, more CO_2 is absorbed by the oceans. Although this is beneficial because it reduces CO_2 concentrations in the atmosphere, it causes harmful effects to the oceans. As shown in **FIGURE 58.2**, when CO_2 molecules move from the atmosphere to the water, a large portion of them quickly react with water molecules to form a compound known as carbonic acid (H_2CO_3). Given that carbonic acid is an acid, it readily releases positively charged hydrogen atoms (H^+) and this causes the water to become more acidic. Since this is an equilibrium reaction, an increase in ocean CO_2 causes more CO_2 to be converted to carbonic acid, which lowers the pH of the water in a process known as **ocean acidification**. An important point to remember is that pH represents the concentration of hydrogen ions on a negative log scale. Thus, the change in ocean pH that has occurred, becoming less basic from 8.2 to 8.1, actually represents a 25 percent increase in the number of hydrogen ions in the water.

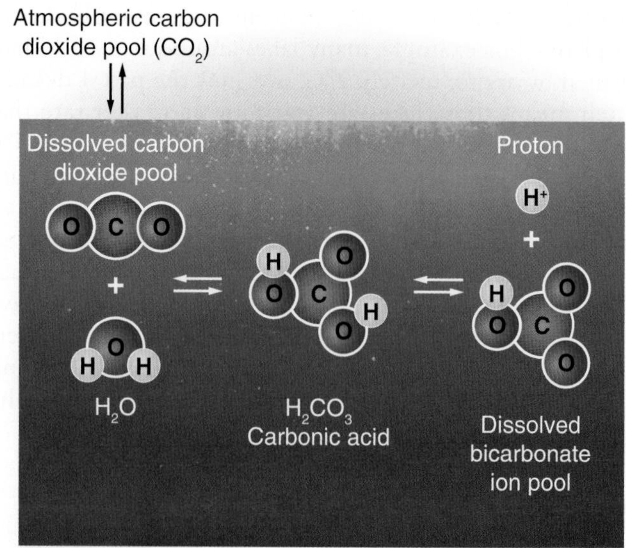

FIGURE 58.2 Ocean acidification. When CO_2 enters the water, it reacts with water to form carbonic acid. Carbonic acid then releases positively charged hydrogen atoms, also known as a proton, which causes the water to become more acidic.

(a)

(b)

FIGURE 58.3 Marine animal responding to elevated pH. (a) When the pencil urchin was raised in ocean water with lower pH caused by increased CO_2 concentrations, it grew much less than individuals grown in modern-day ocean pH. (b) The opposite effect was observed in lobsters, with a decline in pH stimulating the lobsters to grow thicker shells and larger bodies. *(a: © Woods Hole Oceanographic Institution, T. Kleindinst; b: © Woods Hole Oceanographic Institution, Ries, J.B., Cohen, A.L., McCorkle, D.C., 2009, Marine calcifiers exhibit mixed responses to CO_2-induced ocean acidification.* Geology *34: 1131–1134.)*

AP® Exam Tip

Be able to differentiate between coral bleaching and ocean acidification. Coral bleaching is the result of increased sea surface temperatures, pollution, decreasing pH, diseases, and sedimentation. Ocean acidification is the result of increased carbon dioxide in the water, which lowers the pH. Although they both harm coral, they do it in different ways.

Ocean acidification is of particular concern for the wide variety of species that build shells and skeletons made of calcium carbonate, including corals, clams, scallops, oysters, and crustaceans. As pH decreases, the calcium carbonate in these organisms can begin to dissolve and the ocean's saturation point for calcium carbonate declines, which makes it harder for organisms to acquire the material they need to build their shells and skeletons. In fact, more acidic conditions can even cause shells to break down. Corals are of particular concern since their calcium carbonate shell is the foundation of coral reefs. This is why the decline of corals appears to be caused by a combination of increased temperatures (as discussed earlier), declining pH in ocean waters, and coral diseases that may be more deadly when the corals are already experiencing the stress of higher temperatures and lower pH.

Declining pH in the oceans can also affect the behavior of aquatic organisms. For example, some species are less capable of detecting predators with decreases in the pH of water. A fascinating discovery found that species differ in how they respond to declining pH. For example, sea urchins grow smaller under lower pH, while several species of crabs, lobsters, and prawns grow larger and build thicker shells in response to lower pH, as you can see in **FIGURE 58.3**. As a result, some species may perform better than others as oceans become more acidic and this can alter the species composition of communities and potentially alter how these ecosystems function.

Ocean acidification can also affect human health and our food supply. For example, for those species of shellfish that have a hard time building their shells under lower pH conditions, the declining pH of marine ecosystems has the potential to cause major harm to the $1 billion shellfish industry in the United Sates that provides a tremendous amount of food to consumers in the United States and around the world. Such a decline also threatens the jobs of tens of thousands employed in the shellfish industry. In addition, you may recall from Module 5 that marine ecosystems are experiencing harmful algal blooms that are known to produce toxins that can harm people and other animals. Under more acidic conditions, marine algae produce more of these toxins.

58-3 What are the international agreements on global climate change?

International agreements include the Kyoto Protocol and the Paris Agreement

When it comes to global warming and global climate change, the science is clear. Humans have caused an increase in greenhouse gases, these greenhouse gases have caused the planet to slowly warm through the greenhouse effect, and the warming is causing dramatic changes in climate, weather, human health, agriculture, and ecosystems. Many of the adverse effects are expected to be in the developing world, which has received disproportionately fewer benefits from the use of fossil fuels that led to the pollution in the first place. It would be impossible for just one nation to pass legislation and expect that to help it avoid the impacts of climate change. To address the problem of global warming, the nations of the world must work together.

In 1997, representatives of the world's nations convened in Kyoto, Japan, to discuss how best to control the emissions contributing to global warming. At this meeting, they drew up the **Kyoto Protocol**, an international agreement which set a goal for global emissions of greenhouse gases from all industrialized countries to be reduced by 5.2 percent below their 1990 levels by 2012. Due to special circumstances and political pressures, countries agreed to different levels of emission restrictions, including a 7 percent reduction for the United States, an 8 percent reduction for the countries of the European Union, and a 0 percent reduction for Russia. Developing nations, including China and India, did not have emission limits imposed by the protocol. These developing nations argued for different restrictions on developed and developing countries because developing countries are unfairly exposed to the consequences of global warming that in large part came from the developed nations during the past century. Indeed, the poorest countries in the world have only contributed to 1 percent of historic carbon emissions but are still affected by global warming. In modern times, rapidly developing countries such as China are among the largest contributors to CO_2 in the atmosphere. Thus, the approach in the Kyoto Protocol was to have the countries that have historically emitted the most CO_2 into the atmosphere pay most of the costs of reducing CO_2 emissions into the atmosphere.

The main argument for the Kyoto Protocol is grounded in the precautionary principle, which we learned about in Module 53. It states that, in the face of scientific evidence that contains some uncertainty, we should behave cautiously. In the case of climate change, this means that since there is sufficient evidence to suggest human activities are altering the global climate, we should take measures to stabilize greenhouse gas concentrations either by reducing emissions or by removing the gases from the atmosphere. The first option includes trying to increase fuel efficiency or switching from coal and oil to energy sources that emit less or no CO_2 such as natural gas, solar energy, wind-powered energy, or nuclear energy.

The second option includes carbon sequestration—an approach that involves taking CO_2 out of the atmosphere, as discussed in the Pursuing Environmental Solutions feature of Unit 4, "Sequestering Carbon to Reduce Global Warming." Methods of carbon sequestration include storing carbon in agricultural soils or retiring agricultural land and allowing it to become pasture or forest, either of which would return atmospheric carbon to longer-term storage in the form of plant biomass and soil carbon. Researchers are also working on cost-effective ways of capturing CO_2 from the air, from coal-burning power plants, and from other emission sources. This captured CO_2 can then be compressed and pumped into abandoned oil wells or the deep ocean. Such technologies are still being developed, so their economic feasibility and potential environmental impacts are not yet known.

In developed countries, reducing CO_2 emissions would require major changes to manufacturing, agriculture, or infrastructure at significant expense and economic impact. In 1997, before the Kyoto Protocol was finalized, the U.S. Senate voted unanimously (95–0) that the United States should not sign any international agreement that lacked restrictions on developing countries or any agreement that would harm the economy of the United States. Since it was never ratified by the U.S. Senate, the Kyoto Protocol is not legally binding on the United States.

A total of 192 countries have ratified the Kyoto Protocol, including most developed and developing countries, although more than 100 developing countries are exempt from any limits on CO_2 emissions including China and India. The United States is the only developed country that has not yet ratified the agreement.

More recently, the U.S. government has taken stronger steps to regulate CO_2 emissions. In 2007, the U.S. Supreme Court ruled that the U.S. Environmental Protection Agency not only had the authority to regulate greenhouse gases as part of the Clean Air Act, but that it was required to do so. As a result, in 2009 the EPA announced it would begin regulating greenhouse gases for the first time. In 2011, major automakers agreed to substantially increase the average fuel efficiency of new cars and light trucks to reduce the production of CO_2 and other greenhouse gases from vehicles. This would allow the United States to reduce greenhouse gases, invest in new automotive technology, reduce its consumption of fossil fuels, and save money. With new technologies, the fuel efficiency of cars and light trucks has increased by 29 percent from 2005 to 2020.

In response to the Kyoto Protocol, reductions in CO_2 emissions in various countries have been mixed. For example, according to the World Bank, the total emissions of greenhouse gases from 1991 to 2018 have decreased by 29 percent for Germany, 24 percent for Russia, and 21 percent for the United States. Emissions in the United States have been declining sharply since 2007 for a variety of reasons, including increased fuel efficiency and an increase in the use of natural gas rather than coal for power generation. In contrast, 2018 greenhouse gas emissions in Canada have increased 2 percent since 1990.

In 2015, the nations of the world met in Paris to discuss what actions could be taken to combat global climate change. The **Paris Climate Agreement** (also known as the **Paris Climate Accord**) was a pledge by 196 countries to keep global warming less than 2°C above pre-industrial

Kyoto Protocol An international agreement that sets a goal for global emissions of greenhouse gases from all industrialized countries to be reduced by 5.2 percent below their 1990 levels by 2012.

Paris Climate Agreement A pledge by 196 countries to keep global warming less than 2°C above pre-industrial levels. *Also known as* the **Paris Climate Accord**.

FIGURE 58.4 The Paris Climate Agreement. The agreement was a pledge by 196 countries to keep global warming less than 2°C above preindustrial levels. *(Pascal Le Segretain/Getty Images)*

levels (**FIGURE 58.4**). To achieve this goal, each country individually decides how it will help in the global effort to reduce greenhouse gases. The agreement went into effect in 2020; however, there is no enforcement of each country's targets by the United Nations.

In this module we learned that the warming of Earth due to human activities has had a number of effects including melting ice, rising sea levels, changing climates, and altered ecosystems. Because climates and ecosystems are complex, future changes are difficult to predict, but advanced computer models predict increased extremes in heat and cold as well as more intense storms that have the potential to impact human health and economies of nations around the world. In the next module we will focus on the effects of ocean warming and ocean acidification.

Module 58 AP® Review

Preparing for the AP® Exam

Learning Goals Revisited

58-1 How is ocean warming altering ocean ecosystems?

Oceans have absorbed about 90 percent of the excess heat produced through global warming, resulting in a 1.5°C to 2.0°C increase. The warming of the oceans can alter the distribution of marine species as they try to optimize their physiology, growth, and reproduction.

58-2 How does climate change affect ocean pH?

As CO_2 concentrations have increased in the atmosphere, about one-third has diffused into water bodies. As more CO_2 dissolves into the ocean water, much of it gets converted to carbonic acid and the pH of the ocean water decreases. A lower pH makes it harder for some species with shells—such as snails, clams, mussels, corals, and crustaceans—to build their shells and this has the potential to cause declines in coral reefs, cause increasingly harmful algal blooms, and impact the shellfish industry.

58-3 What are the international agreements on global climate change?

Global climate change comes from many nations so the solution has to also involve many nations. The Kyoto Protocol is a 1997 international agreement which set a goal for global emissions of greenhouse gases from all industrialized countries to be reduced by 5.2 percent below their 1990 levels by 2012. Some countries have exceeded this goal while others have not. The Paris Climate Agreement was a pledge by 196 countries in 2015 to keep global warming less than 2°C above preindustrial levels.

AP® Practice Questions

Preparing for the AP® Exam

Multiple-Choice Questions

Use the passage below to answer questions 1 & 2:

As temperatures rise globally due to increased greenhouse gases, some greenhouse gases will have more of an impact than others. With an increase in carbon dioxide, more ice caps will melt, and this will release the methane trapped in those ice caps. Methane is a much more potent greenhouse gas that will speed up climate change.

1. Identify the author's claim from the above passage.
   (a) Climate change will lead to ocean acidification.
   (b) Climate change can lead to global cooling.
   (c) Coral bleaching will be a major consequence of climate change.
   (d) An increase in greenhouse gases can lead to a positive feedback loop.

2. Which of the following is another consequence that can increase climate change with the melting of ice caps?
   (a) Carbon dioxide levels will decrease with melting.
   (b) Fewer ice caps will decrease albedo.
   (c) Soil erosion from melting ice caps will increase.
   (d) Deforestation will reduce a common sink for carbon dioxide

3. In 1997 the United States did not ratify the Kyoto Protocol because
   (a) it required the same emission reductions from all signatories.
   (b) it did not limit the emissions of developing nations.
   (c) current technology was unable to meet the required goals.
   (d) the target emission reductions were seen as impossible to achieve.

4. Which statement is true about elevated CO_2 concentrations?
   (a) CO_2 is elevated in the atmosphere but not in the oceans.
   (b) CO_2 is elevated in the ocean but not in the atmosphere.
   (c) CO_2 is elevated in the atmosphere and in the oceans.
   (d) CO_2 is elevated neither in the atmosphere nor in the oceans.

5. The decline in ocean pH due to human activities is caused by
   (a) a chemical reaction in which dissolved CO_2 is converted to carbonic acid.
   (b) a chemical reaction caused by decomposition in the ocean.
   (c) a chemical reaction caused by shellfish accumulating calcium carbonate.
   (d) a chemical reaction involving a reduction in hydrogen ions in the water.

6. Which is a viable strategy to reduce CO_2 in the atmosphere?
   (a) carbon sequestration
   (b) converting forests to agricultural land
   (c) increased size of sport utility vehicle automobiles
   (d) electrical generation from coal

Free-Response Question

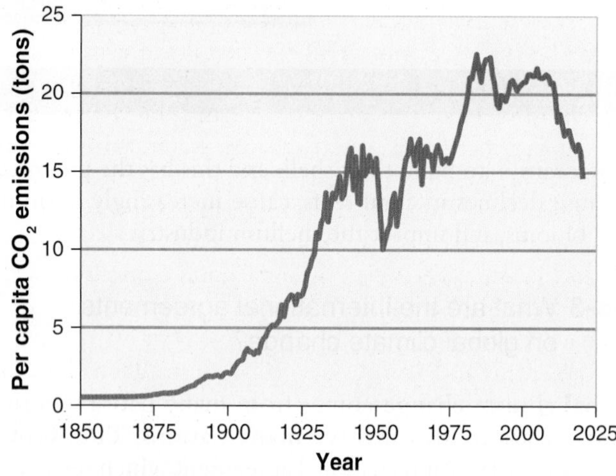

(a) **Identify** the year where CO_2 emissions were lowest after 1920. (1 pt.)
(b) **Describe** the pattern in CO_2 emissions from 1850 to 1920, and from 1921 to 1990. (1 pt.)
(c) **Describe** one reason why there is a difference in the trend of CO_2 emissions from 1850 to 1920 and 1921 to 1990. (1 pt.)
(d) **Describe** the effect that an increase in carbon dioxide would have on oceans. (1 pt.)
(e) **Describe** the primary method to reduce the issue described in part (d). (1 pt.)

As carbon dioxide levels have risen, levels of pH in the ocean have changed. Aquatic ecologists have studied the change in pH levels in two different coastlines. Coastline A has several cities and the pH of the coastal water is 8.0. Coastline B has very few people and the pH of the coastal water is 8.1. Historical data had both coastlines at 8.2 pH.

(f) **Identify** the dependent variable in the study described. (1 pt.)
(g) **Identify** a testable hypothesis for the study described. (1 pt.)
(h) **Describe** why both sites likely had a drop in pH. (1 pt.)
(i) **Explain** what consequences a reduced pH would likely have on coral reefs. (1 pt.)
(j) **Identify** a global agreement that would reduce the source of ocean acidification. (1 pt.)

Module 59

Unit 9 | 55 | 56 | 57 | 58 | 59

Invasive Species, Endangered Species, and Human Impacts on Biodiversity

As we have discussed throughout this unit, a variety of human activities have altered ecosystems and have often caused harm to the abundance and diversity of species in nature. In this module we will examine the magnitude of declines in genetic diversity and species diversity. We will then investigate the major causes of these declines, including habitat loss, invasive species, population growth, pollution, climate change, and overexploitation. With an understanding of the problems, we will then examine the solutions to curb the further loss of biodiversity, including a number of laws and international agreements.

Learning Goals

After reading this module you should be able to

59-1 identify the threats posed by invasive species.

59-2 describe why species are becoming endangered.

59-3 identify how human activities are affecting genetic biodiversity.

59-4 explain the causes of declining biodiversity.

59-5 identify how we can conserve biodiversity.

59-1 What are the threats posed by invasive species?

Invasive species can negatively impact native species, ecosystems, and human activities

You may recall our definitions of native, exotic, and invasive species in Module 1. Native species live within their historical range whereas exotic species live outside their historical range. Some exotic species—such as domesticated farm animals and honeybees—provide benefits to humans. However, when exotic species quickly spread across large areas and cause declines in native species, cause harm to ecosystems, or negatively impact human activities, we call them invasive species. As we have noted in previous modules, the rapid spread of invasive species is possible because invasive species often have no natural enemies in regions where they are introduced.

Many invasive species are r-selected, generalist species. As you might recall, this means they can experience rapid population growth, live in a wide variety of habitats, and consume a wide variety of food. Examples include mosquitoes that have invaded Hawaii and carried deadly avian malaria to Hawaiian birds, Norway rats from Europe that have invaded every continent, and many insects and diseases that have nearly wiped out several species of native trees in the United States.

In the Great Lakes, numerous species have been accidentally introduced from the water that ships from other parts of the world carry inside their hulls to help them maintain proper buoyancy. A high-profile case is the zebra mussel, which is native to the Black Sea and the Caspian Sea in eastern Europe and western Asia. It can achieve very dense populations that filter algae from the water, thus outcompeting native mussels, clams, and other herbivores for available algae. These invasive mussels can achieve such high densities that they can clog water-intake pipes and impede the flow of water on which industries and communities rely. While tremendous efforts have been made to control or eradicate various invasive species, most invasive species are too numerous and widespread to be controlled. The other difficult reality is that new invasive species are arriving on every continent every year from other parts of the world.

While most invasive species arrive by accident, some are intentionally introduced, such as honeybees, which were brought to the United States from Europe. While honeybees are popular for their honey and their pollination of plants, it is less appreciated that North America already had dozens of native bees that were doing these jobs. These native bees have been in decline for decades, in part due to the introduction of competing honeybees.

FIGURE 59.1 The spread of the exotic kudzu vine. The fast-growing kudzu vine, native to Asia, was introduced to the United States to control erosion. It has since spread rapidly, growing over the top of nearly anything that does not move, like this barn in Rose Hill, North Carolina. *(Gilbert S. Grant/Science Source)*

FIGURE 59.2 Silver carp invading the Mississippi River. The silver carp has been introduced into the Mississippi River from Asia and is quickly heading toward an invasion of the Great Lakes. *(Chris Olds—U.S. Fish and Wildlife Service/LaCrosse FWCO)*

Another intentionally introduced species is the kudzu vine, which is native to Japan and southeast China. Kudzu was introduced to the United States in 1876 and farmers in the southeastern states were encouraged to plant kudzu to help reduce erosion in their fields. By the 1950s, it became apparent that the southeastern climate was ideal for kudzu, with growth rates of the vine approaching 0.3 m (1 foot) *per day*. The vine grows up over most wildflowers and trees and shades them from the sunlight, causing the plants to die. Indeed, the vine grows over just about anything that does not move (**FIGURE 59.1**). Kudzu currently covers approximately 3 million hectares (7.4 million acres) in the United States.

A new threat to the Great Lakes is the silver carp (*Hypophthalmichthys molitrix*), a fish that is native to Asia but has been transported around the world. After being brought to the United States, some of the fish escaped and rapidly spread through many of the major river systems, including the Mississippi River. There are two major concerns about this invading fish. First, scientists worry that it will outcompete native species of fish that also consume algae. Second, the silver carp exhibits an unusual behavior: It jumps out of the water when startled by passing boats (**FIGURE 59.2**). Given that the carp can grow to 18 kg (40 pounds) and jump up to 3 m (10 feet) into the air, this poses a major safety issue for boaters.

Around the world, invasive exotic species pose a serious threat to biodiversity by acting as predators, pathogens, or superior competitors to native species. Some of the most complete data exist in Europe. During the past 100 years, Europe has experienced nearly 2,000 exotic species in terrestrial ecosystems. Additional species have been introduced into freshwater and marine ecosystems.

Controlling Invasive Species

It is often quite difficult or impossible to control the spread of invasive species once they have arrived in a region. In some cases, such as aquatic invasive plants, efforts have been made to remove the plants from water bodies when they are first observed and are still low in abundance. Other invasive species can be controlled by introducing an enemy from their native range. For example, when prickly pear cactus was introduced from South America to Australia, it spread quickly and outcompeted many of the native plants that sheep grazed on. To control the cactus, scientists brought in a species of cactus moth from South America. The caterpillar stage of the moth eats the prickly pear cactus and substantially reduced the abundance of the cactus within a few years. Unfortunately, many invasive species are much more difficult to control.

Given the difficulty of controlling invasive species once they arrive in a region, it is more effective to try to prevent invasive species from ever arriving. A number of efforts are used to reduce the introduction of invasive exotic species, including the inspection of goods coming into a country and the prohibition of wooden packing crates made of untreated wood that could contain insect pests.

AP® Exam Tip

Make certain to use the terms "exotic species" and "invasive species" independently. Not all species that are exotic are invasive. Not all invasive species are exotic.

59-2 Why are species becoming endangered?

Plants and animals are becoming endangered due to human activities

In Module 11, we noted that the world has experienced five mass extinctions during the past 500 million years. Many scientists have suggested that we may currently be in the midst of a sixth mass extinction. In a recent assessment, one group of scientists estimated that the world is currently experiencing approximately 1,000 species extinctions per year, although other scientists have estimated much lower rates of extinction. The uncertainty around these estimates depends on the assumption made about how many species (discovered and undiscovered) currently live on Earth and whether the rates of extinction observed in well-documented groups of plants or animals can be applied to all plants and animals. Regardless of the exact estimate of extinction rate, this current sixth mass extinction is unique because it is happening over a relatively short period of time and it is the first mass extinction to occur since humans have been present on Earth.

The Percentage of Endangered Plants and Animals

To understand the current status of species on Earth, we need to first define a few categories. Extinct species are those that were known to exist as recently as the year 1500 but no longer exist today. **Endangered species** are those that are likely to go extinct in the near future. In contrast, threatened species are those that are likely to become endangered in the near future. Thus, threatened species are at a lower risk of extinction than endangered species. Two additional categories are near-threatened (i.e., likely to become threatened in the near future) and least-concern (i.e., species are widespread and abundant). Finally, data-deficient species have no reliable data to assess their status; their populations may be increasing, decreasing, or stable.

Using these categories, the International Union for Conservation of Nature (IUCN) has assembled data on about 50,000 species. Across all groups of organisms that have been assessed, nearly one-third are threatened or endangered. Some of the most complete data are for conifers, birds, reptiles, mammals, amphibians, and fish. As you can see in **FIGURE 59.3**, the percent of species that are threatened or endangered varies substantially among different groups of organisms. For example, only 13 percent of birds and 18 percent of fish are threatened or endangered. However, the percentage is higher for reptiles and mammals and the highest for conifers and amphibians. In fact, amphibians are experiencing the greatest global declines of any assessed group. Among the approximately 5,000 species of amphibians for which reliable data exist, nearly 2,000 are declining around the world.

Many other groups of organisms are also experiencing large declines, but complete assessments have not yet been conducted because of the time and money required for each assessment. From the groups that have been assessed, however, it is clear that the world's biodiversity has experienced a substantial number of extinctions and that a large percentage of species are declining. These results

Endangered species Species that are likely to go extinct in the near future.

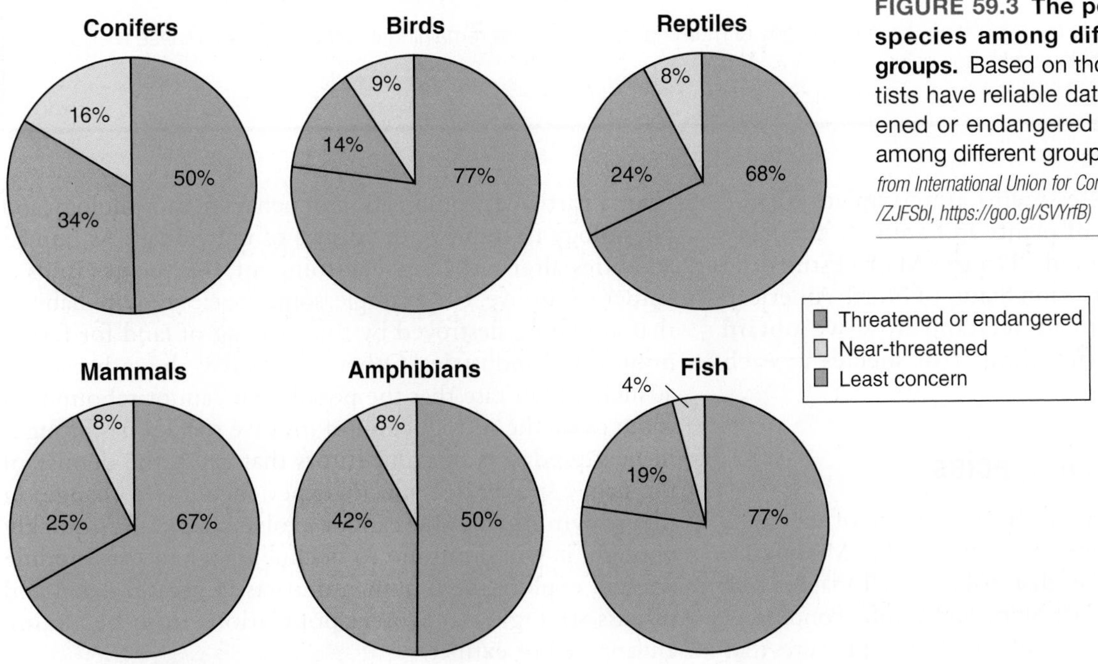

FIGURE 59.3 The percentage of endangered species among different plant and animal groups. Based on those species for which scientists have reliable data, the percentage of threatened or endangered species differ substantially among different groups of plants and animals. *(Data from International Union for Conservation of Nature, 2017, https://goo.gl/ZJFSbl, https://goo.gl/SVYrfB)*

DO THE MATH

Estimating Percentages in the Conservation Status of North American Mammals

Preparing for the AP® Exam

As we have seen, it can be useful to examine both the number of species and the percentage of species that fall within each IUCN category for a given group of plants or animals in the world. We can do the same exercise within a given continent to see the percentages in each category for a given group on a specific continent. For example, in the table provided below we see the number of North American mammal species in each IUCN category. From these numbers we can calculate the percentage of species in each category.

Number of North American mammal species in each IUCN category

IUCN category	Number of species	Percentage of species
Extinct	6	
Threatened or endangered	122	
Near-threatened	33	
Least-concern	546	
Data deficient	37	
Total		

(Data from Smithsonian Museum of Natural History. https://naturalhistory.si.edu/mna/image_menu.cfm)

To calculate the percentages, we first have to determine the total number of mammal species:

$$\text{Total} = 6 + 122 + 33 + 546 + 37 = 744$$

To determine the percentage of species that are extinct, we divide the number of species in that category, by the total number of mammal species, and then multiply the decimal fraction by 100 to make it into a percentage:

$$\text{Percent extinct} = \text{Number of extinct mammal species} \div \text{Total number of mammal species} \times 100$$

$$= 6 \div 744 = 0.008$$

$$0.008 \times 100\% = 0.8\%$$

Complete the table by calculating the percentages of mammal species in the remaining IUCN categories. Please round off to the nearest whole percentage value.

suggest that when the assessments are complete for the remaining major groups of plants and animals, the news will most likely not be good. "Do the Math: Estimating Percentages in the Conservation Status of North American Mammals" will help you understand how we obtain these percentages from the number of species in each conservation category.

Causes of Endangered Species

The reasons for the decline in the abundance of species is connected to human activities. As we learned in Module 13, each species has evolved traits that makes it well adapted to a particular habitat with distinct biotic and abiotic conditions. These traits have evolved as a result of selective pressures that favor a particular combination of behavior, morphology, and physiology to improve the fitness of individuals. As human activities alter a species' environment, the species finds it harder to survive. For example, some species require habitats that are being destroyed by the clearing of land for farms, homes, and industries. Other species have been harvested at such a high rate that the population cannot rebound. In some cases, the introduction of invasive species has brought in new predators or competitors that cause the demise of the native species. If a species experiences these changes in their environment and cannot evolve adaptations quickly enough, it will continue to decline. Later in this module we will explore these human impacts in greater detail and discuss strategies to protect populations from becoming endangered or extinct.

FIGURE 59.4 Declines in genetic diversity. The Florida panther was reduced to such a small population that it suffered severe effects of inbreeding. In recent years the introduction of new genotypes from a Texas population has allowed the Florida panther population to rebound. *(©Tom & Pat Leeson/AGE Fotostock)*

59-3 How are human activities affecting genetic biodiversity?

Human activities are causing declines in genetic biodiversity

In addition to being concerned about the growing number of endangered species and the decline of species diversity, environmental scientists are concerned about the decline of genetic diversity. Populations with low genetic diversity are not well suited to surviving environmental change. High genetic diversity produces a wide range of phenotypes that increases the chances that at least some individuals can survive and reproduce under changing environmental conditions. High genetic diversity also ensures that a wider range of genotypes is present, which reduces the probability that an offspring will receive the same harmful mutation from both parents.

Declines in genetic diversity have natural or human causes. Cheetahs, for example, possess very low genetic diversity. Researchers have determined that this condition is the result of a natural population bottleneck that occurred approximately 10,000 years ago (see Figure 13.7 on page 143). Other declines in genetic diversity have human causes. For example, we discussed in Module 13 that the Florida panther once roamed throughout the southeastern United States (**FIGURE 59.4**). Because of hunting and habitat destruction, the population of the Florida panther shrank to only a small group of 20 individuals. With such a small population, harmful mutations could be present in both parents and this caused a number of offspring defects that caused the population to decline even further. After scientists released 8 panthers from Texas into Florida to increase genetic diversity, the Florida panther population increased to nearly 100 individuals.

Declining Genetic Diversity of Domesticated Species

Although declining genetic variation of plants and animals in the wild is of great concern to scientists, there are also major concerns about declining genetic variation in the domesticated species of crops and livestock on which humans depend. The United Nations notes that while there are 38 species of domesticated animals, the majority of livestock comes from just seven species of mammals (donkeys, buffalo, cattle, goats, horses, pigs, and sheep) and four species of birds (chickens, ducks, geese, and turkeys). Over thousands of years, humans have bred the 38 species of domesticated animals into approximately 8,800 different varieties. Across the world, these species have been bred for a variety of characteristics, including adaptations that allow them to survive local climates. For example, humans have bred for a tremendous diversity of traits in cattle, as illustrated in **FIGURE 59.5**. This wide variety of adaptations, which contains a great deal of genetic variation, could be used in the future for adapting to changing environmental conditions or resisting new diseases.

FIGURE 59.5 The genetic diversity of livestock. Over thousands of years, humans have selected for numerous breeds of domesticated animals to thrive in local climatic conditions and to resist diseases common in their local environments. Modern breeding, which focuses on productivity, has caused the decline or extinction of many of these animal breeds.

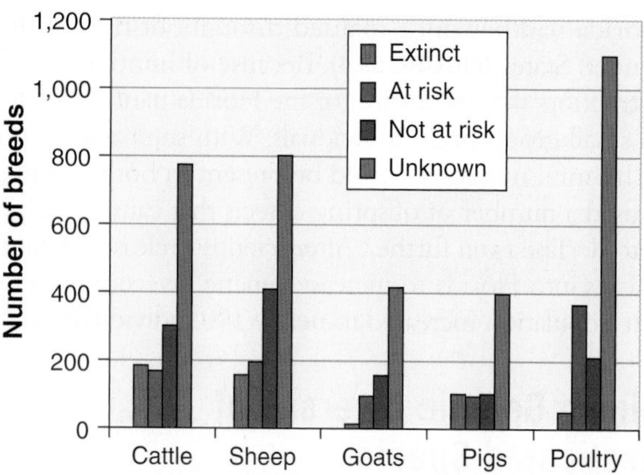

FIGURE 59.6 The global decline in genetic diversity of domesticated animals. Some of the most common species of livestock include cows, sheep, goats, pigs, and chickens. Across all domesticated breeds, more than 400 have gone extinct and 17 percent are at risk of extinction. The status of another 58 percent is unknown due to insufficient data on their population sizes. *(Data from Food and Agriculture Commission of the United Nations (2015). The Second Report on the State of the World's Animal Genetic Resources for Food and Agriculture.)*

Unfortunately, livestock producers have concentrated their efforts on the breeds that are most productive in terms of their economic return in meat or milk production, so much of this genetic variation is being lost. The United Nations recently assessed the state of domesticated animal breeds and concluded that more than 400 breeds have gone extinct and 17 percent of the remaining breeds are at risk of extinction. However, this percentage may be much higher because there are insufficient data for 58 percent of the breeds, which makes it difficult to assess whether any of them are also declining. You can view the global data for cows, sheep, goats, pigs, and chickens in **FIGURE 59.6**.

A similar story exists for crop plants. A century ago, most of the crops that humans consumed were composed of hundreds or thousands of unique genetic varieties. Each variety grew well under specific environmental conditions and was usually resistant to local pests. In addition, each variety often had its own unique flavor. As we saw in Module 25, the green revolution in agriculture focused on techniques that increased productivity. Farmers began planting fewer crop varieties, concentrating on those with higher yields. Fertilizers and irrigation helped humans control many of the abiotic conditions, allowing fewer but higher-yielding varieties to be grown across large regions of the world. According to the United Nations Food and Agriculture Organization, 6,000 crop species have been historically cultivated for food, but today a mere nine species provide two-thirds of all crops consumed by humans. Moreover, we have lost 75 percent of our crop diversity in the last century.

Planting only a few varieties of crops leaves us open to major crop losses if the abiotic or biotic environment changes. For example, in the 1840s farmers in Ireland relied primarily on potatoes as a crop to feed their families and planted just one variety of potato known as the "Irish lumper." In 1845, a fungus appeared that killed about half of the potato crop and it continued to kill potato plants for another 7 years. As a result, more than 1 million people died of starvation and another million left Ireland, with many coming to the United States. This event became known as the Irish Potato Famine.

The nations of the world have recognized the problem of declining crop diversity and have responded by storing seed varieties in specially designed warehouses to preserve genetic diversity for the future. In fact, there are currently more than 1,700 such storage facilities around the world. However, many of these facilities are at risk of being destroyed by war and natural disasters. In the past decade, nations and philanthropists have funded an international storage facility known as the Svalbard Global Seed Vault (**FIGURE 59.7**). This facility

(a)

(b)

FIGURE 59.7 A global seed bank. The Svalbard Global Seed Vault in northern Norway is an international storage area for many varieties of crop seeds from throughout the world. *(a & b: Hemis/Alamy Stock Photo)*

consists of a tunnel built into the side of a frozen mountain on an island above the Arctic Circle of northern Norway. It was designed to resist a wide range of possible calamities, including natural disasters and global warming. Should the environment change in future years, either in terms of abiotic conditions or because of emergent diseases, the seed bank will be available to help scientists address the challenge. The Svalbard facility opened in 2008 with a capacity of 4.5 million seed varieties, with each variety represented by more than 500 seeds. As of 2022, more than 1 million seed samples had been sent to Svalbard for long-term storage.

59-4 What are the causes of declining biodiversity?

Declining biodiversity has a wide variety of causes

Threats to biodiversity exist throughout the world. In this section, we will explore the factors that are causing declines in biodiversity. We can remember the list of threats using the acronym HIPPCO, which stands for habitat destruction, invasive species, human population growth, pollution, climate change, and overexploitation. As we will see, it is the continually increasing population of our species that is the underlying cause of the other drivers of declining biodiversity. Having more people on the planet destroys habitats, introduces invasive species, creates pollution, creates global climate changes, and causes the overexploitation of plants and animals.

Habitat Destruction

For most species, the greatest cause of decline and extinction is habitat loss. In modern times, the primary cause of habitat loss and fragmentation is human development that removes natural habitats and replaces them with homes, industries, agricultural fields, shopping malls, and roads. As we learned previously, many species can only thrive in a particular habitat within a narrow range of abiotic and biotic conditions. Such specialists are particularly prone to population declines, especially when their favored habitat is limited, which restricts their distribution to a specific geographic area suitable only for a small population. For example, for thousands of years the northern spotted owl (*Strix occidentalis caurina*) lived in old-growth forests—those dominated by trees that are hundreds of years old—in the northwestern United States and southwestern Canada. This habitat provided the ideal sites for nesting, roosting, and catching small mammals to eat. The removal of the old trees for lumber and housing developments has transformed much of the former old-growth forest into a different habitat (**FIGURE 59.8**).

(a)

(b)

FIGURE 59.8 Habitat destruction. Habitat destruction caused by humans is the largest threat to biodiversity. (a) Old-growth forests, such as this one in the Mount Rainier National Park in Washington State, serve as habitat for spotted owls. (b) After clear-cutting an old-growth forest for timber, such as this site in the Olympic National Forest in Washington State, the forest provides a very different habitat. It may take hundreds of years before this logged habitat becomes suitable for the spotted owl. *(a: Stephen Matera/DanitaDelimont.com; b: Kent Foster/Science Source)*

This habitat alteration has reduced the number of northern spotted owls by 50 to 75 percent over the past 25 years because they have less old-growth forest in which to nest and find food. As a result, the owl is currently categorized as threatened and may soon be classified as endangered by the U.S. Fish and Wildlife Service.

The map in **FIGURE 59.9** shows the changing forest habitats since 2001. As we learned in Module 24, much of the forest in the United States during the 1700s and 1800s was logged for lumber and cleared for agriculture. In recent decades, forested land has been increasing, although humans have often planted the new forests, which typically have a lower diversity of species than the original forests. At the same time, developing countries in South America, sub-Saharan Africa, and Southeast Asia are clearing their forests much as the United States and Europe did in centuries past. As a result, large declines in forest cover are occurring in developing countries that were once forested. It is currently unclear whether these countries will follow the pattern of Europe and North America and eventually allow their forested areas to increase.

Although deforestation receives a lot of attention, many other habitats are also being lost. According to the most recent Millennium Ecosystem Assessment, approximately 70 percent of the woodland/shrubland ecosystem that borders the Mediterranean Sea has been lost. Similarly, across the globe we have lost nearly 50 percent of grassland habitats and 30 percent of desert habitats. In the United States, we have lost more than half of the wetlands that existed in the 1600s. With this large loss of these natural biomes, we also observe a large decline in the abundance of species that depend on these biomes.

In marine systems, there has been a sharp decline in the amount of living coral in the Caribbean Sea, from a high of 50 percent live coral in the 1970s to a mere 8 percent by 2012. As we have learned in previous modules, living coral provide habitat for thousands of other species, which makes them particularly vital to the persistence of marine habitats. The decline in coral is the result of human impacts including the warming of oceans and increased pollution, and the removal of coral by collectors. Today, the occurrence of coral bleaching at any given site is about five times more frequent than it was in the 1980s.

A species may decline in abundance without complete habitat destruction. As we saw in the case of the Florida panther, a smaller habitat supports a smaller population, which reduces genetic diversity. Less habitat also reduces the variety of physical and climatic options available to individuals during periods of extreme conditions. The presence of cooler, high-altitude areas in a habitat, for example, allows animals a place to move during periods of hot weather. Without such areas, the animals have nowhere to go. Also, loss of habitat can restrict the movement of migratory or highly mobile species. While many species can thrive in small habitats, other species, such as mountain lions, wolves, and tigers, require large tracts of relatively uninhabited, undisturbed land.

When a large habitat is broken up into smaller, more isolated habitat fragments, organisms can experience increased interactions with other harmful species. For example, many songbirds in North America live in forests. When these birds make their nests near the edge of the forests—where the forest meets a field—they often have to contend with the brown-headed cowbird (*Molothrus ater*). The cowbird is a nest parasite—it does not build its own nest but lays its eggs

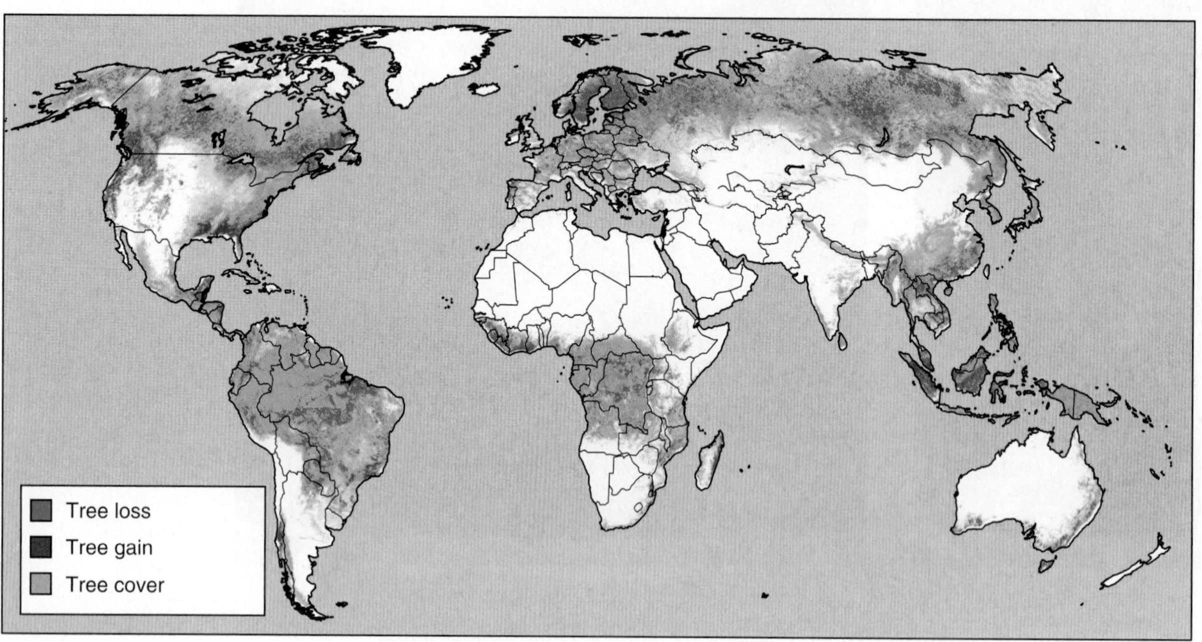

FIGURE 59.9 Changing forests. Some regions of the world experienced large declines in the amount of forested land while other regions have shown little change or have seen increases in forest cover. Tree gain data are from 2001 to 2012; tree loss data are from 2001 to 2020. *(Data from Global Forest Watch, http://www.globalforestwatch.org/map. Source: Hansen/UMD/Google/USGS/NASA, accessed through Global Forest Watch.)*

(a)

(b)

FIGURE 59.10 Habitat fragmentation. (a) Increased fragmentation of forests has caused forest songbirds to come into increasing contact with the brown-headed cowbird. (b) The cowbird does not make its own nest. Instead, it lays its brown, spotted eggs in the nests of other species, such as this nest containing four blue eggs of the chipping sparrow (*Spizella passerina*). *(a: Jim Zipp/Ardea/Animals Animals; b: Paul J. Fusco/Science Source)*

in the nests of several other species of birds (**FIGURE 59.10**). The cowbird tricks forest birds into raising its offspring, which take food away from the forest birds' own offspring. As forests are broken up into smaller fragments, the proportion of forest near the edge increases and, therefore, the number of bird nests that are susceptible to brown-headed cowbirds increases. This has resulted in brown-headed cowbirds causing declines in many species of North American songbirds. The impacts of fragmenting large habitats into smaller habitats will vary among different species.

AP® Exam Tip

When writing about habitat loss on an FRQ, be sure to mention the type of habitat destruction such as deforestation, development, etc. Also, be sure to use the term "habitat fragmentation" when appropriate instead of "habitat destruction."

Invasive Species and Human Population Growth

Two additional threats to biodiversity are invasive species and human population growth. In the previous section we examined the issue of invasive species in detail and noted that invasive species can cause the decline or extinction of native species that have no evolved defenses against the invasive species. At the same time, the human population continues to grow and this has caused an increased demand for space, food, and other resources from nature.

Pollution

In Units 7 and 8, we saw how air and water pollution harm ecosystems. Threats to biodiversity come from toxic contaminants such as pesticides, heavy metals, acids, and oil spills. Other contaminants, such as endocrine disrupters, can prevent or inhibit reproduction. Pollution sources that cause declines in biodiversity also include the release of nutrients that cause algal blooms and dead zones as well as thermal pollution that can make water bodies too warm for species to survive.

In 2010, for example, an oil platform in the Gulf of Mexico owned by BP and named the *Deepwater Horizon* exploded, causing a massive release of oil that lasted for several months. As we discussed in Module 47, the release of oil caused a tremendous amount of death across a wide range of animal species including sea turtles, pelicans, fish, and shellfish. In response to the massive oil spill, BP released hundreds of thousands of liters of oil dispersant, a chemical designed to break up large areas of oil into tiny droplets that can be consumed by specialized species of bacteria. However, the dispersant is also toxic to many species of animals. It is estimated that the oil spill and subsequent dispersant killed tens of thousands of sea turtles; in the northern Gulf it is estimated to have killed 12 percent of brown pelicans (*Pelecanus occidentalis*) and 32 percent of laughing gulls

(*Leucophaeus atricilla*). The total impact of the spilled oil and applied dispersants on the wildlife of the Gulf of Mexico may not be known for many years.

Climate Change

In the previous three modules we discussed the many impacts of climate change on ecosystems. As a threat to biodiversity, the primary concern about climate change is its effect on patterns of temperature and precipitation in different regions of the world. In some regions, a species may be able to respond to warming temperatures and changes in precipitation by relocating to a place where the climate is well suited to the species niche. In other cases, this is not possible and species will experience habitat loss and possibly become endangered or extinct. For example, in southwestern Australia, a small woodland/shrubland peninsula exists on the edge of the continent, surrounded by a much larger area of subtropical desert farther inland (see Figure 11.5 on page 125). Scientists expect conditions on the peninsula to become drier during the next 70 years. If this occurs, many species of plants in this small ecosystem will not have a nearby hospitable environment to which they can relocate, since the surrounding desert ecosystem is already too dry for them. An examination of 100 species of plants in the area (all from the genus *Banksia*) has led scientists to project that 66 percent of the species will decline in abundance and up to 25 percent will become extinct. Many species in the world are expected to be similarly affected by climate change.

FIGURE 59.11 Over exploitation. The dodo was a large flightless bird that served as an easy source of meat for sailors and settlers on the island of Mauritius. Because it evolved on an island with no large predators or humans, the dodo had no instinct to fear humans and was hunted to extinction. *(Stock Montage/Getty Images)*

Overexploitation

Hunting, fishing, and other forms of harvesting are the most direct human influences on wild populations of plants and animals. Most species can be harvested to some degree, but a species is overexploited when individuals are removed at a rate faster than the population can replace them. In the extreme, overharvesting of a species can cause extinction. In the seventeenth century, for example, ships sailing from Europe stopped for food and water at Mauritius, an uninhabited island in the Indian Ocean. On Mauritius, the sailors would hunt the dodo (*Raphus cucullatus*), a large flightless bird that had no innate fear of humans because it had never seen humans during its evolutionary history (**FIGURE 59.11**). The dodos were unable to protect themselves from human hunters and the rats (introduced by humans) that consumed dodo eggs and hatchlings. As a result, they became extinct in just 80 years. This same scenario appears to have taken place with many other large animal species as well. These animals include the giant ground sloths, mammoths, American camels of North and South America, and the 3.7 m (12 feet) tall moa birds of New Zealand. Each species became extinct soon after humans arrived, suggesting that the animals' demise may have been due to overharvesting.

Overexploitation has also occurred in the more recent past. In the 1800s and early 1900s, for example, market hunters slaughtered wild animals to sell their parts on such a scale that many species, including the American bison, declined dramatically. Bison were once abundant on the western plains, with estimates ranging from 60 to 75 million individuals. By the late 1800s fewer than 1,000 were left. This means that 99.999 percent of all bison were killed. Following enactment of legal protections, the bison population today has increased to more than 500,000, including both wild bison and bison raised commercially for meat.

Not all species harvested by market hunters fared as well as the American bison. For instance, the passenger pigeon was once one of the most abundant species of birds in North America. Population estimates range from 3 to 5 billion birds in the nineteenth century. In fact, during annual migrations, people observed continuous flocks of pigeons flying overhead for 3 days straight in densities that blocked out most of the sun. Breeding flocks could cover 40,000 ha (100,000 acres) with 100 nests built into each tree. With such high densities, market hunters could shoot or net the birds in very large numbers and fill train cars with harvested pigeons to be sold for food in eastern cities. This overharvesting, combined with the effects of forest clearing for agriculture, caused the passenger pigeon to decline quickly. Attempts were made to save the species by capturing wild birds and breeding them in captivity, but the birds only lay a single egg and they did not survive well, so these breeding efforts were not successful in increasing the pigeon population. The last passenger pigeon died in 1914 at the Cincinnati Zoo.

During the past century, regulations have been passed to prevent the overexploitation of plants and animals. In the United States, for example, state and federal regulations restrict hunting and fishing of game animals to particular times of the year and limit the number of animals that can be harvested. Such regulations make it a crime to harvest animals during other times of the year or harvest more than the limited number, which is known as poaching. Similar agreements have been reached among countries through international treaties to regulate the trade in pets, plants, and animal body parts. In general, these regulations have proven very successful in preventing species declines caused by overharvesting. In some regions of the world, however, harvest regulations are not enforced and illegal poaching, especially of large, rare animals that include tigers, rhinoceroses, and apes, continues to threaten species with extinction. Harvesting of rare plants, birds, and coral reef species for private collections has also jeopardized these species.

AP® Exam Tip

The AP® Environmental Science Exam may ask how a species may be negatively impacted by a human activity or a change to the ecosystem. Answer with specifics. Describe how it may lead to loss of food sources, loss of nesting habitat, or competition for a specific resource. Clearly describe the impact on the species.

59-5 How do we conserve biodiversity?

To conserve biodiversity, we need to reduce the threats

Now that we know the many threats to biodiversity, we can explore the steps that have been taken to reduce these impacts. There are two general approaches to conserving biodiversity: the single-species approach and the ecosystem approach. In this module we explore each of these approaches.

Conservation of Single Species

The single-species approach to conserving biodiversity focuses our efforts on one species at a time. When a species declines significantly, the natural response is to encourage a population rebound by improving the conditions in which that species exists. This might be accomplished by providing additional habitat, reducing exploitation, or reducing the presence of a contaminant that is impairing survival or reproduction. When the population of a species has declined to extremely low numbers, sometimes the remaining few individuals will be captured and brought into captivity. These captive animals are then bred with the intention of returning the species to the wild. While such attempts do not always work, as we saw in the case of the passenger pigeon, sometimes captive breeding works very well. A well-known example occurred with the California condor. The condor had declined to a mere 22 birds in 1987 and these wild birds were captured and brought into a captive-breeding facility. While condors normally only lay a single egg, the researchers discovered that if they removed the egg, the female condor would lay a second egg. If the second egg was removed, the female would lay a third egg. For those eggs that were removed, the researchers incubated the eggs and then fed the chicks using hand puppets that resembled an adult condor so that the chicks did not become attracted to humans once they were released into the wild. Thanks to this captive breeding and several improvements in the condor's habitat, the condor population grew to more than 500 birds by 2022. Programs such as these are a major function of zoos and aquariums around the world. During the past 50 years, legislation has been passed to help many more species recover. In the United States, the single-species approach to conservation formed the foundation of several important laws to protect plants and animals.

The Lacey Act

For some species, the legal and illegal trade in plants and animals represents a serious threat to their ability to persist in nature. One of the earliest laws in the United States to control the trade of wildlife was the **Lacey Act**. First passed in 1900, the act originally prohibited the transport of illegally harvested game animals, primarily birds and mammals, across state lines. Over the years, a number of amendments have been added so that the Lacey Act today forbids the interstate shipping of all illegally harvested plants and animals.

The Marine Mammal Protection Act

The **Marine Mammal Protection Act** prohibits the killing of all marine mammals in the United States and prohibits the import or export of any marine mammal body parts. Only the U.S. Fish and Wildlife Service and the National Marine Fisheries Service are allowed to approve any exceptions. The act was passed in 1972 in response to declining populations of many marine mammals, including polar

Lacey Act A U.S. act that prohibits interstate shipping of all illegally harvested plants and animals.

Marine Mammal Protection Act A 1972 U.S. law that prohibits the killing of all marine mammals in the United States and prohibits the import or export of any marine mammal body parts.

FIGURE 59.12 Protected marine mammals. The Marine Mammal Protection Act protects marine mammals in the United States such as the sea otter from being killed. *(Hal Beral/Getty Images)*

FIGURE 59.13 Bringing back endangered species. Habitat protection and reduced contaminants in the environment have allowed grizzly populations to increase to the point where they could be taken off the Endangered Species List. These grizzlies live in Yellowstone National Park. *(Sergey Uryadnikov/Shutterstock)*

bears, sea otters, manatees, and California sea lions (*Zalophus californianus*) (**FIGURE 59.12**).

The Endangered Species Act

The Endangered Species Act, first introduced in Module 24, is a law designed to protect species from extinction. This act authorizes the U.S. Fish and Wildlife Service to determine which species can be listed as threatened or endangered and prohibits the harming of such species, including prohibitions on the trade of listed species, their fur, or their body parts.

The Endangered Species Act was first passed in 1973 and has been amended several times since. To assist in the conservation of threatened and endangered species, the act authorizes the government to purchase habitat that is critical to the conservation of these species and to develop recovery plans to increase the population of threatened and endangered species. This is often one of the most important steps in allowing endangered species to persist. For example, grizzly bears in the Greater Yellowstone Ecosystem had declined to 136 individuals in 1975. After being listed as endangered, steps were taken to protect the bear's favored habitat and today it numbers nearly 700 individuals. (**FIGURE 59.13**).

As of 2022, more than 1,600 species of plants and animals have been listed as threatened or endangered in the United States. Although the listing process can take several years, once listed many threatened and endangered species have experienced stable or increasing populations. Indeed, some species have experienced sufficient increases in numbers to be removed from the endangered species list; these include the bald eagle, peregrine falcon, American alligator, and the eastern Pacific population of the gray whale (*Eschrichtius robustus*). Other species are currently increasing in number and may be taken off the list in the future.

Convention on International Trade in Endangered Species of Wild Fauna and Flora (CITES) A 1973 treaty formed to control the international trade of threatened plants and animals.

The Endangered Species Act has occasionally sparked controversy because it can restrict certain human activities in areas where listed species live, including how landowners use their land. For example, some construction projects have been prevented or altered to accommodate threatened or endangered species. Organizations whose activities are restricted by the Endangered Species Act often try to pit the protection of listed species against the jobs of people in the region. During the past decade, several politicians and their constituents have attempted to weaken the Endangered Species Act. However, strong support from the public and scientists has allowed it to retain much of its original power to protect threatened and endangered species. The biggest current challenge is a lack of sufficient funds and personnel required to implement the law.

Convention on International Trade in Endangered Species of Wild Fauna and Flora

At the international level, the United Nations **Convention on International Trade in Endangered Species of Wild Fauna and Flora**, also known as **CITES**, was developed in 1973 to control the international trade of threatened plants and animals. Today, CITES is an international agreement among 182 countries. The IUCN maintains a list of threatened species, known as the Red List. Each member country assigns a specific agency to monitor and regulate the import and export of animals on the list. For example, in the United States, the U.S. Fish and Wildlife Service conducts this oversight.

Despite such international agreements, much illegal plant and animal trade still occurs throughout the world. For example, a report by the Congressional Research Service estimated that illegal trade in wildlife was worth \$5 billion to \$20 billion annually. In 2017, it was estimated that 50,000 illegal shipments of wildlife and wildlife parts had been sent to United States ports over a 10-year period. In some cases, animals are sold for fur or for body parts that are thought to have medicinal value. For instance, 1 kg (2.2 pounds) of rhino horn can sell for $50,000. In other cases, rare animals are in demand as pets.

AP® Exam Tip

Make sure you can discuss legislation that focuses on the protection of biodiversity, including the Endangered Species Act and the Convention on International Trade in Endangered Species of Wild Fauna and Flora (CITES).

Conserving Entire Ecosystems

Awareness of a potential sixth mass extinction has motivated a growing interest in conserving biodiversity of entire ecosystems. This approach recognizes the benefit of preserving particular regions of the world, such as biodiversity hotspots. Protecting entire ecosystems has been one of the major motivating factors in setting aside national parks and marine reserves. In some cases, these areas were originally protected for their aesthetic beauty, but today they are also valued for their communities of organisms. The amount of protected terrestrial and marine habitats has increased throughout the world since 1950, as you can see in **FIGURE 59.14**, with a dramatic increase in protecting marine habitats during the past two decades.

When protecting ecosystems to conserve biodiversity, a number of factors must be taken into consideration including the size and shape of the protected area. We must also consider the amount of connectedness to other protected areas, using habitat corridors, and how best to incorporate conservation while recognizing the need for sustainable habitat use for human needs.

Configuring Protected Areas

A number of questions arise when we consider protecting areas of land or water. For example, how large should the designated area be? Should we protect a single large area or several smaller areas? Does it matter whether protected areas are isolated or near other protected areas? To help us answer these questions, we can return to our discussion of the theory of island biogeography from Module 10.

As you may recall, the theory of island biogeography looks at how the size of islands and the distance between islands and the mainland affect the number of species that are present on different islands. Larger islands generally contain more species because they support larger populations of each species, which makes them less susceptible to extinction. Larger islands also contain more species because they typically contain more habitats and, therefore, provide a wider range of niches for different species to occupy. The distance between an island and the mainland, or between one island and another, is another crucial factor, since more species are capable of dispersing to close islands than to islands farther away.

Although the theory of island biogeography was originally applied to oceanic islands, it has since been applied to islands of protected areas in the midst of less hospitable environments. For example, we can think of all the state and national parks, natural areas, and wilderness areas as islands surrounded by other environments that have high levels of human activity, including agricultural fields, logged forests, housing developments, and cities (**FIGURE 59.15**). These areas provide habitats for species and places to stop and rest for migrating species. Applying the theory of island biogeography from this perspective gives us some idea of the best ways to design and manage protected areas. For example, when protected areas are far apart, it is less likely that species can travel among them. This means that when a species has been lost from one ecosystem, it will be harder for individuals of that species from other ecosystems to recolonize it. So,

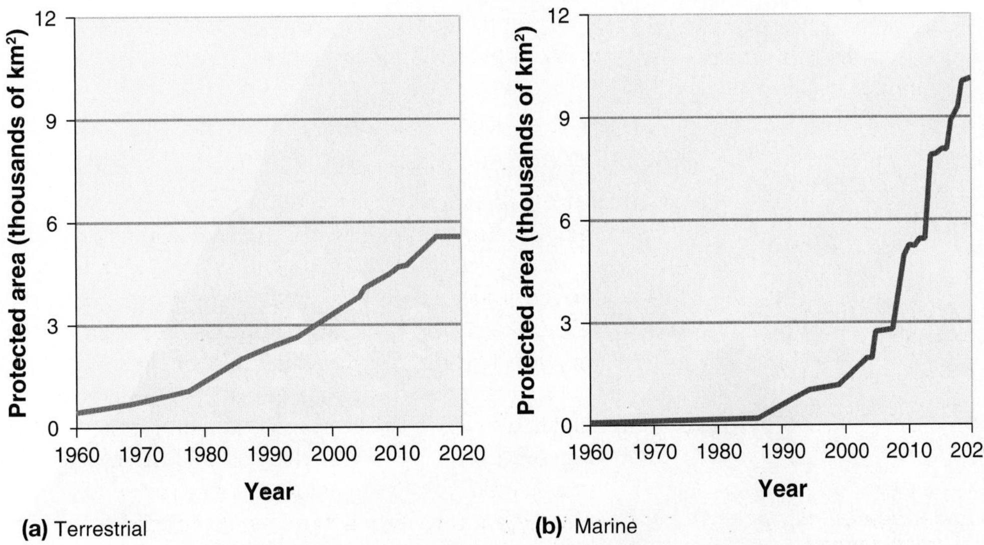

FIGURE 59.14 Changes in protected habitats. Since the 1950s, there has been a large increase in the amount of (a) terrestrial and (b) marine habitat that is protected throughout the world. *(Data from World Database on Protected Areas, https://data.oecd.org/biodiver/protected-areas.htm)*

FIGURE 59.15 Habitat islands. Central Park in New York City is an extreme example of an island of hospitable habitat surrounded by an urban environment that is not hospitable to most species. *(Andrew Thomas/Getty Images)*

when we create smaller areas, they should be close enough for species to move among them easily.

The concepts of island biogeography raise an interesting dilemma for conservation efforts. A single large area would support larger populations, but a species is more likely to survive a disease or natural disaster if it occupies several different areas. The debate over the best approach is known as SLOSS, which is an acronym for "single large or several small." While both approaches have their merits, in reality, human development and other factors often mean that only one of the two strategies is available. For example, due to human development of a region, there may simply not be a single large area available to protect, so the only option is to protect several small areas. A final consideration regarding the size and shape of protected areas is the amount of edge habitat that an area contains. It can be a challenge to protect many small areas with a comparatively larger amount of edge habitat. When we protect several small forests, for example, the proliferation of species such as the cowbird in this larger amount of edge habitat can have a detrimental effect on songbirds that typically live farther inside a forest.

Biosphere Reserves

Managing national parks and other protected areas so they serve multiple users can be a challenge. While we want to make places of great natural beauty available to everyone, when large numbers of people use an area for recreation, some amount of degradation occurs. To address this problem, the United Nations Educational, Scientific and Cultural Organization (UNESCO) developed the innovative concept of biosphere reserves, which are protected areas consisting of zones that vary in the amount of permissible human impact. These reserves protect biodiversity without excluding all human activity. **FIGURE 59.16** shows the different zones in a hypothetical biosphere reserve. The central core is an area that receives minimal human impact and is therefore the best location for preserving biodiversity. A buffer zone encircles the core area. Here, modest amounts of human activity are permitted, including tourism, environmental education, and scientific research facilities. Farther out is a transition

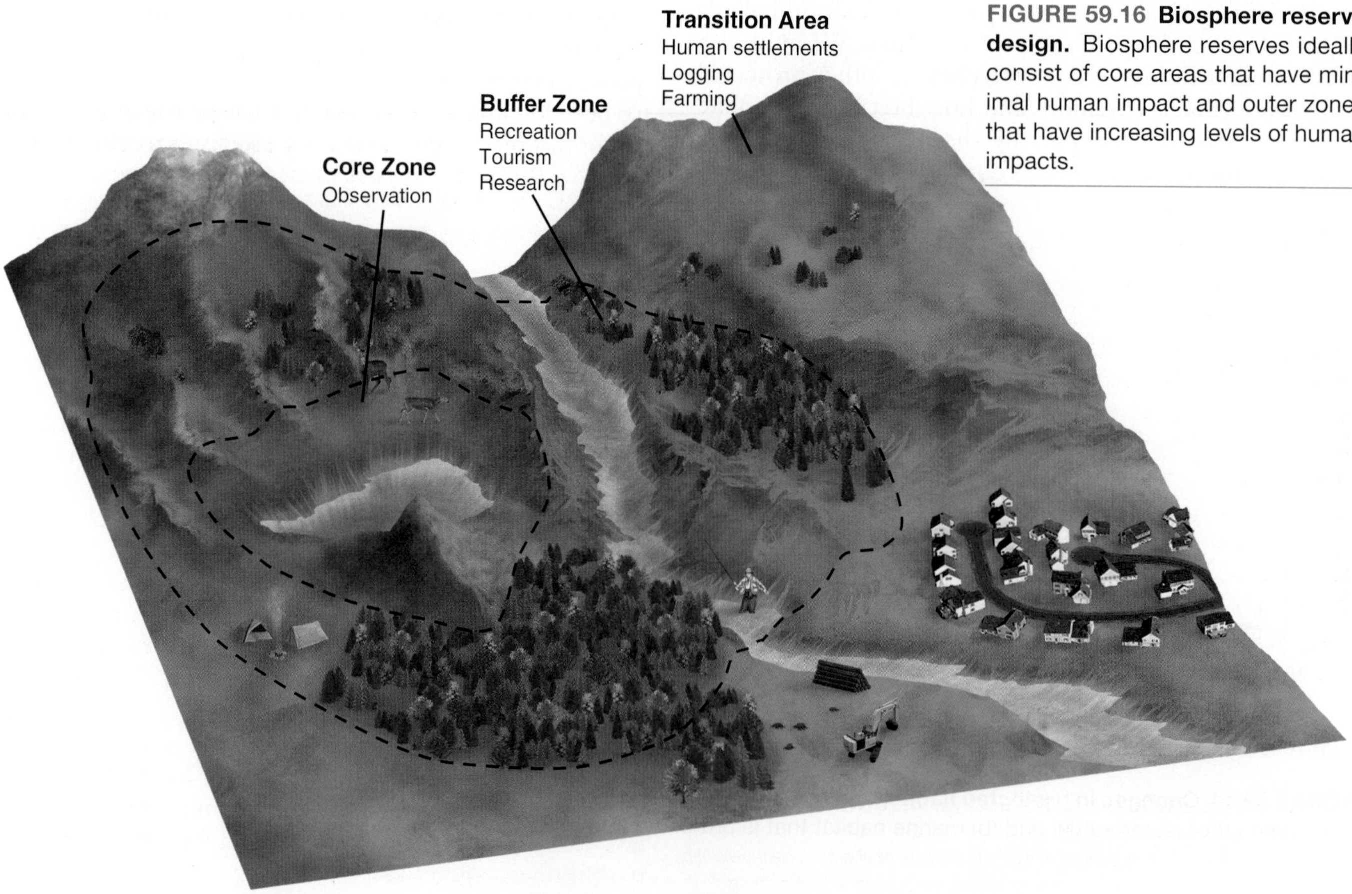

FIGURE 59.16 Biosphere reserve design. Biosphere reserves ideally consist of core areas that have minimal human impact and outer zones that have increasing levels of human impacts.

FIGURE 59.17 Biosphere reserve. Big Bend National Park, located in southwest Texas, serves as a low-impact core area of the Big Bend biosphere reserve. *(Tim Speer/Getty Images)*

area containing sustainable logging, sustainable agriculture, and residences for the local human population.

Designing reserves with these three zones represents an ideal scenario. In reality, biosphere reserves can take many forms depending on their location, though all strive to have low-impact core areas. As of 2022 there are 738 biosphere reserves worldwide, spanning 131 nations. There are 47 biosphere reserves in the United States, including the well-known Big Bend National Park in Texas. The park itself serves as the core area and receives relatively little human impact; here, hikers are permitted to walk through the beautiful desert landscapes and tree-covered mountain peaks where there are several dozen threatened and endangered plants and animals. Outside the boundaries of the park is a region of increased human impact including tourist facilities, human settlements, and agriculture (**FIGURE 59.17**).

The Restoration of Habitats

Many anthropogenic disturbances—for example, housing developments, clear-cutting, or draining of wetlands—are so large that they eliminate an entire ecosystem. In some cases, however, scientists can work to reverse these effects and restore much of the original function of the ecosystem (**FIGURE 59.18**). Growing interest in restoring damaged ecosystems has led to the creation of a scientific discipline called restoration ecology. Restoration ecologists are currently working on two high-profile ecosystem restoration projects, in the Florida Everglades and in the Chesapeake Bay, to restore water flows and nutrient inputs that are closer to historic levels so that the functions of these ecosystems can be restored. At the end of this unit, you can read about a fascinating effort in New Zealand to protect habitat and bring back rare species in Science Applied 9: How Can We Bring Back Biodiversity?

AP® Exam Tip

Be able to identify both the economic drawbacks and benefits of conservation.

FIGURE 59.18 Habitat restoration. (a) Once a forested wetland, this property in Maryland was cleared and drained for agricultural use in the 1970s. (b) Three decades later, efforts began to plug the drainage ditches, remove undesirable trees, and plant wetland plants to restore the property to a wetland habitat. *(a & b: Rich Mason, USFWS)*

In this module we learned about the impacts of invasive species and the growing number of endangered species around the world. We also discussed the decline in genetic diversity of wild and domesticated plants and animals and the major threats that are driving these declines. We have many approaches to slow these declines based on reducing many of the threats and protecting habitats. Conserving the world's biodiversity is a monumental task and it will require people and governments all around the world to work together to provide a sustainable future for coming generations. As you have seen throughout these modules, conserving the biodiversity of our planet highlights the need to understand the full suite of issues in environmental science. For a summarizing visual of how many concepts from this unit are interconnected, including the human causes of several declining species, turn to the Unit 9 Visual Representation feature on pages 710–711.

Preparing for the AP® Exam

Visual Representation 9 **Threats to Biodiversity**

Across the globe biodiversity is being threatened as a result of habitat loss, invasive species, human encroachment, pollution, and climate change. This figure highlights six species that perform critical functions in their ecosystems and are experiencing

Whitebark pine
(Pinus albicaulis)
Endangered

The whitebark pine is native to mountains of the western United States and Canada. This keystone species can live from 500 to 1000 years. Its roots stabilize rocky soil at high elevations. Many animals depend on it for high-protein seeds, including black bears, grizzly bears, red foxes, squirrels, and Clark's nutcrackers.

Survival of this tree is threatened by warmer winter temperatures, fire suppression policies, an invasive pine beetle, and a fungal disease. Conservation efforts include protecting trees from the pine beetle and analyzing the genetics of fungus-resistant trees to help breed more resilient varieties.

Mexican long-nose bat
(Leptonycteris nivalis)
Endangered

This bat migrates from central Mexico to southwest Texas, feeding on agave plants as they bloom. The bats drink nectar from agave flowers and the agave plant depends on the bat for pollination. Many organisms such as woodpeckers, bees, hummingbirds, and mice depend on agave plants for food and shelter.

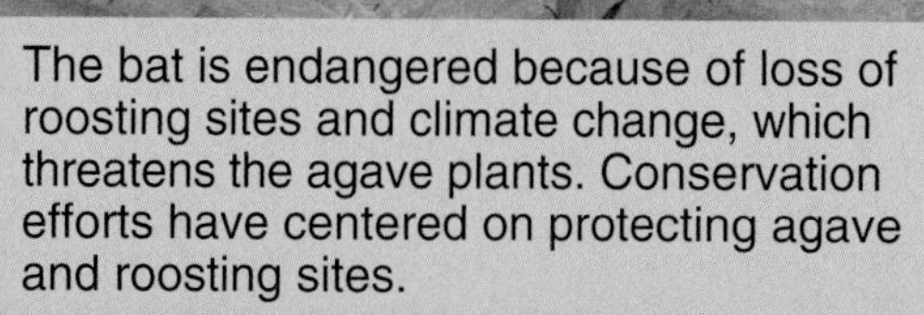

The bat is endangered because of loss of roosting sites and climate change, which threatens the agave plants. Conservation efforts have centered on protecting agave and roosting sites.

Devil's Hole pupfish
(Cyprinodon diabolis)
Critically endangered

Found only in Devils Hole, Nevada, in a single water-filled cavern, these iridescent blue fish have lived in the top 80 feet of the cavern waters for over 10,000 years. They eat snails, beetles, crustaceans, and algae, which helps keep the water clean for other organisms.

Threats to the pupfish include groundwater depletion from agriculture, human intrusion, and climate change. Warmer water has reduced the time for egg hatching so there are fewer new fish each spring. Conservation efforts have focused on protecting the cavern with fencing, addressing the groundwater depletion, and breeding programs. In recent years the number of pupfish have started to climb.

substantial declines due to human impacts. The conservation status of each species (vulnerable, endangered, or critically endangered) reflects the current designation of the International Union for the Conservation of Nature (IUCN).

Elkhorn coral
(Acropora palmata)
Critically endangered

Elkhorn coral was once the dominant coral in the Caribbean. As a key reef builder, it is critical in its ecosystem. The coral forms dense groups in shallow water, which provide habitat for many reef fish and protect beaches and reefs from storm damage.

Unsustainable fishing, pollution, ocean acidification, and ocean warming all threaten the coral. Elkhorn coral is now so reduced in its environment that less than 3 percent of its former amount remains.

Conservation efforts have focused on sustainable fishing, pollution reduction, breeding nurseries, rescuing coral after severe storms and ship groundings, and protecting habitat.

African savanna elephant
(Loxodonta Africana)
Endangered

African savanna elephants are ecosystem engineers. As they migrate, they trample vegetation and larger brush, which provides access to sunlight for smaller plants. In dry periods elephants use their feet and tusks to dig holes in riverbeds, creating watering holes that other animals use and providing habitat for many small animals. They spread plant seed through their dung. Areas with elephants have a greater diversity of reptiles and amphibians.

Since 1965, elephants have declined by 60 percent, due to poaching for their ivory tusks, reduced habitat due to agriculture and human development, and climate change causing reductions in food and water.

Conservation efforts include creating wildlife corridors where elephants can pass safely and encouraging farmers to use more sustainable techniques that take less land. Citizens patrol for poachers and organizations seek to reduce human-elephant conflict by protecting elephants while making sure communities are safe.

Southern cassowary
(Casuarius casuarius)
Vulnerable

The southern cassowary is a large flightless bird that lives in the Australian rainforest. A keystone species, it eats a large variety of plants and clears the way for less competitive plant species to grow. It eats many fruits that are too large for other animals and swallows the seeds whole, excreting them over a wide area, which helps plant species disperse their seeds.

This species is threatened by loss of habitat, fragmented habitat, dogs, wild pigs, and vehicles. Cassowaries can be aggressive, leading to human conflict. Powerful storms, which are becoming more prevalent with global warming, also damage cassowary habitat and food sources.

Conservation efforts include creation of reserves, and posting traffic signs to warn vehicles that cassowaries are in the area. Post-storm feeding stations, rescue centers for injured birds, and population tracking also help maintain populations.

Module 59 AP® Review

Preparing for the AP® Exam

Learning Goals Revisited

59-1 What are the threats posed by invasive species?

Invasive species are those species not living in their historical range, thus causing declines in native species, harm to ecosystems, or negatively impacting human activities. They are often r-selected, generalists species. Given the difficulty in controlling invasive species, it is important to reduce the likelihood of their being introduced.

59-2 Why are species becoming endangered?

Declines in biodiversity are happening at a rate that may indicate the start of a sixth mass extinction. At the species level, we have learned that major groups of organisms, including mammals, fish, and amphibians, have a large percentage of species that are threatened or endangered. The reasons for these declines in the abundance of species is connected to human activities including habitat destruction, overharvesting, and invasive species.

59-3 How are human activities affecting genetic biodiversity?

Populations with low genetic diversity are not well suited to surviving environmental change. Declines in the genetic diversity of wild species is caused by human activities that cause declines in the abundance of individuals of a given species. Domesticated species historically had a large amount of genetic diversity, but in recent decades we have increasingly focused on growing a limited number of animal and plant varieties to maximize food production. However, planting only a few varieties leaves us open to major crop losses if the abiotic or biotic environment changes.

59-4 What are the causes of declining biodiversity?

The biodiversity of our planet is declining for a number of reasons, including habitat loss, invasive species, human population growth, pollution, climate change, and overexploitation.

59-5 How do we conserve biodiversity?

Efforts to conserve biodiversity focus on either single species or entire ecosystems. In the United States, single-species legislation includes the Lacey Act, the Marine Mammal Protection Act, and the Endangered Species Act. Internationally, there are treaties that protect species, including the Convention on International Trade in Endangered Species of Wild Fauna and Flora (CITES). Some conservation efforts have focused on protecting entire ecosystems by considering the concepts of island biogeography and biosphere reserves.

Practice Math and Graphing

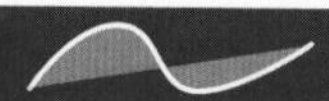

Preparing for the AP® Exam

Answer the following questions. Be sure to show all your work.

1. Practice Math

In the United States, there are endangered species in all of the major vertebrate groups. Given the table below, calculate the percentage of endangered species in each of the five groups of vertebrates.

Vertebrate group	Number of endangered species	Percent of endangered species
Mammals	95	
Birds	102	
Reptiles	45	
Amphibians	36	
Fishes	164	
Total		

2. Practice Graphing

(a) Using the percentage data you calculated in "Practice Math," create a pie graph that shows the percentages of endangered species among the five groups of vertebrates in the United States.

(b) How might the total number of species in each vertebrate group affect the percentage of endangered species among the five groups?

AP® Practice Questions

Multiple-Choice Questions

1. What factor has played the largest role in decreased diversity of domesticated species?
   (a) a focus on increased yields
   (b) the use of genetic engineering
   (c) the adaptation to specific growing environments
   (d) increased seed storage efforts

2. Which group has the highest percentage of threatened and endangered species?
   (a) birds
   (b) amphibians
   (c) reptiles
   (d) fish

3. The most significant cause of species decline and extinction throughout the world is
   (a) habitat loss.
   (b) overharvesting.
   (c) climate change.
   (d) invasive species.

4. Invasive species are
   (a) usually not a threat to biodiversity.
   (b) rare in island habitats.
   (c) successful due to a lack of natural enemies.
   (d) often unable to compete effectively in the new environment.

5. Passenger pigeons were driven extinct primarily by
   (a) habitat loss.
   (b) overexploitation.
   (c) pollution.
   (d) invasive species.

6. The primary impact of climate change on species diversity is expected to be
   (a) an increased variability in weather.
   (b) decreased precipitation worldwide.
   (c) changes in available habitat because of changing temperatures.
   (d) the increased ability of species to disperse.

7. Which of the following is an international agreement forbidding trade of endangered species among involved nations?
   (a) Montreal Protocol
   (b) CITES Treaty
   (c) Kyoto Protocol
   (d) Endangered Species Act

8. In the 1930s, sugar cane crops were being eaten by the Grayback cane beetle. Cane toads had been proven to control beetle populations in Hawaii prior to that. Cane toads did not eat the Grayback cane beetles and instead became an invasive species of their own. Which of the following factors likely contributed to the failure of the cane toad as a biological control?
   (a) Cane toads are specialists, leading them to eat everything but the beetles.
   (b) Cane toads are r-selected species, since they lay 8,000 to 30,000 eggs at a time.
   (c) Cane toad's habitat preferences were not properly studied, which lead to the toads moving away from cane fields to more moist habitats.
   (d) Cane toads were able to outcompete native toads and frogs.

Free-Response Question

At a wildlife refuge in Texas there is an invasive grass known as Guinea grass, which is native to Africa. The habitat is home to a year-round growing season for plants, and supports a diversity of wildlife, birds, and butterflies. Guinea grass was introduced in the 1930s for the purposes of grazing cattle. U.S. Fish and Wildlife Service is proposing a program to mitigate the impact of Guinea grass. Top options are:

Option 1: use herbicides to spray the Guinea grass
Option 2: physical removal of the Guinea grass

(a) **Describe** a disadvantage of using Option 1 for mitigation of this introduced species. (1 pt.)
(b) **Describe** an advantage of using Option 2 for mitigation of this introduced species. (1 pt.)
(c) Some in U.S. Fish and Wildlife have suggested combining Option 1 and Option 2. **Justify** by explaining one advantage of using both options instead of choosing only one. (1 pt.)

The refuge is estimated to produce an amount of 5.6 $kcal/m^2/year$ of Guinea grass. Productivity of native grasses is measured to be 4.1 $kcal/m^2/year$.

(d) **Calculate** the difference between the native grass and the Guinea grass, show your work. (2 pts.)
(e) **Calculate** the percent difference in the two grasses; show your work. (2 pts.)
(f) After implementing a combination of options 1 and 2 above, there is a reduction in the production of Guinea grass to 3.3 $kcal/m^2/year$. **Calculate** the percent reduction in the invasive species. Show your work. (2 pts.)
(g) The ocelot is an endangered species that can be found at the wildlife refuge, but not at a refuge nearby, around 35 km away. **Explain** how the distance between these two refuges can influence the conservation status of the ocelot. (1 pt.)

Preparing for the AP® Exam

UNIT 9
Practice Your Science Skills

Data Analysis

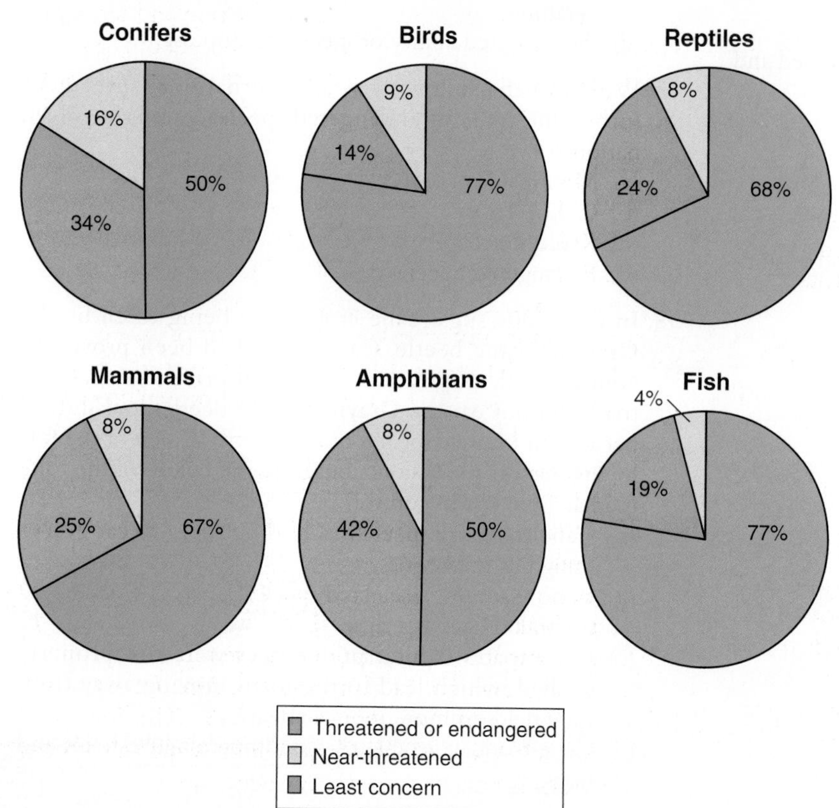

Questions

1. A conifer is a tree that bears cones and has needlelike or scalelike leaves that are typically evergreen. Conifers are of major importance as the source of lumber, resins, and turpentine. There are approximately 615 species of conifer in the world; using the percentages in the figure, **calculate** how many conifer species are threatened or endangered.
2. **Describe** an economic consequence of conifer species becoming extinct.
3. **Make a claim** about which group of animals has the most species that are near-threatened, threatened, and endangered.
4. **Describe** the characteristics of a species that makes it more vulnerable to being threatened or endangered.

Pursuing Environmental Solutions

Swapping Debt for Nature

Preserving biodiversity is expensive. A case in point is the money required to set aside terrestrial or aquatic areas for protection. As an example, if the land is privately owned, it must be purchased. Indirect costs can also be high. Not using the land, water, or other natural resources—such as wood materials, metals, and fossil fuels—results in lost income. Finally, the costs of maintaining the protected area can be prohibitive, ranging from monitoring the biodiversity to hiring guards to prevent illegal activities such as poaching. Given the fact that preserving biodiversity is expensive, how can the developing nations of the world, which contain so much biodiversity but have such little wealth, afford it?

In 1984, Thomas Lovejoy from the World Wildlife Fund came up with an idea that would help protect large areas of land but at the same time improve the economic conditions of developing countries. Lovejoy observed that developing nations possessed much biodiversity but were often deep in debt to wealthier, developed countries. Developing countries borrowed large amounts for the purpose of improving economic conditions and political stability. While the developing countries were slowly repaying their loans with interest, some had fallen so far behind on these payments that it seemed unlikely the loans would ever be repaid in full. These debtor countries had little money left over for investment in an

Swapping debt for nature in Costa Rica. Costa Rica has exceptional biodiversity, including the tropical forest surrounding the Rio Celeste waterfall. For more than a decade, Costa Rica has been forgiven more than $20 million in debt in exchange for conserving its biodiversity. *(Gonzalo Azumendi/Getty Images)*

improved environment after they had paid their loans to developed countries. Lovejoy considered the possibility that the wealthy countries might be willing to let debtor nations swap their debt in exchange for investing in the conservation of the biodiversity of the debtor nations.

The "debt-for-nature" swap has been used several times in Central and South America. In these swaps, the United States government and prominent environmental organizations provide cash to pay down a portion of a country's debt to the United States. The debt is then transferred to environmental organizations within that country with the debtor government making payments to the environmental organizations rather than to the United States. This does not mean that the country is out of debt, just that it now sends its loan payments to the environmental organizations for the purpose of protecting the country's biodiversity. In short, the indebted country switches from sending its money out of the country to investing in its own environmental conservation.

One of the largest debt-for-nature swaps recently happened in the Central American country of Guatemala. The United States government paired with two conservation organizations to provide $17 million to Guatemala. Over a period of 15 years, this amount, with interest, would have grown to more than $24 million, or about 20 percent of Guatemala's debt to the United States. In exchange, Guatemala agreed to pay $24 million over 15 years to improve conservation efforts in four areas of the country, including the purchase of land, the prevention of illegal logging, and future grants to conservation organizations helping to document and preserve the local biodiversity. The four areas include two ecosystems—mangrove forests and tropical forests. Each forms a core area within a biosphere reserve that contains a large number of rare and endangered species including the jaguar (*Panthera onca*). More than twice the size of Yellowstone National Park in the United States, this reserve offers important protection to biodiversity while also preserving historic Mayan temples that are part of Guatemala's cultural heritage and allowing sustainable use of some of the forest by local people.

Since the program began, the United States has used the debt-for-nature swap to protect tropical forests in 15 countries from Central America to the Philippines. To take part in the swap program, the countries are required to have a democratically elected government, a plan for improving their economies, and an agreement to cooperate with the United States on issues related to combating drug trafficking and terrorism. The results of these agreements have been encouraging. In Belize, for example, a debt-for-nature swap allowed 9,300 ha (23,000 acres) to be protected and an additional 109,000 ha (270,000 acres) to be managed for conservation. In Peru, a $10.6 million debt-for-nature swap led to the protection of more than 11 million ha (27 million acres) of tropical forest.

Costa Rica has made tremendous use of the debt-for-nature program with a total debt forgiveness of $26 million since 2007. In 2017, it announced its seventh such arrangement, which involved debt forgiveness of $1 million. As part of its announcement, the Costa Rican government invited conservation groups in the country to submit proposals that were focused on conservation topics, including monitoring the abundance of wildlife species, conducting research on endangered plant species, and recovering degraded areas of tropical rain forest. Although these arrangements were initially applied to tropical forests, this modern-day conservation strategy is now being used by more than 30 nations around the world. For example, in 2021 the government of Pakistan announced that it was planning to enter into an agreement to substantially increase forest cover in the country in exchange for $1 billion in debt relief from Canada, Germany, Italy, and the United Kingdom. The increased forest cover would include mangrove forests, which are critical for resisting the impacts of global climate change.

Critical Thinking Questions

1. In debt-for-nature swaps, why might the United States require that developing countries receiving such assistance have a plan for improving their economies?
2. How might the debt-for-nature program protect biodiversity?

References

Anders, W. 2017. U.S.–Costa Rica debt-for-nature swap will provide $1M for forest conservation. *Costa Rica Star*, September 13. https://news.co.cr/us-costa-rica-debt-for-nature-swap-will-provide-1m-for-forest-conservation/65624/

How debt-for-nature swaps protect tropical forests. The Nature Conservancy. http://www.nature.org/ourinitiatives/regions/centralamerica/guatemala/guatemala-debt-for-nature-swap-is-a-win-for-tropical-forest-conservation.xml

Lacey, M. 2006. U.S. to cut Guatemala's debt for not cutting trees. *New York Times*, October 2. http://www.nytimes.com/2006/10/02/world/americas/02conserve.html

U.S.–Brazil debt for nature swap to protect forests. *BBC News*, August 12, 2010. http://www.bbc.co.uk/news/world-latin-america-10958695

Science Applied 9: Concept Explanation

How Can We Bring Back Biodiversity?

Throughout this book we have discussed the decline of species in their native habitats and the introduction of non-native species, many of which rapidly spread and caused harm in their new surroundings. The island nation of New Zealand is an excellent example of how invasive species can harm native biodiversity. For millions of years, New Zealand had no land mammals, other than three species of bats. This means the island had no mammalian predators, so for millennia the birds of New Zealand evolved no defenses against mammalian predators, including some evolving to be flightless. In fact, many of these bird species are found nowhere else in the world. When humans arrived about 700 years ago, everything changed. They intentionally introduced Australian brushtail possums (*Trichosurus vulpecula*) and several species of weasels—for sources of fur—and accidentally introduced several species of rats. These invasive species have been preying on the birds of New Zealand for centuries, driving one-fourth of the bird species to extinction and many others to the brink of extinction (**FIGURE SA9.1**). In 2016, the New Zealand government announced a bold new initiative to reverse this decline in their unique biodiversity by removing every opossum, weasel, and rat from the country by 2050.

Eliminating invasive species is a difficult task, especially for animals like rats that can rapidly increase their population size in just a few months. Past eradication efforts on small islands have successfully used traps and poisons to remove every last invasive predator, but those islands are much smaller than the two main islands of New Zealand.

FIGURE SA9.1 Invasive mammals. Mammal predators, such as this rat (*Rattus rattus*), have decimated native populations in New Zealand for several centuries. *(Nga Manu Images NZ)*

Moreover, the small islands have few people living on them, making it easier to find and eliminate invasive species.

How can we begin removing invasive predators?

One of the inspirations for the nationwide effort at eradicating the predatory mammals is a sanctuary known as Zealandia. Zealandia is a completely fenced 225-hectare (556-acre) area in which the invasive predators have been removed (**FIGURE SA9.2**). Endangered species of native birds, many of which have persisted only on neighboring predator-free islands, have been re-introduced to this mainland sanctuary and they now thrive. For example, the kaka parrot (*Nestor meridionalis*) is a rare native bird (**FIGURE SA9.3**); 14 individuals were introduced into Zealandia and this population has now expanded to several hundred. With an increased biodiversity of native birds flying and calling throughout the sanctuary—combined with rare species of amphibians, lizards, and plants—Zealandia now offers the public a sense of what New Zealand used to look and sound like before humans arrived.

With this inspiration, the New Zealand government unveiled its plan to expand the effort in what it is calling "Predator-Free 2050." The effort is distributing poisons from the air to cover large areas as well as new technologies using self-resetting traps that can kill large numbers of invasive predators. Researchers are also investigating ways to design poisons that are specifically targeted to the invasive predators to avoid inadvertently poisoning native animals and people's pets. They anticipate having a targeted rat poison developed within a few years. Developing a targeted possum poison is thought to still be a decade away.

Is there opposition to the Predator-Free 2050 plan?

Not everyone is on board with the movement to eliminate the non-native predators from New Zealand. Some people are concerned that controlling the predators sends a message to the public that the way to conserve nature is through widespread killing of thousands of invasive animals. They argue that the predators did not choose to be introduced centuries ago; in fact, many of the predators were intentionally introduced by humans. Changed human preferences, these opponents argue, should not result in the invasive species being exterminated. Proponents counter this argument

FIGURE SA9.2 Zealandia. The 225-hectare (556-acre) sanctuary is surrounded by a fence to keep out invasive predators and facilitate the conservation of New Zealand's declining bird species. The fence is located just inside the road surrounding the preserve. *(Courtesy Rob Suisted)*

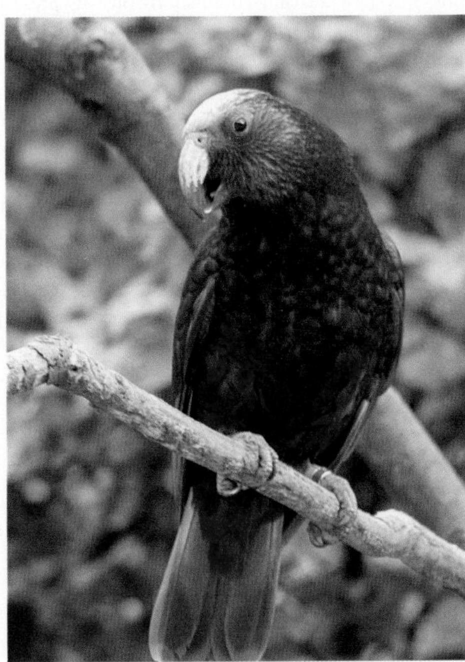

FIGURE SA9.3 Endangered New Zealand birds. Birds such as this kaka parrot are facing extinction due to predation by introduced predators. This bird was photographed inside the Zealandia sanctuary. *(Steve Attwood/Courtesy Zealandia)*

by pointing out that with or without the Predator-Free 2050 effort, thousands of animals will die. If the plan is enacted, the dying animals will be the non-native predators. If the plan is not enacted, the dying animals will be the native birds, including those that are already endangered.

Opponents of the plan have also argued that the people of New Zealand should simply learn to live with the invasive predators. After several centuries of being in the country, they argue that the invasive species are now part of the modern ecosystem and that concept should be embraced. It is also noteworthy that other introduced species—including fallow deer (*Dama dama*), brown trout (*Salmo trutta*), and mallard ducks (*Anas platyrhynchos*)—are not part of the eradication program. These species are desired by hunters and anglers and do not have the same negative stigma as rats, possums, and weasels.

The Predator-Free 2050 plan also does not acknowledge two other common species of non-native mammals in New Zealand: domestic cats and dogs. While the plan does include the removal of feral cats, house cats are also a major cause of bird mortality. In the United States, for example, it is estimated that house cats, which are often allowed to roam outside of people's houses, kill 1.4 to 3.7 billion birds each year.

How do we move forward?

The key to the success of the Predator-Free 2050 plan will be to obtain widespread support from the public to bring back the biodiversity of New Zealand. This is particularly important because homes and businesses are potential places for small numbers of surviving individuals to mount a population rebound following the nationwide extermination effort. In short, the planning for Predator-Free 2050 has to consider social issues as much as it needs to consider biological issues. Fortunately for the proponents of the plan, there has been strong support from the residents. In fact, there are thousands of volunteer groups across the country that are trapping the animals to reduce their numbers.

While the cost of the plan is expensive at $6 billion, it is estimated that the invasive predators currently cost the country $3 billion annually in crop losses. In addition, ecotourism is a major source of revenue for New Zealand, and ecotourism is likely to increase as New Zealand works to improve the populations of its many unique species of birds. Only time will tell whether the monumental effort to remove the invasive predators will be successful and whether the biodiversity of the unique birds will rebound.

Questions

1. Given that declining biodiversity can be caused by both introduced predators and habitat loss, how might the government of New Zealand invest its limited amount of conservation funding?
2. How have proponents and opponents of the Predator-Free 2050 plan invoked arguments for intrinsic versus instrumental values of biodiversity?

Practice AP® Free-Response Question

On the island of Macquarie—a small, uninhabited island between Australia and Antarctica—seal hunters introduced rats, feral cats, and European rabbits (*Oryctolagus cuniculus*) in the 1800s. The rats and cats wreaked havoc on the island's bird species, while the rabbits consumed much of the island's vegetation. From 2011 to 2014, a major effort was undertaken that successfully removed every last rat, cat, and rabbit from the island.

(a) **Explain** how the fact that the island was uninhabited makes the extermination of the invasive species more feasible. (2 pts.)

(b) **Describe** the likely response of the island vegetation after the elimination of the invasive European rabbits. (2 pts.)

(c) In the eradication effort, cats were removed first. During a 15-year period of cat removal, the rabbit population dramatically increased from 10,000 to 100,000. **Describe** what this suggests about the relationship between the cats and rabbits. (2 pts.)

(d) **Explain** why new technologies might be required as successful eradication efforts on small islands such as Macquarie are now being applied to much larger islands such as those of New Zealand. (2 pts.)

(e) **Identify** the potential risk of deploying poisons that are not designed to kill particular non-native species. (2 pts.)

References

Brown, K. V. 2018. The surprising way New Zealand could soon solve its predator problem. *Gizmodo*, January 22. https://gizmodo.com/the-surprising-way-new-zealand-could-soon-solve-its-pre-1822194179

Callabar, R. D. 2021. The nature reserve with a 500-year plan. *BBC Future Planet* https://www.bbc.com/future/article/20210527-zealandia-new-zealands-nature-reserve-with-a-500-year-plan

Greshko, M. 2016. New Zealand announces plan to wipe out invasive predators. *National Geographic*, July 25. https://news.nationalgeographic.com/2016/07/new-zealand-invasives-islands-rats-kiwis-conservation/

Morton, J. 2021. Outgoing boss of crown-owned company optimistic NZ can be predator free. *New Zealand Herald* https://www.nzherald.co.nz/nz/outgoing-boss-of-crown-owned-company-optimistic-nz-can-be-predator-free/DV2FRZTXX6PPXM2NLDBW5GNBGU/

Owens, B. 2017. Behind New Zealand's wild plan to purge all pests. *Nature*, January 11. https://www.nature.com/news/behind-new-zealand-s-wild-plan-to-purge-all-pests-1.21272

UNIT 9

Review

Key Terms to Remember

Chlorofluorocarbons (CFCs) (p. 662)
Montreal Protocol (p. 664)
Global change (p. 667)
Global climate change (p. 668)
Greenhouse effect (p. 668)
Greenhouse warming potential (GWP) (p. 670)
Ocean acidification (p. 690)
Kyoto Protocol (p. 692)
Paris Climate Agreement (Paris Climate Accord) (p. 692)
Endangered species (p. 697)
Lacey Act (p. 705)
Marine Mammal Protection Act (p. 705)
Convention on International Trade in Endangered Species of Wild Fauna and Flora (CITES) (p. 706)

Unit 9 AP® Environmental Science Practice Exam — Preparing for the AP® Exam

Section 1: Multiple-Choice Questions

1. Which statement regarding the decreased concentrations of stratospheric ozone is correct?
   (a) Increased photosynthetic activity has been measured in phytoplankton around Antarctica.
   (b) Significant increases in skin cancers have already occurred.
   (c) Although the Montreal Protocol led to a reduction in the use of CFCs, it will have little effect on stratospheric ozone concentrations in the long term.
   (d) There is no correlation between the incidence of suppressed immune systems and the lower concentrations of stratospheric ozone.

2. Which of these acts is a catalyst and repeatedly breaks down ozone molecules?
   (a) bromine
   (b) chlorine
   (c) fluorine
   (d) serpentine

3. Which of the following is a cause of declining global biodiversity?
   (a) erosion
   (b) desertification
   (c) overexploitation
   (d) migration

4. Which statement about global biodiversity is TRUE?
   (a) Species diversity is decreasing but genetic diversity is increasing.
   (b) Species diversity is decreasing and genetic diversity is decreasing.
   (c) Species diversity is increasing but genetic diversity is decreasing.
   (d) Declines in genetic diversity are occurring in wild plants but not in crop plants.

5. Which species was historically overharvested?
   (a) brown-headed cowbird
   (b) honeybee
   (c) dodo bird
   (d) zebra mussel

6. Corals live in a symbiotic relationship with algae. Coral bleaching is an event that causes the algae in this symbiotic relationship to die or depart from the coral. In the last 20 years, coral bleaching has happened more and more frequently, and though in coral reefs historically only about 2 to 5 percent of the coral permanently die, the coral bleaching events are occurring more frequently and with a higher percentage of permanent loss. These more intense and more frequent bleaching events are most likely due to
   (a) ocean acidification.
   (b) thermohaline circulation.
   (c) El Niño and La Niña events.
   (d) ocean warming due to climate change.

7. Rising concentrations of greenhouse gases are the primary cause of climate change. One of the effects of rising greenhouse gases is the melting of ice sheets. The graph below shows the reduction of ice mass in Antarctica since 2002. The final date of data on this graph was taken in June 2017, when there was −1870 Gt of ice. At the current rate of ice lost per year of −127 Gt of ice, what would be the predicted total loss of ice by June 2020?

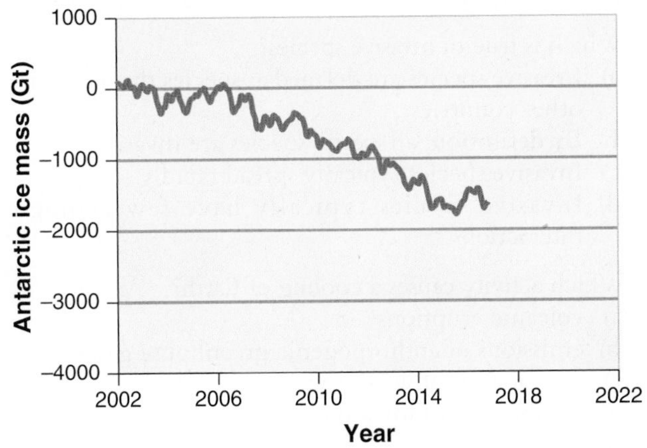

   (a) −381 Gt
   (b) −1997 Gt
   (c) −2124 Gt
   (d) −2251 Gt

8. In a biosphere reserve
   (a) sustainable agriculture and tourism are permitted in different zones.
   (b) human activities are allowed throughout the reserve.
   (c) human activities are restricted to the central core of the reserve.
   (d) sustainable agriculture is permitted, but tourism is not.

9. The Svalbard Global Seed Vault was created to address the problem of
   (a) climate change.
   (b) emerging diseases.
   (c) declining genetic diversity of crop plants.
   (d) disease resistance.

10. The international agreement known as CITES was created to
    (a) monitor populations of endangered species on the Red List.
    (b) determine if a species is endangered or not.
    (c) control the international trade of threatened plants and animals.
    (d) reduce the spread of invasive species.

11. In the "debt-for-nature" swap, the debt of a country is transferred from the ___________ to the ___________.
    (a) UN; Nature Conservancy
    (b) World Bank; local environmental organizations
    (c) U.S. government; local environmental organizations
    (d) IUCN; Nature Conservancy

12. Which is commonly associated with smaller habitat sizes?
    (a) a decrease in genetic diversity
    (b) a decrease in edge habitat
    (c) an increase of habitat diversity
    (d) slower extinction rates

13. The Galápagos Islands, located west of Ecuador, contain a rich diversity of endemic plants and animals. Several of the islands are entirely off-limits to tourists, whereas other islands allow tourists on the outer edges and on designated trails. Which best describes the Galápagos Islands?
    (a) a wildlife refuge
    (b) a biosphere reserve
    (c) multiple-use lands
    (d) a managed resource protection area

14. Which is true of invasive species?
    (a) Invasive species are defined as species that come from other countries.
    (b) By definition, all exotic species are invasive.
    (c) Invasive species typically spread rapidly.
    (d) Invasive species typically have few ecological interactions.

15. Which activity causes a cooling of Earth?
    (a) volcanic eruptions
    (b) emissions of anthropogenic greenhouse gases
    (c) evaporation of water vapor
    (d) combustion of fossil fuels

16. Which is the least potent greenhouse gas?
    (a) carbon dioxide
    (b) methane
    (c) nitrous oxide
    (d) water

17. Which greenhouse gas is correctly paired with one of its sources?
    (a) methane: landfills
    (b) nitrous oxide: melting ice caps
    (c) methane: automobiles
    (d) CFCs: deforestation

18. Carbon sequestration
    (a) is a process to remove CO_2 from the atmosphere.
    (b) is a method for preventing carbon emissions from landfills.
    (c) is a method for emissions reduction that focuses on improved efficiency.
    (d) is the release of carbon from soils due to warming.

19. Which statement about global warming is TRUE?
    (a) The planet is not warming.
    (b) The planet is warming, but humans have not played a role.
    (c) The planet has had many periods of warming and cooling in the past.
    (d) Greenhouse gases compose only a small fraction of the atmosphere, so they cannot be important in causing global warming.

20. Which statement below about feedback loops that occur with climate change is TRUE?
    (a) All feedback loops are positive.
    (b) All feedback loops are negative.
    (c) Increased soil decomposition under warmer temperatures represents a positive feedback loop.
    (d) Increased plant growth under higher CO_2 concentrations represents a positive feedback loop.

21. Which predicted consequence of global warming has not yet occurred?
    (a) melting ice caps
    (b) rising sea levels
    (c) shutting down the thermohaline circulation of the ocean
    (d) altered breeding times and flowering times of animals and plants

22. Which statement regarding the Kyoto Protocol is TRUE?
    (a) All nations agreed to reduce their emission of greenhouse gases.
    (b) All nations agreed to stop emitting greenhouse gases.
    (c) Developed nations agreed to different levels of emission reductions.
    (d) Developing nations agreed to reduce their emission of greenhouse gases.

Use the graph below to answer questions 23–25:

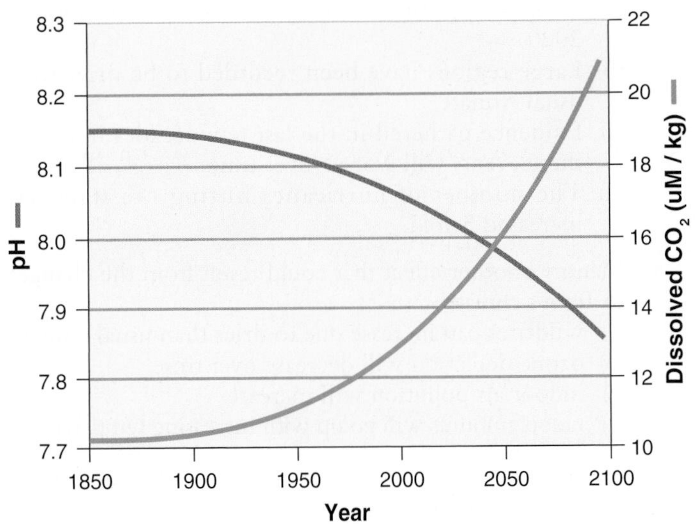

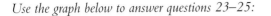

23. Given the data above, approximately how much is dissolved CO_2 expected to rise from 1850 to 2100 in micromoles/Kg?
    (a) 13
    (b) 11
    (c) 0.3
    (d) 9

24. Given the data above, which of the following is likely to occur due to the pattern of CO_2 and pH?
    (a) As the carbon dioxide concentration goes up, ozone in the stratosphere will be depleted.
    (b) As the carbon dioxide concentration goes up, ocean temperatures will drop globally.
    (c) As the pH goes down, coral reefs will be damaged by the increased acidity.
    (d) As the pH goes down, oceanic levels of mercury will rise due to the increased acidity.

25. Given the data above, which of the following is a likely cause in the trends in pH and CO_2?
    (a) Vehicle emissions have increased in the period of 1850 to 2100.
    (b) Deforestation has decreased in the period of 1850 to 2100.
    (c) Burning of fossil fuels has remained the same in the period of 1850 to 2100.
    (d) Tropospheric ozone has increased in the period of 1850 to 2100.

26. Which phenomenon is part of a negative feedback with regard to global warming?
    (a) increased absorption of atmospheric CO_2 by the ocean
    (b) methane production from peatland decomposition
    (c) increased evapotranspiration from plant growth
    (d) CO_2 release from forest fires during dry spells

27. Which country is among the top three emitters of CO_2?
    (a) Germany
    (b) Australia
    (c) India
    (d) Japan

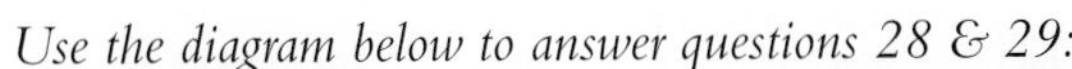
Use the diagram below to answer questions 28 & 29:

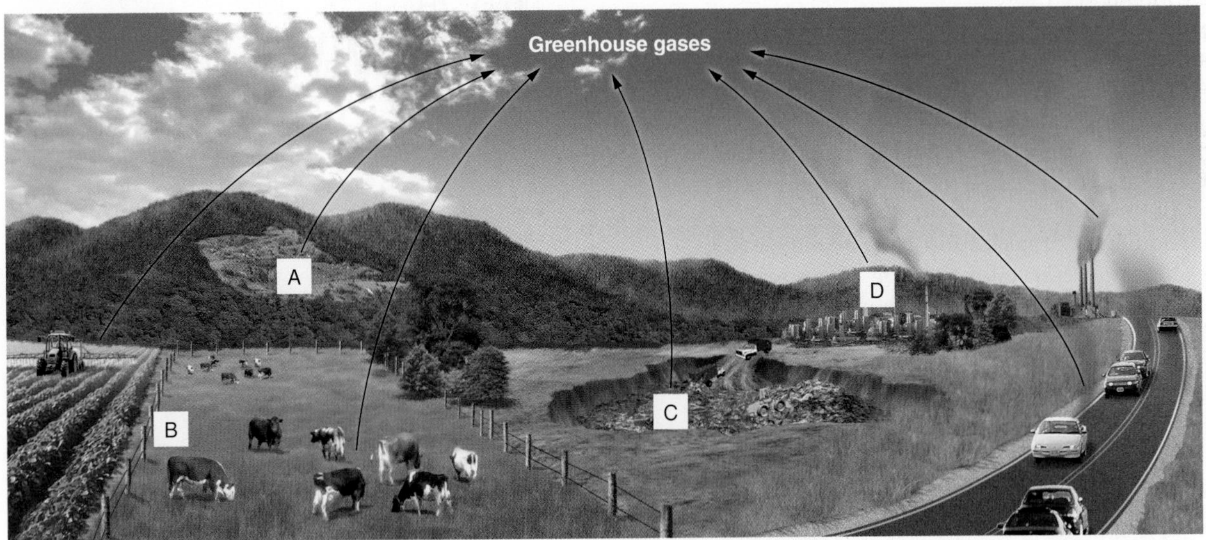

28. Which source of greenhouse gas depicted above adds carbon to the atmosphere via decomposition?
    (a) A - Deforestation
    (b) B - Agriculture
    (c) C - Landfills
    (d) D - Factories

29. Which source of greenhouse gas causes more CO_2 to be in the atmosphere by removing a sink of carbon?
    (a) A - Deforestation
    (b) B - Agriculture
    (c) C - Landfills
    (d) D - Factories

Use the passage below to answer questions 30 & 31:

Several factors have been monitored in the state of Texas over the last few decades, to determine the ways in which climate change is affecting the state:

1. Daily temperatures have risen 2 degrees from 1900 to 2020.
2. The number of hurricanes hitting the state have increased five-fold.
3. Sea levels have risen an average of 4 feet per year, and
4. Large regions have been recorded to be in drier than usual climate, leading to water scarcity.

Evidence gathered in the last few decades only indicates these events will worsen over time unless significant steps are taken to mitigate climate change.

30. Identify the author's claim from the passage above.
    (a) Daily temperatures have risen 2 degrees from 1900 to 2020
    (b) Large regions have been recorded to be drier than usual climate
    (c) Evidence gathered in the last few decades indicates these events will worsen over time
    (d) The number of hurricanes hitting the state has increased 5-fold.

31. Identify another effect that could result from the changes to Texas's climate.
    (a) wildfires can increase due to drier than usual climate
    (b) ozone depletion will decrease over time
    (c) indoor air pollution will increase
    (d) radon amounts will go up with increasing temperatures

Section 2: Free-Response Questions

1. An environmental organization is concerned about protecting a little-studied, endangered species of frog that has recently been listed as endangered by the IUCN. They created habitats to protect the species, setting aside several areas for the species throughout a province.
   (a) As an experiment, the organization has created four different areas around the province, separated by less hospitable habitats.
      (i) **Identify** the independent variable in the experiment. (1 pt.)
      (ii) **Identify** the dependent variable in the experiment. (1 pt.)
      (iii) **Identify** a reasonable hypothesis for the experiment. (1 pt.)
      (iv) **Describe** one variable that was not discussed that could affect the results of the study. (1 pt.)
   (b) **Explain** one likely reason why the frog species has only recently been listed as endangered by the IUCN. (1 pt.)
   (c) The environmental organization has also suggested creating one large reserve for the frog versus several smaller reserves. **Describe** one benefit of each approach. (2 pts.)
   (d) **Describe** one drawback of each approach described in part (c). (2 pts.)
   (e) Suppose the primary reason for the decline in the frog species is an invasive predator that tends to live in edge habitats. **Describe** one method to mitigate the issue of the invasive predator. (1 pt.)

2. Sea ice has been measured and found to be decreasing from historical quantities. At the same time, temperatures have been rising globally, with the largest increases in the Arctic.
   (a) **Identify** one cause for sea ice reduction and temperature increases globally. (1 pt.)
   (b) **Describe** one other issue that has occurred from the cause described in part (a). (1 pt.)

The graph below shows sea ice reduction from 1984 to 2020.

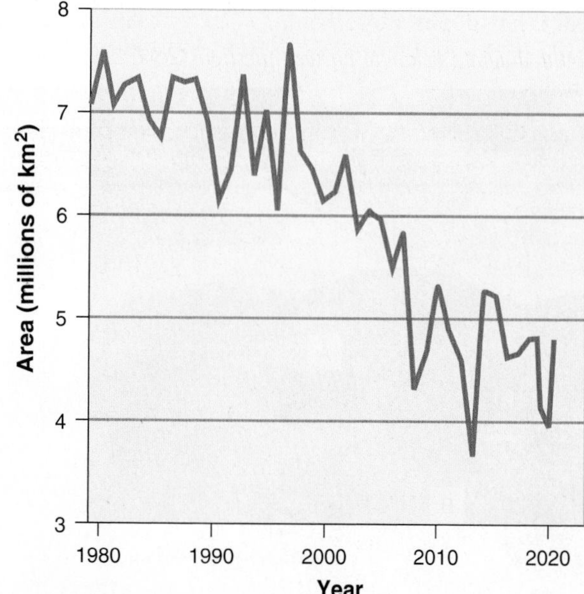

   (c) **Identify** the amount of sea ice in the year 2000. (1 pt.)
   (d) **Identify** the amount of sea ice in the year 1984. (1 pt.)
   (e) **Describe** the effect less sea ice will have on species in the Arctic. (1 pt.)
   (f) **Explain** how a decrease in sea ice will affect the overall climate of the planet. (1 pt.)
   (g) **Describe** TWO methods that an individual could use to help reduce the effects of global climate change. (2 pts.)
   (h) **Propose** one drawback for each method mentioned in part (g). (2 pts.)

3. The map below shows where ozone-depleting substances were produced during the year 2014.

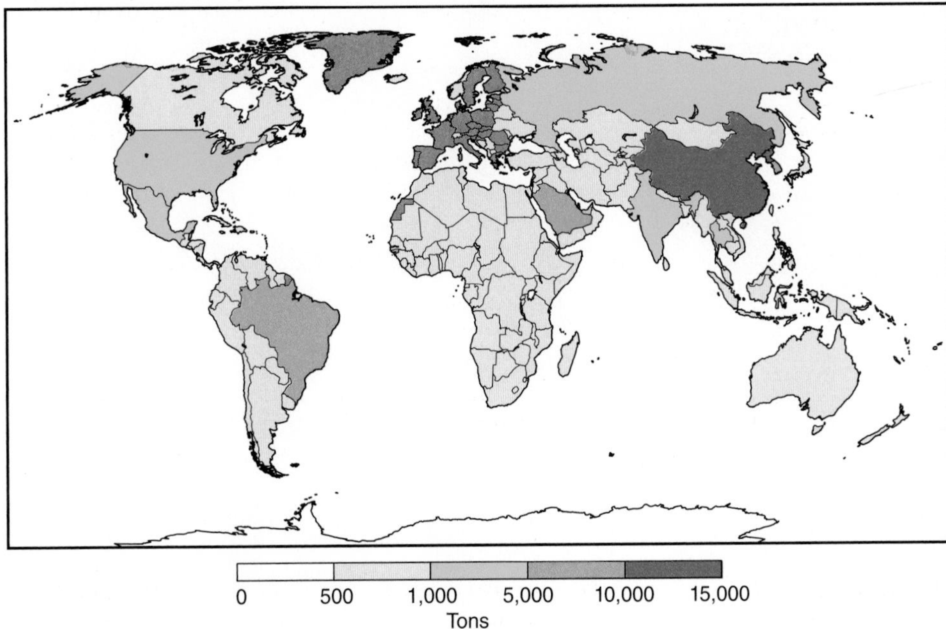

(a) **Identify** the nation using the most ozone-depleting substances in 2014. (1 pt.)

(b) **Describe** what ozone-depleting substances are used for. (1 pt.)

(c) **Explain** one effect reduced ozone has on humans and one effect it has on the environment. (2 pts.)

Ozone has depleted between 5 percent on average each decade in the mid-latitudes, and 11 percent in the southern latitudes.

(d) **Calculate** the difference between the ozone depleted in the mid-latitudes and southern latitudes. Show your work. (2 pts.)

(e) Ozone is measured in Dobson units. It is estimated that the historic ozone concentration was 300 Dobson units. **Calculate** the percentage of depleted ozone in the mid-latitudes in Dobson units, over the span of one decade. Show your work. (2 pts.)

(f) **Calculate** the amount of ozone depleted in the southern latitudes in Dobson units. Show your work. (2 pts.)

Cumulative AP® Environmental Science Practice Exam

Section 1: Multiple-Choice Questions

1. Which biome has the greatest amount of nutrients in the soil?
   (a) tropical rainforest
   (b) temperate seasonal forest
   (c) taiga
   (d) tundra

2. Suppose there is an ecosystem with several plant species and several generalist and specialist consumers that eat those plants. Which of the following best describes what will most likely happen to the ecosystem following a severe drought that kills most of the plant species?
   (a) The specialist species will adapt to consume other plants that they did not originally consume.
   (b) The generalist species in the area will be advantaged over the specialist species and outcompete the specialists.
   (c) The specialist species will be advantaged over the generalist species and outcompete the generalists.
   (d) The generalist species will begin to prey on the specialist species as a way to adapt and survive.

3. The four letters on the map below represent four distinct biomes. Which biome is most likely to develop the thickest soil O-horizon?

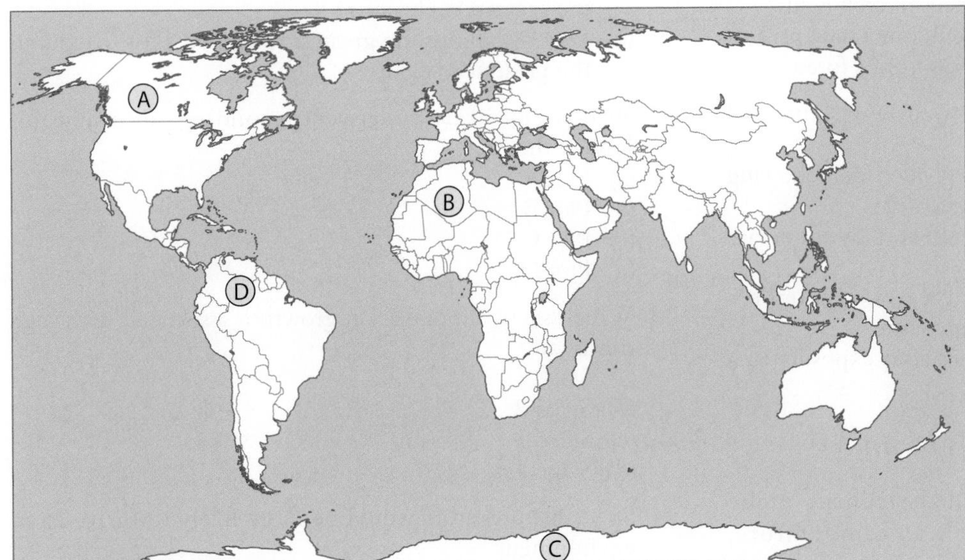

   (a) biome A
   (b) biome B
   (c) biome C
   (d) biome D

4. The graphic below depicts the ways in which phosphorus cycles through the global ecosystem. Which arrow is not part of the phosphorus cycle?

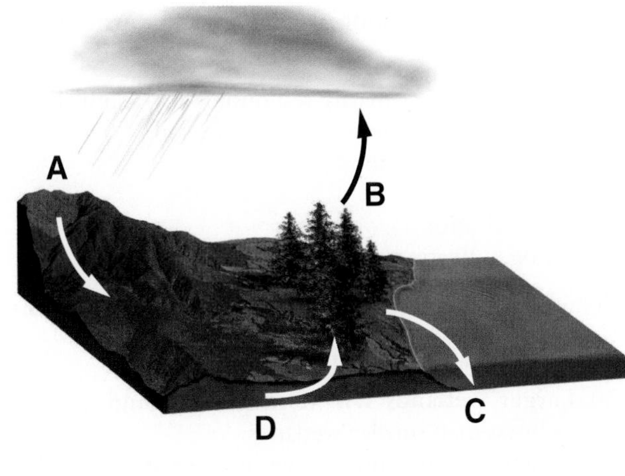

   (a) A
   (b) B
   (c) C
   (d) D

5. Imagine that you place a plant inside a sealed jar along with a sensor that measures the concentration of CO_2 inside the jar. You then place the jar with the plant and sensor near a window exposed to ambient sunlight. You record CO_2 in the jar over 24 hours. How could you estimate the gross primary productivity of the plant?
   (a) Subtract the amount of O_2 lost during the day from the amount of O_2 gained during the night.
   (b) Add the amount of CO_2 lost during the day to the amount of CO_2 gained during the night.
   (c) Determine the difference in average CO_2 concentration during the day and average CO_2 concentration during the night.
   (d) Record the amount of CO_2 gained during the day only.

6. Which process is primarily responsible for assimilating N_2 gas from the atmosphere into water or soil?
   (a) nitrogen assimilation
   (b) nitrification
   (c) denitrification
   (d) nitrogen fixation

7. Which statement accurately describes the theory of island biogeography?
   (a) Larger species generally require more area.
   (b) Larger wetlands will tend to have higher species richness than smaller wetlands.
   (c) Larger islands tend to have greater species richness than smaller islands.
   (d) Greater amounts of habitat diversity are associated with smaller amounts of species diversity.

8. Which activity would lead to an increase in regulatory ecosystem services?
   (a) planting trees in a previously deforested area
   (b) spraying pesticides to remove weeds in crop fields
   (c) erecting an educational visitor center in a state park
   (d) removing all understory vegetation from a forest

9. Which population best responds to an environmental stressor?
   (a) a population of squirrels that sharply declines following the introduction of an invasive predator
   (b) a population of trees that is unaffected by moderate forest fires
   (c) a population of crabs that increases in numbers following a change in water temperature
   (d) a population of fish that collapses after exposure to a novel disease

Use the following passage to answer question 10:

A team of bird researchers examined the climate niche (range of climatic conditions such as temperature, precipitation, humidity) that a variety of bird species occupy. According to their research, birds that have increased in abundance during the previous 50 years now occupy habitats with a wider range of climatic conditions than they did 50 years ago, and declining species now occupy habitats with a smaller range of climate conditions than they did 50 years ago.

10. Based on the passage above, what is the most accurate description of a bird species that has increased in abundance?
    (a) The species is failing to adapt to new climate conditions.
    (b) The species is more of a generalist than it was previously.
    (c) The climate niche of the species is not changing.
    (d) The species is one that does not appear to migrate to South America in winter.

Use the table below to answer questions 11 & 12:

Forest plot	Year 1	Year 2	Year 3	Year 4
Plot A	20	2	19	20
Plot B	15	13	15	16
Plot C	14	0	0	2
Plot D	18	3	5	7

11. Researchers measured species richness along a single transect in four plots of a temperate forest. Measurements in the first year were made before a series of forest fires occurred later that year. Based on the data, which plot exhibits the greatest resistance to fire disturbance and which plot exhibits the greatest ability to recover from fire disturbance?
    (a) Plot A exhibits the greatest resistance; Plot B exhibits the greatest resilience.
    (b) Plot C exhibits the greatest resistance; Plot A exhibits the greatest resilience.
    (c) Plot B exhibits the greatest resistance; Plot A exhibits the greatest resilience.
    (d) Plot D exhibits the greatest resistance; Plot C exhibits the greatest resilience.

12. Which plot likely underwent secondary succession following the fire?
    (a) Plot A
    (b) Plot B
    (c) Plot C
    (d) Plot D

13. Which stage of population growth is associated with high death rate?
    (a) Stage 1
    (b) Stage 2
    (c) Stage 3
    (d) Stage 4

14. Specialist predators would be more likely to thrive in an environment
    (a) with very little primary productivity.
    (b) undergoing early stages of primary succession.
    (c) containing a high diversity of primary consumers.
    (d) with high species evenness.

15. In demographic transition theory, the fourth stage of population growth is generally associated with what type of age structure diagram?
(a) a pyramid
(b) a vertical column
(c) a pyramid with significantly more males than females
(d) an inverted pyramid

Use the graph below to answer questions 16 & 17:

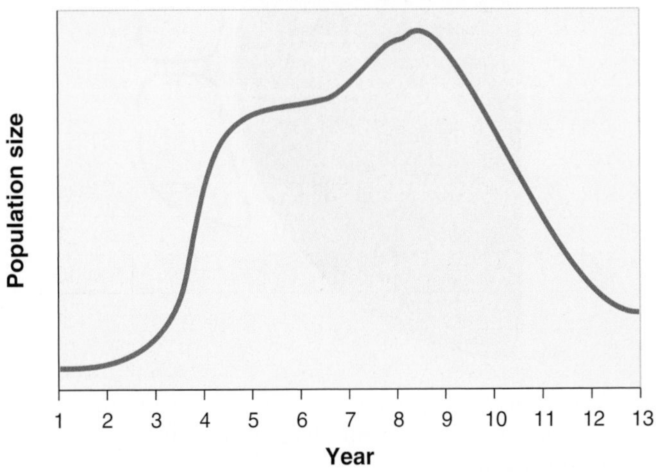

16. Researchers monitored the population size of a species for over a decade. Given the trend shown in the graph, which statement is likely correct?
(a) Available resources increased halfway through the observation period, and then decreased.
(b) The species is most likely *K*-selected.
(c) The species experienced several overshoots and die-offs during the years of observation.
(d) The trend in population growth primarily represents an exponential growth model.

17. When is the growth rate of this population the highest?
(a) Year 1
(b) Year 4
(c) Year 10
(d) Year 13

Use the table below to answer questions 18 & 19:

Country	Crude birth rate	Crude death rate	Net migration rate
Country A	8.8	12.9	1.3
Country B	11.6	11.0	0.1
Country C	33.8	9.3	−5.0
Country D	12.4	8.3	3.0

18. Considering only crude birth rates and crude death rates of the four countries listed in the table, which country has the highest replacement level fertility?
(a) Country A
(b) Country B
(c) Country C
(d) Country D

19. Considering crude birth rates and crude death rates, as well as net migration rates, which country is most likely to be in the second stage of demographic transition?
(a) Country A
(b) Country B
(c) Country C
(d) Country D

20. According to the age structure diagram shown below, which statement best describes this population?

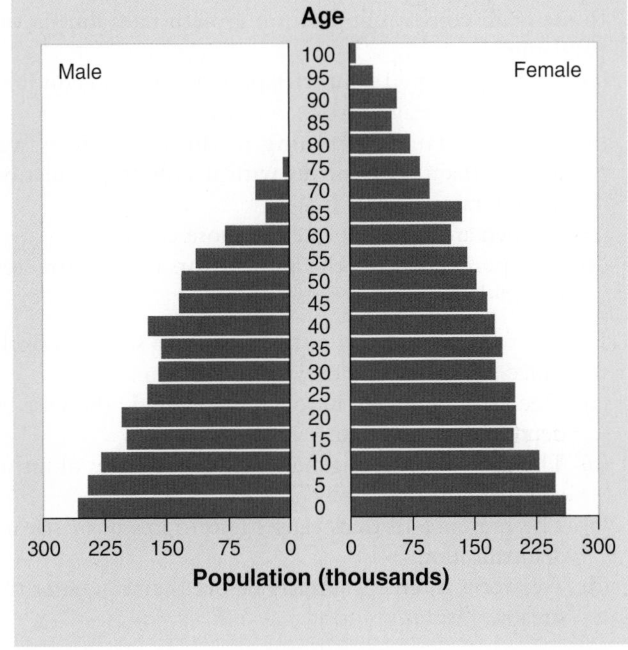

(a) The population experiences a low crude birth rate.
(b) Among newborns, there is a skewed sex ratio toward more females.
(c) The population is exhibiting a decreasing growth rate.
(d) Males have shorter life spans than females.

21. The Malthusian theory describes factors that limit the growth of human population. Which of the following arguments counters the Malthusian theory of human population growth?
(a) Some countries have expanding populations, whereas other countries have contracting populations.
(b) Humans have consistently developed strategies for growing more food to meet increasing demand.
(c) Like all species, human populations are subject to density-independent effects.
(d) The global human population size has been stable for the last century.

Use the following passage to answer question 22:

Throughout human history, pandemics such as the bubonic plague in the Middle Ages have led to increased mortality and declining population growth rates. The current Covid pandemic has also shown this pattern. However, a few countries including Germany, Finland, Denmark, and the Netherlands have not experienced decreased population growth rates.

22. What piece of evidence would support the author's description for why certain developed countries were able to maintain constant population growth rates during the pandemic?
    (a) Some developed countries provided more economic support for health care than others.
    (b) Access to family planning in those countries was greater than in countries with declining population growth rates.
    (c) Women are more educated in those countries.
    (d) The percentage of the population that contracted Covid-19 was less in those countries.

23. Explain why the addition of trees around a stream is likely to reduce aquatic concentrations of nitrogen.
    (a) Tree leaves intercept nitrogen that would otherwise be deposited by acid rain.
    (b) The growth of trees generally increases rates of nitrogen fixation.
    (c) The presence of trees causes faster rates of in-stream denitrification.
    (d) Tree roots intercept nitrates before they can enter the stream.

24. Which property of water leads to a type of physical weathering whereby rocks break up into smaller pieces?
    (a) Water expands when it freezes.
    (b) Water vaporizes at 100°C.
    (c) The polarity of water molecules can dissolve minerals.
    (d) Hydrogen bonding permits strong capillary action.

25. An increase in the amount of impermeable surfaces within a watershed will most likely
    (a) reduce soil porosity.
    (b) increase downstream soil erosion.
    (c) reduce the total amount of water leaving the watershed annually.
    (d) increase the cation exchange capacity of soil under the impermeable surface.

26. Which of the following is a correct description of the formation of global wind patterns?
    (a) Global wind patterns result from increased upwelling in coastal areas.
    (b) Global wind patterns result from high albedo at the poles, leading to rising cold air masses.
    (c) Global wind patterns result from uneven heating of the Earth's surface and changes in the density of air masses.
    (d) Global wind patterns result from seasonal changes in air temperature and varying rates of photosynthesis.

Use the illustration below to answer questions 27 & 28:

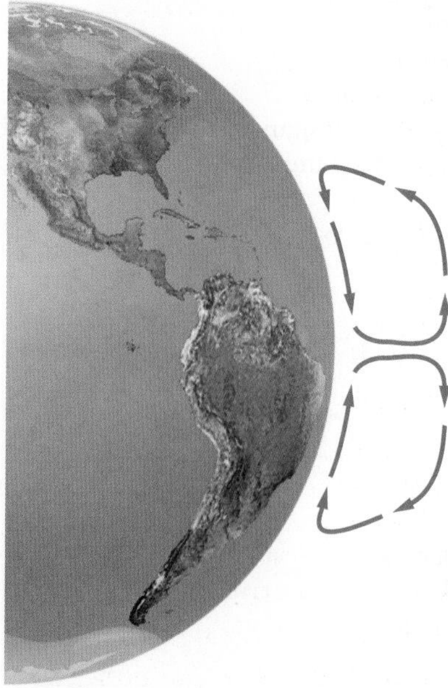

27. Which of the following phenomenon is responsible for the pattern shown in the diagram that illustrates global wind patterns?
    (a) gravitational pull of the moon
    (b) intense solar radiation at the equator
    (c) latent heat release
    (d) water holding capacity of air

28. What feature of Earth's atmosphere does the illustration represent?
    (a) Hadley cells
    (b) rain shadows
    (c) the mesosphere
    (d) Ferrell cells

29. Imagine you are conducting a study to investigate how the invasion of earthworms in North America has affected the amount of organic material in soil. To do this, you identify six 1-acre plots of land in a forest. In half of the plots, you spray an insecticide that targets earthworms. In the other half of the plots, you spray water as a control. After 1 year, you measure the amount of humus in the soil. What is an appropriate hypothesis for this study?
    (a) You hypothesize that the addition of water will have substantial effects on the amount of humus in forest soil.
    (b) You hypothesize that spraying insecticides will increase the amount of humus in forest soil.
    (c) You hypothesize that elimination of earthworms by pesticides will increase the amount of humus in forest soil.
    (d) You hypothesize that earthworms consume humus in forest soil.

30. Relative to other years, what environmental hazard is more likely to occur in the southern United States during El Niño years?
   (a) elevated levels of smog
   (b) increased severity of wildfires
   (c) greater risk of flooding events
   (d) increased risk of acid rain

31. Which is an example of the cost of internalizing an externality associated with environmental systems?
   (a) charging deer hunters for a hunting license
   (b) breathing in fresh air in a forest
   (c) planting trees along roadsides in urban areas
   (d) feeling a sense of calm after hiking up a mountain

32. What U.S. legislation is designed to protect rare and threatened species?
   (a) the Lacey Act
   (b) the Rare and Threatened Species Act
   (c) the Resource Conservation and Recovery Act
   (d) the Endangered Species Act

33. What is an advantage of synthetic fertilizers that is not typically an advantage of organic fertilizers?
   (a) Synthetic fertilizers are less energy intensive to produce than organic fertilizers.
   (b) Synthetic fertilizers are more easily absorbed by plants than organic fertilizers.
   (c) Synthetic fertilizers are less likely to run off farms than organic fertilizers.
   (d) Synthetic fertilizers reduce the need for intensive use of pesticides.

34. Suppose you are attempting to manage the population of a marine fish species for sustainable harvesting. From past research, you know that the species exhibits typical logistic growth, as shown in the graph. What population size is most likely to be the maximum sustainable yield of this species?

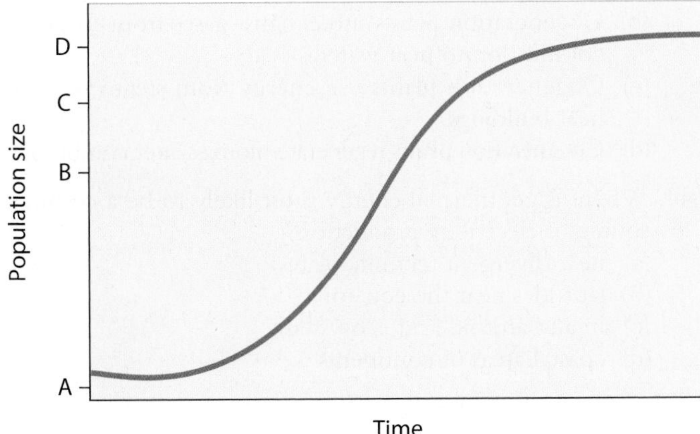

   (a) population size A
   (b) population size B
   (c) population size C
   (d) population size D

Questions 35 & 36 refer to the following excerpt from a Harvard Gazette *article by Robert Paarlberg.*

"The conviction that organic food is a better choice did not become widespread in the United States until the 1980s, when national media reported a number of food safety scares linked to pesticide residues on fresh fruits and vegetables. When worried consumers learned that organic farming methods did not allow the use of any synthetic pesticides (although naturally occurring poisons could be used), they demanded more organic products, along with a credible national system for certifying and labeling those products in the marketplace. Once this system began to operate in 2002, the farmers who had switched to organic methods could capture sizable price premiums for their goods, and this motivated rapid growth in the sector, but only up to a point."

The article continues, "When the new National Organic Program came into full effect in 2002, commercial production and sales began to increase rapidly. Food stores specializing in organic products, such as Whole Foods and Wild Oats, expanded operations by building new outlets and buying up or consolidating existing organic and natural food stores. Like all supermarkets, these retailers sought out suppliers who could deliver a steady volume of high-quality products on time, at a consistent grade, and uniformly packaged. Small, diverse organic farms could not meet these requirements, so it was the highly specialized, industrial-scale operations that expanded to take over. Earthbound Farm in California, for instance, started out with 2.5 acres of organic raspberries in 1984, but now it manages 50,000 acres and has taken over more than half of the national market for organic packaged salad greens."

Information from https://news.harvard.edu/gazette/story/2021/02/author-robert-paarlberg-argues-against-buying-organic/

35. What does the article assert as the reason for why the public began consuming more organic food?
   (a) The public assumed that organically grown products might be less harmful to human health.
   (b) Organically produced foods are more sustainable than conventionally produced foods.
   (c) Organic foods are less processed than conventional foods.
   (d) Buying organic generally supports small and local farmers.

36. Why is it not possible for "small, diverse organic farms" to meet the requirements of large supermarkets that sell organic food?
   (a) Small farms generally produce crops of less quality than larger farms.
   (b) Small farms do not benefit from the economies of scale.
   (c) Small farms often do not produce the types of crops that the larger supermarkets want to sell.
   (d) Small farms often fail to meet the requirements for organic certification.

37. Based on the data in the table, which statement is most accurate?

State	Population	CO_2 emitted per capita (metric tons)
Wyoming	581,075	111
Rhode Island	1,061,509	10.1
New York	19,299,981	8.6
California	39,613,493	9.3

(a) Wyoming has the greatest annual CO_2 emissions of the four states.
(b) Residents of Rhode Island have the lowest carbon footprint of the four states.
(c) California has the greatest annual CO_2 emissions of the four states.
(d) Wyoming emits the least amount of CO_2 of the four states.

38. The table below presents caloric content, land requirements, and CO_2 gas emissions per cup (200 mL) of different milk types commonly produced. Which type of milk provides the least number of calories per mL and which type has the greatest ecological footprint?

Milk type	Energy content in cal/200 mL of milk	Land required per 200 mL (m^2)	CO_2 released per 200 mL (kg CO_2)
Cow milk produced on CAFOs	150	0.24	0.55
Cow milk produced by free-range animals	150	1.79	0.63
Almond milk	25	0.10	0.14
Soy milk	110	0.13	0.20

(a) Free-range cow milk provides the least number of calories per mL and also has the greatest ecological footprint.
(b) Soy milk provides the greatest number of calories per mL; CAFO cow milk has the greatest ecological footprint.
(c) Almond milk provides the least number of calories per mL; free-range cow milk has the greatest ecological footprint.
(d) CAFO cow milk provides the least number of calories per mL; free-range cow milk has the greatest ecological footprint.

39. Farmers operating in the Colorado River Basin rely on diverting water flow from the Colorado River to irrigate their crops. Because so many people rely on the river, the government sells permits that allow farmers to use limited amounts of the water. What does this strategy prevent?
(a) an increase in the use of drip irrigation
(b) tragedy of the commons
(c) cones of depression
(d) aquifer recharging

40. Which farming strategy can prevent erosion?
(a) crop rotation
(b) contour plowing
(c) tilling
(d) monocropping

41. Which statement best explains why developing a sense of place can prevent urban residential area blight?
(a) Developing a sense of place encourages residents to build houses in suburban areas.
(b) Communities with a sense of place tend to attract wealthier residents.
(c) Increasing a sense of place in a community generally reduces the amount of open space in an area.
(d) Areas with a sense of place are less likely to deteriorate over time and more likely to retain residents.

42. Which statement is correct?
(a) All energy sources are ultimately derived from the Sun.
(b) Many forms of renewable energy achieve 100 percent efficiency.
(c) Conversion of energy from one form to another is never 100 percent efficient.
(d) The production of renewable energy does not release any greenhouse gases.

43. Explain how a cogeneration plant increases energy use efficiency.
(a) Cogeneration plants combine biomass and fossil fuel energy to produce electricity.
(b) Cogeneration plants use exhaust gases from fossil fuel combustion to heat water.
(c) Cogeneration plants use energy from steam waste to heat buildings.
(d) Cogeneration plants regenerate biomass after combustion.

44. Where is geothermal energy most likely to be a common source of electricity production?
(a) near divergent tectonic plates
(b) latitudes near the equator
(c) in and around arid ecosystems
(d) coastal areas of continents

Use the figure below to answer questions 45 & 46:

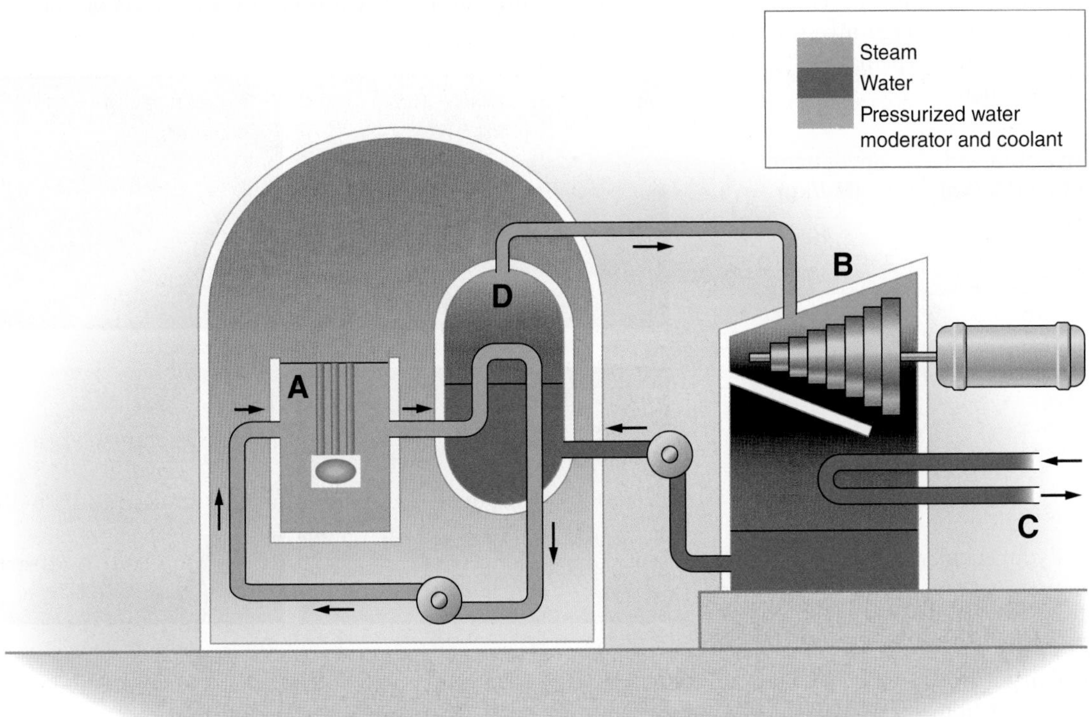

45. The figure represents the basic components of a nuclear energy plant. At what location in this diagram is thermal energy first generated?
    (a) location A
    (b) location B
    (c) location C
    (d) location D

46. What is the purpose of the device located at location C?
    (a) to condense steam into water
    (b) to pump liquid around control rods
    (c) to generate neutrons that will continue the process of nuclear fission
    (d) to convert steam energy into electricity

47. According to the graphs shown below, which is the most accurate statement?

Typical results for residential solar systems in the United States

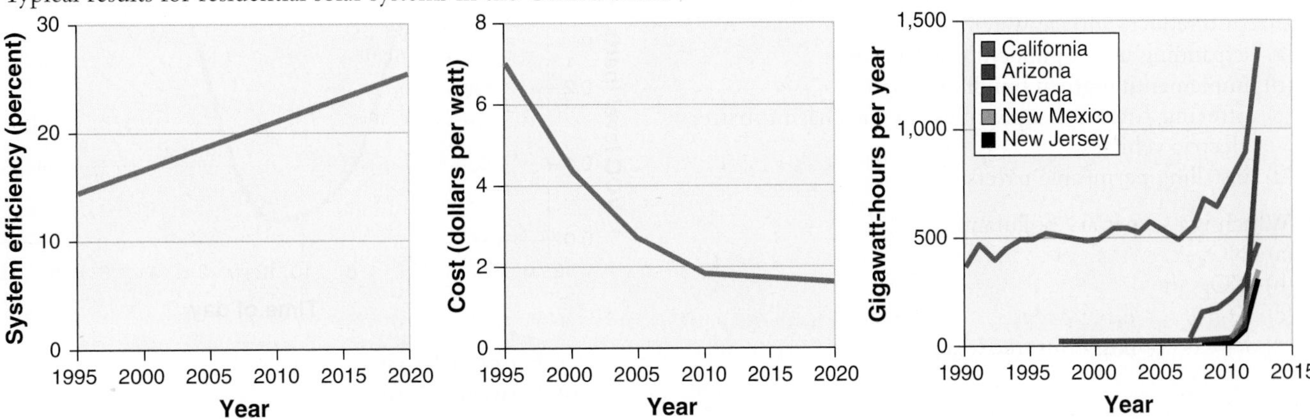

(a) The use of solar energy in the United States increased primarily due to increased system efficiency.
(b) The use of solar energy in the United States began increasing due to a decrease in solar panel cost.
(c) Solar energy was produced in most states prior to 2010, but only became prevalent once efficiency increased and cost decreased.
(d) The cost of solar panels was probably reduced because there was a concurrent increase in panel efficiency and user demand.

48. Suppose that you could use either diesel, gasoline, firewood, or natural gas to run a machine. The table presents data on the energy density and energy investment to obtain each of those fuels. Which fuel type has the greatest energy return on energy investment?

Fuel type	Energy obtained from fuel (MJ/kg)	Investment (MJ/kg)
Diesel	42	10
Gasoline	44	40
Firewood	16	5
Natural gas	50	30

(a) diesel
(b) gasoline
(c) firewood
(d) natural gas

49. Imagine that you replaced your standard washing machine with an Energy Star washing machine. The new washing machine consumes 0.5 kW. Suppose that electricity cost at peak demand is 20 cents per kWh, whereas off-peak cost is 14 cents per kWh. If you ran the machine twice per week and each cycle lasted for 1 hour, what would be the annual difference in electricity cost if you ran the machine during peak demand versus off-peak demand time?
(a) \$1.56
(b) \$3.12
(c) \$3.64
(d) \$6.24

50. What do wood, manure, wind, and geothermal energy sources have in common?
(a) All are nondepletable energy sources.
(b) All generate more air pollution than coal.
(c) All are renewable energy sources.
(d) All use heat energy to rotate a turbine.

51. Which of the following is an energy conservation strategy that also reduces surface water runoff?
(a) expanding use of public transportation
(b) implementing green building design
(c) offering financial incentives for purchasing battery electric vehicles
(d) installing permeable pavers

52. Which is a secondary pollutant?
(a) NO_2
(b) CO
(c) O_3
(d) VOCs coming from tree leaves

53. Which image depicts the most likely city to experience a thermal inversion that will have adverse effects on human health?

(a) image A
(b) image B
(c) image C
(d) image D

54. City managers are trying to reduce concentrations of photochemical smog by reducing production of VOCs from natural and anthropogenic sources. The graph shows VOC concentration from both sources over a 24-hour period. During which time period should city managers attempt to reduce fossil fuel combustion?

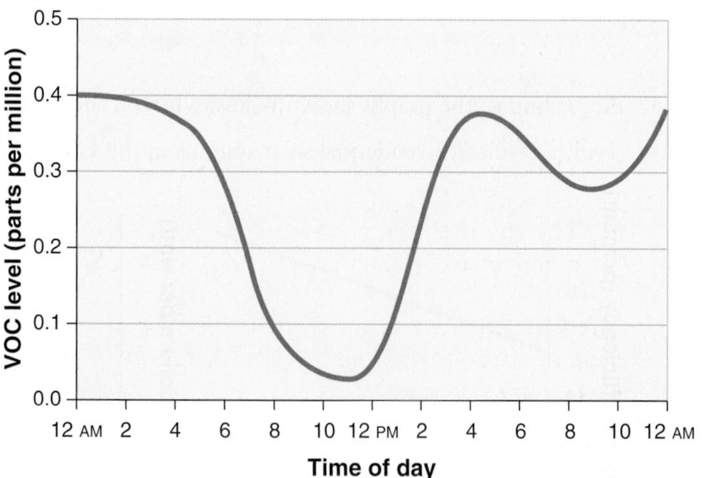

(a) between 10 AM and 1 PM
(b) between 12 PM and 8 PM
(c) between 8 AM and 12 PM
(d) between 10 PM and 4 AM

55. Explain why the increased use of public transportation can reduce atmospheric pollution, even though the engines of public transportation vehicles (e.g., buses, trains) might be much less efficient than single-passenger vehicles.
   (a) Public vehicles are less efficient but tend to produce fewer long-lasting pollutants.
   (b) Public vehicles travel shorter distances than single-passenger vehicles.
   (c) Many forms of public transportation are switching to electric engines, and the generation of electricity emits fewer pollutants than the combustion of gasoline.
   (d) The per-passenger pollutant emissions are lower on public transportation.

56. The graph below shows the concentration of three pollutants for a major city from 2003 to 2010. Based on the changing concentrations of pollutants shown in the graph, what most likely occurred in 2006?

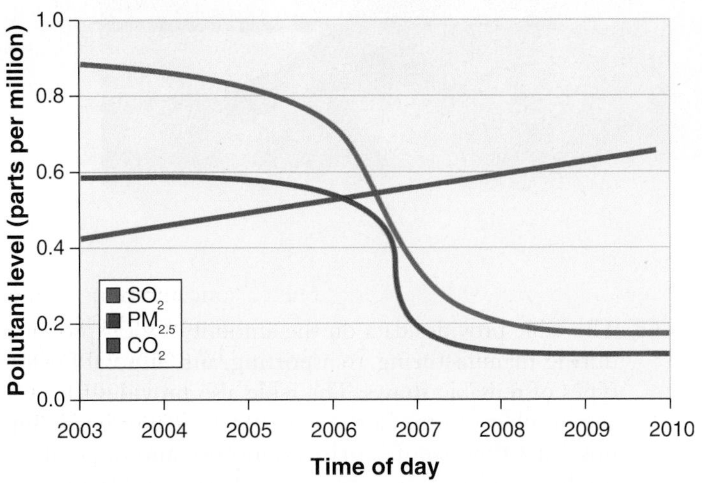

   (a) The city required that power plants install smokestack scrubbers.
   (b) The city began implementing restrictions on single-passenger vehicle transportation.
   (c) The city began seeing a rise in the number of individuals with asthma and other respiratory ailments.
   (d) The city launched a campaign to reduce the amount of indoor air pollutants.

Use the following excerpt from the article "Noise pollution is invading even the most protected natural area," by Ula Chrobak to answer question 57:

Noise pollution—from honking cars to clanging construction equipment—can disturb sleep, cause stress, and impair concentration. Regulators have largely ignored noise in national parks, local parks, wilderness, and other protected areas, which cover 14 percent of the country. Noise pollution can be annoying to people and it can also harm ecosystems. It can also undo the benefits of spending time in nature, like improved mood and memory retention. For plants and animals, the ruckus can disrupt entire communities ... Animals need silence to hear predators approaching or to communicate with their mates: A bird whose song would normally travel 100 meters would, with a 10-fold increase in noise, have its melody stifled to a 10-meter radius. "In so many landscapes, both people and other organisms are living in shrunken perceptual worlds," says Clinton Francis, an ecologist at California Polytechnic State University in San Luis Obispo, who was not involved in the work.

57. Based on the text, what is the most likely future consequence if the level of noise pollution in and around parks continues to grow over the next decade?
   (a) There will be more visitors to the parks.
   (b) Birds will stop migrating to national parks.
   (c) Parks and wilderness areas will lose biodiversity.
   (d) There will be a reduction in the number of invasive species.

58. Sodium chloride (salt) is commonly spread on roadways as a deicing agent during the winter and in this context is called road salt. After rainfall, the road salt washes off the road and into nearby wetlands where amphibians breed. Researchers have found that amphibian populations exposed to road salt tend to have more males than unexposed populations. In this context, what type of pollutant can road salt be categorized as?
   (a) neurotoxin
   (b) endocrine disruptor
   (c) carcinogen
   (d) allergen

59. Which is an example of nonpoint source pollution?
   (a) a network of leaky sewer pipes under a city
   (b) a single toilet that flushes into a sewage system
   (c) a leach field from a single residential home
   (d) the effluent from a wastewater treatment plant

60. What is a common characteristic of water bodies that receive thermal pollution from power plants?
   (a) reduced concentrations of dissolved oxygen
   (b) greater concentrations of nutrients that can lead to algal blooms
   (c) higher amounts of soil erosion
   (d) increased lead concentrations along nearby roads

61. What is the simplest way to measure the persistence of a chemical contaminant in an environment?
   (a) by measuring the solubility of the contaminant in water
   (b) by measuring the rate of bioaccumulation in a single organism
   (c) by measuring the half-life of the chemical under typical environmental conditions
   (d) by measuring the concentration of the chemical at the source of pollution

Use the diagram of a modern wastewater treatment plant below to answer questions 62 & 63:

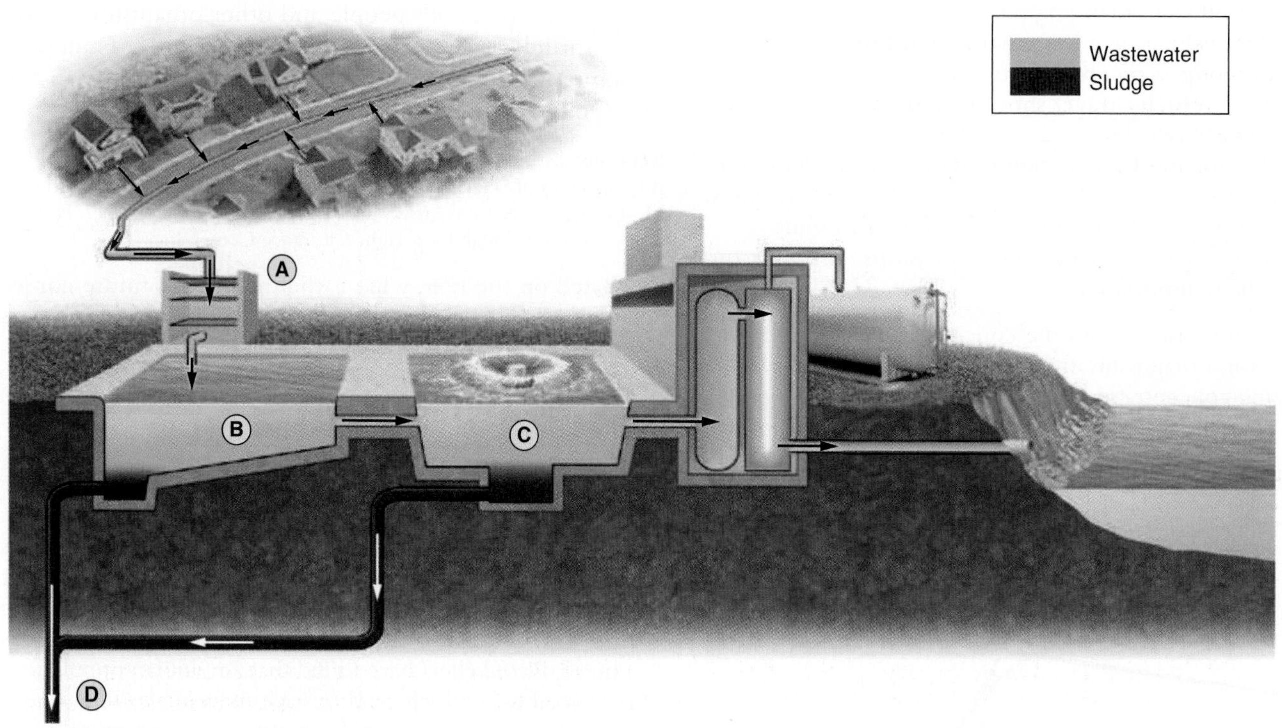

62. During the processing of wastewater at a modern treatment plant, the highest amount of aerobic respiration occurs at
    (a) location A.
    (b) location B.
    (c) location C.
    (d) location D.

63. During the processing of wastewater at a modern treatment plant, where are useful energy and agricultural products harvested?
    (a) location A
    (b) location B
    (c) location C
    (d) location D

64. For 30 years before the EPA banned the production of PCBs, General Electric manufacturing plants discharged waste PCBs into the Hudson River. These PCBs bioaccumulated in fish that were consumed by anglers. General Electric funded the cleanup of PCBs from the river, which was initiated in 2007 and concluded in 2015. A group of scientists would like to conduct a retrospective study on the harmful effects of ingesting PCB contaminated fish. They identified individuals that consumed fish caught from the river prior to 2007 and assessed their overall health. What would be the best control group for this study?
    (a) individuals in the same region that did not consume fish from the river
    (b) individuals in the same region that consumed fish from noncontaminated rivers
    (c) individuals in a region with a similarly sized population but no problems with PCB contamination
    (d) individuals who are currently consuming fish from the Hudson River

65. The table provides data on the amount of CO_2 produced during manufacturing, transporting, and disposal for four types of reusable straws. The table also provides the average number of times a straw is used before disposal. Suppose that the manufacturing, transport, and disposal of a single-use plastic straw requires 1.0, 0.5, and 0.5 g CO_2, respectively. Which type of straw material would result in the greatest CO_2 emissions per use relative to a single-use plastic straw?

Straw type	Manufacturing (g CO_2)	Transport (g CO_2)	Disposal (gg CO_2)	Average # of uses
Metal	50	10	20	90
Silicone	20	5	5	20
Bamboo	1	5	4	7
Glass	20	15	5	10

    (a) metal
    (b) silicone
    (c) bamboo
    (d) glass

66. A municipality of 200,000 people maintains four waste disposal trucks with an annual operating cost of $460,295 each. Calculate the cost, in dollars per year, for each resident to pay for waste disposal services.
    (a) $0.11
    (b) $2.30
    (c) $9.20
    (d) $200.00

67. Which statement is the correct description of an ecological consequence of dam construction?
   (a) The construction of dams increases downstream sedimentation.
   (b) The construction of dams reduces recreational opportunities for outdoor sports enthusiasts.
   (c) The construction of dams reduces fish breeding habitat.
   (d) The construction of dams creates reservoirs that provide drinking water.

68. A reduction in stratospheric ozone concentration is primarily caused by
   (a) insufficient stratospheric O_2 levels for O_3 formation.
   (b) breakdown of O_3 due to higher levels of UV-B and UV-C radiation.
   (c) the reaction of chlorine atoms with O_3 molecules.
   (d) an increase in the average temperature of the Earth's surface at the poles.

69. What is the most abundant greenhouse gas in the atmosphere?
   (a) water
   (b) carbon dioxide
   (c) methane
   (d) chlorofluorocarbons

70. Explain why termites significantly contribute to global warming.
   (a) Anaerobic digestion of wood in termite guts creates methane.
   (b) Termites eat plants faster than they can regenerate, thus creating a net increase in atmospheric CO_2.
   (c) Termites secrete a defensive chemical compound that volatizes to become a potent greenhouse gas.
   (d) The activity of termites typically reduces the amount of primary productivity in an area, leading to atmospheric increases in CO_2 and methane.

71. Coral bleaching is an indicator of
   (a) a reduction in coral food resources.
   (b) ozone pollution.
   (c) ocean warming.
   (d) harmful algal blooms

72. The graph shows annual trends in CO_2 emissions from natural sources throughout a year for a particular ecosystem. Given what you know about natural fluctuations in CO_2, what biome does this data most likely represent?

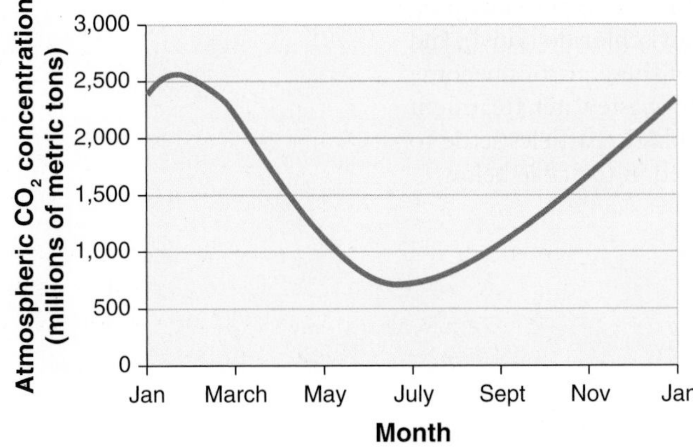

   (a) tundra
   (b) temperate forest
   (c) tropical forest
   (d) desert

Use the following passage entitled "Ohio woman pleads guilty to selling invasive crayfish species across 36 states" to answer questions 73 & 74:

An Ohio woman who sold hundreds of marbled crayfish online has pleaded guilty to raising the freshwater crayfish in a huge tank in her home and selling them to people across 36 different states. The case is the first such enforcement action aimed at stemming the advance of the marbled crayfish, which has already expanded in huge numbers across countries as diverse as Germany and Madagascar. Wildlife officials fear the creature is now threatening to gain a claw hold in the United States, where it is banned in several states but not nationally. The woman was prosecuted after Ohio introduced a state rule in 2020 naming the crayfish as an injurious species.

The marbled crayfish is at first glance relatively unremarkable—measuring around five inches in length, sporting a reddish or blue mottled shell and surviving off an omnivorous diet of algae, plants, and amphibians. But the species is parthenogenetic, meaning that it can asexually reproduce itself. All known marbled crayfish are female—no males have been discovered—and each animal can lay up to 700 unfertilized eggs that develop into genetically identical offspring. This prodigious cloning ability allows the marbled crayfish to quickly dominate any aquatic ecosystem it finds itself in, outcompeting, or consuming, any native species already there.

Source: https://www.theguardian.com/environment/2022/may/12/ohio-woman-marbled-crayfish-invasive-species. Copyright Guardian News & Media Ltd 2022

73. Based on its description in the article, the marbled crayfish is invasive because it
   (a) is an *r*-selected species.
   (b) is a specialist consumer.
   (c) is prohibited under the Lacey Act.
   (d) outcompetes native species.

74. Which of the following is an example of environmental legislation intended to assist people exposed to toxins from hazardous waste disposal?
   (a) the Comprehensive Environmental Response Compensation and Liability Act
   (b) the Endangered Species Act
   (c) the Marine Mammals Protection Act
   (d) the Clean Water Act

75. Why is the conservation of large habitats often a better strategy than the conservation of several small habitats?
   (a) Large habitats are less likely to be exploited by humans.
   (b) Large habitats are generally easier to fund and protect.
   (c) Large habitats support greater genetic and species diversity.
   (d) Large habitats are generally more resistant to climate change.

76. The Irish Potato Famine of 1845 could have been prevented by
   (a) increasing the genetic diversity of potato crops in Ireland.
   (b) occasionally spraying some of the potato crops with fungicide.
   (c) increasing the amount of farmland available for potato growth.
   (d) increasing the connectedness of potato farms in Ireland.

77. The graph shows average permafrost depth per year for a plot of land in Alaskan tundra. Which of the following is an inference researchers can draw from the data?

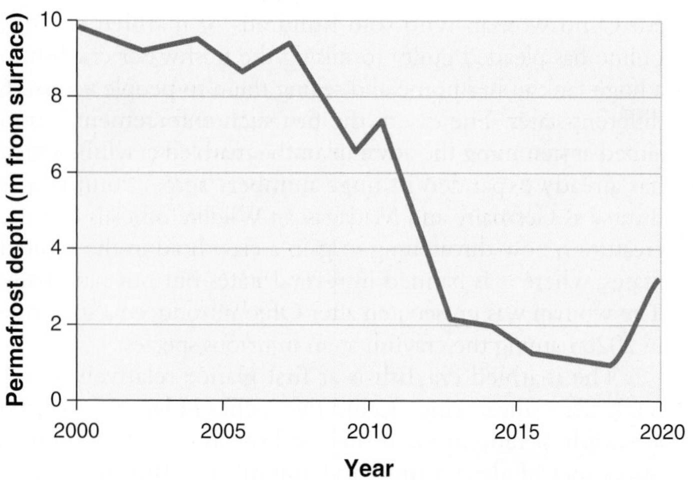

(a) Average permafrost depth exhibits a cyclic pattern over the last 800,000 years.
(b) Methane concentrations likely increased from 2000 to 2020.
(c) A decrease in permafrost depth is associated with negative feedback loops that slow global climate change.
(d) The permafrost depth is unlikely to increase over the next decade.

78. Over the past 50 years, researchers have recorded the dates of first flower and pollinator emergence for a single ecosystem in New England. This is shown in the graph. Given future predictions of a warmer climate for this biome, what do the data in the graph suggest about the future of the ecosystem?

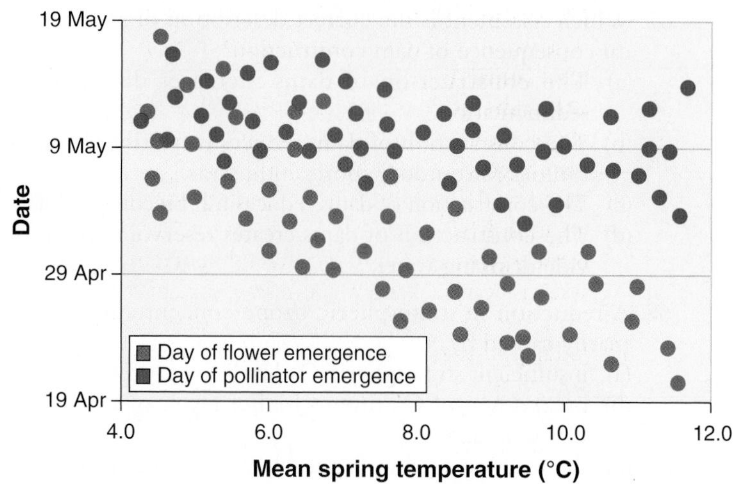

(a) Fewer flowers will be pollinated in the future.
(b) There will be more flowers in the future.
(c) The system will experience more severe heatwaves.
(d) The mean spring temperature will consistently decrease every year.

79. Which statement concerning invasive species is most accurate?
(a) Invasive species are defined as those that were unintentionally brought in from somewhere else and cause ecological harm.
(b) Invasive species are impossible to control.
(c) Invasive species are those that provide no benefit to humans.
(d) Some invasive species can be controlled by introducing a predator or herbivore of that species.

80. Which action is most likely to reduce species extinction as the average temperature of the Earth increases?
(a) the presence of habitat corridors
(b) increased habitat fragmentation
(c) development of edge habitats
(d) elimination of invasive species

Section 2: Free-Response Questions

1. Microplastics are defined as any plastic particle less than 5 mm in length. These particles come from a variety of sources, including fragmentation of larger plastic items, pharmaceutical and personal care products, and glitter. Clothing is also an important source of microplastic particles. Clothing can be made from a variety of plastic types, including polyamide (nylon), polyethylene terephthalate (polyester), polyvinyl chloride (vinyl), and polystyrene (the same material used in Styrofoam). Fragments of these plastics become detached during normal washing cycles and are transported to wastewater treatment plants. Researchers measured the rates at which similarly sized plastic particles settle to the bottom of freshwater ecosystems. Their findings are represented in the table below.

Plastic type	Settling rate (cm/s)
Polyamide	7.8
Polyethylene terephthalate	9.1
Polyvinyl chloride	12.2
Polystyrene	1.2

(Data from Wang et al. 2021. Settling velocity of irregularly shaped microplastics under steady and dynamic flow conditions, Environmental Science and Pollution Research *28:62116–62136)*

(a) **Describe** how microplastic particles are likely to leave wastewater treatment plants and contaminate agricultural lands. (1 pt.)

(b) **Describe** how microplastic particles are likely to leave wastewater treatment plants and contaminate freshwater ecosystems. (1 pt.)

(c) Based on the data in the table, **identify** which plastic types are most likely to end up on agricultural land and which particle types are most likely to end up in freshwater ecosystems near the wastewater treatment plant. (1 pt.)

(d) You are asked to design a study that examines the lethal and sublethal effects of polyethylene terephthalate microplastics on freshwater organisms.

(i) State an appropriate research question for a dose-response study to measure LD_{50} of microplastics on a fish species. **Identify** the dependent and independent variables. (2 pts.)

(ii) **Describe** a dose-response study to measure ED_{50} of microplastics on a fish species. ED_{50} is similar to LD_{50} but it refers to effective dose not lethal dose. (2 pts.)

(e) Some cities are developing a dual-pipe sewer system in which one set of pipes will carry sewage to the wastewater treatment plant and the other set will carry stormwater directly to natural bodies of water.

(i) **Justify** the importance of developing a dual-pipe sewer system. (2 pts.)

(ii) **Explain** how a dual-pipe sewer system might minimize the amount of microplastics contamination in freshwater systems. (1 pt.)

2. Coffee (*Coffea arabica*) is a small tree that produces bright red berries. Each berry has a single bean that is extracted, dried, and roasted to produce the aromatic and flavorful ingredient that is served in coffee shops and residences around the world. In nature, the coffee tree grows under the shade of much larger trees. However, coffee farmers have found that coffee trees produce more berries when they are grown in full sun. Unfortunately, coffee trees that grow in full sun tend to be infested with a beetle known as the coffee berry borer (*Hypothenemus hampei*). When present, populations of this beetle can reach very high numbers and consume all berries before harvest. In addition, infestations of these beetles seem to be greater in regions where more farmers grow their coffee trees in the sun. To better understand and remedy this problem, researchers have examined birds, which are one of the natural predators of insects found on coffee trees.

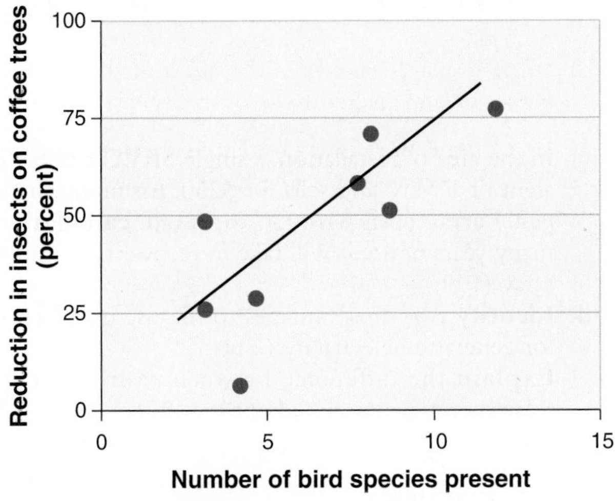

(a) Based on the graph, answer the following questions:

(i) **Identify** the number of bird species present when there is a 50 percent reduction in coffee trees. (1 pt.)

(ii) **Describe** the relationship between the number of bird species and reduction in insects on coffee trees. (1 pt.)

(iii) **Explain** one possible reason for the relationship you described in part a (ii). (1 pt.)

(b) **Identify** the ecosystem service that birds are providing. (1 pt.)

(c) Researchers have found that the diversity of coffee borer beetle predators is greater in shade-grown coffee farms than in sun-grown coffee farms.

(i) **Identify** one reason why researchers found this trend. (1 pt.)

(ii) **Make a claim** for why farmers will gain more profit by growing all coffee in the shade of forests than in open sun. (1 pt.)

(d) Coffee farmers are interested in implementing sustainable agricultural practices to ensure high yields in the future.

(i) **Identify** one strategy the coffee farmers could use to decrease the density of insects infesting their crops (1 pt.)

(ii) **Describe** one unintended consequence of the solution you provided. (1 pt.)

(e) In addition to insect infestations, soil degradation is another problem that could lead to decreased yield.

(i) **Identify** one biotic component of the A horizon in soil that will increase crop yield. (1 pt.)

(ii) **Describe** one soil conservation strategy farmers can implement. (1 pt.)

3. To encourage the creation and use of renewable energy, many states are providing incentives for homeowners to install photovoltaic cells (also known as solar panel arrays) on their roofs. The cost of the panels and installation can be expensive, but the average panel array can provide all the electricity for the home. Consequently, the homeowner will not have to pay any electricity bills to local energy producers. In addition, many states allow homeowners to sell credits based on the amount of energy they produce from their array (known as solar renewable energy credits, or SRECs). Depending on the state, these credits can sell for up to $500 and are purchased by coal- and gas-fired energy producers.

The table below provides energy produced and available solar energy for a given home in New Jersey.

Month	Energy produced per panel in one month (kWh)	Available solar energy per day (kWh per m^2)
January (31 days)	15.4	1.7
February (28 days)	22.0	3.1
March (31 days)	33.1	2.7
April (30 days)	34.1	3.3
May (31 days)	34.9	4.3
June (30 days)	38.7	5.2
July (31 days)	41.1	3.7
August (31 days)	35.9	4.1
September (30 days)	33.0	5.1
October (31 days)	22.0	3.1
November (30 days)	17.5	1.8
December (31 days)	8.9	1.0

(a) Assume each solar panel has a surface area of 2 m^2.
   (i) **Calculate** the efficiency of the solar panels for the months of January and April. (2 pts.)
   (ii) **Describe** how the pattern of monthly energy production would change for a solar panel array located closer to the equator. (1 pt.)

(b) Assume there are 20 panels in the solar panel array.
   (i) **Calculate** the average monthly kWh produced per panel. (1 pt.)
   (ii) Using the average energy produced per panel, **calculate** the total energy produced by the entire array throughout a 365-day year. (2 pts.)

(c) In the area of installation, a single SREC is the equivalent of 1 MW and sells for $250. Assuming the solar panel array costs $10,000 to install, **calculate** how many years of use it will take to recover the cost of the array. (1 pt.)

(d) **Identify** two disadvantages to the use of solar panels for generating electricity. (2 pts.)

(e) **Explain** the difference between passive and active solar energy generation. (1 pt.)

Appendix: Reading Graphs

Environmental scientists often use graphs to display the data they collect. Unlike a table that contains rows and columns of data, graphs help us visualize patterns and trends in the data and they communicate our results more clearly. Because graphs are such a fundamental tool of environmental science as well as most other sciences, we have designed this appendix to help you become familiar with the major types of graphs that environmental scientists use; we also discuss how to create each type of graph and how to interpret the data that are presented in the graphs.

Scientists use graphs to present data and ideas

A graph is a tool that allows scientists to visualize data or ideas. Organizing information in the form of a graph can help us understand relationships more clearly. Throughout your study of environmental science you will encounter many different types of graphs. In this section we will look at the most common types of graphs that environmental scientists use.

Scatter Plot Graphs

Although many of the graphs in this book may look different from each other, they all follow the same basic principles. Let's begin with an example in which researchers investigated a possible relationship between the area of different islands around Malaysia, the Philippines, and New Guinea and the number of bird species living on each island. We can examine this relationship by creating a *scatter plot graph*, as shown in **FIGURE A.1**. In the simplest form of a scatter plot graph, researchers look at two variables; they put the values of one variable on the *x* axis and the values of the other variable on the *y* axis. By convention, the location where the two axes converge in the bottom left corner, called the origin, represents a value of 0 for each variable. The units of measurement get larger as we move from left to right on the *x* axis, and from bottom to top on the *y* axis. In our example, island area is on the *x* axis and the number of bird species is on the *y* axis. As you can see in the graph, as island area increases, the number of bird species increases.

When two variables are graphed using a scatter plot, we can draw a line through the middle of the data points that describes the general trend of these data points. Because such a line is drawn in a way that fits the general trend of the data, we call it *the line of best fit*. The line of best fit allows us to visualize a general trend. In our graph, as we move from low to high income we observe lower fertility. This is known as a negative relationship between the two variables because as one variable gets larger the other variable gets smaller. When graphing data using a scatter plot graph, the line of best fit may be either straight or curved.

Line Graphs

A *line graph* displays data that occur as a sequence of measurements over time or space. For example, scientists have estimated the number of humans living on Earth from 8,000 years ago to the present time. Using all of the

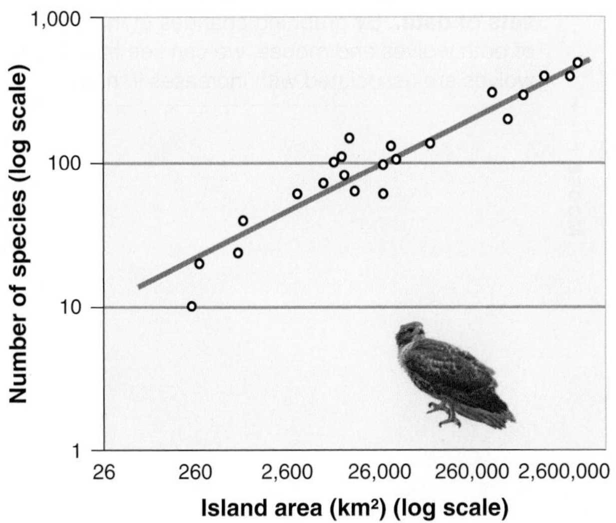

FIGURE A.1 (book Figure 10.2) Scatter plot graph. In this graph, we place data points that coincide with the area of different islands on the *x* axis and the corresponding number of bird species on the *y* axis.

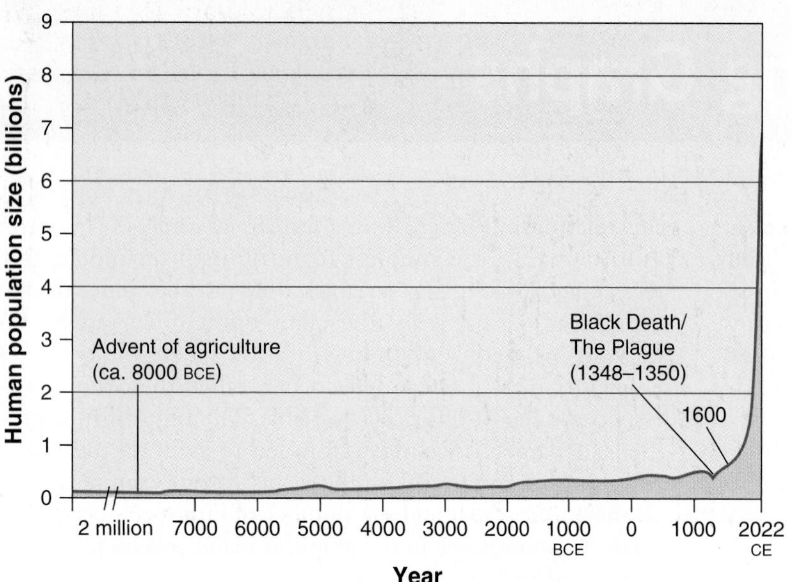

FIGURE A.2 (book Figure 17.1) Line graph. In this graph, a line is used to track the change in human population size over time. Note that this graph extends far back in time and there is only a small increase in population size from 2 million BCE to 7000 BCE, so a break in the *x* axis is used to allow a shorter axis. This allows us to focus on the period of rapid population growth that happened during the past 7,000 years.

available data points, a line graph can be used to connect each data point over time, as shown in **FIGURE A.2**. In contrast to a line of best fit that fits a straight or curved line through the middle of all data points, a line graph connects one data point to another, so it can be straight or curved, or it can move up and down as it follows the movement of the data points.

When a graph includes data points with a very large range of values, the size of the graph can become cumbersome. To keep the graph from becoming too large, we can use a break in the axis. For example in our graph of human population growth, we see that the size of the population varied little from 2 million BCE to 7000 BCE. Showing the data for those years would not provide much additional useful information but it would make the graph a lot wider. The break in the *x* axis between 2 million BCE and 7000 BCE, indicated by the double hatch marks, allows us to shorten the *x* axis. The double hatch marks indicate that we are condensing the middle part of the *x* axis.

Line graphs can also illustrate how several different variables change over time. When two variables contain different units or a different range of values, we can use two *y* axes. For example, **FIGURE A.3** presents data on changes in the population sizes of two different animals on Isle Royale in the years 1955 to 2018. The left *y* axis represents the population changes in the wolf population

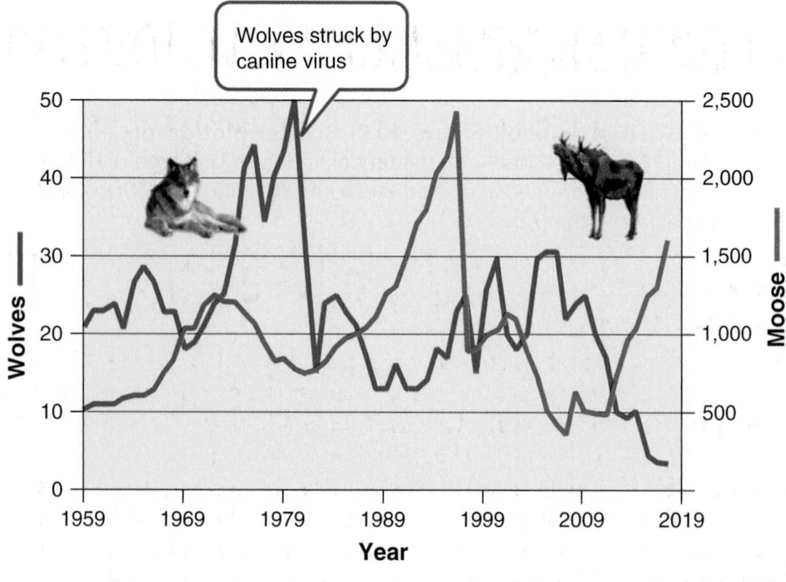

FIGURE A.3 (book Figure 16.8) A line graph with two sets of data. By graphing changes in the populations of both wolves and moose, we can see that declines in wolves are associated with increases in moose.

whereas the right y axis represents the population changes in the moose population during the same time span.

Bar Graphs

A *bar graph* plots numerical values that come from different categories. For example, in **FIGURE A.4** the x axis contains categories that represent regions of the world. The y axis represents a numerical value—the number of people infected with HIV. The visual impact of the different bar heights provides a dramatic comparison of the incidence of HIV in different regions.

A bar graph is a very flexible tool and can be altered in several ways to accommodate data sets of different sizes or even several data sets that a researcher wishes to compare. When scientists measured the net primary productivity of different ecosystems, as shown in **FIGURE A.5**, they put the categories—various ecosystems—on the y axis and the plotted values—net primary productivity—on the x axis. This orientation makes it easier to accommodate the relatively large amount of text needed to name each ecosystem.

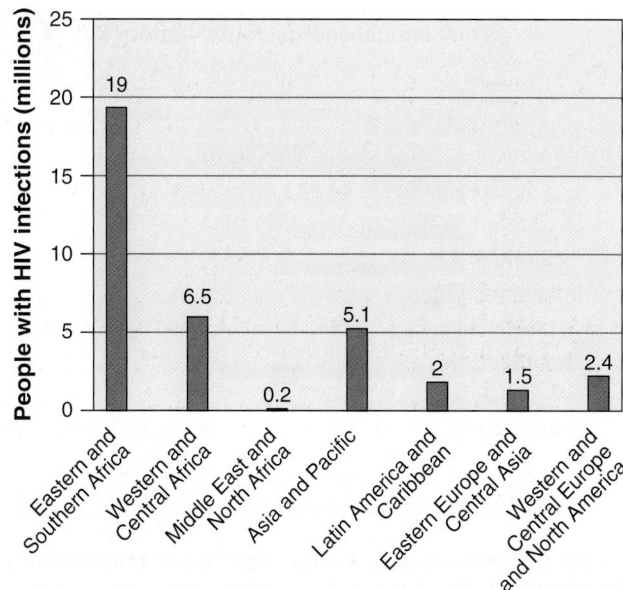

FIGURE A.4 (book Figure 17.6) Bar graph. When we have numerical values that come from different categories we can use a bar graph to plot data. In this example, we can plot the number of people infected with HIV in several areas around the world.

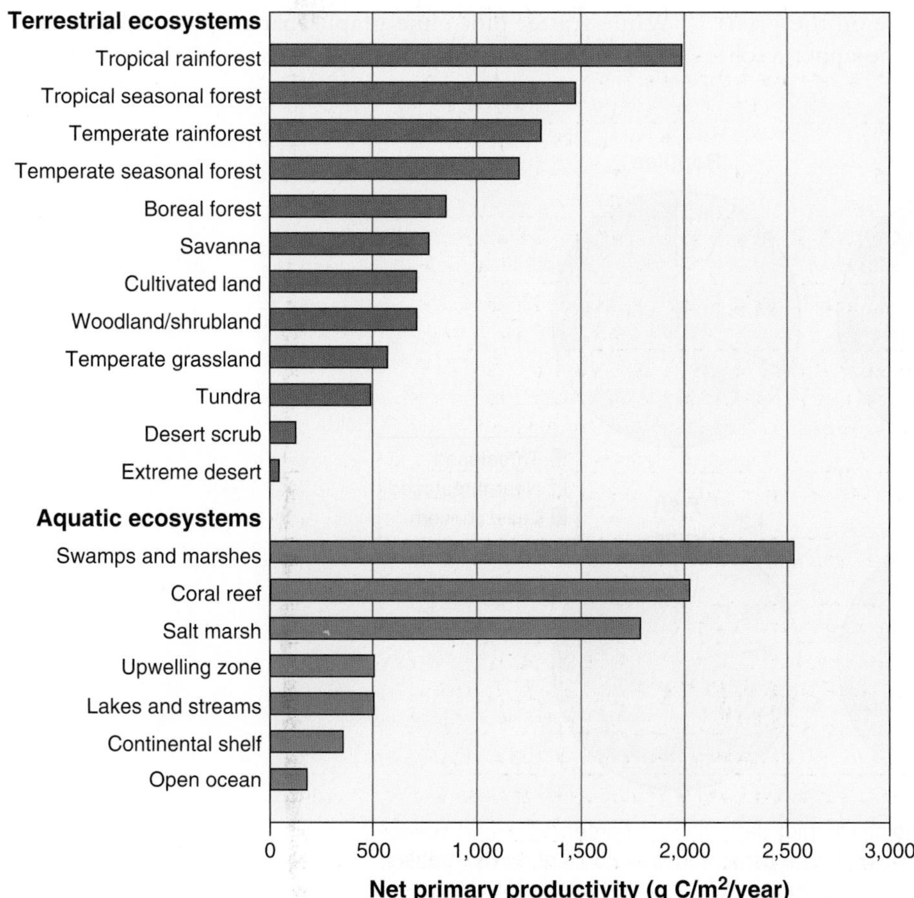

FIGURE A.5 (book Figure 6.5) A rotated bar graph. Bar graphs can place the categories on either the x axis, as in Figure A.4, or on the y axis, as in this figure that plots the net primary productivity of different ecosystems.

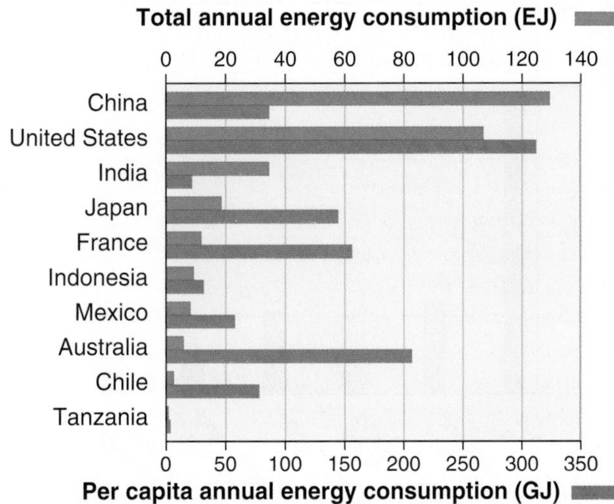

FIGURE A.6 (book Figure 35.3) A rotated bar graph with two sets of data. In this bar graph, we have nations as our categories and two sets of numerical data: the total energy consumed by each country and the per capita annual energy consumed.

FIGURE A.6 shows an example of a bar graph that presents two sets of data for each category. In this example, the bar graph is rotated so the categories are on the *y* axis and the numeric data are on two *x* axes. The upper *x* axis represents total annual energy consumption of each country. The lower *x* axis plots per capita (per person) annual energy consumption. Notice how much information we can gather from this graph; we can compare the total annual energy consumption versus per capita annual energy consumption within each country and also compare the energy consumption among countries.

Pie Charts

A *pie chart* is a graph represented by a circle with slices of various sizes representing categories within the whole pie. The entire pie represents 100 percent of the data and each slice is sized according to the percentage of the pie that it represents. For example, **FIGURE A.7** shows the percentage of conifers, birds, reptiles, mammals, amphibians, and fish from around the world that have been categorized as threatened or endangered, near-threatened, or of least concern from a conservation point of view. For each group of organisms, each slice of the pie represents the percentage of species that fall within each conservation category.

Two special types of graphs used by environmental scientists

While scatter plots, line graphs, bar graphs, and pie charts are used by many different types of scientists, environmental

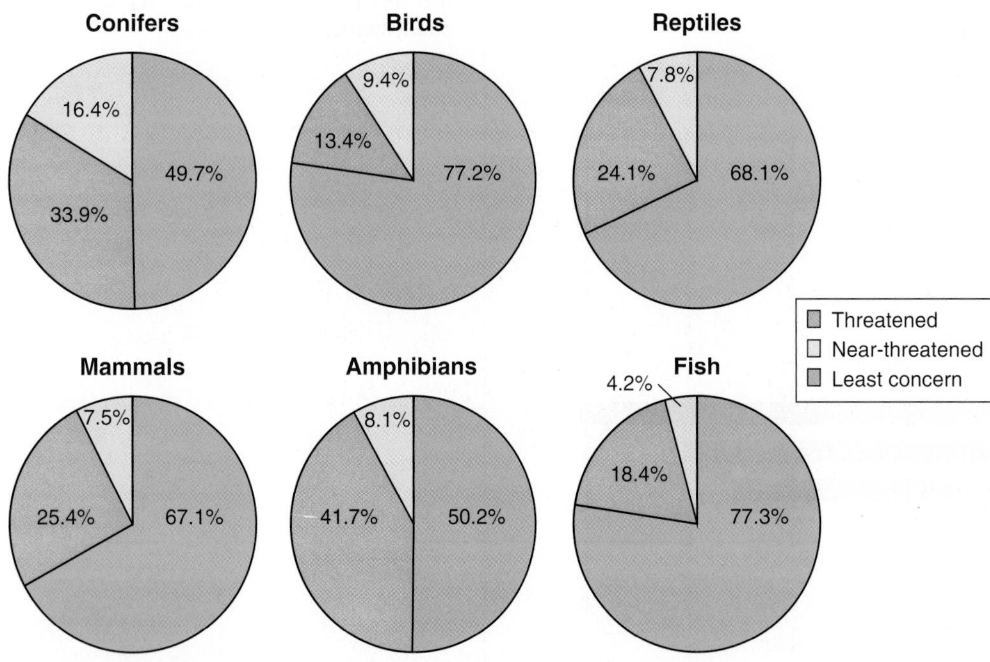

FIGURE A.7 (book Figure 59.3) Pie chart. Pie charts plot data that are percentages and collectively add up to 100 percent. These pie graphs illustrate the percentages of conifers, birds, reptiles, mammals, amphibians, and fish of the world that are categorized as either threatened/endangered, near-threatened, or least concern.

scientists also use two types of graphs that are not common in most other fields of science: *climate diagrams* and *age structure diagrams*. Although these two types of graphs are discussed within the text, we provide them here for review.

Climate Graphs

Climate diagrams are used to illustrate the annual patterns of temperature and precipitation that help to determine the productivity of biomes on Earth. **FIGURE A.8** shows two hypothetical biomes. By graphing the average monthly temperature and precipitation of a biome, we can see how conditions in a biome vary during a typical year. We can also observe the specific time period when the temperature is warm enough for plants to grow. In the biome illustrated in Figure A.8a, the growing season—indicated by the shaded region on the *x* axis—is mid-March through mid-October. In Figure A.8b, the growing season is mid-April through mid-September.

In addition to identifying the growing season, climate diagrams show the relationships among precipitation, temperature, and plant growth. In Figure A.8a, the precipitation line is above the temperature line in every month. This means that water supply exceeds demand, so plant growth is more constrained by temperature than by precipitation throughout the entire year. In Figure A.8b, the precipitation line intersects the temperature line. At this point, the amount of precipitation available to plants equals the amount of water lost by plants through evapotranspiration. When the precipitation line falls below the temperature line from May through September, water demand exceeds supply and plant growth will be constrained more by precipitation than by temperature.

Age Structure Diagrams

Age structure diagrams are visual representations of age distribution for both males and females in a country. **FIGURE A.9** on page APP-6 presents four examples. Each horizontal bar of the diagram represents a 5-year-age group and the length of a given bar represents the number of males or females in that age group.

While every nation has a unique age structure, we can group countries very broadly into three categories. Figure A.9a shows a country with many more young people than older people. The age structure diagram of a country with this population will be in the shape of a pyramid, with its widest part at the bottom, moving toward the smallest at the top. Age structure diagrams with this shape are typical of countries in the developing world.

A country with a smaller difference between the number of individuals in the younger and older age groups has an age structure diagram that looks more like a column. With fewer individuals in the younger age groups, we can deduce that the country has little or no population growth. Figure A.9b shows the age structure of people in the United States, which is similar to the age structure of people in Canada, Australia, Sweden, and many other developed countries. Panels (c) and (d) show countries with a proportionally larger number of older people. This age structure

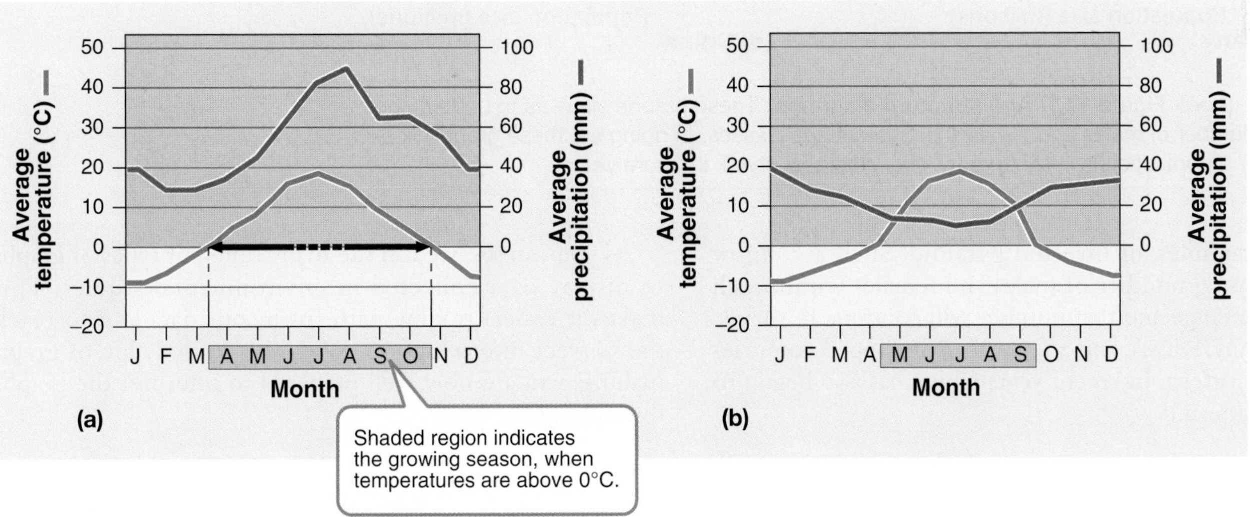

FIGURE A.8 (book Figure 2.4) Climate graph. Climate graphs are a special type of graph that plot monthly temperatures and monthly precipitation in a way that tells us whether plant growth is more limited by temperature or water. These diagrams help us understand the productivity of different biomes.

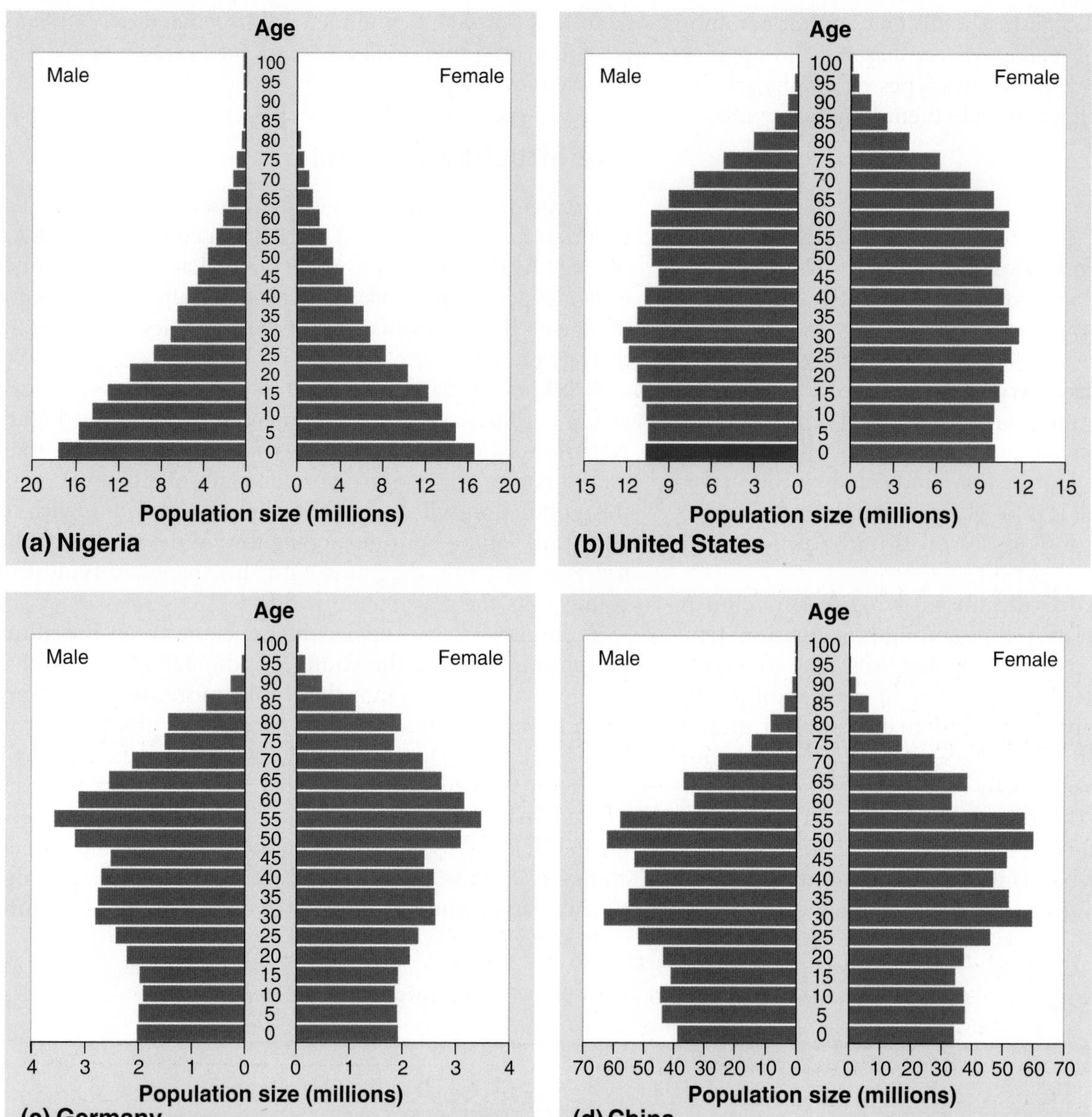

FIGURE A.9 **(book Figure 17.7) Age structure diagrams.** These graphs allow us to understand the relative number of males and females in different age classes. In doing so, these graphs illustrate whether a population is likely to grow, stay stable, or shrink in future years.

diagram resembles an inverted pyramid. Such a country has a decreasing number of males and females within each younger age range and that number will continue to shrink. Italy, Germany, Russia, and a few other developed countries display this pattern. In recent years China has also begun to show this pattern.

As you can see, we can use many different types of graphs to display data collected in environmental studies. This makes it easier to view patterns in our data and to reach the correct interpretation. With this knowledge of graph making, you are now well prepared to interpret the graphs throughout the book.

Glossary / Glosario

Glossary	Glosario
A	
abiotic Nonliving.	**abiótico** Sin vida.
accuracy How close a measured value is to the actual or true value.	**exactitud** Fidelidad que hay entre un valor medido y su valor real.
acid A substance that contributes hydrogen ions to a solution.	**ácido** Sustancia que aporta iones de hidrógeno a una solución.
acid deposition Precipitation high in sulfuric acid and nitric acid. *Also known as* **acid rain**.	**depósitos ácidos** Precipitación con niveles altos de ácido sulfúrico y ácido nítrico. *También se conoce como* **lluvia ácida**.
acid precipitation Precipitation high in sulfuric acid and nitric acid. *Also known as* **acid rain**.	**precipitación ácida** Precipitación con una concentración alta de ácido sulfúrico y ácido nítrico. *También se conoce como* **lluvia ácida**.
acid rain *See* **acid precipitation**.	**lluvia ácida** *Ver* **acid precipitation**.
Acquired Immune Deficiency Syndrome (AIDS) An infectious disease caused by the human immunodeficiency virus (HIV).	**síndrome de inmunodeficiencia adquirida (SIDA)** Enfermedad infecciosa causada por el virus de la inmunodeficiencia humana (VIH).
active solar energy A use of technology that captures and stores the energy of sunlight with electrical equipment and devices.	**energía solar activa** Uso de tecnología que captura y almacena la energía del sol a través de equipos y dispositivos eléctricos.
acute disease A disease that rapidly impairs the functioning of a person's body.	**enfermedad aguda** Enfermedad que rápidamente perjudica el funcionamiento del cuerpo de una persona.
acute study An experiment that exposes organisms to an environmental hazard for a short duration.	**estudio agudo** Experimento que expone a los organismos a un peligro ambiental durante un período de tiempo breve.
adaptation A trait that improves an individual's fitness.	**adaptación** Característica que mejora la adecuación de un individuo.
adiabatic cooling The cooling effect of reduced pressure on air as it rises higher in the atmosphere and expands.	**enfriamiento adiabático** Efecto enfriador al reducir la presión sobre el aire a medida que este asciende en la atmósfera y se expande.
adiabatic heating The heating effect of increased pressure on air as it sinks toward the surface of Earth and decreases in volume.	**calentamiento adiabático** Efecto calentador al aumentar la presión sobre el aire a medida que este desciende hacia la superficie de la Tierra y su volumen disminuye.
aerobic An environment with abundant oxygen.	**aeróbico** Ambiente con oxígeno abundante.
aerobic respiration The process by which cells convert glucose and oxygen into energy, carbon dioxide, and water.	**respiración aeróbica** Proceso mediante el cual las células convierten la glucosa y el oxígeno en energía, dióxido de carbono y agua.
age structure diagram A visual representation of the number of individuals within specific age groups for a country, typically expressed for males and females.	**diagrama de la estructura de edades** Representación visual de la cantidad de individuos dentro de grupos de edades específicos por país; característicamente se expresa por separado para varones y para mujeres.
agribusiness *See* **industrial agriculture**.	**agroindustria** *Ver* **industrial agriculture**.
agroforestry An agricultural technique in which trees and vegetables are intercropped.	**agrosilvicultura** Técnica agrícola en la que se combinan los cultivos de árboles y los de comestibles.
A horizon Frequently the top layer of soil, a zone of organic material and minerals that have been mixed together. *Also known as* **topsoil**.	**horizonte A** A menudo es la capa superficial del suelo, una zona de material orgánico y minerales que se han entremezclado. *También se conoce como* **manto**.
air pollution The introduction of chemicals, particulate matter, or microorganisms into the atmosphere at concentrations high enough to harm plants, animals, and materials such as buildings, or to alter ecosystems.	**contaminación aérea** La introducción de sustancias químicas, materias particuladas o microorganismos en la atmósfera a concentraciones que son lo suficientemente altas como para causar daño a las plantas, los animales o a materiales tales como edificios, o que alteran ecosistemas completos.
albedo The percentage of incoming sunlight reflected from a surface.	**albedo** Porcentaje de luz solar que llega y que se refleja a partir de una superficie.

algal bloom A rapid increase in the algal population of a waterway.	**eflorescencia de algas** Crecimiento acelerado de la población de algas en una vía navegable.
alien species *See* **exotic species**.	**especie introducida** *Ver* **exotic species**.
allergen A chemical that causes allergic reactions.	**alérgeno** Sustancia química que produce reacciones alérgicas.
allopatric speciation The process of speciation that occurs with geographic isolation.	**especiación alopátrida** Proceso de especiación por aislamiento geográfico.
ammonification *See* **mineralization**.	**amonificación** *Ver* **mineralización**.
anaerobic An environment that lacks oxygen.	**anaeróbico** Ambiente que carece de oxígeno.
anaerobic respiration The process by which cells convert glucose into energy in the absence of oxygen.	**respiración anaeróbica** Proceso mediante el cual las células convierten glucosa en energía en la ausencia de oxígeno.
anthracite Contains greater than 90 percent carbon. It has the highest quantity of energy per volume of coal and the fewest impurities. *Also known as* **hard coal**.	**antracita** Contiene más de 90 porciento de carbono. Tiene la cantidad más alta de energía por volumen de carbón y la menor cantidad de impurezas. *También se conoce como* **carbón duro**.
anthropogenic Derived from human activities.	**antropogénico** Derivado de actividades humanas.
aphotic zone The deeper layer of ocean water that lacks sufficient sunlight for photosynthesis.	**zona afótica** Profundidad submarina en la que se carece de suficiente luz solar para que haya fotosíntesis.
aquaculture The farming of fish, shellfish, and seaweed.	**acuacultura** Cultivo de peces, mariscos y algas marinas.
aquatic biome An aquatic region characterized by a particular combination of salinity, depth, and water flow.	**bioma acuático** Región acuática caracterizada por una combinación específica de salinidad, profundidad y flujo acuático.
aqueduct A canal or ditch used to carry water from one location to another.	**acueducto** Canal o trocha que se utiliza para transportar agua de una ubicación a otra.
aqueducts Pipes and canals that move water from where it is abundant to areas where it is scarce.	**acueductos** Tuberías y canales que llevan agua de lugares donde es abundante a lugares donde es escasa.
aquifer Pore spaces found within permeable layers of rock and sediment underneath the soil that store groundwater.	**acuífero** Espacios porosos dentro de capas permeables de roca y sedimento que contienen aguas freáticas.
artesian well A well created by drilling a hole into a confined aquifer.	**pozo artesiano** Pozo que se crea taladrando un agujero en un acuífero cautivo.
asbestos A long, thin, fibrous silicate mineral with insulating properties, which can cause cancer when inhaled.	**asbesto** Mineral silicato fibroso, delgado y largo con propiedades aislantes, que puede causar cáncer si se lo inhala.
ash The residual nonorganic material that does not combust during incineration.	**ceniza** Material residual no orgánico que no se quema durante la incineración.
assimilation A process by which plants and algae incorporate nitrogen into their tissues.	**asimilación** Proceso mediante el cual plantas y algas incorporan nitrógeno en sus tejidos.
asthenosphere The layer of Earth located in the outer part of the mantle, composed of semi-molten rock.	**astenosfera** Capa de la Tierra ubicada en la parte exterior del manto. Está compuesta por piedra semifundida.
atmospheric convection current Global patterns of air movement that are initiated by the unequal heating of Earth.	**corriente de convección atmosférica** Patrones globales de movimientos de aire iniciados por diferencias en el calentamiento de la Tierra.
autotroph *See* **producer**.	**autótrofo** *Ver* **producer**.
B	
base A substance that contributes hydroxide ions to a solution.	**base** Sustancia que aporta iones de hidróxido a una solución.
base saturation The proportion of soil bases to soil acids, expressed as a percentage.	**saturación base** Proporción de bases en el suelo, comparadas con los ácidos en el suelo. Se expresa en forma de porcentaje.
becquerel (Bq) A measurement of the rate at which a sample of radioactive material decays; 1 Bq is equal to the decay of one atom per second.	**becquerelio (Bq)** Medida de la tasa de desintegración de una muestra de un material radiactivo; 1 Bq equivale a la desintegración de un átomo o núcleo por segundo.
benthic zone The muddy bottom of a lake, pond, or ocean beneath the limnetic and profundal zones.	**zona béntica** Fondo de lodo de un lago, una laguna o un océano ubicado debajo de la zona limnética y la zona profunda.
B horizon Commonly known as subsoil, a soil horizon is composed primarily of mineral material with very little organic matter.	**horizonte B** Comúnmente conocido como subsuelo, un horizonte del suelo está compuesto principalmente de material mineral con muy poco material orgánico.
bioaccumulation The selective absorption and concentration of a chemical within an organism over time.	**bioacumulación** Absorción selective y concentración de una sustancia química en un organismo con el paso del tiempo.

biochemical oxygen demand (BOD) The amount of oxygen a quantity of water uses over a period of time at specific temperatures.	**demanda bioquímica de oxígeno** Volumen de oxígeno que una cantidad de agua utiliza durante un período de tiempo a temperaturas específicas.
biocontrol A shortened term for biological control, it uses biological organism to control agricultural pests.	**biocontrol** Forma breve de decir control biológico, método en el que se utilizan organismos biológicos para controlar plagas agrícolas.
biodiesel A diesel substitute produced by extracting and chemically altering oil from plants.	**biodiésel** Sustituto del diésel producido de aceite extraído de plantas, que luego se altera químicamente.
biodiversity The diversity of life forms in an environment.	**biodiversidad** Diversidad de las formas con vida en un entorno.
biodiversity hotspots Isolated areas that are home to so many endemic species, that they contain a high proportion of all the species found on Earth.	**punto caliente de biodiversidad** Zonas aisladas que albergan tantas especies endémicas que contienen una proporción alta de todas las especies que se encuentran en la Tierra.
biofuel Liquid fuel created from processed or refined biomass.	**biocombustible** Combustible líquido creado a partir de biomasa procesada o refinada.
biogeochemical cycle The movements of matter within and between ecosystems involving cycles of biological, geological, and chemical processes.	**ciclo biogeoquímico** Movimientos de materia dentro de ecosistemas y entre ellos, que involucran ciclos de procesos biológicos, geológicos y químicos.
biomagnification The increase in chemical concentration in animal tissues as the chemical moves up the food chain.	**biomagnificación** Aumento en la concentración química en tejidos animales a medida que la sustancia química asciende por la cadena trófica.
biomass Biological material that has mass.	**biomasa** Materia biológica que tiene masa.
biome The plants and animals that are found in a particular region of the world.	**bioma** Plantas y animales que se encuentran en una región específica del mundo.
biosphere The region of our planet where life resides.	**biosfera** Región de nuestro planeta en la que reside la vida.
biotic Living.	**biótico** Con vida.
biotic potential Under ideal conditions with unlimited resources available, every population has a maximum potential for growth.	**potencial biótico** En condiciones ideales y con recursos ilimitados a su disposición, todas las poblaciones tienen el potencial máximo de crecimiento.
bird flu A type of flu caused by the H5N1 virus.	**gripe aviaria** Tipo de influenza causada por el virus H5N1.
bituminous coal A black or dark brown coal that contains bitumen, *also known as* **asphalt**. It typically contains up to 80 percent carbon.	**carbón bituminoso** Carbón negro o café que contiene bitumen, *también conocido como* **asfalto**. Normalmente contiene hasta 80 por ciento de carbono.
boreal forest *See* **taiga**.	**bosque boreal** *Ver* **taiga**.
broad-spectrum pesticide A pesticide that kills many different types of pest.	**plaguicida de espectro amplio** Plaguicida que mata muchos tipos diferentes de plagas.
brownfields Contaminated industrial or commercial sites that may require environmental cleanup before they can be redeveloped or expanded.	**zona contaminada** Sitios comerciales o industriales que podrían precisar una limpieza ambiental antes de que se puedan volver a urbanizar o se puedan ampliar.
brown smog *See* **photochemical smog**.	**esmog marrón** *Ver* **photochemical smog**.
bycatch The unintentional catch of nontarget species while fishing.	**pesca incidental** Captura no intencional de especies que no son las anheladas cuando se pesca.

C

capacity The maximum electrical output of something such as a power plant.	**capacidad** Producción eléctrica máxima de algo, por ejemplo, una central eléctrica.
capacity factor The fraction of time a power plant operates in a year.	**factor de capacidad** Fracción de tiempo que una central eléctrica funciona durante un año.
carbon cycle The movement of carbon around the biosphere among reservoir sources and sinks.	**ciclo de carbono** Movimiento del carbono alrededor de la biosfera entre reservorios y depósitos.
carbon dioxide A by-product of all combustion, carbon dioxide from biofuels contains modern carbon from woody material, rather than fossil carbon from fossil fuels.	**dióxido de carbono** Producto derivado de toda combustión, el dióxido de carbono de biocombustibles contiene carbono moderno de material leñoso en lugar de carbono fósil de combustibles fósiles.
carbon footprint A measure of the total carbon dioxide and other greenhouse gases emissions from the activities, both direct and indirect, of a person, country, or other entity.	**huella de carbono** Medida del total de dióxido de carbono y otras emisiones de gases invernadero de las actividades, directas e indirectas, de una persona, país u otra entidad.

carbon monoxide A colorless, odorless gas that is formed during incomplete combustion of most materials.	**monóxido de carbono** Gas incoloro e inodoro que se forma durante la combustión incompleta de la mayoría de los materiales.
carbon neutral An activity that does not change atmospheric CO_2 concentrations.	**emisión neutra de carbono** Actividad que no cambia las concentraciones atmosféricas de CO_2.
carbon sequestration The capture and long-term storage of carbon dioxide from the atmosphere.	**captura de carbono** Captura y almacenamiento en el largo plazo de dióxido de carbono en la atmósfera.
carcinogen A chemical that causes cancer.	**carcinógeno** Sustancia química que causa cáncer.
carnivore A consumer that eats other consumers.	**carnívoro** Consumidor que se come otros consumidores.
carrying capacity The limit to the number of individuals that can be supported by an existing habitat or ecosystem, and is denoted as *K*.	**capacidad poblacional** Límite del número de individuos que un hábitat o ecosistema existente puede sustentar y se denota como *K*.
catalytic converter A device that uses chemicals to convert pollutants such as nitrogen oxide and carbon monoxide to nitrogen gas and carbon dioxide.	**convertidor catalítico** Dispositivo que utiliza químicos para convertir contaminantes como óxido de nitrógeno y monóxido de carbono en nitrógeno gaseoso y dióxido de carbono.
cation exchange capacity (CEC) The ability of a particular soil to adsorb and release cations.	**capacidad de intercambio de cationes** Capacidad de un suelo dado de adsorber y de liberar cationes.
cellular respiration The process by which cells unlock the energy of chemical compounds.	**respiración celular** Proceso mediante el cual las células liberan la energía de compuestos químicos.
cellulosic ethanol An ethanol derived from cellulose, the cell wall material in plants.	**etanol celulósico** Etanol derivado de celulosa, el material de la pared celular en las plantas.
charcoal Woody material that has been heated in the absence of oxygen so that water and some volatile compounds are driven off.	**carbón** Material leñoso que ha sido calentado en ausencia de oxígeno de tal manera que elimina agua y ciertos compuestos volátiles.
chemical weathering The breakdown of rocks and minerals by chemical reactions, the dissolving of chemical elements from rocks, or both these processes.	**erosión química** Descomposición de rocas y minerales por reacciones químicas; disolución de elementos químicos de rocas, o ambos procesos.
chemosynthesis A process used by some bacteria to generate energy with methane and hydrogen sulfide.	**quimiosíntesis** Proceso que utilizan algunas bacterias para generar energía valiéndose del metano y el ácido sulfhídrico.
child mortality The number of deaths of children under age 5 per 1,000 live births.	**mortalidad infantil** Cantidad de fallecimientos de niños menores de 5 años por millar de partos vivos.
C horizon The least-weathered soil horizon, which always occurs beneath the B horizon and is similar to the parent material.	**horizonte C** Horizonte del suelo menos erosionado, que siempre ocurre debajo del horizonte B y es parecido al material parental.
chlorofluorocarbons (CFCs) Chemical that can be used for cooling refrigerators and air conditioners.	**clorofluorocarburos (CFC)** Químico que sirve para enfriar refrigeradores y sistemas de aire acondicionado.
chronic disease A disease that slowly impairs the functioning of a person's body.	**enfermedad crónica** Enfermedad que lentamente va perjudicando las funciones del cuerpo de una persona.
chronic study An experiment that exposes organisms to an environmental hazard for a long duration.	**estudio crónico** Experimento que expone los organismos a un peligro ambiental durante un período de tiempo prolongado.
Clean Water Act Legislation that supports the "protection and propagation of fish, shellfish, and wildlife and recreation in and on the water" by maintaining and, when necessary, restoring the chemical, physical, and biological properties of surface waters.	**Ley de Agua Limpia** Legislación estadounidense que fomenta la "protección y propagación de peces, moluscos y crustáceos, y vida silvestre, así como recreación en o sobre el agua" mediante el mantenimiento y, cuando se precise, la restauración de las propiedades químicas, físicas y biológicas de las aguas superficiales.
clear-cutting A method of harvesting trees that involves removing all or almost all of the trees within an area.	**tala indiscriminada** Método de cosechar árboles que implica la tala de todos o casi todos los árboles dentro de una zona.
climate The average weather that occurs in a given region over a long period of time.	**clima** Promedio del estado del tiempo que ocurre en una región dada durante un período de tiempo prolongado.
climax community Historically described as the final stage of succession.	**comunidad clímax** Históricamente se le conoce como la etapa final de sucesión.
closed-loop recycling Recycling a product into the same product.	**reciclaje en circuito cerrado** Reciclaje de un producto para formar el mismo producto.
coal A solid fuel formed primarily from the remains of trees, ferns, and other plant materials that were preserved 280 million to 360 million years ago.	**carbón** Combustible sólido formado principalmente de los residuos de árboles, helechos y otros materiales vegetales que fueron preservados hace 280 a 360 millones de años.

cogeneration The use of a fuel to both generate electricity and deliver heat to a building or industrial process. *Also known as* **combined heat and power**.	**cogeneración** Uso de un combustible para generar electricidad y para llevar calor a un edificio o proceso industrial. *También se conoce como* **energía y calor combinados**.
collision zone An area where two continental plates are pushed together and the colliding forces push up the crust to form a mountain range.	**zona de colisión** Área donde dos placas continentales chocan y las fuerzas de colisión elevan la corteza para formar cordilleras.
combined cycle A feature in some natural gas–fired power plants that uses both a steam turbine to generate electricity and a separate turbine that is powered by the exhaust gases from natural gas combustion to turn another turbine to generate electricity.	**ciclo combinado** Característica de ciertas plantas eléctricas de gas natural. En ellas, una turbina de vapor genera electricidad y una turbina separada, que se alimenta de los gases de escape resultantes de la combustión del gas natural, hacen girar otra turbina para generar más electricidad.
combined heat and power *See* **cogeneration**.	**energía y calor combinados** *Ver* **cogeneration**.
commensalism An interaction between two species in which one species benefits and the other species is neither harmed nor helped.	**comensalismo** Interacción entre dos especies en la que una especie se beneficia y a la otra especie no se la ha perjudicado ni ayudado.
commercial energy sources Energy sources that are bought and sold, such as coal, oil, and natural gas.	**fuentes comerciales de energía** Fuentes de energía que se compran y se venden, tales como carbón, petróleo y gas natural.
community ecology The study of interactions among species.	**ecología de la comunidad** Estudio de las interacciones entre las especies.
competition The struggle of individuals, either within or between species, to obtain a shared limiting resource.	**competencia** Batalla entre individuos, dentro de una misma especie o entre especies, por obtener un recurso limitante compartido.
competitive exclusion principle The principle stating that two species competing for the same limiting resource cannot coexist.	**principio de la exclusión competitiva** Principio que explica que dos especies que compiten por el mismo recurso limitante no pueden coexistir.
composting The breakdown of organic materials into organic matter (humus).	**compostaje** Descomposición de materiales orgánicos en materia orgánica (humus).
concentrated animal feeding operation (CAFO) A large indoor or outdoor structure designed for maximum occupancy of animals and maximum output of meat.	**operación concentrada para la alimentación de animales** Estructura grande interior o al aire libre que se ha diseñado para la máxima ocupación de animales y para generar una producción máxima de carne.
cone of depression An area surrounding a well that does not contain groundwater.	**cono de depresión** Área alrededor de un pozo que no contiene aguas subterráneas.
confined aquifer Surrounded by a layer of impermeable rock or clay, which impedes water flow to or from the aquifer.	**acuífero cautivo** Acuífero rodeado por una capa de roca o arcilla impermeable que impide la entrada o salida de agua del acuífero.
consumer An organism that is incapable of photosynthesis and must therefore obtain its energy by consuming other organisms. *Also known as* **heterotroph**.	**consumidor** Organismo que es incapaz de hacer la fotosíntesis y debe, por lo tanto, obtener su energía consumiendo otros organismos. *También se conoce como* **heterótrofo**.
contour plowing Plowing and harvesting parallel to the topographic contours of the land.	**arar en curvas de nivel** Arar y cosechar en paralelo a los contornos topográficos del terreno.
control group In a scientific investigation, a group that experiences exactly the same conditions as the experimental group, except for the single variable under study.	**grupo de control** En una investigación científica, grupo que experimenta exactamente las mismas condiciones que el grupo experimental, con la excepción de una variable que se está estudiando.
control rod A cylindrical device inserted between the fuel rods in a nuclear reactor to absorb excess neutrons and slow or stop the fission reaction.	**vara de control** Dispositivo cilíndrico que se inserta entre las barras de combustible en un reactor nuclear para absorber los neutrones que sobran y para enlentecer o detener la reacción de fisión.
Convention on International Trade in Endangered Species of Wild Fauna and Flora (CITES) A 1973 treaty formed to control the international trade of threatened plants and animals.	**Convenio sobre el Comercio Internacional de Especies Amenazadas de Fauna y Flora Silvestres (CITES)** Convenio promulgado en 1973 con el fin de controlar el comercio internacional de plantas y animales amenazados.
convergent boundary An area where one plate moves toward another plate and collides.	**borde convergente** Zona en la que una placa tectónica se mueve hacia otra y chocan.
coral bleaching A phenomenon in which algae inside corals die, causing the corals to turn white.	**blanqueamiento de corales** Fenómeno en el que las algas dentro de los corales fallecen, motivo por el cual los corales se ponen blancos.

coral reef Represents Earth's most diverse marine biome, and are found in warm, shallow waters beyond the shoreline in tropical regions.	**arrecife de coral** Representa el bioma marino más diverso de la Tierra y se encuentra en aguas tibias y de poca profundidad a cierta distancia del litoral en regiones tropicales.
core The innermost zone of Earth's interior, composed mostly of iron and nickel. It includes a liquid outer layer and a solid inner layer.	**núcleo interno** Zona interior de la Tierra, compuesta en su mayor parte de hierro y níquel. Incluye una capa externa líquida y una capa interna maciza.
Coriolis effect The deflection of an object's path due to the rotation of Earth.	**efecto Coriolis** Desviación de la trayectoria de un objeto debido a la rotación de la Tierra.
cradle-to-grave analysis *See* **life-cycle analysis**.	**análisis de la cuna a la tumba** *Ver* **life-cycle analysis**.
crop rotation A crop-planting strategy in which different types of crop species are planted from season to season or year to year on the same plot of land.	**rotación de cultivos** Estrategia agrícola en la cual se siembran diferentes especies de cultivos entre una temporada y otra sobre la misma parcela de tierra.
crude birth rate (CBR) The number of births per 1,000 individuals per year.	**tasa bruta de nacimientos** Cantidad de nacimientos por millar de individuos por año.
crude death rate (CDR) The number of deaths per 1,000 individuals per year.	**tasa bruta de fallecimientos** Cantidad de fallecimientos por millar de individuos por año.
crude oil A mixture of hydrocarbons such as oil, gasoline, kerosene as well as water and sulfur that exists in a liquid state underground, and when brought to the surface.	**petróleo crudo** Mezcla de hidrocarburos como petróleo, keroseno y agua, así como azufre que existe en estado líquido bajo la tierra y en la superficie.
crustal abundance The average concentration of an element in Earth's crust.	**abundancia en la corteza** Concentración promedio de un elemento en la corteza terrestre.
cultural eutrophication An increase in fertility in a body of water, the result of anthropogenic inputs of nutrients.	**eutrofización cultural** Aumento en la fertilidad de una extensión de agua como resultado del ingreso antropogenético de nutrientes.
curie A unit of measure for radiation, a curie is 37 billion decays per second.	**curio** Unidad de medida de radiación, 1 curio equivale a 37 billones de desintegraciones por segundo.
D	
dam A barrier that runs across a river or stream to control the flow of water.	**presa** Barrera que atraviesa un río o arroyo. Se usa para controlar el flujo de agua.
dead zone When oxygen concentrations become so low that it kills fish and other aquatic animals.	**zona muerta** Cuando las concentraciones de oxígeno son tan bajas que los peces y otras especies acuáticas mueren.
decibel A scale (dBA) A logarithmic scale that measure both the loudness of sound and the frequency.	**escala de decibeles A (dBA)** Escala logarítmica que mide tanto el volumen del sonido como la frecuencia.
decomposers Fungi and bacteria that complete the breakdown process by converting organic matter into small elements and molecules that can be recycled back into the ecosystem.	**saprofitos** Hongos o bacterias que completan el proceso de descomposición al convertir materia orgánica en elementos pequeños y moléculas que se pueden reciclar y devolver al ecosistema.
deductive reasoning The process of applying a general statement to specific facts or situations.	**razonamiento deductivo** Proceso de aplicar una declaración general a actos o situaciones específicas.
Delaney Clause A clause in the Food, Drug, and Cosmetic Act designed to prevent potentially harmful cancer-causing food ingredients.	**cláusula Delaney** Cláusula en la Ley Federal de Alimentos, Medicamentos y Cosméticos de los Estados Unidos diseñada para impedir que haya ingredientes potencialmente dañinos y cancerígenos en los alimentos.
demographer A scientist in the field of demography.	**demógrafo** Científico que trabaja en el campo de la demografía.
demography The study of human populations and population trends.	**demografía** Estudio de las poblaciones humanas y sus tendencias poblacionales.
denitrification The conversion of nitrate (NO_3^-) in a series of steps into the gases nitrous oxide (N_2O) and, eventually, nitrogen gas (N_2), which is emitted into the atmosphere.	**desnitrificación** Conversión, mediante una serie de pasos, de nitrato (NO_3^-) en los gases óxido nitroso (N_2O) y, con el tiempo, nitrógeno gaseoso (N_2), que se emite a la atmósfera.
density-dependent factor A factor that influences an individual's probability of survival and reproduction in a manner that depends on the size of the population.	**factor densodependiente** Factor que influye en la probabilidad de supervivencia y reproducción de una manera que depende del volumen de la población.
density-independent factor A factor that has the same effect on an individual's probability of survival and the amount of reproduction at any population size.	**factor densoindependiente** Factor que tiene el mismo efecto en la probabilidad de supervivencia y cantidad de reproducción de un individuo en una población de cualquier volumen.
dependent variable A variable that is dependent on other factors.	**variable dependiente** Variable que depende de otros factores.

desalination The process for obtaining fresh water by removing the salt from salt water. *Also known* as **desalinization**.	**desalinización** Proceso para obtener agua fresca mediante la eliminación de la sal del agua salobre. *También se conoce como* **desalinización**.
desertification Transformation of arable, productive low-precipitation land to desert or unproductive land due to climate change or destructive land use such as overgrazing and logging.	**desertificación** Transformación de suelos arables, productivos y de baja precipitación a suelos no productivos debido a cambios climáticos o al uso destructivo del suelo, por ejemplo, el pastoreo excesivo y la tala.
detritivore An organism that specializes in breaking down dead tissues and waste products into smaller particles.	**detritívoro** Organismo que se especializa en descomponer tejidos muertos y desechos en partículas más pequeñas.
developed countries Countries that have relatively high levels of industrialization and income.	**país desarrollado** Países que tienen niveles de industrialización e ingresos relativamente altos.
developing countries Countries that have relatively low levels of industrialization and income.	**país en vías de desarrollo** Países que tienen niveles de industrialización e ingresos relativamente bajos.
dieback A rapid decline in a population due to death. *Also known as* **die-off**.	**extinción paulatina** Disminución rápida en una población debido a muertes. *También se conoce como* **muerte regresiva**.
dikes Structure built to prevent ocean waters from flooding adjacent land.	**diques** Estructuras construidas para prevenir que las aguas del océano inunden los terrenos contiguos.
disease Any impaired function of the body with a characteristic set of symptoms.	**enfermedad** Toda función corporal disminuida que presenta un conjunto característico de síntomas.
distillation A process of desalination in which water is boiled and the resulting steam is captured and condensed to yield pure water.	**destilación** Proceso de desalinización en el que el agua se hierve y el vapor resultante se capta y se condensa para producir agua pura.
divergent boundary An area below the ocean where tectonic plates move away from each other.	**borde divergentes** Zona en la profundidad del océano en la que las placas tectónicas se separan y alejan una de la otra.
dose-response study A study that exposes animals or plants to different amounts of a chemical and then looks for a variety of possible responses, including mortality or changes in behavior or reproduction.	**estudio de dosis-respuesta** Estudio que expone a animales o plantas a diferentes cantidades de una sustancia química y luego observa una diversidad de respuestas posibles, las que incluyen la mortalidad o cambios en la conducta o reproducción.
doubling time The number of years it takes a population to double.	**tiempo de duplicación** Número de años que precisa una población para duplicarse.
drip irrigation A form of irrigation where a slowly dripping hose on the ground or buried beneath the soil delivers water directly to the plant roots.	**riego por goteo** Forma de irrigación en la cual una manguera que gotea lentamente sobre el suelo, o enterrada bajo la tierra, lleva agua directamente a las raíces de las plantas.
dysentery An infection of the intestines that causes diarrhea, which results in dehydration and can cause death.	**disentería** Infección de los intestinos que causa diarrea, la cual resulta en deshidratación y puede ser mortal.
E	
earthquake A sudden movement of Earth's crust caused by a release of potential energy from the movement of tectonic plates.	**terremoto** Movimiento repentino de la corteza de la Tierra producido por la liberación de energía potencial como consecuencia del movimiento de las placas teutónicas.
Ebola hemorrhagic fever An infectious disease with high death rates, caused by several species of Ebola virus.	**fiebre hemorrágica del ébola** Enfermedad infecciosa con tasas de mortalidad altas, causada por varias especies del virus del Ébola.
ecological efficiency The proportion of consumed energy that can be passed from one trophic level to another.	**eficiencia ecológica** Proporción de energía consumida que se puede pasar de un nivel trófico a otro.
ecological footprint A measure of the area of land and water an individual, population, or activity requires to produce all the resources it consumes and to process the waste it generates.	**presencia ecológica** Medida del área de tierra que requiere un individuo, población o actividad para producir todos los recursos que consume y procesar los desechos que genera.
ecologically sustainable forestry An approach to removing trees from forests in ways that do not unduly affect the viability of other noncommercial tree species.	**silvicultura ecológicamente sostenible** Estrategia de talar árboles en los bosques de maneras que no afecten indebidamente la viabilidad de otras especies de árboles que no se comercializan.
ecological succession The predictable replacement of one group of species by another group of species over time.	**sucesión ecológica** Reemplazo predecible de un grupo de especies por otro grupo de especies a través del tiempo.
ecological tolerance The suite of abiotic conditions under which a species can survive, grow, and reproduce. *Also known as* **the fundamental niche**.	**tolerancia ecológica** El conjunto de condiciones abióticas bajo las cuales las especies pueden sobrevivir, crecer y reproducirse. *También conocida como* **nicho fundamental**.
economies of scale The observation that average costs of production fall as output increases.	**economías de escala** Observación de que los costos promedio de la producción caen en la medida en que aumenta la producción.

ecosystem A particular location on Earth with interacting biotic and abiotic components.	**ecosistema** Ubicación particular en la Tierra que tiene componentes bióticos y abióticos que interactúan entre sí.
ecosystem diversity The variety of ecosystems that exist in a given region.	**diversidad de ecosistemas** La variedad de ecosistemas que existen en una región determinada.
ecosystem services The processes by which life-supporting resources such as clean water, timber, fisheries, and agricultural crops are produced.	**servicios del ecosistema** Procesos mediante los cuales se producen los recursos que sustentan la vida, tales como el agua limpia, la madera, las pesquerías y la agricultura.
ED_{50} The effective dose of a chemical that causes 50 percent of the individuals in a dose-response study to display a harmful, but nonlethal, effect.	**ED_{50}** Dosis efectiva de una sustancia química que causa que el 50 por ciento de los individuos en un estudio de dosis-respuesta presenten un efecto nocivo, pero no letal.
E horizon A zone of leaching, or eluviation, found in some acidic soils under the O horizon or, less often, the A horizon.	**horizonte E** Zona de lixiviación o eluviación que se encuentra en algunos suelos ácidos debajo del horizonte O o, con menor frecuencia, el horizonte A.
electrical grid A network of interconnected transmission lines.	**red eléctrica** Red de líneas de transmisión interconectadas.
electrolysis The application of an electric current to water molecules to split them into hydrogen and oxygen.	**electrólisis** Aplicación de una corriente eléctrica a moléculas de agua para separarlas en hidrógeno y oxígeno.
electrostatic precipitator A device that removes particulate matter by using an electrical charge to make particles coalesce so they can be removed from the exhaust stream.	**precipitador electrostático** Dispositivo que elimina partículas mediante una carga eléctrica que une las partículas para que puedan ser eliminadas de la corriente de escape.
El Niño–Southern Oscillation (ENSO) A reversal of wind and water currents in the South Pacific.	**El Niño–oscilación austral** Inversión de las corrientes de viento y de agua en el Pacífico Sur.
emergent infectious disease An infectious disease that has not been previously described or has not been common for at least 20 years.	**enfermedad infecciosa emergente** Enfermedad infecciosa que no se ha descrito antes o que no ha sido común durante al menos 20 años.
emigration The movement of people out of a country or region.	**emigración** Desplazamiento de personas que salen de un país o región.
endangered species Species that are likely to go extinct in the near future.	**especie en peligro de extinción** Especies con alta probabilidad de extinguirse en el futuro cercano.
Endangered Species Act A 1973 U.S. law designed to protect plant and animal species that are threatened with extinction, and the habitats that support those species.	**Ley de Especies en Peligro de Extinción** Ley diseñada para proteger especies de plantas y animales que están bajo amenaza de extinción, y los hábitats que las sustentan.
endemic species Species that live in a very small area of the world and nowhere else, often in isolated locations such as the Hawaiian Islands.	**especie endémica** Especies que viven en una zona muy reducida del mundo y en ningún otro sitio, a veces en lugares aislados como las islas de Hawái.
endocrine disruptor A chemical that interferes with the normal functioning of hormones in an animal's body.	**interruptor endocrino** Sustancia química que interfiere con el funcionamiento normal de las hormonas en el organismo de un animal.
energy carrier An energy source such as electricity that can move and deliver energy in a convenient, usable form to end users.	**transportador de electricidad** Fuente de energía, como la electricidad, que puede trasladar y entregar energía de manera conveniente y usable para sus usuarios finales.
energy conservation Methods for finding and implementing ways to use less energy.	**conservación de energía** Métodos para hallar y poner en práctica maneras de consumir menos energía.
energy efficiency The ratio of the amount of energy expended in the form you want to the total amount of energy that is introduced into the system.	**eficiencia energética** Cantidad de energía que se expende en la forma deseada, expresada como proporción de la cantidad de energía que se introduce en el sistema.
energy intensity The energy use per unit of gross domestic product (GDP).	**intensidad energética** La energía por unidad del producto bruto interno (PIB).
energy quality The ease with which an energy source can be used to do work.	**calidad de la energía** Facilidad con la cual una fuente de energía se puede aprovechar para hacer trabajo.
energy return on energy investment (EROEI) The amount of energy we get out of an energy source for every unit of energy expended on its production.	**tasa de retorno energético (TRE)** La cantidad de energía que obtenemos de una fuente de energía por cada unidad de energía gastada en su producción.
energy subsidy The fossil fuel energy and human energy input per calorie of food produced.	**subsidio energético** La energía de combustibles fósiles y la energía humana consumidas por caloría de alimento producida.
environment The sum of all the conditions surrounding us that influence life.	**entorno** Suma de todas las condiciones circundantes que influyen en la vida.

environmental hazard Anything in the environment that can potentially cause harm.	**peligro ambiental** Toda cosa en el medioambiente que potencialmente pueda ser perjudicial.
environmental indicators Describe the current state of an environmental system or the Earth.	**indicador ambiental** Describe el estado actual de un sistema ambiental o de la Tierra.
environmentalism A social movement that seeks to protect the environment through lobbying, activism, and education.	**ecologismo** Movimiento social que busca la protección del medio ambiente mediante el cabildeo político, el activismo y la educación.
environmental justice The study of the disproportionate exposure to environmental hazards experienced by people of color, recent immigrants and people of lower socio-economic backgrounds; and is both an academic field and a social movement.	**justicia ambiental** El estudio de la exposición desproporcionada a riesgos ambientales por personas de color, inmigrantes recientes y personas de niveles socioeconómicos bajos; es un campo académico y un movimiento social.
environmental science The field of study that looks at interactions among human systems and those found in nature.	**ciencia ambiental** Campo en el que se estudian las interacciones entre sistemas humanos y sistemas que se hallan en la naturaleza.
environmental studies The field of study that includes environmental science and additional subjects such as environmental policy, economics, literature, and ethics.	**estudios ambientales** Campo de estudio que incluye la ciencia ambiental y otros temas tales como política ambiental, economía, literatura y ética.
epidemic A situation in which a pathogen causes a rapid increase in disease.	**epidemia** Situación en la que un patógeno causa la propagación rápida de una enfermedad.
episodic disruption Occurring somewhat regularly, such as cycles of high rain and low rain that occur every 5 to 10 years.	**interrupción episódica** Ocurre de manera relativamente regular, como ciclos de alta y baja precipitación que suceden cada 5 a 10 años.
erosion The physical removal of rock fragments from a landscape or ecosystem.	**erosión** Retiro físico de fragmentos de roca de un paisaje o ecosistema.
estuary An area along the coast where the fresh water of rivers mixes with salt water from the ocean.	**estuario** Zona en una costa donde el agua dulce fluvial se mezcla con el agua salada del océano.
ethanol Alcohol made by converting starches and sugars from plant material into alcohol and CO_2.	**etanol** Alcohol que se elabora convirtiendo almidones y azúcares de material vegetal en alcohol y CO_2.
eutrophic Describes a lake with a high level of fertility.	**eutrófico** Se dice de un lago con un gran nivel de fertilidad.
eutrophication Excess nutrients from human activities that make their way into waterbodies; it causes nutrient pollution that alters food webs and harms water quality.	**eutrofización** Nutrientes en exceso que resultan de actividades humanas y llegan a los cuerpos de agua; causan contaminación de nutrientes, alteran redes alimentarias y afectan la calidad del agua.
evaporate The process of converting from liquid to a gas or vapor.	**evaporar** El proceso de conversión de líquido a gas o vapor.
evapotranspiration The combined amount of evaporation and transpiration.	**evapotranspiración** Cantidad combinada de evaporación y transpiración.
evolution A change in the genetic composition of a population over time.	**evolución** Cambio en la composición genética de una población a través del tiempo.
evolution by artificial selection The process in which humans determine which individuals to breed, typically with a preconceived set of traits in mind.	**evolución por selección artificial** Proceso en el que los seres humanos determinan qué individuos se reproducen, característicamente teniendo en cuenta un conjunto de rasgos.
evolution by natural selection The process in which the environment determines which individuals survive and reproduce.	**evolución por selección natural** Proceso en el que el entorno determina qué individuos sobreviven y se reproducen.
evolution by random processes The processes that alter the genetic composition of a population over time, but the changes are not related to differences in fitness among individuals.	**evolución por procesos aleatorios** Los procesos aleatorios que alteran la composición genética de una población a lo largo del tiempo, pero los cambios no están relacionados con diferencias en niveles de aptitud entre individuos.
exosphere The outermost layer of the atmosphere which extends from 600 to 10,000 km (375–6,200 miles) above the surface of Earth.	**exosfera** La capa exterior de la atmósfera que se extiende entre 600 y 10,000 km (375–6,200 millas) por encima de la superficie de la Tierra.
exotic species A species living outside its historical range. *Also known as* **alien species**.	**especie exótica** Especie que vive por fuera de su hábitat histórico. *También se conoce como* **especie introducida**.
exponential growth model A growth model that estimates a population's future size after a period of time based on the biotic potential and the number of reproducing individuals currently in the population.	**modelo de crecimiento exponencial** Modelo de crecimiento que estima el volumen futuro de una población después de un período de tiempo, con base en el potencial biótico y la cantidad de individuos que se pueden reproducir en la actualidad en la población.
externality The cost or benefit of a good or service that is not included in the purchase price of that good or service or otherwise accounted for.	**externalidad** Costo o beneficio de un bien o servicio que no se incluye en el precio de compra de dicho bien o servicio o de lo contrario contabilizado.
exurb An area similar to a suburb, but unconnected to any central city or densely populated area.	**exurbio** Zona parecida a un suburbio, pero desconectada de la ciudad central o de la zona de densidad poblacional alta.

F	
family planning The regulation of the number or spacing of offspring through the use of birth control.	**planificación familiar** Regulación del espaciamiento de la descendencia gracias al uso de un método de control de la natalidad.
fault A fracture in rock caused by a movement of Earth's crust.	**falla** Fractura en la roca causada por un movimiento de la corteza terrestre.
fecal coliform bacteria A group of microorganisms that live in the intestines of humans, other mammals, and birds that serve as an indicator species for potentially harmful microorganisms associated with contaminated sewage.	**bacteria fecal coliforme** Grupo de microorganismos que viven en los intestinos de los humanos, otros mamíferos y aves, que sirven como especies indicadoras de microorganismos potencialmente dañinos asociados cona aguas residuales contaminadas.
fecundity The ability to produce an abundance of offspring.	**fecundidad** La capacidad de producir una abundancia de descendientes.
Ferrell cell A convection current in the atmosphere that lies between Hadley cells and polar cells.	**célula de Ferrell** Corriente de convección en la atmósfera que yace entre las células de Hadley y las células polares.
first law of thermodynamics A theory with no known exception that states that energy is neither created nor destroyed but it can change from one form to another.	**primera ley de termodinámica** Teoría sin excepciones conocidas que establece que la energía no se crea ni se destruye, sino que cambia de una forma a otra.
fishery A commercially harvestable population of fish within a particular ecological region.	**pesquería** Población de peces dentro de una región ecológica dada. Se puede cosechar con fines comerciales.
fishery collapse The decline of a fish population by 90 percent or more.	**colapso de pesquería** La caída en un 90 por ciento o más de una población de peces.
fish ladder A stair-like structure that allows migrating fish to get around a dam.	**escalera para peces** Estructura en forma de escalera que facilita la migración de los peces que ha sido afectada por un embalse.
fission A nuclear reaction in which a neutron strikes a relatively large atomic nucleus, which then splits into two or more parts, releasing additional neutrons and energy in the form of heat.	**fisión** Reacción nuclear en la que un neutrón bombardea un núcleo atómico relativamente grande, lo cual produce su rotura en dos o más partes, liberando otros neutrones y energía en forma de calor.
fitness An individual's ability to survive and reproduce.	**aptitud** Capacidad que tiene un individuo para sobrevivir y reproducirse.
flood irrigation A form of irrigation where an entire field is flooded with water.	**irrigación por inundación** Forma de riego donde se inunda con agua un campo entero.
food chain The sequence of consumption from producers through tertiary consumers.	**cadena trófica** Secuencia de consumo desde productores hasta consumidores terciarios.
food web A model of how energy and matter move through two or more interconnected food chains.	**red trófica** Modelo que explica cómo la energía y la materia se desplazan entre dos o más cadenas alimenticias interconectadas.
forest Land dominated by trees and other woody vegetation and sometimes used for commercial logging.	**bosque** Terreno dominado por árboles y otra vegetación leñosa, que a veces se usa para la explotación forestal comercial.
formaldehyde A naturally occurring compound that is used as a preservative and as an adhesive in plywood and carpeting.	**formaldehído** Compuesto natural que se usa como preservativo y como adhesivo de madera contrachapada y en alfombras.
fossil carbon Old carbon contained in fossil fuels.	**carbono fósil** Carbono viejo contenido en combustibles fósiles.
fossil fuel combustion The chemical reaction between any fossil fuel and oxygen resulting in the production of carbon dioxide, water, and the release of energy.	**combustión de combustibles fósiles** La reacción química entre cualquier combustible fósil y oxígeno que resulta en la producción de dióxido de carbono, agua, y la liberación de energía.
fossil fuels Fuels derived from biological material that became fossilized millions of years ago.	**combustibles fósiles** Combustibles derivados de materiales biológicos que se fosilizaron hace millones de años.
fracking Short for hydraulic fracturing, a method of oil and gas extraction that uses high-pressure fluids to force open existing cracks in rocks deep underground.	**fraqueo** Abreviación de fracturación hidráulica, método para la extracción de petróleo y gas que se vale de líquidos a gran presión que abren a la fuerza fisuras existentes en rocas que se encuentran a grandes profundidades subterráneas.
free range grazing Allowing animals to graze outdoors on grass for most or all of their lifecycle.	**pastoreo a campo abierto** Permite que los animales se alimenten en el césped o al exterior durante la mayoría de su ciclo de vida.
freshwater biome Categorized as streams and rivers, lakes and ponds, or freshwater wetlands.	**bioma de agua dulce** Categorizado como arroyos y ríos, lagos y estanques o humedales de agua dulce.
freshwater wetland An aquatic biome that is submerged or saturated by water for at least part of each year, but shallow enough to support emergent vegetation.	**humedales de agua dulce** Bioma acuático sumergido o saturado de agua al menos una parte de cada año, pero suficientemente pando para dar sustento a vegetación emergente.

fuel cell An electrical-chemical device that converts fuel, such as hydrogen, into an electrical current.	**celda de combustible** Dispositivo electroquímico que convierte un combustible, por ejemplo, el hidrógeno, en una corriente eléctrica.
fuel rod A cylindrical tube that encloses nuclear fuel within a nuclear reactor.	**barra de combustible** Tubo cilíndrico que contiene el combustible nuclear dentro de un reactor nuclear.
fundamental niche *See* **ecological tolerance**.	**nicho fundamental** *Ver* **tolerancia ecológica**.
fungicide A pesticide that specifically targets fungi (the plural of fungus).	**fungicida** Pesticida dirigido específicamente a hongos.
furrow irrigation A form of irrigation where the farmer digs trenches, or furrows, along the crop rows, and fills them with water.	**irrigación en surcos** Forma de irrigación donde los agricultores excavan trincheras, o surcos, junto a las hileras de cultivos y los llenan de agua.

G

generalists Species that can live under a wide range of biotic or abiotic conditions.	**generalistas** Especies que viven bajo una amplia variedad de condiciones bióticas o abióticas.
genetic diversity A measure of the genetic variation among individuals in a population.	**diversidad genética** Medida de la variación genética entre individuos en una población.
genetically modified organism (GMO) An organism produced by copying genes from a species with some desirable trait and inserting them into other species of plants, animals, or microbes.	**organismo transgénico** Organismo producido al copiar genes de una especie con algún rasgo deseado e insertarlos en otras especies de plantas, animales o microbios.
geographic range Areas of the world in which a species lives.	**rango geográfico** Áreas del mundo en las que vive una especie.
geothermal energy Heat energy that comes from the natural radioactive decay of elements deep within Earth.	**energía geotérmica** Energía térmica que proviene de la descomposición natural radiactiva de elementos en lo profundo de la Tierra.
global change Change that occurs in the chemical, biological, and physical properties of the planet.	**cambio global** Cambio que se da en las propiedades químicas, biológicas y físicas del planeta.
global climate change A type of global change that is focused on changes in the average weather that occurs in an area over a period of years or decades.	**cambio climático global** Tipo de cambio global cuyo enfoque son los cambios en el tiempo promedio que se presenta en una región durante un período de años o décadas.
global warming The increase in global temperatures due to humans producing more greenhouse gases.	**calentamiento global** Aumento de las temperaturas globales debido a la producción humana de más gases invernadero.
gray smog *See* **sulfurous smog**.	**esmog gris** *Ver* **sulfurous smog**.
greenhouse effect Absorption of infrared radiation by atmospheric gases and radiation of the energy back toward Earth.	**efecto invernadero** Absorción de radiación infrarroja por gases atmosféricos y la consecuente radiación de energía de regreso a la Tierra.
greenhouse gases Gases in Earth's atmosphere that trap heat near the surface.	**gases de efecto invernadero** Gases de la atmósfera terrestre que atrapan el calor y lo mantienen cerca de la superficie.
greenhouse warming potential (GWP) An estimate of how much a molecule of any compound can contribute to global warming over a period of 100 years relative to a molecule of CO_2.	**potencial de calentamiento por gases de efecto invernadero** Estimación de cuánto una molécula de cualquier compuesto puede contribuir al calentamiento global durante un lapso de 100 años con relación a una molécula de CO_2.
green manure Plant material deliberately grown in a field with the intention of plowing it under at the end of the season.	**Estiércol verde** Material vegetal que se cultiva deliberadamente en un campo con la intención de ararlo al final de la temporada.
Green Revolution A shift in agricultural practices in the twentieth century that included new management techniques, mechanization, fertilization, irrigation, and improved crop varieties, that resulted in increased food output.	**Revolución verde** Modificación en el cultivo agrícola que se dio en el siglo XX, la cual incluyó nuevas técnicas de gestión, mecanización, fertilización, riego, así como cultivos mejorados, con lo cual se obtuvo un aumento en la producción alimentaria.
gross primary productivity (GPP) The total amount of solar energy that producers in an ecosystem capture via photosynthesis over a given amount of time.	**productividad bruta primaria (PBP)** Monto total de energía solar que los productores de un ecosistema captan por fotosíntesis durante un lapso de tiempo dado.
ground source heat pump A technology that transfers heat from the ground to a building.	**bomba térmica de origen terrestre** Tecnología mediante la cual se transfiere el calor de la tierra al interior de un edificio.
groundwater recharge The process by which water from precipitation percolates through the soil into groundwater.	**recarga de aguas freáticas** Proceso mediante el cual el agua de la precipitación se filtra a través del suelo y llega al agua subterránea.
gyre A large-scale pattern of water circulation that moves clockwise in the Northern Hemisphere and counterclockwise in the Southern Hemisphere.	**giro oceánico** Sistema a gran escala de la circulación del agua que se desplaza en el sentido de las manecillas del reloj en el hemisferio septentrional (Norte) y en el sentido contrario al de las manecillas del reloj en el hemisferio austral (Sur).

H	
habitat An area where a particular species lives in nature.	**hábitat** Lugar en la naturaleza donde vive una especie en particular.
habitat diversity The variety of habitats that exist in a given ecosystem.	**diversidad de hábitats** La variedad de hábitats que existen en un ecosistema dado.
Hadley cell A convection current in the atmosphere that cycles between the equator and 30° N and 30° S.	**célula de Hadley** Corriente de convección en la atmósfera entre el ecuador y 30° N y 30° S.
half-life The time it takes for one-half of an original radioactive parent atom to decay.	**hemivida** Tiempo necesario para que se desintegre la mitad de un átomo radiactivo original.
hazardous waste Liquid, solid, gaseous, or sludge waste material that is harmful to humans, ecosystems, or materials.	**residuos nocivos** Material líquido, sólido, gaseoso o fangoso que es perjudicial para los seres humanos, ecosistemas o materiales.
haze Reduced visibility.	**bruma** Visibilidad reducida.
herbicide A pesticide that targets plant species that compete with crops.	**herbicida** Plaguicida que se enfoca en especies de plantas que compiten con los cultivos.
herbivore A consumer that eats producers. *Also known as* **primary consumer**.	**herbívoro** Consumidor que come productores. *También se conoce como* **consumidor primario**.
herbivory An interaction in which an animal consumes plants or algae.	**herbivorio** Interacción en la que un animal consume plantas o algas.
heterotroph *See* **consumer**.	**heterótrofo** *Ver* **consumer**.
homeostasis The ability to experience relatively stable internal conditions in their bodies.	**homeostasis** La capacidad de experimentar condiciones internas relativamente estables dentro del cuerpo.
horizon A horizontal layer in a soil defined by distinctive physical features such as color and texture.	**horizonte** Capa horizontal en un suelo que se define por sus rasgos físicos característicos tales como color y textura.
hot desert A biome located at roughly 30° N and 30° S, and characterized by hot temperatures, extremely dry conditions, and sparse vegetation.	**desierto cálido** Bioma localizado en 30° N and 30° S, caracterizado por temperaturas cálidas, condiciones extremadamente secas y vegetación escasa.
hot spot In geology, a place where molten material from Earth's mantle reaches the lithosphere.	**punto caliente** En geología, sitio en el que el material en estado líquido del manto terrestre alcanza la litosfera.
Hubbert curve A graph that represents oil use and projects both when world oil production will reach a maximum and when world oil will be depleted.	**curva de Hubbert** Gráfica que representa el uso de petróleo y que proyecta tanto el momento en el cual se llegará al nivel máximo de producción mundial y cuando el petróleo del mundo se agotará.
Human Immunodeficiency Virus (HIV) A type of virus that causes Acquired Immune Deficiency Syndrome (AIDS).	**virus de la inmunodeficiencia humana (VIH)** Tipo de virus que causa el síndrome de inmunodeficiencia adquirida (sida).
humus The most fully decomposed organic matter in the lowest section of the O horizon.	**humus** La sustancia orgánica en mayor descomposición en la parte más baja del horizonte O.
hydrocarbons Pollutant compounds that contain carbon-hydrogen bonds, such as gasoline and other fossil fuels, lighter fluid, dry-cleaning fluid, oil-based paints, and perfumes.	**hidrocarburos** Compuestos contaminantes que contienen enlaces de hidrógeno y carbono, como la gasolina y otros combustibles fósiles, líquido para encendedores, líquido para lavar al seco, pinturas de aceite y perfumes.
hydroelectricity Electricity generated by the kinetic energy of moving water.	**hidroelectricidad** Electricidad generada por la energía cinética del agua en movimiento.
hydrologic cycle The movement of water around the biosphere among reservoir sources and sinks.	**ciclo hidrológico** Movimiento del agua alrededor de la biosfera entre reservorios y depósitos.
hypothesis A testable conjecture about how something works.	**hipótesis** Conjetura comprobable acerca de cómo funciona algo.
hypoxic Low in oxygen.	**hipóxico** De poco oxígeno.

I	
igneous rock Rock formed directly from magma.	**roca ígnea** Roca que se formó directamente del magma.
immigration The movement of people into a country or region, from another country or region.	**inmigración** Desplazamiento de personas a un país o región, provenientes de otro país o región.
impervious surface Pavement or other surfaces that do not allow water penetration.	**superficie impermeable** Pavimento u otras superficies que no permiten que el agua penetre.
incineration The process of burning waste materials to reduce volume and mass, and sometimes to generate electricity or heat.	**incineración** Proceso de quemar desechos para reducir su volumen y masa y que a veces se usa para generar electricidad o calor.

independent variable A variable that is not dependent on other factors.	**variable independiente** Variable que no depende de otros factores.
indicator species A species that demonstrates a particular characteristic of an ecosystem.	**especie indicadora** Especie que demuestra una característica particular de un ecosistema.
indoor air pollutants Compounds that adversely affect the quality of air in buildings and structures.	**contaminantes del aire interior** Compuestos que afectan negativamente la calidad del aire en edificios y estructuras.
inductive reasoning The process of making general statements from specific facts or examples.	**razonamiento inductivo** El proceso de hacer declaraciones generales a partir de datos o ejemplos específicos.
industrial agriculture Agriculture that applies the techniques of mechanization and standardization to the production of food. *Also known as* **agribusiness**.	**agricultura industrial** Agricultura que aplica las técnicas de mecanización y normalización a la producción de alimentos. *También se conoce como* **agroindustria**.
industrial smog *See* **sulfurous smog**.	**esmog industrial** *Ver* **sulfurous smog**.
infant mortality The number of deaths of children under 1 year of age per 1,000 live births.	**mortalidad infantil** Número de muertes de niños menores de un año por millar de partos vivos.
infectious disease A disease caused by a pathogen.	**enfermedad infecciosa** Enfermedad causada por un patógeno.
innocent-until-proven-guilty principle A principle based on the belief that a potential hazard should not be considered an actual hazard until the scientific data definitively demonstrate that it actually causes harm.	**principio de ser inocente hasta que se compruebe la culpabilidad** Principio que se basa en la convicción de que un posible peligro no debe ser considerado un peligro real sino hasta que se cuente con datos científicos que demuestren de manera definitiva que es perjudicial.
inorganic fertilizer *See* **synthetic fertilizer**.	**fertilizante inorgánico** *Ver* **synthetic fertilizer**.
insecticide A pesticide that targets species of insects and other invertebrates that consume crops.	**insecticida** Plaguicida que se enfoca en especies de insectos y otros invertebrados que consumen cultivos.
insolation Incoming solar radiation, which is the main source of energy on Earth.	**insolación** Radiación solar entrante, que es la mayor fuente de energía en la Tierra.
integrated pest management (IPM) An agricultural practice that uses a variety of techniques designed to minimize pesticide inputs.	**gestión integrada de plagas** Técnica agrícola que emplea una diversidad de técnicas diseñadas para minimizar el uso de plaguicidas.
integrated waste management An approach to waste disposal that employs several waste reduction, management, and disposal strategies in order to reduce the environmental impact of MSW.	**gestión integrada de desechos** Manera de enfocar la disposición de desechos que emplea varias estrategias para la reducción, la gestión y la eliminación de desechos con el fin de mitigar el impacto de los desechos sólidos municipales.
intercropping An agricultural technique that calls for physical spacing of different crops growing at the same time, in close proximity to one another, to promote biological interaction.	**cultivos asociados** Técnica agrícola que, para promover interacción biológica, requiere espacio físico entre cultivos diferentes que crecen de manera cercana al mismo tiempo.
intermediate disturbance hypothesis The hypothesis that ecosystems experiencing intermediate levels of disturbance will favor a higher level of diversity of species than those with high or low disturbance levels.	**hipótesis de la perturbación intermedia** Hipótesis que explica que los ecosistemas experimentan niveles intermedios de perturbaciones favorecerán un nivel más alto de diversidad de especies que aquellos con niveles de disturbios altos o bajos.
intertidal zone The narrow band of coastline that exists between the levels of high tide and low tide.	**zona intermareal** Banda delgada en el litoral que existe entre los niveles conocidos de marea alta y marea baja.
intertropical convergence zone (ITCZ) The latitude that receives the most intense sunlight, which causes the ascending branches of the two Hadley cells to converge.	**zona de convergencia intertropical** Latitud que recibe la luz solar más intensa, lo que lleva a que las ramas ascendentes de las dos células de Hadley se junten.
intrinsic growth rate (*r*) *See* **population growth rate**.	**tasa de crecimiento intrínseco (*r*)** *Ver* **tasa de crecimiento poblacional**.
invasive species A species that spreads rapidly across large areas and causes harm.	**especie invasora** Especie que se disemina rápidamente a través de zonas extensas y causa daño.
inversion layer The layer of warm air that traps emissions in a thermal inversion.	**capa de inversión** Capa de aire cálido que atrapa las emisiones en una inversión térmica.
IPAT equation A conceptual representation of the three major factors that influence environmental **I**mpact: **P**opulation of humans, **A**ffluence, **T**echnology.	**ecuación IPAT** Representación conceptual de los tres grandes factores que influyen en el **I**mpacto ambiental: **P**oblación de humanos, **A**fluencia, **T**ecnología.
island arc A chain of islands formed by volcanoes as a result of two tectonic plates coming together and experiencing subduction.	**arco insular** Cadena de islas formada por volcanes como resultado de dos placas tectónicas que se unen y experimentan subducción.
island biogeography The study of how species are distributed and interacting on islands.	**biogeografía de islas** Estudio de cómo las especies se distribuyen e interaccionan en las islas.

J	
J-shaped curve The curve of the exponential growth model when graphed.	**curva con forma de J** Curva del modelo de crecimiento exponencial cuando se representa de manera gráfica.

K	
keystone species A species that is not very abundant but has large effects on an ecological community.	**especie clave** Especie que no es muy abundante, pero tiene grandes efectos en una comunidad ecológica.
***K*-selected species** A species with a low intrinsic growth rate that causes the population to increase slowly until it reaches the carrying capacity of the environment.	**especie de selección *K*** Especie con una tasa de crecimiento intrínseco baja que es motivo de que la población crezca lentamente hasta que llegue a la capacidad poblacional del medio ambiente.
Kyoto Protocol An international agreement that sets a goal for global emissions of greenhouse gases from all industrialized countries to be reduced by 5.2 percent below their 1990 levels by 2012.	**protocolo de Kyoto** Convenio internacional que fija una meta para el total de emisiones mundiales de gases de efecto invernadero provenientes de todos los países industrializados. Para el año 2012, dichos gases se habrían de reducir por 5.2 por ciento a los niveles que tenían en 1990.

L	
Lacey Act A U.S. act that prohibits interstate shipping of all illegally harvested plants and animals.	**Ley Lacey** Ley federal estadounidense que prohíbe el despacho interestatal de plantas y animales obtenidos de manera ilícita.
La Niña Following an El Niño event, trade winds in the South Pacific reverse strongly, causing regions that were hot and dry to become cooler and wetter.	**La Niña** Después de un suceso de el Niño, los vientos alisios en el Pacífico Sur se revierten con fuerza y causan que regiones cálidas y secas sean más cálidas y húmedas.
latent heat release The release of energy when water vapor in the atmosphere condenses into liquid water.	**liberación térmica latente** Liberación de energía que se produce cuando el vapor de agua en la atmósfera se condensa y forma agua líquida.
LD_{50} The lethal dose of a chemical that kills 50 percent of the individuals in a dose-response study.	**LD_{50}** Dosis letal de una sustancia química que mata el 50 por ciento de los individuos en un estudio de dosis-respuesta.
leachate Liquid that can contain elevated levels of pollutants as a result of having passed through the solid waste of a landfill.	**lixiviado** Líquido que contiene niveles elevados de contaminantes como resultado de haber pasado por desechos sólidos o por suelos contaminados.
leach field A component of a septic system, made up of underground pipes laid out below the surface of the ground.	**campo de lixiviación** Componente de un sistema séptico compuesto por tuberías subterráneas dispuestas en un diseño planificado.
leaching A process in which dissolved molecules are transported through the soil via groundwater.	**lixiviación** Proceso en el que moléculas disueltas son transportadas a través de la tierra por medio de aguas freáticas.
lead (Pb) A trace metal that occurs naturally in rocks and soils, is present in small concentrations in coal and oil and is a neurotoxin.	**plomo (Pb)** Metal cuyas trazas se presentan de manera natural en piedras y en la tierra, que se presenta en concentraciones pequeñas en el carbón y el petróleo, y es una neurotoxina.
levee An enlarged bank built up on each side of a river.	**dique de contención** Orilla agrandada que se amplía a cada costado de un río.
life-cycle analysis A systems tool that looks at the materials used and released throughout the lifetime of a product—from the procurement of raw materials through their manufacture, use, and disposal. *Also known as* **cradle-to-grave analysis**.	**análisis de ciclo de vida** Recurso en sistemas que estudia los materiales utilizados y emanados durante la vida completa de un producto (es decir, desde que se compra la materia prima hasta que se fabrica, utiliza y tira). *También se conoce como* **análisis de la cuna a la tumba**.
life expectancy The average number of years that an infant born in a particular year in a particular country can be expected to live, given the current average life span and death rate in that country.	**expectativa de vida** Cantidad promedio de años que se puede prever que viva un bebé nacido en un año dado y en un país específico, dada la media de vida o la tasa de fallecimientos en dicho país.
lignite A brown coal that is a soft sedimentary rock that sometimes shows traces of plant structure; it typically contains 60 to 70 percent carbon.	**lignito** Carbón marrón que es una roca sedimentaria y que a veces muestra rastros de estructuras de plantas; normalmente contiene entre 60 y 70 por ciento de carbono.
limestone A calcium carbonated sedimentary rock that has been ground up or crushed for easy application as fertilizer.	**caliza** Roca sedimentaria de carbonato de calcio que ha sido molida o aplastada para su aplicación como fertilizante.
limiting nutrient A nutrient required for the growth of an organism but available in a lower quantity than other nutrients.	**nutriente limitante** Nutriente que se precisa para el crecimiento de un organismo, pero que está disponible en cantidad menor que otros nutrientes.

limiting resource A resource that a population cannot live without and that occurs in quantities lower than the population would require to increase in size.	**recurso limitante** Recurso sin el que una población no puede vivir y que existe en cantidades menores de las que la población precisaría para aumentar en volumen.
limnetic zone A zone of open water in lakes and ponds as deep as the sunlight can penetrate.	**zona limnética** Espacio de aguas abiertas en lagos y lagunas cuya profundidad llega hasta donde llegue la luz del sol.
lithosphere The outermost layer of Earth, including the solid upper mantle and crust.	**litosfera** Capa exterior de la Tierra que incluye el manto sólido superior y la corteza.
littoral zone The shallow zone of soil and water in lakes and ponds near the shore where most algae and emergent plants such as cattails grow.	**zona litoral** El espacio pando de tierra y agua en lagos y lagunas cercanos a la orilla, en los que crece la mayoría de las algas y las plantas emergentes, como las aneas.
logistic growth model A growth model that describes a population whose growth is initially exponential, but slows as the population approaches the carrying capacity of the environment.	**modelo de crecimiento logístico** Modelo de crecimiento que describe una población cuyo crecimiento inicialmente es exponencial, pero que pierde velocidad a medida que se aproxima a la capacidad poblacional del entorno.
London-type smog *See* **sulfurous smog**.	**esmog tipo londinense** *Ver* **sulfurous smog**.
Los Angeles–type smog *See* **photochemical smog**.	**esmog tipo angelino o de Los Ángeles** *Ver* **photochemical smog**.
Lyme disease A disease caused by a bacterium (*Borrelia burgdorferi*) that is transmitted by ticks.	**la enfermedad de Lyme** es una enfermedad causada por una bacteria (*Borrelia burgdorferi*) transmitida por garrapatas.

M

macroevolution Evolution that gives rise to new species, genera, families, classes, or phyla.	**macroevolución** Evolución de la que surgen nuevas especies, géneros, familias, clases o filos.
mad cow disease A disease in which prions mutate into deadly pathogens and slowly damage a cow's nervous system.	**enfermedad de las vacas locas** Enfermedad en la que priones mutan en patógenos mortales que lentamente perjudican el sistema nervioso de las vacas.
magma Molten rock.	**magma** Roca fundida.
malaria An infectious disease caused by one of several species of protists in the genus *Plasmodium*.	**malaria** Enfermedad infecciosa causada por una o varias especies de protistas del género *Plasmodium*.
mangrove swamp A swamp that occurs along tropical and subtropical coasts, and contains salt-tolerant trees with roots submerged in water.	**manglar** Pantano que se da en las costas tropicales y subtropicales, que contiene árboles que toleran la sal y cuyas raíces permanecen sumergidas en el agua.
mantle The layer of Earth above the core, containing magma, the asthenosphere, and the solid upper mantle.	**manto** Capa de la Tierra que está sobre el núcleo interno y que contiene el magma, la astenosfera y el manto sólido superior
manure lagoon Human-made pond lined with rubber built to handle large quantities of manure produced by livestock.	**lago anaeróbico o de estiércol** Estanque artificial con forro de caucho o hule construido para manejar grandes cantidades de estiércol producido por ganado.
Marine Mammal Protection Act A 1972 U.S. law that prohibits the killing of all marine mammals in the United States and prohibits the import or export of any marine mammal body parts.	**Ley de Protección de los Mamíferos Marinos** Ley federal estadounidense promulgada en 1972 que prohíbe la importación o exportación de cualquier parte del cuerpo de mamíferos marinos.
mass extinction A large number of species that went extinct over a relatively short period of time.	**extinción masiva** Gran número de especies que se extinguen en un periodo de tiempo relativamente corto.
maximum contaminant level (MCL) The standard for safe drinking water established by the EPA under the Safe Drinking Water Act.	**nivel máximo de contaminantes** La norma que define el agua potable, establecida por la EPA conforme a la Ley de Agua Potable Segura.
maximum sustainable yield (MSY) The largest quality of a renewable resource that can be harvested indefinitely.	**producción sostenida máxima** Calidad máxima de un recurso renovable que se puede cosechar de manera indefinida.
MERS-CoV A coronavirus that causes the disease known as Middle Eastern Respiratory Syndrome.	**MERS-CoV** Coronavirus que causa la enfermedad conocida como síndrome respiratorio de Oriente Medio.
mesosphere The layer of the atmosphere above the stratosphere, extending roughly 50 to 85 km (31–53 miles) above the surface of Earth.	**mesosfera** La capa de la atmósfera que se encuentra arriba de la estratósfera y se extiende aproximadamente entre 50 y 85 km (31 a 53 millas) por encima de la superficie de la Tierra.
mesotrophic Describes a lake with a moderate level of fertility.	**mesotrópico** Se dice de un lago con un nivel moderado de fertilidad.
metal An element with properties that allow it to conduct electricity and heat energy and to perform other important functions.	**metal** Elemento con propiedades que le permiten conducir energía eléctrica y térmica, y realizar otras funciones importantes.

metamorphic rock Rock that forms when sedimentary rock, igneous rock, or other metamorphic rock is subjected to high temperature and pressure.	**roca metamórfica** Roca que se forma cuando una roca sedimentaria, una roca ígnea o cualquier otra roca metamórfica se somete a temperatura y presión alta.
microevolution Evolution at the population level.	**microevolución** Evolución al nivel de población.
mineralization The process by which fungal and bacterial decomposers break down the organic matter found in dead bodies and waste products and convert it into inorganic compounds, such as inorganic ammonium (NH_4^+). *Also known as* **ammonification**.	**mineralización** El proceso a través del cual los descomponedores fúngicos y bacterianos descomponen la materia orgánica que se halla en cadáveres y productos de desecho para convertirla en compuestos inorgánicos, como el amonio inorgánico (NH_4^+). *También se conoce como* **amonificación**.
mine tailings Unwanted waste material created during mining; chemical compounds and rock residues that are left behind after the desired metal or ore is removed.	**relaves** Materiales de desecho que sobran del proceso de minería; de compuestos químicos y residuos de roca que quedan luego de que se extrae el metal deseado, o mena.
modern carbon Carbon in biomass that was recently in the atmosphere.	**carbono moderno** Carbono en forma de biomasa que hace poco estuvo en la atmósfera.
monocropping An agricultural method that utilizes large plantings of a single species or variety.	**monocultivo** Método agrícola en el que se siembran grandes plantaciones de una sola especie o variedad.
Montreal Protocol A commitment by 24 nations to reduce CFC production by 50 percent by the year 2000.	**Protocolo de Montreal** Compromiso de 24 países de reducir la producción de clorofluorocarburos en 50 por ciento antes del año 2000.
mountaintop removal A mining technique in which the entire top of a mountain is removed with explosives.	**remoción de cumbres** Técnica de minería en la que la cumbre completa de una montaña se elimina con explosivos.
municipal solid waste (MSW) Solid waste collected by municipalities from households, small businesses, and institutions such as schools, prisons, municipal buildings, and hospitals.	**desechos sólidos municipales** Desechos sólidos que las municipalidades recogen de hogares, negocios pequeños e instituciones como escuelas, cárceles, edificios municipales y hospitales.
mutagen A type of carcinogen that causes damage to the genetic material of a cell.	**mutágeno** Tipo de agente carcinógeno que causa daños a los materiales genéticos de una célula.
mutualism An interaction between two species that increases the chances of survival or reproduction for both species.	**mutualismo** Interacción entre dos especies que aumenta las posibilidades de supervivencia o reproducción para ambas especies.
N	
narrow spectrum pesticide *See* **selective pesticide**.	**pesticida estrecho del espectro** *Ver* **narrow spectrum pesticide**.
native species A species that lives in its historical range, typically where it has lived for thousands or millions of years.	**especies nativas** Especie que viven dentro de su hábitat histórico, generalmente donde ha vivido durante miles o millones de años.
natural experiment A natural event that acts as an experimental treatment in an ecosystem.	**experimento natural** Acontecimiento natural que hace las veces de tratamiento experimental en un ecosistema.
natural gas A relatively clean fossil fuel containing 80 to 95 percent methane (CH_4) and 5 to 20 percent ethane, propane, and butane.	**gas natural** Combustible fósil relativamente limpio que contiene entre 80 y 95 por ciento metano (CH_4) y entre 5 a 20 por ciento etano, propano y butano.
natural predators Predators that occur naturally in the environment.	**depredadores naturales** Depredadores que se presentan de manera natural en el ambiente.
net migration rate The difference between immigration and emigration in a given year per 1,000 people in a country.	**tasa neta de migración** Diferencia entre inmigración y emigración en un año dado por cada millar de personas que viven en un país.
net primary productivity (NPP) The energy captured by producers in an ecosystem minus the energy producers respire.	**productividad primaria neta** Energía captada por productores en un ecosistema menos la energía que respiran los productores.
neurotoxin A chemical that disrupts the nervous systems of animals.	**neurotoxina** Sustancia química que altera el sistema nervioso de los animales.
nitrification The conversion of ammonia (NH_4^+) into nitrite (NO_2^-) and then into nitrate (NO_3^-).	**nitrificación** Conversión de amonio (NH_4^+) a nitrito (NO_2^-) y luego a nitrato (NO_3^-).
nitrogen cycle The movement of nitrogen around the biosphere among reservoir sources and sinks.	**ciclo de nitrógeno** Movimiento del nitrógeno alrededor de la biosfera entre reservorios y depósitos.
nitrogen fixation The process that converts nitrogen gas in the atmosphere (N_2) into forms of nitrogen that plants and algae can use.	**fijación de nitrógeno** Proceso por el que el gas nitrógeno de la atmósfera (N_2) se convierte en formas de nitrógeno que las plantas y algas pueden utilizar.

nitrogen oxides A by-product of combustion of any fuel in the atmosphere (which contains 78 percent nitrogen).	**óxidos de nitrógeno** Producto derivado de la combustión de cualquier combustible en la atmósfera (que contiene 78 por ciento nitrógeno).
noise pollution Unwanted sound that interferes with normal activities that is loud enough to cause health issues including hearing loss.	**contaminación acústica** Sonido no deseado que interfiere con actividades normales y cuyo volumen alto causa problemas de salud, incluida la pérdida de capacidad auditiva.
nomadic grazing The feeding of herds of animals by moving them to seasonally productive feeding grounds, often over long distances.	**pastoreo nómada** Alimentación de manadas de animales mudándolas cada temporada de un terreno de alimentación productiva a otro, a menudo cubriendo grandes distancias.
nondepletable An energy source that cannot be used up.	**inagotable** Fuente de energía que no se puede agotar ni consumir en su totalidad.
nonpersistent pesticide Pesticide that breaks down relatively rapidly, usually in weeks or months, and have fewer long-term effects but because they must be applied more often their overall environmental impact is not always lower than that of persistent pesticides.	**pesticida no persistente** Pesticida que se descompone de manera relativamente rápida, generalmente en semanas o meses, y que tiene menos efectos a largo plazo pero que debe aplicarse con mayor frecuencia por lo cual su impacto ambiental no es inferior al de pesticidas persistentes
nonpoint source A diffuse area that produces pollution.	**fuente sin punto** Zona indefinida que produce contaminación.
nonrenewable energy resource An energy source with a finite supply, primarily the fossil fuels and nuclear fuels.	**recurso energético no renovable** Fuente de energía con un abastecimiento finito, primordialmente combustibles fósiles y combustibles nucleares.
no-observed-effect level (NOEL) The highest concentration of a chemical that causes no lethal or sublethal effects.	**nivel sin efecto observable (NOEL, por sus siglas en inglés que significan, No-observed-effect level)** La concentración más alta de un químico que no causa efectos letales o subletales.
no-till agriculture An agricultural method used in fields of annual crops where farmers do not till or plow the soil between seasons.	**agricultura sin arar** Método agrícola empleado en cultivos anuales donde los agricultores no aran la tierra entre estaciones.
nuclear power Electricity generated from the nuclear energy contained in nuclear fuel.	**energía nuclear** Electricidad generada de la energía nuclear contenida en el combustible nuclear.
null hypothesis A prediction that there is no difference between the groups or conditions that are being compared.	**hipótesis nula** Predicción de que no hay diferencia entre los grupos o las condiciones, o declaración o concepto que puede ser falsificado o que se puede demostrar que es erróneo.

O

ocean acidification A process in which an increase in ocean CO_2 causes more CO_2 to be converted to carbonic acid, which lowers the pH of the water.	**acidificación de los océanos** Proceso en el que un aumento en el CO_2 oceánico provoca que se convierta más CO_2 en ácido carbónico, lo cual reduce el pH del agua.
O horizon The organic horizon at the surface of many soils, composed of organic detritus in various stages of decomposition.	**horizonte O** Horizonte orgánico en la superficie de muchos suelos, compuesto por desperdicios en diversas etapas de descomposición.
oil sands *See* **tar sands**.	**arenas petrolíferas** *Ver* **arenas de alquitrán**.
oligotrophic Describes a lake with a low level of phytoplankton due to low amounts of nutrients in the water.	**oligotrópico** Se dice de un lago con un nivel bajo de fitoplancton debido a la baja cantidad de nutrientes en el agua.
open ocean Deep-ocean water, located away from the shoreline where sunlight can no longer reach the ocean bottom.	**mar abierto** Aguas profundas del océano, ubicadas a una distancia del litoral donde la luz solar ya no puede llegar hasta el fondo.
open-loop recycling Recycling one product into a different product.	**reciclaje en circuito abierto** Reciclaje de un producto para terminar con otro producto diferente.
open-pit mining A mining technique that creates a large visible pit or hole in the ground.	**minería a cielo abierto** Técnica de minería en la que crea un tajo o agujero grande y visible en el suelo.
ore A concentrated accumulation of minerals from which economically valuable materials can be extracted.	**mena** Acumulación concentrada de minerales de los que se pueden extraer materiales de valor económico.
organic agriculture The production of crops in a way that sustains or improves the soil, without the use of synthetic pesticides or fertilizers.	**agricultura orgánica** Producción de cultivos sin usar plaguicidas ni fertilizantes sintéticos.
organic fertilizer Fertilizer composed of organic matter from plants and animals.	**fertilizante orgánico** Fertilizante hecho de materia orgánica tomada de plantas y animales.
overgrazing Excessive grazing that can reduce or remove vegetation and erode and compact the soil.	**sobrepastoreo** Pastoreo excesivo que reduce o elimina la vegetación y erosionar o comprimir el suelo.

overshoot When a population becomes larger than the environment's carrying capacity.	**extralimitación** Cuando una población crece más que la capacidad poblacional del entorno.
oxygenated fuel A fuel with oxygen as part of the molecule.	**combustible oxigenado** Combustible que incluye oxígeno como parte de la molécula.
oxygen sag curve The relationship of oxygen concentrations to the distance from a point source of decomposing sewage or other pollutants.	**curva de hundimiento de oxígeno** Relación de concentraciones de oxígeno con la distancia desde una fuente de aguas residuales en descomposición u otros contaminantes.
ozone A pale blue gas composed of molecules made up of three oxygen atoms (O_3).	**ozono** Gas azul claro compuesto por moléculas de tres átomos de oxígeno (O_3).

P

pandemic An epidemic that occurs over a large geographic region, such as an entire continent.	**pandemia** Epidemia que cuando ocurre abarca una región geográfica amplia, como un continente.
parasitism An interaction in which one organism lives on or in another organism, referred to as the host.	**parasitismo** Interacción en la que un organismo vive sobre o en otro organismo, conocido como el anfitrión.
parasitoid A specialized type of predator that lays eggs inside other organisms—referred to as its host.	**parasitoide** Tipo especializado de depredador que pone huevos dentro de otros organismos, a los que se los designa huéspedes.
parent material The underlying rock material from which the inorganic components of a soil are derived.	**material parental** Material de rocoso subyacente a partir del cual se derivan los componentes inorgánicos del suelo.
Paris Climate Agreement A pledge by 196 countries to keep global warming less than 2°C above pre-industrial levels. *Also known as* the **Paris Climate Accord**.	**Acuerdo de Paris sobre el cambio climático** Compromiso de 196 países de mantener el aumento de la temperatura global por debajo de 2°C con respecto a los niveles preindustriales.
particles *See* **particulate matter**.	**partículas** *Ver* **particulate matter**.
particulate matter (PM) Solid or liquid particles suspended in the air. *Also known as* **particulates; particles**.	**materia particulada** Partículas sólidas o líquidas suspendidas en el aire. *También se conoce como* **particulados; partículas**.
particulates *See* **particulate matter**.	**particulados** *Ver* **particulate matter**.
passive solar A use of energy from the sun that takes advantage of solar radiation without active technology.	**solar pasiva** Uso de la energía del sol que aprovecha la radiación solar sin tecnología activa.
passive solar design Construction technique designed to take advantage of solar radiation without active technology.	**diseño solar pasivo** Técnica de construcción diseñada para aprovechar la radiación solar sin emplear tecnología activa.
pathogen A parasite that causes disease in its host.	**patógenos** Parásitos que pueden enfermar a su huésped (u hospedador).
peak demand The greatest quantity of energy used at any one time.	**demanda pico** Cantidad máxima de energía que se utiliza simultáneamente.
peak oil The point at which oil extraction and use would increase steadily until roughly half the supply had been used up.	**pico petrolero** Punto en el cual la extracción y uso del petróleo aumentaría de manera estable hasta agotar aproximadamente la mitad del abasto.
peat A precursor to coal, made up of partly decomposed organic material, including mosses.	**turba** Precursora del carbón, compuesta principalmente de material orgánico descompuesto, incluidos los musgos.
perennial plants Plants that live for multiple years and do not need to be replanted at the beginning of each growing season.	**planta perenne** Plantas que viven múltiples años y que no es necesario volver a sembrar al inicio de cada estación
periodic disruption Occurring regularly, such as the cycles of day and night or the daily and nightly cycle of the moon's effects on ocean tides.	**interrupción periódica** Ocurren regularmente, por ejemplo, el ciclo de día y noche o el ciclo diurno y nocturno de los efectos de la luna sobre las mareas oceánicas.
permafrost An impermeable, permanently frozen layer of soil.	**permafrost** Capa del suelo impermeable y permanentemente congelada.
permeability The ability of water to move through the soil.	**permeabilidad** La capacidad del agua para moverse por el suelo.
persistence The length of time a chemical remains in the environment.	**persistencia** Tiempo que una sustancia química permanece en el medio ambiente.
persistent organic pollutants (POPs) Synthetic, carbon-based molecules that break down very slowly in the environment.	**contaminantes orgánicos persistentes (COPs)** Moléculas sintéticas a base de carbono que se descomponen muy lentamente en el medio ambiente.
persistent pesticide A pesticide that remains in the environment for years to decades.	**plaguicida persistente** Plaguicida que permanece en el medio ambiente años o décadas.
pesticide A substance, either natural or synthetic, that kills or controls organisms that people consider pests.	**plaguicida** Sustancia, sea natural o sintética, que mata o controla organismos que las personas consideran plagas.

pesticide resistance A trait possessed by certain individuals that are exposed to a pesticide and survive.	**resistencia a plaguicidas** Rasgo de ciertos individuos que se exponen a un plaguicida y sobreviven.
pH The relative strength of acids and bases in a substance. It is a logarithmic scale, meaning that each number on the scale represents a change by a factor of 10.	**pH** La fuerza relativa de ácidos y bases en una sustancia. Es una escala logarítmica, es decir que cada número en la escala representa un cambio por un factor de 10.
phantom loads Electrical demand by a device that draws electrical current, even when it is turned off.	**carga fantasma** Demanda eléctrica por parte de un dispositivo que atrae extrae corriente eléctrica, incluso cuando está apagado.
phosphorus cycle The movement of phosphorus around the biosphere among reservoir sources and sinks.	**ciclo de fósforo** Movimiento de fósforo alrededor de la biosfera entre reservorios y depósitos.
photic zone The upper layer of ocean water in the ocean that receives enough sunlight for photosynthesis.	**zona fótica** Capa superior del agua del océano en la que el océano recibe suficiente luz solar para la fotosíntesis.
photochemical oxidant A class of air pollutants formed as a result of sunlight acting on compounds such as nitrogen oxides and sulfur dioxide.	**oxidante fotoquímico** Clase de contaminantes aéreos que se forman como resultado de la acción de la luz solar sobre compuestos tales como óxidos de nitrógeno y dióxido de azufre.
photochemical smog Smog that is dominated by oxidants such as ozone. *Also known as* **Los Angeles–type smog; brown smog**.	**Esmog fotoquímico** Esmog dominado por oxidantes tales como ozono. *También se conoce como* **esmog angelino o de Los Ángeles** o como **esmog marrón**.
photosynthesis The process by which plants and algae use solar energy to convert carbon dioxide (CO_2) and water (H_2O) into glucose ($C_6H_{12}O_6$) and oxygen (O_2).	**fotosíntesis** Proceso mediante el cual las plantas y algas aprovechan la energía solar para convertir dióxido de carbono (CO_2) y agua (H_2O) en glucosa ($C_6H_{12}O_6$) y oxigeno (O_2).
photovoltaic solar cell A use of energy from the Sun as light, not heat, and converting it directly into electricity.	**celda solar fotovoltaica** Uso de energía del sol como luz, no calor, para convertirla directamente en electricidad.
physical weathering The mechanical breakdown of rocks and minerals.	**desgaste físico** Descomposición mecánica de rocas y minerales.
phytoplankton Floating algae.	**fitoplancton** Algas flotantes.
pioneer species In primary succession, species that can survive with little to no soil.	**especie pionera** En la sucesión primaria, especies que sobreviven con poco nada de suelo.
placer mining The process of looking for minerals, metals, and precious stones in river sediments.	**minería de arenal** Proceso mediante el cual se buscan minerales, metales y piedras preciosas en los sedimentos de ríos.
plague An infectious disease caused by a bacterium (*Yersinia pestis*) that is carried by fleas.	**peste** Enfermedad infecciosa causada por una bacteria (*Yersinia pestis*) transportada por pulgas.
plate tectonics The theory that the lithosphere of Earth is divided into plates, most of which are in constant motion.	**tectónica de placas** Teoría que explica que la litosfera de la Tierra está dividida en placas, la mayoría de las cuales está en constante movimiento.
plowing The process of digging deep into the soil and turning it over.	**arar** Proceso de excavar profundamente y voltear la tierra.
PM_{10} Particles smaller than 10 μm are called Particulate Matter-10 and are not filtered out by the nose and throat and can be deposited deep within the respiratory tract.	**PM_{10}** A las partículas menores de 10 μm se les conoce como material particulado -10 y no se filtran por la nariz ni la garganta, por lo que pueden depositarse en el fondo de las vías respiratorias.
$PM_{2.5}$ Particles of size 2.5 μm and smaller can travel further within respiratory tract and are of even greater health concern.	**$PM_{2.5}$** Las partículas de 2.5 μm o menos pueden viajar más lejos en el tracto respiratorio y presentan un riesgo aún más grande para la salud.
point source A distinct location from which pollution is directly produced.	**fuente de punto** Ubicación definida en la que se produce la contaminación directamente.
polar cell A convection current in the atmosphere, formed by air that rises at 60° N and 60° S and sinks at the poles, 90° N and 90° S.	**célula polar** Corriente de convección en la atmósfera, formada por aire que asciende a 60° N y 60° S y se hunde en los polos, 90° N y 90° S.
polychlorinated biphenyls (PCBs) A group of industrial compounds compounds that were once used to manufacture plastics and insulate electrical transformers.	**bifenilos policlorados** Grupo de compuestos industriales que se usaban para fabricar plásticos y aislar transformadores eléctricos.
population bottleneck When a large population declines in number, the amount of genetic diversity carried by the surviving individuals is greatly reduced.	**cuello de botella de población** Cuando los números de una población decaen, la cantidad de diversidad genética que llevan los individuos sobrevivientes se reduce en gran medida.
population growth models Mathematical equations that can be used to predict population size at any moment in time.	**modelos de crecimiento poblacional** Ecuaciones matemáticas que se pueden utilizar para predecir el volumen poblacional en un momento dado.

population growth rate The number of offspring an individual can produce in a given time period, minus the deaths of the individual or its offspring during the same period. *Also known as* **intrinsic growth rate**.	**tasa de crecimiento poblacional** La cantidad de descendientes que un individuo puede producir en un período de tiempo dado, menos los fallecimientos del individuo o su descendencia durante ese mismo período de tiempo. *También se conoce como* **tasa de crecimiento intrínseco**.
population momentum Continued population growth after growth reduction measures have been implemented.	**momento poblacional** Continuación del crecimiento de la población después de implementarse medidas para reducir la población.
population pyramid An age structure diagram that is widest at the bottom and smallest at the top, typical of developing countries.	**pirámide poblacional** Diagrama de estructuras de edad que es más ancho en la parte inferior y más pequeño en la cima, aspecto característico de los países en vías de desarrollo.
porosity The size of the air spaces between particles.	**porosidad** El tamaño de los espacios de aire entre partículas.
potentially renewable An energy source that can be regenerated indefinitely as long as it is not overharvested.	**potencialmente renovable** Fuente de energía que se puede regenerar indefinidamente siempre y cuando no sea sobreexplotada.
precautionary principle A principle based on the belief that when a hazard is plausible but not yet certain, we should take actions to reduce or remove the hazard.	**principio precautorio** Principio basado en la convicción de que cuando algo es un riesgo plausible sin ser un riesgo confirmado, debemos tomar medidas para retirar o reducir el riesgo.
precision How close the repeated measurements of a sample are to one another.	**precisión** Cercanía que existe entre las reiteradas mediciones de una muestra.
predation An interaction in which one animal typically kills and consumes another animal.	**depredación** Interacción en la que un animal característicamente mata y consume otro animal.
prescribed burn When a fire is deliberately set under controlled conditions, thereby decreasing the accumulation of dead biomass on the forest floor.	**quema prescrita** Cuando se inicia un incendio a propósito y en condiciones controladas, y por lo tanto se reduce la acumulación de biomasa muerta en el suelo forestal.
primary consumer *See* **herbivore**.	**consumidor primario** *Ver* **herbivore**.
primary pollutant A polluting compound that comes directly out of a smokestack, exhaust pipe, or natural emission source.	**contaminante primario** Compuesto contaminante que proviene directamente de una chimenea, un tubo de escape o una fuente de emisión natural.
primary productivity The rate of converting solar energy into organic compounds over a period of time.	**productividad primaria** Tasa de conversión de energía solar en compuestos orgánicos durante un periodo de tiempo.
primary succession Ecological succession occurring on surfaces with bare rock and no soil.	**sucesión primaria** Sucesión ecológica que se da en las superficies de roca expuesta que no tienen suelo.
prion A small, beneficial protein that occasionally mutates into a pathogen.	**prión** Proteína pequeña y benigna que a veces muta en un patógeno.
producers Plants, algae, and some bacteria that use the Sun's energy to produce usable forms of energy, such as sugars. *Also known as* **autotrophs**.	**productor** Plantas, algas y bacterias que usan la energía del Sol para producir formas utilizables de energía, como azúcares. *También se conocen como* **autótrofos**.
profundal zone A region of water where sunlight does not reach, below the limnetic zone in very deep lakes.	**zona de profundidad** Región del agua a la que no llega la luz solar, debajo de la zona limnética en lagos muy profundos.
prospective study A study that monitors people who might become exposed to an environmental hazard, such as a harmful chemical in the future.	**estudio prospectivo** Estudio en el que se vigila a personas que en un futuro podrían quedar expuestas a un peligro medioambiental como las sustancias nocivas.
provision A good produced by an ecosystem that humans can use directly.	**provisión** Bien producido por un ecosistema que los seres humanos pueden aprovechar directamente.

R

radioactive decay When a parent radioactive isotope emits alpha or beta particles or gamma rays.	**descomposición radiactiva** Cuando un isótopo radiactivo padre emite partículas alfa o beta o rayos gama.
radioactive waste Nuclear fuel that can no longer produce enough heat to be useful in a power plant but continues to emit radioactivity.	**residuos radiactivos** Combustible nuclear que ya no puede producir suficiente calor para ser útil en una planta generadora pero que sigue emitiendo radiactividad.
radioactivity The emission of ionizing radiation or particles caused by the spontaneous disintegration of atomic nuclei.	**radiactividad** Emisión de radiación o partículas ionizantes causada por la desintegración espontánea de núcleos atómicos.
radon-222 A radioactive gas that occurs naturally from the decay of uranium and is an indoor air pollutant.	**radón-222** Gas radiactivo que se produce naturalmente al descomponerse el uranio que es, además, un contaminante del aire al interior.

rain shadow A region with dry conditions found on the leeward side of a mountain range as a result of humid winds from the ocean causing precipitation on the windward side.	**sombra pluviométrica** Región de condiciones secas que se encuentra en el lado de sotavento de una cordillera como resultado de los vientos húmedos que provienen del océano que producen precipitación en el lado de barlovento.
random disruption Occurring with no regular pattern such as volcanic eruptions or hurricanes.	**interrupción aleatoria** Se producen sin seguir un patrón regular, por ejemplo, erupciones volcánicas o huracanes.
rangelands Dry, open grassland primarily used for grazing cattle.	**pastizales** Campos de pasto abiertos y secos usados principalmente para pastoreo de ganado.
REACH A 2007 agreement among the nations of the European Union about regulation of chemicals; the acronym stands for registration, evaluation, authorization, and restriction of chemicals.	**REACH** Convenio logrado en el año 2007 entre los países de la Unión Europea en el que se controlan las sustancias químicas. La sigla representa en inglés registro, evaluación, autorización y restricción de las sustancias químicas.
realized niche The range of abiotic and biotic conditions under which a species actually lives.	**nicho realizado** Gama de condiciones abióticas y bióticas bajo las cuales vive una especie.
recycling The process by which materials destined to become municipal solid waste (MSW) are collected and converted into raw materials that are then used to produce new objects.	**reciclaje** Proceso en el que se recogen los materiales destinados a convertirse en desechos sólidos municipales y se convierten en materia prima que luego se aprovecha para producir objetos nuevos.
reduce, reuse, recycle A popular phrase promoting the idea of diverting materials from the waste stream. *Also known as* **the three Rs**.	**reducir, reutilizar, reciclar** Frase de gran acogida que fomenta el concepto de extraer materiales del flujo de desechos. *También se conoce como* **las tres erres**.
reforestation The natural or intentional restocking of trees after clear-cutting to repopulate the forest reduce erosion, and begin the process of removing carbon dioxide from the atmosphere.	**reforestación** Restauración natural o intencional de árboles después de la tala para reestablecer el bosque, reducir erosión y comenzar el proceso de retirar dióxido de carbono de la atmósfera.
renewable energy resources Sources of energy that are infinite.	**fuentes renovables de energía** Fuentes de energía que son infinitas.
replacement-level fertility The total fertility rate required to offset the average number of deaths in a population in order to maintain the current population size.	**fertilidad a nivel de reemplazo** Tasa de fertilidad total requerida para compensar la cantidad promedio de fallecimientos en una población a fin de mantener el volumen poblacional actual.
replication The data collection procedure of taking repeated measurements.	**replicación** Procedimiento de recolección de datos en el que se repiten las mediciones.
reserve In resource management, the known quantity of a resource that can be economically recovered.	**reserva** En la gestión de recursos, la cantidad conocida de un recurso que se puede recuperar en términos económicos.
reservoir The water body created by damming a river or stream.	**estanque** Extensión de agua creada cuando se le coloca una presa a un río o arroyo.
reservoirs The components of the biogeochemical cycle that contain matter, including air, water, and organisms.	**reservorio** Componentes del ciclo biogeoquímico que contienen materia, incluyendo el agua, el aire y organismos.
resilience The rate at which an ecosystem returns to its original state after a disruption.	**resiliencia** Ritmo al cual un ecosistema regresa a su estado original después de una perturbación.
resistance In an ecosystem, a measure of how much a disruption can affect flows of energy and matter.	**resistencia** En un ecosistema, medida del efecto que una perturbación puede tener respecto de los flujos de energía y materia.
resource partitioning When two species divide a resource based on differences in their behavior or morphology.	**partición de recursos** Cuando dos especies se dividen un recurso según diferencias en su comportamiento o morfología.
retrospective study A study that monitors people who have been exposed to an environmental hazard such as a harmful chemical at some time in the past.	**estudio retrospectivo** Estudio en el que se vigila a personas que en algún momento fueron expuestas a un peligro medioambiental como las sustancias nocivas.
reuse Using a product or material that would otherwise be discarded.	**reutilizar** Hacer uso de algún producto o material que en otras circunstancias sería descartado.
reverse osmosis A process of desalination in which water is forced through a thin semipermeable membrane at high pressure.	**ósmosis inversa** Proceso de desalinización en el que el agua se pasa por una membrana delgada y semipermeable bajo gran presión.
rock cycle The geologic cycle governing the constant formation, alteration, and destruction of rock material that results from tectonics, weathering, and erosion, among other processes.	**ciclo rocoso** Ciclo geológico que dirige la constante formación, alteración y destrucción de material rocoso que es resultado de la tectónica, la desintegración y la erosión, entre otros procesos.
rodenticide A pesticide that specifically targets rodents.	**rodenticida** Pesticida dirigido específicamente a los roedores.
rotational grazing The rotation of farm animals to different pastures and fields to prevent overgrazing.	**pastoreo rotacional** Rotación de animales de granja a diferentes pastizales y campos para prevenir el sobrepastoreo.

route of exposure The way in which an individual might come into contact with an environmental hazard, such as a chemical.	**ruta de exposición** Manera en que un individuo puede entrar en contacto con un peligro medioambiental como un producto químico.
***r*-selected species** A species that has a high intrinsic growth rate, and their population typically increases rapidly.	**especies de selección *r*** Especie que tiene una alta tasa de crecimiento intrínseco, y su población suele crecer rápidamente.
rule of 70 A method which dictates that by dividing the number 70 by the percentage population growth rate we can determine a population's doubling time.	**ley de 70** Método que establece que al dividir el número 70 por la tasa de crecimiento porcentual de la población se puede determinar el tiempo que tarda una población en duplicarse.
runoff Water that moves across the land surface and into streams and rivers.	**escorrentía** Agua que se desplaza a través de la superficie terrestre y desemboca en ríos y arroyos.
run-of-the-river Hydroelectricity generation in which water is retained behind a low, small dam or no dam.	**flujo del agua** Generación hidroeléctrica en la que el agua se retiene detrás de una presa pequeña, de poca altura, o se aprovecha sin presa.

S

Safe Drinking Water Act Legislation that sets the national standards for safe drinking water.	**Ley de Agua Potable Segura** Legislación federal estadounidense que fija las normas para el agua potable segura.
salinization A form of soil degradation that occurs when the small amount of salts in irrigation water becomes highly concentrated on the soil surface through evaporation.	**salinización** Forma de degradación del suelo que ocurre cuando una cantidad reducida de sales en el agua de riego termina con una concentración alta en la superficie del suelo por medio de la evaporación.
salt marsh Found along the coast in temperate climates, a marsh containing nonwoody emergent vegetation.	**marisma salina** Ubicada a lo largo de la costa en climas templados, marisma que contiene vegetación emergente no leñosa.
saltwater intrusion An infiltration of salt water in an area where groundwater pressure has been reduced as a result of a cone of depression from extensive pumping of wells.	**intrusión de agua salada** Infiltración de agua salada en una zona en la que la presión de las aguas freáticas se ha visto reducida como resultado de un cono de depresión después de bombear un pozo de manera extensa.
sample size (*n*) The number of times a measurement is replicated in data collection.	**tamaño de la muestra (*n*)** Cantidad de veces que se replica una medición en una recolección de datos.
sanitary landfill An engineered ground facility designed to hold municipal solid waste (MSW) with as little contamination of the surrounding environment as possible.	**basurero sanitario** Instalación terrestre diseñada para retener desechos sólidos municipales con el mínimo posible de contaminación del entorno ambiental.
SARS-CoV-2 A coronavirus that causes the disease known as Covid-19.	**SARS-CoV-2** Coronavirus que causa la enfermedad conocida como Covid-19.
saturation point The maximum amount of water vapor in the air at a given temperature.	**punto de saturación** Cantidad máxima de vapor de agua en el aire a una temperatura dada.
savanna A biome marked by warm temperatures and distinct wet and dry seasons. *Also known as* **tropical seasonal forest**.	**sabana** Bioma que se distingue por sus temperaturas cálidas y sus estaciones de lluvia y sequía claramente marcadas. *También conocido como* **bosque tropical estacional**.
scavenger An organism that consumes dead animals.	**carroñero** Organismo que consume animales muertos.
scientific method An objective method to explore the natural world, draw inferences from it, and predict the outcome of certain events, processes, or changes.	**método científico** Método objetivo para explorar el mundo natural, trazar inferencias del mismo y predecir el resultado de ciertos eventos, procesos o cambios.
scrubber A device that uses a combination of lime and or water to separate and remove particles from industrial exhaust streams.	**depurador** Dispositivo que combina caliza y/o agua para separar y eliminar partículas de corrientes de escape industriales.
seafloor spreading Caused by a divergent boundary, in which rising magma forms new ocean crust on the seafloor at the boundaries between those plates.	**ensanchamiento del fondo marino** Causado por un límite divergente, en el que el magma ascendente forma una nueva corteza marítima en el lecho marino sobre el límite entre las dos placas.
secondary consumer A carnivore that eats primary consumers.	**consumidor secundario** Carnívoro que se come consumidores primarios.
secondary pollutant A primary pollutant that has undergone transformation in the presence of sunlight, water, oxygen, or other compounds.	**contaminante secundario** Contaminante primario que se ha transformado ante la presencia de luz solar, agua, oxígeno u otros compuestos.
secondary succession The succession of plant life that occurs in areas that have been disturbed but have not lost their soil.	**sucesión secundaria** Sucesión de flora que ocurre en áreas que han sido perturbadas pero que no han perdido su suelo.

second law of thermodynamics The physical law stating that when energy is transformed, the quantity of energy remains the same, but its ability to do work diminishes.	**segunda ley de termodinámica** Ley física que manifiesta que cuando la energía se transforma, la cantidad de energía sigue igual pero su capacidad de hacer trabajos disminuye.
sedimentary rock Rock that forms when sediments such as muds, sands, or gravels are compressed by overlying sediments.	**roca sedimentaria** Roca que se forma cuando sedimentos tales como lodos, arenas o gravas se comprimen sobre sedimentos sobreyacentes.
selective cutting The method of harvesting trees that involves the removal of single trees or a relatively small number of trees from the larger forest.	**tala selectiva** Método de tala de árboles que implica retirar un solo árbol o una cantidad relativamente reducida de árboles entre los muchos que hay en un bosque o selva.
selective pesticide A pesticide that targets a narrow range of organisms. *Also known as* **narrow-spectrum pesticide**.	**plaguicida selectivo** Plaguicida que se enfoca en una gama estrecha de organismos. *También conocido como* **pesticida de espectro estrecho**.
sense of place The feeling that an area has a distinct and meaningful character.	**sentido de sitio** Sentido de que una zona tiene un carácter distintivo y significativo.
septage A layer of fairly clear water found in the middle of a septic tank.	**residuos sépticos** Capa de agua relativamente clara que se encuentra en el centro del tanque séptico.
septic system A relatively small and simple sewage treatment system, made up of a septic tank and a leach field, often used for homes in rural areas.	**sistema séptico** Sistema relativamente pequeño y sencillo para el tratamiento de aguas residuales. Consta de un tanque séptico y un campo de lixiviación. A menudo se utilizan en hogares en zonas rurales.
septic tank A large container that receives wastewater from a house as part of a septic system.	**tanque séptico** Recipiente grande que forma parte de un sistema séptico y que recibe aguas negras del hogar.
severe acute respiratory syndrome (SARS) A type of flu caused by a coronavirus.	**síndrome respiratorio agudo grave (SRAG)** Tipo de gripe causada por uno de los virus de la corona.
shifting agriculture *See* **slash-and-burn agriculture**.	**agricultura de desplazamiento** *Ver* **agricultura de roza y quema**.
shrubland A biome characterized by hot, dry summers and mild, rainy winters. *Also known as* **woodland**.	**terreno de árboles y arbustos** Bioma caracterizado por veranos calurosos y secos, e inviernos templados y lluviosos.
sick building syndrome A buildup of toxic pollutants in weatherized spaces, such as newer buildings in the developed world.	**síndrome de edificio enfermo** Acumulación de contaminantes tóxicos en espacios climatizados, por ejemplo, en edificios del mundo desarrollado.
siltation Sediments from moving water that accumulate on the bottom of a reservoir.	**sedimentación** Sedimentos de agua en movimiento que se acumulan en el fondo de un reservorio.
slash-and-burn agriculture An agricultural method in which land is cleared and farmed for only a few years until the soil is depleted of nutrients. *Also known as* **shifting agriculture**.	**agricultura de roza y quema** Método de agricultura en el cual se despeja la tierra y se usa para cultivos durante pocos años hasta que la tierra se queda sin nutrientes. *También conocida como* **agricultura itinerante**.
sludge Solid waste material from wastewater.	**fangos residuales** Material de desecho sólido que proviene de aguas negras.
smart grid An efficient, self-regulating electricity distribution network that accepts any source of electricity and distributes it automatically to end users.	**red inteligente** Red de distribución eléctrica eficiente, que se autorregula, y que recibe cualquier fuente de electricidad y la distribuye automáticamente a los usuarios finales.
smog A type of air pollution that is a mixture of oxidants and particulate matter.	**esmog** Tipo de contaminación del aire que es una mezcla de oxidantes y materias particuladas.
soil conservation The prevention of soil erosion while simultaneously increasing soil depth and increasing the nutrient content and organic matter content of the soil.	**conservación del suelo** Prevención de la erosión del suelo mientras se aumenta al mismo tiempo la profundidad del suelo, los nutrientes y el contenido de material orgánico del suelo.
solid waste The waste produced by humans as discarded materials that is not in liquid or gas form and do not pose a toxic hazard to humans and other organisms.	**desechos sólidos** Desechos producidos por humanos que descartan materiales que no están en estado líquido o gaseoso y que no presentan un peligro tóxico para humanos u otros organismos.
solubility How well a chemical dissolves in a liquid.	**solubilidad** La facilidad con que una sustancia química se disuelve en un líquido.
source reduction An approach to waste management that seeks to cut waste by reducing the use of potential waste materials in the early stages of design and manufacture.	**reducción de fuentes** Manera de enfocar la gestión de desechos mediante la cual se procura reducir el uso de materiales que posiblemente se tengan que desechar en las etapas iniciales de diseño y fabricación.

specialists Species that only live under a narrow range of biotic or abiotic conditions.	**especialistas** Especies que solo viven en un rango angosto de condiciones bióticas y abióticas.
species-area curve A description of how the number of species on an island increases with the area of the island.	**curva especie-área** Descripción de cómo el número de especies en una isla aumenta con el área de la isla.
species diversity The number of species in a region or in a particular ecosystem.	**diversidad de especies** Número de especies en una región o en ecosistema en particular.
species evenness The relative proportion of individuals within the different species in a given area.	**uniformidad de especies** Proporción relativa de individuos dentro de especies diferentes en un área dada.
species richness The number of different species in a given area.	**riqueza de especies** Cantidad de especies diferentes en un área dada.
spray irrigation A form of irrigation where water is pumped into an apparatus that contains a series of spray nozzles.	**riego por aspersión** Forma de riego donde se bombea agua a un aparato que contiene una serie de aspersores.
spring Water that naturally percolates up to the surface.	**manantial** Agua que se filtra naturalmente hasta la superficie.
S-shaped curve The shape of the logistic growth model when graphed.	**curva en forma de S** Representación gráfica de la forma del modelo de crecimiento logístico.
standing crop The amount of biomass present in an ecosystem at a particular time.	**cultivo activo** Cantidad de biomasa presente en un ecosistema a una hora dada.
steady state When a system's inputs equal outputs, so that the system is not changing over time.	**régimen estacionario** Cuando las entradas son equivalentes a las salidas de un sistema de manera que con el pasar del tiempo el sistema no cambia.
Stockholm Convention A 2001 agreement among 127 nations concerning 12 chemicals to be banned, phased out, or reduced.	**Convenio de Estocolmo** Convenio suscrito en 2001 por 127 países referente al tratamiento de 12 sustancias químicas que se han de prohibir, eliminar de manera paulatina o reducir.
stratosphere The layer of the atmosphere above the troposphere, extending roughly 16 to 50 km (10–31 miles) above the surface of Earth.	**estratosfera** Capa de la atmósfera encima de la troposfera, que se extiende aproximadamente 16 a 50 km (de 10 a 31 millas) sobre la superficie terrestre.
strip cropping An agricultural method of planting crops with different spacing and rooting characteristics in alternating sets of rows to prevent soil erosion.	**cultivos en franjas** Método agrícola de sembrar cultivos con características de espaciamiento y enraizamiento en conjuntos de franjas alternas para evitar la erosión del suelo
strip mining The removal of overlying vegetation and "strips" of soil and rock to expose underlying ore.	**minería a cielo abierto** Remoción de la vegetación suprayacente y de "tajos" de suelo y roca para dejar expuesto un mineral subyacente.
subduction The process in which the edge of an oceanic plate moves downward beneath the continental plate and is pushed toward the center of Earth.	**subducción** Proceso en el que la orilla de una placa oceánica se mueve hacia debajo de la placa continental y es empujada hacia el centro de la Tierra.
sublethal effect The effect of an environmental hazard that does not kill an organism but which may impair an organism's behavior, physiology, or reproduction.	**efecto subletal** El efecto de un riesgo ambiental que no mata a un organismo, pero puede afectar su comportamiento, fisiología o reproducción.
sublimate The process of converting from a solid to a gas or vapor.	**sublimar** Proceso de conversión de sólido a gas o vapor.
subsistence energy sources Energy sources gathered by individuals for their own immediate needs including straw, sticks, and animal dung.	**fuentes de energía de subsistencia** Fuentes de energía reunida por individuos para satisfacer sus necesidades inmediatas, por ejemplo, paja, varas y heces animales.
subsistence farming Farming for consumption by the farming family and maybe a few neighbors.	**agricultura de subsistencia** Agricultura para el consumo de la familia agricultora y posiblemente algunos vecinos.
subsurface mining Mining techniques used when the desired resource is more than 100 m (328 feet) below the surface of Earth.	**minería subsuperficial** Técnicas de minería que se emplean cuando el recurso deseado se encuentra a más de 100 m (328 pies) por debajo de la superficie terrestre.
suburbs Areas that surround metropolitan centers.	**suburbios** Zonas circundantes de un centro metropolitano.
sulfur dioxide (SO_2) A corrosive gas that comes primarily from combustion of fuels such as coal and oil, including diesel fuel from trucks.	**dióxido de azufre (SO_2)** Gas corrosivo que resulta principalmente de la combustión de combustibles como carbón y petróleo, incluyendo diésel para camiones.
sulfurous smog Smog dominated by sulfur sulfur dioxide, sulfate compounds, and particulate matter. *Also known as* **London-type smog**; **gray smog**; **industrial smog**.	**esmog sulfuroso** Esmog dominado por el dióxido de azufre, compuestos sulfatados y partículas. *También se conoce como* **esmog londinense**; **esmog gris**; **esmog industrial**.

Superfund Act The common name for the Comprehensive Environmental Response, Compensation, and Liability Act (CERCLA); a 1980 U.S. federal act that imposes a tax on the chemical and petroleum industries, funds the cleanup of abandoned and nonoperating hazardous waste sites, and authorizes the federal government to respond directly to the release or threatened release of substances that may pose a threat to human health or the environment.	**Ley *Superfund*** Nombre popular de la Ley Integral de Respuesta, Indemnización y Responsabilidad Medioambiental (CERCLA); ley federal estadounidense promulgada en 1980 que impone un gravamen a las industrias químicas y petroleras, fondos para la limpieza de sitios con desechos nocivos que se hayan abandonado o que hayan dejado de funcionar, y que autoriza al gobierno federal a responder directamente a la liberación o posible liberación de sustancias que podrían atentar contra la salud humana o contra el medioambiente.
survivorship curve A graph that represents the distinct patterns of species survival as a function of age.	**curva de supervivencia** Representación gráfica de los distintos patrones de supervivencia de las especies como función de la edad.
sustainable agriculture Fulfills the need for food and fiber while enhancing the quality of the soil, minimizing the use of nonrenewable resources, and allowing economic viability for the farmer.	**agricultura sostenible** Satisface la necesidad de alimento y fibra al mismo tiempo que mejora la calidad del suelo, minimizando el uso de recursos no renovables y facilitando la viabilidad económica del agricultor.
sustainable development Development that balances current human well-being and economic advancement with resource management for the benefit of future generations.	**desarrollo sostenible** Desarrollo en el que se equilibran el bienestar y el progreso económico de los seres humanos de hoy con la gestión de los recursos por el bien de las generaciones futuras.
sustainable forestry A methodology for managing forests so they provide wood while also providing clean water, maximum biodiversity, and maximum carbon sequestration in both trees and soil.	**silvicultura sostenible** Metodología de gestión de bosques para que proporcionen leña y al mismo tiempo agua limpia, máxima biodiversidad y captura máxima de carbono tanto en árboles como el suelo.
sustainability Living on Earth in a way that allows humans to use its resources without depriving future generations of those resources.	**sostenibilidad** Vivir en la Tierra de una manera que les permite a los humanos aprovechar sus recursos sin privar a generaciones futuras de tales recursos.
swine flu A type of flu caused by the H1N1 virus.	**gripe porcina** Tipo de gripe causada por el virus H1N1.
symbiosis Two species living in close and long-term association with one another in an ecosystem.	**simbiosis** Dos especies que viven en asociación cercana y duradera en un ecosistema.
sympatric speciation The evolution of one species into two, without geographic isolation.	**especiación simpátrida** Evolución de una especie en dos, sin que medie aislamiento geográfico.
synergistic interaction A situation in which two risks together cause more harm than expected based on the separate effects of each risk alone.	**interacción sinérgica** Situación en la que dos riesgos se unen para producir más perjuicios que lo previsto con base en los efectos separados de cada riesgo por sí solo.
synthetic fertilizer Fertilizer produced commercially, normally with the use of fossil fuels. *Also known as* **inorganic fertilizer**.	**fertilizante sintético** Fertilizante producido de manera comercial, generalmente a partir de combustibles fósiles. *También se conoce como* **fertilizante inorgánico**.
T	
taiga A forest biome made up primarily of coniferous evergreen trees that can tolerate cold winters and short growing seasons. *Also known as* **boreal forest**.	**taiga** Bioma forestal compuesto principalmente de árboles coníferos perennes que toleran inviernos fríos y temporadas cortas de crecimiento. *También conocido como* **bosque boreal**.
tar sands Slow-flowing, viscous deposits of bitumen or asphalt, mixed with sand, water, and clay; *also known as* **oil sands**.	**arenas bituminosas** Depósitos viscosos de flujo lento de bitumen mezclado con arena, agua y barro; *también conocidas como* **arenas petrolíferas**.
temperate grassland A biome characterized by cold, harsh winters, and hot, dry summers. *Also known as* **cold desert**.	**pradera templada** Bioma caracterizado por inviernos fríos y agrestes, y veranos calientes y secos. *También conocido como* **desierto frío**.
temperate rainforest A coastal biome typified by moderate temperatures and high precipitation.	**bosque lluvioso templado** Bioma litoral caracterizado por temperaturas moderadas y precipitación abundante.
temperate seasonal forest A biome with warm summers and cold winters with over 1 m (39 inches) of annual precipitation.	**bosque templado estacional** Bioma con veranos cálidos e inviernos fríos con más de 1 metro (39 pulgadas) de precipitación anual.
teratogen A chemical that interferes with the normal development of embryos or fetuses.	**teratógeno** Sustancia química que interfiere con el desarrollo normal del embrión o feto.
terracing An agricultural technique where farms shape sloping land into step-like terraces that are flat.	**bancal** Técnica agrícola en la que a las tierras inclinadas se les da forma de terrazas planas escalonadas.

terrestrial biome A geographic region of land categorized by a particular combination of average annual temperature, annual precipitation, and distinctive plant growth forms.	**bioma terrestre** Región geográfica de tierra categorizada por una combinación dada de los promedios de temperatura anual y de precipitación anual, así como por las plantas terrestres características que se distinguen en la zona.
tertiary consumer A carnivore that eats secondary consumers.	**consumidor terciario** Carnívoro que come consumidores secundarios.
the 10% rule Of the total biomass available at a given trophic level, only about 10 percent can be converted into energy at the next highest trophic level.	**regla del 10 %** Del total de biomasa disponible en cierto nivel trófico, apenas 10 por ciento puede ser convertida en energía en el siguiente nivel trófico más alto.
the three Rs *See* **reduce, reuse, recycle**.	**las tres erres** *Ver* **reduce, reuse, recycle**.
theory A hypothesis that has been repeatedly tested and confirmed by multiple groups of researchers and has reached wide acceptance.	**teoría** Hipótesis que múltiples grupos de investigadores han comprobado y confirmado en repetidas ocasiones y que ha logrado amplia aceptación.
theory of demographic transition A theory that states that a country moves from high to lower birth and death rates as development occurs and that country moves from a preindustrial to an industrialized economic system.	**teoría de transición demográfica** Teoría que establece que un país pasa de tasas de natalidad y mortalidad altas a tasas más bajas a la par del desarrollo y que ese país transita de un sistema económico preindustrial a uno industrializado.
thermal inversion An atmospheric condition in which a relatively warm layer of air at mid-altitude covers a layer of cold, dense air below.	**inversión térmica** Condición atmosférica en la que una capa de aire relativamente cálida a media altura cubre una capa de aire frío y denso más abajo.
thermal mass A property of a building material that allows it to maintain heat or cold.	**masa térmica** Propiedad de un material de construcción que le permite mantenerse caliente o frío.
thermal pollution Occurs when humans cause a substantial change in the temperature of a water body.	**contaminación térmica** Sucede cuando los seres humanos causan un cambio sustancial en la temperatura de un cuerpo de agua.
thermal shock A dramatic change in temperature that can kill many species.	**choque térmico** Cambio drástico en la temperatura del agua que puede matar muchas especies.
thermohaline circulation An oceanic circulation pattern that drives the mixing of surface water and deep water.	**circulación termohalina** Patrón de circulación oceánica en el que se mezclan las aguas superficiales con las aguas profundas.
thermosphere The layer of the atmosphere above the mesosphere, extending 85 to 600 km (53–375 miles) above the surface of Earth.	**termósfera** Capa de la atmósfera que está encima de la mesósfera, que se extiende de 85 a 600 km (53 a 375 millas) por encima de la superficie de la Tierra.
tidal energy Energy that comes from the movement of water driven by the gravitational pull of the Moon.	**energía mareomotriz** Energía que se genera del movimiento del agua impulsado por la atracción gravitatoria de la Luna.
tilling The preparation of soil through a variety of activities including plowing but also including stirring, digging, and cultivating.	**labranza** Preparación del suelo a través de una variedad de actividades que incluyen no solo el arado sino también remover, excavar y cultivar.
tipping fee A fee charged for trucks that deliver and tip solid waste into a landfill or incinerator.	**tarifa de carga** Tarifa que se cobra a camiones que entregan y depositan desechos sólidos en un basurero o incinerador.
topsoil *See* **A horizon**.	**manto** *Ver* **A horizon**.
total fertility rate (TFR) An estimate of the average number of children that each woman in a population will bear throughout her childbearing years.	**tasa de fertilidad total** Estimación del número promedio de niños que cada mujer en una población dará a luz durante sus años en edad reproductiva.
tragedy of the commons The tendency of a shared, limited resource to become depleted if it is not regulated in some way.	**tragedia de los comunes** Tendencia de un recurso compartido y limitado si no está regulado de alguna manera.
transform boundary An area where tectonic plates move sideways past each other.	**borde de transformación** Zona en la que las placas tectónicas se desplazan lateralmente, una pasando la otra.
transpiration The release of water from leaves into the atmosphere during photosynthesis.	**transpiración** Liberación de agua de las hojas a la atmósfera durante la fotosíntesis.
tree plantation A large area typically planted with a single fast-growing tree species.	**plantación de árboles** Zona extensa en la que característicamente se han sembrado especies de árboles de crecimiento rápido.
trophic levels The successive levels of organisms consuming one another.	**niveles tróficos** Niveles sucesivos de organismos que se consumen unos a otros.
trophic pyramid A representation of the distribution of biomass, numbers, or energy among trophic levels.	**pirámide trófica** Representación gráfica de la distribución de biomasa, cifras o energía entre los niveles tróficos.

tropical rainforest A warm and wet biome found between 20° N and 20° S of the equator, with little seasonal temperature variation and high precipitation.	**bosque lluvioso tropical** Bioma cálido y húmedo que se encuentra entre 20° N y 20° S del ecuador, con poca variación estacional de temperatura y niveles altos de precipitación.
tropical seasonal forest *See* **savanna**.	**bosque tropical estacional** *Ver* **sabana**.
troposphere A layer of the atmosphere closest to the surface of Earth, extending up to approximately 16 km (10 miles).	**troposfera** Capa de la atmósfera contigua a la superficie de la Tierra que se extiende en forma ascendente unos 16 km (10 millas).
tsunami A series of waves in the ocean caused by seismic activity or an undersea volcano that causes a massive displacement of water.	**tsunami** Series de olas en el océano causadas por actividad sísmica o un volcán submarino que genera un gran desplazamiento de agua.
tuberculosis A highly contagious disease caused by the bacterium *Mycobacterium tuberculosis* that primarily infects the lungs.	**tuberculosis** Enfermedad muy contagiosa causada por la bacteria *Mycobacterium tuberculosis.* Primordialmente afecta los pulmones.
tundra A cold and treeless biome with low-growing vegetation.	**tundra** Bioma frío y carente de vegetación arbórea con vegetación de poco crecimiento.
turbine A device that can be turned by water, steam, or wind to produce power such as electricity.	**turbina** Dispositivo que puede ser girado por agua, vapor o viento para producir energía como electricidad.
type I survivorship curve A pattern of survival over time in which there is high survival throughout most of the life span, but then individuals start to die in large numbers as they approach old age.	**curva de supervivencia tipo I** Patrón de supervivencia temporal en el que existe un nivel de supervivencia alto durante la mayor parte de la duración de la vida, pero luego los individuos comienzan a fallecer en gran volumen a medida que se aproximan a la vejez.
type II survivorship curve A pattern of survival over time in which there is a relatively constant decline in survivorship throughout most of the life span.	**curva de supervivencia tipo II** Patrón de supervivencia temporal en el que existe una disminución relativamente constante en la supervivencia durante la mayoría de la duración de la vida.
type III survivorship curve A pattern of survival over time in which there is low survivorship (a high death rate) early in life with few individuals reaching adulthood.	**curva de supervivencia tipo III** Patrón de supervivencia temporal en el que existe un bajo nivel de supervivencia (tasa de mortalidad elevada) temprano en la vida y son pocos los individuos que llegan a la adultez.

U

uncertainty An estimate of how much a measured or calculated value differs from a true value.	**incertidumbre** Estimación de la diferencia que hay entre un valor medido o calculado y el valor real.
unconfined aquifer Porous rock covered by soil.	**acuífero no cautivo** Roca porosa cubierta de suelo.
upwelling The upward movement of ocean water toward the surface as a result of diverging currents.	**surgencia** Movimiento hacia arriba del agua oceánica hacia la superficie como resultado de corrientes divergentes.
urban area An area that contains more than 385 people per square kilometer (1,000 people per square mile).	**zona urbana** Extensión que contiene más de 385 personas por kilómetro cuadrado (1,000 personas por milla cuadrada).
urban blight The lack of support for and deterioration of urban communities.	**deterioro urbano** Falta de apoyo y degradación de comunidades urbanas.
urbanization The process of making an area more urban, which means increasing the density of people per unit area of land.	**urbanización** Proceso de hacer un área más urbana, lo cual significa aumentar la densidad de personas por unidad de área de tierra.
urban runoff Runoff, water that does not evapotranspire or infiltrate the soil, that occurs in an urban area.	**escorrentía urbana** Escorrentía, agua que no se evapotranspira ni infiltra el suelo, que se produce en zonas urbanas.
urban sprawl Urbanized areas that spread into rural areas.	**expansión urbana** Zonas urbanizadas que se desparraman hacia las zonas rurales.

V

variable Any categories, conditions, factors, or traits that differ in the natural world or in experimental situations.	**variable** Cualquier categoría, condición, factor o rasgo que difiere en el mundo natural o en situaciones experimentales.
vapor recovery nozzle A device that prevents VOCs from escaping into the atmosphere while a person is fueling their vehicle.	**boquilla de recuperación de vapor** Dispositivo que impide que compuestos orgánicos volátiles escapen a la atmósfera mientras una persona carga combustible a su vehículo.
volatile organic compounds (VOCs) An organic compound that evaporates at typical atmospheric temperatures.	**compuestos volátiles orgánicos** Compuesto orgánico que se evapora a las temperaturas atmosféricas características.
volcano A vent in the surface of Earth that emits ash, gases, or molten lava.	**volcán** Escape en la superficie terrestre por el que se emiten cenizas, gases o lava fundida.

W	
waste stream The flow of solid waste that is recycled, incinerated, placed in a solid waste landfill, or disposed of in another way.	**flujo de desechos** El flujo de desechos sólidos que se reciclan, incineran, tiran en un basurero o eliminan de alguna otra manera.
waste-to-energy A system in which heat generated by incineration is used as an energy source rather than released into the surrounding environment.	**desecho a energía** Sistema en el que se genera calor mediante incineración. Dicho calor se utiliza como fuente de energía en lugar de liberarlo en el medio ambiente.
wastewater The water produced by livestock operations and human activities, including human sewage from toilets and gray water from bathing and washing of clothes and dishes.	**aguas residuales** Aguas producidas por operaciones de ganado y actividades humanas, incluidas aguas negras humanas de inodoros y aguas grises de baños y del lavado de ropas y trastes.
water footprint Total daily per capita use of fresh water for a country or the world.	**consumo de agua** Uso total diario per cápita de agua dulce en un país o a nivel mundial.
water holding capacity The amount of water a soil can hold against the draining force of gravity.	**capacidad de retención de agua** Cantidad de agua que puede contener el suelo frente a la fuerza de drenaje de la gravedad.
water impoundment The storage of water in a reservoir behind a dam.	**embalse de agua** Almacenamiento de agua en un reservorio, detrás de una presa.
waterlogging A form of soil degradation that occurs when soil remains under water for prolonged periods.	**saturación** Forma de degradación del suelo que ocurre cuando este permanece bajo agua por lapsos de tiempo prolongados.
watershed All land in a given landscape that drains into a particular stream, river, lake, or wetland.	**cuenca hidrográfica** Todo el terreno en un paisaje dado que desemboca en un arroyo, río, lago o humedal dado.
water table The uppermost level at which the groundwater in a given area fully saturates the rock or soil.	**nivel freático** El nivel más alto en el que el agua satura totalmente la roca o el suelo.
weather The short-term conditions of the atmosphere in a local area, which include temperature, humidity, clouds, precipitation, and wind speed.	**estado del tiempo** Condiciones de la atmósfera a corto plazo en una zona local; incluye la temperatura, la humedad, la nubosidad, la precipitación y la velocidad del viento.
West Nile virus A virus that lives in hundreds of species of birds and is transmitted among birds by mosquitoes.	**virus del Nilo Occidental** Virus que vive en centenares de especies de aves y que es transmitido entre aves por mosquitos.
windbreaks An agricultural technique that literally plants tall objects that "break" the wind and prevent soil erosion.	**cortavientos** Técnica de agricultura que literalmente siembra objetos altos que «cortan» el viento y evitan la erosión del suelo.
wind energy Energy generated from the kinetic energy of moving air.	**energía eólica** Energía generada a partir de la energía cinética del aire en movimiento.
wind turbine A turbine that converts the kinetic energy of moving air into electricity.	**turbina aerogeneradora** Turbina que convierte la energía del aire en movimiento en electricidad. *También se conoce como* **torre eólica**.
Z	
Zika virus disease A disease caused by a pathogen that causes fetuses to be born with unusually small heads and damaged brains.	**enfermedad por el virus del Zika** Enfermedad causada por un patógeno que causa microcefalia (reducción del tamaño de la cabeza) y daños cerebrales en los fetos.

Index

Note: Page numbers in **boldface** indicate definitions; those followed by f indicate figures; those followed by t indicate tables.

Math Review

Metric Prefixes A prefix in front of a unit of measure indicates a multiple or fraction of that unit. The table below lists the most frequently used prefixes of the metric system, which you should know for the AP® Environmental Science Exam.

	Prefix	Equivalence	Scientific notation	Abbreviation
Largest	mega	1,000,000	10^6	M
	kilo	1,000	10^3	k
	hecto	100	10^2	h
	deca	10	10^1	da
Base unit	e.g., meter, gram, liter, joule, watt	1	10^0	m, g, L, J, W
	deci	0.1	10^{-1}	d
	centi	0.01	10^{-2}	c
	milli	0.001	10^{-3}	m
Smallest	micro	0.000001	10^{-6}	µ

Scientific Notation *(See pages 298, 343.)*
Use scientific notation when dealing with very large or very small numbers. Here are some examples of numbers written in scientific notation:

$34{,}670 = 3.467 \times 10^4$
$0.0053 = 5.3 \times 10^{-3}$
$259 = 2.59 \times 10^2$

The first part of this notation (e.g., 2.59) is called the digit term. The second part of the notation (e.g., 10^2) is called the exponential term. When adding or subtracting numbers that are written in scientific notation, first make sure that both numbers use the same exponent. Then add or subtract the digit terms and retain the original exponential term.

Example

$$1.23 \times 10^5 + 1.51 \times 10^6 = (0.123 + 1.51) \times 10^6 = 1.633 \times 10^6$$

When multiplying in scientific notation, multiply the two digit terms and then add the two exponential terms.

Example

$$(2.4 \times 10^2) \times (3.0 \times 10^4) = (2.4 \times 3.0) \times (10^{2+4}) = 7.2 \times 10^6$$

When dividing in scientific notation, divide the two digit terms and then subtract the two exponential terms.

Example

$$(3.6 \times 10^4) \div (1.2 \times 10^2) = (3.6 \div 1.2) \times (10^{4-2}) = 3.0 \times 10^2$$

Converting Units *(See pages 38, 100, 298, 314, 367, 417, 561, 590.)*
To convert a value into different units, multiply by one or more ratios that are equal to 1, where the numerator and denominator are equivalent. The units in the numerator and denominator should cancel each other out until only the desired units are left.

Example

Convert 6.85 km into inches. (1 km = 1,000 m, 1 m = 3.28 feet, and 1 foot = 12 inches)

$$6.85\ \cancel{km} \times \frac{1{,}000\ \cancel{m}}{1\ \cancel{km}} \times \frac{3.28\ \cancel{foot}}{1\ \cancel{m}} \times \frac{12\ \text{inches}}{1\ \cancel{foot}} = 269{,}616\ \text{inches}$$

Calculating Averages *(See pages 13, 515.)*
To find the average, or mean, of a set of values, add all the given values and divide the total by the number of given values.

average (mean) $= (x + y + z) \div n$, *where* $n =$ *the number of given values*

Example

The average (mean) of 107, 53, 100, 114, and 236 = (107 + 53 + 100 + 114 + 236) ÷ 5 = 610 ÷ 5 = 122

Percent of a Total Value *(See pages 54, 62, 144, 590, 602, 698.)*
To calculate the percent of a total value, divide the subset value by the total value, then multiply by 100%.

percent of a total value = (subset value ÷ total value) × 100%

Example

A group of rabbits consists of 10 brown, 15 white, and 25 black rabbits.
The percentage of brown rabbits = [10 ÷ (10 + 15 + 25)] × 100% = (10 ÷ 50) × 100% = 0.2 × 100% = 20%

Percent Change *(See pages 62, 144, 298, 525, 683.)*
To calculate percent change, take the difference between the final and initial values and divide this difference by the initial value. Then multiply by 100%.

percent change = [(final value − initial value) ÷ initial value] × 100%

Example

Annual U.S. meat consumption in 1960 = 60 kg per person

Annual U.S. meat consumption in 2012 = 75 kg per person

The percent change from 1960 to 2012 = [(75 kg − 60 kg) ÷ 60 kg] × 100% = (15 ÷ 60 kg) × 100% = 0.25 × 100 = 25%

Annual Rate of Change *(See pages 235, 525, 677.)*
To calculate the annual rate of change for a given measurement, first find the difference between the final value and the initial value. Then divide that difference by the total number of years considered, which is the difference between the final year and the initial year.

annual rate of change = (final value − initial value) ÷ (final year − initial year)

Example

Atmospheric CO_2 = 320 ppm in 1960

Atmospheric CO_2 = 390 ppm in 2010

Average annual rate of change = (390 ppm − 320 ppm) ÷ (2010 − 1960) = 70 ppm ÷ 50 years = 1.4 ppm/year

pH Scale *(See page 530.)*
The pH scale is logarithmic, meaning that each number on the scale changes by a factor of 10 .

Example

A substance with a pH = 5 is 10 times more acidic than a substance with a pH = 6, and 100 times more acidic than a substance with a pH = 7.

Primary Productivity *(See pages 70–71.)*
The gross primary productivity (GPP) of an ecosystem is a measure of the total amount of solar energy captured by producers via photosynthesis. To determine net primary productivity (NPP) of an ecosystem, take the gross primary production (GPP) and subtract the respiration rate of the producers.

GPP – respiration rate = NPP

Example

A forest with a GPP of 3.8 kg C/m^2/year and a respiration rate of 2.4 kg C/m^2/year has an NPP = 3.8 – 2.4 = 1.4 kg C/m^2/year.

Global and National Population Growth Rates *(See pages 195, 207.)*
To calculate the global population growth rate as a percentage, start with the crude birth rate (CBR), which is the number of births per 1,000 individuals per year, and subtract the crude death rate (CDR), which is the number of deaths per 1,000 individuals per year. Then divide by 10.

global population growth rate (%) = crude birth rate (CBR) – crude death rate (CDR) ÷ 10

Example

In 2014, there were 20 births and 8 deaths per 1,000 people worldwide.

Global growth rate = (20 – 8) ÷ 10 = 1.2%

To calculate the population growth rate for a single nation, we also need to account for immigration and emigration.

national population growth rate = [(CBR + immigration) – (CDR + emigration)] ÷ 10

Example

A population of 100,000 people has 2,000 births, 500 deaths, 200 emigrants, and 100 immigrants over a 1-year period. Growth rate = [(20 + 1) – (5 + 2)] ÷ 10 = (21 – 7) ÷ 10 = 1.4%

Doubling Time *(See pages 206–207.)*
The time that it takes for a population to double can be calculated using the "rule of 70."

doubling time (years) = 70 ÷ growth rate (%)

(Note that when dividing by the growth rate, ignore the percent.)

Example

If a population grows at 2% per year, doubling time = 70 ÷ 2 = 35 years.

Population Density *(See pages 181–183.)*
Population density is the number of individuals living in a given area divided by the size of that area.

population density = number of individuals ÷ area

Example

2,500 individuals live on 10,000 hectares of land

Population density = 2,500 ÷ 10,000 = 0.25 individual/hectare

Per Capita *(See pages 324, 367.)*
To calculate the per capita value for a given measurement of a population, divide the measurement by the number of people in the population.

per capita value = population measurement ÷ population size

Example

A population of 50,000 people consumes a total of 3.75 million liters of milk per year.

Per capita milk consumption = $(3.75 \times 10^6) \div (5.0 \times 10^4)$
$= 0.75 \times 10^2 = 75$ liters/person/year

Total Energy Used *(See pages 417, 451, 472.)*
A watt (W) is a unit that measures the rate of energy use.
1,000 W = 1 kilowatt (kW).

Total energy used over a given period of time = kW × time (hours) = kWh

Example

If an electrical appliance that uses 500 W operates for 6 hours, total amount of energy used = 0.5 kW × 6 hours = 3 kWh

Energy Efficiency *(See page 432.)*
To calculate energy efficiency, multiply the individual efficiencies in each step of the system.

overall energy efficiency = (efficiency of step 1) × (efficiency of step 2) × (efficiency of step 3) × (efficiency of step 4 [if needed])

Example

Efficiency of coal to electricity = 0.35
Efficiency of electricity transport = 0.9
Light bulb efficiency = 0.05

The efficiency of converting coal energy to electricity in a light bulb = (efficiency of coal to electricity) × (efficiency of electricity transport) × (light bulb efficiency) = 0.35 × 0.90 × 0.05 = 0.016 (rounding to two significant figures) = 1.6% overall efficiency

Half-life *(See pages 438–439, 577.)*
The half-life of an element is the time it takes for one-half of the original radioactive parent atoms to decay. For a given element with a half-life = n, the element loses half of its radioactivity every n years.

amount radioactivity remaining = (original amount) × (0.5^x), where x = the number of half-lives

Example

Calculate when carbon-14 will have $\frac{1}{8}$ of its radioactivity. Amount remaining ÷ original amount = $(0.5^x) = \frac{1}{8}$

$$\frac{1}{8} = \left(\frac{1}{2}\right)^x$$

$$x = 3$$

time for a given reduction in radioactivity = $x \times n$, where n = the half-life of an element

Example

Carbon-14's half-life = 5,730 years = n
Time for carbon-14 to have $\frac{1}{8}$ of its radioactivity = $x \times n$, = 3 × 5,730 years = 17,190 years

Hydrogen 1 H 1.008 — Element name / Atomic number / Symbol / Atomic mass

Elements that are abundant on Earth

Elements that are less abundant on Earth but prominent in the study of environmental science.

Hydrogen 1 **H** 1.008																	Helium 2 **He** 4.003
Lithium 3 **Li** 6.941	Beryllium 4 **Be** 9.012											Boron 5 **B** 10.81	Carbon 6 **C** 12.01	Nitrogen 7 **N** 14.01	Oxygen 8 **O** 16.00	Fluorine 9 **F** 19.00	Neon 10 **Ne** 20.18
Sodium 11 **Na** 22.99	Magnesium 12 **Mg** 24.31											Aluminum 13 **Al** 26.98	Silicon 14 **Si** 28.09	Phosphorus 15 **P** 30.97	Sulfur 16 **S** 32.07	Chlorine 17 **Cl** 35.45	Argon 18 **Ar** 39.95
Potassium 19 **K** 39.10	Calcium 20 **Ca** 40.08	Scandium 21 **Sc** 44.96	Titanium 22 **Ti** 47.87	Vanadium 23 **V** 50.94	Chromium 24 **Cr** 52.00	Manganese 25 **Mn** 54.94	Iron 26 **Fe** 55.85	Cobalt 27 **Co** 58.93	Nickel 28 **Ni** 58.69	Copper 29 **Cu** 63.55	Zinc 30 **Zn** 65.39	Gallium 31 **Ga** 69.72	Germanium 32 **Ge** 72.64	Arsenic 33 **As** 74.92	Selenium 34 **Se** 78.96	Bromine 35 **Br** 79.90	Krypton 36 **Kr** 83.80
Rubidium 37 **Rb** 85.47	Strontium 38 **Sr** 87.62	Yttrium 39 **Y** 88.91	Zirconium 40 **Zr** 91.22	Niobium 41 **Nb** 92.91	Molybdenum 42 **Mo** 95.94	Technetium 43 **Tc** (98)	Ruthenium 44 **Ru** 101.1	Rhodium 45 **Rh** 102.9	Palladium 46 **Pd** 106.4	Silver 47 **Ag** 107.9	Cadmium 48 **Cd** 112.4	Indium 49 **In** 114.8	Tin 50 **Sn** 118.7	Antimony 51 **Sb** 121.8	Tellurium 52 **Te** 127.6	Iodine 53 **I** 126.9	Xenon 54 **Xe** 131.3
Cesium 55 **Cs** 132.9	Barium 56 **Ba** 137.3		Hafnium 72 **Hf** 178.5	Tantalum 73 **Ta** 180.9	Tungsten 74 **W** 183.8	Rhenium 75 **Re** 186.2	Osmium 76 **Os** 190.2	Iridium 77 **Ir** 192.2	Platinum 78 **Pt** 195.1	Gold 79 **Au** 197.0	Mercury 80 **Hg** 200.6	Thallium 81 **Tl** 204.4	Lead 82 **Pb** 207.2	Bismuth 83 **Bi** 209.0	Polonium 84 **Po** (209)	Astatine 85 **At** (210)	Radon 86 **Rn** (222)
Francium 87 **Fr** (223)	Radium 88 **Ra** (226)		Rutherfordium 104 **Rf** (267)	Dubnium 105 **Db** (268)	Seaborgium 106 **Sg** (271)	Bohrium 107 **Bh** (272)	Hassium 108 **Hs** (277)	Meitnerium 109 **Mt** (276)	Darmstadtium 110 **Ds** (281)	Roentgenium 111 **Rg** (280)	Copernicium 112 **Cn** (285)	Nihonium 113 **Nh** (286)	Flerovium 114 **Fl** (289)	Moscovium 115 **Mc** (289)	Livermorium 116 **Lv** (292)	Tennessine 117 **Ts** (293)	Oganesson 118 **Og** (294)

Lanthanum 57 **La** 138.9	Cerium 58 **Ce** 140.1	Praseodymium 59 **Pr** 140.9	Neodymium 60 **Nd** 144.2	Promethium 61 **Pm** (145)	Samarium 62 **Sm** 150.4	Europium 63 **Eu** 152.0	Gadolinium 64 **Gd** 157.3	Terbium 65 **Tb** 158.9	Dysprosium 66 **Dy** 162.5	Holmium 67 **Ho** 164.9	Erbium 68 **Er** 167.3	Thulium 69 **Tm** 168.9	Ytterbium 70 **Yb** 173.0	Lutetium 71 **Lu** 175.0
Actinium 89 **Ac** (227)	Thorium 90 **Th** 232.0	Protactinium 91 **Pa** 231.0	Uranium 92 **U** 238.0	Neptunium 93 **Np** (237)	Plutonium 94 **Pu** (244)	Americium 95 **Am** (243)	Curium 96 **Cm** (247)	Berkelium 97 **Bk** (247)	Californium 98 **Cf** (251)	Einsteinium 99 **Es** (252)	Fermium 100 **Fm** (257)	Mendelevium 101 **Md** (258)	Nobelium 102 **No** (259)	Lawrencium 103 **Lr** (262)

The value in parentheses is the average atomic mass of the longest lasting isotope of the element known at the time of publication.

The Periodic Table

Commonly Used Conversions for Metric and U.S. Units of Measurement

Measurement	Unit Abbreviation	Metric Equivalent	Metric to U.S.	U.S. to Metric
Area	1 square meter (m^2) 1 square centimeter (cm^2)	= 10,000 square centimeters = 100 square millimeters	1 m^2 = 1.1960 square yards 1 m^2 = 10.764 square feet 1 cm^2 = 0.155 square inch	1 square yard = 0.8361 m^2 1 square foot = 0.0929 m^2 1 square inch = 6.4516 cm^2
Length	1 kilometer (km) 1 meter (m) 1 centimeter (cm) 1 millimeter (mm)	= 1,000 meters = 100 centimeters = 1,000 millimeters = 0.01 meter = 0.001 meter	1 km = 0.62 mile 1 m = 1.09 yards 1 m = 3.28 feet 1 m = 39.37 inches 1 cm = 0.394 inch 1 mm = 0.039 inch	1 mile = 1.61 km 1 yard = 0.914 m 1 foot = 0.305 m 1 inch = 2.54 cm
Mass/Weight	1 metric ton (t) 1 kilogram (kg) 1 gram (g) 1 milligram (mg)	= 1,000 kilograms = 1,000 grams = 1,000 milligrams = 0.001 gram	1 t = 1.102 U.S. tons 1 kg = 2.205 pounds 1 g = 0.0353 ounce	1 U.S. ton = 0.907 t 1 pound = 0.4536 kg 1 ounce = 28.35 g
Volume of solids	1 cubic meter (m^3) 1 cubic centimeter (cm^3 or cc) 1 cubic millimeter (mm^3)	= 1,000,000 cubic centimeters = 0.000001 cubic meter = 1 milliliter = 0.000000001 cubic meter	1 m^3 =1.3080 cubic yards 1 m^3 = 35.315 cubic feet 1 cm^3 = 0.0610 cubic inch	1 cubic yard = 0.7646 m^3 1 cubic foot = 0.0283 m^3 1 cubic inch = 16.387 cm^3
Volume of liquids and gases	1 liter (L) 1 kiloliter (kl) 1 milliliter (ml)	= 1,000 milliliters = 1,000 liters = 0.001 liter = 1 cubic centimeter	1 L = 0.264 gallon 1 kL = 264.17 gallons 1 L = 1.057 quarts 1 ml = 0.034 fluid ounce 1 ml = approximately 1/5 teaspoon	1 quart = 0.946 L 1 gallon = 3.785 L 1 quart = 946 ml 1 pint = 473 ml 1 fluid ounce = 29.57 ml 1 teaspoon = approx. 5 ml
Time	1 millisecond (ms)	= 0.001 second		
Temperature	Degrees Celsius (°C)		°C = 5/9 (°F – 32)	°F = 9/5 °C + 32
Energy and Power	1 kilowatt-hour 1 watt 1 calorie 1 horsepower 1 joule	= 3,412 BTUs = 860,421 calories = 3.412 BTU/hr = 14.34 calorie/min = amount of energy needed to raise the temperature of 1 gram (1 cm^3) of water by 1 degree Celsius = 746 watts = 9.478×10^{-4} BTU = 0.239 cal = 2.778×10^{-7} kilowatt-hour		

Math Review

Metric Prefixes A prefix in front of a unit of measure indicates a multiple or fraction of that unit. The table below lists the most frequently used prefixes of the metric system, which you should know for the AP® Environmental Science Exam.

	Prefix	Equivalence	Scientific notation	Abbreviation
Largest	mega	1,000,000	10^6	M
	kilo	1,000	10^3	k
	hecto	100	10^2	h
	deca	10	10^1	da
Base unit	e.g., meter, gram, liter, joule, watt	1	10^0	m, g, L, J, W
	deci	0.1	10^{-1}	d
	centi	0.01	10^{-2}	c
	milli	0.001	10^{-3}	m
Smallest	micro	0.000001	10^{-6}	μ

Scientific Notation *(See pages 298, 343.)*
Use scientific notation when dealing with very large or very small numbers. Here are some examples of numbers written in scientific notation:

$$34,670 = 3.467 \times 10^4$$
$$0.0053 = 5.3 \times 10^{-3}$$
$$259 = 2.59 \times 10^2$$

The first part of this notation (e.g., 2.59) is called the digit term. The second part of the notation (e.g., 10^2) is called the exponential term. When adding or subtracting numbers that are written in scientific notation, first make sure that both numbers use the same exponent. Then add or subtract the digit terms and retain the original exponential term.

Example

$$1.23 \times 10^5 + 1.51 \times 10^6 = (0.123 + 1.51) \times 10^6 = 1.633 \times 10^6$$

When multiplying in scientific notation, multiply the two digit terms and then add the two exponential terms.

Example

$$(2.4 \times 10^2) \times (3.0 \times 10^4) = (2.4 \times 3.0) \times (10^{2+4}) = 7.2 \times 10^6$$

When dividing in scientific notation, divide the two digit terms and then subtract the two exponential terms.

Example

$$(3.6 \times 10^4) \div (1.2 \times 10^2) = (3.6 \div 1.2) \times (10^{4-2}) = 3.0 \times 10^2$$

Converting Units *(See pages 38, 100, 298, 314, 367, 417, 561, 590.)*
To convert a value into different units, multiply by one or more ratios that are equal to 1, where the numerator and denominator are equivalent. The units in the numerator and denominator should cancel each other out until only the desired units are left.

Example

Convert 6.85 km into inches. (1 km = 1,000 m, 1 m = 3.28 feet, and 1 foot = 12 inches)

$$\frac{6.85\ \text{km}}{} \times \frac{1,000\ \text{m}}{1\ \text{km}} \times \frac{3.28\ \text{foot}}{1\ \text{m}} \times \frac{12\ \text{inches}}{1\ \text{foot}} = 269,616\ \text{inches}$$

Calculating Averages *(See pages 13, 515.)*
To find the average, or mean, of a set of values, add all the given values and divide the total by the number of given values.

average (mean) $= (x + y + z) \div n$, *where* n = *the number of given values.*

Example

The average (mean) of 107, 53, 100, 114, and 236 = (107 + 53 + 100 + 114 + 236) ÷ 5 = 610 ÷ 5 = 122

Percent of a Total Value *(See pages 54, 62, 144, 590, 602, 698.)*
To calculate the percent of a total value, divide the subset value by the total value, then multiply by 100%.

percent of a total value = (subset value ÷ total value) × 100%

Example

A group of rabbits consists of 10 brown, 15 white, and 25 black rabbits.
The percentage of brown rabbits = [10 ÷ (10 + 15 + 25)] × 100% = (10 ÷ 50) × 100% = 0.2 × 100% = 20%

Percent Change *(See pages 62, 144, 298, 525, 683.)*
To calculate percent change, take the difference between the final and initial values and divide this difference by the initial value. Then multiply by 100%.

percent change = [(final value − initial value) ÷ initial value] × 100%

Example

Annual U.S. meat consumption in 1960 = 60 kg per person

Annual U.S. meat consumption in 2012 = 75 kg per person

The percent change from 1960 to 2012 = [(75 kg − 60 kg) ÷ 60 kg] × 100% = (15 ÷ 60 kg) × 100% = 0.25 × 100 = 25%

Annual Rate of Change *(See pages 235, 525, 677.)*
To calculate the annual rate of change for a given measurement, first find the difference between the final value and the initial value. Then divide that difference by the total number of years considered, which is the difference between the final year and the initial year.

annual rate of change = (final value − initial value) ÷ (final year − initial year)

Example

Atmospheric CO_2 = 320 ppm in 1960

Atmospheric CO_2 = 390 ppm in 2010

Average annual rate of change = (390 ppm − 320 ppm) ÷ (2010 − 1960) = 70 ppm ÷ 50 years = 1.4 ppm/year

pH Scale *(See page 530.)*
The pH scale is logarithmic, meaning that each number on the scale changes by a factor of 10 .